HENDERSON'S DICTIONARY OF

Biology

HENDERSON'S DICTIONARY OF

Biology

Fifteenth Edition

Edited by
Eleanor Lawrence

Benjamin Cummings
is an imprint of

Harlow, England • London • New York • Boston • San Francisco • Toronto
Sydney • Tokyo • Singapore • Hong Kong • Seoul • Taipei • New Delhi
Cape Town • Madrid • Mexico City • Amsterdam • Munich • Paris • Milan

Pearson Education Limited
Edinburgh Gate
Harlow
Essex CM20 2JE
England

and Associated Companies throughout the world

Visit us on the World Wide Web at:
www.pearsoned.co.uk

First published 1920
Eighth edition 1963
Ninth edition 1979
Tenth edition 1989
Eleventh edition 1995
Twelfth edition 2000
Thirteenth edition 2005
Fourteenth edition 2008
Fifteenth edition 2011

We are grateful to the authors and Pearson Education for permission to reproduce
Appendix 8: The Geological Timescale, from Lawrence, E., Jackson, R.W. and
Jackson, J.M. (1998) *Longman Dictionary of Environmental Science*,
Pearson Education: Harlow.

Pearson Education is not responsible for the content of third party internet sites.

ISBN 978-1-4082-3430-3

British Library Cataloguing-in-Publication Data
A catalogue record for this book is available from the British Library

Library of Congress Cataloging-in-Publication Data
Henderson's dictionary of biology / [edited by] Eleanor Lawrence.. -- 15th ed.
      p. cm.
   ISBN 978-1-4082-3430-3 (pbk.)
   1. Biology--Dictionaries.   I. Lawrence, Eleanor, 1949-   II. Title:
Dictionary of biology.
   QH302.5.H45 2010
   570.3--dc22

                                                                                    2010044452

10   9   8   7   6   5   4   3   2   1
14   13   12   11   10

Typeset in 8/9pt Times NR by 35
Printed and bound in Malaysia (CTP-PJB)

CONTENTS

PREFACE TO THE FIFTEENTH EDITION

Since the publication of the Fourteenth Edition of *Henderson's Dictionary of Biological Terms* new words have entered the biological vocabulary and old ones have acquired new or more precise usages. Progress in genetics, stem-cell biology and biotechnology continues to be rapid, driven by the genome-sequencing programmes, bioinformatics, the use of gene manipulation and advances in experimental cell biology.

The overall classification of the living world followed in this edition reflects the division of all living things into three domains or superkingdoms – Eukaryotes, Bacteria and Archaea (the latter making up the prokaryotes). Entries in the body of the dictionary are given for all the main phyla, divisions and classes of plants, fungi, animals, protists and some prokaryotes, with some orders being included for groups such as the insects, birds, mammals and flowering plants. The formal classifications of the Bacteria and the Archaea have now been revised to reflect relationships that rely on DNA sequence data. There are also entries under many common names of organisms. The appendices at the back give a fuller outline of the various kingdoms. Viruses are covered by entries for the main groups and in an appendix. Appendix 8 shows the geological timescale.

Terms are arranged in strict alphabetical order, disregarding hyphenation and spaces between words, with abbreviations and acronyms included in their appropriate place within the body of the dictionary. Numbers, Greek letters and configurational letters at the beginning of chemical names are ignored for alphabetization purposes. Within an entry, different meanings of a term are numbered and separated by semicolons. The abbreviations (*bot.*), (*zool.*), etc. have been used in some cases to indicate more clearly which subject area a definition refers to. Almost all the technical terms that may be used as part of a definition are defined within the dictionary; to avoid complicating the text with excessive cross-referencing, such terms are not generally indicated by (*q.v.*) within the body of an entry.

Common suffixes and prefixes derived from Latin and Greek are entered in the body of the dictionary, along with their usual meanings, and Appendix 9 gives the etymological origins of some common word elements.

I should like to thank the staff at Pearson Education for their help and encouragement throughout the project. Comments concerning errors or omissions in this edition will be greatly appreciated, so that they may be rectified in future reprints or editions.

Eleanor Lawrence
London, 2010

ABBREVIATIONS

a. adjective
adv. adverb
alt. alternative (synonym)
anat. anatomy
appl. applies or applied to
biochem. biochemistry
bioinf. bioinformatics
bot. botany
ca. circa (approximately)
cf. compare
dev. biol. developmental biology
EC Enzyme Commission number
  (1978)
e.g. for example
esp. especially
et al. and others
etc. and so forth
evol. evolution
genet. genetics
Gk Greek
i.e. that is

immunol. immunology
L. Latin
microbiol. microbiology
mol. biol. molecular biology
mycol. mycology
n. noun
n haploid no. of chromosomes
neurobiol. neurobiology
pert. pertaining to
plu. plural
p.p.m. parts per million
q.v. see
r.n. Enzyme Commission
  recommended name, where it differs
  from that used as the headword
sing. singular
sp. species (*sing.*)
spp. species (*plu.*)
v. verb
virol. virology
zool. zoology

UNITS AND CONVERSIONS

Basic SI units
ampere, A (electric current)
candela, cd (luminous intensity)
kelvin, K (temperature)
kilogram, kg (mass)
metre, m (length)
mole, mol (amount of substance)
second, s (time)

Units and Conversions

Some common derived SI units

joule, J (energy)	kg m^2 s^{-2}
molar, M (concentration)	mol dm^{-3} (mol l^{-1})
newton, N (force)	kg m s^{-2}
pascal, Pa (pressure)	kg m^{-1} s^{-2}
volt, V (electrical potential)	
watt, W (power)	kg m^2 s^{-3}

Conversions from SI or metric

centimetre, cm (10^{-2} m)	0.394 inches (in.)
degree Celsius (centigrade), °C*	(9/5) °F
gram, g (10^{-3} kg)	0.035 ounces (oz.)
hectare, ha (10^4 m^2)	2.471 acres
joule, J (kg m^2 s^{-2})	0.239 calories (cal.)
kilogram, kg	2.20 pounds (lb)
litre, l (dm^3)	1.76 pints (pt)
metre	39.37 in.
millimetre, mm (10^{-3} m)	0.039 in.
tonne (metric ton), t (1 Mg)	0.984 tons

Conversions to SI or metric

acre (4840 sq. yd)	4046.86 m^2
ångström unit, Å	10^{-10} m
atmosphere, standard, atm (14.72 p.s.i.)	101,325 Pa
bar	10^5 Pa
British thermal unit, Btu	1.055 kJ
British thermal unit/hour, Btu h^{-1}	0.293 W
bushel, bu	0.0364 m^3
bushel (US), bu	0.0352 m^3
calorie, thermochemical	4.184 J
cubic foot, cu. ft, ft^3	0.0283 m^3
cubic inch, cu. in., in^3	16.387 cm^3
cubic yard, cu. yd, yd^3	0.7645 m^3
degree Fahrenheit, °F*	(5/9) °C
dram (avoirdupois), dr	1.772 g
fathom (6 ft)	1.829 m
fluid ounce, fl. oz	28.413 cm^3
foot, ft	0.3048 m
gallon, gal	4.546 dm^3
gallon (US), US gal	3.785 dm^3
grain, gr	64.799 mg
hundredweight, cwt	50.802 kg
inch, in.	25.4 mm
kilocalorie/h, kcal h^{-1}	1.163 W
mile	1.6093 km
millibar, mbar	100 Pa
millimetre of mercury, mm Hg	133.32 Pa
millimetre of water	9.807 Pa
ounce (avoirdupois), oz	0.0283 kg

pint, pt	0.568 dm^3
pound (avoirdupois), lb	0.4536 kg
square foot, sq. ft, ft^2	0.0929 m^2
square inch, sq. in., in^2	645.16 mm^2
square mile, sq. mile (640 acres)	2.590 km^2
square yd, sq. yd, yd^2	0.836 m^2
ton (long) (2240 lb)	1016.05 kg
yard, yd	0.9144 m

* To convert temperature in °C to °F, multiply by 9/5 and add 32; to convert °F to °C subtract 32, then multiply by 5/9.

SI PREFIXES

The following prefixes may be used to construct decimal multiples of units.

Multiple	Prefix	Symbol	Multiple	Prefix	Symbol
10^{-1}	deci	d	10	deca	da
10^{-2}	centi	c	10^2	hecto	h
10^{-3}	milli	m	10^3	kilo	k
10^{-6}	micro	μ	10^6	mega	M
10^{-9}	nano	n	10^9	giga	G
10^{-12}	pico	p	10^{12}	tera	T
10^{-15}	femto	f	10^{15}	peta	P
10^{-18}	atto	a	10^{18}	exa	E
10^{-21}	zepto	z	10^{21}	zetta	Z
10^{-24}	yocto	y	10^{24}	yotta	Y

GREEK ALPHABET

Name	Greek letter		English equivalent	Name	Greek letter		English equivalent
alpha	A	α	a	nu	N	ν	n
beta	B	β	b	xi	Ξ	ξ	x
gamma	Γ	γ	g	omicron	O	o	o
delta	Δ	δ	d	pi	Π	π	p
epsilon	E	ε	e	rho	P	ρ	r
zeta	Z	ζ	z	sigma	Σ	σ	s
eta	H	η	e or h	tau	T	τ	t
theta	Θ	θ	th or q	upsilon	Y	υ	u
iota	I	ι	i	phi	Φ	ϕ	ph or f
kappa	K	κ	k	chi	X	χ	ch or c
lambda	Λ	λ	l	psi	Ψ	ψ	ps or y
mu	M	μ	m	omega	Ω	ω	w

COMMON LATIN AND GREEK
NOUN ENDINGS*

sing.	*plu.*
-a	-ae (L.)
-a	-ata (Gk)
-is	-es (L.)
-on	-a (Gk)
-um	-a (L.)
-us	-i (L.)

* Some familiar exceptions to these rules include genus (*sing.*), genera (*plu.*) and opus (*sing.*), opera (*plu.*).

A

α- *for headwords with prefix α- refer also to alpha or to headword itself.*

α+β name given to a class of protein fold or protein domain composed of separate regions of α-helix and β-sheet.

α/β name given to a class of protein fold or protein domain composed of α-helices connected by β-strands.

a- prefix denoting lacking, without. Derived from Gk *a*, not.

A (1) absorbance *q.v.*; (2) adenine *q.v.*; (3) alanine *q.v.*

Å ångström *q.v.*

AAV adeno-associated virus *q.v.*

ab- prefix derived from Gk *ab*, from.

Ab antibody *q.v.*

ABA abscisic acid *q.v.*

abactinal *a.* situated on the part of the echinoderm body not bearing tube-feet. *alt.* **abambulacral.**

A-band dark band seen in longitudinal sections of striated muscle myofibrils, repeating alternately with light bands (I-bands), and representing thick filaments interdigitating with thin filaments at their ends. *see* Fig. 28 (p. 425).

abapical *a.* (1) *pert.* or situated at lower pole; (2) away from the apex.

abaxial *a. pert.* that surface of any structure which is farthest from, or turned away from, the axis, e.g. the under surface of a leaf is known as the abaxial surface because during leaf formation it faces away from the shoot tip. *cf.* adaxial.

abbreviated *a.* (1) shortened; (2) curtailed.

abcauline *a.* outwards from, or not close to, the stem.

ABC transporters large superfamily of ATP-dependent proteins, in prokaryotes and eukaryotes, that transport a wide variety of substances across cellular membranes. Examples include the multidrug-resistance proteins that pump hydrophobic drugs out of mammalian cells and the type I secretion systems of bacteria.

abdomen *n.* (1) in vertebrates, the lower part of the body cavity, containing the digestive organs, reproductive organs, kidneys and liver. In mammals it is separated from the thorax by the diaphragm; (2) in arthropods and certain polychaete worms, the posterior part of the body, behind the head and thoracic regions; (3) in tunicates, the section of the body containing stomach and intestines. *a.* **abdominal.**

abdominal histoblast nests *see* histoblast nests.

abdominal pores single or paired openings leading from coelom to exterior, in certain fishes.

abdominal reflex contraction of abdominal wall muscles when skin over the side of the abdomen is stimulated.

abdominal regions nine regions into which the human abdomen is divided by two horizontal and two vertical imaginary lines, comprising hypochondriac (two regions), lumbar (two), inguinal (two), epigastric, umbilical, hypogastric.

abdominal ribs bony structures occurring in fibrous tissue in abdominal region between skin and muscles in certain reptiles.

abdominal ring one of two openings in the connective tissue sheath of the abdominal muscles, through which passes the spermatic cord in males and the round ligament in females.

abducens *n.* sixth cranial nerve, supplying the rectus externus muscle of the eyeball.

abduction *n.* movement away from the median axis. *cf.* adduction.

abductor *n.* muscle that draws a limb or part outwards.

abequose *n.* 3,6-dideoxyhexose sugar found in lipopolysaccharide in outer membrane of some enteric bacteria.

aberrant *a.* with characteristics not in accordance with type, *appl.* species, etc.

abhymenial *a.* on or *pert.* the side of the gill opposite that of the hymenium in agaric fungi.

abience *n.* (1) retraction from stimulus; (2) avoiding reaction. *a.* **abient.**

abiocoen *n.* non-living parts of the environment in total.

abiogenesis *n.* production of living from non-living matter, as in the origin of life. Also sometimes refers to the theory of spontaneous generation, held in the 19th century and before, which stated that microorganisms or higher organisms could arise directly from non-living material.

abiology *n.* study of non-living things in a biological context.

abioseston *n.* non-living material floating in the plankton.

abiosis *n.* apparent suspension of life.

abiotic *a.* non-living.

abiotic environment that part of an organism's environment consisting of non-biological factors such as topography, geology, climate and inorganic nutrients.

abjection *n.* shedding of spores from a spore-bearing organ, usually with some force.

abjunction *n.* delimitation of fungal spores by septa at the tip of a hypha.

ablactation *n.* (1) cessation of milk secretion; (2) weaning.

ablation *n.* destruction or removal of a particular structure, piece of tissue or individual cell.

abomasum *n.* in ruminants, the fourth chamber of the stomach, into which acid and digestive enzymes are secreted, and in which the final stages of digestion take place.

aboral *a.* away from, or opposite to, the mouth.

abortion *n.* (1) premature birth of a foetus, whether spontaneous or deliberately induced that results in its death. In humans, it is technically the expulsion of a foetus from the time of fertilization up to 3 months gestation; (2) arrest of development of an organ (in plants and animals). *v.* to abort.

abortive infection virus infection in which no new infectious viral particles are produced.

ABO system the main human blood group system used for matching blood for transfusion. There are four blood groups (A, B, AB and O) which result from the presence or absence of either or both of two antigens (A or B) on the red blood cells and the presence of naturally occurring antibodies against the absent antigen(s) in the serum (e.g. anti-A in B individuals and anti-B in A individuals). This results in agglutination of red blood cells and 'transfusion shock' if blood is given to an individual of an incompatible blood group. AB individuals possess A and B antigens and no antibodies and are therefore universal recipients, being able to receive A, B, AB or O blood. O individuals lack either antigen on their red cells and have antibodies against both A and B antigens, and can thus only receive O blood but are universal donors.

abranchiate *n.* without gills.

abrin *n.* lectin isolated from *Abrus precatorius*, specific for D-galactose.

abrupt *a.* appearing as if broken, or cut off, at extremity, *appl.* leaves.

abruptly-acuminate *a.* having a broad extremity from which a point arises, *appl.* leaf.

abrupt speciation formation of a species as a result of a sudden change in chromosome number or constitution.

abscise *v.* (1) to become separated; (2) to fall off, as leaves or fruit.

abscisic acid (ABA) sesquiterpene plant hormone discovered through its promotion of leaf senescence and leaf fall, but whose major functions are in the adaptation of plants to environmental stresses such as drought. It induces stomatal closure and synthesis of proteins that protect against the effects of desiccation. It is also involved in the maintenance of dormancy in seeds and buds.

abscission *n.* separation of a part from the rest of the plant.

abscission zone region at the base of a leaf, flower, fruit, or other part of the plant.

It consists of an abscission layer of weak cells whose breakdown separates the part from the rest of the plant, and a protective layer of corky cells formed over the wound when the part falls.

absenteeism *n.* practice of certain animals of nesting away from their offspring and visiting them from time to time to provide them with food and a minimum of care.

absolute age age of a rock, fossil or archaeological specimen in years before present (BP).

absolute configuration the designation D- or L-, which denotes the handedness of a structural isomer and is based on the atomic configuration around the asymmetric carbon atom. It applies to molecules that contain a centre of asymmetry and occur in mirror-image structural isomers, e.g. many sugars and amino acids.

absolute refractory phase *see* refractory.

absorbance (A) *n.* spectrophotometric measurement of the absorption of light at a particular wavelength by a substance in solution. It can be used to determine the concentration of a substance and to follow conversion of substrate to product in enzyme reactions. *alt.* extinction, optical density.

absorption *n.* (1) uptake of fluid and solutes by living cells and tissues; (2) passage of nutritive material through living cells; (3) of light, when neither reflected nor transmitted by a substance.

absorption spectrum the characteristic plot of wavelength versus intensity of electromagnetic radiation that has been absorbed by a given substance.

absorptive cell *see* brush-border cell.

abstriction *n.* detachment of spores or conidia from a fungal hypha by rounding off and constriction of the tip.

abterminal *a.* going from the end inwards.

abundance *n.* (1) the total number of individuals of a species or type present in a given area; (2) of mRNA, the average number of molecules of a particular mRNA per cell.

abyssal *a.* (1) *appl.* or *pert.* the deep sea below 2000 m and to the organisms and material found there; (2) *appl.* the depths of a lake where light does not penetrate.

abyssobenthic *a. pert.* or found on the ocean floor in the depths of the ocean, in the abyssal zone.

abyssopelagic *a. pert.* or inhabiting the ocean depths of the abyssal zone, but floating, not on the ocean floor.

abzyme *n.* antibody with enzymatic activity.

acanaceous *a.* bearing prickles.

acanthaceous *a.* bearing thorns or prickles.

acanthella *n.* larval stage of acanthocephalans.

acanthion *n.* the most prominent point on the nasal spine.

acanthocarpous *a.* having fruit covered in spines or prickles.

Acanthocephala, acanthocephalans *n., n.plu.* phylum of pseudocoelomate animals, commonly called thorny-headed worms. As adults they are intestinal parasites of vertebrates and as larvae have an arthropod host.

acanthocephalous *a.* with a hooked proboscis.

acanthocladous *a.* having spiny branches.

acanthocyst *n.* sac containing lateral or reserve stylets in nemertine worms.

Acanthodii, acanthodians *n., n.plu.* extinct group of bony fishes, among the earliest jawed vertebrates known, with bony spines projecting in front of their fins and very small, diamond-shaped scales, present from the Silurian to the Permian.

acanthoid *a.* resembling a spine or prickle.

Acanthomorpha *n.* large group of bony fishes, including the Acanthopterygii, Paracanthopterygii, Polymixiiformes and the Lampridiformes.

Acanthopterygii, acanthopterygians *n., n.plu.* large group of bony fishes, including, among many others, perches, mackerel, wrasses, sea basses and dories.

acanthor *n.* the first larval stage of acanthocephalans.

acanthosis *n.* thickening of the inner layer of cells of the epidermis.

acanthozooid *n.* tail part of proscolex of cestodes (tapeworms).

acapnia *n.* condition of low carbon dioxide level in blood.

Acari, Acarina *n.* very large and varied order of arachnids, commonly called mites and ticks. The adults usually have a rounded body carrying four pairs of legs. Ticks are relatively large and parasitic, living as ectoparasites on mammals and sucking their blood. They are vectors of several serious diseases, including Rocky Mountain spotted fever and tick-borne encephalitis. Mites are smaller, inhabiting both plants and animals and are common in soil. They include both parasites and non-parasites, and are the vectors of scrub typhus in humans.

acaricide *n.* chemical compound used to kill ticks and mites (Acarina).

acarocecidium *n.* plant gall caused by gall mites.

acarology *n.* study of mites and ticks.

acarophily *n.* symbiosis of plants and mites.

acarpous *a.* not fruiting.

acaryote, acaryotic *see* akaryote, akaryotic.

acaudate *a.* lacking a tail.

acaulescent *a.* having a shortened stem.

acauline, acaulous *a.* having no stem or stipe.

accelerator *n.* muscle or nerve that increases rate of activity.

accepted name the name adopted as the correct name for a taxon.

acceptor *n.* substance that receives and unites with another substance in a chemical reaction, as in oxidation–reduction processes where the oxygen acceptor is the substance oxidized, the hydrogen acceptor the substance reduced.

acceptor splice region, acceptor splice site the conserved sequence AG|G, comprising the last two nucleotides (AG) at the 3′ end of an intron in a primary RNA transcript (or the corresponding DNA) and the first nucleotide (G) of the next exon. During the RNA splicing reaction, the intron is removed by a final cleavage at this site and the 5′ end of the exon is joined to the 3′ end of the preceding exon. *alt.* 3′ splice site. *cf.* donor splice site.

accessorius *n.* (1) muscle aiding in the action of another; (2) accessory nerve *q.v.*

accessory bud (1) additional axillary bud; (2) bud formed on a leaf.

accessory cell (1) (*bot.*) subsidiary cell; (2) in the immune system, macrophages and other non-lymphoid cells which are involved in immune responses.

accessory chromosomes supernumerary chromosomes *q.v.*

accessory nerve 11th cranial nerve, supplying muscle of soft palate and pharynx and the sternomastoid and trapezius muscles.

accessory pigment photosynthetic pigment that captures light energy and transfers it to chlorophyll *a*, e.g. chlorophyll *b*, carotenoids and phycobilins.

accessory pulsatory organs sac-like structures in insects, variously situated, pulsating independently of the heart.

accessory species in plant ecology, a species that is found in a quarter to a half of the area of a stand. *cf.* accidental species.

accidental species in plant ecology, a species that is found in less than a quarter of a stand. *cf.* accessory species.

accipiters *n.plu.* hawks, medium-sized birds of prey with rounded wings and long tails, part of the family Accipitridae.

acclimation *n.* physiological habituation of an organism to a change in a particular environmental factor, for example the onset of winter. *cf.* acclimatization, adaptation.

acclimatization *n.* physiological and/or behavioural habituation of an organism to a different climate or environment. *cf.* acclimation, adaptation.

accommodation *n.* in the eye, the rapid change in the shape of the lens and thus in its focal length that enables the eye to receive clear images of objects at different distances away.

accrescence *n.* (1) growth through addition of similar tissues; (2) continued growth after flowering. *a.* **accrescent.**

accrete *a.* (1) grown or joined together; (2) formed by accretion.

accretion *n.* growth by external addition of new matter.

accumbent *a.* (*bot.*) reclining, lying against each other; *appl.* embryo having cotyledons with edges turned towards radicle, as in dicot plants of the family Cruciferae.

accumulator *n.* plant that accumulates relatively high concentrations of certain chemical elements, such as heavy metals, in its tissues.

ACE angiotensin-converting enzyme *q.v.*

A cell α-cell of islets of Langerhans in the pancreas, which secretes the hormone glucagon.

acellular *a.* not divided into cells.

acellular slime moulds plasmodial slime moulds *q.v.*

acelomate acoelomate *q.v.*

acelous acoelous *q.v.*

acentric *a. appl.* chromosome or chromosome fragment lacking a centromere and which therefore does not segregate correctly at mitosis or meiosis.

acentrous *a.* with no vertebral centra, but with a persistent notochord, as in some fishes.

acephalocyst *n.* hydatid stage of some tapeworms.

acephalous *a.* having no structure comparable to a head, *appl.* some molluscs, *appl.* larvae of certain Diptera, *appl.* ovary without terminal stigma.

Acerales Sapindales *q.v.*

acerate *a.* (1) needle-shaped; (2) pointed at one end.

acerose *a.* narrow and slender, with sharp point, as leaf of pine.

acerous *a.* (1) hornless; (2) without antennae; (3) without tentacles.

acervate *a.* (1) heaped together; (2) clustered.

acervuline *a.* irregularly heaped together, *appl.* shape of foraminiferal tests.

acervulus *n.* small cluster of spore-bearing hyphae. *a.* **acervulate**.

Acetabularia genus of large unicellular green algae of which *A. mediterranea* has been used for experiments in developmental biology.

acetabulum *n.* (1) cup-shaped socket in pelvic girdle for head of femur, forming the hip joint in tetrapod vertebrates; (2) in insects, thoracic cavity in which leg is inserted; (3) socket of coxa in arachnids; (4) sucker used for attachment to host in flukes, tapeworms and leeches; (5) sucker on arm of cephalopod. *a.* **acetabular**.

acetic acid bacteria bacteria (e.g. *Acetobacter*, *Acetomonas*) that partially oxidize ethyl alcohol to produce acetic acid, a reaction used in manufacturing vinegar. *alt.* acetobacters.

acetoacetate *n.* compound produced in liver which can be used as a source of energy by cells, esp. heart muscle and kidney cells, and is used by the brain instead of glucose during periods of starvation. *see also* ketone body.

acetobacters acetic acid bacteria *q.v.*

acetotrophic *a. appl.* methanogenic bacteria that break down acetate into methane and carbon dioxide.

acetyl chemical group, $-COCH_3$, formed by removal of $-OH$ from acetic acid.

acetylase *see* acetyltransferase.

acetylation *n.* addition of an acetyl group to a molecule.

acetylcholine (ACh) *n.* neurotransmitter secreted by motor neurons that stimulate skeletal muscle, where it acts as an excitatory transmitter, and by certain neurons in the peripheral nervous system and in the brain. It acts as an inhibitory transmitter between the vagus nerve and heart muscle. Neurons secreting acetylcholine are known as cholinergic. It acts at several different types of receptor.

acetylcholine receptor cell-surface protein that binds and is activated by the neurotransmitter acetylcholine. There are two main types, each with several subtypes. The nicotinic receptor (nAChR) is an ion channel and is found on the postsynaptic terminals of some neurons and on skeletal muscle at neuromuscular junctions. The muscarinic acetylcholine receptor (mAChR) is a G-protein-coupled receptor and is mainly present on neurons, heart muscle and in the gastrointestinal tract.

acetylcholinesterase (AChE) cholinesterase *q.v.*

acetyl-CoA pathway pathway of autotrophic carbon dioxide fixation found in obligate anaerobic bacteria such as methanogens and sulphate-reducing bacteria.

acetyl-coenzyme A (acetyl-CoA) acetyl thioester of coenzyme A, produced during the aerobic breakdown of sugars in cells. It is formed in mitochondria from pyruvate and coenzyme A and the acetyl group is subsequently oxidized to CO_2 in the tricarboxylic acid cycle. Also formed during the breakdown of some amino acids and esp. in fatty acid oxidation. It is an important carrier of activated acetyl groups in metabolic reactions.

acetylene reduction assay assay for the enzyme nitrogenase, in which acetylene is

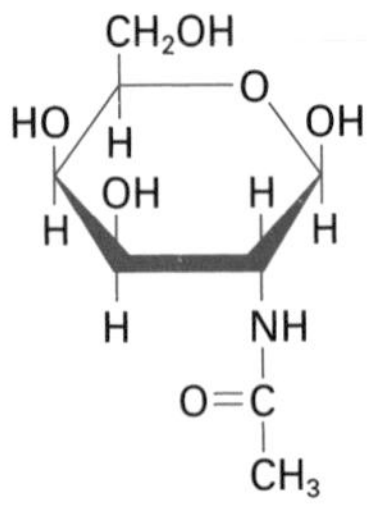

Fig. 1 *N*-acetylgalactosamine.

substituted for its normal substrate N_2 and is reduced to ethylene or ethane.

***N*-acetylgalactosamine (GalNAc)** *n.* acetyl derivative of the amino sugar galactosamine, with the acetyl group carried on the amino nitrogen. It is a common constituent of glycoproteins, certain heteropolysaccharides (e.g. chondroitin) and glycolipids. *see* Fig. 1.

***N*-acetylglucosamine (GlcNAc)** *n.* acetyl derivative of the amino sugar glucosamine, with the acetyl group carried on the amino nitrogen. It is a common constituent of glycoproteins, certain heteropolysaccharides (e.g. hyaluronic acid) and glycolipids and is the subunit of the polysaccharide chitin.

***N*-acetylmuramic acid (NAM)** monosaccharide present in the peptidoglycans of bacterial cell walls, comprising *N*-acetylglucosamine condensed with lactic acid.

***N*-acetylneuraminic acid (NANA)** nine-carbon sugar acid, a component of gangliosides and of the carbohydrate side chains of some glycoproteins. *alt.* sialic acid.

acetylsalicylic acid aspirin, a compound with analgesic and anti-inflammatory properties, the latter being due to its inhibition of prostaglandin synthesis.

acetyltransferase *n.* any of a group of enzymes catalysing the transfer of acetyl groups from acetyl-CoA (included in EC 2.3.1). *alt.* transacetylase.

ACh acetylcholine *q.v.*

achaetous *a.* without chaetae.

AChE acetylcholinesterase *q.v.*

acheilary *a.* having an undeveloped labellum, as some orchids.

achelate *a.* without claws or chelae.

achene *n.* one-seeded, dry, indehiscent fruit formed from one carpel, with the seed usually not fused to the fruit wall.

achiasmate, achiasmatic *a.* lacking chiasma in meiosis, as some Diptera (e.g. in spermatogenesis in male *Drosophila*).

Achilles tendon the tendon of the heel, the united strong tendon of the gastrocnemius and solaeus muscles. *alt.* tendo calcaneus.

achlamydate *a.* lacking a mantle, as some gastropods.

achlamydeous *a.* without calyx or corolla. *alt.* gymnanthous.

achlorophyllous *a.* lacking chlorophyll.

acholeplasma *n.* mycoplasma-like microorganism.

achondroplasia *n.* dominantly inherited form of dwarfism characterized by disturbance of ossification of the long bones of the limbs and of certain facial bones during development.

AChR (1) acetylcholine receptor *q.v.*; (2) **mAChR** muscarinic acetylcholine receptor; (3) **nAChR** nicotinic acetylcholine receptor.

achroglobin *n.* colourless respiratory pigment found in some molluscs and tunicates.

achroic *a.* colourless.

achromasie *n.* emission of chromatin from nucleus.

achromatic *a.* (1) colourless; (2) *appl.* threshold, the minimal stimulus inducing sensation of luminosity or brightness.

achromatopsia *n.* lack of colour vision but not black-and-white vision.

achromatous, achromic *a.* colourless, unpigmented.

A chromosomes the normal chromosomes of a diploid set, as opposed to the B or supernumerary chromosomes.

acicle *n.* small needle-like bristle, spine or crystal. *alt.* **acicula**. *plu.* **aciculae**.

aciculate *a.* having acicles.

aciculilignosa *n.* evergreen forest and bush made up of needle-leaved coniferous trees and shrubs.

aciculum *n.* thick central bristle (chaeta) in tuft of chaetae on parapodia of polychaete worms. *plu.* **acicula**.

acid *n.* substance that releases H^+ ions (protons) in solution and thus causes a rise in proton concentration in the solution. Acid solutions have a pH < 7. *cf.* base.

acidaemia *n.* abnormally high acidity (low pH) of the blood.

acid-alcohol fast remaining stained with aniline dyes after treatment with acid alcohol.

acid–base balance the correct ratio of acids to bases in blood for maintaining a suitable pH.

acid–base catalysis type of enzymatic catalysis in which a proton is transferred to or from an acid or base at some stage of the catalytic reaction.

acid deposition rain (acid rain) or other form of precipitation, or dry deposition, which contains acids and acid-forming compounds and has a pH < 5.6. It can cause acidification of lakes, with harmful effects on the aquatic flora and fauna, and damage to terrestrial vegetation. Acid deposition is caused mainly by atmospheric sulphur dioxide produced by the burning of coal and other fossil fuels, which is precipitated as sulphuric acid and sulphates, and by nitrogen oxides emitted from fossil fuel burning and vehicle exhausts, which form nitric acid and nitrogen dioxide.

acid-fast *a.* remaining stained with aniline dyes on treatment with acid.

acid gland (1) acid-secreting gland of ants, bees and wasps; (2) acid-secreting oxyntic cells of mammalian stomach.

acid hydrolase any of a range of hydrolytic enzymes active at acid pH (around pH 5), found esp. in lysosomes and including proteinases, phosphatases, nucleases, glycosidases, lipases, phospholipases and sulphatases.

acidic *a.* (1) having the properties of an acid; (2) *appl.* stains such as eosin that react with basic components of protoplasm such as cytoplasm and collagen.

Acidobacteria *n.* a phylum of the Bacteria distinguished on DNA sequence criteria, currently containing only three species.

acid peptidases proteolytic enzymes that require relatively acid pH to be active.

acidophile *n.* (1) plant that grows best on acid soils; (2) microorganism that grows best in acidic (pH < 5) conditions, and can be isolated on acidic media. *a.* **acidophilic, acidophilous.**

acidophobic *a.* unable to tolerate acid conditions.

acidosis *n.* abnormally high acidity of body tissues and fluids due to failure of normal regulation of acid–base balance.

acid phosphatase enzyme, found esp. in lysosomes, that catalyses the hydrolysis of an orthophosphate monoester to an alcohol and orthophosphate at acid pH. EC 3.1.3.2.

acid precipitation *see* acid deposition.

acid proteases acid peptidases *q.v.*

acid rain *see* acid deposition.

acid shock the biological disruption due to rapid acidification of aquatic ecosystems.

acid tide transient increase in acidity of body fluids that follows the decrease in acidity (alkaline tide) after eating.

aciduria *n.* condition in which pH of urine is lowered.

aciduric *a.* tolerating acid conditions.

aciform *a.* needle-shaped.

acinaciform *a.* shaped like a sabre or scimitar, *appl.* leaf.

acinar *a. pert.* acinus.

acinar cell pancreatic secretory cell, characterized by sac-like terminations.

acinarious *a.* having globose vesicles, as some algae.

aciniform *a.* grape- or berry-shaped.

acinus *n.* (1) cluster of cells forming the inner secretory region of a gland, usually a branched or compound gland. *alt.* alveolus. *plu.* **acini**; (2) (*bot.*) drupel *q.v.*

acleidian *a.* with clavicles vestigial or absent.

acoelomate *a. appl.* animals not having a true coelom (i.e. sponges, sea anemones and corals, nematodes, rotifers, platyhelminths and nemertean worms). *alt.* acoelous.

acoelous *a.* (1) acoelomate *q.v.*; (2) *appl.* vertebrae with flattened centra.

acondylous *a.* without nodes or joints.

acone *a. appl.* insect compound eye without crystalline or liquid secretion from cone cells.

aconitase *n. see* aconitic acid. EC 4.2.1.3, *r.n.* aconitate hydratase.

aconitic acid, aconitate *n.* six-carbon intermediate (as *cis*-aconitate) in the tricarboxylic acid cycle, formed from citrate and converted into isocitrate by the enzyme aconitase.

acontium *n.* thread-like process armed with stinging cells, borne on the mesenteries of some sea anemones. *plu.* **acontia**.

acorn worms Enteropneusta *q.v.*

acotyledonous *a.* lacking cotyledons.

acoustic *a. pert.* organs or sense of hearing, *appl.* e.g. nerve.

acoustico-lateralis system sensory system concerned with movement detection and avoidance of obstacles in fish and amphibians. It consists of vibration-sensitive hair cells located in the ear and/or on the external surface (the lateral line system in fishes), which detect vibrations in the surrounding liquid.

acoustic reflex adjustment of the muscles that regulate the positions of the ossicles in the mammalian ear. It is a protective mechanism against damage by too loud a noise.

ACP acyl carrier protein *q.v.*

acquired behaviour behaviour brought on by conditioning or learning.

acquired character modification or permanent structural change brought on during the lifetime of an individual by use or disuse of a particular organ, disease, trauma or other environmental influence, and which is not heritable.

acquired immune deficiency syndrome (AIDS) fatal T-cell deficiency disease caused by infection with the human immunodeficiency virus (HIV), a retrovirus, which results in severe depletion of the CD4 class of T lymphocytes and consequent immunodeficiency. The immunodeficient patient is susceptible to infection by opportunistic pathogens (e.g. *Candida*, *Pneumocystis*) and to the development of unusual cancers (Kaposi's sarcoma). HIV is transmitted sexually or by infected blood and blood products, and can be transmitted from mother to child at birth or by breast feeding.

acquired immunity immunity to infection produced by previous exposure to a pathogen or by immunization against it. *alt.* adaptive immunity. *cf.* innate immunity, passive immunity.

acral *a. pert.* extremities.

acrandrous *a.* having antheridia borne at the tips, *appl.* bryophytes. *n.* **acrandry**.

Acrania *n.* group of chordates, sometimes considered as a subphylum, that includes the urochordates and the cephalochordates and excludes the craniates.

Acraniata *n.* invertebrates *q.v.*

acranthous *a.* having the inflorescence borne at the tip of the main stem.

acrasids *n.plu.* group of eukaryotic microorganisms resembling cellular slime moulds, but thought not to be members of the Mycetozoa. They form multicellular sporangia from a small cluster of amoebae.

Acrididae *n.* family of grasshoppers (order Orthoptera) with antennae shorter than their bodies, and which includes the locust, *Locusta migratoria*. *alt.* short-horned grasshoppers.

acridines *n.plu.* group of dyes (e.g. acridine orange, acriflavin, ethidium bromide, proflavin) which intercalate into DNA, causing addition or deletion of a single base during DNA replication and producing frameshift mutations.

acroblast *n.* body in spermatid which gives rise to the acrosome.

acrocarpic, acrocarpous *a.* bearing fructifications at the tips of main stem or branches.

acrocentric (1) *a. appl.* chromosome with the centromere nearer one end than the other; (2) *n.* an acrocentric chromosome.

acrochordal *a. appl.* a chondrocranial unpaired frontal cartilage in birds.

acrocoracoid *n.* process at the dorsal end of the coracoid bone in birds.

acrodont *a. appl.* teeth attached to the summit of a parapet of bone, as in lizards.

acrodromous *a. appl.* leaf with veins converging at its point.

acrogenous *a.* increasing in growth at summit or tip.

acromion *n.* ventral prolongation of the spine of the scapula. *a.* **acromial**.

acron *n.* pre-oral region of insects.

acronematic *a. appl.* smooth whip-like flagella.

acropetal *a.* ascending, *appl.* leaves, flowers, roots or spores developing successively along an axis so that the youngest are at the apex. *cf.* basipetal.

acropodium *n.* digits of a limb collectively, the fingers or toes.

acrosarc *n.* pulpy berry arising from the union of ovary and calyx.

acrosomal process long actin-containing process that projects from the head of some invertebrate sperm on fertilization and which helps penetrate the egg.

acrosomal reaction release of hydrolytic enzymes from the acrosome of sperm when it contacts the egg.

acrosome *n.* organelle at apex of sperm, containing hydrolytic enzymes that digest coating of egg, enabling sperm to penetrate. *alt.* acrosomal vesicle.

acrospire *n.* first shoot, being spiral, at the end of germinating seed.

acrospore *n.* spore at the apex of a sporophore or hypha.

acrostichoid *a. appl.* fern sporangia produced all over surface of frond and not grouped in sori over a vein.

acroteric *a. pert.* outermost points, as tips of fingers, ears, nose.

acrotonic, acrotonous *a.* having anther united at its apex with rostellum.

acrotroch *n.* circlet of cilia at the extreme anterior end of the trochophore larva of some marine invertebrates.

acrotrophic *a. appl.* ovariole having nutritive cells at apex joined to oocytes by nutritive cords. *alt.* telotrophic.

act *n.* in psychology and behavioural science denotes a complex behaviour, as opposed to a simple movement. *alt.* action pattern.

ACTH adrenocorticotropic hormone *q.v.*

actin *n.* globular protein, G-actin, found in all eukaryotic cells but not present in prokaryotes. It can polymerize end-to-end into a fibrous form, F-actin, which forms the thin filaments of muscle and the microfilaments of the cytoskeleton. F-actin consists of two chains of actin monomers wound round each other in a helix and, with myosin, forms a contractile complex, actomyosin.

actinal *a.* (1) *appl.* area of echinoderm body bearing tube feet, *alt.* ambulacral; (2) star-shaped.

actin-binding proteins large and disparate group of proteins that bind to the cytoskeletal protein actin, having diverse effects on actin filaments (microfilaments) which include bundling of filaments into fibres and severing of filaments into actin subunits.

actin filament *see* actin, microfilament.

actinians sea anemones *q.v.*

actinic *a. pert.* radiation of wavelength between that of visible violet and X-rays.

actiniform *a.* star-shaped.

α-actinin *n.* actin-binding protein present in Z-lines of muscle fibrils, and at the site of non-muscle cell contact with substrate.

Actinobacteria *n., n.plu.* a phylum of the Bacteria distinguished on DNA sequence criteria containing mainly spore-forming bacteria that grow as slender branched filaments (hyphae). They are found in soil, river muds and lake bottoms and include many species (e.g. *Streptomyces*) that produce antibiotics. Formerly known as the Actinomycetes.

actinoblast *n.* mother cell from which a spicule develops, as in sponges.

actinocarpous, actinocarpic *a.* having flowers and fruit radially arranged.

actinodromous *a.* palmately veined.

actinoid *a.* (1) rayed; (2) star-shaped.

actinology *n.* (1) term formerly used for the study of the action of radiation; (2) homology of successive regions or parts radiating from a common central region.

actinomorphic, actinomorphous *a.* (1) radially symmetrical; (2) regular.

Actinomycetes, actinomycetes *n., n.plu.* large group of Gram-positive, spore forming, prokaryotic microorganisms belonging to the Bacteria, which grow as slender branched filaments (hyphae). They are found in soil, river muds and lake bottoms and include many species (e.g. *Streptomyces*) that produce antibiotics.

actinomycins *n.plu.* antibiotics produced by species of the actinomycete *Streptomyces*, which block transcription in both bacterial and eukaryotic cells by binding to DNA. Actinomycin D (dactinomycin) is used as an anticancer drug.

actinophage *n.* bacteriophage that infects actinomycetes.

Actinopoda, actinopods *n., n.plu.* phylum of non-photosynthetic protists (protozoa) characterized by long slender cytoplasmic projections (axopodia) that are stiffened by a bundle of microtubules. They include the marine radiolarians and the mainly freshwater heliozoans. The radiolarians are divided into three groups: acantharians, which have a radially symmetrical skeleton of rods of strontium sulphate, and polycystids and phaeodarians, which have silica spicules. *alt.* **Actinopodea**.

Actinopterygii, actinopterygians *n.*,
n.plu. the ray-finned fishes, a subclass of
bony fishes (Osteichthyes) which includes
all extant bony fishes except the lungfishes (Dipnoi) and the coelacanth.

actinorrhiza *n.* nitrogen-fixing nodule-like
structure formed on the roots of some
non-legumes (e.g. alders, *Alnus* spp.) by
nitrogen-fixing actinomycetes of the
genus *Frankia*.

actinost *n.* basal bone of fin-rays in teleost
fishes.

actinostele *n.* column of vascular tissue in
plant stem lacking pith, the xylem being
star-shaped in cross-section.

actinostome *n.* mouth of sea anemone or
starfish.

actinotrichia *n.plu.* unjointed horny rays at
edge of fins in many fishes.

actinotrocha *n.* free-swimming larval
form of Phoronida *q.v.*

Actinozoa Anthozoa *q.v.*

actinula *a.* larval stage in some jellyfish,
which develops into a medusa.

action pattern act *q.v.*

action potential self-propagating electrical potential difference produced across
the plasma membrane of nerve or muscle
cells when they are stimulated, reversing the membrane potential from about
−70 millivolts (mV) to about +30 mV, and
being an easily measured manifestation of
a nerve impulse. *cf.* resting potential.

action spectrum range of wavelengths of
light or other electromagnetic radiation
within which a certain reaction or process
takes place.

activated carrier molecule such as ATP,
acetyl-CoA or NADH, which can donate
high-energy chemical groups for participation in biochemical reactions. *alt.*
coenzyme, cofactor.

activated sludge process sewage treatment process in which a mixture of protozoa and bacteria (activated sludge) is
added to aerated sewage to break down the
organic matter. As the microorganisms
use the organic matter for food they multiply, producing more activated sludge.

activation energy free energy of activation
q.v.

activation-energy barrier *see* free
energy of activation *q.v.*

activation hormone hormone secreted by
the brain in insects and which stimulates the
prothoracic gland to produce the steroid hormone ecdysone, which triggers moulting.

**activation-induced cytidine deaminase
(AID)** enzyme that deaminates cytidine
residues in DNA and is involved in
somatic hypermutation and isotype switching in activated B lymphocytes.

activator *n.* (1) any substance that stimulates
a given process; (2) of genes, a protein that
acts as a positive regulator of transcription.
cf. repressor.

active centre active site *q.v.*

active immunity immunity acquired after
immunization with an antigen. *cf.* passive
immunity.

active site region of an enzyme molecule
at which substrates are bound and activation and chemical reaction of substrate
takes place. *alt.* active centre.

active space of a pheromone or other
chemical signal, the space within which
the chemical is above the threshold concentration for its detection by another
individual.

active transport movement of substances
across biological membranes into cells
or organelles other than by passive diffusion or passive transport, often occurring
against concentration or electrochemical
gradients. It involves carrier proteins and
requires energy.

activin *n.* protein growth factor involved in
mesoderm induction in early amphibian
development.

activity *n.* (1) of enzymes, the rate at which
they catalyse a chemical reaction; (2) in
ecology, the total flow of energy through a
system in unit time.

actomyosin *n.* complex of myosin and
actin, threads of which contract on addition of ATP, K^+ and Mg^{2+}. *see also* muscle.

actual evapotranspiration (AET) total
water loss over a given area with time by
the removal of water from soil and other
surfaces by evaporation and from plants by
transpiration.

actual vegetation in ecology, the vegetation existing at the time of observation.

aculeate *a.* having prickles, sharp points or
a sting.

aculeiform *a.* prickle-shaped.

aculeus *n.* (1) prickle growing from bark, as in roses; (2) sting or hair-like projection; (3) sting of an insect.

acuminulate *a.* having a very sharp tapering point.

acute *a.* (1) ending in a sharp point; (2) temporarily severe, not chronic; (3) *appl.* toxicity: causing severe poisoning, leading to severe illness, damage or death, within 24 to 96 hours of exposure to the toxic substance; (4) *appl.* medical conditions or diseases that progress rapidly to a crisis. *cf.* chronic.

acute-phase proteins proteins synthesized by the liver, such as C-reactive protein and mannose-binding lectin, which are present in blood of healthy people in small amounts but whose levels are greatly increased after infection and other traumas and which help to combat infection.

acute-phase response a part of the innate immune system that is activated soon after infection. It involves the increased production of proteins that circulate in the blood and help combat infection. *see also* acute-phase proteins.

acyclic *a.* flowers with parts arranged in a spiral.

acyclovir *n.* antiviral drug active that is used to treat cold sores and other herpes infections. *alt.* acycloguanosine, Zovirax.

acyl chemical group, –RCO⁻, where R is any alkyl or aryl group.

acylation *n.* addition of an acyl group to a molecule.

acyl carrier protein (ACP) small protein with a pantothenate prosthetic group that carries acyl groups in metabolic cycles concerned with fatty acid synthesis.

acyl-CoA synthetase enzyme catalysing formation of activated acyl-coenzyme A (CoA) compounds, involved e.g. in the activation of fatty acids before their oxidation in mitochondria. EC 6.2.1.3. *r.n.* long-chain-fatty-acid-CoA ligase. *alt.* fatty acid thiokinase.

acyl-coenzyme A (acyl-CoA) any acyl thioester of coenzyme A, e.g. fatty acid esters of CoA and succinyl-CoA.

acyltransferase *n.* any of a group of enzymes catalysing the transfer of acyl groups, including e.g. acetyltransferases, succinyltransferases. *alt.* transacylase.

ADA adenosine deaminase *q.v.*

adamantoblast enamel cell, ameloblast *q.v.*

adambulacral *a. appl.* structures adjacent to ambulacral areas in echinoderms.

adaptation *n.* (1) evolutionary process involving genetic change by which a population becomes fitted to its prevailing environment; (2) structure or habit fitted for some special environment or activity; (3) process by which a cell, organ or organism becomes habituated to a particular level of stimulus and ceases to respond to it, a more intense stimulus then being needed to produce a response; (4) in the eye, increasing sensitivity of retina to the available light.

adaptin *n.* protein that binds clathrin to proteins in the membrane of coated vesicles.

adaptive *a.* (1) capable of fitting different conditions; (2) adjustable; (3) inducible, *appl.* enzymes synthesized only when their specific substrates are available, *cf.* constitutive; (4) *appl.* control of metabolism, changes in rates of synthesis and degradation of enzymes in response to the organism's requirements; (5) *appl.* any trait that confers some advantage on an organism and thus is maintained in a population by natural selection. Traits can only be defined as adaptive with reference to the environment pertaining at the time, as a change in environment can render a previously adaptive trait non-adaptive, and vice versa.

adaptive immune response a selective response mounted by the immune system of vertebrates in which specific antibodies and/or cytotoxic cells are produced against an invading microorganism, parasite, transplanted tissue or other substance recognized as foreign by the body (antigens). This type of immune response often results in long-lived and specific immunity against the pathogen. *cf.* innate immunity.

adaptive immunity long-lasting and specific protection against re-infection that follows infection with a pathogen or immunization against it. It is the result of an adaptive immune response. *cf.* innate immunity, passive immunity.

adaptive landscape three-dimensional graphical representation which plots the mean fitness for all possible combinations of allele frequency for a set of given genes in a population. The surface produced resembles a topographic map and shows adaptive peaks of high fitness and valleys of low fitness. *alt.* adaptive surface.

adaptive peak(s) with respect to a particular combination of genes, the allele frequency (or frequencies) at which fitness is highest. Selection will tend to drive the population towards an adaptive peak. *see* adaptive landscape.

adaptive radiation evolution of a number of divergent species from a common ancestor, each species becoming adapted to occupy a different ecological niche.

adaptive surface adaptive landscape *q.v.*

adaptor protein intracellular protein that forms part of a signalling pathway leading from a cell-surface receptor. It has at least two different binding sites, one of which, in many cases, binds to phosphorylated tyrosine residues, usually on the receptor tail, while the other(s) bind to other proteins and recruit them to the pathway.

adaxial *a. appl.* the surface facing the main axis of a structure, e.g. the upper surface of a leaf is known as the adaxial surface because during leaf formation it faces towards the shoot tip. *cf.* abaxial.

adcauline *a.* towards or nearest the stem.

ADCC antibody-dependent cell-mediated cytotoxicity *q.v.*

ad-digital *n.* primary wing quill feather connected with phalanx of 3rd digit.

addition reactions chemical reactions that add an atom or chemical group to either end of a double bond in an organic molecule, converting it into a single bond.

additive alleles alleles that interact in such a way that the phenotype of the heterozygote is the average of the phenotypes of the two corresponding homozygotes.

additive mortality total mortality in a population due to all factors (e.g. predation, disease, accident) over a given period.

additive tree in phylogenetic analysis, a phylogenetic tree in which the lengths of the branches represent a measure of evolution, such that the evolutionary distance between any two taxa is given by the sum of the lengths of the branches connecting them.

additive variance in quantitative genetics, that part of the genotypic variance in a trait that can be attributed to the additive effects of alleles.

addressin *n.* sialyl-rich cell adhesion molecule expressed on blood vessel endothelium that binds selectins expressed on lymphocytes, enabling the cells to migrate out of the blood vessel. Different addressins enable lymphocytes to enter different types of tissue.

adduction *n.* movement towards the median axis. *cf.* abduction.

adductor *n.* muscle that brings one part of the body towards another.

adeciduate *a.* not falling or coming away.

adecticous *a.* without functional mandibles to escape from puparium or cocoon, *appl.* pupa of some insects.

adelomorphic, adelomorphous *a.* indefinite in form.

adelophycean *a. appl.* stage or generation of many seaweeds when they appear as prostrate thalli.

adelphogamy *n.* mating between two individuals derived vegetatively from the same parent.

adelphous *a.* joined together in bundles, *appl.* filaments of stamens that are joined together.

adendritic *a.* without dendrites or cellular processes, *appl.* cells.

adendroglia *n.* type of neuroglia lacking cellular processes.

adenine (A) *n.* purine base, one of the four nitrogenous bases in DNA and RNA, in which it pairs with thymine (T) and uracil (U) respectively (*see* Fig. 5, p. 69). It is also a component of the nucleoside adenosine, the nucleotides AMP, ADP and ATP, the nicotinamide cofactors NAD and NADP, and the flavin nucleotide cofactors FAD and FMN.

adenine deaminase enzyme that catalyses the deamination of adenine with the formation of hypoxanthine and ammonia. EC 3.5.4.2.

adeno-associated viruses (AAV) group of DNA viruses of the Parvoviridae, whose multiplication is dependent on the presence of adenovirus. They are being developed as DNA vectors for gene therapy.

adenoblast *n.* embryonic glandular cell.

adenocarcinoma *n.* malignant tumour of glandular epithelium.

adenocheiri *n.plu.* elaborate accessory copulatory organs, outgrowths of atrial wall in turbellarians. *alt.* adenodactyli.

adenocyte *n.* secretory cell of gland.

adenodactyli adenocheiri *q.v.*

adenohypophysis *n.* the glandular, non-neural portion of the pituitary body, from which many pituitary hormones are secreted. *alt.* anterior pituitary. *cf.* neurohypophysis.

adenoid *a. pert.* or resembling a gland. *n.* mucosal-associated lymphoid tissue in nasopharynx.

adenoma *n.* generally benign tumour of glandular epithelium.

adenophore *n.* stalk of nectar gland.

adenose *a.* glandular.

adenosine *n.* nucleoside made up of the purine base adenine linked to the sugar ribose.

adenosine deaminase (ADA) enzyme (EC 3.5.4.4) that catalyses the conversion of adenosine to inosine. A deficiency of the enzyme, owing to a genetic defect, leads to a build-up of toxic nucleosides, causing the death and non-development of lymphocytes, especially T cells, and resulting in severe combined immunodeficiency.

adenosine diphosphate (ADP) nucleotide made of adenosine linked to two phosphate groups in series through C atom 5 (5′) on the ribose ring, important in all living cells in energy transfer reactions, where it is converted to ATP (e.g. during oxidative phosphorylation and photosynthesis) or formed from the hydrolysis of ATP.

adenosine monophosphate (AMP) nucleotide made of adenosine linked to a phosphate group through C atom 5 (5′) on the ribose ring. It is one of the four types of nucleotide subunit in RNA. *alt.*

adenylic acid, adenylate, adenosine 5′-phosphoric acid. *see also* cyclic AMP, deoxyadenylate acid (dAMP).

adenosine triphosphatase (ATPase) enzyme activity that catalyses the hydrolysis of ATP to ADP and orthophosphate (P_i, phosphate(V)) with the release of free energy, which is used to drive mechanical work (as in muscle), enzyme reactions and ion transport across membranes. Many proteins possess ATPase activity, including motor proteins such as myosin and dynein associated with the cytoskeleton, and transport proteins such as the sodium and proton pumps in cellular membranes. An ATPase is also part of mitochondrial and chloroplast coupling factors, in which it catalyses the reverse reaction, $ADP + P_i = ATP$, and is often known as ATP synthase. *see* F-type ATPases, P-type ATPases, V-type ATPase.

adenosine triphosphate (ATP) an activated carrier molecule, a nucleotide made of adenosine linked to three phosphate groups in series through C atom 5 (5′) on the ribose ring. It is one of the main sources of energy and phosphate groups for metabolic reactions in all living cells when it is hydrolysed to ADP with release of free energy and phosphate, or to AMP with release of free energy and pyrophosphate (PP_i). It is chiefly regenerated from ADP during photosynthesis and oxidative phosphorylation. *see* Fig. 2.

***S*-adenosylmethionine (SAM)** *n.* activated carrier molecule that carries methyl groups for participation in biochemical reactions.

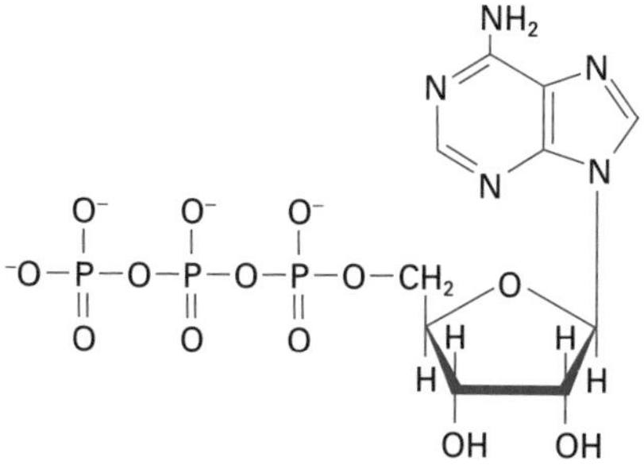

Fig. 2 Adenosine triphosphate.

Adenoviridae, adenoviruses *n.*, *n.plu.* family of double-stranded DNA viruses that cause respiratory infections in mammals. It includes adenovirus type 2, which causes a respiratory infection in various mammals including humans, but which is tumorigenic in newborn hamsters.

adenylate adenosine monophosphate *q.v.*

adenylate cyclase membrane-associated enzyme in animal cells that converts ATP into the second messenger cyclic AMP. One of the first steps in the actions of many hormones and other chemical messengers on cells is to stimulate or inhibit the activity of adenylate cyclase. *see* G-protein-coupled receptors. EC 4.6.1.1. *alt.* adenylylcyclase.

adenylate energy charge *see* energy charge.

adenylate kinase enzyme catalysing the interconversion of adenosine tri-, di- and monophosphates. EC 2.7.4.3. *alt.* (formerly) myokinase.

adenylylcyclase adenylate cyclase *q.v.*

adenylic acid adenosine monophosphate *q.v.*

adenylyltransferase *see* nucleotidyl-transferase.

adequate *a. appl.* stimulus appropriate to a given sensory receptor or sensory organ and producing the appropriate sensation.

adesmic *a. appl.* scales of some extinct fish that grow from margin outwards, and which are made up of separate small tooth-like units.

adesmy *n.* a break or division in an organ usually entire.

ADH (1) alcohol dehydrogenase *q.v.*; (2) antidiuretic hormone. *see* vasopressin.

adherens junction, adhaerens junction intermediate junction, belt desmosome. *see* desmosome.

adherent *a.* (1) touching but not growing together; (2) attached to substratum.

adhesin *n.* surface protein on bacteria by which they adhere to host cells.

adhesion *n.* (1) condition of touching but not growing together of parts normally separate; (2) the binding together of cells, esp. animal cells, by means of intercellular junctions and/or cell-surface molecules.

adhesion molecule, adhesion receptor any of a large number of different cell-surface glycoproteins that are involved in the binding of cells to one another and to the extracellular matrix. Some types of adhesion molecule bind to an identical molecule on the other cell whereas others bind to different receptor molecules. *see* addressin, cadherins, cell adhesion molecule, integrin, selectins.

adhesion plaque area of a cell contacting substrate and at which it makes connections with the extracellular matrix via membrane proteins.

ADI aerobic dependence index *q.v.*

adiabatic *a.* without losing or gaining heat.

adience *n.* (1) advance towards a stimulus; (2) approaching reaction. *a.* **adient.**

adipocellulose *n.* cellulose with a large amount of suberin, as in cork tissue.

adipocyte *n.* animal cell specialized for fat storage, containing large globules of fat (triacylglycerols) in the cytoplasm. *alt.* adipose cell, fat cell.

adipoleukocyte *n.* blood cell containing fat or wax droplets, in insects.

adipolysis lipolysis *q.v.*

adipose *a.* fatty, *pert.* animal fat.

adipose body fat body *q.v.*

adipose cell adipocyte *q.v.*

adipose fin modified rayless posterior dorsal fin in some fishes.

adipose tissue type of connective tissue in animals made up of cells (adipocytes) filled with fat droplets. Colloquially known as fat. *see* brown fat, white fat.

aditus *n.* anatomical structure forming approach or entrance to a part, e.g. to larynx.

adjacent segregation in cells with a reciprocal translocation, the segregation of a translocated and a normal chromosome to each of the poles at mitosis.

adjuvant *n.* substance that helps stimulate an immune response when injected together with an antigen.

adlacrimal *n.* the lacrimal bone of reptiles, not homologous with that of mammals.

admedial *a.* (1) near the middle; (2) near the median plane.

ADMET, ADME-Tox absorption, distribution, excretion, metabolism and toxicity tests carried out in the development of a new drug.

ADMR average daily metabolic rate *q.v.*

A-DNA *see* DNA.

adnate *a.* (1) fused to another organ of a different kind; (2) *appl.* leaves and stipules that are closely attached to petiole or stalk; (3) *appl.* anther that is attached throughout its length to the filament; (4) *appl.* gills of agaric fungi which are fused with the stalk for the whole of their width. *alt.* conjoined, connate.

adnexa *n.plu.* (1) structures or parts closely associated with an organ; (2) extraembryonic membranes, such as foetal membranes or placenta.

adnexed *a. appl.* gills of agaric fungi which are fused to stalk for only part of their width.

adolescence *n.* stage in human and animal development from the onset of puberty to full sexual maturity. *a.* **adolescent**.

adoptive immunity passive immunity *q.v.*

adoptive transfer transfer of cell-mediated immunity from an immunized to an unimmunized animal by the transfer of lymphocytes.

adoral *a.* near or *pert.* mouth.

ADP adenosine diphosphate *q.v.*

ADPR adenosine diphosphate ribosyl. *see* ADP-ribosylation.

adpressed appressed *q.v.*

ADP-ribosylation addition of a ribosyl group derived from the ADP moiety of NAD to a protein. The effects of diphtheria toxin and cholera toxin are due to ADP-ribosylation and consequent inactivation of protein synthesis elongation factor 2 and the GTPase activity of G_S protein, respectively.

adradius *n.* in coelenterates, the radius midway between perradius and interradius, a radius of the third order.

adrectal *a.* near to or closely connected with rectum.

adrenal *a.* (1) situated near kidneys; (2) *pert.* the adrenal glands *q.v.*

adrenal glands, adrenals paired bodies adjacent to kidneys in mammals, consisting of an inner medulla secreting adrenaline and noradrenaline, and an outer cortex secreting various steroid hormones (corticosteroids). In some vertebrates the two types of secretory tissue are segregated into separate glands.

adrenaline *n.* catecholamine secreted by adrenal medulla and by nerve endings in

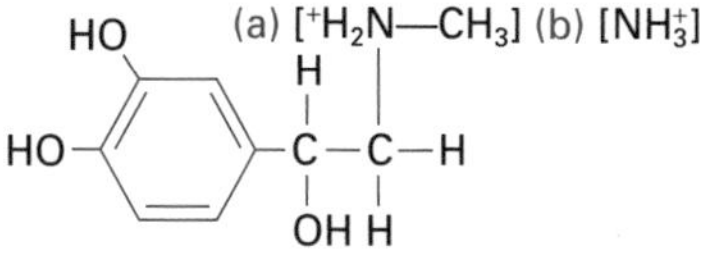

Fig. 3 Adrenaline (a) and noradrenaline (b).

the sympathetic nervous system. It acts via specific adrenergic receptors on a wide variety of tissues and has many effects, e.g. speeding up heartbeat, stimulating breakdown of glycogen to glucose in muscle and liver. It is also a neurotransmitter. *alt.* (US) epinephrine. *see* Fig. 3.

adrenergic *a. appl.* nerve endings, of the sympathetic nervous system, that secrete adrenaline or noradrenaline.

adrenergic receptor any of several types of receptor specific for adrenaline and noradrenaline, found on the surface of a variety of cells. β-adrenergic receptors and α-adrenergic receptors are the two main classes and act through different intracellular signalling pathways. *alt.* **adrenoceptor**.

adrenocortical *a.* produced by the adrenal cortex, *appl.* hormones.

adrenocorticotropic hormone, adrenocorticotropin (ACTH) polypeptide hormone synthesized by anterior pituitary and which acts on adrenal cortex, stimulating its growth and the synthesis and release of adrenocortical steroids. Also has effects on other tissues, stimulating lipid breakdown and release of fatty acids from fat cells.

adrenodoxin *n.* non-haem iron protein component of the cytochrome P_{450} system which is important in detoxification of foreign compounds.

adrenogenital syndrome type of pseudohermaphroditism in genetically female (XX) humans which is caused by a hereditary defect in the adrenal glands. This results in accumulation of the hormone progesterone and its breakdown to aldosterone, which has similar but weaker effects to testosterone, with consequent partial virilization.

adrostral *a.* near to or closely connected with beak.

adsere *n*. that stage in a plant succession that precedes its development into another at any time before the climax stage is reached.

adsorption *n*. adhesion of molecules to the surface of a solid or liquid.

adtidal *a. appl.* organisms living just below the low-tide mark.

adult *n*. the fully sexually mature form of an animal.

adult neurogenesis generation of new neurons in the brains of adults. It occurs in fish and other cold-blooded vertebrates, in birds, and to a much more limited extent in mammals.

adult stem cell any stem cell that is the source of new cells in the tissues of young or adult animals. Examples are the haematopoietic stem cells in bone marrow, and the epithelial stem cells that renew the lining of the gut. *cf.* embryonic stem cells.

adultoid *a. appl.* nymph having imaginal characters differentiated further than in normal nymph.

adult-onset diabetes type 2 diabetes *q.v.*

aduncate *a.* crooked, bent in the form of a hook.

adustous *a.* browned, as if scorched.

advanced *a.* of more recent evolutionary origin. *cf.* primitive.

advehent *a.* carrying to an organ, afferent.

adventitia *n.* external connective tissue layer of blood vessels.

adventitious *a.* (1) accidental; (2) found in an unusual place; (3) *appl.* tissues and organs arising in abnormal positions; (4) secondary, *appl.* dentine.

adventive *a.* (1) not native; (2) *appl.* organism in a new habitat but not completely established there.

AEC adenylate energy charge. *see* energy charge.

aecidiospore, aeciospore *n.* binucleate spore of rust fungi, produced in an aecidium.

aecidium, aecium *n.* cup-shaped structure containing chains of spores (aecidiospores) in rust fungi. *plu.* **aecidia, aecia**.

aedeagus *n.* copulatory organ of male insects. *alt.* intromittent organ.

aegithognathous *a. appl.* type of palate characteristic of passerine birds (e.g.

sparrow), with maxillopalatines separate, vomers forming a wedge in front and diverging behind.

aeolian *a.* wind-borne, *appl.* deposits.

Aepyornithiformes *n.plu.* order of very large Pleistocene birds of the subclass Neornithes from Madagascar, known as elephant birds.

aequorin *n.* calcium-binding protein from the jellyfish *Aequorea*, which emits a flash of light when it binds Ca^{2+}.

aerenchyma *n.* (1) parenchyma with large intercellular spaces; (2) air-storing tissue in cortex of some aquatic plants.

aerial *a.* (1) *appl.* roots growing above ground; (2) *appl.* small bulbs appearing in leaf axils.

aerial plankton spores, bacteria and other microorganisms floating in the air.

aeroallergen *n.* any substance present in the air, such as pollen, which causes an allergic immune reaction when inhaled by those sensitized to it.

aeroaquatic *a. appl.* fungi living in water and liberating spores into the air.

aerobe *n.* any organism capable of living in the presence of oxygen, obligate aerobes being unable to live without oxygen and facultative aerobes being able to live in either aerobic or anaerobic conditions. All animals and higher plants are obligate aerobes. *cf.* anaerobe. *a.* **aerobic**.

aerobic *a. pert.*, containing, involving or requiring oxygen: *appl.* reactions, organisms.

aerobic dependence index (ADI) a measure of the relative use of aerobic and anaerobic metabolism in muscle during exercise in animals. It is calculated using the lactate content of resting muscles after exercise as a measure of anaerobic metabolism, and maximum oxygen consumption as a measure of aerobic metabolism. Higher ADIs tend to indicate more active animals.

aerobic respiration respiration requiring oxygen. *see also* oxidative phosphorylation.

aerobiology *n.* study of airborne organisms and their distribution.

aerobiosis *n.* existence in the presence of oxygen.

aerobiotic *a.* living mainly in the air.

aerocyst *n.* air vesicle of algae.

Aeroendospora *n.* the aerobic endospore-forming Gram-positive bacteria (e.g. *Bacillus* spp.).

aerogenic *a.* gas-producing, *appl.* certain bacteria.

aerolae *n.plu.* large depressed box-like structures in the walls of diatoms.

aeromycology *n.* study of airborne fungi.

aerophore *n.* aerating outgrowth in ferns.

aerophyte *n.* epiphyte attached to the aerial portion of another plant.

aeroplankton aerial plankton *q.v.*

aerosol *n.* suspension of particles in airborne water droplets.

aerostat *n.* air sac in insect body or bird bone.

aerostatic *a.* containing air spaces.

aerotaxis *n.* movement towards or away from oxygen. *a.* **aerotactic.**

aerotolerant *a. appl.* an anaerobic organism that is not inhibited by molecular oxygen, although it cannot use it.

aerotropism *n.* reaction to gases, generally to oxygen, particularly the growth curvature of roots and other parts of plants in response to changes in oxygen tension. *a.* **aerotropic.**

aesthesis *n.* (1) sensibility; (2) sense perception.

aesthetasc *n.* type of chemoreceptor found in the antennule of crustaceans.

aestidurilignosa *n.* mixed evergreen and deciduous broadleaf forest.

aestilignosa *n.* broadleaf deciduous forest which loses its leaves in winter.

aestival *a.* (1) produced in, or *pert.*, summer; (2) *pert.* early summer.

aestivation *n.* (1) (*bot.*) the mode in which different parts of a flower are arranged in the bud; (2) (*zool.*) torpor during heat and drought during summer in some animals.

AET actual evapotranspiration *q.v.*

aethalium *n.* reproductive stage of some acellular slime moulds in which the plasmodium forms a large cushion-shaped sporangium. *plu.* **aethalia.**

aetiological *a.* causal.

aetiology *n.* cause, esp. of a disease.

AFDW ash-free dry weight.

affectional *a. appl.* behaviour concerned with social relationships, important in development and maintenance of social cohesion and organization.

afferent *a.* bringing towards, *appl.* sensory nerves bringing impulses towards the central nervous system, *appl.* vessels carrying blood or lymph to an organ or set of organs. *cf.* efferent.

affinity *n.* the strength of binding of a molecule for its ligand. It is defined for antibodies and other molecules with more than one binding site as the strength of binding of the ligand at one specified binding site. *cf.* avidity.

affinity chromatography technique for purifying proteins or other macromolecules by their affinity for particular chemical groups, by passage through a column containing those groups, to which the molecule binds. *see also* immunoaffinity column chromatography.

affinity constant *see* association constant, equilibrium constant.

affinity maturation in an immune response, the increase in binding strength (affinity) of the antibodies for the immunizing antigen as the response progresses. It is the result of selection of antibody-producing cells producing antibodies of higher affinity.

afforestation *n.* the production of forest over an area, either by planting or by allowing natural regeneration or colonization.

aFGF acidic fibroblast growth factor.

aflagellate *a.* lacking flagella.

aflatoxin *n.* carcinogenic mycotoxin formed by some species of the mould *Aspergillus*.

AFLP amplified fragment length polymorphism *q.v.*

AFP α-foetoprotein *q.v.*

African subkingdom subdivision of the Palaeotropical floristic kingdom, comprising Madagascar, Ascension Island and St Helena and all of the African continent except the southernmost tip. It is divided into the following regions: Ascension and St Helenan, East African Steppe, Madagascan, North African Desert, Northeast African Highland and Steppe, South African, Sudanese Park Steppe, West African Rainforest.

afterbirth *n.* placenta and foetal membranes expelled after offspring's birth.

afterpotential *n.* slow change in membrane potential in nerve cells following trains of impulses.

after-ripening period after dispersal in which a seed cannot germinate, even if conditions are favourable.

after-sensation persistent sensation due to continued activity in sense receptor after cessation of external stimulation.

Ag antigen *q.v.*

agameon *n.* species comprising only apomictic individuals.

agamete *n.* young form or gamete which develops directly into an adult without fusing with another gamete.

agametoblast *n.* one of the cells into which an agamont or schizont divides by multiple fission, and which gives rise to merozoites.

agamic *a.* (1) asexual *q.v.*; (2) parthenogenetic *q.v.*

agamic complexes group of apomictic plants that are usually allopolyploids and consist of many different biotypes, forming a taxonomically difficult group.

agammaglobulinaemia *n.* deficiency of immunoglobulin in the blood.

agamodeme *n.* small assemblage of closely related individuals consisting predominantly of apomictic plants or asexual organisms.

agamogenesis *n.* (1) any reproduction without participation of a male gamete; (2) asexual reproduction. *a.* **agamogenetic**.

agamogony *n.* asexual reproduction.

agamohermaphrodite *a.* with neuter and hermaphrodite flowers on the same plant, usually in the same inflorescence.

agamont *n.* stage in protozoan life cycle when it divides by multiple asexual fission, giving rise to agametes.

agamospecies *n.* species that only reproduces asexually.

agamospermy *n.* any form of apomixis in which embryos and seeds are produced asexually.

agamotropic *a. appl.* flowers which having opened do not close again.

agamous *a.* lacking sex organs.

agar *n.* gelatinous substrate used for bacterial cultures and a constituent of some gels used for electrophoresis. It is prepared from agar, a galactose/galacturonic acid polysaccharide extracted from red algae.

agarase *n.* enzyme that degrades agar into its constituent monosaccharides.

Agaricales, agarics *n., n.plu.* large group of basidiomycete fungi typified by mushrooms and 'toadstools' and the boletes. Agarics have conspicuous fleshy fruiting bodies, usually composed of a stalk and cap, with the spores borne on the lining of gills or pores on the underside of the cap.

agarose *n.* polysaccharide prepared from agar and which is used to make gels for electrophoresis.

agarose gel *see* agarose, electrophoresis.

age distribution *see* age structure.

ageotropism *n.* (1) not responding to gravity; (2) negative geotropism.

age polyethism in social animals, the changing of roles by members of a society as they age.

age ratio of a population, the relative frequencies of different age groups within it.

age-specific death rate mortality rate for different age groups within a population.

age-specific reproductive value an index of the extent to which the members of a given age group contribute to the next generation between now and the time they die.

age structure of a population, the percentage of the population in each age group, or the number of individuals of each sex in each age group. *alt.* age distribution.

agg. abbreviation of aggregate (3).

agglomerate *a.* clustered, as a head of flowers.

agglutinate *v.* to cause agglutination or to undergo agglutination. *n.* the mass formed by agglutination. *a.* stuck together.

agglutination *n.* formation of clumps or floccules of cells, esp. pollen, bacteria, red blood cells, spermatozoa and some protozoans, either spontaneously or after treatment with an antibody or other agent.

agglutinin *n.* any substance causing agglutination of cells (e.g. antibodies, lectins, viral haemagglutinins).

agglutinogen *n.* cell-surface antigen, such as the ABO antigens of human red cells, that interacts with antibodies (agglutinins) resulting in agglutination of the cells.

aggregate *a.* (1) formed in a cluster; (2) *appl.* a fruit formed from a flower with several separate carpels, such as raspberry; (3) *appl.* a group of species or hybrids that are difficult to distinguish from each other morphologically.

aggregation *n.* (1) group of individuals of the same species gathered in the same place but not socially organized or engaged in cooperative behaviour; (2) the assembling of amoebae of cellular slime moulds to form a fruiting body.

aggregation chimaera animal (usually a mouse) that has developed from an embryo formed by the mixing *in vitro* of cells from two genetically different embryos.

aggresome intracellular structure into which misfolded and aggregated proteins are collected for destruction.

aggression *n.* an act or threat of action by one individual that limits the freedom of action of another individual, often shown by animals in ritualized trials of strength to gain mates, territory, etc.

aglomerular *a. appl.* kidneys that have no glomeruli, as in certain fishes.

aglossate *a.* having no tongue.

aglutone, aglycone *n.* non-carbohydrate portion of a glycoside, produced from the glycoside by hydrolysis.

Agnatha, agnathans *n., n.plu.* taxon of primitive jawless vertebrates, including the lampreys (Monorhina), hagfishes (Diplorhina) and their extinct relatives. *see* Appendix 3.

agnathous, agnathostomatous *a.* having no jaws.

agnosia *n.* inability to recognize objects, in spite of being able to describe them in terms of form and colour.

agonist *n.* (1) hormone, neurotransmitter or drug responsible for triggering a response in a cell by stimulating a receptor, *cf.* antagonist; (2) any substance that mimics the function of a natural ligand such as a hormone or neurotransmitter by binding to its receptor and triggering a normal response; (3) muscle directly responsible for a change in position of a part of body.

agonistic *a.* (1) in animal behaviour, *pert.* any activity related to fighting, whether aggressive or conciliatory; (2) *pert.* an agonist.

agranular *a.* (1) lacking granules; (2) without a conspicuous layer of granular cells, *appl.* motor areas in brain.

agraphia *n.* inability to write.

agrestal *a. appl.* uncultivated plants growing on arable land.

Agricultural Revolution the gradual shift from a hunter-gatherer way of life to settled agriculture, in which animals and wild plants were domesticated, which began between 10,000 and 12,000 years ago.

Agrobacterium tumefaciens plant pathogenic bacterium, the cause of crown gall in numerous dicotyledonous plants. It carries a plasmid, the Ti plasmid, which becomes integrated into the chromosomes of infected tissue. *Agrobacterium* and its plasmid have been extensively modified by genetic engineering and are widely used to introduce novel genes into plant cells to produce transgenic plants.

agroecosystem *n.* ecosystem that develops on farmed land, and which includes the indigenous microorganisms, plants and animals, and the crop species.

agroinfection, agroinoculation *n.* infection or inoculation of plants with genetically engineered *Agrobacterium* spp. as a means of introducing novel genes.

agrostology *n.* that part of botany dealing with grasses.

A-horizon upper, or leached, soil layers. *alt.* eluvial layer.

AI artificial intelligence *q.v.*

AI, AID artificial insemination by donor.

AID activation-induced cytidine deaminase *q.v.*

AIDS acquired immune deficiency syndrome *q.v.*

air bladder (1) swim bladder in fishes; (2) hollow dilation of thallus in bladderwrack.

air cells (1) air-filled cavities in bone; (2) air spaces in plant tissue.

air chamber (1) gas-filled compartment of *Nautilus* shell, previously occupied by the animal; (2) accessory respiratory organ or respiratory sac in some air-breathing teleost fishes.

air pollution any gaseous or particulate matter in the air that is not a normal constituent of air or not normally present in such large amounts. It may be the result of human activity, such as sulphur dioxide from burning of coal, and carbon monoxide and nitrogen oxides from exhaust emissions, or can result from natural causes, such as desert dust, methane and hydrogen sulphide from microbial activity in bogs and volcanic debris in the atmosphere.

air quality level of pollution in the air, which may be judged by a variety of criteria such as chemical and physical analysis, medical symptoms, damage to plants and damage to buildings. Air quality is deemed to be high when pollution is low.

air sacs (1) spaces filled with air and connected with lungs in birds; (2) dilatations of tracheae in many insects; (3) sacs representing tracheal system in some insect larvae and having hydrostatic function.

air sinuses cavities in various facial bones with passages to the nasal cavities.

akaryote *n.* cell lacking a nucleus. *a.* **akaryotic.**

akinesia *n.* (1) absence or arrest of motion; (2) inability to initiate movement, usually as the result of brain damage. *a.* **akinetic.** *alt.* **akinesis.**

akinete *n.* non-motile spore.

ala *n.* (1) wing-like projection or structure, as on fruit, leaf-stalk or bone; (2) lateral petal of flowers of the pea family. *plu.* **alae.**

Ala alanine *q.v.*

alanine (Ala, A) α-amino propionic acid, simple non-polar amino acid with a methyl side chain, a constituent of protein. Non-essential in the human diet as it can be synthesized in the body by the reaction of pyruvate and glutamate to give alanine and α-ketoglutarate.

alar *a.* (1) wing-like; (2) *pert.* wings or alae; (3) axillary *q.v.*

alarm behaviour types of behaviour shown by animals when disturbed and which are intended to distract predators or to hide the animal from view.

alarm pheromone chemical substance released in minute amounts by an animal which induces a fright response in other members of the species.

ala spuria bastard wing *q.v.*

alate *a.* (1) winged; (2) having a wing-like extension, as of leaf-stalk or stem; (3) broad-lipped, *appl.* shells.

albedo *n.* (1) ratio of the amount of light reflected by a surface to the amount of incident light, surfaces with the higher albedo having the greater reflectivity; (2) (*bot.*) mesocarp, the white tissue of rind of orange, lemon, etc.

albescent *a.* growing whitish.

albinism *n.* (1) genetically determined or environmentally induced absence of pigmentation in animals that are normally pigmented, leading to lack of pigmentation in hair, skin and eyes. In humans, albinism is generally an autosomal recessive trait in which there is a deficiency of the enzyme tyrosinase; (2) (*bot.*) absence of green or other colour in plants due to genetically determined lack or non-development of chloroplasts or other chromoplasts. Organisms showing albinism are known as albinos.

albino *n. see* albinism.

albomaculus *a. appl.* variegation in plants consisting of irregular green and white patches. It is the result of mitotic segregation of chloroplasts or of the genes directing their development and function into certain cells, with their absence from others, during development.

albuginea *n.* dense white connective tissue surrounding testis, ovary, spleen or eye.

albumen *n.* (1) white of egg, containing proteins such as ovalbumin, ovomucoid and conalbumin; (2) endosperm *q.v.*

albuminous cell *see* phloem.

albumins *n.plu.* general name for a class of small globular proteins found in blood (serum albumin), synovial fluid, milk and other mammalian secretions, and as storage proteins in plant seeds. They were characterized originally as proteins that are soluble in water and dilute buffers at neutral pH.

alburnum *n.* sapwood or splintwood, the young wood of dicotyledons.

alcaptonuria, alkaptonuria *n.* inherited metabolic deficiency in which the enzyme homogentisate oxidase is absent, leading to accumulation of homogentisate in urine, which turns dark on exposure to air.

alcohol *n.* any organic compound that has a structure equivalent to that of a hydrocarbon with one or more of its hydrogen atoms each replaced by a hydroxyl (−OH) group. In non-technical usage often refers to ethanol (ethyl alcohol).

alcohol dehydrogenase (ADH) any one of several enzymes catalysing the conversion of acetaldehydes or ketones to alcohols (e.g. in alcoholic fermentation in yeast) and the reverse reaction. EC 1.1.1.1–2.

alcoholic fermentation fermentation of hexose sugar to ethyl alcohol and CO_2.

Alcyonaria Octocorallia *q.v.*

aldehyde group –CHO.

alder flies *see* Neuroptera.

aldolase *n.* (1) enzyme that catalyses the conversion of fructose 1,6-bisphosphate into dihydroxyacetone phosphate and glyceraldehyde 3-phosphate in glycolysis, and the reverse reaction in photosynthesis, EC 4.1.2.13, *r.n.* fructose-bisphosphate aldolase; (2) name used generally for any enzyme that catalyses aldol condensations involving aldose and ketose monosaccharides. *alt.* aldehyde-lyases. EC 4.1.2.

aldose *n.* any monosaccharide containing an aldehyde (–CHO) group. *cf.* ketose.

aldosterone *n.* steroid hormone produced by adrenal cortex, involved in regulation of mineral and water balance in the body by its action on kidneys.

alecithal *a.* with little or no yolk.

alepidote *a.* without scales.

aleuriospore, aleurospore *n.* thick-walled terminal fungal spore.

aleurone grains storage granules in endosperm and cotyledon of seeds, consisting of proteins, phytin and hydrolytic enzymes. They form the aleurone layer in cereal seeds.

aleurone layer in cereal seeds, a protein-rich layer of cells found immediately under the testa, containing aleurone grains.

aleuroplast *n.* protein storage body in plant cells.

alexia *n.* inability to read.

alfalfa *n.* lucerne, *Medicago sativa*, leguminous plant widely grown for forage and green manure.

algae *n.plu.* general name for a heterogeneous group of unicellular, colonial or multicellular eukaryotic photosynthetic organisms of simple structure. The different groups of algae are classified as divisions of the kingdom Protoctista or Protista. They are aquatic or live in damp habitats on land, and include unicellular organisms such as *Chlamydomonas* and diatoms, colonial forms such as *Volvox*, the multicellular green, red and brown seaweeds, and freshwater multicellular algae

such as *Spirogyra*. The algal body is known as a thallus and in multicellular forms is generally filamentous or flattened into a thin sheet or ribbon. *sing.* **alga**. *see* Bacillariophyta, Charophyta, Chlorophyta, Chrysophyta, Cryptophyta, Dinoflagellata, Euglenophyta, Eustigmatophyta, Gamophyta, Glaucophyta, Haptophyta, Phaeophyta, Pyrrophyta, Rhodophyta, Xanthophyta, zoochlorellae, zooxanthellae. For bluegreen algae, *see* cyanobacteria.

algal bloom exceptional growth of algae or cyanobacteria in lakes, rivers or oceans, which may occur in particular climatic conditions or as a result of excess nutrients in the water. In some cases, the microorganisms produce toxic compounds.

algesis *n.* the sense of pain.

algicide *n.* any substance that kills algae.

algicolous *a.* living on algae.

alginate *n.* gel-like polysaccharide derived from the cell walls of brown algae, which is used as a food stabilizing and texturing agent and in dental moulding materials.

algoid *a. pert.*, resembling or of the nature of an alga.

Algonkian *a. pert.* the late Proterozoic era.

alien *n.* plant species thought to have been introduced by humans but now more or less naturalized.

aliform *a.* wing-shaped.

alignment *n.* the process of matching up two nucleotide or amino acid sequences (a pairwise alignment) or more than two sequences (a multiple alignment) to assess their degree of similarity and thus to assess whether they are homologous to each other. In various forms alignment is the basis for the search algorithms used to find homologous sequences in database searches. Gaps may be introduced into the aligned sequences to optimize the number of matching positions.

alima *n.* a larval stage of certain crustaceans.

alimentary *a. pert.* nutritive functions.

alimentary canal the tube running from mouth to anus in animals into which food is ingested and in which it is digested, in vertebrates comprising the oesophagus, stomach and intestines. *alt.* digestive tract, gastrointestinal tract.

alimentation *n.* the process of nourishing or being nourished.

aliphatic *a. appl.* organic compounds in which the carbon atoms are joined in straight or branched chains. *cf.* aromatic.

aliquot *n.* one of a number of equal parts of a sample.

Alismales *n.* order of herbaceous monocots, aquatic or partly aquatic, and including the families Alismaceae (water plantain), Butomaceae (flowering rush) and Limnocharitaceae. *alt.* **Alismatales**.

alisphenoid *n.* wing-like portion of the sphenoid bone forming part of the cranium.

alkaline *a. pert.* substances that release hydroxyl ions (OH⁻) in solution. Alkaline solutions have a pH greater than 7. *see also* base.

alkaline gland Dufour's gland *q.v.*

alkaline soils temperate-region soils rich in calcium compounds, with a pH > 7.5 and up to 8 or 9, which develop over chalk or limestone. *alt.* calcareous soils. *cf.* alkali soils.

alkaline tide transient decrease in acidity of body fluids after taking food.

alkaliphile *n.* organism that grows best at high pH (e.g. pH > 9).

alkaliphilic *a. appl.* organism that grows best at high pH, or enzyme with activity optimum at high pH.

alkali soils soils with a very high surface content of mineral salts such as sodium chloride, sodium sulphate, sodium carbonate and borax. They are formed in dry regions where evaporation is much greater than rainfall. *see* solonchak, solonetz.

alkaloid *n.* any of a group of nitrogen-containing organic bases found in plants and which are toxic or physiologically active in vertebrates, such as caffeine, morphine, nicotine, strychnine.

alkalosis *n.* rise in the pH of blood associated with some diseases.

alkanes *n.plu.* saturated hydrocarbons thought to be chemical fossils indicating the existence of life, which have been found in Precambrian rocks.

alkaptonuria alcaptonuria *q.v.*

alkyl a chain of CH_2 groups of any length, terminating in a methyl (CH_3) group.

alkylating agents highly reactive organic chemicals (e.g. mustard gas) that attach alkyl groups to bases in DNA, and to other molecules. They are potent mutagens as well as causing damage to tissues.

all-α, all-β names given to classes of protein folds composed entirely of α-helices or β-strands, respectively.

allaesthetic *a. appl.* characters effective when perceived by another organism.

allantoate (allantoic acid) hydration product of allantoin, nitrogenous excretion product of teleost fish and also found in other animals and plants.

allantochorion *n.* foetal membrane formed of outer wall of allantois and the primitive chorion.

allantoicase *n.* enzyme catalysing the hydrolysis of allantoate to ureidoglycolate and urea, which is found in amphibians, certain fishes and invertebrates. EC 3.5.3.4.

allantoin *n.* an end-product of purine and pyrimidine metabolism. It occurs in allantoic fluid and urine of mammals other than primates, in some gastropods and insects, and in plants.

allantoinase *n.* enzyme catalysing the conversion of allantoin to allantoate. EC 3.5.2.5.

allantois *n.* sac-like outgrowth from posterior part of alimentary canal in embryos of reptiles, amphibians, birds and mammals, which acts as an organ of respiration and/or nutrition and/or excretion. *a.* **allantoic**.

allatectomy *n.* excision or removal of corpora allata.

Allee effect the decrease in reproduction and survival of individuals in smaller populations, e.g. when population gets too thinly spread and individuals cannot find a mate.

Alleghany subregion subdivision of the Nearctic zoogeographical region, comprising eastern North America south of the Great Lakes to Florida and east of the central plains.

allele *n.* an alternative form of a gene. For example, a hypothetical gene, C, could exist in three variant forms within a population – the alleles C, c and c^1. Each allele represents a DNA sequence with slight differences from the others. A diploid organism carries two alleles for each gene locus, one on each homologous chromosome. The two alleles may be identical (e.g. genotype CC), or different (e.g. genotype Cc), and it is the particular combination of alleles that determines the phenotype of the organism. *see also* dominant (1), recessive allele.

allele frequency a measure of the commonness of an allele in a population, being the proportion of a given allele in the population with respect to all alleles of that gene.

allelic *a.* (1) *pert.*, or the state of being, alleles; (2) *appl.* two or more mutations mapping to the same area and which do not complement each other in the heterozygous state, showing that they are affecting the same genetic locus.

allelic complementation *see* complementation.

allelic exclusion the situation in any particular B lymphocyte or T lymphocyte that antibody or T-cell receptor synthesis, respectively, is specified by the genes on only one of the chromosomes in the relevant pair of homologous chromosomes.

allelomimetic *a. appl.* animal behaviour involving imitation of another animal, usually of the same species.

allelomorph *n.* (1) the characteristic specified by an allele; (2) formerly also used for allele *q.v.*

allelopathic *a. pert.* the influence or effects (sometimes inhibitory or harmful) of a living plant on other nearby plants or microorganisms. *n.* **allelopathy**.

allelotype *n.* the frequency of different alleles in a population.

Allen's rule rule that in a widely distributed species of endothermic animal (an animal that generates its own body heat), the extremities (e.g. ears, feet, tail) tend to be smaller in the colder regions of the species' range than in the warmer regions.

allergen *n.* substance to which an individual is hypersensitive and which causes an inappropriate immune response, or allergy.

allergic *a.* (1) *appl.* immune reaction due to hypersensitivity to an antigen; (2) *appl.* person who produces a hypersensitive immune response to an otherwise innocuous antigen.

allergic asthma constriction of the bronchial tubes due to an allergic reaction to an inhaled antigen.

allergic conjunctivitis, allergic rhinitis *n.* allergic conditions which affect the mucous membranes of the eyelids and nose,

respectively, caused by a reaction to pollen and other airborne antigens. *alt.* hayfever.

allergy *n.* hypersensitive immune reaction exhibited by certain individuals on exposure to an otherwise innocuous antigen (e.g. pollen, certain drugs, dust-mite faeces, certain foods). The response is generally a local inflammatory reaction but severe systemic shock (anaphylactic shock) may occur in some individuals. *see also* anaphylaxis, delayed hypersensitivity.

alliaceous *a. pert.* or resembling garlic or onion.

alloantibody *n.* antibody produced in response to immunization with an antigen from another member of the same species.

alloantigen *n.* an antigen, e.g. an MHC molecule, which is present in different forms in different individuals of the same species. An alloantigen from one individual provokes an immune response in a genetically dissimilar individual of the same species.

alloantiserum *n.* antiserum raised in one animal against the antigens of a genetically dissimilar animal of the same species.

allobiosis *n.* the changed reactivity of an organism in a changed internal or external environment.

allocarpy *n.* production of fruit after cross-fertilization.

allocheiral *a.* (1) having right and left sides reversed; (2) *pert.* reversed symmetry.

allochroic *a.* (1) able to change colour; (2) showing colour variation.

allochronic *a.* (1) not contemporary, *appl.* species in evolutionary time; (2) *appl.* species or populations that have non-overlapping breeding seasons or flowering periods.

allochthonous *a. appl.* material or species that has originated elsewhere.

allocortex *n.* the most primitive cortical areas. *cf.* neocortex.

allodynia *n.* the stimulation of pain by normally innocuous stimuli.

alloenzyme allozyme *q.v.*

allogamy *n.* cross-fertilization *q.v.* *a.* **allogamous**.

allogeneic *a.* genetically different, when *appl.* individuals of the same species.

allogenic *a. appl.* plant successions caused by external factors such as fire or grazing.

allogenous *a.* persisting from an earlier environment, *appl.* floras.

allograft *n.* tissue or organ transplanted from one individual to another genetically dissimilar individual of the same species.

allogrooming *n.* grooming directed at another individual of the same species. *alt.* social grooming.

alloheteroploid *n.* heteroploid derived from genomes of different species.

allokinesis *n.* reflex or passive movement.

allokinetic *a.* moving passively or drifting, as plankton.

allometric *a.* (1) differing in growth rate; (2) *pert.* allometry.

allometry *n.* (1) study of relative growth; (2) change of proportions with increase in size; (3) growth rate of a part differing from standard growth rate or from growth rate as a whole.

allomixis *n.* cross-fertilization *q.v.* *a.* **allomictic.**

allomone *n.* chemical secreted by one individual which causes an individual of another species to react to it, such as scent given out by flowers that attracts pollinating insects.

allomorphosis *n.* evolution with rapid increase in specialization.

alloparental care assistance in the care of the young by individuals other than the parents (alloparents).

allopatric *a.* having separate and mutually exclusive areas of geographical distribution, *appl.* populations, species.

allophenic *a.* (1) chimaeric (*appl.* animals); (2) *appl.* a phenotype not due to a mutation in the cells showing the characteristic, but due to the influence of other cells.

allophore *n.* cell or chromatophore containing red pigment in skin of fishes, amphibians and reptiles.

allophycocyanin *n.* red protein pigment in the phycobilisomes of red algae and cyanobacteria which acts as a light-harvesting pigment for photosynthesis.

allopolyploid, alloploid *n.* polyploid produced from a hybrid between two or more different species and therefore possessing two or more unlike sets of chromosomes. *alt.* amphiploid.

alloreactive *a. appl.* antibodies and T cells that react against antigens or cells from a genetically dissimilar individual of the same species. *n.* **alloreactivity.**

all-or-none principle principle that a response to a stimulus is either completely effected or is absent, first observed in heart muscle. *alt.* Bowditch's Law.

allosematic *a.* having markings or coloration imitating warning signs in other, usually dangerous, species. *cf.* aposematic.

allosomal *n. pert.* inheritance of characters controlled by genes carried on an allosome.

allosome *n.* a chromosome other than an autosome, e.g. a sex chromosome.

allosteric *a. pert.* allostery.

allosteric activator a ligand that binds to a site on an allosteric protein and induces a conformational change that increases the protein's activity.

allosteric inhibitor a ligand that binds to a site on an allosteric protein and induces a conformational change that decreases the protein's activity.

allostery *n.* property displayed by many proteins such that the binding of a molecule or a covalent modification (e.g. phosphorylation) at one site induces a conformational change in the protein that results in a change in the properties of another, distant, site on the protein. The sites affected may be on the same subunit or on a different subunit of the same protein. The activity of allosteric proteins is usually regulated, either negatively or positively, by such binding of a molecule, or a covalent modification, at regulatory sites that leads to a change in conformation at the active, or functional, site. Proteins whose activity is controlled by allosteric regulation include many enzymes, haemoglobin, motor proteins, DNA-binding proteins and receptors.

allosynapsis, allosyndesis *n.* pairing of homologous chromosomes from opposite parents in a polyploid.

allotetraploid *n.* organism that results from the doubling of chromosome number in a hybrid between two species. *alt.* amphidiploid.

allotherm *n.* organism with body temperature dependent on environmental temperature.

allotopic *a. appl.* sympatric populations occupying different habitats within the same geographical range of distribution.

allotopic gene expression expression of a gene in a cell or organelle in which it is not normally expressed.

allotriploid *n.* organism whose somatic cells contain three sets of chromosomes, one of which differs from the others.

allotrophic heterotrophic *q.v.*

allotropous *a.* (1) *appl.* insects not limited to or adapted to visiting special kinds of flowers; (2) *appl.* flowers whose nectar is available to all kinds of insects.

allotropy *n.* (1) tendency of certain cells or structures to approach each other; (2) mutual attraction, as between gametes.

allotype *n.* allelic differences in light or heavy chains of immunoglobulins in individuals of the same species. *cf.* isotypes, idiotype.

allozygous *a. appl.* alleles at the same locus that are different or, if identical, are different origins, i.e. their identity is not due to common descent.

allozyme *n.* one of a number of different forms of the same enzyme encoded by different alleles of the same gene locus.

alluvial *a. pert.* soils composed of sediment transported and deposited by flowing water.

alpestrine *a.* growing high on mountains but not above the tree line.

alpha- *for headwords with prefix α- or alpha- refer also to headword itself, e.g. for α-globin look under globin.*

alpha *n.* the highest-ranking individual within a dominance hierarchy.

alpha diversity ecological diversity within a habitat due to the coexistence of many subtly different ecological niches, each occupied by a different species. This type of diversity results from competition between species that tends to reduce the variation within species as each species becomes more finely adapted to its niche. *alt.* niche diversification. *cf.* beta diversity, gamma diversity.

alpha motor neuron motor neuron that innervates the main contractile fibres of a muscle, outside the spindle. *cf.* gamma efferent motor neuron.

alpha-proteobacteria group of the Proteobacteria that includes the photoautotrophic non-sulphur purple bacteria, which contain bacteriochlorophyll and various carotenoid pigments (e.g. *Rhodopseudomonas*, *Rhodomicrobium*), related non-phototrophic Gram-negative bacteria (e.g. *Rhizobium*, *Agrobacterium*), and most of the budding and/or stalked bacteria (e.g. *Hyphomicrobium*). *alt.* **alpha purple bacteria**.

alpha rhythm brain electrical potential, frequency 8–12 Hz, that occurs in a state of relaxed wakefulness.

alphaviruses *n.plu.* group of animal viruses in the Togaviridae, including Semliki Forest virus.

alpine *a. appl.* the part of a mountain above the tree line and below the permanent snow line, and to species mainly restricted to this zone.

alpine grassland grassland found above the tree line on high mountains.

alpine tundra zone of tundra-like vegetation found on high mountains above the alpine grasslands and below the permanent snow line.

alternate *a.* not opposite, *appl.* leaves, branches, etc., occurring at different levels successively on opposite sides of stem.

alternate host in the life cycle of heteroecious rust fungi, the host on which spermogonial and aecial stages are produced.

alternate segregation in a cell with a reciprocal translocation, the movement of both normal chromosomes to one pole and both translocated chromosomes to the other.

alternation of generations alternation of haploid and diploid stages which occurs in the life cycle of sexually reproducing eukaryotic organisms. In some organisms, e.g. mosses, the haploid phase is predominant; in others, e.g. flowering plants and many animals, the diploid phase is dominant and the haploid phase is much reduced or represented only by the gametes; and in others, e.g. hydrozoan coelenterates, diploid and haploid organisms alternate.

alternation of parts general rule that the parts of the different whorls of a flower alternate in position with each other, sepals with petals, stamens with petals.

alternative pathway of complement activation, pathway triggered by cell surfaces possessing certain properties and by endotoxins of Gram-negative bacteria. It involves complement component C3 and serum factors B, D and P (properdin) and results in formation of C3b which then follows the classical pathway. *see also* C3, complement system, Factor B, Factor D, properdin.

alternative splicing the production of different RNAs from the same primary RNA transcripts by different patterns of splicing.

alterne *n.* vegetation exhibiting disturbed zonation due to abrupt change in environment or to interference with normal plant succession.

alternipinnate *a. appl.* leaflets or pinnae arising alternately on each side of midrib.

altrices *n.plu.* birds whose young are hatched in a very immature condition.

altricial *a.* requiring care or nursing after hatching or birth.

altruism *n.* any act or behaviour which results in an individual increasing the genetic fitness of another at the expense of its own, e.g. by devoting large amounts of time and resources to caring for another individual's offspring at the expense of producing its own. *a.* **altruistic**. *see also* reciprocal altruism.

alula *n.* small lobe of a wing.

Alu sequences family of repetitive interspersed DNA sequences of around 300 bp, present in up to 1 million copies in the human genome. They were named after the restriction enzyme *Alu*I, which was originally used to identify them in restriction maps. Similar sequences are present in other mammals. *see also* LINE, SINE.

alveolar *a.* (1) *pert.* an alveolus; (2) *pert.* tooth socket, *appl.* nerve, artery in connection with jaw bone.

Alveolata, alveolates *n., n.plu.* name given to a group of protists comprising the ciliates, apicomplexans and dinoflagellates, which have a system of sacs beneath the cell membrane.

alveolate *a.* deeply pitted or honeycombed.

alveolus *n.* (1) air cavity in lungs; (2) tooth socket; (3) a cavity; (4) small pit or depression.

Alzheimer's disease degenerative neurological disease relatively common in elderly people, with atrophy of neurons in certain parts of the forebrain and consequent disturbance of brain function. It is accompanied by fibrillar amyloid deposits in the brain. Symptoms include profound confusion, memory loss and often changes in personality. Formerly called senile dementia. A rare inherited form of the disease, familial Alzheimer's disease (FAD), causes symptoms at a much earlier age than usual.

amacrine *a.* having no conspicuous axon, *appl.* type of neuron in retina which forms a layer with bipolar and horizontal cells and makes lateral connections.

α-amanitin *n.* cyclic octapeptide toxin from the fungus *Amanita phalloides*, a potent inhibitor of RNA polymerases II and III.

Amastigomycota *n.* in some classifications, a major group of the Fungi comprising the zygomycetes, ascomycetes, basidiomycetes and deuteromycetes. They are terrestrial fungi, usually with a well-developed mycelium, and do not have motile flagellate zoospores or gametes. They include moulds (e.g. *Mucor, Penicillium*), mushrooms and toadstools.

amastigote *n.* stage in the life cycle of trypanosomatid protozoa. The oval or round cell has a nucleus, kinetoplast and basal body but the flagellum is either very short or entirely absent.

amatoxins *n.plu.* cyclic octapeptide toxins found in some mushrooms, e.g. α-amanitin.

Amazon region that part of the Neotropical floristic kingdom consisting of the Amazon basin.

amb ambulacra *q.v.*

amber *n.* (1) translucent yellow or brown material, known from the Cretaceous onwards, which is the fossilized resin of coniferous trees; (2) the UAG termination (stop) codon.

ambergris *n.* secretion of the sperm whale, formerly used as a musk fragrance in perfumery, now superseded by synthetic compounds.

amber mutation mutation generating an amber termination codon and resulting in premature termination of synthesis of the protein product of the mutated gene.

amber suppressor a mutant gene producing a tRNA which overcomes the effects of an amber mutation by inserting an amino acid at UAG.

ambiens *n.* thigh muscle in some birds, whose action causes the toes to maintain grasp on perch.

ambient *a.* surrounding.

ambilateral *a. pert.* both sides.

ambiparous *a. appl.* buds containing the beginnings of both flowers and leaves.

ambiquitous *a. appl.* enzymes for which the degree to which they are associated with subcellular particulate structures is dependent on metabolic activity.

ambisexual *a.* (1) *pert.* both sexes; (2) monoecious *q.v.*

ambitus *n.* (1) the outer edge or margin; (2) outline of echinoid shell viewed from apical pole.

amblyopia *n.* reduced visual acuity.

ambon *n.* ring of fibrous cartilage surrounding the socket of a joint.

ambosexual *a.* (1) common to, or *pert.* both sexes; (2) activated by both male and female hormones.

ambrosial *a. appl.* class of odours typified by ambergris and musk.

ambulacra *n.plu.* (1) region containing the tube-feet of echinoderms; (2) the bands of tube-feet themselves. *sing.* **ambulacrum**, *alt.* amb.

ameba *alt.* spelling of amoeba *q.v.*

ameiosis *n.* (1) occurrence of only one division in meiosis instead of two; (2) the absence of pairing of chromosomes in meiosis.

ameiotic *a. appl.* parthenogenesis in which meiosis is suppressed.

amelification *n.* formation of tooth enamel.

ameloblast *n.* columnar or hexagonal epithelial cell that secretes enamel and is part of enamel organ in tooth. *alt.* enamel cell.

amensalism *n.* form of competition between two species in which one is inhibited and the other is not. *alt.* antagonism.

amentaceous, amentiferous *a.* bearing catkins.

amentum *n.* catkin. *plu.* **amenta**.

ameristic *a.* (1) not divided into parts; (2) unsegmented; (3) undifferentiated or undeveloped.

Ames test simple *in vitro* test devised by the American biochemist Bruce Ames to screen compounds for potential mutagens and carcinogens by their ability to cause mutations in bacteria.

Ametabola *n.* the ametabolous (*q.v.*) insects.

ametabolous *a.* not changing form, *appl.* the orders of primitive wingless insects (the Ametabola) in which the young hatch from the egg resembling young adults, and comprising the Diplura (two-pronged bristletails), Thysanura (three-pronged bristletails), Collembola (springtails) and Protura (bark-lice).

amethopterin *n.* folate analogue that blocks regeneration of tetrahydrofolate and dTMP synthesis and is used as an anticancer drug to inhibit rapidly dividing cells. *alt.* methotrexate.

ametoecious *a.* parasitic on one host during one life cycle. *alt.* autoecious.

AMH anti-Müllerian duct hormone. *see* Müllerian-inhibiting substance.

amicronucleate *a. appl.* fragments of certain protozoans in which there is no micronucleus.

amictic *a.* (1) *appl.* eggs that cannot be fertilized and which develop parthenogenetically into females; (2) *appl.* females producing such eggs.

amidase *n.* (1) enzyme catalysing the hydrolysis of a monocarboxylic acid amide to a monocarboxylic acid and ammonia. EC 3.5.1.4; (2) any of a group of enzymes hydrolysing non-peptide C–N linkages of amides and including asparaginase, urease and glutaminase. EC 3.5.1–2.

amide *n.* compound that contains the group $-CO.NH_2$, biological amides being derived from carboxylic acids and amino acids by replacement of the –OH of the carboxyl group with $-NH_2$. An amide bond is formed by the condensation of a carboxylic acid with an amino group.

amidinase *n.* any of a group of enzymes hydrolysing non-peptide C–N linkages in amidines, and including arginase. EC 3.5.3–4.

amidine *n.* compound that contains the group $-CNH.NH_2$.

amine *n.* any compound containing the functional group $-NH_2$.

amino acid *n.* any of a class of compounds of the general formula $RCH(NH_2)COOH$ (α-amino acids) where R is a distinctive side chain. Side chains vary from the single hydrogen atom of glycine to the aromatic side chains of tryptophan and phenylalanine and the sulphur-containing side chains of cysteine and methionine. Proteins are composed of amino acids covalently linked together through their amino and carboxyl groups into a polypeptide chain with the side chains projecting from the covalently linked backbone. Amino acids can occur as optically active D- and L-isomers, of which only L-isomers are found in proteins. Around 20 different amino acids are present in proteins, all of which can be synthesized by autotrophs but which in heterotrophs are chiefly obtained by breakdown of dietary protein. Amino acids are also biosynthetic precursors of many important molecules such as purines, pyrimidines, histamine, thyroxine, adrenaline, melanin, serotonin, the nicotinamide ring and porphyrins among others. *see* Fig. 4 and individual entries for each amino acid.

(a)

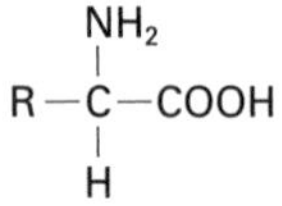

$$R - \overset{\displaystyle NH_2}{\underset{\displaystyle H}{C}} - COOH$$

(b)

Amino acid	Abbreviations	
Glycine	Gly	G
Alanine	Ala	A
Valine	Val	V
Leucine	Leu	L
Isoleucine	Ile	I
Serine	Ser	S
Threonine	Thr	T
Lysine	Lys	K
Arginine	Arg	R
Histidine	His	H
Aspartic acid	Asp	D
Asparagine	Asn	N
Glutamic acid	Glu	E
Glutamine	Gln	Q
Proline	Pro	P
Tryptophan	Trp	W
Phenylalanine	Phe	F
Tyrosine	Tyr	Y
Methionine	Met	M
Cysteine	Cys	C

Fig. 4 (a) General structure of an amino acid. R = side chain. (b) Table of three-letter and single-letter abbreviations for the amino acids found in proteins.

D-**amino acid** type of amino acid found in peptides in bacterial cell walls and a few other materials but never in proteins.

amino acid analysis determination of the amino acid composition of a protein.

amino acid neurotransmitters the amino acids glycine and glutamate, which act as inhibitory and excitatory neurotransmitters respectively in the central nervous system.

amino acid notation the abbreviations for the amino acids found in proteins, which are used in displaying protein sequences and which may be of one or three letters.

amino acid racemization conversion of L-amino acids to D-amino acids, which occurs at a very slow rate in nature. It can be used to date certain fossils by measuring the amount of D-amino acids in the sample and calculating the time taken for them to form from the original L-amino acids (only L-amino acids are found in living tissue to any appreciable extent). The method can date fossils between 15,000 and 100,000 years old.

amino acid sequence order of amino acids in a polypeptide or protein.

aminoaciduria *n.* presence of abnormally large amounts of amino acids in the urine.

aminoacylase *n.* enzyme catalysing the hydrolysis of an acyl-amino acid to carboxylate and an amino acid. EC 3.5.1.14.

aminoacyltransferase *n.* any of a group of enzymes that transfer aminoacyl groups and including peptidyltransferase. EC 2.3.2.

aminoacyl-tRNA *n.* tRNA carrying an activated amino acid covalently attached to its 3′ end (e.g. methionyl-tRNA, seryl-tRNA, tyrosyl-tRNA).

aminoacyl-tRNA synthetase *n.* any of a large group of enzymes catalysing the attachment of an amino acid to tRNA, each aminoacyl-tRNA synthetase being specific for a particular amino acid and one or more acceptor tRNAs.

p-**aminobenzoic acid (PBA)** substance necessary as a vitamin in rats, but does not seem to be required in humans.

γ-**aminobutyric acid (GABA)** amino acid that acts as an inhibitory neurotransmitter in the central nervous system.

aminoglycoside *n.* any of a group of antibiotics such as streptomycin and kanamycin that consist of several amino-substituted sugars and/or sugar alcohols linked by glycosidic bonds.

amino group –NH_2.

δ-aminolevulinate synthase (DALA synthetase) the first enzyme in the pathway through which haem is synthesized and through which haem synthesis is regulated by feedback inhibition.

aminopeptidase *n.* any of a group of enzymes that remove amino-terminal amino acid residues from a protein or peptide. EC 3.4.11.

aminopterin *n.* folate analogue that blocks the regeneration of tetrahydrofolate and dTMP synthesis and is used as an anticancer drug.

aminopurine (2-aminopurine, AP) *n.* analogue of adenine which pairs with cytosine rather than thymine and is therefore mutagenic.

4-aminopyridine (4-AP) compound used to selectively block potassium conductance channels in neurons.

amino sugar monosaccharide in which a hydroxyl group has been replaced by an amino group, e.g. galactosamine and glucosamine.

amino terminus (N terminus) the end of a protein chain that bears the free α-amino group. A protein is synthesized starting at the amino terminus. *a.* **amino terminal (N-terminal)**.

aminotransferase *n.* any of a class of enzymes that catalyse the transfer of an α-amino group, usually from an α-amino acid to a α-keto acid. EC 2.6.1. *alt.* transaminase.

amitosis *n.* division of the nucleus by constriction without the condensation of mitotic chromosomes or formation of a spindle and without breakdown of the nuclear membrane, e.g. in the macro-nucleus of ciliates.

amixia *n.* cross-sterility between members of the same species as a result of morpho-logical, geographical or physiological iso-lating mechanisms.

amixis *n.* (1) absence of fertilization, *a.* **amictic**; (2) sometimes used for the absence of gonads; (3) apomixis in haploid organisms.

ammocoete *n.* larva of lamprey.

ammonia (NH3) *n.* chemical compound that either directly or in the form of the ammonium ion (NH_4^+) is a source of nitrogen for plants. It is produced by biological nitrogen fixation, by the action of soil microorganisms that break down protein in dead organic matter, and as an excretion product of e.g. teleost fishes. *see also* ammonification, ammonotelic.

ammonia-oxidizing bacteria bacteria that oxidize ammonia to nitrite (e.g. *Nitrosomonas, Nitrosococcus*). *alt.* nitrosifiers.

ammonification *n.* the production of ammonia (as ammonium ion) from organic nitrogenous compounds, carried out by a variety of heterotrophic microorganisms (ammonifiers).

ammonifiers *n.plu.* bacteria that can produce ammonia from organic nitrogenous compounds.

ammonites *n.plu.* an extinct group of cephalopods, familiar from their coiled shells, similar to the nautiloids but probably had a calcareous larval shell.

ammonitiferous *a.* carrying fossil remains of ammonites.

ammonium ion *see* ammonia.

ammonotelic *a.* excreting nitrogen mainly as ammonia, e.g. most aquatic invertebrates, tadpoles and some teleost fish.

amniocentesis *n.* clinical procedure in which cells from the amniotic fluid surrounding the foetus are withdrawn for chromosomal analysis to detect genetically determined defects.

amnion *n.* (1) innermost of the embryonic membranes surrounding the developing embryo of reptiles, birds and mammals, *alt.* amniotic sac; (2) inner embryonic membrane of insects; (3) membrane like an amnion found in other invertebrates.

Amniota, amniotes *n., n.plu.* the land-living vertebrates, e.g. reptiles, birds and mammals, which have an amnion around the developing embryo.

amniotic *a. pert.* amnion, *appl.* folds, sac, cavity, fluid.

amniotic egg egg of birds, reptiles and prototherians (egg-laying mammals) within which extra-embryonic membranes are formed during embryonic development.

amoeba *n.* (1) unicellular non-photosynthetic wall-less protist whose shape is subject to constant change due to formation and retraction of broad, finger-like pseudopodia. The amoebas are classified in the large grouping Amoebozoa (*q.v.*), or the protist phylum Rhizopoda, and were formerly classified in the protozoan class Rhizopodea; (2) myxamoeba *q.v. plu.* **amoebae, amoebas.** *alt.* ameba.

amoebic *a. pert.*, or caused by, amoebae, *appl.* dysentery (amoebiasis) caused by the parasitic amoeba *Entamoeba histolytica.*

amoebiform amoeboid *q.v.*

amoebism *n.* amoeboid form or behaviour, as of some cells.

amoebocyte *n.* (1) any cell having the shape or properties of an amoeba; (2) a cell in coelomic fluid of echinoderms.

amoeboflagellate *n.* protist that can change from an amoeboid to a flagellate form.

amoeboid *a.* resembling an amoeba in shape, in properties or mode of movement.

Amoebozoa *n.* proposed major monophyletic taxonomic grouping, based on molecular phylogenetics, which includes non-photosynthetic unicellular eukaryotes that move via broad finger-like pseudopodia (amoebae), the cellular slime moulds (e.g. *Dictyostelium*), and the plasmodial slime moulds (e.g. *Physarum*).

amoebula *n.* the amoeboid swarm spore of various protists.

amorph *n.* mutation in which no active gene product is formed. *alt.* null mutation.

amorphous *a.* (1) of indeterminate or irregular form; (2) with no visible differentiation in structure.

AMP adenosine monophosphate *q.v.*

AMPA receptor type of glutamate receptor in the nervous system.

ampheclexis sexual selection *q.v.*

ampherotoky amphitoky *q.v.*

amphetamine *n.* sympathomimetic drug chemically related to adrenaline and which is a powerful stimulant of the central nervous system.

amphi- Gk prefix denoting both, on both sides.

amphiapomict *n.* a group of genetically identical individuals reproduced from facultatively sexual forms.

amphiarthrosis *n.* a slightly moveable joint.

amphiaster *n.* a sponge spicule star-shaped at both ends.

Amphibia, amphibians *n.*, *n.plu.* vertebrate class including the extant subclass Lissamphibia, comprising the frogs and toads (order Anura), newts and salamanders (order Urodela) and the worm-like caecilians (order Apoda). There are also a number of extinct subclasses dating from the Devonian onwards, of which the ichthyostegalians are the earliest amphibian fossils found. Amphibians are ectothermic anamniote tetrapod vertebrates that typically return to the water for reproduction and pass through an aquatic larval stage with gills. Adults generally have lungs, are carnivorous, and may be at least partly land-living, and modern amphibians have a moist skin without scales, which is permeable to water and gases. *see also* anthracosaurs, ichthyostegalians, lepospondyls, temnospondyls.

amphibian, amphibious *a.* adapted for life either in water or on land.

amphibivalent *n.* a ring of chromosomes arising in the metaphase and anaphase of the 1st meiotic division as a result of the reciprocal translocation of chromosome segments between two chromosomes.

amphiblastic *a. appl.* telolecithal ova with complete but unequal segmentation.

amphibolic *a.* capable of turning backwards or forwards, such as the outer toe of some birds.

amphicarpous *a.* producing two kinds of fruit.

amphicoelous, amphicelous *a.* concave on both surfaces.

amphicondylous *a.* having two occipital condyles.

amphicribral *a.* with the phloem surrounding the xylem, *appl.* some concentric vascular bundles. *alt.* amphiphloic. *cf.* amphivasal.

amphid *n.* one of a pair of anterior sense organs in nematodes, possibly detecting chemical stimuli.

amphidetic *a.* extending behind and in front of umbo, *appl.* hinge ligaments of some bivalve shells.

amphidiploid allotetraploid *q.v.*

amphidromous *a.* going in both directions, *appl.* animal migration.

amphigenous *a*. (1) borne or growing on both sides of a structure; (2) borne or growing on all sides of an organism or structure, *appl*. fungi in which hymenium covers the whole surface of the spore-bearing body. *alt*. perigenous.

amphigony *n*. sexual reproduction involving two individuals.

amphigynous *a*. *appl*. antheridium surrounding the base of oogonium, as in some Peronosporales.

amphihaploid *n*. a haploid arising from an amphidiploid species.

amphimict *n*. (1) a group of individuals resulting from sexual reproduction; (2) an obligate sexual organism.

amphimixis *n*. reproduction by sexual reproduction involving fusion of gametes from two organisms. *a*. **amphimictic**. *cf*. apomixis.

Amphineura *n*. class of marine molluscs, commonly called chitons, having an elongated body and a mantle bearing calcareous plates.

amphioxus *n*. the lancelet, a cephalochordate of the genus *Branchiostoma*.

amphipathic *a*. possessing both a hydrophobic (non-polar) and a hydrophilic (polar) portion in the same molecule, *appl*. proteins or parts of proteins, and to molecules such as phospholipids. The amphipathic nature of phospholipids allows them to orient in aqueous solution into micelles with hydrophobic groups together at the centre of the sphere and the polar groups on the surface, and to form bilayered biological membranes with the hydrophobic groups pointing towards the centre. *alt*. **amphiphilic**.

amphiphloic *a*. with phloem both external and internal to xylem, *appl*. stems, vascular bundles.

amphiphyte *n*. an amphibious plant, one that can live on land or in water.

amphiploid allopolyploid *q.v.*

amphipneustic *a*. (1) having both gills and lungs throughout life history; (2) with only anterior and posterior pairs of spiracles functioning, as in most dipteran larvae.

amphipod *n*. member of the Amphipoda, an order of terrestrial, marine and freshwater malacostracan crustaceans, having a later-ally compressed body, elongated abdomen and no carapace, e.g. sand-hopper.

amphiprotic *a*. *appl*. a molecule that can either donate or receive a proton.

amphirhinal *a*. having, or *pert*. two nostrils.

amphisarca *n*. a superior, many-seeded fruit with pulpy interior and woody exterior.

Amphisbaenia, amphisbaenians, amphisbaenids *n*., *n.plu*., *n.plu*. group of worm-like, burrowing, generally limbless reptiles with inconspicuous eyes and rounded tails.

amphispermous *a*. having seed closely surrounded by pericarp.

amphisporangiate *a*. (1) having sporophylls bearing both megasporangia and microsporangia; (2) hermaphrodite, *appl*. flowers.

amphispore *n*. (1) reproductive spore which functions as a resting spore in certain algae; (2) uredospore modified to withstand dry conditions.

amphisternous *a*. *appl*. type of sternum structure in some sea urchins.

amphistomatous *a*. having stomata on both surfaces, *appl*. some leaves.

amphistomous *a*. having a sucker at each end of body, as leeches.

amphistylic *a*. primitive type of jaw suspension, having the mandibular arch articulated with the hyoid arch, and the palatoquadrate (upper jaw) making three connections with the chondrocranium, as in some sharks, *appl*. skulls.

amphitelic *a*. *appl*. orientation of chromosomes on the spindle equator at metaphase of mitosis, with centromeres exactly equidistant from each pole.

amphithecium *n*. in bryophytes, peripheral layer of cells in sporogonium.

amphitoky *n*. parthenogenetic reproduction of both males and females.

amphitrichous *a*. with a flagellum at each pole, *appl*. bacteria.

amphitroph *n*. normally autotrophic organism which adapts itself to heterotrophic nutrition if placed in the dark for long periods.

amphitropical *a*. occurring on both sides of the tropics.

amphitropous *a*. having the ovule inverted, with hilum in middle of one side.

amphiumas *n.plu.* small family of wholly aquatic eel-like amphibians from southeastern United States, comprising a single genus with three species.

amphivasal *a.* with the xylem surrounding the phloem, *appl.* some concentric vascular bundles. *alt.* **amphixylic.**

amphogenic *a.* producing offspring consisting of males and females.

ampholyte *n.* molecule that contains both acidic and basic groups.

amphoteric *a.* (1) possessing both acidic and basic properties, e.g. amino acids; (2) with opposite characters.

amphotericin B a polyene antibiotic effective against fungal infections.

ampicillin *n.* a semi-synthetic aminophenylacetyl penicillin active against Gram-positive and some Gram-negative bacteria.

amplectant *a.* clasping or winding tightly round a support, as tendrils.

amplexicaul *n.* clasping or surrounding the stem, as base of sessile leaf.

amplexus *n.* mating embrace in frogs and toads during which eggs are shed into the water and fertilized.

ampliate *a.* having outer edge of wing prominent, as in some insects.

amplicon *n.* a stretch of DNA that has become copied many times to form an array of repeated sequences.

amplification *n.* (1) changes towards increased structural or functional complexity in ontogeny or phylogeny; (2) of genes or DNA, multiplication of a gene or DNA sequence to produce numerous copies within the chromosomes.

amplified fragment length polymorphism (AFLP) type of polymorphic DNA marker used for genetic mapping, esp. in plants. It is detected by restriction digestion of DNA followed by amplification of fragments by polymerase chain reaction.

ampulla *n.* (1) a membranous vesicle; (2) dilated portion at one end of each semicircular canal of ear; (3) dilated portion of various ducts and tubules; (4) internal reservoir on ring canal of water vascular system in echinoderms; (5) terminal vesicle of sensory canals in elasmobranch fishes; (6) (*bot.*) submerged bladder of bladderwort, *Utricularia.* *plu.* **ampullae.**

ampullaceous *a.* flask-shaped.

ampullae of Lorenzini jelly-filled tubes in the head of selachian fish (sharks, rays, etc.) opening to the exterior and terminating in sensory cells that detect changes in temperature or salinity of water or changes in electrical potential in the tissue.

amygdala *n.* (1) almond, or almond-shaped structure; (2) subcortical structure in the medial anterior part of temporal lobes of cerebral hemispheres, thought to be involved in memory.

amygdalin *n.* cyanogenic glycoside found in kernels of bitter almond, peach, cherry.

amylase *n.* α-amylase (EC 3.2.1.1), which randomly hydrolyses α-1,4 linkages in starch, glycogen and other glucose polysaccharides, or β-amylase (EC 3.2.1.2), which successively removes maltose units, or a mixture of these two enzymes.

amyliferous *a.* containing or producing starch.

amylogenesis *n.* starch formation.

amyloid *n.* (1) complex proteinaceous fibrillar material deposited in heart, liver, spleen and other organs in various diseases. It is formed from a variety of proteins that lose their normal folded structure and form fibrous aggregates rich in β-sheet structure that are resistant to proteolytic degradation. Amyloid deposits in brain are characteristic of Alzheimer's disease; (2) *a.* starch-like.

amylolytic *a.* starch-digesting.

amylopectin *n.* branched polymer of glucose, found in starch, with a structure similar to glycogen.

amyloplast *n.* colourless plastid in which starch accumulates.

amylose *n.* unbranched polymer of glucose, found in starch.

ANA antinuclear antibody, autoantibodies against various components of the cell nucleus.

anabiosis *n.* a condition of apparent death or suspended animation produced in certain organisms by e.g. desiccation, and from which they can be revived.

anabolism *n.* the constructive biochemical processes in living organisms involving the formation of complex molecules from simpler ones and the storage of energy. *a.* **anabolic.**

anabolite *n.* any substance involved in anabolism.

anacanthous *a.* without spines or thorns.

anachoresis *n.* the phenomenon of living in holes or crevices.

anadromous *a. appl.* fishes which migrate from salt to fresh water annually.

anaemia *n.* blood disorder characterized by a lack of red cells, which may be due to a variety of causes.

anaerobe *n.* any organism that can live in the absence of gaseous oxygen (O_2), obligate anaerobes being unable to live in even low oxygen concentrations, facultative anaerobes being able to live in low or normal oxygen concentrations as well. *cf.* aerobe.

anaerobic *a.* (1) *appl.* organisms that live in the absence of gaseous oxygen (O_2); (2) *appl.* environments or conditions lacking in oxygen.

anaerobic decomposition the breakdown of organic matter under anaerobic conditions by microorganisms.

anaerobic respiration (1) respiration involving electron transport and reduction–oxidation reactions, leading to generation of a protonmotive force, but which uses a final electron acceptor other than O_2; (2) glycolysis; (3) fermentation.

anaerobiotic *a.* lacking or depleted in oxygen, *appl.* habitats.

anaerogenic *a.* not producing gas during fermentation, *appl.* microorganisms.

anagenesis *n.* (1) progressive evolution within a lineage by the gradual change of one type into another without any branching or splitting of the lineage. *a.* **anagenetic**. *cf.* cladogenesis; (2) regeneration of tissues.

anal *a.* (1) *pert.* or situated at or near the anus; (2) *appl.* posterior median ventral fin of fishes.

analgesic (1) *a.* reducing or abolishing pain; (2) *n.* pain-killing drug.

analogous *a. appl.* structures that are similar in function but not in structure and developmental and evolutionary origin, e.g. the wings of insects and birds.

analogue *n.* (1) any organ or part similar in function to one in a different plant or animal, but of unlike origin; (2) any compound chemically related to but not identical with another. Analogues of natural metabolites compete with them for binding sites on enzymes, receptors, etc., often blocking the normal reaction; (3) DNA or protein sequences that are similar by virtue of convergent evolution rather than by evolution from a common ancestor.

analogy *n.* resemblance in function though not in structure or origin. *a.* **analogous**.

analysis of variance (ANOVAR) a statistical method by which the variance of a set of data can be apportioned to different causes.

anamestic *a. appl.* small variable bones filling spaces between larger bones of more fixed position.

anamnestic *a.* in immunology, *appl.* secondary immune responses.

anamniotes *n.plu.* fishes, amphibians and the Agnatha (lampreys and hagfishes), characterized by the absence of an amnion around the embryo.

anamorph *n.* the asexual stage in the life cycle of a fungus. *cf.* holomorph, teleomorph.

anamorpha *n.plu.* larvae hatched with an incomplete number of segments.

anamorphic *a. appl.* fungi that have no known sexual stage in the life cycle.

anamorphosis *n.* (1) evolution from one type to another through a series of gradual changes; (2) excessive or abnormal structure of plant origin.

anandrous *a.* without anthers.

anangian *a.* without a vascular system.

ananthous *a.* (1) not flowering; (2) without an inflorescence.

anaphase *n.* stage of mitosis or meiosis which follows metaphase. In mitosis, duplicated chromosomes split lengthways, the two chromatids moving to opposite poles of the spindle. In meiosis, homologous chromosomes move to opposite poles of the spindle in anaphase I (1st meiotic division), and sister chromatids separate and move to opposite poles in anaphase II (2nd meiotic division).

anaphase-promoting complex (APC) protein complex that promotes the separation of sister chromatids during the transition from metaphase to anaphase in mitosis.

anaphylactic shock, anaphylaxis severe and sometimes fatal systemic shock symptoms produced by exposure to an antigen to which an individual is hypersensitive (allergic).

anaphylatoxin *n.* any of the complement fragments C3a, C4a or C5a, which can initiate inflammatory reactions that mimic some of the features of anaphylaxis. They are produced from their parent complement proteins – C3, C4 and C5 – by enzymatic cleavage during activation of the complement system by infection.

anaphysis *n.* (1) an outgrowth; (2) sterigma-like filament in apothecium of certain lichens.

anaplasia *n.* reversion to a less differentiated structure.

anaplast(id) leucoplast *q.v.*

anaplerotic reaction a replenishing reaction in intermediary metabolism, such as the carboxylation of pyruvate to oxaloacetate to replenish oxaloacetate in the tricarboxylic acid cycle after its withdrawal for amino acid synthesis.

anapleurite *n.* upper thoracic pleurite, as in some bristletails.

anapophysis *n.* small dorsal projection rising near transverse process in lumbar vertebrae.

anapsid *a.* with skull completely roofed over, with no temporal fenestrae, the only gaps in the dorsal surface being the nares, eye orbits and the parietal foramen.

anapsids *n.plu.* tortoises and turtles and extinct members of the reptilian subclass Anapsida, characterized by a sprawling gait and a skull with no temporal opening.

anarthous *a.* having no distinct joints.

anaschistic *a. appl.* tetrads which divide twice longitudinally in meiosis.

anaspids *n.plu.* order of extinct jawless vertebrates (agnathans) with slender flexible bodies covered in thin bony scales.

anastomosis *n.* formation of network or meshwork, e.g. the union of fine ramifications of leaf veins, the union of fungal hyphae, the union of blood vessels arising from a single trunk.

anastomosis group isolates of a fungal species whose hyphae can fuse with each other.

anastral *a. appl.* type of mitosis without aster formation.

anatomical *a. pert.* the structure of a plant or animal.

anatomically modern humans the species *Homo sapiens*, to which all living humans belong, which arose around 200,000 years ago. *alt.* modern humans.

anatomy *n.* study of the structure of plants and animals as determined by dissection.

anatoxin toxoid *q.v.*

anatriaene *n.* a trident-shaped spicule with backwardly directed branches.

anatropous *a.* inverted, *appl.* ovule bent over so that hilum and micropyle are close together and chalaza is at other end.

anautogenous *a. appl.* adult female insect that must feed if her eggs are to mature.

anaxial *a.* (1) having no distinct axis; (2) asymmetrical.

anchorage-dependent growth the requirement of many mammalian cells, such as fibroblasts, for a suitable surface, such as glass or plastic, on which to grow and divide in culture.

anchorage-independent growth the property shown by some transformed mammalian cells in culture, which can divide in semi-solid agar, no longer requiring a surface such as glass or plastic on which to grow.

ancient forest, ancient woodland native forest or woodland, which may be virgin forest or old secondary forest developed by secondary succession and that has been continuously present on a site for hundreds of years. It may have been managed but has never undergone extensive clear-felling, and can be recognized by its characteristic flora. *cf.* old-growth forest, secondary forest.

ancipital *a.* flattened and having two edges.

anconeal *a. pert.* the elbow.

anconeus *n.* small extensor muscle situated over elbow.

Andean region that part of the Neotropical floristic kingdom consisting of the Andes and their western coastline.

Andreaeidae granite mosses *q.v.*

andric *a.* male.

androconia *n.plu.* modified wing scales producing a sexually attractive scent in certain male butterflies. *sing.* **androconium**.

androdioecious *a.* having male and hermaphrodite flowers on different plants.

androecium *n.* male reproductive organs of a plant. In flowering plants, the stamens collectively.

androgamone *n.* any substance produced by a male gamete which acts on a female gamete.

androgen *n.* any of various male steroid sex hormones, e.g. androsterone and testosterone, that are concerned with development of male reproductive system and production and maintenance of secondary sexual characteristics. They are secreted chiefly by the testis.

androgenesis *n.* (1) development from an embryo that contains paternal chromosomes only; (2) development from a male gamete, i.e. male parthenogenesis. *a.* **androgenetic.**

androgenic *a.* (1) stimulating male characters, masculinizing, *appl.* hormones; (2) *appl.* tissue capable of making androgenic hormones.

androgenone *n.* artificially produced embryo containing only male-derived chromosomes.

androgenous *a.* producing only male offspring.

androgonidia *n.plu.* male sexual individuals produced after repeated divisions of reproductive individuals of the colonial protistan, *Volvox*.

androgonium *n.* cell in antheridium (male reproductive organ of cryptogams) which gives rise to antherozoid mother cell.

androgyne *a.* hermaphrodite *q.v.*

androgynism hermaphroditism *q.v.*

androgynous *a.* (1) hermaphrodite *q.v.*; (2) bearing both staminate and pistillate flowers in the same inflorescence; (3) with antheridium and oogonium on the same hypha.

androgyny *a.* having the characteristics of both male and female.

andromerogony *n.* development of an egg fragment with only paternal chromosomes.

andromonoecious *a.* having male and hermaphrodite flowers on the same plant.

andromorphic *a.* having a morphological resemblance to males.

andropetalous *a.* having petaloid stamens.

androphore *n.* stalk or hypha carrying male reproductive organs (e.g. antheridia or androecium).

androsporangium *n.* sporangium containing androspores.

androspore *n.* an asexual zoospore which gives rise to a male dwarf plant.

androstenedione *n.* steroid sex hormone, an androgen, synthesized in gonads and adrenals.

androsterone *n.* a male steroid sex hormone, produced chiefly by testis, and which is less active than testosterone.

androtype *n.* type specimen of the male of a species.

anellus *n.* a small ring-shaped or triangular plate supported by valves and vinculum, in Lepidoptera.

anelytrous *a.* without elytra.

anemo- prefix derived from Gk *anemos*, wind.

anemochory *a.* dispersal of seeds by the wind. *a.* **anemochorous,** *appl.* seeds dispersed by the wind and *appl.* plants having seeds dispersed by the wind.

anemophily *n.* wind pollination or any other type of fertilization brought about by wind. *a.* **anemophilous.**

anemoplankton *n.* wind-borne microorganisms, spores and pollen.

anemosporic *a.* having spores or seeds dispersed by air currents.

anemotaxis *n.* movement in response to air currents.

anemotropism *n.* orientation of body, or plant curvature, in response to air currents.

anencephaly *n.* condition of having no brain. *a.* **anencephalous.**

anenterous, anenteric *a.* lacking a gut.

aner *n.* insect male, especially of ants.

anergy *n.* state of non-responsiveness to antigen.

aneuploid *a.* (1) having more or fewer than an exact multiple of the haploid number of chromosomes or haploid gene dosage; (2) *appl.* chromosomal abnormalities that disrupt relative gene dosage, such as deletions, *alt.* unbalanced. *n.* **aneuploidy.**

aneuronic *a.* without innervation, *appl.* chromatophores controlled by hormones.

aneusomic *a. appl.* organisms whose cells have varying numbers of chromosomes.

ANF atrial natriuretic factor *q.v.*

angienchyma *n.* vascular tissue.

angioblast *n.* cell from which endothelial lining of blood vessels is derived.

angiocarpic, angiocarpous *a*. (1) having fruit enclosed in a covering; (2) having spores enclosed in some kind of receptacle. *n*. **angiocarpy.**

angiogenesis *n*. development of new blood vessels by sprouting from preexisting vessels.

angiogenesis factors cytokines that promote the formation of blood vessels. *see also* tumour angiogenesis factors.

angiogenic *a*. (1) stimulating the formation of new blood vessels; (2) giving rise to blood vessels.

angiogenic clusters blood islands *q.v.*

angiogenin a tumour angiogenesis factor *q.v.*

angiography *n*. technique for visualizing the vascular system within an organ by injecting blood vessels with a dye that shows up on an X-radiograph. The picture produced is known as an angiogram.

angiology *n*. anatomy of blood and lymph systems.

Angiospermae, angiosperms Anthophyta *q.v.*

angiosporous *a*. having spores contained in a theca or capsule.

angiostomatous *a*. narrow-mouthed, *appl.* molluscs and snakes with non-distensible mouth.

angiotensin *n*. either of the two short polypeptides angiotensin I (inactive) or II (active hormone) released in the blood by action of renin on angiotensinogen, angiotensin II being formed from angiotensin I. Angiotensin II acts on blood vessels, causing constriction and a rise in blood pressure. It also causes contraction of the uterus and stimulates aldosterone secretion from the adrenal cortex.

angiotensinogen *n*. protein formed in liver and released into the blood where it may be split by the enzyme renin to produce angiotensin I, the inactive precursor of the vasopressor angiotensin II.

ångström (Å) *n*. unit of ultramicroscopic measurement, 10^{-10} m, 0.1 nm.

angular *a*. (1) having or *pert.* an angle; (2) *appl.* leaf originating at forking of stem, as in many ferns; (3) *n*. membrane bone in lower jaw of most vertebrates.

angulosplenial *n*. bone forming most of lower and inner part of mandible in amphibians.

angustifoliate *a*. with narrow leaves.

angustirostral *a*. with narrow beak or snout.

angustiseptate *a*. having a silicula laterally compressed with a narrow septum.

anholocyclic *a*. (1) *pert.* alternation of generations with suppression of sexual part of cycle; (2) permanently parthenogenetic.

animal *see* Animalia, Metazoa.

animal cap in the *Xenopus* and similar blastulas, the region of small rapidly cleaving cells at and around the animal pole.

animal hemisphere that half of an egg or blastula containing the animal pole.

Animalia *n*. the animal kingdom. In modern classifications it comprises all multicellular eukaryotic organisms with wall-less, non-photosynthetic cells. Animals are holozoic feeders, taking in solid organic material. All multicellular animals except the sponges possess some form of nervous system and contractile muscle or muscle-like cells, and most can move about. In older classifications protozoa were also included in the animal kingdom. *alt.* animals, Metazoa, metazoans. *see* Appendix 3.

animal pole (1) in yolky eggs, that part free of yolk and which cleaves more rapidly than the opposite, vegetal, pole; (2) the end of a blastula at which the smaller cleavage products (micromeres) collect.

animal traps (1) a group of zygomycete fungi (Zoopagales) that capture and parasitize small soil animals; (2) a group of deuteromycete fungi (e.g. *Arthrobotrys*) that form hyphal loops and branches to which soil nematodes adhere and become entangled.

animal–vegetal axis the axis running from animal pole to vegetal pole in amphibian eggs.

animal viruses viruses that infect animals.

anion *n*. negatively charged ion (e.g. Cl⁻) which moves towards the anode, the positive electrode. *cf.* cation.

anion channel protein spanning cell membrane and allowing the passive transport of anions across the membrane.

aniridia *n*. absence of the iris of the eye.

aniso- prefix from Gk *anisos*, unequal.

anisocarpous *a*. having number of carpels less than the number of parts in the other floral whorls.

anisocercal *a*. with lobes of tail-fin unequal.

anisocytic *a. appl.* stomata of a type in which three subsidiary cells, one distinctly smaller than the other two, surround the stoma. Formerly called cruciferous.

anisocytosis *n*. excessive variation in size of red blood cells.

anisodactylous *a*. having unequal toes, three toes forward, one backward.

anisodont heterodont *q.v.*

anisogamete *n*. one of two conjugating motile gametes differing in form or size. *a*. **anisogamous**.

anisogamy *n*. the union of morphologically unlike motile gametes. *cf.* isogamy.

anisognathous *a*. (1) with jaws of unequal width; (2) having teeth in upper and lower jaws unlike.

anisomerous *a*. having unequal numbers of parts in floral whorls. *n*. **anisomery**.

anisomorphic *a*. differing in size, shape or structure.

anisophylly heterophylly *q.v.*

anisopleural *a*. asymmetrical bilaterally.

anisoploid *a*. with an odd number of chromosome sets in somatic cells.

anisopterans *n.plu.* dragonflies, members of the suborder Anisoptera of the order Odonata.

anisopterous *a*. unequally winged, of seeds.

anisospore *n*. anisogamete *q.v.*

anisostemonous *a*. (1) having number of stamens unequal to number of parts in other floral whorls; (2) having stamens of unequal size.

anisotropic *a*. (1) doubly refracting, *appl.* dark bands of striated muscle fibres; (2) *appl.* eggs with predetermined axis or axes; (3) *appl.* crystals, displaying properties with different values when measured in different directions.

ankistroid *a*. (1) like a barb; (2) barbed.

ankyloblastic *a*. with a curved germ band.

ankylosis *n*. union of two or more bones or hard parts to form one, e.g. bone to bone, tooth to bone. *a*. **ankylosing**.

ankyroid *a*. hook-shaped.

anlage *n*. the first structure or cell group indicating development of a part or organ. *alt.* primordium.

annealing *n*. reconstitution of a double-stranded nucleic acid from single strands.

annectant *a*. linking, *appl.* intermediate species or genera.

Annelida, annelids *n*., *n.plu.* phylum of segmented coelomate worms, commonly called ringed worms. They have a soft elongated body with a muscular body wall, divided into many similar segments, usually separated by septa, and covered with a thin, flexible collagenous cuticle. They possess a blood system, nephridia and a central nervous system. The Annelida contains three main classes: Polychaeta (ragworms, lugworms), Oligochaeta (e.g. earthworms) and Hirudinea (leeches).

annelide *n*. type of fungal cell that produces conidia basipetally.

annexin *n*. any of a structurally related group of proteins with calcium-binding and phospholipid-binding properties and a wide range of biochemical functions including inhibition of blood coagulation.

annidation *n*. situation in which a mutant organism survives in a population because an ecological niche exists which the normal individual cannot use.

Annonales Magnoliales *q.v.*

annotation *n*. the determination of the positions of genes and other features in a genomic DNA sequence, and the addition of information about function where available.

annotinous *a. appl.* growth during the previous year.

annual *a*. (1) *appl.* structures or growth features that are marked off or completed yearly; (2) living for a year only; (3) completing life cycle in a year from germination; (4) *n*. plant that completes its life cycle in a year.

annual increment the increase in biomass of an organism or community in a year.

annual ring growth ring *q.v.*

annular *a*. (1) ring-like; (2) *appl.* certain ligaments in wrist and ankle; (3) *appl.* orbicular ligament encircling head of radius and attached to radial notch of ulna; (*bot.*) (4) *appl.* certain vessels in xylem having ring-like thickenings in their interior; (5) *appl.* bands formed on inner surface of cell wall.

Annulata, annulates *n*., *n.plu.* a group of invertebrates including the annelid worms, arthropods and some related forms, having bilateral symmetry and true metameric segmentation.

annulate *a.* (1) ring-shaped; (2) composed of ring-like segments; (3) with ring-like constrictions; (4) having colour arranged in ring-like bands.

annulate lamellae stacks of membranes containing structures like nuclear pore complexes, seen in the cytoplasm of some eukaryotic cells.

annulus *n.* (1) any ring-like structure, such as segment of annelid; (*bot.*) (2) remains of veil forming ring around stalk in mushrooms and toadstools; (4) in ferns, a row of specialized cells in a sporangium; (*zool.*) (3) growth ring of fish scale; (4) 4th digit of hand. *plu.* **annuli.**

anococcygeal *a. pert.* region between coccus and anus.

anoestrus *n.* (1) the non-breeding period; (2) period of absence of sexual receptiveness in females.

anoikis *n.* apoptosis of cells that lose contact with the extracellular matrix.

anomaly *n.* any departure from type characteristics.

anomer *n.* either of two isomers of a monosaccharide differing only in the arrangement of atoms around the carbonyl carbon atom, as α-D-glucopyranose and β-D-glucopyranose. *a.* **anomeric.**

anomerism *n.* the existence of anomers of a molecule.

anomia *n.* inability to name objects or persons easily.

anomocytic *a. appl.* stomata which have no subsidiary cells. Formerly called ranunculaceous.

anomophyllous *a.* with irregularly placed leaves.

anopheline *a. appl.* mosquitoes of the genus *Anopheles*, vectors of malaria and some other diseases.

Anoplura *n.* order of insects, the sucking or body lice, which are ectoparasites of mammals. *alt.* Siphunculata.

anorexia *n.* absence of appetite.

anorthogenesis *n.* evolution showing changes in direction of adaptations.

anorthospiral paranemic *q.v.*

anosmic *a.* having no sense of smell. *n.* **anosmia.**

ANOVAR analysis of variance *q.v.*

anoxia *n.* lack of oxygen.

anoxic *a.* devoid of molecular oxygen, *appl.* habitats.

anoxygenic *n.* not producing oxygen, *appl.* photosynthesis carried out by bacteria other than cyanobacteria.

anoxygenic photosynthesis use of light energy to synthesize ATP in an electron-transport-based cyclic photophosphorylation reaction without the production of oxygen by splitting of water. Carried out by some photosynthetic bacteria, e.g. purple bacteria. The electron donors used in this type of photosynthesis include hydrogen, reduced sulphur compounds and organic compounds.

ANP atrial natriuretic polypeptide. *see* atrial natriuretic factor.

ANS autonomic nervous system *q.v.*

ansa *n.* loop, as of certain nerves.

Anseriformes *n.* a large order of birds, the waterfowl, including ducks, geese and swans. *a.* **anseriform.**

anserine *a. pert.* a goose.

ansiform *a.* loop-shaped, or looped.

antagonism *n.* (1) the effect of a hormone, etc., that counteracts the effects of another; (2) the inhibitory action of one species on another, such as the action of certain substances secreted by plant roots which inhibit other plants nearby. *alt.* amensalism. *a.* **antagonistic.**

antagonist *n.* (1) a muscle working against the action of another; (2) any substance that interferes with or prevents the action of a hormone, neurotransmitter or drug.

Antarctic kingdom one of the five generally recognized floristic kingdoms, comprising the Antarctic, the islands in the Southern Ocean, New Zealand, and the southern tip of South America (below 40°S). It is divided into regions: New Zealand, Patagonian, South Temperate Islands.

ante- prefix from the L. *ante*, before, in front of.

anteclypeus *n.* the anterior portion of clypeus when it is differentiated by a suture.

antecosta *n.* internal ridge of tergum for attachment of intersegmental muscles in insects, extended to phragma in segments that bear wings.

antecubital *a.* in front of the elbow, *appl.* fossa.

antedorsal *a.* situated in front of the dorsal fin in fishes.

antefurca *n.* forked process of anterior thoracic segment in some insects.

antemarginal *a. appl.* sori of ferns when they lie within margin of frond.

antenatal *a.* before birth, *appl.* tests for genetic defects performed on the foetus in the womb.

antenna *n.* (1) one of a pair of jointed feelers on head of arthropod; (2) feeler of rotifers; (3) in some fish, a modified flap on dorsal fin which attracts prey; (4) group of chlorophyll and other pigment molecules involved in light capture in photosynthesis. *plu.* antennae. *a.* antennal, antennary.

Antennapedia complex (ANT-C) cluster of homeotic genes in the fruitfly *Drosophila melanogaster* which control the specification of particular segments in the embryo.

antennary *a. pert.* antenna, *appl.* nerve, artery, etc.

antennation *n.* touching with the antenna, serving as tactile communication signal or an exploratory probing.

antennifer *n.* socket of antenna.

antennule *n.* small antenna or feeler, especially the 1st pair of antennae in Crustacea.

anterior *a.* (1) nearer head end; (2) ventral in human anatomy; (3) facing outwards from axis; (4) previous. *cf.* posterior.

anterior commissure tract of fibres connecting the two cerebral hemispheres, anterior and ventral to the anterior end of corpus callosum.

anterior horn of the spinal cord, that part of grey matter containing cell bodies of motor neurons. *cf.* posterior horn.

anterior pituitary adenohypophysis (*q.v.*), excluding the pars intermedia.

anterior–posterior antero-posterior *q.v.*

anterograde *a.* (1) *appl.* transport of movement of material in axons of neurons away from cell body; (2) *appl.* degeneration of a neuron due to injury to the distal end. *alt.* Wallerian degeneration.

anterolateral system somatosensory nerve system which carries pain information to the brain. *alt.* spinothalamic tract.

antero-posterior *a. appl.* axis, from head to tail of the animal body.

antesternite *n.* anterior sternal sclerite of insects. *alt.* basisternum.

anthela *n.* the cymose inflorescence of the rush family, Juncaceae.

anther *n.* terminal part of stamen, which produces the pollen. *see* Fig. 18 (p. 239).

antheridiophore *n.* structure bearing antheridia.

antheridium *n.* organ or receptacle in which male gametes are produced in many cryptogams (ferns, mosses and liverworts), algae and fungi. *plu.* antheridia.

antherophore *n.* stalk of a stamen bearing many anthers, in male cone of some gymnosperms.

antherozoid *n.* motile male gamete produced from an antherozoid mother cell in an antheridium. *alt.* sperm, spermatozoid.

anthesis *n.* (1) stage or period at which flower bud opens; (2) flowering.

anthoblast *n.* young sessile polyp of a stony coral.

anthocarp *n.* collective, composite or aggregated fruit formed from an entire inflorescence, such as pineapple and fig.

Anthocerophyta, Anthoceratopsida, Anthocerotae, Anthocerotales *n.* the hornworts, a group of small spore-bearing non-vascular green plants with a thalloid gametophyte and rosette-like habit of growth. The cells typically contain a single large chloroplast with a pyrenoid. Hornworts often carry symbiotic photosynthetic cyanobacteria in the intercellular spaces. The sporophyte is typically an upright elongated sporangium on a stalk growing from the gametophyte. *see also* bryophytes.

anthoclore *n.* yellow pigment dissolved in cell sap of corolla, as of primrose.

anthocodia *n.* distal portion of polyp of soft corals (Alcyonaria) bearing mouth and tentacles.

anthocyanidins *see* anthocyans.

anthocyanins *see* anthocyans.

anthocyans *n.* water-soluble red, blue or purple flavonoid pigments found in the cell vacuoles of plants and comprising the anthocyanidins and the flavonoid glycoside anthocyanins. They are also found in some insects, absorbed with plant food.

anthodium *n.* head of florets, as in Compositae.

anthogenesis *n.* in some aphids, production of both males and females by asexual forms.

anthophilous *a.* (1) attracted by flowers; (2) feeding on flowers.

anthophore *n.* elongation of receptacle between calyx and corolla.

Anthophyta *n.* the flowering plants, one of the five main divisions of extant seed-bearing plants. Reproductive organs (stamens and ovary) are carried in flowers, in which the sporophylls (stamens and carpels) are typically surrounded by sterile leaves (petals and sepals). After pollination and fertilization the closed ovary containing the seeds develops into a fruit. The haploid gametophyte generation is much reduced, being restricted to the male and female gametes and cells giving rise to them. *alt.* angiosperms, Magnoliophyta. *see* Appendix 1.

anthostrobilus *n.* strobilus (cone) of certain cycads.

anthotaxis *n.* arrangement of flowers on an axis.

anthoxanthins *n.plu.* sap-soluble flavone glycoside flower pigments giving colours from ivory to deep yellow, also found in insects, having been absorbed from the plant on which the insect feeds.

Anthozoa, anthozoans *n.*, *n.plu.* class of coelenterates of the phylum Cnidaria, comprising the soft corals, sea pens and sea fans (subclass Alcyonaria), and the sea anemones and stony corals (subclass Zoantharia). The soft corals, sea fans and stony corals are generally colonial, with individual polyps connected by living tissue and, in the case of the stony corals, embedded in a calcium carbonate matrix. Sea anemones are generally solitary.

anthracobiontic *a.* growing on burned soil or scorched material.

Anthracosauria, anthracosaurs *n.*, *n.plu.* order of Carboniferous to Permian labyrinthodont amphibians, among whose members were the ancestors of the reptiles.

anthraquinone *n.* any of a class of orange or red pigments found in lichens, fungi, higher plants and insects such as the cochineal beetle and lac insect.

anthropic zone that area of the Earth's surface that is under the influence of humans.

Anthropocene *n.* the most recent period in the Earth's history, the time when human activity began to have a significant impact on the Earth's climate and ecosystems – considered by some to have begun with the Industrial Revolution, and by some much earlier, with the rise of farming.

anthropocentrism *n.* an exclusively human-centred view that human activities are paramount and need take no account of non-human species. *a.* **anthropocentric**.

anthropochory *n.* accidental dispersal by humans (via spores, pollen, etc.).

anthropogenesis *n.* the evolutionary descent of humans. *a.* **anthropogenetic**.

anthropogenic *a.* produced or caused by humans.

anthropoid *a.* resembling or related to humans, as the anthropoid apes (family Pongidae): orang utan, chimpanzee and gorilla. Gibbons (family Hylobatidae) are sometimes also included.

Anthropoidea *n.* the suborder of primates consisting of monkeys, apes and humans.

anthropology *n.* the scientific study of human beings and human societies, especially differences in social organization, racial differences, physiological differences and social and religious development.

anthropometry *n.* study of proportional measurements of parts of the human body.

anthropomorphism *n.* ascribing human emotions and behaviour to animals.

anthropomorphous *a.* resembling humans.

anthropozoonotic *n.* a human infection that can be transmitted to a variety of animal hosts, which act as reservoirs of infection.

anti- prefix derived from Gk *anti*, against, or L. *ante*, before.

antiae *n.plu.* feathers at base of bill in some birds.

antiauxin *n.* any compound that regulates or inhibits growth stimulation by auxins.

antibiosis *n.* antagonistic association of organisms in which one produces compounds, known as antibiotics, which are harmful to the other(s).

antibiotic (1) *n.* any of a diverse group of organic compounds produced by microorganisms which selectively inhibit the growth of or kill other microorganisms. Many antibiotics are used therapeutically against bacterial and fungal infections in humans and animals; (2) *a.* killing or inhibiting growth.

antibiotic resistance *see* drug-resistance factors, R plasmid, transposon.

antibiotin avidin *q.v.*

antibody (Ab) *n.* immunoglobulin which is secreted by plasma cells, derived from B lymphocytes after activation by an antigen, and which specifically recognizes that antigen, binding selectively to it and aiding its elimination by other components of the immune system. The body can make an almost unlimited variety of different antibodies, each B lymphocyte being genetically programmed early in its development to produce antibody of a single antigen specificity. Antibodies fall into several different classes or isotypes, which vary in their function in the immune response. *see also* adaptive immune response, B lymphocyte, immunoglobulin, immunoglobulin A, immunoglobulin D, immunoglobulin E, immunoglobulin G, immunoglobulin M.

antibody class *see* immunoglobulin A, immunoglobulin D, immunoglobulin E, immunoglobulin G, immunoglobulin M.

antibody combining site the site on an antibody that binds the antigen.

antibody-dependent cell-mediated cytotoxicity (ADCC) cell killing by various types of white blood cell, especially natural killer cells and eosinophils, which requires the target cell to be coated with specific antibody.

antibody diversity the production of an almost unlimited repertoire of antibodies of different specificities by an individual immune system. Different antibodies are produced by different B cells. Each B cell is programmed, by gene rearrangment at the immunoglobulin loci early in its development, to produce an antibody of a single specificity. *see* immunoglobulin genes.

antibody engineering the production by genetic engineering of antibodies with new properties (e.g. abzymes), or hybrid antibodies in which sequences from two dif-

ferent species (e.g. mouse and human) have been combined. One product of antibody engineering is human antibodies bearing specific antigen-binding sites derived from mouse antibodies. In this way, antibodies against a particular antigen, which can be obtained more easily in mice, are rendered non-immunogenic to humans and thus less likely to be inactivated than the unmodified animal antibody.

antibody genes *see* immunoglobulin genes.

antibody repertoire the total variety of antibodies that an individual can produce.

antiboreal *a. pert.* cool or temperate regions of the Southern Hemisphere.

antibrachial *a. pert.* the forearm or corresponding portion of a forelimb.

antical *a. appl.* the upper or front surface of a thallus, leaf or stem, esp. in liverworts.

anticlinal *a.* (1) *appl.* plane of cell division at right angles to surface of apex of a growing point; (2) in quadrupeds, *appl.* one of the lower thoracic vertebrae with upright spine towards which those on either side incline.

anticoagulant *n.* any substance which prevents coagulation or clotting of blood, such as dicoumarol, warfarin, heparin.

anticoding strand antisense strand *q.v.*

anticodon *n.* group of three consecutive bases in tRNA complementary to a codon in mRNA.

anticryptic *a. appl.* protective coloration facilitating attack.

antidiuretic *a.* (1) reducing the volume of urine; (2) *appl.* hormone (vasopressin) that controls water reabsorption by kidney tubules.

antidiuretic hormone vasopressin *q.v.*

antidromic *a.* (1) contrary to normal direction; (2) (*bot.*) *appl.* stipules with fused outer margins.

antifertilizin *n.* a protein in cytoplasm of spermatozoa which reacts with fertilizin produced by ovum.

antifreeze compounds (1) compounds such as glycerol, sorbitol and mannitol, which lower the freezing point of body fluids and protect against freezing, found in the haemolymph of some insects; (2) glycoproteins found in the blood of some polar fish, and which depress the freezing point of the blood by enveloping small ice crystals that would otherwise form ice nuclei and cause the blood to freeze.

antigen (Ag) *n.* any substance capable of binding specifically to an antibody or a T-cell receptor. An antigen may be unable to induce a specific immune response when administered on its own, but will do so if attached to a suitable carrier. *a.* **antigenic**. *cf.* immunogenicity.

antigen–antibody complex complex of antibody with its specific antigen non-covalently bound to the antigen-binding site. Such complexes are formed when antigen and antibody come together, and are the form in which foreign antigens are most effectively scavenged by phagocytic cells and thus removed from the body. *see also* Fc receptors, immune complex.

antigen-binding site the part of an immunoglobulin molecule that specifically binds an antigen. It is composed of the variable regions of one light and one heavy chain, each different antigen-binding site recognizing a single antigenic determinant. Each immunoglobulin monomer has two identical antigen-binding sites.

antigenic determinant site on an antigen molecule that elicits the formation of a specific antibody or activates a specific T cell, and against which the antibody or T-cell activity is directed. *alt.* epitope.

antigenic drift a gradual change in the antigens carried by some viruses, esp. influenza viruses, as a result of small genetic changes.

antigenicity *n.* the property possessed by a substance that can bind specifically to the antigen receptors on B or T lymphocytes, and thus, in principle, is able to stimulate a specific immune response. *see* antigen. *cf.* immunogenicity.

antigenic shift a substantial change in the antigens carried by some viruses, esp. influenza viruses, caused by the recombination between two virus strains, which manifests itself as the sudden appearance of a new virus type.

antigenic variation the ability of African trypanosomes (e.g. *Trypanosoma brucei*) and some other microorganisms to change the cell-surface antigens that they synthes-

ize in successive generations. In trypanosomes, this is due to DNA rearrangements which bring a different gene into a position where it is expressed.

antigen presentation the process by which a foreign protein antigen is prepared for recognition by the T lymphocytes of the immune system. Cells take up incoming protein antigens and break them down into peptide fragments intracellularly. The peptides form complexes with MHC molecules, and these complexes are displayed on the cell surface. In this form the antigen can be recognized and responded to by T lymphocytes specific for the peptide–MHC combination. *see also* antigen-presenting cell, cytotoxic T lymphocyte, helper T lymphocyte, MHC molecules, T-cell receptor.

antigen-presenting cell (1) any cell displaying a complex of peptide antigen and MHC molecule on its surface; (2) professional antigen-presenting cell *q.v.*

antigen processing the partial breakdown of protein antigens by cells. The resulting peptide fragments are then displayed on the cell surface in complexes with MHC molecules.

antigen receptor cell-surface protein on lymphocytes that binds antigen. The antigen receptors on B cells are immunoglobulins and on T cells they are structurally similar molecules known as T-cell receptors.

antigiberellin *n.* any compound (e.g. phosphon, maleic hydrazide) with action on plant growth opposite to that of giberellins, causing plants to grow with short thick stems.

antihaemorrhagic *a. appl.* agents that stop bleeding, *appl.* vitamin: vitamin K *q.v.*

antihelix *n.* the curved prominence in front of helix of ear.

antihelminthic *n.* drug effective against parasitic flatworms or roundworms.

anti-idiotypic antibody antibody specific for an antigenic determinant located in the variable region of another antibody.

anti-immunoglobulin an antibody against an immunoglobulin, most usually against a determinant on the constant region.

Antillean subregion subdivision of the Neotropical zoogeographical region, comprising the islands of the Caribbean.

antimeres *n.plu.* (1) corresponding parts, such as left and right limbs, of a bilaterally symmetrical animal; (2) a series of equal radial parts, or actinomeres, of a radially symmetrical animal.

antimetabolite *n.* any substance that blocks a metabolic reaction, e.g. by competing with the natural substrate for enzyme active sites.

antimitotic *a.* inhibiting or preventing mitosis.

antimorph *n.* a mutant allele that has an opposite effect to the normal allele, competing with the normal allele when in the heterozygous state.

anti-Müllerian duct hormone (AMH) Müllerian-inhibiting substance *q.v.*

antimutagen *n.* any substance or other agent that slows down the mutation rate or reverses the action of a mutagen.

antimycin *n.* compound used experimentally as an inhibitor of cellular respiration.

antineuritic *a. appl.* vitamin: thiamine, lack of which causes polyneuritis.

antinociceptive *a. appl.* any agent that can lessen or prevent the generation or transmission of a painful or injurious sensation.

antiparallel *a.* describes two similar structures arranged in opposite orientations (e.g. the two strands of the DNA double helix).

antiparallel β-sheet type of β-sheet secondary structure in proteins in which adjacent strands of the sheet run in opposite directions. It can be formed by the folding of contiguous regions of the protein chain. *cf.* parallel β-sheet.

antiperistalsis *n.* peristalsis in the posterior–anterior direction.

antipetalous *a.* inserted opposite the insertion of the petals.

antipodal *a.* in plants, *appl.* group of three cells at the end of embryo sac opposite the micropyle.

antiport, antiporter *n.* membrane protein that transports a solute across the membrane, the transport depending on the simultaneous or sequential transport of another solute in the opposite direction. *see also* coupled transport, symport.

antiprostate *n.* bulbo-urethral gland *q.v.*

antipyretic *a. appl.* drugs that lower body temperature.

antirachitic *a.* preventing rickets, *appl.* vitamin: vitamin D.

anti-reductionism *see* reductionism.

Antirrhinum majus snapdragon, an ornamental dicot plant widely used in experimental plant genetics.

antiscorbutic *a.* preventing or counteracting scurvy, *appl.* vitamin: vitamin C (ascorbic acid).

antisense RNA RNA complementary to the normal RNA transcript of a gene, and which can block its expression by hybridizing to the DNA or by binding to the RNA transcript and preventing its translation.

antisense strand in double-stranded DNA, the DNA strand used as the template strand for transcription and thus complementary to the sense, or coding, strand. *alt.* anti-coding strand, template strand.

antisepalous *a.* inserted opposite the insertion of the sepals.

antiseptic (1) *n.* a substance that destroys harmful microorganisms; (2) *a.* preventing putrefaction.

antiserum *n.* blood serum containing specific antibodies produced after immunization or natural infection.

antisocial factor any selection pressure that tends to reverse social evolution.

antispadix *n.* a group of four modified tentacles in internal lateral lobes of *Nautilus*.

antitermination *n.* the continuation of transcription by RNA polymerase past the usual termination point in a gene, caused by the interaction of proteins known as antitermination factors with the enzyme.

antithesis, principle of *see* principle of antithesis.

antithrombin *n.* general name for any substance that neutralizes the action of thrombin and thus limits blood coagulation. Antithrombin III is an α-globulin that inactivates thrombin.

antitoxin *n.* substance which neutralizes a toxin by combining with it.

antitragus *n.* prominence opposite tragus of external ear.

antitrochanter *n.* in birds, an articular surface on ilium, against which trochanter of femur abuts.

antitrope *n.* any structure that forms a bilaterally symmetrical pair with another.

antitropic *a.* (1) turned or arranged in opposite directions; (2) arranged to form bilaterally symmetrical pairs.

antitropous *a.* (1) inverted; (2) *appl.* plant embryos with radicle directed away from hilum.

α₁-antitrypsin deficiency familial emphysema, an inherited defect leading to emphysema (overinflation and distension of air sacs in the lungs causing shortness of breath), caused by a genetic defect resulting in the production of inactive antitrypsin.

antitrypsins *n.plu.* protein inhibitors of the enzyme trypsin, produced in various animals and plants. Some are members of the serpin protein family (e.g. α₁-antitrypsin). The genes for some plant antitrypsins have been transferred to crop species that lack them in order to make them resistant to insect pests, which cannot digest the plant material as a result of the antitrypsins it contains and so starve to death.

antiviral *a. appl.* antibodies, drugs or other agents that destroy or neutralize a virus or prevent its replication.

antlers *n.plu.* paired bony growths, projections from the skull, on the heads of members of the deer family, which are often branched, are shed annually and are usually confined to males.

antlia *n.* the spiral sucking proboscis of Lepidoptera.

ant lions a group of insects in the order Neuroptera *q.v.*

antral follicle immature fluid-filled ovarian follicle.

antrorse *a.* directed forwards or upwards.

antrum *n.* (1) a cavity or sinus; (2) fluid-filled cavity in developing ovarian follicle. *a.* **antral**.

ants *n.plu.* social insects of the superfamily Formicoidea of the order Hymenoptera, which live in colonies composed of a queen, with male, worker and, in some cases, soldier castes.

anucleate *a.* without a nucleus.

anucleolate *a.* without a nucleolus.

Anura, anurans *n., n.plu.* one of the three orders of extant amphibians, comprising the frogs and toads. In some classifications called the Salientia.

anural, anurous *a.* tailless.

anus *n.* the opening of the alimentary canal (usually posterior) through which undigested food is voided. *a.* **anal**.

anvil incus *q.v.*

aorta *n.* (1) in mammals, the great trunk artery that carries blood from the heart to the arterial system of the body; (2) in other animals, major blood vessel carrying oxygenated blood. *see* dorsal aorta, ventral aorta.

aortic *a. pert.* aorta, *appl.* e.g. hiatus, isthmus, lymph glands, semilunar valves.

aortic arches paired arteries in vertebrate embryos, which connect dorsal and ventral arteries, running between gill slits on either side.

aortic bodies two small masses of chromaffin cells in a capillary plexus, one on each side of foetal abdominal aorta, being part of system for controlling the oxygen content and acidity of blood. *alt.* Zuckerkandl's bodies.

AP aminopurine *q.v.*

AP-1 a transcriptional regulatory protein in mammalian cells, a dimer of the proteins Jun and Fos.

4-AP 4-aminopyridine *q.v.*

apandrous *a.* (1) without functional male sex organs; (2) without antheridia; (3) parthenogenetic, such as oospores in certain oomycetes.

apatetic *a. appl.* misleading coloration.

APC anaphase-promoting complex *q.v.*

ape *n.* any member of the primate superfamily Hominoidea (*q.v.*), comprising human, chimpanzee, gorilla, orangutan and gibbon species and their extinct relatives back to the divergence of this lineage from that of the Old World monkeys. *see* great apes, lesser apes.

AP endonuclease any of a group of DNA repair endonucleases that make a single-stranded cut in DNA to the 5′ side of a nucleotide from which the purine or pyrimidine base has been removed.

aperispermic *a. appl.* seeds without nutritive tissue.

apertura piriformis anterior nasal aperture of skull.

apes *see* Primates.

apetalous *a.* without petals.

apex *n.* tip or summit, as of wing, heart, lung, root, shoot. *plu.* **apices**.

aphagia *n.* refusal to eat.

Aphaniptera Siphonaptera *q.v.*

aphanipterous *a.* apparently without wings.

aphanoplasmodium *n.* plasmodium consisting of a network of fine strands of protoplasm, resembling lace.

aphasia *n.* impairment in the understanding or production of language as the result of an injury to the brain.

Aphasmidia *n.* class of nematode worms with no phasmids, and whose amphids open on to the posterior part of the head capsule.

apheliotropism *n.* tendency to turn away from the sun.

aphid *n.* insect of the family Aphididae (Aphidae) of the Hemiptera with mouthparts adapted for piercing and sucking plants. Aphids are of economic importance as plant pests and vectors of plant virus diseases, and have a parthenogenetic and a sexual reproductive phase.

aphidicolin *n.* fungal antibiotic that inhibits DNA replication in eukaryotes and DNA polymerase α *in vitro*.

aphins *n.plu.* red and yellow fat-soluble pigments extracted from various aphids, probably arising after death from protaphin.

aphlebia *n.* lateral outgrowth from base of frond stalk in certain ferns.

aphodus *n.* short tube leading from internal chamber lined with flagellate cells to the excurrent canal system in sponges. *a.* **aphodal**.

aphotic *a.* *pert.* absence of light, *appl.* zone of deep sea where daylight fails to penetrate.

aphototropism *n.* tendency to turn away from light.

Aphragmabacteria *n.* in some classifications the name for the group of prokaryotes comprising the mycoplasmas and similar organisms, small bacteria lacking the typical bacterial cell wall.

aphthous *a.* producing blisters.

aphthoviruses *n.plu.* group of animal viruses in the Picornaviridae, e.g. foot-and-mouth disease virus.

Aphyllophorales *n.* lineage within the basidiomycetes that contains the polypores, chanterelles, tooth fungi, coral fungi and corticoids.

aphyllous *a.* without foliage leaves. *n.* **aphylly**.

aphytic *a.* without plant life, *appl.* zone of coastal waters below *ca.* 100 m or to the bottoms of deep lakes.

Apiales Cornales *q.v.*

apical *a.* (1) at the tip of any cell, structure or organ; (2) *pert.* distal end; (3) *appl.* cell at tip of growing point; (4) (*bot.*) *appl.* style arising from summit of ovary.

apical–basal axis the axis running from the tip of a plant shoot to the tip of the root.

apical cell in development of plant embryo, the small cell that is produced by unequal division of the fertilized egg and gives rise to the embryo.

apical dominance phenomenon common in plant development in which the bud at the tip of the shoot, the apical bud, suppresses the development of lateral buds that have formed further down the stem. If the apical bud is removed, the lateral buds then develop. The apical bud produces a growth-inhibitory hormone that is carried back down the stem.

apical ectodermal ridge ridge on end of limb bud in mammalian and avian embryos which produces signals that influence the development of the limb. *alt.* apical ridge.

apical membrane in an epithelium lining an internal cavity (e.g. gut, lungs, glands), the face of an epithelial cell which is adjacent to the cavity.

apical meristem dividing tissue at tip of developing shoot and young root, at which growth occurs.

apical placentation in plant ovary, placentation where ovule is at the apex of ovary.

apical ridge apical ectodermal ridge *q.v.*

apices *plu.* of apex.

Apicomplexa, apicomplexans *n.*, *n.plu.* phylum of non-photosynthetic heterotrophic protists parasitic in animals, comprising the sporozoan protozoans, e.g. gregarines, coccidians, *Plasmodium* and piroplasms. They are transmitted from host to host in the form of 'spores', small infective bodies produced by schizogony. *alt.* Sporozoa.

apiculate *a.* coming abruptly to a small tip, as in some leaves.

apilary *a.* having upper lip missing or suppressed in corolla.

apileate *a.* having no pileus.

apitoxin *n.* main toxic fraction of bee venom.

apivorous *a.* feeding on bees.

aplacental *a.* having no placenta, *appl.* mammals: the monotremes.

aplanetic *a.* non-motile, *appl.* spores.

aplanetism *n.* absence of motile spores or gametes.

aplanogamete *n.* a non-motile gamete.

aplanospore *n.* a non-motile resting spore.

aplasia *n.* (1) arrested development; (2) non-development; (3) defective development.

aplastic *a.* (1) *pert.* aplasia; (2) without change in development or structure.

aplastic anaemia lack of red blood cells as a result of their non-development in the bone marrow.

aplerotic *a.* not entirely filling a space.

aploperistomatous *a.* having a peristome with one row of teeth, as do mosses.

aplostemonous *a.* with a single row of stamens.

Aplysia genus of opisthobranch molluscs known as sea hares, usually referring to *Aplysia californica*, which is used as an experimental animal in neurobiology.

apneustic *a.* with spiracles closed or absent, *appl.* aquatic larvae of certain insects.

apocarp *n.* individual carpel of a composite fruit.

apocarpous *a.* having separate or partially united carpels. *n.* **apocarpy.**

apocentric *a.* diverging or differing from the original type.

apocratic *a.* opportunistic, *appl.* species.

apocrine *a.* *appl.* glands, e.g. mammary glands, whose secretion accumulates beneath the surface and is released by breaking away of the outer part of the cells.

apocytium *n.* multinucleate mass of naked cytoplasm.

Apoda *n.* (1) an order of limbless burrowing amphibians, commonly known as caecilians, having a reduced or absent larval stage and minute calcified scales in the skin. In some classifications called the Gymnophiona; (2) the name has also been given to orders of parasitic barnacles (crustaceans) and burrowing sea cucumbers (echinoderms).

apodal *a.* (1) having no feet; (2) having no ventral fin; (3) stemless.

apodeme *n.* an internal skeletal projection in arthropods.

apoderma *n.* enveloping membrane secreted during resting stage between instars by certain ticks and mites.

Apodiformes *n.* an order of birds including the swifts.

apoenzyme *n.* inactive protein part of an enzyme remaining after removal of the prosthetic group. *cf.* holoenzyme.

apogamy *n.* a type of apomixis in which the embryo is produced from the unfertilized female gamete or from an associated cell. *see also* generative apogamy.

apogeotropism ageotropism *q.v.*

apogynous *a.* lacking functional female reproductive organs.

apoinducer *n.* regulatory protein that activates a gene by binding to control regions in the DNA and allowing transcription to take place.

apolegamic *a.* *appl.* mating associated with sexual selection.

apolipoprotein *n.* the protein component of a lipoprotein.

apolysis separation of old cuticle from epidermis that occurs in insect larvae at the start of moulting.

apomeiosis *n.* sporogenesis without meiosis. *a.* **apomeiotic.**

apomict *n.* an organism reproducing by apomixis.

apomixis *n.* (1) asexual reproduction without meiosis or fertilization in plants, akin to parthenogenesis but including development from cells other than ovules. *see* apogamy *q.v.*, apospory *q.v.*; (2) vegetative apomixis *q.v.* *a.* **apomictic.** *cf.* amphimixis.

apomorphy *n.* in cladistic phylogenetics, a novel character evolved from a preexisting character. The original character and the derived character form a homologous pair termed an evolutionary transformation series. *see also* synapomorphy.

aponeurosis *n.* flattened tendon for insertion of a muscle.

apopetalous polypetalous *q.v.*

apophyllous *a.* having free perianth leaves.

apophysate *a.* having an apophysis.

apophysis *n.* (1) (*zool.*) a projecting process on bone or other skeletal material, usually for muscle attachment; (2) (*bot. &*

mycol.) various small protuberances, as on hyphae or on capsule of mosses; (3) small protuberance at base of seed-bearing scales in pine cones.

apoplasmodial *a.* not forming a typical plasmodium.

apoplast *n.* the cell walls collectively of a tissue or a complete plant. *a.* **apoplastic**.

apoplastic pathway in plant tissue, the movement of ions and other solutes across stems or roots via cell walls.

apoplastid *n.* a plastid lacking chromatophores.

apoprotein *n.* protein lacking its prosthetic group. *see also* apoenzyme.

apoptosis *n.* cell death as a result of activation of an intracellular 'suicide' programme. It is a normal and essential event during development generally and within the immune system. Apoptosis does not lead to lysis of cells and thus avoids damage to neighbouring tissue. *alt.* programmed cell death. *cf.* necrosis.

apopyle *n.* exhalent pore of sponges.

aporogamy *n.* entry of pollen tube into ovule by some method other than through the micropyle.

aporrhysa *n.plu.* exhalent canals in sponges.

aposematic *a. appl.* warning coloration or markings that signal to a predator that an organism is toxic, dangerous or distasteful. *cf.* parasematic.

aposporogony *n.* absence of sporogony.

apospory *n.* a type of apomixis in which a diploid gamete is produced from the sporophyte without spore formation.

apostasis *n.* condition of abnormal growth of axis that causes separation of perianth whorls from one another.

apostatic *a.* differing markedly from the normal.

apostatic selection type of frequency-dependent selection in which a predator selects the most common morph in the population.

apostaxis *n.* abnormal or excessive exudation.

apostrophe *n.* arrangement of chloroplasts along lateral walls of leaf cells in bright light.

apothecium *n.* (1) open cup-shaped fruiting body of Discomycetes (cup fungi, morels and truffles) bearing asci on the inner surface; (2) fruiting body of some lichens. *plu.* **apothecia**.

apothelium *n.* a secondary tissue derived from a primary epithelium.

apotracheal *a.* with xylem parenchyma independent of vessels, or dispersed, *appl.* wood.

apotropous *a. appl.* an anatropous ovule with a ventrally situated line of fusion with funicle.

apotypic *a.* diverging from a type.

apparent free space that part of a tissue lying outside the plasma membranes of its constituent cells. It includes the intercellular spaces and, in plant tissues, the cell wall as well.

apparent mortality a measure of mortality in a population at a given developmental stage (e.g. age group) expressed as the percentage of the number alive at the beginning of the stage. *alt.* percentage successive mortality.

appeasement *n. appl.* behaviour which ends the attack of one animal on another of the same species by the loser's adopting a submissive posture or gesture.

appendage *n.* organ or part attached to a trunk, as limb or branch.

appendical *a. pert.* appendix.

appendices *plu.* of appendix.

appendicular *a.* (1) *pert.* appendages, *appl.* skeleton of limbs; (2) *pert.* appendix.

appendiculate *a.* (1) having a small appendage, as a stamen or filament; (2) having an appendiculum.

appendiculum *n.* remains of the partial veil on rim of cap of some agaric fungi.

appendix *a.* an outgrowth, esp. the vermiform appendix of human intestine. *plu.* **appendices**.

appendix colli the hanging tuft of hairs on the neck of an animal, e.g. goat.

appetitive *a. appl.* behaviour at the beginning of a fixed behaviour pattern. It can be very variable, ranging from unoriented wanderings to apparently purposeful behaviour.

applanate *a.* flattened.

application factor a factor used to determine the maximum safe concentration of a substance for an organism. It is the ratio of the concentration of the substance that produces a certain long-term response in the organism to the concentration causing death in 50% of the population within a given time period.

apposition *n.* laying down of material on a preformed surface, as in growth of cell wall or bone.

appressed *a.* pressed together without being united.

appressorium *n.* (1) adhesive disc, as of sucker or haustorium; (2) modified hyphal tip of parasitic fungi which may form a haustorium or penetrate the substrate.

apraxia *n.* impairment in the ability to carry out skilled voluntary movements although there is no muscle paralysis.

aproterodont *a.* having no premaxillary teeth.

AP site apurinic or apyrimidinic site. A site on DNA at which a purine or pyrimidine base has been removed.

aptamer *n.* small DNA molecule that has been selected for its binding properties for a particular molecule.

apteria *n.plu.* naked or down-covered surfaces between feather tracts on bird skin.

apterous *a.* (1) (*zool.*) wingless; (2) (*bot.*) having no wing-like extensions on stems or petioles.

apterygial *a.* (1) wingless; (2) without fins.

Apterygiformes *n.* an order of flightless birds including the kiwis.

apterygote, apterygotous *a.* wingless, *appl.* a group of insects, the subclass Apterygota, that have no wings, little or no metamorphosis, and abdominal appendages in the adult. It comprises the orders Thysanura, Diplura, Protura and Collembola.

apurinic *a. appl.* a nucleotide in DNA that has lost its purine base.

apyrene *a.* seedless, *appl.* certain cultivated fruits.

apyrimidinic *a. appl.* a nucleotide in DNA that has lost its pyrimidine base.

aquaculture *n.* raising of algae, fish and shellfish for human use in artificial or natural freshwater ponds, lakes, irrigated fields and irrigation ditches, and, for marine organisms, in enclosures in coastal inlets and estuaries.

aquaporins *n.plu.* family of channel-forming membrane proteins which allow the passage of water into cells through the plasma membrane.

aquatic *a.* living in or near water.

aquatic ecosystems any ecosystem of which the principal component is water, such as ponds, lakes, rivers, streams and oceans. *cf.* wetland.

aqueduct *n.* fluid-filled channel or passage, as that of cochlea and of vestibule of ear.

aqueduct of Sylvius channel running through midbrain, connecting 3rd and 4th ventricles. *alt.* aqueduct, cerebral aqueduct.

aqueous *a.* watery, *appl.* humour: fluid between lens and cornea.

aqueous solution solution of a solute in water.

Aquificae *n.* a phylum of Bacteria, distinguished on DNA sequence data. It includes hyperthermophilic chemolithotrophs that oxidize H_2 or reduced sulphur compounds. *Aquifex* is thought to be the closest known relative of the universal ancestor of all Bacteria. Previously known as the Aquifex–Hydrogenobacter group.

aquiherbosa *n.* herbaceous vegetation growing submerged in water.

Arabidopsis thaliana thale-cress, a small annual plant of the Cruciferae, widely used as a model organism in plant molecular genetic and developmental research because of its ease of culture and small simple genome.

arabinan *n.* any of a group of poly-saccharides composed predominantly of arabinose subunits, found in some plant cell walls.

arabinogalactan *n.* polysaccharide found in plant cell walls, composed of arabinose and galactose subunits.

arabinose *n.* five-carbon aldose sugar found esp. in plant gums, pectins and cell-wall polysaccharides.

arable *a. appl.* land that is cultivated, and to crops grown on cultivated land, except trees. *cf.* pasture, rangeland.

arachidonic acid a fatty acid, a precursor of prostaglandins and prostacyclins.

Arachnida, arachnids *n.*, *n.plu.* class of mainly terrestrial, carnivorous arthropods, included in the subphylum Chelicerata, comprising spiders (order Araneae), scorpions (Scorpiones), mites and ticks (Acari), false scorpions (order Pseudoscorpiones), palpigrades (order Palpigrada), solifugids (order Solifugae) and harvestmen (Opiliones). They have a body usually divided into a prosoma of eight fused segments and a posterior opisthosoma of 13 fused segments. The prosoma is not differentiated into a head and thorax and bears the clawed and prehensile chelicerae, the pedipalps, and four pairs of walking legs.

arachnidium *n.* the spinning apparatus of a spider, including spinning glands and spinnerets.

arachniform *a.* arachnoid, stellate.

arachnodactyly *n.* having abnormally long and slender fingers.

arachnoid *a.* (1) *pert.* or resembling a spider; (2) like a cobweb; (3) consisting of fine entangled hairs; (4) *n.* the middle of the three membranes surrounding the brain and spinal cord.

arachnoidal *a. pert.* the arachnoid membrane.

arachnoidal granulations Pacchionian bodies *q.v.*

aragonite *n.* a crystalline form of calcium carbonate, one of the constituents of mollusc shells.

Arales *n.* an order of herbaceous monocots comprising the families Araceae (arum) and Lemnaceae (duckweed).

Araliales Cornales *q.v.*

Araneida spiders *q.v.*

araneose, araneous *a.* covered with or consisting of fine entangled filaments.

arbacioid *a.* of sea urchins, having one primary pore plate, with a secondary on either side.

arboreal *a.* (1) living in trees; (2) *pert.* trees.

arborescence arborization *q.v.*

arborescent *a.* branched like a tree.

arboretum *n.* a collection of species of trees.

arboriculture *n.* the cultivation of trees.

arborization *n.* tree-like branching, as of dendrites and axon on nerve cells.

arboroid *a.* tree-like.

arbovirus *n.* arthropod-borne virus, a virus that replicates in an arthropod as its intermediate host and in a vertebrate as its definitive host. Examples are yellow fever virus, which is transmitted by mosquitoes, and the tick-borne encephalitis viruses. Previously known as arborviruses.

arbuscular *a.* shrub-like.

arbuscular mycorrhiza common type of endomycorrhiza characterized by the occurrence of arbuscules (discrete masses of branched hyphae) in infected tissues and hyphae ramifying within and between the cells of the root cortex.

arbuscule *n.* small tree-like shrub or dwarf tree.

arbutoid *a. appl.* endomycorrhizas formed on members of the tribe Arbutoideae (family Ericaceae), with a well-defined fungal sheath and Hartig net and extensive penetration of the cells of the root cortex.

arcade *n.* (1) an arched channel or passage; (2) a bony arch.

arch-, arche- prefix derived from Gk *arch*, beginning.

archae- prefix derived from Gk *archaios*, primitive.

Archaea, archaea *n., n.plu.* (1) superkingdom or domain of prokaryotic microorganisms previously known as the archaebacteria. It is distinguished from the other prokaryotic domain, the Bacteria, on genetic and biochemical grounds. Its members are known as **archaea** (*sing.* **archaeon**) and include the methanogens, the extreme thermophiles, the extreme halophiles, and *Thermoplasma*. Archaea differ from bacteria in having no muramic acid in their cell walls and in having ether-linked membrane lipids that contain side chains of repeating isoprene units instead of fatty acids. *see* Appendix 6 for the main groups within the Archaea, which all have individual entries within the body of the dictionary; (2) the name has been used in the past for microorganisms found in the oldest Precambrian rocks, i.e. rocks of the Archean era.

archaeal, archaean *a. pert.* the Archaea, or a member of the Archaea.

Archaean Archean *q.v.*

archaebacteria *see* Archaea.

archaeocyte *n.* one of the three types of somatic cells in sponges. It can differentiate into all of the other cell types. *alt.* archeocyte.

archaeon *n.* a member of the Archaea (*q.v.*). *plu.* **archaea**.

Archaeopteryx* and *Archaeornis genera of fossil birds from the Jurassic, which show many reptilian features.

Archaeornithes *n.* subclass of primitive reptile-like fossil birds that includes *Archaeopteryx.*

archaeostomatous *a.* having the blastopore persistent and forming the mouth.

Archaeozoic *see* Archaean.

arch-centra *n.plu.* vertebral centra formed by fusion of basal growths of primary dorsal and ventral cartilaginous outgrowths (arcualia) from centrum external to chordal sheath. *a.* **archecentrous, archicentrous, archocentrous**.

arche- *see* archae-, archi-, arch-.

Archea, archea Archaea *q.v.*

Archean *n.* earliest geological era, the earlier aeon or division of the Precambrian, beginning about 4000 million years ago with the formation of Earth's crust and lasting to 2500 million years ago. The time during which life originated (first traces found at around 3800 million years ago) and was solely unicellular. *a.* **Archean**.

archecentric *a.* conforming more or less with the original type.

archedictyon *n.* intervein network in the wings of some primitive insects.

archegoniophore *n.* branches of bryophytes, or parts of fern prothallus, that bear archegonia.

archegonium *n.* female sex organ on gametophyte of liverworts, mosses, ferns and related plants. It consists of a multicellular, flask-shaped structure containing one ovum (oosphere) in the base (venter). *adj.* **archegonial**. *plu.* **archegonia**.

archencephalic *a. pert.* forebrain.

archencephalon forebrain *q.v.*

archenteron *n.* cavity formed at gastrulation which develops into the gut of the embryo.

archespore sporoblast *q.v.*

archesporium *n.* a cell or mass of cells dividing to form spore mother cells. In anthers it generates the pollen mother cells.

Archezoa name proposed for a class of single-celled eukaryotic microorganisms that lack mitochondria.

archi- prefix derived from Gk *archi*, first. *see also* archae-, arche-, arch-.

Archiascomycetes *n.* lineage of ascomycetes that contains some yeasts and yeast-like fungi: *Schizosaccharomyces, Taphrina, Protomyces, Saitoella, Pneumocystis*.

archibenthic *a. pert.* sea bottom from edge of continental shelf to upper limit of abyssal zone, at depths of *ca.* 200–1000 m.

archicarp *n.* (1) spirally coiled region of thallus or stalk bearing female sex organ in certain fungi; (2) cell which gives rise to a fruiting body.

archichlamydeous *a.* having no petals, or having petals entirely separate from one another.

archicortex *n.* evolutionarily old cortex, such as the hippocampus in mammals. *alt.* archipallium. *cf.* palaeocortex.

archinephridium *n.* excretory organ of certain larval invertebrates, usually a solenocyte.

archipallium archicortex *q.v.*

archipterygium *n.* type of fin in which the skeleton consists of an elongated segmented central axis and a row of jointed rays on either side.

archisternum *n.* cartilaginous elements in ligaments joining muscle blocks in ventral region of thorax, as in tailed amphibians.

archistriatum *n.* region of forebrain in birds involved in organization of motor function.

architomy *n.* asexual reproduction by fission with subsequent regeneration, in certain annelids.

architype *n.* an original type from which others may be derived. *alt.* archetype.

Archosauria, archosaurs *n., n.plu.* subclass of diapsid reptiles, the 'ruling reptiles', that included the dinosaurs, mainly extinct but including the living crocodilians, with specializations of the skeleton showing trend towards bipedalism.

arciform *n.* shaped like an arch or bow.

arcocentrum *n.* vertebral centrum formed from parts of neural and haemal arches.

Arctic and Subarctic region that part of the Boreal kingdom that extends from the far north south to central Alaska, Labrador, central Scandinavia and northern Siberia.

Arctic Circle latitude 66°30′N, to the north of which there is at least one period of 24 hours in summer in which the Sun does not set, and at least one period of 24 hours in winter in which the Sun does not rise.

arcualia *n.* small cartilaginous pieces, dorsal and ventral, fused or free, on vertebral column of fishes.

arcuate *a.* (1) shaped like an arch or bow; (2) *appl.* nucleus, an arc-shaped nucleus in the hypothalamus.

arculus *n.* arc formed by the two wing veins of certain insects.

ardellae *n.* small apothecia of certain lichens, having appearance of dust.

area centralis area corresponding to the fovea in some animal eyes.

area opaca outer, dark area of the chick blastoderm.

area pellucida central lighter area of the chick blastoderm.

Arecales *n.* order of tree-like or climbing monocots with feather- or fan-like leaves, comprising the family Arecaceae (Palmae) (e.g. palms, yuccas).

arena *n.* area used for communal courtship displays.

arenaceous *a.* having properties or appearance of sand.

Arenaviridae, arenaviruses *n.*, *n.plu.* family of RNA viruses with single-stranded genomes in two parts. It includes Lassa fever virus and lymphocytic choriomeningitis virus.

arenicolous *a.* living or growing in sand.

areola *n.* (1) small coloured circle round nipple; (2) part of iris bordering pupil of eye; (3) a small pit. *a.* **areolar**.

areolar glands sebaceous glands on areola of nipple.

areolar tissue type of connective tissue consisting of cells (macrophages, fibroblasts and mast cells) embedded in a matrix of glycoproteins and proteoglycans in which are embedded collagen and elastin fibres.

areolate *a.* divided into small areas by cracks or other margins.

areole *n.* (1) areola *q.v.*; (2) (*bot.*) space occupied by a group of hairs or spines, as in cacti; (3) small area of mesophyll in leaf delimited by intersecting veins.

arescent *a.* becoming dry.

Arg arginine *q.v.*

argentaffin *a.* staining with silver salts, *appl.* certain cells in gastric glands and crypts of Lieberkühn which secrete digestive enzymes.

argenteous *a.* like silver.

argenteum *n.* silvery reflecting layer of iridocytes in the skin of fishes.

Argentinian Floral Region part of the Austral Realm that includes Argentina, Paraguay, southern Chile and the offshore islands including the Falklands (Malvinas).

argentophil *a.* staining with silver salts. *alt.* argyrophil.

argillaceous *a.* (1) having clay-sized particles, *appl.* soil; (2) having the properties of clay.

arginase *n.* enzyme hydrolysing the amino acid arginine to urea and ornithine, important in the urea cycle and also found in some plants. EC 3.5.3.1.

arginine (Arg, R) *n.* amino acid with a basic side chain that is positively charged at physiological pH. It is a constituent of protein, and is essential in the human diet. It is hydrolysed to ornithine and urea in the urea cycle.

arginine–urea cycle urea cycle *q.v.*

arginosuccinate *n.* compound formed from arginine and succinate, intermediate in the urea cycle of vertebrates.

argyrophil *a.* staining with silver salts. *alt.* argentophil.

ARIAs neuregulins *q.v.*

ariboflavinosis *n.* condition of skin cracking and lesions caused by a deficiency of the vitamin riboflavin.

arid *a. appl.* climate or habitat with less than 250 mm annual rainfall, very high evaporation and sparse vegetation.

aridity index a measure of the aridity of an area which takes into account both rainfall and evaporation.

arid zone regions extending from latitudes 15° to 30° in both hemispheres in which rainfall is very low and either evaporates in the high daytime temperatures or drains away rapidly so that it is unavailable to vegetation. This zone contains most of the world's deserts. Parts of the zone support vegetation and some cultivation but are subject to overgrazing and overcultivation in times of drought, which can lead to desertification.

aril *n.* additional covering formed on some seeds after fertilization, and which may be spongy, fleshy (as the red aril of yew berries), or a tuft of hairs.

arillate *a.* having an aril.

arillode *n.* a false aril arising from region of micropyle as an expansion of the opening in outer wall of ovule.

arista *n.* (1) (*bot.*) awn, long-pointed process as in many grasses; (2) (*zool.*) bristle borne on antenna of some dipteran flies.

aristate *a.* with awns, or with a well-developed bristle.

Aristolochiales *n.* an order of dicots of the subclass Magnoliidae, containing one family of herbs and climbing plants, the birthworts (Aristolochiaceae).

Aristotle's lantern calcareous structure around mouth of sea-urchin supporting five long teeth.

aristulate *a.* having a short awn or bristle.

arithmetic growth linear growth *q.v.*

armature *n.* any structure that serves as a defence, e.g. hairs, prickles, thorns, spines, stings.

armilla *n.* (1) bracelet-like fringe; (2) superior ring on stalk of certain fungi.

armillate *a.* fringed.

arm-palisade palisade tissue in which the chloroplast-bearing surface is enlarged by infolding of cell walls beneath the epidermis.

arms race the sequence of evolutionary changes seen in e.g. a predator and its prey, as each advantageous adaptation in one species is countered by a further adaptation in the other.

aroid *n.* plant of the family Araceae, the arum family.

arolium *n.* soft hairy pad at extreme tip of insect leg.

aromatic *a. appl.* organic compounds that contain one or more benzene rings. *cf.* aliphatic.

aromatic amino acids amino acids with an aromatic side chain: phenylalanine, tryptophan, tyrosine.

aromatization *n.* chemical reaction that converts testosterone to oestradiol and other androgens to other oestrogens.

aromorphosis *n.* evolution with an increase in degree of organization without much increase in specialization.

arousal *n.* level of responsiveness to a stimulus in an animal.

array *n.* (1) arrangement in order of magnitude; (2) *see* DNA microarray; protein array.

arrect *a.* upright or erect.

Arrhenius principle the principle that a relatively small percentage change in the average kinetic energy of a population of molecules may result in a relatively large change in the fraction of molecules having energy greater than the activation energy.

arrhenogenic *a.* producing offspring preponderantly or entirely male. *n.* **arrhenogeny**.

arrhenotoky *n.* type of parthenogenesis where males are formed from unfertilized eggs and are haploid.

arrhizal *a.* without true roots, as some parasitic plants.

arrow worms Chaetognatha *q.v.*

ARS autonomously replicating sequence *q.v.*

arsenate *n.* arsenate-containing salt that is toxic as a result of its similarity to phosphate, for which it substitutes, esp. inhibiting the energy-generating reactions of cellular respiration.

artefact, artifact *n.* (1) apparent structure or experimental result obtained as a result of the method of preparing the specimen or the experimental conditions; (2) human-made object.

arterial *a. pert.* an artery, or to the system of vessels which carries blood from the heart to the rest of the body.

arteriole *n.* small artery.

Arteriviridae *n.* family of single-stranded positive-strand RNA viruses that infect vertebrates, e.g. equine arteritis virus.

artery *n.* vessel that conveys blood from heart to rest of body.

arthral, arthritic *a. pert.* or at joints.

arthrobranchiae *n.plu.* joint gills, arising at junction of thoracic appendage with trunk, in some arthropods.

arthrocyte *n.* (1) large resorptive cell in nephridium of bryozoans; (2) type of coelomocyte in nematodes.

arthrocytosis *n.* the capacity of cells to selectively absorb and retain solid particles in suspension, such as dyes.

arthrodia *n.* joint admitting only of gliding movements.

arthrodial *a. appl.* articular membranes connecting appendages with thorax in arthropods.

arthrogenous *a.* formed as a separate joint, *appl.* spores developed from separated portions of a plant.

arthromere *n.* an arthropod body segment or somite.

Arthropoda, arthropods *n., n.plu.* very large phylum of segmented invertebrate animals with heads, jointed appendages (feelers, mouth-parts and legs), and a thickened chitinous cuticle forming an exoskeleton. The main body cavity is a haemocoel. The phylum is generally divided into several different groups, most commonly the Chelicerata, Atelocerata, Crustacea and the extinct Trilobita. In this classification, the Chelicerata includes the spiders, ticks, mites, scorpions, pycnogonids, horseshoe crabs and the extinct eurypterids, the Atelocerata (sometimes known as the Uniramia) includes the insects and myriapods (centipedes and millipedes), and the Crustacea includes the crustaceans (e.g. crabs, shrimps, barnacles). The velvet worms (Onychophora) are sometimes placed in a separate phylum. *see* Appendix 3.

arthropterus *a.* having jointed fin-rays, as fishes.

arthrosis *n.* articulation, as of joints.

arthrospore *n.* (1) in some cyanobacteria, thick-walled resting cell formed by segmentation of filament; (2) a cell, e.g. an ooidium, formed by fragmentation of a fungal hypha.

arthrostraceous *a.* having a segmented shell.

arthrous articulate *q.v.*

Arthus reaction local inflammatory response evoked by injection into the skin of an antigen to which the subject already has antibodies.

articulamentum *n.* in chitons, the lower part of each of the body plates.

articular *a. pert.* or situated at a joint, *appl.* e.g. cartilage, surface, capsule.

articular(e) *n.* bone of lower jaw in reptiles and birds, articulating with the quadrate bone of skull. In mammals it is modified to form the malleus, an auditory ossicle of the inner ear.

Articulata *n.* class of brachiopods with shells joined by a hinge joint in which two teeth on one shell move in sockets on the other. *cf.* Inarticulata.

articulate *a.* (1) jointed; (2) separating easily at certain points.

articulation *n.* joint between bones or between segments of a stem or fruit.

artifact *alt.* spelling of artefact.

artificial chromosomes small chromosomelike structures constructed by genetic engineering and which include a centromere, origin of replication and telomeres. They replicate and segregate like chromosomes when introduced into a eukaryotic cell. Large pieces of additional DNA can be introduced into these chromosomes. *see also* YAC.

artificial classification, artifical key classification that groups organisms or objects together on the basis of a few convenient characteristics rather than on the basis of evolutionary relationships. *cf.* natural classification.

artificial formation a pattern of vegetation caused by human activity.

artificial insemination the introduction of sperm collected from a male into the female reproductive tract in mammals.

artificial intelligence (AI) the study of machine 'intelligence' and that branch of computer science that aims to create machines that would be considered truly intelligent by a human being interacting with them.

artificial neural network a technique of machine learning that attempts to simulate a natural neural network. Composed of layers of processing elements ('neurons') with multiple interconnections ('synapses') whose transmitting properties can be adjusted, it can be trained to recognize particular types of patterns in data. Neural networks have been used to model biological learning, and are also used e.g. in bioinformatics in statistical pattern-recognition applications such as secondary structure prediction from sequence.

artificial selection the selection of particular forms as a result of environmental pressures deliberately imposed, either in plant or animal breeding or in *in vitro* cell cultures.

artiodactyl *a.* having an even number of digits.

Artiodactyla, artiodactyls *n., n.plu.* even-toed ungulates, including pigs, sheep, cattle and camels, which have a complex stomach for dealing with plant food and in which the 3rd and 4th digit of the limb forms a cloven hoof.

aryl *n.* chemical group that results when a hydrogen atom is removed from an aromatic hydrocarbon such as benzene.

arytaenoid *a.* pitcher-like, *appl.* two cartilages at back of larynx.

ascending *a.* curving or sloping upwards.

ascending aestivation arrangement of petals, where each petal overlaps the edge of the one posterior to it.

Ascension and St Helenan region part of the African subkingdom of the Palaeotropical floristic kingdom that comprises the islands of Ascension and St Helena in the South Atlantic.

ascertainment artefact data that appear to demonstrate some finding, but which do not do so because they are collected from a population that is selected in a biased fashion.

aschelminths *n.plu.* the phyla Gastrotricha, Kinorhyncha, Nematoda, Nematomorpha and Rotifera, which contain pseudocoelomate, mainly worm-like animals.

asci *plu.* of ascus.

Ascidiacea, ascidians *n., n.plu.* class of marine tunicates (urochordates), commonly called sea squirts, in which the adults are generally colonial and fixed to a substrate.

ascidium *n.* a pitcher-leaf or part of leaf as in *Nepenthes* and other pitcher plants.

ascigerous *a.* sexual, ascus-bearing, reproductive phase of ascomycete fungi.

ascites *n.* accumulation of watery fluid and cells in abdominal cavity. *a.* **ascitic**.

ascocarp *n.* sexual fruiting body of ascomycete fungi, containing asci surrounded by a protective covering, and which may be an apothecium, cleistocarp or perithecium. *alt.* ascoma.

ascogenous *a.* producing asci, *appl.* hyphae.

ascogonium *n.* female sex organ in ascomycete fungi, which gives rise to asci.

Ascolichenes *n.* lichens in which the fungal partner is an ascomycete.

ascoma ascocarp *q.v. plu.* **ascomata**.

Ascomycota, Ascomycotina, ascomycetes *n., n., n.plu.* phylum (division) of terrestrial fungi, commonly called sac fungi, which have a septate mycelium and develop their spores in sac-like structures called asci. They include the yeasts, leaf-curl fungi, black and green moulds,

powdery mildews, cup fungi, morels and truffles.

ascophore *n.* structure bearing cell from which ascus develops.

ascorbic acid vitamin C *q.v.*

ascospore *n.* haploid spore of ascomycete fungi, produced in an ascus.

ascostroma *n.* type of ascocarp characteristic of the Loculoascomycetidae, in which the asci arise in locules within the stroma. *plu.* **ascostromata**.

Ascoviridae *n.* family of double-stranded DNA viruses infecting invertebrates (e.g. Spodoptera frugiperda virus).

ascus *n.* club-shaped or cylindrical sac-like structure containing ascospores in ascomycete fungi. The ascus and the (usually) eight ascospores it contains are formed from a single multinucleate cell. *plu.* **asci**.

-ase suffix denoting an enzyme, usually joined to a root denoting the substance acted on or the type of reaction, e.g. proteinase, lipase, glucosidase, asparaginase, hydrolase, oxidase.

asemic *a.* without markings.

asepsis *n.* sterile conditions.

aseptate *a.* without a septum or septa, *appl.* fungi.

aseptic *a.* (1) sterile; (2) *appl.* diseases usually caused by a bacterium but in which no bacterial agent can be found, e.g. aseptic meningitis, which may be due to a viral infection or to a non-infectious cause.

asexual reproduction reproduction which does not involve formation and fusion of gametes and results in progeny with an identical genetic constitution to the parent and to each other. Reproduction may occur by binary fission, budding, asexual spore formation or vegetative propagation. In asexual division in eukaryotic organisms, all cell divisions are by mitosis.

Asfarviridae *n.* family of double-stranded DNA viruses that infect vertebrates (e.g. African swine fever virus).

asialoglycoprotein *n.* glycoprotein that has lost the terminal sialic acid residues from its carbohydrate side chains. Such damaged glycoproteins are removed from the circulation by the liver.

Asiatic subregion subdivision of the Euro-Siberian region of the Boreal floristic kingdom, consisting of central Asia between latitudes 50° and 70°N.

A-site site on the ribosome at which the next codon on mRNA is exposed and incoming aminoacyl-tRNAs attach.

Asn asparagine *q.v.*

Asp aspartic acid *q.v.*

asparaginase *n.* enzyme hydrolysing asparagine via aspartate to glutamate and fumarate. EC 3.5.1.1.

asparagine (Asn, N) *n.* amino acid with a hydrophilic amine side chain, an uncharged derivative of aspartic acid. It is a constituent of proteins, was first discovered in asparagus, and is important in nitrogen metabolism in plants. Required in human diet.

aspartase *n.* enzyme that hydrolyses aspartic acid to yield fumaric acid and ammonia, present in some bacteria and higher plants. EC 4.3.1.1. *r.n.* aspartate ammonia-lyase.

aspartate, aspartic acid (Asp, D) *n.* amino acid with an acidic (carboxylic acid) side chain. It is a constituent of proteins, and also important in transamination reactions. Required in human diet.

aspartate transcarbamoylase enzyme involved in control of pyrimidine nucleotide biosynthesis. EC 2.1.3.2. *r.n.* aspartate carbamoyltransferase.

aspect *n.* (1) direction in which a surface faces; (2) appearance or look; (3) seasonal appearance.

asperate *a.* having a rough surface.

aspergilliform *a.* tufted like a brush.

aspergillosis *n.* disease caused by species of *Aspergillus* in humans.

Aspergillus genus of ascomycete fungi.

asperity *n.* roughness, as on a leaf.

aspirin *see* acetylsalicylic acid.

asplanchnic *a.* without an alimentary canal.

asporocystid *a. appl.* oocyst of certain sporozoan protozoans when zygote divides into sporozoites without sporocyst formation.

asporogenous *a.* not producing spores.

asporous *a.* having no spores.

assay *n.* a procedure for measurement or identification.

assembly *n.* the smallest community unit of plants or animals, e.g. a colony of aphids on a stem.

assimilate *n.* any of the first organic compounds produced during assimilation in autotrophs. *v. see* assimilation.

assimilated energy in an animal's energy budget, assimilated energy (A) = consumption (C) − faeces (F) and is equivalent to production (P) + energy lost as heat during respiration (R) + energy lost by small metabolites voided in urine (U).

assimilate stream the movement of sugars out of the leaves where they are manufactured during photosynthesis to other parts of the plant via the phloem.

assimilation *n.* (1) in autotrophic organisms, the uptake of elements and simple inorganic compounds such as CO_2, N_2, H_2O from the environment and their incorporation into complex organic compounds, *a.* **assimilative, assimilatory**; (2) in heterotrophs, the conversion of digested food material into complex biomolecules. *v.* to assimilate.

assimilation efficiency in animal physiology and ecophysiology, a measure of the efficiency of utilization of food. It is expressed as a ratio of assimilated energy (A) divided by energy consumed (C) and is mainly influenced by the nature of the food consumed, carnivores having a greater assimilation efficiency than herbivores.

assimilative, assimilatory *a.* (1) *appl.* metabolism in which inorganic compounds are reduced for use as nutrient sources, *cf.* metabolism; (2) *pert.* or used for assimilation *q.v.*; (3) *appl.* growth preceding reproduction.

association *n.* (1) *see* associative learning; (2) (*bot.*) plant community forming a division of a larger unit of vegetation and characterized by a dominant species.

association centres, association cortex areas of cerebral cortex where different aspects and types of incoming sensory information are integrated and interpreted.

association constant (K_a) measure of the strength of reversible binding between two molecules, being the concentration of the bound form [AB] divided by the product of the concentrations of the free forms of each molecule [A][B] at equilibrium. It is the reciprocal of the dissociation constant (*q.v.*). *alt.* affinity constant. *see also* equilibrium constant.

association fibres nerve fibres connecting the white matter of the interior of the brain with the cortex.

associative learning learning by forming an association between a stimulus (the cause) and a particular outcome (the response) or by associating two stimuli.

associative nitrogen fixation non-symbiotic nitrogen fixation by bacteria (e.g. *Azospirillum*, *Azotobacter*) associated with the rhizosphere of certain grasses and cereals.

associes *n.* an association representing a stage in the process of succession.

assortative mating non-random mating within a population where individuals tend to mate with individuals resembling themselves. In human populations for example, mating tends to be random for certain characteristics such as blood groups and assortative for others such as height and ethnic group.

astely *n.* absence of a central vascular cylinder or stele in stem.

aster *n.* star-shaped system of microtubules radiating from the centriole, present at either end of the spindle in many cells during cell division, but not found in plants.

Asteraceae in some classifications the name for the Compositae *q.v.*

Asterales *n.* an order of herbaceous dicots, rarely trees or woody climbers, with flowers usually crowded into closely packed heads and comprising the family Asteraceae (Compositae) (e.g. daisy).

astereognosis *n.* inability to recognize objects by touch and feel.

asterigmate *a.* not borne on sterigma, *appl.* spores.

asternal ribs false ribs *q.v.*

asteroid *a.* (1) star-shaped; (2) *pert.* starfish.

Asteroidea, asteroids *n.*, *n.plu.* class of echinoderms, commonly called starfish or sea stars, having a star-shaped body with five radiating arms not sharply marked off from the central disc.

asterophysis *n.* a rayed cystidium-like hair in hymenium of certain fungi.

asterospondylous *a.* *appl.* vertebra with centrum of radiating calcified cartilage.

asthenic *a.* (1) weak; (2) tall and slender, *appl.* physical constitutional type.

astichous *a.* not set in a row or rows.

astigmatous *a.* (1) (*bot.*) without a stigma; (2) (*zool.*) without spiracles.

astipulate exstipulate *q.v.*

astogeny *n.* the development of a siphonophore colony from a single egg. Individual zooids bud from a larva that develops from the zygote.

astomatous *a.* (1) lacking a mouth or cytostome; (2) (*bot.*) without stomata.

astomous *a.* (1) without a stomium or line of dehiscence; (2) bursting irregularly.

astragalus *n.* one of the bones of the vertebrate ankle (tarsus).

astrobiology exobiology *q.v.*

astrocyte *n.* type of neuroglial cell forming large part of the supporting non-neuronal tissue in the central nervous system. Fibrous astrocytes are found mainly in white matter, and bear fine branched processes some of which abut on blood vessels. Protoplasmic astrocytes are found in grey matter and bear thick branched processes similar to pseudopodia.

astroglia *n.* astrocytes collectively.

astropodia *n.plu.* fine unbranched radiating pseudopodia, as in some protozoans.

astropyle *n.* chief aperture of central capsule in some radiolarian protozoans.

astrosclereid *n.* a multiradiate sclereid or stone cell.

Astroviridae *n.* family of single-stranded positive-strand RNA viruses that infect vertebrates, e.g. human astrovirus.

Asx aspartic acid or asparagine.

asymmetrical *a.* (1) *pert.* lack of symmetry; (2) having two sides unlike or disproportionate; (3) *appl.* structures that cannot be divided into similar halves by any plane.

asymmetric division in development, cell division that produces two daughter cells that are different from each other because of unequal distribution of cytoplasmic determinants between them.

asymptomatic *a.* having no symptoms, whether disease is present or not.

asynapsis *n.* absence of pairing of homologous chromosomes in meiosis. *alt.* **asyndesis**.

atactostele *n.* a complex stele, having vascular bundles scattered in the ground tissue, as in monocotyledons.

atavism *n.* presence of an ancestral characteristic not observed in more recent progenitors. *a.* **atavistic.**

ataxia *n.* impairment of muscular movement. *a.* **atactic.**

ATCase aspartate transcarbamoylase *q.v.*

ateleosis *n.* dwarfism where individual is a miniature adult.

atelia *n.* (1) the apparent uselessness of a character of unknown biological significance; (2) incomplete development. *a.* **atelic.**

Atelocerata *n.* in some classifications the name of the group of arthropods that includes the insects and myriapods.

Atherinomorpha *n.* small group of advanced teleost fish including the toothcarps (guppies and swordtails).

athermopause *n.* dormancy of animals due to lack of water or food.

atherogenesis *n.* formation of fatty plaques or scar tissue on the walls of arteries, eventually causing atherosclerosis.

atheroma *n.* fatty deposits in the wall of an artery.

atherosclerosis *n.* the disease caused by deposition of fatty plaques and degeneration of walls in arteries, narrowing vessel lumen and restricting blood flow.

Atlantic North American region part of the Boreal floristic kingdom consisting of North America east to the Rockies, south to the Gulf of Mexico and north to the Arctic Circle.

atlanto-occipital occipito-atlantal *q.v.*

atlas *n.* the 1st cervical vertebra in humans.

atocia *n.* sterility in the female. *a.* **atokous, atokal** without offspring.

atoll *n.* coral reef surrounding a central lagoon.

atomic coordinates the positions in three-dimensional space of the atoms in a molecule relative to each other.

atomic force microscope type of scanning probe microscope with resolution down to fractions of a nanometre.

atopy *n.* an idiosyncratic sensitivity or allergy to a particular chemical compound or to an antigen. *a.* **atopic.**

ATP adenosine triphosphate *q.v.*

ATPase adenosinetriphosphatase *q.v.*

ATP synthases F-type ATPases in inner mitochondrial and thylakoid membranes and in bacterial membranes that synthesize ATP from ADP and inorganic phosphate using energy derived from a proton (H^+) gradient across the membrane. They couple the protonmotive force generated by aerobic respiration or photosynthesis to ATP synthesis. ATP synthases can also perform the reverse reaction of hydrolysing ATP and pumping protons across the membrane. Formerly known as chloroplast factor, coupling factor (in mitochondria).

atractoid *a.* spindle-shaped.

atretic *a.* (1) having no opening; (2) imperforate; (3) *appl.* vesicles resulting from degeneration of Graafian follicles.

atrial *a.* *pert.* atrium.

atrial natriuretic factor, atrial natriuretic peptide (ANF, ANP) polypeptides (α, β, γ) isolated from atrium of heart which act on the kidney to regulate salt and water balance and dilate blood vessels. Similar polypeptides (brain natriuretic peptide) are produced by the brain.

atrichous *a.* having no flagella or cilia.

atriopeptin atrial natriuretic factor *q.v.*

atrioventricular *a.* (1) *pert.* atrium and ventricle of heart; (2) *appl.* node: mass of cardiac muscle fibres in the wall of the right auricle.

atrioventricular bundle His' bundle *q.v.*

atrium *n.* (1) the main chamber of the auricle of the heart, sometimes refers to the whole auricle; (2) various chambers or cavities, e.g. the tympanic cavity in the ear, chamber from which tracheae extend into body in insects, chamber surrounding pharynx in tunicates and cephalochordates.

atrochal *a.* *appl.* trochophore larvae in which the preoral circlet of cilia is absent and surface is uniformly ciliated.

atropal atropous *q.v.*

atrophy *n.* diminution in size and function.

atropine *n.* alkaloid obtained from the deadly nightshade, *Atropa bella-donna* and other plants of the Solanaceae, used medically as a muscle relaxant.

atropous *a.* *appl.* ovule which is not inverted.

attached chromosome isochromosome *q.v.*

attenuated *a.* (1) gradually tapering to a point; (2) reduced in density, strength or pathogenicity; (3) *appl.* vaccines prepared from live strains of virus that have become non-pathogenic by mutation during long-term growth in culture.

attenuation *n.* (1) loss of virulence in a pathogenic microorganism while still retaining its capacity to immunize; (2) regulatory process occurring in some bacterial biosynthetic operons in which translation of a leader sequence in mRNA terminates transcription at a site called the attenuator.

attic *n.* the recess in bone above the eardrum.

auditory *a.* (1) *pert.* sense of hearing; (2) *pert.* hearing apparatus, as auditory organ, auditory canal, auditory meatus, auditory ossicle, auditory capsule, *see* ear; (3) *appl.* nerve: 8th cranial nerve, connecting inner ear with hindbrain. It carries signals of sound and pitch from cochlea for relay to auditory area of cerebral cortex, and postural information from semicircular canals to cerebellum.

Auerbach's plexus myenteric plexus *q.v.*

augmentation *n.* in plants, an increase in the number of whorls in a flower.

aulostomatous *a.* having a tubular mouth or snout.

aural *a. pert.* ear or hearing.

auricle *n.* (1) the anterior chamber of the heart; (2) the external ear; (3) any ear-shaped appendage.

auricula auricle *q.v.*

auricular (1) *a. pert.* an auricle; (2) *n.* feathers covering apertures of ears in birds.

auricularis *n.* superior, anterior, posterior, extrinsic: muscles of the external ear.

auriculate *a.* (1) eared, having auricles; (2) *appl.* leaf with expanded bases surrounding stem; (3) *appl.* leaf with lobes separate from rest of blade.

auriculotemporal *a.* (1) *pert.* external ear and temples (temporal regions); (2) *appl.* nerve: a branch of the mandibular nerve.

auriculoventricular *a. pert.* or connecting auricle and ventricle of heart.

auriform *a.* ear-shaped.

aurones *n.plu.* a group of yellow flavonoid plant pigments.

Australian kingdom one of the five generally recognized floristic kingdoms, comprising Australia and Tasmania. It is divided into regions: North and East Australian, Southwest Australian, Central Australian.

Australian region zoogeographical region in Wallace's classification that includes Australia, New Zealand, the Melanesian, Micronesian and Polynesian islands of the Pacific, Papua New Guinea, Sulawesi and other islands south-east of Wallace's line. It is divided into subregions as follows: Australian, Austro-Malayan, New Zealand, Polynesian.

Australian subregion subdivision of the Australian zoogeographical region, comprising Australia and Tasmania.

australopithecines, *Australopithecus* *n.plu, n.* genus of fossil hominids, believed to have lived from at least 4 million until 1 million years ago, found in southern and eastern Africa. They include the 'gracile' australopithecines (*Australopithecus africanus*) at around 2.5 million years, the 'robust' australopithecines (*A. robustus* and *A. boisei*) at around 2 to 1 million years, and an older species, *A. afarensis*, at around 3.5 million years, found in Kenya and Ethiopia. They had an upright posture and a relatively small brain compared with *Homo*. Fossils now accepted as australopithecine were also formerly known as *Paranthropus* and *Zinjanthropus*. *A. africanus* and *A. robustus* are offshoots from the direct line of human evolution and *A. robustus* is contemporaneous with early species of *Homo*.

Austro-Malayan subregion subdivision of the Australian zoogeographical region, comprising Papua New Guinea and the islands of Indonesia east of Wallace's line.

autacoid *n.* an internal secretion, such as a hormone or growth factor, which has a physiological effect.

autapomorphy *n.* unique, derived character state found only in a single taxon.

autapses *n.plu.* synapses formed between one part of a nerve cell and another part of the same cell, typically between axon and dendrite or axon and soma.

autecious, autecism *alt.* spelling of autoecious, autoecism *q.v.*

autecology *n.* the ecology of individual organisms or species. *cf.* synecology.

auto- prefix derived from Gk *autos*, self.

autoagglutination *n*. the clumping of an individual's cells by its own serum.

autoantibody *n*. antibody made against an antigen of one's own body. Autoantibodies are responsible for causing damage to tissues in some autoimmune diseases.

autoantigen *n*. any molecule of an individual's own body which is recognized by that individual's own immmune system, provoking an autoimmune response.

autobasidium *n*. (1) basidium having sterigmata bearing spores laterally; (2) a non-septate basidium.

autocarp *n*. fruit resulting from self-fertilization.

autocatalysis *n*. catalysis of a reaction by one of its own products. *a*. **autocatalytic**.

autochory *n*. self-dispersal of spores or seeds by an explosive mechanism.

autochthonous *a*. (1) aboriginal; (2) indigenous; (3) *appl*. indigenous soil microflora which is normally active; (4) inherited or hereditary; (5) *appl*. characteristics originating within an organ, as the pulsating of an excised heart; (6) formed where found.

autoclave *n*. equipment used for sterilization of glassware and media by steam heat.

autocoprophagy refection *q.v.*

autocrine *a*. *appl*. effects of a substance secreted by a cell on that cell itself.

autocyst *n*. thick membrane formed by some sporozoan parasites which separates them from host tissues.

autodeliquescent *a*. becoming liquid as the result of self-digestion, as the cap and gills of fungi of the genus *Coprinus* (inkcaps).

autodont *a*. designating or *pert*. teeth not directly attached to jaws, as in cartilaginous fishes.

autoecious *a*. *appl*. fungi and other parasites, parasitic on one host only during a complete life cycle. *n*. **autoecism**.

autogamy *n*. (1) self-fertilization in which the two nuclei that fuse are products of different meioses during gametogenesis; (2) conjugation of nuclei within a single cell; (3) conjugation of two protozoans originating from division of the same cell. *a*. **autogamous**. *cf*. automixis.

autogenesis *n*. origin, production or reproduction within the same organism.

autogenic *a*. (1) *appl*. plant successions caused by interactions between the members of the community itself; (2) *appl*. organisms that complete their life cycles within the confines of a particular system.

autogenous *a*. (1) produced in same organism, *cf*. exogenous; (2) *appl*. adult female insect that does not need to feed for her eggs to mature.

autogenous control situation in which the expression of a gene is controlled by its own product, either at the level of transcription or translation.

autogeny, autogony autogenesis *q.v.*

autograft *n*. tissue transplanted from one site to another in the same individual. *alt*. **autotransplant**.

autoheteroploid *a*. heteroploid derived from a single genome.

autoimmune *a*. *appl*. (1) immune responses against the constituents of one's own cells; (2) diseases caused by an immune response against a normal constituent of the body.

autoimmunity *n*. condition where the body mounts an immune response against its own components, leading e.g. to chronic inflammation and tissue destruction.

autoinfection *n*. re-infection from a host's own parasites or body flora.

autointoxication *n*. intoxication by poisonous substances formed within the body.

autokinetic *a*. moving by its own action.

autologous *a*. *appl*. (1) tissue graft from one part of body to another part in same individual; (2) immunization of animal of one species with antigen from another individual of the same species.

autolysin *n*. any enzyme that causes autolysis.

autolysis *n*. self-digestion of a cell by its own hydrolytic enzymes.

autolytic *a*. causing or *pert*. autolysis, *appl*. enzymes.

automimicry *n*. (1) imitation of a communication signal used by a particular sex or age-group in a species, by a member of the opposite sex or another age-group; (2) intraspecific mimicry, as when some members of a species are unpalatable, and palatable members of the same species mimic them.

automixis *n.* self-fertilization in which the two nuclei that fuse to form the zygote are the products of the same meiosis. *a.* **automictic.**

autonarcosis *n.* the state of being poisoned, rendered dormant or arrested in growth, owing to self-produced carbon dioxide.

autonomic *a.* (1) autonomous, self-governing, spontaneous; (2) *appl.* nervous system: the involuntary nervous system as a whole, comprising sympathetic and parasympathetic systems supplying smooth and cardiac muscles and glands, and whose actions are not under conscious control.

autonomously replicating sequence (ARS) DNA sequence that enables any DNA containing it to replicate in the yeast *Saccharomyces cerevisiae*, and which represents a yeast DNA origin of replication.

autonomous phenotype *see* cell-autonomous (2).

autopalatine *n.* in some fish, an ossification at anterior end of pterygoquadrate.

autoparasite *n.* parasite subsisting on another parasite.

autoparthenogenesis *n.* development from unfertilized eggs activated by a chemical or physical stimulus.

autophagic *a.* involved in or *pert.* autophagy.

autophagosome *n.* intracellular membrane-bounded vesicle containing intracellular membranes and organelles in the process of digestion. *alt.* **autophagic vacuole.**

autophagous *a. appl.* birds capable of running about and securing food for themselves when newly hatched.

autophagy *n.* (1) self-digestion, sometimes seen in cells, apparently mediated by lysosomes; (2) subsistence by self-absorption of products of cellular metabolism, such as the consumption of their own glycogen by yeasts.

autophilous *a.* self-pollinating.

autophosphorylation *n.* phosphorylation of a protein by the protein's own enzymatic activity.

autophyllogeny *n.* growth of one leaf on or out of another.

autophyte *n.* an autotrophic plant. *a.* **autophytic.**

autoploid autopolyploid *q.v.*

autopodium *n.* hand or foot.

autopolyploid, autoploid *a. appl.* (1) organism having more than two sets of homologous chromosomes; (2) a polyploid in which chromosome sets are all derived from a single species.

autoproteolysis *n.* the cleavage of a protein chain by its own enzymatic activity.

autoradiography *n.* technique by which large molecules, cell components or body organs are radioactively labelled and their image recorded on photographic film, producing an **autoradiograph** or **autoradiogram.**

autoreactivity *n.* self-reactivity, the ability to mount an immune response against an antigen of one's own body.

autoreceptor *n.* receptor that is activated by a compound secreted by the cell bearing the receptor.

autoregulation *n.* the regulation of initiation of transcription of a gene by its own protein product.

autoshaping *n.* Pavlovian conditioned response which may be induced in learning experiments in which animals are exposed to repeated pairings of stimulus and reward and begin to exhibit the behaviour appropriate to the reward on presentation of the stimulus without any training or reinforcement.

autoskeleton *n.* a true skeleton formed from elements secreted by the animal itself.

autosomal *a. appl.* to a gene carried on an autosome, i.e. any chromosome other than a sex chromosome.

autosomal dominant inheritance pattern of inheritance characteristic of a phenotypic trait determined by a dominant allele carried on an autosome. The trait is displayed by individuals who carry only one copy of the allele so that each offspring of a heterozygous individual has a 50% chance of inheriting the allele and showing the trait. *cf.* autosomal recessive inheritance, sex-linked.

autosomal recessive inheritance typical pattern of inheritance of a phenotypic trait determined by a recessive allele carried on an autosome. The trait is displayed by individuals who carry two copies of the allele. The offspring of parents both heterozygous for the recessive allele will each have a 1 in 4 chance of being homozygous for the allele and displaying the trait. *cf.* autosomal dominant inheritance, sex-linked.

autosome *n.* any chromosome other than the sex chromosomes.

autospore *n.* (1) a non-motile spore resembling the parent cell; (2) the protoplast formed by longitudinal division of a diatom, and which forms new valves.

autostoses cartilage bones *q.v.*

autostylic *n. appl.* jaw suspension, having the mandibular arch self-supporting, with the palatoqudrate (upper jaw) essentially fused with the cranium. Present in most modern tetrapods.

autosynapsis, autosyndesis *n.* in a polyploid organism, pairing of chromosomes from the same parent at meiosis.

autotetraploid *n.* organism whose nuclei contain four sets of chromosomes of identical origin.

autotomy *n.* shedding of a body part, as in some worms, arthropods and lizards.

autotransplantation *n.* grafting of tissue from one part of the body to another in the same individual.

autotriploid *n.* organism with nuclei containing three sets of chromosomes of identical origin.

autotroph *n.* organism able to utilize carbon dioxide as its sole source of carbon, and inorganic sources of nitrogen (e.g. nitrates, ammonium salts) and other elements as its sole starting materials for biosynthesis. *a.* **autotrophic.** *cf.* heterotroph.

autotropism *n.* (1) tendency to grow in a straight line, *appl.* plants unaffected by external influences; (2) tendency of organs to resume original form, after bending or straightening due to external factors.

autoxenous autoecious *q.v.*

autozooid *n.* fully formed polyp of an alcyonarian colony, which can feed and digest nutrients.

autozygote *n.* homozygote in which the two homologous genes have a common origin.

autozygous *a. appl.* alleles at the same locus that are identical by virtue of common descent.

auxesis *n.* growth owing to increase in cell size rather than in cell numbers. *a.* **auxetic.**

auxiliaries *n.plu.* female social insects that associate with other females of the same generation and become workers.

auxiliary cells (1) two or more modified epidermal cells adjoining guard cells, or surrounding stomata; (2) cells formed in the soil by some endomycorrhizal fungi.

auxin *n.* the plant growth hormone indole-3-acetic acid (IAA). Auxin is involved in plant embryonic development, cellular elongation and differentiation, root growth, the development of vascular tissue, phototropism, the development of fruits, phyllotaxis, and the normal suppression of the growth of lateral buds. It is responsible for the curvature of plant shoots towards the light by causing a differential elongation of cells on the side away from light where a greater concentration of auxin accumulates. Auxin produced by the tip of the shoot suppresses the development of lateral buds.

auxoautotroph *n.* organism that can synthesize all the growth substances needed for its development.

auxoheterotroph *n.* organism that cannot synthesize all the growth substances needed for its development.

auxospore *n.* the zygote in diatoms.

auxotonic *a.* (1) (*bot.*) induced by growth, *appl.* movements of immature plants; (2) (*zool.*) *appl.* contractions against an increasing resistance in muscle.

auxotroph *n.* organism that has acquired a requirement for the external supply of a particular nutrient as a result of mutation. *see also* auxoheterotroph. *a.* **auxotrophic.**

average daily metabolic rate (ADMR) a measure of the daily energy requirements of an animal engaging in its normal activities. It is calculated from the energy content of the food eaten in a day.

average heterozygosity a measure of genetic variation in a population, being the average frequency of heterozygotes at a selected group of genetic loci.

average life expectancy the number of years a newborn individual can be expected to live, averaged over the population.

aversive *a. appl.* stimuli which decrease the strength of a response if applied several times, and evoke fear and avoidance behaviour.

Aves *n.plu.* birds, a class of bipedal homo-iothermic vertebrates having the body clothed in feathers and front limbs modified as wings, and the skin of the jaw forming a horny bill (beak). They are descended from the extinct archosaurian reptiles. The earliest known fossil bird is *Archaeopteryx* (subclass Archaeornithes) from the upper Jurassic. Modern birds belong to the subclass Neornithes. *see* Appendix 3.

Aviales *n.* taxonomic grouping proposed to contain all modern birds, *Archaeopteryx*, and those feathered dinosaurs apparently capable of flight and considered as bird ancestors. The exact range and validity of this grouping is disputed at this time. *alt.* **avialan.**

avian *a. pert.* birds.

avicularium *n.* type of zooid in certain bryozoans.

avidin *n.* protein, abundant in egg white, which binds strongly to the vitamin biotin.

avidity *n.* the total binding strength of a multivalent ligand bound to a receptor with more than one binding site, e.g. an antibody with two or more binding sites binding to a repeated antigenic determinant on the surface of a pathogenic microorganism. *cf.* affinity.

avifauna *n.* all the bird species of a region or period.

avirulence gene gene found in some bacterial plant pathogens that determines their ability to cause disease on a host plant containing a corresponding resistance gene. *see* gene-for-gene resistance.

avirulent *a. appl.* a strain of bacterium, virus or other potential pathogen that does not cause disease.

avitaminosis *n.* vitamin deficiency.

Avogadro's number number of atoms (6×10^{23}) in 1 gram of hydrogen, or in the molecular weight equivalent in grams of any element or molecule.

avoidance behaviour a wide range of defensive behaviour in animals (e.g. freezing posture, running for cover, giving warning signals) by which they minimize exposure to apparently harmful situations, and which may be innate or learned.

avoidance reaction movement away from stimulus.

awn *n.* stiff bristle-like projection from the tip or back of the lemma or glumes in grasses, or from a fruit, or from tip of leaf.

axenic *a. appl.* pure cultures of micro-organisms in which the organism is grown in the absence of any other contaminating microorganism or, in the case of parasites or symbionts, of its plant or animal host.

axes *plu.* of axis.

axial *a.* (1) *pert.* axis or stem; (2) *appl.* filaments and other structures running longitudinally in a stem or the axon of a nerve cell; (3) *appl.* mesoderm in vertebrate embryos that gives rise to the notochord. *cf.* paraxial.

axial sinus a nearly vertical canal of the water vascular system in echinoderms, opening into an internal division of oral ring sinus, and communicating with the stone canal.

axial skeleton skeleton of head and trunk.

axil *n.* the angle between leaf or branch and the axis from which it springs.

axile *a.* (1) *pert.*, situated in, or belonging to the axis; (2) *appl.* placentation in which ovules are situated in middle of ovary in the angles formed by the meeting of the septa.

axilla *n.* armpit. *plu.* **axillae.**

axillary *a.* (1) (*bot.*) *pert.* axil; (2) growing in axil, as buds; (3) (*zool.*) *pert.* armpit.

axipetal *a.* travelling in direction of cell body, *appl.* nerve impulses.

axis *n.* (1) the central line of a structure; (2) any of several directions along which body structure is organized, e.g. apical–basal axis, antero-posterior axis, dorso-ventral axis, medio-lateral axis; (3) the main stem of plant or central cylinder of plant stem; (4) the trunk and head of a vertebrate; (5) the 2nd cervical vertebra. *plu.* **axes.**

axis cylinder the axon of a myelinated nerve fibre.

axoaxonic *a. appl.* synapse between axon terminal and axon of another neuron.

axodendritic *a. appl.* synapse in which the axon terminal contacts a dendrite.

axolotl *n.* sexually mature form of some salamanders which retains larval features such as gills and remains aquatic.

axon *n*. cytoplasmic process which carries nerve impulses away from the cell body of neuron. It may be up to several metres long in the case of motor neurons of large animals. The end of the axon makes contact with the dendrites or cell body of other nerve cells. *alt.* nerve fibre.

axonal *a. pert.* axon or axons.

axonal transport active transport of material in small cytoplasmic vesicles between cell body and axon terminus in nerve cells. Transport may be anterograde (from cell body to axon terminus) or retrograde (in the other direction) and is believed to occur along tracks formed from microtubules.

axoneme *n*. central core of eukaryotic cilium or flagellum, composed of a regular array of microtubules emanating from the basal body.

axon hillock small clear area of nerve cell at point of exit of axon, at which conducted impulses often start.

axon terminal the (usually) branched tip of an axon, each branch making a synapse on the dendrites or cell body of another nerve cell and secreting neurotransmitter in response to the arrival of a nerve impulse down the axon.

axoplast *n*. filament extending from kinetoplast to end of body in some trypanosomes.

axopod *n*. spike-like, usually permanent pseudopodium with a strengthening axial filament, in heliozoan protozoans. *alt.* **axopodium**.

axosomatic *a. appl.* synapses in which axon terminal contacts nerve cell body.

axospermous *a. appl.* carpels with axile placentation.

axostyle *n*. a supporting axial filament in flagellate protozoans.

axotomy *n*. cutting of an axon.

5-azacytidine (azaC) nucleoside analogue which is incorporated into DNA in place of cytidine and which, unlike cytidine, cannot be methylated. *see* DNA methylation.

azathioprine *n*. an immunosuppressive drug which acts by killing rapidly dividing cells.

azoic *a*. (1) uninhabited; (2) without remains of organisms or their products.

azonal *a. appl.* soils without definite horizons.

azotodesmic *a. appl.* lichens in which the algal partner is nitrogen-fixing.

AZT the deoxyribonucleoside analogue 3′-azido-3′-deoxythymidine, which inhibits the enzyme reverse transcriptase and is used in the chemotherapy of HIV infection. *alt.* zidovudine.

azurophilic *a*. staining readily with blue aniline dyes.

azygomatous *a*. without a cheek bone arch.

azygomelous *a*. having unpaired appendages.

azygospore *n*. spore developed directly from a gamete without conjugation.

azygote *n*. organism resulting from haploid parthenogenesis.

azygous *a*. unpaired.

B

β- *for headwords with prefix β- refer to beta- or to headword itself.*

B (1) either asparagine or aspartic acid in the single-letter code for amino acids; (2) factor B *q.v.*

BAC bacterial artificial chromosome *q.v.*

bacca *n.* berry, esp. if formed from an inferior ovary.

baccate *a.* (1) pulpy, fleshy, as a berry; (2) bearing berries.

bacciferous *a.* berry-bearing.

bacciform *a.* berry-shaped.

Bacillariophyta *n.* the diatoms, a phylum of unicellular photosynthetic protists, which have a silicified wall in two halves, chlorophyll and carotenoid pigments, and store oils and leucosin instead of starch. Now considered as members of the kingdoms Stramenopila or Chromista. There are two main groups: centric diatoms, which are radially symmetrical, and pennate diatoms, which have bilateral symmetry *alt.* **Bacillariophyceae**.

bacillary *a.* (1) rod-like; (2) *appl.* layer, the layer of rods and cones in the retina; (3) *pert.* or caused by bacilli.

bacillus *n.* (1) formerly much used, esp. in medical bacteriology, for any rod-shaped bacterium; (2) specifically, any member of the genus *Bacillus*, aerobic spore-forming rods widely distributed in the soil. *plu.* **bacilli**.

back cross cross between the heterozygous F_1 generation and the homozygous recessive parent, which allows the different genotypes present in the F_1 to be distinguished.

background *a. appl.* the rate at which spontaneous mutations occur in any particular organism.

back mutation mutation that reverses the effects of a previous mutation in the same gene, either by an exact reversal of the original mutation or by a compensatory mutation elsewhere in the gene (also known as intragenic suppression). *alt.* reverse mutation, reversion.

bacteraemia *n.* the transient presence of bacteria in the blood.

Bacteria *n.* superkingdom or domain of microorganisms containing all prokaryotes that are not members of the Archaea. Its members are known as **bacteria** (*sing.* **bacterium**). Bacteria are distinguished from archaea on the basis of genetics and biochemistry. They are an extremely metabolically and ecologically diverse group of unicellular microorganisms, free-living in soil and water or as parasites and saprophytes of plants and animals, the parasitic forms causing many familiar infectious diseases. Bacteria typically possess cell walls, and reproduce by binary fission or asexual endospores and also transfer genetic material by sexual processes (conjugation) and by virus (bacteriophage)-mediated transfer (transduction). *see* Appendix 5 for the main groups distinguished within the Bacteria, which have individual entries in the dictionary.

bacterial *a. pert.* or caused by bacteria.

bacterial artificial chromosome (BAC) an F plasmid modified to act as a cloning vector that can carry large inserts of 'foreign' DNA.

bacterial envelope all outer layers of a bacterial cell, including plasma membrane, outer membrane (if present), cell wall, and capsule (if present). *alt.* cell envelope.

bactericidal, bacteriocidal *a.* causing death of bacteria.

bacteriochlorophylls *n.plu.* green pigments related to chlorophyll, which act as light-collecting pigments in photosynthetic bacteria other than cyanobacteria. There are several types: *a*, *b*, *c*, *d*, *e*, *g*. They have absorption maxima in the infrared and violet.

bacteriocidal bactericidal *q.v.*

bacteriocins *n.plu.* proteins produced by bacteria that kill bacteria of another strain or species.

Bacteroidetes *n.* a bacterial phylum comprising Gram-negative, anaerobic, rod-shaped bacteria found in the guts of animals and in the soil, and can be opportunistic pathogens. Members of the Bacteroidetes are normally a major component of the beneficial intestinal microbiota of humans.

Bacteriological Code International Code of Nomenclature of Bacteria.

bacteriology *n.* the study of bacteria. *a.* **bacteriological**.

bacteriolysin *n.* substance that causes the breakdown or dissolution of a bacterial cell.

bacteriolysis *n.* disintegration or dissolution of bacteria.

bacteriophage *n.* a virus that infects bacteria, e.g. lambda, T2, T4. *alt.* phage. *see also* lysogeny, lytic infection. *see* Appendix 7.

bacteriophagous bacterivorous *q.v.*

bacteriorhodopsin *n.* purple protein found in the purple membrane of some salt-loving Archaea (halobacteria), which acts as a light-driven transmembrane proton pump.

bacteriostatic *a.* inhibiting the growth of, but not killing, bacteria.

bacterium *n.* individual member of the Bacteria. The term is also in general use for any prokaryotic microorganism, including members of the Archaea. *plu.* **bacteria**.

bacterivorous *a.* feeding on bacteria.

bacteroid *n.* irregularly shaped bacterial cell, the form in which rhizobia are found in root nodules of legumes.

bacteroidal *a. appl.* rod-shaped uric acid crystals, in certain annelids.

Bacteroides–Flavobacteria group *see* cytophagas.

baculiform *a.* rod-shaped.

Baculoviridae, baculoviruses *n.*, *n.plu.* family of enveloped double-stranded rod-shaped DNA viruses infecting arthropods.

baeocyte *n.* cell showing gliding motility, produced in some cyanobacteria.

bag cell secretory neuron in *Aplysia*, involved in regulation of egg-laying.

balanced lethal existence of two non-allelic recessive lethal genes on different chromosomes of a homologous pair, so that the double heterozygote is viable. The lethal genes are maintained in the population, because on interbreeding half the offspring are homozygous for one or the other gene and die, but the other half are heterozygous for both genes and survive.

balanced polymorphism stable coexistence of two or more distinct types of individual, forms of a character or different alleles of a gene in a population.

balanced translocation mutual exchange of material between two non-homologous chromosomes in which none is lost. Two monocentric chromosomes are produced which segregate normally at mitosis, so preserving the normal diploid number of chromosomes in the somatic cells. *alt.* **balanced reciprocal translocation**.

balancers *n.plu.* (1) halteres *q.v.*; (2) paired larval head appendages functioning as props until forelegs are developed in some salamanders.

balance trials methodology for determining quantitative nutritional requirements of animals, in which total intake and total loss of a particular nutrient are measured over a long period.

balancing selection natural selection that maintains a balanced polymorphism in a population.

balanoid *a.* (1) acorn-shaped; (2) *pert.* barnacles.

Balanopales *n.* order of dicot trees and shrubs comprising the family Balanopaceae, sometimes placed in the Fagales.

balausta *n.* many-celled, many-seeded, indehiscent fruit with tough outer rind, such as a pomegranate.

Balbiani body specialized organelle found in the developing oocytes of many species, characterized by large numbers of mitochondria. In some oocytes it is known to be involved in transporting and localizing maternal RNAs. *alt.* mitochondrial cloud.

Balbiani rings type of very large puff seen on certain dipteran polytene chromosomes, esp. in the gnat *Chironomus tentans* in which they were first discovered by E.G. Balbiani in 1881.

baleen *n.* whalebone. The horny plates attached to upper jaw in some whales, used to filter plankton from water. Whales using this method of feeding are known as the baleen whales.

baler scaphognathite *q.v.*

ball-and-socket *appl.* joints in which the hemispherical end of one bone fits into a cup-shaped socket on another, allowing movement in several planes, as in shoulder and hip joints.

ball-and-stick format a way of depicting chemical structures by means of coloured balls, representing atoms, joined by rods representing the bonds between them.

ballast *n.* elements present in plants and which are not apparently essential for growth, such as aluminium or silicon.

ballistic *a. appl.* fruits that explode when ripe, forcibly discharging seeds.

ballistospore *n.* spore that is forcibly discharged.

balsamic *a. appl.* class of odours typified by vanilla, heliotrope, resins.

balsamiferous *a.* producing balsam.

balsams *n.plu.* complex fragrant substances found in some plants, consisting of resin acids, esters and terpenes, often exuded from wounds.

BALT bronchus-associated lymphoid tissue *q.v.*

band *n.* (1) in gel electrophoresis, the region of a gel in which molecules of a particular size class have accumulated, visualized by various means; (2) in mitotic and polytene chromosomes, densely staining regions which make up an invariant pattern for each chromosome, by which the chromosome can be recognized.

band III protein protein spanning the red blood cell membrane, forming an anion channel through which HCO_3^- is exchanged for Cl^- in the lungs as red cells dispose of carbon dioxide.

banner standard *q.v.*

baraesthesia *n.* sensation of pressure.

barb *n.* (1) one of the delicate threadlike structures extending from the shaft of a feather and forming the vane; (2) hooked hair-like bristle; (3) type of fish.

barbate *a.* bearded or tufted.

barbel *n.* whisker-like tactile extension arising on the lower jaw in some fishes.

barbellate *a.* with stiff hooked hair-like bristles.

Barbeyales *n.* order of dicot trees with one family, Barbeyaceae, including one genus *Barbeya*.

barbicel *n.* one of the tiny hooked processes on barbules of feather which interlock to hold the barbs together.

barbule *n.* lateral projection from barb of feather, serving to hold barbs together to form an unbroken vane.

bare lymphocyte syndrome any of a number of rare genetic defects in humans that result in a lack of expression of MHC class II molecules on cells.

baresthesia *alt.* spelling of baraesthesia *q.v.*

bark *n.* the layer of tissue external to the vascular cambium in woody plants, comprising the secondary phloem, cortex and outermost periderm.

bark lice common name for some members of the Psocoptera, small wingless insects with a globular abdomen and incomplete metamorphosis, inhabiting the bark of trees.

barnacle *n.* common name for a member of the Cirripedia *q.v.*

baroceptor baroreceptor *q.v.*

barognosis *n.* capacity to detect changes in pressure.

barophile *n.* organism that grows best under high pressures, e.g. deep-sea species of bacteria.

baroreceptor *n.* sensory receptor responding to stretch in the wall of heart and large blood vessels that signals changes in blood pressure. *alt.* baroceptor.

barotaxis *n.* directed movement or orientation in response to a pressure stimulus.

barotolerant *a. appl.* organism that can tolerate high pressures.

barrage reaction reaction between the mycelia of two incompatible strains of fungi.

Barr body densely staining body seen in the nuclei of somatic cells from female mammals. It is the inactivated X chromosome. Individuals of abnormal genetic constitution, e.g. 3X, 4X, have two and three such bodies respectively. *alt.* sex-chromatin body.

β-barrel element of protein architecture consisting of a β-sheet curved into a barrel-shaped structure.

barrier forest forest in the mountains that holds back snow from the lower slopes.

Bartholin's duct the larger duct of the sublingual gland.

Bartholin's glands glands secreting vaginal lubricant, situated on either side of the vagina.

basad *adv.* towards the base.

basal *a. pert.*, at, or near the base.

basalar *a. appl.* sclerites below the base of wing in insects.

basal body cylindrical structure at base of the axoneme of eukaryotic flagella and cilia, composed of nine microtubules. It organizes the assembly and arrangement of the microtubules of the axoneme. *alt.* basal granule, kinetosome.

basal cell (1) cell in lowest layer of a stratified epithelium, such as epidermis, and from which that tissue is renewed; (2) (*bot.*) in development of plant embryo, the lower, larger cell produced by unequal division of the fertilized egg; (3) (*mycol.*) uninucleate cell which supports the dome and tip of a hyphal crozier; (4) (*zool.*) contractile epithelial cell, as in coelenterates.

basal disc (1) in corals, the area of ectoderm that secretes the calcareous skeleton; (2) in hydra, lower end of body by which it attaches to substratum.

basale *n.* bone of variable structure supporting fish fins.

basal ganglia paired masses of grey matter in midbrain, which connect with other brain centres and are involved in motor control. Each is composed of globus pallidus, putamen (together forming the lentiform nucleus), and caudate nucleus.

basal knobs swellings or granules at points of emergence of cilia in ciliated epithelial cells.

basal lamina (1) specialized thin mat-like layer of collagen-containing extracellular matrix that underlies all epithelia, forming part of the basement membrane that separates the epithelial layer from underlying tissues; (2) layer of extracellular matrix lying between nerve terminal and muscle membrane and surrounding muscle and nerve terminals.

basal leaf leaf produced near base of stem, often different in shape from the leaves on the rest of the stem.

basal metabolic rate minimum metabolic rate required for survival, measured in humans at complete rest in a thermally neutral environment after fasting for 12 hours.

basal metabolism normal state of metabolic activity of organism at rest.

basal placentation condition where ovules are situated at the base of the ovary in the flower.

basal ridge ridge around base of crown in a tooth.

basapophysis *n.* transverse process arising from the ventrolateral side of a vertebra.

base *n.* (1) substance that accepts an H^+ ion (proton) in solution, *cf.* acid; (2) in molecular genetics, refers to the nitrogenous bases, which are the purine and pyrimidine constituents of nucleotides and nucleic acids. *see* adenine, cytosine, guanine, thymine, uracil.

base analogue substance chemically similar to one of the nucleotide bases and which is incorporated into DNA, often causing mutations.

base exchange capacity the extent to which exchangeable cations can be held in a soil. *alt.* cation exchange capacity.

base excision repair type of DNA repair used to remove individual damaged bases and restore the original DNA sequence.

basement membrane layer separating an epithelium from the underlying tissue. It consists of the basal lamina and a fine fibrous meshwork of collagen fibrils from the underlying tissue that interacts with the lower face of the basal lamina. *alt.* basilemma.

base pair (bp) single pair of complementary nucleotides from opposite strands of the DNA double helix. The number of base pairs is used as a measure of length of a double-stranded DNA.

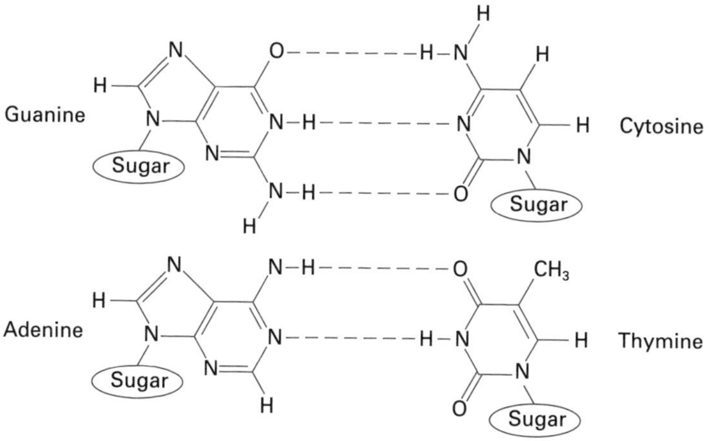

Fig. 5 Base pairing.

base pairing non-covalent bonding between purine and pyrimidine bases within nucleic acids, adenine pairing with thymine (in DNA) or uracil (in RNA) and cytosine with guanine (in DNA and RNA). *see* Fig. 5.

base ratio the ratio of the bases (A+T)/(C+G) in DNA, which varies widely from species to species.

base-rich *a. appl.* soils containing a relatively large amount of free basic ions such as magnesium or calcium.

base sequence *see* nucleotide sequence.

base sequencing *see* DNA sequencing.

base substitution in DNA, replacement of one type of nucleotide with another.

basibranchial *n.* central ventral or basal skeletal portion of branchial arch.

basibranchiostegal urohyal *q.v.*

basic *a.* (1) having the properties of a base; (2) *appl.* stains which act in general on the nuclear contents of the cell; (3) *appl.* number: (i) the minimum haploid chromosome number occurring in a series of euploid species in a genus, (ii) chromosome number in gametes of diploid ancestor of a polyploid organism; (4) of soils, rich in alkaline minerals.

basic helix-loop-helix (bHLH) the DNA-binding structural motif in many DNA-binding proteins.

basicranial *a.* situated or relating to base of skull.

basidia *plu.* of basidium *q.v.*

basidiocarp *n.* the fruiting body of basidiomycete fungi, which bears the basidia. *alt.* basidioma, basidiome.

basidiole, basidiolum *n.* sterile basidium-like structure in some basidiomycetes.

basidiolichen *n.* lichen in which the fungus partner is a basidiomycete.

basidioma basidiocarp *q.v.*

Basidiomycota, Basidiomycotina, basidiomycetes *n., n., n.plu.* phylum (division) of terrestrial fungi that have septate hyphae and bear their spores on the outside of spore-producing bodies (basidia). These are often borne on or in conspicuous fruiting structures. Basidio-mycetes include the rusts, smuts, jelly fungi, mushrooms and toadstools, puffballs, stinkhorns, bracket fungi and bird's nest fungi.

basidiospore *n.* spore formed by meiosis in basidiomycetes, borne on the outside of a basidium.

basidium *n.* in basidiomycete fungi, a usually club-shaped cell on the surface of which the basidiospores, usually four in number, are formed by meiosis. *plu.* basidia. *a.* **basidial**.

basifugal *a.* (1) growing away from base; (2) acropetal *q.v.*

basifuge *n.* a plant unable to tolerate basic soils. *alt.* calcifuge.

basigamic, basigamous *a.* having ovum and synergids at the far end of the embryo sac, away from the micropyle.

basihyal *n.* basal or ventral portion of hyoid arch.

basilabium *n.* sclerite formed by fusion of labiostipites in insects.

basilar *a. pert.*, near, or growing from base.

basilar artery artery formed by fusion of the vertebral arteries and whose branches supply blood to the brain stem and posterior cerebral hemispheres.

basilar membrane basal membrane of organ of Corti in ear.

basilemma basement membrane *q.v.*

basilic *a. appl.* large vein on inner side of biceps of arm.

basilingual *a. appl.* broad cartilaginous plate, the body of the hyoid, in crocodiles, turtles and amphibians.

basimandibula *n.* small sclerite on insect head, at base of mandible.

basimaxilla *n.* small sclerite on insect head, at base of maxilla.

basin peat fen peat *q.v.*

basinym, basionym *n.* name on which new names of species have been based.

basion *n.* the middle of the anterior margin of the foramen magnum.

basiophthalmite *n.* basal joint of eyestalk in crustaceans.

basiotic mesotic *q.v.*

basipetal *a.* (1) descending; (2) developing from apex to base, i.e. with youngest at base, *appl.* leaves on a stem, flowers in a spike, spores in a chain.

basipharynx *n.* in insects, the epipharynx and hypopharynx united.

basipodite *n.* 2nd joint of certain limbs of crustaceans.

basipodium *n.* wrist or ankle.

basiproboscis *n.* membranous portion of proboscis of some insects.

basipterygium *n.* bone or cartilage in pelvic fin of fishes.

basirostral *a.* situated at, or *pert.*, the base of beak or rostrum.

basisphenoid *n.* cranial bone in middle of base of skull.

basistyle *n.* base of clasper in male mosquitoes.

basitarsus *n.* 1st segment of tarsus, usually the largest.

basitonic *a.* having anther united at its base with rostellum.

basivertebral *a. appl.* veins within bodies of vertebrae and communicating with vertebral plexuses.

basket cell (1) one of the myoepithelial cells surrounding the base of certain glands; (2) type of interneuron found in cerebellum with cell body situated just above Purkinje cell layer, acts on Purkinje cells.

basolateral *a. pert.* sides and base of any cell, structure or organ.

basophil *n.* (1) type of white blood cell classed as a granulocyte or polymorphonuclear leukocyte. It contains granules that stain strongly with basic dyes and which release histamine and serotonin. It is involved in inflammatory reactions and responses to infection; (2) type of secretory cell of the anterior lobe of pituitary gland that stains with basic dyes.

basophilic *a.* staining strongly with basic dyes.

bass (1) (*bot.*) bast *q.v.*; (2) (*zool.*) type of fish.

bassorin *n.* mucilaginous carbohydrate food store in orchids, used to make saloop.

bast *n.* (1) inner fibrous layer of some trees; (2) phloem fibres.

bastard wing group of three quill feathers on the 1st digit of bird's wing.

BAT brown adipose tissue, brown fat *q.v.*

batch culture microbial culture being grown in a fixed volume of medium, such that as the culture grows the nutrients become exhausted.

Batesian mimicry resemblance of one animal (the mimic) to another (the model) to the benefit of the mimic, as when the model is dangerous or unpalatable. It was first described by the English naturalist H.W. Bates.

bathyaesthesia *n.* sensation of stimuli within the body.

bathyal *a. appl.* or *pert.* zone of seabed between the edge of the continental shelf and the abyssal zone at a water depth of 2000 m.

bathylimnetic *a.* living or growing in the depths of lakes or marshes.

bathymetric *a. pert.* vertical distribution of organisms in space.

bathypelagic *a.* inhabiting the deep sea (1000–3000 m).

bathyplankton *n.* plankton that undergo a daily migration, moving up towards the surface at dusk, and down to lower depths at dawn.

bathysmal *a. pert.* deepest depths of the sea.

batrachians *n.plu.* frogs and toads.

batrachosaurs *n.plu.* group of labyrinthodonts of the Carboniferous and Permian, which may include the ancestors of reptiles.

batrachotoxin *n.* alkaloid poison from skin of certain frogs that acts on the nervous system.

bauplan *n.* generalized, idealized, archetypal body plan of a particular group of animals.

B cell (1) B lymphocyte *q.v.*; (2) β-cell of pancreas; (3) β-cell of pituitary.

B-1 cell, B-2 cell *see* B lymphocyte.

B-cell co-receptor cell-surface receptor on B lymphocytes which is activated by certain complement components and which cooperates with the B-cell receptor in activating the B lymphocyte in response to encounter with antigen.

B-cell growth and differentiation factors *see* interleukin.

B-cell receptor the antigen receptor on B lymphocytes, which is a membrane-bound immunoglobulin.

BCG Bacille Calmette-Guérin, a modified variant of a bovine strain of *Mycobacterium tuberculosis* used as a vaccine against human tuberculosis.

B chromosomes extra heterochromatic chromosomes present in some organisms above the normal number for the species.

bdelloid *a.* having the appearance of a leech.

B-DNA *see* DNA.

BDNF *see* neurotrophins.

beak *n.* (1) bill (*q.v.*) of birds; (2) elongated jaws or mandibles of other animals, as the elongated jaw of a dolphin; (3) long, angled projections on certain fruits, as those of cranesbills (Geraniales).

beard *n.* (*bot.*) (1) barbed or bristly outgrowths on seed or fruit; (2) awn *q.v.*

beard worms Pogonophora *q.v.*

Becker muscular dystrophy (BMD) mild heritable form of muscular dystrophy, caused by a defect at the DMD (Duchenne muscular dystrophy) locus.

becquerel (Bq) *n.* derived SI unit for expressing the activity of a radionuclide, which is equal to 2.7×10^{-11} curies (Ci).

bedeguar *n.* moss-like outgrowth produced on rosebushes by gall wasps.

bees *n.plu.* insects of the superfamily Apoidea of the order Hymenoptera, some of which are social and some solitary. They include the honey bees (*Apis*), bumblebees (*Bombus*) and flower bees (*Anthophora*). Bees feed themselves and their young on pollen and nectar gathered from flowers, and are important plant pollinators. The social bees form colonies with a single queen, males (drones) and workers.

beetle *n.* common name for a member of the Coleoptera *q.v.*

beets *n.plu.* various cultivated forms of *Beta vulgaris*, such as sugar-beet, beetroot, chard.

Begoniales *n.* order of mostly succulent dicot herbs, but also shrub-like herbs and some large trees, comprising the families Begoniaceae (begonia) and Datiscaceae.

behavioural ecology study of the relationship of animals to their environment and to other animals, which also includes the effects of their behaviour and the way it may be modified by environmental factors and by interactions with members of their own species.

behavioural genetics study of the genetic basis of behaviour, which includes the study of the contribution of environment and nurture and heritable traits to behaviour.

behavioural silence condition in which an animal is thought to learn even though no change in behaviour is observed.

behaviourism *n.* mechanistic psychological theory of behaviour. In its most extreme form it postulated that animal and human behaviour may be explained purely in terms of muscular and other physiological (e.g. hormonal) reactions to external and internal stimuli.

belemnoid *a.* shaped like a dart.

Bellini's ducts tubes opening at apex of kidney papilla, formed by union of smaller collecting tubules.

belonoid *a.* shaped like a needle.

belt desmosome intermediate junction *q.v.*

Beltian bodies, Belt's bodies small nutritive organs containing oils and proteins, borne at the tips of leaves of swollen-thorn acacias. They provide food for ants that live on the plant.

belt transect the recording of organisms within a strip of ground, typically up to a metre wide, along a predetermined line.

Bence-Jones protein immunoglobulin light chains found in urine of patients with multiple myeloma.

β-bend β-turn *q.v.*

Benedict's solution solution containing sodium citrate, sodium carbonate and copper sulphate, which forms a rust-brown cuprous oxide precipitate on boiling with reducing sugars.

benefit *n.* in animal behaviour, the quantity that is maximized by the behavioural choices made. *alt.* negative cost. *cf.* cost.

benign *a. appl.* tumours that do not become invasive and spread to other sites in the body.

Bennettitales *see* Cycadeoidophyta.

Benson–Calvin cycle Calvin cycle *q.v.*

benthal, benthic *a. pert.*, or living on, the bottom of sea, lake, river or other water body.

benthophyte *n.* bottom-living plant.

benthos *n.* flora and fauna of sea or lake bottom from high water mark down to the deepest levels.

benzo[a]pyrene, benzpyrene *n.* carcinogenic polycyclic aromatic hydrocarbon produced by burning fossil fuels. Also present in coal tar.

benzodiazepines *n.plu.* class of compounds used as minor tranquillizers. They interact with receptors for the excitatory neurotransmitter glycine in the central nervous system, inhibiting its action.

benzyladenine *n.* synthetic plant growth hormone.

Bergmann glia glial cells that span the wall of the developing cerebellum and provide guidance for migrating neurons.

Bergmann's rule idea that geographically variable warm-blooded animal species have smaller body sizes in the warmer parts of their range than in the colder.

beriberi *n.* disease resulting in degeneration of the nerves or polyneuritis, caused by a deficiency of the vitamin thiamine (vitamin B$_1$) in the diet, or to an inability to absorb thiamine.

berry *n.* (1) several-seeded indehiscent fruit with a fleshy covering and without a stony layer surrounding the seeds; (2) dark knob-like structure on bill of swan.

Bertin's columns renal columns *q.v.*

beta- *for headwords with prefix beta- or β- refer also to headword itself, e.g. for β-globin look under globin.*

betacyanin *n.* any of a group of complex flavonoid pigments which give a reddish colour to some flowers.

beta diversity biological diversity resulting from competition between species that produces a finer adaptation of a species to the complete habitat, thus narrowing its range of tolerance to other environmental factors. Beta diversity is represented by the extent of change of species composition along an environmental gradient, e.g. of altitude, from one habitat to another. The greater the change, the higher the beta diversity. *alt.* habitat diversification. *cf.* alpha diversity.

betaine *n.* glycine derivative, intermediate in choline synthesis.

beta-proteobacteria group of proteobacteria (purple Bacteria) distinguished on the grounds of their DNA sequence. It includes the spirilla, *Neisseria*, non-fluorescent pseudomonads, *Bordetella* and other genera. *alt.* **beta purple bacteria**.

Betulales *n.* in some classifications an order of dicot trees including the families Betulaceae (birch) and Corylaceae (hazel).

Betz cells giant pyramidal cells in motor area of cerebral cortex.

bFGF basic fibroblast growth factor. *see* fibroblast growth factor.

BFU-E erythrocytic burst-forming unit *q.v.*

bHLH basic helix-loop-helix *q.v.*

B-horizon layer of deposition and accumulation below the topmost layer in soils. *alt.* **illuvial layer**.

bi- prefix from L. *bis*, twice, often indicating having two of.

biacuminate *a.* having two tapering points.

biarticulate *a.* two-jointed.

biaxial *a.* (1) with two axes; (2) allowing movement in two planes, as condyloid and ellipsoid joints.

bicapsular *a.* (1) having two capsules; (2) having a capsule with two chambers.

bicarinate *a.* with two keel-like processes.

bicarpellary, bicarpellate *a.* with two carpels.

bicaudal, bicaudate *a.* possessing two tail-like processes.

bicellular *a.* composed of two cells.

bicentric *a. pert.* two centres, *appl.* e.g. to a discontinuous distribution of a species.

biceps *n.* muscle with two heads or origins, esp. the large muscle in upper arm or leg.

biciliate *a.* having two cilia.

bicipital *a.* (1) *pert.* biceps; (2) *pert.* groove: the intertubercular sulcus, on upper part of humerus; (3) *appl.* ridges: the crests of the greater and lesser tubercles of the humerus; (4) *appl.* a rib with dorsal tuberculum and ventral capitulum; (5) divided into two parts at one end.

bicollateral *a.* having two sides similar.

biconjugate *a.* with two similar sets of pairs.

Bicornes Ericales *q.v.*

bicornute *a.* with two horn-like processes.

bicrenate *a.* (1) doubly crenate, as crenate leaves with notched toothed margins; (2) having two rounded teeth.

bicuspid *a.* (1) having two longitudinal ridges or ribs, as leaf; (2) having two cusps or points, *appl.* premolar teeth; (3) *appl.* valve: mitral valve of heart between left auricle and ventricle, *alt.* **bicuspidate**.

bicyclic *a.* arranged in two whorls.

bidentate *a.* having two teeth or tooth-like processes.

bidenticulate *a.* with two small teeth or tooth-like processes, as some scales.

bidirectional replication DNA replication which proceeds in both directions from an origin of replication, as in bacterial and eukaryotic chromosomes.

bidiscoidal *a.* consisting of two disc-shaped parts, *appl.* type of placenta.

biennial *n.* plant living for two years and fruiting only in the second.

bifacial *a.* flattened, and having upper and lower surface of distinctly different structure.

bifarious *a.* arranged in two rows, one on each side of axis.

bifid *a.* (1) forked; (2) divided nearly to middle line.

biflagellate *a.* having two flagella.

biflex *a.* twice curved.

biflorate *a.* having two flowers.

biflorous *a.* flowering in both spring and summer.

bifoliar *a.* having two leaves.

bifoliate *a.* (1) *appl.* palmate compound leaves with two leaflets; (2) having two leaves.

biforate *a.* having two foramina or pores.

bifurcate *a.* forked.

bigeminal *a.* (1) with structures arranged in double pairs; (2) *pert.* optic lobes of vertebrate brain.

bigeminate *a.* twin-forked.

bigeneric *a. appl.* hybrids between two different genera.

bijugate *a.* with two pairs of leaflets.

bilabiate *a.* two-lipped.

bilamellar *a.* formed of two plates or scales.

bilaminar, bilaminate *a.* having two layers.

bilateral *a. pert.* or having two sides.

bilateral symmetry type of symmetry in which a central axis divides the body into two sides that are mirror images of each other, as in most animals. *cf.* radial symmetry.

bilateria *n.* collective name for all animals with bilateral symmetry.

bilayer lipid bilayer *q.v.*

bile *n.* secretion of liver cells which collects in gall bladder and passes via the bile duct to the duodenum. It contains emulsifying bile salts, bile pigments (derived from breakdown of haemoglobin), cholesterol and lecithin, and some other substances. *alt.* gall.

bile duct *see* bile.

bile salts sodium glycocholate and sodium taurocholate, which are found in bile and aid the emulsification of dietary fat in the gut.

bilharzia *n.* schistosomiasis. *see* schistosome.

biliary *a. pert.* or conveying bile.

bilicyanin *n.* blue pigment resulting from oxidation of biliverdin and bilirubin.

bilin *n.* the chromophore in biliproteins.

biliprotein *n.* any of a group of protein pigments present in some groups of algae, including phycoerythrin and phycocyanin.

bilirubin *n*. red pigment, reduction product of biliverdin. Excreted in bile. It is also found in spleen and seen in bruising.

biliverdin *n*. green pigment, breakdown product of the haem group of haemoglobin. Excreted in bile. Also found in spleen and seen in bruising.

bill *n*. the beak of a bird, formed from outgrowth of cornified skin at the corners of the jaws.

bilobate, bilobed *a*. having two lobes.

bilobular *a*. having two lobules.

bilocellate *a*. divided into two compartments.

bilocular *a*. having two cavities or compartments.

bilophodont *a*. *appl*. molar teeth of tapir, which have ridges joining the anterior and posterior cusps.

bimaculate *a*. having two spots.

bimanous *a*. having two hands, *appl*. certain primates.

bimodal distribution statistical distribution having two modes.

bimolecular sheet *see* bilayer.

bimuscular *a*. having two muscles.

binary *a*. (1) composed of two units; (2) *appl*. compounds of only two chemical elements.

binary fission in prokaryotic organisms, the chief mode of division, in which a cell divides into two equal daughter cells, each containing a copy of the chromosome.

binary vector genetic engineering vector composed of two separate plasmids, one of which carries the foreign DNA and the other some function required for its transfer or maintenance.

binate *a*. (1) growing in pairs; (2) *appl*. leaf composed of two leaflets.

binaural *a*. *pert*. hearing involving both ears.

bindin *n*. protein isolated from sea urchin sperm which is responsible for the species-specific adherence of sperm to egg.

binding constant *see* association constant.

binding energy the free energy (*q.v.*) of association of a molecule (such as a protein) and its ligand. The free energy of the complex will be lower than the total free energy of the unbound components.

binemic *a*. two-stranded.

binervate *a*. having two veins, *appl*. insect wing, leaf.

binocular *a*. (1) *pert*. both eyes; (2) stereoscopic, *appl*. vision.

binodal *a*. having two nodes.

binomial *n*. name consisting of two parts.

binomial nomenclature the system of double Latin names given to plants and animals, consisting of a generic name followed by a specific name, e.g. *Felis* (genus) *tigris* (species).

binotic binaural *q.v.*

binovular *a*. *pert*. two ova, *appl*. twins arising from two eggs.

binuclear, binucleate *a*. having two nuclei.

-bio- word element derived from Gk *bios*, life, indicating living, *pert*. living organisms.

bioaccumulation *n*. build-up of a pollutant in the body of an aquatic organism by uptake in food and directly from the surrounding water. *cf*. bioconcentration, biomagnification, although all these terms are often used synonymously.

bioaccumulator *n*. plant or animal species that accumulates heavy metals or other environmental contaminants (e.g. fat-soluble pesticides) in its tissues by uptake from its environment. It can be used as an indicator of the presence of chronic pollution by these compounds, especially where amounts of pollutant in the environment are too low to be easily detectable directly.

bioamplification biomagnification *q.v.*

bioassay *n*. (1) determination of the biological activity of a substance; (2) use of a living organism or tissue for the purpose of assaying the presence or amount of something.

bioastronomy exobiology *q.v.*

bioavailability *n*. the availability of nutrients and/or pollutants to a living organism.

biobank *n*. a collection of tissue samples, e.g. from different individuals and/or diseases, usually kept as reference material for research and population studies.

biocenosis biocoenosis *q.v.*

biochemical evolution *see* molecular evolution.

biochemical oxygen demand (BOD) measurement of the amount of organic pollution in water, measured as the amount of oxygen taken up from a sample containing a known amount of dissolved oxygen kept at 20°C for 5 days. A low BOD indicates little pollution, a high BOD indicates increased activity of heterotrophic microorganisms and thus heavy pollution. *alt.* biological oxygen demand.

biochemistry *n.* the chemistry of living organisms and its study.

biochip general term for any small device such as a DNA microarray, a protein array, or a microfluidics device, which is used to study biological molecules and processes.

biochore *n.* (1) boundary of a floral or faunal region; (2) climatic boundary of a floral region.

biochrome *n.* any biological pigment.

biocide *n.* any agent that kills living organisms.

bioclimatology *n.* study of the relationship of living organisms and climate.

biocoen *n.* (1) living parts of an environment; (2) biosphere *q.v.*

biocoenosis *n.* the community of different organisms that live in a particular type of habitat.

bioconcentration *n.* uptake of a pollutant by an aquatic organism by direct uptake from the surrounding water, *cf.* bioaccumulation, biomagnification, although all these terms are often used synonymously.

biocontrol biological control *q.v.*

bioconversion *n.* use of microorganisms to convert one substance into another.

biocycle *n.* one of the three main divisions of the biosphere: marine, freshwater or terrestrial habitat.

biodegradable *a. appl.* materials that can be broken down by microorganisms into small molecules such as water and carbon dioxide. *n.* **biodegradation**.

biodemography *n.* the science dealing with the integration of ecology and population genetics.

biodiesel *n.* replacement 'renewable' fuel for diesel that can be manufactured from vegetable oils and animal fats, including recycled cooking oils.

biodiversity *see* biological diversity.

bioelectric *a. appl.* electric currents produced in living organisms.

bioenergetics *n.* (1) energy flow in an ecosystem; (2) study of energy transformation in living organisms.

bioengineering *n.* (1) use of artificial replacements for body organs; (2) use

of technology in the biosynthesis of economically important compounds. *see also* genetic engineering.

bioethanol *n.* ethyl alcohol (ethanol) manufactured from corn (maize) and other carbohydrate-rich feedstocks and from cellulose biomass by microbial fermentation followed by distillation, which can be used as a fuel in modified petrol engines.

biofilm *n.* adhesive material enclosing colonies of microorganisms attached to a surface.

bioflavonoid *n.* any of a group of flavonoids present in citrus and other fruits such as paprika, which have activity in animals due to their reducing and chelating properties, e.g. citrin.

biofuel *n.* gas such as methane or liquid fuel such as ethanol (ethyl alcohol) made from organic waste material, usually by microbial action. *see* biodiesel, bioethanol.

biogas *n.* gas with a high methane content which is produced by microbial fermentation of organic wastes.

biogenesis *n.* the theory that living organisms must be generated from living organisms, as opposed to the 19th century theory of spontaneous generation.

biogenic amines 5-hydroxytryptophan (serotonin), adrenaline, noradrenaline and dopamine.

biogenic, biogenetic *a.* originating from living organisms, *appl.* deposits such as coal, oil and chalk.

biogeochemical cycle closed cycle described by elements found in living material, such as carbon, nitrogen, oxygen and sulphur, as they pass from within organisms (biotic phase) into the non-living environment (abiotic phase) and back again. *alt.* nutrient cycle.

biogeochemistry *n.* study of the distribution and movement of elements present in living organisms in relation to their geographical environment, and the movement of elements between living organisms and their non-living environment.

biogeocoenosis *n.* community of organisms in relation to its special habitat.

biogeographical province area of the Earth's surface defined by the endemic species it contains.

biogeographical realms, biogeographical kingdoms major geographical divisions of the terrestrial environment characterized by their overall flora and fauna. *see* floristic kingdoms, zoogeographical regions.

biogeography *n.* that part of biology dealing with the geographical distribution of plants (phytogeography) and animals (zoogeography).

bioinformatics *n.* the science of the analysis, handling and storage of biological information, esp. the large amounts of data emerging from genome-sequencing and proteomics work.

biolistics *n.* technique of introducing DNA into a cell by firing minute (*ca.* 1.6 μm) DNA-coated particles (e.g. of gold) into the cell using a device powered by pressurized helium – the biolistic gun or 'gene gun'.

biological *a. pert.* living organisms and their products.

biological amplification biomagnification *q.v.*

biological clocks regular metabolic and behavioural rhythms seen in many cells and organisms.

biological containment in genetic engineering, the use of non-infectious, enfeebled and nutritionally fastidious strains of microorganism, which cannot survive outside the laboratory, as vehicles in which to clone recombinant DNA in order to minimize risk in case of accident.

biological control control of pests and weeds by other living organisms, usually other insects, bacteria or viruses, or by biological products such as hormones.

biological diversity as defined by the United Nations Convention on Biological Diversity: 'the variability among living organisms from all sources, including, *inter alia*, terrestrial, marine, and other aquatic ecosystems and the ecological complexes of which they are part. This includes diversity within species, between species and of ecosystems.' The number of different living species is estimated at between 40 and 80 million, most of them still undiscovered and uncharacterized, each species containing yet further genetic diversity. Commonly known as biodiversity.

see also alpha diversity, beta diversity, gamma diversity, genetic diversity, species diversity, ecological diversity.

biological indicator *see* indicator species.

biological oxygen demand biochemical oxygen demand *q.v.*

biological races strains of a species which are alike morphologically but differ in some physiological way such as host requirement or food or habitat preference.

biological rhythm internally generated, periodic daily or seasonal behaviour or metabolic change in living organisms. It will continue, for some time at least, even when the environmental rhythm (e.g. cycle of light and dark) to which it is entrained is absent. *see* circadian rhythm, circannual, diurnal. *alt.* biorhythm.

biological species population of individuals that can interbreed, i.e. a true species.

biology *n.* the science dealing with living organisms, a term coined by J.B. de Lamarck in 1802.

bioluminescence *n.* production of light by living organisms. It occurs at normal temperatures and is caused by an enzyme-catalysed biochemical reaction in which an inactive precursor is converted into a light-emitting chemical.

biomagnification *n.* increase in concentration of pollutants, e.g. fat-soluble pesticides such as DDT, in the bodies of living organisms at successively higher levels in the food chain. *cf.* bioaccumulation, bioconcentration, although all these terms are often used synonymously.

biomanipulation *n.* deliberate manipulation of the species composition of an ecosystem, e.g. to try to regenerate a hypereutrophic lake after the organic pollution itself has been ameliorated.

biomarker *n.* biological marker: (1) in geology, a molecule whose presence in a rock sample is indicative of the presence of life, either now or in the past. *alt.* biosignature; (2) in medicine and biomedical research, any molecule (usually a protein) whose presence or level indicates a particular disease or physiological state; (3) a protein, usually a cell-surface protein, that distinguishes a particular cell type.

biomass *n.* (1) total weight, volume or energy equivalent of organisms in a given area; (2) plant materials and animal wastes used as a source of fuel or other industrial products; (3) in biotechnology, the microbial matter in the system.

biome *n.* climatically controlled group of plants and animals of a characteristic composition and distributed over a wide area, such as tropical rainforest, tundra, temperate grassland, desert, savanna, taiga and northern coniferous forest.

biomedical *a. pert.* or *appl.* basic research, e.g. in cell biology, molecular biology or genetics, that is intended eventually to have an application in medicine.

biometeorology *n.* study of the effects of the weather on plants and animals.

biometrics, biometry *n.* statistical study of living organisms and their variations.

biomineralization *n.* production of partly or wholly mineralized internal or external structures by living organisms.

biomolecule *n.* any molecule that is produced in nature only by a living cell, esp. large organic molecules such as nucleic acids, proteins and polysaccharides.

biophage *n.* organism feeding upon other living organisms.

biopharmaceuticals *n.plu.* drugs or therapies consisting of proteins, peptides or nucleic acids or their derivatives, as opposed to small organic molecules.

biophysics *n.* (1) study of biological processes in terms of their underlying physical principles; (2) physics as applied to biology.

biophyte *n.* parasitic plant.

bioplastic *see* biopolymer.

biopolymer *n.* biodegradable polymer produced by living organisms, e.g. polysaccharide gums (xanthans) and bioplastics such as poly-β-hydroxybutyric acid and other poly-β-hydroxyalkanoates produced by bacteria.

biopsy *n.* examination of living tissue.

bioreactor *n.* vessel or other device in which living cells or microorganisms are grown, either as a means of large-scale production of the cells, or in order to carry out a particular biochemical process or produce a particular product.

bioregion *n.* unique area with distinctive soils, landforms, climates and indigenous plants and animals.

bioremediation *n.* recovery of a contaminated site by the use of living organisms, usually microorganisms, to break down the pollutants.

biorhythm *see* biological rhythm.

bios *n.* living organisms.

bioscaffold *n.* biocompatible material, such as certain polymers, that can be used as an inert matrix on which to grow cells, esp. in relation to implanting the scaffold and living cells in the body to regrow damaged tissues.

biosensor *n.* any organism, microorganism, enzyme system or other biological structure used as an assay or indicator.

bioseries *n.* succession of changes of any single heritable character.

bioseston *n.* plankton or particulates of organic matter suspended in water.

biosignature biomarker (1), *q.v.*

biospecies biological species *q.v.*

biospeleology *n.* the study of the biology of cave-dwelling organisms.

biosphere *n.* the part of the planet containing living organisms: the living world.

biostasis *n.* ability of living organisms to withstand environmental changes without being changed themselves.

biostatics *n.* study of the structure of living organisms in relation to function.

biostratigraphic unit rock stratum or collection of strata that is distinguishable by the fossils it contains, without recourse to geological or physical features, and which is established by worldwide correlations. *alt.* **biostratigraphic zone.**

biosynthesis *n.* formation of organic compounds by living organisms.

biosynthetic *a. appl.* any reaction or process that results in the synthesis of an organic compound in a biological system.

biosystem ecosystem *q.v.*

biosystematics *n.* study of the variation and evolution of taxa.

biota *n.* (1) the total fauna and flora of a region; (2) the population of living organisms in general.

biotechnology *n.* the industrial use of living cells, usually microorganisms, or of their isolated enzymes to carry out chemical processes. Examples are brewing and wine-making, food processing and manufacture, sewage treatment, and the manufacture of some organic chemicals and drugs. Nowadays the term esp. refers to the industrial, agricultural and pharmaceutical use of genetically modified cells and organisms, and to the application of genomics and proteomics approaches to the discovery of new drugs.

bioterrorism *n.* the deliberate release of harmful biological agents, such as bacteria and viruses, that cause disease in humans, animals or plants, as a weapon of terrorism.

biotic *a. pert.* life and living organisms.

biotic climax plant community that is maintained in the climax state by a biotic factor such as grazing.

biotic community the whole community of plants and animals that share a particular habitat or region.

biotic environment the part of an organism's environment produced by its interaction with other organisms.

biotic factors the influence of organisms and their activities on other organisms and the environment.

biotic index a measure of the ecological quality of the environment, usually in regard to organic pollution, using assessments of the number and abundance of key indicator species present. A high biotic index, representing high diversity and the presence of pollution-sensitive species, indicates a high-quality environment. A low biotic index, reflecting the presence of only a few pollution-tolerant species, indicates heavy organic pollution. *cf.* diversity index.

biotic potential highest possible rate of population increase (r_{max}), resulting from maximum rate of reproduction and minimum mortality.

biotic province major ecological region of a continent.

biotic pyramid ecological pyramid *q.v.*

biotic succession that part of a plant succession that is controlled by the activities and interactions of the species present rather than by the physical environment.

biotin *n.* vitamin B_4, a water-soluble vitamin which is required as the prosthetic group of enzymes involved in the incorporation of carbon dioxide into organic compounds. Liver, egg yolk and yeast are rich sources.

biotin–streptavidin methodology for detecting nucleic acids and proteins by labelling them with biotin, which is in turn bound by streptavidin, a protein isolated from bacteria.

biotope *n.* area or habitat of a particular type, defined by the organisms (plants, animals, microorganisms) that typically inhabit it, e.g. grassland, woodland. On a smaller scale denotes a microhabitat.

biotransformation *n.* the metabolic conversion of toxic compounds into non-toxic compounds in the body.

biotroph *n.* (1) any organism that feeds on other living organisms; (2) in plant pathology, a parasitic fungus that requires living host cells. *a.* **biotrophic.**

biotrophism *n.* type of parasitism in which the parasite requires living host cells.

biotype *n.* group of organisms of similar genetic constitution.

bipaleolate *a.* furnished with two small bracts (paleae).

bipalmate *a.* divided into lobes with each lobe divided again.

biparietal *a.* connected with the two parietal lobes of brain.

biparous *a.* bearing two young at a time.

bipectinate *a. appl.* structures having the two edges furnished with teeth like a comb.

biped *n.* a two-footed animal. *a.* **bipedal** walking on two legs.

bipeltate *a.* having, or consisting of, two shield-like structures.

bipenniform *a.* shaped like a feather, with sides of vane of equal size.

bipetalous *a.* with two petals.

bipinnaria *n.* starfish larva with two bands of cilia.

bipinnate *a. appl.* compound pinnate leaf in which leaflets grow in pairs on paired stems.

bipinnatifid *a. appl.* pinnate leaf whose segments are again divided.

bipinnatipartite *a.* bipinnatifid, but with divisions extending nearly to midrib.

bipinnatisect *a.* bipinnatifid, but with divisions extending completely to midrib.

biplicate *a.* having two folds, having two distinct wavelengths, *appl.* flagellar movement in certain bacteria.

bipolar *a.* having two distinct ends.

bipolar cell type of nerve cell with one process leaving cell body at either end, present in retina between photoreceptors and ganglion cells.

bipolarity *n.* the condition of having two distinct ends or poles.

biradial *a.* symmetrical both radially and bilaterally, as some coelenterates.

biradiate *a.* two-rayed.

biramous, biramose *a.* dividing into two branches.

bird lice common name for an order of insects, the Mallophaga or biting lice, which are ectoparasites of birds.

birds Aves *q.v.*

bird's nest fungus common name for a member of the Nidulariales, an order of gasteromycete fungi which have fruiting bodies resembling minute bird's nests (the peridium) full of eggs (the peridioles containing the spores).

Birnaviridae *n.* family of non-enveloped RNA viruses infecting fish and birds.

birostrate *a.* having two beak-like processes.

birth pore uterine pore of trematodes and cestodes.

birth rate number of live births within a population over a set period, usually a year. The crude birth rate is calculated for human populations as the annual number of live births per 1000 population in a given geographical area, with the population number usually taken at the midpoint of the year in question. A more accurate measure is the annual number of live births per 1000 females of reproductive age (15–44 years), known as the standardized birth rate or fertility rate.

biscotiform *a.* biscuit-shaped, *appl.* spores.

biscuspid *a. appl.* tooth: premolar *q.v.*

bisect (1) *n.* a stratum transect chart with root system as well as shoot included; (2) *v.* to divide into two equal halves.

bisegmental *a.* involving or *pert.* two segments.

biseptate *a.* with two partitions.

biserial, biseriate *a.* arranged in two rows or series.

biserrate *a.* having marginal teeth which are themselves notched.

bisexual *a.* having both male and female organs.

bisporangiate *a.* having both micro- and megasporangia.

bispore *n.* a paired spore, as in certain red algae.

bisporic, bisporous *a.* having two spores.

bistipulate *a.* having two stipules.

bistrate *a.* having two layers.

bistratose *a.* with cells arranged in two layers.

bisulcate *a.* (1) having two grooves; (2) cloven-hoofed.

biternate *a.* divided into three with each part again divided into three.

bithorax complex (BX-C) cluster of homeotic genes in the fruitfly *Drosophila melanogaster* that are involved in controlling the identity of different segments.

biting lice *see* bird lice.

bitubercular *a.* with two tubercles or cusps, *appl.* biscuspid premolar teeth.

bitunicate *a. appl.* structures, e.g. asci, with distinct and often separable outer and inner walls.

Biuret reaction simple test for the presence of proteins and peptides which is based on a reaction with the peptide bond. Solutions of copper sulphate and sodium hydroxide are added to the test sample, and a purple copper complex is formed if a compound containing a peptide bond is present.

bivalent *n.* (1) chromosome that has duplicated to form two sister chromatids still held together at the centromere; (2) pair of duplicated homologous chromosomes held together by chiasmata at meiosis, *alt.* tetrad; (3) *a.* (*immunol.*) having two identical binding sites, *appl.* antibody or antigen.

bivalve *a.* (1) *appl.* shells consisting of two hinged parts or valves; (2) *appl.* molluscs having such shells; (3) (*bot.*) *appl.* seed capsule of similar structure.

Bivalvia, bivalves *n., n.plu.* class of bilaterally symmetrical molluscs which are laterally flattened and have a shell made of two hinged valves, e.g. clams, mussels, scallops, cockles.

biventer cervicalis the spinalis capitis muscle, or central part of the spinalis, a muscle of the neck, consisting of two fleshy ends with a narrow portion of tendon in the middle.

biventral *a. appl.* muscles of the biventer type.

bivoltine *a*. having two broods in a year.

black coral common name for a member of an order of stony corals which are colonial and have a black or brown skeleton.

black earth chernozem *q.v.*

bladder *n*. (1) general term for a membranous sac filled with air or fluid; (2) in humans, the urinary bladder, the structure in which urine collects before release from the body.

bladder cell (1) (*mycol.*) a globular modified hyphal cell in integument or stalk of fruit body in fungi; (2) (*zool.*) in tunicates, a large vacuolated cell in outer layer of tunic.

bladderworm *n*. larval stage of some Cestoda (tapeworms) in the intermediate host, which is in the form of a bladder containing an inverted scolex. *see* cysticercoid.

blade *n*. flat part of leaf or bone.

Blandin's glands anterior lingual glands.

blanket bog, blanket mire, blanket peat acid peat bog covering large stretches of country which develops in very wet climates on upland regions where drainage is impeded and the soil is acid. It is composed of a layer of undecomposed organic matter (peat) overlying waterlogged, acidic, nutrient-poor ground.

BLAST Basic Local Alignment Search Tool, a computerized algorithm used to search electronic databases of DNA and protein sequences for sequences similar to each other.

-blast, -blastic word elements from Gk *blastos*, bud, signifying a cell or structure that can produce new cells.

blast cell cell that divides to give rise to two different progeny cells.

blastema *n*. mass of undifferentiated cells that develops on end of amputated limb of some amphibians, reptiles and insects, and from which the limb is regenerated.

blastic *a*. *pert*. or stimulating growth by cell division.

blastocarpous *a*. developing while still surrounded by pericarp, *appl*. fruits.

blastocele, blastocoel *n*. hollow fluid-filled cavity inside the blastula.

blastocyst *n*. hollow sphere of cells that develops from the morula in mammalian embryogenesis and which implants in the uterine wall. It consists of trophoblast cells, which will develop into the chorion, and the inner cell mass, from which the embryo itself and other extra-embryonic membranes will develop.

blastocyte *n*. any undifferentiated embryonic cell.

blastoderm *n*. (1) the layer of syncytial protoplasm that forms at the periphery of the fertilized insect egg and which develops into a layer of cells, the cellular blastoderm, which forms the embryo; (2) blastodisc *q.v.*

blastodisc *n*. disc-shaped cellular structure that is formed by cleavage of the zygote in large yolky eggs such as those of birds and reptiles. *alt*. germinal disc.

blastogenesis *n*. reproduction by budding.

blastokinesis *n*. movement of embryo in the egg, as in some insects and cephalopods.

blastomere *n*. any one of the cells formed by the first few divisions of a fertilized egg.

blastopore *n*. slit-like or circular invagination on the surface of the blastula in some animal embryos at which internalization of endoderm and mesoderm occurs at gastrulation.

blastospore *n*. spore developed by budding and itself capable of budding, as in yeast cells.

blastostyle *n*. specialized structure in which medusae develop in some colonial hydrozoans.

blastozooid *n*. an individual or zooid formed by budding.

blastula *n*. hollow ball of cells that develops from the morula in embryogenesis in many animals.

blastulation *n*. development of a blastula.

blending inheritance the idea, now known to be mistaken but current in the 19th century, that the intermediate and novel characteristics seen in hybrids were due to the physical blending of fluid-like carriers of parental characteristics. Gregor Mendel showed, on the contrary, that the individual 'factors' that determine hereditary characteristics remain unchanged during inheritance.

blennoid *a*. resembling mucus.

blephara *n*. peristome tooth in mosses.

blepharal *a*. *pert*. eyelids.

blight *n.* insect or fungal infestation of plants.

blind spot region of retina where optic nerve leaves and which is devoid of rods and cones.

blocky *a. appl.* soil crumbs of squarish or rectangular shape.

blood *n.* fluid circulating in the vascular system of animals, distributing e.g. nutrients, oxygen, metabolites and hormones, and taking up waste products for excretion. In vertebrates it contains cells specialized for oxygen transport (red blood cells) and white blood cells (leukocytes), which are concerned with protection against infection and scavenging dead cells.

blood–brain barrier structural and physiological barriers which prevent the movement of most blood components into brain tissue and cerebrospinal fluid or vice versa.

blood cell any cell forming a normal part of the blood. *see* erythrocyte, leukocyte, lymphocyte, haemocyte.

blood clotting response to any wound that causes bleeding, in which liquid blood is converted into a solid clot that plugs the wound. Damaged blood vessels release chemicals that cause the narrowing of blood vessels and adherence of platelets to vessel walls. Activation of plasma proteins called coagulation factors results in the conversion of soluble plasma fibrinogen to insoluble fibrin, which forms a meshwork with platelets, red blood cells and other plasma proteins, forming a permanent clot. A genetically determined deficiency of coagulation factor VIII is the cause of classical haemophilia, in which the blood cannot clot.

blood corpuscle red or white blood cell. *see* erythrocyte, leukocyte.

blood gills delicate blood-filled sacs functioning in uptake of salts, in certain insects.

blood groups immunological classification of genetically different types of blood in human populations, used for matching blood for transfusion. Blood groups are determined by antigens carried on the red blood cells, the main system being the ABO system (*q.v.*). Other blood-group systems in humans include the Rh system (Rh positive and Rh negative), and the MN system.

blood islands groups of mesodermal cells in early embryo from which immature erythrocytes develop.

blood plasma fluid part of the blood including the soluble blood proteins, which is left as a clear yellow fluid after the red and white cells have been removed by centrifugation. *cf.* blood serum.

blood platelet *see* platelet.

blood serum clear fluid which can be expressed from clotted whole blood or clotted blood plasma, and which comprises the fluid portion of the blood from which fibrin has been removed. It contains many of the soluble blood proteins. *see also* antiserum, serum.

blood sugar glucose *q.v.*

blood vessel any vessel or space in which blood circulates (e.g. artery, vein or capillary), strictly used only in regard to special vessels with well defined walls.

bloodworm *n.* (1) reddish thread-like aquatic larva of the chironomid midge, *Chironomus riparius*, which is tolerant of heavy organic pollution; (2) reddish oligochaete worm living in river mud; (3) a red bristleworm found on muddy shores.

bloom *n.* (1) waxy layer on surface of certain fruits such as grapes; (2) blossom or flower; (3) seasonal dense growth of algae or phytoplankton.

Bloom syndrome inherited condition characterized by low birth weight, short stature, extreme sensitivity to sunlight, and an enhanced frequency of sister-chromatid exchange during mitosis.

BLOSUM matrix type of amino-acid matrix used in scoring alignments. *see* substitution matrix.

blotting *see* immunoblot, Northern blotting, Southern blotting, Western blot.

blubber *n.* insulating layer of fat between skin and muscle layer in aquatic mammals such as whales and seals.

blue coral common name for a member of the genus *Heliopora*, corals having a solid calcareous skeleton with vertical tubular cavities containing polyps.

blue-green algae common and inaccurate name for the cyanobacteria *q.v.*

blue-light receptor any of a class of flavoproteins in plants that act as receptors for light in the blue region of the spectrum (wavelength 400–500 nm), and which are involved in phototropism, inhibition of stem growth and promotion of leaf expansion.

blunt-end ligation technique used in the construction of recombinant DNAs in which any two DNA molecules may be joined.

B lymphocyte, B cell type of lymphocyte that gives rise to antibody-producing cells in the adaptive immune response to infection or immunization. B cells develop in bone marrow (in mammals) and mature B cells are present in large numbers in lymph nodes, spleen and other lymphoid tissues, and in the blood. During development in the bone marrow an individual B cell undergoes rearrangement of its immunoglobulin genes to produce genes encoding immunoglobulin of a single antigen specificity, which is expressed on the B-cell surface. After encounter with its corresponding antigen, a B cell proliferates and differentiates into plasma cells producing antibodies of the same specificity. There are two lineages of B cell, the majority **B-2 cells** and a minority population of **B-1 cells**, which appear to have a more limited range of antigen specificity. *see also* antibody, plasma cell. *cf.* T lymphocyte.

BMD Becker muscular dystrophy *q.v.*

BMI body mass index *q.v.*

BMP bone morphogenetic protein(s) *q.v.*

BOD biochemical oxygen demand *q.v.*

body cavity internal cavity in many animals in which various organs are suspended and which is bounded by the body wall. *alt.* coelom.

body cell somatic cell, as opposed to a germ cell or reproductive cell.

body fluids the fluid components of the animal body and fluids secreted or excreted by the body, including blood, lymph, urine, semen, sweat and secretions from mucous membranes.

body lice common name for members of the insect order Anoplura which are ecto-parasitic on mammals (e.g. body louse, bed bug).

body mass index (BMI) a measure of the amount of body fat in humans, calculated from the body weight (in kilograms) divided by the height (in metres) squared.

'Normal' values for adults range from 18.5 to 24.9, with a value of greater than 30 being classified as clinical obesity.

body plan of an animal embryo, the general disposition of the main axes of symmetry and the germ layers. *see also* bauplan.

body stalk band of mesodermal tissue connecting tail end of embryo and chorion.

bog *n.* (1) characteristic plant community developing on wet, very acid peat, containing e.g. sundews (*Drosera* spp.), sphagnum moss and bog myrtle (*Myrica gale*). *see* blanket bog, raised bog; (2) sometimes also refers to alkaline bogs developing in valleys. *cf.* fen.

Bohr effect decrease in the affinity of haemoglobin for oxygen which occurs when pH is lowered and carbon dioxide concentration is increased. It results in the release of oxygen from oxyhaemoglobin, as in metabolically active tissues.

boletes *n.plu.* basidomycete fungi of the genus *Boletus* and related genera, similar in general form to agarics, but with pores and not gills on the underside of the cap.

boletiform *a.* shaped like an elliptic spindle, *appl.* spores of some boletes.

boll *n.* fruit in the form of a capsule or spherical pericarp, as in cotton plant.

bolus *n.* (1) rounded mass; (2) lump of chewed food.

bone *n.* calcified connective tissue forming the skeleton of some fishes and of all reptiles, amphibians, birds and mammals. It consists of an organic phase, mostly the protein collagen, and an inorganic phase composed of calcium phosphate as hydroxyapatite, and other minerals. The organic matrix is secreted by cells called osteoblasts, which persist in mature bone as non-dividing osteocytes. Bone may be formed directly or by calcification of cartilage. *see also* compact bone, Haversian system, membrane bone, ossification, osteoblast, osteoclast, osteocyte.

bone-beds fossil deposits formed largely by remains of bones of fishes and reptiles, such as the Liassic bone-beds.

bone cell *see* osteoblast, osteocyte.

bone marrow connective tissue filling the central cavities of the long bones, from which all blood cells are produced. It is also a site of antibody production, from plasma cells that migrate to it after differentiation in lymph nodes.

bone marrow chimaera animal in which the bone marrow has been killed by irradiation and then replaced with transplanted bone marrow from another, genetically dissimilar, animal.

bone morphogenetic proteins (BMP) family of proteins that provide developmental signals in embryogenesis in many animals. They are members of the TGF-β family of signalling proteins.

bonitation *n.* evaluation of the numerical distribution of a species in a particular locality or season, esp. in relation to its agricultural, veterinary or medical implications.

bonobo *Pan paniscus*, formerly known as the pygmy chimpanzee and one of the two species of living chimpanzees, the other being the common chimpanzee *Pan troglodytes.*

bony fishes common name for the Osteichthyes, a class of fishes with bony skeletons, usually possessing a swimbladder or lung and gills covered by an operculum. They include teleosts, lungfishes and crossopterygians.

book lice common name for some members of the Psocoptera, small wingless insects with a globular abdomen and incomplete metamorphosis, often living on paper.

book lung lung book *q.v.*

booster *a. appl.* a second or subsequent injection of a vaccine or other antigen.

booted *a.* equipped with raised horny plates of skin, as the feet of some birds.

bootstrapping *n.* procedure involved in building phylogenetic trees, in which the accuracy of the tree is checked by repeatedly taking random samples of the original data, constructing new trees and analysing them.

bordered pit form of pit developed on walls of xylem vessels with overarching border of secondary cell wall.

boreal *a.* (1) *appl.* the northern coniferous forest growing in a band across the Northern Hemisphere, e.g. in North America and Siberia; (2) *pert.* post-glacial age with a continental type of climate. *see also* Boreal kingdom.

Boreal kingdom major phytogeographical kingdom comprising the Northern Hemisphere north of latitude 30°N. It is divided

into the following regions: Atlantic North American, Arctic and Subarctic, Euro-Siberian, Macaronesian, Mediterranean, Pacific North American, Sino-Japanese, Western and Central Asiatic. *alt.* Holarctic kingdom.

bosset *n.* the beginning of antler formation in deer in 1st year.

Botallo's duct ductus arteriosus, a small blood vessel representing the 6th gill arch and connecting pulmonary with systemic arch.

botany *n.* branch of biology dealing with plants.

bothrenchyma *n.* plant tissue formed of pitted ducts.

bothridium *n.* muscular cup-shaped outgrowth from scolex of some tapeworms, used for attachment to host.

bothrium *n.* sucker.

bothrosome *n.* indentation in the wall of cell of net slime moulds, from which the extracellular matrix is secreted.

botryoid(al) *a.* in the form of a bunch of grapes.

botryose racemose *q.v.*

bottleneck *n.* sudden decrease in population density with a resulting decrease in genetic variability within a population.

botuliform *a.* sausage-shaped.

botulinum toxin protein toxin produced by the bacterium *Clostridium botulinum* under anaerobic conditions, e.g. in inadequately sterilized canned or bottled food. It is a powerful poison which affects transmission of nerve impulses between nerve and muscle, causing botulism.

botulism *n.* respiratory and muscular paralysis caused by botulinum toxin, which can be fatal if untreated.

boundary elements insulator elements *q.v.*

bouquet *n.* bunch of muscles and ligaments connected with the styloid process of the temporal bone.

bourrelet *n.* poison gland associated with sting in ants.

bouton *n.* enlarged terminal of an axon branch where it forms a synapse.

Bovidae, bovids *n., n.plu.* mammalian family of the Artiodactyla comprising cattle, sheep, goats and bison, which have four-chambered stomachs and horns that are not shed. *a.* **bovine.**

bovine serum albumin (BSA) small globular protein isolated commercially from bovine blood and used in biochemistry and molecular biology as a carrier protein, as a stabilizing agent in enzymatic reactions, and as a blocking agent in nucleic acid hybridizations, among other uses. *see also* albumins.

bovine spongiform encephalopathy (BSE) *see* transmissible spongiform encephalopathies.

Bowditch's Law all-or-none principle *q.v.*

Bowman's capsule the cup-like dilated end of a vertebrate kidney tubule, surrounding a glomerulus.

Bowman's glands glands in nasal mucous membranes secreting a watery fluid to wash over olfactory epithelium.

box-jellies the Cubozoa, a class of Cnidaria with free-swimming cuboidal medusae with venomous tentacles at each corner, which can deliver a sting fatal to humans. *alt.* cubozoans, sea wasps.

bp base pair (in a nucleic acid) *q.v.*; used as the unit of length in DNA.

bp, b.p. before present.

b.p. boiling point.

Bq becquerel *q.v.*

braccate *a.* having additional feathers on legs or feet, *appl.* birds.

brachia *n.plu.* (1) arms; (2) two spirally coiled structures, one at each side of mouth in brachiopods. *sing.* **brachium.**

brachial *a. pert.* arms, arm-like.

brachialis *n.* flexor muscle of the forearm.

brachiate *a.* (1) branched; (2) having arms; (3) having opposite, widely spread, paired branches on alternate sides.

brachiating, brachiation *n.* moving along by swinging the arms from one hold to another, as in the gibbon. *v.* to brachiate.

brachidia *n.plu.* calcareous skeleton supporting brachia in some brachiopods.

brachiocephalic *pert.* arm and head, *appl.* artery, vein.

brachiocubital *a. pert.* arm and forearm.

brachiolaria *n.* larval stage in some starfish.

Brachiopoda, brachiopods *n., n.plu.* small phylum of shelled coelomate animals, the lamp shells, superficially resembling the bivalve molluscs but different in symmetry of the shell and internal structure. A characteristic structure is the lophophore, consisting of coiled tentacles (brachia) surrounding the mouth.

brachioradialis *n.* supinator longus muscle in forearm, one of the muscles used in turning the palm of hand upwards.

brachium *n.* (1) arm, or branching structure; (2) forelimb of vertebrate; (3) bundle of nerve fibres connecting cerebellum to cerebrum or pons.

brachy- prefix from Gk *brachys*, short.

brachyblast brachyplast *q.v.*

brachyblastic *a.* with a short germ band.

brachycephalic *a.* short-headed.

Brachycera *n.* short-horned flies, a suborder of Diptera with short stout antenna, which include the blood-sucking horse-flies and gadflies, the metallic-coloured soldier-flies, the slender, long-legged snipe-flies, the beeflies, which have a strong superficial resemblance to bees and hover and feed in flowers, the robber-flies, and other families.

brachycerous *a.* (1) short-horned; (2) with short antennae.

brachydactyly *n.* condition where fingers or toes are abnormally short.

Brachydanio rerio the zebrafish, an important model animal in developmental biology.

brachydont, brachyodont *a. appl.* molar teeth with low crowns. *cf.* hypsodont.

brachyelytrous *a.* having short elytra.

brachyism *n.* dwarfism in plants caused by shortening of internodes.

brachyplast *n.* short spur bearing tufts of leaves, occurring with normal branches on the same plant.

brachypleural *a.* with short pleura or side plates.

brachypodous *a.* with short legs or stalk.

brachypterous *a.* with short wings, *appl.* insects.

brachysclereid(e) stone cell *q.v.*

brachystomatous *a.* with a short proboscis, *appl.* certain insects.

brachytic *a.* dwarfish, *appl.* plants in which dwarfism is caused by shortening of the internodes.

brachyural, brachyurous *a.* having short abdomen usually tucked in below thorax, *appl.* certain crabs.

brachyuric *a.* short-tailed.

brackish water water with a salinity of 0.5–30 parts per thousand total dissolved solids. *cf.* freshwater.

bract *n.* (1) modified leaf in whose axil an inflorescence or flower arises; (2) floral leaf; (3) leaf-like structure.

bracteate *a.* having bracts.

bracteiform *a.* like a bract.

bracteolate *a.* possessing bracteoles.

bracteole, bractlet *n.* secondary bract at the base of an individual flower.

bracteose *a.* having many bracts.

bractlet bracteole *q.v.*

bract scales small scales developed directly on axis of cones in conifers, not bearing ovules.

bradyauxesis *n.* (1) relatively slow growth; (2) growth of a part at a slower rate than that of the whole.

bradycardia *n.* slowing of heart (and pulse) rate.

bradykinin *n.* polypeptide of nine amino acids, circulating in the blood and causing dilation of blood vessels and contraction of smooth muscle. Derived by post-translational cleavage from kininogen by plasma kallikrein. *alt.* kinin.

bradytelic *a.* evolving at a rate slower than the standard rate.

bradyzoite *n.* a slowly growing merozoite.

brain *n.* the coordinating centre of the nervous system, present as a distinct part of the nervous system in many animals, and most highly developed in vertebrates, where it forms an anterior continuation of the spinal cord and is enclosed in the bony cranium. The vertebrate brain is a highly organized mass of billions of interconnected nerve cells and supporting tissues, and is divided into three main regions: the forebrain, the midbrain and the hindbrain. The brain analyses and integrates incoming sensory information and generates output that is sent to muscles and glands. In invertebrates lacking a distinct brain the supraoesophageal or suprapharyngeal ganglia are the main coordinating centres.

brain-derived neurotrophic factor (BDNF) a neurotrophin *q.v.*

brainstem part of brain at its base, before it becomes spinal cord, and which consists of the midbrain, pons and medulla oblongata.

brain vesicles three dilations at the anterior end of the embryonic neural tube which give rise to the forebrain, midbrain and hindbrain.

branch *n.* (1) taxonomic group used in different ways by different specialists but usually referring to a level between subphylum and class; (2) in a phylogenetic tree, the line joining two nodes.

branch gaps gaps in the vascular cylinder of a main plant stem, each subtending a branch trace.

branchiae *n.plu.* gills of aquatic animals. *sing.* **branchia**.

branchial *a. pert.* gills.

branchial arches series of ectodermal ridges developed in pharyngeal region in vertebrate embryos. They develop into the gill arches in fishes and into a variety of structures in other vertebrates. *alt.* visceral arches.

branchial basket framework of cartilaginous bars around the gill region in lampreys and cartilaginous fish.

branchial clefts infolding of ectoderm between branchial arches in pharyngeal region of vertebrate embryos. They develop into the gill slits in fishes and into a variety of structures in other vertebrates. *alt.* **branchial grooves**.

branchial siphon in molluscs, the siphon through which water is drawn in over the gills.

branchiate *a.* having gills.

branchicolous *a.* parasitic on fish gills.

branchiform *a.* gill-like.

branching enzyme enzyme that catalyses the transfer of a segment of 1,4-α D-glucan chain to a primary hydroxyl group in a similar chain, forming a branchpoint in a polysaccharide chain. EC 2.4.1.18.

branchiocardiac *a.* (1) *pert.* gills and heart; (2) *appl.* vessel leading off ventrally from ascidian heart; (3) *appl.* vessels conveying blood from gills to pericardial cavity in certain crustaceans.

branchiomeric *a. appl.* muscles derived from gill arches.

branchiopallial *a. pert.* gill and mantle of molluscs.

branchiopneustic *a. appl.* insects having spiracles replaced functionally by gills.

Branchiopoda, branchiopods *n., n.plu.* the water fleas, brine shrimps and their allies, a subclass of mainly freshwater crustaceans whose carapace, if present, forms a dorsal shield or bivalve shell, and which have broad lobed trunk appendages fringed with hairs.

branchiostegal *a*. with, or *pert*. a gill cover, *appl*. membrane, rays.

branchiostegal ray one of a group of dermal bones ventral to gill cover in bony fishes.

branchiostegite *n*. expanded lateral portion of carapace forming gill cover in some crustaceans.

Branchiura *n*. class of crustaceans, commonly called fish lice, that are ectoparasites on fish and amphibians.

branch migration in DNA recombination, the extension of strand pairing by movement of the DNA crossover point along the DNA, resulting in a crossed-over DNA strand pairing with its complementary strand in the opposite DNA molecule as it displaces the resident strand.

branch trace vascular bundle extending from stem into base of lateral bud.

brand *n*. burnt appearance of leaves, caused by rust and smut fungi.

brand fungi a name for the rust and smut fungi *q.v.*

brand spore teleutospore *q.v.*

brassica *n*. any plant of the family Brassicaceae (such as mustard, oilseed rape, cabbage).

brassinosteroids *n.plu*. class of plant steroids that act as growth regulators during development and are involved in mediating developmental responses to darkness, such as etiolation.

Brazilian subregion subdivision of the Neotropical zoogeographical region, comprising northern South America to southern Brazil, excluding only the Andes to the west.

BrdU, BrU bromouracil *q.v.*

breeding season for animals that do not breed all the year round, the period each year in which animals court, mate and rear their young.

breeding size number of individuals in a population actually involved in reproduction, per generation.

breeding system extent and mode of interbreeding within a species or group of closely related species.

bregma *n*. (1) that part of skull where frontals and parietals meet; (2) intersection of sagittal and coronal sutures.

brevi- prefix from L. *brevis*, short.

brevicaudate *a*. with a short tail.

brevilingual *a*. with a short tongue.

brevipennate *a*. with short wings, *appl*. birds.

brevirostrate *a*. with a short beak.

brevissimus oculi the obliquus inferior, the shortest of the eye muscles.

bridge *n*. structure produced from a dicentric chromosome at meiosis, in which the two centromeres in the same chromosome are segregating to opposite poles, and which may lead to chromosome breakage.

brigalow forest acacia forest covering large areas in Australia, in which the dominant species is the brigalow (*Acacia harpophylla*).

bright-field microscopy technique of optical microscopy in which a living cell is viewed by direct transmission of light through it.

brightness *n*. one of the three basic dimensions of perception of visible light by humans. Refers to the gradient from dark to light. *see also* hue, saturation.

brille *n*. transparent covering over the eyes of snakes.

brine shrimp small marine crustacean of the subclass Branchiopoda, usually refers to *Artemia*.

bristletails *n.plu*. common name for certain insects of the orders Thysanura and Diplura, small wingless insects characterized by a single or two-pronged bristle at the tail end.

bristle worm common name for a member of the Polychaeta, a class of annelid worms.

brittle star common name for a member of the Ophiuroidea, a class of echinoderms with five slender arms clearly marked off from the central disc.

broad heritability (H^2) the proportion of total phenotypic variance at the population level that is contributed by the genotypic variance.

broadleaf, broadleaved *a*. (1) *appl*. the angiosperm trees of temperate climates, which bear thin flat leaves (as opposed to the needle-bearing conifers); (2) *appl*. dicot weeds with broad flat leaves, as opposed to grasses.

broad-spectrum antibiotic antibiotic that is active against both Gram-negative and Gram-positive bacteria.

Broca's area region of left frontal cerebral cortex involved in speech production. *alt.* area 44.

Brodman areas small areas into which the human cerebral cortex is conventionally divided, each being numbered. Areas 17 and 18, for example, comprise the primary visual cortex. The division is based on the cytology and detailed morphology of the cortex.

bromatium *n.* hyphal swelling on a fungus cultivated by ants, and which they use as food.

bromelain *n.* proteolytic and milk-clotting enzyme found in pineapple. EC 3.4.22.4. *alt.* bromelin.

Bromeliales, bromeliads *n.*, *n.plu.* order of terrestrial and epiphytic monocots with a reduced stem and a rosette of fleshy water-storing leaves, and comprising the family Bromeliaceae (pineapple).

bromodeoxyuridine, bromouracil (BrdU, BrU) thymidine analogue in which the methyl group of thymine has been replaced by a bromine atom, and which causes mutations when incorporated into DNA because it can pair with guanine as well as adenine. Used as an anticancer drug to preferentially kill rapidly dividing cancer cells.

bromovirus group plant virus genus named after the type member, brome mosaic virus, which is a small isometric single-stranded RNA virus causing mosaic symptoms (mottling of leaves). They are multicomponent viruses in which four genomic RNAs are encapsidated in three different virus particles.

bronchi *n.plu.* tubes connecting the trachea (windpipe) and the lungs. *sing.* **bronchus.**

bronchia *n.plu.* the subdivisions or branches of each bronchus.

bronchial *a. pert.* bronchi.

bronchiole *n.* small terminal branch of a bronchus.

bronchopulmonary *a. pert.* bronchi and lungs.

bronchus *sing.* of bronchi.

bronchus-associated lymphoid tissue (BALT) lymphoid tissues in the connective tissue of the walls of the respiratory tract.

brood *n.* (1) offspring of a single birth or clutch of eggs; (2) any young animals being cared for by adults.

brood cell chamber built to house immature stages of insects.

brood parasite animal that lays its eggs in the nest of another member of the same species (intraspecific brood parasitism) or of a different species (interspecific brood parasitism), who then rears them.

brood pouch sac-like cavity in which eggs or embryos are placed.

brown alga common name for a member of the Phaeophyta, mainly large marine algae (seaweeds), such as the kelps, containing the brown pigment fucoxanthin.

brown earths, brown forest soils dark brown friable soils associated with areas of the Earth's land surface originally covered with deciduous forest.

brown fat highly vascularized adipose tissue rich in mitochondria, the cytochromes of which help to give it a brown colour. Involved in thermoregulation in hibernators and in young mammals generally. Typically occurring around the shoulder blades, neck, heart, large blood vessels and lungs. It is specialized for heat generation as a result of uncoupling of fatty acid oxidation and electron transfer in mitochondria from ATP synthesis. *alt.* brown adipose tissue (BAT).

Brownian movement movement of small particles such as pollen grains or bacteria when suspended in a colloidal solution, due to their bombardment by molecules of the solution.

brown podzolic soil acid forest soil with a layer of litter over a greyish-brown organic and mineral layer and a pale leached layer below.

brown soils soils similar to chernozems (*q.v.*) but found in warmer and drier areas and supporting short grassland.

browse (1) *n.* tender parts of woody plants such as leaves and shoots; (2) *v.* to eat such material. *cf.* grazing.

Bruch's membrane thin basal membrane forming the inner layer of the choroid.

Brunner's glands branched glands in the submucosa of the duodenum which open into the crypts of Lieberkühn and which secrete an alkaline fluid containing glycoproteins. *alt.* duodenal glands.

brush border dense covering of minute finger-like projections (microvilli) on the surface of the absorptive cells in the epithelium lining the lumen of the small intestine, and on renal epithelium.

brush-border cell cell type in the epithelium lining the small intestine that is specialized for absorption of nutrients and has the apical surface formed into numerous microvilli. *alt.* absorptive cell.

Bryidae true mosses *q.v.*

bryocole *n.* animal living among moss.

bryology *n.* branch of botany that deals with mosses and liverworts.

Bryophyta, bryophytes *n.*, *n.plu.* division of the plant kingdom containing the mosses (Musci), liverworts (Hepaticae) and hornworts (Anthocerotae). They are small non-vascular plants either thalloid in form (hornworts and some liverworts) or differentiated into stems and leaves (mosses and some liverworts) and attached to the substrate by rhizoids. They have a well-marked alternation of generations, the independent plant being the gametophyte, which produces motile male gametes that fertilize single egg cells contained in flask-shaped archegonia. The sporophyte (the capsule) grows out from the fertilized egg and produces spores from which new plants develop.

Bryopsida Bryophyta *q.v.*

Bryozoa, bryozoans *n.*, *n.plu.* moss animals. *see* Ectoprocta.

BSA bovine serum albumin *q.v.*

BSE bovine spongiform encephalopathy. *see* transmissible spongiform encephalopathies.

Bt toxin toxic protein produced by the soil bacterium *Bacillus thuringiensis* that kills lepidopteran larvae. The gene for the toxin has been engineered into plants to confer resistance to lepidopteran pests.

buccae *n.plu.* the cheeks.

buccal *a. pert.* the cheek or mouth.

buccal cavity part of the alimentary tract between mouth and pharynx.

buccinator *n.* broad thin muscle of the cheek.

buccolabial *a. pert.* mouth cavity and lips.

buccolingual *a. pert.* cheeks and tongue.

bucconasal *a.* (1) *pert.* cheek and nose; (2) *appl.* membrane closing posterior end of nasal pit.

buccopharyngeal *a. pert.* mouth and pharynx.

bud *n.* (1) in plants, structure from which shoot, leaf or flower develops; (2) incipient outgrowth, as limb buds in animal embryo, from which limbs develop.

budding *n.* (1) production of buds; (2) (*zool.*) method of asexual reproduction common in sponges, coelenterates and some other invertebrates, in which new individuals develop as outgrowths of the parent organism, and may eventually be set free; (3) (*bot.*) artificial vegetative propagation by insertion of a bud within the bark of another plant; (4) (*mycol.*) cell division by the outgrowth of a new cell from the parent cell; (5) (*virol.*) release of certain animal viruses from the host cell by their envelopment in a piece of plasma membrane which subsequently pinches off from the cell.

budding and appendaged bacteria large and heterogeneous group of bacteria that form cytoplasmic extrusions such as stalks, hyphae and appendages. Cell division is often unequal.

budding yeast *Saccharomyces cerevisiae* and related species, which multiply by budding a small daughter cell off the parent cell. *cf.* fission yeast.

budget *see* energy budget, time-energy budget.

buffer *n.* (1) salt solution that minimizes changes in pH when an acid or alkali is added; (2) any factor that reduces the impact of external changes on a system.

buffer species species that is usually only of secondary importance as a food source but which becomes the primary food source in adverse conditions.

bufonin *n.* toxin present in toads.

bufotoxin *n.* toxin present in toads.

bug *n.* common name for an insect of the order Hemiptera *q.v.*

bulb *n.* (1) (*bot.*) specialized underground reproductive organ consisting of a short stem bearing a number of swollen fleshy leaf bases or scale leaves, the whole enclosing next year's bud; (2) (*general*) any part or structure resembling a bulb, a bulb-like swelling.

bulbar *a. pert.* bulb or a bulb-like part.

bulbiferous *a.* bearing bulbs or bulbils.

bulbil *n.* (1) fleshy axillary bud which may fall and produce a new plant, as in some lilies; (2) any small bulb-shaped structure or swelling.

bulbonuclear *a. pert.* medulla oblongata and nuclei of cranial nerves.

bulbo-urethral *a. appl.* two branching glands opening into the bulb of the male urethra. *alt.* antiprostate, Cowper's or Mery's glands.

bulbous *a.* (1) like a bulb; (2) developing from a bulb; (3) having bulbs.

bulbus *n.* knob-like or bulb-like structure or swelling.

bulla *n.* (1) rounded prominence, such as the bony projection of skull encasing the middle ear in many tetrapod vertebrates; (2) bubble-shaped structure.

bullate *a.* (1) appearing blistered; (2) puckered like a savoy cabbage leaf.

bulliform *a.* (1) bubble-shaped; (2) *appl.* large thin-walled epidermal cells in grasses which cause rolling, folding or opening of leaves by changes in turgor, *alt.* motor cell.

bundle scar traces of the vascular bundle remaining on a leaf scar after leaf fall.

bundle sheath one or more layers of large parenchyma or sclerenchyma cells surrounding a vascular bundle.

bundle sheath cells in some tropical plants, the cells in photosynthetic tissues in which carbon dioxide incorporated into aspartate and malate in the C4 pathway is released and enters the Calvin cycle.

α-bungarotoxin protein toxin from venom of snakes of the genus *Bungarus*, binds specifically to acetylcholine receptors.

bunodont *a. appl.* molar and premolar teeth with low crowns and cusps, as those of e.g. pigs, monkeys and humans.

bunoid *a.* low and conical, *appl.* cusps of molar teeth.

bunolophodont *a.* between bunodont and lophodont (having transverse ridges) in structure, *appl.* molar teeth.

bunoselenodont *a. appl.* molar teeth having internal cusps bunoid, external cusps crescent-shaped.

bunt fungi smut fungi *q.v.*

Bunyaviridae *n.* enveloped, spherical single-stranded, segmented RNA viruses, including the Bunyamwera virus and Rift Valley fever viruses.

burden *n.* of parasites, the total number or mass of parasites infecting an individual.

Burkitt's lymphoma childhood B-cell lymphoma, rare except in certain regions of Africa, where Epstein–Barr virus infection and malarial infection are together thought to increase the risk of the disease.

bursa *n.* (1) pouch or sac-like cavity; (2) sac containing a viscid fluid that prevents friction at joints; (3) bursa of Fabricius *q.v. a.* **bursal**.

bursa copulatrix (1) region of female genitalia in insects that receives the adeagus and sperm during copulation; (2) genital pouch of various animals.

bursa of Fabricius pouch of lymphoid tissue opening into the cloaca in birds, the site of maturation of B lymphocytes in birds.

bursa seminalis fertilization chamber of female genital ducts, as in turbellarians.

bursicle *n.* in orchid flowers, a flap- or purse-like structure surrounding the sticky disc at the base of stalk of pollinium, and containing a sticky liquid to prevent the disc drying up.

bursicule *n.* small sac.

burster neuron neuron that typically produces rythmic bursts of impulses when activated.

bush *n.* (1) small shrub; (2) vegetation cover composed of grassland and shrubs.

bush layer shrub layer *q.v.*

2,3-butanediol fermentation a basic fermentation pattern in enteric bacteria in which butanediol, ethanol, CO_2 and H_2 are the main products along with smaller amounts of the organic acids. *cf.* mixed-acid fermentation.

butterfly *n.* common name for a member of the Lepidoptera which has clubbed antennae. *cf.* moth.

butterfly bone sphenoid *q.v.*

buttress root root coming off stem above ground, arching away from stem before entering the soil and forming additional support for trunk.

by abbreviation for 1 billion (10^9) years.

byssaceous byssoid *q.v.*

byssal *a. pert.* a byssus.

byssoid *a.* formed of fine threads, resembling a byssus.

byssus *n.* (1) tuft of strong filaments secreted by the byssogenous gland of certain bivalve molluscs, by which they become attached to substrate; (2) the stalk of certain fungi.

C

C (1) symbol for the element carbon *q.v.*; (2) cysteine *q.v.*; (3) cytosine *q.v.*; (4) Calorie (= 1000 cal); (5) Simpson dominance index *q.v.*

C$_\gamma$ Morisita's similarity index *q.v.*

C$_H$ constant region of immunoglobulin heavy chain.

C$_L$ constant region of immunoglobulin light chain.

C1 complement protein that binds to antibody–antigen complexes and initiates the classical pathway of complement activation. Formed of three protein subunits, C1q, C1r and C1s. *see also* complement system.

C1 inhibitor (C1INH) protein that inhibits the activity of the C1 component of complement. A genetic deficiency of C1NH causes the disease hereditary angioneurotic oedema. *see also* complement system.

C2 complement protein of the classical pathway that is cleaved by C1s to a larger fragment C2a and a smaller fragment C2b on activation of the pathway, C2a forming the active protease of the enzyme C3 convertase. (In most complement cleavages the larger fragment is named 'b', C2a is an exception.) *see also* complement system.

C3 complement protein that is central in all pathways of complement activation. It is cleaved by C3 convertase to give C3a, which is an anaphylatoxin, and the larger fragment C3b, which binds covalently to bacteria and other antigens. Antigens coated with C3b are taken up more readily by phagocytes. C3b also forms one component of C3 and C5 convertases. *see also* complement system.

C3a, C4a, C5a *see* anaphylatoxin.

C3 convertase complex of activated complement components (C4b,2a in the classical pathway, C3b,Bb in the alternative pathway) that catalyses the conversion of C3 to C3a and C3b, with the consequent binding of large amounts of C3b to the cell surface. *see also* anaphylatoxin. *see also* complement system.

C3 pathway (*bot.*) carbon dioxide fixation in plants via the Calvin cycle.

C3 plant any plant in which carbon dioxide fixation is solely via the Calvin cycle as in most temperate plants. *cf.* C4 plant.

C4 complement protein of the classical pathway that is cleaved to give the anaphylatoxin C4a and the larger fragment C4b, which bonds covalently to the bacterial surface and facilitates its uptake by phagocytes. With C2a, C4b forms a C3 convertase. *see also* complement system.

C4 pathway (*bot.*) pathway of carbon dioxide fixation present in many tropical plants. CO$_2$ is incorporated via phosphoenolpyruvate into 4-carbon compounds, first oxaloacetate, then malate and aspartate, which are transported to chloroplasts of bundle sheath cells. Here CO$_2$ is released and enters the Calvin cycle. The pathway ensures an adequate and continuous supply of CO$_2$ for photosynthesis under tropical conditions. *alt.* Hatch–Slack pathway.

C4 plant any plant that possesses the C4 pathway (*q.v.*) of carbon dioxide fixation. In such plants metabolic pathways concerned with photosynthesis are compartmented between mesophyll cells and bundle sheath cells in the leaf. C4 plants are chiefly tropical plants adapted to high temperatures and low humidity. *cf.* C3 plant.

C5, C6, C7, C8, C9 the terminal components of the complement system (*q.v.*), which form a pore in the cell membrane, leading to lysis of the cell.

C5 convertase complex of activated complement components (C4b,2b,3b in the classical pathway, C3b₂,Bb in the alternative pathway) that catalyses the conversion of C5 to C5a and C5b, with the deposition of C5b molecules on the cell surface. *see also* anaphylatoxin, complement system, terminal complement components.

Ca symbol for the chemical element calcium *q.v.*

Ca²⁺-activated K⁺ channel calcium-gated ion channel in neuronal membranes that is sensitive to the intracellular build-up of Ca²⁺ and is involved in regulating the timing between action potentials.

CAAT box conserved sequence in the promoter region of some eukaryotic genes about 70–80 base pairs upstream from the start-point of transcription, and which is involved in control of initiation of transcription.

Ca²⁺-ATPase membrane transport protein in eukaryotic cells that actively transports calcium ions across the membrane using the energy of ATP hydrolysis.

Ca²⁺/calmodulin-dependent protein kinase (CaM kinase) a protein serine/threonine kinase whose phosphorylating activity is regulated by the binding of a Ca²⁺/calmodulin complex. Examples are phosphorylase kinase, myosin light-chain kinase and the multifunctional CaM kinases abundant in the brain.

cacao *n. Cacao theobroma*, tree whose seeds provide the raw material for cocoa and chocolate.

cachectin *see* tumour necrosis factor.

cachexia *n.* wasting of the body, seen in some cancers and other diseases. *a.* **cachetic.**

cacogenesis *n.* inability to hybridize.

Cactales, cacti *n., n.plu.* order of succulent dicot plants, found in arid or semiarid regions, mainly in tropical America, and adapted to hot, dry conditions. Leaves are absent or much reduced and the fleshy stems often bear clusters of spines. One family, the Cactaceae.

cacuminous *a.* with a pointed top, *appl.* trees.

cadastral *n. appl.* genes involved in flower development that set boundaries for expression of the genes that specify the identities of the different parts of the developing flower.

cadavericole *n.* animal that feeds on carrion. *alt.* carrion feeder.

cadaverine *n.* foul-smelling toxic polyamine, formed by decarboxylation of lysine and produced during bacterial breakdown of protein, e.g. in putrefying meat.

caddis fly common name for a member of the Trichoptera, an order of insects somewhat resembling moths, with weak flight and mouth parts adapted for licking, with aquatic larvae (caddis worms) that construct protective cases.

cadherins *n.plu.* one of the main families of cell adhesion molecules. They are cell-surface proteins that can bind an identical cadherin molecule on another cell in a calcium-dependent interaction, causing cells to bind strongly to each other. The family includes the E-cadherins, N-cadherins, P-cadherins and VE-cadherins.

cadophore *n.* dorsal outgrowth bearing buds in certain tunicates.

caducous *a. pert.* parts that fall off early.

caecal *a.* (1) ending without outlet, *appl.* to stomach with cardiac part prolonged into blind sac; (2) *pert.* caecum.

caecilians *see* Apoda.

caecum *n.* blind-ended diverticulum or pouch from part of the alimentary canal or other hollow organ. *plu.* **caeca.**

Caenorhabditis elegans species of small soil nematode, used as an experimental subject in developmental biology and genetic research. Its genome has been fully sequenced.

caespitose *a.* (1) having low, closely matted stems; (2) growing densely in tufts.

caffeine *n.* 1,3,7-trimethylxanthine, a purine with a bitter taste, found in coffee, tea, matè and kola nuts, which is a stimulant of the central nervous system and a diuretic.

caged molecule a light-sensitive inactive form of a small molecule such as ATP, which can be introduced into cells and activated as desired by illumination.

caino- *alt.* caeno-, ceno-, kaino-.

Cainozoic Cenozoic *q.v.*

caisson *n.* box-like arrangement of longitudinal muscle fibres in earthworms.

Cajal bodies intranuclear structures proposed to be sites where snRNPs are being recycled.

cal calorie *q.v.*

calamistrum *n.* comb-like structure on metatarsus of certain spiders.

calamus *n.* (1) hollow reed-like stem without nodes; (2) the quill of a feather.

calcaneus *n.* the heel or heel-bone.

calcar *n.* (1) (*bot.*) hollow prolongation or tube at base of petal or sepal; (2) (*zool.*) spur-like process on leg or wing of birds; (3) bony process on heel-bone that supports web between wing and tail in bats.

calcarate *a.* spurred, *appl.* petal, corolla.

calcar avis protuberance in posterior part of lateral ventricle of brain, the hippocampus minor.

calcareous *a.* (1) composed chiefly of calcium carbonate (lime); (2) growing on limestone or chalky soil; (3) *pert.* limestone.

calcareous sponges sponges of the class Calcarea, with a skeleton of 1-, 3- or 4-rayed spicules composed chiefly of calcite (calcium carbonate).

calcariform *a.* spur-shaped.

calcarine *a. appl.* fissure extending to the hippocampal gyrus, on medial surface of cerebral hemisphere.

calceolate *a.* slipper-shaped, *appl.* flowers such as those of some orchids.

calcicole *n.* plant that thrives in soil rich in lime or other calcium salts. *a.* calcicolous, *appl.* grassland.

calciferol *n.* vitamin D_2, a sterol ($C_{28}H_{43}OH$) that can be made by irradiation of ergosterol, and which is used as a dietary supplement.

calciferous *a.* containing or producing calcium salts.

calcification *n.* (1) deposition of calcium salts in tissue; (2) accumulation of calcium salts in soil. *a.* calcified.

calcifuge *n.* plant that thrives only in soils poor in lime and usually acid. *a.* calcifugous, *appl.* grassland.

calcigerous calciferous *q.v.*

calcineurin *n.* a protein serine/threonine phosphatase (protein phosphatase IIB) that is involved in signalling from the T-cell receptor. The immunosuppressive drugs cyclosporin A and tacrolimus (FK506) act by interfering with its activity and preventing T-cell activation.

calciphile calcicole *q.v.*

calciphobe calcifuge *q.v.*

calciphyte calcicole *q.v.*

calcite *n.* crystalline form of calcium carbonate, one of the constituents of mollusc shells and the skeletons of calcareous sponges.

calcitonin *n.* peptide hormone secreted by the thyroid and/or parathyroid gland in mammals and by the ultimobranchial bodies in other vertebrates. It lowers the level of calcium in the blood by reducing release of calcium from bone, opposing the activity of parathyroid hormone.

calcium (Ca) *n.* metallic element that occurs in many rocks and in seawater and is an essential macronutrient. As Ca^{2+} it is an important regulatory ion in living cells, involved e.g. in regulation of enzyme activity, stimulation of muscle contraction, and control of secretion. As calcium salts (e.g. calcium phosphates), it is a major constituent of bone. *see also* calcite.

calcium-binding protein any of a large number of proteins with specific binding sites for calcium ions (Ca^{2+}). Calcium binding generally causes a conformational change, which leads to a change in activity of the protein. Calcium-binding proteins associate with many other proteins and regulate their activity, and are the main agents mediating the effects of calcium in living cells.

calcium carbonate *see* aragonite, calcite.

calcium channel ion channel in a biological membrane which allows the passage of calcium ions.

calcium cycle the movement of calcium from inorganic sources in the soil and water, first into plants and microorganisms, and then through the food chain, eventually returning to the inorganic environment.

calcium ion Ca^{2+}.

calcium pump Ca^{2+}-ATPase *q.v.*

calice, calicle calycle *q.v.*

Caliciviridae, caliciviruses *n., n.plu.* family of icosahedral, single-stranded RNA viruses that includes vesicular exanthema of swine.

Californian subregion subdivision of the Nearctic zoogeographical region, consisting of the western Pacific seaboard of southern British Columbia and Oregon, California, and the Baja California peninsula of Mexico.

caligate *a.* (1) sheathed; (2) veiled.

callosal

callosal *a. pert.* corpus callosum *q.v.*

callose (1) *a.* having hardened thickened areas on skin or bark; (2) *n.* amorphous polysaccharide of glucose, usually found on sieve plates in phloem but also in parenchyma cells after injury.

callow workers in colonies of social insects, newly emerged adult workers, having exoskeletons that are still rather soft and lightly pigmented.

callunetum *n.* plant community dominated by the heather *Calluna vulgaris*.

callus *a.* (1) small hard outgrowth or swelling; (2) mass of hard tissue that forms over cut or damaged plant surface; (3) mass of undifferentiated cells that initially arises from plant cell or tissue in artificial culture.

calmodulin *n.* calcium-binding protein abundant in eukaryotic cells. It undergoes a conformational change on binding Ca^{2+} and binds to regulatory sites on many proteins, or acts as a regulatory subunit, and is the means by which their activity is regulated by Ca^{2+}.

calnexin *n.* protein chaperone involved in the assembly of MHC class I–peptide complexes in the endoplasmic reticulum.

calobiosis *n.* in social insects, when one species lives in the nest of another and at its expense.

Calorie (C) 1000 calories or 1 kilocalorie (kcal).

calorie (cal) *n.* non-SI unit of heat quantity. It is variously defined. The thermochemical calorie is equivalent to 4.184J, The 15° calorie is the quantity of heat needed to warm 1 g water from 14.5 to 15.5°C and is equivalent to 4.1855J.

calorific *a.* heat-producing.

calorigenic *a.* promoting oxygen consumption and heat production.

calorimetry *n.* measurement of heat. In animal physiology calorimetry is used to determine metabolic rate by measuring heat production.

calotte *n.* lid or cap, anterior end of a dicyemid parasite.

calpain *n.* calcium-dependent protease that is involved in activation of fertilized eggs and regulation of the cell cycle.

calreticulin *n.* protein chaperone involved in the assembly of MHC class I–peptide complexes in the endoplasmic reticulum.

calsequestrin *n.* calcium-binding protein from sarcoplasmic reticulum of muscle.

calvaria *n.* dome of the skull.

Calvin cycle cycle of reactions in stroma of chloroplasts in which the ATP and NADPH produced during the light reaction of photosynthesis provide energy and reducing power for the incorporation of carbon dioxide into carbohydrate. The first reaction is that of ribulose-1,5-bisphosphate with carbon dioxide to form 3-phosphoglycerate. This is converted in several stages to reform ribulose-1,5-bisphosphate, producing in the process the 3-carbon sugar glyceraldehyde-3-phosphate, which is the precursor of starch, amino acids, fatty acids and sucrose.

Calycerales *n.* order of herbaceous dicots comprising the South American family Calyceraceae (calyceras).

calyces *plu.* of calyx.

calyciflorous *a. appl.* flowers in which stamens and petals are attached along their length to the calyx.

calyc- word element denoting cup-like or calyx-like, from L. *calyx*.

calycle *n.* (1) small cup-shaped cavity or structure; (2) epicalyx of a flower.

calypter *n.* modified wing sheath covering haltere in certain flies.

calyptobranchiate *a.* with gills not visible from outside.

calyptopsis *n.* larva of some crustaceans.

calyptra *n.* (1) enlarged archegonial wall surrounding the developing sporophyte in bryophytes, which in some cases persists as a protective covering to the spore capsule of the sporophyte; (2) root cap *q.v.*

calyptrate *a.* (1) *appl.* a calyx that falls off, separating from its lower portion; (2) having a lid, *appl.* capsules, seed pods; (3) (*zool.*) *appl.* certain flies which have halteres hidden by scales.

calyptrogen *n.* root meristem cells giving rise to the root cap.

calyptron calypter *q.v.*

calyx *n.* (1) the sepals collectively, forming the outer whorl of the flower; (2) various structures resembling the calyx of a flower, e.g. the cup-like body of a crinoid. *plu.* **calyces**.

CAM (1) cell adhesion molecule (*q.v.*), used esp. for those of the immunoglobulin superfamily; (2) crassulacean acid metabolism *q.v.*

cambial *a. pert.* cambium.

cambiform *a.* similar to cambial cells.

cambiogenetic *a. appl.* cells that produce cambium.

cambium *n.* meristematic tissue from which radial secondary growth occurs in roots and shoots, producing xylem from one face and phloem from the other.

Cambrian *n.* geological period lasting from about 542 to 488 million years ago and during which many phyla of multicellular animals first arose.

camera eye non-compound eye like those of vertebrates and cephalopods, where light entering the eye is diffracted through a lens and falls on a light-sensitive retina.

cameration *n.* division into a large number of separate chambers.

CaM kinase Ca^{2+}/calmodulin-dependent protein kinase *q.v.*

cAMP cyclic AMP *q.v.*

campaniform *a.* bell- or dome-shaped.

Campanulales *n.* order of mainly herbaceous dicots, often with latex vessels, and including the families Campanulaceae (bellflower), Goodeniaceae and Lobeliaceae (lobelia). *alt.* **Campanulatae.**

campanulate *a.* bell-shaped, *appl.* flowers.

campodeiform *a.* flattened and elongated with well-developed legs and antennae, *appl.* lacewing and certain beetle larvae.

cAMP phosphodiesterase cyclic AMP phosphodiesterase *q.v.*

camptodactyly *n.* crookedness of little finger due to a congenital shortness of tendon, often due to inheritance of a dominant allele.

camptodrome *a. pert.* leaf venation in which secondary veins bend forward and join together before reaching end of leaf.

camptotrichia *n.plu.* jointed dermal fin-rays in some primitive fishes.

campylodrome *a. appl.* leaf with veins converging at its point.

campylospermous *a. appl.* seeds with groove along inner face.

campylotropous *a. appl.* ovules bent so that the funicle appears attached to the side halfway between the chalaza and micropyle.

Canadian subregion subdivision of the Nearctic zoogeographical region, comprising northern North America south to the Great Lakes in the east, to central British Columbia in the west and to southern Saskatchewan in the centre.

canaliculus *n.* (1) minute channel, e.g. containing bile in liver; (2) small channel for passage of nerves through bone. *plu.* **canaliculi.** *a.* **canalicular.**

canaliform *a.* canal-like.

canalizing selection selection for phenotypic characters which is largely unaffected by environmental fluctuations and genetic variability.

cancellated, cancellous *a.* consisting of slender fibres and lamellae, which join to form a meshwork, *appl.* inner spongy portion of bone.

cancer *n.* malignant, ill-regulated proliferation of cells, causing either a solid tumour or other abnormal conditions. Usually fatal if untreated. Cancer cells are abnormal in many ways, esp. in their ability to multiply indefinitely, to invade underlying tissue and to migrate to other sites in the body and multiply there (metastasis).

cancerous *a. appl.* cells that have undergone certain changes which enable them to divide indefinitely and invade underlying tissue. *alt.* malignant.

cancrisocial *a. appl.* organisms living as commensals with crabs.

candidate gene gene of unknown function that has been isolated and sequenced and which subsequently becomes a candidate for a particular function.

candidiasis *n.* infection caused by the yeast-like fungus *Candida albicans*, a normal inhabitant of the body. It commonly takes the form of thrush (infection of mucous membranes of either mouth or vagina) or, in immunosuppressed or otherwise debilitated patients, a more serious systemic infection.

cane sugar sucrose *q.v.*

canid *n.* member of the mammalian family Canidae, which includes the dogs, wolves, foxes, jackals and coyotes.

canine (1) *a. pert.* a dog, or the genus *Canis*; (2) *n.* the tooth next to the incisors.

caninus *n.* muscle from canine fossa to angle of mouth, which lifts the corner of the mouth.

canker *n.* general term for any plant disease that causes decay of bark and wood.

cannabinoid *n.* any substance that binds to the receptor for tetrahydrocannabinol. *see Cannabis sativa.*

Cannabis sativa hemp, varieties of which are cultivated for the drug cannabis or marijuana and for fibre. The active principle is tetrahydrocannabinol.

cannibalistic *a.* eating the flesh of one's own species.

cannon bone bone supporting limb from hock to fetlock.

canoids *n.plu.* mammals of the dog, hyena, bear, panda and related families.

canonical sequence consensus sequence *q.v.*

canopy *n.* the cover formed by the branches and leaves of trees in a wood or forest.

canopy cover the amount of ground shaded by trees and shrubs in a forest.

canthus *n.* angle where upper and lower eyelids meet. *a.* **canthal.**

cap *n.* in eukaryotic mRNA, a structure found at the 5′ end of the molecule, comprising a terminal methylguanosine residue and sometimes methylations at other sites, and which is formed after transcription. It is involved in the proper processing and export of mRNAs and in their translation.

CAP catabolite gene activator protein *q.v.*

capacitation *n.* final stage of maturation of mammalian sperm which occurs in contact with the secretions of the female genital tract. Without this stage sperm are not capable of fertilization.

cap-binding proteins (*mol. biol.*) proteins that bind to the 5′ cap region of mRNA.

Cape region the single region within the South African floristic kingdom, consisting of the very southernmost tip of Africa.

capillary *a.* (1) hair-like; (2) *appl.* moisture held in and around particles of soil; (3) *appl.* spontaneous creeping movement of water in very fine tubes; *n.* (4) the finest, thin-walled blood vessel, which forms networks in tissues; (5) any similar small vessel, e.g. those that carry lymph or bile.

capillary bed network of fine, thin-walled blood vessels in tissues that delivers oxygenated blood to tissues from the arterial system and returns deoxygenated blood to the venous system.

capillary DNA sequencing rapid automated fluorescence-based method of DNA sequencing in which fluorescent-tagged nested DNA fragments are separated in gel-filled capillaries.

capillitium *n.* (1) system of sterile, thread-like structures present among the spores in fruiting bodies of many myxomycetes; (2) system of skeletal hyphae in the gleba of gasteromycetes.

capitate *a.* (1) enlarged or swollen at tip; (2) gathered into a mass at tip of stem, *appl.* inflorescence.

capitatum *n.* the third carpal bone.

capitellum capitulum *q.v.*

capitular *a. pert.* a capitulum.

capitulum *n.* (*zool.*) (1) knob-like swelling on end of a bone, e.g. on humerus; (2) anterior part of body in ticks and mites carrying the mouthparts; (3) part of cirriped body enclosed in mantle; (4) swollen end of hair or tentacle; (5) enlarged end of insect proboscis or antenna; (6) (*bot.*) inflorescence forming a head of stalkless florets crowded together on a receptacle and often surrounded by an involucre, as in dandelion and certain other Compositae.

Capparales *n.* order of dicot herbs, shrubs, small trees and lianas, including the families Brassicaceae (Cruciferae) (mustard, etc.), Capparaceae (caper) and Resedaceae (mignonette).

capping *n.* (1) the clustering of membrane proteins at one end of an animal cell to form a 'cap' after treatment with lectins or specific antibody; (2) RNA capping, *see* cap; (3) addition of a protein to the end of an actin filament, stabilizing the filament.

capreolate *a.* having tendrils.

caprification *n.* pollination of flowers of fig trees by chalcid wasps.

Caprimulgiformes *n.* order of birds including the nightjars.

capsaicin *n.* pungent compound found in chilli peppers.

capsid *n.* (1) external protein coat of a virus particle; (2) common name for a bug of the family Capsidae.

capsomere *n.* one of the protein units of which a virus capsid is made.

capsula glomeruli Bowman's capsule *q.v.*

capsular *a.* like or *pert.* a capsule.

capsule *n.* (*zool.*) (1) sac-like membrane enclosing an organ; (2) membrane surrounding the nerve cells of sympathetic ganglia; (*bot.*) (3) any closed box-like vessel containing spores, seeds or fruits; (4) one- or many-celled, many-seeded dehiscent fruit; (5) in bryophytes, the portion of the sporogonium containing the spores; (*bact.*) (6) thick compact layer of polysaccharide outside the cell wall of some bacteria.

capsuliferous, capsuligerous, capsulogenous *a.* with, or forming, a capsule.

captacula *n.plu.* extrudable filamentous tactile organs with sucker-like ends near mouth of tusk-shells (Scaphopoda).

Captorhinida, captorhinids *n.*, *n.plu.* extinct group of primitive anapsid reptiles of the Upper Carboniferous to Triassic, possibly the ancestral reptiles from which many later forms developed. *alt.* **cotylosaurs**.

capture–recapture method method of estimating population size by marking and releasing a sample of individuals, allowing them to mingle with the population, then taking another sample. The ratio of marked to unmarked animals in this sample – the Lincoln ratio – is used to estimate total population size.

caput *n.* (1) head; (2) knob-like swelling at apex of a structure.

carapace *n.* (1) hard chitinous covering in crustaceans starting behind the head and covering the whole or part of the trunk; (2) protective covering of body of tortoises and other chelonians, composed of bony plates covered with a horny shell.

carbohydrate *n.* any member of a class of biological molecules of general formula $C_x(H_2O)_y$. Carbohydrates include sugars and their derivatives and polysaccharides such as starch and cellulose.

carbon (C) *n.* non-metallic element which occurs as various isotopes, of which ^{12}C is by far the most common. Carbon atoms can bond to each other and to other elements to form rings and chains, producing a vast variety of large carbon-based molecules. All life on Earth is based on carbon compounds (organic compounds) in which carbon is combined mainly with hydrogen, oxygen, nitrogen

and phosphorus. *see also* carbon isotope ratio, fixed carbon, radiocarbon.

carbon-14 *see* radiocarbon.

carbon capture and storage (CCS) the removal of CO_2 from waste gases from fossil-fuel-burning power stations and other industries before they reach the atmosphere, and the long-term storage of the CO_2 underground.

carbon cycle the various processes by which carbon from atmospheric carbon dioxide (CO_2) enters the biosphere, circulates within it as organic carbon, and is eventually returned to the atmosphere as CO_2. Carbon enters the biosphere (both terrestrial and marine) by fixation of CO_2 into organic compounds by photosynthetic organisms and is returned to the atmosphere as CO_2 formed chiefly as a waste product of the respiration of living organisms, but also by burning of wood and fossil fuels. *see* Fig. 6 (p. 98). *see also* greenhouse effect.

carbon dating *see* radiocarbon.

carbon dioxide CO_2, a gas present in the atmosphere at a current global average concentration of about 385 ppm. The main sources of atmospheric CO_2 are biological respiration, of which it is a waste product, combustion of organic material, and outgassing of dissolved CO_2 from the oceans. Relatively small amounts are also released into the atmosphere by volcanic activity. Carbon dioxide is mainly removed from the atmosphere by photosynthesis and by solution in seawater. Human activity is increasing the atmospheric concentration of CO_2 by the burning of fossil fuels (coal, oil and natural gas) and biomass (e.g. burning of tropical rain forest). It is a major greenhouse gas, trapping heat in the lower atmosphere, and is thus a major potential contributor to climate change *q.v.*

carbon dioxide compensation point ambient concentration of carbon dioxide at which, when light is not limiting, photosynthesis just compensates for respiration. At 25°C and 21% O_2 this value is about 45 ppm for C3 plants. *see also* compensation point. *alt.* **carbon dioxide compensation concentration**.

carbon fixation conversion of carbon dioxide into organic compounds by living

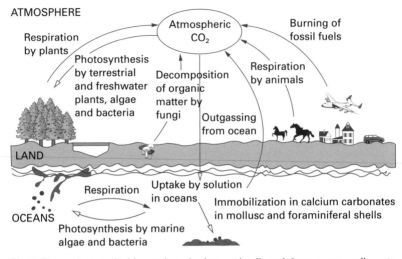

Fig. 6 The carbon cycle. Vegetation, detritus and soil, and deep-ocean sediments act as carbon sinks, locking up more carbon than they release as carbon dioxide (CO_2). Animals, decomposer fungi, and the burning of fossil fuels and biomass are net releasers of CO_2 to the atmosphere.

organisms, e.g. the conversion of carbon dioxide into carbohydrates that occurs in the stroma of chloroplasts during photosynthesis.

carbonic anhydrase enzyme of red blood cells that catalyses the formation of bicarbonate from carbon dioxide in the reaction $CO_2 + H_2O \rightarrow H^+ + HCO_3^-$. EC 4.2.1.1. *r.n.* carbonic dehydratase.

carbonicole, carbonicolous *a.* living on burnt soils or burnt wood.

Carboniferous *n.* period of the late Paleozoic era, following the Devonian and preceding the Permian, lasting from *ca.* 359 million years ago to 299 million years ago, and during which the Coal Measures were formed. Typical vegetation was swamp forest dominated by seed ferns and lycophytes such as *Lepidodendron*, with the earliest relatives of the conifers appearing later in the period. The first amniote fossils are found in the Carboniferous, in the form of small lizard-like reptiles.

carbon isotope ratio the ratio of ^{12}C to ^{13}C in a material, which is evidence of whether or not it has originated from living matter. The difference arises because photosynthetic carbon-fixing enzymes discriminate between the two isotopes of carbon, which occur in a stable ratio in atmospheric carbon dioxide, preferentially incorporating ^{12}C into living matter. Thus organic carbon is 'lighter' than carbon in atmospheric carbon dioxide or geological deposits. *see also* isotope fractionation.

carbonization *n.* form of fossilization, esp. with respect to plant material, in which the structure is turned into coal or into a thin film of carbon within a rock.

carbon monoxide (CO) poisonous gas which when inhaled in large amounts binds tightly to the haemoglobin of red blood cells, making the haemoglobin unable to bind and transport oxygen. It is also produced by some cells as a signal molecule and acts by stimulating guanylate cyclase.

carbon sink any part of the biosphere in which carbon is absorbed and immobilized faster than it is released, e.g. ocean sediments and tropical rain forests.

carbon source (1) any carbon-containing compound that can be utilized as a source of carbon by an organism; (2) in the carbon cycle, a source of carbon dioxide, such as respiration.

carbonyl group the –C=O group, found in aldehydes and ketones.

carboxydotrophic *a. appl.* bacteria that can use carbon monoxide (CO) as an energy source, oxidizing it to CO_2.

carboxyhaemoglobin, carbomonoxyhaemoglobin (HbCO) compound of carbon monoxide and haemoglobin formed in the blood following carbon monoxide poisoning.

carboxylase *n.* any of a group of enzymes, mostly containing a biotin prosthetic group, that catalyse the addition of a carboxyl group (–COOH, –COO⁻) to a compound. EC 6.4.

carboxyl group –COOH.

carboxylic acid organic acid containing one or more carboxyl (–COOH, –COO⁻) groups.

carboxyl terminus *see* carboxy terminus.

carboxypeptidase *n.* enzyme that removes the carboxy-terminal amino acid from a peptide by hydrolysing a peptide bond. EC 3.4.16–18. Carboxypeptidases A and B (EC 3.4.17.1 and 2) are mammalian digestive enzymes synthesized in the pancreas as inactive precursors and enzymatically activated in the small intestine.

carboxysome *n.* polyhedral inclusion composed of crystalline ribulose-bisphosphate carboxylase, found in some bacteria that are able to fix atmsopheric carbon into carbohydrate.

carboxy terminus (C terminus) the end of a protein chain that normally bears the free carboxyl group. It is the last part of a protein chain to be synthesized. *a.* **carboxy-terminal (C-terminal).**

carcerulus *n.* superior, dry, capsular fruit composed of several cells, each containing a single seed, which split off as separate nutlets when ripe (e.g. fruit of dead-nettle).

carcinoembryonic antigen (CEA) cell-surface protein found on embryonic tissues and also on the surface of tumour cells derived from the gastrointestinal tract.

carcinogen *n.* any agent capable of causing cancer in humans or animals. *a.* **carcinogenic.**

carcinogenesis *n.* process by which a cancerous cell arises from a normal cell.

carcinology *n.* study of crustaceans.

carcinoma *n.* malignant tumour of epithelial tissues. *cf.* sarcoma.

cardenolides *n.plu.* class of plant steroids that includes cardiac glycosides such as digitalin.

cardia *n.* (1) opening between oesophagus and stomach; (2) in sucking insects, the enlarged anterior part of the digestive chamber in front of the stomach.

cardia bifida condition in which the two embryonic heart primordia do not fuse.

cardiac *a.* (1) *pert.*, near, or supplying the heart; (2) *pert.* anterior part of the stomach.

cardiac glycosides plant glycosides such as digitalin that have stimulatory effects on the human heart and are thus often highly toxic.

cardiac muscle specialized muscle tissue of vertebrate heart, formed of muscle fibres made up of cylindrical muscle cells joined end to end. Cardiac muscle cells have orderly arrays of actin and myosin filaments and a single nucleus.

cardiac sphincter thick ring of muscle around opening between oesophagus and stomach.

cardiac valve in insects, a valve at the junction of the fore-gut and mid-gut, probably serving to prevent or reduce regurgitation of food.

cardinal *a.* (1) *pert.* that upon which something depends or hinges; (2) *pert.* hinge of bivalve shell; (2) *appl.* points for plant growth: maximum, minimum and optimal temperatures or temperature ranges.

cardinal gene in *Drosophila*, any of the zygotically expressed pattern-formation genes that respond to the antero-posterior and dorso-ventral gradients of positional information set up in the embryo by maternal gene products.

cardinal sinuses and veins veins uniting in Cuvier's duct, persistent in most fishes, embryonic in other vertebrates.

cardines *plu.* of cardo.

cardiobranchial *a. appl.* enlarged posterior basibranchial cartilage below the heart in elasmobranch fishes.

cardiogenic *a.* arising in the heart.

cardiolipin *n.* a phospholipid (diphosphatidylglycerol) found in heart tissue.

cardiovascular *a. pert.* heart and blood vessels.

cardo *n*. (1) basal segment of maxilla or secondary jaw in insects; (2) hinge of bivalve shell. *plu.* **cardines**.

Caribbean region region of the Neotropical floristic kingdom comprising the Caribbean islands, Central America and the southern tip of Florida.

carina *n*. (*zool.*) (1) keel-like ridge on certain bones, as on breast bone of birds; (2) median dorsal plate of a barnacle; (*bot.*) (3) the two joined lower petals (the keel) of a leguminous flower; (4) ridge on bracts of certain grasses.

carinate *a*. having a ridge or keel.

cariniform *a*. keel-shaped.

carnassial *a. pert.* cutting teeth of carnivora, 4th premolar above and 1st molar below.

carneous *a*. flesh-coloured.

carnitine *n*. methyl-substituted amino acid that carries long-chain acyl-CoA molecules (derived from fatty acids) across the inner mitochondrial membrane for oxidation.

Carnivora *n*. order of flesh-eating mammals containing the suborders Fissipedia, comprising the terrestrial carnivores, and Pinnipedia, the aquatic carnivores (seals, walruses and sea-lions).

carnivore *n*. animal that feeds on other animals, esp. the flesh-eating mammals (Carnivora) such as dogs, cats, bears and seals, which feed on meat or fish.

carnivorous *a*. (1) flesh-eating; (2) (*bot.*) *appl.* certain plants that trap and digest insects and other small animals.

carnose *a*. fleshy or pulpy.

carnosine *n*. the dipeptide β-alanyl L-histidine.

carotene *n*. any of several orange or yellow, fat-soluble, hydrocarbon pigments with the formula $C_{40}H_{56}$. They are synthesized in plants, are accessory pigments in photosynthesis (β-carotene), and precursors of the A vitamins such as retinol, to which they are converted in animals.

carotenoid *n*. any of a group of widely distributed orange, yellow, red or brown fat-soluble plant pigments. They are involved in photosynthesis as accessory pigments and are also found in flowers and fruits. There are two main groups: carotenes (e.g. β-carotene) and xanthophylls (e.g. lutein, violaxanthin, fucoxanthin).

carotid *a. pert.* main arteries in the neck.

carotid body one of two small masses of tissue associated with the carotid sinus in neck, involved in control of oxygen content and acidity of blood. They are composed of sensory cells and chromaffin cells secreting catecholamines.

carotid sinus small dilation inside carotid artery containing baroceptors that sense changes in arterial pressure and are involved in regulating heart rate and vasodilation.

carpal *n*. wrist bone.

carpel *n*. female reproductive structure in angiosperm flowers, composed of a stigma, a style and an ovary containing one or more ovules. A flower may have more than one carpel, which collectively make up the pistil or gynoecium. *see* Fig. 18 (p. 239).

carpellate *a. appl.* flower containing carpels but no functional stamens. *alt.* **pistillate**.

carpogenous *a*. growing in or on fruit, *appl.* fungi.

carpogonium *n*. female sex organ of red algae and some fungi, whose basal portion functions as the female gamete.

carpolith *n*. fossil fruit.

carpometacarpus *n*. bone in bird wing which corresponds to three metacarpals and some carpals fused together.

carpophagous *a*. feeding on fruit.

carpophore *n*. (1) part of flower axis to which carpels are attached; (2) stalk of a fruiting body containing spores.

carposporangium *n*. sporangium formed after fertilization of carpogonium in red algae, which produces diploid spores (carpospores) from which the tetrasporophyte arises.

carpotropism *n*. movements of fruit stalk, esp. after fertilization, to place fruit in a good position for ripening and dispersal.

carpus *n*. (1) the wrist; (2) region of forelimb between forearm and metatarsus.

carr *n*. fen woodland, usually dominated by alder or willow.

carrier *n*. (1) individual infected with a pathogenic microorganism, who does not suffer disease but can transmit the pathogen to others, causing outbreaks of disease; (2) (*genet.*) individual heterozygous for a recessive allele, esp. one responsible for a genetic disease, who shows no symptoms of disease but can pass the allele on to their offspring; (3) (*immunol.*) protein to which a hapten is attached to render the hapten immunogenic.

carrier-mediated transport transport of ions and other solutes across cell membranes with the aid of a protein in the membrane.

carrier molecule *see* activated carrier, electron carrier.

carrier protein membrane protein that binds molecules and transports them across membranes, either by facilitated diffusion or by active transport.

carrying capacity (*K*) (1) maximum number of individuals of a particular species that can be supported indefinitely by a given part of the environment; (2) number of grazing animals a piece of land can support without deterioration; (3) level of use an environment or resource can sustain without being destroyed or suffering unacceptable deterioration.

cartilage *n.* firm, elastic, skeletal tissue in which the cells, chondrocytes, are embedded in a matrix of the proteoglycan chondrin and collagen fibres. There are three main types: smooth bluish-white hyaline cartilage, fibrous cartilage, and yellow elastic cartilage, containing elastin fibres. Hyaline cartilage is the main skeletal material of the cartilaginous fishes such as sharks and dogfish. In other vertebrates it forms the embryonic skeleton, to be mainly replaced later by bone.

cartilage bone bone formed by the ossification of cartilage.

cartilaginous *a.* (1) gristly, consisting of or *pert.* cartilage; (2) resembling consistency of cartilage.

cartilaginous fishes common name for fishes of the class Selachii, which includes the sharks and their allies. They have a skeleton of cartilage, a spiral valve in the gut and no lungs or air bladder.

caruncle *n.* naked fleshy excrescence.

Caryoblastea *n.* phylum of protists containing one species, the giant multinucleate amoeba *Pelomyxa palustris*, which lacks mitochondria and other organelles characteristic of eukaryotic cells and whose nuclei divide without mitosis.

caryophyllaceous *a. pert.* flowers of the family Caryophyllaceae, e.g. the pinks and campions, whose flowers have long clawed petals. Also *appl.* any flower with long clawed petals.

Caryophyllales *n.* order of herbaceous dicots, rarely shrubs or trees, containing betalain pigments and including the families Amaranthaceae (amaranth), Caryophyllaceae (pink), Chenopodiaceae (goose-foot), Phytolaccaceae (pokeweed) and others.

caryopsis *n.* achene with pericarp and tests inseparably fused, as in grasses.

cascade *n.* series of enzymatic reactions in which the activated form of one enzyme catalyses the activation of the next, greatly amplifying the initial response.

casein *n.* protein found in milk, synthesized in mammary glands in response to the hormone prolactin.

Casparian strip zone of corky cells in the endodermis of plant roots.

caspases family of cysteine proteases that cleave target proteins at specific aspartic acid residues. They are involved in apoptosis (programmed cell death).

casque *n.* helmet-like structure in animals, such as the horny outgrowth from beak in hornbill.

cassava *n. Manihot esculenta*, tropical plant of the family Euphorbiaceae, whose starchy roots are used as food (tapioca) in parts of the world. The roots have to be processed before being turned into flour to remove toxic compounds.

cassette model description of the DNA rearrangements underlying switching of mating type in the yeast *Saccharomyces cerevisiae*, in which the mating-type locus, MAT, can be occupied by either of two genes (**a** or α, designating two different mating types) transposed from sites on either side of the locus.

cassette mutagenesis method of causing a mutation in a gene by insertion of a DNA fragment containing an easily selectable marker. Such an insertion usually destroys the function of the gene, creating a gene knock-out *q.v.*

caste *n.* one of the distinct forms found among social insects, e.g. worker, drone, queen.

caste polyethism the division of labour between different castes in social insects.

castrate(d) *a. appl.* animal from which gonads, esp. male gonad, have been removed.

casual *n.* non-native plant which has been introduced but has not yet become established as a wild plant, although occurring uncultivated.

casual society temporary and highly unstable group formed by individuals within a society, e.g. for play or feeding.

Casuariformes *n.* order of flightless birds including the cassowaries and emus.

Casuarinales *n.* order of dicot trees or shrubs with branches in whorls and which contains one family, the Casuarinaceae (she-oak).

CAT (1) chloramphenicol acetyltransferase *q.v.*; (2) computerized axial tomography *q.v.*

catabolism *n.* the breaking down of complex molecules by living organisms with release of energy. *a.* **catabolic.**

catabolite *n.* any substance that is the product of catabolism.

catabolite gene activator protein (CAP, CRP) regulatory protein in bacteria which, in a complex with cyclic AMP, can activate some catabolic operons.

catabolite repression phenomenon seen in bacteria when glucose is available as a substrate. The genes involved in the uptake and metabolism of alternative energy sources are repressed.

catacorolla *a.* secondary corolla.

catadromous *a.* (1) tending downwards; (2) *appl.* fishes which migrate from fresh to salt water for spawning.

catalase *n.* iron-containing enzyme found in all aerobic organisms, which catalyses the breakdown of hydrogen peroxide into water and molecular oxygen. EC 1.11.1.6.

catalepsis, catalepsy *n.* shamming dead reflex, as in spiders.

catalysis *n.* acceleration of a reaction due to the presence of a substance (catalyst) which can be recovered unchanged after the reaction. *a.* **catalytic.**

catalyst *n.* substance that accelerates a chemical reaction, usually by forming temporary complexes with intermediates and reducing the free energy of activation, but which can be recovered unchanged at the end of the reaction. Enzymes are the catalysts of most biochemical reactions.

catapetalous *a.* having the petals united with the base of stamens.

cataphyllary *a. appl.* rudimentary scale-like leaves which cover buds.

cataplasis *n.* regression or decline after maturity.

cataplasmic *a. appl.* irregular plant galls caused by parasites or other factors.

cataplexy *n.* (1) condition of an animal feigning death; (2) maintenance of a postural reflex induced by restraint or shock, without loss of consciousness.

catarrhines *n.plu.* group of primates comprising the Old World monkeys, apes and humans.

catastrophism *n.* the idea, held in the 18th and 19th centuries, that the fossil record represented a series of discrete creations, each terminated by a catastrophic mass extinction. For the modern view of this aspect of the fossil record *see* mass extinction events.

catch muscles muscles such as those that close the shells of bivalve molluscs, which can remain in the contracted state for long periods with little expenditure of energy.

catechins *n.plu.* group of colourless flavonoids of plant origin.

catecholamines *n.plu.* amine derivatives of catechol (2-hydroxyphenol), which act as neurotransmitters or hormones. They include adrenaline, noradrenaline and dopamine.

catena *n.* sequence of soil types which is repeated in a corresponding sequence of topographical sites, as between ridges and valleys of a region.

catenane *n.* two interlocked circles of duplex DNA, produced by the action of DNA topoisomerases.

catenation *n.* production of a catenane *q.v.*

catenin *n.* intracellular protein that acts as an accessory adhesion molecule, linking the cytoplasmic tail of the cell adhesion molecule cadherin to the cytoskeleton. β-catenin has a dual role, acting as a transcription factor in the Wnt signalling pathway in some circumstances and as an accessory adhesion molecule in others.

catenoid *a.* chain-like, *appl.* certain protozoan colonies.

catenular *a.* chain-like, *appl.* e.g. colonies of butterflies, and to colour markings on butterfly wings and on shells.

caterpillar *n.* fleshy thin-skinned larva, esp. of Lepidoptera, having segmented body, true legs and also prolegs on abdomen, and no cerci.

catfishes *n.plu.* group of mainly tropical, mainly freshwater bony fish (the order Siluriformes) often with long whisker-like barbels from which they take their name. (The marine fish *Anarhichas*, commonly called the catfish in English, is not a member of this group.)

cathepsins a family of proteolytic enzymes present in eukaryotic cells, mainly in lysosomes. Some are involved in initiating apoptosis.

cation *n.* positively charged ion which moves towards cathode, or negative pole, e.g. K⁺, Na⁺, Ca²⁺. *cf.* anion.

cation exchange capacity base exchange capacity *q.v.*

catkin *n.* inflorescence consisting of a hanging spike of small unisexual flowers interspersed with bracts, as in willows, poplars, hazel.

cauda *n.* tail or tail-like appendage.

caudal *a.* (1) of or *pert.* a tail, e.g. caudal fin of fishes; (2) towards the tail end of the body.

Caudata Urodela *q.v.*

caudate *a.* (1) having a tail; (2) *appl.* to a lobe of the liver.

caudate nucleus semicircular region of grey matter in brain lying on either side of the lateral ventricles.

caudatolenticular *a. appl.* caudate and lenticular (lentiform) nucleus of corpus striatum in brain.

caudex *n.* stem or trunk of tree ferns and palms.

caudicle *n.* stalk of pollinium in orchids.

Caudofoveata *n.* class of shell-less wormlike molluscs.

caudotibialis *n.* muscle connecting caudal vertebrae and tibia, as in some seals.

caul *n.* (1) amnion *q.v.*; (2) an enclosing membrane.

caulescent *a.* with leaf-bearing stem above ground.

caulicolous *a.* growing on the stem of a plant, usually *appl.* fungi.

cauliflory *n.* condition of having flowers arising from axillary buds on the main stem or older branches.

cauliform *a.* stem-like.

cauligenous *a.* borne on the stem.

Caulimoviridae plant virus family containing double-stranded DNA viruses with a reverse transcription step in their replication, e.g. cauliflower mosaic virus.

cauline *a.* (1) *pert.* stem; (2) *appl.* leaves growing on upper portion of a stem; (3) *appl.* vascular bundles not passing into leaves.

caulocarpous *a.* (1) with fruit-bearing stems; (2) fruiting repeatedly.

caulome *n.* the shoot structure of a plant as a whole.

caulonema *n.* profusely branched portion of protonema with relatively few chloroplasts, present in some genera of mosses.

caulotrichome *n.* hair-like or filamentous outgrowths on a stem.

caveolae *n.plu.* invaginations in the plasma membrane, formed at the site of lipid rafts, that become pinched off as pinocytotic vesicles.

cavernicolous *a.* cave-dwelling.

cavernosus *a.* hollow or honeycombed, *appl.* tissue.

cavicorn *a.* hollow-horned, *appl.* certain ruminants.

cavitation *n.* formation of a cavity by enlargement of intercellular space within a cell mass.

cavum *n.* (1) hollow or chamber; (2) in helical shells, the lower division of the internal cavity caused by origin of the helix.

C-banding technique for staining heterochromatin which generates dense staining at centromeric regions.

C4-binding protein (C4BP) plasma protein that acts as a complement-regulatory protein. It binds to C4b, making it more susceptible to cleavage and inactivation by Factor I. *see also* complement system.

CCL when followed by a number denotes a particular chemokine.

CCK cholecystokinin *q.v.*

CCR when followed by a number denotes a particular receptor for a CCL chemokine.

¹²C/¹³C ratio *see* carbon isotope ratio.

CCS carbon capture and storage *q.v.*

CD 'cluster of differentiation' system of nomenclature for cell-surface proteins on white blood cells. It derives from the identification of cell-surface proteins by particular sets of monoclonal antibodies. Individual polypeptides are designated CD followed by a unique number, e.g. CD1, CD2, CD3, CD4. *see also* cluster of differentiation.

CD3 complex proteins associated with the T-cell receptor at the lymphocyte surface and which are essential for the signalling function of the receptor.

CD4 membrane glycoprotein that characterizes the Th1, Th2, Th17 subtypes of effector T cells and some regulatory T cells. By binding to MHC class II molecules on the antigen-presenting cell it acts as a co-receptor to enhance the signal received through the T-cell receptor.

CD40 membrane protein produced by B cells and other cells, which interacts with CD40 ligand on activated T cells. This interaction is required for T cells to activate B cells and to stimulate isotype switching in B cells.

CD40 ligand (CD40L) *see* CD40.

CD4 T cells T lymphocytes (*q.v.*) that carry the co-receptor protein CD4. They recognize peptide antigens bound to MHC class II molecules on an antigen-presenting cell, and after initial activation by antigen differentiate into Th1, Th2 or Th17 effector cells (*all q.v.*).

CD5 B cell B-1 cell. *see* B lymphocyte.

CD8 membrane glycoprotein that characterizes cytotoxic T lymphocytes. By binding to MHC class I molecules on the antigen-presenting cell it acts as a co-receptor to enhance the signal received through the T-cell receptor.

CD8 T cells T lymphocytes (*q.v.*) that carry the co-receptor protein CD8. They recognize peptide antigens bound to MHC class I molecules on an antigen-presenting cell and after initial activation by antigen differentiate into effector cytotoxic T lymphocytes (*q.v.*).

cDC conventional dendritic cell, *see* dendritic cell.

Cdk, CDK cyclin-dependent kinase *q.v.*

cDNA complementary DNA *q.v.*

C-DNA form of DNA occurring *in vitro* in the presence of lithium ions, with fewer base pairs per turn than B-DNA.

cDNA clone DNA clone derived from amplification of a complementary DNA (cDNA) transcript of an mRNA.

cDNA library collection of cDNA clones. *cf.* genomic library.

CDP cytidine diphosphate *q.v.*

CDR complementarity-determining region *q.v.*

CDS coding sequence *q.v.*

CDV canine distemper virus, a paramyxovirus affecting dogs and other canids.

CEA carcinoembryonic antigen *q.v.*

Ceboidea *n.* superfamily of primates comprising the New World monkeys.

cecal, cecum *see* caecal, caecum.

cecidium *see* gall.

Celastrales *n.* order of dicot trees, shrubs or vines, rarely herbs, and including the families Aquifoliaceae (holly), Celastraceae (staff tree), and others.

celiac coeliac *q.v.*

cell *n.* the basic structural building block of living organisms, consisting of protoplasm delimited by a cell membrane. In plants, bacteria and fungi, the cell is also surrounded by a non-living rigid cell wall. Some organisms consist of a single cell (bacteria, archaea, protozoans and some algae and fungi), others of more than one cell. Multicellular organisms generally contain cells specialized for different functions. Prokaryotic cells (archaea and bacteria) have a relatively simple internal structure in which the DNA is not enclosed in a discrete nucleus and the cytoplasm is not differentiated into specialized organelles. Cells from all other living organisms (Eukarya) are called eukaryotic cells. They are in general larger than prokaryotic cells, the DNA is enclosed in a nucleus and is organized into chromosomes, and the cytoplasm contains a cytoskeleton and specialized membrane-bounded organelles such as mitochondria and (in photosynthetic plants) chloroplasts. *see* Fig. 7; (2) small cavity or hollow; (3) space between veins of insect wing.

α-cell (1) cell in islets of Langerhans of pancreas that secretes glucagon; (2) oxyphilic cell in the anterior pituitary. *alt.* A cell.

β-cell (1) cell in islets of Langerhans of pancreas that secretes insulin; (2) secretory basophil cell in the anterior pituitary. *alt.* B cell.

cell adhesion the binding of a cell to another cell, as in tissues, or to extracellular matrix, mediated by cell-surface molecules.

cell adhesion molecule (CAM) any of a large and heterogenous group of cell-surface glycoproteins that promote adhesion

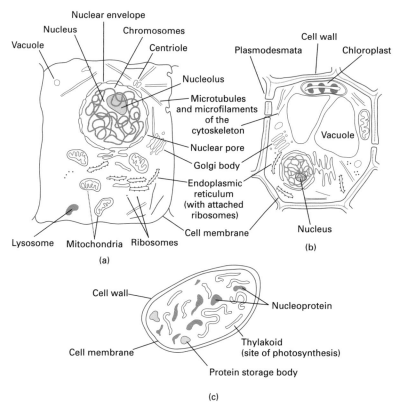

Fig. 7 Schematic diagram of (a) a generalized animal cell; (b) a generalized plant cell; (c) a prokaryotic cell (the example illustrated here is a cyanobacterium).

between animal cells or between cells and the extracellular matrix, by binding to each other or to other receptor molecules. Cell adhesion molecules fall into several different families on the basis of structure: proteins of the immunoglobulin superfamily (*q.v.*), cadherins (*q.v.*), integrins (*q.v.*) and selectins (*q.v.*).

cell-autonomous *a.* (1) *appl.* development of a blastomere or embryonic cell entirely determined by factors within the cell itself, and not by interactions with its neighbours; (2) *appl.* a genetic trait in a multicellular organism in which only genotypically mutant cells exhibit the mutant phenotype. *alt.* autonomous phenotype.

cell body that part of a neuron containing the nucleus and most organelles. *alt.* soma.

cell–cell interaction any interaction between one cell and another, whether involving direct contact or action via secreted proteins or other molecules.

cell–cell junction cell junction *q.v.*

cell centre centrosome *q.v.*

cell coat carbohydrate-rich peripheral layer at the surface of most eukaryotic cells. It is composed of the carbohydrate side chains of membrane glycoproteins and also includes secreted and adsorbed polysaccharides and proteoglycans. *alt.* glycocalyx.

cell cortex *see* cortex.

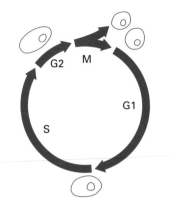

Fig. 8 The stages of the cell cycle of a eukaryotic cell. G1 is gap phase 1; S phase is DNA synthesis; G2 is gap phase 2; M phase is mitosis and cell division.

cell coupling chemical and electrical communication between adjacent animal cells via gap junctions.

cell culture cells growing outside the organism in nutrient medium in the laboratory. *see also* primary culture, secondary cell culture.

cell cycle the period between the formation of a cell as one of the products of cell division and its own subsequent division. During this period all cells undergo replication of the DNA. In eukaryotic cells the cell cycle is divided into phases termed G1, S, G2 and M. G1 is the period immediately after mitosis and cell division, when the newly formed cell is in the diploid state and at which growth takes place. S is the phase of DNA synthesis, and is (sometimes) followed by G2, at which time the cell is in a tetraploid state and further cell growth may take place. Mitosis (M) follows to restore the diploid state, accompanied by cell division. The interphase of the cell cycle comprises the G1, S and G2 phases. *see* Fig. 8.

cell death *see* apoptosis, necrosis, programmed cell death.

cell differentiation *see* differentiation.

cell division splitting of a cell into two complete new cells. It occurs by binary fission in bacteria and other prokaryotes, and by divison of nucleus (mitosis) and cytoplasm in eukaryotic cells.

cell doctrine cell theory *q.v.*

cell envelope *see* bacterial envelope.

cell fate in development, what a given embryonic cell will develop into in the usual circumstances.

cell fractionation process of breaking up cells and then subjecting the extract to centrifugation to separate out the different cellular components of different densities.

cell-free system any mixture of cell components reconstituted *in vitro* in which processes such as translation, transcription and DNA replication can be studied.

cell fusion the coming together of two cells to form one cell, not necessarily accompanied by fusion of the two cell nuclei. *see* cell hybrid.

cell hybrid cell produced by induced fusion *in vitro* of two somatic cells of different genetic constitution (often from different species). They are used for genetic and other studies. Plant cell hybrids can sometimes be regenerated into whole plants of novel genetic constitution.

cell junction point at which the plasma membrane of one cell is closely apposed to that of its neighbouring cell. *see* desmosome, gap junction, intermediate junction, septate junction, tight junction.

cell line clone of plant or animal cells that can be propagated indefinitely in culture. *cf.* primary culture.

cell lineage (1) the family tree of cells in a tissue or part of a developing embryo, tracing their mitotic line of descent; (2) any pedigree of cells related by asexual division.

cell lineage restriction the situation when the descendants of a group of founder cells remain within a given boundary and do not mix with adjacent cells. *see also* compartment.

cell-mediated immune response, cell-mediated immunity *see* immune response.

cell membrane plasma membrane *q.v.*

cell memory (1) the phenomenon that differentiated eukaryotic cells give rise only to cells of the same type; (2) the phenomenon that in eukaryotic development, determined cells give rise to similarly determined cells, even though the extracellular influences that resulted in determination no longer are acting. *see also* determination.

cellobiase *n.* enzyme that hydrolyses cellobiose to glucose. EC 3.2.1.21, *r.n.* β-D-glucosidase.

cellobiose *n.* disaccharide produced on partial hydrolysis of cellulose, made up of two glucose units joined by a β1,4 linkage.

cell of Boettcher granular epithelial cell found between cells of Claudius and the basal membrane in the ear.

cell of Claudius columnar or cuboid epithelial cell found in the lining of the endolymphatic space in the ear.

cell plate material laid down across middle of dividing plant cell, and from which the partitioning cell wall is formed.

cell proliferation multiplication of cells by division.

cell sap fluid in vacuoles in plant cells, being a solution of small organic molecules in water.

cell senescence ageing of cells. The phenomenon that most cells in higher animals can only go through a certain number of cell divisions, after which they become moribund.

cell signalling the communication between cells by means of chemical signals, or the stimulation of cells by physical stimuli such as light, and the intracellular biochemical machinery cells use to respond to these signals.

cell sorting (1) *see* flow cytometry; (2) the ability of different cell types from a disaggregated tissue to sort out by aggregating with their own cell type when recombined.

cell-surface receptor a membrane-bound receptor protein that is exposed at the cell surface to the extracellular environment of a cell.

cell theory the idea that plant and animal bodies are made up of cells, and that the cell is the unit structure of an organism, first proposed by Matthias Schleiden for plants and by Theodore Schwann for animals in 1838–40. *alt.* cell doctrine.

cell type the differentiated state and function of a cell, *e.g.* fibroblast, skeletal muscle cell, cardiac muscle cell and neuron are different cell types.

cellular *a. pert.* or consisting of cells.

cellular blastoderm in insect embryonic development, the stage at which the syncytial blastoderm has become divided into cells.

cellular differentiation *see* differentiation.

cellular immune response, cellular immunity *see* immune response.

cellular oncogene proto-oncogene *q.v.*

cellular respiration the biochemical processes at the cellular level by which molecules such as glucose are oxidized to generate ATP.

cellular slime moulds group of eukaryotic heterotrophic soil microorganisms (the Dictyostelia or dictyostelids) which are classified as protists. Free-living unicellular amoebae aggregate to form a multicellular fruiting body differentiated into stalk and sporehead (a compound sporangium). *cf.* plasmodial slime moulds, protostelids.

cellulase *n.* enzyme catalysing the degradation of cellulose by hydrolysis of internal β1,4 linkages, found in some invertebrates, plants, fungi and bacteria but lacking in most animals. EC 3.2.1.4.

cellulin granules granules containing chitin, present in some Oomycota.

cellulolytic *a.* able to degrade cellulose.

cellulose *n.* linear polysaccharide made up of glucose residues joined by β1,4 linkages. Considered to be the most abundant organic compound in the biosphere as it comprises the bulk of plant and algal cell walls, where it occurs as cellulose microfibrils. It is also found in certain tunicates. *see* Fig. 9 (p. 108).

cellulose synthase enzyme complex bound to the plasma membrane of a plant cell that synthesizes cellulose.

cellulosic *a.* composed of cellulose.

cell wall non-living rigid structure surrounding the plasma membrane of algal, plant, fungal and most prokaryotic cells. Depending on the type of organism, it

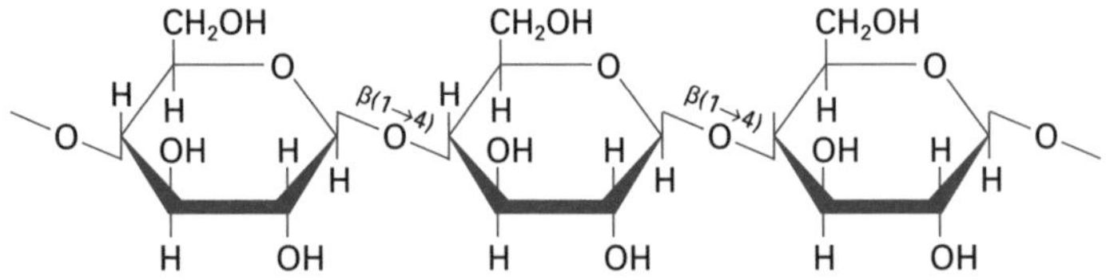

celo-

Fig. 9 Cellulose.

is composed of polysaccharides such as cellulose (plants, algae and some fungi) or chitin (fungi), or of peptidoglycans (prokaryotes). *see* Fig. 7 (p. 105).

celo- *alt.* spelling of coelo-.

cement, cementum *n.* (1) bone-like material covering root of tooth; (2) material that sticks cells together or anchors organism or part of organism to its substrate.

cementoblast, cementocyte *n.* cell secreting the bone-like cementum that covers the root of a tooth.

cen- *alt.* spelling of coen-.

Cenozoic *n.* geological era following the Mesozoic, commencing *ca.* 65 million years ago and lasting until the present. It is subdivided into the Tertiary and Quaternary sub-eras.

censer mechanism method of seed dispersal by which seeds are shaken out of fruit by wind action.

census *n.* complete counting of a whole population with respect to the variable under study.

centimorgan (cM) *n.* unit for the relative distances between genetic loci on a map of the chromosome as determined by the frequency of recombination between them. 1 cM = 1% recombination between two gene loci on the same chromosome. *alt.* map unit.

centipede *n.* common name for a member of the Chilopoda, a group of arthropods having numerous and similar body segments each with one pair of walking legs, except the 1st segment which bears a pair of poison claws.

centra *plu.* of centrum *q.v.*

central *a.* (1) situated in, or *pert.* the centre; (2) *pert.* a vertebrate centrum.

Central Australian region that part of the Australian floristic kingdom consisting of the central Australian desert and west

to the western Australian coast, excepting the small southwestern tip.

central cylinder stele *q.v.*

central dogma principle that the transfer of genetic information from DNA to RNA to protein goes only one way. Now modified to take into account the formation of DNA from an RNA template by reverse transcription.

central lymphoid organs primary lymphoid organs *q.v.*

central mother cells large cells found under the superficial layer in shoot apical meristem.

central nervous system (CNS) that part of the nervous system that serves an integrating and coordinating function. In vertebrates it consists of the brain and spinal cord. In those invertebrates with a central nervous system it consists of a brain or cerebral ganglia and a nerve cord, which may be dorsal or ventral, single or double. *cf.* peripheral nervous system.

central sulcus major transverse groove that divides the frontal lobes from the parietal lobes in brain.

central tolerance self-tolerance that is developed in the lymphocyte population as it develops in the primary, or central, lymphoid organs. *cf.* peripheral tolerance.

central vacuole large fluid-filled cavity that takes up most of the volume of many plant and algal cells, and which maintains the turgor of plant cells, and also contains digestive enzymes.

centrarch *a. appl.* protostele with central protoxylem.

centres of diversity regions identified originally by the Russian botanist N. Vavilov, in which large numbers of different strains or races of a particular cultivated plant occur. In some cases this diversity may indicate the geographical place of origin of the crop plant.

centric *a.* (1) *appl.* leaves that are cylindrical or nearly so; (2) *appl.* chromosomes having a centromere; (3) *appl.* diatoms radially symmetrical when viewed face on.

centrifugal *a.* (1) turning or turned away from centre or axis; (2) (*bot.*) *appl.* radicle, *appl.* compact cymose inflorescences having youngest flowers towards outside, *appl.* xylem differentiating from centre towards edge of stem or root, *appl.* thickening of cell wall when material is deposited on outside of wall, as in pollen grains; (3) (*zool.*) away from centre, *appl.* transmission of nerve impulses from a nerve centre towards parts supplied by nerve.

centrifugation *n.* procedure for separating a mixture of particles into components of different sizes and densities by spinning the mixture at high speed in a centrifuge. *see* differential centrifugation, velocity sedimentation, density-gradient sedimentation, ultracentrifuge.

centrifuge *see* ultracentrifuge.

centriole *n.* one of a pair of cylindrical organelles formed of short microtubules, found just outside the nuclear envelope in animal cells but absent from plant cells.

centripetal *a.* (1) turning or turned towards centre or axis; (2) (*bot.*) *appl.* radicle, *appl.* racemose inflorescences having youngest flowers at apex, *appl.* xylem differentiating from edge towards centre of stem or root, *appl.* thickening of cell wall where material is deposited on the inside of wall, as most cells; (3) (*zool.*) towards centre, *appl.* transmission of nerve impulses from periphery to nerve centres.

centroblast *n.* large, rapidly dividing lymphoblast that develops from an activated B cell within a germinal centre.

centrocyte *n.* non-dividing cell into which a centroblast develops in a germinal centre.

centrogenous *a. appl.* skeleton of spicules which meet in a common centre and grow outwards.

centrolecithal *a. appl.* eggs with yolk aggregated in centre.

centromere *n.* region on a chromosome at which the kinetochore assembles and at which a chromosome becomes attached

to the microtubules of the spindle during mitosis or meiosis. It is composed of highly repeated DNA. It becomes visible in metaphase chromosomes as a constriction at which the sister chromatids of the duplicated chromosome are still held together. Normal chromosomes have only one centromere.

centroplast *n.* extranuclear spherical body forming division centre of mitosis, as in some radiolarians.

centrosome *n.* organelle in plant and animal cells, situated near the nucleus, and which organizes the microtubule cytoskeleton.

centrosome matrix pericentriolar material *q.v.*

Centrospermae Caryophyllales *q.v.*

centrum *n.* (1) main body of a vertebra, from which neural and haemal arches arise; (2) bone situated between proximal and distal rows in wrist or ankle; (3) (*mycol.*) all structures enclosed by an ascocarp wall.

cephalad *a.* towards head region or anterior end.

cephalanthium *n.* flowerhead of closely packed florets, as in Compositae.

cephalic *a.* (1) *pert.* head; (2) in the head region.

cephalic index one hundred times maximum breadth of skull divided by maximum length.

cephalization *n.* increasing differentiation and importance of anterior end in animal evolution.

cephalo-, -cephalic word elements derived from Gk *kephal*, head.

Cephalocarida *n.* class of minute marine crustaceans living in sand and having a horseshoe-shaped carapace.

Cephalochordata, cephalochordates *n., n.plu.* subphylum of small cigar-shaped aquatic chordates commonly called lancelets and including the amphioxus (*Branchiostoma*). They have a persistent notochord in the adult and a large sac-like pharynx with gill slits for food collection and respiration.

cephalodium *n.* gall-like outgrowth developing on or within lichens with both cyanobacterial and algal phycobionts and containing nitrogen-fixing cyanobacteria. *plu.* **cephalodia**.

cephalogenesis *n.* development of the head region.

cephalon *n.* the head of some arthropods.

cephalopedal *a. appl.* the haemocoel cavity in the head and foot of snails and their relatives.

Cephalopoda, cephalopods *n., n.plu.* class of marine molluscs including octopus, squid and nautiloids. They have a well-developed head, large brain, and eyes somewhat resembling those of vertebrates in structure. The head is surrounded by prehensile tentacles and the animals can move very rapidly by jet propulsion, squirting water out of the large mantle cavity that communicates with the exterior by a siphon or funnel.

cephalopodium *n.* head and arms constituting the head region in cephalopods.

cephalosporins β-lactam antibiotics based on a compound isolated originally from the fungus *Cephalosporium acremonium*. They interfere with bacterial cell-wall synthesis.

cephalosporium *n.* globular mucilaginous mass of spores.

cephalostegite *n.* anterior part of cephalothoracic shield in certain arthropods.

cephalothorax *n.* (1) body region formed by fusion of head and thorax in crustaceans; (2) in arachnids, the prosoma *q.v.*

cephalula *n.* free-swimming embryonic stage in certain brachiopods.

ceraceous *a.* waxy, wax-like.

ceral *a. pert.* the cere in birds.

ceramide *n.* *N*-acyl sphingosine, widely distributed in plant and animal tissues, and a constituent of sphingolipids such as cerebrosides and gangliosides.

cerata *n.plu.* lobes or leaf-like processes functioning as gills on the back of nudibranch molluscs (e.g. sea-slugs). *sing.* **ceras.**

ceratium *n.* type of siliqua.

ceratobranchial *n.* ventral skeletal element of the gill arch.

ceratohyal *n.* ventral skeletal element of the hyoid arch in fishes, next below the epihyal.

ceratomorphs *n.plu.* suborder of mammals that contains the rhinoceroses and tapirs.

ceratotrichia *n.plu.* thin rods of collagen (elastoidin) which form sheets between rays of fins of elasmobranch fishes (e.g. sharks, dogfishes) and stiffen them.

cercal *a.* (1) *pert.* the tail; (2) *pert.* a cercus, *appl.* hairs, nerve.

cercaria *n.* heart-shaped, tailed, larval stage of a trematode (fluke) produced in the snail host. It is released from the snail, sometimes then encysting, and subsequently infects a vertebrate host. *plu.* **cercariae.**

cerci *plu.* of cercus.

Cercidiphyllales *n.* order of dicot trees comprising the family Cercidiphyllaceae, with a single genus *Cercidiphyllum.*

cercoid *n.* one of paired appendages on the 9th or 10th abdominal segment of certain insect larvae.

cercopithecoid *a. appl.* monkeys of the superfamily Cercopithecoidea, the Old World monkeys (e.g. baboons) which together with apes and humans are the only primates with a fully opposable thumb.

cercopod cercus *q.v.*

Cercozoa *n.* in some classifications a phylum of non-photosynthetic amoeba-like protozoa with thread-like pseudopodia, some groups having shells of siliceous scales. The Foraminifera, the Radiolaria and the Cercozoa together make up the Rhizaria.

cercus *n.* (1) jointed appendage at end of abdomen in many arthropods; (2) appendage bearing acoustic hairs in some insects. *plu.* **cerci.**

cere *n.* swollen fleshy patch at base of bill in birds.

cereal *n.* any plant of the family Gramineae, the grasses, whose seeds are used as food.

cerebellospinal tract tract of motor nerve fibres running from cerebellum in brain to anterior horn cells of the spinal cord, concerned with automatic regulation of muscle tone and posture.

cerebellum *n.* pair of finely convoluted rounded hemispherical masses of tissue situated behind the midbrain and forming part of the hindbrain. It is concerned with regulation of muscle tone and posture and coordination of movement in relation to sensory signals received in other parts of the brain. *a.* **cerebellar.**

cerebral *a.* (1) *pert.* the brain, more particularly *pert.* hemispheres of forebrain; (2) *appl.* arteries supplying the frontal and middle parts of cerebral hemispheres.

cerebral aqueduct aqueduct of Sylvius *q.v.*

cerebral cortex superficial layer of grey matter of the cerebral hemispheres, some 2 mm deep and consisting of several layers of nerve cell bodies and their complex interconnections. In humans and anthropoid apes the highly convoluted surface of the brain provides a much enlarged area of cortex compared with other animals. The cortex is the site of analysis and interpretation of sensory information, of generation of all voluntary motor action, and of higher cognitive functions such as learning and memory and conscious perception. It receives input from cranial nerves and from peripheral sensory receptors via various nuclei in other parts of the brain. Different functions, such as the analysis of visual information, auditory information and generation of motor signals, are at least partly localized to different areas of the cortex.

cerebral ganglia the supraoesophageal ganglia, or 'brain', of invertebrates.

cerebral hemispheres pair of symmetrical rounded masses of convoluted tissue forming the bulk of the human brain, separated longitudinally by a deep fissure but connected at the base by a bridge of fibres, the corpus callosum. *alt.* cerebrum.

cerebral peduncles twin short pillars of tissue in the brain, supporting and carrying fibres to and from the cerebral hemispheres, and which together with the corpora quadrigemina and the aqueduct of Sylvius form the midbrain.

cerebrifugal *a. appl.* nerve fibres which pass from brain to spinal cord.

cerebropedal *a. appl.* nerve fibres connecting cerebral and pedal ganglia in molluscs.

cerebrose *a.* resembling the convolutions of the brain, *appl.* e.g. surface of spores.

cerebroside *n.* any of a group of glycosphingolipids containing one sugar residue, glucose or galactose, as headgroup. They are found chiefly in nerve cell membranes.

cerebrospinal *a. pert.* brain and spinal cord.

cerebrospinal fluid (CSF) in vertebrates, the fluid filling the cavity in the brain and spinal cord and the subarachnoid space surrounding the brain and spinal cord. It is secreted by the choroid plexuses and reabsorbed by veins on the brain surface. *alt.* neurolymph.

cerebrovisceral *a. appl.* nerve fibres connecting cerebral and visceral ganglia in molluscs.

cerebrum cerebral hemispheres *q.v.*

cereous *a.* waxy.

ceriferous *a.* wax-producing.

cernuous *a.* drooping or pendulous.

ceroma cere *q.v.*

cerous *a. appl.* a structure resembling a cere (fleshy patch at base of bird's bill).

ceruloplasmin *n.* blue copper-containing protein present in blood plasma.

cerumen *n.* (1) ear wax, secreted by the ceruminous glands of the external auditory meatus of the ear; (2) wax secreted by scale insects; (3) wax of the nests of certain bees.

cervical *a.* (1) *appl.* or *pert.* structures connected with the neck, such as nerves, bones, blood vessels; (2) *appl.* cervix or neck of an organ; (3) *appl.* groove across carapace of certain crustaceans that appears to delimit a 'head'.

cervids *n.plu.* members of the mammalian family Cervidae: the deer.

cervix *n.* neck or narrow mouth of an organ, such as the uterine cervix, the neck of the uterus just above the vagina.

cespitose caespitose *q.v.*

Cestoda, cestodes *n., n.plu.* tapeworms, a class of platyhelminths (flatworms) that are internal parasites of humans and animals. They have a long flattened ribbon-like body lacking a gut or mouth, usually divided into many identical segments (proglottids), and attach themselves to the wall of the gut through an attachment organ at the anterior end. Reproduction is by detachment of mature proglottids, which each form a new reproductive unit. The complex life cycle involves two or more hosts.

Cetacea, cetaceans *n., n.plu.* order of wholly aquatic placental mammals comprising the whales and dolphins. Their bodies are steamlined and highly adapted for swimming, with the forelimbs modified as flippers, a broad flattened tail (fluke) and the hind limbs hardly developed and invisible externally.

cetology *n.* study of whales and dolphins.

Ceylonese subregion subdivision of the Oriental zoogeographical region, comprising southern India and the island of Sri Lanka.

CF cystic fibrosis *q.v.*

CF$_1$, CF$_0$ chloroplast factor *q.v.*

CFU colony-forming unit *q.v.*

CFU-E erythrocyte colony-forming unit *q.v.*

C gene gene coding for the constant region of an immunoglobulin polypeptide chain.

CGH comparative genomic hybridization *q.v.*

cGMP cyclic GMP *q.v.*

CGN *cis* Golgi network *q.v.*

chaeta *n.* (1) retractable bristle partly made of chitin projecting from the body wall in oligochaete worms and on parapodia of polychaete worms; (2) chitinous sensory bristle on body and appendages of insects. *plu.* **chaetae.**

chaetiferous, chaetigerous *a.* bristle-bearing.

Chaetognatha *n.* arrow-worms, a phylum of marine coelomate animals found in swarms in plankton, having an elongated transparent body with head, trunk and tail.

chaetophorous chaetiferous *q.v.*

chaetopods *n.plu.* the annelid worms that bear chaetae (bristles): the Polychaeta and Oligochaeta.

chaetotaxy *n.* the arrangement of bristles or chaetae on an insect.

chain behaviour series of actions, each being induced by the antecedent action and being an integral part of a unified performance.

chain-termination method one of the basic methods of DNA sequencing, which involves controlled interruption of synthesis of a new DNA strand on a template DNA. *alt.* Sanger method.

chalaza *n.* (1) (*zool.*) one of two spiral bands attaching yolk to membrane in a bird's egg; (2) (*bot.*) region of ovule or seed where the stalk joins the nucellus and integuments.

chalazogamy *n.* fertilization in which the pollen tube pierces the chalaza of the ovule.

chalcone *n.* any of a group of yellow flavonoid plant pigments.

chalcone synthetase key enzyme in flavonoid biosynthesis in plants. EC 2.3.1.74, *r.n.* naringenin-chalcone synthase.

chalice *n.* (1) arms and disc of a crinoid; (2) simple gland cell or goblet cell.

chalice cell goblet cell *q.v.*

chalicotheres *n.plu.* extinct family of ungulates which had clawed feet.

chamaephyte, chamaeophyte *n.* perennial woody plant having overwintering buds at or just above ground level.

chambered organ in sea lilies, an aboral cavity with five compartments, which occupies the body region enclosed by the thecal plates and which sends branches to the cirri.

chameleon sequence an amino acid sequence that can adopt different stable secondary structures (i.e. α-helix or β-strand) in different proteins.

channel protein membrane protein forming a water-filled pore (a channel) in a cellular membrane. This channel enables the diffusion of specific solutes and ions across membranes that are otherwise impermeable to them. *see* ion channel.

chaparral *n.* type of vegetation found in areas with a mediterranean climate, dominated by evergreen shrubs with broad hard leaves.

chaperone protein involved in facilitating the folding or assembly of newly synthesized proteins. There are several classes of chaperone, which act by different mechanisms.

chaperonin *n.* member of one class of molecular chaperone *q.v.*

characins *n.plu.* group of tropical freshwater bony fish (the Characinoidei) that includes the tetras and piranhas. They have complex teeth bearing 5–7 cusps, strong jaws and a scaly body.

character *n.* (1) any physical feature or attribute of an organism that can be used to compare it with another organism; (2) in genetics, any heritable feature present in a taxon or in a population, such as flower or eye colour, which can exist in two or more different states.

character convergence in evolution, the condition in which two newly evolved species interact in such a way that one or both converges in one or more traits towards the other.

character displacement, character divergence in evolution, the condition in which two newly evolved species interact in such a way that both diverge further from each other.

characteristic species plant species that are almost always found within a particular association.

character state (1) any of the range of values or qualitatively different states of a particular character; (2) a character as expressed in a specific taxon.

Charadriiformes *n.* large diverse group of shore and wading birds including the gulls and terns, auks, oystercatchers, plovers, curlews, snipes and waders.

Charophyta *n.* taxon of chlorophyte algae that are encrusted with calcium carbonate and are commonly called stoneworts. They have a filamentous or thalloid body bearing lateral branches in whorls. *alt.* **Charophyceae.**

chartaceous *a.* like paper.

chasmocleistogamy *n.* the condition of having both chasmogamous (opening) and cleistogamous (never opening) flowers.

chasmogamy *n.* opening of a mature flower in the normal way to ensure fertilization. *cf.* cleistogamy.

chasmophyte *n.* plant which grows in rock crevices.

cheating *n.* in animal behaviour, any behaviour (e.g. exaggeration of body size) intended to mislead a rival or potential mate into an incorrect estimate of the animal's strength or genetic fitness.

checkpoint *n.* any point in the course of a developmental or intracellular process at which successful completion of the previous steps in the pathway is checked before the pathway is allowed to proceed. The term is used mostly to denote such points in the eukaryotic cell cycle.

cheiropterygium *n.* the pentadactyl limb, as of all vertebrates except fish.

chela *n.* large claw borne on certain limbs of arthropods, such as the pincer of a crab. *plu.* **chelae.**

chelate (1) *a.* claw-like or pincer-like; (2) *v.* to combine with a metal iron to form a stable compound, a chelate. *n.* **chelation.**

chelating agent compound that can react with a metal ion and form a stable compound, a chelate, with it.

chelicera *n.* one of the pair of prehensile appendages at the extreme anterior end of arachnids and horseshoe crabs. In arachnids often modified into fangs used to inject poison into prey. *plu.* **chelicerae.** *see also* Chelicerata.

Chelicerata, chelicerates *n., n.plu.* a class or subphylum of arthropods with a body generally in two parts, a prosoma bearing the paired chelicerae (poison jaws) and sensory pedipalps, and a posterior opisthsoma bearing usually four pairs of walking legs. The Chelicerata include the arachnids (e.g. spiders, ticks, mites, scorpions), pycnogonids (sea spiders), horseshoe crabs and the extinct eurypterids. *see also* Arachnida, Merostomata, Pycnogonida.

cheliferous *a.* having chelae or claws.

cheliform *see* chelate (*a.*).

cheliped *n.* claw-bearing appendage in crustaceans.

Chelonia, chelonians *n., n.plu.* turtles and tortoises, an order of reptiles having a short broad trunk protected by a dorsal shield (carapace) and ventral shield (plastron) composed of bony plates overlain by epidermal plates of tortoiseshell.

chemical defences unpalatable or toxic chemicals, such as astringent tannins and toxic alkaloids, produced by plants in their tissues to deter herbivores.

chemical ecology study of the secondary chemical compounds produced by plants and animals (e.g. antibiotics, alkaloids, unpalatable and/or toxic compounds) and their effect on the interaction of the organism with other animals and plants in the ecosystem, especially in respect of plants' defences against herbivores and animals' defences against predators.

chemical fossils supposed chemical traces of life, such as alkanes and porphyrins, found in rocks older than the earliest true fossil-bearing rocks.

chemical genomics the systematic screening of large numbers of small molecules (chemical libraries) for their potential actions on cells, using information from the human genome to identify possible interactions between the small molecules and cellular protein targets, with the aim of developing new drugs.

chemical group a set of covalently linked atoms, such as an amino group $(-NH_2)$ or a methyl group $(-CH_3)$, with specific chemical properties.

chemical mediator any molecule, esp. a small non-protein molecule, that influences a physiological process.

chemical oxygen demand (COD) chemical test for the degree of organic pollution of water. It measures the amount of oxygen taken up from a sample of water by the organic matter in the sample.

chemical potential activity or free energy of a substance.

chemical synapse see synapse.

chemiluminescence n. light production during a chemical reaction at ordinary temperature.

chemiosmosis n. mechanism by which energy derived from aerobic respiration or from sunlight can be used to power ATP synthesis. Electrons resulting from aerobic respiration or light capture are passed along an electron-transport chain in a membrane to a final electron acceptor. This leads to the generation of a proton gradient across the membrane. The energy stored in the proton gradient drives ATP synthesis as protons flow back down the gradient through ATP synthases located in the membrane, which is otherwise impermeable to protons and other ions. alt. chemiosmotic theory. a. chemiosmotic.

chemiosmotic coupling see chemiosmosis.

chemoaffinity hypothesis the theory that neurons recognize their correct targets in the developing nervous system by a system of chemical identification tags on neurons and target cells.

chemoattractant n. chemical that attracts cells or organisms to move towards it.

chemoautotroph chemolithotroph q.v.

chemoceptor chemoreceptor q.v.

chemogenomics n. the large-scale integration of chemical information about chemical compounds that might be potential drugs with information about protein structure and function gained from the study of fully sequenced genomes (genomics), with the aim of speeding up the discovery of possible new drugs and their target proteins in the body or in microorganisms.

chemoheterotroph chemoorganotroph q.v.

chemoinformatics the creation, organization, management, retrieval, analysis, dissemination, visualization and use of chemical information, esp. in relation to the design and identification of new drugs. alt. chemometrics, computational chemistry, chemical informatics, cheminformatics.

chemokine n. any of a large number of small proteins that act as chemoattractants for cells, esp. white blood cells entering tissues during inflammation. There are two main structural classes denoted by CCL or CXCL. Individual chemokines are designated either CCL or CXCL followed by a number.

chemokinesis n. movement in response to the intensity of a chemical stimulus, including that of scent.

chemolithotroph n. organism able to oxidize inorganic chemicals as its source of energy. a. chemolithotrophic. alt. chemoautotroph. cf. chemoorganotroph, phototroph.

chemometrics chemoinformatics q.v.

chemonasty n. nastic movement in response to diffuse or indirect chemical stimuli.

chemoorganotroph n. organism that uses the oxidation of organic chemicals as its sole source of energy. a. chemoorganotrophic. alt. heterotroph, a. heterotrophic. cf. chemolithotroph, phototroph.

chemoreceptor n. (1) sense organ or cell receiving chemical stimuli, such as taste bud, olfactory neuron; (2) receptor protein on such cells.

chemoreflex n. reflex caused by chemical stimulus.

chemorepellant n. any chemical that causes cells or organisms to move away from it.

chemosensory a. pert. sensing of chemical stimuli.

chemostat n. (1) organ concerned in maintaining constancy of chemical conditions, such as the carotid body, which regulates pH of blood; (2) continuous culture device for maintaining constant culture conditions, controlled by the concentration of limiting nutrient and dilution rate.

chemosynthesis n. (1) biosynthesis of organic compounds; (2) oxidation of inorganic compounds as energy source for biosynthesis, in some bacteria. *a.* **chemosynthetic.**

chemotaxis *n.* reaction of motile cells or microorganisms to chemical stimuli by moving towards or away from source of chemical. *a.* **chemotactic.**

chemotaxonomy, chemosystematics *n.* the classification of organisms according to similarities and differences in the organic compounds (other than DNA and RNA) they produce, esp. secondary metabolites in the case of plants.

chemotherapeutic agent antimicrobial or anti-cancer drug that can be taken internally.

chemotherapy *n.* treatment of disease with chemical agents.

chemotroph *n.* organism that uses chemicals as its source of energy. These may be inorganic chemicals as in chemolithotrophs, or organic compounds as in chemoorganotrophs. *a.* **chemotrophic.** *cf.* phototroph.

chemotropism *n.* curvature of plant or plant organ in response to chemical stimuli.

Chenopodiales Caryophyllales *q.v.*

chernozem *n.* black soil, formed under continental climatic conditions and characteristic of subhumid to temperate grasslands. *alt.* black earth.

chersophilous *a.* thriving on dry wasteland.

chersophyte *n.* plant which grows on wasteland or on shallow soil.

chestnut soils dark-brown soils of semiarid steppe-lands, fertile under adequate rainfall or when irrigated.

cheta *see* chaeta.

chevron *n.* V-shaped bones articulating with ventral surface of spinal column in caudal region of many vertebrates.

chiasma *n.* (1) partial crossing-over of nerve fibres from either side of the body, as at optic chiasma in brain, *alt.* **chiasm;** (2) X-shaped structure formed by homologous chromatids in prophase of meiosis and which represents the site of crossing-over and exchange of segments of DNA between chromatids during homologous recombination. *plu.* **chiasmata.**

chigger *n.* larval form of a trombiculid mite, which is an ectoparasite on the skin of terrestrial and aquatic vertebrates.

chilaria *n.plu.* pair of processes between 6th pair of appendages in the king crab, *Limulus.*

Chilean subregion subdivision of the Neotropical zoogeographical region, comprising the Andes and the rest of South America south of Brazil.

chilidium *n.* shelly plate covering the opening between hinge and beak in brachiopods.

Chilopoda, chilopods *n., n.plu.* in some classifications, a class of arthropods comprising the centipedes, which have numerous and similar body segments each with one pair of walking legs, except the first segment which bears a pair of poison claws. Considered as a subclass or order of class Myriapoda in some classifications.

chimaera *n.* (1) organism developing from an embryo composed of cells from two different individuals and therefore composed of cells of two different genotypes; (2) organism composed of cells of two different genotypes as a result of grafting tissue from one individual to the other.

chimaeric *a.* (1) *appl.* animals or plants consisting of some cells with one genetic constitution and some with another, *see* chimaera; (2) *appl.* DNA or proteins consisting of a mixture of sequences or subunits from different sources. *alt.* hybrid.

chimeric, chimera *see* chimaeric *q.v.*

chimpanzee *n.* common name for the great apes of the genus *Pan,* generally referring to the common chimpanzee *Pan troglodytes* (now rare and threatened in its native African habitat). *see also* bonobo.

chionophyte *n.* snow-loving plant.

ChIP, ChIP-chip, ChIP-seq *see* chromatin immunoprecipitation.

chirality *n.* property of molecules with handedness in their chemical structure (i.e. their mirror image cannot be exactly superimposed on the 'real' image in any orientation) and possessing optical activity. *a.* **chiral.** *cf.* prochirality.

chiro- cheiro-.

chironomid *a. appl.* midge of the genus *Chironomus,* whose aquatic larvae are used as indicator species in assessing biotic indices for freshwater habitats. Abundance of the larvae of *C. riparius* (the bloodworm) on their own indicates heavy organic pollution. *n.* **chironomid**.

Chiroptera, chiropterans *n., n.plu.* order of flying placental mammals comprising the echolocating bats (microchiroptera) and the 'flying foxes' and Old World fruit bats (megachiroptera). They have the digits of the forelimbs greatly elongated to support a wing membrane stretched between fore- and hindlimb and attached to the body.

chiropterophilous *a.* pollinated by the agency of bats.

chiropterygium cheiropterygium *q.v.*

chisel teeth chisel-shaped incisors of rodents.

chi sequence short DNA sequence that occurs many times in the bacterial (*E. coli*) chromosome and at which RecA-mediated recombination is stimulated.

chi-square test statistical test used to determine the probability of obtaining the observed results by chance, given a specified hypothesis.

chi (χ) structure X-shaped structures derived from figure-of-eight DNA molecules isolated from bacterial cells undergoing genetic recombination and which are presumed to represent recombination intermediates of circular molecules.

chitin *n.* long-chain polymer of *N*-acetylglucosamine. It is the chief polysaccharide in fungal cell walls and in the exoskeletons of arthropods.

chitinase *n.* enzyme that hydrolyses chitin. EC 3.2.1.14.

chitinous *a.* composed of, or containing chitin.

chitons *n.plu.* common name for the Polyplacophora, a class of marine molluscs with an elongated body bearing a shell of calcareous plates and a muscular foot on which they crawl about.

chitosan *n.* fully non-acetylated glucoaminoglycan polymer, present in some fungal cell walls.

chitosome *n.* small vesicle containing chitin-synthesizing enzyme, in fungi.

chlamydate *a.* (1) having a mantle; (2) ensheathed, enclosed in a cyst.

chlamydeous *a.* having a perianth.

chlamydiae *n.plu.* Gram-negative bacteria of the genus *Chlamydia*, forming a distinct phylogenetic group within the Bacteria. They are obligate intracellular parasites in which many metabolic and structural features have been lost. They have muramic acid but not diaminopimelic acid in their cell walls and no ATP-generating capacity. Chlamydiae are responsible for a variety of diseases of humans and animals, including trachoma and psittacosis.

chlamydocyst *n.* encysted zoosporangium, as in certain saprobic fungi.

chlamydospore *n.* thick-walled resting spore of certain fungi and protozoans.

chloragogen cells *n.* cells in the heart and the roof of annelid intestine containing stores of oil and glycogen. *alt.* **chloragen**.

chloramphenicol *n.* antibiotic produced by the actinomycete *Streptomyces venezuelae*, which blocks translation in bacteria and mitochondria by attaching to the 50S ribosomal subunit and preventing the addition of an amino acid to the polypeptide chain by inhibiting the peptidyltransferase reaction.

chloramphenicol acetyltransferase (CAT) bacterial enzyme (EC 2.3.1.28) that confers resistance to the antibiotic chloramphenicol. The bacterial *cat* gene encoding the enzyme is widely used as a marker in gene transfer experiments, as cells carrying it can be selected by their survival when cultured in the presence of chloramphenicol.

chloranthy *n.* reversion of petals, sepals and other flower parts to ordinary green leaves.

chlorenchyma *n.* plant tissue containing chlorophyll.

chloride cell columnar cell of gill epithelium, specialized for excretion of chloride, in certain fishes.

chloride channel anion channel in a membrane through which chloride ions (Cl^-) pass.

chloride shift movement of chloride ions into red blood cells and bicarbonate ions out.

chlorinated hydrocarbons compounds such as DDT and PCBs, which are persistent in the environment, entering food chains and often accumulating in organisms.

chlorocruorin *n.* green, haem-containing, oxygen-carrying protein found in the blood of certain polychaete worms.

Chloroflexi *n.* a phylum of the Bacteria distinguished on DNA sequence criteria.

chloronema *n.* in mosses, a type of protonemal branch which grows along the surface of the substrate or into the air for a short distance and contains many conspicuous chloroplasts.

chlorophore *n.* chlorophyll granule in protists.

Chlorophyceae Chlorophyta *q.v.*

chlorophycophilous *a. appl.* lichens in which the photosynthetic partner is a green alga.

chlorophyll *n.* principal light-capturing pigment of plants, algae and cyanobacteria, consisting of a porphyrin (tetrapyrrole) ring with a magnesium atom at the centre and esterified to a long-chain aliphatic alcohol (phytol), different chlorophylls having different side chains. In plants and algae, chlorophyll is located in the thylakoid membranes of chloroplasts. It absorbs light best in the red and violet-blue regions of the spectrum, chemically distinct chlorophylls having different absorption maxima. Chlorophylls *a* and *b* are found in higher plants and green algae, chlorophyll *a* in cyanobacteria, chlorophylls *c* and *d* in algae. *see also* bacteriochlorophylls.

chlorophyllose cells elongated, very narrow living cells containing chloroplasts, separated from each other by large empty cells, in *Sphagnum* moss leaves.

Chlorophyta, chlorophytes *n., n.plu.* the green algae, a large and varied group of unicellular and multicellular eukaryotic photosynthetic organisms. Mostly freshwater, but some marine (the green seaweeds), they have chlorophylls and carotenoids similar to those of plants, carry out oxygenic photosynthesis, store food as starch and have cellulose cell walls.

chloroplast *n.* green organelle found in the cytoplasm of the photosynthetic cells of plants and algae, and in which the reactions of photosynthesis take place. A chloroplast is bounded by a double membrane and contains a system of internal membranes (thylakoids) embedded in the matrix or stroma. The thylakoid membranes contain the green pigment chlorophyll and other pigments involved in light collection, electron-transport chains and ATP synthase, and are the site of the light reactions of photosynthesis in which ATP and NADPH are generated. The dark reactions of photosynthesis, in which carbohydrate is synthesized, take place in the stroma. A chloroplast also possesses a small DNA genome which specifies rRNAs, tRNAs and some chloroplast proteins. *alt.* plastid. *see also* chemiosmotic theory, endosymbiont hypothesis, photosynthesis, stroma, thylakoids. *see* Fig. 7 (p. 105).

chloroplast DNA *see* chloroplast.

chloroplast factor *see* ATP synthases.

chloroplast pigments pigments present in plant cell chloroplasts: chlorophylls, carotene and xanthophyll.

chloroquine *n.* antimalarial drug that raises the pH of cellular organelles such as lysosomes, endosomes and Golgi apparatus.

chlorosis *n.* abnormal condition characterized by lack of green pigment in plants, owing to lack of light, or to magnesium or ion deficiency, or to genetic deficiencies in chlorophyll synthesis. *a.* **chlorotic.**

chlorosome *n.* photosynthetic structure containing chlorophyll in green bacteria.

Chloroxybacteria prochlorophytes *q.v.*

choana *n.* (1) funnel-shaped opening; (2) opening of the nostrils into the pharynx or throat.

Choanichthyes Sarcopterygii *q.v.*

choanocyte *n.* flagellated cell, with protoplasmic collar around base of the flagellum, lining body cavity of sponges. Involved in uptake of food particles from water drawn into cavity by the beating of the flagella. *alt.* collar cell.

choanoflagellates *n.plu.* group of flagellates which have a protoplasmic collar around the base of the flagella.

cholecalciferol *n.* vitamin D_3, a natural form of vitamin D that is formed in the skin on exposure to sunlight. It is a sterol ($C_{27}H_{43}OH$) and is found especially in fish, egg yolk and fish-liver oil. It controls levels of calcium and phosphorus in the body and prevents rickets.

cholecyanin bilicyanin *q.v.*

cholecystokinin (CCK) *n.* peptide hormone produced by duodenal cells and acting on pancreatic acinar cells to induce secretion of digestive enzymes. It also

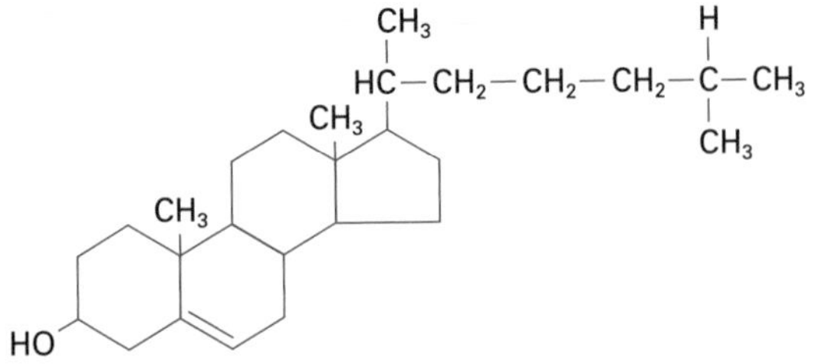

Fig. 10 Cholesterol.

induces contraction of gall bladder and relaxation of sphincter around the duodenal end of the common bile duct. Also present in the central nervous system.

choleic *a. appl.* a bile acid.

cholera toxin protein toxin produced by the cholera bacterium, which inhibits the GTPase activity of G proteins, leading to their permanent activation.

cholesterol *n.* a sterol present in animal cell membranes, where it influences membrane fluidity. Synthesized from acetyl-CoA as the starting point, it is itself the precursor of many biologically active steroids such as the steroid hormones and vitamin D. The main site of synthesis in humans is the liver, and cholesterol is transported in the blood mainly in the form of lipoprotein particles. In certain conditions it is deposited from plasma onto blood vessel walls forming atherosclerotic plaques. *see* plasma lipoproteins. *see* Fig. 10.

cholic *a. pert.*, present in, or derived from, bile.

cholic acid *n.* an acid, $C_{24}H_{40}O_5$, present in bile, helps in fat digestion.

choline *n.* $(CH_3)_3NCH_2CH_2OH$, which is acetylated to form the neurotransmitter acetylcholine, and which is also a moiety of the membrane phospholipid phosphatidylcholine. Choline is sometimes considered as a vitamin, one of the components of the B vitamin complex. Glandular tissue, nervous tissue, egg yolk and some vegetable oils are particularly rich in choline.

choline acetyltransferase enzyme which synthesizes the neurotransmitter acetylcholine from choline. EC 2.3.1.6.

cholinergic *a. appl.* nerve fibres that liberate acetylcholine from their terminals.

cholinesterase *n.* enzyme found in the synaptic cleft at cholinergic synapses, which hydrolyses acetylcholine to choline and acetic acid. EC 3.1.1.8. *alt.* acetylcholinesterase.

chondral *a. pert.* cartilage.

chondric *a.* gristly, cartilaginous.

Chondrichthyes *n.* class of fishes known from the Devonian to the present day, commonly known as the cartilaginous fishes, having a cartilaginous skeleton, spiral valve in the gut and no lungs or air bladder, and including the rays, skates and sharks. *cf.* Osteichthyes.

chondrin *n.* gelatinous bluish-white substance forming the ground substance of cartilage, having a firm elastic consistency.

chondrioclast chondroclast *q.v.*

chondroblast, chondrocyte *n.* cartilage cell. It secretes chondrin, the extracellular matrix of cartilage.

chondroclast *n.* large multinuclear cell that breaks down cartilage matrix. *alt.* chondrioclast.

chondrocranium *n.* the skull when in a cartilaginous condition, either temporarily as in embryos, or permanently as in some fishes.

chondrogenesis *n.* the formation of cartilage.

chondroid *a.* cartilage-like, *appl.* tissue: undeveloped cartilage or pseudocartilage, serving as support in certain invertebrates and lower vertebrates.

chondroitin *n.* glycosaminoglycan part of a proteoglycan found in cartilage and other connective tissues. It contains repeating sulphated disaccharide units composed of D-glucuronic acid and N-acetylgalactosamine.

chondroma *n.* generally benign tumour of cartilage.

chondrophore *n.* structure that supports the inner hinge cartilage in a bivalve shell.

chondrophores *n.plu.* colonial hydrozoans of the order Chondrophora, showing a degree of division of labour and cooperation between zooids approaching that of siphonophores.

chondroseptum *n.* cartilaginous part of the septum of the nose.

chondroskeleton *n.* a cartilaginous skeleton.

Chondrostei, chondrosteans *n., n.plu.* group of primitive actinopterygian bony fishes including the extant bichirs of the Nile, which have lungs, and paddlefishes and sturgeons, as well as many extinct groups. The bony skeleton has largely been substituted by cartilage. They usually have a spiral valve in the gut, and retain the spiracle, and paddlefishes and sturgeons have a heterocercal tail. *see also* palaeoniscids.

chondrosteosis *n.* conversion of cartilage to bone.

chondrosteous *a.* having a cartilaginous skeleton.

chondrosternal *a. pert.* rib cartilages and sternum.

chorda *n.* any cord-like structure. *plu.* **chordae.**

chordacentra *n.plu.* vertebral centra formed by conversion of notochordal sheath into a number of rings.

chordae tendineae tendons connecting papillary muscles with valves of heart.

chordal *a.* (1) *pert.* a chorda or chordae; (2) *pert.* the notochord.

chordamesoderm *n.* mesoderm cells that will form the notochord in vertebrate embryos.

Chordata, chordates *n., n.plu.* phylum of coelomate animals having a notochord and gill clefts in the pharynx at some point in their life history, and a hollow nerve cord running dorsally, with the anterior end usually dilated to form a brain. The chordates include the vertebrates, the cephalochordates (e.g. amphioxus) and the urochordates (e.g. sea squirts). *see* Appendix 3.

chordotonal *a. appl.* sensilla or organ: in insects, a rod-like or bristle-like structure for reception of mechanical or sound vibrations.

chore *n.* area showing a unity of geographical or environmental conditions.

chorioallantoic *a. appl.* placenta when chorion is lined by allantois and allantoic vessels convey blood to embryo, as in some marsupials and all eutherian mammals.

choriocapillaris *n.* innermost vascular layer of the chorion.

chorion *n.* (1) embryonic membrane external to and enclosing the amnion and yolk sac; (2) allantochorion *q.v.*; (3) tough shell covering the eggs of insects; (4) (*bot.*) outer membrane of a seed.

chorion frondosum villous placental part of chorion.

chorionic *a.* (1) *pert.* the chorion; (2) *appl.* hormone. *see* gonadotropins.

chorionic villus sampling (CVS) method of antenatal diagnosis of genetic disease by analysing foetal cells taken from chorionic villi, the embryo-derived part of the placenta. Can be carried out earlier than amniocentesis.

chorioretinal *a. pert.* choroid and retina.

choriovitelline *a. appl.* placenta when part of chorion is lined with yolk sac, vitelline blood vessels being connected with uterine wall, as in some marsupials.

chorisis *n.* increase in parts of floral whorl due to division of its primary members.

chorismate (chorismic acid) *n.* aromatic carboxylic acid, intermediate in synthesis of aromatic amino acids in microorganisms.

C-horizon lowest layer of a soil profile above the bedrock.

choroid (1) *a.* infiltrated with many blood vessels, *appl.* delicate and highly vascular membranes; (2) *n.* intermediate pigmented layer of eye, underlying the retina, and forming the diaphragm of the iris.

choroid plexus (1) highly vascularized network of interlaced capillaries and nerves between retina and the tough opaque fibrous layer of the eyeball; (2) vascular tissue lining the ventricles of the brain, which secretes cerebrospinal fluid.

chorology *n.* study of the geographical distribution of plants and animals.

choronomic *a.* external, *appl.* influences of geographical or regional environment.

chorotypes *n.plu.* local types.

Christmas disease *see* haemophilia.

Christmas factor Factor IX *q.v.*

chroma *n.* the hue and saturation of a colour.

chromaffin cell catecholamine-secreting cell type of neural origin found in carotid body and in medulla of adrenal gland in vertebrates, and in some invertebrates such as annelids.

Chromalveolata *n.* proposed major monophyletic taxonomic grouping, based on molecular phylogenetics, which includes the Chromista (*q.v.*) and the Alveolata (*q.v.*) together.

chromatic *a.* of, produced by, or full of bright colour.

chromatic threshold minimal stimulus, varying with the wavelength of the light, which induces a sensation of colour.

chromatid *n.* one of the copies of a chromosome present after DNA and chromosome replication. Chromatids become visible in the microscope at prophase and metaphase of mitosis and meiosis as the two halves of a double structure held together at the centromere. A chromatid essentially becomes a separate new chromosome at anaphase of mitosis and in the 2nd meiotic division.

chromatid conversion *see* gene conversion.

chromatin *n.* the complex of DNA and histone proteins that makes up the basic material of eukaryotic chromosomes. It is composed of molecules of DNA associated with histones and packaged into nucleosomes. Although the term chromatin strictly refers to the complex of DNA and histone, it is often used to refer to the chromosomal material as a whole, which includes other proteins also. *see also* heterochromatin, euchromatin.

chromatin immunoprecipitation (ChIP) technique that isolates a protein-bound DNA fragment from a mixture of genomic DNA by means of an antibody specific for the protein. After isolation, the DNA fragment is identified either by DNA microarray analysis (**ChIP-chip**) or DNA sequencing (**ChIP-seq**). The technique is used particularly to identify the DNA motifs targeted by specific gene regulatory proteins.

chromatin remodelling the alteration of the structure and functional state of chromatin by processes such as the repositioning of nucleosomes (nucleosome remodelling), histone modification and DNA methylation.

chromatin-remodelling complex *see* nucleosome-remodelling complex.

chromatocyte *n.* pigment-containing cell.

chromatography *n.* separation of compounds in a mixture on the basis of their affinity for, and migration with, a nonpolar solvent such as water on a polar support such as paper or through a column of various materials. *see also* affinity chromatography, column chromatography, partition chromatography, gel-filtration chromatography.

chromatophil(ous) chromophilous *q.v.*

chromatophore *n.* (1) organelle containing pigment; (2) pigment cell or group of pigment cells which, under the control of the nervous system or of hormones, can be altered in shape or colour.

chromatotropism *n.* orientation in response to stimulus consisting of a particular colour.

Chromista *n.* a proposed taxonomic kingdom containing the Hyphochytriomycota, Labyrinthulomycota, Oomycota (water moulds), Phaeophyta (brown algae), Chrysophyta (golden algae), Bacillariophyta (diatoms), Haptophyta (coccolithophorids) and Xanthophyta (yellow-green algae). A very similar grouping has also been given the name Stramenopila.

chromo-argentaffin *a.* staining with bichromates and silver nitrate, *appl.* flask-shaped cells in epithelium of crypts of Lieberkühn.

chromoblast *n.* embryonic cell giving rise to a pigment cell.

chromocentre *n.* (1) region at which polytene chromosomes appear to be attached together; (2) granule of heterochromatin, part of interphase chromosomes, many of which show up on staining interphase nuclei.

chromocyte *n.* pigment-containing cell.

chromogen *n.* substance that is converted into a pigment.

chromogenesis *n.* production of colour or pigment.

chromogenic *a.* colour-producing.

chromomere *n*. one of numerous bead-like structures along meiotic and mitotic chromosomes in prophase.

chromonema *n*. thread of chromatin (chromosome) as detected during interphase, when the chromosomes are extended and dispersed in the nucleus. *plu.* **chromonemata.**

chromophane *n*. (1) pigmented oil globule (red, yellow or green) found in retina of birds, reptiles, fishes, marsupials; (2) any retinal pigments.

chromophilic, chromophilous *a*. staining readily.

chromophore *n*. the group of atoms in a molecule responsible for absorbing light energy and to whose presence colour in a compound is due.

chromoplast *n*. coloured plastid containing no chlorophyll but usually containing some red or yellow pigment.

chromoprotein *n*. protein that is combined with a pigment, such as haemoglobin, chlorocruorin, cytochromes.

chromosomal *a*. of or *pert*. a chromosome or chromosomes.

chromosomal incompatibility failure to interbreed due to differences in chromosome composition.

chromosomal mosaic an individual who has somatic cells of more than one type of chromosome constitution, usually arising from an error in mitosis during development.

chromosome *n*. (1) in eukaryotic cells, a structure composed of DNA and associated proteins, and which carries the genetic information. Chromosomes become visible under the light microscope at mitosis and meiosis as small, rod-shaped, deeply staining structures. At other stages of the cell cycle the chromosomes are thread-like structures enclosed within the nucleus. In the interphase stage of the cell cycle, before DNA replication has taken place, each chromosome consists of a single very long molecule of DNA associated with histones and other proteins. Chromosomes entering mitosis and meiosis have undergone DNA replication and are composed of two identical new chromosomes (chromatids) which are still joined together. Each species has a char-

acteristic set of chromosomes which can often be distinguished from those of other species in shape, size and number. *see also* chromatin, diploid, DNA, gene, haploid; (2) the single DNA molecule in prokaryotic cells.

chromosome 21 the chromosome found in a trisomic state in the cells of persons with Down syndrome.

chromosome aberration any departure from the normal in gross chromosomal structure or number, such as trisomy, deletions, duplications, inversions and translocations.

chromosome banding the characteristic pattern of transverse bands that is revealed on a metaphase chromosome by staining with dyes such as Giemsa, and which can be used to identify chromosomes.

chromosome complement karyotype *q.v.*

chromosome map a plan showing the position of genes on a chromosome. *see also* genetic map.

chromosome mutation chromosomal aberration *q.v.*

chromosome number the usual constant number of chromosomes in a somatic cell that is characteristic of a particular species.

chromosome painting the delineation of a particular chromosome or set of chromosomes in a karyotype by staining with fluorescent dyes attached to a collection of DNA probes specific for that chromosome or chromosomes.

chromosome pair the pair of homologous chromosomes, one derived from the maternal, one from the paternal parent, which become associated at meiosis.

chromosome puff *see* puffs.

chromosome races races of a species differing in number of chromosomes, or in number of sets of chromosomes.

chromosome spread preparation of stained metaphase chromosomes, separated so that each is visible.

chromosome theory of inheritance the theory stating that patterns of inheritance can be explained by assuming that chromosomes are the carriers of genetic information and that a given gene is located at a particular site on a chromosome.

chromosome translocation *see* translocation.

chromosome walking technique for mapping chromosomes from a collection of overlapping restriction fragments. Starting from a known DNA sequence, the overlapping sequences can be detected in other restriction fragments and a map of a particular area gradually built up.

chromotropic *a. appl.* any agent or factor controlling pigmentation.

chronaxia, chronaxie, chronaxy *n.* (1) latent time between electrical stimulation and muscle contraction; (2) minimal time required between successive firings of a nerve cell under a prolonged stimulus.

chronic *a. appl.* medical conditions or diseases of slow progress and long duration. *cf.* acute.

chronotropic *a.* affecting the rate of action.

chrysalis *n.* the pupa of insects with complete metamorphosis, enclosed in a protective case, which is sometimes itself called the chrysalis.

Chrysiogenetes *n.* a phylum of the Bacteria distinguished on DNA sequence criteria, currently including a single species.

chryso- prefix derived from Gk *chrysos*, gold.

chrysocarpous *a.* with golden-yellow fruit.

chrysolaminarin *n.* storage polysaccharide of golden-brown algae.

chrysophanic *a.* having a golden or brown colour, *appl.* an acid formed in certain lichens and in leaves.

chrysophyll *n.* a yellow colouring matter in plants, a decomposition product of chlorophyll.

Chrysophyta, chrysophytes *n., n.plu.* phylum of unicellular or colonial protists commonly known as the golden algae, formerly included in the algae but now classified as members of the kingdom Stramenopila or Chromista. Mainly freshwater, but the marine groups are an important constituent of the marine nanoplankton. At some stage of the life cycle the cells bear two unequal undulipodia. Chrysophytes contain large amounts of carotenoid pigments, reserves of oil and the polysaccharide chrysolaminarin. Some possess siliceous or organic skeletons or scales, *alt.* **Chrysophyceae.**

chyle *n.* lymph containing globules of emulsifed fat, found in the lymphatic vessels of small intestine during digestion.

chylifaction *n.* formation of chyle.

chyliferous *a. appl.* tubes and vessels conveying chyle.

chylific *a.* chyle-producing, *appl.* ventricle or true stomach of insects.

chylocaulous *a.* with fleshy stems.

chylomicrons *n.plu.* small lipoprotein particles in plasma and other body fluids, which transport cholesterol, triacylglycerols and other lipids from intestine to adipose tissue.

chylophyllous *a.* with fleshy leaves, *appl.* succulent plants adapted to dry conditions, such as stonecrops and sedums.

chyme *n.* partially digested food leaving the stomach.

chymosin *n.* peptidase enzyme in gastric juice. Clots milk by limited cleavage of bonds in casein. EC 3.4.23.4. *alt.* rennin.

chymotrypsin *n.* protein-digesting enzyme (peptidase) in pancreatic juice. It is formed in the pancreas by enzymatic cleavage of an inactive precursor, chymotrypsinogen, by trypsin. EC 3.4.21.1.

Chytridiomycota, chytrids *n., n.plu.* phylum of small coenocytic saprobic or parasitic fungi living in soil or fresh water. They produce motile asexual zoospores and some species have motile gametes. Motile cells bear a single posteriorly directed whiplash flagellum. *alt.* **Chytridiomycetes.**

chytridium *n.* structure containing spores in certain fungi.

Ci curie *q.v.*

cibarium *n.* part of buccal cavity anterior to pharynx in insects.

cicatrice, cicatrix *n.* (1) scar; (2) small scar in place of previous attachment of organ.

cicatricial tissue newly-formed fine fibrous connective tissue that closes and draws together the edges of a wound.

Ciconiiformes *n.* order of wading birds with long necks, legs and bill, including herons, bitterns, storks and flamingos.

cilia *plu.* of cilium.

ciliaris *n.* muscle forming a ring outside anterior part of choroid in the eye, and which is responsible for regulating the convexity of the lens.

ciliary *a.* (1) *pert.* cilium *q.v.*; (2) *pert.* eyelashes, *appl.* sweat glands; (3) in the vertebrate eye, *appl.* ligament that regulates the size of the pupil, and muscle that regulates the convexity of the lens; (4) *appl.* branches of the nasociliary nerve and ganglion.

ciliated *a.* bearing cilia.

ciliated epithelium epithelium composed of cells with cilia projecting from the free surface. It lines various internal passages, such as the nasal and bronchial passages, and the beating action of the cilia draws fluid over the epithelial surface, helping to clear bacteria and other unwanted material.

ciliates Ciliophora *q.v.*

ciliograde *a. appl.* movement due to cilia.

Ciliophora, ciliates *n.*, *n.plu.* phylum of protozoan protists comprising the ciliates (e.g. *Paramecium, Stentor, Tetrahymena*). They possess cilia, at least when immature, and often have structurally complex cells. Ciliates are common in marine and fresh water, and in the rumen of cattle and other ruminants, where they digest cellulose. Almost all ciliates possess two nuclei, a macronucleus that directs vegetative growth, and a diploid micronucleus involved in sexual reproduction. *alt.* **Ciliata, Ciliatea.**

ciliospore *n.* a ciliated protozoan swarm spore.

cilium *n.* (1) motile hair-like outgrowth present on the surface of many eukaryotic cells, which makes whip-like beating movements. The synchronized beating of cilia propels free-living unicells (e.g. protozoans) or, in stationary cells (e.g. of nasal epithelium), produces a flow of material over the cell surface. A cilium is composed of a central core of microtubules (the axoneme) anchored to the cell by a basal body, the whole enclosed in plasma membrane. In multicellular organisms cilia chiefly occur on epithelial cells lining various internal passages; (2) various other hair-like structures, esp. eyelash. *plu.* **cilia.**

cinchonine *n.* alkaloid found in various dicots of the family Rubiaceae.

cinereous *a.* ash-grey.

cingula *n.* (1) ring formed on stalk of some mushroooms and toadstools at which the edge of the cap was attached to the stalk, *plu.* **cingulae**; (2) *plu.* of cingulum.

cingulate *a.* shaped like a girdle, *appl.* a gyrus (ridge) and sulcus (groove) above the corpus callosum on the medial surface of cerebral hemisphere.

cingulum *n.* (1) any structure that is like a girdle; (2) part of plant between root and stem; (3) ridge round base of crown of a tooth; (4) tract of fibres connecting callosal and hippocampal convolutions of brain. *plu.* **cingula.**

circadian rhythm, circadian clock metabolic or behavioural rhythm with a cycle of about 24 hours.

circannual *a.* (1) occurring on an approximately annual basis; (2) *appl.* rhythm, a rhythm or cycle of behaviour of approximately one year.

circaseptan *a. appl.* biological rhythm of around 7 to 8 days.

circinate, circinnate *a.* rolled up on the axis so that apex is the centre, as in young fern fronds.

circle of Willis structure at base of brain that is formed by the joining of carotid and basilar arteries.

Circoviridae *n.* family of single-stranded DNA viruses found in vertebrates (e.g. chicken anaemia virus).

circuit *n.* in the nervous system, a group of interconnected neurons which performs a particular limited function.

circulation *n.* (1) the regular movement of any fluid within definite channels in the body; (2) the blood circulatory system.

circulatory system (1) in vertebrates, the heart, blood vessels and blood; (2) any system with a similar function in other organisms.

circulus *n.* a ring-like arrangement.

circum- prefix derived from L. *circum,* around, surrounding.

circumduction *n.* form of movement exhibited by a bone describing a conical space with the joint cavity as the apex.

circumesophageal circumoesophageal *q.v.*

circumferential *a.* (1) around the circumference; (2) *appl.* primary lamellae parallel to circumference of bone.

circumfila *n.plu.* looped or wreathed filaments on antennae of some insects.

circumflex *a.* bending round.

circumgenital *a.* (1) surrounding the genital pore; (2) *appl.* glands secreting a waxy powder in oviparous species of coccid bugs.

circumnutation *n.* irregular elliptical or spiral movement exhibited by a growing stem, shoot or tendril.

circumoesophageal *a.* structures or organs surrounding the gullet.

circumoral *a.* surrounding the mouth, *appl.* e.g. cilia, tentacles, nerve ring.

circumorbital *a.* surrounding the orbit of the eye, *appl.* bones of the skull.

circumpolar *a.* (1) *appl.* flora and fauna of Polar regions; (2) *appl.* to distributions of plants and animals in northerly parts of Northern Hemisphere that extend through Asia, Europe and North America.

circumscissile *a.* splitting along a circular line, *appl.* dehiscence of a pyxidium.

circumvallate *a.* (1) surrounded by a wall, as of tissue; (2) *appl.* papillae, one type of small projections on tongue that contain tastebuds.

circumventricular organ organ situated in the wall of a cerebral ventricle.

cirral *a. pert.* cirrus or cirri.

cirrate *a.* having cirri.

cirrhosis *n.* formation of fibrous tissue in the liver.

cirrhus *n.* ribbon-like string of spores held together by mucus, issuing from a sporocarp.

cirri *plu.* of cirrus *q.v.*

Cirripedia, cirriped(e)s *n., n.plu.* subclass of aquatic crustaceans, commonly called barnacles, which as adults are stalked or sessile sedentary animals with the head and abdomen reduced and the body enclosed in a shell of calcareous plates.

cirrose, cirrous *a.* (1) with cirri or tendrils; (2) *appl.* leaf with prolongation of midrib forming a tendril.

cirrus *n.* (1) tendril or tendril-like structure; (2) feathery feeding appendage of barnacle; (3) bundle of cilia on a ciliate protozoan; (4) respiratory and copulatory appendage of annelids; (5) projection from the stalk of sea lilies, which anchors them to the substratum; (6) organ having the function of penis in some molluscs and flatworms; (7) hair-like structure on appendages of insects. *plu.* **cirri**.

cis (1) in genetics, two different mutations at the same locus in a diploid organism are said to be in *cis* if they are both on the same chromosome, *alt. cis* configuration, *cis* conformation, *cf. trans*; (2) *appl.* molecular configuration, one of two configurations of a molecule caused by the limitation of rotation around a double bond, the alternative configuration being the *trans*-configuration.

***cis*-acting, *cis*-regulatory** *a. appl.* control elements in the regulatory region of a gene. It refers to their capacity to act only on the adjacent gene. *cf. trans*-acting, *trans*-regulatory.

***cis* configuration, *cis* conformation** *cis q.v.*

***cis*-dominance** with respect to an allele of a gene, the ability of the allele to control adjacent genes irrespective of the presence of other alleles of the gene in the cell, but not to control corresponding genes on other DNA molecules or chromosomes. *alt. cis*-acting.

***cis* Golgi network (CGN)** a system of interconnected tubules and cisternae at the *cis* face (the face away from the plasma membrane) of the Golgi apparatus. *alt.* intermediate compartment.

Cistales Violales *q.v.*

***cis–trans* complementation test** complementation test *q.v.*

cisterna *n.* (1) closed space containing fluid, such as subarachnoid space; (2) flattened membrane-bounded fluid-filled structures in cell, as in the Golgi apparatus and endoplasmic reticulum. *plu.* **cisternae**.

cistern epiphyte an epiphyte lacking roots and gathering water between leaf bases.

cistron *n.* a gene, a DNA sequence coding for a single polypeptide or structural RNA.

citric acid, citrate six-carbon tricarboxylic acid, a component of the tricarboxylic acid cycle, where it is converted to isocitrate by the enzyme aconitase. It is also responsible for the sour taste of oranges and grapefruit.

citric acid cycle tricarboxylic acid cycle *q.v.*

citrulline *n.* amino acid, first isolated from water melon. It is an intermediate in the urea cycle in vertebrates.

CJD Creutzfeldt–Jakob disease *q.v.*

Cl⁻ chloride ion.

cladanthous *a.* having terminal archegonia on short lateral branches.

clade *n.* branch of a phylogenetic tree containing the set of all organisms descended from a particular common ancestor which is not an ancestor of any non-member of the group.

cladistics *n.* method of classification of living organisms that makes use of lines of descent only, rather than phenotypic similarities, to deduce evolutionary relationships, and which groups organisms strictly on the relative recency of common ancestry. Cladistic methods of classification only permit taxa in which all the members share a common ancestor who is also a member of the taxon and which include all the descendants of that common ancestor.

cladocarpous cladanthous *q.v.*

cladode *n.* green flattened lateral shoot, arising from the axil of a leaf and resembling a foliage leaf.

cladodont *a.* having or *appl.* teeth with prominent central and small lateral cusps.

cladogenesis *n.* (1) branching of evolutionary lineages so as to produce new types; (2) evolutionary change as a result of multiplication of species at any one time and their subsequent evolution along different lines. *cf.* anagenesis.

cladogenous *a.* (1) borne on stem, *appl.* certain roots; (2) borne on branches.

cladogram *n.* tree-like diagram showing the evolutionary descent of any group of organisms or set of protein or nucleic acid sequences.

cladophyll(um) cladode *q.v.*

cladoptosis *n.* annual or other shedding of twigs.

cladose *a.* branched.

cladus *n.* a branch, as of a branched spicule.

clamp connection swelling on hyphae of certain basidiomycete fungi, through which a daughter nucleus has passed and at which a septum forms to make two binucleate cells.

clan phratry *q.v.*

clandestine *a. appl.* evolutionary change which is not apparent in adult forms.

Clara cell cell type in the epithelium lining the bronchioles of the lung that secretes various products that protect the epithelium.

claspers *n.plu.* modified organs or parts of various types enabling the two sexes to clasp one another during mating, e.g. the rod-like processes on pelvic fins of some fishes.

claspettes harpagones *q.v.*

class *n.* (1) taxonomic group into which a phylum or a division is divided, and which is itself divided into orders; (2) (*immunol.*) of antibodies, *see* isotype.

class I MHC molecules *see* MHC class I molecules.

class II-associated invariant-chain peptide (CLIP) peptide fragment cleaved from the invariant chain and which remains bound to the MHC class II molecule during its maturation, preventing peptide from binding.

class II MHC molecule *see* MHC class II molecules.

classical conditioning experimental behavioural technique whereby a response (the unconditioned response) elicited by a natural stimulus such as food (the unconditioned stimulus) becomes a response (the conditioned response) to an unrelated stimulus (the conditioned stimulus) by repeated association of the conditioned and unconditioned stimuli. Positive conditioning increases an animal's response to the stimulus, negative conditioning results in increased avoidance of the stimulus. *alt.* Pavlovian conditioning.

classical pathway of complement activation, pathway triggered by antibody binding to antigen on cell surfaces. Complement component C1 binds to antigen and the subsequent series of reactions leads to covalent binding of C3b to pathogen surfaces. This either facilitates phagocytosis of the pathogen, or leads to activation of the terminal complement components and bacterial lysis. *see also* C1, C2, C3, C4, C5, etc., complement system.

classification *n.* arrangement of living organisms into groups on the basis of observed similarities and differences.

Modern classifications of plants and animals attempt wherever possible to reflect degrees of evolutionary relatedness. The smallest group in classification is usually the species, although subspecies, races and varieties (in cultivated plants) below the level of the species are recognized. Species are grouped into genera, genera into families, families into orders, orders into classes, classes into phyla (for animals and protists) or divisions (for plants), phyla and divisions into kingdoms, and kingdoms into superkingdoms or domains. There are also intermediate categories such as superfamilies and infraclasses. *see* Appendices 1–7.

class switching, class switch recombination DNA rearrangement that occurs in the functional heavy-chain gene of activated B lymphocytes which results in the cells switching from making the IgM class of immunoglobulin to IgG, IgA or IgE. A recombination event joins the rearranged heavy-chain V-region sequence to a heavy-chain C-region gene other than C_{mu} (i.e. C_{gamma}, C_{alpha} or $C_{epsilon}$). It enables clonal lineages of B cells to successively make antibodies with different effector functions but the same antigen specificity during the immune response. *alt.* isotype switching.

clathrate, clathroid *a.* lattice-like.

clathrin *n.* fibrous protein that forms a characteristic polyhedral coat on the cytosolic surface of some coated pits and coated vesicles.

claustrum *n.* thin strip of grey matter in hemispheres of brain, located just under the cortex at each side.

clava *n.* club-shaped structure, e.g. club-shaped spore-bearing branch in certain fungi, the knob-like end of antenna in some insects. *plu.* **clavae.**

clavate *a.* club-shaped, thickened at one end.

clavicle *n.* collar bone, forming anterior or ventral portion of the shoulder girdle. *a.* **clavicular.**

clavicularium *n.* in turtles and tortoises, one of a pair of anterior bony plates in shell.

claviform *a.* club-shaped.

clavula *n.* club-shaped fruiting body of certain fungi.

clavus *n.* club-like structure at the end of the antenna in some insects.

claw *n.* (*bot.*) the stalk of a petal.

clay *n.* soil in which most of the mineral particles are less than 0.002 mm in diameter. Typically, soils of this type have high moisture- and nutrient-holding capacities but tend to be poorly drained and slow to warm in the spring.

cleavage *n.* the series of mitotic cell divisions, usually occurring with no increase in cytoplasmic mass, that first transform the single-celled zygote into a multicellular morula or blastula.

cleavage furrow the deep groove that runs round the surface of a zygote at each of the early cleavage divisions, dividing the cytoplasm.

cleavage nucleus (1) nucleus of fertilized egg of zygote produced by fusion of male and female pronuclei; (2) the egg nucleus of parthenogenetic eggs.

cleidoic *a.* having or *pert.* eggs enclosed within a shell or membrane.

cleisto- prefix derived from the Gk *kleistos,* closed.

cleistocarp cleistothecium *q.v.*

cleistocarpous, cleistocarpic *a.* having closed fruiting bodies, as in certain fungi and mosses.

cleistogamy *n.* condition of having flowers that never open and are self-pollinated, and are often small and inconspicuous. *a.* **cleistogamic, cleistogamous.**

cleistothecium *n.* closed fruiting body (ascocarp) characteristic of some ascomycete fungi, in which spores are produced internally and are released by breakdown of the wall.

cleithrum *n.* clavicular element in some fishes.

cleptobiosis *n.* robbing of food stores or scavenging in the refuse piles of one species by another that does not live in close association with it. *alt.* kleptobiosis.

cleptoparasitism kleptoparasitism *q.v.*

Cl⁻-HCO₃⁻ exchanger (1) sodium-driven chloride–bicarbonate exchanger in animal cells in which an influx of Na^+ and HCO_3^- is coupled to an outflow of Cl^- and H^+; (2) sodium-independent chloride–bicarbonate exchanger in which HCO_3^- flows out of the cell down its electrochemical gradient and Cl⁻ flows in. Both are important in regulating intracellular pH.

climacteric *n.* (1) a critical phase, or period of changes, in living organisms; (2) *appl.* change associated with menopause or with recession of male function; (3) (*bot.*) *appl.* phase of increased respiratory activity at ripening of fruit.

climactic *a. pert.* a climax.

climate change any permanent alteration in climate, generally refers to the potential alterations in world climates that will occur as a result of the warming of the atmosphere (global warming) due to increased atmospheric carbon dioxide and the greenhouse effect (*q.v.*).

climatype *n.* biotype resulting from selection in particular climatic conditions.

climax *n.* the mature or stabilized stage in a successional series of plant communities, when the dominant species are completely adapted to environmental conditions.

clinandrium *n.* cavity in the column between anthers in orchids.

clinanthium *n.* dilated floral receptacle, as in the flowerhead of Compositae.

cline *n.* graded series of different forms of the same species, usually distributed along a spatial dimension.

clinidium *n.* spore-producing filament in a pycnidium.

CLIP class II-associated invariant-chain peptide *q.v.*

clisere *n.* succession of communities that results from a changing climate.

Clitellata *n.* a taxonomic group comprising the earthworms and leeches.

clitellum *n.* swollen glandular portion of skin of certain annelids, such as earthworm, which secretes the cocoon in which an embryonic worm develops.

clitoris *n.* erectile organ, homologous with penis, at upper part of vulva in women.

clivus *n.* (1) shallow depression in sphenoid, behind dorsum sellae; (2) posterior sloped part of the monticulus of the cerebellum.

cloaca *n.* (1) chamber into which intestinal, genital and urinary canals open in vertebrates, except most mammals; (2) posterior end of intestinal tract in some invertebrates. *a.* **cloacal**.

clock gene gene that regulates a rhythmic or periodic activity or behaviour.

clonal *a. pert.* a clone.

clonal deletion the deletion of self-reactive lymphocytes from the repertoire during lymphocyte development in the primary lymphoid organs.

clonal expansion the proliferation of a lymphocyte after encounter with its corresponding antigen.

clonal selection in an immune response, the proliferation, in response to stimulation by an antigen, of clones of lymphocytes of the corresponding specificities.

clone *n.* (1) group of genetically identical individuals or cells derived from a single cell by repeated asexual divisions; (2) DNA clone *q.v.*; (3) animal or plant derived from a single somatic cell or cell nucleus is termed a clone of the individual from which the cell or nucleus came; (4) (*bot.*) an apomict strain; (5) *v.* to isolate a single cell or DNA molecule, and multiply it.

cloning *n.* (1) the isolation of a single cell and the growth of a clone of identical cells from it; (2) DNA cloning *q.v.*; (3) the development of an animal or plant from a single somatic cell or cell nucleus. Such an organism is termed a clone of the individual from which the cell was taken. Mammals have been cloned by substituting a somatic cell nucleus for the nucleus of an unfertilized or fertilized egg.

cloning vector specially modified plasmid or phage into which 'foreign' genes can be inserted for introduction into bacterial or other cells for multiplication.

clonotype *n.* specimen of an asexually propagated part of a type specimen or holotype.

clonotypic *a.* characteristic of a particular clone of cells.

clonus *n.* series of muscular contractions in which individual contractions are discernible.

closed community ecological community into which further colonization is prevented as all niches are occupied.

closed forest forest where the crowns of the trees touch and produce a closed canopy for all or part of the year.

Closteroviridae plant virus family containing very long thread-like single-stranded RNA viruses, e.g. beet yellows virus.

Clostridium, clostridia *n.*, *n.plu.* bacteria of the genus *Clostridium*, strictly anaerobic, spore-forming rods widely distributed in soil, some species of which produce powerful toxins. They include *Cl. tetani*, the causal agent of tetanus, *Cl. botulinum*, whose toxin formed in contaminated food causes botulism, *Cl. histolyticum*, the cause of gas gangrene, and *Cl. difficile*, which causes severe intestinal infection in hospitals.

clotting (1) blood clotting *q.v.*; (2) coagulation of milk by the action of enzymes such as rennin, which converts soluble proteins to insoluble forms.

clotting factor coagulation factor *q.v.*, *see* blood clotting.

clover-leaf structure description of secondary structure common to all transfer RNAs in which base pairing within the RNA chain forms three loops resembling a clover-leaf.

club moss common name for a member of the division Lycophyta, a group of seedless vascular plants.

clunes *n.plu.* buttocks.

cluster-crystals globular aggregates of calcium oxalate crystals in plant cells.

cluster-cup aecidium *q.v.*

cluster of differentiation (CD) set of monoclonal antibodies that distinguishes a particular cell-surface antigen (CD antigen) on white blood cells. Many cell-surface proteins have been characterized in this way and given a CD number, e.g. CD3 is a protein complex associated with the T-cell receptor, CD4 and CD8 are glycoprotein adhesion molecules on T cells, CD25 is the receptor for interleukin-2, CD45 is a tyrosine phosphatase involved in signal transduction in white blood cells.

clutch *n.* number of eggs laid by a female at one time.

clypeal *a. pert.* clypeus *q.v.*

clypeate *a.* round or shield-like.

clypeus *n.* (1) shield-shaped plate of exoskeleton in the centre front of insect head; (2) strip of cephalothorax between eyes and bases of chelicerae in spiders.

CML chronic myelogenous leukaemia.

CMP cytidine monophosphate *q.v.*

CMV (1) cowpea mosaic virus; (2) cytomegalovirus.

cnemial *a.* (1) *pert.* tibia; (2) *appl.* ridge along dorsal margin of tibia.

cnemidium *n.* lower part of bird's leg devoid of feathers and usually scaly.

cnemis *n.* shin or tibia.

Cnidaria *n.* phylum of simple, aquatic, mostly marine, invertebrate animals containing corals, sea fans and sea anemones (class Anthozoa), the hydroids and milleporine corals (Hydrozoa), the jellyfishes (Scyphozoa) and the box-jellies (Cubozoa). They include both colonial and solitary forms. Individuals are generally radially symmetrical, with only one opening (mouth) to the gut and a simple two-layered body with a primitive nerve net between the two layers. Cnidaria have hydroid (polyp) and/or medusa forms, and bear stinging cells (cnidoblasts) on the tentacles fringing the mouth. With the phylum Ctenophora, the Cnidaria form the large group of animals known as coelenterates.

cnidoblast *n.* stinging cell of sea anemone, jellyfish and other Cnidaria, containing a coiled thread which is discharged oncontact with prey. *alt.* nematoblast, nematocyst.

cnidocil *n.* minute process projecting from a cnidoblast (stinging cell), whose stimulation causes discharge of a nematocyst. *alt.* **cnidocyte.**

cnidophore *n.* modified zooid in a hydrozoan colony that bears cnidoblasts.

Cnidosporidia Myxozoa *q.v.*

CNS central nervous system *q.v.*

CO carbon monoxide *q.v.*

CO₂ carbon dioxide *q.v.*

CoA coenzyme A *q.v.*

coacervate *n.* inorganic colloidal particle, e.g. clay, on which organic molecules have been adsorbed.

co-action *n.* the reciprocal activity of organisms within a community.

coactivator gene regulatory protein that binds to a DNA-bound protein to help activate transcription.

co-adaptation *n.* correlated variation and adaptation displayed in two mutually dependent organs or organisms.

co-adapted *a. appl.* a set of genes all involved in effecting a complex process.

coagulase *n.* fibrin-clotting enzyme produced by pathogenic *Staphylococcus aureus.*

coagulation *n.* (1) curdling or clotting; (2) the change from a liquid to a viscous or solid state by chemical reaction.

coagulation factor (1) any of a group of blood proteins involved in blood clotting, such as Factors VIII (antihaemophilia factor), XII (Hageman factor), IX (Christmas factor), kallikrein, fibrinogen and prothrombin (*see individual entries*); (2) vitamin K *q.v.*

coagulocyte *n.* granular blood cell in insects.

coagulum *n.* (1) coagulated mass or substance; (2) a clot.

coal ball more-or-less spherical aggregate of petrified plant structures found in certain coal measures.

coalescence *n.* union of floral parts of the same whorl.

co-ancestry coefficient of kinship *q.v.*

coarctate *a.* (1) compressed; (2) closely connected; (3) with abdomen separated from thorax by a constriction.

coarctate larva, coarctate pupa larval stage in certain Diptera in which the larval skin is retained as a protective puparium.

coastal zone the zone that extends from the high-tide mark on land to the edge of the continental shelf. It comprises the flora and fauna of the beach and rocks and of the relatively warm, nutrient-rich shallow waters over the continental shelf.

coated pit *see* coated vesicle.

coated vesicle type of vesicle found in most eukaryotic cells, which is enclosed in a protein coat formed of either clathrin or COP. Coated vesicles are formed from depressions known as coated pits which form on plasma membrane and Golgi membranes.

cobalamin *n.* cobalt-containing vitamin, vitamin B_{12}, synthesized by microorganisms, a prosthetic group of certain mammalian enzymes. *alt.* cobalamine.

cobamide coenzyme coenzyme form of cobalamin *q.v.*

coca *n.* the dried leaves of the coca plant *Erythroxylon coca*, which contains cocaine.

cocaine *n.* alkaloid obtained from coca leaves, which has been used as a local anaesthetic, is taken as a stimulant drug of abuse, and can result in addiction.

cocci *plu.* of coccus.

Coccidia *see* coccidiosis.

coccidioidomycosis *n.* fungal disease of humans and other animals caused by *Coccidioides immitis*, marked by fever and granulomas in lung.

coccidiosis *n.* disease of animals caused by sporozoan protozoan parasites of the order Coccidia, which infect the lining of the gut. *plu.* **coccidioses**.

coccids *n.plu.* scale insects, minute bugs with winged males, scale-like females that are attached to the infested plant, and young that suck the sap.

coccogone *n.* reproductive cell in certain algae.

coccoid *a.* spherical or globose, like or *pert.* a coccus.

coccolith *see* coccolithophoroids.

coccolithophorids *n.plu.* resting form of some species of small motile golden unicellular eukaryotic organisms (Haptophyta), commonly known as the golden algae, having calcareous plates (coccoliths) covering the cells.

coccospheres *n.plu.* remains of hard parts of certain algae and radiolarians.

coccus *n.* roughly spherical bacterial cell. *plu.* **cocci**.

coccygeal *a. pert.* or in region of coccyx.

coccygeomesenteric *a. appl.* a branch of the caudal vein, in birds.

coccyges *plu.* of coccyx.

coccyx *n.* the terminal part of the vertebral column beyond the sacrum. *a.* **coccygeal**.

cochlea *n.* the part of the inner ear concerned with hearing. It is a hollow structure spirally coiled like a snail's shell and containing sensory cells that respond to vibrations of fluid inside the cochlea caused by the transmission of sound from the outer ear. The cochlea is maximally stimulated at different points along its length by sound of different frequency.

cochlear *a.* (1) (*bot.*) *appl.* a mode of disposition of parts of flower in the bud where a wholly internal leaf is next but one to a wholly external leaf; (2) (*zool.*) *pert.* cochlea *q.v.*

cochleariform *a.* screw- or spoon-shaped.

cochleate *a.* (1) screw-like; (2) like a snail's shell.

cockroaches *n.plu.* common name for many members of the Dictyoptera *q.v.*

cocoon *n.* (1) protective case of many larval forms before they become pupae; (2) silky or other covering formed by many animals for their eggs.

COD chemical oxygen demand *q.v.*

codeine *n.* alkaloid present in opium and having effects similar to but weaker than morphine.

coding *n.* in the nervous system, the rules by which action potentials in a sensory system reflect a physical stimulus.

coding region, coding sequence (CDS) that part of a gene which encodes an amino acid sequence.

coding strand in double-stranded DNA, the DNA strand that carries the same base sequence as the RNA transcribed from the DNA, i.e. the strand that is not used by the RNA polymerase as the template for transcription. *alt.* sense strand.

co-dominance *n.* (1) case where alleles present in a heterozygous state produce a different phenotype from that produced by either allele in the homozygous state; (2) case where each of a pair of heterozygous alleles is expressed equally and makes an equal contribution to the phenotype.

co-dominant *a.* (1) (*genet.*) *appl.* alleles showing co-dominance; (2) (*bot.*) *appl.* two species being equally dominant in climax vegetation.

codon *n.* group of three consecutive bases in RNA which specifies an amino acid or a signal for termination of translation. *alt.* triplet (in DNA). *cf.* anticodon. *see* Fig. 19 (p. 258).

codon bias preferential use of a particular codon for an amino acid, which may vary between different genes or between different species.

codon family set of codons differing only in the third base and which all specify the same amino acid.

coefficient of coincidence ratio of the observed number of recombinants to the expected number of recombinants.

coefficient of community measurement of the degree of resemblance of two plant communities, based upon their species compositions. It may be calculated in various ways using the ratios of number of common species to the total number of species in each. *alt.* similarity index. *see* Gleason's index, Jaccard index, Kulezinski index, Morisita's similarity index, Simpson's index of floristic resemblance, Sørensen similarity index.

coefficient of consanguineity coefficient of kinship *q.v.*

coefficient of genetic relatedness coefficient of relationship *q.v.*

coefficient of inbreeding *see* inbreeding coefficient.

coefficient of kinship (*f*) of two individuals, the probability that two gametes taken at random, one from each individual, carry alleles at a given locus that are identical by virtue of common descent.

coefficient of relationship (*r*) probability that a gene in one individual will be identical by virtue of common descent to a gene in a particular relative, e.g. for monozygotic twins $r = 1$, for parents and offspring $r = 0.5$, for full siblings $r = 0.5$, for grandparents and grandchildren $r = 0.25$, etc. *r* also gives the fraction of genes identical by common descent between two individuals.

coefficient of selection (*s*) a measure of the strength of natural selection, calculated as the proportional reduction in contribution of one genotype to the gametes compared with a standard genotype. It may have any value from zero to one.

coefficient of variation standard deviation expressed as a percentage of the mean value.

coelacanth *see* Crossopterygii.

coelenterates *n.plu.* the animal phyla Cnidaria (corals, sea anemones, jellyfish and hydroids) and Ctenophora (sea combs or sea gooseberries) collectively. They have radial symmetry, a single body cavity (coelenteron) opening in a mouth, and a simple body wall of endoderm (gastrodermis) and ectoderm (epidermis) separated by usually noncellular gelatinous mesogloea, and include both solitary and colonial (e.g. corals) forms. Contractile musculo-epithelial cells and a simple nerve net are present.

coelenteron *n.* the single body cavity of coelenterates.

coeliac *a.* (1) *pert.* the abdominal cavity, *appl.* arteries, veins, nerves; (2) *appl.* plexus: solar plexus *q.v.*

coelo- *alt.* spelling of celo-.

coelom *n.* body cavity in animals that houses gonads and excretory organs and opens to the exterior. Animals possessing a coelom are known as coelomates.

coelomates *n.plu.* animals possessing a true coelom, comprising the phyla Mollusca, Annelida, Arthropoda, Phoronida, Bryozoa, Brachiopoda, Echinodermata, Chaetognatha, Hemichordata and Chordata.

coelomic *a. pert.* a coelom.

coelomocyte *n.* any of various types of cell found in the coelom in annelid worms, in the body cavity of nematodes, and in the coelomic, water and blood circulation systems in echinoderms.

coelomoduct *n.* channel leading from the body cavity to the exterior.

coelomostome *n.* external opening of a coelomoduct.

coelomyarian *a.* having a longitudinal row of muscle cells bulging into the pseudocoel, *appl.* some nematodes.

coelomycete *n.* informal term indicating a fungus that produces conidia in pycnidia or acervuli.

coelozoic *a. appl.* a trophozoite when parasitizing some cavity of the body.

coen- *alt.* spelling of cen-.

coenangium *n.* coenocytic sporangium.

coenanthium *n.* inflorescence with a nearly flat receptacle having upcurved margins.

coenenchyme *n.* the common tissue that connects individual polyps in a coral.

coenobium *n.* colony of unicellular organisms, as in some green algae, having a definite form and organization, which behaves as an individual and reproduces to give daughter coenobia. *plu.* **coenobia**.

coenocyte *n.* fungal or algal tissue in which constituent protoplasts are not separated by cell walls. *a.* **coenocytic**. *alt.* aseptate.

coenoecium *n.* the common ground material of a bryozoan colony.

coenogametangium *n.* coenocytic gametangium, as in zygomycete fungi.

coenogamete *n.* multinuclear gamete.

coenogamy *n.* union of coenogametangia, as in some zygomycete fungi.

coenosarc *n.* thin outer layer of tissue that unites all the members of a coral colony.

coenosis *n.* random assemblage of organisms with similar ecological preferences. *cf.* community.

coenosite *n.* organism habitually sharing food with another.

coenospecies *n.* a group of taxonomic units such as species, ecospecies or varieties, which can intercross to form hybrids that are sometimes fertile, the group being equivalent to a subgenus or superspecies.

coenosteum *n.* the common colonial skeleton in corals and bryozoans.

coenozygote *n.* zygote formed by coenocytic gametes, as in some fungi.

coenurus *n.* metacestode with large bladder, from whose walls many daughter cysts arise, each with one scolex.

coenzyme cofactor *q.v.*

coenzyme A (CoA) carrier of activated acyl groups in many metabolic reactions, made up of phosphoadenosine diphosphate linked to a pantothenate unit linked to a β-mercaptoethylamine unit, to which an acyl group can be linked to form an acyl-CoA (often acetyl-CoA). *see* Fig. 11.

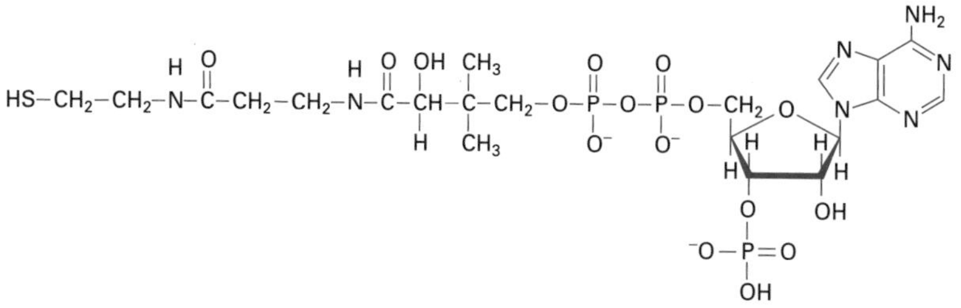

Fig. 11 Coenzyme A.

coenzyme M 2-mercaptoethanesulphonic acid, a cofactor that carries the methyl group that is reduced to methane in the final step of methanogenesis.

coenzyme Q ubiquinone *q.v.*

coevolution *n.* parallel evolution of two species or genes, in which changes in one tend to produce changes in the other.

cofactor *n.* any non-protein substance required by a protein for biological activity, such as prosthetic groups and compounds such as NAD, NADP, flavin nucleotides, coenzyme A.

cognate *a.* (1) *appl.* a tRNA recognized by its corresponding aminoacyl-tRNA synthetase; (2) *appl.* antigen recognized by its corresponding antibody or T-cell receptor; (3) *appl.* B cells and T cells that recognize the same antigen (via different epitopes).

cognition *n.* those higher mental processes in humans and animals, such as the formation of associations, concept formation and insight, whose existence can only be inferred and not directly observed. *a.* **cognitive.**

coherent *a.* with similar parts united but capable of separating with a little tearing.

cohesin multisubunit protein complex that holds sister chromatids together until anaphase of mitosis.

cohesion *n.* (*bot.*) condition of union of separate parts of floral whorl.

cohesive ends the complementary single-stranded ends of DNA treated with certain restriction enzymes, which can pair to form a circular molecule or join to other similar cohesive ends. *alt.* sticky ends.

cohort *n.* group of individuals of the same age in a population.

coiled-coil type of protein structural element in which two α-helices are wound around each other.

co-immunoprecipitation *n.* technique for detecting proteins that interact with each other in cells. A labelled antibody specific for one of the proteins is used to pull out the protein complex from a tissue extract, and the interacting proteins can then be identified.

co-integrate *n.* product of the integration of one DNA into another.

coition, coitus *n.* sexual intercourse, copulation.

colchicine *n.* toxic drug isolated from the autumn crocus *Colchicum* which arrests cells in metaphase by disrupting microtubule organization. Derivatives such as colcemid are widely used in cell biology.

cold-blooded poikilothermal *q.v.*

cold-sensitive mutation mutation resulting in a gene product that is functional at a normal or higher temperature but not at a lower temperature.

coleogen *n.* meristematic tissue in plants that gives rise to the endodermis.

Coleoptera *n.* very large order of insects, commonly called beetles. In the adult the forewings are modified as hard wing-cases (elytra) that cover membranous hindwings, which may be reduced or absent.

coleoptile *n.* protective sheath surrounding the germinating shoot of some monocotyledonous plants such as grasses.

coleorhiza *n.* protective sheath surrounding the germinating root of some monocotyledons such as grasses.

colic *a. pert.* the colon.

colicin *n.* protein produced by certain bacteria which kills or inhibits the growth of other bacteria.

colicinogenic *a.* (1) producing colicins; (2) *appl.* plasmids: Col plasmid *q.v.*

coliform (1) *a.* sieve-like; (2) *n.* any of a group of non-sporing Gram-negative rod-like bacteria typified by *Escherichia coli.* Their presence is used as a standard indicator of faecal pollution of water.

Coliiformes *n.* order of birds including the colies or mousebirds.

co-linear *a.* having a linear correspondence to each other, e.g. the correspondence between a DNA sequence and the protein sequence it encodes. *n.* **co-linearity.**

coliphage *n.* bacteriophage that attacks *Escherichia coli.*

colitose *n.* 3,6-dideoxyhexose sugar present in the lipopolysaccharide in the outer membrane of some enteric bacteria.

collagen *n.* fibrous protein secreted by connective tissue cells in animals and forming a major component of the extracellular matrix of connective tissues. There are around 20 different types of collagen. Individual collagen molecules are composed of a triple helix of three polypeptide chains wound around each other. In some types of collagen (e.g. type I, the commonest form), individual molecules bind together to form insoluble fibrils which then aggregate into thicker fibres of high tensile strength. Other types (e.g. type IV collagen of basal laminae) form a sheet-like network. Collagen is rich in glycine and proline and contains the unusual amino acids hydroxyproline and hydroxylysine. *see also* procollagen, tropocollagen.

collagenase *n.* any of a group of enzymes that break down collagen, found in certain bacteria, e.g. *Clostridium histolyticum*, the bacterium responsible for gas gangrene, and in developing and metamorphosing animal tissue. *alt.* κ-toxin (in clostridia).

collagenous *a.* containing collagen.

collaplankton *n.* plankton organisms rendered buoyant by a mucilaginous or gelatinous envelope.

collar *n.* any structure comparable to or resembling a collar, such as the fleshy rim projecting beyond the edge of a snail's shell, the junction between root and stem in a plant, the junction between blade and leaf sheath in grasses.

collar cell choanocyte *q.v.*

collateral *a.* (1) side by side; (*bot.*) (2) *appl.* vascular bundles with xylem and phloem on the same radius, the phloem to the outside of the xylem; (3) *appl.* bud at side of axillary bud; (*zool.*) (4) *appl.* fine lateral branches from the axon of a nerve cell; (5) *appl.* circulation established through anastomosis with other parts when the chief vein is obstructed; (6) (*genet.*) *appl.* inheritance of a trait from a common ancestor in individuals not lineally related.

collateral ganglia ganglia of the autonomic nervous system. In mammals they lie e.g. at the roots of the coeliac, anterior mesenteric and posterior mesenteric arteries. They are not joined together by nerve fibres. *cf.* lateral ganglia.

collectins *n.plu.* a family of calcium-dependent sugar-binding proteins or lectins containing collagen-like sequences. Examples are mannose-binding lectin and complement component C1q.

collecting ducts ducts that convey urine from kidney tubules in the cortical region

of the kidney to the ureter. *alt.* collecting tubules.

collecting hair one of the pollen-retaining hairs on stigma or style of certain flowers.

collecting tubules collecting ducts *q.v.*

collective fruit anthocarp *q.v.*

Collembola *n.* order of insects containing the springtails, small wingless insects with only six segments in the abdomen and with two long projections from the abdomen that spring out from a folded position to make the animal jump. Common in soil and leaf litter.

collenchyma *n.* (1) in plants, peripheral parenchymal supporting tissue with cells more-or-less elongated and thickened; (2) the middle layer of sponges. *a.* **collenchymal**.

collet *n.* root zone of hypocotyl, where cuticle is absent.

colleter *n.* glandular hair on resting bud.

colleterium *n.* mucus-secreting gland in the female reproductive system of insects, which is involved in production of egg case or in cementing eggs to substrate. *alt.* **colleterial glands**.

collicular *a. pert.* a colliculus.

collicular commissure tract of fibres connecting the two cerebral hemispheres at base of occipital lobes.

colliculate, colliculose *a.* having small elevations.

colliculus *n.* (1) any of various structures forming a rounded eminence; (2) in the brain. *see* inferior colliculi, superior colliculi; (3) in the eye, the slight elevation formed by optic nerve at entrance to retina. *plu.* **colliculi**.

colligation *n.* the combination of persistently discrete units.

colloblast *n.* cell on tentacles and pinnae of ctenophores, which carries small globules of adhesive material. *alt.* lasso cell.

colloid *n.* (1) gelatinous substance found in some diseased tissues; (2) heterogeneous material composed of submicroscopic particles of one substance dispersed in another substance.

collum *n.* neck, collar, or any similar structure.

colon *n.* (1) in vertebrates, a portion of the large intestine preceding the rectum and whose main function is reabsorption of water; (2) in insects, the second portion of the intestine.

colonial *a. appl.* organisms that live together in large numbers, esp. where the individual organisms form part of a larger structure. Examples are some algae, soft corals, reef-building corals, gorgonians and other anthozoans, and siphonophores.

colonization *n.* (1) invasion of a new habitat by a species; (2) occupation of bare ground by seedlings.

colony *n.* (1) group of individuals of the same species living together in close proximity. A colony may be an aggregate structure with organic connections between the individual members (e.g. sponges, corals) or, as in the social insects (e.g. bees, ants, termites), large numbers of free-living individuals which habitually live together in a strictly organized social group; (2) group of animals or plants living together and somewhat isolated, or recently established in a new area; (3) coenobium *q.v.*; (4) (*microbiol.*) aggregate of microorganisms (esp. bacteria and fungi) formed when growing on solid media, and which is visible to the naked eye. A bacterial colony comprises the progeny of a single cell. A fungal or actinomycete colony comprises the mycelium arising from a single spore or single fragment of hypha.

colony fission production of new colonies by departure of some members while leaving the parent colony intact.

colony-forming unit (CFU) haematopoietic cell which has the capacity to multiply.

colony-stimulating factor (CSF-1) a cytokine that stimulates production of macrophages from bipotential macrophage-granulocyte precursor cells.

colorectal *n. pert.* colon and rectum, *appl.* e.g. cancer.

colostrum *n.* clear fluid secreted from mammary glands at the end of pregnancy and differing in composition from the milk secreted later.

colour blindness genetically determined inability in humans to distinguish all or certain colours. The most common form, red–green colour blindness, is determined by various X-linked recessive genes and is thus usually found in males.

colour phase unusual but regularly occurring colour variety of a plant or animal, e.g. the white colour forms of many plants.

colour vision the ability to distinguish colours within the visible light spectrum, which is due to colour-sensitive cone cells in the retina. Many vertebrates, including humans, possess three types of cone each with a distinctive spectral sensitivity, and are said to have trichromatic vision. Some animals possess only two types of cone and have dichromatic vision, in which some wavelengths will be confused and be perceived as the same colour.

colpate *a.* furrowed, *appl.* pollen grains.

Col plasmid colicinogenic plasmid, any of a group of plasmids found in certain bacteria, especially strains of *Escherichia coli*, which carry genes for colicins.

Columbiformes *n.* order of birds including the doves and pigeons.

columella *n.* (*bot.*) (1) column of sterile cells in centre of capsule in mosses; (2) central core in root cap; (3) sterile structure within a fruiting body in fungi; (*zool.*) (4) central pillar in skeleton of some corals; (5) the central pillar in some gastropod shells; (6) small bone in skull of some reptiles; (7) partly bony and partly cartilaginous rod connecting tympanum with inner ear in birds, reptiles and amphibians; (8) the axis of the cochlea; (9) lower part of nasal septum. *a.* **columellar.**

column, columna *n.* (1) any structure like a column, such as spinal column, cylindrical body of hydroids up to the tentacles, stalk of a crinoid; (2) (*bot.*) structure formed of united stamens or of stamens and style.

columnar *a.* (1) *pert.*, or like, a column, *appl.* cells longer than broad; (2) *appl.* epithelium composed of such cells.

column chromatography separation of proteins by running in a solvent through a column of a permeable solid-state matrix to which the proteins bind with different affinities. After flushing with solvent, proteins will emerge from the end of the column in different fractions. According to the choice of matrix, proteins can be separated on the basis of their charge, hydrophobicity or ability to bind to particular chemical groups.

Columniferae Malvales *q.v.*

coma *n.* (1) terminal cluster of bracts, as in pineapple; (2) hair tuft on seed; (3) compact head of clustered leaves or branches in mosses such as *Sphagnum*.

combination colours colours produced by structural features of a surface in conjunction with pigment.

combining site site on an antibody to which specific antigen binds.

comb jellies common name for the Ctenophora, a phylum of marine coelenterates, also called sea gooseberries.

comb-rib *see* Ctenophora.

comes *n.* blood vessel that runs alongside a nerve. *plu.* **comites.**

co-metabolism *n.* metabolic transformation of a substance while a second substance serves as the primary energy or carbon source.

comitalia *n.plu.* small two- or three-rayed spicules in sponges.

Commelinales *n.* order of herbaceous monocots including the family Commelinaceae (tradescantia) and others.

commensal *n.* a partner, usually the one that benefits, in a commensalism. *a. pert.* commensalism.

commensalism *n.* association between two organisms of different species that live together and share food resources, one species benefiting from the association and the other not being harmed.

commissure *n.* (1) joining line between two parts; (2) connecting band of nerve tissue. *a.* **commissural.**

commitment *n.* the case when cells follow their normal course of development if no external factors intervene. *a.* **committed.**

community *n.* well-defined assemblage of plants and/or animals, clearly distinguishable from other such assemblages.

community biomass total weight per unit area of the organisms in a community.

community effect induction of cell differentiation only when a sufficient number of responding cells are present.

community production (PP) primary production, the quantity of dry matter formed by the vegetation covering a given area.

comose *a.* (1) hairy; (2) having a tuft of hairs.

Comoviridae plant virus family containing small isometric single-stranded RNA viruses, e.g. cowpea mosaic virus. They are multicomponent viruses in which two genomic RNAs are encapsidated in three different virus particles, one of which lacks nucleic acid.

compact bone osteon bone *q.v.*

compaction *n.* flattening of blastomeres against each other in morula of mouse embryo, with development of cell polarity.

companion cell *see* phloem.

comparative genomic hybridization (CGH) a molecular cytogenetic method of screening a tumour for genetic changes, by comparing the pattern of hybridization of tumour DNA and normal DNA to a set of normal metaphase chromosomes. The method picks up changes such as translocations and deletions at both chromosomal and subchromosomal levels.

comparative genomics the comparison of different genomes.

comparative modelling homology modelling *q.v.*

compartment *n.* (1) in development, an invariant, precisely delimited area that contains all the descendants of a small group of founder cells and which tends to act as a discrete developmental unit. Cells in compartments do not enter adjacent compartments; (2) metabolic compartments *q.v.*

compartmentation *n.* segregation of different metabolic and cell biological processes into particular membrane-bounded areas or organelles inside a eukaryotic cell.

compass *n.* curved forked ossicle, part of Aristotle's lantern in echinoderms.

compass plants plants with a permanent north and south direction of their leaf edges.

compatibility group group of plasmids whose members are unable to coexist within the same bacterial cell.

compatible *a.* (1) able to cross-fertilize; (2) having the capacity for self-fertilization; (3) able to coexist; (4) in plant pathology, *appl.* an interaction between host and pathogen that results in disease; (5) in tissue transplantation, *appl.* tissue graft that will not be rejected.

compensation point (1) the point at which respiration and photosynthesis are balanced at a given temperature, as determined by the intensity of light or the concentration of carbon dioxide. *see* carbon dioxide compensation point; (2) limit of lake or sea depth at which green plants and algae lose more by respiration than they gain by photosynthesis. *alt.* **compensation depth, compensation level.**

competence *n.* (1) state in which a cell or organism is able to respond to a stimulus, refers esp. to the ability of embryonic cells to respond to a specific developmental stimulus; (2) ability of a bacterium to take up DNA and become transformed. *a.* **competent.**

competition *n.* (1) active demand by two or more organisms for a material or condition, so that both are inhibited by the demand, e.g. plants competing for light and water, *cf.* amensalism; (2) active demand by two or more substances for the same binding site on an enzyme or receptor. *alt.* **competitive binding.**

competitive exclusion principle the principle that two different species cannot indefinitely occupy the same ecological niche, one eventually being eliminated. *alt.* Gause's principle.

competitive inhibitor *see* inhibitor.

complement (1) *n.* the proteins of the complement system (*q.v.*) collectively; (2) *v. see* complementation.

complement activation *see* complement system.

complemental air volume of air which can be taken in in addition to that drawn in during normal breathing.

complemental male pygmy male *q.v.*

complementarity-determining regions (CDRs) three highly variable loops of polypeptide chain, CDR1, CDR2 and CDR3, which are part of the variable region of an immunoglobulin or T-cell receptor chain. They are involved in binding antigen and are responsible for the antigen specificity of the antigen-binding site. *cf.* framework regions.

complementary *a.* (1) *appl.* two strands of DNA or RNA that can base-pair with each other; (2) *appl.* non-suberized cells loosely arranged in cork tissue and forming air passages in lenticels; (3) *n.* coronoid bone in mandible of reptiles.

complementary DNA (cDNA) DNA synthesized *in vitro* on an RNA template by reverse transcriptase.

complementary male pygmy male *q.v.*

complementary RNA synthetic RNA transcribed from a single-stranded DNA template.

complementation *n.* the production of a wild-type phenotype when two mutations with the same mutant phenotype are combined *in trans* in a diploid organism or heterokaryon. The two mutations are said to complement each other.

complementation group group of mutations which fail to complement each other *in trans* in a complementation test, formally defining a cistron (a DNA sequence specifying a single polypeptide chain).

complementation test test used to determine whether two recessive mutations giving the same phenotype lie in the same or different genes. A cross is made between two organisms, each homozygous for one of the mutations. If the heterozygote progeny have recovered wild-type function, the mutations are located in different genes. *alt.* cis–trans complementation test.

complement fixation *see* complement system.

complement receptors *see* CR1, CR2, CR3, CR4.

complement system a system of blood proteins, some of which are proteases, which becomes activated on infection and aid in clearing bacteria and other pathogens from the body. It is part of innate immunity. Certain complement proteins coat the pathogen (complement fixation), facilitating its removal by phagocytic cells, which bear receptors for complement proteins, while others form a complex in the cell membrane that causes lysis of the pathogen. Complement can act on its own or in conjunction with antibodies. The main proteins of the complement system are C1, C2, C3, C4, C5, C6, C7, C8, C9, factor B, factor D, ficolin, mannose-binding lectin, properdin (*see individual entries*). *see also* anaphylatoxin, alternative pathway, C3 convertase, C5 convertase, classical pathway, lectin-mediated pathway, lytic complex, terminal complement components.

complete *a. appl.* flowers containing sepals, petals, stamens and carpels.

complete metamorphosis insect metamorphosis in which the young are morphologically different from the adult and are called larvae, and go through a resting stage (the pupa), before reaching the adult stage (the imago). The wings and other adult organs are developed inside the body during the pupal stage. *cf.* incomplete metamorphosis.

complex *n.* (1) two or more molecules held together, usually by non-covalent bonding, so that they are quite easily separable; (2) (*ecol.*) the meeting of several distinct communities related to each other by certain shared species.

complexity *n.* amount of sequence information in a DNA, measured as the total length of different sequences present.

complex locus genetic locus which encodes an apparent single function, but which is divisible by various genetic tests into several subloci, each acting in some respects as a separate locus. Complex loci usually turn out to be composed of a cluster of genes involved in determining different aspects of the same function.

complex medium culture medium whose precise chemical composition is unknown. *cf.* defined medium.

complex tissue tissue composed of more than one type of cell.

complex trait genetically determined character that is not inherited as a simple Mendelian trait. Complex traits are usually determined by a large number of genes acting independently.

complexus *n.* (1) aggregate formed by a complicated interweaving of parts; (2) *appl.* a muscle, the semispinalis capitis.

complicant *a.* folding over one another, *appl.* wings of certain insects.

complicate *a.* (1) folded, *appl.* leaves or insect wings folded longitudinally so that right and left halves are in contact; (2) in fungi, *appl.* fruit bodies of some hymenomycetes composed of several caps with stalks fusing to form a central stalk.

component population a population containing all the individuals of a specified life-history phase at a given place and time.

Compositae, composites *n., n.plu.* large family of dicotyledonous flowering plants, typified by dandelions, daisies, thistles, in which the inflorescence is a capitulum made up of many tiny florets.

composite *a.* (1) closely packed, as the florets in a capitulum; (2) *appl.* fruits. *see* sorosis, syconus, strobilus; (3) *appl.* member of the plant family Compositae.

composite transposon a transposon with a central region flanked by insertion sequences.

compound *a.* made up of several elements, *appl.* e.g. flowerheads made up of many individual flowers, leaves made up of several leaflets.

compound chromosome isochromosome *q.v.*

compound eye eye characteristic of insects and most crustaceans, made up of many identical visual units. *see* ommatidium.

compound microscope the compound light microscope, in which the image is magnified via a series of two lenses, the ocular and objective lenses. The upper limit of magnification with a modern compound light microscope is *ca.* 1500×.

compound nest nest containing colonies of two or more species of social insects, in which the galleries of the nest run together. Adults sometimes intermingle but the broods are kept separate.

compressed *a.* flattened transversely.

compression wood in conifers, reaction wood formed on the lower sides of crooked stems or branches, and having a dense structure and much lignification.

compressor *n.* muscle that serves to compress.

computerized axial tomography (CAT) non-invasive technique for examining brain structure by computer analysis of X-radiographs taken at different positions around the head.

Con A concanavalin A *q.v.*

conalbumin *n.* iron-binding glycoprotein found in egg white, also found in plasma where it transports iron. *alt.* ovotransferrin.

conarium *n.* transparent deep-sea larva of certain coelenterates.

concanavalin A (con A) lectin that binds α-mannosyl and glucose residues. It is used in experimental cell biology as a marker for glycoproteins on cell membranes. It also has mitogenic activity.

concatamer *n.* DNA molecule composed of two or more separate molecules linked end to end.

concatenate *a.* forming a chain, as spores.

concentration factor the factor by which the level of a toxic pollutant, e.g. a heavy metal or pesticide, accumulates in the tissue of a living organism compared with its concentration in the environment, or with its concentration in the previous organism in the food chain.

concentric *a.* (*bot.*) having a common centre, *appl.* vascular bundles with one kind of tissue surrounding another.

conceptacle *n.* depression in the thallus of certain algae in which gametangia are borne.

conceptus *n.* the fertilized ovum, in mammals.

concerted evolution coevolution *q.v.*

concha *n.* cavity of the external ear, opening into the external auditory meatus.

conchiform, conchoid *a.* shell-shaped (like a conch shell).

conchiolin *n.* the protein component of ligament and external layer of mollusc shells.

conchology *n.* branch of zoology dealing with molluscs and their shells.

concolorate, concolorous *a.* (1) similarly coloured on both sides or throughout; (2) of the same colour as a specified structure.

concordant *a. appl.* a trait that is present in both of a pair of identical twins or a pair of siblings. *cf.* discordant.

Concorde effect in animal behaviour, the case where future behaviour is determined by past investment rather than future prospects.

concrescence *n.* the growing together of parts.

concrete *a.* grown together to form a single structure.

condensation *n.* (1) compaction or aggregation of a structure out of a less organized tissue; (2) of chromosomes, the coiling up of the chromatin thread to form a shorter more compact structure that takes place as chromosomes enter mitosis or meiosis; (3) (*biochem.*) type of chemical reaction in which two organic molecules are covalently joined to each other with the elimination of water.

condensed *a.* (*bot.*) *appl.* inflorescence with short-stalked or sessile flowers crowded together.

condensin protein involved in chromosome codensation, possibly by coiling the DNA.

condensing vacuoles large immature secretory vesicles associated with the Golgi apparatus.

conditional *a. appl.* dominance owing to influence of modifying genes.

conditional learning form of learning in which the individual is taught that a particular response to a particular stimulus is appropriate in one circumstance but not in another.

conditional lethal mutation mutation that causes the death of the embryo only under certain conditions, e.g. of temperature or nutritional status.

conditional mutation mutation that only produces an effect under certain conditions, as e.g. of temperature or nutritional status.

conditioned reflex conditioned response. *see* classical conditioning.

conditioned response (CR), conditioned stimulus (CS) *see* classical conditioning.

conditioning *see* classical conditioning, operant conditioning, pseudoconditioning, second-order conditioning.

conductance *n.* a measure of the permeability of a biological membrane or a single ion channel to ions. It is the inverse of electrical resistance or impedance and is measured in siemens (S).

conducting *a.* conveying, *appl.* tissues or structural elements that convey material or a signal from one place to another.

conduction *n.* (1) (*bot.*) transfer of soluble material from one part of a plant to another; (2) (*zool.*) movement of an electrical impulse along a nerve fibre.

conduction aphasia language disorder in which comprehension is unimpaired but the subject is unable to repeat spoken language.

conduplicate *a.* (1) *appl.* cotyledons folded to embrace the radicle; (2) *appl.* leaves in which one half is folded longitudinally upon the other.

condyle *n.* (1) a process on a bone at which it articulates at a joint; (2) rounded structure adapted to fit into a socket; (3) (*bot.*) antheridium of stoneworts. *a.* **condylar.**

condyloid *a.* shaped like, or situated near, a condyle.

cone *n.* (1) cone- or flask-shaped light-sensitive sensory cell in the retina, responsible for colour vision and vision in good light, individual cones being sensitive to blue, green or red wavelengths; (2) (*bot.*) reproductive structure in certain groups of plants. *see* strobilus.

conferted *a.* closely packed, crowded.

conflict *n.* situation in which two motivations compete for dominance in the control of behaviour, as when an animal is deciding which of two objects to approach (or avoid) or whether to approach or run away from an object.

confluence *n.* (1) angle of union of superior sagittal and transverse sinuses at occipital bone; (2) of cell cultures, the point at which cells have formed a continuous sheet over the dish, at which point they usually stop dividing.

confocal microscopy type of fluorescence microscopy that uses a laser to scan across at a given depth in an object, excluding light from other parts of the object, and thus produce a sharp image, or optical section, of the scanned plane.

conformation *n.* three-dimensional arrangement of atoms in a structure.

conformational change in respect of a protein or other macromolecule, any change in the shape or the disposition of the structure in three-dimensional space.

conformer *n.* animal in which features of the internal environment such as temperature and salt concentration of internal fluids are alterable by the external conditions. *cf.* regulator.

congeneric *a.* belonging to the same genus.

congenetic *a.* having the same origin.

congenic *a. appl.* specially bred strains of animals in which a particular allele or block of alleles from one strain is superimposed on the genetic background of another.

congenital *a.* present at birth, *appl.* physiological or morphological defects, not necessarily inherited.

congestin *n.* toxin of sea anemone tentacles.

conglobate *a.* ball-shaped.

conglomerate *a.* bunched or crowded together.

congression *n.* the movement of sister chromatids or paired homologous chromosomes to the equator of the mitotic or meiotic spindle, respectively, during mitosis or meiosis.

coni (1) *plu.* of conus, *appl.* various cone-shaped structures; (2) coni vasculosi: lobules forming head of epididymis.

conidia *plu.* of conidium.

conidial *a. pert.* a conidium.

conidioma *n.* specialized conidium-bearing structure in fungi. *see* acervulus, pycnidium, sporodochium, synnema. *plu.* **conidiomata.**

conidiophore *n.* hypha with sterigmata that bear conidia.

conidiosporangium *n.* conidium which may produce zoospores or germinate directly.

conidium, conidiospore *n.* asexual fungal spore produced by the constriction of a tip of a hypha or sterigma and not enclosed in a sporangium. *plu.* **conidia.**

conifer *n.* cone-bearing tree of the Coniferophyta *q.v.*

Coniferales, Coniferae *n.* order of gymnosperm trees with reproductive organs usually as separate male and female cones, and usually needle-shaped leaves (e.g. pines, larches, cypresses). *alt.* conifers.

Coniferophyta *n.* a division of gymnosperm trees and shrubs, commonly called conifers. One of the five main divisions of extant seed-bearing plants. They have simple, often needle-like leaves, and bear their megasporangia usually in compound strobili (cones). *alt.* **Coniferopsida.**

coniferous *a.* cone-bearing, *pert.* conifers or other cone-bearing plants.

conjoined adnate *q.v.*

conjugate(d) *a.* (1) united, esp. united in pairs; (2) *appl.* protein when it is united with a non-protein.

conjugation *n.* (1) fusion of two gametes; (2) the pairing of chromosomes; (3) in unicellular organisms, the transfer of genetic material from one cell to another in contact with it.

conjugation canal tube formed from fused outgrowths from opposite cells of parallel algal filaments, for passage of male gametes to the other filament.

conjugation tube tube formed between two bacterial cells at conjugation through which the genetic material passes. *alt.* sex pilus.

conjunctiva *n.* delicate mucous membrane lining eyelids, reflected over cornea and sclera and constituting the corneal epithelium.

conjunctive *a. appl.* symbiosis in which the partners are organically connected.

Connarales *n.* order of dicots comprising the family Connaraceae.

connate *a. appl.* similar parts that are united or fused, e.g. petals united into a corolla tube.

connate-perfoliate *a.* joined together at base so as to surround stem, *appl.* opposite sessile leaves.

connective *n.* (1) connecting band of nerve fibres between two ganglia; (2) tissue separating two lobes of anther.

connective tissue supporting tissues of the animal body, including bone, cartilage, adipose tissue and the fibrous tissues supporting and connecting internal organs. It is derived from the embryonic mesoderm.

connexon *n.* unit composed of two protein particles from opposing membranes at a gap junction.

connivent *a.* (1) converging; (2) arching over so as to meet.

conodonts *n.plu.* minute tooth-like fossils abundant from the Paleozoic up to the early Triassic, and thought to be part of the feeding apparatus of an extinct early vertebrate, placed in a separate phylum, Conodonta.

conoid *a.* cone-like, but not quite conical.

conoid tubercle small rough eminence on posterior edge of clavicle, to which the conoid ligament is attached.

conopodium *n.* conical receptacle or thalamus of a flower.

consanguineous *a.* (1) related; (2) *appl.* matings between relatives.

consciousness *n.* (1) awareness of one's actions or intentions, and having a purpose and intention in one's actions; (2) sometimes defined as the presence of mental images and their use by an animal to regulate its behaviour, although the question of whether animals possess consciousness is controversial.

conscutum *n.* dorsal shield formed by united scutum and alloscutum in certain ticks.

consensual *a.* (1) *appl.* involuntary action correlated with voluntary action; (2) relating to excitation of a corresponding organ; (3) *appl.* contraction of both pupils when only one retina is directly stimulated.

consensus sequence the 'ideal' form of a DNA sequence, in which the base present in a given position is the base most often found in comparisons of experimentally determined sequence. *alt.* canonical sequence.

conservation *n.* (1) management of the environment and its natural resources with the aim of protecting it from the damaging effects of human activity; (2) the maintenance of a nucleotide or protein sequence relatively unchanged over evolutionary time and in different species.

conservative *a.* (1) *appl.* characters that change little during evolution; (2) *appl.* taxa retaining many ancestral characters.

conservative recombination breakage and reunion of pre-existing strands of DNA without any synthesis of new stretches of DNA. *see also* site-specific recombination.

conservative substitution the replacement of one amino acid in a protein sequence by another with similar physicochemical properties.

conserved *a.* (1) *appl.* structures, proteins, genes, DNA sequences that are identical or very similar in different organisms; (2) *appl.* structure or function that is identical or very similar in different molecules.

conserved synteny the preservation of large blocks of genes in the same order in different species.

consimilar *a.* (1) similar in all respects; (2) with both sides alike, as some diatoms.

consociation *n.* climax community characterized by a single dominant species.

consocies *n.* consociation representing a stage in the process of plant succession.

consolidation *n.* in memory formation, that stage in which short-term or intermediate-term memory is transferred to long-term memory.

consortes *n.plu.* associate organisms other than symbionts, commensals, or hosts and parasites. *sing.* **consors.**

consortium *n.* kind of symbiosis or assemblage involving two or more species in which all partners gain benefit from each other.

conspecific *a.* belonging to the same species.

consperse *a.* densely scattered, *appl.* dot-like markings, pores.

constancy *n.* in ecology, the frequency with which a particular species occurs in different samples of the same association.

constant *n.* in ecology, a species that occurs in at least 95% of samples taken at random within a community.

constant region (C region) the carboxy-terminal region of an immunoglobulin heavy or light chain, or of the two chains of a T-cell receptor, that has a relatively constant sequence from molecule to molecule. The individual protein domains in the constant region are known as constant domains. *cf.* variable region. *see also* isotype.

constitutive *a.* (1) *appl.* enzymes synthesized by the cell in the absence of any specific stimulus; (2) *appl.* genes which are expressed all the time and do not require a specific external stimulus for initiation of transcription; (3) *appl.* heterochromatin: chromosomal regions that form heterochromatin in all cells. *cf.* inducible.

constricted *a.* narrowed, compressed at regular intervals.

constrictor *n.* muscle that compresses or constricts.

consumer *n.* heterotrophic organism, i.e. one that must consume resources provided by autotrophic organisms. *see also* primary consumer, secondary consumer.

consummatory behaviour in classical ethology, a piece of behaviour, such as drinking, which is considered to be the end result of a physiological need (in this case thirst) which drives a search for a suitable stimulus (in this case water).

consute *a.* with stitch-like markings.

contact guidance guidance of developing neurons in development by contact with molecules of the underlying extracellular matrix.

contact hypersensitivity type of allergic response caused by direct contact with certain substances and producing severe inflammation at site of contact.

contact inhibition cessation of cell growth and division in cultured cells when they come into contact with each other.

contact receptor sensory receptor in dermis or epidermis.

contact sensitivity contact hypersensitivity *q.v.*

contagious *n.* transmissible, *appl.* disease.

contest competition type of competition in which the successful competitor gains sufficient resources for survival and reproduction whereas the unsuccessful competitors gain nothing or insufficient for survival.

context *n.* (*mycol.*) fibrous tissue between hymenium and true mycelium in the cap of some basidiomycete fungi.

contig *n.* a set of overlapping DNA clones that have been put in order so that they represent the complete DNA sequence of a region of a chromosome.

contiguous *a.* touching each other at the edges but not actually united.

contiguous gene syndrome clinical syndrome with a complex pathology associated with the deletion of a set of adjacent genes.

continental *a. appl.* climate characterized by some or all of the features of weather associated with continental interiors, namely hot summers and cold winters with a wide temperature range between extremes, short spring and autumn seasons, and marked rainy and dry seasons.

continental drift the movement of the continents over the surface of the Earth.

Continental Southeast Asiatic region floristic region consisting of China south of 30°N and the remainder of South-East Asia except the Malaysian peninsula.

continuous culture method of growing microbial cultures so that nutrients and space do not become exhausted and the culture is always in the rapidly multiplying phase of growth. *cf.* batch culture.

continuous variation variation in a character among individuals of a population in which differences are slight and grade into each other. Continuously variable phenotypic characters are determined by a number of different genes and/or considerable environmental input.

continuum *n.* (*bot.*) form of vegetation cover in which one type passes almost imperceptibly into another and no two types are repeated exactly.

contorted *a. appl.* arrangement of floral parts in which one petal overlaps the next with one margin and is overlapped by the previous on the other.

contortuplicate *a. appl.* bud with contorted and folded leaves.

contour *n.* outline of a figure or body, *appl.* outermost feathers that cover the body of a bird.

contractile *a.* capable of contracting.

contractile bundle small transient assemblages of myosin and actin present in many types of animal cell. Their contraction is involved in cell movement and changes in cell shape.

contractile fibre cells elongated spindle-shaped more-or-less polyhedral muscle cells, containing a central bundle of myofibrils.

contractile ring bundles of actin filaments immediately beneath the plasma membrane which are involved in forming the constriction that separates the two new cells in animal cell division.

contractile vacuole small spherical vesicle found in the cytoplasm of many freshwater protozoans, by which surplus water is expelled.

contractility *n.* (1) power by which muscle fibres are able to contract; (2) the capacity to change shape.

contracture *n.* contraction of muscles persisting after stimulus has been removed.

contra-deciduate *a. appl.* foetal placenta and distal part of allantois that are absorbed by maternal tissues at birth.

contralateral *a. pert.* or situated on the opposite side. *cf.* ipsilateral.

contranatant *a.* swimming or migrating against the current.

control *n.* an experiment or test carried out to provide a standard against which experimental results can be evaluated.

controlling element DNA sequence in maize which can become inserted at various sites on the chromosomes during somatic cell division, and may influence the expression of nearby genes. *alt.* transposable genetic element.

control region that region of a gene involved in controlling its expression. It does not encode an amino acid sequence but contains nucleotide sequences to which regulatory proteins bind.

control sequence short nucleotide sequence in the control region of a gene at which a gene-regulatory protein or protein complex binds specifically.

conuli *n.plu.* tent-like projections on surface of some sponges, caused by principal skeletal elements.

conus *n.* (1) any cone-shaped structure; (2) diverticulum of right ventricle from which pulmonary artery arises.

conus arteriosus funnel-shaped structure between ventricle and aorta in fishes and amphibians.

conus medullaris tapering end of the spinal cord.

conventional behaviour any behaviour by which members of a population reveal their presence and allow another organism to assess their numbers.

conventional dendritic cell *see* dendritic cell.

convergence *n.* coordinated movement of eyes when focusing on a near point.

convergent evolution similarity between two organisms, structures or molecules due to independent evolution along similar lines rather than descent from a common ancestor.

convergent extension in development, a change in shape of a sheet of cells in which the sheet extends in one direction and narrows (converges) in the direction at right angles to the extension. It is due to cell movements within the sheet.

convolute *a.* (1) rolled together, *appl.* leaves and cotyledons; (2) *appl.* shells in which outer whorls overlap inner; (3) coiled, *appl.* parts of renal tubule.

convolution *n.* a coiling or twisting.

Coombs test test used to determine the presence of anti-Rhesus blood group antibodies in a Rhesus-negative mother. If present, these antibodies can attack the blood cells of a Rhesus-positive foetus, causing haemolytic disease of the newborn.

cooperation *n.* in animal behaviour, the sharing of a task between different animals, as in hunting by a wild dog pack, the sharing of infant care by relatives of the mother in chimpanzees, feeding of nestlings by other members of the community in some communal birds. *see also* symbiosis.

cooperative binding, cooperativity case where binding of a ligand to one site changes the affinity of another site(s) for either the same or another ligand. Positive cooperativity results in an increase in the affinity of the second site for its ligand, making binding easier. Negative cooperativity is when the affinity of the second site is lowered, making binding more difficult. *see also* allostery.

cooperative breeding breeding system in which parents are assisted in the care of their offspring by other adults.

coordinate regulation of gene expression, the expression of a particular set of genes at the same time.

COP coat protein, the protein that forms the coat on some coated vesicles.

copal *n.* resin exuding from various tropical trees and hardening to a colourless, yellow, red or brown mass. Used in varnishes.

Copepoda, copepods *n., n.plu.* subclass of free-living or parasitic small crustaceans that form a large part of the marine zooplankton and are also found in fresh water. They have no carapace and have one median eye in the adult.

copepodid *n.* juvenile stage after the nauplius stage in copepods.

coppicing *n.* woodland management practice in which trees such as willow, hazel and sweet chestnut are regularly cut back almost to the ground every few years so that they develop a low 'stool' from which many long straight shoots arise. *cf.* pollarding.

coprodaeum *n.* the part of the cloaca that receives the rectum.

coprolite, coprolith *n.* fossilized faeces.

coprophage *n.* animal that feeds on dung. *a.* **coprophagous**.

coprophagy *n.* (1) habitual feeding on dung; (2) reingestion of faeces.

coprophil, -ic, -ous *a.* growing in or on dung.

coprophyte *n.* plant that grows on dung.

coprosterol *n.* sterol produced by bacterial reduction of cholesterol, found in faeces.

coprozoic *a.* living in faeces, as some protozoans.

coprozoite *n.* any animal that lives in or feeds on dung.

copula *n.* (1) basihyal in certain reptiles; (2) fused basibranchial and basihyal in birds; (3) any bridging or connecting structure.

copularium *n.* cyst formed around two associated gametocytes in gregarines.

copulation *n.* (1) sexual union; (2) in protozoans, complete fusion of two individuals.

copy number the number of plasmids of a particular type that can accumulate in a bacterial cell.

CoQ coenzyme Q, ubiquinone *q.v.*

coracidium *n.* ciliated embryo of certain cestodes (tapeworms), developing into a procercoid within first intermediate host.

Coraciformes *n.* order of birds including the kingfishers, rollers and hornbills.

coracoid *a.* (1) *appl.* or *pert.* bone or part of the pectoral girdle between shoulder-blade and breastbone; (2) *appl.* ligament which stretches over the suprascapular notch.

coracoid process rudimentary coracoid bone fused to the shoulder-blade in most mammals.

coral *n.* common name for a wide variety of colonial cnidarians, which comprise colonies of individual polyps joined together by living tissue. Many species live within a calcium carbonate exoskeleton (the stony or true corals) and can form extensive underwater reefs. *see* Octocorallia, Hexacorallia.

coralliferous *a.* coral-forming or containing coral.

coralliform *a.* resembling a coral or branching like a coral.

coralligenous *a.* coral-forming.

coralline *a.* (1) resembling a coral, *appl.* to some lime-encrusted red algae; (2) composed of or containing coral; (3) *appl.* zone of coastal waters at about 30–100 m.

corallite *n.* cup of an individual coral polyp.

coralloid *a.* (1) resembling or branching like a coral; (2) *appl.* negatively geotropic roots of cycads (gymnosperms of the family Cycadaceae) which arise from the swollen hypocotyl and taproot, often deep in the soil, and are infected with nitrogen-fixing cyanobacteria.

corallum *n.* skeleton of a compound coral.

corbiculum *n.* the pollen basket on the hind legs of many bees, formed by stout hairs on the borders of the tibia.

Cordaitales *n.* order of fossil conifers, being mostly tall trees with slender trunks and a crown of branches, with spirally arranged simple grass-like or paddle-like leaves and with mega- and microsporangia in compound strobili.

cordate, cordiform *a.* heart-shaped.

cordycepin *n.* 3'-deoxyadenosine, a nucleotide analogue that inhibits polyadenylation of eukaryotic RNA.

core area that part of the home range of an animal in which it spends most of its time.

co-receptor *n.* a cell-surface receptor whose activity enhances the effect of another receptor.

core enzyme enzyme containing a functional catalytic site but lacking one or more of the polypeptide subunits necessary for all its normal regulated functions.

coregonid *n.* freshwater fish of the genus *Coregonus*, e.g. houting, powan and vendace.

coremata *n.plu.* accessory copulatory organ in moths, composed of paired sacs bearing hairs, located between 7th and 8th abdominal segments. *sing.* **corema**.

coremiform *a.* shaped like a broom or a sheaf.

coremiospore *n.* one of the series of spores on top of a coremium.

coremium *n.* a sheaf-like aggregation of conidiophores or hyphae.

co-repressor *n.* (1) in bacteria, a small molecule that prevents the expression of the operon specifying the enzymes involved in synthesizing the molecule. It acts by binding to a gene-regulatory protein to cause this to bind to DNA and block the initiation of transcription; (2) in eukaryotes, regulatory protein that binds to a DNA-bound gene regulatory protein to suppress gene expression.

core temperature temperature of an animal's body measured at or near the centre.

Cori cycle conversion into glucose in the liver of the lactate formed by glycolysis in contracting muscle. The glucose is then transported to muscle in the blood.

coriaceous, corious *a.* leathery, *appl.* leaves.

corium *n.* (1) dermis *q.v.*; (2) central division of an insect elytron (wing-case); (3) the main part of the front wing of a hemipteran bug.

cork *n.* external layer of plant tissue composed of dead cells filled with suberin, forming a seal impermeable to water. Present on woody stems and derived from the cork cambium. *alt.* phellem.

cork cambium secondary meristematic layer in woody stems that gives rise to the corky layer of the bark.

corm *n.* enlarged, solid underground stem, rounded in shape, composed of two or more internodes, and covered externally with a few thin scales or leaves.

cormel *n.* secondary corm produced by an old corm.

cormidium *n.* group of individuals of a siphonophore colony that can separate from the colony and live a separate existence.

cormoid *a.* like a corm.

cormophyte *n.* any plant differentiated into roots, shoots and leaves and adapted for life on land.

cormous *a.* producing corms.

cormus *n.* (1) body of a plant that is developed into root and shoot systems; (2) body or colony of a compound animal.

corn *n.* wheat (UK), maize (US).

Cornales *n.* order of dicot trees, shrubs and herbs, with leaves often much divided, and including the families Umbelliferaceae (carrot, etc.), Araliaceae (ginseng), Cornaceae (dogwood), Davidiaceae (dove tree) and others.

cornea *n.* transparent layer covering the front of the eye or of each element of a compound eye.

corneosclerotic *a. pert.* cornea and sclera.

corneoscute *n.* an epidermal scale.

corneous *a.* horny, *appl.* sheath covering bill of birds.

cornicle *n.* (1) wax-secreting organ of aphids; (2) any small horn or horn-like process. *alt.* corniculum, cornule.

corniculate *a.* having little horns.

corniculate cartilages two small conical elastic cartilages articulating with apices of arytaenoids in larynx.

corniculum *see* cornicle.

cornification *a.* formation of outer horny layer of epidermis. *alt.* keratinization.

cornified *a.* (1) keratinized, *appl.* epithelium; (2) transformed into horn.

cornified layer outer keratinized layer of vertebrate epidermis, consisting of flattened dead cells filled with keratin, which flake off the surface. *alt.* stratum corneum. *see* Fig. 35 (p. 622).

cornifying keratinizing *q.v.*

cornua *n.plu.* (1) horns; (2) horn-like prolongations, as of bones, nerve tissues, cavities; (3) the dorsal, lateral and ventral columns of grey matter in the spinal cord. *sing.* **cornu**.

cornule *see* cornicle.

cornute *a.* with horn-like processes.

corolla *n.* the petals of a flower collectively. *a.* **corollaceous** having petals.

corolliferous *a.* having a corolla.

corona *n.* (1) (*bot.*) frill at mouth of corolla tube formed by union of scales on petals, as in the trumpet of a daffodil; (*zool.*) (2) the theca and arms of a crinoid; (3) ciliated disc or circular band of certain animals such as rotifers; (4) the head or upper portion of any structure.

coronal *a.* (1) *pert.* a corona; (2) *appl.* suture between frontal and parietal bones of skull; (3) *appl.* plane of section through brain, being a vertical section at right-angles to the long axis.

corona radiata layer of cells surrounding mammalian egg.

coronary *a.* (1) crown-shaped or crown-like; (2) encircling; (3) *n.* coronary bone *q.v.*

coronary arteries/veins arteries/veins supplying tissue of heart.

coronary bone (1) small conical bone in mandible of reptiles; (2) small pastern bone in horses.

coronary sinus channel receiving the output of most of the cardiac veins and opening into the right auricle of the heart.

coronary vessels coronary arteries/veins *q.v.*

coronate *a.* (1) having a corona; (2) having a row of tubercles encircling a structure, or mounted on whorls of spiral shells.

Coronaviridae, coronaviruses *n., n.plu.* family of medium-sized, enveloped, single-stranded RNA viruses covered with protein petal-like projections, and which include avian infectious bronchitis virus and the SARS (severe acute respiratory syndrome) virus.

coronoid *a.* shaped like a beak.

coronula *n.* (1) group of cells forming a crown on oosphere, as in green algae of the Charophyceae; (2) circle of pointed processes around frustule of certain diatoms.

corpora *plu.* of corpus.

corpora albicantia white bodies or scars formed in ovarian follicle after disintegration of luteal cells.

corpora allata endocrine glands in insects situated just behind the brain, which secrete juvenile hormone. They may be paired ovoid whitish structures or may fuse during development to form a single median structure, the corpus allatum.

corpora amylacea spherical bodies composed of nucleic acid and protein, present in alveoli of prostate gland and becoming more numerous with age.

corpora bigemina optic lobes of vertebrate brain, corresponding to the superior colliculi of corpora quadrigemina in mammals.

corpora cardiaca neurosecretory bodies between cerebral ganglia and corpora allata, in some insects.

corpora cavernosa (1) erectile mass of tissue forming anterior part of penis; (2) erectile tissue of clitoris.

corpora mamillaria *see* mamillary bodies.

corpora quadrigemina four rounded eminences which form the dorsal part of the midbrain in mammals.

corpus *n.* (1) body; (2) any fairly homogeneous structure which forms part of an organ. *plu.* **corpora**. *a.* **corporal**.

corpus allatum *see* corpora allata.

corpus callosum wide band of axons connecting the two cerebral hemispheres and involved in the transfer of information from one hemisphere to the other.

corpuscle *n.* (1) cell *q.v.*, esp. red blood cell; (2) any of various small multicellular structures, e.g. Malpighian corpuscle.

corpuscles of Ruffini Ruffini's organs *q.v.*

corpuscular *a.* compact or globular.

corpus luteum glandular mass of yellow tissue that develops from a Graafian follicle in the mammalian ovary after the ovum is extruded, and which secretes progesterone.

corpus spongiosum mass of erectile tissue forming posterior wall of penis.

corpus striatum in cerebral hemisphere of brain, mass of grey matter containing white nerve fibres and consisting of the caudate nucleus and putamen, and situated on the outer side of each lateral ventricle.

correlation centres regions in the brain where information from various sense organs is integrated and the resultant response determined.

correlation coefficient *r*, statistical measure of the likelihood that the value of one variable is dependent on the value of another.

corrin *n.* cobalt atom surrounded by four pyrrole units (corrin ring).

corrugator *n.* muscle that causes wrinkling.

cortex *n.* (1) cerebral cortex *q.v.*; (2) outer layer of a structure or organ; (3) in vascular plants, the tissue in stem and root surrounding but not part of the vascular tissue; (4) in eukaryotic cells, the region just under the cell membrane; (5) in bacterial spores, the layer between spore wall and the outer spore coat. *plu.* **cortices.**

cortical *a. pert.* the cortex.

cortical barrel barrel-shaped portion of somatosensory cortex that receives input from a single whisker, in e.g. rats.

cortical column one of the vertical columns of neurons that make up the basic organization of the neocortex.

cortical granules in ovum, granules lying just beneath the plasma membrane which release their contents when a sperm enters the egg, changing the composition of the outer coat of the egg and preventing entry of other sperm.

cortical rotation phenomenon seen in amphibian eggs immediately after fertilization in which the egg cortex rotates with respect to the underlying cytoplasm.

corticate *a.* having a special outer covering.

cortices *plu.* of cortex.

corticiferous *a.* forming or having a bark-like cortex.

corticoid corticosteroid *q.v.*

corticolous *a.* inhabiting, or growing on, bark.

corticospinal *a. pert.* or connecting cerebral cortex and spinal cord.

corticospinal system pyramidal system *q.v.*

corticosteroid *n.* any of a group of steroids secreted by the cortex of the adrenal glands, some of which are hormones. They include the glucocorticoids (*q.v.*) and the mineralocorticoids (*q.v.*). *see also* corticosterone, cortisone, cortisol, aldosterone. *a.* **corticosteroid.**

corticosterone *n.* steroid hormone produced by the cortex of the adrenal glands in response to adrenocorticotropic hormone, and which has effects on lipid, carbohydrate and protein metabolism.

corticostriate *a. appl.* nerve fibres that join corpus striatum to cerebral cortex.

corticotrop(h)ic adrenocorticotrop(h)ic *q.v.*

corticotropin adrenocorticotropic hormone (ACTH) *q.v.*

corticotropin-releasing factor (CRF) small peptide secreted by the hypothalamus which stimulates the release of adrenocorticotropic hormone (ACTH) from the anterior pituitary. *alt.* **corticotropin-releasing hormone (CRH).**

cortina *n.* the veil in some agaric fungi.

cortinate *a.* (1) of a cobwebby texture; (2) having a veil, of mushrooms and toadstools, *alt.* craspedote.

cortisol hydrocortisone *q.v.*

cortisone *n.* steroid hormone secreted in small amounts by the human adrenal cortex, and which has many effects, including suppression of inflammation and the promotion of carbohydrate formation. Drugs based on cortisone are used as anti-inflammatory drugs.

Corti's organ, Corti's membrane, Corti's rods *see* organ of Corti.

coruscation *n.* twinkle, rapid fluctuation in a flash or oscillation in light emission, as of fireflies.

corvine *a. appl.* birds of the Corvidae or crow family.

corymb *n.* inflorescence in the form of a raceme with lower flower stalks elongated so that the top is nearly flat. *a.* **corymbose.**

corymbose cyme flat-topped cyme which therefore resembles a corymb in appearance but not in mode of development.

corynebacteria *n.plu.* bacteria of the family Corynebacteriaceae, which are characterized by irregularly shaped cells dividing by 'snapping fission', and which include the human pathogen *Corynebacterium diphtheriae*, the cause of diphtheria. *sing.* **corynebacterium.**

coryneform *a. appl.* bacteria characterized by irregular, often club-shaped, Gram-positive cells.

coscinoid *a.* sieve-like.

cosere *n.* series of plant successions on the same site.

cosmid *n.* type of cloning vector consisting of a bacterial plasmid into which the *cos* sequences of phage lambda have been inserted. This both allows the vector to grow as a plasmid in bacterial cells and enables subsequent purification of vector DNA by packaging into phage particles *in vitro*.

cosmine *n.* type of dentine with tiny branched canals, found in the scales of some fishes.

cosmoid scale type of scale found in typical crossopterygian fishes. It has an outer layer of enamel, a layer of cosmine (a type of dentine), and then bone, growth in thickness being by addition of inner layers only. *cf.* ganoid scales.

cosmopolitan *a.* world-wide in distribution.

cosmopolite *n.* species with a worldwide distribution.

co-speciation *n.* the formation of new species in one species in response to the speciation of another.

cost *n.* the decrement in an animal's inclusive fitness that results from a particular behaviour. *cf.* benefit.

costa *n.* (1) rib; (2) anything rib-like in shape, as a ridge on a shell; (3) anterior vein or margin of insect wing; (4) swimming plate of sea gooseberries; (5) structure at base of undulating membrane in certain protozoans. *plu.* **costae.**

costaeform *a.* (1) rib-like; (2) *appl.* unbranched parallel leaf veins.

costal *a.* (1) *pert.* ribs or rib-like structures; (2) *appl.* bony shields of turtles and tortoises; (3) *pert.* costa of insect wing.

costalia *n.plu.* the supporting plates in test of echinoderms.

costate *a.* (1) with one or more longitudinal ribs; (2) with ridges or costae.

cost function the combination of the various costs of an animal's behaviour, which is used to evaluate all aspects of the animal's state and behaviour.

co-stimulatory signal in immunology, a stimulatory signal required for activation of a naive T cell, in addition to the binding of antigen. Such signals are delivered by professional antigen-presenting cells.

Cot a measure of reassociation of double-stranded DNA from single strands. It is the product of single-stranded DNA concentration at time 0 (C_0) and the time of incubation (t), $Cot_{1/2}$ being the value when reassociation is half complete. The larger and more complex the genome, the greater the Cot value.

coterie *n.* social group which defends a common territory against other coteries.

co-terminous *a.* (1) of similar distribution; (2) bordering on; (3) having a common boundary.

co-transduction *n.* transduction of two different genes in a single event.

co-transfection *n.* simultaneous transfection of a cell with two different genes.

co-transformation *n.* (1) the neoplastic transformation of cells by the simultaneous introduction of two different proteins, genes or transforming viruses, neither of which can completely transform the cells on their own; (2) the simultaneous transformation of a bacterial cell by two different genes.

co-translational transfer, co-translational translocation process in eukaryotic cells in which proteins destined for cell membranes or for secretion start their passage across the membrane of the endoplasmic reticulum before translation is complete.

co-transport membrane transport in which the transport of one solute depends on the simultaneous transport of another.

cotyle acetabulum *q.v.*

cotyledon *n.* (1) (*bot.*) part of plant embryo in which food is stored and which may form a leaf or be left below ground when seed germinates; (2) (*zool.*) patch of villi on mammalian placenta. *a.* **cotyledonary.**

cotyloid, cotyliform *a.* cup-shaped.

cotylophorous *a.* with a cotyledonary placenta.

cotylosaurs Captorhinida *q.v.*

cotype syntype *q.v.*

coumarin *n.* substance found in many plants, esp. clover, having an odour of new-mown hay. It is used in perfumery and to make the anticoagulant dicoumarin (dicumarol).

counteracting selection the operation of selection pressures on two or more levels of organization, e.g. individual, family or population, in such a way that certain genes are favoured at one level but disfavoured at another. *cf.* reinforcing selection.

counterevolution *n.* the evolution of traits in one population in response to adverse interactions with another population, as between prey and predator.

countershading *n.* condition of an animal being dark dorsally and pale ventrally. When lighting is from above, the ventral shadow is obscured and the animal appears evenly coloured and inconspicuous.

countertranscript *n.* RNA that exerts some function because of its partial complementarity to and presumed base pairing with another RNA coded in the same region of DNA.

coupled reaction chemical reaction in which an energy-requiring reaction is linked to an energy-releasing reaction.

coupled transport transport of a solute across a cellular membrane in which the transport of one solute against its concentration gradient is coupled to the transport of another down its concentration gradient. *see* antiport, symport.

coupling factor *see* ATP synthases.

court *n.* a small area of a lek used by an individual male for display.

courtship *n.* behaviour pattern preceding mating in animals, often elaborate and ritualized.

covalent bond chemical bond formed by the sharing of electrons between two atoms. *cf.* non-covalent bond.

covariance *n.* statistical measure used in calculating the correlation coefficient between two variables. For two variables x and y the covariance (cov) is related to the variance (var) by the following expression: $\text{var}(x + y) = \text{var}(x) + (\text{var})y + 2\text{cov}(xy)$.

coverts *n.plu.* feathers covering bases of quills.

Cowper's glands bulbo-urethral glands *q.v.*

coxa *n.* joint of leg nearest body in insects, arachnids and some other arthropods. *a.* **coxal**.

coxite *n.* one of paired lateral plates of exoskeleton next to insect sternum.

coxocerite *n.* basal joint of an insect antenna.

coxopleurite catapleurite *q.v.*

coxopodite *n.* the joint nearest the body in crustacean limbs.

Coxsackie virus a picornavirus which multiplies chiefly in the intestinal tract, but which can cause aseptic meningitis and other conditions.

cpDNA chloroplast DNA *q.v.*

CPE cytopathic effect *q.v.*

CpG island cluster of CG sequences present in the 5′ control regions of many constitutively expressed genes. Methylation of the C in such sequences is associated with inactivation of the gene. *see also* DNA methylation.

cpm, c.p.m. counts per minute, a measure of radioactivity.

CR conditioned response. *see* classical conditioning.

CR1, CR2, CR3, CR4 cell-surface receptors for complement components. They are present on phagocytes, such as macrophages, and facilitate the phagocytosis of complement-coated microorganisms.

crampon *n.* an aerial root, as in ivy.

craniad *a.* towards the head.

cranial *a. pert.* skull, or that part which encloses the brain.

cranial nerves nerves arising from the brain, and concerned mainly with sensory and motor systems associated with the head. Cranial nerves are one of the three main subdivisions of the peripheral nervous system.

Craniata, craniates Vertebrata *q.v.*

cranihaemal *a. appl.* anterior lower portion of a sclerotome.

cranineural *a. appl.* anterior upper portion of a sclerotome.

craniology *n.* study of the skull.

craniometry *n.* science of the measurement of skulls.

craniosacral *a. appl.* nerves, the parasympathetic system.

cranium *n.* the skull, more particularly that part enclosing the brain.

craspedium *n.* a bordered seed-pod.

craspedote *a. appl.* fungi having a veil or cortina.

Crassulacean acid metabolism (CAM) metabolic pathway in some succulent plants (Crassulaceae and Cactaceae) in which carbon dioxide is first fixed non-photosynthetically into carboxylic acids (chiefly malate) which are then mobilized in the light and used as CO_2 sources for the Calvin cycle.

crateriform *a.* bowl-shaped.

craticular *a.* bowl-shaped, *appl.* stage in life history of diatom where new valves are formed before old ones are lost.

CRE cyclic AMP response element. *see* cyclic AMP response element binding protein.

C-reactive protein blood protein, synthesized in the liver, which is involved in general resistance to bacterial infection.

creatine *n.* amino acid present in vertebrate muscle, which as creatine phosphate is a high-energy compound donating phosphoryl groups to ADP during muscle contraction to recover ATP.

creatinine *n.* compound formed from creatine by dehydration, present in muscles, blood and urine.

Creationism *n.* doctrine that the different types of living organisms have all arisen independently by divine creation. It has now been superseded as a serious doctrine in biology by the overwhelming evidence for evolution.

CREB cyclic AMP response element binding protein *q.v.*

C region constant region *q.v.*

Cre/lox system a technique for producing targeted gene deletions in experimental animals such as the mouse, involving the recombinase Cre and its target sequences *loxP*. Cre will cut out of the genome any sequence flanked by two *loxP* sequences. The mice in which the mutation is to be made carry two *loxP* sequences in the desired positions and also a gene encoding Cre, which can be activated when wished by the experimenter.

cremaster *n.* (1) cluster of hooks on the hind end of butterfly pupae, from which it is suspended; (2) abdominal spine in subterranean insect pupae; (3) thin muscle along the spermatic cord.

cremocarp *n.* fruit composed of two bilocular carpels, as in umbellifers, in which the two carpels separate into one-seeded indehiscent capsules which remain attached to the supporting axis before dispersal.

crena *n.* (1) cleft, as anal cleft; (2) deep groove, such as the longitudinal groove of heart.

Crenarchaeota *n.* kingdom within the Archaea, defined on the basis of DNA sequence data, which contains sulphur-reducing hyperthermophiles, e.g. *Sulfolobus* and *Thermococcus* and the methanogen *Methanopyrus. cf.* Euryarchaeota.

crenate *a.* with scalloped margin.

crenation *n.* (1) scalloped margin or rounded tooth, as of leaf; (2) notched or wrinkled appearance.

crenulate *a.* with minutely scalloped margin.

creodonts *n.plu.* group of extinct placental mammals of the Cretaceous and Pliocene, which were probably archaic carnivores.

crepitaculum *n.* (1) stridulating organ as in some Orthoptera; (2) rattle in rattlesnake's tail.

crepitation *n.* in insects, the discharge of fluid with an explosive sound for defence.

crepuscular *a. pert.* dusk, *appl.* animals flying before sunrise or at twilight. *alt.* vespertine.

crescentic, crescentiform *a.* crescent-shaped.

Cretaceous *n.* last period of the Mesozoic era, from *ca.* 146 million years ago to 65 million years ago, during which chalk was being laid down, occurring after the Jurassic period and before the Tertiary sub-era. Comprises two epochs, Lower Cretaceous (146–99 million years ago) and Upper Cretaceous (99–65 million years ago). The end of the Cretaceous saw the extinction of the dinosaurs and many other groups. *see* K–T boundary, mass extinction event.

Creutzfeldt–Jakob disease (CJD) a rare fatal neurodegenerative disease of humans, a prion-caused spongiform encephalopathy. A variant form of the disease, vCJD, with younger onset and more rapid progression, is thought to be caused by the consumption of material from cows infected with bovine spongiform encephalopathy (BSE).

CRF, CRH corticotropin-releasing factor *q.v.*

cribellum *n.* (1) in spiders, a plate perforated by openings of silk ducts; (2) in insects, a perforated chitinous plate.

cribrellate *a.* having many pores, as certain spores.

cribriform *a.* sieve-like.

cribriform plate portion of ethmoid or mesethmoid bone perforated by many foramina for exit of olfactory nerves.

cribrose *a.* having sieve-like pitted markings.

cricket *n.* common name for many members of the Orthoptera *q.v.*

crico-thyroid *a. pert.* cricoid and thyroid cartilages, *appl.* tensor muscle of vocal chord.

cricoid *a.* (1) ring-like; (2) *appl.* cartilage in larynx, articulating with thyroid and arytaenoid cartilages; (3) *appl.* placenta lacking villi on central part of the disc, as in certain Edentata.

crinite *a.* with hairy or hair-like structures or tufts. *n.* a fossil crinoid.

Crinoidea, crinoids *n., n.plu.* a class of echinoderms, commonly called sea lilies and feather stars, present since the Cambrian. They have a cup-shaped body with feathery arms, and are attached to the substratum by a stalk in the case of sea lilies.

crinose *a.* with long hairs.

crisped, crispate *a.* curled, frizzled.

crissal *a. pert.* the crissum.

criss-cross inheritance pattern of inheritance of an X-linked gene, i.e. from male parent to female child to male grandchild.

crissum *n.* (1) the region around the cloaca in birds; (2) vent feathers or lower tail coverts. *a.* **crissal**.

crista *n.* (1) crest or ridge; (2) a single fold of the inner membrane in mitochondria. *plu.* **cristae**.

cristate *a.* (1) crested; (2) shaped like a crest.

crithidial *a. appl.* long slender form of trypanosome with a partial undulating membrane and kinetoplast anterior to the nucleus.

critical frequency (1) maximum frequency of successive stimuli at which they can produce separate sensations; (2) minimum frequency for a continuous sensation.

critical group taxonomic group that cannot be divided into smaller groups, such as an apomictic species.

critical links organisms in a food chain that are responsible for primary energy capture and nutrient assimilation. They are critical in the transformation of nutrients into forms that can be used by organisms at higher trophic levels in the chain.

critical period that period during the development of any organ when its development may be disturbed, e.g. by drugs or viral infection.

cRNA complementary RNA *q.v.*

crochet *n.* (1) larval hook used in locomotion in insects; (2) terminal claw of chelicerae in arachnids.

Crocodylia, crocodilians *n., n.plu.* order of reptiles found first in the Triassic, typified by the present-day crocodiles and alligators, which are armoured, have front limbs shorter than hind and a body elongated for swimming.

crop *n.* (*zool.*) (1) sac-like dilation of gullet of bird in which food is stored; (2) similar structure in alimentary canal in insects.

crop-milk secretion of epithelium of crop in pigeons, stimulated by prolactin, for nourishment of nestlings.

crosier crozier *q.v.*

cross (1) *n.* organism produced by mating parents of different genotypes, strains or breeds; (2) *v.* to hybridize.

cross-fertilization the fusion of male and female gametes from different individuals, especially of different genotypes. *alt.* allogamy, allomixis.

crossing-over exchange of genetic material between homologous chromosomes at DNA recombination during meiosis. The structure formed during recombination is called a chiasma or crossover and is visible under the light microscope. *see also* unequal crossing-over.

Crossopterygii, crossopterygians *n., n.plu.* group of mainly extinct bony fishes, of which the coelacanth (*Latimeria*) is the only living member, also called tassel-finned or lobe-finned fishes.

crossover chiasma *q.v. see also* crossing-over.

crossover fixation model proposed mechanism for maintenance of fidelity of multiple repeated DNA sequences (such as the rRNA genes, and some satellite DNAs in arthropods) in which the entire cluster is continually rearranged by the mechanism of unequal crossing-over during recombination.

cross-pollination transfer of pollen from the anther of a flower on one plant to the stigma of a flower on another plant.

cross-presentation (*immunol.*) the presentation of an extracellularly derived antigen by MHC class I molecules (which normally present intracellularly derived antigens). *see also* antigen presentation.

cross-priming (*immunol.*) the activation of naive CD8 T cells by antigen presented to them by cross-presentation.

cross-reacting (1) *appl.* antibodies that react with more than one antigen; (2) *appl.* antigen that interacts with antibodies that were raised against a different antigen.

cross-reflex the reaction of an effector on one side of the body to stimulation of a receptor on the other side.

crotchet *n.* a curved bristle, notched at the end.

crown *n.* (1) crest; (2) top of head; (3) the exposed part of a tooth, especially the grinding surface; (4) head, cup and arms of a crinoid; (*bot.*) (5) leafy upper part of a tree; (6) short rootstock with leaves.

crown gall disease plant tumour caused by the bacterium *Agrobacterium tumefaciens*.

crozier *n.* (1) (*bot.*) the coiled young frond of a fern; (2) (*mycol.*) hook formed by the terminal cells of ascogenous hyphae.

CRP cyclic AMP receptor protein, catabolite gene activator protein *q.v.*

crucial, cruciate *a.* in the form of a cross.

crucifer *n.* member of a large family of dicotyledons, the Cruciferae, whose flowers have four petals in the form of a cross, e.g. cabbage, wallflower, mustard, cress. *a.* **cruciferous**.

cruciform *a.* (1) arranged like the points of a cross; (2) in the shape of a cross.

crumena *n.* a sheath for retracted stylets in Hemiptera.

cruor *n.* the clots in coagulated blood.

crura *n.plu.* columnar structures, in various organs and organisms. *sing.* **crus**.

crura cerebri cerebral peduncles, two cylindrical masses forming the sides and floor of the midbrain.

crural *a. pert.* the thigh. *alt.* femoral.

crureus *n.* parts of the quadriceps muscle in thigh.

crus *sing.* of crura.

Crustacea, crustaceans *n.*, *n.plu.* subphylum of arthropods, considered as a class in older classifications. They are mainly aquatic, gill-breathing animals, such as crabs, lobsters and shrimps. The body is divided into a head bearing five pairs of appendages (two pairs of pre-oral sensory feelers and three pairs of postoral feeding appendages) and a trunk and abdomen bearing a variable number of often biramous appendages which serve as walking legs and gills. Crustacea often have a hard carapace or shell. *see* Appendix 3.

crustaceous *a.* (1) with characteristics of a crustacean; (2) thin or brittle; (3) forming a thin crust.

crustose *a.* forming a thin crust on the substratum, *appl.* certain lichens and sponges.

cryobiology *n.* study of the freezing of cells and tissues.

cryo-electron microscopy electron microscopy carried out on unstained thin frozen films of material, used to investigate the structure of large molecules and molecular complexes.

cryoenzymology *n.* study of properties of enzymes at very low temperatures (e.g. −50°C) at which the catalytic process is slowed down.

cryoglobulin *n.* any immunoglobulin that spontaneously precipitates, gels or crystallizes when cooled. Present in small amounts in normal human blood serum, and in larger amounts in some diseases.

cryoglobulinaemia *n.* disease caused by accumulation of immune complexes of cryoglobulins.

cryophil(ic) *a.* thriving at low temperature.

cryophylactic *a.* resistant to low temperature, *appl.* bacteria.

cryophyte *n.* plant, alga, bacterium or fungus that lives in snow and ice.

cryoplankton *n.* plankton found around glaciers and in the polar regions.

crypsis *n.* camouflage that makes the organism resemble part of the environment.

crypt *n.* (1) simple glandular tube or cavity, e.g. the narrow epithelium-lined cavities in the wall of the small intestine; (2) (*bot.*) pit of stoma.

cryptdins *n.plu.* antibacterial proteins of the defensin family that are secreted by the Paneth cells of the gut epithelium.

cryptic *n.* hidden; (1) *appl.* protective colouring making concealment easier; (2) *appl.* genetic variation due to the presence of recessive genes; (3) *appl.* species extremely similar in external appearance but which do not normally interbreed.

crypto- prefix derived from the Gk *kryptos*, hidden.

cryptocarp cystocarp *q.v.*

cryptochrome *n.* a type of blue-light receptor in plants.

cryptococcosis *n.* human disease affecting lungs or central nervous system, caused by the yeast-like fungus *Cryptococcus neoformans* (*Lipomyces neoformans*).

cryptofauna *n.* organisms living in concealment in protected situations, such as crevices in coral reefs.

cryptogams *n.plu.* plants reproducing by spores, such as the mosses and ferns. The term has also been used for plants without flowers, or without true stems, roots or leaves.

cryptogram *n.* method of expressing in standard form a collection of data used in taxonomic classification.

cryptomonad *see* Cryptophyta.

cryptonema *n.* filamentous outgrowth in cryptostomata of brown algae.

Cryptophyta, cryptophytes *n., n.plu.* phylum of mostly free-living but some parasitic unicellular protists with an ovoid flattened cell and a gullet (crypt) from which arise two unequal flagella. The phylum contains both photosynthetic and non-photosynthetic organisms. Cryptophytes are distinguished from similar protists such as euglenoids by their mode of cell division. Reproduction is by asexual division only. Found in a wide range of moist habitats including the intestinal tracts of some animals. *alt.* cryptomonads.

cryptophyte *n.* (1) perennial plant persisting by means of rhizomes, corms or bulbs underground, or by underwater buds;

(2) member of the protist phylum Cryptophyta *q.v.*

cryptoptile *n.* a feather filament, developed from a papilla.

cryptorchid *a.* having testes abdominal in position.

cryptosphere *n.* the habitat of the cryptozoa.

cryptostomata *n.plu.* non-sexual apparent conceptacles in some large brown algae, bearing only sterile hairs. *sing.* **cryptostoma.**

cryptozoa *n.* (1) (*ecol.*) the small terrestrial animals that live on the ground but above the soil in leaf litter and twigs, among and under pieces of bark or stones; (2) (*zool.*) animals that live in crevices. *a.* **cryptozoic.**

cryptozoite *n.* stage of the sporozoite of parasitic protozoa when living in tissues before entering blood.

crypts of Lieberkühn tubular exocrine glands in the intestines that secrete digestive enzymes.

crystal cells cells found in the ceoelom of echinoderms, containing rhomboid crystals.

crystal-containing body microbody *q.v.*

crystallin *n.* any member of a family of small globular proteins which are the principal components of the lens in the eye.

crystalline *a.* transparent or translucent.

crystalline cone cone-shaped extracellular jelly-like structure in ommatidium of a compound eye.

crystalline style translucent proteinaceous rod containing carbohydrases and involved in carbohydrate digestion in the alimentary canal of some molluscs.

crystallization *n.* (1) formation of crystals from a solution, e.g. of a protein; (2) the final stage in formation of the fully adult song of a bird.

crystallography *see* X-ray crystallography.

crystalloid *n.* (1) substance which in solution readily diffuses through a semipermeable membrane, *cf.* colloid; (2) crystal composed of protein, found in some plant cells.

crystal structure the three-dimensional structure of a protein or other molecule obtained by X-ray crystallography.

CS conditioned stimulus *q.v.*

CSF cerebrospinal fluid *q.v.*

CSF-1 colony-stimulating factor *q.v.*

ctene *see* Ctenophora.

ctenidium *n.* (1) feathery or comb-like structure, esp. respiratory apparatus in molluscs; (2) rows of fused cilia forming the swimming plate of sea gooseberries (Ctenophora); (3) row of spines forming a comb in some insects. *plu.* **ctenidia.**

cteniform, ctenoid, ctenose *a.* comb-shaped, with comb-like margin.

ctenocyst *n.* sense organ of Ctenophora, borne on the aboral surface.

Ctenophora, ctenophores *n.*, *n.plu.* phylum of coelenterates comprising the sea gooseberries or comb jellies. They are free-living and biradially symmetrical with eight meridional rows of ciliated ribs (known as ctenes, comb-ribs or swimming plates) by which they propel themselves. They have no nematoblasts (stinging cells).

C-terminal, C terminus *see* carboxy terminus.

CTL cytotoxic T lymphocyte *q.v.*

CTP cytidine triphosphate *q.v.*

CT scan *see* computerized axial tomography.

C-type lectins large family of calcium-dependent lectins found in animals.

C-type viruses class of retroviruses whose virions contain two copies of a single-stranded RNA genome and the enzyme reverse transcriptase. DNA is transcribed from the viral RNA in the infected cell and becomes integrated into the host genome as proviral DNA. Here it is transcribed to produce new viral genomes and viral mRNA. Replication-defective C-type viruses in which part of the viral genome has been replaced by an oncogene and which can induce transformation and tumour formation have arisen by the acquisition of modified host cell sequences during infection. *see also* RNA tumour viruses.

cubical *a. appl.* cells as long as broad.

cubital *a.* (1) *pert.* elbow; (2) *pert.* ulna or cubitus.

cubitus *n.* (1) ulna or the forearm generally; (2) the main vein in an insect wing.

cuboid *n.* bone in the ankle.

Cubozoa *n.* class of Cnidaria comprising the box-jellies *q.v.*

Cuculiformes *n.* order of birds including the cuckoos.

cucullate *a.* (1) hooded; (2) (*bot.*) with hood-like petals or sepals; (3) (*zool.*) with prothorax hood-shaped, in insects.

cucurbit *n.* member of the family of dicotyledonous plants, the Cucurbitaceae, the marrows, cucumbers and gourds.

Cucurbitales *n.* order of dicots, herbs and small trees, often climbing by tendrils and comprising the family Cucurbitaceae (e.g. gourds, melon, cucumber).

cuiller *n.* spoon-like terminal portion of male insect clasper.

cuirass *n.* bony plates or scales arranged like a cuirass.

culm *n.* flowering stem of grasses and sedges.

culmen *n.* (1) median longitudinal ridge of a bird's beak; (2) part of superior vermis of cerebellum.

culmicole, culmicolous *a.* living on grass stems.

culmination *n.* stage in life cycle of dictyo-stelid slime moulds which begins when the pseudoplasmodium stops migrating and ends with fruiting body formation.

cultellus *n.* sharp knife-like organ, one of the mouthparts of certain blood-sucking flies.

cultivar *n.* plant variety found only in cultivation, denoted by the species name followed by the abbreviation cv. followed by the cultivar name, e.g. *Rosa foetida* cv. Persian Yellow.

cultural eutrophication eutrophication as a result of human activity such as agriculture, urbanization and sewage discharge.

cultural inheritance transmission of particular traits and behaviours from generation to generation by learning rather than by genetic inheritance.

culture (1) *n.* microorganisms, cells or tissues growing in nutrient medium in the laboratory; (2) *v.* to isolate and grow microorganisms, cells or tissues.

culture collection a reference collection of different species and strains of microorganisms or cultured cells.

culture medium solution of nutrients required for the growth of microorganisms, cells or tissues in culture.

cumarin coumarin *q.v.*

cumulose *a. appl.* deposits consisting mainly of plant remains, e.g. peat.

cumulus *n.* the mass of epithelial cells bulging into the cavity of an ovarian follicle and in which the ovum is embedded.

cuneate *a.* wedge-shaped; (1) (*bot.*) *appl.* leaves with broad abruptly pointed apex and tapering to the base; (2) (*neurobiol.*) *appl.* nucleus of grey matter situated in anterior medulla oblongata at which spinal neurons carrying sensory information from skin and other tissues first make connections with neurons that transmit the information onwards to other regions of the brain.

cuneiform *a.* wedge-shaped.

cuneiform nucleus area in midbrain, implicated in control of locomotion.

cup fungi common name for the Discomycetes *q.v.*

cupula *n.* (1) the bony apex of the cochlea; (2) jelly-like cup over a group of hair cells (a neuromast) in acoustico-lateralis system of fish and amphibians and vestibular system of mammals; (3) cupule *q.v.*

cupulate *a.* (1) cup-shaped; (2) having a cupule.

cupule *n.* (1) (*bot.*) the green involucre of bracts round the fruit of some trees, e.g. acorns; (2) (*zool.*) a small sucker in various animals.

curare *n.* alkaloid extracted from cinchona bark that blocks neuromuscular transmission causing paralysis.

curie (Ci) *n.* unit of radiation corresponding to an amount of radioactive material producing 3.7×10^{10} disintegrations per second, which is the activity of radium. It is being replaced by the derived SI unit, the becquerel (Bq). 1 Ci = 3.7×10^{10} Bq.

currant gall type of small spherical gall found on oak leaves or catkins caused by larvae of the gall wasp *Neuoterus quercus baccarum*.

cursorial *a.* having limbs adapted for running, *appl.* e.g. birds, bipedal dinosaurs.

cusp *n.* (1) a prominence, as on molar teeth; (2) a sharp point. *alt.* tubercle.

cuspidate *a.* terminating in a sharp point.

cut-and-paste transposition type of transposition in which the transposable element excises itself from the DNA and reinserts elsewhere in the genome.

cutaneous *a. pert.* the skin.

cuticle *n.* (1) an outer skin or pellicle, sometimes referring to the epidermis as whole, esp. when impermeable to water; (2) (*zool.*) an outer protective layer of material, of various composition, produced by the epidermal cells, that covers the body of many invertebrates; (3) (*bot.*) layer of waxy material, cutin, on the outer wall of epidermal cells in many plants, making them fairly impermeable to water. *a.* **cuticular**.

cuticularization *n.* (1) formation of a cuticle; (2) the laying down of cutin in the outer layers of plant epidermis.

cuticulin *n.* lipoprotein secreted by epidermal cells and forming the outer layer of insect cuticle.

cutin *n.* wax-like mixture of fatty substances impregnating walls of epidermal plant cells, and also forming a separate layer, the cuticle, on the outer surface.

cutis *n.* (1) dermis of skin; (2) outer layer of cap and stalk of mushrooms and toadstools.

cuttlebone *n.* shell of the cuttlefish, *Sepia*, which acts as a buoyancy organ.

cuttlefish *n.* common name for a group of cephalopod molluscs, e.g. *Sepia*, characterized by a shell of unusual structure (cuttlebone), and having eight short arms around the mouth and long tentacles.

Cuvierian ducts short paired veins opening into the sinus venosus, and formed by the union of anterior and posterior cardinal veins, in fish and in tetrapod embryos.

Cuvierian organs tubular organs in sea cucumbers (Holothuroidea) which secrete collagen and polysaccharide which is released as slime via the cloaca when the animal is harassed.

C value total amount of DNA in the haploid genome of a species, either measured directly in picograms or expressed in base pairs or daltons.

C-value paradox the fact that C values for related species can differ widely and also appear not to be closely related to the relative organizational complexity of different organisms, and that the total amounts of DNA in higher multicellular organisms seem to be much more than is needed to encode the required number of genes.

CVS chorionic villus sampling *q.v.*

CXCL when followed by a number denotes a particular chemokine.

CXCR when followed by a number denotes a particular receptor for a CXCL chemokine.

cyanelle *n.* DNA-containing photosynthetic organelle in glaucophyte algae such as *Cyanophora paradoxa.* It has some features of cyanobacteria and may be a relic of a bacterial endosymbiont.

cyanidin *n.* violet flavonoid pigment present in many flowers.

Cyanobacteria *n.* major phylum of photosynthetic Bacteria. Although they are prokaryotic, they have an oxygen-evolving type of photosynthesis resembling that of green plants and contain chlorophyll *a* and phycobilin pigments. Unicellular, filamentous and colonial types are found. They live in aquatic and terrestrial environments, either free-living or in symbiotic associations, as with fungi in lichens. Some species can fix atmospheric nitrogen. Some cyanobacteria produce toxins which can become a health hazard in conditions where cyanobacterial 'algal blooms' appear. Known also as the blue-green algae from their previous classification in the plant kingdom as the Cyanophyta or Cyanophyceae. *sing.* **cyanobacterium.**

cyanocobalamin *n.* common commercial form of cobalamin (vitamin B_{12}) with CN substituted at the 6th position on the cobalt atom.

cyanogenesis *n.* production of hydrocyanic acid, as in some plants. *a.* **cyanogenic.**

cyanogenic glycosides plant glycosides that, when fully hydrolysed, liberate hydrogen cyanide, which is toxic to most cells. Found in species of *Sorghum*, *Prunus* and *Linum.*

cyanophil(ic) *a.* with special affinity for blue or green stains.

cyanophycophilous *n. appl.* lichens in which a cyanobacterium is the photosynthetic partner.

cyathiform *a.* cup-shaped.

cyathium *n.* inflorescence of the spurges (*Euphorbia*), consisting of a cup-shaped involucre of bracts surrounding staminate flowers each with a single stamen, with a central pistillate flower.

cyathozooid *n.* the primary zooid in certain tunicates.

cyathus *n.* small cup-shaped organ or structure.

cybernetics *n.* science of communication and control, as by nervous system and brain.

Cycadeoidophyta, cycadeoids *n. n.plu.* division of fossil cycad-like plants that had massive stems, pinnate leaves and were usually monoecious, with sporophylls arranged in a flower-like structure. They have been considered as possible ancestors of angiosperms.

Cycadophyta, cycads *n.* one of the five main divisions of extant seed-bearing plants, commonly called cycads or sago palms. They are palm-like in appearance with massive stems which may be short or tree-like. Microsporangia and megasporangia are borne on sporophylls arranged in cones, male and female cones being borne on separate plants. In some classifications called the Cycadopsida or Cycadales.

Cyclanthales *n.* order of often palm-like monocots, also climbers and large herbs, comprising the family Cyclantheraceae.

cyclic *a.* having parts of flowers arranged in whorls. *alt.* cyclical.

cyclic AMP (cAMP) adenosine 3′,5′-cyclic monophosphate, in which an oxygen of the phosphate is bonded to a carbon in the ribose. It is synthesized from ATP by the enzyme adenylate cyclase, and acts as an intracellular signalling molecule in both eukaryotic and prokaryotic cells. In eukaryotic cells it is produced in response to a variety of hormonal and other chemical signals that act at cell-surface receptors to activate adenylate cyclase; it then acts as a second messenger, binding to and regulating the activity of a variety of enzymes or ion channels. It is rapidly converted to AMP by cyclic AMP phosphodiesterase. In some cellular slime moulds cyclic AMP acts as the chemoattractant that causes the aggregation of amoebae. *see* Fig. 12 (p. 156).

cyclic AMP-dependent protein kinase protein kinase A *q.v.*

cyclic AMP phosphodiesterase (cAMP phosphodiesterase) enzyme, present constitutively in the cell, that converts

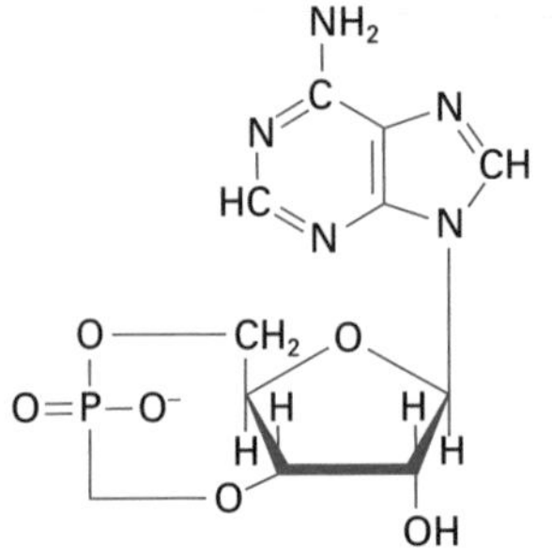

Fig. 12 Cyclic AMP.

cyclic AMP into AMP, thereby limiting cyclic AMP's regulatory activity.

cyclic AMP response element binding protein (CREB) gene-regulatory protein that is phosphorylated as a result of a rise in the second messenger cyclic AMP within the cell. CREB then binds to the cAMP response element (CRE) in DNA and activates transcription of the associated genes, thus mediating the effect of cyclic AMP on gene expression.

cyclic GMP small molecule analogous to cyclic AMP, formed from guanosine triphosphate by the enzyme guanylate cyclase, and which functions as a second messenger in some intracellular signalling pathways.

cyclic GMP-dependent protein kinase protein kinase G *q.v.*

cyclic GMP phosphodiesterase enzyme that converts cyclic GMP to GMP, thereby limiting cyclic GMP's regulatory activity.

3′,5′-cyclic-nucleotide phosphodiesterase *see* cyclic AMP phosphodiesterase, cyclic GMP phosphodiesterase. EC 3.1.4.17.

cyclic photophosphorylation type of light-dependent phosphorylation of ADP involving photosystem I only, in which ATP is generated without concomitant NADPH generation. *alt.* **cyclic phosphorylation**.

cyclin-dependent kinases (CDKs) family of protein kinases that are activated by cyclins and other regulatory proteins and which by their actions initiate particular stages of the eukaryotic cell cycle.

cyclins *n.plu.* family of proteins whose concentrations rise and fall at specific times during the cell cycle of eukaryotic cells, and which are involved in controlling progression through the cycle.

cyclocoelic *a.* with the intestines coiled in one or more distinct spirals.

cyclogenous *a. appl.* a stem growing in concentric circles.

cyclogeny *n.* production of a succession of different morphological types in a life cycle.

cycloheximide *n.* antibiotic produced by the actinomycete *Streptomyces griseus*, which inhibits protein synthesis in eukaryotic cells only, by inhibiting translation.

cycloid *a. appl.* scales with an evenly curved free border.

cyclomorial *a. appl.* scales growing by apposition at margin only.

cyclomorphosis *n.* cycle of changes in form, usually seasonal, esp. in marine zooplankton, possibly in response to changes in salinity.

cyclopean *a. appl.* a single median eye developed under certain artificial conditions or as a mutation, instead of the normal pair.

cyclophilin *see* immunophilin.

cyclopia *n.* abnormal development of a single central eye instead of two bilateral ones.

cyclopoid naupliiform *q.v.*

cyclops *n.* larval stage in some copepods, having many characteristics of the adult.

cyclosis *n.* the streaming of cytoplasm within a cell.

cyclospermous *a.* with embryo coiled in a circle or spiral.

cyclospondylic, cyclospondylous *a. appl.* vertebral centra in which the calcified material forms a single cylinder surrounding the notochord.

cyclosporin A immunosuppressive drug isolated from the fungus *Tolypocladium*, used clinically to suppress unwanted immune responses that lead to organ rejection after transplantation. *alt.* **cyclosporine**.

Cyclostomata *n.* (1) cyclostomes *see* Agnatha; (2) order of Ectoprocta in which the zooids are completely fused, the case enclosing them is completely calcified and the pore has no lid.

cyclostomes *n.plu.* lampreys and hagfish, primitive jawless fishes. *see* Agnatha.

cydippid *n.* a ctenophore larva.

cygneous *a.* shaped like a swan's neck.

cylindrical *a. appl.* leaves rolled on themselves, or to solid cylinder-like leaves.

cymba *n.* upper part of cavity of external ear.

cymbiform *a.* boat-shaped.

cymbium *n.* boat-shaped tarsus of pedipalp in some spiders.

cyme *n.* repeatedly branching determinate inflorescence in which each growing point ends in a flower, with the oldest flowers at the end of the branch.

cymose *a. appl.* inflorescence formed by successive growths of axillary shoots after growth of main shoot in each branch has stopped.

cymotrichous *a.* having wavy hair.

cynarrhodium, cynarrhodon *n.* fruit of the rose-hip type.

Cynodontia, cynodonts *n., n.plu.* group of extinct mammal-like reptiles, found from the late Permian to the middle of the Jurassic, possessing a secondary bony palate, complex crowns on the cheek teeth, and other mammalian features, and which are believed to be the direct ancestors of mammals.

cynopodous *a.* with non-retractile claws, as a dog.

Cyperales *n.* the sedges, an order of herbaceous monocots with rhizomes and solid stems triangular in cross-section and comprising the family Cyperaceae.

cyprid, cypris *n.* larval stage that follows the nauplius stage in cirripedes.

cyprinids, cyprinoids *n.plu.* group of freshwater fish (the Cyprinoidei) widespread in Europe, Asia, Africa and North America, and including the carps and minnows.

cyprinodonts *n.plu.* order of small, mainly tropical fishes, the Cyprinidontiformes, including the toothed carps.

cypsela *n.* an inferior achene composed of two carpels, as in Compositae.

Cys cysteine *q.v.*

cyst *n.* bladder-like or sac-like structure; (1) resting cell surrounded by a protective coat, which is formed in dry conditions by some microorganisms; (2)

bladder or air vesicle in certain seaweeds; (3) abnormal fluid-filled sac developing in tissues.

cystacanth *n.* infective juvenile stage in acanthocephalans.

cysteine (Cys, C) *n.* sulphur-containing amino acid with a thiol group on the side chain. It is a constituent of many proteins. Two cysteines can form a disulphide bond cross-linking and stabilizing the folded protein chain.

cystic *a. pert.* gall bladder or urinary bladder.

cysticercoid, cysticercus *n.* larval stage in tapeworm, consisting of a fluid-filled sac containing a scolex, which in the appropriate host attaches to the gut wall and develops into an adult. *a.* **cysticercoid**.

cystic fibrosis (CF) inherited condition in humans caused by a recessive genetic defect. It is characterized by malfunction of pancreas and abnormal secretion from the lungs, leading to serious secondary effects such as infection, and is fatal in early childhood if the symptoms are untreated. The primary defect is in the transport of chloride ions.

cysticolous *a.* living in a cyst.

cystidium *n.* inflated hair-like cell in the hymenium of some fungi. *plu.* **cystidia**.

cystine (Cys–Cys) *n.* amino acid residue formed in proteins by oxidation of the sulphydryl groups of two cystine residues to form a disulphide bridge cross-linking the protein chain.

cystoarian *a. appl.* gonads when enclosed in coelomic sacs, as in most teleost fishes.

cystocarp *n.* a small spore-bearing structure in red algae.

cystochroic *a.* having pigment in cell vacuoles.

cystocyte *n.* any of the interconnected cells that are the product of division of the oogonium in *Drosophila* oogenesis.

cystogenous *a.* (1) cyst-forming; (2) *appl.* large nucleated cells that secrete the cyst in cercaria.

cystolith *n.* mass of calcium carbonate, occasionally of silica, formed on ingrowths of epidermal cell walls in some plants.

cystozooid *n.* body portion of a metacestode.

cyt cytochrome *q.v.*

cytidine *n.* nucleoside made up of the pyrimidine base cytosine linked to ribose.

cytidine diphosphate (CDP) cytosine nucleotide containing a diphosphate group.

cytidine monophosphate (CMP) nucleotide composed of cytosine, ribose and a phosphate group, product of the partial hydrolysis of RNA and also found in the activated intermediate CMP-*N*-acetylneuraminate in ganglioside synthesis. *alt.* cytidylate, cytidylic acid, cytidine 5′-phosphate.

cytidine triphosphate (CTP) cytosine nucleotide containing a triphosphate group, one of the ribonucleotides required for RNA synthesis, also acts as an energy donor in metabolic reactions in a manner analogous to ATP, esp. in triacylglycerol synthesis where it forms activated CDP-acylglycerols.

cytidylate (cytidylic acid) cytidine monophosphate *q.v.*

cytidyl transferase *see* nucleotidyltransferase.

cytoarchitectonics *n.* study of brain anatomy based on the types and spacing of nerve cells and the distribution of their axons.

cytocentrum centrosome *q.v.*

cytochalasin *n.* fungal toxin that inhibits the polymerization of actin into microfilaments.

cytochemistry *n.* study of the chemical composition and chemical processes in cells by specific staining and microscopical examination. *alt.* histochemistry.

cytochimaera *n.* tissue or organism in which cells have different chromosome numbers.

cytochroic *a.* having pigmented cytoplasm.

cytochrome *n.plu.* any of a group of haem-containing electron-transporting proteins. Cytochromes are components of the respiratory and photosynthetic electron transport chains.

cytochrome *a* and *a₃* components of cytochrome *c* oxidase. They are the terminal electron carriers in the respiratory chain, electrons being transferred from a copper prosthetic group of cytochrome *a₃* to molecular oxygen to form water.

cytochrome *b-c₁* complex transmembrane proton-pumping complex of cytochromes in respiratory electron chain. Receives electrons from the NADH dehydrogenase complex via ubiquinone and donates electrons to cytochrome *c* oxidase via cytochrome *c*. *alt.* QH₂-cytochrome *c* reductase.

cytochrome *b₆-f* complex transmembrane proton-pumping complex of cytochromes which is part of the photosynthetic electron-transport chain in green plants. It receives electrons from photosystem II via plastoquinone and donates electrons to photosystem I via plastocyanin.

cytochrome *c* mobile component of the respiratory electron transport chain, transfers electrons from the cytochrome *b-c₁* complex to the cytochrome *c* oxidase complex.

cytochrome oxidase complex cytochrome *c* oxidase, an enzyme complex catalysing the terminal transfer of electrons to oxygen in the respiratory chain, with the production of water. It contains cytochromes *a* and *a₃*. EC 1.9.3.1.

cytochrome P₄₅₀ specialized cytochrome that is the terminal component of an electron-transport chain in adrenal mitochondria and liver microsomes that is involved in hydroxylation reactions.

cytocidal *a.* cell-destroying.

cytocuprein superoxide dismutase *q.v.*

cytocyst *n.* envelope formed by remains of host cell within which a protozoan parasite multiplies.

cytode *n.* non-nucleated protoplasmic mass.

cytodeme *n.* local interbreeding unit of a taxon which differs cytologically, usually in chromosome number, from the rest of the taxon.

cytofluorimetric *a. appl.* the identification and isolation of specific cell types by means of immunological or other specific staining with fluorescent antibodies or dyes.

cytogamy *n.* cell conjugation.

cytogenesis *n.* development or formation of cells.

cytogenetic *a.* (1) *pert.* cytogenetics; (2) *pert.* cytogenesis.

cytogenetics *n.* study of the microscopic structure of chromosomes.

cytogenic *a. appl.* reproduction by cell division, as in a clone.

cytogenous *a.* producing cells.

cytohet *n.* cell containing two genetically distinct types of an organelle such as mitochondria or chloroplasts.

cytokine *n.* any protein or polypeptide produced by a cell and which affects the growth or differentiation of the same or another cell. The term usually excludes the endocrine hormones. Examples are chemokines, differentiation factors, growth factors, interleukins.

cytokinesis *n.* cytoplasmic division into two daughter cells after mitosis or meiosis.

cytokinin *n.* plant growth hormone that acts in concert with IAA (an auxin) to promote rapid cell division. Cytokinins are derivatives of adenine and include the naturally occurring zeatin and i⁶Ade, and kinetin, which probably does not occur in nature. They are used to induce the formation of plantlets from callus tissue in culture. *alt.* kinin, phytokinin.

cytology *n.* study of the structure, functions and life history of cells.

cytolysin *n.* substance causing cell lysis.

cytolysis *n.* cell lysis or disintegration.

cytolysosome *n.* large lysosome apparently involved in autolysis.

cytolytic *a.* causing cell lysis.

cytomegaloviruses *n.plu.* group of DNA viruses of the herpesvirus family which cause the formation of large inclusion bodies in infected glandular cells. Cytomegalovirus infections can be fatal in very young children.

cytomere *n.* cell formed by schizont and giving rise to merozoites.

cytometry *n.* counting cells.

cytomorphosis *n.* (1) series of structural modifications of a cell or successive generations of cells; (2) cellular change, as in senescence.

cytonemes *n.* long protrusions put out by some cells.

cytopathic effect (CPE) the destruction of cells caused by infection with a virus.

cytophagas *n.plu.* major group of the Bacteria, distinguished on DNA sequence data. It contains a mixture of physiological types, including *Bacteroides*, *Cytophaga* and *Flexibacter*. *alt.* Bacteroides–Flavobacteria group.

cytopharynx *n.* tube-like structure leading from 'mouth' into cell interior in certain protozoans. *alt.* gullet.

cytophilic *a.* (1) having an affinity for cells; (2) binding to cells; (3) staining cells.

cytophilic antibody antibody that becomes adsorbed to cells by its constant region, leaving the antigen-binding site free to subsequently bind antigen, as IgE.

cytoplasm *n.* all the living part of a cell inside the cell membrane and excluding the nucleus. *cf.* cytosol.

cytoplasmic *a. pert.* or found in the cytoplasm.

cytoplasmic determinant in developmental biology, a substance laid down in cytoplasm of egg during its formation that directs the early development of the zygote.

cytoplasmic inheritance inheritance of genes carried in organelles such as chloroplasts and mitochondria, or in other cytoplasmic particles. Such genes behave in a non-Mendelian fashion as they are usually inherited only from the maternal parent.

cytoplasmic localization in development, the non-uniform distribution of a cytoplasmic determinant within a cell, so that when the cell divides, the substance is unequally distributed to the daughter cells.

cytoplasmic male sterility form of male sterility in plants that is determined by cytoplasmic factors, usually mitochondrial DNA.

cytoplasmic streaming the rapid movement of cytoplasm within eukaryotic cells, seen most clearly in plant and algal cells, some protozoa and in the plasmodia of slime moulds.

cytoplasmic tail that portion of a transmembrane receptor protein that projects into the cytoplasm.

cytoplast *n.* a cell containing only cytoplasm, e.g. after removal of the nucleus.

cytoproct, cytopyge *n.* the 'anus' of a unicellular organism, as certain protozoans.

cytosine (C) *n.* pyrimidine base, one of the four nitrogenous bases in DNA and RNA, in which it pairs with guanine (G). It is the base in the nucleoside cytidine and its derivatives. Cytosine in DNA is sometimes methylated to form methylcytosine.

see DNA methylation. *see* Fig. 5 (p. 69).

cytosine arabinoside arabinosyl cytosine, an antiviral drug.

cytoskeleton *n.* internal system of protein fibres and tubules that extends throughout the cytoplasm of a eukaryotic cell. It is composed of actin microfilaments, intermediate filaments and microtubules. The cytoskeleton gives shape to a cell and provides support for cell extensions such as villi and axons of nerve cells. Elements of the cytoskeleton form the mitotic and meiotic spindles and are involved in intracellular transport and cell movement. *see also* intermediate filament, microfilament, microtubule.

cytosol *n.* the fluid part of the cytoplasm outside the membrane-bounded organelles.

cytosome microbody *q.v.*

cytostatic *a. appl.* any substance suppressing cell growth and multiplication.

cytostome *n.* specialized region acting as a 'mouth' of a unicellular organism, as in some protozoans.

cytotaxis *n.* movement of cells to or away from a stimulus. *a.* **cytotactic.**

cytotaxonomy *n.* classification based on characteristics of chromosome structure and number.

cytotoxic *a.* attacking or destroying cells.

cytotoxic T lymphocyte, cytotoxic T cell (T$_c$, CTL) effector T lymphocyte distinguished by the cell-surface protein CD8. It can kill cells and is the means of destroying virus-infected cells in an adaptive immune response.

cytotoxin *a.* substance that can poison or destroy cells.

cytotrophoblast *n.* inner layer of trophoblast.

cytotropism *n.* mutual attraction of two or more cells.

cytotype *n.* genetically determined characteristic carried by cytoplasmic part of cell, different cytotypes usually being revealed by their effects on the expression of the nuclear genotype.

cytozoic *a.* living inside a cell, *appl.* sporozoan trophozoite.

D

δ- *for headwords with prefix δ-, refer to delta- or to headword itself.*

d prefix attached to abbreviations for nucleotides to indicate that they contain the sugar deoxyribose, e.g. dATP, dCTP.

D (1) aspartic acid *q.v.*; (2) dalton *q.v.*; (3) deuterium *q.v.*; (4) Simpson diversity index *q.v.*

D decimal reduction time *q.v.*

D- prefix denoting a particular molecular configuration, defined according to convention, of certain chiral compounds esp. monosaccharides and amino acids. The alternative configuration, L, is a mirror image of D.

2,4-D 2,4-dichlorophenoxyacetic acid, a synthetic auxin used as a herbicide for controlling broad-leaved weeds.

Da, dal dalton *q.v.*

dacryocyst *n.* pouch in the angle of the eye between upper and lower eyelids, which receives tears from the lacrimal ducts.

dacryoid *a.* tear-shaped.

dactyl *n.* digit, a finger or toe.

dactyloid *a.* like a finger or fingers.

dactylopatagium *n.* the part of the flying membrane of bats that is carried on the metacarpals and phalanges.

dactylopodite *n.* (1) joint furthest away from the body in limbs of certain crustaceans; (2) tarsus and metatarsus of spiders.

dactylozooid palpon *q.v.*

DAF decay-accelerating factor *q.v.*

DAG diacylglycerol *q.v.*

dal dalton *q.v.*

DALA synthetase δ-aminolevulinate synthase *q.v.*

dalton (D, Da, dal) *n.* mass unit equal (by definition) to one-twelfth of the mass of a single atom of carbon-12, which is ca. 1.66×10^{-27} kg. *alt.* atomic mass unit.

dammar *n.* resin obtained from several Malaysian trees.

damselflies *see* Odonata.

dance, of bees series of movements performed by honeybees on their return to the hive after finding a food source, which informs other bees in the hive of the location of the food. *see also* waggle dance.

DAP diaminopimelic acid *q.v.*

daphnid *n.* any of various small water fleas, esp. of the genus *Daphnia*.

DAPI a blue fluorescent stain used in cell biology to visualize DNA in live cells.

dark adaptation the visual adaptations that occur in the eye for vision in dim light compared with bright light. The threshold of just-visible light is lowered as a result of e.g. the switch from the use of the retinal cones to the more sensitive rod cells, dilation of the pupil, and (in many fish, amphibians, reptiles and birds) the migration of choroid pigment away from the outer segments of the photoreceptor.

dark-field microscopy type of high-resolution light microscopy used to study living cells, in which the lighting is modified to produce an illuminated object on a dark background.

dark reactions in photosynthesis, reactions occurring in the stroma of chloroplasts, for which light is not required, in which carbon dioxide is reduced to carbohydrate. *alt.* Calvin cycle, carbon dioxide fixation, photosynthetic carbon reduction cycle (PCR cycle).

Darlington's rule the fertility of an allopolyploid is inversely proportional to the fertility of the original hybrid.

dart *n.* (1) crystalline structure in molluscs, used in copulation; (2) in nematodes, a sharp point used to penetrate the host.

dart sac small sac, containing a calcareous dart, attached to vagina near its opening in some gastropods.

Darwinian evolution the theory of evolution by means of natural selection, put forward by Charles Darwin in his book *Origin of Species* published in 1859 but formulated some years earlier. The theory was based on his observation of the genetic variability that exists within a species and the fact that organisms produce more offspring than can survive. Under any particular set of environmental pressures, those heritable characteristics favouring survival and successful reproduction would therefore be preferentially passed on to the next generation (natural selection). Selection for particular aspects of life-style or in different environmental conditions could therefore eventually lead to two populations differing in many ways from the original and to the development of complex adaptations to a particular mode of life or environment. A very similar theory was proposed independently by Alfred Russel Wallace. In the 1930s and 1940s the theory of Mendelian genetics was incorporated into Darwin's original theory to produce a modern version, the neo-Darwinian synthesis. *see also* neo-Darwinism, natural selection.

Darwinian fitness the fitness of a genotype measured by its proportional contribution to the gene pool of the next generation.

Darwinian tubercle slight prominence on the helix of external ear, near the point where it bends downwards.

Darwinism Darwinian evolution *q.v.*

Darwin's finches the 14 species of finches found on the Galapagos Islands, which all possess features adapting them to a different mode of life, which were studied by Charles Darwin in the course of his voyage on the HMS *Beagle*, and which are said to have helped to stimulate his ideas on evolution and the origin of species.

dasypaedes *n.plu.* birds whose young are downy at hatching. *a.* **dasypaedic**.

dasyphyllous *a.* with thickly haired leaves.

data *plu.* factual information such as measurements and statistics, DNA and protein sequences, etc., produced e.g. from experiments, assays and screens. *sing.* **data point, datum** a single piece of data.

database *n.* in genomics or proteomics, usually refers to an electronic archive containing large numbers of DNA sequences, protein sequences or protein structures. Large publicly accessible databases of this type are GenBank for genomic DNA sequences, Swiss-Prot for protein sequences, dbEST for partial cDNAs derived from cellular mRNA, and the Protein Data Bank (PDB) for protein structures.

data mining the extraction of useful biological information from very large amounts of data.

Datiscales Begoniales *q.v.*

dauer larva larval state in nematode worms that is produced when food is short and which can last for several months. In this state the larva neither eats nor grows.

daughter *n.* progeny of cell or nucleus arising from a mitotic division, as daughter cells, daughter nuclei.

Davson–Danielli model one of the first models proposing a lipid bilayer structure for biological membranes – now superseded by the fluid mosaic model.

Dayhoff matrix a type of amino-acid substitution matrix used in scoring alignments. *see* substitution matrix.

day-neutral *a. appl.* plants in which flowering can be induced by either a long or a short photoperiod or by neither.

DBA lectin isolated from *Dolichos biflorus*.

DBH diameter at breast height (1.4 m), standard measurement of a trunk of a tree.

DC dendritic cell *q.v.*

D-DNA form adopted by synthetic DNA molecules lacking guanine, with eight base pairs per turn.

DDT dichlorodiphenyltrichloroethane, a persistent organochlorine insecticide.

de- prefix from L. *de*, away from, denoting removal of.

deacetylase *n.* enzyme that catalyses the removal of acetyl groups. *see also* histone deacetylase.

dealation *n.* the removal of wings, as by female ants after fertilization, or by termites.

deamination *n.* removal of an amino group (–NH$_2$) from a molecule.

death point temperature, or other environmental variable, above or below which organisms cannot exist.

death rate number of deaths within a population over a set period, usually a year. The crude death rate is calculated for human populations as the annual number of deaths per 1000 population in a given geographical area, with the population number usually taken at the midpoint of the year in question. *alt.* mortality rate.

Débove's membrane layer between tunica propria and epithelium of tracheal, bronchial and intestinal membranes.

debranching enzyme enzyme that hydrolyses the α-1,6-glycosidic bonds at the branch points in glycogen molecules. EC 3.2.1.10, *r.n.* oligo-1,6-glucosidase.

deca- prefix derived from Gk *deka*, ten, denoting having ten of, divided into ten, etc.

decagynous *a.* having ten pistils.

decalcify *n.* to treat with acid to remove calcareous parts.

decamerous *a.* with the various parts arranged in tens, *appl.* flowers.

decandrous *a.* having ten stamens.

decaploid *a.* having ten times the normal chromosome set.

Decapoda, decapods *n., n.plu.* (1) order of freshwater, marine and terrestrial crustaceans having five pairs of legs on the thorax and a carapace completely covering the throat, and including the prawns, shrimps, crabs and lobsters; (2) order of cephalopods having two retractile tentacles as well as eight normal tentacles, including the squids and cuttlefish.

decarboxylase *n.* enzyme that removes CO$_2$ from organic compounds.

decarboxylation *n.* removal of CO$_2$ from organic compounds.

decay *n.* (1) decomposition of dead organic matter by the action of microorganisms; (2) radioactive decay. *see* radioactivity.

decay-accelerating factor (DAF) membrane glycoprotein on mammalian cells that binds activated complement components C3b and C4b and inhibits further action of complement. It protects the host cells against the destructive effects of complement.

decem- prefix derived from the L. *decem*, ten, denoting having ten of, divided into ten, etc.

decemfid *a.* cut into ten segments.

decemfoliate *a.* ten-leaved.

decemjugate *a.* with ten pairs of leaflets.

decempartite *a.* having ten lobes.

deception *n.* any behaviour or feature that deceives a predator or other animal. It can range from physical mimicry to apparently cognitive deceptions. *cf.* honest behaviour.

decidua *n.* mucous membrane lining the pregnant uterus, cast off after giving birth. *a.* **decidual**.

deciduate *a.* (1) characterized by having a decidua; (2) partly formed by the decidua.

deciduous *a.* (1) falling at the end of growth period or at maturity; (2) *appl.* teeth: milk teeth; (3) *appl.* placenta, such as that of most mammals, where maternal endometrium and foetal chorion cannot be separated without damage to both mother and foetus; (4) *appl.* trees whose leaves fall all at the same time. *cf.* evergreen.

decimal reduction time (*D*) time required for a tenfold reduction in the population density of a bacterium or other microorganism at a given temperature.

declarative memory conscious memory consisting of learned facts and information that one is aware of accessing. *cf.* non-declarative memory.

declinate *a.* bending aside in a curve.

decline phase of population growth *see* S-shaped growth curve *q.v.*

declivis *n.* part of superior vermis of cerebellum, continuous laterally with lobulus simplex of cerebellar hemispheres.

decollated *a.* with apex of spire wanting.

decomposed *a.* (1) not in contact; (2) not adhering, *appl.* barbs of feather when separated; (3) decayed; (4) rather shapeless and gelatinous, *appl.* cortical hyphae in lichens.

decomposer *n.* any organism that feeds on dead plant and animal matter, breaking it down physically and chemically and recycling elements and organic and inorganic compounds to the environment. Decomposers are chiefly microorganisms and small invertebrates.

decomposition *n.* decay of organic material, mediated by microorganisms.

decompound *a. appl.* compound leaf whose leaflets are also compound.

decondensation *n.* the uncoiling and extension of the mitotic chromosomes into long fine thread-like structures, which occurs after mitosis has been completed. *cf.* condensation.

deconjugation *n.* separation of paired chromosomes, as before end of meiotic prophase.

deconvolution *n.* the deblurring of an optical image by computer-aided techniques.

decorticate *a.* with bark (of trees) or cortex (of brain, other organs) removed.

decumbent *a.* lying on ground but rising at tip, *appl.* stems.

decurrent *a.* (1) having leaf base prolonged down stem as a winged expansion of rib; (2) prolonged down stalk, *appl.* gills of agarics.

decurved *a.* curved downwards.

decussate *a.* (1) crossed; (2) having paired leaves, succeeding pairs crossing at right angles.

decussation *n.* crossing of nerves (from either side of body) with interchange of fibres, as in optic and pyramidal tracts.

dedifferentiation *n.* loss of characteristics of a specialized cell and its regression to an undifferentiated state. This may be followed by redifferentiation to a different cell type.

dediploidization *n.* in basidiomycete and ascomyete fungi, the production of haploid cells or hyphae from a dikaryotic cell or mycelium.

deep-sea hydrothermal vent *see* hydrothermal vent community.

default state the developmental state that a cell or cells will adopt in the absence of the action of a regulatory switch.

defaunation *n.* removal of animal life from an area.

defective viruses viruses that can infect cells but not reproduce within them, as they lack genes for essential viral components. They can reproduce if a related non-defective helper virus is also present to direct the synthesis of the missing viral components. *alt.* replication-defective viruses.

defensins *n.plu.* large family of antimicrobial peptides, produced by animals and plants.

deferent *a.* carrying away from.

deferred *a. appl.* shoots arising from dormant buds.

Deferribacteres *n.* a phylum of the Bacteria distinguished on DNA sequence criteria.

deficiency *n.* in genetics, the inactivation or absence of a gene or segment of chromosome. *alt.* deletion.

deficiency diseases pathological conditions in plants and animals due to lack of some vitamin, trace element or other minor nutrient, e.g. crown rot in sugar beet due to boron deficiency, scurvy in humans due to lack of vitamin C.

defined medium culture medium whose composition is chemically defined. *cf.* complex medium.

definite *a.* (1) fixed; (2) constant; (3) (*bot.*) *appl.* inflorescences with primary axis terminating in a flower. *cf.* indefinite.

definitive *a.* (1) defining or limiting; (2) complete; (3) fully developed; (4) final, *appl.* host of adult parasite.

deflorate *a.* after the flowering stage.

defoliate *v.* to strip off leaves.

defoliation *n.* removal of all leaves from a plant.

deforestation *n.* complete and permanent removal of forest or woodland and its associated undergrowth.

degeneracy *n. pert.* the genetic code, the fact that one type of amino acid can be specified by more than one codon. The code is thus said to be degenerate.

degeneration *n.* (1) breakdown in structure; (2) change to a less specialized or functionally less active form; (3) evolutionary change resulting in change from a complex to a simpler form. *a.* **degenerate**.

degenerative disease disease caused by deterioration of organs or tissues, rather than by infection, e.g. osteoarthritis, Alzheimer's disease, cardiovascular disease.

deglutition *n.* the process of swallowing.

degranulation *n.* release of granules from a cell, as in mast cells during an allergic reaction.

degree of relatedness coefficient of relationship *q.v.*

dehiscence *n.* spontaneous opening of an organ or structure, such as a seed-pod, along certain lines or in a definite direction. *a.* **dehiscent**.

dehydration *n.* removal of water from a substance.

dehydrogenase *n.* any enzyme catalysing the transfer of hydrogen from a donor to an acceptor compound, classified amongst the oxidoreductases in EC class 1.

dehydrogenation *n.* removal of hydrogen from a compound, which is an oxidation reaction. *cf.* hydrogenation.

deimatic display frightening behaviour consisting of adoption of a posture by one animal to intimidate another.

Deinococcus-Thermus a phylum of the Bacteria, distinguished on DNA sequence data. It includes the radiation-resistant *Deinococcus* and the thermophile *Thermus aquaticus*.

Deiter's cells non-neural supporting cells between rows of outer hair cells in organ of Corti in cochlea.

Deiters' nucleus lateral vestibular nucleus, the area in the medulla of brain involved in relaying sensory information relating to balance to motor pathways.

delamination *n.* (1) the splitting off of cells to form a new layer; (2) splitting of a layer.

delayed hypersensitivity immunological hypersensitivity reactions arising 24–48 hours after introduction of the antigen and induced by activated T cells. *alt.* delayed-type hypersensitivity. *cf.* immediate hypersensitivity.

delayed implantation reproductive phenomenon seen in many mammals in which the fertilized blastocyst does not implant in the uterine wall for some months after fertilization.

deletion *n.* mutation involving loss of part of a chromosome, or loss of a base or a stretch of bases in a DNA sequence.

deliquescent *a.* (1) becoming liquid; (2) (*bot.*) having lateral buds more vigorously developed, so that the main stem seems to divide into a number of irregular branches.

delomorphic *a.* with a definite form.

delphinidin *n.* blue flavonoid pigment present in many flowers.

delphinology *n.* the study of dolphins.

delta- *for headwords with prefix delta- also refer to headword itself.*

delta agent a replication-defective virus that requires hepatitis B virus for its coat proteins.

deltaG (Δ*G*) Gibbs' free energy change. *see* free energy.

delta-proteobacteria group of the proteobacteria (purple Bacteria) distinguished on DNA sequence data. It consists of chemoorganotrophs, e.g. *Desulfovibrio* and other sulphate-reducing bacteria, *Bdellovibrio*, *Myxococcus* and *Desulfuromonas*. *alt.* delta purple bacteria.

deltidium *n.* plate covering the opening between hinge and beak, where the stalk emerges in many brachiopods.

deltoid *a.* more-or-less triangular in shape.

demanian *a. appl.* complex system of paired efferent tubes connecting the intestine and uteri in nematodes, and associated with secretion of gelatinous material to protect the eggs.

deme *n.* (1) (*bot.*) assemblage of individuals of a given taxon, usually qualified by a prefix, e.g. ecodeme, gamodeme, topodeme (*all q.v.*); (2) (*zool.*) a gamodeme, a local population unit of a species within which breeding is completely random.

dementia *n.* chronic disorder in mental processes as a result of brain disease. Typical symptoms include memory loss, confusion and personality changes.

demersal *a.* living on or near the bottom of sea or lake.

demersed *a.* growing under water, *appl.* parts of plants.

demethylation *n.* the loss of a methyl group.

demibranch hemibranch *q.v.*

demicyclic *a. appl.* rust fungus that lacks a uredinial stage.

demographics *n.* the changes in population numbers, or of particular age or sex classes within the population, over time.

demographic society a society that is relatively stable throughout time, being relatively closed to newcomers and whose composition is therefore the result largely of the demographic processes of birth and death.

demography *n.* the study of numbers of organisms in a population and their variation over time.

Demospongia *n.* class of sponges (Porifera, *q.v.*) which often have the body wall strengthened by a tangled mass of spongin fibres (e.g. in the bath sponge *Spongia*). They may have silica spicules, in the form of simple needles or a four-armed spicule whose points describe a tetrahedron, or may have no spicules. They are found on shores, and down to depths of more than 5000 m.

denatant *a.* swimming, drifting or migrating with the current.

denaturation *n.* alteration in the structural properties of a macromolecule such as a protein or a nucleic acid, leading to loss of function, as a result of heating, change in pH, irradiation, etc. In most cases denaturation refers to the disruption of non-covalent bonding leading to loss of secondary structure (e.g. unfolding of a protein chain or separation of the two strands of a DNA double helix).

dendriform *a.* branched like a tree.

dendrite *n.* fine cytoplasmic process on a neuron at which the neuron receives signals from other neurons.

dendritic *a.* (1) much branched; (2) resembling, *pert.*, or having, dendrites or dendrons.

dendritic cell (DC) type of non-lymphoid leukocyte that is the main cell type responsible for presenting antigens to naive T cells and providing lymphocyte-activating signals. Immature dendritic cells are found in most tissues of the body. They pick up invading microorganisms and their products by phagocytosis and macropinocytosis and are stimulated to mature and migrate to secondary lymphoid tissues, where they encounter T cells. Two main lineages of dendritic cell are generally distinguished: conventional dendritic cells (cDCs), whose role is antigen presentation, and plasmacytoid dendritic cells (pDCs), which can also secrete large amounts of interferon in response to stimulation by pathogens. *see also* follicular dendritic cell.

dendrochronology *n.* study of the age of trees and timber, generally by counting tree-rings, and the study and analysis of tree-rings in relation to changes in climate over time.

Dendrogaea *n.* biogeographical region including all the neotropical region except temperate South America.

dendrogram *n.* any branching tree-like diagram illustrating the relationship between organisms or objects.

dendroid *a.* tree-like, much branched.

dendrology *n.* study of trees.

dendrometer *a.* device for measuring small changes in the diameter of a tree trunk, such as the minute amounts of shrinkage and swelling that accompany the daily fluctuations in transpiration.

dendron *n.* cytoplasmic process on nerve cells that conveys impulses towards the nerve cell body. It is usually much branched into dendrites.

denervation *n.* removal of the nerve supply to an organ, muscle, etc.

denitrification *n.* (1) conversion of nitrate to nitrite, and nitrite to molecular nitrogen, leading to the loss of nitrogen from the biosphere. These reactions are carried out by a few genera of anaerobic bacteria. *see* denitrifiers; (2) the reduction of nitrates to nitrites and ammonia, as in plant tissues.

denitrifiers, denitrifying bacteria diverse anaerobic bacteria capable of dissimilative conversion of nitrate to nitrite and nitrite to molecular nitrogen, e.g. *Pseudomonas*, *Achromobacter*, *Thiobacillus* and *Micrococcus* spp. *see also* denitrification.

de novo from new, *appl.* e.g. synthesis of a compound from its basic constituents.

dens *n.* tooth or tooth-like process. *plu.* **dentes**.

density-dependent *a.* (1) *appl.* factors limiting the growth of a population which are dependent on the existing population density. They are generally the effects of competition from other species; (2) *appl.* selection that either favours or disfavours the rarer forms of individual within a population.

density gradient centrifugation procedure for separating cell components or macromolecules by centrifugation in a sucrose or caesium chloride solution of graded density. A component of a particular density will collect at the band of identical density within the solution. *alt.* equilibrium sedimentation.

density-independent *a. appl.* factors limiting the growth of a population that are independent of the existing density of the population. They are usually abiotic factors such as temperature and light intensity.

dental *a.* (1) *pert.* teeth; (2) *appl.* pulp: the inner soft tissue of tooth, innervated and supplied with blood vessels; (3) *appl.* papilla: the small mass of undifferentiated tissue from which a tooth forms.

dental formula method of representing the number of each type of tooth in a mammal. It consists of a series of fractions, the numerators representing the number of each type of tooth in one half of the upper jaw, and the denominators the number in the corresponding lower jaw.

dental plaque bacterial cells surrounded by extracellular polysaccharides, present on teeth.

dentary *n.* bone in lower jaw of many vertebrates. *plu.* **dentaries**.

dentate *a.* (1) toothed; (2) with large saw-like teeth on the margin, *appl.* leaves.

dentate-ciliate with teeth and hairs on the margin, *appl.* leaves.

dentate-crenate with rounded teeth on the margin, *appl.* leaves.

dentate gyrus part of the hippocampal formation in brain, the other part being the hippocampus.

dentes *plu.* of dens.

denticidal *a. appl.* dehiscent fruit with tooth-like formation around top of capsule, as in the dicot family Caryophyllaceae.

denticle *n.* (1) small tooth-like process; (2) small outgrowth of the epidermis on some insects; (3) type of scale present in many fish and covering the whole of the body surface in elasmobranchs, having a shape and structure similar to a small tooth.

denticulate *a.* (1) (*zool.*) having denticles; (2) (*bot.*) with tiny teeth on the margin.

dentin dentine *q.v.*

dentinal *a.* (1) *pert.* dentine; (2) *appl.* tubules: the minute canals in dentine of teeth.

dentine *n.* hard elastic substance, also called ivory, with same constituents as bone (collagen and calcium salts), constituting the interior hard part of vertebrate teeth and outer layer of denticles and dermal bone.

dentition *n.* the type, number and arrangement of teeth.

deoxyadenosine, deoxycytidine, deoxyguanosine nucleosides consisting of

the relevant purine or pyrimidine base linked to the sugar deoxyribose. *see also* thymidine.

deoxyadenosine triphosphate (dATP) deoxyribonucleotide of adenine, one of the four deoxyribonucleotides needed for DNA synthesis.

deoxyadenylate *n.* deoxyadenosine monophosphate (dAMP), a product of partial hydrolysis of DNA.

deoxycytidine triphosphate (dCTP) deoxyribonucleotide of cytosine, one of the four deoxyribonucleotides needed for DNA synthesis.

deoxycytidylate *n.* deoxycytidine monophosphate (dCMP), a product of partial hydrolysis of DNA.

deoxyguanosine triphosphate (dGTP) deoxyribonucleotide of guanine, one of the four deoxyribonucleotides needed for DNA synthesis.

deoxyguanylate *n.* deoxyguanosine monophosphate (dGMP), a product of partial hydrolysis of DNA.

deoxynucleoside triphosphates (dNTPs) general name for the four deoxyribonucleotides needed for DNA synthesis.

deoxyribonuclease (DNase) *n.* any of various enzymes that cleave DNA into shorter oligonucleotides or degrade it completely into its constituent deoxyribonucleotides. *alt.* nuclease. *see* endonuclease, exonuclease.

deoxyribonucleic acid *see* DNA and associated entries.

deoxyribonucleoside *n.* nucleoside containing the sugar deoxyribose.

deoxyribonucleotide *n.* nucleotide containing the sugar deoxyribose.

deoxyribose *n.* pentose sugar similar to ribose but lacking an oxygen atom, present in DNA. *see* Fig. 13.

deoxythymidine thymidine *q.v.*

Fig. 13 D-deoxyribose.

deoxythymidylic acid, deoxythymidylate thymidine monophosphate *q.v.*

deperulation *n.* the pushing apart or throwing off of bud scales.

dephosphorylation *n.* removal of a phosphate group. Often refers to the dephosphorylation of proteins by protein phosphatases, an important mechanism of control of cellular activities.

depigmentation *n.* loss of pigment by a cell.

deplanate *a.* (1) levelled; (2) flattened.

deplasmolysis *n.* re-entry of water into a plant cell after plasmolysis with the reversal of shrinkage of protoplasm.

deplumation *n.* moulting, in birds.

depolarization *n.* reduction in electrical potential difference across a membrane. With respect to living cells, indicates the inside becoming less negative with respect to the outside. *cf.* hyperpolarization.

depolymerization *n.* the disassembly of a multisubunit structure into its constituent subunits, usually referring to the disassembly of structures such as an actin filament into its constituent proteins.

deposit feeder aquatic organism that swallows sediments in order to feed on the microorganisms they contain.

depressant *a.* anything that lowers activity.

depressed *a.* flattened dorsoventrally.

depressomotor *a. appl.* any nerve that lowers muscular activity.

depressor (1) *n.* muscle that lowers or depresses any structure; *a.* (2) *appl.* nerve that lowers the activity of an organ; (3) *appl.* compound that slows down metabolic rate.

depurination *n.* loss of a purine base from DNA.

derepressed *a. appl.* genes which have been switched on from a previously repressed state. *alt.* induced.

derived *a. appl.* character or character state not present in the ancestral stock.

dermal *a. pert.* dermis or, more generally, to the skin.

dermal bone bone derived from the dermis of skin.

dermalia *n.plu.* small hard plates in the surface layer of certain sponges.

dermal tissue system of higher plants, the epidermis and associated tissues.

dermamyotome *n.* the part of a somite that develops into the skeletal muscle cells

of trunk and limbs and into the connective tissue of the dermis.

Dermaptera *n.* order of insects, commonly called earwigs, having cerci modified as forceps, small leathery forewings and membranous hindwings. They undergo a slight metamorphosis.

dermatan sulphate sulphated glycosaminoglycan containing *N*-acetylgalactosamine. It is a constituent of the extracellular matrix in skin, blood vessels and other organs.

dermatic dermal *q.v.*

dermatitis *n.* inflammation of the surface of the skin.

dermatogen *n.* (1) young epidermis in plants; (2) tissue giving rise to the epidermis in plants.

dermatoglyph *n.* pattern of whorls and lines on skin of palm, finger, toe or sole of foot.

dermatoid *a.* (1) resembling a skin; (2) functioning as a skin.

dermatology *n.* the study and treatment of diseases of the skin.

dermatome *n.* (1) the part of a somite that develops into the connective tissue of the dermis; (2) area of skin supplied by a single spinal nerve.

dermatomycosis *n.* fungal infection of human or animal skin. *plu.* **dermatomycoses**.

dermatophyte *n.* any fungal parasite of skin.

dermatopsy *n.* having a skin sensitive to light.

dermatosis *n.* disease of the skin.

dermatosome *n.* unit of cellulose in plant cell wall.

dermatozoon *n.* any animal parasite of the skin.

dermethmoid *n.* a dermal bone above the ethmoid bone of the nasal cavity in fish skull. *alt.* supraethmoid.

dermis *n.* the deeper layers of vertebrate skin underlying the epidermis, from which it is separated by a basement membrane. It is composed of connective tissues and derived from the mesoderm. *see* Fig. 35 (p. 622).

dermo-, dermato-, -derm, -dermis, -dermal word elements from Gk *dermis*, skin.

dermo-ossification *n.* a bone formed in the skin.

dermopharyngeal *n.* superior or inferior plate of membrane bone supporting pharyngeal teeth in some fishes.

dermophyte dermatophyte *q.v.*

dermosclerite *n.* mass of spicules found in tissues of some alcyonarians.

dermosphenotic *n.* bone located around the orbit of eye, between supraorbitals and suborbitals, in teleost fishes.

dermotrichia *n.plu.* fin-rays made of dermal bone.

dermotropic *n.* having a predilection for colonizing dermis, *appl.* parasites.

derris *n.* broad-spectrum, non-persistent insecticide and acaricide, extracted from the roots of certain legumes. It is toxic to fish but not to mammals or birds. *alt.* rotenone.

descending *a.* (1) directed downwards, or towards caudal region, *appl.* blood vessels or nerves; (*bot.*) (2) growing or hanging downwards; (3) *appl.* arrangement of parts of a flower where each petal overlaps the one in front of it.

desegmentation *n.* fusion of segments previously separate.

desensitization *n.* (1) loss of the capacity to make a response; (2) loss of regulatory sites by an enzyme while retaining full catalytic activity.

desert *n.* biome where the average amount of precipitation is erratic and less than 25 cm per annum, and evaporation exceeds precipitation. Such areas have sparse, highly adapted, vegetation, e.g. cacti, succulents and spiny shrubs. Hot deserts such as the Sahara have very high daytime temperatures. Cold deserts such as the Gobi and the northern Californian desert have very low winter temperatures.

deserticolous *a.* living in the desert.

desertification *n.* conversion of semi-arid pastureland and crop land into desert, or the gradual enlargement and encroachment of deserts into formerly marginal arid lands. It is caused by climatic factors such as prolonged drought and by overgrazing and overcultivation.

desiccation *n.* dehydration, drying.

desmergate *n.* type of ant intermediate between a worker and a soldier.

desmids *n.plu.* group of unicellular or colonial freshwater green algae whose cells are typically almost divided in two by a narrow constriction of the cell wall.

desmin *n.* component protein of intermediate filaments in muscle cells.

desmoid *a.* (1) band-like; (2) forming a chain or ribbon; (3) resembling a desmid.

desmology *n.* the anatomy of ligaments.

desmoneme *n.* nematocyst in which the distal end of the thread or closed tube coils around prey when discharged.

desmosome *n.* cell junction that anchors an animal cell strongly to its neighbours by joining the intermediate filaments in one cell to those of neighbouring cells. Belt desmosomes are continuous bands of contact extending around the cell, spot desmosomes are roughly circular points of contact, and hemidesmosomes resemble spot desmosomes but join the cell to the basal lamina. All desmosomes are characterized by thickening of the apposed cell membranes separated by an enlarged intercellular space filled with various types of filamentous material connecting the two cells.

desmotubule *n.* fine tubular structure running through a plasmodesma joining two plant cells and connecting the endoplasmic reticulum of the cells.

despotism *n.* social system in animals in which one individual dominates the rest of the flock which are all equally subservient to him and of equal rank with each other. *see also* dominance systems.

desquammation *n.* shedding of cuticle or epidermis in flakes.

desulphurylation *n.* the removal of a sulphuryl group ($-SH_2$) from a compound.

desynapsis, desyndesis *n.* (1) the normal dissociation of synapsed chromosomes in meiosis, which starts at diplotene; (2) failure of synapsis, caused by disjunction of homologous chromosomes.

detergent *n.* compound that disperses lipids. Detergents are used to release membrane proteins from lipid membranes and thus to solubilize them.

determinant *see* antigenic determinant, cytoplasmic determinant.

determinate *a.* (1) with certain limits, *appl.* growth that stops when an organism or part of an organism reaches a certain size and shape; (2) with a well-marked edge; (3) *appl.* inflorescence with primary axis that terminates early with a flower bud; (4) *appl.* development in which the fate of each cell is already determined by the time it is formed. *cf.* regulative.

determination *n.* the irreversible commitment of an embryonic cell to a particular developmental pathway, which occurs in many cases long before visible differentiation.

determined *a. appl.* embryonic cells once their fate has been irrevocably established.

detoxification *n.* rendering a toxic substance harmless. *alt.* **detoxication.**

detritiphage, detritivore *n.* organism that feeds on detritus. *alt.* detritus feeder.

detritus *n.* (1) small pieces of dead and decomposing plants and animals. *see also* detritus layer; (2) detached and broken down fragments of a structure. *a.* **detrital.**

detritus feeder soil animal that extracts nutrients from detritus, e.g. earthworm. *see also* decomposer.

detritus layer layer on the surface of the soil composed of dead and decaying fragments of organic material.

detrusor *n.* outer of three layers of the muscular coat of urinary bladder, or may refer to all three layers.

detumescence *n.* subsidence of swelling.

deuterencephalic *a. pert.* hindbrain.

deuterencephalon hindbrain *q.v.*

deuterium *n.* isotope of the element hydrogen with mass number 2 (^{2}H), which is the hydrogen isotope in heavy water, deuterium oxide, D_2O.

deuterocerebrum *n.* that portion of crustacean brain from which antennular nerves arise.

deuterocoel coelom *q.v.*

deuterocone *n.* cusp on mammalian premolar teeth corresponding to the protocone of molar teeth.

deuteroconidium *n.* a conidium produced by division of protoconidium in dermatophytes (fungal parasites of skin).

deuterogamy *n.* (1) secondary fertilization; (2) pairing substituting for the union of gametes, as in fungi.

deuterogenesis *n.* second phase of embryonic development, involving growth in length and consequent bilateral symmetry.

Deuterolichenes *n.* class of lichens in which a deuteromycete is the fungal partner.

Deuteromycota, Deuteromycotina, deuteromycetes *n., n., n.plu.* large group of fungi, known only in the asexual, conidia-bearing form, but which display strong affinities to the ascomycetes. They may be ascomycetes that have lost their ascus stage or in which the sexual phase has not yet been discovered. *alt.* Fungi Imperfecti.

deuterostome *n.* mouth formed secondarily in embryonic development, as distinct from the gastrula mouth. *alt.* **deuterostoma.**

deuterostomes *n.plu.* animals with a true coelom, radial cleavage of the egg, and in which the blastopore becomes the anus. They comprise pogonophorans, hemichordates, echinoderms, urochordates and chordates.

deuterotoky *n.* parthenogenesis where both sexes are produced.

deuterotype *n.* the specimen chosen to replace the original type specimen for designation of a species.

deutocerebrum *n.* portion of insect brain derived from fused ganglia of antennary segment of head.

deutomerite *n.* posterior division of certain gregarines. *cf.* primite.

deutonymph *n.* second nymphal stage or instar, either chrysalis-like or motile, in the development of mites and ticks.

deutoscolex *n.* a secondary scolex produced by budding, in bladderworm stage of certain tapeworms.

deutosporophyte *n.* second sporophyte phase in life cycle of red algae.

deutosternum *n.* sternite of segment bearing pedipalps in mites and ticks.

development *n.* in biology, the changes that occur as a multicellular organism develops from a single-celled zygote, from the first cleavage of the fertilized ovum until maturity. *alt.* ontogeny.

developmental *a. pert.* or involved in development, *appl.* e.g. genes or hormones active at a particular time during development and which have a specific effect on development.

Devonian *n.* geological period lasting from *ca.* 416 million years ago to 359 million years ago, ending in an extinction event. Lobe-finned fishes, bony fishes and ammonites first appeared. On land, the first insect fossils are found in the mid-Devonian and the first amphibian fossils in the later Devonian. Vegetation included lycophytes, ferns and progymnosperms (including the earliest trees), and, by the end of the period, the first seed-bearing plants (seed ferns).

dexiotropic *a.* turning from right to left, as whorls, *appl.* shells.

dextral *a.* on or *pert.* the right.

dextran *n.* branched polysaccharide made of glucose residues joined by α-1,6 linkages. Dextrans are found in capsules and slime layers of bacteria.

dextranase *n.* enzyme that degrades dextrans to glucose.

dextrin *n.* any of a group of small soluble polysaccharides, partial hydrolysis products of starch.

dextro-rotatory *a. appl.* optically active molecules that rotate a beam of plane-polarized light in a clockwise (+) direction.

dextrorse *a.* (1) growing in a spiral which twines from left to right; (2) clockwise.

dextrose glucose *q.v.*

Df in genetics the abbreviation for deficiency, the absence or inactivation of a given gene.

D-factor double-stranded RNA virus that causes disease in some fungi.

D gene segment any of the 'diversity' gene segments found in immunoglobulin heavy-chain loci and in T-cell receptor β and δ loci. They are short DNA sequences, of which one is spliced to a V gene segment and a J gene segment to produce the DNA sequence coding for the variable region of the immunoglobulin polypeptide chain. *alt.* **D segment, D region**, which may also designate the corresponding region of the polypeptide chain. *see also* gene rearrangement, J gene segment, variable region, V gene segment.

DHAP dihydroxyacetone phosphate *q.v.*

DHFR dihydrofolate reductase *q.v.*

diabetes mellitus condition characterized by abnormally high levels of blood glucose due to inability of cells to take it up from the blood. It may have various causes, e.g. non-production of the hormone

insulin (which stimulates glucose uptake from the blood), inability of cells to respond to insulin, or excessive production of hormones that have opposing effects to those of insulin. *see* type 1 diabetes, type 2 diabetes.

diabetogenic *a.* causing diabetes.

diachaenium *n.* each part of a cremocarp.

diachronous *a.* dating from different periods, *appl.* fossils of different ages occurring in the same geological formation.

diacoel(e) *n.* third ventricle of the brain.

diacranteric *a.* with a distinct space between front and back teeth, as in snakes.

diactinal *a.* having two rays pointed at each end.

diacylglycerol (DAG) *n.* small membrane-bound lipid molecule that acts as an intracellular signalling molecule and which is generated as the result of stimulation of various cell-surface receptors. It is produced by the action of phospholipase C on membrane phosphatidylinositol phosphate and, together with Ca^{2+}, activates the enzyme protein kinase C.

diacytic *a. appl.* stomata of a type with one pair of subsidiary cells, with their common axis at right angles to the long axis of the guard cells, surrounding the stoma.

diadelphous *a.* having stamens in two bundles due to fusion of filaments.

diadematoid *a.* of sea urchins, having three primary pore plates with occasionally a secondary between aboral and middle primary.

diadromous *a.* (1) migrating between fresh and sea water; (2) having veins radiating in a fan-like manner, *appl.* leaves.

diaene *n.* a two-pronged spicule.

diagenesis *n.* changes (due e.g. to pressure and temperature) that occur in the course of fossilization of an organism. *a.* **diagenetic**.

diageotropism *n.* growth movement in a plant organ so that it assumes a position at right angles to the direction of gravity.

diagnosis *n.* (1) concise description of the distinctive characters of an organism; (2) identification of a physiological or pathological condition by its distinctive signs.

diagnostic *a.* (1) distinguishing; (2) *appl.* characters that differentiate e.g. a species or genus from others similar.

diakinesis *n*. last stage of meiotic prophase, in which the nuclear membrane breaks down.

dialect *n*. local variant of e.g. bird songs, mating calls, bee waggle dances.

diallelic *a. pert.* or involving two alleles, *appl.* polyploid with two different alleles at a locus.

dialysate *n*. any substance which passes through a semipermeable membrane during dialysis. *alt.* diffusate.

dialysis *n*. separation of large molecules such as proteins from small molecules and ions by the inability of the larger molecules to pass through a semipermeable membrane.

dialystelic *a*. having the steles in the stem remaining more or less separate. *n*. **dialystely**.

diamine *n*. compound containing two amino groups, such as cadaverine or putrescine.

diaminopimelic acid (DAP) amino acid structurally similar to lysine, found in the peptidoglycans of bacterial cell walls.

diandrous *a*. (1) having two free stamens; (2) in a moss, having two antheridia each surrounded by a bract.

dianthovirus group plant virus group containing isometric single-stranded RNA viruses, type member carnation ringspot virus.

diapause *n*. spontaneous state of dormancy occurring in the life cycles of many insects, esp. in larval stages.

diapedesis *n*. the crawling of white blood cells between the walls of endothelial cells as they migrate from capillaries into surrounding tissue.

Diapensales *n*. order of dicot trees and herbs including the families Ebenaceae (ebony), Sapotaceae (sapote), Styracaceae (styrax) and others.

diaphototropism *n*. (1) growth movement in plant organs to assume a position at right angles to rays of light. *alt.* **diaheliotropism**, when the light is sunlight.

diaphragm(a) *n*. (1) sheet of muscle and tendon separating chest cavity from abdominal cavity in mammals and whose movement aids breathing; (2) various other partitions in other organisms, such as the fibromuscular abdominal septum enclosing perineural sinus in some insects and the transverse septum separating cephalothorax from abdomen in some arachnids.

diaphysis *n*. (1) shaft of limb bone, *a*. **diaphyseal**; (2) abnormal growth of an axis or shoot.

diaplexus *n*. choroid plexus of 3rd ventricle of brain.

diapophysis *n*. lateral or transverse process of neural arch of a vertebra.

diapsid *a*. having skull with both dorsal and ventral temporal fenestrae on each side.

diapsids *n.plu.* group of reptiles with diapsid skulls, known from the late Carboniferous, to which most modern reptiles belong and also including extinct forms such as the dinosaurs.

diarch *a*. (1) with two xylem and two phloem bundles; (2) *appl.* root in which protoxylem bundles meet and form a plate of tissue across cylinder with phloem bundle on each side.

diarthric biarticulate *q.v.*

diarthrosis *n*. an articulation allowing considerable movement, a moveable joint.

diaschistic *a. appl.* type of tetrad produced by one transverse and one longitudinal division of mother cell during meiosis.

diaspore *n*. any spore, seed, fruit or other part of plant or fungus when being dispersed and able to produce a new organism.

diastase *n*. original name for a mixture of α- and β-amylase. For a time in the 19th century the term was used for any enzyme. Later used to refer to α- or β-amylase.

diastasis *n*. (1) rest period preceding systole in heart contraction; (2) abnormal separation of parts that are usually joined together.

diastema *n*. a toothless space, usually between two types of teeth.

diaster *n*. stage in mitosis where daughter chromosomes are grouped near spindle poles ready to form new nuclei.

diastereoisomers *n*. non-mirror-image structural isomers.

diastole *n*. (1) rhythmical relaxation of the heart; (2) rhythmical expansion of a contractile vacuole. *cf.* systole.

diastomatic *a*. (2) through stomata or pores; (2) giving off gases from spongy parenchyma through stomata.

diatom *n.* common name for a member of the phylum Bacillariophyta, a group of algae characterized by delicately marked thin double shells of silica.

diatomaceous *a.* containing the shells of diatoms, *appl.* earth.

diatomin phycoxanthin *q.v.*

diatropism *n.* tendency of organs or organisms to place themselves at right angles to line of action of stimulus.

diauxic growth growth occurring in two phases separated by a period of inactivity.

diauxy *n.* adaptation of a microorganism to the use of an additional sugar by the induction of the relevant enzymes.

diaxon *a.* with two axes.

diaxonic *a. appl.* neurons with two axons.

diazotroph *n.* organism able to convert elemental nitrogen into ammonia.

dibranchiate *a.* with two gills.

DIC disseminated intravascular coagulation *q.v.*

dicaryo- *see* dikaryo-.

dicentral *a. appl.* central canal in fish vertebral column.

dicentric *a. appl.* chromosome possessing two centromeres, which often leads to chromosome breakage at mitosis as the two centromeres are pulled towards opposite poles. *n.* **dicentric**.

dicerous *a.* (1) with two horns; (2) with two antennae.

dichasium *n.* cymose inflorescence in which two lateral branches occur at about same level. *alt.* dichasial cyme.

dichlamydeous *a.* having both calyx and corolla.

dichocarpous *a.* with two forms of fructification, *appl.* certain fungi.

dichogamy *n.* maturing of sexual elements at different times, ensuring cross-fertilization. *a.* **dichogamous**.

dichophysis *n.* rigid, dichotomous hypha, as in hymenium and trama of some fungi.

dichoptic *a.* with eyes quite separate. *cf.* holoptic.

dichotomy *n.* (1) branching that results from division of growing point into two equal parts; (2) repeated forking. *a.* **dichotomous**.

dichroism *n.* property displayed by some substances of showing two colours, as one

by transmitted and one by reflected light. *a.* **dichroic**.

dichromatic *a.* (1) showing dichromatism *q.v.*; (2) seeing only two colours.

dichromatism *n.* condition in which members of a species show one of only two distinct colour patterns. *a.* **dichromatic**.

diclesium *n.* multiple fruit or anthocarp formed from an enlarged and hardened perianth.

diclinous *a.* (1) with stamens and pistils on separate flowers; (2) with staminate and pistillate flowers on same plant; (3) with antheridia and oogonia on separate hyphae.

dicots dicotyledons *q.v.*

Dicotyledones, dicotyledons *n., n.plu.* class of flowering plants having an embryo with two cotyledons (seed leaves), parts of the flower usually in twos or fives or their multiples, leaves with net veins, and vascular bundles in the stem in a ring surrounding a central pith. In some classifications called the Magnoliopsida. *a.* **dicotyledonous**. *see* Appendix 1.

dicratic *a.* with two spores of a meiotic tetrad being of one sex and the other two of the opposite sex.

dictyate *a. appl.* stage in oogenesis in human females, the prolonged diplotene stage of the first meiotic prophase, in which oocyte development is arrested.

dictyodromous *a.* net-veined, when the smaller veins branch and anastomose freely.

Dictyoptera *n.* order of insects including the cockroaches and praying mantises. They are winged but often non-flying, with long antennae, biting mouthparts, tough narrow forewings and broad membranous hindwings.

dictyosome Golgi body *q.v.*

dictyospore *n.* spore with both vertical and horizontal septa.

dictyostele *n.* (1) network formed by leaf traces; (2) stele having large overlapping leaf gaps which dissect the vascular system into strands, each having phloem surrounding xylem.

dictyostelid slime moulds cellular slime moulds *q.v.*

Dictyostelium discoideum a cellular slime mould, widely used as a model system for developmental and genetic studies.

dictyotene *n. appl.* stage of meiosis in which diplotene is prolonged, as in oocytes during yolk formation.

dicyclic *a.* (1) with two whorls; (2) biennial *q.v., appl.* herbaceous plants.

Dicyemida(e), dicyemids *n., n.plu.* class of Mesozoa having a body that is not annulated. They are parasites of the kidneys of cephalopod molluscs. *see also* Rhombozoa.

dicystic *a.* with two encysted stages.

didactyl *a.* having two fingers, toes or claws.

didelphic *a.* (1) having a paired uterus, as in certain nematodes; (2) having two uteri, as in marsupials.

dideoxy method the Sanger method of DNA sequencing, which employs dideoxynucleotides.

dideoxynucleotide *n.* nucleotide analogue lacking the 3′-hydroxyl group on the deoxyribose sugar. When incorporated during DNA replication it terminates extension of the DNA chain. Used in DNA sequencing by the Sanger method.

diductor *n.* muscle running posterior to axis of hinge and joining the two valves of the shell in articulate brachiopods, and which opens shell as it contracts.

Didymelales *n.* order of dicot trees comprising the family Didymelaceae with a single genus *Didymeles*.

didymous *a.* growing in pairs.

didynamous *a.* with four stamens, two long and two short.

dieback *n.* (1) population crash; (2) death of stems of woody plants from the tip backwards.

diecdysis *n.* in the moulting cycle of arthropods, a short period between proecdysis and metecdysis.

diecious *alt.* spelling of dioecious *q.v.*

diel *a.* (1) during or *pert.* 24 hours; (2) occurring at 24-hour intervals.

diencephalon *n.* that part of the developing forebrain that will form the thalamus and hypothalamus and other structures.

diestrus *alt.* spelling of dioestrus.

differential centrifugation the successive centrifugation of a cell extract at different speeds so that different cellular components are sedimented at each speed.

differential-interference-contrast microscopy *see* **Nomarski differential-interference-contrast microscopy**.

differentiated *a. appl.* cells that have developed their final specialized structure and function, e.g. muscle cell, nerve cell, red blood cell.

differentiation *n.* (1) in a general sense, the increasing specialization of organization of the different parts of an embryo as a multicellular organism develops from the undifferentiated fertilized egg; (2) of cells, the development of cells with specialized structure and function from unspecialized precursor cells.

differentiation antigens cell-surface antigens that are specific to different cell types and tissues.

differentiation factor name given to cytokines that induce cell differentiation.

diffluence *n.* disintegration by vacuolation.

diffraction colours colours produced not by pigment but by the unevenness of the surface of an organism causing the diffraction of reflected light.

diffusate dialysate *q.v.*

diffuse *a.* (1) widely spread; (2) not localized; (3) not sharply defined at the margin.

diffuse-porous *appl.* wood in which vessels of approximately the same diameter tend to be evenly distributed in a growth ring. *cf.* ring-porous.

diffusion *n.* the passive movement of molecules or ions from a region of high concentration to a region of low concentration.

diffusion pressure deficit (DPD) suction pressure *q.v.*

digalactosyl diacylglycerol glycolipid found in plant cell membranes.

digametic heterogametic *q.v.*

digastric *a.* (1) *appl.* muscles fleshy at ends with a central portion of tendon; (2) *appl.* one of the suprahyoid muscles; (3) *appl.* a branch of the facial nerve; (4) *appl.* a lobule of cerebellum.

digenean *a. appl.* parasitic flatworms of the order Digenea. They include liver, blood and gut flukes, such as *Schistosoma*, the cause of schistosomiasis in humans. As adults they are endoparasites of many vertebrates. They have complex life cycles with larval stages in molluscs and sometimes also in several other different hosts. *see also* cercaria, metacercaria, miracidium.

digenesis *n.* alternation of sexual and asexual generations.

digenetic *a.* (1) *pert.* digenesis; (2) requiring an alternation of hosts, *appl.* parasites.

digenic *a. pert.* or controlled by two genes.

digeny *n.* sexual reproduction.

digestion *n.* (1) the process whereby nutrients are rendered soluble and capable of being absorbed by the organism or cell. It involves the action of hydrolytic enzymes that break down macromolecules such as proteins, polysaccharides and fats to their small-molecule constituents; (2) the similar enzymatic breakdown of molecules in organelles such as lysosomes.

digestive *a. pert.* digestion, or having power of aiding in digestion.

digestive gland sac-like portion of intestine in molluscs and other invertebrates, which produces digestive enzymes and in which food is digested.

digestive system, digestive tract the canal leading from mouth to anus in which food is digested, comprising the oesophagus, stomach and intestines, and associated glands secreting digestive enzymes, such as the pancreas.

digestive vacuole type of lysosome resulting from phagocytosis of large particles in animal cells.

digger wasps solitary insects of the superfamilies Pompiloidea and Sphecoidea of the order Hymenoptera, somewhat resembling the true wasps in appearance, and including species that nest in burrows they dig in the ground.

digit *n.* terminal division of limb as finger or toe, in vertebrates other than fishes.

digital *n.* the terminal joint of pedipalp of spider.

digitaliform *a.* finger-shaped, *appl.* flowers that are shaped like the fingers of a glove, e.g. foxglove.

digitalin, digitonin, digitoxin glycosides from leaves of foxglove, *Digitalis purpurea*.

digitate *a.* (1) having parts arranged like fingers of a hand; (2) having fingers.

digitiform *a.* finger-shaped.

digitigrade *a.* walking with only the digits touching the ground.

digitinervate *a.* having veins radiating out from base like the fingers of a hand, with usually five or seven veins, *appl.* leaves.

digitonin, digitoxin *see* digitalin.

digitule *n.* any small finger-like process.

digitus digit *q.v.*

diglycerol tetraether *n.* type of membrane lipid found only in Archaea, in which two glycerol headgroups are joined by ether linkages to two long phytanyl chains composed of isoprene units. This type of lipid spans the membrane as a single molecule.

diglyphic *a.* having two siphonoglyphs.

digoneutic *a.* breeding twice a year.

digynous *a.* with two carpels.

dihaploid *n.* organism arising from a tetraploid but only containing half a normal tetraploid chromosome complement.

dihybrid *n.* (1) progeny of a cross in which the parents differ in two distinct characters; (2) organism heterozygous at two distinct loci.

dihydrofolate reductase (DHFR) enzyme catalysing the regeneration of tetrahydrofolate from dihydrofolate. It is inhibited by the folate antagonists methotrexate (amethopterin) and aminopterin.

dihydrotachysterol *n.* vitamin D_4, an irradiation product of the dihydro derivative of ergosterol, which counteracts impaired parathyroid function.

dihydrouridine *n.* unusual nucleotide found in tRNA, formed by addition of hydrogen to positions 5 and 6 of uracil, saturating the double bond.

dihydroxyacetone phosphate (DHAP) three-carbon ketose monosaccharide phosphate intermediate in photosynthetic carbon dioxide fixation and glycolysis, important in cellular respiratory metabolism as part of the glycerol phosphate shuttle.

diisopropylphosphofluoridate (DIPF) compound that inhibits some enzymes, including acetylcholinesterase, which is the basis for its use in insecticides and nerve gas.

Dikarya *n.* name sometimes given to those fungi (the Ascomycota and Basidiomycota) that have hyphae with paired nuclei.

dikaryon *n.* (1) pair of nuclei situated close to one another and dividing at the same time, as in some fungal hyphae; (2) the hypha containing such a pair of nuclei. *a.* **dikaryotic**.

dikinetid *a.* a kinetid having two kineto-somes.

dikont biflagellate *q.v.*

dilambodont *a. appl.* insectivores having molar teeth with W-shaped ridges.

dilatation *n.* dilation.

dilatator dilator *q.v.*

dilate *a.* flattened, *appl.* structure with a wide margin.

dilation *n.* expansion or swelling of a hollow structure such as a blood vessel.

dilator *n.* any muscle that expands or dilates an organ.

Dilleniales *n.* order of woody, often climbing dicots, comprising the families Crossosomataceae and Dilleniaceae.

diluvial *a.* produced by a flood, *appl.* soil deposits.

dimastigote biflagellate *q.v.*

dimegaly *n.* condition of having two sizes, *appl.* sperm and ova.

dimer *n.* protein made up of two subunits.

dimeric *a.* (1) of a protein, having two subunits; (2) having two parts, bilaterally symmetrical.

dimerous *a.* (1) in two parts; (2) having each whorl of two parts, of flowers; (3) with a two-jointed tarsus.

dimitic *a. appl.* basidiocarp composed of generative hyphae and either binding or skeletal hyphae.

dimixis *n.* fusion of two kinds of nuclei, in heterothallism.

dimorphic *a.* (1) having or *pert.* two different forms; (2) *appl.* fungi producing two morphologically different zoospores.

dimorphism *n.* condition of having two distinct forms within a species.

dinergate *n.* soldier ant.

dineuronic *a.* with double innervation, *appl.* chromatophores with two sets of nerve fibres, one directing dispersion and the other concentration of pigment.

dinitrogenase *n.* the iron- and molybdenum-containing component of bacterial nitrogenase (Mo-Fe protein). It reduces N_2.

dinitrogenase reductase the iron-containing component of bacterial nitrogenase. It transfers electrons from ferredoxin to the Mo-Fe protein component of nitrogenase.

dinitrophenol (DNP) small lipid-soluble organic molecule with various uses in experimental biology. It is used as an uncoupler of oxidative phosphorylation, as a marker for certain amino acids in protein sequencing, and as a hapten in experimental immunology.

Dinoflagellata, dinoflagellates *n., n.plu.* phylum of unicellular protists having two flagella, one pointing forwards, the other forming a girdle around the body. Sometimes considered as part of the superkingdom Alveolata. A major component of marine and freshwater plankton, some are autotrophic and photosynthetic, some heterotrophic. The nuclear organization of dinoflagellates differs from that of a typical eukaryote or prokaryote: dinoflagellate DNA is not complexed with histones and is condensed into microscopically visible chromosomes throughout the life cycle even though it is still transcriptionally active. The stages of mitosis are absent. Formerly class Phytomastigophorea in animal classification or Pyrrophyta (Dinophyta) in plant classification.

dinokaryotic *a.* term sometimes applied to the nuclear organization of a dinoflagellate *q.v. alt.* mesokaryotic.

Dinornithiformes *n.* order of flightless birds from New Zealand, comprising the extant kiwis and extinct species such as the moa.

dinosaur *n.* member of either of two orders of reptiles that flourished during the Mesozoic: the Saurischia, the lizard-hipped dinosaurs, or the Ornithischia, the bird-hipped dinosaurs. The Saurischia included both bipedal carnivores and very large quadrupedal herbivores. The Ornithischia were mostly quadrupedal and all herbivorous.

dinucleotide *n.* two nucleotides linked together by a $3',5'$-phosphodiester bond.

diocoel *n.* cavity of the diencephalon, particularly in embryo.

dioecious *a.* (1) having the sexes separate; (2) having male and female flowers on different individuals. *n.* **dioecism**.

dioestrus *n.* quiescent period between periods of sexual receptivity in female animals with more than one period of fertility each year.

dionychous *a.* having two claws, as tarsi of certain spiders.

dioptrate *a.* having eyes or ocelli separated by a narrow line.

dioptric *a. pert.* transmission and refraction of light.

diorchic *a.* having two testes.

diosgenin *n.* complex steroid obtained from certain species of yam and which can be converted into 16-dehydropregnenolone, one of the main active ingredients in oral contraceptives.

dioxygenase *n.* enzyme that incorporates both atoms of an oxygen molecule (O_2) into a compound. *cf.* monooxygenase.

dipeptidase *n.* an exopeptidase that catalyses the cleavage of terminal dipeptides from oligopeptides or polypeptides. EC 3.4.13.

dipeptide *n.* two amino acids linked by a peptide bond.

dipetalous *a.* having two petals.

DIPF diisopropylphosphofluoridate *q.v.*

diphasic *a.* having two distinct states, *appl.* e.g. life cycle.

diphosphatidyl glycerol any phosphoglyceride with glycerol as the alcohol group, found chiefly in plants, cardiolipin being one of the few examples known from animals.

2,3-diphosphoglycerate (DPG) compound present in red blood cells in equimolar amounts with haemoglobin. It binds to deoxyhaemoglobin, reducing its affinity for oxygen and allowing oxygen unloading in tissues.

diphtheria toxin toxic protein from the bacterium *Corynebacterium diphtheriae* that inhibits the action of elongation factor eEF-2 in eukaryotic cells by catalysing ADP-ribosylation of the factor. It thus blocks protein synthesis.

diphycercal *a.* with a caudal fin in which the vertebral column runs straight to tip, dividing the fin symmetrically.

diphygenetic *a.* producing embryos of two different types.

diphyletic *a. pert.* or having origin in two separate lines of descent.

diphyllous *a.* having two leaves.

diphyodont *a.* with deciduous and permanent sets of teeth, i.e. two successive sets of teeth.

diplanetic *a.* with two distinct kinds of zoospores.

diplanetism *n.* condition of having two periods of motility in one life history, as of zoospores in some fungi.

dipleurula *n.* bilaterally symmetrical larva of echinoderms.

diplobiont *n.* organism characterized by at least two kinds of individual in its life cycle, such as sexual and asexual. *cf.* haplobiont.

diplobivalent *n.* bivalent containing two anomalous doubly duplicated chromosomes and hence eight chromatids.

diploblastic *a.* having only two germ layers, e.g. endoderm and ectoderm, as in coelenterates and sponges. *cf.* triploblastic.

diplocardiac *a.* with the two sides of the heart quite distinct.

diplocaryon diplokaryon *q.v.*

diplocaulescent *a.* with secondary stems and branches.

diplochromosome *n.* anomalous chromosome, having four chromatids instead of two, attached to centromere.

diplocyte *n.* a cell having conjugate nuclei.

diplogangliate *a.* with ganglia arranged in pairs.

diplogenesis *n.* development of two parts instead of a single part.

diplohaplont *n.* organism with alternation of diploid and haploid generations.

diploic *a.* occupying channels in cancellous tissue of bone.

diploid (1) *a. appl.* organisms whose cells (apart from the gametes) have two sets of chromosomes, and therefore two copies of the basic genetic complement of the species. Designated $2n$. *cf.* haploid; (2) *n.* a diploid organism or cell.

diploid arrhenotoky the production of diploid males from unfertilized eggs, which is known only in the scale insect *Lecanium putnami*.

diploidization *n.* (1) doubling of chromosome number in haploid cells; (2) the restoration of the diploid state.

diploidy *n.* having two sets of chromosomes in the somatic cells.

diplokaryon *n.* nucleus with two diploid sets of chromosomes.

diplomonads *n.plu.* group of parasitic freshwater flagellate protozoa, which includes the human parasite *Giardia*.

diplomycelium *n.* a diploid or dikaryotic mycelium.

diploneural *a.* supplied with two nerves.

diplont *n.* organism having diploid somatic nuclei. *alt.* diploid.

diploperistomous *a.* having a double projection or peristome.

diplophase *n.* (1) stage in life history of an organism in which nuclei are diploid, *alt.* sporophyte phase; (2) diplotene phase in meiosis.

diplophyll *n.* a leaf having palisade tissue on upper and lower side with intermedial spongy parenchyma tissue.

Diplopoda, diplopods *n., n.plu.* in some classifications, a class of arthropods commonly called millipedes, having numerous similar apparent segments each in fact made up of two segments and therefore bearing two pairs of legs. In some classifications it is considered as a subclass or order of class Myriapoda.

diploptile *n.* double down feather, without a shaft, formed by precocious development of barbs of the adult feather.

diplosome *n.* (1) double centrosome lying outside the nuclear membrane; (2) a paired heterochromosome.

diplosomite *n.* body segment consisting of two annular parts, prozonite and metazonite, in diplopods.

diplospondyly *n.* the condition of having two centra to each myotome, or with one centrum and well-developed intercentrum. *a.* **diplospondylic**.

diplospory *n.* type of apomixis in which a diploid megaspore mother cell gives rise directly to the embryo.

diplostemonous *a.* (1) with two whorls of stamens in regular alternation with perianth leaves; (2) with stamens double the number of petals.

diplostichous *a.* arranged in two rows or series.

diplotegia *a.* an inferior fruit with dry dehiscent pericarp.

diplotene *n.* stage in 1st division of meiosis at which bivalent chromosomes appear split longitudinally.

Diplura *n.* order of wingless insects with a pair of cerci and two 'tails' on last segment, sometimes called two-pronged bristletails. Minute white insects with no eyes, found in soil and under stones.

dipnoan *a.* breathing by lungs and gills.

Dipnoi, dipnoans *n., n.plu.* group of bony fishes, commonly called lungfish, known from the Devonian, possessing lungs and broad, crushing toothplates. The three genera of modern lungfish (found in Australia, South America and Africa) are air-breathing and live in tropical areas with a dry season, and have a reduced skeleton.

diprotodont *a.* having two anterior incisors large and prominent, the rest of incisors and canines being smaller or absent.

Dipsacales *n.* order of dicot herbs and shrubs, rarely small trees, comprising the families Adoxaceae (moschatel), Caprifoliaceae (honeysuckle), Dipsacaceae (teasel) and Valerianaceae (valerian).

Diptera, dipterans *n., n.plu.* large order of insects including the housefly and other two-winged (true) flies, mosquitoes and fruitflies (*Drosophila*). They have one pair of functional wings only, the second pair being reduced to small halteres. There is complete metamorphosis.

dipterocecidium *n.* gall caused by a dipteran insect.

dipterous *a.* (1) with two wings, or winglike expansions; (2) *pert.* the Diptera (true flies).

direct development type of heterochronous development in which the embryo abandons the larval stages of development and proceeds directly to adult stage.

directed mutagenesis the alteration of an isolated DNA at some specified position and its reintroduction into an organism.

direct germination germination of a fungal spore by means of a germ tube.

directional selection (1) selection that changes the frequency of an allele in a constant direction; (2) selection that acts on one extreme of the range of variation in a particular character, and therefore tends to shift the entire population towards the opposite end of the range.

dirhinic *a.* (1) having two nostrils; (2) *pert.* both nostrils.

disaccharide *n.* any of a group of carbohydrates which are the condensation products of two monosaccharide units with the elimination of a molecule of water, and which include sucrose, lactose and maltose.

disarticulate *a.* separated at a joint or joints.

disassortative mating mating between organisms of unlike phenotype.

disc *n.* (*bot.*) (1) middle portion of capitulum in Compositae; (2) adhesive end of tendril; (3) base of seaweed thallus, by which it adheres to rocks; (4) (*zool.*) area around mouth in many animals, as in starfish.

discal *n.* large cell at base of wing of Lepidoptera completely enclosed by veins, also in some Diptera.

disc florets inner florets borne on much reduced stalks in many flower-heads of the Compositae type, e.g. the yellow florets in a daisy. *cf.* ray florets.

disciflorous *a.* having flowers in which the receptacle is large and disc-like.

disciform discoid *q.v.*

disclimax *n.* subclimax stage in plant succession replacing or modifying true climax, usually the result of animal and human activity.

discoblastic *a. pert.* blastula formed from egg with a disc-like blastoderm.

discocarp *n.* (1) special enlargement of thalamus below calyx in certain flowers; (2) (*mycol.*) disc-shaped apothecium *q.v.*

discocellular vein cross vein between 3rd and 4th longitudinal veins of insect wing.

discoctasters *n.plu.* sponge spicules with eight rays terminating in discs, each disc corresponding in position to the corners of a cube.

discodactylous *a.* with a sucker at end of digit.

discohexactine *n.* sponge spicule with six equal rays meeting at right angles.

discohexaster *n.* hexactine sponge spicule with rays ending in discs.

discoid *a.* (1) flat and circular; (2) disc-shaped.

discoidal *a.* (1) disc-like; (2) *appl.* type of incomplete cleavage in which blastoderm initially forms a one-layered disc or cap of cells on top of yolk, as in eggs of birds and reptiles.

discoidins *n.* family of lectins produced by the slime mould *Dictyostelium discoideum*.

discolichen *n.* lichen in which the fungal partner is a discomycete.

Discomycetes *n.* group of ascomycete fungi, commonly called cup fungi, and including also the earth tongues, truffles and morels, in which the fruiting body is in the form of an apothecium, usually black or brightly coloured. The ascothecium may be open and cup-shaped or disc-like, or closed and subterranean, as in truffles.

discontinuity *n.* (1) occurrence (e.g. of a species) in two or more separate areas of geographical regions; (2) *appl.* layer: thermocline.

discontinuous distribution pattern of geographical distribution where the same or similar species are found in widely separated parts of the world, which is taken to indicate that the species was formerly distributed over the whole area but has become extinct in the intervening regions.

discontinuous variation variation between individuals of a population in which differences are marked and do not grade into each other, brought about by the effects of different alleles at a few major genes. *alt.* qualitative inheritance. *cf.* continuous variation.

disconula *n.* eight-rayed stage in larval development of certain coelenterates.

discoplacenta *n.* placenta with villi on a circular cake-like disc.

discous discoid *q.v.*

discrimination learning situation in which although an animal is capable of discriminating between two stimuli, it only does so after it has learned to tell them apart.

discus proligerus granular zone in a Graafian follicle, the mass of cells in which the ovum is embedded.

disease gene the mutant allele responsible for an inherited disease.

dishabituation *n.* the abolition of habituation to a particular stimulus, seen e.g. after administration of a strong generalized stimulus of a different type.

disinhibit *v.* to remove inhibition. *n.* **disinhibition**.

disjunct *a.* (1) with body regions separated by deep constrictions; (2) *appl.* distribution in which potentially interbreeding populations are separated by a sufficient distance to preclude gene flow.

disjunction *n.* (1) separation of paired chromosomes in anaphase of mitosis or meiosis; (2) discontinuity (1) *q.v.*

disjunction mutants mutants in which chromosomes are divided unequally between daughter cells at meiosis.

disjunctive symbiosis mutually helpful condition of symbiosis although there is no direct connection between the partners.

disjunctor *n.* zone of separation between successive conidia.

disk *alt.* spelling of disc.

disomic *a. pert.* or having two homologous chromosomes or genetic loci. *n.* **disomy.**

disoperation *n.* (1) co-actions resulting in disadvantage to individual or group; (2) indirectly harmful influence of organisms upon each other.

dispermous *a.* having two seeds.

dispermy *n.* penetration of an ovum by two sperm.

displacement ·activity performance of a piece of behaviour, usually in moments of frustration or indecision, that is not directly relevant to the situation at hand.

display *n.* series of stereotyped movements or sounds that cause a specific response in another animal, usually of the same species. Often deployed in courtship or territorial defence.

disporocystid *n. appl.* oocyst of sporozoans when two sporocysts are present.

disporous *a.* with two spores.

disruptive coloration colour patterns that obscure the outline of an animal and so act as camouflage and protection against predators.

disruptive selection selection that operates against the middle range of variation in a particular character, tending to split a population into two populations showing the extreme phenotype at either end of the range.

dissected *a.* (1) having leaf blade cut into lobes, with incisions nearly reaching the midrib; (2) with parts displayed.

dissection *n.* the cutting open of an animal or plant such that the internal structures are clearly displayed.

disseminated *a.* distributed widely, e.g. throughout the whole body.

disseminated intravascular coagulation (DIC) massive systemic blood clotting that occurs in septic shock.

disseminule diaspore *q.v.*

dissepiment *n.* (1) partition in a compound plant ovary; (2) calcareous partition in corals; (3) trama *q.v.*

dissilient *a.* springing open, *appl.* capsules of various plants which dehisce explosively.

dissimilation *n.* (1) the breakdown of nutrients to provide energy and simple compounds for intermediary metabolism; (2) dissimilative metabolism *q.v.*

dissimilative metabolism *a. appl.* metabolism in which large amounts of inorganic compounds are reduced for energy metabolism, with the excretion of the reduced compound to the environment as a waste product. *alt.* **dissimilatory mechanism.** *cf.* assimilative.

dissociation constant (K_d) the inverse of the association constant (K_a) *q.v.* It is a measure of the rate at which two molecules dissociate from each other and thus of their strength of binding.

dissogeny, dissogony *n.* condition of having two sexually mature periods in the same animal, one in larva and the other in adult.

dissolved organic carbon (DOC) the fraction of carbon bound in organic compounds in water in particles smaller than 0.45 μm.

dissolved organic matter (DOM) in oceans and water on land, soluble organic molecules derived from degradation of dead organisms or excretion of molecules synthesized by organisms.

dissolved oxygen level (DO) amount of oxygen gas (O_2) dissolved in a given volume of water at a given temperature and pressure, and usually expressed as a concentration of oxygen in parts per million of water. *see also* biological oxygen demand.

distad *adv.* (1) towards or at a position away from centre or from point of attachment; (2) in a distal direction.

distal *a.* (1) *pert.* the end of any structure furthest away from middle line of organism or point of attachment; (2) *appl.* region of a gene furthest away from the promoter; (3) far apart, distant, *appl.* e.g. bristles. *cf.* proximal.

distalia *n.plu.* the distal or 3rd row of carpal or tarsal bones.

distance-matrix method a method of generating phylogenetic trees from rDNA sequences.

distance receptor sense organ that responds to stimuli emanating from distant objects, e.g. an olfactory, visual or auditory receptor.

distemonous *a.* having two stamens.

distichous *a.* (1) arranged in two rows; (2) *appl.* alternate leaves arranged so that 1st is directly below 3rd and so on.

distractile *a.* widely separated, *appl.* long-stalked anthers.

distraction display behaviour in female birds which distracts an enemy from the eggs or chicks, and often takes the form of feigning injury to entice the predator away.

distribution *n.* geographical range of a species or group of species.

disturbance *n.* in ecology, any perturbation (either natural or caused by humans) experienced by an ecosystem.

disturbance climax disclimax *q.v.*

disulphide bond covalent S–S bond formed between the sulphydryl groups (–SH) of two cysteine residues in proteins. It links two portions of the polypeptide chain in the same or different subunits, and contributes to the stabilization of tertiary and quaternary structure.

dithecal *a.* two-celled.

ditocous, ditokous *a.* producing two eggs or two young at one time.

ditrematous *a.* with genital and anal openings separate.

ditrochous *a.* with a divided trochanter.

ditypism *n.* (1) occurrence or possession of two types; (2) sexual differentiation, represented by + and −, of two apparently similar haplonts.

diuresis *n.* increased or excessive secretion of urine.

diuretic *a.* increasing the secretion of urine. *n.* any agent causing the above.

diurnal *a.* (1) occurring every day, with a cycle of 24 hours; (2) active in the daytime; (3) opening in the daytime.

diurnal rhythm metabolic or behavioural rhythm with a cycle of about 24 hours.

divaricate *a.* (1) widely divergent; (2) forked.

divergence *n.* the evolution of two species, two DNAs or two proteins from a common ancestor such that they come to differ from each other.

divergency *n.* the fraction of stem circumference, usually constant for a species, which separates two consecutive leaves in a spiral.

divergent *a.* (1) separated from another, having tips further apart than the bases; (2) *appl.* evolutionary change tending to produce differences between e.g. two species or two genes.

diversion behaviour (1) distraction display *q.v.*; (2) behaviour likely to confuse an enemy, e.g. squids ejecting a cloud of black 'ink'.

diversity *n.* (1) (*ecol.*) *see* alpha diversity, beta diversity, biodiversity, ecological diversity, gamma diversity, species composition, species diversity, species richness; (2) (*immunol.*) the variety of antigen specificities displayed within the total repertoire of antigen receptors, i.e. the immunoglobulins and T-cell receptors.

diversity gene segment D gene segment *q.v.*

diversity gradient geographical gradient (e.g. in altitude or latitude) along which a change in species diversity is found.

diversity index a measure of the biological diversity (generally the species diversity) within an environment. There are various types of diversity index, which are calculated in various ways from the number of species present and their relative abundance (*see* e.g. Simpson diversity index). Such indices can be used to detect ecological changes due e.g. to stress on an environment. *cf.* biotic index.

diverticulate *a.* (1) having a diverticulum; (2) having short offshoots approximately at right angles to axis.

diverticulum *n.* blind-ended tube or sac opening off a canal or cavity.

divided *a.* with leaf blade cut by incisions reaching to midrib.

division *n.* major taxonomic grouping in plants, corresponding to a phylum in animals, e.g. Bryophyta (mosses and liverworts), Pterophyta (ferns, etc.),

Spermatophyta (seed-bearing plants: the gymnosperms and angiosperms).

dixenous *a.* parasitizing or able to parasitize two host species.

dizygotic *a.* originating from two fertilized ova, *appl.* non-identical or fraternal twins (DZ twins). *cf.* monozygotic.

D loop (1) single-stranded 'displacement loop' seen in replication of mitochondrial and chloroplast DNA, where replication proceeds for a short length along one strand of parental DNA only, displacing the other parental strand; (2) single-stranded loop seen in duplex DNA when invaded by a homologous single strand of DNA or RNA which pairs with one strand of the duplex, displacing the other to form a loop.

DM dry matter *q.v.*

DMD Duchenne muscular dystrophy *q.v.*

DNA deoxyribonucleic acid, a very large linear molecule which acts as the store of genetic information in all cells. It contains carbon, oxygen, hydrogen, nitrogen and phosphorus. A DNA molecule is composed of two chains of covalently linked deoxyribonucleotide subunits, and a single molecule may be millions of nucleotides in length. The two chains are wound round each other to form the Watson–Crick right-handed 'double helix'. The four types of deoxyribonucleotide subunit in DNA contain the bases adenine (A), thymine (T), cytosine (C) and guanine (G) respectively. Each chain is composed of nucleotides covalently linked through regularly repeating sugar (deoxyribose) phosphate ester bonds between the hydroxyl group on C5 of one deoxyribose and the phosphate group on C3 of another. The two chains are exactly complementary to each other in base sequence and are held together by specific hydrogen bonding between A on one chain and T on the other, and between C one one chain and G on the other. Genetic information is encoded in the sequence of the bases along the polynucleotide chains, which forms a genetic code directing the synthesis of RNAs and proteins. In eukaryotic cells, DNA is present in a complex with histones, forming chromatin, which is packaged with other proteins into discrete chromosomes.

DNA replicates by semi-conservative replication, in which the two strands of the helix separate and two new complementary strands are synthesized using the old ones as templates. DNA can occur in several configurations: **B-DNA** is the classical right-handed double helix with 10–10.4 nucleotide residues in one complete turn and is the form generally found *in vivo*; **A-DNA** consists of a more tightly wound right-handed helix with around 11 residues per turn and is formed by dehydration of B-DNA; **Z-DNA** is a left-handed double helix proposed on the basis of the crystal structure of the duplex trinucleotide d(CG)$_3$, and containing 12 residues per turn. Other DNA helices are possible but only A-DNA, B-DNA and Z-DNA have been found in biological systems. *see* Figs 5 (p. 69), 14. *see other DNA entries. see also* chromosome, complementary DNA, gene, genetic code, nucleic acid, recombinant DNA.

DNA amplification the production of many copies of a particular DNA sequence, usually by the polymerase chain reaction.

DNA array *see* DNA microarray.

DNA-binding protein any protein that binds to a specific sequence in DNA.

DNA blotting *see* Southern blotting *q.v.*

DNA chip DNA microarray *q.v.*

DNA clone piece of DNA incorporated into a bacterial plasmid or phage such that many identical copies can be made by replication in an appropriate host cell.

DNA cloning isolation and multiplication of a piece of DNA by incorporating it into a specially modified phage or plasmid and introducing it into a bacterial cell. The DNA of interest is replicated along with the phage or plasmid DNA and can subsequently be recovered from the bacterial culture in large amounts. *see also* polymerase chain reaction.

DNA-directed RNA polymerase RNA polymerase *q.v.*

DNA fingerprinting, DNA profiling, DNA typing method of ascertaining individual identity, family relationships, etc., by means of DNA analysis. The DNA fingerprint consists of a pattern of DNA fragments obtained on analysis of certain highly variable repeated DNA sequences

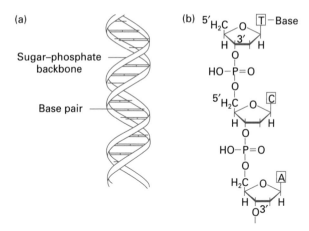

Fig. 14 DNA. (a) The double-helical structure of DNA. (b) Polydeoxyribonucleotide chain of a single DNA strand showing the sugar–phosphate backbone. See Fig. 5 (p. 69) for details of the pairing between bases.

within the genome, whose number and arrangement are virtually unique to each individual. DNA fingerprints can be obtained from a tiny quantity of blood, semen or hair, and are widely used in forensic work and also in ecological studies.

DNA footprinting method of determining the extent and sequence of the DNA sequence bound by a particular protein, by digesting away the unbound DNA in the protein–DNA complex and then isolating and sequencing the DNA that remains.

DNA glycosylases enzymes that remove damaged bases from DNA by splitting the bond linking the base to deoxyribose.

DNA gyrase *see* DNA topoisomerases.

DNA helicase helicase *q.v.*

DNA hybridization (1) technique for determining the similarity of two DNAs (or DNA and RNA) by reassociating single strands from each molecule and determining the extent of double-helix formation. *alt.* DNA renaturation; (2) general method involving reassociation of complementary DNA or RNA strands, used to identify and isolate particular DNA or RNA molecules from a mixture. *see* DNA probe. *see also in situ* hybridization.

DNA immunization, DNA vaccination the induction of immunity by injection of DNA encoding the required protein antigen into a muscle. The protein is expressed and elicits an immune response.

DNA library collection of cloned DNAs. *see* cDNA library, genomic library.

DNA ligase enzyme that joins breaks in the sugar–phosphate backbone in double-stranded DNA. It is involved in DNA replication and repair and is also used in DNA recombination work to join two DNAs end to end. EC 6.5.1.1 and 6.5.1.2. *r.n.* polydeoxyribonucleotide synthetase.

DNA marker a DNA sequence of known location, but usually of no function, in the genome that can be used as a reference point e.g. in genetic mapping.

DNA melting the dissociation of the two strands of a double-stranded DNA when heated.

DNA methylation addition of a methyl group by DNA methyltransferases (DNA methylases) to cytosine at CG sequences. It is found in the DNAs of plants, mammals and some other eukaryotes. Methylation is thought to be associated with genes that are not being transcribed, such as inactivated X chromosomes and imprinted genes (*see* genomic imprinting).

In bacteria, methylation of DNA at C and A residues is used to protect the DNA against the bacterium's own restriction endonuclease, and is known as modification. *see also* CpG island.

DNA microarray hundreds to thousands of 'spots' of different cDNAs or synthetic oligonucleotide probes, usually representing a selection of genes, arranged in a predetermined array on a glass microscope slide or other similar small-sized format. Microarrays are commonly used to measure the simultaneous expression of thousands of genes, by hybridization of the array with isolated cellular mRNA (or the cDNAs derived from it). They can also be used to detect single-nucleotide polymorphisms or other genomic features by hybridizing genomic DNA to an array of appropriate oligonucleotides. *alt.* DNA chip.

DNA modification modification *q.v.*

DNA-only transposon type of transposable element that exists as DNA throughout its life cycle. Examples are the IS elements of bacteria and the P element of *Drosophila*. *cf.* retrotransposon.

DNA polymerase (pol) enzyme that adds deoxyribonucleotides to the 3′ end of a DNA strand using DNA or (in the retroviruses) RNA as template. DNA-dependent DNA polymerases carry out DNA replication and are involved in DNA repair. In eukaryotic cells, DNA polymerases α, δ and ϵ are responsible for nuclear DNA synthesis, β for repair functions, and γ for mitochondrial DNA replication. Some DNA polymerases have a proofreading and editing activity with which they can recognize and excise mispaired nucleotides during DNA replication. Three main classes of replicative DNA polymerases are defined on the basis of sequence: type A (e.g. bacterial PolI), type B (Polα and archaeal DNA polymerases) and type C (bacterial PolIII). *see also* reverse transcriptase.

DNA primase *see* primase.

DNA probe *see* probe.

DNA profiling DNA fingerprinting *q.v.*

DNA puff chromosome puff generated in the salivary glands of sciarid insects which contains locally amplified sequences of DNA as well as RNA.

DNA rearrangement *see* antigenic variation, gene rearrangement, locus.

DNA recombination *see* recombination.

DNA renaturation *see* DNA hybridization.

DNA repair various biochemical processes by which DNA damaged by the action of chemicals or irradiation can be restored. Altered and incorrectly matched bases are recognized by enzymes that excise them and new DNA is then synthesized by reference to the undamaged strand. Double-strand breaks are repaired by homologous or non-homologous endjoining. The repair of certain types of damage may result in alteration of the base sequence of the DNA and thus in mutation.

DNA replication the process by which a new copy of a DNA molecule is made. The two strands of the double helix are separated and each acts as a template for the synthesis of a new complementary strand by the enzyme DNA polymerase, resulting in two new identical double-stranded DNA molecules. This type of replication is termed semi-conservative. *see also* Okazaki fragments.

DNase, DNAse deoxyribonuclease *q.v.*

DNase I endonuclease that makes single-stranded nicks in duplex DNA.

DNA sequence the order of nucleotides in a DNA molecule. In protein-coding DNA this determines the order of amino acids in the protein specified. *alt.* base sequence, nucleotide sequence.

DNA sequencing determination of the order of nucleotides in DNA. The first rapid methods were developed independently by Sanger and associates and by Maxam and Gilbert in the 1970s, and various automated methods have since been developed. *alt.* nucleotide sequencing, base sequencing. *see also* massively parallel DNA sequencing, pyrosequencing.

DNase-sensitive sites, DNasehypersensitive sites sites in chromosomal DNA which are susceptible to cleavage by DNase I. They are thought to indicate a more open and extended conformation of chromatin.

DNA splicing the rearrangement of DNA sequences into different combinations. This occurs naturally, e.g. in the generation of functional immunoglobulin genes in somatic cells, and artificially, in DNA cloning and genetic engineering procedures.

DNA supercoiling *see* supercoiled DNA.

DNA synthesis *see* DNA replication.

DNA topoisomerases enzymes that can untwist tangled or supercoiled DNA by making transient single-strand breaks (type I topoisomerases), around which the rest of the DNA molecule can swivel, or transient double-strand breaks (type II topoisomerases). Type II topoisomerases can also separate two interlinked DNA molecules.

DNA transcription *see* transcription.

DNA tumour viruses a group of unrelated DNA viruses that can cause cancers by various means. They include Epstein–Barr virus, hepatitis B and certain human papilloma viruses, and adenovirus, SV40 (simian virus 40) and polyoma, which are only tumorigenic in certain highly susceptible newborn animals.

DNA-uracil glycosidase uracil-DNA glycosidase *q.v.*

DNA vaccine plasmid DNA encoding a protein antigen. The DNA is injected into muscle, where it is transcribed and translated into protein, which elicits an immune response.

DNA viruses viruses containing DNA as the genetic material and including the Adenoviridae, Herpesviridae, Poxviridae, Papovaviridae, Parvoviridae and Iridoviridae amongst vertebrate viruses, and the caulimoviruses and geminiviruses among the plant viruses. *see* Appendix 7.

DNP dinitrophenol *q.v.*

dNTPs deoxynucleoside triphosphates *q.v.*

DO dissolved oxygen level *q.v.*

DOC dissolved organic carbon *q.v.*

dodecagynous *a.* having 12 pistils.

dodecamer *n.* a structure (e.g. a protein) composed of 12 subunits.

dodecamerous *a.* having each whorl composed of 12 parts.

dodecandrous *a.* having at least 12 stamens.

dolabriform, dolabrate *a.* axe-shaped.

dolichocephalic *a.* (1) long-headed; (2) with a cephalic index of under 75.

dolichofacial *a.* long-faced.

dolichohieric *a.* having a sacral index below 100.

dolichol phosphate very long-chain lipid, carrier of activated oligosaccharides for attachment to glycoproteins in lumen of endoplasmic reticulum.

dolichostylous *a. pert.* long-styled anthers in dimorphic flowers.

dolioform *a.* barrel-shaped.

doliolaria *n.* in some sea cucumbers, a barrel-shaped larva which develops from an auricularia.

dolipore septum septum which flares out in the middle of the hypha, forming a barrel-shaped structure with open sides.

Dollo's rule or law that evolution is irreversible and that structures and functions once lost are not regained.

dolphin *n.* member of the marine families of mammals Delphinidae or Platanistidae (river dolphins) of the suborder Odontoceti (toothed whales) of the order Cetacea. They are slim and fast-moving, with a prominent elongated snout or 'beak'.

DOM dissolved organic matter *q.v.*

domain *n.* (1) the highest level of biological classification. The three domains of organisms are the Bacteria, the Archaea and the Eukarya. *alt.* superkingdom; (2) (*mol. biol.*) structurally defined compact globular portion of a protein molecule. Some protein chains fold into a single domain, others into two or more domains joined by less structured portions of polypeptide chain; (3) a length of looped chromatin in a chromosome; (4) region of cell membrane of particular lipid and protein composition.

α domain protein domain consisting entirely or predominantly of α-helices.

β domain protein domain consisting entirely or predominantly of β-sheet.

domain shuffling the joining together of protein domains in new combinations that has occurred during evolution, leading to the presence of the same domain in a variety of different proteins.

domain swapping in a protein of more than one subunit, the replacement of a secondary structural element or domain in one subunit with the same element from another subunit, and vice versa.

domatium *n.* crevice or hollow in some plants that serves as a lodging for insects or mites.

dome cell penultimate cell of a hyphal crozier, containing two nuclei which fuse. It is the first stage in ascus formation.

dominance *n.* (1) (*genet.*) property possessed by some alleles of determining the phenotype when present in one copy in a cell. As one member of a heterozygous pair they mask the effects of the other allele (the recessive allele) to give a phenotype identical to that when the dominant allele is present as two copies. This phenomenon is known as complete dominance. Incomplete dominance is exhibited when the effects of the other allele are not completely masked. *see also* co-dominance; (2) (*ecol.*) the extent to which a particular species predominates in a community and affects other species; (3) (*behav.*) *see* dominance systems.

dominance frequency in ecology, proportion of samples in which a particular species is predominant.

dominance systems social systems in which certain individuals aggressively dominate others. In the case where one individual dominates all the others with no intermediate ranks it is known as a despotism. In the more common dominance hierarchies or social hierarchies there are distinct ranks, with individuals of any rank dominating those below them and submitting to those above them.

dominance variation genetic variance at a single locus that is attributable to dominance of one allele over the other.

dominant *a.* (1) (*genet.*) *appl.* a phenotypic character state or an allele that masks an alternative character state or allele when both are present in a hybrid. *see also* heterozygote. *cf.* recessive allele; (*ecol.*) (2) *appl.* plants which by their numbers and extent determine the biotic conditions in an area; (3) *appl.* species most prevalent in a particular community, or at a given period; (4) (*behav.*) *appl.* an individual which is high ranking in the social hierarchy or pecking order.

dominant-negative *a. appl.* mutation that inactivates a particular cellular function by the production of a defective RNA or protein that blocks the normal function of the gene product.

Domin scale scale (1–10) used to indicate the approximate percentage cover of individual plant species in a given area, with 1 corresponding to insignificant cover, 3 corresponding to 1–5% cover, 8 to 50–75% cover, and 10 equal to 100% cover. + indicates organisms occurring singly.

Donnan free space fraction of a tissue available for ion-exchange reactions, i.e. the cell wall in plant tissue.

donor splice region, donor splice site the conserved sequence comprising the first two nucleotides (GU) at the 5′ end of an intron in a primary RNA transcript and the last two nucleotides (AG) of the preceding exon. During the RNA splicing reaction that removes the intron, this site is the first to be cleaved. The 3′ end of the exon is finally joined to the 5′ end of the next exon. *alt.* 5′ splice site. *cf.* acceptor splice site.

DOPA, dopa (L-dopa) 3,4-dihydroxyphenylalanine, formed from tyrosine in the adrenal medulla, brain and sympathetic nerve terminals by the enzyme tyrosine hydroxylase (*r.n.* tyrosine-3-monooxygenase, EC 1.14.16.2). It is a biosynthetic precursor of noradrenaline, adrenaline and dopamine. Also oxidized by dopa-oxidase to a melanin precursor, e.g. in the basal layers of skin. L-dopa is used in the treatment of Parkinson's disease.

dopamine *n.* a catecholamine that acts as a neurotransmitter in the central nervous system. Also produced in small amounts by adrenal medulla. A deficiency of dopamine as a result of destruction of dopamine-producing neurons in forebrain is an underlying cause of Parkinson's disease. *see* Fig. 15.

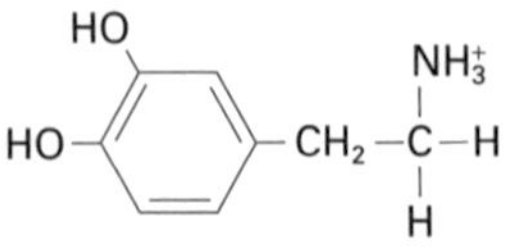

Fig. 15 Dopamine.

dopaminergic *a. appl.* neurons that release dopamine from their terminals.

dormancy *n.* resting or quiescent condition with reduced metabolism, as in seeds or spores in unfavourable conditions for germination. *a.* **dormant**.

dorsad *a.* towards the back or dorsal surface.

dorsal *a.* (1) *pert.* or nearer the back (not hind end) of an animal, which is usually the upper surface; (2) *appl.* upper surface of leaf or wing; (3) *appl.* placentation of ovule, ovules attached to midrib of carpels. *cf.* ventral.

dorsal aorta major artery carrying oxygenated blood to the rest of the body in vertebrates and cephalochordates. In mammals the dorsal aorta is formed from the left branch of the 4th (systemic) aortic arch.

dorsal fin the large fin on the back of most fish and cetaceans.

dorsalis *n.* artery which supplies the dorsal surface of any organ.

dorsalized *a.* in experimental embryology, *appl.* embryo that develops an increased dorsal region at the expense of the ventral region, as a result of some treatment. *n.* **dorsalization**.

dorsal lip of blastopore *see* organizer.

dorsal root dorsal branch of a spinal nerve as it enters the spinal cord. It carries sensory information from the body to the spinal cord. *cf.* ventral root.

dorsal root ganglion mass of cell bodies lying on the dorsal root, just outside the spinal cord. It contains the cell bodies of sensory neurons of both a spinal nerve and a visceral sympathetic nerve.

dorsal–ventral dorso-ventral *q.v.*

dorsiferous *a.* bearing sori on back of leaf.

dorsifixed *n.* having filament attached to back of anther.

dorsigerous *a.* carrying young on back.

dorsigrade *a.* having back of digit on the ground when walking.

dorsilateral *a.* of or *pert.* back and sides.

dorsispinal *a. pert.* or referring to back and spine.

dorsiventral *a.* flattened and having upper and lower surface of distinctly different structure, *appl.* leaves.

dorsobronchus *n.* one of a set of tubes in lungs of birds which branch off the bronchi and are connected with the posterior air sacs.

dorsocentral *a.* (1) *pert.* mid-dorsal surface; (2) *pert.* aboral surface in echinoderms.

dorsolumbar *a. pert.* lumbar region of back.

dorso-ventral *a.* (1) *appl.* axis, from back to belly of the animal body, from upper to lower surface of a limb, leaf, etc.; (2) *pert.* structures, axis, gradients, etc., stretching from dorsal to ventral surface.

dorsum *n.* (1) tergum and notum of insects and crustaceans; (2) inner margin of insect wing; (3) the back of higher animals; (4) upper surface, as of tongue.

dorylaner *n.* exceptionally large male ant of driver-ant group.

dosage compensation regulation of the number of copies of a gene expressed from the sex chromosomes, so that the level of expression is the same in the two sexes even though the number of sex chromosomes is not the same. In mammals for example, one copy of the X chromosome is permanently inactivated in the somatic cells of the female and its genes are not expressed. The 'dosage' of X-linked genes in males and females is thereby made equal.

dose equivalent standardized measure of the effect of ionizing radiation on tissue, which is the measured absorbed dose multiplied by weighting factors for particular tissues and types of radiation. It represents the risk to health from that amount of radiation if it had been absorbed uniformly throughout the body. It is expressed in sieverts (SI unit).

dose–response curve for any assay, the relation between the concentration of active agent (virus, hormone, enzyme, etc.) in the sample, and the quantitative response in that particular assay.

dot blot type of nucleic acid hybridization technique in which the nucleic acid sample is 'dotted' directly onto a nylon or nitrocellulose membrane, and a labelled DNA probe is then applied.

double bond covalent bond that consists of two pairs of electrons shared between the bonded atoms. Represented by a double line between the bonded atoms, e.g. $O=O$.

double crossover two chiasmata occurring in a given chromosomal region.

double fertilization characteristic feature of flowering plants, in which one male haploid nucleus fuses with the polar nuclei to form the triploid primary endosperm nucleus, and the other fuses with the egg-cell nucleus to produce the diploid zygote.

double helix the typical conformation of double-stranded DNA in which the two strands are wound around each other with base pairing between the strands. Also found in double-stranded RNA. *see* Fig. 14 (p. 183).

double-minute chromosomes self-replicating extrachromosomal DNA elements lacking centromeres. They are found in eukaryotic cells after certain treatments that result in selective gene amplification.

double-positive thymocyte immature developing T cell that produces both CD4 and CD8 cell-surface proteins.

double recessive cell or organism homozygous for the recessive allele at two different genes.

double-strand break a break in a DNA molecule across both strands.

doubling time the time it takes for the numbers of something growing exponentially (e.g. a population of living organisms) to double. *alt.* generation time.

down, down feathers the first fluffy feathers of young birds, with a short quill and with barbules not interlocking to form a flat vane. Some birds retain a down layer under the adult plumage. *alt.* prepenna.

down, downland *n.* grassland vegetation typical of the chalk downs of southern England, which is generated and maintained by continuous grazing (by sheep and rabbits), and which is short turf rich in small flowering plants.

down mutation mutation in which transcription of a particular gene(s) is much reduced but the gene product is unaltered, usually due to a mutation in DNA regions involved in regulation of gene expression.

downregulation *n.* a decrease in e.g. the rate of transcription of a gene or the number of receptors on a cell surface.

Down syndrome condition due to the presence of three copies of chromosome 21 (trisomy 21), characterized by some degree of mental retardation, short stature and poor muscle tone. *alt.* **Down's syndrome**.

downstream *a.* (1) *appl.* sequences to the 3' side of any given point on a DNA coding strand or in the RNA corresponding to it; (2) *appl.* events occurring after a given stage in a metabolic or signalling pathway. *cf.* upstream.

2D-PAGE *see* polyacrylamide gel electrophoresis.

Dp in genetics, the abbreviation for a duplication of a given gene or chromosomal segment.

DPD diffusion pressure deficit. *see* suction pressure.

DPG 2,3-diphosphoglycerate *q.v.*

dpm, d.p.m. disintegrations per minute, a measure of radioactivity.

dragonflies *see* Odonata.

drepanium *n.* helicoid cyme with secondary axes developed in a plane parallel with that of main peduncle and its first branch, so that the whole inflorescence is sickle-shaped.

drepanoid *a.* sickle-shaped.

drift *see* genetic drift, continental drift.

drive *n.* the motivation of an animal that results in its achieving a goal or satisfying a need.

dromaeognathous *a.* having a palate in which palatines and pterygoids do not articulate, owing to intervention of vomer.

dromotropic *a.* bent in a spiral.

drone *n.* male social insect, esp. honeybee.

drop mechanism the mechanism by which a pollen grain, trapped in a drop of liquid in gymnosperm ovules, is drawn into the micropyle.

dropper *n.* downward outgrowth of a bulb which may form a new bulb.

Drosophila melanogaster a species of fruitfly, a member of the Diptera, a favoured subject for experimental genetics and developmental biology.

drosopterin *n.* red pteridine pigment in eyes and other organs of some insects, including *Drosophila*.

drought *n.* situation in which a region does not receive its normal amount of water because of decreased rainfall, increased evaporation due to higher than normal temperatures or a combination of both.

drug-resistance factors, drug-resistance plasmids (R factors) plasmids in enterobacteria and other medically important bacteria that carry genes for resistance to various commonly used antibiotics.

drupaceous *a.* bearing drupes.

drupe *n.* a more-or-less fleshy fruit with one compartment and one or more seeds, having the pericarp differentiated into a thin epicarp, a fleshy mesocarp, and a hard stony endocarp. Plums and cherries are drupes. *alt.* stone fruit.

drupel *n.* one of a collection of small drupes forming an aggregate fruit, e.g. raspberry, blackberry. *alt.* **drupelet**.

dry mass *see* dry weight.

dry matter (DM) material left after removal of water from organic matter, such as plant biomass or soil, obtained by heating to constant weight in an oven at 90–95°C.

dryopithecids, dryopithecines *n.plu.* a group of Miocene ape-like fossils from India and Africa, including the genera *Dryopithecus* and *Proconsul*, dating from *ca.* 20 to 8 million years ago, and which are thought to include the ancestors of modern apes.

dry weight (DW) the weight or mass of organic matter or soil after removal of water by heating to constant weight. *alt.* dry mass.

dry weight rank method technique for estimating the contribution each plant species makes to the total yield of a pasture.

dsDNA double-stranded DNA.

D segment D gene segment *q.v.*

dsRNA double-stranded RNA.

dual-specificity protein kinase protein tyrosine kinase that can phosphorylate proteins on both tyrosine and serine/threonine residues.

dual-specificity protein phosphatase protein phosphatase that can remove phosphate groups from phosphorylated tyrosine and serine/threonine residues in proteins.

Duchenne muscular dystrophy (DMD) X-linked heritable disease leading to muscular atrophy and eventual death, affecting around 1 in 4000 newborn males. It is due to a defect in the gene encoding dystrophin, a large protein associated with the sarcolemma of muscle fibres.

duct *n.* any tube that conveys fluid or other material.

ductless glands glands such as endocrine glands that do not release their secretions into a duct.

ductule *n.* minute duct.

ductulus efferens one of the ductules leading from the testis to the vas deferens, known collectively as the vasa efferentia.

ductus *n.* a duct.

ductus arteriosus the connection between the pulmonary arch and dorsal aorta in mammalian foetus.

ductus venosus the connection between the umbilical vein and vena cava in mammalian foetus.

duetting *n.* rapid antiphonal calling back and forth between mated pairs in some birds, presumed to be a recognition and bonding device.

Dufour's gland gland that leads into the poison sac at the base of the sting in certain Hymenoptera. *alt.* alkaline gland.

dulosis *n.* slavery among ants, in which those of one species are captured by another species and work for them, an extreme example of social parasitism.

dumose *a.* shrub-like in appearance.

duodenum *n.* first part of the small intestine, lying between the stomach and jejunum.

duplex *a.* (1) double; (2) compound, *appl.* flowers; (3) consisting of two distinct structures; (4) having two distinct parts.

duplex DNA double-stranded DNA. *see* DNA.

duplication *n.* situation in which a chromosome, segment of a chromosome, or gene is present in more than the normal number of copies.

duplicident *a.* with two pairs of incisors in upper jaw, one behind the other.

duplicity *n.* condition of being twofold.

duplicocrenate *a.* with scalloped margin, and each rounded tooth again notched, *appl.* leaf.

duplicodentate *a.* with marginal teeth on leaf bearing smaller teeth-like indentations.

duplicoserrate *a.* with marginal saw-like teeth and smaller teeth directed towards leaf tip.

dural *a. pert.* dura mater.

dura mater tough membrane forming the outermost covering of brain and spinal cord.

duramen *n.* hard darker central wood of a tree trunk. *alt.* heartwood.

dura spinalis tough membrane lining the spinal canal.

dust lice *see* Psocoptera.

DW dry weight *q.v.*

dwarf male small, usually simply formed male individual in many classes of animal, either free-living or carried by the female. *alt.* pygmy male.

dyad *n.* (1) one member of a pair of homologous replicated chromosomes synapsed at meiosis; (2) *appl.* symmetry about a twofold axis.

dynamic instability the property displayed by actin filaments and microtubules that are in a constant state of assembly and disassembly.

dynamin *n.* protein involved in the formation of coated vesicles in eukaryotic cells.

dynein *n.* motor protein with ATPase activity that can attach to microtubules and travel along them. Dynein is also a permanent component of eukaryotic cilia and flagella, where its activity causes their bending movement.

dynorphins *n.plu.* endorphin-like peptides found in brain and gut. Together with the endorphins and enkephalins they are known as the endogenous opioids.

dysgenesis *n.* infertility of hybrids in matings between themselves, although fertile with individuals of either parental stock.

dysgenic *a. appl.* traits inimical to the propagation of the organism, such as sterility, chromosomal aberrations, mutations, abnormal segregation at meiosis.

dysmerogenesis *n.* segmentation resulting in unlike parts.

dysphotic *a.* (1) dim; (2) *appl.* zone: waters at depths between 80 and 600 m, between euphotic and aphotic zones.

dysplasia *n.* abnormal development of a tissue. *a.* **dysplastic**.

dyspnoea *n.* difficulty in breathing.

dystrophic *a.* (1) wrongly or inadequately nourished; (2) inhibiting adequate nutrition; (3) *pert.* faulty nutrition; (4) *appl.* lakes rich in undecomposed organic matter so that nutrients are scarce.

dystrophin *n.* protein associated with the muscle cell membrane, and whose absence or abnormality is responsible for Duchenne muscular dystrophy.

DZ twins dizygotic twins *q.v.*

E

ε- *for all headwords with prefix ε-, refer to epsilon or to headword itself.*

E glutamic acid *q.v.*

E_0' redox potential *q.v.*

E_D evolutionary distance *q.v.*

EAE experimental allergic encephalomyelitis *q.v.*

e- prefix derived from L. *ex*, out of or *ex* without, often denoting a lack of, e.g. ebracteolate, lacking bracts, or ecaudate, without a tail.

ear *n.* (1) sense organ in vertebrates concerned with hearing and gravity detection. In mammals it consists of an external ear (pinna) surrounding the auditory canal or auditory meatus leading to the membranous eardrum. A row of small bones transmits vibrations of the eardrum caused by sound waves across the air-filled space of the middle ear to the fluid-filled inner ear, which contains the cochlea and the semicircular canals. The cochlea contains sensory cells (hair cells) that detect sound-induced vibrations of the surrounding fluid and transmit signals encoding tone and pitch to the brain via the auditory nerve. The semicircular canals are concerned largely with gravity detection and maintenance of balance. *see also* acousticolateralis system, and individual entries for cochlea, semicircular canals, etc.; (2) (*bot.*) the spike of grasses, usually *appl.* cereals.

eardrum tympanic membrane *q.v.*

early *a.* (1) *appl.* the first phase of the lytic cycle of phage or virus infection, from the entry of virus DNA (or RNA) to the start of its replication; (2) *appl.* phage or viral genes expressed early in infection, before replication.

early wood wood formed in the first part of the growing season, which shows as a ring of fewer and larger cells in the annual ring.

ear sand otoconia *q.v.*

earth ball common name for basidiomycete fungi of the genus *Scleroderma* and their relatives. They are gasteromycetes with hard tuberous unstalked fruiting bodies that crack open to release a mass of spores.

earthworm *n.* common name for a number of terrestrial oligochaete worms (*Lumbricus* spp. and others) inhabiting the soil, which contribute to soil aeration through their tunnels and to soil fertility by bringing humus-containing soil to the surface.

earwig *n.* common name for an insect of the order Dermaptera *q.v.*

East African Steppe region part of the Palaeotropical floristic kingdom comprising Africa east of the Rift Valley from northern Uganda to southern Mozambique and below Lake Tanganyika extending westwards to the Atlantic coast of Angola above the Namibian desert.

East African subregion subdivision of the Ethiopian zoogeographical region, comprising sub-Saharan Africa south to South Africa with the exception of the equatorial rain forest belt of West and Central Africa.

eavesdropping *n.* behavioural strategy in which rival males are attracted to a female by another male's courtship display.

Ebner's gland von Ebner's gland *q.v.*

ebracteate, ebracteolate *a.* without bracts, without bracteoles.

EBV Epstein–Barr virus *q.v.*

EC effective concentration *q.v.*

ecad *n.* (1) plant or animal form modified by the environment; (2) habitat form *q.v.*

ecalcarate *a.* having no spur or spur-like process, *appl.* petals.

ecardinal, ecardinate, *a.* having no hinge, *appl.* shells.

ecarinate *a.* having no keel or keel-like ridge.

ecaudate *a.* without a tail.

eccentric excentric *q.v.*

eccentric cell nerve cell in ommatidium of compound eye of *Limulus* which receives signals from retinula cells and transmits them to the central nervous system.

eccrine *a. appl.* glands that secrete without disintegration of secretory cells. *cf.* apocrine.

eccritic *a.* (1) causing or *pert.* excretion; (2) preferred, *appl.* temperature or other environmental state.

ecdemic *a.* not native.

ecdysial *a. pert.* ecdysis or moulting.

ecdysiotropic hormone insect protein hormone produced in the 'brain' that acts on the resting ovary in concert with juvenile hormone, resulting in the secretion of the steroid hormone ecdysone.

ecdysis *n.* moulting, the periodic shedding of cuticular exoskeleton in insects and some other arthropods, to allow for growth.

ecdysone, ecdysterone *n.* steroid hormone produced by the prothoracic gland of insects and the Y-organs of crustaceans, which stimulates growth and moulting.

Ecdysozoa, ecdysozoans *n., n.plu.* one of the two major divisions of the protostomes (the other being the Lophotrochozoa), comprising the animal phyla Arthropoda, Cephalorhyncha, Nematoda, Nematomorpha, Onychophora, Rotifera and Tardigrada. The body is supported by an external organic cuticle, which is periodically moulted (ecdysis) and regrown as the animal grows in size.

ecesis *n.* invasion of organisms into a new habitat.

ECG electrocardiogram. *see* electrocardiography.

echidna *n.* the spiny anteater, a monotreme.

echinate *a.* bearing spines or bristles.

echinidium *n.* marginal hair with small pointed or branched outgrowths on cap of fungi.

echinococcus *n.* vesicular metacestode developing a number of daughter cysts, each with many heads. *see* polycercoid.

Echinodermata, echinoderms *n., n.plu.* phylum of marine coelomate animals that are bilaterally symmetrical as larvae but show five-rayed symmetry as adults and have a calcareous endoskeleton and a water vascular system. It includes the classes Crinoidea (sea lilies and feather stars), Asteroidea (starfish), Ophiuroidea (brittle stars), Echinoidea (sea urchins) and Holothuroidea (sea cucumbers).

echinoid *a. pert.* or like a sea urchin.

Echinoidea *n.* class of echinoderms, commonly called sea urchins, having a typically globular body with skeletal plates fitting together to form a rigid outer case.

echinopluteus *n.* the pluteus larva of sea urchins.

echinulate *a.* (1) having small spines; (2) *appl.* bacterial colonies, having pointed outgrowths.

Echiura, echiurans *n., n.plu.* phylum of unsegmented coelomate marine worms, with soft plump bodies, which live in U-shaped tubes or in rock crevices down to abyssal depths. They have an extensible but not eversible proboscis with a ciliated groove for collecting food. *alt.* spoon worms.

echolocation *n.* locating objects by sensing the echoes returned by very high frequency sounds emitted by the animal, as used e.g. by bats.

echoviruses, ECHO viruses group of picornaviruses infecting the intestinal tract and which may also cause respiratory illnesses and meningitis.

eclipse *a.* (1) *appl.* the less colourful plumage assumed after moulting at the end of the breeding season by the males of some bird species; (2) *appl.* period immediately after entry into the host cell when a virus is not easily detectable.

eclosion *n.* hatching from egg or pupa case.

ecobiotic *a.* adaptation to a particular mode of life within a habitat.

ecoclimatic *a. appl.* adaptation to the physical and climatic conditions in a particular region.

ecocline *n.* continuous gradient of variation of ecotypes in relation to variation in ecological conditions.

ecodeme *n.* deme occupying a particular ecological habitat.

E. coli the bacterium *Escherichia coli q.v.*

ecological *a. pert.* or concerned with ecology.

ecological diversity the diversity of ecosystems (e.g. forest, desert, grassland, oceans) within a given region.

ecological efficiency the efficiency of use of energy by an ecosystem. It is described in terms of several energy coefficients: the energy coefficient of the first order (production, P, divided by consumption, C), the energy coefficient of the second order (P divided by assimilated energy, A) and the assimilation efficiency (A/C).

ecological niche the role of an organism in a community in terms of the habitat it occupies, its interactions with other organisms and its effect on the environment. A given niche, e.g. the small herbivore niche, may be occupied by different species in different ecosystems and different parts of the world. *alt.* niche. *see also* realized niche.

ecological pyramid diagram showing the biomass, numbers or energy levels of individuals of each trophic level in an ecosystem, starting with the primary producers (e.g. green plants) at the base. *see* Fig. 16.

ecological succession the natural process whereby communities of plant and animal species are replaced by others, usually more complex, over time as a mature ecosystem develops. *see* climax, primary succession, secondary succession, succession.

ecology *n.* (1) the interrelationships between organisms and their environment and each other; (2) the study of these interrelationships.

economic density of a population, the number of individuals per unit of inhabited area.

ecoparasite *a.* (1) parasite that can infect a healthy and uninjured host; (2) parasite restricted to a specific host or small group of host species.

ecophysiology *n.* the study of animal physiology in relation to life-style and adaptation to environment.

ecorticate *a.* without a cortex, as certain lichens.

ecospecies *n.* group of individuals associated with a particular ecological niche and behaving like a species, but capable of interbreeding with neighbouring species.

ecosphere *n.* the planetary ecosystem, consisting of the living organisms of the world and the components of the environment with which they interact.

ecostate *a.* without ribs or costae.

ecosystem *n.* community of different species interdependent on each other, together with their non-living environment, which is relatively self-contained in terms of energy flow, and is distinct from neighbouring communities. Different types of ecosystem are defined by the collection of organisms found within them, e.g. forest, soil, grassland. Continuous ecosystems covering very large areas, such as the northern coniferous forest or the steppe grassland, are known as biomes.

ecotone *n.* zone where two ecosystems overlap, and which supports species from both ecosystems as well as species found only in this zone.

ecotope *n.* (1) a particular kind of habitat within a region; (2) the total relationship of an organism with its environment, being the interaction of niche, habitat and population factors.

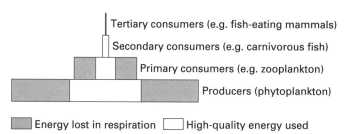

Fig. 16 Typical pyramid of energy flow through a river ecosystem.

ecotype *n*. subspecific form within a true species, resulting from selection within a particular habitat and therefore adapted genetically to that habitat, but which can interbreed with other members of the species.

ecsoma *n*. retractile posterior part of body in certain trematodes.

ect-, ecto- word elements derived from Gk *ektos*, outside.

ectad *adv*. (1) towards the exterior; (2) outwards.

ectadenia *n.plu.* ectodermal accessory genital glands in insects.

ectal *a*. (1) outer; (2) external.

ectamnion *n*. ectodermal thickening in proamnion, beginning of head-fold.

ectangial *a*. (1) outside a vessel; (2) produced outside a primary sporangium.

ectendotrophic *a*. partly ectotrophic and partly endotrophic, *appl.* mycorrhizas.

ectethmoid *n*. lateral ethmoid bone of skull.

ectoascus *n*. outer membrane of ascus wall in certain ascomycete fungi. *cf.* endoascus.

ectoblast epiblast *q.v.*

ectobronchus *n*. lateral branch of main bronchus in birds.

ectochondrostosis *n*. ossification beginning in perichondrium and gradually invading cartilage.

ectochroic *a*. having pigmentation on surface of cell or hypha.

ectocommensal *n*. commensal living on the surface of another organism.

ectocrine *a. appl.* or *pert.* substances or decomposition products in the external medium that inhibit or stimulate plant growth.

ectoderm *n*. (1) primary germ layer in embryo of multicellular animals that gives rise to epidermis and nervous system; (2) in coelenterates, the outer layer of epithelium of body wall. *a.* **ectodermal**.

ectoenzyme exoenzyme *q.v.*

ectogenesis *a*. (1) embryonic development outside the maternal organism; (2) development in an artificial environment. *a.* **ectogenetic**.

ectogenic *a*. of external origin, not produced by organisms themselves.

ectogenous *a*. (1) able to live an independent life; (2) originating outside the organism.

ectoglia *n*. an outer layer of glia in the nervous system.

ectolecithal *a. appl.* eggs having yolk around the periphery.

ectoloph *n*. the ridge stretching from paracone to metacone in a lophodont molar tooth.

ectomeninx *n*. outer membrane covering embryonic brain and giving rise to the dura mater.

ectomere *n*. a blastomere that gives rise to ectoderm.

-ectomy suffix derived from Gk *ek*, out, and *temnein*, to cut, signifying an excision, e.g. thyroidectomy, gonadectomy.

ectomycorrhiza *n*. type of mycorrhiza in which the fungal hyphae form a superficial covering and do not extensively penetrate the root. It is found on both coniferous and broadleaved forest trees, the infecting fungi being chiefly higher basidiomycetes.

ectoneural *a. appl.* system of oral ring, radial and subepidermal nerves in echinoderms.

ectoparasite *n*. parasite that lives on the surface of an organism.

ectopatagium *n*. the part of the wing-like flying membrane of bats that is carried on the metacarpals and phalanges.

ectophagous *a*. feeding on the outside of a food source.

ectophloeodeic *a*. growing on plants, *appl.* lichens.

ectophloic *a*. with phloem outside xylem.

ectophyte *n*. any external plant parasite of plants and animals. *a.* **ectophytic**.

ectopic *a*. not in normal position, *appl.* e.g. organs, pregnancy, gene expression.

ectoplasm *n*. (1) external layer of cytoplasm in a cell, next to the cell membrane and usually clear and non-granular; (2) ectosarc *q.v. cf.* endoplasm.

ectoplast plasma membrane *q.v.*

Ectoprocta, ectoprocts *n., n.plu.* phylum of small marine and freshwater colonial animals, which superficially resemble mosses, hence the common name of moss animals. A colony is composed of zooids each bearing a crown of ciliated tentacles (a lophophore), and living in a horny, calcareous or gelatinous case. *alt.* (formerly) Bryozoa, bryozoans, Polyzoa.

ectopterygoid *n.* ventral membrane bone behind palatine and extending to quadrate.

ectosarc *n.* external layer of cytoplasm, esp. in protozoans. *alt.* ectoplasm.

ectospore *n.* spore formed at end of each sterigma in basidiomycete fungi.

ectosporous *a.* with spores borne exteriorly.

ectostosis *n.* formation of bone in which ossification begins under the perichondrium and either surrounds or replaces the cartilage.

ectostracum *n.* outer primary layer of exocuticle of exoskeleton in ticks and mites.

ectostroma *n.* fungal tissue penetrating cortical tissue of host and bearing conidia.

ectothermic *a. appl.* regulation of body temperature by animals using heat from the surrounding environment, e.g. as in fish, lizards, insects. Such animals are known as ectotherms. *n.* **ectothermy.** *see also* poikilothermic. *cf.* endothermic.

ectotrophic *a.* (1) finding nourishment from outside; (2) *appl.* mycorrhiza in which the fungal hyphae form a superficial covering on the host roots and do not extensively penetrate the root itself.

ectotropic *a.* tending to curve or curving outwards.

ED effective dose *q.v.*

edaphic *a. pert.* or influenced by conditions of soil or substratum.

edema oedema *q.v.*

Edentata, edentates *n., n.plu.* order of placental mammals known from the Paleocene, extant members of which include the armadillo, anteater and two-toed sloth, having reduced teeth and often an armoured body.

edentate *a.* without teeth or tooth-like projections.

edge *n.* in a phylogenetic tree, a line joining two branch points (nodes).

edge effect tendency to have greater variety and density of organisms in the boundary zone between communities.

edge species species living primarily or most frequently or numerously at junctions of communities.

Ediacaran fauna soft-bodied animals of possible coelenterate or problematic affinities present in late Precambrian strata, and considered by some to bear little resemblance to any later organisms. *alt.* Vendian biota.

editing *n.* in DNA synthesis, the exonuclease functions of DNA polymerase which correct mistakes (mismatched bases). *see also* proofreading, RNA editing.

Edman degradation method of determining the amino-acid sequence of peptides by sequentially removing amino acids for identification from the N-terminal end of the peptide.

E-DNA form of DNA taken up by synthetic molecules lacking guanine, with $7\frac{1}{2}$ base pairs per turn.

edriophthalmic *a.* having sessile eyes, *appl.* to certain crustaceans.

EDTA ethylenediaminetetraacetate, a chelating agent that binds magnesium and calcium ions. Used widely in experimental biochemistry to study e.g. the role of metal ions in the structure of proteins and protein assemblies.

eEF general designation for eukaryotic protein synthesis elongation factors.

EEG electroencephalogram. *see* electroencephalography.

eelworms *n.plu.* group of soil nematode worms, some of which cause serious damage to crop plants.

EF general designation for bacterial protein synthesis elongation factors *q.v.*

EFA essential fatty acids *q.v.*

E face in freeze-fractured membranes, the face representing the interior side of the outer half of the lipid bilayer.

effective concentration (EC) the concentration of a toxic substance that is sufficient to cause adverse symptoms (in cases where effects other than death are being studied) within a given period, and which is expressed as e.g. 48-hour EC_{50}, the concentration required to cause symptoms in 50% of the animals tested within 48 hours.

effective dose (ED) usually expressed as ED_{50}, the dose of a drug that produces the desired effect in 50% of the subjects.

effective population size (N_e) the number of individuals in an ideal, randomly breeding population with a 1/1 sex ratio that would have the same rate of heterozygosity decrease as the actual population under consideration.

effector *n.* any organ or cell that reacts to a stimulus by producing something, carrying out a specific set of functions, or doing mechanical work. Examples are muscle, electric and luminous organs, chromatophores, glands, and cells such as the plasma cells and activated T cells of the immune system. *a.* **effector**.

effector T cell a T lymphocyte once it has been activated, or primed, by contact with antigen on a professional antigen-presenting cell. On recognizing antigen on a target cell an effector T cell is stimulated to carry out its particular function. *see* cytotoxic T lymphocyte, helper T lymphocytes, regulatory T cell, Th1 cell, Th2 cell, Th17 cell.

efferent *a.* (1) conveying from, *appl.* motor nerves carrying impulses outwards from central nervous system, *appl.* vessels conveying blood from an organ or lymph from a lymph node; (2) *appl.* ductules from rete testis opening into epididymis. *cf.* afferent.

effigurate *a.* having a definite shape or outline.

efflorescence *n.* (1) flowering; (2) time of flowering; (3) bloom, as on surface of grapes and other fruits.

effodient *a.* having the habit of digging.

effoliation *n.* shedding of leaves.

effuse *a.* (1) spreading loosely, *appl.* flowerheads; (2) spreading thinly, *appl.* bacterial colonies.

effused-reflexed *a. appl.* flattened basidiocarp whose edge rises from the substrate and forms a small shelf-like structure.

eft *n.* juvenile phase in life cycle of a newt.

egesta *n.plu.* (1) the sum total of material and fluid discharged from the body; (2) material passed out of the body in egestion (2).

egestion *n.* (1) process of ridding the body of any waste material, as by defecation and excretion; (2) specifically, the excretion of material that has never been taken out of the gut, as defecation.

EGF epidermal growth factor *q.v.*

egg *n.* (1) ovum *q.v.*; (2) in certain animals, e.g. reptiles, birds, amphibians and insects, a structure composed of the fertilized ovum and nutritive and protective tissues, surrounded by a protective shell, which is laid by the female, and from which the young animal hatches.

egg apparatus egg cell and two cellular synergids that develop at the micropylar end of megagametophyte in flowering plants.

egg cell the female gamete in animals and plants. *alt.* ovum.

egg coat outer glycoprotein coat of ovum, the inner layer of which is the zona pellucida or vitelline layer.

egg membrane (1) vitelline layer *q.v.*; (2) chorion *q.v.*; (3) layer of tough tissue lining an egg shell.

egg nucleus the female pronucleus *q.v.*

egg sac thin-walled sac-like structure from which eggs are released in e.g. brown algae.

egg tooth small structure on tip of upper jaw or beak with which hatchling breaks out of shell.

eglandular *a.* without glands.

egocentric *a. appl.* behaviour that benefits the survival of the individual that exhibits it.

EGTA a chelating agent.

EI energy index. *see* phosphorylation potential.

EIA environmental impact assessment *q.v.*

Eichler's rule groups of hosts with more variation are parasitized by more species than taxonomically uniform groups.

eicosanoid *n.* any member of a class of compounds derived from the fatty acid arachidonic acid. Eicosanoids include prostaglandins, prostacyclins, thromboxanes, leukotrienes and lipoxins. They can act as intracellular messengers and be secreted as local chemical signals in e.g. inflammatory and hypersensitivity reactions, and have many different effects on a variety of cells.

eIF followed by a numeral, indicates a eukaryotic protein synthesis initiation factor.

Eimer's organs organs in the snout of moles, probably tactile organs.

einkorn *n.* a primitive cultivated diploid wheat (*Triticum monococcum*) first cultivated in the Near East and South-west Asia around 11,000 years ago, and which is derived from the wild *T. boeoticum*.

ejaculate *a.* the emitted seminal fluid.

ejaculatory *a.* throwing out, *appl.* certain ducts that discharge their secretions with some force.

ejaculatory sac organ pumping ejaculate from vas deferens through ejaculatory duct to penis, in certain insects.

elaborate *v.* to form complex organic substances from simple materials.

Elaeagnales *n.* order of shrubs, often with leathery leaves and thorns, comprising the family Elaeagnaceae (oleaster).

elaeoblast *n.* mass of nutrient material at posterior end of body in some tunicates.

elaeocyte *n.* cell containing lipid droplets, in coelomic fluid of annelid worms.

elaeodochon *n.* oil gland in birds.

elaio- elaeo-.

elaioplankton *n.* planktonic organisms rendered buoyant by oil globules in their cells.

elaioplast *n.* colourless organelle in plant cells in which oils or fats are formed and stored.

elaiosome *n.* fleshy oil-containing appendage present on seeds that are to be dispersed by ants, such as those of castor oil plant.

elaiosphere *n.* oil globule in a plant cell.

elasmobranchs *n.plu.* group of cartilaginous fishes including the sharks, dogfishes, skates and rays, which have an outer covering of bony tooth-like scales, a spiracle and no covering over the gill openings.

elastase *n.* proteolytic enzyme secreted in the pancreas, by leukocytes and by some bacteria, and acting on elastin and other proteins. EC 3.4.21.36–37.

elastic fibres yellow fibres *q.v.*

elastin *n.* connective tissue protein, major component of elastic fibres as in blood vessels and ligaments. It is formed from a soluble precursor, proelastin.

elater *n.* (1) (*bot.*) cell in spore capsule of liverworts with spiral thickening in its wall which is sensitive to changes in humidity, expanding or twisting and aiding the dispersal of spores; (2) flattened appendage of spore of horsetails, which unfolds in dry conditions and aids in dispersal by wind; (3) (*zool.*) springing organ in Collembola (springtails).

electrical coupling the passive flow of electric current from one animal cell to its neighbour through gap junctions, as in

cells of heart muscle where it mediates the synchronous contraction of individual fibres.

electrically active cells muscle and nerve cells.

electrical synapse junction between cells, including those of the nervous system, at which electric current flows directly from one cell to another, identified with gap junction in many instances. *cf.* chemical synapse. *see* synapse.

electric organ modifications of muscle or epithelium which discharge electric energy, found chiefly in certain fishes, e.g. electric eel.

electric receptors receptors present in some fish, sensitive to potential differences across the skin, used to detect distortions of the electric field that the fish sets up around itself.

electroblast *n.* modified muscle cell that gives rise to an electroplax.

electrocardiography *n.* recording of electrical activity in the heart, from electrodes placed on the body surface over the heart, the resulting trace being known as an electrocardiogram (ECG).

electrochemical gradient the gradient of an ion or other charged solute across a membrane, which is made up of both the concentration gradient of the solute and the gradient of electrical charge across the membrane.

electrochemical potential electrical potential difference across a membrane due to the presence of an electrochemical gradient of a solute across the membrane.

electrocyte *n.* modified muscle or nerve cell in electric organ of gymnotoid fish (American knife-fish) that produces an electric discharge.

electroencephalography *n.* recording of gross electrical activity in the brain from electrodes placed on the scalp, the resulting trace being known as an electroencephalogram (EEG).

electrogenic *a.* generating an electrical potential across a membrane, *appl.* certain ion pumps in cell membrane.

electrolemma *n.* membrane surrounding an electroplax.

electromagnetic senses senses that detect electric fields and which are used by some animals, e.g. some fish, to detect distortions in the Earth's electric field caused by objects in the locality. *see also* magnetotaxis.

electromyography *n.* measurement of electrical activity of muscle, by inserted electrodes, the resultant trace being known as an electromyogram (EMG).

electron acceptor in an oxidation–reduction reaction, the chemical partner that accepts electrons and becomes reduced.

electron carrier (1) any of the proteins and other molecules that transport electrons in an electron-transport chain; (2) molecule that carries electrons between the electron donor and electron acceptor in oxidation and reduction reactions, itself undergoing a cycle of oxidation and reduction, e.g. quinone in the respiratory electron-transport chain, NADH and NADPH in numerous metabolic pathways. *alt.* coenzyme.

electron crystallography the determination of molecular structure by the computer-aided analysis of multiple electron microscopic images of a crystalline array of molecules.

electron-dense *a. appl.* material that does not allow the passage of electrons and thus appears dark on an electron micrograph.

electron density map a contour plot of the distribution of electrons around the atoms of a molecule. Such a map is the end result of the X-ray crystallographic determination of a protein's atomic structure, and can be interpreted to give the spatial coordinates of each non-hydrogen atom in the protein.

electron donor in an oxidation–reduction reaction, the chemical partner that loses electrons and becomes oxidized.

electron microscope microscope in which images are produced by a beam of electrons passing through or impinging on an object, the beam being focused by magnetic lenses. *see* scanning electron microscope, transmission electron microscope, scanning-tunnelling electron microscope, tunnelling electron microscope.

electron-transfer chain, electron-transport chain general term for a series of electron carriers as found in the membranes of mitochondria and chloroplasts, along which electrons are transferred in a series of redox reactions. A typical chain consists of cytochromes, quinones, ferredoxin (in chloroplasts), flavoproteins (in mitochondria) and other components. The energy released during electron transport is used to pump protons across the membrane. *see also* respiratory chain.

electrophoresis *n.* technique for separating molecules such as proteins or nucleic acid fragments on the basis of their net charge and mass, by their differential migration through paper, or through a polyacrylamide or agarose gel (gel electrophoresis) in an electric field. *a.* **electrophoretic.**

electrophysiology *n.* study of physiological processes in relation to electrical phenomena.

electroplax *n.* one of the constituent plates of electric organ of e.g. electric eel.

electroporation *n.* the entry of DNA and other large molecules into animal cells under the influence of a strong electric field that makes temporary pores in the membrane.

electroreceptor *n.* sensory receptor sensitive to an electric field. *see* electrosensory.

electroretinography *n.* recording of changes in the electrical potential in the retina, after stimulation with light. The resulting trace is known as an electroretinogram (ERG).

electrosensory *a. appl.* sensory systems in animals, such as the ampullae Lorenzini in dogfish (*Scyliorhinus*), that detect changes in electric fields.

electrostatic bond bond formed by electrostatic interaction between oppositely charged ions or chemical groups. It is the bond formed, for example, between two ions in a salt, and is one of the types of bonds involved in the three-dimensional structure of proteins. *alt.* ionic bond, ion pair, salt bridge, salt linkage.

electrotaxis *n.* orientation of movement of a cell or organism within an electric field.

electrotonic *a.* (1) *appl.* localized potentials produced by subthreshold ionic currents in nerve cell membranes, determined by the passive electrical properties of cell; (2) *appl.* passive propagation of an electrical impulse in nerve cell membrane which gradually diminishes and dies away over a distance of a few millimetres. *cf.* action potential.

electrotonus *n.* modified condition of a nerve when subjected to a constant current of electricity.

electrotropism *n.* plant curvature in an electric field.

elementary body one of the forms taken by chlamydias during their life cycle. It is a small dense cell, relatively resistant to drying and is the dispersal and infectious form of the microorganism. *cf.* reticulate body.

eleocyte fat-containing cell in annelids that originates from the chlorogogen tissue.

elephant-tooth shells common name for the Scaphopoda *q.v.*

elicitor *n.* in plant pathology, a compound that induces a defence response to damage or infection in the plant. It may either be derived from a plant pathogenic microorganism (a biotic elicitor) or may be an inorganic material such as mercury and other heavy metals.

elimination reaction chemical reaction that removes an atom or a chemical group from an organic molecule, forming a double bond.

ELISA enzyme-linked immunosorbent assay *q.v.*

elittoral *a. appl.* zone out from the coast where light ceases to penetrate to the sea bottom.

elliptical *a. appl.* leaves of about the same breadth at equal distances from apex and base, which are themselves slightly acute.

elongation factor (EF) any of several accessory proteins (e.g. EF-Tu, EF-Ts, EF-G in *E. coli*) required for translation to proceed. They are involved in loading aminoacyl-tRNAs onto the mRNA–ribosome complex. Eukaryotic elongation factors are abbreviated eEF.

eluvial *a.* (1) *appl.* layer in soils which is impoverished and leached, above the illuvial layer, *alt.* A-horizon; (2) *appl.* gravels formed by breakdown of rocks *in situ*.

elytra *plu.* of elytron *q.v.*

elytroid *a.* resembling an elytron.

elytron *n.* the forewing of beetles, which is hard and stiff and is not flapped in flight, but serves at rest as a protective covering (wing case or wing sheath) for the membranous hindwing. *plu.* **elytra**.

elytrum elytron *q.v.*

emarginate *a.* with a distinct notch or indentation.

emasculation *n.* (1) removal of anthers to prevent self-pollination; (2) removal of testes.

Embden–Meyerhof pathway (EMP) glycolysis.

embedding *n.* method used in preparing permanent microscope slides or specimens for electron microscopy. For optical microscopy the prepared material is impregnated with molten paraffin wax before sectioning. For electron microscopy the specimen is embedded in a colourless resin before sectioning.

embiid *n.* common name for a member of the Embioptera *q.v.*

Embioptera *n.* order of insects with soft flattened bodies, and incomplete metamorphosis, commonly called embiids or foot spinners, which live in groups in silken tunnels and have wingless females and winged males.

embolism *n.* a blood clot blocking a blood vessel.

embolium *n.* outer or costal part of forewing in certain insects.

embolomerous *a. appl.* type of vertebra having two vertebral rings in each segment, due to union of hypocentra with neural arch and union of two pleurocentra below notochord.

embolus *n.* (1) projection closing foramen of an ovule; (2) apical division of palp in some spiders; (3) the core of horn of ruminants; (4) embolism *q.v.*

emboly invagination *q.v.*

embryo *n.* multicellular animal or plant before it is fully formed and capable of independent life. It develops from a fertilized egg (zygote) and develops into a freeliving miniature adult or larva, in animals, or germinates into a seedling, in plants. In human development, the conceptus is technically known as an embryo either from the moment of conception, or sometimes from the blastocyst stage, until the main parts of the body and main internal organs have started to take shape at around the 7th week of gestation. After this it is called a foetus. *see also* proembryo.

embryo cell in some plants, the one of the two cells formed from the 1st division of the fertilized egg that becomes the embryo, the other developing into the suspensor.

embryogenesis *n.* development of an embryo from a fertilized ovum.

embryogeny *n.* formation of the embryo, used esp. in botany.

embryoid bodies partly differentiated masses of tissue that are formed when cultured embryonic carcinoma cells are treated with retinoic acid.

embryology *n.* the study of the formation and development of embryos.

embryonal carcinoma cells stem cells derived from a teratocarcinoma. They can differentiate into a wide variety of tissues.

embryonated *a. appl.* eggs in which embryo has developed.

embryonic *a. pert.* an embryo.

embryonic membranes the various membranes surrounding the developing embryo, including the amnion. *see also* extra-embryonic.

embryonic stem cells (ES cells) cultured cell lines of pluripotent cells isolated from the inner cell mass of the very early mammalian embryo. On introduction into another early embryo they will develop to populate all tissues of the developing animal, including the germline. Genetic manipulation of cultured mouse ES cells and introduction of these cells into a mouse blastocyst is now a common method of producing transgenic mice with a desired mutation. Human ES cells are being investigated for their therapeutic potential in replacement cell therappy.

embryophore *n.* ciliated mantle enclosing embryo in many tapeworms, and formed from superficial blastomeres of embryo.

embryo sac in flowering plants, the mature female gametophyte (megagametophyte) which develops in the ovule, comprising an egg cell and accessory cells.

embryotega *n.* small hardened portion of testa which marks micropyle in some seeds and separates like a little lid at germination.

embryo transfer reproductive technology in which very early embryos produced by *in vitro* fertilization or artificial insemination are transferred into a surrogate mother for further development, used in cattle and sheep breeding to produce many more offspring from a prize female than she could produce naturally.

emergence *n.* (1) the hatching of adult insects, e.g. butterflies or moths, from their pupae; (2) outgrowth from subepidermal tissue; (3) epidermal appendage.

emergent *a. appl.* plants that are rooted under the water but which partly rise above the surface. *alt.* emersed.

emergent disease, emerging disease infectious disease that suddenly becomes prevalent in a human population, previously having been unknown or present at only a very low level.

emersed emergent *q.v.*

EMG electromyogram. *see* electromyography.

eminence *n.* ridge or projection on surface of bones.

emissary *a.* (1) coming out; (2) *appl.* veins passing through apertures in cranial wall and establishing connection between sinuses inside and veins outside.

emmenophyte, emmophyte *n.* water plant without any floating parts.

emmer *n.* a primitive cultivated wheat (*Triticum dicoccum*) first domesticated in the Near East and South-west Asia around 11,000 years ago, and which is thought to be derived from wild emmer (*T. dicoccoides*), which is a hybrid between *T. monococcum* and goat-grass (*Aegilops*).

EMP Embden–Meyerhoff–Parnas pathway. *see* Embden–Meyerhoff pathway.

empirical *a.* derived from experiment or observation.

empodium *n.* outgrowth between the claws on the last tarsal segment in flies.

EMS ethane methanesulphonate, a compound often used as an experimental mutagen.

enamel *n.* hard material containing over 90% calcium and magnesium salts which forms cap over dentine or may form complete tooth or scale.

enamel cell ameloblast *q.v.*

enameloid *a. appl.* enamel-like material in fish.

enamel organ complex structure of tall columnar epithelium (ameloblasts or enamel cells) forming the surface of the dental papilla, from which tooth enamel is developed.

enantiomer, enantiomorph *n.* either one of two structural isomers of a molecule that are mirror images of each other and rotate the plane of polarized light equally but in opposite directions. *a.* **enantiomeric.** *see also* chirality, optical isomer.

enantiomorphic *a.* (1) *appl.* or *pert.* a molecule, crystal or compound that exists as two enantiomers; (2) similar but contraposed, as mirror image; (3) deviating from normal symmetry.

enantiomorphism *n.* the existence of a molecule in two enantiomeric forms. *alt.* **enantiomerism**

enantiostylous *a.* having flowers whose styles protrude right or left of the axis, with the stamens on the other side.

enarthrosis ball-and-socket joint *q.v.*

enation *n.* outgrowth from a previously smooth surface.

encapsulated *a. appl.* bacteria that are surrounded by a capsule. Capsulated bacteria resist phagocytosis by the scavenger cells of the immune system unless coated with antibodies or complement.

encephalic *a. pert.* the brain.

encephalitis *n.* inflammation of the brain.

encephalization *n.* evolutionary tendency towards formation of a distinct brain.

encephalization factor measure of brain capacity in relation to body size. It is given by brain weight divided by (body weight)$^{0.69}$.

encephalomyelic *a. pert.* brain and spinal cord.

encephalomyelitis *n.* any disease that involves inflammation of the brain and spinal cord, may be viral or autoimmune in origin.

encephalon *n.* the brain.

enchytracheids *n.plu.* group of small oligochaete worms, living in soil.

encoding *n.* process of memory formation in which information entering memory pathways is passed into short-term memory.

encyst *v.* of a cell or small organism, to surround itself with an outer tough coat or capsule. *n.* **encystment.**

endangered *a.* IUCN definition *appl.* species or higher taxa whose numbers have become so low, or whose habitats have been so drastically reduced, that they

are thought to be in immediate danger of extinction in the wild in the foreseeable future if there is no change in circumstances. *see also* rare, rarity, vulnerable.

endangium *n.* innermost lining or tunica intima of blood vessels.

endarch *a.* with central protoxylem, or several protoxylem groups surrounding pith, produced when xylem matures centrifugally so the oldest protoxylem is closest to the centre of the axis.

end-bulbs minute oval or cylindrical bodies, representing the terminals of sensory neurons in mucous and serous membranes, in skin of genitalia, and in synovial layer of certain joints.

endemic *a.* (1) restricted to a certain region or part of region; (2) *appl.* disease, present at relatively low levels in the population all the time. *n.* **endemicity, endemism.**

endergonic *a.* absorbing or requiring energy, *appl.* metabolic reactions.

endexine *n.* the inner membranous layer of extine (exine) of a spore.

endites *n.plu.* offshoots from the border of certain appendages of arthropods.

endo- prefix derived from Gk *endon,* within, signifying within, inside, acting inside, opening to the inside, etc.

endoascus *a.* inner coat of ascus wall, protruding after rupture of the ectoascus, typical of certain ascomycete fungi.

endobasal *a. appl.* body in nucleus which acts as a centrosome in certain protozoans.

endobiotic *a.* living within a substratum or within another living organism.

endoblast hypoblast *q.v.*

endobronchus entobronchus *q.v.*

endocardiac, endocardial *a.* situated within the heart.

endocarditis *n.* inflammation of the endocardium and heart valves.

endocardium *n.* membrane lining internal cavities of heart.

endocarp *n.* the innermost layer of pericarp of fruit, usually fibrous, hard or stony, as the 'stone' enclosing the seed in plums and cherries.

endocarpy angiocarpy *q.v.*

endocast *n.* cast of the cranial cavity of a skull.

endochondral *a. appl.* ossification in cartilage beginning inside and working outwards.

endochorion *n.* inner lamina of chorion of insect eggs.

endochroic *a.* having pigment within a cell or hypha.

endochrome *n.* any pigment within a cell, esp. other than chlorophyll.

endocoel *n.* (1) coelom *q.v.*; (2) the cavities in proboscis, collar and trunk of certain hemichordates.

endocoelar *a. pert.* inner wall of coelom.

endoconidium *n.* conidium formed within a conidiophore.

endocranium neurocranium *q.v.*

endocrine *a.* (1) *appl.* ductless glands secreting hormones directly into blood; (2) *pert.* such glands. *cf.* exocrine.

endocrine system the system of endocrine glands secreting a variety of hormones, which are controlled by peptide hormones released from the pituitary and by direct neural input. *see* adrenal glands, neuro-endocrine system, ovary, steroid hormones, testis, thyroid gland.

endocrinology *n.* study of hormones produced in endocrine glands and their effects.

endocuticle *n.* innermost layer of the cuticle (exoskeleton) of arthropods.

endocycle *n.* layer of tissue separating internal phloem from endodermis.

endocyclic *a.* (*zool.*) (1) with the mouth remaining in axis of coil of gut, *appl.* crinoids; (2) having an apical system with double circle of plates surrounding anus, *appl.* echinoids; (3) (*bot.*) *pert.* endocycle; (4) (*mycol.*) *appl.* a rust fungus in which the teliospores resemble aeciospores.

endocyst *n.* (1) soft body wall of a poly-zoan zooid; (2) membranous inner lining of a protozoan cyst.

endocytic *a. pert.* or *appl.* endocytosis.

endocytic vesicle membrane-bounded vesicle in eukaryotic cells that has been formed as a result of endocytosis.

endocytosis *n.* process by which eukaryotic cells take up extracellular material by invagination of the plasma membrane to form vesicles enclosing the external material. *see also* pinocytosis, phago-cytosis, receptor-mediated endocytosis. *a.* **endocytotic, endocytic.**

endodeme *n.* gamodeme composed of pre-dominantly inbreeding dioecious plants or bisexual animals.

endoderm *n.* (1) primary germ layer in embryo of multicellular animals that gives rise to digestive tract, respiratory tract and associated glandular epithelium; (2) in coelenterates, the inner epithelial layer of body wall, surrounding the coelenteron. *a.* **endodermal.**

endodermal *a.* (1) (*zool.*) *pert.* endoderm *q.v.*; (2) (*bot.*) *pert.* endodermis *q.v.*

endodermis *n.* innermost layer of cortical cells in stems and roots of vascular plants, the layer surrounding the pericycle. *a.* **endodermal.**

endoenzyme *n.* any intracellular enzyme.

endogamy *n.* (1) zygote formation within a cyst by the fusion of two of the products of preceding division; (2) self-pollination; (3) inbreeding.

endogastric *a.* (1) having curvature of body with enclosing shell towards ventral side; (2) within the stomach.

endogenote *n.* in bacterial sexual processes, the part of the recipient cell's chromosome that is homologous to the incoming DNA.

endogenous *a.* (1) originating within the organism, cell or system being studied; (2) originating from a deep-seated layer, *appl.* lateral roots; (2) *appl.* metabolism, biosynthetic and degradative processes in tissues. *cf.* exogenous.

endogenous opioids peptides that are the body's natural ligands for the opiate receptors. *see* endorphins, enkephalins, dynorphins.

endogenous pyrogen fever-causing substance released by host cells when stimulated by bacterial endotoxin as a result of infection, and which acts on the temperature-control centres in the brain.

endogenous rhythm metabolic or behavi-oural rhythm that originates within the organism and persists even when external conditions are kept constant. There may be some slight change in the periodicity in constant conditions, when it is said to be free-running, such as a circadian rhythm changing to a periodicity of 23 or 25 hours. Endogenous circadian rhythms are main-tained at the regular 24-hour cycle by an external stimulus such as light or tem-perature, the *zeitgeber*.

endogenous virus virus (e.g. C-type retroviruses) carried permanently in an inactive proviral state in the genome and inherited from generation to generation, and which may become activated on infection with another virus.

endognath, endognathite *n.* inner branch of oral appendages of crustaceans.

endogonidium *n.* the colony-forming cell in colonial protists such as *Volvox*.

endolaryngeal *a.* within the larynx.

endolithic *a.* living or penetrating into rock, as some algae and fungi.

endolymph *n.* fluid filling the semi-circular canals and cochlea of the ear. *a.* **endolymphatic.**

endolymphangial *a.* situated in a lymphatic vessel.

endolymphatic *a. pert.* lymphatic system, or labyrinth of ear.

endolysis *n.* destruction of a cell or other material within another cell that has engulfed it.

endomembrane system the intracellular membrane system of a eukaryotic cell, comprising the endoplasmic reticulum, Golgi apparatus, lysosomes and plasma membrane. These are all connected by a flow of membrane from one to the other by means of small membrane vesicles. Mitochondria and chloroplasts, with their double membranes, are not part of this endomembrane system.

endomeninx *n.* single inner membrane covering embryonic brain, giving rise to pia mater and arachnoid.

endomere *n.* a blastomere that gives rise to endoderm.

endomesoderm embryonic cells that can give rise to both endoderm and mesoderm.

endometrium *n.* mucous epithelium lining the uterus.

endomitosis *n.* multiplication and separation of interphase chromosomes without subsequent nuclear division, leading to endopolyploidy.

endomixis *n.* self-fertilization in which a male and female nucleus from the same individual fuse.

endomycorrhiza *see* mycorrhiza.

endomysium *n.* connective tissue binding muscle fibres.

endoneurium *n.* delicate connective tissue holding nerve fibres together in a bundle within a nerve.

endonuclease *n.* nuclease that cleaves a nucleic acid at internal sites.

endoparasite *n.* parasite that lives inside another cell or organism.

endopeptidase *n.* peptidase that splits a protein molecule into peptides by hydrolysing internal peptide bonds. EC 3.4.21–25, 3.4.99. *alt.* protease.

endoperidium *n.* inner layer of peridium.

endophagous *a.* feeding inside a food source.

endophragm *n.* an internal septum formed from projections from the cephalic and thoracic regions in crustaceans.

endophyllous *a.* (1) sheathed by a leaf; (2) living within a leaf.

endophyte *n.* bacterium, fungus or alga living inside the body or cells of an organism to which they cause no apparent damage. *alt.* endosymbiont.

endoplasm *n.* (1) the inner part of the cytoplasm of a cell, which is often granular; (2) endosarc *q.v. cf.* ectoplasm.

endoplasmic reticulum (ER) extensive, convoluted internal membrane-bounded space in eukaryotic cells, continuous with the nuclear envelope. It is involved in the synthesis, glycosylation and transport of membrane proteins and of proteins destined for secretion from the cell, which are delivered into the ER from their site of synthesis in the cytosol. It is also the site of new membrane lipid synthesis and stores calcium ions. The cytosolic face of rough endoplasmic reticulum (RER) is studded with protein-synthesizing ribosomes. Areas lacking ribosomes are known as smooth endoplasmic reticulum (SER). *see* Fig. 7 (p. 105).

endopleura *n.* inner seed coat.

endopodite *n.* inner or mesial branch of a two-branched crustacean limb, or if not branched, the only part of the limb remaining.

endopolyploidy *n.* polyploidy resulting from repeated doubling of chromosome number without normal mitosis.

Endoprocta Entoprocta *q.v.*

Endopterygota, endopterygotes *n.,* *n.plu.* a division of insects having complete metamorphosis with wings developing internally and larvae different from adults. Includes the Coleoptera, Neuroptera,

Mecoptera, Trichoptera, Lepidoptera, Diptoptera, Siphonaptera, Hymenoptera and other orders. *cf.* Exopterygota.

endorachis *n.* layer of connective tissue lining canal of vertebral column and skull. *alt.* endorhachis.

endoral *a. pert.* structures situated within the 'mouth' of certain protozoans.

endoreplication *n.* continued replication of DNA without separation of new chromosomes or cell division, which produces giant polytene chromosomes.

end organ a structure at the end of a nerve, such as a sensory receptor or motor end plate.

endorhizal *a.* with the radicle enclosed, as in seed of monocotyledons.

endorhizosphere *n.* the epidermis and cortex of normal healthy roots which is invaded by soil microorganisms.

endorphins *n.plu.* family of peptides produced in gut, brain and pituitary, which can mimic the narcotic effects of morphine by binding to opiate receptors, and which together with the enkephalins and the dynorphins are known as the endogenous opioids. β-Endorphin is a peptide of 31 amino acids, generated from proopiomelanocorticotropin, which produces effects such as analgesia, rigidity and changes in mood and behaviour.

endosarc *n.* internal layer of cytoplasm, esp. in protozoans.

endosclerite *n.* any sclerite of the endoskeleton of arthropods.

endosiphuncle *n.* tube leading from the protoconch to siphuncle in certain cephalopods.

endoskeleton *n.* any internal skeleton or supporting structure.

endosmosis *n.* osmosis in an inward direction, such as into a cell.

endosome *n.* vesicle formed in animal cells by fusion of endocytic vesicles.

endosperm *n.* nutritive tissue surrounding plant embryo in most angiosperm seeds. It is triploid and develops from fusion of two haploid female nuclei with a haploid male nucleus. *alt.* albumen.

endospore *n.* (1) spore formed within a sex organ or sporangium; (2) (*bact.*) thickwalled heat- and drought-resistant asexual spore formed within the cells of certain

bacteria, e.g. *Bacillus* and *Clostridium* spp., in unfavourable growing conditions, and which can remain dormant for many years.

endosporium *n.* inner coat of spore wall.

endosporous *a.* (1) having spores borne inside an organ such as a sporangium, *appl.* fungi and plants; (2) producing endospores, *appl.* bacteria.

endosteal *a. pert.* endosteum.

endosternite *a.* internal skeletal plate for muscle attachment.

endosteum *n.* membrane lining cavities in bones.

endostosis *n.* ossification that begins in cartilage.

endostracum *n.* the inner layer of mollusc shell.

endostyle *n.* longitudinal groove in ventral wall of pharynx in urochordates and some primitive chordates, involved in mucus secretion.

endosymbiont *see* endosymbiosis.

endosymbiont hypothesis the idea, now widely accepted, that the mitochondria and chloroplasts of eukaryotic cells are the relics of bacterial endosymbionts, and that the eukaryotic cell evolved from symbiotic associations of prokaryotic cells.

endosymbiosis *n.* symbiosis in which one partner (the endosymbiont) lives inside the cells of the other, e.g. photosynthetic cyanobacteria living in the cells of non-photosynthetic dinoflagellates. *a.* **endosymbiotic**.

endotergite *n.* infolding from tergite of insects, for muscle attachment.

endotheca *n.* system of membranes lining and joining polyp cavities in a coral.

endothecium *n.* (1) in bryophytes, inner tissue of sporogonium, formed from central region of embryo; (2) inner lining of an anther. *a.* **endothecial**.

endothelial *a. pert.* or part of an endothelium.

endothelin *n.* polypeptide produced by endothelial cells that stimulates contraction of the underlying smooth muscle of blood vessel walls.

endotheliochorial *a. appl.* placenta with chorionic epithelium in contact with endothelium of uterine capillaries, as in carnivores and some other mammals.

endothelium *n.* single layer of flattened cells, of mesodermal origin, bound tightly together, that lines internal body surfaces such as blood and lymphatic vessels, heart and other fluid-filled cavities.

endothermic *a.* (1) *appl.* chemical reaction that absorbs heat, *cf.* exothermic; (2) *appl.* regulation of body temperature by an animal by producing heat within its own tissues, as do birds and mammals. Such animals are known as endotherms, *cf.* ectothermic. *see also* homoiothermic. *n.* **endothermy**.

endotoky *n.* the development of larvae within the body of the mother, which consume her and kill her.

endotoxin *n.* bacterial toxin that is an integral component of the bacterial cell and is usually only released when the cell is degraded. Generally refers to the lipopolysaccharides of outer cell membranes of Gram-negative bacteria, which induce fever, and in large quantities can also cause septic shock and tissue destruction. *cf.* exotoxin.

endotrophic *a. appl.* mycorrhizas in which the fungal hyphae extensively penetrate the cells of the host roots.

endotunica *n.* inner layer of ascus wall.

endozoic *a.* living within an animal or involving passage through an animal, as in the distribution of some seeds.

endozoochore *n.* any spore, seed or organism dispersed by being carried within an animal.

endozoochory *n.* dispersal of seeds by ingestion by animals and passage through them.

end-piece filament forming the tail of a sperm.

end plate the contact of a nerve fibre terminal with a muscle fibre.

end-plate potential (EPP) change in potential in muscle cell membrane seen on stimulation of its associated nerve fibre.

end-product inhibition feedback control *q.v.*

energid *n.* nucleus in syncytial blastoderm of insect embryo.

energy budget balance of energy input and use in a biological system. It is expressed as consumption (C) = production (P) + respiration (R) + rejecta (faeces and urine) (F + U). *see also* assimilated energy.

energy charge (EC) index of the energy status of a cell in terms of proportions of ATP, ADP and AMP: EC = ([ATP] + [ADP])/([ATP] + [ADP] + [AMP]). *see also* phosphorylation potential.

energy flow the transfer of energy from one organism to another through the trophic levels in an ecosystem. About 90% of the chemical energy is lost at each transfer.

energy of activation *see* free energy of activation.

energy source any substance, organic or inorganic, that can be taken in by a living organism and used as a source of energy. For photosynthetic organisms, the primary energy source is light.

enervose *a.* having no veins, *appl.* certain leaves.

engram *n.* a memory trace, a supposed permanent change in the brain accounting for the existence of memory.

engraved *a.* with irregular linear grooves on the surface.

enhalid *a.* (1) containing salt water, *appl.* soils; (2) growing in saltings or on loose soil in salt water, *appl.* plants.

enhancer *n.* type of control sequence present in many eukaryotic genes, and also found in prokaryotes, whose binding by gene-regulatory proteins dramatically increases the level of transcription. Enhancers generally control the tissue-specific expression of their associated gene.

enhancer mutation mutation that intensifies the phenotypic effect of other mutations.

enhancer protein gene regulatory protein that acts at an enhancer sequence.

enhancer trap artificial DNA construct that is used to locate and isolate tissue-specific genes. It contains a weak promoter and a marker gene such as *lacZ* which will only be expressed strongly if the enhancer trap inserts into the chromosome near an enhancer.

enkephalins *n.plu.* pentapeptides found in brain, gut and adrenal gland which bind to opiate receptors in brain, mimicking the effects of morphine, producing effects such as analgesia, rigidity and changes in mood and behaviour. Two main types, Leu-enkephalin and Met-enkephalin.

Together with endorphins and dynorphins they are known as the endogenous opioids.

enneagynous *a.* having nine pistils.

enneandrous *a.* having nine stamens.

enolase *n.* enzyme catalysing the formation of phosphoenolpyruvate from 2-phosphoglycerate in glycolysis and other metabolic pathways. EC 4.2.1.11. *r.n.* phosphopyruvate hydratase.

Enopla *n.* class of proboscis worms having the mouth anterior to the brain, and usually an armed proboscis.

enphytotic *a.* afflicting plants, *appl.* diseases restricted to a locality.

enrichment culture technique for isolating a particular microbial species from a natural mixed population by culture in conditions particularly favourable for its growth, so that it becomes the predominant form in the culture.

ensiform *a.* sword-shaped.

entactin nidogen *q.v.*

entad *adv.* towards the interior.

ental *a.* (1) inner; (2) internal.

entangial *a.* (1) within a vessel; (2) produced inside a sporangium.

entelechy *n.* in vitalist theories, a vital principle or influence, distinct from physicochemical forces, inherent in living organisms and directing their vital processes.

enteral *a.* (1) within intestine; (2) *appl.* the parasympathetic portion of the autonomic nervous system.

enteric *a.* (1) *pert.* alimentary canal; (2) *pert.* intestines or gut; (3) *appl.* parasympathetic ganglia.

enteric bacteria large group within the Bacteria, comprising Gram-negative, rod-shaped, non-sporulating, oxidase-negative, heterotrophic bacteria with a facultatively aerobic metabolism and able to ferment glucose. They include many plant, animal and human pathogens. Some species live in soil or water, some are normal inhabitants of the mammalian intestine, and some cause disease. Examples are *Escherichia, Erwinia, Proteus, Salmonella, Shigella, Yersinia.* The group is part of the gamma-proteobacteria. *alt.* enterobacteria.

entero-, -enteron word elements derived from Gk *enteron*, gut.

enterobacteria enteric bacteria *q.v.*

enterococcus *n.* streptococcus present in the intestinal tract, esp. *Streptococcus faecalis.*

enterocoel *n.* coelom arising as a pouch-like outgrowth of the archenteron, or as a series of such outgrowths.

enterocoelomate *a.* having an enterocoel.

enterocyte *n.* intestinal epithelial cell.

enteroendocrine cells cells in the epithelium lining the lumen of the small intestine that secrete serotonin and a variety of peptide hormones.

enterogastrone *n.* duodenal hormone which inhibits secretion and motility of stomach.

enterohaemorrhagic *a. appl.* strains of bacteria and viruses that cause bloody diarrhoea.

enterokinase enteropeptidase *q.v.*

enteron *n.* (1) alimentary tract; (2) coelenteron *q.v.*; (3) any structure corresponding to the archenteron of gastrula.

enteronephric *a.* with nephridia opening into gut, *appl.* oligochaetes.

enteropathogenic *a.* producing disease in the intestinal tract.

enteropeptidase *n.* enzyme produced by duodenal cells that activates trypsinogen by cleavage of a peptide bond to produce trypsin. EC 3.4.21.9.

Enteropneusta *n.* class of solitary, worm-like burrowing hemichordates, having many gill slits and no lophophore, commonly called acorn worms.

enteroproct *n.* opening from the endodermal gut into the posterior (ectodermally derived) gut.

enterostome *n.* (1) in coelenterates, the aboral opening of the actinopharynx, leading to the coelenteron; (2) the posterior opening of the stomodaeum into endodermal gut.

enterosympathetic *a. appl.* that part of the sympathetic nervous system supplying the gut.

enterotoxic *a.* having a toxic effect on the intestine.

enterotoxin *n.* any toxin secreted by bacteria infecting food or growing in the intestine, and which has effects on the intestine.

enteroviruses *n.plu.* picornaviruses typically found in the gut, and which include e.g. Coxsackie viruses, echoviruses, poliovirus.

enterozoon *n.* any animal parasite inhabiting the intestines.

enthalpy *n.* thermodynamic term describing the energy lost as heat to the environment in an exothermic chemical reaction or in an open thermodynamic system such as a living organism.

enthetic *a.* (1) introduced; (2) implanted.

entire *a.* (1) unimpaired; (2) with continuous margin, *appl.* e.g. leaves, bacterial colonies.

Entner–Doudoroff pathway metabolic pathway in some bacteria in which glucose is converted to 2-ketogluconic acid.

ento- endo- *q.v.*

entobranchiate *a.* having internal gills.

entobronchus *n.* the dorsal secondary branch of bronchus in birds.

entochondrite *n.* carapace or endosternum of the king crab *Limulus*.

entocoel *n.* space enclosed by a pair of mesenteries in Anthozoa.

entoconid *n.* the postero-internal cusp of a lower molar.

entocuneiform *n.* the most internal of distal tarsal bones.

entoectad *a.* from within outwards.

entoglossal *a.* lying within tissue of tongue.

entoglossum *n.* extension of basihyal into tongue in some fishes.

entomochory *n.* dispersal of seeds or spores through the agency of insects. *a.* **entomochoric.**

entomofauna *n.* the insects of a particular environment or region.

entomogenous *a.* growing in or on insects, as certain fungi.

entomology *n.* the study of insects.

entomophagous insectivorous *q.v.*

entomophilous *a.* pollinated by insects. *n.* **entomophily.**

Entomophthorales *n.* order (in some classifications considered as a class) of zygomycete fungi chiefly parasitic on insects and also known as fly fungi, which reproduce by forcibly discharged conidia, and have a mycelium with septa and a tendency to fragment.

entomophyte *n.* any fungus growing on or in an insect.

entomostracans *n.plu.* large group of small crustaceans including branchiopods, *Branchiura* (fish lice), copepods, barnacles and ostracods.

entoneural *a.* *appl.* system of aboral ring and genital nerves in enchinoderms.

entopic *a.* in the normal position. *cf.* ectopic.

entoplastron *n.* anterior median plate in the shell of turtles or tortoises, often called the episternum.

Entoprocta, entoprocts *n.*, *n.plu.* phylum of solitary or colonial small marine pseudocoelomate animals, in some classifications included in the Bryozoa. They have a U-shaped gut with anus opening to the exterior within a circle of ciliated tentacles.

entopterygoid *n.* dorsal membrane bone behind palatine in some fishes.

entoptic *a.* (1) within the eye; (2) *appl.* visual sensations caused by eye structures and processes, not by light.

entorhinal cortex area of grey matter in the brain that provides an interface between the hippocampus and higher cortical areas and is thought to have roles in spatial processing and memory.

entosternum *n.* (1) entoplastron *q.v.*; (2) an internal process of sternum of numerous arthropods.

entotympanic *n.* a separate tympanic bone in some mammals.

entovarial *a.* inside the ovary.

entozoic *a.* living inside the body of another animal or plant.

entrainment *n.* process by which a free-running endogenous biological rhythm is synchronized to an environmental stimulus.

entropy *n.* thermodynamic term describing the disorder or randomness of a system.

enucleate (1) *n.* lacking a nucleus; (2) *v.* to remove the nucleus from a cell.

env gene of C-type retroviruses specifying envelope protein of the virus particle.

envelope *n.* (1) of certain viruses, layer of lipid and protein surrounding capsid, the lipid being derived from host cell membrane as the virus is discharged from the cell, the protein being virus-encoded. Such viruses are known as **enveloped viruses**; (2) bacterial envelope *q.v. see also* floral envelope, nuclear envelope.

environment *n.* the external influences acting on an organism. *a.* **environmental.**

environmental audit a survey of all aspects of an environment, including e.g. the geology, hydrology, and habitats and species present, and human influences, often undertaken as part of an environmental impact assessment.

environmental impact assessment (EIA) an evaluation of the likely environmental consequences of a proposed development and of the measures to be taken to minimize adverse effects, which is now a legal requirement in some countries.

environmental resistance the factors limiting the population growth of a species in a given environment.

environmental science the study of how humans and other species interact with their non-living and living environment, which includes the study of ecology, resource management and conservation, population dynamics, economics, politics and ethics.

environmental variance that variance in a phenotype which is due to environmental influences.

enzootic *a*. (1) afflicting animals; (2) *appl.* disease in animals restricted to a locality. *cf*. epizootic.

enzymatic *a. pert.* enzymes, enzyme action, or reactions catalysed by enzymes. *alt*. **enzymic**.

enzyme *n*. any of a large and diverse group of (mainly) proteins that function as catalysts in virtually all biochemical reactions, essential in all cells, different enzymes being highly specific for a particular chemical reaction and reactants. Enzymes are classified and named according to the type of chemical reaction catalysed and substrate acted on, e.g. hydrolases, carboxylases, oxido-reductases, nucleases, proteinases. Certain RNAs can also act as enzymes. *see* ribozyme. *see also* immobilized enzyme.

enzyme kinetics the progress of an enzymatic reaction over time, the exact course of which is dependent on substrate concentration, temperature and the molecular properties of the enzyme (e.g. allostery).

enzyme-linked immunosorbent assay (ELISA) a type of serological assay in which the antibodies used to detect a particular substance are labelled by linkage to an enzyme. The test substance is immobilized on a plastic surface and a positive reaction, i.e. antibody binding to the

surface, is detected by the action of the enzyme on a colourless substrate to produce a coloured product.

enzyme-linked receptor any of a large class of receptors that either have intrinsic enzymatic activity in their cytoplasmic tails or become associated with cytoplasmic enzymes when the receptor is stimulated. They include the receptor tyrosine kinases and the kinase-associated receptors.

enzyme–substrate complex intermediate in the action of an enzyme on its substrate, when the substrate becomes non-covalently or covalently bound to the enzyme at the active site.

enzymic enzymatic *q.v.*

enzymology *n*. study of enzymes and their action.

enzymopathy *n*. disease resulting from a failure to produce an enzyme, due to a genetic defect.

eobiogenesis *n*. formation of living matter from prebiotic macromolecular systems.

Eocene *n*. early epoch of the Tertiary sub-era and the Paleogene period, between Paleocene and Oligocene, lasting from *ca.* 55-34 million years ago to 34 million years ago. The first fossil evidence of most modern mammalian orders appears.

Eogaea *n*. zoogeographical division including Africa, South America and Australasia.

eolian *alt*. spelling of aeolian.

eosin *n*. red/brown acidic dye, sodium or potassium salt of eosin ($C_{20}H_8Br_4O_5$).

eosinophil *n*. type of white blood cell classed as a granulocyte. It contains granules that stain with the red acidic dye eosin and which contain substances that can induce inflammation when the granules are secreted by exocytosis. It is involved in destruction of parasites and in inflammatory reactions.

eosinophilia *n*. abnormal increase in number of eosinophils, seen in certain allergic states.

eosinophilic *a*. staining readily with eosin.

epacme *n*. (1) the stage in the phylogeny of a group just before its highest point of development; (2) stage in development of an individual just before adulthood.

epactal *a*. (1) supernumerary; (2) intercalary; (3) *appl.* bones situated in sutures between two other bones.

epalpate *a.* lacking palps.

epanthous *a.* living on flowers, *appl.* certain fungi.

epapillate *a.* lacking papillae.

epapophysis *n.* median process arising from centre of vertebral neural arch.

eparterial *a.* situated above an artery.

epaulettes *n.plu.* (1) branched or knobbed processes projecting from oral arms of some jellyfish; (2) tegula, in Hymenoptera.

epaxial *a.* above the axis, dorsal, usually with respect to axis formed by the vertebral column.

ependyma *n.* layer of cells lining cavities of brain and spinal cord. *a.* **ependymal**.

ephedrine *n.* sympathomimetic alkaloid obtained from *Ephedra* spp., having the same effects as adrenaline and used as a nasal decongestant and to treat asthma.

ephemeral *a.* (1) short-lived; (2) taking place once only, *appl.* plant movements as expanding buds; (3) completing life cycle within a brief period; (4) *n.* a short-lived plant or animal species.

Ephemeroptera, ephemerids *n.*, *n.plu.* order of insects including the mayflies, whose adult life is very short, sometimes less than a day. They have an aquatic nymph (larval) stage. Adults have three appendages, 'tails', projecting from the end of the abdomen, and the hindwing is smaller than the forewing.

ephippial *a. appl.* winter eggs, as of rotifers and daphnids.

ephippium *n.* (1) the pituitary fossa, the cavity in which the pituitary body lies in the sphenoid bone; (2) saddle-shaped modification of cuticle detached from carapace and enclosing winter eggs in daphnids.

Eph receptors *see* ephrins.

ephrins *n.plu.* membrane proteins that regulate cell adhesion and repulsion responses in the guidance of migrating cells and axons in animal development, by binding to receptors (Eph receptors) on other cells.

ephyra *n.* immature medusa in some jellyfish, formed by strobilization from a polyp. *plu.* **ephyrae**.

epi- prefix derived from Gk *epi*, upon, signifying above or upon.

epibasal *n.* upper segment of plant zygote or embryo, ultimately giving rise to the shoot.

epibasidium *n.* upper part of cell of septate basidium from which sterigmata and basidiospores are produced.

epibenthos *n.* fauna and flora of sea bottom between low-water mark and 200-metre line.

epibiotic *a.* (1) *appl.* endemic species that are relics of a former flora or fauna; (2) growing on the exterior of a living organism, with reproductive organs at surface and part or all of soma within the substrate; (3) living on a surface, as of sea bottom.

epiblast *n.* in mammalian and avian embryos that part of the blastoderm that gives rise to the embryo proper. *a.* **epiblastic**.

epiblem(a) *n.* outermost layer of tissue in roots, which may be the piliferous layer, or the exodermis in an older root where the piliferous layer has worn away.

epiboly *n.* extension of one part over another, as in the extension of ectoderm over mesoderm in amphibian gastrulation.

epibranchial *n.* dorsal skeletal element of gill arch.

epicalyx *n.* calyx-like structure formed outside but close to the true calyx, formed from stipules fused in pairs, or by aggregation of bracts, bracteoles or small sepal-like structures.

epicanthus *n.* a prolongation of upper eyelid over inner angle of eye. *alt.* **epicanthic fold**.

epicardia *n.* (1) that part of oesophagus running from diaphragm to stomach; (2) *plu.* of epicardium *q.v.*

epicardium *n.* that part of the pericardium that is reflexed over the viscera (internal organs) including the heart. *plu.* **epicardia**.

epicarp exocarp *q.v.*

epicentral *a.* attached to or arising from vertebral centra, *appl.* intermuscular bones.

epicerebral *a.* situated above the brain.

epichilium *n.* in orchid flowers, the outer part of the lip where there are two distinct parts.

epichondrosis *n.* formation of cartilage on the fibrous membrane surrounding a bone, as in formation of antlers.

epichordal *a.* (1) upon the notochord; (2) *appl.* vertebrae in which the ventral cartilaginous portions are almost completely suppressed; (3) *appl.* upper lobe of caudal fin in fishes.

epichroic *a.* discolouring, as after injury.

epiclinal *a.* situated on the receptacle or torus of a flower.

epicoel *n.* (1) cavity of mid-brain in lower vertebrates; (2) cerebellar cavity; (3) a perivisceral cavity formed by invagination.

epicondyle *n.* (1) a medial and a lateral protuberance at distal end of humerus and femur. *a.* **epicondylar, epicondylic.**

epicone *n.* the part anterior to the girdle in dinoflagellates.

epicoracoid *a. pert.* a skeletal element, usually cartilaginous, at sternal end of coracoid in amphibians, reptiles and monotremes.

epicormic *a.* growing from a dormant bud.

epicortex *n.* an outer layer, as of filaments covering cortex of some fungi.

epicotyl *n.* the stem-like axis above the cotyledons in a plant embryo or young seedling, terminating in an apical meristem and sometimes bearing one or more young leaves.

epicranial *a.* (1) *pert.* the cranium; (2) *pert.* the epicranium of insects.

epicranium *n.* (1) region between and behind eyes in insect head; (2) scalp; (3) the structures covering the cranium in vertebrates.

epicritic *a. appl.* stimuli and nerves concerned with delicate touch and other special sensations in skin.

epicuticle *n.* outer waxy layer of the exoskeleton of arthropods.

epicutis *n.* outer layer of cutis in agarics.

epicyst *n.* the external resistant cyst of an encysted protozoan.

epicytic *a.* on the surface of a cell.

epideictic display a suggested display by which members of a population reveal their presence and allow others to assess the density of the population.

epidemic *a. appl.* an outbreak of a disease affecting large numbers of individuals at the same time over quite a wide geographical area. *n.* **epidemic.** *cf.* endemic, pandemic.

epidemiology *n.* (1) the study of the occurrence of infectious diseases, their origins and pattern of spread through the population; (2) the cause and pattern of spread of a disease; (3) the study of the

incidence of non-infectious disease (e.g. cancer) with a view to finding causes and contributing factors, e.g. the causal link between smoking and lung cancer was found by epidemiological studies.

epidermal *a. pert.* epidermis *q.v.*

epidermal growth factor (EGF) cytokine that stimulates the division of epidermal and other cells, and has been used to promote wound healing.

epidermis *n.* (1) outer layer or layers of the skin, derived from embryonic ectoderm. In vertebrates a non-vascular stratified tissue, often keratinized, *see* Fig. 35 (p. 622); (2) outer epithelial covering of roots, stems and leaves in plants. *a.* **epidermal.**

epidermoid *a.* resembling epidermis.

epididymis *n.* mass at back of testicle composed mainly of ductules leading from testis to vas deferens. *a.* **epididymal.** *plu.* **epididymides.**

epidural *a.* (1) *pert.* dura mater; (2) *appl.* space between dura mater and wall of vertebral canal around spinal cord.

epifauna *n.* (1) animals living on the surface of the ocean floor; (2) any encrusting fauna.

epigamic *a.* (1) *appl.* any trait related to courtship and sex other than the essential sexual organs and copulatory behaviour; (2) tending to attract the opposite sex, e.g. *appl.* colour displayed in courtship.

epigaster *n.* that part of embryonic intestine that later develops into the colon.

epigastric *a.* (1) *pert.* anterior wall of abdomen; (2) middle region of upper zone of artificial divisions of abdomen.

epigastrium *n.* the middle region of upper part of abdomen above the navel.

epigeal, epigean, epigeous *a.* (1) above ground; (2) (*bot.*) *appl.* type of seed germination in which cotyledons are carried above ground as shoot grows; (3) (*zool.*) *appl.* insects, living near the ground.

epigenesis *n.* the accepted central concept of embryonic development, that an embryo is formed by the gradual differentiation and organization of its parts from an undifferentiated single fertilized egg cell. This was originally opposed to the earlier theory of preformation, which held that the sperm (or the egg) contained a fully formed miniature individual, and that development consisted only in an increase in size.

epigenetic *a.* (1) *appl.* or *pert.* epigenesis *q.v.*; (2) *appl.* processes by which modifications in gene function that can be inherited by a cell's progeny occur without a change in DNA nucleotide sequence. Such modifications include DNA methylation, heterochromatin formation, genomic imprinting, paramutation, X-chromosome inactivation.

epigenous *a.* developing or growing on a surface.

epigeous epigeal *q.v.*

epiglottis *n.* moveable flap of fibrocartilage which bends over the opening of the windpipe in throat when food is being swallowed.

epignathous *a.* having upper jaw longer than lower.

epigonial *a. appl.* sterile posterior portion of genital ridge.

epigonium *n.* (1) cover over the young sporogonium of a liverwort; (2) calyptra *q.v.*

epigyne, epigynium, epigynum *n.* external female genitalia in arachnids.

epigynous *a. appl.* flowers in which sepals, petals and stamens are attached on top of the ovary. *cf.* hypogynous. *n.* **epigyny**.

epilepsy *n.* brain disorder characterized by sudden massive waves of electrical activity in the brain, which produce seizures or fits. *a.* **epileptic**.

epilimnion *n.* upper water layer, above thermocline in lakes, rich in oxygen.

epilithic *a.* attached on rocks, *appl.* algae, lichens.

epimandibular *a. pert.* a bone in the lower jaw of vertebrates.

epimeletic *a. appl.* animal behaviour relating to the care of others.

epimer *n.* any one of two or more molecules, esp. monosaccharides, that differ only in the arrangement of atoms around a single carbon atom other than the carbonyl carbon atom. *a.* **epimeric**.

epimerase *n.* any enzyme that can convert one epimer into the other, included in EC 5.1.

epimere *n.* the upper part of a somite in vertebrate embryos, giving rise to muscle.

epimerism *n.* the existence of epimers of a molecule.

epimerite *n.* prolongation of protomerite of certain gregarines for attachment to host.

epimerization *n.* conversion of one epimer into another.

epimeron *n.* posterior part of the side wall of any segment in insects.

epimorphic *a.* maintaining the same form in successive stages of growth.

epimorphosis *n.* type of regeneration in which regenerated structures are formed by new growth by cell proliferation.

epimyocardium *n.* outer wall of embryonic heart tube, which will form the heart muscle.

epimysium *n.* sheath of areolar tissue which invests the entire muscle.

epinasty *n.* the more rapid growth of upper surface, compared with lower surface, of a dorsoventral organ, e.g. a leaf, thus causing unrolling or downward curvature.

epinekton *n.* nekton that are incapable of actively swimming themselves but are attached to actively swimming organisms.

epinephrin(e) *alt.* name for adrenaline esp. in North America.

epineural *a.* (1) arising from vertebral neural arch; (2) *pert.* canal external to radial nerve in some echinoderms; (3) *appl.* sinus between embryo and yolk, beginning of body cavity in insects.

epineurium *n.* fibrous connective tissue sheath around a nerve.

epineuston *n.* those animals living at the surface of water, in the air.

epinotum propodeon *q.v.*

epiostracum *n.* thin cuticle or epicuticle covering exocuticle or exostracum in mites and ticks.

epiotic *a. pert.* upper element of bony capsule of ear.

epiparasite ectoparasite *q.v.*

epipelagic *a.* (1) *pert.* deep-sea water between surface and bathypelagic zone; (2) inhabiting oceanic water at depths not exceeding around 200 m, i.e. above mesopelagic zone.

epipelon *n.* algal community living in or on the surface of sediments in shallow waters where light penetrates.

epipetalous *a.* having stamens inserted upon petals.

epipharynx *n.* the upper lip of dipteran insects (e.g. flies), adapted as a piercing organ in some cases.

epiphenomenon *n.* something produced as a side-effect of a process.

epiphloem outer bark, periderm *q.v.*

epiphragm *n.* membrane or plate that closes an opening, e.g. the shell of certain molluscs, or capsule of certain mosses.

epiphyllous *a.* (1) growing upon leaves; (2) united to perianth, *appl.* stamens.

epiphysis *n.* (1) any part or outgrowth of a bone which is formed from a separate centre of ossification and later fuses with the main part of the bone, becoming its terminal portion. *a.* **epiphyseal**; (2) pineal body *q.v.*

epiphyte *n.* plant that lives on the surface of another plant but does not derive nourishment from it.

epiphyton *n.* community of 'plants' living attached to other plants, e.g. algae living on plants in aquatic environments.

epiphytotic *n.* epidemic disease in plants.

epiplankton *n.* that portion of plankton from surface to about 200 m.

epiplasm *n.* cytoplasm within an ascus that is excluded from the ascospore when that is delimited.

epiplastron *n.* in turtles and tortoises, one of a pair of anterior bony plates in shell.

epipleura *n.* (1) rib-like structure in teleost fishes which is not preformed in cartilage; (2) hooked process on rib in birds; (3) turned down outer margin of elytra in some beetles.

epiploic *a. pert.* omentum *q.v.*

epiploic foramen opening between bursa omentalis and large sac of peritoneum.

epipodite *n.* process arising from basal joint of crustacean limb, modified for various functions.

epiproct *n.* the central 'tail' of silverfish and other members of the insect order Thysanura.

epipteric *a.* (1) winged at tip, *appl.* certain seeds; (2) *pert.*, or shaped like, or placed above, wing.

epipterygoid *n.* small bone extending nearly vertically downwards from pro-otic to pterygoid in skull of some reptiles.

epipubic *a. pert.* or borne upon the pubis.

epipubis *n.* unpaired cartilage or bone borne anteriorly on pubis.

epirhizous *a.* growing upon a root.

episclera *n.* connective tissue between sclera and conjunctiva.

episematic *a.* aiding in recognition, *appl.* coloration, markings. *cf.* aposematic, parasematic, sematic.

episeme *n.* a marking or colour aiding in recognition.

episepalous *a.* adnate to sepals.

episkeletal *a.* outside the skeleton.

episodic memory memory of a particular incident or of a particular time or place.

episodic selection strong directional selection, often occurring in newly disturbed habitats.

episome *n.* (1) autonomous self-replicating DNA in bacteria which can integrate into the bacterial chromosome semipermanently, e.g. F factor, some plasmids; (2) often used interchangeably with the term plasmid. *a.* **episomal**.

episperm *n.* the outer coat of seed.

epistasis, epistasy *n.* the suppression or masking of the effect of a gene by another, different, gene.

epistatic *a. appl.* the gene whose action masks or suppresses. *cf.* dominance.

episternalia *n.plu.* two small elements preformed in cartilage frequently intervening in development between clavicles and sternum, and ultimately fusing with sternum.

episternum *n.* (1) bone between the clavicles; (2) anterior part of the side wall of any of the thoracic segments in insects.

epistome *n.* (1) small lobe overhanging mouth in Polyzoa and containing part of body cavity; (2) region between antenna and mouth in Crustacea; (3) portion of insect head immediately behind labrum; (4) portion of rostrum in some dipterans.

epistrophe *n.* the position assumed by chloroplasts alongside the walls of a plant cell when exposed to diffuse light.

epistropheus *n.* the 2nd cervical or axis vertebra.

epithalamus *n.* part of brain comprising habenula, pineal body and posterior commissure.

epithalline *a.* growing upon the thallus.

epithallus *n.* cortical layer of hyphae in lichens.

epitheca *n.* (1) external layer surrounding theca in many corals; (2) older half of frustule in diatoms.

epithecium *n.* layer of tissue on the surface of hymenium of an apothecium.

epithelia *plu.* of epithelium.

epithelial *a. pert.* or *appl.* an epithelium *q.v.*

epithelial–mesenchymal induction type of induction in which mesenchyme is induced to differentiate by the influence of an adjacent epithelium. Occurs e.g. in development of kidney tubules and in tooth development.

epitheliochorial *a. appl.* placenta with apposed chorionic and uterine epithelia, and villi pitting the uterine wall, as in marsupials and ungulates.

epithelioid *a.* resembling epithelium.

epithelium *n.* sheet of cells tightly bound together, lining all external surfaces and internal surfaces continuous with the external environment in multicellular animals, e.g. the epidermis, surfaces of mucous membranes, the lining of the gut, and the linings of secretory ducts and glands. Epithelia variously serve protective, secretory or absorptive functions. An epithelium may be composed of a single layer of cells, or be composed of several layers (stratified epithelium). *plu.* **epithelia**. *a.* **epithelial**. *cf.* endothelium.

epitoke *n.* the sexually reproductive form of marine polychaete worms, which develops from the posterior part of the animal and separates from the anterior asexual part. *n.* **epitoky** reproduction by this mode.

epitope antigenic determinant *q.v.*

epitreptic *a. appl.* animal behaviour causing another animal of the same species to approach.

epitrichium *n.* an outer layer of foetal epidermis of many mammals, usually shed before birth.

epitrochlea *n.* inner condyle at distal end of humerus.

epixylous *a.* occurring on wood, *appl.* fungi, lichens.

epizoic *a.* (1) living on or attached to the body of an animal; (2) having seeds or fruits dispersed by being attached to the surface of an animal.

epizoite *n.* organism that lives on the shell or surface of another animal but is not parasitic on it.

epizoochory *n.* dispersal of seeds by being carried on the body of an animal.

epizoon *n.* animal that lives on the body of another animal.

epizootic *n.* epidemic disease amongst animals.

eplicate *a.* not folded.

epoch *n.* in geological time, the subdivision of a period.

eponychium *n.* thin fold of cuticle which overlaps the lunula (half-moon) of nail.

eponym *n.* name of a person used in designation of e.g. a species, organ, law or disease.

EPP end-plate potential *q.v.*

epsilon-proteobacteria group of the proteobacteria (purple Bacteria) distinguished on DNA sequence data. It consists of chemoorganotrophs, e.g. *Campylobacter*, *Heliobacter*, *Thiovulvum* and *Wolinella*. *alt.* **epsilon purple bacteria**.

EPSP excitatory postsynaptic potential *q.v.*

Epstein–Barr virus (EBV) a herpes virus, the cause of infectious mononucleosis and also associated with Burkitt's lymphoma and nasopharyngeal carcinoma. Infects primarily B cells and epithelial cells and can transform cells on prolonged latent infection.

equatorial furrow cleavage furrow around equator of fertilized egg undergoing the first cleavage division.

equatorial plate (1) metaphase plate *q.v.*; (2) in plant cells, site of new cell wall formation during cell division.

equiaxial *a.* having axes of equal length.

equibiradiate *a.* with two equal rays.

equicellular *a.* composed of equal number of cells, or composed of cells of equal size.

equifacial *a.* having equivalent surfaces or sides, as vertical leaves.

equifinality *n.* arrival at a common endpoint in behavioural development by different routes.

equilateral *a.* (1) having the sides equal; (2) *appl.* shells symmetrical about a transverse line drawn through umbo.

equilibrium constant (K) number that characterizes the equilibrium state for a reversible chemical reaction. It is given by the ratio of forward and reverse rate constants.

equilibrium dialysis technique for measuring the affinity of a protein for a small-molecule ligand that can diffuse across a dialysis membrane.

equilibrium potential value of membrane potential at which a given ion can pass easily in either direction across the membrane.

equilibrium sedimentation density gradient centrifugation *q.v.*

equimolecular *see* isotonic.

equinoctial *a. appl.* flowers that open and close at definite times.

equipotent *a.* (1) totipotent *q.v.*; (2) able to perform the function of another cell, part or organ.

equipotential *a.* of equal developmental potential.

Equisetales horsetails. *see* Sphenophyta.

equitant *a.* overlapping like a saddle, as leaves in leaf bud, leaves on stem of e.g. iris.

equivalence group group of embryonic cells with the same developmental potential.

equivalve *a.* having two halves of shell alike in form and size.

ER endoplasmic reticulum *q.v.*

era *n.* a main division of geological time, such as Paleozoic, Mesozoic, Cenozoic, and divided into periods.

eradication *n.* the extinction of a species in a particular area.

erect *a.* upright; (1) *appl.* ovule, directed towards summit of ovary; (2) *appl.* plants, growing upright, not decumbent.

erectile *a.* capable of being raised or erected, as of a penis or crest.

erector *n.* a muscle which raises an organ or part.

Eremian *a. appl.* or *pert.* part of the Palaearctic region including deserts of North Africa and Asia.

eremic *a. pert.*, or living in, deserts.

eremobic *a.* (1) growing or living in isolation; (2) having a solitary existence.

eremophyte *n.* a desert plant.

eRF eukaryotic release factor. *see* release factor.

ERG electroretinogram. *see* electroretinography.

ergaloid *a.* having adults sexually capable though wingless, *appl.* insects.

ergastic substance storage or waste product produced by a cell.

ergatandrous *a.* having worker-like males.

ergataner *n.* male ant resembling worker.

ergate(s) *n.* worker ant.

ergatogyne *n.* female ant resembling a worker, intermediate between queen and worker.

ergatoid *a.* resembling a worker, *appl.* male ant.

ergocalciferol *see* calciferol.

ergonomics *n.* the anatomical, physiological and psychological study of humans in their working environment.

ergonomy *n.* (1) the differentiation of functions; (2) physiological differentiations associated with morphological specialization.

ergosterol *n.* sterol present in yeasts, moulds and certain algae, and in animal tissues, and which is converted to vitamin D_2 (ergocalciferol) on irradiation.

ergot *n.* the hardened mycelial mass (sclerotium) of the fungus *Claviceps purpurea*, which replaces the grain of infected rye and some other grasses. It contains poisonous alkaloids that cause ergotism – abortion, hallucinations and sometimes death – in animals and humans who eat the infected grain.

ergotism *see* ergot.

ericaceous *a.* of or *pert.* the Ericaceae, the heather family, which includes the heaths, heathers and rhododendrons.

Ericales *n.* order of dicot shrubs, rarely trees or herbs, and including the families Actinidiaceae, Empetraceae (crowberry), Ericaceae (heaths and rhododendrons), Monotropaceae (Indian pipe) and Pyrolaceae (wintergreen).

ericoid *a. appl.* endomycorrhizas formed on members of the Ericaceae, lacking a fungal sheath and with extensive penetration of the cell of the root cortex, with formation of intracellular hyphal coils.

erineum *n.* outgrowth of abnormal hairs on leaves, produced by certain gall mites.

Eriocaulales *n.* order of monocot herbs comprising the family Eriocaulaceae (pipe-wort).

eriocomous *a.* (1) having woolly hair; (2) fleecy.

eriophyid *n.* any of a large family (Eriophyidae) of minute plant-eating mites that have two pairs of legs at front and no respiratory system.

eriophyllous *a.* having leaves with a cottony appearance.

erose *a.* having margin irregularly notched, *appl.* leaf, bacterial colony.

E-rosette red blood cell surrounded by T cells bound to it.

erostrate *a.* having no beak, *appl.* anthers.

errantia *n.* mobile organisms. *cf.* sedentaria.

error-prone repair DNA repair that introduces mutations in the process of restoring the structural integrity of the DNA.

eruca *n.* (1) a caterpillar; (2) an insect larva having the shape of a caterpillar.

erucic acid unusual unsaturated fatty acid ($C_{22}H_{42}O_2$) found in large quantities in rapeseed (*Brassica campestris*) and some other related plants.

eruciform *a.* (1) *appl.* insect larvae with a more-or-less cylindrical body and stumpy legs on the abdomen as well as the true thoracic legs, as caterpillars and grubs; (2) having the shape of a caterpillar or grub.

erumpent *a.* breaking through suddenly, *appl.* fungal hyphae.

erythema *n.* abnormal reddening of the skin due to dilation of blood capillaries.

erythrism *n.* abnormal prescence, or excessive amount, of red colouring matter, as in petals, feathers, hair, eggs.

erythritol *n.* sugar alcohol found in raised concentrations in placenta and which stimulates growth of the bacterium *Brucella abortus.*

erythroaphins *n.plu.* red pigments formed by the postmortem enzymatic transformation of yellow plant pigments in aphids.

erythroblast *n.* nucleated cell of bone marrow which gives rise to erythrocytes.

erythroblastosis foetalis a severe form of haemolytic anaemia in the foetus and newborn. It is caused by a Rh-negative mother making anti-Rh antibodies which destroy the foetus's Rh-positive red blood cells.

erythrocruorin *n.* name formerly used for the haemoglobin respiratory pigments of annelids and molluscs.

erythrocuprein superoxide dismutase *q.v.*

erythrocyte *n.* predominant type of cell in vertebrate blood. It is small and discshaped, and lacks most internal organelles (including the nucleus in mature mammalian erythrocytes). It contains large amounts of the oxygen-binding protein haemoglobin, which transports oxygen and gives the cell its red colour. *alt.* red blood cell, red blood corpuscle.

erythrocyte colony-forming unit (CFU-E) erythrocyte precursor cell which forms colonies of erythrocytes in culture if erythropoietin is present.

erythrocytic *a. appl.* phase of the malaria parasite life cycle in humans in which merozoites invade red blood cells.

erythrocytic burst-forming unit (BFU-E) bone marrow cell which in culture forms large colonies of erythrocytes if stimulated with erythropoietin, presumed to be a precursor of the erythrocyte colony-forming unit.

erythrogenic *a.* producing reddening, as in inflammation or a rash.

erythrogenic toxin toxin produced by *Streptococcus pyogenes* which causes the red rash in scarlet fever.

erythrolabe *n.* the red-sensitive pigment of the human eye.

erythromycin *n.* antibiotic synthesized by the actinomycete *Streptomyces erythreus*, which inhibits bacterial protein synthesis.

erythron *n.* the red cells of blood and bone marrow collectively.

erythrophilous *a.* having an affinity for red stains.

erythrophore xanthophore *q.v.*

erythrophyll *n.* a red anthocyanin found in some leaves and red algae.

erythropoiesis *n.* production of red blood cells.

erythropoietic *a.* (1) generating red blood cells; (2) stimulating the production of red blood cells.

erythropoietin *n.* glycoprotein hormone produced chiefly by the kidney and which stimulates the final differentiation of red blood cells from precursor cells.

erythrose *n.* four-carbon sugar, involved esp. in carbon fixation in green plants.

ESC embryonic stem cell *q.v.*

escape (1) *n.* plant or animal originally domesticated and now established in the wild; (2) *a. appl.* behaviour in which an animal moves away from an unpleasant stimulus.

ES cells embryonic stem cells *q.v.*

Escherichia coli *E. coli*, a usually harmless bacterial inhabitant of the colon of humans and some other mammals. *E. coli* is widely used as an experimental subject in bacterial genetics and as a host bacterium in recombinant DNA work. Some strains of *E. coli* are pathogenic, causing serious gastrointestinal disease in humans and animals.

escutcheon *n.* area on rump of many quadrupeds which is either variously coloured or has the hair specially arranged.

eseptate *a.* without septa.

eserine *n.* plant alkaloid, specific inhibitor of cholinesterase. *alt.* physostigmine.

E site site on ribosome at which a tRNA, having discharged its amino acid, binds immediately before it leaves.

esophageal, esophagus *alt.* spelling of oesophageal, oesophagus.

esoteric *a.* arising within the organism.

espathate *a.* having no spathe.

esquamate *a.* having no scales.

ESS evolutionarily stable strategy *q.v.*

essential amino acids amino acids which cannot be synthesized by the body, or only in insufficient amounts, and must be supplied in the diet: for humans these are Arg, His, Ile, Leu, Lys, Met, Phe, Thr, Trp, Val.

essential elements trace elements *q.v.*

essential fatty acids (EFA) fatty acids that cannot be synthesized by mammals and must be present in the diet, e.g. linoleate and linolenate.

essential oils mixtures of various volatile oils derived from benzenes and terpenes found in plants and producing characteristic odours, and having various functions such as attracting insects or warding off fungal attacks.

EST expressed sequence tag *q.v.*

ester *n.* organic product of a condensation reaction between an alcohol and an acid (e.g. a carboxylic acid), the bond formed by this reaction being known as an ester bond ($-CH_2-O-CO-$).

esterase *n.* any hydrolytic enzyme that attacks an ester, splitting off the acid, EC 3.1.

esterification *n.* formation of an ester.

esthesia *alt.* spelling of aesthesia.

estipulate *a.* having no stipules.

estival *alt.* spelling of aestival.

estivation *alt.* spelling of aestivation.

estradiol *alt.* spelling of oestradiol.

estriate *a.* (1) not marked by narrow parallel grooves or lines; (2) not streaked *cf.* striate.

estrin, estriol, estrogen, estrone, estrous, estrus *alt.* spellings of oestrin, oestriol, oestrogen, oestrone, oestrous, oestrus.

estuarine *a.* living in the lower part of a river or estuary where freshwater and seawater meet.

etaerio *n.* aggregate fruit composed of achenes, berries, drupels, follicles or samaras.

ethanol *n.* ethyl alcohol, CH_3CH_2OH, which is produced by yeasts and other microorganisms from sugars by a fermentation in which pyruvate is converted into acetaldehyde and then into ethanol. Large-scale industrial biotechnological processes for producing ethanol include brewing and wine-making as well as the production of fuel alcohol from organic wastes.

ethanolamine *n.* small hydrophilic molecule that forms the headgroup in the phospholipid phosphatidylethanolamine.

ether *n.* organic compound consisting of an oxygen atom bonded to two hydrocarbon radicals: $-CH_2-O-CH_2-$. The type of linkage found in membrane lipids of Archaea. *cf.* ester.

ethereal *a.* (1) *appl.* class of odours including those of ethers and fruits; (2) *appl.* fragrant oils in many seed plants.

ethidium bromide reagent showing orange fluorescence on binding to double-stranded DNA.

Ethiopian region zoogeographical region in Wallace's classification consisting of Africa south of the Sahara, Madagascar, and the south-western part of the Arabian peninsula. It is divided into subregions as follows: East African, Malagasy, South African, West African.

ethmohyostylic *a.* with mandibular suspension from ethmoid region and hyoid bar.

ethmoid *a. pert.* bones that form a considerable part of the nasal cavity.

ethmoidal *a. pert.* ethmoid bones or region.

ethmoidal notch a quadrilateral space separating the two orbital parts of the frontal bone.

ethmopalatine *a. pert.* ethmoid and palatine bones or their region.

ethmoturbinals *n.plu.* cartilages or bones in the nasal cavity which are folded so as to increase area of olfactory epithelium.

ethmovomerine *a.* (1) *pert.* ethmoid and vomer regions; (2) *appl.* the cartilage that forms the nasal septum in early embryo.

ethnobotany *n.* the study of the use of plants by humans.

ethnography *n.* the description and study of human races.

ethnology *n.* science dealing with the different human races, their distribution, relationship and activities.

ethnozoology *n.* the study of the use of animals by humans.

ethogram *n.* catalogue of the natural behaviours of an animal and the contexts in which they occur.

ethological isolation prevention of interbreeding between species as the result of behavioural differences.

ethology *n.* study of the behaviour of animals in their natural habitats. *a.* **ethological**.

ethylene *n.* C_2H_4, a gas produced by plants in minute amounts and which has various developmental effects as a plant hormone, including regulation of fruit ripening. It acts at specific ethylene receptors in the plant cell membrane.

ethylenediaminetetraacetate EDTA *q.v.*

etiolation *n.* the appearance of plants grown in the dark, having no chlorophyll, chloroplasts not developing, internodes being greatly elongated so the plant is tall and spindly, and having small, rudimentary leaves. *a.* **etiolated**.

etiolin protochlorophyll *q.v.*

etiological *alt.* spelling of aetiological.

etiology *alt.* spelling of aetiology.

etioplast *n.* (1) chloroplast formed in the absence of light, found in etiolated leaves. It lacks thylakoid membranes and chlorophyll, and will develop into a functional chloroplast on illumination; (2) chloroplast precursor.

-etum in ecology, suffix used to indicate a plant community dominated by a particular species, e.g. a callunetum, a community dominated by heather (*Calluna vulgaris*).

euapogamy *n.* development of a diploid sporophyte from one or more cells of the gametophyte without fusion of gametes. For haploid apogamy, *see* meiotic apogamy.

Euascomycetae *n.* large class of ascomycete fungi including the black moulds, blue moulds, powdery mildews, discomy-cetes, tar spots, morels and truffles, in which the asci are enclosed in an ascocarp.

euaster *n.* aster in which rays meet at a common centre.

Eubacteria Bacteria *q.v.*

eucarpic *a.* having the fruiting body arising from only part of the thallus, while the rest of the thallus continues to carry out its somatic functions, *appl.* fungi. *cf.* holo-carpic, monocentric.

Eucarya *alt.* spelling of Eukarya.

eucaryote, eucaryotic *alt.* spellings of eukaryote, eukaryotic.

eucentric pericentric *q.v.*

eucephalous *a.* with well-developed head, *appl.* certain insect larvae.

euchroic *a.* having normal pigmentation.

euchromatin *n.* chromatin that is in a structural state such that the DNA can be transcribed. *alt.* active chromatin. *a.* **euchromatic**. *cf.* heterochromatin.

Eucommiales *n.* order of dicot trees comprising a single family Eucommiaceae with one genus *Eucommia*.

eucone *a.* having crystalline cones fully developed in each ommatidium, *appl.* compound eyes.

eudominant *n.* a dominant species that is more or less restricted to a particular climax vegetation.

eudoxome *n.* free-swimming stage of a siphonophore lacking the nectocalyx.

eugamic *a. appl.* mature period rather than youthful or senescent.

eugenic *a. pert.* or able to increase the fitness of a race or breed.

eugenics *n.* a pseudoscientific philosophy at its height in Europe and the United States in the early 20th century which aimed to 'improve' the genetic quality of the human population, and which eventually led to abuses such as compulsory sterilization of those deemed 'unfit' and persecution of racial minorities.

euglenoid *a.* (1) *pert.* Euglenophyta *q.v.*; (2) *appl.* movement, resulting from a change in shape, as in *Euglena*.

Euglenophyta, euglenoids *n., n.plu.* phylum of mainly unicellular freshwater flagellate protists typified by *Euglena*, which have no rigid cell wall and store food as fat or the polysaccharide par-amylon. They have a single locomotory undulipodium and a smaller pre-emergent flagellum. Most, but not all, are photosynthetic. In older botanical classifications they are treated as a division of the algae. In zoological classifications they are included in the Mastigophora.

eugonic *a.* (1) prolific; (2) growing profusely, *appl.* bacterial colonies.

euhaline *a.* (1) *appl.* seawater or water of comparable salinity, i.e. *ca.* 35 parts per thousand of sodium chloride (salt); (2) living only in saline waters.

euhyponeuston *n.* organisms living in the top 5 cm of water for the whole of their lives.

Eukarya *n.* one of the three domains or superkingdoms into which all living organisms are classified, the other two being the Bacteria and the Archaea. The Eukarya includes all organisms with cells possessing a membrane-bounded nucleus in which the DNA is complexed with histones and organized into chromosomes. Eukaryotic cells also have an extensive cytoskeleton of protein filaments and tubules, and many cellular functions are sequestered in membrane-bounded organelles in the cytoplasm, such as mitochondria, chloroplasts, endoplasmic reticulum and Golgi apparatus. The eukaryotes comprise protozoans, algae, fungi, slime moulds, plants and animals. *a.* **eukaryotic.** *see* Appendices 1–4.

eukaryote *n.* a member of the Eukarya *q.v.*

eukaryotic cell any cell from a member of the Eukarya *q.v.*

eulamellibranch *a. appl.* gills of bivalve molluscs whose filaments are attached to adjacent ones by bridges of tissue.

Euler–Lotka equation an equation used to compute the intrinsic rate of increase, r, of a population using survivorship and fertility data, where l_x is the probability of females surviving from birth to the beginning of age class x, and m_x is the number of offspring per female in class x.

eumelanin *n.* black melanin.

eumerism *n.* an aggregation of like parts.

eumeristem *n.* meristem composed of isodiametric thin-walled cells with dense cytoplasm and large nuclei.

eumerogenesis *n.* segmentation in which the units are similar for at least some time.

eumetazoa *n.* the multicellular animals excluding the sponges.

eumitosis *n.* typical mitosis, as occurs in the cells of most multicellular plants and animals.

euphausiid *n.* member of the order Euphausiacea, small, usually luminescent shrimplike crustaceans forming an important part of the marine plankton.

Euphorbiales *n.* order of dicot trees, shrubs and occasionally herbs, including the families Buxaceae (boxwood), Euphorbiaceae (spurge), Simmondsiaceae (jojoba) and others.

euphotic *a.* (1) well-illuminated, *appl.* zone of surface waters to depth of *ca.* 80–100 m; (2) upper layer of photic zone.

euphotometric *a. appl.* leaves oriented to receive maximum diffuse light.

euplankton *n.* the plankton of open water.

euploid *a.* (1) having an exact multiple of the haploid number of chromosomes, e.g. being diploid, triploid, tetraploid, etc., *n.* **euploidy;** (2) *appl.* chromosomal abnormalities that do not disrupt relative gene dosage, *alt.* balanced; (3) *n.* organism having cells with an exact multiple of the haploid number of chromosomes.

eupotamic *a.* thriving both in streams and in their backwaters, *appl.* plankton.

Euptales *n.* order of dicot trees and shrubs comprising the family Eupteleaceae with a single genus *Euptelea.*

European subregion (1) area within the Euro-Siberian floristic region. It consists of the whole of Europe from southern Scandinavia to northern Spain (and excluding the Mediterranean), and east to the Ural mountains; (2) zoogeographical area, a subdivision of the Palaearctic region, consisting of Northern and Western Europe south to the Pyrenees and the southern Alps and east to the Caspian Sea and the Ural Mountains. It also includes Iceland and Greenland.

Euro-Siberian region floristic region within the Boreal kingdom, consisting of the whole of Europe from southern Scandinavia to northern Spain, and Asia between *ca.* 65°N and *ca.* 50°N.

Euryarchaeota *n.* kingdom within the Archaea, defined on the basis of DNA sequence data, which contains the extreme halophiles such as *Halobacterium*, most methanogens, e.g. *Methanobacterium*,

Methanococcus and *Methanosarcina*, and *Thermoplasma* and *Archaeoglobus*. *cf.* Crenarchaeota.

eurybaric *a. appl.* animals adaptable to great differences in altitude or pressure.

eurybathic *a.* having a wide range of vertical distribution.

eurybenthic *a. pert.* or living within a wide range of depths in the sea, of organisms that live on the ocean floor.

eurychoric *a.* widely distributed.

euryhaline *a. appl.* marine organisms adaptable to a wide range of salinity.

euryhygric *a. appl.* organisms adaptable to a wide range of atmospheric humidity.

euryoecious *a.* having a wide range of habitat selection.

euryphotic *a.* adaptable to a wide range of illumination.

Eurypterida, eurypterids *n., n.plu.* extinct subclass (or order) of giant (2 m long) predatory fossil aquatic arthropods of the class Merostomata, present in the Ordovician, having a short non-segmented prosoma and a long segmented opisthosoma, and resembling scorpions.

eurythermic *a. appl.* organisms adaptable to a wide range of temperature. *alt.* **eurythermous**.

eurytopic *a.* having a wide range of geographical distribution.

euryxerophilous *a. appl.* plants adaptable to a wide range of dry conditions within a temperate climate.

eusocial *a. appl.* social insects which display cooperative care of the young, reproductive division of labour with more-or-less sterile individuals working on behalf of those engaged in reproduction, and an overlap of at least two generations able to contribute to colony labour. They include all ants, termites, and some bees and wasps.

eusporangiate *a. appl.* ferns in which a sporangium develops from several initial cells, which form a sporangium with a wall of more than one layer of cells (a eusporangium). *cf.* leptosporangiate.

eusporangium *see* eusporangiate.

Eustachian tube canal connecting middle ear and pharynx (throat).

Eustachian valve valve guarding orifice of inferior vena cava in atrium of heart.

eustele *n.* type of stele in which strands of vascular tissue (vascular bundles) surround a central pith and are separated from each other by parenchymatous ground tissue. Present in stems of horsetails and of dicotyledons and some gymnosperms.

Eustigmatophyta *n.* phylum of unicellular yellow-green photosynthetic protists, possessing a distinctive eyespot composed of drops of carotenoid pigments, and usually a single flagellum.

eustomatous *a.* having a distinct mouth or mouth-like opening.

eustroma *n.* in lichens, stroma formed of fungal cells only.

Eutheria, eutherians *n., n.plu.* an infraclass of mammals, including all mammals except the monotremes and marsupials, which are viviparous with an allantoic placenta, and have a long period of gestation, after which the young are born as immature adults. *alt.* placental mammals.

euthycomous *a.* straight-haired.

eutrophic *a. appl.* water bodies rich in plant nutrients and therefore usually highly productive, with very large numbers of plankton, often dominated by cyanobacteria, and often with turbid water in summer. Eutrophic waters suffer frequent algal blooms. Coarse fish (e.g. perch, roach and carp) are dominant. Larger aquatic plants may be absent as the water can become depleted of dissolved oxygen through the decay of large amounts of organic matter. *n.* **eutrophy**. *see* eutrophication.

eutrophication *n.* the enrichment of bodies of fresh water by inorganic plant nutrients (e.g. nitrate, phosphate). It may occur naturally but can also be the result of human activity (cultural eutrophication from fertilizer runoff and sewage discharge) and is particularly evident in slow-moving rivers and shallow lakes. The biomass of phytoplankton and herbivorous zooplankton increases, and species diversity decreases. The water becomes turbid in summer, the growth of the larger aquatic plants may eventually become suppressed and algal blooms are frequent. The water may become anoxic through the decay of large amounts of organic matter. Increased sediment deposition can eventually raise the level of the lake or river bed, allowing land plants to colonize the edges, and eventually converting the area to dry land.

eutropic *a.* (1) turning sunward; (2) dextrorse *q.v.*

eutropous *a.* (1) *appl.* insects adapted to visiting special kinds of flowers; (2) *appl.* flowers whose nectar is available to only a restricted group of insects.

euxerophyte *n.* plant that shows adaptations to and thrives in very dry conditions.

evaginate *v.* (1) to evert from a sheathing structure; (2) to protrude by eversion.

evagination *n.* (1) the process of unsheathing, or the product of this process; (2) the process of turning inside out.

E-value false-positive expectation value, a statistical measure that is used as a cut-off point to filter the matches found in a search of a sequence database against a query sequence, to try to ensure that only homologous sequences will be found. In its simplest form it represents the number of alignments with scores at least equal to a chosen value, *S*, that would be expected by chance alone, given the size of the database.

evanescent *a.* (1) disappearing early; (2) *appl.* flowers that fade quickly.

evaporative water loss (EWL) (1) (*bot.*) *see* evapotranspiration; (2) (*zool.*) in mammals and birds, the loss of heat from the body through the evaporation of water, which may occur from the body surface (sweating) in some mammals and/or from the respiratory tract (thermal panting) in some mammals and birds. Although EWL is actively employed by some birds and animals for cooling, in small mammals it could lead to dehydration and is minimized.

evapotranspiration *n.* loss of water from the soil by evaporation from the surface and by transpiration from the plants growing thereon.

evelate *a.* lacking a veil, *appl.* certain agaric fungi.

even-toed ungulates artiodactyls *q.v.*

event-related potential large change in electrical activity in the brain that is elicited by a sensory or motor event. *alt.* evoked potential.

evergreen *a. appl.* vascular plants that do not shed all their leaves at the same time and therefore appear green all the year round.

eviscerate *v.* (1) to disembowel; (2) to eject the internal organs, as do holothurians (sea cucumbers) on capture.

evocation *n.* in developmental biology, the ability of an inducer to elicit a particular pathway of differentiation in the induced tissue.

evoked potential event-related potential *q.v.*

evolute *a.* (1) turned back; (2) unfolded.

evolution *n.* the development of new types of living organisms from pre-existing types by the accumulation of genetic differences over long periods of time. It is studied by reference to the fossil record and to the anatomical, physiological and genetical differences between extant organisms. Present-day views on the process of evolution are based largely on the theory of evolution by natural selection formulated by Charles Darwin and Alfred Russel Wallace in the 19th century. Darwin's theory has undergone certain modifications to incorporate the principles of Mendelian genetics, unknown in his day, and the more recent discoveries of molecular biology, but still remains a basic framework of modern biology. *see also* Creationism, Darwinism, natural selection, neo-Darwinism, macroevolution, microevolution, molecular evolution.

evolutionarily stable strategy (ESS) in evolutionary theory, a behaviour pattern or strategy which, if most of the population adopt it, cannot be bettered by any other strategy and will therefore tend to become established by natural selection. Using games theory the results of various different strategies (e.g. in contests between males) can be worked out and a theoretical ESS determined and compared with actual behaviour.

evolutionary clock molecular clock *q.v.*

evolutionary distance (*ED*) a measure of the evolutionary relatedness of two taxa. It is obtained by comparing e.g. rDNA sequences and using the number of positions in the sequence at which the two taxa differ (after correction for various factors). *alt.* genetic distance.

evolutionary grade the level of development of a structure, physiological process or behaviour occupied by a species or group of species that are not necessarily related.

evolutionary psychology an approach to psychology in which knowledge and principles from evolutionary biology are used in research on the structure of the human mind and human behaviour.

evolutionary taxonomy taxonomic philosophy and method that utilizes both phenotypic characters and lines of descent in the classification of organisms. One difference from the cladistic method of classification is the acceptance of taxa which do not contain all the descendants of the common ancestor, e.g. the class Reptilia, which does not contain the birds, who are descendants of a reptile ancestor. *cf.* cladistics.

evolutionary transformation series pair of homologous characters, one of which is derived from the other.

evolvability *n.* the capacity to undergo evolutionary change.

evolvate *a.* lacking a volva, *appl.* certain agaric fungi.

evolve *v.* to undergo evolution.

EWL evaporative water loss *q.v.*

ex- prefix derived from Gk *ex*, without.

exafferent *a. appl.* stimulation that results solely from factors outside the body.

exalate *a.* wingless.

exalbuminous *a.* (1) lacking albumen; (2) *appl.* seeds without endosperm or perisperm.

exannulate *a.* having sporangia not furnished with an annulus, *appl.* certain ferns.

exanthema *n.* a skin rash, or a disease in which such a rash appears, e.g. measles.

exaptation *n.* an evolutionary process in which a characteristic that evolved under natural selection for a particular function is placed under selection for a different function: *e.g.* feathers first developed for heat insulation were later adapted for flight.

exarate *a. appl.* insect pupae in which all the appendages are free.

exarch *n.* stele with protoxylem strands to the outside of the metaxylem, produced when xylem matures centripetally so that the oldest protoxylem is farthest from the centre of the axis.

exarillate *a.* lacking an aril.

exasperate *a.* furnished with hard, stiff points.

Excavata, excavates *n., n.plu.* proposed major monophyletic taxonomic grouping, based on molecular phylogenetics, which includes a diverse range of heterotrophic, mostly flagellate, unicellular eukaryotes such as diplomonads, trypanosomes, euglenids and also some amoeba-like protozoa and the acrasid slime moulds.

excavate *a.* hollowed out.

excentric *a.* (1) one-sided; (2) having the two portions of a lamina or pileus unequally developed. *alt.* eccentric.

exchange diffusion *see* antiport *q.v.*

exciple, excipulum *n.* outer covering of apothecium.

excision repair DNA repair process in which abnormal or mismatched nucleotides are enzymatically cut out of one strand of a DNA molecule and the correct nucleotides replaced by enzymatic synthesis using the remaining intact strand as template.

excitability *n.* capability of a living cell or tissue to respond to an environmental change or stimulus.

excitation *n.* (1) act of producing or increasing stimulation; (2) immediate response of a cell, tissue or organism to a stimulus.

excitation–contraction coupling the process by which the contractile fibrils of a muscle are stimulated to contract by excitation by a neuron.

excitatory *a.* (1) tending to excite, *appl.* e.g. stimuli, cells; (2) *appl.* neuron, neurotransmitter or synapse whose activity tends to cause generation of an action potential in a postsynaptic neuron. *cf.* inhibitory.

excitatory cells motor neurons in the sympathetic nervous system.

excitatory postsynaptic potential (EPSP) electrical potential generated in a postsynaptic neuron by the action of neurotransmitter liberated at a synapse, and which tends to produce an action potential.

excitotoxicity *n.* refers to the death of neurons when over-stimulated with large amounts of the excitatory neurotransmitter glutamate, which causes a prolonged depolarization of the neuronal membrane.

exclusive species species that is confined to one community.

exconjugant *n.* (1) microorganism which is leading an independent life after conjugation with another; (2) female bacterial cell that has been in conjugation with a male cell and received DNA from the male.

excorticate decorticate *q.v.*

excreta *n.plu.* (1) waste material eliminated from body or any tissue thereof; (2) harmful substances formed within a plant.

excretion *n.* the elimination of waste material from the body of a plant or animal, specifically the elimination of waste materials produced by metabolism.

excretophores *n.plu.* in invertebrates, cells of coelomic epithelium in which waste substances from blood accumulate, for discharge into coelomic fluid.

excretory *a. pert.* or functioning in excretion, *appl.* e.g. organs, ducts.

excurrent *a.* (1) *pert.* ducts, channels or canals in which there is an outgoing flow; (2) with undivided main stem; (3) having midrib projecting beyond apex, *appl.* leaves.

excurvate *a.* curved outwards from centre.

excystation *n.* emergence from a cyst.

exendospermous *a.* without endosperm.

exergonic *a.* releasing energy, *appl.* metabolic reactions.

exflagellation *n.* process of microgamete formation by microgametocyte in protozoan blood parasites.

exfoliation *n.* (1) the shedding of leaves or scales from a bud; (2) peeling in flakes, as of bark or skin.

exhalant, exhalent *a.* carrying from the interior outwards.

exine *n.* tough and durable outer layer of wall of pollen grain, often intricately sculptured, composed mainly of sporopollenin.

exinguinal *a.* (1) occurring outside the groin; (2) *pert.* 2nd joint of arachnid leg.

exites *n.plu.* offshoots on outer lateral border of axis of certain arthropod limbs.

exo- prefix derived from Gk *exo*, without, signifying outside, acting outside, opening to the outside, etc.

exobiology *n.* the search for and study of life originating outside Earth, and the study of the effects of extraterrestrial environments on living organisms. *alt.* astrobiology.

exobiotic *a.* living on the exterior of a substrate or the outside of an organism.

exocardiac *a.* situated outside the heart.

exocarp *n.* outermost layer of pericarp of fruit, the skin. *alt.* epicarp.

exoccipital *a. pert.* a skull bone on each side of the foramen magnum.

exochorion *n.* outer layer of membrane secreted by follicular cells surrounding the egg in ovary of insects.

exocoel *n.* (1) space between adjacent mesenteries in sea anemones and their relatives; (2) exocoelom *q.v.*

exocoelar *a. pert.* parietal wall of coelom.

exocoelom *n.* an extra-embryonic cavity of embryo.

exocone *a. appl.* insect eye with cones of cuticular origin.

exocrine *a.* (1) *appl.* glands whose secretion is drained by ducts; (2) *pert.* such glands. *cf.* endocrine.

exocuticle *n.* the main layer of the cuticle (exoskeleton) of arthropods, which in crustaceans often contains calcium salts.

exocytosis *n.* process by which proteins and some other molecules are secreted from eukaryotic cells. They are packaged in membrane-bounded vesicles which then fuse with the plasma membrane, releasing their contents to the outside of the cell. *a.* **exocytotic, exocytic**.

exodermis *n.* specialized cell layer in root immediately underneath the epidermal layer. It produces root hairs.

exoenzyme *n.* any enzyme secreted by a cell and which acts outside the cell, i.e. an extracellular enzyme.

exo-erythrocytic *a.* outside red blood cells, *appl.* phase of the malaria parasite life cycle in humans in which merozoites produced from schizonts reinvade tissue cells.

exogamy *n.* (1) outbreeding *q.v.*; (2) cross-pollination *q.v.*; (3) disassortative mating *q.v.*

exogastric *a.* having the shell coiled towards dorsal surface of body.

exogastrula *n.* artificially induced abnormal amphibian gastrula in which the mesoderm remains on the outside.

exogastrulation *n.* the formation of an exogastrula.

exogenote *n.* in bacterial conjugation, the chromosome fragment that passes from donor to recipient to form part of the merozygote.

exogenous *a.* (1) originating outside the organism, cell or system being studied; (2) developed from superficial tissue, the superficial meristem; (3) growing from parts that were previously ossified; (4) *appl.* metabolism concerned with motor and sensory activities, hormone production and action, temperature control, etc. *cf.* endogenous.

exogenous rhythm metabolic or behavioural rhythm which is synchronized by some external factor and which ceases to occur when this factor is absent.

exognath, exognathite *n.* outer branch of oral appendages of crustaceans.

exogynous *a. appl.* flower with style longer than corolla and projecting above it.

exo-intine *n.* middle layer of a spore covering, between exine and intine.

exomixis *n.* union of gametes derived from different sources.

exon *n.* block of DNA sequence encoding part of a polypeptide chain (or of tRNA or rRNA), which forms part of the coding sequence of a eukaryotic gene, and which is separated from the next exon by a noncoding region of DNA (an intron).

exonephric *a.* with nephridia opening to exterior, *appl.* oligochaetes.

exon shuffling in evolution, the formation of new genes by the linking together of different combinations of exons encoding different protein sequences.

exon trapping cloning technique for isolating protein-coding sequences using specialized vectors in which only DNA containing exons can be maintained.

exonuclease *n.* any of various enzymes that degrade DNA or RNA by progressively splitting off single nucleotides from one end of the chain. *alt.* nuclease, deoxyribonuclease, ribonuclease.

exopeptidase *n.* peptidase that successively cleaves off terminal amino acids or dipeptides. EC 3.4.11–19. *see also* aminopeptidase, carboxypeptidase, omega peptidase.

exoperidium *n.* the outer layer of spore covering (peridium) in certain fungi.

exophytic *a.* on, or *pert.* exterior of plants.

exopodite *n.* the outer branch of a typical two-branched (biramous) crustacean limb.

Exopterygota *n.* major division of the insects including those with only slight metamorphosis and no pupal stage. Includes the Anoplura, Dermaptera, Dictyoptera, Embioptera, Ephemeroptera, Hemiptera, Isoptera, Mallophaga, Orthoptera, Odonata, Phasmida, Plectoptera, Psocoptera, Thysanoptera.

exopterygote, exopterygotous *a. appl.* insects in which the wings develop gradually on the outside of the body and there is no pupal stage, e.g. dragonflies, and whose young are called nymphs. *see* Exopterygota.

exoskeleton *n.* hard supporting structure secreted by and external to the epidermis, such as the calcareous exoskeletons of some sponges and the chitinous exoskeleton of arthropods.

exosmosis *n.* osmosis in an outward direction, e.g. out of a cell.

exosporium *n.* outer wall of spore coat.

exostosis *n.* (1) formation of knots on surface of wood; (2) formation of knob-like outgrowths of bone or of dental tissue at a damaged portion.

exoteric *a.* produced or developed outside the organism.

exothecium *n.* the outer specialized dehiscing cell layer of anther.

exothermic *a.* (1) ectothermic *q.v.*, *n.* **exotherm**; (2) *appl.* chemical reactions that release heat.

exotic *n.* foreign plant or animal which has not acclimatized or naturalized.

exotoxin *n.* toxin secreted by bacteria. *cf.* endotoxin.

exotunica *n.* outer layer of ascus wall.

experimental allergic encephalomyelitis (EAE) autoimmune disease affecting the nervous system that can be induced in experimental animals by immunization with myelin basic protein and other neural antigens in a stong adjuvant.

experimental extinction blotting out an acquired behavioural response.

expiration *n.* (1) the act of emitting air or water from the respiratory organs; (2) emission of carbon dioxide by plants and animals.

explanate *a.* having a flat extension.

explant *n.* small fragment of tissue taken from a living organism and grown in culture.

explosive *a.* (1) *appl.* flowers in which pollen is suddenly discharged on decompression of stamens by alighting insects, as of broom and gorse; (2) *appl.* fruits with sudden dehiscence, seeds being discharged to some distance; (3) *appl.* evolution, rapid formation of numerous new types; (4) *appl.* speciation, rapid formation of new species from a single species in one locality.

exponential growth type of growth in which numbers increase by a fixed percentage of the total population in a given time period, and that gives a J-shaped curve when numbers are plotted over time. *alt.* logarithmic growth. *cf.* linear growth.

exponential growth phase phase of maximum population growth.

expressed sequence tag (EST) in genomics, a short cDNA sequence that has been derived from cellular mRNA, and which thus represents part of a protein-coding gene.

expression *see* gene expression.

expression library library of DNA clones in which the vector enables the DNA insert to be expressed as mRNA.

expression vector DNA vector which has been constructed in such a way that the 'foreign' gene it contains will be expressed.

expressivity *n.* the degree to which a particular genotype produces a phenotypic effect.

exsculptate *a.* having the surface marked with more-or-less regularly arranged raised lines with grooves between.

exscutellate *a.* having no scutellum, *appl.* certain insects.

exserted *a.* (1) protruding beyond; (2) *appl.* stamens that protrude beyond corolla.

exsertile *a.* capable of extrusion.

exsiccata *n.plu.* dried specimens, as in an herbarium.

exstipulate *a.* without stipules.

exsuccate, exsuccous *a.* sapless, without juice.

exsufflation *n.* forced expiration from lungs.

extant *a.* still in existence.

extein *see* intein.

extended phenotype the concept of the phenotype of an organism extended to include its behaviour, and its relations with its family group, who share some of

its genes, and other members of its own species.

extensor *n.* any muscle which extends or straightens a limb or part. *cf.* flexor.

exterior *a.* situated on side away from axis or definitive plane.

external *a.* (1) outside or near the outside; (2) away from the mesial plane.

external fertilization fertilization of eggs by sperm that takes place outside the female's body.

external respiration respiration considered in terms of the gaseous exchange between organism and environment and the transport of gases to and from cells.

exteroceptor *n.* sense organ or receptor that detects stimuli originating outside the body. *cf.* interoceptor.

extinct *a.* no longer in existence.

extinction *n.* (1) the complete disappearance of a species from the Earth; (2) (*behav.*) process by which learned behaviour patterns cease to be performed when they are not reinforced; (3) (*phys.*) absorbance *q.v.*

extinction point minimum level of illumination below which a plant is unable to survive in natural conditions.

extine exine *q.v.*

extra- prefix derived from L. *extra*, outside, signifying located outside, etc.

extrabranchial *a.* arising outside the branchial arches.

extracapsular *a.* (1) arising or situated outside a capsule; (2) *appl.* ligaments at a joint; (3) *appl.* protoplasm lying outside the central capsule in some protozoans.

extracellular *a.* (1) occurring outside the cell; (2) secreted by or diffused out of the cell.

extracellular fluid the fluid in the spaces between cells and in the vascular system (blood and lymph in mammals).

extracellular matrix macromolecular ground substance of connective tissue, secreted by fibroblasts and other connective tissue cells, and which generally consists of proteins, polysaccharides and proteoglycans.

extracellular space the spaces between cells, which may be filled with extracellular matrix in some tissues.

extrachorion *n.* in certain insect eggs, an outermost layer external to the exochorion.

extrachromosomal *a. appl.* DNA molecules such as plasmids and episomes in bacteria, and the rRNA genes in certain animals, which replicate and are expressed independently of the chromosomes.

extracortical *a.* not within the cortex, *appl.* part of brain.

extra-embryonic *a.* situated outside the embryo proper, e.g. the various foetal membranes (amnion, chorion, allantois, yolk sac) that are involved in the nutrition and protection of the embryo.

extra-embryonic ectoderm in a mammalian embryo, ectoderm derived from trophectoderm and which contributes to the placenta.

extra-embryonic endoderm in a mammalian embryo, endoderm derived from the inner cell mass and which contributes to the yolk sac.

extra-embryonic mesoderm in a mammalian embryo, mesoderm that forms the blood vessels of the placenta.

extra-enteric *a.* situated outside the alimentary tract.

extrafloral *a.* outside the flower, *appl.* nectaries.

extrafoveal *a. pert.* yellow spot surrounding the fovea centralis.

extrafusal *a. appl.* muscle fibres outside muscle spindle, which provide most of the force for contraction.

extrahepatic *a. appl.* cystic duct and common bile duct.

extranuclear *a.* outside the nucleus; (1) *pert.* or *appl.* processes occurring outside the nucleus, or to structures situated outside the nucleus; (2) *appl.* genes: mitochondrial and chloroplast genes.

extraocular *a.* exterior to the eye; (1) *appl.* antennae of insects; (2) *appl.* muscle: one of the muscles attached to the eyeball that controls its position and movement. *n.* **extraocular**.

extraperitoneal subperitoneal *q.v.*

extrapulmonary *a.* external to the lungs, *appl.* bronchial system.

extrapyramidal system motor system in brain which includes the basal ganglia and some related brainstem structures.

extrastriate cortex visual cortex outside primary visual cortex (striate cortex).

extraterrestrial *a. appl.* or *pert.* to planets other than Earth, or to anything originating in space or on another planet.

extravaginal *a.* forcing a way through the sheath, as shoots of many plants.

extravasation *n.* the movement of fluid, cells and molecules from blood into surrounding tissues, which occurs in conditions of inflammation.

extraxylary *a.* on the outside of the xylem, *appl.* fibres.

extreme barophile microorganism that lives in the deep sea and is an obligate barophile, being unable to grow at pressures below 400 atm and often growing best at much higher pressures (*ca.* 1000 atm).

extreme halophile microorganism that requires very high salt concentrations to live, at least 1.5 M (9%) NaCl and most species requiring 2–4 M (12–23%) NaCl. All those known are members of the Archaea.

extremeophile extremophile *q.v.*

extremeozyme extremozyme *q.v.*

extreme psychrophile microorganism that requires very low temperatures to live. For some deep-sea archaea, optimal growth temperature is 2°C, and they grow poorly above 10°C.

extreme thermophile hyperthermophile *q.v.*

extremophile *n.* prokaryotic microorganism that grows in extreme conditions (e.g. very high or very low temperature, high salt concentration, very acid pH) that other microorganisms cannot survive in.

extremozyme *n.* any enzyme from an extremophile that is adapted to function in the relevant extreme conditions.

extrinsic *a.* (1) acting from the outside; (2) *appl.* muscles not entirely within the part or organ on which they act; (3) *appl.* membrane proteins which are embedded in the outer layer of a biological membrane.

extrinsic isolating mechanism an environmental barrier which isolates potentially interbreeding populations.

extrinsic pathway series of reactions in blood leading to clot formation, triggered by trauma to tissue or addition of tissue extracts to plasma *in vitro*, the final stages being the same as in the intrinsic pathway (*q.v.*).

exudate *n.* any substance released from a cell, organ or organism by exudation, e.g. sweat, gums, resins.

exudation *n.* discharge of material from a cell, organ or organism through a membrane, incision, pore or gland.

exumbral *a. pert.* the rounded upper surface of a jellyfish.

exuviae *n.plu.* the cast-off skin, shells, etc., of animals.

exuvial *a. appl.* insect glands whose secretions facilitate moulting.

eye *n.* (1) light-sensitive organ in animals, the organ of sight or vision, which takes various forms in different groups of animals. Insects and most crustaceans have compound eyes, made up of many separate visual units or ommatidia, as well as simple single eyes in some cases. The vertebrate eye consists of a jelly-filled ball, the back of which is lined with a photosensitive layer, the retina. Light is focused onto the retina through a single transparent lens. The amount of light entering the eye is regulated by varying the size of the pupil. Cephalopods also have a similar type of eye. *see also* aqueous humour, cornea, iris, lens, sclera, vitreous humour; (2) (*bot.*) the bud of a tuber. *see also* eye-spot.

eye-spot *n.* (1) small cup-shaped pigmented spot of sensory tisssue in invertebrates, and also in some vertebrates, which has a light-detecting or visual function; (2) orange carotenoid-containing structure in some flagellates; (3) eye-like marking on wings of some butterflies and moths, or on the bodies of other animals, and which is exposed to distract predators when the animal is disturbed. *alt.* stigma.

eye teeth upper canine teeth.

F

f coefficient of kinship *q.v.*

F phenylalanine *q.v.*

F$_1$, F$_0$ *see* F-type ATPases.

F$_1$, F1 denotes 1st filial generation, hybrids arising from a cross between two pure-breeding parents. The generation arising from intercrossing the F$_1$ generation is denoted F$_2$, and so on. P$_1$ denotes the parents of the F$_1$ generation, P$_2$ the grandparents.

F$_2$, F2 generation. *see* F$_1$.

Fab 'fragment antigen-binding', that portion of an antibody molecule containing an antigen-binding site, obtained by proteolytic cleavage of antibody by papain. It comprises one light chain and its paired portion of heavy chain, held together by an intra-chain disulphide bond.

Fabales *n.* order of dicots, also known as legumes, whose fruit is a pod, and whose roots contain nitrogen-fixing bacteria of the genus *Rhizobium*. They include the families Mimosaceae, Cesalpinaceae and Papilionaceae (Fabaceae), and are also known as the Leguminosae.

fabiform *a.* bean-shaped.

facet *n.* (1) smooth, flat or rounded surface for articulation; (2) surface of an ommatidium (one of the units making up a compound eye), or the ommatidium itself.

facial *a.* (1) *pert.* face, *appl.* e.g. arteries, bones; (2) *appl.* nerve: 7th cranial nerve, which supplies facial muscles, activates some salivary glands and conveys taste sensation from front of tongue.

faciation *n.* (1) formation or character of a facies; (2) grouping of dominant species within an association; (3) geographical differences in abundance or proportion of dominant species in a community.

facies *n.* (1) a surface, in anatomy; (2) aspect, e.g. superior or inferior; (3) grouping of dominant plants in the course of a successional series; (4) one of different types of deposit in a geological series or system, and the palaeontological and lithological character of a deposit.

facies fossils fossils characterizing the environment prevailing in a particular area of sedimentation. *cf.* index fossil.

facilitated diffusion transport of molecules or ions across a membrane down their concentration gradient by a carrier system without the expenditure of energy.

facilitation *n.* (1) in neurophysiology, the process whereby the amount of neurotransmitter liberated at an axon terminal, and therefore the magnitude of the postsynaptic potential, increases with the frequency of stimulation of the presynaptic nerve cell; (2) in animal behaviour, an improvement in a pre-existing capability in response to a particular stimulus; (3) social facilitation, the initiation or increase in an ordinary behaviour pattern by the presence or actions of another animal.

faciolingual *a. pert.* or affecting face and tongue.

FACS fluorescence-activated cell sorter. A piece of equipment for counting and separating cells in a mixed population after staining for distinctive cell-surface molecules using antibodies linked to different coloured fluorescent dyes. A stream of labelled cells is run through a fluorescence detector, which counts the cells of different types and which can also deflect appropriately labelled cells from the main stream, thus separating them from the mixture. *see also* flow cytometry.

F actin *see* actin.

factor *n.* (1) any agent (e.g. biological, climatic, nutritional) contributing to a result or effect; (2) in physiology and cell biology, a name given to any endogenous substance that appears to have a physiological effect. Many 'factors' have subsequently been characterized as proteins and polypeptides, e.g. Factor VIII, nerve growth factor, epidermal growth factor.

Factor B plasma protein which is a component of the alternative pathway of complement activation. It binds C3b, and in this form is cleaved by Factor D to form the complex C3bBb. This is an active C3 convertase, cleaving C3 to produce C3a and C3b.

Factor D plasma protein which is a component of the alternative pathway of complement activation. It cleaves Factor B.

Factor H plasma protein that acts as a complement-regulatory protein. It binds to C3b, displacing C2b or Bb and making C3b susceptible to cleavage and inactivation by Factor I.

Factor I plasma protein that acts as a complement-regulatory protein. It cleaves and inactivates C3b and C4b.

Factor IX a blood clotting factor, a serine protease activated by Factor XI to give the active protease (Factor IXa), and whose deficiency is the cause of haemophilia B (Christmas disease). Together with Factor VIII it activates Factor X. EC 3.4.21.22, *r.n.* coagulation factor IXa. *alt.* Christmas factor.

Factor P properdin *q.v.*

Factor V a blood clotting factor, a modifier protein that stimulates the conversion of prothrombin to thrombin by the enzyme Factor X.

Factor VII a blood clotting factor, a serine protease which, together with tissue factor, activates Factor X. EC 3.4.21.21, *r.n.* coagulation factor VIIa. *alt.* proconvertin.

Factor VIII non-enzyme protein involved in blood clotting, accelerating the conversion of Factor X to its active form by the serine protease Factor IXa, and whose deficiency causes haemophilia A. *alt.* antihaemophilia factor.

Factor X a blood clotting factor, a protease activated by Factors IX and VIII, which converts prothrombin to thrombin.

EC 3.4.21.6, *r.n.* coagulation factor Xa. *alt.* Stuart factor (formerly known as thrombokinase or thromboplastin).

Factor XI a blood clotting factor, a serine protease activated by Factor XII, which cleaves and activates Factor IX. EC 3.4.21.27, *r.n.* coagulation factor XIa. *alt.* plasma thromboplastin antecedent.

Factor XII a blood clotting factor, a precursor serine protease circulating in the blood which, on contact with an abnormal surface such as wounded tissue, is converted to the active protease, Factor XIIa, which can cleave and activate Factor XI and Factor VII. EC 3.4.21.38. *r.n.* coagulation factor XIIa. *alt.* Hageman factor.

Factor XIII plasma protein involved in blood clotting, cross-linking fibrin and stabilizing clots.

facultative *a.* having the capacity to live under different conditions or to adopt a different mode of life. When *appl.* e.g. to aerobe, anaerobe, symbiont, parasite, it indicates organisms that can live in this way but are not obliged to, and may under certain conditions adopt another mode of life. *cf.* obligate.

facultative heterochromatin *see* heterochromatin.

FAD (1) oxidized form of flavin adenine dinucleotide *q.v.*; (2) familial Alzheimer's disease. *see* Alzheimer's disease.

FADH$_2$ reduced form of flavin adenine dinucleotide *q.v.*

faeces *n.* excrement from alimentary canal.

Fagales *n.* order of dicots including many deciduous forest trees such as beech, birch, oak, sweet chestnut, hazel and hornbeam. *see also* Betulales.

fairy ring a circle of fruit bodies (mushrooms or toadstools) of certain agaric fungi, representing the outer edge of the underground mycelium.

falcate *a.* (1) sickle-shaped; (2) hooked.

falces *n.plu.* (1) chelicerae (poison claws) of arachnids; (2) *plu.* of **falx**.

falcial *a. pert.* falx, esp. the falx cerebri.

falciform *a.* sickle- or scythe-shaped.

Falconiformes *n.* the birds of prey, an order of birds with strong clawed feet and hooked beaks, with a well-developed ability to soar. It includes eagles, hawks, buzzards, kites, kestrels, falcons and vultures.

falcula *n.* curved scythe-like claw.

fallopian tube one of a pair of narrow ducts in mammals each leading from an ovary to the uterus, into which ova are released on ovulation, and in which fertilization normally takes place.

false foot pseudopodium *q.v.*

false fruits fruits formed from the receptacle or other part of the flower in addition to the ovary, or from complete inflorescences.

false ribs those ribs whose cartilaginous ventral ends do not join the breastbone directly. *alt.* floating ribs.

false scorpions pseudoscorpions *q.v.*

falx *n.* (1) a sickle-shaped structure; (2) falx cerebri, a sickle-shaped fold in the dura matter. *plu.* **falces**. *a.* **falcate, falcular, falciform**.

familial *a.* (1) *pert.* family; (2) *appl.* traits that tend to occur in several members and subsequent generations of a family. They are not necessarily genetically based but may be due to the shared family environment; (3) *appl.* disease, usually signifies a genetically based heritable condition, as opposed to the spontaneous and sporadic occurrence of the same condition. *see* genetic disease.

familial cancers cancers in which there is an underlying inheritable predisposition to develop the disease, usually much earlier than is normal. *see also* tumour suppressor genes.

familial emphysema α_1-antitrypsin deficiency *q.v.*

familial hypercholesterolaemia *see* hypercholesterolaemia.

familiality *n.* the clustering of a trait in families. *see* familial.

family *n.* taxonomic group of related genera, related families being grouped into orders. Family names usually end in -aceae in plants and -idae in animals.

farina *n.* (1) flour or meal; (2) fine mealy powder present on the surface of some plants, e.g. certain primulas, and insects.

farinaceous *a.* (1) containing flour or starch; (2) mealy; (3) covered with a fine powder or dust; (4) covered with fine white hairs that can be detached like dust. *alt.* **farinose**.

farnesyl anchor *see* prenylation.

far red light light of wavelength between 700 and 800 nm. It reverses the effect of red light on phytochrome in plants.

Fas membrane protein present on some mammalian cell types including activated T cells. Stimulation by its ligand, which is another cell-surface protein, Fas ligand, signals the Fas-bearing cell to undergo apoptosis.

fascia *n.* (1) ensheathing band of connective tissue; (2) transverse band of different colour, as in some plants; (3) any band-like structure.

fasciated *a.* (1) banded; (2) arranged in bundles or tufts; (3) *appl.* stems or branches malformed and flattened.

fasciation *n.* (1) the formation of bundles; (2) the coalescent development of branches of a shoot system, as in cauliflower; (3) abnormal development of flattened, malformed fused stems or branches.

fascicle *n.* (1) bundle, as of pine needles; (2) small bundle, as of muscle fibres, nerve fibres; (3) tuft, as of leaves. *alt.* **fasciculus**.

fascicular *a.* (1) *pert.* a fascicle; (2) arranged in bundles or tufts; (3) (*bot.*) *appl.* cambium or tissue within vascular bundle. *alt.* intrafascicular.

fasciculation *n.* the bundling together of the axons of nerve cells to form a nerve.

fasciola *a.* narrow band of colour.

fasciole *n.* ciliated band on certain sea urchins for sweeping water over surrounding parts.

Fas ligand *see* Fas.

FASTA a computer program for sequence alignment.

fast block to polyspermy, the change in the electrical potential of the egg membrane that occurs immediately after sperm entry and which prevents penetration by other sperm.

fastigiate *a.* with branches close to stem and erect.

fast muscle fibres muscle fibres that contract rapidly and strongly but tire quickly. *alt.* **fast-twitch muscle fibres**. *see* FG fibres, FOG fibres.

fat *n.* general name for any of the triacylglycerols that contain highly saturated fatty acids and are solid at 20°C. They contain carbon, oxygen and hydrogen, but no nitrogen. Fats are stored in animal and plant cells where they provide a concentrated source of energy. Adipose tissue in animals, also known as fat, consists of cells filled with globules of fats. When energy is required, fats are hydrolysed by lipases to fatty acids and glycerol and the fatty acids are metabolized in the mitochondria. *cf.* oils.

fat body (1) diffuse gland dorsal to gut in insects, with function analogous to that of liver in vertebrates. It stores fats, glycogen and protein and is a major site of metabolism; (2) structure filled with fat globules and associated with gonads in amphibians; (3) other fat-storage organs in animals and plants.

fat cell adipocyte *q.v.*

fate *n.* of an embryonic cell, the type of cell or structure it will develop into in normal circumstances.

fate map a 'map' of the surface of the fertilized egg or early embryo predicting which regions will form various tissues or parts of the body.

fatigue *n.* effect produced by unduly prolonged stimulation on cells, tissues or other structures so that they are less responsive to further stimulation.

fat index ratio of dry weight of total body fat to that of non-fat.

fat-soluble *appl.* hydrophobic organic compounds, such as hydrocarbons, that are soluble in lipids. Such compounds can be taken up into fatty tissue and remain there. When *appl.* vitamins, refers to vitamins A, D and K.

fatty acid long-chain organic acid of the general formula $CH_3(C_nH_x)COOH$, where the hydrocarbon chain is either saturated ($x = 2n$) (e.g. palmitate, $C_{15}H_{31}COO^-$) or unsaturated (e.g. oleate, $C_{17}H_{33}COO^-$). They are a constituent of many lipids, including the phospholipids of cell membranes, and are also used as a fuel molecule for respiration.

fatty acid synthetase multifunctional protein complex containing the seven enzyme activities involved in fatty-acid synthesis from acetyl-CoA, found in eukaryotes.

fatty acid thiokinase acyl-CoA synthetase *q.v.*

fauces *n.plu.* (1) upper or anterior portion of throat between palate and pharynx; (2) mouth of a spirally coiled shell.

fauna *n.* the animals peculiar to a country, area, specified environment or period. Microscopic animals are usually called the microfauna.

faunal collapse local extinction of an animal or a number of animal species.

faunal region (1) area characterized by a special group or groups of animals; (2) zoogeographical region *q.v.*

favella *n.* conceptacle of certain red algae.

faveolate *a.* honeycombed.

faveolus *n.* small depression or pit.

favoid *a.* resembling a honeycomb.

favus *n.* ringworm, when it affects the scalp.

Fc 'fragment crystallizable', that portion of an antibody molecule comprising the paired carboxy-terminal constant regions of the two heavy chains up to the hinge region, held together by disulphide-bonding in the hinge region. It is obtained on digestion of antibody with papain.

Fc receptor any of several receptors on macrophages, neutrophils and some other immune system cells which bind the constant region of an antibody (the Fc region) when it is part of an antigen–antibody complex. Binding to Fc receptors stimulates phagocytosis of the antigen–antibody complex.

F'-duction *n.* transfer of chromosomal genes from one bacterium to another by the agency of an F factor, which becomes attached to the chromosome.

Fe symbol for the chemical element iron *q.v.*

feather *n.* keratinous epidermal structure forming the body cover (plumage) of birds. Each feather consists of a midrib (rachis) from which project on either side many delicate thread-like barbs. In the fluffy down feathers of nestlings, the barbs do not interlock, whereas in the outer feathers and flight feathers of older birds, the barbs interlock to form a flat flexible wind-resistant surface (the vane).

feather epithelium epithelium on inner surface of nictitating membrane in birds and reptiles, whose cells each have a process with numerous lateral filaments, and which acts to clean the eye surface.

feather follicle epithelium surrounding the base of a feather, and from which the feather has developed.

feature detection column column or slab of nerve cells in visual cortex, perpendicular to the surface, which all respond to a particular type of visual stimulus, such as a line in a particular orientation.

feces faeces *q.v.*

fecundity *n.* (1) the number of eggs produced by an individual; (2) fertility *q.v.*

feedback control, feedback inhibition, feedback regulation type of metabolic regulation in which the first enzyme in a metabolic pathway (usually a biosynthetic pathway) is inhibited by reversible binding of the final product of the pathway. *alt.* end-product inhibition.

feedback mechanism general mechanism operative in many biological and biochemical processes, in which once a product or result of the process reaches a certain level it inhibits (negative feedback) or promotes (positive feedback) further reaction.

felid *n.* member of the mammalian family Felidae, the cats.

feloid *n.* member of the cat (Felidae) or mongoose (Viverridae) families.

female *n.* individual whose gonads contain only female gametes, symbol ♀.

female pronucleus the haploid nucleus of the egg.

feminization *n.* the production (e.g. by hormones) of secondary female sexual characteristics in genetic males.

femoral *a. pert.* femur.

femur *n.* (1) the thigh bone, the large bone in the upper part of the hindlimb of vertebrates; (2) 3rd segment of insect, spider and myriapod leg, counting from the body. *plu.* femora. *a.* femoral.

fen *n.* plant community on alkaline, neutral or slightly acid wet peat, characterized by tall herbaceous plants, e.g. reeds, reed canary grass. *cf.* bog.

fenestra *n.* (1) an opening or small hole in a structure; (2) transparent spot on wings of insects. *plu.* fenestrae.

fenestra ovalis oval window *q.v.*

fenestra pseudorotunda opening covered by the endotympanic membrane in birds, the fenestra rotunda in mammals having a different embryonic origin.

fenestra rotunda round window *q.v.*

fenestra tympani fenestra rotunda *q.v.*

fenestra vestibuli fenestra ovalis *q.v.*

fenestrate(d) *a.* (1) having small perforations or fenestrae; (2) having transparent spots, *appl.* insect wings.

fenestrated membrane (1) close network of elastic connective tissue resembling a membrane with perforations, as in inner tunic of arterial wall; (2) basal membrane of compound eye, perforated by ommatidial nerve fibres.

fen peat type of peat (alkaline, neutral or slightly acidic) formed where depressions allow the accumulation of drainage water and the rate of growth of grasses, sedges and trees exceeds that of plant decomposition, which is usually substantial. *alt.* basin peat.

feral *a.* wild, or escaped from domestication and reverted to wild state.

fermentation *n.* (1) glycolysis *q.v.*; (2) anaerobic breakdown of carbohydrates by living cells, esp. by microorganisms, often with the production of heat and waste gases (as in alcoholic fermentation in yeasts) and a variety of end-products (e.g. ethanol, lactic acid).

fern *n.* common name for a member of the Pterophyta *q.v.*

ferralitic soils deep red soils, acid in reaction, found on freely drained sites in humid tropical regions.

ferredoxin *n.* iron–sulphur protein that acts as an electron carrier and as a biological reducing agent in its reduced form. It is a component of the photosynthetic electron-transport chain and an electron donor in nitrogen fixation in microorganisms.

ferrihaemoglobin methaemoglobin *q.v.*

ferritin *n.* protein to which iron binds, forming an iron-storage complex. Found in spleen, liver and bone marrow.

ferrocyte *n.* iron-containing cell in ascidians, apparently concerned with the production of the cellulose-like polysaccharide tunicin.

ferruginous *a.* (1) having the appearance of iron rust; (2) rust-coloured.

fertile *a.* (1) producing viable gametes; (2) capable of producing living offspring; (3) of eggs or seeds, capable of developing; (4) *appl.* a soil containing the necessary nutrients for plant growth.

fertilin *n.* transmembrane protein in sperm that helps the sperm bind to the egg plasma membrane.

fertilis- *alt.* fertiliz-.

fertility *n.* (1) the ability to reproduce; (2) the reproductive performance of an individual or population, measured as the number of surviving offspring produced per unit time. *alt.* fecundity.

fertility factor F factor *q.v.*

fertility rate *see* birth rate.

fertility schedule demographic data giving the average number of female offspring that will be produced by a female at each particular age. *alt.* fecundity schedule.

fertilization *n.* the union of male and female gametes, e.g. sperm and egg, to form a zygote. *alt.* syngamy.

fertilization cone protuberance on ovum at point of contact and entry of spermatozoon before fertilization.

fertilization envelope the vitelline envelope after it has been raised off the surface of the fertilized egg by the release of cortical granule contents.

fertilization membrane membrane formed by the ovum in response to penetration by a sperm, which grows rapidly from the point of penetration and covers the ovum, excluding other sperm.

fertilization tube antheridial structure that grows through the oogonial wall, for passage of male gamete in certain fungi.

fertilize *v.* to bring about fertilization.

fetal, fetus *alt.* spelling of foetal, foetus.

Feulgen stain histological stain that shows up DNA as purple.

fever a rise in body temperature that often accompanies infection, and which is caused both by bacterial components (exogenous pyrogens) and molecules produced in response to the infection (endogenous pyrogens).

FFA free fatty acid *q.v.*

F factor, F plasmid transmissible plasmid in the bacterium *Escherichia coli* that acts as a sex factor. It directs synthesis of the sex pilus (F pilus), conjugation, and transfer of itself, and chromosomal genes, from an F^+ to an F^- bacterium. It can exist as a free element or integrated into the bacterial chromosome, in which state it mediates the transfer of chromosomal genes at greater frequency. An F factor into which bacterial genes have been incorporated is known as an **F′ factor**. *alt.* fertility factor, sex factor. *see also* Hfr strain.

F_1 generation, F_2 generation *see* F_1.

FGF fibroblast growth factor *q.v.*

FG fibres fast glycolytic muscle fibres in muscle of mammalian limbs. They are white muscle fibres adapted for mainly anaerobic metabolism, and are used only when the animal is running fast. *cf.* FOG fibres, SO fibres.

F_1 hybrid in horticulture and experimental genetics, the first cross between two pure-breeding lines.

fiber *see* fibre.

Fibonacci series the unending sequence 1, 1, 2, 3, 5, 8, 13, 21, 34 . . . , where each term is defined as the sum of its two predecessors.

fibre *n.* (1) an elongated cell or aggregation of cells forming a strand of e.g. muscle, nerve or connective tissue; (2) protein filament, such as the filaments made of keratin in wool and hair; (*bot.*) (3) a delicate root; (4) a tapering elongated thick-walled sclerenchyma cell providing mechanical strength in plant stem.

fibril *n.* (1) small thread-like structure or fibre; (2) individual strand of a fibre; (3) root hair.

fibrilla *n.* minute muscle-like thread found in some ciliates. *plu.* **fibrillae.**

fibrillar *a. pert.*, or like, fibrils or fibrillae.

fibrillar flight muscles wing muscles of dipteran (flies) and hymenopteran (wasps and bees) insects, which, unlike ordinary striated muscle, do not need an action potential to initiate every new contraction.

fibrillate *a.* possessing fibrillae or hair-like structures.

fibrillation *n.* (1) spontaneous, asynchronous contractions of muscle fibres seen after nerve supply has been cut; (2) in heart, spontaneous arrhythmia due to asynchronous muscle fibre contractions that can result from a variety of causes.

fibrillose *a.* furnished with fibrils, *appl.* mycelia of certain fungi.

fibrin *n.* insoluble protein produced from fibrinogen in the blood by the proteolytic action of thrombin, and forming a mesh of fibres (a clot) in which platelets and blood cells are caught.

fibrinogen *n*. soluble blood plasma protein, precursor to fibrin.

fibrinolysin plasmin *q.v.*

fibrinolysis *n*. the dissolving of blood clots as a result of fibrin degradation.

fibrinolytic *a*. able to break down fibrin.

fibrinopeptide *n*. peptide that is cleaved from the protein fibrinogen when it is activated to form fibrin during blood clotting.

fibrin-stabilizing factor Factor XIII *q.v.*

fibroblast *n*. flattened, irregular-shaped connective tissue cell, ubiquitous in fibrous connective tissue. It secretes components of the extracellular matrix, including type I collagen and hyaluronic acid.

fibroblast growth factor (FGF) secreted cytokine produced by many cell types. There are at least 10 different types in mammals. It stimulates proliferation of fibroblasts and also has many activities in embryonic development in vertebrates, including roles in the induction of the spinal cord and limb development.

fibrocartilage *n*. type of cartilage whose matrix is mainly composed of fibres similar to connective tissue fibres, found at articulations, cavity margins and grooves in bones.

fibrocyte *n*. type of white blood cell resembling a fibroblast in its ability to secrete large quantities of extracellular matrix proteins such as collagen. It can enter tissues from the blood and participate in wound repair.

fibrohyaline chondroid *q.v.*

fibroin *n*. fibrous protein found in silk fibres, produced by proteolysis from a precursor protein, fibroinogen.

fibroinogen *n*. a protein secreted by the silk glands of certain insects, and which is cleaved to form fibroin.

fibroma *n*. generally benign tumour of fibrous connective tissue.

fibronectin *n*. glycoprotein of extracellular matrix, to which animal cells can bind by means of integrins in their plasma membranes. It is involved in interactions of animal cells with extracellular matrix.

fibrosarcoma *n*. tumour of fibrous connective tissue.

fibrosis *n*. scarring or thickening of connective tissue, e.g. of lungs.

fibrous *a*. (1) composed of or resembling fibres, *appl.* e.g. tissue, mycelium; (2) forming fibres, *appl.* proteins such as collagen, elastin, keratin, fibrin, fibroin.

fibrous astrocyte type of astrocyte found mainly in white matter of brain, having thick branched cytoplasmic processes. Some have foot-like projections that abut on blood vessels.

fibrous root system root system in which the roots form a mass of fibres without a tap root.

fibula *n*. (1) in tetrapod vertebrates, the bone posterior to the tibia in the shank of the hindlimb; (2) in humans, the outer and smaller shin bone; (3) in some insects, a structure holding fore- and hindwings together. *a*. **fibulate**.

fibulare *n*. outer bone of proximal row of tarsus.

fibularis peroneus *q.v.*

ficolins *n.plu.* proteins of innate immunity present in vertebrates and urochordates. They bind to *N*-acetylglucosamine-containing carbohydrates on the surfaces of pathogens, initiating complement fixation via the lectin-mediated pathway and thus facilitating phagocytosis of the pathogen.

fidelity *n*. (1) (*ecol.*) the degree of limitation of a species to a particular habitat; (2) (*mol. biol.*) of DNA replication, transcription and translation, the probability of an error being made during the copying of DNA into DNA or RNA, or during the translation of RNA into protein.

field layer herb layer *q.v.*

field metabolic rate metabolic rate as measured in a freely ranging animal, most commonly by the doubly labelled water method. In this method, water labelled with either deuterium (^{2}H) or tritium (^{3}H) and the oxygen isotope ^{18}O is injected at the start of the experiment. The decline of ^{18}O and labelled hydrogen in the blood after a period of days or weeks is measured and the concomitant rate of CO_2 production calculated.

filament *n*. (1) slender thread-like structure, e.g. a fungal or actinomycete hypha, a chain of bacterial or algal cells, very fine fibres in cells, *a*. **filamentous**; (2) the stalk of a stamen. *see* Fig. 18 (p. 239); (3) the rachis of a down feather; (4) slender apical end of egg tube of insect ovary. *see also* thick filament, thin filament.

filamentous actinomycetes *see* actinomycetes.

filamentous bacteria bacteria that form long thin cells or chains of cells.

filamentous bacteriophages group of bacteriophages with individual phage particles in the form of long filaments, with helically arranged coat protein subunits and a circular single-stranded DNA genome. An example is M13, which is used to clone DNA for DNA sequencing.

filamin *n.* cytoplasmic protein which converts a solution of actin filaments from a viscous fluid to a solid gel.

filaria *n.* parasitic nematode worm. *plu.* **filariae.**

filasome *n.* small vesicle, coated with dense filamentous material, seen in electron micrographs of tips of developing fungal hyphae.

filator *n.* (1) part of the spinning organ of silkworms that regulates the size of the silk fibre; (2) the spinnerets of other caterpillars.

filial generation F$_1$, F$_2$ *q.v.*

filial imprinting imprinting resulting in attachment of an offspring to parents or foster parents.

filibranch *a. appl.* gills of bivalve molluscs whose filaments are attached to adjacent ones by cilia.

Filicales *n.* group of ferns (Pterophyta) that includes most extant species. They are mainly terrestrial, with generally large compound leaves and rhizomatous roots, and produce spores on the undersides of the leaves.

filicauline *a.* with a thread-like stem.

filiciform *a.* (1) shaped like the frond of a fern; (2) fern-like.

Filicinophyta name sometimes given to the Pterophyta *q.v.*

filiform *a.* thread-like.

filiform papillae papillae on tongue ending in numerous minute slender processes.

filigerous *a.* with thread-like outgrowths or flagella.

filipendulous *a.* thread-like, with tuberous swelling at middle or end, *appl.* roots.

filoplume *n.* delicate hair-like feather with long axis and a few free barbs at apex.

filopodia *n.plu.* (1) fine stiff cytoplasmic protrusions that are put out by the leading edge

of an animal cell as it moves over a surface; (2) stiff thread-like pseudopodia of some protozoans. *sing.* **filopodium.**

Filoviridae, filoviruses *n., n.plu.* family of enveloped single-stranded negative-strand RNA animal viruses, with long thread-like particles. It comprises Marburg fever and Ebola fever viruses, which cause often fatal haemorrhagic fevers.

filose *n.* slender and thread-like.

filter feeders organisms that feed on small organisms in water or air, straining them out of the surrounding medium by various means.

filter sterilization sterilization of a liquid or gas by passing it through a filter with pores small enough to hold back microorganisms.

filtrate *n.* clear liquid obtained by filtration.

filtration *n.* separation process in which a liquid is passed through a porous material (the filter) to separate out any particles or solid material.

fimbria *n.* (1) any fringe-like structure; (2) (*bact.*) one of numerous filaments, smaller than flagella but with similar structure, fringing certain bacteria; (3) (*neurobiol.*) posterior prolongation of fornix to hippocampus. *plu.* **fimbriae.**

fimbriate(d) *a.* bordered with fine hairs.

fimbrin *n.* protein associated with actin filaments in the supporting cytoskeleton of intestinal microvilli.

fin *n.* (1) fold of skin supported by bony or cartilaginous rays in fishes and used for e.g. locomotion, balancing, steering, display. Most fishes have an upright dorsal fin on the back, a caudal fin at the end of the tail, an anal fin on the underside just anterior to the anus, a pair of pelvic fins on the underside and a pair of pectoral fins just behind the gills. The pectoral and pelvic fins represent the fore- and hindlimbs of other vertebrates; (2) any similarly shaped structure in other aquatic animals.

fingerprinting *see* DNA fingerprinting, protein fingerprinting.

fin-rays stiff rods of connective tissues, generally cartilage or bone, which support the fins.

fire climax plant community maintained as climax vegetation by natural or human-made fires which destroy the plants that would otherwise become dominant.

Firmibacteria *n.* class of Bacteria that includes most Gram-positive bacteria, e.g. staphylococci, bacilli, streptococci.

Firmicutes *n.* phylum of the Bacteria distinguished on DNA sequence criteria. It includes two large classes of Gram-positive bacteria – the Clostridia and the Bacilli – and the Mollicutes, wall-less obligate parasites such as *Mycoplasma*.

first-set rejection rejection of tissue when transplanted to a recipient of incompatible tissue type who has never before received a tissue transplant.

FISH fluoresence *in situ* hybridization *q.v.*

Fisher's sex-ratio theory theory that states that when daughters and sons are of equal reproductive value, the ratio of sons to daughters produced should be adjusted so that the average fitness costs of sons and daughters are equal.

fishes *n.plu.* group of aquatic limbless vertebrates, breathing mainly by means of gills, with streamlined bodies and fins and with the body covered in scales (in bony fishes), and comprising the Chondrichthyes (cartilaginous fishes) and the Osteichthyes (bony fishes). The Agnatha are also sometimes called fishes.

fish lice *see* Branchiura.

fissile *a.* (1) tending to split; (2) cleavable.

fissilingual *a.* with a forked tongue.

fission *n.* asexual reproduction by division of a unicellular organism into two or more equal-sized cells. *see also* binary fission.

fission yeast *Schizosaccharomyces* and related species, whose cells multiply by division into two equal-sized daughter cells. *cf.* budding yeast.

fissiparous *a.* reproducing by fission.

fissiped *a.* with digits of feet separated, as in the terrestrial carnivorous mammals (e.g. cats, dogs, bears). *cf.* pinnipeds.

Fissipedia *n.* name given to an order that contains the terrestrial carnivorous mammals (e.g. cats, dogs, bears), the marine carnivores (e.g. seals) being placed in the order Pinnipedia.

fissirostral *a.* with deeply cleft beak.

fissure *n.* (1) deep groove or furrow dividing an organ into lobes, or subdividing and separating certain areas of the lobes; (2) sulcus *q.v.*

fistula *n.* (1) a pathological or artificial pipe-like opening; (2) a water-conducting vessel.

fistular *a.* hollow and cylindrical, *appl.* stems of umbellifers, *appl.* leaves surrounding the stem in some monocotyledons.

FITC fluorescein isothiocyanate *q.v.*

fitness *n.* the fitness of an individual is defined as the relative contribution of its genotype to the next generation relative to the contributions of other genotypes, and is determined by the number of offspring it manages to produce and rear successfully. *alt.* Darwinian fitness. *see also* inclusive fitness.

fixation *n.* (1) of carbon and nitrogen, *see* carbon dioxide fixation, nitrogen fixation; (2) treatment of specimens to preserve structure, for microscopy; (3) (*genet.*) of an allele, its spread throughout a population until it is the only allele found at that locus; (4) (*behav.*) in experimental psychology, a stereotyped response shown by an animal regardless of whether the stimulus is accompanied by positive or negative reinforcement, and often shown in an insoluble problem situation.

fixation index in population genetics, a measure of genetic differentiation between subpopulations, being the proportionate reduction in average heterozygosity compared with the theoretical heterozygosity if the different subpopulations were a single randomly mating population.

fixative *n.* chemical such as ethanol or formaldehyde which is used to preserve cells and cellular structure.

fixed-action pattern stereotyped and fixed response found in animal behaviour where learning has not occurred.

fixed allele an allele that is present in the homozygous state in all members of the population.

fixed anions the negatively charged organic molecules in a cell.

fixed carbon carbon that has been incorporated into organic matter by photosynthesis.

flabellate *a.* with projecting flaps on one side, *appl.* to certain insect antennae.

flabellum *n.* fan-shaped organ or structure.

flaccid *a.* limp, *appl.* leaves that do not have enough water and are wilting or about to wilt.

flagella *plu.* of flagellum.

flagellate(d) *a.* bearing flagella.

flagellated chambers in sponges the central cavities lined with choanocytes, which are flagellated cells specialized for uptake of food particles from water.

flagellates *n.plu.* diverse group of unicellular eukaryotic microorganisms, including photosynthetic and non-photosynthetic species, and classified in various schemes as protozoans, protists or algae. They are motile in the adult stage, swimming by means of flagella. They include both free-living marine and freshwater species and some important commensals such as those living in the guts of ruminants, and human parasites such as trypanosomes. For the photosynthetic flagellates *see* Chrysophyta, Euglenophyta, Eustigmatophyta, Pyrrophyta, Prymnesiophyta and Xanthophyta. For the non-photosynthetic flagellates *see* Zoomastigina.

flagellation *n.* the arrangement of flagella on a cell. *see* lophotrichous, peritrichous, polar.

flagellin *n.* protein constituent of bacterial flagella.

flagellum *n.* (1) long whip-like or feathery structure borne either singly or in groups on a cell and which propels it through a liquid medium. Flagella are borne by the motile cells of many bacteria and unicellular eukaryotes and by the motile male gametes of many eukaryotes. Bacterial and eukaryotic flagella differ in internal structure and mechanism of action. A bacterial flagellum is made up of a single protein fibre and rotates, while a eukaryotic flagellum has a structure and beating action like a cilium; (2) in insects, the distal part of the antenna, beyond the 2nd segment. *plu.* **flagella**.

flame cell *see* protonephridial system.

flash colours the sudden flash of colour displayed by some species during an attempt to escape from a predator, which may startle and distract the predator or deceive it into thinking the prey has gone.

flask fungi common name for the Pyrenomycetes *q.v.*

flatworms Platyhelminthes *q.v.*

flavedo *n.* the outer layer, or rind, of pericarp in citrus fruits.

flavescent *a.* growing or turning yellow.

flavin *n.* any of a group of yellowish pigments to which the vitamin riboflavin belongs.

They contain a nitrogenous base, usually isoalloxazine, and show greenish-yellow fluorescence. They occur free in higher animals and plants and as part of flavin nucleotides.

flavin adenine dinucleotide (FAD) a derivative of riboflavin that is tightly bound as a prosthetic group to certain oxidative enzymes, which are thus known as flavoproteins. Can exist in an oxidized form (FAD) and a reduced form ($FADH_2$) and acts as an electron carrier in many metabolic reactions. *see also* flavin mononucleotide.

flavin adenine mononucleotide flavin mononucleotide *q.v.*

flavin mononucleotide (FMN) riboflavin 5′-phosphate, an electron-carrying prosthetic group in several enzymes, including NADH dehydrogenase in the mitochondrial respiratory chain. It is reduced to $FMNH_2$. *see* Fig. 17.

Flaviviridae, flaviviruses *n., n.plu.* family of enveloped single-stranded positive-strand RNA animal viruses with icosahedral particles, including the viruses of yellow fever and dengue haemorrhagic fever.

flavodoxin *n.* flavin-containing protein that is an electron carrier in some bacterial electron-transport chains.

flavone *n.* any of a group of pale yellow flavonoid plant pigments, with the C_3 part of the molecule forming an oxygen-containing ring.

flavonoid *n.* any of various compounds containing a $C_6C_3C_6$ skeleton, the C_6 parts being benzene rings and the C_3 part varying in different compounds, and which include many water-soluble plant pigments.

flavonol *n.* any of a group of pale yellow flavonoid plant pigments.

flavoprotein *n.* any protein that contains a flavin prosthetic group (FAD or FMN) which can exist in oxidized or reduced form. They are yellow when oxidized but colourless when reduced. Flavoproteins are involved in electron transfers and oxidation reactions.

flavoxanthin *n.* yellow carotenoid pigment found in plants.

flea *n.* common name for a member of the Siphonaptera *q.v.*

flexor *n.* any muscle that bends a limb or its part by its contraction. *cf.* extensor.

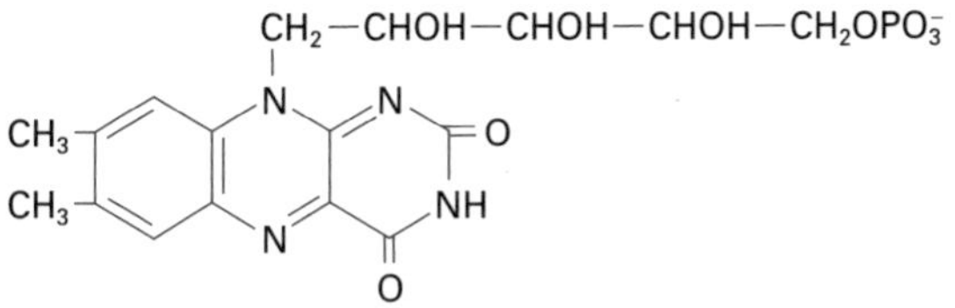

$$CH_2-CHOH-CHOH-CHOH-CH_2OPO_3^-$$

Fig. 17 Flavin mononucleotide.

flexuose, flexuous *a*. curving in a zig-zag manner.

flexure *n*. a curve or bend.

flimmer *n*. minute hairs borne on some types of eukaryotic flagella, giving them the appearance of tinsel.

FLIP fluorescence loss in photobleaching *q.v.*

flippase *n*. enzyme that transfers a phospholipid molecule from inner to outer leaflet of a biological membrane during membrane synthesis.

floating ribs false ribs *q.v.*

floccose *a*. covered with wool-like tufts.

flocculation *n*. clumping of small particles in the disperse phase of a colloidal system, such as the clumping of clay particles which can be brought about by lime.

flocculence *n*. adhesion in small flakes, as of a precipitate.

flocculent *a*. covered in a soft waxy substance, giving appearance of wool.

flocculus *n*. (1) small lobe on each lateral lobe of the cerebellum; (2) posterior hairy tuft in some Hymenoptera.

floccus *n*. (1) the tuft of hair terminating a tail; (2) downy plumage of young birds; (3) any tuft-like structure; (4) a mass of hyphal filaments, in fungi or algae.

floor plate ventral region of neural tube.

flor *n*. covering of yeasts, bacteria and other microorganisms which forms on the surface of some wines during fermentation.

flora *n*. (1) the plants peculiar to a country, area, specified environment or period; (2) a book giving descriptions of these plants; (3) the microorganisms that naturally live in and on animals and plants, e.g. gut flora, skin flora.

floral *a*. (1) *pert*. flora of a country or area; (2) *pert*. flowers.

floral axis receptacle *q.v.*

floral diagram conventional way of representing a flower, indicating the position of the parts relative to each other.

floral envelope the perianth, or calyx and corolla considered together.

floral formula numerical expression summarizing the number and position of parts of each whorl of a flower.

floral kingdom *see* floristic kingdoms.

floral leaf petal or sepal.

floral meristem region of dividing cells at tip of a shoot that gives rise to a flower.

floral organ a part of a flower, such as petal, sepal, stamen, carpel.

floral realm (1) large geographical area characterized by the plant species it contains; (2) specifically, one of the major divisions of the world according to its flora, *see* floristic kingdoms.

floral tube cup or tube formed by fusion of the bases of sepals, petals and stamens.

floret *n*. (1) one of the small individual flowers of a crowded inflorescence such as a capitulum of Compositae; (2) individual flower of grasses.

floridean starch type of starch found in red algae which gives a brown reaction with iodine instead of blue and is diagnostic for that group.

florigen *n*. a leaf-produced stimulus that causes flowering, now identified in *Arabidopsis* as the protein FT.

floristic *a*. (1) *pert*. the species composition of a plant community; (2) *pert*. or *appl*. flora.

floristic kingdoms major geographical divisions of the world distinguished according to their flora: Antarctic, Australian, Holarctic (Boreal), Neotropical, Palaeotropical (Paleotropical) and South African (Cape).

floristic province geographical area within a floristic kingdom that has a relatively uniform composition of plant species, merging with adjacent provinces via a transitional region containing species from both provinces.

floristics *n.* the study of an area of vegetation in terms of the species of plants in it.

flow cytometry technique for counting cells and distinguishing different types of cells in a mixed cell population. Cells are usually stained with different fluorescently labelled antibodies to distinctive cell-surface molecules. A stream of labelled cells is then run through laser beams that detect fluorescence and other optical parameters, and which count the numbers of cells of each type. *see also* FACS.

flower *n.* the reproductive structure of angiosperms (Anthophyta *q.v.*), being derived evolutionarily from a leafy shoot in which leaves have become modified into petals, sepals and calyx, and into the carpels and stamens in which the gametes are formed. Although flowers can take many different forms, they can all be represented by concentric whorls of different parts inserted on a base (the receptacle). The outermost whorl of sepals (often green) forms the calyx, inside that is a whorl of often brightly coloured petals, next is a ring of stamens (the male reproductive organs), and in the centre are the carpels (the female reproductive organs). *see* Fig. 18.

flowering plant *see* Anthophyta.

flow sorting *see* FACS.

fluctuation test test that distinguishes whether a phenotypic change in bacteria is due to physiological adaptation or to mutation. It was devised in 1943 by Delbrück and Luria to determine the mutant origin of phage-resistant bacteria.

fluid mosaic model accepted model for the structure of biological membranes. A membrane is composed of a lipid bilayer in which proteins are embedded. Proteins and lipids are free to diffuse laterally within the plane of the membrane but cannot flip vertically. *see* Fig. 26 (p. 392).

flukes *n.plu.* group of parasitic flatworms (platyhelminths). It includes the Monogenea, which are ectoparasitic on skin and gills of fishes, and the Trematoda.

The latter includes endoparasitic blood, liver and gut flukes, such as *Fasciola*, the common liver fluke of sheep, and the blood fluke *Schistosoma*, the cause of schistosomiasis in humans. *see also* digenean, Platyhelminthes, Trematoda.

fluorescamine *n.* a dye that reacts with amino acids to give a fluorescent product, used in analysis of amino acid composition of proteins.

fluorescein isothiocyanate (FITC) fluorescent compound widely used as a label for proteins.

fluorescence *n.* the phenomenon when light of one optical wavelength is emitted by a material in response to illumination with light of another wavelength. The dyes used as stains in fluorescence microscopy are usually those that emit light of a particular colour, e.g. red, green, blue, on activation by ultraviolet light.

fluorescence-activated cell sorter *see* FACS.

fluorescence *in situ* hybridization (FISH) method of determining the location of a DNA sequence on a chromosome by hybridization of mitotic chromosomes with a fluorescently labelled DNA probe specific for that sequence.

fluorescence loss in photobleaching (FLIP) technique used to study the lateral diffusion rates of fluorescently labelled membrane proteins by continuously illuminating a small area of membrane such that all proteins entering it become bleached, and measuring the rate of loss of fluorescence from the surrounding membrane.

fluorescence microscopy type of light microscopy used to visualize objects that fluoresce if light of a particular wavelength is shone on them. Fluorescence may be due to the presence of naturally fluorescent materials such as chlorophyll, or can be added by staining with fluorescent dyes. *see also* immunofluorescence microscopy.

fluorescence recovery after photobleaching (FRAP) technique used to study the lateral diffusion rates of fluorescently labelled membrane proteins by transiently illuminating a small area of membrane such that the proteins within it become bleached, and measuring the rate

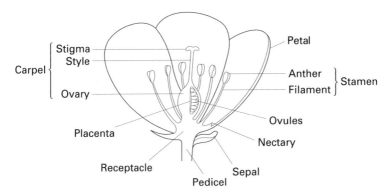

Fig. 18 Cross-section of a generalized angiosperm flower.

at which new fluorescent proteins move into the area.

fluorescence resonance energy transfer (FRET) technique for detecting protein–protein interactions by measuring the transfer of energy of one fluorescently labelled protein to another when they come into close contact.

fluorescent *a.* having the ability to emit light of one colour when activated by light of another wavelength.

fluorescyanine *n.* mixture of pterins with a yellow or blue fluorescence, found in the eyes, eggs and luminous organs of some insects.

fluoroacetate *n.* a poison that is converted *in vivo* to fluorocitrate which inhibits the enzyme aconitase and thus the tricarboxylic acid cycle.

fluorochrome *n.* any fluorescent compound.

fluorouracil (FU) *n.* fluorine-substituted analogue of uracil. As the deoxynucleotide, FUdR, it blocks dTMP synthesis and is used as an anticancer drug.

flush *n.* (1) patch of ground where water lies but does not run into a channel; (2) period of growth, esp. in a woody plant; (3) rapid increase in the size of a population.

flushing *n.* the washing of dissolved substances upwards in the soil so that they are deposited near the surface. *cf.* leaching.

fluviatile *a.* (1) growing in or near streams; (2) inhabiting and developing in streams, *appl.* certain insect larvae; (3) caused by rivers, *appl.* deposits.

fluvioterrestrial *a.* found in streams and in the land beside them.

fly *n.* (1) common name for a member of the Diptera *q.v.*; (2) in developmental biology, genetics and molecular biology, usually refers to the fruit fly *Drosophila melanogaster*.

F-mediated transduction sexduction *q.v.*

fMet formylmethionine *q.v.*

fMet-tRNA^f formylmethionyl-tRNA, the first aminoacyl-tRNA to be bound to the initiation complex in bacterial protein synthesis.

FMN flavin mononucleotide *q.v.*

FMNH$_2$ reduced form of flavin mononucleotide *q.v.*

fMRI functional magnetic resonance imaging *q.v.*

focal adhesion area of cell membrane that is bound strongly to the extracellular matrix by interactions between integrins in the cell membrane and extracellular matrix proteins, and through which signals from the extracellular matrix can be transmitted to the interior of the cell.

fodrin *n.* structural protein that is part of the cell cortex, present in neurons. It is identical to the spectrin of red blood cells.

foet- *alt.* fet-.

foetal *a. pert.* a foetus.

foetal haemoglobin (HbF) type of haemoglobin present in foetal blood. It consists of two α and two γ subunits and possesses a higher oxygen affinity than adult haemoglobin under foetal conditions.

foetal membranes membranes that protect or nourish the foetus, such as the chorion, amnion, allantois and yolk sac in mammals. *alt.* extra-embryonic membranes.

α-foetoprotein (α-FP, AFP) protein secreted by yolk sac and embryonic liver epithelial cells. In adults, it is produced by proliferating liver cells and can be used as an indicator of a liver cancer. Raised blood levels of α-foetoprotein in a pregnant woman may indicate a foetus with spina bifida.

foetus *n.* mammalian embryo after the stage at which it becomes recognizable. Technically, it is a human embryo from 7 weeks after fertilization.

fog-basking behaviour of some small insects in desert environments, in which they maximize the condensation of fog on their body surface and collect the water for drinking.

FOG fibres fast oxidative glycolytic muscle fibres in muscle of mammalian limbs. They are red muscle fibres adapted for aerobic metabolism and are used when the animal is walking or running. *cf.* FG fibres.

folate *see* folic acid.

fold *n.* the three-dimensional conformation of a protein chain. Different folds are defined in terms of the secondary structure elements present and their arrangement.

foldback DNA DNA in which inverted repeats on the same strand have formed internal base-paired regions.

foliaceous *a.* (1) having the texture of a foliage leaf; (2) thin and leaf-like; (3) bearing leaves.

foliage *n.* the leaves of a plant or plant community, collectively.

Folian process anterior projection on malleus, one of the auditory ossicles of mammalian middle ear.

foliar *a.* (1) *pert.* or consisting of leaves; (2) bearing leaves (as opposed to flowers), *appl.* side spurs.

foliate papilla one type of small projection on tongue that contains taste buds.

folic acid, folate pteroylglutamic acid, a water-soluble vitamin, sometimes considered as part of the vitamin B complex, found esp. in yeast, liver and green vegetables. It is important in nucleic acid metabolism because tetrahydrofolate is a coenzyme for purine and pyrimidine biosynthesis. A deficiency causes megaloblastic anaemia. *alt.* PGA.

folicole, folicolous, foliicolous *a.* growing on leaves, *appl.* certain fungi.

foliobranchiate *a.* possessing leaf-like gills.

foliolate *a. pert.*, having, or like, leaflets.

foliole *n.* (1) small leaf-like organ or appendage; (2) leaflet of a compound leaf.

foliose *a.* (1) with many leaves; (2) having leaf-like lobes, *appl.* the thallus of some lichens and liverworts.

folium *n.* (1) flattened structure in the cerebellum, expanding laterally into superior semilunar lobes; (2) fold on side of tongue.

folivorous *a.* leaf-eating.

follicle *n.* small sac-like structure; (*zool.*) (1) in the ovary, a discrete structure containing the developing oocyte. *see also* Graafian follicle; (2) hair follicle or feather follicle, sheath of epithelium surrounding hair or feather root. *see* Fig. 35 (p. 622); (3) (*immunol.*) area rich in B cells in the cortex of secondary lymphoid tissues such as lymph nodes; (4) (*bot.*) dry dehiscent fruit consisting of a single carpel and opening along one side only.

follicle-stimulating hormone (FSH) glycoprotein hormone secreted by the anterior pituitary. It acts on the gonads to stimulate growth of Graafian follicles and oestrogen secretion in the ovaries, and spermatogenesis in the testes.

follicular *a.* (1) *pert.*, like, or consisting of follicles, *appl.* first stage of human menstrual cycle, in which a follicle begins to mature; (2) *appl.* an ovarian hormone: oestrone *q.v.*

follicular dendritic cell type of stromal cell found in the follicles of lymph nodes and other secondary lymphoid tissues. Antigens from microorganisms infecting the body become bound to its surface and it is thought to be involved in stimulating activated B cells to develop into plasma cells. It is distinct from the dendritic cells found elsewhere in the lymphoid tissue.

folliculate *a.* containing, consisting of, or enclosed in follicles.

folliculose *a.* having follicles.

following response the innate response shown by the young of many species (e.g. chicks, ducklings), which will indiscriminately follow moving objects.

fomite *n.* inanimate object that can transmit disease when contaminated by a viable pathogenic microorganism.

fontanel(le) *n.* gap or space between bones of cranium, closed only by a membrane.

food bodies Beltian bodies *q.v.*

food chain a sequence of organisms within an ecosystem in which each is the food of the next member in the chain. A chain starts with the primary producers, which are photosynthetic organisms (e.g. algae, plants, bacteria) or chemolithotrophic bacteria. These are eaten by herbivores (primary consumers) which are in turn eaten by carnivories (secondary consumers). Small carnivores may be eaten by larger carnivores. *see also* food web.

food infection gastrointestinal disease caused by the ingestion of food contaminated with certain bacteria, e.g. salmonella, and subsequent growth of the bacteria in the intestine. *cf.* food poisoning.

food poisoning strictly, any disease caused by the ingestion of food containing toxins produced by bacteria growing in the food, e.g. botulism. Disease caused by the growth of bacteria from contaminated food in the intestine, e.g. salmonellosis, is also commonly known as food poisoning but is more strictly termed a food infection.

food pollen pollen present in flowers to provide food for visiting insects, instead of or as well as nectar, and which may be sterile and produced in special anthers.

food vacuole small vacuole enclosing fluid and ingested food particles, in many heterotrophic protists.

food web the interconnected food chains in an ecosystem.

foot-jaws poison claws or 1st pair of legs in centipedes. *alt.* maxillipeds.

foot plate terminal enlargement of cytoplasmic process of astrocyte in contact with a small blood vessel.

footprinting *n.* technique for determining the area of DNA to which a protein binds. The DNA–protein complex is subjected to digestion by an endonuclease so that only the area of DNA protected by the protein remains intact.

foot spinners common name for the Embioptera *q.v.*

forage (1) *n.* the vegetation eaten by grazing and browsing animals; (2) *v.* to search for food.

foraging *n.* the collection of food by animals.

foraging strategy method by which animals seek out food. *see also* optimal (2).

foramen *n.* (1) any small perforation; (2) aperture through a shell, bone or membrane. *plu.* **foramina**.

foramen magnum the opening in back of skull for passage of the spinal cord.

foramen of Monro passage between third and lateral ventricles of brain.

foramen ovale (1) aperture in great wing of sphenoid bone, passage for mandibular nerve; (2) opening between atria in foetal heart, which closes at birth.

foramen Panizzae an opening at point of contact between left and right systemic arteries in crocodilians.

foramen rotundum aperture in great wing of sphenoid bone, passage for maxillary nerve.

foramina *plu.* of foramen *q.v.*

foraminate *a.* pitted, having foramina or perforations.

Foraminifera, foraminifers, foraminiferans, forams *n.*, *n.plu.* phylum of mainly marine unicellular protists (classified as protozoans of class Sarcodina in older zoological classifications), having a highly vacuolated outer layer of cytoplasm and a calcareous, siliceous or composite shell through which project fine pseudopodia. Chalk is largely composed of foraminifer shells and they are also major components of many deep-sea oozes. *see* globigerina ooze. *a.* **foraminiferal**.

foraminiferous *a.* (1) having foramina; (2) containing shells of foraminiferans.

forb *n.* herbaceous plant, esp. a pasture plant other than grasses.

forceps *n.* (1) the clasper-shaped anal cercus of some insects; (2) the large fighting or seizing claws of crabs and lobsters; (3) (*neurobiol.*) fibres of corpus callosum curving into frontal and occipital lobes.

forcipate *a.* forked like forceps.

forebrain *n.* the telencephalon (cerebral hemispheres) and diencephalon (thalamus and hypothalamus). *alt.* prosencephalon.

fore-gut stomodaeum *q.v.*

foreign DNA DNA from another organism.

fore-kidney pronephros *q.v.*

forelimbs *n.plu.* anterior pair of limbs in tetrapod vertebrates.

foreshore *n.* the zone of beach between the high- and low-water marks. *see also* seashore.

forest *n.* biome consisting of continuous or semi-continuous tree cover, which may develop in areas where the average annual precipitation is sufficient (>75 cm) to support the growth of trees and shrubs. Broadleaved forest, coniferous forest, pine forest, etc., describe forest in which the named types of tree comprise at least 80% of the canopy. *see also* ancient forest, mixed forest, monsoon rain forest, old-growth forest, rain forest, secondary forest, taiga, temperate rain forest, tropical rain forest. *cf.* desert, grassland.

forest floor (1) ground level in a forest; (2) the surface of the soil, including its litter covering, in a forest ecosystem.

forficiform, forficulate *a.* scissor-shaped.

form *n.* (1) taxonomic unit consisting of individuals that differ from those of a larger unit by a single character, therefore being the smallest category of classification. *alt.* **forma**; (2) one of the kinds of a polymorphic species; (3) taxonomic group whose status is not clear but may be species or subspecies; (4) the concealed resting place of a hare.

formaldehyde *n.* HCHO, a gas at room temperature and pressure, with an irritating odour. *see also* formalin.

formalin *n.* aqueous solution of formaldehyde, used as a preservative for biological specimens.

forma specialis form of a plant pathogenic fungus that is very similar to other members of the species but differs in host range. Abbreviated *f. sp. plu.* **formae speciales**.

formation *n.* the vegetation proper to a definite type of habitat over a large area, e.g. tundra, coniferous forest, prairie, tropical rain forest.

form genus a genus whose constituent species may not be related by common ancestry.

formic acid HCOOH, an organic acid present in some ants and other insects, and in some plants.

formicarian *a.* (1) *pert.* ants; (2) *appl.* plants which attract ants by means of sweet secretions.

formicarium *n.* an ants' nest, particularly an artificial arrangement for purposes of study.

form species a member of a form genus.

formylmethionine (fMet) *n.* a modified form of methionine, with an added formyl group, which is the first amino acid of a polypeptide chain in bacteria.

fornical *a.* *pert.* fornix.

fornicate(d) *a.* concave within, convex without, arched.

fornix *n.* (1) an arched recess, as between eyelid and eyeball, or between vagina and cervix; (2) arched sheet of white fibres beneath corpus callosum in brain. *plu.* **fornices**.

forward mutation mutation that inactivates a gene, a mutation from normal wild type.

fossa *n.* pit or trench-like depression.

fossette *n.* small pit or depression.

fossil *n.* the remains or trace of an organism that lived in the past. Fossils are found embedded in rock either as petrified hard parts of the organism or as moulds, casts or tracks.

fossil assemblage group of different fossils found together, representing organisms that lived together in a particular environment and time.

fossil diagenesis *see* diagenesis.

fossil fuels refers to coal, oil and natural gas (methane), which have been formed from the remains of dead organisms over long periods of time.

fossiliferous *a.* containing fossils.

fossilization *n.* the process of fossil formation, e.g. by carbonization, permineralization or recrystallization of the remains of organisms.

fossil record the record of past life on Earth as it can be obtained from fossils.

fossorial *a.* adapted for digging.

fossulate *a.* with slight grooves and hollows.

fossulet *n.* long narrow depression.

founder effect genetic differences between an original population and an isolated offshoot due to alleles in the small number of founder members of the new population being unrepresentative of the alleles in the original population as a whole.

fourth ventricle the internal cavity of hindbrain in vertebrates.

fovea *n.* (1) small pit, fossa or depression; (2) shallow pit in centre of retina that contains only cones and is the point of greatest acuity of vision. It is present in diurnal birds, lizards and primates. *alt.* **fovea centralis**; (*bot.*) (3) small hollow at leaf base in quillworts (Isoetales), containing a sporangium; (4) pollinium base in orchids. *plu.* **foveae**.

foveal *a.* (1) *pert.* fovea; (2) *pert.* fovea centralis, *appl.* cone vision.

foveate *a.* pitted.

foveola *n.* (1) small pit; (2) shallow cavity in bone. *plu.* **foveolae**.

foveolate *a.* having regular small depressions.

FP flavoprotein *q.v.*

α-FP α-foetoprotein *q.v.*

F pilus filamentous appendage produced by strains of bacteria carrying F factors and through which the F factor is transmitted to another bacterium at conjugation. *alt.* sex pilus.

F plasmid F factor *q.v.*

fractionation *see* cell fractionation.

fragile X syndrome a form of heritable mental retardation which is associated with abnormal fragility of a section of the X-chromosome.

fragmentation *n.* (1) type of asexual reproduction in which the organism breaks up into smaller pieces, each of which can develop into a new individual, as in some algae; (2) (*ecol.*) *see* habitat fragmentation, species–area effect.

fragmentin granzyme *q.v.*

frame *see* reading frame.

frameshift *n.* mutation that causes a change in translational reading frame as a result of the insertion or deletion of non-multiples of three consecutive nucleotides in a DNA sequence. Frameshift mutations usually lead to no or a truncated gene product being produced, as a result of the generation of a premature termination codon in the altered reading frame.

frameshift suppression the case where a previous frameshift mutation is apparently abolished. Suppression may be due to a subsequent insertion or deletion mutation in the same gene, which restores the original reading frame, or to the action of certain tRNAs (extragenic frameshift suppressors) that can recognize four-base 'codons'.

frame silk one type of silk produced by spiders, which forms the supporting frame and radii of a typical orb web. *see also* viscid silk.

framework regions the less variable parts of the variable regions of immunoglobulin and T-cell receptor chains. They provide the structural framework of the antigen-binding site. *cf.* complementarity-determining regions.

frankincense *n.* a balsam obtained from plants of the genus *Boswellia*.

FRAP fluorescence recovery after photobleaching *q.v.*

Fraser Darling effect the stimulation of reproductive activity by the presence and activity of other members of the species in addition to the mating pair.

fraternal *a. appl.* twins produced from two separately fertilized eggs.

free central *appl.* placentation of plant ovary, ovules borne on a free-standing central column of tissue.

free energy (G) (Gibbs free energy) a thermodynamic function used to describe the ability of a system to undergo change, in that a system cannot change spontaneously to a state of higher free energy. For a chemical reaction, the change in free energy of the reaction, ΔG, is given by the equation $\Delta G = \Delta E - T\Delta S$ for a system undergoing change at constant temperature (T) and pressure, where ΔE is the change in internal energy of the system and ΔS is the change in entropy of the system. A reaction can only occur spontaneously if ΔG is negative.

free energy of activation (ΔG‡) in a chemical reaction A → B, the difference in free energy between A and the unstable transition state of highest free energy, on which the rate of reaction depends. The state of highest energy is known as the activation-energy barrier, and has to be attained for the reaction to occur. Enzymes act as catalysts by binding the transition state more tightly than the substrate and thus lowering the free energy of activation of the reaction.

free-living *a. appl.* any cell or organism that can support life independently of any other organism. *cf.* colonial, parasitic.

freemartin *n.* sterile female or intersex twin born with a male, the abnormality being due to sharing of blood circulation in the uterus and consequent masculinization of the female by male hormones.

free nuclear division division of nucleus not followed by formation of new cell walls, as in the endosperm of angiosperm seeds.

free radical a molecule or fragment of a molecule that contains one or more unpaired electrons, and is thus highly reactive.

free-running *appl.* an endogenous rhythm unaffected by any external influence.

freeze-etching an extension of the freeze-fracture method of preparing specimens for electron microscopy. Before platinum shadowing, freeze-fractured material is subjected to freeze-drying which exposes the cytoplasmic and exterior faces of membranes and structures in the interior of the cell.

freeze-fracture method of preparing specimens for electron microscopy which enables the interior of cell membranes to be visualized. A frozen block of cells is split, often fracturing along the middle of the lipid bilayer of cell membranes, a platinum replica of the exposed surface is prepared by shadowing and the organic material dissolved away.

frenulum *n.* a projection on hindwing of Lepidoptera for attachment to forewing.

frenum *n.* fold of integument at junction of mantle and body of barnacles, carrying eggs in some species.

frequency-dependent fitness case where the fitness of a genotype varies with its relative frequency in the population.

frequency-dependent selection selection occurring when the fitness of particular genotypes is related to their frequency in the population. When rare, the particular genotype is at an advantage compared with the other possible genotypes, but when it is common it is at a disadvantage.

freshwater *a. pert.* or living in water containing less than 0.5 parts per thousand dissolved salt (sodium chloride), such as that in rivers, ponds and lakes. *cf.* brackish water.

FRET fluorescence resonance energy transfer *q.v.*

Freund's complete adjuvant an adjuvant used for experimental immunization in animals. It consists of an emulsion of killed mycobacteria in mineral oil.

Freund's incomplete adjuvant Freund's complete adjuvant lacking the mycobacterial component.

friable *a.* crumbly, easily powdered.

frilled organ organ at anterior end of some cestodes, for attachment to host.

frog *n.* (1) common name for a member of the amphibian order Anura, *see* Amphibia; (2) a triangular mass of horny tissue in the middle of sole of the foot of a horse and related animals.

frond *n.* (1) a leaf, esp. of fern or palm; (2) flattened thallus of certain seaweeds or liverworts; (3) leaf-like outgrowth from thallus; (4) any leaf-like structure.

frondescence *n.* development of leaves.

frons *n.* forehead or comparable part of head in other animals.

frontal *n.* (1) frontal scale in reptiles; (2) frontal bone in vertebrates; *a.* (3) in region of forehead; (4) *appl.* plane of section of brain at right angles to median longitudinal or sagittal plane.

frontal cortex cortex of the frontal lobes of the brain, which in humans occupies almost one-third of the entire cerebral cortical surface. It contains the primary motor and premotor cortex and prefrontal areas involved in higher-order aspects of cognitive and emotional functions. *see also* prefrontal cortex.

frontal lobe of brain, the front part of cerebral hemisphere, anterior to the central sulcus.

frontocerebellar fibres nerve fibres passing from frontal lobes of cortex to cerebellum.

frontoclypeus *n.* frons and clypeus, fused, in insects.

fronto-ethmoidal *a. pert.* frontal and ethmoidal bones, *appl.* suture.

frontoparietal *a.* (1) *pert.* frontal and parietal bones; (2) *pert.* suture: the coronal suture.

frontosphenoidal *a. pert.* frontal and sphenoidal bones, *appl.* process of zygomatic bone articulating with frontal.

fructicole *a.* inhabiting fruits, *appl.* fungi.

fructification *n.* fruiting body, or any morphologically complex spore-producing structure.

fructokinase phosphofructokinase *q.v.*

fructosan *n.* polysaccharide, such as inulin, which is made of condensed fructose units.

fructose *n.* hexose sugar which gives many fruits their sweet taste and is sweeter than sucrose. As various fructose phosphates it is an intermediate in gluconeogenesis and photosynthetic carbon fixation. Fructose 1,6-bisphosphate is an intermediate in glycolysis, where it is converted to pyruvate.

fructose 1-phosphate pathway metabolic pathway in liver for the conversion of dietary fructose into glyceraldehyde phosphate for glycolysis.

frugivore *n.* a fruit-eater. *a.* **frugivorous**.

fruit *n.* the developed ovary of a flower, containing the ripe seeds, and any associated structures. In mosses the spore-containing capsule is often called the fruit.

fruit body in fungi, slime moulds and algae, any specialized structure that produces spores or gametes, e.g. a mushroom in fungi. *alt.* fruiting body.

fruit fly any member of the dipteran genus *Drosophila*, especially *Drosophila melanogaster*, a favoured organism for experimental genetics and developmental biology.

fruiting body fruit body *q.v.*

fruit wall outer part of a fruit, either the pericarp derived from the ovary wall, or a structure derived from the ovary wall together with receptacle or other parts of the flower.

frustose *a.* (1) cleft into polygonal pieces; (2) covered with markings resembling cracks.

frustration *n.* situation where an animal cannot make an appropriate response to a stimulus.

frustule *n.* the two-part silica wall of a diatom.

frutescent *a.* becoming shrub-like.

frutex *n.* shrub. *plu.* **frutices**.

fruticose *a.* shrubby, *appl.* lichens that grow in the form of tiny shrubs.

fry *n.* newly hatched fish.

FSH follicle-stimulating hormone *q.v.*

F-type ATPases proton-transporting proteins with ATPase activity, found in inner mitochondrial, thylakoid and bacterial membranes. When protons are transported in one direction these proteins act as ATP synthases whereas ATPase activity transports protons in the other direction. They are multisubunit proteins composed of a spherical head (F_1), which contains the active site for ATP synthesis, joined by a short 'stalk' to a transmembrane portion which contains the proton channel (F_0). Formerly called chloroplast factor, coupling factor (mitochondria).

FU fluorouracil *q.v.*

fucoid *a. pert.* or resembling a seaweed.

fucosan *n.* polysaccharide composed of fucose units, found in vesicles (fucosan vesicles) in cells of brown algae where it may be a storage polysaccharide or a waste metabolic product.

fucose *n.* five-carbon sugar, constituent of some plant polysaccharides.

fucoxanthin *n.* brown xanthophyll carotenoid pigment found in brown algae, diatoms and golden-brown algae.

fugacious *a.* withering or falling off very rapidly.

fugitive species a species typical of newly disturbed habitats, which has a high ability to disperse, and which is usually eliminated from established habitats by interspecific competition.

fugu *n.* the pufferfish *Takifugu (Fugu) rubripes*.

fulcrum *n.* (1) a supporting structure such as a tendril or stipule; (2) the pivot of a lever, *appl.* points of articulation of some bones; (3) the hinge-line in brachiopod shells.

fulvic acid fraction remaining in a solution of humus in weak alkali after removal of humin and humic acid.

fulvous *a.* deep yellow, tawny.

fumaric acid, fumarate four-carbon dicarboxylic acid which participates in the tricarboxylic acid cycle, where it is hydrated by fumarase to form malate. It is also involved in the urea cycle in vertebrates.

functional complementation the restoration of wild-type phenotype to a mutant by the introduction of a cloned gene.

functional genomics the determination of the function of the proteins encoded in a genome, using a variety of techniques.

functional magnetic resonance imaging (fMRI) non-invasive imaging technique that allows visualization of brain regions and detection of increases in oxygen consumption by active brain regions. This technique is used to locate areas of the brain involved in a particular task.

fundament primordium *q.v.*

fundamental niche the largest niche an organism could occupy in the absence of competition or other interacting species.

fundamental tissue ground tissue *q.v.*

fundatrix *n.* a female founding a new colony by oviposition, *appl.* aphids.

fundic *a. pert.* a fundus, *appl.* cells of the stomach.

fundiform *a.* looped.

fundus *n.* the base of an organ. *a.* **fundic**.

fungal *a.* of or *pert.* fungi.

Fungi, fungi *n., n.plu.* kingdom of heterotrophic, non-motile, non-photosynthetic and chiefly multicellular organisms that absorb nutrients from dead or living organisms. Multicellular terrestrial fungi comprise four main groups, the Zygomycota (zygomycetes) (e.g. bread moulds), the Ascomycota (ascomycetes) (the unicellular yeasts and multicellular sac fungi), the Basidiomycota (basidiomycetes) (e.g. mushrooms and toadstools), and the form group Deuteromycota (Fungi Imperfecti), which lack a sexual stage. The slime moulds and water moulds are now classed as protists. Many fungi are serious plant pathogens and there are also some human pathogens. Multicellular fungi grow vegetatively as a mycelium, a mat of thread-like hyphae from which characteristic fruiting bodies arise (e.g. the blue or black spore-heads of many common moulds, the mushrooms and toadstools of agarics, and the brackets of bracket fungi). In some fungi the hyphae are divided into uninucleate or binucleate cells by transverse partitions (septa). Hyphae possess rigid cell walls that differ in composition from those of plants, chitin rather than cellulose being a main constituent in most fungi. *see* Appendix 2.

fungicidal *a.* able to kill fungi.

fungicolous *a.* living in or on fungi.

fungiform *a.* mushroom-shaped, *appl.* certain rounded papillae scattered irregularly on the tongue and containing a taste bud.

Fungi Imperfecti Deuteromycota *q.v.*

fungistasis mycostasis *q.v.*

fungistatic *a.* inhibiting the growth of, but not killing, fungi.

fungivore *n.* organism that feeds on fungi. *a.* **fungivorous**.

fungoid, fungous *a.* (1) *pert.* fungi; (2) with the character or consistency of a fungus.

fungus *sing.* of fungi *q.v.*

funicle *n.* (1) small stalk, e.g. of an ovule; (2) small cord or band, as of nerve fibres.

funicular *a. pert.* a funicle or funiculus.

funiculose *a.* rope-like.

funiculus *n.* (1) funicle *q.v.*; (2) one of the ventral lateral or dorsal columns of white matter in the spinal cord.

funnel *n.* siphon of cephalopods.

furanose *n.* a monosaccharide in the form of a five-membered ring with four carbon and one oxygen atoms. *cf.* pyranose.

furca *n.* any forked structure. *a.* **furcal, furcate**.

furcula *n.* (1) the forked 'spring' of springtails (Collembola); (2) the united clavicles of birds, the wishbone.

furfuraceous *a.* scurfy, covered with scurf-like or bran-like particles.

fuscin *n.* brown pigment in retinal epithelium.

fuscous *a.* of a dark, almost black, colour.

fuseau *n.* spindle-shaped thick-walled spore divided by septa, in some fungi.

fusi *n.plu.* in spiders, organs composed of two retractile processes, which form the threads of silk.

fusiform *a.* spindle-shaped, tapering gradually at both ends, *appl.* innermost layer of cerebral cortex, *appl.* a gyrus of temporal lobe.

fusiform initial in vascular cambium, a spindle-shaped cell that gives rise to secondary xylem or phloem.

fusimotor *a. appl.* motor nerve fibres which cause contraction of muscle fibres within muscle spindles.

fusion biotrophism type of parasitism found in some fungi that parasitize other fungi, in which protoplasm of host and parasite becomes fused at one location.

fusion gene gene composed of the 3′ part of one gene and the 5′ part of another, and which may be generated spontaneously by mutation or constructed by recombinant DNA techniques.

fusion nucleus central nucleus of embryo sac of ovule, formed by fusion of odd nuclei from each end.

fusion protein (1) the protein product of a fusion gene *q.v.*; (2) protein that induces fusion of two cells, or of an enveloped virus and cell membrane.

fusocellular *a.* having, or *pert.* spindle-shaped cells.

fusogenic *a.* promoting fusion.

fusoid *a.* somewhat fusiform.

fusulae *n.plu.* minute tubes of spinnerets of spiders.

futile cycle substrate cycle *q.v.*

Fv 'fragment variable', truncated form of an antibody Fab fragment, containing the antigen-binding site, obtained by chemical cleavage or by genetic engineering.

fynbos *n.* South African term for temperate evergreen scrubland, a type of chaparral.

G

γ- for headwords with prefix γ-, see gamma, or refer to headword itself.

Γ *see* gamma distribution *q.v.*

G (1) Gibbs free energy. *see* free energy; (2) glycine *q.v.*; (3) guanine *q.v.*

ΔG‡ Gibbs free energy of activation. *see* free energy of activation.

ΔG° standard free energy change *q.v.*

G₀, G0 phase of eukaryotic cell cycle in which progress through the cycle is arrested. A cell may enter G_0 at some point in G_1 and remain in it for some length of time. Some cells, such as nerve cells, that do not divide after they are fully differentiated remain in G_0 for the rest of their life.

G₁, G1 gap 1 phase of the eukaryotic cell cycle. It follows cell division and lasts to the start of DNA synthesis, and is a phase at which cell growth takes place. *see* Fig. 8 (p. 106).

G₂, G2 gap 2 phase of the eukaryotic cell cycle, between the end of DNA replication and the start of mitosis, and is a phase at which cell growth can take place. *see* Fig. 8 (p. 106).

Gα the alpha subunit of a heterotrimeric G protein.

Gβγ the beta and gamma subunits of a heterotrimeric G protein, which remain together as a dimer when the alpha subunit dissociates.

G₀₁, G₀₂, Gₘ₁, Gₘ₂, Gₜ₁, Gₜ₂, etc. gangliosides containing one (G_M), two (G_D) and three (G_T) N-acetylneuraminic acid residues. The number represents the number of other sugar residues subtracted from 5.

G₁ heterotrimeric G proteins that inhibit adenylate cyclase, leading to a suppression of cyclic AMP production.

G₀ heterotrimeric G protein that activates K^+ channels and inactivates Ca^{2+} channels by means of the dissociated α subunit, and couples receptors to the phosphoinositide second messenger pathway by activating phospholipase C by means of the βγ complex.

G₀ₗf olfactory-specific heterotrimeric G protein that activates adenylate cyclase in olfactory sensory neurons in response to odorants.

G_q heterotrimeric G protein that couples receptors to the phosphoinositide second messenger pathway by activating phospholipase C.

G_s heterotrimeric G proteins that stimulate adenylate cyclase, leading to a rise in cyclic AMP production.

G_t transducin *q.v.*

Ga 10^9 years.

GA (1) gibberellic acid or gibberellin *q.v.*; (2) Golgi apparatus *q.v.*

GABA γ-aminobutyric acid *q.v.*

GABA receptors receptors on neurons and other cells for the neurotransmitter γ-aminobutyric acid (GABA). They are ion channels.

G actin *see* actin.

gadfly *see* Brachycera.

Gadidae, gadids *n., n.plu.* large and economically important family of marine bony fishes including the cod, whiting and haddock.

GAG glycosaminoglycan *q.v.*

gain-of-function *appl.* mutation that confers a new function on a gene and produces a new phenotype. *cf.* loss-of-function.

Gal galactose *q.v.*

galactan *n.* any polysaccharide composed wholly or largely of galactose.

galactolipid *n.* any glycolipid in which the sugar is galactose, such as cerebrosides.

galactosaemia *n.* genetic disease that results in the inability to metabolize dietary galactose. There are three types, each due to a deficiency in a different enzyme. Illness, and sometimes death, of affected infants is the result of accumulation of toxic compounds.

galactosamine *n.* amino derivative of the sugar galactose, substituted at carbon 2.

galactosan galactan *q.v.*

galactose *n.* six-carbon aldose sugar, a constituent, with glucose, of the disaccharide lactose. Also found in various complex carbohydrates such as the pectins of plant cell walls and in some glycolipids and glycoproteins.

galactosemia *alt.* spelling of galactosaemia *q.v.*

β-galactosidase *n.* enzyme that hydrolyses the glycosidic bond in lactose and some other galactosides, producing galactose as one of the products. EC 3.2.1.23.

galactoside *n.* any compound in which galactose is linked via a glycosidic bond to another sugar or a non-sugar alcohol, and including the disaccharide lactose and the galactolipids such as cerebrosides.

galactotropic *a.* stimulating milk secretion.

galacturonate (galacturonic acid) sugar acid derived from galactose by oxidation, a component of polysaccharides of plant cell walls.

galbulus *n.* a closed globular female cone with shield-shaped scales which are fleshy or thickened, as in the false-cypresses.

galea *n.* (1) (*bot.*) helmet-shaped petal; (2) (*zool.*) outer branch of maxilla in insects.

Galen, veins of internal cerebral veins and great cerebral vein formed by their union.

galericulate *a.* bearing or covered by a small cap.

galeriform *a.* shaped like a cap.

gall *n.* (1) abnormal outgrowth from plant stem or leaf caused by the presence of young insects (e.g. gall wasps or gall mites) in the tissues (as e.g. oak-apple) or by infection by certain fungi or bacteria (e.g. crown gall caused by *Agrobacterium*); (2) bile *q.v.*

gall bladder pear-shaped or spherical sac that stores bile. In humans it drains via the bile duct into the duodenum.

gall flower in fig trees, an infertile female flower in which the fig wasp lays its eggs.

gallic acid organic acid obtained from hydrolysis of tannin, and present in tea, galls and some plants.

gallicolous *a.* living in plant galls.

Galliformes *n.* order of heavy-bodied, chicken-like land birds, including grouse, partridges, pheasants, quails and domestic fowl.

gallinaceous *a.* resembling the domestic fowl, *appl.* birds of the same family.

gall midges common name for cecidomyiid flies that cause galls on plants.

gallotannin *n.* a tannin found in many types of galls, esp. on oak, and which is a glucoside of glucose and digallic acid.

gall wasps minute hymenopteran insects belonging to the superfamily Cynipoidea, which lay their eggs in the leaf and stem tissue of oak and some other plants (e.g. roses), inducing the formation of a gall, inside which the grub develops.

GalNAc *N*-acetylgalactosamine *q.v.*

GALT gut-associated lymphoid tissue *q.v.*

Galton's law of filial regression the tendency of offspring of outstanding parents to revert to the average for species.

galvanotaxis, galvanotropism *n.* movement in response to an electrical stimulus.

gametal *a.* (1) *pert.* a gamete; (2) reproductive.

gametangiogamy *n.* union of gametangia.

gametangium *n.* any structure containing gametes. *plu.* **gametangia**. *a.* **gametangial**.

gamete *n.* haploid reproductive cell produced by meiosis that then fuses with another of opposite sex or mating type to produce a diploid zygote. *see also* antherozoid, spermatozoon, ovum, egg. *cf.* spore.

game theory mathematical theory concerned with determining the optimal strategy in situations of competition or conflict. This theory can be applied to the relationships within a community, and the computer simulation of such relationships to determine winning strategies can help to throw light on ecological and social relationships and their evolution.

gametic *a. pert.* a gamete.

gametic imprinting genomic imprinting *q.v.*

gametic meiosis meiosis that gives rise to a gamete. *cf.* zygotic meiosis, sporic meiosis.

gametic number the haploid number of chromosomes (*n*) for a species, which is the number present in a gamete.

gametocyst *n.* cyst surrounding two associated individuals in which sexual reproduction takes place, as in some gregarine protozoans.

gametocyte *n.* stage in the life cycle of the malaria parasite which is released from red blood cells in humans and reinfects mosquitoes. Gametocytes mature into gametes in the mosquito.

gametogamy *n.* the union of gametes.

gametogenesis *n.* gamete formation.

gametogenetic *a.* stimulating gamete formation.

gametogenic variation variation arising from mutations in gametes.

gametogony *n.* that part of a parasite life cycle in which gametes are formed.

gametoid *n.* a structure behaving like a gamete, such as the multinucleate masses of protoplasm that fuse to form a zygotoid in some fungi.

gametophore *n.* (1) in some gametophytes, that part on which gametangia are borne, such as the upright leafy shoot of a moss; (2) in fungi, a hyphal outgrowth that fuses with a similar neighbouring outgrowth to form a zygospore.

gametophyll *n.* (1) modified leaf bearing sexual organs; (2) microsporophyll (*q.v.*) or megasporophyll (*q.v.*).

gametophyte *n.* the haploid, gameteforming phase in the alternation of generations in plants. *cf.* sporophyte.

gametothallus *a.* thallus that produces gametes. *cf.* sporothallus.

gametropic *a.* movement of plant organs before or after fertilization.

gamic *a.* fertilized.

gamma *for headwords with prefix γ- or gamma- refer also to headword itself, e.g. for γ-aminobutyric acid look under A, for γδT cells, look under T.*

gamma distribution (Γ) a type of continuous probability distribution that can be used to make corrections for the variation in substitution rates among different sites in a nucleotide or amino acid sequence

when developing models of sequence evolution for use in phylogenetic analysis.

gamma diversity the diversity of species within a given geographical area. *cf.* alpha diversity, beta diversity.

gamma efferent motor neuron motor neuron that innervates muscle fibre within a muscle spindle, and is involved in coordinating movement.

gammaglobulins *n.plu.* blood serum proteins with a particular range of electrophoretic mobility. They include the immunoglobulins as well as some non-immunoglobulins.

gamma-proteobacteria large and physiologically heterogeneous group of the proteobacteria (purple Bacteria) distinguished on DNA sequence data. It includes the phototrophic purple sulphur bacteria (e.g. *Chromatium*), the enteric bacteria (e.g. *Escherichia*, *Salmonella*), *Legionella*, *Vibrio*, fluorescent pseudomonads and many other genera. *alt.* **gamma purple bacteria**.

gammopetalous *a.* having petals joined into a tube, at least at the base. *alt.* sympetalous. *cf.* monopetalous, polypetalous.

gammosepalous *a.* having sepals joined into a tube, at least at the base. *alt.* monosepalous. *cf.* polysepalous.

gamobium *n.* the sexual generation in alternation of generations, i.e. the gametophyte.

gamodeme *n.* a deme forming a relatively isolated interbreeding community.

gamodesmic *a.* having the vascular bundles fused together instead of separated by parenchyma tissue.

gamogenesis *n.* sexual reproduction.

gamogenetic, gamogenic *a.* sexual, produced from union of gametes.

gamogony *n.* formation of gametes or gametocytes from a gamont in protozoans.

gamont *n.* in some protozoa, an individual that produces gametes which then unite in pairs to form the zygote or sporont.

Gamophyta *n.* phylum of green pigmented photosynthetic protists comprising: (i) those multicellular filamentous green algae that lack flagella at all stages of the life cycle and engage in sexual reproduction by conjugation between haploid vegetative cells (e.g. *Spirogyra*), and (ii) the desmids (*q.v.*).

gamostely *n.* the condition in stems with several steles when the separate steles are fused together and surrounded by pericycle and endodermis.

ganglia *plu.* of ganglion.

gangliar *a. pert.* a ganglion or ganglia.

gangliform *a.* in the shape of a ganglion.

ganglioid *a.* like a ganglion.

ganglion *n.* structure formed of a mass of nerve cell bodies outside the central nervous system. *plu.* **ganglia**.

ganglionated *a.* having ganglia.

ganglion cells in mammalian retina, nerve cells forming the outermost layer of nervous tissue, and whose fibres feed into the optic nerve.

ganglioneural *a. appl.* a system of nerves, consisting of a series of ganglia connected by nerve fibres.

ganglionic *a. pert.*, consisting of, or near a ganglion, *appl.* layer of retina containing ganglion cells.

ganglioplexus *n.* a diffuse ganglion.

ganglioside *n.* glycosphingolipid containing one or more *N*-acetylneuraminic acid residues in the oligosaccharide head-group. Gangliosides are found chiefly in nerve cell membranes but are also present in other cell types. Ganglioside G_{M1} in the membranes of intestinal epithelial cells acts as a receptor for cholera toxin. *see also entry under* G_{D1}.

ganoid scales rhomboidal scales, found in primitive fish, with many outer layers of enamel, below which is dentine then layers of bone, growth in thickness being by layers above and below.

GAP (1) glyceraldehyde 3-phosphate *q.v.*; (2) GTPase-activating protein *q.v.*

gape *n.* the distance between the open jaws of e.g. birds, fishes.

gap genes developmental genes acting early in *Drosophila* embryogenesis which, if mutant, result in an embryo lacking chunks of contiguous segments. They code for transcription factors that subdivide the embryo into regions along the antero-posterior axis.

gap junction type of cell junction in which the opposing plasma membranes contain protein-lined channels that connect the cytoplasm of the two adjacent cells. Gap junctions allow the passage of ions and

small molecules up to 1000–1500 molecular weight.

gap penalty in a alignment, a penalty deducted from the overall 'score' of the alignment for the introduction of a gap.

gap phases of the cell cycle, *see* G_1 and G_2.

Garryales *n.* in some classifications, an order of woody dicots with opposite evergreen leaves and flowers in hanging catkin-like panicles.

gas bladder swimbladder *q.v.*

gas chromatography (GC) *see* gas–liquid chromatography.

gaseous exchange the exchange of gases between an organism and its surroundings, including uptake of oxygen and release of carbon dioxide in respiration in animals and plants, and the uptake of carbon dioxide and release of oxygen in photosynthesis in plants.

gas gangrene tissue destruction and bloating produced by infection of wounds with certain species of *Clostridium*.

gas gland glandular portion of air bladder of certain fishes which secretes gas into the bladder.

gas–liquid chromatography (GLC) column chromatographic technique for analysing the composition of a solute, which depends on partioning of the solute between gas and liquid phases. Compounds with the lowest solubility in the solvent, or the highest volatility, leave the column first.

gaster *n.* (1) an abdomen, esp. a swollen one; (2) swollen portion of abdomen of hymenopterans, which lies behind the waist.

Gasteromycetes *n.plu.* basidiomycete fungi in which the hymenium is completely enclosed in a basidiocarp and never exposed. Basidiospores mature within the basidiocarp and are not forcibly discharged from their basidia. Mature spores are released when the basidiocarp ruptures. Examples are puffballs, earth balls, earth stars, stinkhorns and bird's nest fungi.

Gasteropoda, gasteropods Gastropoda, gastropods *q.v.*

gastral *a. pert.* stomach.

gastralia *n.plu.* abdominal ribs, as in some reptiles.

gastral layer in sponges the layer of flagellated cells (choanocytes) lining the internal cavities which take up food particles from the water.

gas transport transport of gases between respiratory surface and tissues.

gastric *a. pert.* or in region of stomach.

gastric filaments in some jellyfish, endodermal filaments lined with nematocysts which kill any live prey entering the stomach.

gastric gland tubular gland at the base of the stomach that secretes gastric juice containing the digestive peptidases pepsin and rennin, as well as hydrochloric acid and mucus.

gastric intrinsic factor intrinsic factor *q.v.*

gastric juice *see* gastric gland.

gastric mill in decapod crustaceans, the hard lining of the gizzard and its associated muscles, which grinds and strains food.

gastric shield in bivalve molluscs, a hard structure in the stomach against which the crystalline style rubs and is worn away, releasing amylase.

gastrin *n.* peptide hormone produced by the stomach. It stimulates secretion of gastric juice. Also found in the central nervous system.

gastrocnemius *n.* the large calf muscle of the leg.

gastrocolic *a. pert.* stomach and colon.

gastrodermis *n.* single layer of epithelium lining the gut cavity in simple animals such as coelenterates, flatworms, nematodes. *a.* **gastrodermal**.

gastroepiploic *a. pert.* stomach and the main peritoneal fold, *appl.* arteries, veins.

gastrohepatic *a. pert.* stomach and liver, *appl.* a portion of lesser omentum, a mesentery connecting liver and stomach in reptiles.

gastrointestinal *a. pert.* stomach and intestines.

gastrointestinal tract (GI tract) the digestive tract or alimentary canal, the tube through which food enters the body and solid waste is excreted in animals, leading from the mouth to the anus.

gastrolienal *a. pert.* stomach and spleen.

gastrolith *n.* mass of calcareous matter found on each side of gizzard of crustaceans before a moult.

Gastromycetes Gasteromycetes *q.v.*

gastroparietal *a. pert.* stomach and body wall.

gastrophrenic *a. pert.* stomach and diaphragm, *appl.* ligament.

Gastropoda, gastropods *n., n.plu.* class of molluscs including the winkles and whelks, sea slugs, water snails, and land snails and slugs. They are characterized by a large flat muscular foot on which they crawl about. When a shell is present, it is a spirally coiled shell in one piece.

gastropores *n.plu.* in milleporine corals, the larger pores in the surface of the colony, through which protrude polyps with four knobbed tentacles.

gastropulmonary *a. pert.* lungs and stomach.

gastrosplenic *a. pert.* stomach and spleen.

Gastrotricha *n.* phylum of marine and freshwater microscopic pseudocoelomate animals which have an elongated body and move by ventral cilia.

gastrozooid *n.* individual specialized for feeding in siphonophore colony.

gastrula *n.* animal embryo when undergoing gastrulation *q.v.*

gastrulation *n.* early stage in animal embryogenesis involving extensive cell movements and reshaping of the embryo. Endoderm and mesoderm move inside the embryo and the gut cavity is formed.

gas vacuoles, gas vesicles gas-filled vacuoles in cyanobacteria and other aquatic prokaryotes which confer buoyancy on the cells and through which their position in the water column can be regulated. They are spindle-shaped structures with a surrounding membrane composed only of protein.

gated *a.* (1) *appl.* ion channels whose opening is dependent on a specific stimulus, e.g. binding by a specific molecule (ligand-gated) or a change in membrane potential to some threshold level (voltage-gated); (2) *appl.* any activity that requires a stimulus at a certain level to allow it to occur.

gating current ionic current associated with the opening or closing of gated ion channels in membranes.

Gause's principle ecological principle stating that usually only one species may occupy a particular niche in a habitat. *alt.* competitive exclusion principle.

Gaussian curve the symmetrical bell-shaped curve representing the frequency distribution of a normally distributed population.

Gaussian distribution a symmetrical distribution about the mean. *alt.* normal distribution.

Gaviiformes *n.* order of birds that includes divers and loons.

G-banding *n.* staining of mitotic chromosomes with Giemsa, which produces a specific pattern of dark and light bands on each chromosome.

GC box short DNA sequence rich in guanine and cytosine, typically found in the promoters of many eukaryotic genes transcribed by RNA polymerase II.

GC gas chromatography, *see* gas–liquid chromatography.

GC ratio the ratio of G + C to A + T in the DNA of a species.

G-CSF granulocyte colony-stimulating factor *q.v.*

GDP guanosine diphosphate *q.v.*

GEFs guanine-nucleotide exchange factors *q.v.*

gel *n.* in electrophoresis, a thin slab of jelly-like material (agarose or polyacrylamide) through which the mixture to be separated is run.

gelatin *n.* jelly-like substance obtained from animal tissue on heating, and which is denatured collagen.

gelatinous *a.* jelly-like in consistency.

gel electrophoresis *see* electrophoresis, polyacrylamide gel electrophoresis.

gel-filtration chromatography technique for separating molecules on the basis of size by passage through a column of beads of an insoluble, highly hydrated polymer (e.g. Sephadex). Larger molecules are unable to enter the spaces within the beads and thus are more rapidly eluted from the column.

gel mobility shift assay an assay that detects the sequence-specific binding of proteins to DNA by a change in the rate of migration of the relevant DNA fragment on gel electrophoresis compared with the unbound DNA. *alt.* **gel retardation assay, gel shift assay**.

gelsolin *n.* cytoplasmic protein that can fragment actin filaments by inserting between actin subunits.

GEM genetically engineered microorganism. *see* genetically modified organisms.

gemellus *n.* either of two muscles, superior and inferior, running from ischium to greater trochanter and to trochanteric foassa of femur respectively.

geminate *a.* (1) growing in pairs; (2) *appl.* species or subspecies: corresponding forms in similar but separate regions, as reindeer and caribou.

geminiflorous *a. appl.* a plant whose flowers are arranged in pairs.

gemini of coiled bodies (GEMS) intranuclear structures proposed to consist of snRNPs undergoing recycling.

Geminiviridae *n.* family of plant viruses with single-stranded circular DNA genomes. They have 'twinned' particles composed of two isometric virus particles attached to each other. There are three subgroups. A, leafhopper-transmitted viruses with genomes composed of a single type of DNA, infecting monocotyledonous plants, type member maize streak virus. B, whitefly-transmitted viruses with two types of DNA, infecting dicotyledonous plants, type member African cassava mosaic virus. C, leafhopper-transmitted viruses with genomes composed of a single type of DNA, infecting dicots, type member beet curly top virus.

geminous *a.* in pairs, paired.

gemma *n.* (1) an outgrowth from a plant or animal that develops into a new organism; (2) thick-walled cell similar to a chlamydospore, often found in oomycetes. *plu.* **gemmae**.

gemmaceous *a. pert.* gemmae or buds.

gemma cup cup-shaped or crescent-shaped structure in which gemmae are produced in some liverworts.

gemmate *a.* having or reproducing by buds or gemmae.

Gemmatimonadetes *n.* phylum of the Bacteria distinguished on DNA sequence criteria.

gemmation *n.* (1) budding; (2) the development of new independent individuals by budding off from the parent; (3) the arrangement of buds on a twig or branch.

gemmiform *a.* bud-shaped.

gemmiparous *a.* reproducing by gemmae or buds.

gemmule *n*. any small asexual propagative unit.

GEMS gemini of coiled bodies *q.v.*

gena *n*. cheek or side part of head. *a*. **genal**.

GenBank the main publicly accessible electronic database of DNA sequences, containing many millions of sequences.

gene *n*. the basic unit of inheritance, by which hereditary characteristics are transmitted from parent to offspring. At the molecular level a single gene consists of a length of DNA (or in some viruses, RNA) which exerts its influence on the organism's form and function by encoding and directing the synthesis of a protein, or a tRNA, rRNA or other structural RNA. Each living cell carries a full complement of the genes typical of the species, borne in linear order on the chromosomes. Cells from diploid organisms carry two copies (alleles) of each gene. *see also* allele.

gene amplification the repeated duplication of a gene, producing a number of identical copies.

gene bank gene library. *see* DNA library. *see also* germplasm, seed bank.

gene centres geographical regions in which certain species of cultivated plant are represented in their greatest number of different varieties or forms. In some, but not all, cases such a region corresponds to the crop's centre of origin and initial domestication.

gene cloning *see* DNA cloning.

gene cluster two or more contiguous genes of related or identical nucleotide sequence.

genecology *n*. ecology studied in relation to the population genetics of the organisms concerned.

gene complex a group of genes whose combined effects determine a phenotypic character.

gene conversion the 'correction' of heteroduplex DNA to one genotype or the other during genetic recombination, before segregation of the products of crossing-over. It can be detected by abnormal ratios of a pair of parental alleles in the recombinants. Gene conversion in somatic cells is also used in birds and some other animals to diversify immunoglobulin genes after the initial gene rearrangement.

gene diminution the routine reduction or elimination of certain genes during development which occurs in certain organisms, such as the nematode *Parascaris equorum* and ciliates.

gene disruption *see* cassette mutagenesis, gene knock-out, homologous recombination.

gene dosage the number of copies of a particular gene in a nucleus or cell.

gene duplication the generation of additional copies of a gene during normal cellular processes such as recombination. This is thought to be the origin of families of related genes such as the globin genes.

gene evolution *see* molecular evolution.

gene expression the activation of a gene, and the consequent production of the protein or RNA that it encodes.

gene family a set of genes with related nucleotide sequences, and which are descended by duplication and divergence from an ancestral gene. The proteins encoded by a gene family are called a protein family. *see also* gene superfamily.

gene flow the spread of particular alleles within a population and between populations resulting from outbreeding and subsequent intercrossing.

gene-for-gene resistance in plants, a type of resistance against fungal infection in which a so-called avirulence gene in the pathogen is matched by a resistance gene in the plant. The outcome of an infection – resistance or disease – depends on which alleles of each gene are present.

gene frequency the frequency of a particular allele of a gene in the population.

gene fusion the joining of two genes, as can occur in a translocation, so that the new sequence is read as one gene. *see also* fusion gene.

gene gun colloquial term for the equipment used to introduce DNA into a cell using biolistics.

gene knock-in the substitution of one gene or part of a gene with another functioning sequence in an experimental animal such as the mouse, using genetic engineering techniques.

gene knock-out destruction of the function of a particular gene. It usually involves the insertion of a large DNA fragment into the gene, thus disrupting its function. When the technique is applied to mouse embryonic stem cells, the mutant cells can then be used to generate transgenic mice lacking a specific gene function. *see* cassette mutagenesis, homologous recombination. *alt.* gene disruption.

gene library *see* DNA library.

gene locus the site on a chromosome occupied by a given gene. In a diploid cell there are two copies of each locus, each occupied by an allele of the gene in question. *alt.* genetic locus.

gene manipulation (1) alteration of genes by mutation *in vitro*; (2) genetic engineering *q.v.*

gene mapping determination of the location of a gene on a chromosome, which may be carried out by genetic mapping and/or physical mapping.

gene pair the two copies of a gene in a diploid organism.

gene pool all the genes, and their different alleles, present in an interbreeding population.

gene product the protein, tRNA, rRNA or other structural RNA encoded by a gene, and which is produced when the gene is activated.

genera *plu.* of genus.

general fertility rate demographic measurement of fertility rate that takes the age structure of the population into account. For human populations it is calculated from the crude birth rate multiplied by the percentage of women in the age group 15–45 years.

generalist, generalist species organism or species with a very broad ecological niche, which can tolerate a wide range of environmental conditions and eat a variety of foods. *cf.* specialist species.

generalized *a.* (1) combining characteristics of two groups, as in many fossils; (2) not specialized.

generalized transduction transduction in which any part of the host chromosome becomes part of the transducing phage particle. *cf.* specialized transduction.

general recombination genetic recombination between homologous DNA sequences. This is the typical form of reciprocal recombination that occurs at crossing-over at meiosis during the formation of gametes.

general transcription factor any of a group of gene regulatory proteins that are

required for the initiation of transcription at most eukaryotic genes. They assemble around the TATA box and aid the correct positioning and activation of RNA polymerase.

generation (1) in a line of descent, individuals that share a common ancestor and are all the same number of broods away from that ancestor; (2) in human populations often refers to all individuals in the same age group.

generation of diversity in immunology, the various processes that combine in the development of lymphocytes to generate the highly variable repertoire of immunoglobulins and T-cell receptors.

generation time (1) of cells, the time between formation of a somatic cell by mitosis and its own division; (2) of human and animal populations, the average span of time between the birth of parents and the birth of their offspring; (3) of microorganisms, time taken for one cell to divide and produce two new cells, often expressed as the time taken for the population to double.

generative *a.* concerned in reproduction.

generative apogamy the condition where the sporophyte plant is developed from the ovum or another haploid cell of the gametophyte, with no fertilization. *alt.* haploid apogamy, meiotic apogamy, reduced apogamy.

generative cell (1) in many gymnosperms, the cell of the male gametophyte that divides to form the sterile and sperm-producing cells; (2) in flowering plants, the cell of the male gametophyte that divides to form two sperm.

generative hyphae in a basidiocarp, the hyphae that give rise to basidia.

generator potential electrical potential difference across the cell membrane which is produced in a sensory receptor in response to a stimulus. It is graded in strength proportionally to the strength of the stimulus and triggers an action potential when it reaches a certain threshold.

gene rearrangement *see* antigenic variation, immunoglobulin genes, mating-type locus, T-cell receptor.

gene regulatory network (GRN) set of genes that interact directly with each other to accomplish a particular function or process. Such gene networks controlling certain aspects of early embryonic development in animals such as sea urchin and zebrafish are starting to be identified.

gene regulatory proteins proteins that control the expression of a gene by interacting with a control site in DNA and influencing the initiation of transcription. *see also* enhancer protein, activator, repressor, transcription factor.

generic *a.* (1) common to all species of a genus; (2) *pert.* a genus.

gene segment *see* D gene segment, immunoglobulin genes, J gene segment, V gene segment, T-cell receptor.

gene sequencing *see* DNA sequencing.

gene silencing the shutdown of gene expression. The term usually refers to cases in which a particular gene or group of genes is inactive by virtue of the state of the chromatin in which they are located, a state that is heritable from one generation of cells to the next. The term is also used to describe the shutdown of gene expression by RNA interference *q.v.*

genesis *n.* formation, production or development of a cell, organ, individual or species.

gene superfamily genes that are related in sequence and derive from a single ancestral gene, but which have diverged to such an extent that they now encode proteins with many different functions and roles. An example is the immunoglobulin superfamily, which contains immunoglobulins, cell adhesion molecules and numerous other proteins. The proteins encoded by a gene superfamily are known as a protein superfamily.

genet *n.* unit or group of individuals deriving by asexual reproduction from a single zygote.

gene targeting the replacement or mutation of a particular gene, using recombinant DNA techniques.

gene therapy *n.* amelioration of a genetic disease by the replacement or supplementation of affected cells with genetically corrected cells, or by introduction of correct copies of the gene directly into affected cells. Trials of somatic gene therapy (*q.v.*) are in progress for a small number of human genetic diseases.

genetic *a.* (1) *appl.* anything involving, caused by, or *pert.* genes; (2) *pert.* genetics; (3) *pert.* genesis. *see also* gene.

genetic adaptation adaptation to a new habitat or changed environmental conditions as a result of genetic change.

genetically modified crops (GMCs, GM crops) commercial varieties of crop plants such as maize, sugar-beet, cotton and soybean that have been produced by genetic engineering rather than conventional crop breeding.

genetically modified organisms (GMOs) plants, animals or microorganisms that have been genetically altered by means other than conventional breeding, usually by the introduction of a gene from another species. *alt.* genetically engineered organisms.

genetic code rules by which the amino acid sequence of a protein chain is determined by the order of the bases in the DNA that specifies the protein. A group of three consecutive bases (triplets if referring to DNA, codons if referring to the mRNA) specifies one of the 20 amino acids of which proteins are composed. The code is redundant, or degenerate, in that almost all the amino acids can be specified by more than one codon. *see* Fig. 19 (p. 258).

genetic disease, genetic defect heritable disease or condition, such as haemophilia or sickle-cell anaemia, which is caused by a specific defect in a single gene.

genetic distance a measure of the difference between two DNAs from different species, which is used in the construction of phylogenetic trees. In its crudest form it is the percentage of nucleotide differences between the two DNAs. *see also* p-distance. *alt.* evolutionary distance.

genetic diversity variability within a species due to genetic differences between individuals.

genetic drift (1) random changes in gene frequency in small isolated populations owing to factors other than natural selection, such as sampling of only a small number of gametes in each generation. *alt.* Sewall Wright effect; (2) random nucleotide changes in a gene not subject to natural selection; (3) small changes in the genome of the influenza virus that

		Second base			
		U	C	A	G
First base	U	UUU ⎫Phe UUC ⎭ UUA ⎫Leu UUG ⎭	UCU ⎫ UCC ⎬Ser UCA ⎪ UCG ⎭	UAU ⎫Tyr UAC ⎭ UAA Stop UAG Stop	UGU ⎫Cys UGC ⎭ UGA Stop UGG Trp
	C	CUU ⎫ CUC ⎬Leu CUA ⎪ CUG ⎭	CCU ⎫ CCC ⎬Pro CCA ⎪ CCG ⎭	CAU ⎫His CAC ⎭ CAA ⎫Gln CAG ⎭	CGU ⎫ CGC ⎬Arg CGA ⎪ CGG ⎭
	A	AUU ⎫ AUC ⎬Ile AUA ⎭ AUG Met	ACU ⎫ ACC ⎬Thr ACA ⎪ ACG ⎭	AAU ⎫ AAC ⎬Asn AAA ⎫ AAG ⎭	AGU ⎫Ser AGC ⎭ AGA ⎫Arg AGG ⎭
	G	GUU ⎫ GUC ⎬Val GUA ⎪ GUG ⎭	GCU ⎫ GCC ⎬Ala GCA ⎪ GCG ⎭	GAU ⎫Asp GAC ⎭ GAA ⎫Glu GAG ⎭	GGU ⎫ GGC ⎬Gly GGA ⎪ GGG ⎭

Fig. 19 The genetic code.

cause annual outbreaks but not serious epidemics or pandemics. *cf.* genetic shift.

genetic element (1) any structure or entity that has a genetic function, e.g. chromosome, plasmid, virus, transposable element; (2) any DNA sequence that has a particular property or function.

genetic engineering the deliberate alteration of the genetic constitution of a living organism or cell by artificial means (i.e. not by conventional breeding), such as the introduction of a gene from another species or the introduction of a mutation into a specific gene. Bacteria, cultured cells, plants and some animals can be altered in this way. *see also* embryonic stem cells, microinjection, transfection, transformation, transgene, transgenic, recombinant DNA.

genetic equilibrium condition where gene frequencies stay constant from generation to generation in a population.

genetic fingerprinting *see* DNA fingerprinting.

genetic fitness *see* fitness.

genetic immunization *see* DNA immunization, DNA vaccine.

genetic information the information for synthesizing RNAs and proteins, which is contained in DNA or RNA.

genetic instability the increased rate of mutation and/or chromosomal abnormalities seen in cells (e.g. cancer cells) when normal mechanisms of DNA repair and maintenance of chromosomal integrity break down.

genetic isolation lack of interbreeding between groups of a population or between different populations due to geographic isolation or cultural preferences.

genetic load all the mutant alleles with deleterious effects that are carried by an individual and which reduce its fitness when compared with the theoretical optimum.

genetic locus *see* gene locus.

genetic manipulation any deliberate alteration made to the genome of an organism by genetic engineering techniques.

genetic map a schematic diagram showing the relative positions of gene loci on a chromosome.

genetic mapping determining the position of a gene on the chromosomes by means of recombination frequencies, and other purely genetic means. *cf.* physical mapping.

genetic map unit *see* centimorgan.

genetic marker a gene or other piece of DNA whose properties, and sometimes position on the chromosome, are known and which may be used to identify particular cells or organisms, or as a point of reference in a genetic mapping experiment.

genetic modification *see* genetic engineering, recombinant DNA.

genetic polymorphism the stable, long-term existence of multiple alleles of a gene in the population. Technically a gene is said to be polymorphic if the most common homozygote occurs at a frequency of less than 90% in the population.

genetic ratios usually refers to the classic Mendelian ratios of different genotypes of progeny arising from various standard genetic crosses.

genetic recombination *see* recombination.

genetics *n.* (1) that part of biology dealing with inherited variation and its physical basis in DNA, the genetic material; (2) of an organism, the physical basis of its inherited characteristics, i.e. the sequence and arrangement of its genes.

genetic screen the detection of mutations on a large scale in a population of experimental organisms (e.g. *Drosophila*, yeast), usually after a treatment that increases the rate of mutation.

genetic screening the screening of humans for specific mutations that are known to cause disease or increase the risk of disease.

genetic shift major change in the genome of the influenza virus that causes serious epidemics or pandemics.

genetic spiral in a spiral arrangement of leaves around an axis, the imaginary spiral line following points of insertion of successive leaves.

genetic transformation *see* transformation.

genetic variation heritable variation in a population. It is the result of the presence of different variants (alleles) of genes and, in eukaryotes, the shuffling of alleles into new combinations by sexual reproduction and recombination.

gene transcription *see* transcription.

gene transfer the introduction of genes from one species into another, using recombinant DNA techniques. *see* genetic engineering, recombinant DNA, transgenic, transgene.

genial *a. pert.* chin.

genic *a. pert.* genes.

genicular *a.* (1) *pert.* region of the knee; (2) *pert.* geniculum, *appl.* a ganglion of the facial nerve; (3) *appl.* bodies, the lateral and medial corpora geniculata, comprising part of the thalamus. *see* lateral geniculate nucleus.

geniculate *a.* bent like a knee. *see also* lateral geniculate nucleus.

geniculum *n.* (1) sharp bend in a nerve; (2) part of facial nerve in temporal bone where it turns abruptly towards the stylomastoid foramen.

genioglossal, geniohyoglossal *a.* connecting chin and tongue, *pert.* muscle that moves tongue in vertebrates.

geniohyoid *a. pert.* chin and hyoid, *appl.* muscles running from hyoid to lower jaw.

genital *a.* (1) *pert.* reproductive organs; (2) *pert.* external accessory sexual organs, e.g. penis.

genital bursae in brittle stars, sacs into which the gonads open, and which open on each side of the base of each arm, also concerned with respiration and sometimes with the brooding of larvae.

genital canals in crinoids, the canals in the arms carrying genital cords that enlarge into gonads when they reach the pinnule.

genital coelom in cephalopods, a coelom at the apex of the visceral hump.

genital cord the cord formed by the posterior ends of the Müllerian and Wolffian ducts in the mammalian embryo.

genital disc in *Drosophila* and other insects, a disc of cells formed in the embryo and which gives rise to the external genitalia and reproductive system of adult at morphogenesis.

genital duct duct leading from gonads to the exterior.

genitalia, genitals *n.plu.* the organs of reproduction, the gonads and accessory organs, esp. the external organs.

genital operculum in some arachnids, a soft rounded median lobe divided by a cleft on the sternum of the 1st preabdominal segment, with the opening of the genital duct at its base.

genital plates the plates in sea urchins bearing the opening of the gonads.

genital pleurae in some hemichordates, a pair of lateral ridges or folds in the region of the gills in which the gonads lie.

genital rachis in many echinoderms, a ring of genital cells on which the gonads are borne.

genital ridge embryonic mesoderm from which gonads develop.

genital sinus (1) fused male and female genital ducts in some trematodes; (2) in cartilaginous fish, a paired sinus opening into the posterior cardinal sinus.

genital tubercle tissue in mammalian embryo that develops into the clitoris in females and the penis in males.

genito-anal *a.* in the region of the genitalia and anus.

genitocrural *a.* in the region of the genitalia and thigh, *appl.* a nerve originating from the 1st and 2nd lumbar nerves.

genitofemoral genitocrural *q.v.*

genito-urinary urinogenital *q.v.*

Gennari's band layer of white nerve fibres in middle layer of cerebral cortex in occipital lobe.

genocline *n.* a gradual reduction in the frequency of various genotypes within a population in a particular spatial direction.

genocopy *n.* production of the same phenotype by different mimetic non-allelic genes.

genodeme *n.* a deme differing from others genotypically, but not necessarily phenotypically.

genoholotype *n.* a species defined as typical of its genus.

genome *n.* (1) the complete set of genes carried by an organism or virus. If specified quantitatively in respect of a diploid organism, it usually refers to the haploid set of genes; (2) sometimes refers to the total DNA content of a nucleus.

genome annotation the addition of information that interprets the DNA sequence of a genome, such as the identification of genes, protein functions, and sequence elements such as promoters.

genome browser internet-based servers and bioinformatics tools that can be used to find and interpret a wide range of information about fully sequenced genomes.

genome library genomic library *q.v.*

genome sequence the complete DNA sequence of the genome of a species.

genomic *a.* (1) *pert.* genome; (2) *appl.* chromosomal DNA (as opposed to cDNA synthesized on a messenger RNA template).

genomic hybridization method of determining the taxonomic relatedness of two organisms by measuring the similarity of their genomes by means of DNA hybridization.

genomic imprinting situation where expression of either the maternally derived or the paternally derived allele of a gene is suppressed in the embryo. Imprinting occurs for only some genes, and is established during formation of the gametes. Gene inactivation in the gamete is correlated with increased DNA methylation of the gene. *alt.* parental genomic imprinting.

genomic library collection of DNA clones isolated directly from chromosomal DNA. *alt.* genome library, genomic DNA library. *cf.* cDNA library.

genomics *n.* the study of complete genomes, and of genes and their expression on a large scale rather than as individual genes.

genospecies *n.* group of bacterial strains capable of gene exchange.

genosyntype *n.* a series of species together defined as typical of their genus.

genotype *n.* the precise genetic constitution of an organism. For a diploid organism it refers to the particular pair of alleles present for any given gene. *a.* genotypic. *cf.* phenotype.

genotypic *a. pert.* genotype. *cf.* phenotypic.

genotypic variance a statistical measure used in determining the heritability of a particular trait. It is that part of the phenotypic variance for the trait that can be attributed to differences in genotype between individuals.

genotyping *n.* the determination of the detailed genetic make-up, or genotype, of an individual, usually in respect to particular genes or sets of genes.

gens *n.* (1) clan or tribe; (2) those members of a parasitic species that parasitize a particular host. *plu.* **gentes**.

gentamicin *n.* aminoglycoside antibiotic related to streptomycin, which interferes with bacterial protein synthesis.

gentes *plu.* of gens.

Gentianales *n.* order of dicot trees, shrubs or herbs that includes the families Asclepiadaceae (milkweed), Gentianaceae (gentian), Rubiaceae (madder) and others.

genu *n.* (1) the knee; (2) a joint between femur and tibia in some ticks and mites; (3) a knee-like bend in an organ or part; (4) anterior end of corpus callosum.

genus *n.* taxonomic group of closely related species, similar and related genera being grouped into families. Generic names are italicized in the scientific literature, e.g. *Homo* (human), *Quercus* (oaks), *Canis* (wolves and dogs), *Salmo* (salmon and relatives). *plu.* **genera**. *a.* **generic**.

geobiotic terrestrial *q.v.*

geobotany phytogeography *q.v.*

geocarpic *a.* having fruits maturing underground due to young fruits being pushed underground by curvature of stalk after fertilization.

geochronology *n.* science dealing with the measurement of time in relation to the Earth's evolution.

geocline *n.* gradual and continuous change in a phenotypic character over a considerable area as a result of adaptation to changing geographical conditions.

geocoles *n.plu.* organisms that spend part of their life in the soil and affect its aeration and drainage.

geocryptophyte *n.* plant with dormant parts hidden underground.

geographical isolation the separation of two populations originally of the same species by a physical barrier such as mountains, oceans or rivers. Eventually this may lead to such differences evolving in one or both of the isolated populations that they are unable to interbreed and new species are formed.

geographical race a population separated from other populations of the same species by geographical barriers such as mountain ranges or oceans, and showing little or no differences from the rest of the population. Members of this population can usually interbreed with the rest of the species to produce fertile offspring, if brought into contact.

geographical speciation evolution of a new species from a geographically isolated population.

geology *n.* science dealing with the physical structure, activity and history of Earth.

geonasty *n.* a curvature of plant stem, usually a growth curvature, in response to gravity. *a.* **geonastic**.

geophilous *a.* (1) living in or on the soil; (2) having leaves borne at soil level on short stout stems.

geophyte *n.* (1) a land plant; (2) a perennial herbaceous plant with dormant parts (tubers, bulbs, rhizomes) underground.

geoplagiotropic *n.* growing at right angles to the surface of the ground, in response to the stimulus of gravity.

geosere *n.* (1) series of climax plant formations developed through geological time; (2) the total plant succession of the geological past.

geosmins *n.plu.* volatile terpene derivatives that contribute to the 'earthy' smell of soil, produced by actinomycetes.

geotaxis *n.* movement in response to gravity. *a.* **geotactic**.

geotonus *n.* normal position in relation to gravity.

geotropism *n.* movement or growth in relation to gravity, either in the direction of gravity (positive geotropism, as plant roots) or away from the ground (negative geotropism, as plant shoots). *alt.* gravitropism.

Geraniales *n.* order of dicot trees, shrubs or herbs that includes the families Balsaminaceae (balsam), Erythroxylaceae (coca), Geraniaceae (geranium), Linaceae (flax), Tropaeolaceae (nasturtium) and others.

geranylgeranyl anchor *see* prenylation.

germ *n.* (1) vernacular for a microorganism, esp. one that causes disease; (2) the embryo of a seed.

germ band (1) primitive streak *q.v.*; (2) in a developing insect embryo, that part of the blastoderm that will give rise to the embryo. *see* long-germ development, short-germ development.

germ cell (1) a diploid reproductive cell that gives rise to a gamete; (2) gamete *q.v.*

germicidal *a. appl.* any substance that can kill microorganisms, esp. bacteria. *n.* **germicide**.

germiduct *n.* oviduct of trematodes.

germinal

germinal *a. pert.* a seed, a germ cell or reproduction.

germinal centre focus of dividing and differentiating B cells that forms in a follicle of a secondary lymphoid organ after immunization or infection.

germinal crescent region of chick blastoderm forming a crescent of primordial germ cells partially surrounding anterior end of primitive streak.

germinal epithelium (1) epithelial cells covering stroma of an ovary; (2) layer of epithelial cells lining the testis which give rise to the spermatogonia and Sertoli cells.

germinal mutation mutation occurring in a cell destined to give rise to gametes.

germinal pore germ pore *q.v.*

germinal ridge mesodermal tissue in embryo that gives rise to gonads.

germinal spot the nucleolus of the germinal vesicle or ovum.

germinal streak primitive streak *q.v.*

germinal vesicle the nucleus of an oocyte before the formation of polar bodies.

germination *n.* the beginning of growth from a spore, seed, or similar reproductive structure.

germ layers regions of the very early animal embryo that will give rise to distinct types of tissues. Triploblastic animals have three germ layers: ectoderm, mesoderm and endoderm. Diploblastic animals have two: endoderm and ectoderm. Ectoderm gives rise mainly to epidermis and nervous system, mesoderm to muscle and connective tissues, and endoderm to the gut. *alt.* primary germ layer.

germ line in animals, the lineage of cells from which gametes (e.g. ova and sperm) are produced and through which hereditary characteristics (genes) are passed on from generation to generation in animals. *a.* **germline.**

germline configuration of immunoglobulin and T-cell receptor genes, the state of such genes before DNA rearrangement has occurred.

germline diversity that diversity among immunoglobulins and T-cell receptors that is due to the presence of multiple different gene segments in the germline genome. *see* immunoglobulin genes.

germline stem cell (GSC) type of stem cell that gives rise to the germ cells (sperm and ova).

germline theory a theory of antibody diversity that proposed that each antibody was encoded in a separate germline gene. Now known not to be true.

germ nucleus egg or sperm nucleus.

germplasm originally a term coined by the 19th-century biologist, A. Weismann, to denote the idea of protoplasm that was transmitted unchanged from generation to generation in the germ cells (as opposed to the inheritance of acquired characteristics). Nowadays often denotes cells from which a new plant or animal can be regenerated, as in collections of plant seeds in seed banks. *alt.* **germ plasm.**

germ pore (1) thin region in pollen grain wall through which the pollen tube emerges; (2) similar area in a spore wall for exit of a germ tube.

germ tube short filament put forth by a germinating spore.

gerontology *n.* the study of senescence and ageing.

gestalt *n.* (1) an organized or fixed response to an arrangement of stimuli; (2) coordinated movements or configuration of motor reactions; (3) a mental process considered as an organized pattern, involving explanation of parts in terms of the whole; (4) a pattern considered in relation to background or environment, *appl.* morphology irrespective of taxonomic or phylogenetic relationships.

gestation *n.* the period between conception and birth in animals that give birth to live young.

GFAP glial fibrillary acidic protein *q.v.*

GFP green fluorescent protein *q.v.*

GH growth hormone *q.v.*

ghosts red cell ghosts *q.v.*

GHRF growth-hormone releasing factor, somatoliberin *q.v.*

giant axon large-diameter axons of nerve cells, found in some invertebrates. Squid giant axons were used to discover the mechanism of impulse conduction in nerve cells.

giant cells large nerve fibres in annelids.

giant chromosomes *see* polytene chromosome.

giant fibres nerve fibres of very large diameter running longitudinally through ventral nerve cord of many invertebrates.

giant-grass community grassland with grasses up to 4 m high.

gibberellins (GA) *n.plu.* plant growth hormones that increase both cell division and cell elongation in stems and leaves. They also induce stem elongation and flowering in some plants and are able to break dormancy in some seeds. In germinating seeds, gibberellin stimulates growth by inducing the synthesis of enzymes that break down the starch stored in the endosperm into sugars.

gibbose, gibbous *a.* inflated, pouched.

Gibbs free energy *see* free energy.

Gibbs free energy of activation *see* free energy of activation.

Gibbs standard free energy the free energy of a chemical reaction under a specified set of standard conditions.

Giemsa banding *see* G-banding.

Giemsa stain stain that produces dark bands in regions of chromosome that are rich in AT base pairs.

gigantism *n.* growth of an organ or organism to an abnormally large size.

gill *n.* (1) respiratory organ of many aquatic animals (e.g. crustaceans, fishes, amphibians), plate-like or filamentous outgrowth well supplied with blood vessels at which gas exchange between water and blood occurs; (2) (*mycol.*) thin plate radiating out from the centre on the underside of cap of agaric fungi, and which bears the spore-forming cells.

gill arch the bony or cartilaginous skeleton forming an arch supporting the tissues, blood vessels and nerves of an individual gill.

gill bar in chordates, the tissue separating the gill slits from each other and containing blood vessels, nerves and skeletal support.

gill basket branchial basket *q.v.*

gill book type of gill in horseshoe crabs, consisting of a large number of leaf-like structures between which the water circulates.

gill cavity in agaric fungi, the cavity in the developing fruit body in which the gills are formed.

gill clefts gill slits *q.v.*

gill cover operculum covering the gills in bony fish.

gill filaments fine lateral processes of gills, increasing the area for gas exchange.

gill fungi members of the basidiomycete fungi whose hymenium is borne on gills, including most mushrooms and toadstools.

gill heart in cephalopods, the auricle attached to each gill and which connects to a common ventricle.

gill plume the ctenidium in gastropod molluscs.

gill pouches (1) oval pouches containing gills and communicating with the exterior, as in lampreys and hagfish; (2) outpushings of side wall of pharynx in all chordate embryos which develop into gill slits in fish and some amphibians.

gill rakers bristle-like processes on gill arches.

gill rays slender skeletal structures on outer margin of gill arches, which stiffen gills.

gill remnants epithelial, postbranchial or suprapericardial bodies arising in pharynx of vertebrates other than fish.

gill rods gelatinous bars supporting the pharynx in cephalocordates.

gill slits series of perforations on each side of pharynx to exterior, persistent in fish and some amphibia, embryonic only in reptiles, birds and mammals. When functional they allow the passage of water over the gill tissue.

gingivae *n.plu.* the gums. *a.* **gingival**.

gingivitis *n.* inflammation of the gums.

ginglymus *n.* a hinge joint. *a.* **ginglymoid**.

Ginkgophyta *n.* one of the five main divisions of extant seed-bearing plants, containing only one living genus, *Ginkgo*, characterized by its distinctive fan-shaped leaves. Ovules are borne in pairs on long stalks and microsporophylls in cones on separate plants.

girdle bundles leaf trace vascular bundles which girdle the stem and converge at the leaf insertion, as in cycads.

girdle scar a series of scale scars on axis of twig where bud scales have fallen off.

girdling *n.* removal of a complete ring of bark from around the trunk of a tree, thus preventing the flow of solutes from leaves to roots through the phloem, which is removed with the bark. It eventually kills the tree.

GI tract gastrointestinal tract.

gizzard *n.* muscular chamber in alimentary canal in which food is ground up in some animals, e.g. birds and insects, posterior to the crop.

glabrous *a.* (1) with a smooth even surface; (2) without hairs.

glacial *a.* in the biological literature usually refers to those periods during the Pleistocene in which large parts of the Northern hemisphere were covered by ice, the ice ages. *cf.* interglacial.

gladiate *a.* sword-like.

gladiolus mesosternum *q.v.*

gladius *n.* the pen or chitinous shell in colonial hydroids of the Chondrophora.

gland *n.* (1) single cell or organized structure specialized to secrete substances such as hormones or mucus; (2) lymph gland. *see* lymph node; (3) small vesicle containing oil, resin, or other liquid on any part of a plant.

glandiform *a.* acorn-shaped.

glands of Brunner Brunner's glands *q.v.*

glands of Nuhn Nuhn's glands *q.v.*

glandula *n.* gland *q.v.*

glandular *a.* (1) *pert.* a gland or glands; (2) with a secretory function.

glandular epithelium the epithelium lining a gland, composed of polyhedral, columnar or cubical cells, which contain or manufacture the material to be secreted.

glandular tissue (1) in plants, parenchymatous tissue adapted for secretion of aromatic or other substances; (2) the tissue making up a gland.

glandulose-serrate *a.* having the serrations tipped with glands, of leaves.

glans *n.* gland *q.v.*

glans clitoridis the small rounded mass of erectile tissue at the end of the clitoris.

glans penis the bulbous tip of the penis in mammals.

glareal *a. pert.*, or growing on, dry gravelly ground.

Glaserian fissure a fissure in temporal bone of mammals that holds the Folian process of malleus of middle ear.

glass sponges hexactinellid sponges *q.v.*

Glaucophyta *n.* a small group of freshwater photosynthetic unicellular eukaryotes carrying unusual chloroplasts called cyanelles, which still retain some features of the ancestral cyanobacterium from which they are thought to derive.

glaucous *a.* (1) bluish-green; (2) covered with a pale green bloom.

Glc glucose *q.v.*

GLC gas–liquid chromatography *q.v.*

GlcN glucosamine *q.v.*

GlcNAc *N*-acetylglucosamine *q.v.*

Gleason's index an index of the similarity between two communities (*see* coefficient of community) that is a modified version of the Jaccard index. It takes account of the dominance or importance of species common to both communities and of unique species, rather than simply measuring the number of common and unique species. *cf.* Jaccard index, Kulezinski index, Morisita's similarity index, Simpson's index of floristic resemblance, Sørensen similarity index.

gleba *n.* in gasteromycete fungi, the spore-bearing tissue in the interior of the basidiocarp. It develops into a mass of mature spores.

glenoid *a.* like a socket.

glenoid cavity, glenoid fossa (1) cavity in pelvic girdle into which the head of femur fits in tetrapods; (2) cavity on the squamosal for articulation of the lower jaw in mammals. *alt.* mandibular fossa.

gley *n.* a soil formed under conditions of poor drainage and which is waterlogged all or part of the time.

glia neuroglia *q.v.*

gliadin *n.* seed storage protein found in wheat.

glial cell any of the various non-neuronal cell types making up the neuroglia (*q.v.*) of the central nervous system, and similar cells associated with the peripheral nervous system, e.g. Schwann cells.

glial fibrillary acidic protein (GFAP) component of intermediate filaments in glial cells.

gliding bacteria diverse group of the Bacteria that have no flagella but are able to move by a gliding motion when in contact with a surface. They include the myxobacteria, which can form multicellular fruiting bodies, and filamentous bacteria such as the sulphur-oxidizing bacteria that form white streamers in polluted water and sewage sludge.

glioblast *n.* immature glial cell.

gliogenesis *a.* the generation of glial cells.

glioma *n.* tumour of glial cells in central nervous system.

Gln glutamine *q.v.*

global aphasia the complete loss of the ability to understand language, to speak, read or write.

global sequence alignment alignment of two sequences along their whole lengths.

global warming increase in the temperature of the lower atmosphere that has occurred over the past 100 years and which is thought to be principally due to increased emissions of carbon dioxide into the atmosphere as a result of fossil fuel burning and deforestation. *see also* greenhouse effect.

globigerina ooze mud formed largely from the shells of foraminiferans, esp. the calcareous shells of *Globigerina*.

globin *n.* (1) the protein constituent of haemoglobin, each haemoglobin molecule consisting of four globin subunits of two different types, e.g. ααββ in adult human haemoglobin; (2) generally, any member of the large globin superfamily of related proteins which also includes myoglobin and leghaemoglobin.

α-globin one of several types of globin found in animals, and one of the constituents of foetal and adult human haemoglobins.

α-globin genes gene cluster encoding α-globin and related globins. In humans it consists of the genes for ζ-globin, foetal and adult α-globins, and θ1-globin.

β-globin one of several types of globin found in animals, and a constituent of adult human haemoglobin (ααββ).

β-globin genes gene cluster encoding β-globin and related globins. In humans it consists of the genes for ε-globin, γ-globins, δ-globin and β-globin.

γ-globin type of globin, related to β-globin, a constituent of human embryonic and foetal haemoglobins (ζζγγ and ααγγ).

δ-globin type of globin, related to β-globin, a constituent of a minor type of human adult haemoglobin (ααδδ).

ε-globin type of globin, related to β-globin, a constituent of human embryonic haemoglobins (ααεε and ζζεε).

θ1-globin globin encoded by the θ1 gene in the α-globin gene cluster in primates.

ζ-globin type of globin, related to α-globin, a constituent of human embryonic haemoglobin (ζζεε).

globin genes the genes specifying the various types of globin subunits. In mammals they comprise two clusters of genes on different chromosomes. The α-globin cluster contains genes for ζ-globin, α-globin and δ-globin. The β-globin cluster contains those for ε-globin, γ-globin, θ1-globin and β-globin.

globoid *n.* spherical body in aleurone grains, made up of double phosphate of magnesium and calcium and protein.

globose *a.* spherical or globular.

globular protein any protein with a compact, approximately rounded shape. *cf.* fibrous (2).

globule *n.* any small spherical structure.

β-globulin fraction of blood protein that contains some immunoglobulins.

γ-globulin gammaglobulin *q.v.*

globulins *n.plu.* large group of compact globular proteins which are water-insoluble but soluble in dilute salt solution, from which they can be salted out. They are found in blood serum and other secretions, and as storage proteins in plant seeds.

globulose *a.* (1) spherical; (2) consisting of, or containing, globules.

globus *n.* a globe-shaped structure.

globus pallidus central portion of lentiform nucleus in the basal ganglia.

glochidiate *a.* covered with barbed hairs.

glochidium *n.* (1) hairs bearing barbed processes; (2) the parasitic larva of freshwater mussels such as *Unio* and *Anodon*.

gloea *n.* adhesive secretion of some protozoans.

gloeocystidium *n.* a cystidium containing a slimy or sticky substance.

Glomeromycota *n.* proposed fungal phylum containing the arbuscular–mycorrhizal fungi.

glomerular *a. pert.* or like a glomerulus.

glomerulate *a.* (1) arranged in clusters; (2) bearing a glomerulus.

glomerulonephritis *n.* disease of kidney due to inflammation of glomerular capillaries.

glomerulus *n.* (1) network of blood capillaries; (2) in the kidney, the knot of capillaries surrounded by the dilated end (Bowman's capsule) of a kidney tubule, the whole enclosed in a basement membrane, *alt.* renal glomerulus; (3) complex tree of dendrites arising from a group of olfactory cells; (4) (*bot.*) compact mass of almost sessile flowers; (5) compact mass of spores. *plu.* **glomeruli**.

glomus *n.* a number of glomeruli run together.

glossa *n.* (1) the tongue in vertebrates; (2) in insects, one of a pair of small lobes at tip of labium (lower lip) which are long in honeybees and bumblebees and used to suck nectar; (3) any tongue-like structure. *plu.* **glossae**. *a.* **glossal**.

glossarium *n.* the slender-pointed glossa of some Diptera.

glossate *n.* having a tongue or tongue-like structure.

glosso-epiglottic *a. pert.* tongue and epiglottis, *appl.* folds of mucous membrane.

glossohyal *n.* extension of the basihyal bone into the mouth of some fishes.

glossopalatine *a.* connecting tongue and soft palate.

glossopalatine nerve branch of facial nerve that supplies the tongue and the palate.

glossopalatinus, glossopalatine muscle a thin muscle which arises on each side of the soft palate and is inserted into the tongue.

glossophagine *a.* securing food by means of the tongue.

glossopharyngeal *a.* (1) *pert.* tongue and pharynx; (2) *appl.* nerve: 9th cranial nerve, chiefly conveying sensation, including taste, from pharynx, tonsil and back of tongue.

glottis *n.* the opening from the trachea (windpipe) into the throat.

Glu (1) glucose *q.v.*; (2) glutamic acid *q.v.*

glucagon *n.* polypeptide hormone synthesized by α-cells of pancreas when blood glucose level is low, and which stimulates breakdown of glycogen to glucose in liver and thus raises blood glucose level.

glucan *n.* general term for a polysaccharide composed wholly or chiefly of glucose residues, e.g. starch, cellulose, glycogen.

glucocerebroside *n.* lipid found in cell membranes, a cerebroside containing glucose rather than galactose.

glucocorticoid *n.* any of the steroid hormones produced by the adrenal cortex that influence carbohydrate metabolism, such as the formation of carbohydrate from fat and protein. They include corticosterone and cortisone, but not aldosterone, which is a mineralocorticoid.

glucocorticoid receptor *see* nuclear receptor.

glucocorticoid response element (GRE) control site in DNA at which glucocorticoid hormone/receptor complexes bind to activate gene expression.

glucogenic *a.* (1) *appl.* amino acids that are degraded to tricarboxylic acid cycle intermediates which can be converted to glucose and thus enter the carbohydrate metabolism of the body, e.g. glycine, alanine, aspartic acid, glutamic acid, arginine, ornithine; (2) stimulating glucose formation.

glucokinase *n.* enzyme catalysing the formation of glucose 6-phosphate from glucose. EC 2.7.1.2.

glucokinin *n.* insulin-like protein found in plants, which can reduce blood sugar.

glucomannans *n.plu.* hemicellulose polysaccharides composed of mannose and glucose residues linked together, usually in random order, found esp. in cell walls of conifers.

gluconeogenesis *n.* metabolic pathway by which glucose is synthesized from non-carbohydrates such as lactate, some amino acids and glycerol. It occurs chiefly in liver and kidney and in plants esp. in seeds.

glucosamine *n.* amino derivative of glucose, substituted at carbon 2. *see* Fig. 20.

glucosan glucan *q.v.*

glucose *n.* hexose sugar found in all living cells, and in plant sap and in the blood and tissue fluids of animals. It is the end-product of breakdown of starch, glycogen and cellulose, and is also a constituent of many other polysaccharides. It is the chief fuel molecule of most living cells, being converted to glucose 6-phosphate and

Fig. 20 Glucosamine.

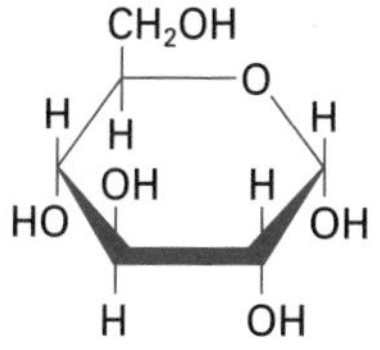

CH₂OH

Fig. 21 Glucose.

oxidized by glycolysis. *alt.* dextrose, grape sugar, starch sugar, blood sugar. *see* Fig. 21.

glucose-6-phosphatase enzyme from liver and kidney that converts glucose 6-phosphate into glucose, enabling it to be released into the blood. EC 3.1.3.9.

glucose 6-phosphate (G6P) compound to which glucose is converted by the enzyme hexokinase in the first reaction of glycolysis.

glucose-6-phosphate dehydrogenase (G6P dehydrogenase, GPDH) enzyme catalysing the formation of phosphogluconate from glucose 6-phosphate. EC 1.1.1.49.

glucose-6-phosphate dehydrogenase deficiency lack of the enzyme glucose-6-phosphate dehydrogenase, due to a mutation in the gene. It is generally symptomless but can lead to severe anaemia if the individual is exposed to certain drugs (e.g. the antimalarial drug chloroquine) or a lack of oxygen. The deficiency appears to confer some resistance to falciparum malaria. It is an X-linked recessive genetic defect and is thus seen most commonly in males.

glucosephosᵢ hate isomerase enzyme catalysing the conversion of glucose 6-phosphate to fructose 6-phosphate. EC

5.3.1.9. *alt.* phosphoglucose isomerase, phosphohexose isomerase.

glucosidases *n.plu.* enzymes that split off terminal glucose units from oligosaccharides. There are two types: (i) α-D-glucosidase (EC 3.2.1.20) which hydrolyses 1,4-α-D-glucosidic bonds and includes maltase, and (ii) β-D-glucosidase (EC 3.2.1.21) which hydrolyses β-D-glucosides, and which includes gentiobiase and cellobiase.

glucoside *n.* any glycoside yielding a sugar, usually glucose, on hydrolysis.

glucosinolate *n.* sulphur-containing compound found in some crucifers (e.g. brassicas, horseradish, mustard, oilseed rape) and responsible for the pungent flavour of these plants.

glucostat *n.* sensory cell that monitors levels of glucose circulating in the blood.

glucuronic acid, glucuronate sugar acid derived from glucose by oxidation, a component of polysaccharides in e.g. plant cell walls, bacterial capsules.

glumaceous *a.* dry and scaly-like glumes, formed of glumes.

glume *n.* dry chaffy bract, present in a pair at the base of spikelet in grasses.

glumella palea *q.v.*

glumiferous *a.* bearing or producing glumes.

Glumiflorae Poales (*q.v.*), the grasses.

glumiflorous *a.* having flowers with glumes or bracts at their base.

glutaeal, glutaeus gluteal, gluteus *q.v.*

glutamate *see* glutamic acid.

glutamate dehydrogenase enzyme involved in amino acid metabolism which catalyses the conversion of glutamate into a keto acid and ammonium ion, and which is also involved in the incorporation of ammonium into glutamate in nitrogen assimilation in plants. EC 1.4.1.2–4.

glutamate receptors cell-surface receptors for the neurotransmitter glutamate, present on neurons and other cells. There are two main structural types: ion channels (e.g. the NMDA receptor) and G protein-coupled receptors (the metabotropic glutamate receptors).

glutamate synthase (GOGAT) enzyme involved in nitrogen assimilation in bacteria, algae and plants, which catalyses the reductive transfer of an amino group from glutamine to 2-oxoglutarate to form glutamate. There are different forms of the enzyme using different electron donors (i.e. ferredoxin, NADH and NADPH). EC 1.4.7.1, 1.4.1.13, 1.4.1.14.

glutamic acid, glutamate (Glu, E) amino acid with an acidic side chain, negatively charged form of α-aminoglutaric acid. It is a constituent of protein and also acts as an excitatory neurotransmitter in the central nervous system.

glutaminase *n.* enzyme catalysing the conversion of glutamine to glutamate and ammonium ion. EC 3.5.1.2.

glutamine (Gln, Q) *n.* amino acid, uncharged monoamide derivative of glutamic acid, a constituent of protein, and a key component in control of nitrogen metabolism in bacteria and plants.

glutamine synthetase (GS) key enzyme in regulation of nitrogen metabolism and nitrogen assimilation in bacteria and plants. It catalyses the incorporation of ammonium ion into glutamine. EC 6.3.1.2.

glutaredoxin *n.* protein cofactor acting with glutathione in the synthesis of deoxyribonucleotides from ribonucleotides.

glutathione (GSH, GSSG) *n.* the tripeptide Glu-Cys-Gly, found in many tissues, esp. red blood cells where it acts as a buffer for haemoglobin. It exists in reduced form (GSH, the tripeptide with the SH group on the cysteine) or oxidized form (GSSG, two molecules of glutathione linked through a disulphide bond at the cysteines). GSH acts as a coenzyme in certain redox reactions.

gluteal *a. pert.* or in region of buttocks or hindquarters.

glutelins *n.plu.* class of small storage proteins in plant seeds. Examples are glutenin and oryzenin.

glutinous *a.* having a sticky, slimy surface.

Glx glutamic acid or glutamine.

Gly glycine *q.v.*

glycan *n.* (1) general term for a carbohydrate polymer such as a glycosaminoglycan or polysaccharide; (2) the oligosaccharide or polysaccharide portions of any macromolecule containing a considerable amount of carbohydrate, such as proteoglycans and glycoproteins.

glycation *n.* direct non-enzymatic addition of a free sugar to a compound.

glyceraldehyde 3-phosphate (GAP, G3P) triose phosphate intermediate in the Calvin cycle of photosynthesis, in glycolysis and in gluconeogenesis.

glyceride *n.* an ester of glycerol.

glycerine *n.* pure glycerol, a sweet viscous liquid.

glycerol *n.* three-carbon sugar alcohol. When combined with fatty acids it forms triacylglycerols (fats and oils) and membrane lipids.

glycerol diether type of membrane lipid found only in Archaea, in which two phytanyl chains composed of isoprene units are joined by ether linkages to a glycerol molecule.

glycerolipid *n.* any lipid based on glycerol. Membrane glycerolipids (e.g. phosphatidylcholine, phosphatidylinositol) have two long-chain fatty acids attached to the glycerol along with another group (e.g. phosphocholine, phosphoinositol). The glycerol diethers and tetraethers of archaeal cell membranes have chains of isoprene units attached to glycerol by ether linkages. Triacylglycerols (triglycerides), in which three long-chain fatty acids are attached to glycerol, are not found in membranes but are storage compounds in many plant seeds (as oils) and in animal fatty tissue.

glycerol phosphate shuttle pathway in which electrons from NADH formed in the cytoplasm by glycolysis are carried into mitochondria and transferred to the respiratory electron chain. Glycerol 3-phosphate formed by reduction of dihydroxyacetone phosphate in the cytoplasm is transported into mitochondria. There it is oxidized to dihydroxyacetone phosphate, which diffuses back to the cytoplasm.

glycine (Gly, G) *n.* the simplest amino acid, having a hydrogen atom as its side chain. It is a constituent of protein and also acts as an inhibitory neurotransmitter in the central nervous system.

glycinin *n.* protein from soya beans.

glycobiology *n.* the study of carbohydrates.

glycocalyx *n.* general term for carbohydrate-rich layer outside the cell wall of bacteria or the cell membrane of animal cells. *see* capsule, cell coat, slime layer.

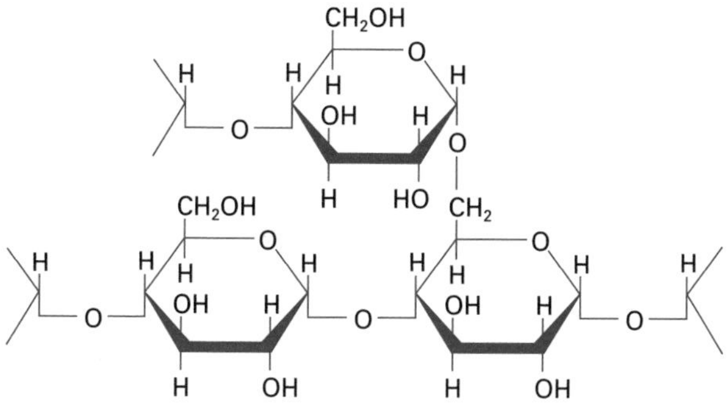

Fig. 22 Glycogen.

glycocholate *n.* the major bile salt, formed by the breakdown of cholesterol in the liver.

glycoconjugate *n.* any molecule composed of a glycan portion or portions attached to a non-carbohydrate.

glycogen *n.* branched-chain polysaccharide of glucose. It is the glucose storage compound in vertebrate liver and muscle and in animals generally. Also found in bacteria and fungi. Glycogen is broken down by phosphorolysis to glucose 1-phosphate, and can also be hydrolysed via dextrins and maltose to glucose. *see* Fig. 22.

glycogenase *n.* general name for two enzymes that catalyse synthesis of glycogen in liver: glycogen (starch) synthetase, EC 2.4.1.11, and glucose-1-phosphate uridylyltransferase, EC 2.7.7.9.

glycogenesis *n.* the synthesis of glycogen from glucose.

glycogen granules granules in cytoplasm of liver and muscle cells which contain glycogen and also contain the enzymes involved in its synthesis and degradation.

glycogenolysis *n.* the breakdown of glycogen to produce glucose 1-phosphate, stimulated in mammalian cells by adrenaline and glucagon, and inhibited by insulin.

glycogenosis *n.* disease caused by disturbance of glycogen metabolism.

glycogen phosphorylase enzyme that catalyses the cleavage of glycogen by inorganic phosphate (phosphorolysis) to glucose 1-phosphate in liver and muscle. EC 2.4.1.1.

glycogen synthase enzyme catalysing the transfer of glucose from UDP-glucose to a growing glycogen chain. EC 2.4.1.11, *r.n.* glycogen (starch) synthase. *alt.* **glycogen synthetase**.

glycolate *see* glycolic acid.

glycolate cycle complex metabolic cycle underlying photorespiration in plants, in which glycolate is used as a substrate, producing serine and glycine via glyoxylate (in peroxisomes).

glycolic acid, glycolate organic acid, formerly known as glycollate (glycollic acid), found chiefly in plants where it is a substrate for photorespiration, and also gives a sour taste to some unripe fruit.

glycolipid *n.* any complex lipid containing one or more carbohydrate residues, and including cerebrosides, gangliosides, sulphatides (from animal brain) and sulpholipids (from plants). They are components of cell membranes.

glycollic acid glycolic acid *q.v.*

glycolysis *n.* conversion of glucose to pyruvate in living cells, with the accompanying synthesis of ATP and NADH. It does not require the presence of oxygen. *adj.* **glycolytic**. *alt.* Embden–Meyerhof pathway.

glycome *n.* collectively, all the carbohydrate-containing molecules (e.g. glycosaminoglycans, polysaccharides, proteoglycans, the oligosaccharide side chains of glycoproteins) that can be produced by a particular organism, tissue or cell type.

glycomics *n.* the study of glycomes.

glyconeogenesis *n.* synthesis of glycogen starting from non-carbohydrate compounds.

glycophorins *n.plu.* proteins rich in sialic acid spanning the red blood cell membrane, and which carry the Gerbich blood group antigens.

glycophyte *n.* a plant unable to tolerate saline conditions.

glycoprotein *n.* any protein that contains carbohydrate in the form of chains of sugars attached to specific amino acid residues. Most membrane proteins and secreted proteins are glycoproteins. Carbohydrate is added to glycoproteins during their passage through the endoplasmic reticulum and Golgi apparatus.

glycosaminoglycan (GAG) *n.plu.* polysaccharide made up of repeating disaccharide units of amino sugar derivatives. Glycosaminoglycans include hyaluronate, chondroitin sulphate, keratan sulphate and heparin, and are found in the proteoglycans of connective tissue. *alt.* (formerly) mucopolysaccharide.

glycoside *n.* any of a class of compounds which on hydrolysis give a sugar and a nonsugar (aglutone) residue, e.g. glucosides give glucose, galactosides give galactose.

glycosidic bond type of covalent chemical linkage formed by condensation between sugar residues. It is the chemical bond between sugar subunits in oligo- and polysaccharides, and often between sugar residues and non-carbohydrates, such as purines and pyrimidine bases in nucleosides.

glycosome *n.* an organelle in some protozoans, including trypanosomes, containing the enzymes concerned with glycolysis.

glycosphingolipid *n.* membrane glycolipid containing the long-chain amino alcohol sphingosine, a fatty acid, and a head-group, and no glycerol.

glycosylation *n.* covalent addition of oligosaccharide side chains to proteins to form glycoproteins. It takes place in the lumen of the endoplasmic reticulum and Golgi apparatus. The oligosaccharide chain can be *N*-linked to the protein by attachment via the amino group of asparagine or *O*-linked by attachment via a hydroxyl group of serine or threonine.

glycosylphosphatidylinositol anchor (GPI anchor) a glycolipid attached to some proteins after translation and which links them to the internal face of the plasma membrane.

glycosyltransferases *n.plu.* enzymes that add monosaccharides to the sugar side chains of glycoproteins and glycolipids in the endoplasmic reticulum and Golgi apparatus.

glyoxalase system enzyme system converting methylglyoxal to lactic acid (the glyoxalate reaction) in animal tissues, and comprising two enzymes, lactoyl-glutathione lyase (EC 4.4.1.5) and hydroxyacyl-glutathione hydrolase (EC 3.1.2.6) formerly known as glyoxalase I and glyoxalase II, and utilizing glutathione as a cofactor.

glyoxylate (glyoxylic acid) cycle a modified version of the tricarboxylic acid cycle found in higher plants (especially germinating seeds) and bacteria. Isocitrate is converted to glyoxylate and succinate, glyoxylate being converted to malate. *alt.* Kornberg cycle.

glyoxysome *n.* plant cell organelle, found esp. in germinating seeds, which is involved in the breakdown and conversion of fatty acids to acetyl-CoA for the glyoxylate cycle. *alt.* microbody.

GM crop genetically modified crop *q.v.*

GM-CSF granulocyte-macrophage colony-stimulating factor *q.v.*

GMM genetically modified microorganism. *see* genetically modified organisms.

GMO genetically modified organism *q.v.*

GMP guanosine monophosphate *q.v.*

gnathal, gnathic *a. pert.* the teeth and jaws.

gnathism *n.* shape of jaw with reference to the degree of projection.

gnathites *n.plu.* mouth appendages of arthropods.

gnathopod maxilliped *q.v.*

gnathosoma *n.* the mouth region, including oral appendages, of some arachnids.

Gnathostomata, gnathostomes *n., n.plu.* (1) the jawed vertebrates, a subphylum of Chordata comprising the fishes, amphibians, reptiles, birds and mammals; (2) group of irregularly shaped sea urchins.

gnathostomatous *a.* with jaws at the mouth.

Gnathostomulida *n.* phylum of tiny acoelomate marine worms living in sediments and on plant and algal surfaces in shallow waters.

gnathothorax *n.* that part of the cephalothorax posterior to protocephalon in crabs and lobsters, etc.

Gnetophyta *n.* one of the five divisions of extant seed-bearing plants, comprising only three genera, *Gnetum*, *Ephedra* and *Welwitschia*, which differ considerably in form and reproduction. They are considered as gymnosperms although having some angiosperm features.

gnotobiosis *n.* the rearing of laboratory animals in a germ-free state or containing only a known prespecified flora of microorganisms.

gnotobiotics *n.* (1) the study of organisms or species when other organisms or species are absent; (2) germ-free culture; (3) study of germ-free animals.

GnRH gonadotropin-releasing hormone *q.v.*

goblet cell vase-shaped epithelial cell that secretes mucus.

GOGAT glutamate synthase *q.v.*

goitre *n.* hypertrophy of the thyroid gland due to lack of iodine in the diet.

golden algae common name for the Chrysophyta *q.v.*

golden-yellow algae common name for the Xanthophyta *q.v.*

Golgi apparatus (GA) organelle in eukaryotic cells which is formed of one or more stacks of flattened sacs of membrane (Golgi stacks). It receives material from the endoplasmic reticulum and is involved in directing membrane lipids and proteins and secretory proteins to their correct destination. It is functionally linked with the endoplasmic reticulum, lysosomes and plasma membrane by membrane transport vesicles. The Golgi apparatus is also the site where sugar side chains of glycoproteins are trimmed and remodelled. The side of a stack facing away from the plasma membrane and towards the nucleus is known as the *cis* face, while the side facing away from the nucleus and towards the plasma membrane is known as the *trans* face. Material moves through the Golgi stacks in a *cis* to *trans* direction. A simple form of Golgi apparatus present in fungi is sometimes known as the Golgi equivalent. *alt.* **Golgi bodies, Golgic complex**, dictyosomes. *see* Fig. 7 (p. 105).

Golgi body one of the stacks of the Golgi apparatus. *alt.* (plants) dictyosome.

Golgi cell type of interneuron found in cerebellum with cell body in granular layer, acts on granule cells.

Golgi staining staining of brain tissue by a silver impregnation method, devised by the neuroanatomist Camillo Golgi in the 19th century, which allows visualization of a whole neuron and its axon and dendrites, and which is still in use.

Golgi stack *see* Golgi apparatus.

Golgi tendon organ tendon organ *q.v.*

Golgi vesicles small membrane vesicles derived from the cisternae of the Golgi stacks. Some contain material being exported to the exterior of the cell or to the plasma membrane, others will become lysosomes.

gomphosis *n.* articulation by insertion of a conical process into a socket, as of roots of teeth into their sockets.

gonad *n.* organ in which reproductive cells are produced, such as the ovary and testis, or in lower animals an ovotestis.

gonadal *n.* of, *pert.*, or *appl.* gonads.

gonadectomy *n.* excision of gonad, castration in male, spaying in female.

gonadotropic *a.* affecting the gonads, *appl.* hormones: the gonadotropins *q.v.*

gonadotropin-releasing hormone (GnRH) hypothalamic peptide that stimulates the release of gonadotropins (follicle-stimulating hormone and luteinizing hormone) from the pituitary. *alt.* **gonadotropin-releasing factor**, luteinizing hormone-releasing hormone.

gonadotropins *n.plu.* hormones that stimulate gonadal function. They are follicle-stimulating hormone and luteinizing hormone, which are produced by the anterior pituitary, and chorionic gonadotropin, produced by the placenta, which is similar but not identical to luteinizing hormone. Prolactin also has gonadotropic activity.

gonal *a.* giving rise to a gonad, *appl.* middle section of germinal ridge which alone forms a functional gonad.

gonapophyses *n.plu.* (1) chitinous outgrowths aiding copulation in insects; (2) the component parts of a sting; (3) any genital appendages.

Gondwana *n.* southern land mass or supercontinent composed of the land that is now South America, Africa, India, Australia and Antarctica, existing from before 650 million years ago up to *ca.* 130 million years ago, after which it broke up and continental drift moved these continents to their present positions. For some time during its history it formed part of the massive supercontinent Pangaea, which separated into the northern supercontinent Laurasia and a reformed Gondwana in the Jurassic. *alt.* (formerly) Gondwanaland. *see also* Laurasia, Pangaea.

gongylidia *n.plu.* hyphal swellings or modifications in fungi cultivated by ants.

gongylus *n.* a round reproductive body in certain algae and lichens.

gonia, gonial(e) *n.plu.* primitive sex cells, spermatogonia or oogonia.

gonidangium *n.* a structure producing or containing gonidia.

gonidiophore *n.* aerial hypha supporting a gonidangium.

gonidiophyll *n.* gametophyte leaf bearing gonidia.

gonidium *n.* asexual non-motile reproductive cell, in some algae and cyanobacteria. *plu.* **gonidia**. *a.* **gonidial**.

gonimium *n.* bluish-green spore in the cyanobacterial cells of some lichens.

gonimoblasts *n.plu.* filamentous outgrowths of a fertilized carpogonium of red algae.

goniocyst *n.* in lichens, a cluster of gonidia.

gonion *n.* the angle point on the lower jaw.

gonocalyx *n.* the bell of a reproductive medusoid individual in a siphonophore colony.

gonochoristic *a. appl.* species having sex chromosomally determined as either female or male.

gonococcus *n.* the bacterium *Neisseria gonorrhoeae*, the cause of gonorrhoea.

gonocoel *n.* the body cavity containing the gonads.

gonocytes *n.plu.* in sponges, the mother cells of ova and spermatozoa or the gametes themselves.

gonoduct *n.* a duct leading from gonad to the exterior.

gonoecium *n.* a reproductive individual in a polyzoan colony.

gonomery *n.* separate grouping of maternal and paternal chromosomes during cleavage stages in development of some organisms.

gononephrotome *n.* embryonic segment containing primordia of the urinogenital system.

gonophore *n.* a reproductive zooid in a hydrozoan colony.

gonopod(ium) *n.* (1) modified anal fin serving as copulatory organ in some male fishes; (2) clasper of male insects, centipedes and millipedes.

gonopore *n.* a reproductive aperture.

gonosome *n.* the reproductive zooids of a coelenterate colony collectively.

gonostyle *n.* (1) in hydrozoans, a columnar zooid with or without mouth and tentacles, bearing gonophores; (2) sexual siphon of siphonophores; (3) bristle-like process on base of clasper of some male insects; (4) part of sting in some hymenopterans.

gonotheca *n.* transparent protective expansion of perisarc around blastostyle in some colonial hydrozoans.

gonotome *n.* the embryonic segment containing primordia of the gonad.

gonotrema, gonotreme *n.* genital aperture of arachnids.

gonozooid *n.* a reproductive individual of a hydrozoan colony.

Goodpasture's syndrome autoimmune disease of kidney caused by attack on glomerular basement membrane.

gordian worms Nematomorpha *q.v.*

gorgonians *n.plu.* the sea fans, horny corals with large, delicate, fan-shaped, branching colonies anchored by a 'stalk' to the substrate.

gorgonin *n.* fibrous protein in the mesoglea of sea fans (gorgonians) which forms the stiff skeleton of the colony.

gp as a prefix, denotes a glycoprotein.

G3P glyceraldehyde 3-phosphate *q.v.*

G6P glucose 6-phosphate *q.v.*

GPCRs G-protein-coupled receptors *q.v.*

G phases of cell cycle. *see* G_0, G_1, G_2.

GPI anchor glycosylphosphatidylinositol anchor, a lipid linkage that attaches some membrane proteins to the lipid bilayer.

GPP gross primary production *q.v.*

G protein *see* guanine-nucleotide binding proteins, heterotrimeric G proteins, small G proteins.

G-protein-coupled receptors (GPCRs) large class of cell-surface receptors, each composed of a single polypeptide chain with seven transmembrane regions. On stimulation of the receptor, the cytoplasmic portion associates with and activates a heterotrimeric G protein on the cytoplasmic face of the membrane. Examples are rhodopsin, β-adrenergic receptors, muscarinic acetylcholine receptors. *alt.* serpentine receptor, seven-span receptor.

Graafian follicle in the mammalian ovary, a spherical vesicle containing a developing ovum and a liquid, the liquor folliculi, surrounded by numerous follicle cells, and from which the ovum is released on ovulation.

gracile *a.* lightly built, small and slender, *appl.* australopithecines: e.g. *Australopithecus africanus*, *A. afarensis*.

gracile nucleus area in medulla of brain involved in relaying sensory information (touch, pressure and vibration) from skin, deep tissues and joints.

Gracilicutes *n.* in some classifications, a division of the Bacteria that includes all Gram-negative bacteria (e.g. enterobacteria, pseudomonads, Gram-negative cocci, spirilla, Gram-negative rods) and the photosynthetic bacteria (green and purple photosynthetic bacteria and the cyanobacteria).

gracilis *n.* (1) a slender muscle on the inside of the thigh; (2) fasciculus gracilis: a bundle of fibres in the medulla oblongata; (3) nucleus of grey matter ventral to clava in medulla oblongata.

grade *n.* a taxonomic category representing a level of morphological organization. A group of organisms forming a grade have a number of characteristics in common but may not owe them to a common ancestor, e.g. the protozoa.

graded *a. appl.* potential, an electrical potential that can vary continuously in strength, produced across a postsynaptic membrane. *alt.* postsynaptic potential.

gradient *see* positional information.

graduated *a.* becoming longer or shorter in discrete steps.

graduate sorus in ferns, a sorus in which sporangia develop from the tip towards the base.

graft (1) *n.* tissue, organ or part of an organism inserted into and uniting with a larger part of the same or another organism; (2) *v.* to insert scion into stock, in plants, or to transfer tissue from donor to recipient, as in an organ transplant in animals.

graft chimaera, graft hybrid an individual into which tissue from another individual of different genotype has been grafted or transplanted, and which shows characteristics of both individuals.

graft rejection the destructive immune reaction that occurs in vertebrates against a tissue or organ transplanted from another individual, due to the recognition of different MHC molecules on the transplanted tissue.

graft-versus-host reaction (GVH, GVHR) the reaction seen when immunocompetent cells in a graft attack and destroy host tissue in a cell-mediated immune response.

grain caryopsis *q.v.*

gramicidin A linear peptide antibiotic produced by species of *Bacillus*, and which is not synthesized by the mRNA/tRNA/ ribosome system. It dimerizes to form an ion channel in membranes.

gramicidin S cyclic peptide antibiotic produced by species of *Bacillus*, and which is not synthesized by the mRNA/tRNA/ ribosome system.

graminaceous *a.* (1) *pert.* grasses, *appl.* members of the grass family (Gramineae) such as cereals; (2) grass-coloured, *appl.* insects.

Gramineae *n.* the grasses, a large and ubiquitous family of monocotyledonous plants, mostly herbaceous but with some woody species (e.g. bamboos). The linear leaves arise from the nodes of jointed stems. Flowering stems bear clusters of windpollinated or cleistogamous flowers. The Gramineae contain many important crop plants including wheat (*Triticum* spp.), barley (*Hordeum vulgare*), oats (*Avena sativa*), maize (*Zea mays*), rice (*Oryza sativa*) and sugar-cane (*Saccharum officinale*).

Pasture grasses include cocksfoot (*Dactylis*), fescues (*Festuca* spp.) and ryegrasses (*Lolium* spp.). In many grasses, the growing shoot remains near or under the ground with the leaves arising mainly from ground level. Leaves can therefore be continually cropped to near the ground by grazing without damaging the plant.

graminicolous *a.* living on grasses.

graminifolious *a.* with grass-like leaves.

graminivorous *a.* grass-eating.

graminoid *a. appl.* grasses and grass-like forms.

graminology *n.* the study of grasses.

grammate *a.* (1) striped; (2) marked with lines or slender ridges.

Gram-negative *appl.* bacteria with a complex cell wall which is composed of a thin inner peptidoglycan cell wall and an outer membrane of lipopolysaccharide, lipoprotein and other complex macromolecules. They do not retain purple Gram stain on alcohol treatment.

Gram-positive *appl.* bacteria with a cell wall composed chiefly of peptidoglycan and lacking an outer membrane. They retain purple Gram stain on alcohol treatment. Gram-positive bacteria form a distinct phylogenetic lineage within the Bacteria, containing bacteria as physiologically and morphologically diverse as mycobacteria, streptomycetes, clostridia and mycoplasmas.

Gram staining staining procedure for bacteria devised by the Danish physician H.C.J. Gram, which distinguishes cell walls of different structures and is used in identification. It consists of staining with crystal violet and subsequent treatment with alcohol. Those bacteria that are decolorized by alcohol are termed Gram-negative and those that retain the stain are termed Gram-positive.

grana *n.plu.* in chloroplasts, groups of stacked disc-shaped structures (thylakoids) in whose membranes are located the photosynthetic pigments and photosynthetic electron-transport chains. *sing.* **granum**.

grandifoliate, grandifolious *a.* large-leaved, particularly when the leaves are the dominant organ, as in water-lilies.

grand postsynaptic potential the sum of all the synaptic potentials received by a neuron, and whose magnitude determines whether the neuron will generate an impulse.

Grandry's corpuscle a touch receptor in skin of beak and tongue of birds.

granellae *n.plu.* refractile granules consisting chiefly of barium sulphate in the tubes of certain protozoans (Sarcodina).

granite mosses a small group of mosses, class Andreaeidae, sometimes called rock mosses, small blackish-green or olive-brown tufted plants, in which the gametophyte arises from a plate-like protonema and spores are shed from slits in the capsule wall.

granivorous *a.* feeding on grain.

granose *a.* in appearance like a chain of grains, as some insect antennae.

granular *a. appl.* soil crumbs which are rounded and rather small.

granular, granulate, granulose *a.* appearing as if made up of, or covered with, granules.

granular cell granule cell *q.v.*

granular layer (1) of epidermis, layer of cells containing keratin granules and which develop into keratinocytes. *see* Fig. 35 (p. 622); (2) of cerebellum, layer containing granule cell bodies.

granule cell type of small interneuron found in cerebellum, with cell body in granular layer and axon projecting into molecular layer where it forms parallel fibres synapsing with dendrites of Purkinje cells.

granulocyte colony-stimulating factor (G-CSF) cytokine that stimulates formation of granulocytes from bipotential macrophage-granulocyte precursor cells.

granulocyte-macrophage colony-stimulating factor (GM-CSF) cytokine required for the growth and differentiation of myeloid (e.g. granulocyte) and monocyte (e.g. macrophage) lineages of white blood cells, and which increases in response to infection.

granulocytes *n.plu.* general term for those white blood cells that characteristically contain numerous secretory vesicles or granules and an irregularly shaped nucleus. They comprise the basophils, eosinophils and neutrophils. *alt.* polymorphonuclear leukocytes.

granuloma *n.* nodule caused by chronic inflammation around a focus of infection that cannot be cleared (e.g. mycobacterial infection) or a non-degradable foreign body.

granulopoiesis *n.* the formation of granulocytes.

granulosa cells cells in mammalian ovarian follicles that produce oestrogens and eventually form the corpus luteum.

granulysin *n.* cytotoxic protein present in the granules of cytotoxic T cells and NK cells. It also has antimicrobial activity.

granum *sing.* of grana.

granzyme *n.* any of several types of proteolytic enzyme present in inactive form in the granules of cytotoxic T cells. They are released and activated when the T cell makes contact with a target cell and are involved in inducing apoptosis (cell suicide) in the target cell.

grape sugar glucose *q.v.*

Graptolita, graptolites *n., n.plu.* group of fossil invertebrates from the Paleozoic, of doubtful affinity but thought to be allied with the hemichordates.

grasses *n.plu.* common name for members of the large monocot family Gramineae *q.v.*

grasshoppers *n.plu.* common name for many members of the Orthoptera *q.v.*

grassland *n.* biome found in regions where the average annual precipitation (*ca.* 25–76 cm) is sufficient to support the growth of grasses and other herbaceous plants but generally insufficient to support continuous tree cover. In wetter regions grassland is maintained as a result of grazing by herbivores. *see* alpine grassland, downland, giant-grass community, prairie, savanna, short-grass community, steppe, tall-grass community, veld.

graveolent *a.* having a strong or offensive smell.

Graves' disease autoimmune disease involving the thyroid, in which antibodies against the receptor for thyroid-stimulating hormone cause overstimulation of the thyroid and overproduction of thyroid hormones. *cf.* Hashimoto's disease.

gravid *a.* (1) *appl.* female with eggs; (2) *appl.* pregnant uterus.

graviperception *n.* the sensation or perception of gravity.

graviportal *a.* with the legs adapted to supporting great weights, as in elephants.

gravitational *a. appl.* water in excess of soil requirements, which sinks under action of gravity and drains away.

gravitaxis *n.* geotaxis *q.v.*

gravitropism geotropism *q.v.*

gray *see* grey.

gray (Gy) the SI unit of the dose of ionizing radiation absorbed by living tissue, being equal to 1 J of energy imparted to 1 kg of mass, and which replaces the non-SI unit, the rad. 1 Gy = 100 rad.

grazing *n.* the consumption of green plant material or algae by animals and micro-organisms such as protozoa.

GRE glucocorticoid response element *q.v.*

great apes members of the primate family Hominidae (*q.v.*): humans, chimpanzees, gorillas and orangutans.

greater omentum a fold of peritoneum attached to the colon and stomach and hanging over the small intestine.

green algae common name for the Chlorophyta *q.v.*

green bacteria diverse group of bacteria with anoxygenic photosynthesis. They have bacteriochlorophylls *c*, *d*, *e*, and small amounts of *a*, enclosed in chlorosomes, and one photosystem. They include the photoautotrophic anaerobic green sulphur bacteria, which oxidize reduced sulphur compounds, and the photoheterotrophic thermophilic green non-sulphur bacteria.

green deserts areas of little ecological interest despite their 'green' appearance, e.g. extensive areas of crop monocultures, heavily fertilized pastureland replanted with alien grass species, and large areas of closely mown grass in urban parks.

green fluorescent protein (GFP) fluorescent protein used as a marker protein in cell biology.

green glands paired excretory organs at base of antennae in some crustaceans.

greenhouse effect the trapping of a proportion of the heat radiated from Earth's surface by water vapour, carbon dioxide, methane and other compounds in the lower atmosphere, and its re-radiation back to the surface. If the levels of e.g. carbon dioxide in the atmosphere continue to increase as a result of the burning of fossil fuels, it is thought that this natural effect will result in a continuing rise in the temperature of the lower atmosphere, eventually leading to widespread alteration in climates throughout the world and melting of polar ice.

green non-sulphur bacteria *see* green bacteria.

Green Revolution the introduction of scientifically bred high-yielding varieties of staple crop plants such as wheat, rice and maize from the 1960s onwards into areas of the world where lower-yielding traditional varieties adapted to local conditions were formerly grown.

green sulphur bacteria *see* green bacteria.

gregaloid *a. appl.* protozoan colony of indefinite shape, usually with gelatinous base, formed by incomplete division of individuals or partial union of adults.

gregarines *n.plu.* group of parasitic protozoans of invertebrates, in which only the adults live outside the host cell, and in which the male and female gametes are smaller than the normal cell.

gregariniform *a.* (1) *pert.* gregarine protozoans; (2) *appl.* spores moving with the gliding motion characteristic of gregarines.

gregarious *a.* (1) tending to herd together; (2) growing in clusters; (3) colonial *q.v.*

gressorial *a.* adapted for walking, *appl.* certain insects and birds.

grex *n.* the 'slug' stage formed by aggregating cells of cellular slime moulds.

grey crescent lightly pigmented crescent-shaped band visible on side of fertilized eggs of some amphibians, which forms opposite the site of sperm entry. The first cleavage plane cuts through it, delimiting right and left sides of the body.

grey matter regions of brain and spinal cord that appear grey, consisting chiefly of cell bodies of neurons. *cf.* white matter.

grinding teeth molars *q.v.*

griseofulvin *n.* antibiotic produced by some species of *Penicillium*, esp. *P. griseofulvum*, which is toxic to some fungi by inhibiting mitotic metaphase.

grooming *n.* the cleaning of fur or feathers by an animal (generally called preening in birds), performed either by itself (self grooming) or to another member of the same species (social grooming). Grooming not only functions to keep the animal clean, but as a displacement activity or to improve the social cohesion of the group.

gross efficiency a measure of the efficiency of an animal in converting food consumed into body substance.

gross primary production (GPP) the total assimilation of inorganic nutrients in a plant community per unit time, *cf.* net production.

ground layer moss layer *q.v.*

ground meristem the primary meristem that gives rise to ground tissue. It is derived from the apical meristem in the developing plant embryo.

ground substance non-living matrix of connective tissue.

ground tissue of plants, the parenchyma, collenchyma and sclerenchyma, i.e. all tissues other than the epidermis, reproductive tissues and vascular tissues. *alt.* fundamental tissue.

groundwater *n.* (1) water that sinks down through soil and rock and collects in underground aquifers; (2) underground water in the zone of saturation, below the water table. *cf.* soil water.

group I introns, group II introns *see* intron.

group selection selection that operates on two or more members of a lineage group as a unit, such that characters that benefit the group rather than the individual may be selected for.

group translocation in some bacteria, type of membrane transport in which molecules entering the cell are chemically modified during transport.

growing point part of plant body at which cell division is localized, generally terminal, and composed of meristematic cells.

growth *n.* (1) increase in mass and size of a tissue or organism by cell division and/or cell enlargement; (2) *appl.* cell: increase in the mass or size of the cell. *cf.* proliferation.

growth cone flattened triangular leading end of a developing axon or dendrite. It is involved in movement along the substrate and in guidance of movement in response to chemical signals in the extracellular matrix.

growth curvature the curved shape imparted to a plant organ by the difference in the rates of growth of its sides.

growth curve plot of log numbers of a population of living organisms against time. For a culture of microorganisms with limiting nutrients the growth phase is typically bell-shaped. A short lag phase of little or no increase in numbers is followed by a steep rise in numbers (exponential growth or log phase) to a plateau, followed by a rapid fall in growth rate as nutrients become exhausted, waste products accumulate and the available 'habitat' becomes overcrowded.

growth factor (1) any organic compound, other than those required as carbon and energy sources, which is needed by an organism for its proper growth and development, and which may include e.g. vitamins, amino acids, purines, pyrimidines; (2) biomolecules, usually proteins, that stimulate cell growth (an increase in cell mass).

growth factor receptors cell-surface receptors for certain types of cytokines. They fall into several different structural classes and include the tyrosine protein kinase receptors for growth factors such as insulin, epidermal growth factor and platelet-derived growth factor.

growth hormone (GH) (1) in mammals, a growth-promoting protein hormone produced by the anterior pituitary. It stimulates the production of somatomedins in the liver, which in turn stimulate bone and muscle growth, *alt.* somatotropin; (2) in animals generally, any of various growth-promoting hormones; (3) in plants, any of the growth-promoting phytohormones, e.g. giberellic acid, indole acetic acid (IAA).

growth hormone-releasing hormone (GHRH) somatoliberin *q.v.*

growth layer area of radial growth in the secondary xylem or secondary phloem.

growth rate (r) increase in the size of a population of organisms or cells per unit of time. Often expressed as the generation time *q.v.*

growth regulators the general name given to both natural growth hormones and artificial compounds that control growth in plants.

growth rings (1) rings seen on the cut trunk of some trees, each representing the radial growth of wood in one year. The width of each ring reflects climatic conditions (e.g. temperature and rainfall), wider rings being produced in favourable years. *see also* dendrochronology; (2) (*zool.*) layers of shell laid down in each growth period in some molluscs; (3) distinct layers of growth on a fish scale.

growth substance generally refers to any compound, natural or artificial, that when present in small amounts has a marked effect on the growth and/or development of a plant. *alt.* plant growth substance.

grub *n.* legless larva of insects of the Diptera, Coleoptera and Hymenoptera.

Gruiformes *n.* order of birds that includes the cranes, bustards, rails, gallinules and coots, and hemipodes.

Gruinales Geraniales *q.v.*

GS glutamine synthetase *q.v.*

GSC germline stem cell *q.v.*

GSH, GSSG *see* glutathione.

GTP guanosine triphosphate *q.v.*

GTPase guanosine triphosphatase *q.v. see also* guanine-nucleotide binding proteins.

GTPase-activating protein (GAP) protein that interacts with the GTP-bound forms of guanine-nucleotide binding proteins to stimulate their intrinsic GTPase activity.

GTP-binding protein guanine-nucleotide binding protein *q.v.*

GTP-exchange factor guanine-nucleotide exchange factor *q.v.*

guaiacyl coniferyl alcohol. *see* lignin.

guanidine *n.* a base produced by oxidation of guanine, whose metabolism is regulated by the parathyroids.

guanine (G) *n.* a purine base, one of the four bases in DNA and RNA. It is the base in the nucleoside guanosine and in the guanine nucleotides (GMP, GDP, GTP) and is also found as iridescent granules or crystals in certain chromatophores. *see* Fig. 5 (p. 69).

guanine-nucleotide binding proteins (GTP-binding proteins) large family of proteins that bind the guanine nucleotides GDP and GTP and act as molecular switches in a variety of cellular processes. They include the heterotrimeric G proteins and the single-subunit 'small G proteins' (e.g. Ras, Rac) and the protein synthesis elongation factors. They can bind either GDP or GTP and in the inactive form are usually complexed with GDP. The replacement of GDP with GTP activates the protein, which remains active until the GTP is hydrolysed to GDP by the G protein's intrinsic GTPase activity. The heterotrimeric G proteins are involved in relaying signals from cell-surface receptors (G-protein-coupled receptors). Small G proteins have many functions, including participation in intracellular signalling pathways and conveying signals for changes in the cytoskeleton. *see also* elongation factor, G_i, G_q, G_s, heterotrimeric G proteins, small G proteins, transducin.

guanine-nucleotide exchange factors (GEFs) proteins that catalyse the exchange of GDP for GTP on guanine-nucleotide binding proteins.

guano *n.* (1) deposits of seabird droppings rich in nitrates and phosphates, formerly collected from islands off the west coast of South America for use as a fertilizer; (2) bat dung.

guanophore *n.* chromatophore containing guanine, either as pale granules usually giving a yellow colour, or as iridescent crystals as in the skin of some reptiles and fishes.

guanosine *n.* nucleoside made up of the purine base guanine linked to the pentose sugar ribose.

guanosine diphosphate (GDP) guanine nucleotide containing a diphosphate group.

guanosine monophosphate (GMP) nucleotide composed of guanine, ribose and a phosphate group, a product of the partial hydrolysis of RNA, and synthesized *in vivo* from xanthylate by amination. *alt.* guanylate, guanylic acid, guanosine 5′-phosphate.

guanosine pentaphosphate (pppGpp) unusual nucleotide with a 5′ triphosphate and a 3′ diphosphate, accumulated by bacterial cells in conditions of amino acid starvation.

guanosine tetraphosphate (ppGpp) unusual nucleotide with diphosphate at the 3′ and 5′ positions, accumulated by bacterial cells in conditions of amino acid starvation.

guanosine triphosphatase (GTPase) enzyme catalysing the hydrolysis of GTP to GDP with the release of free energy. GTPase activity is part of several groups of proteins, notably the guanine-nucleotide binding proteins, in which an intrinsic GTPase activity hydrolyses bound GTP to GDP, thus converting the protein from an activated to an inactive state.

guanosine triphosphate (GTP) nucleotide made of guanosine linked to three phosphate groups in series.

guanylate cyclase enzyme catalysing the formation of cyclic GMP from guanosine triphosphate (GTP). *alt.* guanylyl cyclase.

guanylic acid, guanylate guanosine monophosphate *q.v.*

guanyltransferase *see* nucleotidyltransferase.

guard cells two specialized epidermal cells that surround the central pore of a stoma and which contain chloroplasts. Changes in the turgor of these cells open and close the stomatal opening.

guard polyp nematocalyx *q.v.*

gubernaculum *n.* (1) cord stretching from epididymis to scrotal sac and supporting testis; (2) tissue between gum and dental sac of permanent teeth.

Guérin's glands racemose mucous glands of the female urethra. *alt.* paraurethral glands, Skene's glands.

guest *n.* an animal living and breeding in the nest of another, esp. an insect.

guide RNAs (1) any small RNA or RNA sequence that base-pairs specifically with another RNA or DNA, localizing it in a site in which it can be modified, e.g. the RNA in the enzyme telomerase, which pairs with single-stranded telomeric DNA and enables the enzyme to replicate this DNA; (2) *see* RNA editing.

guild *n.* (1) those species in a community that have similar requirements and foraging habits; (2) those microorganisms in a microbial community that carry out similar metabolic reactions.

gula *n.* the upper part of the throat. *a.* **gular**.

gular *n.* an unpaired anterior horny shield in shell of tortoises and turtles.

gullet *n.* (1) oesophagus *q.v.*; (2) in protozoans, the cytopharynx *q.v.*; (3) any cavity by which food may be taken into the body.

gummiferous *a.* gum-producing or exuding.

gummosis *n.* condition of plant tissue when cell walls become gummy, caused by certain bacteria.

gums *n.plu.* (*bot.*) (1) various materials, composed largely of polysaccharides, resulting from breakdown of plant cell walls and exuding from wounds; (2) trees of the genus *Eucalyptus*; (3) (*zool.*) fibrous tissue covering the jaws around the base of the teeth.

gustatory *a. pert.* sense of taste.

gustducin *n.* heterotrimeric G protein associated with some taste receptors.

gut *n.* the intestines.

gut-associated lymphoid tissue (GALT) secondary lymphoid tissue associated with the gastrointestinal tract and including the tonsils in the throat, the appendix, and Peyer's patches in the small intestine. It is a site of production of IgA antibodies.

gutta *n.* (1) small spot of colour on e.g. insect wing; (2) oil drop, as in fungal hypha; (3) latex of various Malaysian trees, including gutta percha and balata.

guttation *n.* (1) formation of drops of water on leaves, forced out of leaves through special pores (hydathodes) by root pressure. It is seen, e.g., on tips of leaves of many grasses in early morning; (2) formation of drops of water on surface of plant from moisture in air; (3) exudation of aqueous solutions, e.g. by sporangiophores and nectaries.

guttiferous *a.* exuding a resin or gum.

guttiform *a.* in the form of a drop.

guttula *n.* a small drop-like spot.

Gy abbreviation for the gray (*q.v.*), the SI unit of ionizing radiation absorbed by tissue.

gymnetrous *a.* without an anal fin.

gymno- prefix derived from Gk *gymnos*, naked.

gymnoarian *a. appl.* gonads when naked and not enclosed in coelomic sacs.

gymnoblastic *a.* without hydrothecae and gonothecae, as certain coelenterates.

gymnocarpic, gymnocarpous *a.* (1) having the fruit naked, not covered by some kind of masking structure; (2) having hymenium exposed during maturation of spores.

gymnogynous *a.* with exposed ovary.

Gymnolaemata *n.* class of marine Ectoprocta (Bryozoa) in which individual zooids have a circular crown of tentacles (lophophore) and are enclosed in a box-like calcareous exoskeleton. They encrust seaweeds, stones and shells.

Gymnophiona Apoda *q.v.*

gymnopterous *a.* having bare wings, without scales.

gymnosomatous *a.* having no shell or mantle, such as certain molluscs.

gymnosperms *n.plu.* seed-bearing woody plants in which the seeds are not enclosed in an ovary but are borne on the surfaces of the sporophylls. They comprise the cycads (Cycadophyta), *Ginkgo* (Ginkgophyta), the conifers (Coniferophyta) and the gnetophytes (Gnetophyta). Extinct gymnosperms include the seed ferns, cycadeoids (Bennettitales) and cordaites (Cordaitales). *alt.* (formerly) Gymnospermae.

gymnospore *n.* a naked spore not enclosed in a protective coat.

gymnostomatous, gymnostomous *a.* having no peristome, *appl.* mosses.

gymnotoids *n.plu.* small group of slender freshwater bony fish from Central and South America (the Gymnotodei) with very long anal fins, and an electric organ that sets up an electric field around the fish, commonly called American knifefishes, and including the electric eel *Electrophorus*.

-gyn- word element derived from Gk *gyne*, woman, indicating female.

gynaecaner *n.* a male ant resembling a female.

gynaecium gynoecium *q.v.*

gynaecoid *n.* an egg-laying worker ant.

gynaecophore *n.* a canal or groove of certain trematodes, formed by inrolling of the sides, in which the female is carried.

gynander gynandromorph *q.v.*

gynandrism hermaphroditism *q.v.*

gynandromorph *n.* individual exhibiting a spatial mosaic of male and female characters, such as one side female the other male. *alt.* gynander, sex mosaic.

gynandromorphism *n.* condition of being a gynandromorph *q.v.*

gynandrosporous *a.* with androspores adjoining the oogonium, as in some algae.

gynandrous *a.* having stamens fused with pistils, as in some orchids.

gynantherous *a.* having stamens converted into pistils.

gyne *n.* female ant, esp. a queen.

gynecium gynoecium *q.v.*

gynecophoric canal external grooved surface of male schistosomes in which female resides during mating.

gynic *a.* female.

gynobasic *a. appl.* a style arising from the base of the ovary.

gynodioecious *a.* having female and hermaphrodite flowers on different plants.

gynoecious *a.* having female flowers only.

gynoecium *n.* the female reproductive organ of a flower, consisting of one or more carpels, each forming an ovary, together with their stigmas and styles. *alt.* pistil, gynaecium.

gynogenesis *n.* development from eggs penetrated by sperm but not fertilized.

gynomerogony *n.* the development of an egg fragment obtained before fusion with male nucleus, and containing maternal chromosomes only.

gynomonoecious *a.* having female and hermaphrodite flowers on the same plant.

gynomorphic *a.* having a morphological resemblance to females.

gynosporangium *n.* (1) female sporangium; (2) megasporangium *q.v.*

gynostemium *n.* the column composed of united pistil and stamens in orchids.

gypsophil(ous) *a.* thriving in soils containing chalk or gypsum.

gyral, gyrate *a.* (1) *pert.* a gyrus; (2) *pert.* spiral or circular movement.

gyrase DNA gyrase *q.v.*

gyration *n.* (1) rotation, as of a swimming bacterial cell; (2) whorl of a spiral shell.

gyre *n.* circular movement.

gyrencephalic, gyrencephalous *a.* having a convoluted surface to the cerebral hemispheres.

gyri *plu.* of gyrus.

gyrodactyloid *n.* ciliated larva of certain ectoparasitic flukes which swims to find new host.

gyromitrin *n.* toxin present in fungi of the genus *Gyromitra*, the false morels, and which is lethal to some people.

gyrose *a.* (1) with undulating lines; (2) sinuous; (3) curving.

gyrus *n.* (1) a convolution of the surface of the brain; (2) a ridge winding between two grooves. *plu.* **gyri**.

H

H (1) symbol for the chemical element hydrogen *q.v.*; (2) histidine *q.v.*

H⁺ symbol for the proton (hydrogen ion).

H-2 the major histocompatibility complex (*q.v.*) in the mouse.

HA (1) hyaluronic acid *q.v.*; (2) haemagglutinin *q.v.*

habenula *n.* band-like structure. *a.* **habenular.**

habilines *n.plu.* fossil hominids from East Africa, dated at *ca.* 2 million years ago, assigned by some palaeoanthropologists to the species *Homo habilis.*

habit *n.* (1) the external appearance or way of growth of a plant, e.g. climbing, bushy, erect, shrubby; (2) the normal or regular behaviour of an animal.

habitat *n.* the environment within which an organism is normally found. A habitat is characterized by the physical characteristics of the environment and/or the dominant vegetation or other stable biotic characteristics. Examples of habitats can be as general as lakes, woodland or soil, or more specific, such as mudflats, the bark of an oak tree, chalk downland. *cf.* niche. *see also* law of minimum, law of tolerance.

habitat breadth the range of different types of habitat an organism can inhabit.

habitat diversification beta diversity *q.v.*

habitat form the way a plant grows resulting from conditions in that particular habitat.

habitat 'island' isolated area or fragment of a particular habitat completely surrounded by a different habitat.

habitat space the habitable part of a space or area available for establishing a population.

habitat type group of plant communities that produce similar habitats.

habit formation trial-and-error learning *q.v.*

habit fragmentation the splitting up to a hitherto continuous area habit for an organism into smaller isolated areas surrounded by habitats they cannot live in or which impede their free movement.

habituation *n.* (1) adjustment, by a cell or an organism, by which subsequent contacts with the same stimulus produce diminishing effects; (2) form of learning in which reflex behaviour is extinguished when the animal finds it has no adaptive value.

habitus *n.* the external appearance of an individual.

hackle *n.* long erectile feather on the necks of some birds.

hadal *a. appl.* or *pert.* deep sea below 6000 m.

haem *n.* a porphyrin (protoporphyrin IX) with an iron atom in its centre. The iron is in either the ferrous or the ferric state. Haem is the oxygen-carrying prosthetic group in haemoglobin, with a ferrous iron atom binding one molecule of O_2, and acts as an electron-carrying group in other haem proteins. *alt.* **heme.**

haemacyte *n.* a blood cell in insects and other invertebrates. *alt.* haemocyte.

haemadsorption *n.* the binding of red blood cells by some virus-infected cells, used in diagnosis and infectivity assays.

haemagglutination *n.* clumping together of red blood cells, either spontaneously or in response to treatment with antibody or other agent. The reaction is used as a diagnostic assay for viruses that contain haemagglutinating proteins.

haemagglutination inhibition (HI) an immunological assay for certain viruses, in which antibody binding to the virus prevents it agglutinating red blood cells.

haemagglutinin *n.* any substance that causes the agglutination of red blood cells.

haemal *a.* (1) *pert.* blood or blood vessels; (2) situated on the same side of the vertebral column as the heart.

haemal arches lateral extensions from tail vertebrae which fuse ventrally to form the haemal canal, enclosing an artery and a vein.

haemal ridge haemapophysis of vertebra.

haemal strands strands of spongy tissue in echinoderms which are part of the blood vascular system and may be concerned with the phagocytosis and destruction of invading microorganisms.

haemangioblast blood island *q.v.*

haemapophysis *n.* one of the plate-like or spine-like extensions projecting out to each side from the lower edge of the centrum of a vertebra.

haematin *n.* hydroxide of haemin (*q.v.*), a decomposition product of haemoglobin.

haemato- *see also* haem, haema-, haemo-.

haematobium *n.* an organism that lives in blood. *a.* **haematobic.**

haematoblast, haemoblast *n.* precursor cell to an erythroblast.

haematochrome *n.* carotenoid pigment in some red algae.

haematogenous *a.* (1) formed in blood; (2) derived from blood.

haematology *n.* the study of the blood and its formation.

haematolymphoid *a. pert.* the lymphatic system and the blood.

haematolysis haemolysis *q.v.*

haematophagous *a.* feeding on or obtaining nourishment from blood.

haematopoiesis *n.* formation of blood, development of blood cells from stem cells.

haematopoietic *a.* (1) *pert.* haematopoiesis or blood formation; (2) *appl.* cytokines (growth factors) involved in haematopoiesis; (3) *appl.* cells derived from bone marrow (i.e. blood cells).

haematopoietic stem cell (HSC) type of stem cell present in bone marrow that produces all the cellular elements of the blood.

haematosis haematopoiesis *q.v.*

haematoxylin *n.* pink dye that preferentially stains DNA and RNA in a cell.

haematozoon *n.* any animal parasitic in blood.

haemerythrin haemoerythrin *q.v.*

haemic haemal *q.v.*

haemin *n.* haem with the iron atom in the oxidized (ferric) state, usually found as the chloride.

haemobilirubin *n.* breakdown product of haemoglobin which is converted to bilirubin and biliverdin in the liver.

haemoblast haematoblast *q.v.*

haemochorial *a. appl.* placenta with branched chorionic villi penetrating blood sinuses after breaking down uterine tissues, as found in insectivores, rodents and primates.

haemoclastic *a.* breaking down blood cells.

haemocoel *n.* blood-filled cavity consisting of spaces between organs, which is the main body cavity in molluscs.

haemoconia *n.plu.* minute particles of red blood cells taken up by phagocytes of the reticuloendothelial system.

haemocyanin *n.* copper-containing protein, blue in the oxidized form, that acts as a respiratory pigment in the blood of many annelids and arthropods.

haemocyte *n.* a blood cell in insects and other invertebrates.

haemocytometer *n.* chamber of known volume used for counting numbers of cells in a sample of blood under the microscope.

haemodynamics *n.* study of the principles of blood flow.

haemoerythrin *n.* red, iron-containing, non-haem protein that acts as a respiratory pigment in the blood of sipunculids, various molluscs and crustaceans. *alt.* hemerythrin.

haemogenesis haematopoiesis *q.v.*

haemoglobin (Hb) *n.* oxygen-carrying haem protein consisting of two pairs of different globin subunits ($\alpha\alpha\beta\beta$ in adult human haemoglobin), each subunit associated with one haem prosthetic group. It is the oxygen carrier in vertebrate blood, oxygen binding to the iron atom in the haem group. It is contained in the red blood cells and gives blood its red colour. *see also* carboxyhaemoglobin, foetal haemoglobin, haemoglobinopathies, leghaemoglobin, methaemoglobin, sickle-cell haemoglobin, thalassaemia. *alt.* **hemoglobin.**

haemoglobinopathies *n.plu.* inherited diseases, such as sickle-cell anaemia, in which a genetic defect affects the structure, function or production of haemoglobin.

haemoglobin S mutant haemoglobin that causes sickle-cell disease.

haemoid *a.* resembling blood.

haemolymph *n.* fluid in the coelom of some invertebrates, regarded as equivalent to the blood and lymph of vertebrates.

haemolysin *n.* any agent capable of lysing red blood cells.

haemolysis *n.* the lysis of red blood cells.

haemolytic *a. pert.*, caused by, or causing haemolysis.

haemolytic anaemia anaemia caused by destruction of blood cells.

haemolytic disease of the newborn erythroblastosis foetalis *q.v.*

haemoparasite *n.* blood parasite, a parasitic organism that lives in the blood of its host.

haemopathic *a.* affecting the blood circulatory system.

haemophilia *n.* genetically determined condition characterized by excessive bleeding due to inability of blood to clot normally. Haemophilia A is caused by a deficiency of the blood-clotting protein Factor VIII. The rarer haemophilia B or Christmas disease is caused by a deficiency of Factor IX.

haemopoiesis haematopoiesis *q.v.*

haemopoietins *n.plu.* growth factors involved in the proliferation and differentiation of blood cells.

haemoprotein *n.* a protein that contains haem.

haemorrhagic fever disease characterized by fever accompanied by a rash and/or internal bleeding.

haemorrhoidal *a.* rectal, *appl.* blood vessels, nerve.

haemosiderin *n.* a complex of protein and ferric hydroxide, a yellow pigment of blood and found in most tissues, stored as large granules esp. in liver, spleen and bone marrow.

haemostasis *n.* arrest of bleeding, by the natural process of blood clotting or by medical intervention.

haemostatic *a. appl.* an agent that stops bleeding.

haemotoxin *n.* toxin that lyses red blood cells.

haemotropic *a.* affecting or acting upon blood.

haerangium *n.* in some ascomycete fungi, an adhesive droplet containing spores.

Hageman factor Factor XII *q.v.*

hagfishes *n.plu.* common name for the Myxiniformes, a small order of bottom-living jawless marine fish, eel-like in shape, lacking pectoral or pelvic fins and with no scales.

hair *n.* (*bot.*) (1) trichome *q.v.*; (2) root hair *q.v.*; (3) (*zool.*) in mammals, a thread-like epidermal structure consisting of cornified epithelial cells which grows by cell division from a hair follicle at its base.

hair cell (1) sensory cell in the organ of Corti in the ear, involved in hearing; (2) sensory cell in the vestibular apparatus of the ear, involved in sense of balance; (3) similar cell in the acoustico-lateralis system of fishes and amphibia.

hair follicle *see* follicle.

hairpin *n.* region of double-stranded structure in RNA and DNA formed by base-pairing between complementary sequences on the same strand, the unpaired bases between the sequences forming a single-stranded loop (hairpin loop) at the end of the hairpin.

hairpin turn β-turn *q.v.*

hairy root disease abnormal proliferation of root hairs caused by the bacterium *Agrobacterium rhizogenes*.

halarch succession halosere *q.v.*

Haldane's rule a rule stating that when offspring of one sex produced from a cross are inviable or infertile, it is always the heterogametic sex.

half-inferior *a.* having ovary only partly adherent to calyx.

half-life *n.* (1) time required for the disappearance of half the original quantity of a given substance, e.g. from the circulation, assuming that the substance disappears at a regular rate; (2) of radioactive elements, time required for the radioactive decay of half the original amount of material.

half-sibs siblings having only one parent in common.

half-spindle one half of a mitotic or meiotic spindle, comprising the microtubules arising from one pole as far as the equator.

half-terete *a.* rounded on one side, flat on the other.

hallucinogen *n.* any drug capable of altering sensory perception and evoking hallucinations, e.g. LSD, cannabis, psilocybins.

hallux *n.* first digit of hindlimb in vertebrates, the big toe.

halobacteria *n.plu.* group of Archaea that require high salt concentrations and live in habitats such as salt lakes.

halobenthos *n.* marine benthos.

halobios *n.* (1) sum total of organisms living in sea; (2) animals living in sea or any salt water. *a.* **halobiotic**.

halodrymium *n.* a mangrove association.

halolimnic *a. appl.* marine organisms adapted to live in fresh water.

halomorphic *a. appl.* soils containing an excess of salt or an alkali.

halophile *n.* organism that requires salt for growth.

halophilic *a.* (1) salt-loving; (2) thriving in the presence of salt.

halophilic bacteria halobacteria *q.v.*

halophobe *n.* plant intolerant of salt.

halophyte *n.* (1) sea-shore plant; (2) plant that can thrive on soils impregnated with salt. *a.* **halophytic**.

haloplankton *n.* the organisms drifting in the sea.

Haloragales Hippuridales *q.v.*

halorhodopsin *n.* light-driven transport protein in cell membrane of halobacteria that transports Cl⁻ into the cell.

halosere *n.* plant succession originating in a saline area, as in salt marshes.

halotolerant *a.* capable of growing in the presence of salt but not requiring it.

haloxene *a.* tolerating salt water.

halteres *n.plu.* pair of small rounded wing-like structures borne on metathorax of flies and representing rudimentary posterior wings.

Hamamelidales *n.* order of woody dicots that includes the families Hamamelidaceae (witch hazel), Myrothamnaceae and Platanaceae (plane).

hamate *a.* hooked, or hook-shaped at the tip.

hamathecium *n.* all the sterile cells and hyphae that are interspersed among asci in an ascocarp.

hamatum *n.* the hooked bone in the carpus (wrist).

hamiform *a.* hook-shaped.

hamirostrate *a.* having a hooked beak.

hammer malleus *q.v.*

hamose hamate *q.v.*

hamstrings *n.plu.* tendons of insertion of the posterior femoral muscles of fore- and hindlimbs.

Ha-MSV Harvey mouse sarcoma virus.

hamular *a.* hooked or hook-like.

hamulate *a.* having small hook-like processes.

hamulose hamate *q.v.*

hamulus *n.* hooklet or hook-like process on bone, feathers and other structures, such as one of the minute hooks on front edge of hindwing of Hymenoptera which link fore- and hindwings together. *plu.* **hamuli**.

handling time the time it takes a predator to catch and eat its prey.

hanger *n.* a wood situated on a hillside.

H antigen (1) flagellar antigen of salmonella; (2) histocompatibility antigen.

hapanthous *a.* reproducing only once, towards the end of a plant's life.

hapaxanthic, hapaxanthous *a.* with only a single flowering period.

haplobiont *n.* organism with only one type of individual in its life cycle. *cf.* diplobiont.

haplocaulescent *a. appl.* plants with a simple axis, i.e. capable of producing seed on the main axis.

haplocheilic *a. appl.* type of stomata in gymnosperms in which the subsidiary cells are not related developmentally to the guard cells.

haplochlamydeous *a.* having only one whorl of perianth segments.

haplodioecious heterothallic *q.v.*

haplodiploid *a. appl.* species in which sex is determined by the male being haploid, the female diploid, as in bees and wasps. *n.* **haplodiploidy**.

haplodiplontic *a. appl.* life-cycles with multicellular haploid and diploid phases.

haplogroup *n.* a group of similar haplotypes, in particular of mitochondrial or Y-chromosomal DNA, that derive from a common ancestral sequence characterized by a particular single nucleotide polymorphism.

haploid (1) *a. appl.* cells having one set of chromosomes representing the basic genetic complement of the species, usually designated *n*; (2) *n.* a haploid organism or cell.

haploid apogamy/apogamety generative apogamy *q.v.*

haploidization *n.* an event occurring in the parasexual cycle in certain fungi during which a diploid cell becomes haploid by loss of one chromosome after another by non-disjunction.

haploidy *n.* the state of being haploid.

haploinsufficient *a. appl.* gene loci in diploid organisms where the presence of one copy of the normal functional allele is not sufficient to produce a normal phenotype. *n.* **haploinsufficiency**.

haplomonoecious homothallic *q.v.*

haplomycelium *n.* a haploid mycelium.

haploneme *a.* having threads of uniform diameter.

haplont *n.* any organism having haploid somatic nuclei or cells.

haploperistomic, haploperistomatous *a.* (1) having a single peristome; (2) having a peristome with a single row of teeth, *appl.* mosses.

haplopetalous *a.* with a single row of petals.

haplophase *n.* stage in life history of an organism when nuclei are haploid.

haplophyte *n.* a haploid plant or gametophyte.

haploptile *n.* a down feather without a shaft, formed by precocious development of the barbs of the adult feather.

haplostele *n.* a simple stele having a cylindrical core of xylem surrounded by phloem.

haplostemonous *a.* having one whorl of stamens.

haplosufficient *a. appl.* genetic loci in diploid organisms where the presence of one copy of the normal functional allele is sufficient to produce a normal phenotype. *n.* **haplosufficiency**.

haplotype *n.* the set of alleles, or DNA sequence, or set of single-nucleotide polymorphisms borne on one of a pair of homologous chromosomes, or some portion of this, e.g. for a region such as the major histocompatibility complex (MHC).

haploxylic *a.* possessing only one vascular bundle.

hapten *n.* small molecule capable of eliciting an immune response only when attached to a larger macromolecule. *a.* **haptenic**.

hapteron *n.* (1) holdfast *q.v.*; (2) functionally similar adhesive disc-like structure in other organisms, e.g. fungi. *plu.* **haptera**.

haptic *a. pert.* touch, *appl.* stimuli and reactions.

haptoglobin *n.* serum protein that combines with haemoglobin, having the function of ridding serum of free haemoglobin.

haptomonad *n.* an attached form of certain parasitic flagellates.

haptonasty *n.* plant movement elicited by touching, as the drooping of the leaves of mimosa.

haptonema *n.* distinctive, very long flagellum in certain algae, having no locomotory function and consisting of three concentric membranes surrounding a central space.

haptophore *n.* part of a protein molecule that carries the site at which it binds to other molecules or cells.

Haptophyta, haptomonads *n., n.plu.* group of small motile golden unicellular protists, mainly marine, bearing a coiled thread-like haptoneme between the two flagella, which acts as a holdfast, and a cell surface covered with thin scales. Formerly considered as algae, they are now generally classified in the kingdoms Stramenopila or Chromista. Many species have a resting stage, when the surface becomes covered with elaborate calcareous scales (coccoliths), when the organisms are known as coccolithophoroids. *alt.* Prymnesiophyta.

haptor *n.* (1) attachment organ in skin flukes (flatworms); (2) mass of adhesive hyphae at base of cord that attaches a peridiole to the cup in bird's nest fungi.

haptospore *n.* an adhesive spore.

haptotropism *n.* in plants, response by curvature to a contact stimulus, as in tendrils or stems that twine around a support. *alt.* thigmotropism.

Harderian gland accessory lacrimal gland of 3rd eyelid or nictitating membrane.

hard pan hard layer developed in the B-horizon of the soil, consisting of deposited salts. It restricts drainage and root growth.

hard-wired *appl.* neuronal circuits whose connections and properties are fixed during development, as opposed to those whose properties can be modified by experience.

hardwoods *n.plu.* broadleaved trees, although the wood from some broadleaved trees is not as hard as some softwoods (wood from coniferous trees).

Hardy–Weinberg Law in a large randomly mating population, in the absence of migration, mutation and selection, allele frequencies stay the same from generation to generation according to the following rule (the Hardy–Weinberg rule). This states that if the frequency of one of the alleles at a locus is p and that of the other is q, then the frequencies of the two homozygotes and the heterozygote are given by p^2, q^2 and $2pq$ respectively.

harem *n.* group of breeding females which a dominant male mates with and guards from rival males.

harlequin chromosome duplicated chromosome composed of sister chromatids that stain differently.

harmonic suture an articulation formed by apposition of edges or surfaces as between palatine bones.

harp *n.* region on wings of e.g. grasshoppers and crickets whose vibration helps produce and amplify the characteristic sounds produced by these insects.

harpagones *n.plu.* claspers of certain male insects.

harpes *n.plu.* (1) claspers of male lepidoptera; (2) chitinous processes between claspers of mosquito. *sing.* **harpe.**

Hartig net network of fungal hyphae within the epidermis and outer cortex of roots of plants with mycorrhizas. The hyphae do not penetrate into the endodermis and only rarely enter the root cells.

harvestman *n.* common name for a member of the Opiliones, an order of arachnids having very long slender legs and with the prosoma and opisthosoma forming a single structure.

Hashimoto's disease autoimmune disease in which thyroid tissue appears to become converted to lymphoid tissue that produces antibodies against thyroid proteins, thus leading to the underproduction of thyroid hormones. *cf.* Graves' disease.

Hassall's corpuscles structures in medulla of thymus containing epithelial cells, macrophages and cell debris.

hastate *a.* spear-shaped, more or less triangular with the two basal lobes divergent, *appl.* leaves.

HAT histone acetyltransferase *q.v.*

Hatch–Slack pathway C4 pathway *q.v.*

H⁺-ATPase P-type ATPase present in the membranes of lysosomes and similar vesicles in animal cells and in plasma membrane of plants, fungi and bacteria. It pumps protons (H^+) out of the cytoplasm against their concentration gradient, using the energy of ATP hydrolysis. The H⁺-ATPases in lysosomal membranes are responsible for acidification of the vesicle contents.

HAT selection technique for identifying cell hybrids in somatic cell genetics, based on selection of HGPRT⁺, TK⁺ cells in a HAT medium containing hypoxanthine, thymidine and aminopterin, when the major pathway of DNA synthesis is blocked by aminopterin.

haulm *n.* stem of peas, and of grasses.

haustellate *a.* having a haustellum, a proboscis adapted for sucking liquids.

haustorium *n.* an outgrowth of stem, root or hyphae of certain parasitic plants or fungi, through which they obtain food from the host plant. *plu.* **haustoria.** *a.* **haustorial.**

Haversian canals small canals in bone in which run blood capillaries, nerves and lymph.

Haversian system the unit of concentric rings of bone (Haversian lamellae) surrounding a central canal (Haversian canal) with the bone cells and canaliculi, which forms the basic structural unit of compact bone. *alt.* osteon.

Hawaiian region that part of the Palaeotropical floristic kingdom comprising the Hawaiian islands.

hayfever allergic rhinitis *q.v.*

Hb haemoglobin *q.v.*

HbA adult human haemoglobin.

H band a singly refracting light band in the centre of the sarcomere of striated muscle under the microscope, representing thick filaments only. *alt.* H disc, H line.

HbCO carboxyhaemoglobin *q.v.*

HbF foetal haemoglobin *q.v.*

HbH disease type of α-thalassaemia in which only one α-globin gene is functional in the diploid cell, which causes the formation of an abnormal β-globin tetramer from the excess β-globin chains.

HbO₂ oxyhaemoglobin *q.v.*

hCG human chorionic gonadotropin *q.v.*

H chain (1) heavy chain of immunoglobulin; (2) the heavy DNA strand in mitochondrial DNA.

HDAC histone deacetylase *q.v.*

H disc H band *q.v.*

HDL high-density lipoprotein *q.v.*

HD protein helix-destabilizing protein. *see* single-strand DNA-binding protein.

head-cap acrosome of sperm.

head-case the hard outer covering of insect head.

head fold in gastrulating avian and mammalian embryos, a fold that separates the future head region from the surface of the blastoderm.

head-group *n.* small hydrophilic chemical group, e.g. choline or ethanolamine, present in membrane phospholipids and which gives them their amphiphilic property.

head process in early gastrulating avian and mammalian embryos, the notochord as it starts to form anterior to the node.

heart *n.* hollow muscular organ which by rhythmic contractions pumps blood around the body. The mammalian heart consists of four chambers, an upper thin-walled atrium and lower thick-walled ventricle on either side. Venous blood from the body enters the right atrium and leaves for the lungs from the right ventricle. Oxygenated blood re-enters the heart through the left atrium and leaves the heart from the left ventricle. The muscular contraction of the heart is self-sustaining and is synchronized by pacemaker cells located in the sinuatrial node.

heart muscle, heart muscle cell cardiac muscle *q.v.*

heart rot fungal decay of the wood in the centre of a tree.

heart-stage embryo in dicotyledons, a heart-shaped embryonic stage preceding cotyledon development.

heartwood *n.* the darker, harder, central wood of trees, containing no living cells.

heath, heathland *n.* ecosystem developing on poor, usually acid, sandy or gravelly soils in the lowlands and dominated by gorse (*Ulex*), heathers (*Calluna* and *Erica*) and other narrow-leaved plants.

heat shock in microbiology and cell biology, a short period of exposure of

a cell to temperatures above its normal range, which results in synthesis of proteins that protect the cell against the effects of the heat stress.

heat-shock protein a protein whose synthesis is induced by heat shock. Some heat-shock proteins are molecular chaperones.

heavy chain (H chain) (1) one of the two larger identical polypeptide chains in an immunoglobulin molecule, each containing a variable and a constant region. The variable region contributes to an antigen-binding site, the constant region determines antibody class and effector function; (2) *see* myosin.

heavy-chain switching *see* isotype switching.

Hebbian synapse a synapse whose transmitting properties can be modulated by the input it receives.

Hebb rule rule stating that external stimuli that cause two neurons to be active at the same time or in quick succession favour the making or strengthening of synapses between them.

hebetate *a.* blunt-ended.

hebetic *a. pert.* adolescence.

hectocotylus *n.* arm of male cephalopod specialized for transfer of spermatophore to female.

hederiform *a.* shaped like an ivy-leaf, *appl.* nerve endings such as pain receptors in skin.

Hedgehog signalling pathway important and ubiquitous intracellular signalling pathway in animals which is stimulated by members of the Hedgehog family of secreted signalling proteins (e.g. Sonic hedgehog in humans, Hedgehog in *Drosophila*). It results in changes in gene expression in the responding cells and is of particular importance during embryonic development, where Hedgehog signalling can determine cell fate.

hedonic *a. appl.* skin glands of certain reptiles, which secrete a musk-like substance and are specially active at mating season.

Heidelberg man type of primitive human known from fossils found near Heidelberg in Germany, which is now considered as a subspecies of *Homo erectus*.

hekistotherm *n.* a plant that grows well in generally cold conditions (e.g. arctic and alpine plants).

HeLa cells an aneuploid line of human epithelial cells originating from a cervical carcinoma and which have been propagated in tissue culture since 1952.

heleoplankton *n.* plankton of marshy ponds and lakes.

helical *a.* (1) *appl.* long rod-shaped virions in which the protein subunits of the coat and associated nucleic acid are arranged in helical spirals, as in tobacco mosaic virus; (2) *appl.* arrangement of myofibrils in some smooth muscle; (3) *appl.* a type of cell wall thickening in xylem.

helicase *n.* enzyme that can unwind the DNA double helix, e.g. Rep protein. *alt.* unwindase. *see also* RNA helicase.

helices *plu. of* **helix**.

helicine *n.* (1) spiral, convoluted; (2) *pert.* outer rim of external ear.

helicoid *a.* (1) spiral; (2) shaped like a snail's shell.

helicoid cyme inflorescence produced by helicoid dichotomy, so that the blooms are on only one side of the axis.

helicoid dichotomy a type of branching in which there is repeated forking but with the branches on one side uniformly more vigorous than on the other.

helicone *n.* in gastropods, a shell coiled in a helical spiral.

helicorubin *n.* red pigment in gut of pulmonates (e.g. snails and slugs) and in liver of certain crustaceans.

helicospore *n.* a convoluted or spirally twisted spore.

heliobacteria *n.plu.* group of Gram-positive green anoxygenic photosynthetic bacteria unrelated to the other green non-sulphur and green sulphur bacteria. They contain bacteriochlorophyll *g* and have no chlorosomes.

heliophil, heliophilic, heliophilous *a.* adapted to a relatively high intensity of light.

heliophobe *n.* plant that thrives in shade.

heliophyll *n.* plant with leaves of similar structure on both sides and arranged vertically.

heliophyte *n.* plant requiring full sunlight to thrive.

heliosis *n.* production of discoloured spots on leaves through sunlight.

heliotaxis *n.* movement in response to the stimulus of sunlight.

heliothermism *n.* method of regulating body heat adopted by some ectothermic animals, which adjust their body orientation throughout the day to change the amount of solar radiation received.

heliotropism *n.* plant growth movement in response to the stimulus of sunlight.

helioxerophil *n.* plant that thrives in full sunlight and in arid conditions.

heliozoan *n.* member of the Heliozoa, an order of mostly freshwater protozoans of the Sarcodina, having a radially symmetrical body, stiff slender pseudopodia and often a skeleton of spicules.

helitron *n.* type of DNA transposon found in eukaryotes that is thought to transpose via a rolling-circle type of DNA replication rather than the 'cut-and-paste' mechanism usual for eukaryotic DNA transposons.

helix *n.* (1) the outer rim of external ear in humans; (2) double helix. *see* DNA; (3) α-helix *q.v. a.* helical.

α-helix regular periodic secondary structure common in proteins. The polypeptide backbone is twisted in a right-handed spiral to form a rigid rod-like structure held together by regular hydrogen bonding.

helix-destabilizing protein single-strand DNA-binding protein *q.v.*

helix-loop-helix (HLH) DNA-binding structural motif present in some gene-regulatory proteins, in which two α-helices are joined by an unstructured loop. Proteins of this type bind DNA as dimers.

helix-turn-helix (HTH) DNA-binding structural motif present in some gene-regulatory proteins, in which two α-helices are joined by a β-turn. Proteins of this type bind to DNA as dimers.

helmet *n.* (1) thick structure on top of bill of hornbills; (2) (*bot.*) helmet-shaped petal or galea of certain flowers.

helminth *n.* parasitic flatworm (fluke and tapeworm) or roundworm.

helminthoid *a.* shaped like a worm.

helminthology *n.* (1) study of worms; (2) study of parasitic flatworms and roundworms.

helobious *a.* living in marshes.

helophyte *n.* marsh plant, esp. a perennial herbaceous plant of marshes with the perennating parts lying in the mud.

helotism *n.* symbiosis in which one organism enslaves another and forces it to labour on its behalf, e.g. in some species of ants.

helper phage *see* helper virus.

helper T cells, helper T lymphocytes effector T lymphocytes distinguished by the cell-surface protein CD4 and which interact in an antigen-specific fashion with other cells to stimulate those cells' effector functions. There are several functional types, which are distinguished by the spectrum of cytokines they produce. Th1 cells act on macrophages persistently infected with microorganisms, enabling the macrophage to clear the infection, and also interact with activated B cells to cause isotype switching. Th2 cells act on B cells, stimulating their differentiation and proliferation into antibody-secreting plasma cells and are particularly involved in causing isotype switching to the IgE class of antibodies. Th17 cells help recruit neutrophils to sites of infection early in the adaptive immune response.

helper virus a virus needed for the multiplication of related viruses which have lost the capacity to replicate. Helper viruses assist the replication of defective viruses by providing the necessary viral components.

hem-, hema-, hemo- *see also* haem-, haema-, haemo-.

hemelytron *n.* forewing of heteropteran bug, with a distal membranous section.

hemeranthic, hemeranthous *a.* flowering by day.

hemi- prefix derived from Gk *hemi*, half.

Hemiascomycetae *n.* class of unicellular and simple mycelial ascomycete fungi including the budding yeasts and leaf-curl fungi, which bear naked asci not enclosed in an ascocarp. *alt.* **Hemiascomycetidae**.

hemi-autophyte *n.* parasitic plant that produces its own chlorophyll.

hemibasidium *n.* a septate basidium, as found in some basidiomycete fungi. *alt.* heterobasidium.

hemibiotroph *n.* parasitic fungus that initially requires living host cells, but also lives off them after they are dead. *cf.* biotroph.

hemibranch *n.* gill with gill filaments on only one side.

hemicellulase *n.* any of a group of enzymes that hydrolyse hemicelluloses.

hemicellulose *n.* any of a diverse group of polysaccharides that contain a mixture of sugars such as xylose, arabinose, mannose, glucose, glucuronic acid and galactose. They are found in plant cell walls and as storage carbohydrates in some seeds and include xylans, glucomannans, arabinoxylans, xyloglucans.

hemicephalous *a. appl.* insect larvae with a reduced head.

Hemichordata, hemichordates *n., n.plu.* phylum of marine, worm-like, coelomate invertebrate animals, which have pharyngeal gill slits and a body divided into three regions.

hemichordate *a.* possessing a rudimentary notochord.

hemicryptophyte *n.* herbaceous perennial plant in which the perennating parts are at soil level, often protected by the dead leaves.

hemicyclic *a.* (1) with some floral whorls cyclic, some spiral; (2) (*mycol.*) lacking summer stages, in life cycle of rust fungi.

hemidesmosome *see* desmosome.

hemielytron hemelytron *q.v.*

hemiepiphyte *n.* plant that does not spend its whole life cycle as a complete epiphyte. It may either be a plant whose seeds germinate on another plant, but which later sends roots to the ground, or a plant that begins life rooted but later becomes an epiphyte.

hemigamy *n.* activation of ovum by male nucleus without nuclear fusion.

hemignathous *a.* having one jaw shorter than the other, as some fishes and birds.

hemimetabolous *a. appl.* the orders of insects having an incomplete metamorphosis, with no pupal stage in the life history. They are the Orthoptera (crickets, locusts), Dictyoptera (cockroaches), Plecoptera (stoneflies), Dermaptera (earwigs), Ephemeroptera (mayflies), Odonata (dragonflies), Embioptera (foot spinners), Isoptera (termites), Psocoptera (booklice), Anoplura (biting and sucking lice), Thysanoptera (thrips) and Hemiptera (bugs) (*see individual entries*).

hemimethylated *a. appl.* DNA in which a CG sequence is methylated (at the C residue) on only one strand.

hemiparasite *n.* (1) individual which is partly parasitic but which can survive in the absence of its host; (2) parasitic plant that develops from seeds germinating in the soil rather than in host body; (3) parasite that can exist as a saprophyte.

hemipenis *n.* one of paired grooved copulatory structures present in males of some reptiles. *plu.* **hemipenes.**

hemipneustic *a.* with one or more pairs of spiracles closed.

hemipodes *n.plu.* order of small quail-like birds of the order Gruiformes, related to cranes and rails.

Hemiptera, hemipterans *n.*, *n.plu.* order of sucking insects commonly known as bugs. It includes water boatmen and pond skaters, blood-sucking bugs parasitic on mammals, the sap-sucking aphids, scale insects, leaf-hoppers, mealy bugs and cicadas. Some blood-sucking bugs transmit disease, and the aphids are important vectors of plant viral diseases.

hemisaprophyte *n.* (1) plant living partly by photosynthesis, partly by obtaining food from humus; (2) saprophyte that can also survive as a parasite.

hemisphere cerebral hemisphere *q.v.*

hemisystole *n.* contraction of one ventricle of heart.

hemitropous *a.* (1) turned half round, having an ovule with hilum on one side and micropyle opposite in plane parallel to placenta; (2) *appl.* flowers restricted to medium-length tongued insects for pollination; (3) *appl.* insects with medium-length tongues visiting such flowers.

hemixis *n.* fragmentation and reorganization of macronucleus without involving micronucleus, in the ciliate *Paramecium*.

hemizygous *a. appl.* gene locus present in only one copy in a diploid organism. This may be a sex-linked gene in the heterogametic sex (e.g. X-linked genes in human males), or a gene in a segment of chromosome whose homologous partner has been deleted.

Henle's layer outermost layer of nucleated cubical cells in inner root sheath of hair follicle.

Henle's loop loop of Henle *q.v.*

Henle's sheath perineurium, or its prolongation surrounding branches of nerve.

Hensen's cells columnar supporting cells on basal membrane in ear.

Hensen's node thickened region at the anterior end of the primitive streak in avian and mammalian embryos, through which cells can pass into the interior of the embryo. It is the functional equivalent of the dorsal lip of blastopore in amphibian embryos.

Hepadnaviridae, hepadnaviruses *n.*, *n.plu.* family of single-stranded DNA viruses which includes the hepatitis B virus. They have an unusual mode of replication, in which viral DNA is synthesized from an RNA template.

heparan *n.* glycosaminoglycan composed of repeating disaccharides of *N*-acetylglucosamine and glucuronic acid or iduronic acid. Heparan sulphate is a component of extracellular matrix.

heparin *n.* glycosaminoglycan with anticoagulant activity, secreted by mast cells. It inhibits blood clotting by increasing the rate of inactivation of thrombin by antithrombin III.

hepatectomy *n.* removal of the liver.

hepatic *a.* (1) *pert.*, like, or associated with the liver; (2) (*bot.*) *pert.* liverworts; (3) *n.* a liverwort. *see* Hepatophyta.

Hepaticae, Hepaticopsida Hepatophyta *q.v.*

hepatic duct tube leading from liver to duodenum, which serves as drainage duct for liver.

hepatic portal system in vertebrates, the part of the vascular system carrying blood to the liver. It consists of the hepatic portal vein, which carries blood from gut to liver, and the hepatic artery which carries blood away from liver.

hepatitis *n.* inflammation of the liver.

hepatitis viruses viruses that cause acute or chronic inflammation of the liver. They belong to several different virus families.

hepatobiliary *a.* applied to the channels in liver for passage of bile.

hepatocystic *a.* liver and gall bladder.

hepatocyte *n.* liver epithelial cell, specialized for the synthesis, degradation and storage of a large number of substances, and which secretes bile.

hepatocyte growth factor (HGF) cytokine secreted by liver and other cells, which stimulates liver cells to proliferate after wounding. *alt.* scatter factor.

hepatoduodenal *a. pert.* liver and duodenum.

hepatoenteric *a.* of or *pert.* liver and intestine.

hepatogastric *a. pert.* liver and stomach.

hepatoma *n.* malignant tumour of the liver.

hepatopancreas *n.* in many invertebrates, a gland that secretes digestive enzymes, and which is presumed also to perform a function similar to the liver.

Hepatophyta, hepatophytes *n., n.plu.* division of non-vascular spore-bearing green plants commonly called liverworts, which with the hornworts and mosses are known as the bryophytes. Liverworts are small, generally inconspicuous plants, growing in low clumps on the ground or on rocks or tree-bark. The photosynthetic gametophyte is thalloid or leafy, most liverworts having stems with three rows of leaves. The sporophyte is a stalked capsule growing from the gametophyte. The plant is anchored to the ground by fine unicellular rhizoids. Liverworts lack specialized conducting tissue (with a few possible exceptions), a cuticle and stomata, and are the simplest of multicellular plants. *alt.* hepatics.

hepatoportal system hepatic portal system *q.v.*

hepatorenal *a. pert.* liver and kidney.

hepatoumbilical *a.* joining liver and umbilicus.

hepta- prefix derived from Gk *hepta*, seven, and denoting having seven of, or arranged in sevens.

heptad *n.* a group of seven.

heptagynous *a.* with seven pistils.

heptamer *n.* a seven-nucleotide DNA sequence.

heptarch *a. appl.* stele having seven initial groups of xylem.

heptastichous *a.* arranged in seven rows, *appl.* leaves.

heptose *n.* any sugar having the formula $(CH_2O)_7$, e.g. sedoheptulose.

herb *n.* any seed plant with non-woody green stems.

herbaceous *a.* (1) *appl.* seed plants with non-woody green stems; (2) soft, green, with little woody tissue, *appl.* plant organs.

herbarium *n.* collection of dried or preserved plants, or of their parts, and the place where they are kept.

herbicide *n.* chemical that kills plants.

herbivore *n.* animal that feeds exclusively on plants. *a.* **herbivorous**.

herbivory *n.* feeding on plants.

herb layer a horizontal ecological stratum of a plant community comprising the herbaceous plants. *alt.* field layer.

herbosa *n.* vegetation composed of herbaceous plants.

hercogamy *n.* the condition in which self-fertilization is impossible.

herd *n.* group of animals, esp. large herbivores, that feed and travel together.

herd immunity resistance of a population to a pathogen as a result of the immunity of a large proportion of the group to the pathogen.

hereditary *a. appl.* characteristics that can be transmitted from parent to offspring, i.e. traits that are genetically determined.

hereditary angioneurotic oedema inherited condition in which a lack of the C1 inhibitor causes spontaneous activation of the complement system, with fluid leakage from the blood and tissue swelling (oedema).

hereditary persistence of foetal haemoglobin (HPFH) type of β-thalassaemia in which δ- and β-globins are absent but there are no clinical symptoms because of the continued synthesis of foetal haemoglobin (ααγγ) into adulthood.

heredity *n.* (1) the genetic constitution of an individual; (2) the transmission of genetically based characteristics from parents to offspring.

heritability *n.* (1) capacity for being transmitted from one generation to the next; (2) in quantitative genetics can be used in two senses: (i) broad heritability (H^2), which is the proportion of total phenotypic variation in a particular trait at the population level that is attributable to variation in genotype; (ii) narrow heritability (h^2), the proportion of phenotypic variance that can be attributed to additive genetic variance, and which can be used to predict the response of the population to natural selection or selective breeding.

heritable *a.* able to be inherited, *appl.* character, trait, disease.

hermaphrodite *n.* (1) animal with both male and female reproductive organs, *alt.* bisexual; (2) plant in which male and female organs are borne in the same flower; (3) in mammals and some other groups of animals, an individual with a mixture of male and female organs arising as a result of a developmental abnormality, more properly called a pseudohermaphrodite. *a.* **hermaphrodite, hermaphroditic.** *alt.* androgynous, bisexual.

hermaphroditism *n.* condition of being a hermaphrodite.

heroin *n.* an addictive alkaloid obtained from morphine by acetylation, and which acts as a narcotic.

Herpesviridae, herpesviruses *n., n.plu.* family of double-stranded DNA viruses that includes the various herpesviruses that cause cold sores and genital herpes, and the Epstein–Barr virus which causes glandular fever and is also involved in Burkitt's lymphoma in children in Africa and nasopharyngeal carcinoma in China and South-East Asia.

herpetology *n.* that part of zoology dealing with the study of reptiles.

hES cell, hESC human embryonic stem cell *q.v.*

hesmosis *n.* the splitting of some ant and termite colonies by the departure of reproductive individuals with an attendant group of sterile workers.

hesperidin *n.* a flavone derivative, the active constituent of citrin, and which affects the permeability and fragility of blood capillaries.

hesperidium *n.* type of indehiscent fruit exemplified by oranges and lemons.

hesthogenous *A.* covered with down at hatching.

heteracanthous *A.* having the spines in dorsal fin asymmetrically turning alternately to one side then the other.

heteractinal *a. pert.* sponge spicules having a disc of six to eight rays in one plane and a stout ray at right angles to these.

heterandrous *A.* with stamens of different length or shape.

heterauxesis *n.* (1) irregular or asymmetric growth of organs; (2) relative growth rate of parts of an organism.

heteraxial *a.* with three unequal axes.

heterecious, heterecism *alt.* spelling of heteroecious, heteroecism.

hetero- prefix from Gk *heteros*, other. Indicates e.g. difference in structure, from different sources, of different origins, containing different components.

heteroagglutinin *n.* agglutinin of ova which reacts with sperm of a different species.

heteroallelic *a. appl.* mutant alleles which have mutations at different sites so that intragenic recombination can yield a functional gene.

heteroantibody *n.* heterologous antibody, i.e. antibody produced in one species in response to an antigen from another species. *alt.* xenoantibody.

heteroantigen *n.* heterologous antigen, i.e. antigen that is antigenic in a species other than that from which it was obtained. *alt.* xenoantigen.

heterobasidiomycetes *n., n.plu.* basidiomycete fungi producing their basidiospores from septate basidia (heterobasidia). They include the rusts, smuts and jelly fungi. *cf.* Hymenomycetes.

heterobasidium *n.* basidium that is divided by a septum, the upper half only giving rise to basidiospores. *see* heterobasidiomycetes.

heteroblastic *a.* arising from dissimilarcells.

heterocarpous *a.* bearing more than one distinct kind of fruit.

heterocaryo- *see* heterokaryo-.

heterocellular *a.* composed of cells of more than one sort.

heterocephalous *a.* having pistillate flowers and staminate flowers on separate heads.

heterocercal *a.* having vertebral column terminating in upper lobe of tail fin, which is usually larger than lower lobe, as in dogfish and other sharks.

heterochlamydeous *a.* having a calyx differing from corolla in e.g. colour, texture.

heterochromatin *n.* chromatin that is inactive in interphase cells, i.e. it is not generally transcribed and usually silences any otherwise active genes placed within it. It comprises regions that are not transcribed in any cell (e.g. highly repeated DNA such as that at centrosomes and telomeres), regions that are not expressed in one cell lineage but are in another (e.g. one or other of the X chromosomes in female mammals), and chromosomal regions present in different states in different cells. It is generally characterized by the presence of particular proteins and particular covalent modifications to histones and DNA. *a.* **heterochromatic.** *cf.* euchromatin.

heterochromosome *n.* chromosome composed mainly of heterochromatin.

heterochromous *a.* differently coloured, *appl.* disc and marginal florets of some composite flowers.

heterochronic *a. appl.* mutations or genes that affect the timing of a developmental process.

heterochrony *n.* (1) departure from ancestral sequence in timing of different stages of life cycle. *see* direct development, neoteny, progenesis; (2) departure from typical timing and sequence of formation of organs. *a.* **heterochronous.**

heterochrosis *n.* abnormal coloration.

heteroclitic *a. appl.* antibody raised against one antigen but having a higher affinity for another antigen not used in the original immunization.

heterocoelous *a. pert.* vertebrae with saddle-shaped articulatory centra.

heterocotyledonous *a.* having cotyledons unequally developed.

heterocyst *n.* rounded, thick-walled cell found at intervals in filaments of some cyanobacteria. It lacks the photosynthetic apparatus and is the site of nitrogen fixation.

heterodactylous *a.* with the 1st and 2nd toes turned backwards.

heterodimer *n.* protein composed of two different subunits.

heterodont *a.* having teeth differentiated for various purposes.

Heterodontiformes *n.* order of selachians, including the hornsharks, having the notochord only partially replaced in extant species.

heterodromous *a.* having genetic spiral of stem leaves turning in different direction to that of branch leaves.

heteroduplex DNA DNA duplex comprising two strands from two different DNA molecules, and, sometimes, small differences in sequence. Stretches of heteroduplex DNA are formed naturally by DNA strands from two homologous chromosomes during genetic recombination in meiosis. *alt.* hybrid DNA.

heteroecious *a.* (1) passing different stages of life history in different hosts, *n.* **heteroecy**; (2) requiring two hosts to complete its life cycle, *appl.* some rust fungi, *n.* **heteroecism.**

heterofacial *a.* showing regional differentiation.

heterofermentation *n.* fermentation of glucose or another sugar to a mixture of reduced products.

heterogameon *n.* species consisting of races which, when selfed, produce a morphologically stable population, but when crossed may produce several types of viable and fertile progeny.

heterogametangia *n.plu.* gametangia that are morphologically different.

heterogametangic *a.* having more than one kind of gametangium.

heterogametes *n.plu.* male and female gametes that are morphologically different.

heterogametic *a.* having unlike gametes.

heterogametic sex the sex possessing a pair of non-homologous sex chromosomes (e.g. the male, which is XY, in mammals, and the female, which is WZ, in birds). The heterogametic sex produces two different types of gametes, one having one type of sex chromosome and one having the other.

heterogamous *a.* having two or more types of flower, e.g. male and female.

heterogamy *n.* (1) alternation of generations *q.v.*; (2) alternation of two sexual generations, one being true sexual, the other parthenogenetic.

heterogangliate *a.* with widely spaced and asymmetrically placed ganglia.

heterogeneous *a.* consisting of dissimilar parts or composed of a mixture of different components. *n.* **heterogeneity.** *cf.* heterogenous, homogeneous.

heterogeneous nuclear RNA (hnRNA) unstable RNA of very broad size distribution found in the nucleus of eukaryotic cells. It consists of the primary transcripts of nuclear genes.

heterogeneous summation, law of rule that the different independent features of an environmental stimulus (e.g. shape, size and coloration) are additive in their effect on an animal's behaviour.

heterogenetic *a.* (1) descended from different ancestral stocks; (2) *appl.* induction or stimulation by a complex of stimuli of different sorts.

heterogenous *a.* (1) formed of particles of different sizes or a mixture of different ingredients; (2) having a different origin, not originating in the body.

heterogenic *a. pert.*, due to, or having different genes.

heterogenic incompatibility in fungi, the inability of genetically dissimilar individuals to fuse.

heterogeny *n.* having several different distinct generations succeeding each other in a regular series.

heterograft *n.* tissue graft originating in a donor of a different species from the recipient. *alt.* xenograft.

heterogynous *a.* with two types of females.

heteroimmune *a.* (1) displaying immunity to an antigen from another species; (2) *appl.* sera, containing antibodies raised in one species to an antigen from another species.

heterokaryon *n.* cell or mycelium containing genetically different nuclei. *a.* **heterokaryote, heterokaryotic**.

heterokaryosis *n.* the presence of genetically dissimilar nuclei within the same cell or mycelium.

heterokont *a.* bearing different kinds of flagella.

heterolactic fermentation fermentation of a hexose sugar resulting in lactic acid, ethanol and CO_2.

heterolecithal *a. appl.* eggs with the yolk distributed unevenly.

heterologous *a.* (1) of different origin; (2) derived from a different species; (3) differing morphologically, *appl.* alternation of generations; (4) *appl.* various substances, such as agglutinins that affect cells from species other than their own; (5) cross-reacting, *appl.* antibody.

heterologous anti-immunoglobulin antibody against the constant region of a class of immunoglobulin (e.g. IgG), produced by immunizing one species with immunoglobulin from another species, used in many types of immunoassay to detect the binding of a specific antibody to the antigen being assayed. *alt.* secondary antibody.

heterology *n.* non-correspondence of parts owing to different origin. *alt.* non-homology.

heterolysis *n.* the dissolution of cells or tissue by the action of exogenous enzymes or other agents. *a.* **heterolytic**.

heteromallous *a.* spreading in different directions.

heteromastigote *a.* having two different kinds of flagella.

heteromeric *a.* (1) *pert.* another part; (2) *appl.* neuron with axon extending to other side of spinal cord.

heteromerous *a.* (1) having, or consisting of, an unequal number of parts; (2) in insects, having unequal numbers of tarsal segments on the three pairs of legs; (3) *appl.* lichen thallus in which the algal cells form a distinct layer.

heterometabolous *a. appl.* insects having incomplete metamorphosis. *alt.* hemimetabolous.

heteromixis *n.* the union of genetically different nuclei, as in heterothallism.

heteromorphic *a.* (1) having different forms at different times; (2) *appl.* chromosome pairs in which homologues differ in size or some other feature; (3) *appl.* species in which haploid and diploid generations are morphologically dissimilar.

heteromorphosis *n.* (1) development of a part in an abnormal position; (2) regeneration when the new part is different from that which was removed.

heteromorphous *a. pert.* an irregular structure, or departure from the normal.

heteromultimer *n.* a protein composed of different types of subunits. *alt.* hetero-oligomer.

heteronomous *a.* (1) subject to different laws of growth; (2) specialized along different lines; (3) *appl.* segmentation into dissimilar segments.

hetero-oligomer *n.* a protein composed of different types of subunits. *alt.* heteromultimer.

heteropetalous *a.* with dissimilar petals.

heterophagous *a.* having very immature young.

heterophil *a. appl.* non-specific antigens and antibodies present in an organism, affording natural immunity.

heterophilic *a.* binding like-to-unlike, *appl.* certain cell adhesion molecules that bind to different receptor molecules on other cells. *cf.* homophilic.

heterophyllous *a.* bearing foliage leaves of different shape on different parts of the plant. *n.* **heterophylly**.

heterophyte *n.* saprophytic or parasitic plant.

heterophytic *a.* with two kinds of spores, borne by different sporophytes.

heteroplanogametes *n.plu.* motile gametes that are unlike one another.

heteroplasia *n.* the development of a tissue from another of a different kind.

heteroplasm *n.* tissue formed in abnormal places.

heteroplasmy *n.* the presence of organelles (e.g. mitochondria) of more than one genetic type in a single cytoplasm, indicating a cell with cytoplasm derived from different sources. *a.* **heteroplasmic**. *cf.* homoplasmy.

heteroplastic *a.* (1) *appl.* grafts of unrelated material; (2) *appl.* grafts between individuals of different species or genera.

heteroploid *a.* (1) having an extra chromosome through non-disjunction of a pair in meiosis; (2) not having a multiple of the basic haploid number of chromosomes; (3) *n.* an organism having heteroploid nuclei.

heteropolymer *n.* polymer composed of different types of subunit.

heteropolysaccharide *n.* any polysaccharide made up of different types of monosaccharide subunit.

Heteroptera *n.* in some classifications, an order of insects including the water boatmen, capsids and bed bugs.

heteropycnotic *a. appl.* regions of chromosomes that remain compact and densely staining, even in interphase nuclei, when the remainder of the chromatin is more dispersed.

heterorhizal *a.* with roots coming from no determinate point.

heterosexual *a.* of, or *pert.* the opposite sex, *appl.* e.g. hormones.

heterosis *n.* (1) cross-fertilization *q.v.*; (2) hybrid vigour *q.v.*

heterosomal *a.* (1) occurring in, or *pert.*, different bodies; (2) *appl.* rearrangements in two or more chromosomes.

heterosporangic *a.* bearing two kinds of spores in separate sporangia.

heterosporous *a. appl.* plants that produce two kinds of spores – megaspores and microspores – by meiosis. Applies to all seed plants, some ferns and club mosses. Megaspores give rise to the female gametophyte, microspores to the male gametophyte, both much reduced in these plants. *n.* **heterospory**.

heterostemonous *a.* with unlike stamens.

heterostrophic *a.* coiled in a direction opposite to normal.

heterostyly *n.* condition in which individuals within a species differ in the length of style in their flowers, as in primroses with their pin-eyed (long-styled) and thrum-eyed (short-styled) flowers. Anthers in one type of flower are at the same level as stigmas in the other, thus ensuring cross-pollination. *a.* **heterostylic, heterostylous**.

heterosynapsis *n.* pairing of two non-homologous chromosomes.

heterosynaptic facilitation *see* presynaptic faciliation.

heterotaxis *n.* abnormal or unusual arrangement of organs or parts.

heterotetramer *n.* a protein composed of four subunits and having more than one type of subunit.

heterothallic *a. appl.* cell, thallus or mycelium of alga or fungus which can only undergo sexual reproduction with a member of a physiologically different strain. *cf.* homothallic. *see also* mating type. *n.* *hetereothallism*.

heterotic *a. pert.* cross-fertilization, *appl.* vigour: hybrid vigour.

heterotopic *a.* in a different or unusual place, *appl.* transplantation of tissue or organ.

heterotopy *n.* (1) displacement; (2) abnormal habitat.

heterotrichous *a.* (1) having two types of cilia; (2) having a thallus consisting of prostrate and erect filaments, as in certain algae.

heterotrimer *n.* a protein composed of three subunits and having more than one type of subunit.

heterotrimeric G proteins class of guanine-nucleotide binding proteins composed of three subunits: α, β and γ. They are associated with the cytoplasmic face of the plasma membrane of mammalian cells and are involved in transmitting signals from certain cell-surface receptors to intracellular pathways. In the inactive state, the α subunit has GDP bound. On activation, the α subunit binds GTP in place of GDP and dissociates from the βγ subunit. The GTP-bound α subunit is usually, but not always, the effector subunit. The α subunit contains intrinsic GTPase activity which converts the bound GTP to GDP, eventually inactivating the subunit, which then reforms the heterotrimeric G protein. *see also* G_i, G_q, G_s, G protein-coupled receptors, GTPase-activating protein, guanine-nucleotide exchange factors, gustducin, small G proteins, transducin.

heterotroph *n.* organism requiring organic compounds as a carbon source. *a.* **heterotrophic**. *cf.* autotroph. *see also* chemoorganotroph, mixotroph, photoheterotroph.

heterotropous *a. pert.* ovule with micropyle and hilum at opposite ends in a plane parallel with that of placenta.

heterotypic *a.* (1) *pert.* mitotic division in which daughter chromosomes remain united and form rings; (2) *pert.* fusion of membranes from different intracellular compartments; (3) heterophilic *q.v.*

heterotypical *a. appl.* genus comprising species that are not truly related.

heteroxylous *a. appl.* wood containing vessels and fibres as well as tracheids.

heterozygosis *n.* (1) formation of a zygote by two genetically different gametes; (2) the condition of being heterozygous.

heterozygosity *n.* proportion of heterozygotes for a given locus in a population.

heterozygote *n.* heterozygous organism or cell. *alt.* hybrid. *cf.* homozygote.

heterozygote advantage the case where the heterozygote for a given pair of alleles is of superior fitness than either of the two homozygotes.

heterozygous *a. appl.* diploid organism, or cell or nucleus, that has two different alleles at a given gene locus. *alt.* hybrid. *cf.* homozygous.

heuristic *a. appl.* methods of problemsolving that use past experience and a trial-and-error approach.

HEV high endothelial venule *q.v.*

hexa- prefix derived from Gk *hex*, six, signifying having six of, arranged in sixes.

hexacanth *a.* having six hooks.

Hexacorallia *n.* subclass of corals with a 6-fold radial symmetry, including the stony reef-forming corals and the sea anemones. *alt.* Zoantharia. *cf.* Octocorallia.

hexactinal *a.* with six rays.

hexactine *n.* a sponge spicule with six equal and similar rays meeting at right angles.

Hexactinellida *n.* class of Porifera, the glass sponges or hexactinellid sponges, typically radially symmetrical with a skeleton of large six-rayed spicules of silica, often fused to form a three-dimensional network.

hexactinian *a.* with tentacles or mesenteries in multiples of six, *appl.* certain coelenterates.

hexacyclic *a.* having floral whorls consisting of six parts.

hexaene *n.* sponge spicule like a trident but with six branches.

hexagynous *a.* (1) having six pistils or styles; (2) with six carpels to a gynoecium.

hexamer *n.* (1) a protein with six subunits; (2) a sequence of six nucleotides or six amino acids.

hexamerous *a.* occurring in sixes, or arranged in sixes.

hexandrous *a.* having six stamens.

hexapetaloid *a.* with petaloid perianth of six parts.

hexapetalous *a.* having six petals.

hexaphyllous *a.* having six leaves.

hexaploid (1) *a.* having six sets of chromosomes; (2) *n.* an organism having six times the haploid chromosome number.

hexapod *a.* having six legs.

hexapterous *a.* having six wings or winglike expansions.

hexarch *a.* (1) *appl.* stele having six alternating xylem and phloem groups; (2) having six vascular bundles.

hexasepalous *a.* having six sepals.

hexaspermous *a.* having six seeds.

hexasporous *a.* having six spores.

hexastemonous *a.* having six stamens.

hexaster *n.* a hexactine spicule in which the rays branch and produce star-shaped structures.

hexastichous *a.* having the parts arranged in six rows.

hexokinase *n.* enzyme that catalyses the phosphorylation of glucose and some other hexose sugars. EC 2.7.1.1.

hexosamine *n.* amino sugar in which the sugar is a hexose. Examples are galactosamine, glucosamine.

hexosaminidase *n.* enzyme which catalyses the cleavage of a terminal hexose amino sugar from compounds such as gangliosides.

hexosan *n.* polysaccharide made of linked hexose subunits. Examples are starch, glycogen, inulin, cellulose.

hexose *n.* monosaccharide containing six carbon atoms (formula $C_6H_{12}O_6$). Examples are glucose, fructose, galactose, mannose.

hexose monophosphate shunt pentose phosphate pathway *q.v.*

Hfr strain bacterial strain in which the F factor is integrated into the chromosome, leading to an increased frequency of transfer of chromosomal genes.

Hg symbol for the chemical element mercury *q.v.*

HGF hepatocyte growth factor *q.v.*

HGH human growth hormone.

HGPRT hypoxanthine guanine phospho ribosyltransferase *q.v.*

HI haemagglutination inhibition *q.v.*

hiatus *n.* any large gap or opening.

hibernaculum *n.* a winter bud.

hibernal *a.* of the winter.

hibernating glands former term for brown adipose tissue *q.v.*

hibernation *n.* the condition of passing the winter in a resting state of deep sleep, when metabolic rate and body temperature drop considerably. Only a few small mammals, e.g. some rodents, hedgehogs, bats, and other small insectivores, undergo a 'true' hibernation. Obligate hibernators enter hibernation spontaneously as the result of a circannual behavioural rhythm. Facultative hibernators enter hibernation when food becomes scarce and temperatures drop below a certain level. Related conditions include winter torpor in reptiles and winter lethargy in larger mammals, e.g. bears, badgers, skunks and racoons. *see also* aestivation.

hidden Markov model (HMM) type of statistical model that is used in bioinformatics e.g. to construct large multiple alignments that include very distantly related sequences.

hidrosis *n.* sweating, perspiration.

hiemal *a. pert.* winter, *appl.* aspect of a community.

hiemilignosa *n.* monsoon forest composed of small-leaved trees and shrubs which shed their leaves in the dry season.

hierarchy *n.* (1) *see* dominance systems; (2) a natural classification system in which organisms are grouped according to the number of characteristics they have in common and ranked one above another.

high-density lipoproteins (HDL) group of lipoproteins found in blood plasma, which are rich in phospholipids and cholesterol. The protein component is synthesized in the liver.

high endothelial venule (HEV) blood capillary found in lymphoid tissues, whose walls are composed of high endothelial cells.

high-energy bond a misleading term denoting a chemical bond whose breakage releases a large amount of free energy, such as the bond between the two terminal phosphate groups in ATP.

highly repetitive DNA *see* repetitive DNA.

high mobility group proteins HMG proteins *q.v.*

high-performance liquid chromatography (HPLC) type of column chromatography using small-particle media and a mobile phase pumped through at a constant rate, used for analytical separations.

hilar *a.* of or *pert.* a hilum.

hiliferous *a.* having a hilum.

Hill coefficient (nH) a number obtained from equilibrium binding experiments that gives information on the number of binding sites for a ligand present on a protein and about whether they show cooperativity.

Hill reaction the reaction showing that isolated chloroplasts could, on illumination, cause the reduction of suitable electron acceptors such as ferricyanide to ferrocyanide and generate oxygen, first demonstrated by Robert Hill in 1939.

hilum *n.* (*bot.*) (1) scar on ovule or seed where it was attached to ovary; (2) nucleus of a starch grain; (3) (*zool.*) notch, opening or depression in an organ, usually where a blood vessel or nerve enters. *alt.* **hilus.**

hindbrain *n.* cerebellum, pons and medulla oblongata. That part of the brain concerned with basic body activities independent of conscious control, such as regulation of muscle tone, posture, heartbeat, respiration and blood pressure. *alt.* rhombencephalon.

hind-gut (1) outgrowth of the yolk sac extending into tail-fold in human embryo; (2) posterior portion of alimentary tract; (3) proctodaeum *q.v.*

hind-kidney metanephros *q.v.*

hinge cells large epidermal cells which, by changes in turgor, control rolling and unrolling of a leaf.

hinge ligament tough elastic substance that joins the two parts of a bivalve shell.

hinge region flexible part of an IgG or IgA antibody molecule, joining the antigen-binding arms to the stem.

hinge tooth one of the projections found on the hinge line, or line of articulation, of a bivalve shell.

hinoid *a.* with parallel veins at right angles to midrib, *appl.* leaves.

hip *n.* (1) (*bot.*) common name for the type of pome fruit produced by some members of the Rosaceae (e.g. roses); (2) (*zool.*) region of articulation of vertebrate hindleg with trunk. *see also* coxa.

hip girdle pelvic girdle *q.v.*

hippocampal commissure tract of fibres connecting the two cerebral hemispheres in the region of the hippocampal areas.

hippocampal formation brain region consisting of hippocampus and dentate gyrus.

hippocampal gyrus subiculum *q.v.*

hippocampus *n.* area in centre of cerebral hemisphere, lying around the thalamus and just above the corpus callosum. It is thought to be important for learning and memory. Damage to the hippocampus is associated with amnesia.

hippomorphs *n.plu.* group of the Perissodactyla including the extinct brontotheres and the horses (family Equidae).

hippuric acid benzoyl glycine, a constituent of the urine of herbivorous animals.

Hippuridales *n.* order of dicots, land, marsh or water plants, comprising the families Gunneraceae (gunnera), Haloragaceae and Hippuridaceae (mare's tail).

hirsute *a.* hairy; (1) *appl.* birds, covered with hair-like feathers; (2) having stiff, hairy bristles or covering.

hirsutidin *n.* a blue anthocyanin pigment.

hirudin *n.* protein obtained from buccal secretions of leech, which inhibits action of thrombin on fibrinogen, preventing clotting of blood.

Hirudinea *n.* class of carnivorous or ectoparasitic annelids, commonly called leeches, which have 33 segments, cirumoral and posterior suckers and usually no chaetae.

His histidine *q.v.*

His' bundle band of muscle fibres, with nerve fibres, connecting auricles and ventricles of heart. *alt.* atrioventricular bundle.

hispid *a.* having stiff hairs, spines or bristles.

histamine *n.* amine synthesized from histidine by decarboxylation and involved in producing inflammation. It is produced by mast cells and is responsible for many of the symptoms of allergies. It causes contraction of smooth muscle of airways, and dilation of blood vessels, causing them to become leaky. Histamine is also produced by some neurons and acts as a neurotransmitter in the central nervous system.

histidine (His, H) *n.* amino acid with a basic side chain, constituent of protein, possibly essential in the human diet, precursor of histamine.

histidine kinase *see* protein histidine kinase.

histioblast *n.* an embryonic cell of sponges.

histiogenic histogenic *q.v.*

histioid *a.* like a web.

histiotypic *a.* according to cell type, *appl.* reaggregation of dissociated cells.

histoblast nests small groups of cells (histoblasts) in abdomen of dipteran larvae which develop into the adult epidermal structures of the abdomen at metamorphosis.

histochemistry *n.* the study of cells and tissues esp. in respect of their staining properties. *a.* histochemical. *see also* immunohistochemistry.

histocompatibility antigens cell-surface proteins that determine the acceptance or rejection of a tissue when transplanted into another individual of the same or another species. *see* MHC molecules, minor histocompatibility antigen.

histocompatible *a. appl.* tissue that is not rejected if transplanted into another individual. *n.* **histocompatibility**.

histogen *n.* zone of tissue in apical meristems in plants from which new tissue develops.

histogenesis *n.* development of tissues.

histogenic *a.* producing tissues.

histogram *n.* type of graphical representation in which data are grouped in some way and represented as a set of columns, the height of each column being the amount or frequency of the data item in the group.

histoid histioid *q.v.*

histoincompatible *a. appl.* tissue that is rejected if transplanted into another individual. *n.* **histoincompatibility**.

histology *n.* the study of the detailed structure of living tissue, by staining and microscopy.

histolysis *n.* the dissolution of tissues.

histone *n.* any one of a set of simple basic proteins (H2A, H2B, H3, H4), rich in arginine and lysine, which are bound to DNA in eukaryote chromosomes to form nucleosomes *q.v.* Histone H1 binds to the linker DNA between nucleosomes.

histone acetylation the post-translational covalent addition of acetyl groups to specific lysine residues in the amino-terminal tails of histone proteins. Acetylation of histone tails in chromatin modifies the functional state of the chromatin by attracting additional proteins to the modified site. Histones can also be acetylated on certain lysine residues before their incorporation into nucleosomes. This acetylation is only temporary and helps to mark newly synthesized chromatin.

histone acetyltransferase (HAT) enzyme that carries out histone acetylation, using acetyl-CoA as the acetyl donor. *see also* histone deacetylase.

histone deacetylase (HDAC) any of a class of enzymes that removes acetyl groups from acetylated lysines on histone proteins in chromatin, thus altering the functional state of the chromatin. *see also* histone acetyltransferase.

histone-like proteins small basic proteins present in prokaryotes, some of which are associated with bacterial DNA.

histone methylation the covalent post-translational addition of one or more methyl groups to lysine residues in the amino-terminal tails of histone proteins in chromatin. Additional proteins are attracted to the modified site, altering the functional state of the chromatin.

histone methyltransferase enzyme that carries out histone methylation, using *S*-adenosylmethionine as the methyl donor. *alt.* histone methylase.

histone modification covalent modification of histones in chromatin by the addition of acetyl groups, methyl groups or phosphoryl groups, which changes the properties of the chromatin in ways that affect the expression of the genes it contains.

histoplasmosis *n.* fungal disease of humans caused by systemic infection with the yeast *Histoplasma capsulatum.*

histotrophic *a. pert.* or connected with tissue formation or repair.

histotypic *a.* according to cell type, *appl.* reaggregation of dissociated cells.

histozoic *a.* living within tissue, *appl.* the trophozoite stage of certain sporozoan parasites.

HIV human immunodeficiency virus *q.v.*

hives urticaria *q.v.*

HIV receptor the CD4 cell-surface protein on T cells, which acts as a receptor for entry of the human immunodeficiency virus (HIV).

HLA complex human leukocyte antigen complex. *see* major histocompatibility complex, MHC molecules. **HLA-A, HLA-B, HLA-C** are the human MHC class I molecules, **HLA-DP, HLA-DQ, HLA-DR** are the human MHC class II molecules.

HLH helix-loop-helix *q.v.*

H line H band *q.v.*

HMG proteins high mobility group proteins, the non-histone proteins in chromatin that have high mobility on electrophoresis.

HMM heavy meromyosin. *see* meromyosin.

hnRNA heterogeneous nuclear RNA *q.v.*

hoary *n.* greyish-white, having a frosted appearance.

hock *n.* in horses and other hoofed mammals, the joint on hindleg corresponding to the tarsal joint.

Hogness box TATA box *q.v.*

holandric *a.* transmitted from male to male through Y chromosomes, *appl.* sex-linked characters.

holandry *n.* having the full number of testes, as two pairs in oligochaete worms.

Holarctica *n.* zoogeographical region comprising the Nearctic and Palaearctic regions.

holcodont *a.* having the teeth in a long continuous groove.

holdfast *n.* adhesive region by which an organism can attach itself to a surface, used esp. for the adhesive disc by which members of the brown seaweeds attach themselves to rocks.

holistic *a. appl.* explanations that attempt to explain complex phenomena in terms of the properties of the system as a whole. *n.* **holism.** *cf.* reductionism.

Holliday junction the four-armed structure formed between two double-stranded DNA molecules at the point of crossing-over during recombination. *alt.* **Holliday intermediate.**

holobasidium *n.* a single-celled basidium. *cf.* heterobasidium, phragmobasidium.

holobenthic *a.* living on sea bottom or in depths of sea throughout life.

holoblastic *a. pert.* cleavage of fertilized egg in which cleavage furrow extends throughout the whole egg. *cf.* discoidal, meroblastic, superficial.

holobranch *n.* a gill in which gill filaments are borne on both sides.

holocarpic *a.* (1) having fruit body formed by entire thallus; (2) *appl.* parasitic fungi without rhizoids or haustoria, living in host cell.

Holocene *n.* recent geological epoch following Pleistocene, began *ca.* 10,000 years ago. *alt.* Recent.

holocentric *a. appl.* chromosomes having a 'diffuse centromere' so that when fragmented each part of the chromosome behaves at mitosis as though it possesses a centromere.

holocephalian *a. pert.* cartilaginous fishes of the subclass Holocephali, the rabbit fishes, with crushing teeth, a whip-like tail and an operculum covering the gills. *alt.* chimaeras, rat-fish.

holocephalous *a. appl.* a rib with a single head.

holochroal *a.* having eyes with globular or biconvex lenses closely crowded together, so that cornea is continuous over whole eye.

holocrine *a. appl.* glands whose secretion is accompanied by complete breakdown of the secretory cells, e.g. sebaceous glands.

holocyclic *a. pert.* or completing alternation of sexual or parthenogenetic generations.

holoenzyme *n.* complete, fully functional enzyme molecule, consisting of the enzymatic subunit (apoenzyme) and any prosthetic group, cofactor, or regulatory or accessory protein subunits required for full regulated function.

hologamodeme *n.* group of freely interbreeding individuals of the same taxon in a local area.

hologamy *n.* (1) condition of having gametes similar to somatic cells; (2) fusion between mature individuals as in some protozoans.

holognathous *a.* having jaw in a single piece.

hologynic *a.* transmitted directly from female to female, *appl.* sex-linked characters.

holomastigote *a.* having one type of flagellum scattered evenly over the body.

holometabolous *a. appl.* the orders of insects that undergo a full metamorphosis, with a four-stage life history (egg, larva, pupa, adult). They are the Neuroptera (alderflies, lacewings), Mecoptera (scorpion flies), Trichoptera (caddis flies), Lepidoptera (butterflies and moths), Coleoptera (beetles), Strepsiptera, Hymenoptera (ants, bees and wasps), Diptera (two-winged flies) and Siphonaptera (fleas) (*see individual entries*).

holomictic *a. appl.* lakes that are stratified seasonally, because of thermal differences. *cf.* meromictic.

holomorph *n.* all the possible forms of a particular fungus. *cf.* anamorph, teleomorph.

holomorphosis *n.* regeneration in which the entire part is replaced. *a.* **holomorphic.**

holoparasite *n.* parasite that cannot exist independently of its host or on a dead host.

holoplankton *n.* organisms that complete their life cycle in the plankton.

holopneustic *a.* with all spiracles open for respiration.

holoptic *a.* with eyes touching or almost touching on top of head.

holoschisis *n.* division of the nucleus by constriction without the formation of chromosomes or a spindle and without the breakdown of the nuclear membrane.

holosericeous *a.* (1) completely covered with silky hairs; (2) having a silky lustre or sheen.

holostean, holosteous *a.* having a bony skeleton, *appl.* fishes.

Holostei, holosteans *n.*, *n.plu.* group of bony fishes present from the Mesozoic but now represented only by the garpike and bowfin.

holostomatous *a.* with mouth of aperture entire.

holostylic *a. appl.* type of jaw suspension in which the palatoquadrate is fused with the cranium without involving the hyoid arch, typical of rabbit-fishes.

holosystolic *a. pert.* a complete systole.

Holothuroidea, holothurians *n.*, *n.plu.* class of sausage-shaped echinoderms commonly called sea cucumbers. They have minute skeletal plates embedded in the fleshy body wall.

Holotrichia, holotrichans *n.*, *n.plu.* group of ciliate protozoans having no obvious zone of composite cilia around the mouth, and swimming by cilia distributed all over the body.

holotype type specimen *q.v.*

holozoic *a.* obtaining food in the manner of animals, by ingesting food material and then digesting it.

holozygote *n.* zygote containing the entire genomes of both uniting cells.

homaxial *a.* built up around equal axes.

homeobox *n.* nucleotide sequence first identified in homeotic genes in *Drosophila* and present in developmental genes in a wide range of other organisms. It encodes a DNA-binding sequence, the homeodomain.

homeodomain *n.* DNA-binding protein domain which is encoded by the homeobox sequence, and which is found in many gene-regulatory proteins involved in development.

homeologous chromosomes homoeologous chromosomes *q.v.*

homeoprotein *n.* protein containing a homeodomain.

homeosis *n.* the transformation of one part into another, as in the modification of antenna into a leg in the *Drosophila* mutant *Antennapedia*, or of petal into stamen in some plant mutants. *alt.* homoeosis, metamorphy, metamorphosis.

homeostasis *n.* (1) maintenance of the constancy of internal environment of the body or part of body; (2) maintenance of equilibrium between organism and environment; the balance of nature. *a.* **homeostatic.** *alt.* homoeostasis.

homeotely *n.* evolution from homologous parts, but with less close resemblance.

homeothermic homoiothermic *q.v.*

homeotic *a.* (1) *appl.* mutations that transform part of the body into another part; (2) *appl.* genes identified by these mutations. *alt.* homoeotic.

homeotic selector gene *see* Hox genes.

home range territory *q.v.*

homing *n.* the selective entry of leukocytes into different sites in the body, mediated by interactions between homing receptors on the leukocyte and cell adhesion molecules on the vascular endothelial cells at these sites.

Hominidae, hominids *n.*, *n.plu.* taxonomic classification used by palaeoanthropologists in various ways at different times. Currently it is proposed to designate the great apes, a family of primates comprising humans (*Homo* spp.), chimpanzees, gorillas and orangutans, and their extinct ancestors and relatives back to the divergence of this lineage from the rest of the primates; (2) **hominid** *n.* often used in a non-technical sense for a member of a human (*Homo* spp.) or human-like (e.g. *Australopithecus*) fossil species, which are characterized by upright posture and other features distinguishing them from the apes (pongids). *a.* **hominid.** *cf.* Homininae, Hominini, Hominoidea.

Homininae *n.plu.* subfamily of the Hominidae, comprising humans, gorillas and chimpanzees and their extinct close relatives back to the time of the divergence of this lineage from the orangutan lineage. *cf.* Hominini, Hominoidea.

hominine *a.* having the characteristics of a human.

Hominini, hominins *n.*, *n.plu.* tribe of the primate family Homininae that includes humans and chimpanzees and their extinct close relatives. *cf.* Hominidae, Hominoidea.

Hominoidea, hominoids *n.*, *n.plu.* superfamily of primates that includes the families Hominidae (humans, human-like hominids, and the great apes) and Hylobatidae (gibbons). *a.* **hominoid.**

homiothermic homoiothermic *q.v.*

Homo the genus of true humans, including extinct forms (e.g. *H. habilis, H. erectus, H. neanderthalensis*) and modern humans, *H. sapiens*, who are or were primates characterized by completely erect stature, bipedal locomotion, reduced dentition, and above all by a large brain.

homo- prefix from Gk *homo*, the same, indicating e.g. similarity of structure, from the same source, of similar origins, containing similar components.

homoacetogen *n.* bacterium that produces acetate as the sole product of sugar fermentation or from $H_2 + CO_2$.

homoacetogenic fermentation fermentation of fructose to acetic acid.

homoallelic *a. appl.* allelic mutant genes which have mutations at the same site, so that intragenic recombination cannot yield a functional gene.

homobasidiomycetes *n., n.plu.* basidiomycete fungi producing their basidiospores on typically club-shaped, non-septate basidia (homobasidia or holobasidia). They comprise the mushrooms and toadstools, bracket fungi, coral fungi, puffballs, earthstars, stinkhorns and bird's nest fungi. *alt.* Hymenomycetes. *cf.* heterobasidiomycetes.

homoblastic *a.* arising from similar cells.

homocarnosine *n.* a dipeptide, ala-γ-aminobutyric acid, found chiefly in brain.

homocarpous *a.* bearing only one kind of fruit.

homocellular *a.* composed of cells of one type only.

homocercal *a. appl.* type of tail fin in which vertebral column ends before it, and the upper and lower lobes are more-or-less equal.

homochlamydeous *a.* having the outer and inner perianth whorls alike, not distinguishable as calyx and corolla.

homochromous *a.* of one colour, *appl.* florets of a composite flowerhead.

homochronous *a.* occurring at the same age or period, in successive generations.

homocysteine *n.* an amino acid, not found in protein, an intermediate in the biosynthesis of methionine.

homocytotropic antibody IgE *q.v.*

homodimer *n.* protein composed of two identical subunits.

homodont *a.* having teeth all alike, not differentiated.

homodromous *a.* (1) having the genetic spiral alike in direction in stem and branches; (2) moving or acting in the same direction.

homoecious *a.* occupying the same host or shelter throughout the life cycle.

homoeo- homeo- *q.v.*

homoeo box homeobox *q.v.*

homoeologous *a.* partially homologous, *appl.* genetically and evolutionarily related chromosomes from different genomes within a heterogenomic polyploid or from related species.

homoeostasis homeostasis *q.v.*

homoeotic homeotic *q.v.*

homofermentation *n.* fermentation of glucose or other sugar resulting in lactic acid as the sole product.

homogametic sex the sex possessing a pair of homologous sex chromosomes and therefore producing gametes all of one sex. In mammals it is the female, which is XX. In birds, reptiles and lepidopterans the homogametic sex (ZZ) is the male.

homogamy *n.* (1) inbreeding due to some type of isolation; (2) condition of having flowers all alike; (3) having stamens and pistils mature at same time.

homogangliate *a.* having ganglia symmetrically arranged.

homogenate *n.* cell extract obtained by breaking open cells and releasing their contents.

homogeneous *a.* composed of identical or similar components. *alt.* homogenous.

homogenetic *a.* having the same origin.

homogenic *a. pert.*, or having different genes.

homogenic incompatibility in fungi, the inability of genetically similar individuals to fuse.

homogenization *n.* mechanical disruption of tissue so that cells are ruptured and their contents released.

homogenous (1) *appl.* parts or organs that are similar due to descent from a common ancestral type; (2) homogeneous *q.v.*

homogentisate *n.* intermediate compound in the degradation of the amino acids phenylalanine and tyrosine. In the inherited enzyme deficiency alcaptonuria, it accumulates in urine and turns it black on exposure to air.

homogeny *n.* correspondence between parts due to common descent.

homoiomerous *a. appl.* lichens in which the algal cells are fairly evenly distributed throughout the thallus.

homoiosmotic *a. appl.* organisms with constant internal osmotic pressure.

homoiothermic *a. appl.* animals that maintain a more-or-less constant body temperature regardless of external temperature variations, e.g. birds and mammals. Although virtually all homoiotherms are also endothermic, the two terms are not synonyms and describe different aspects of thermoregulation. *n.* **homoiotherm**. *cf.* poikilothermic.

homokaryon *n.* hypha or mycelium having more than one haploid nucleus of identical genetic constitution.

homokaryotic *a.* having genetically identical nuclei in a multinucleate cell, or in different cells of a hypha.

homolactic fermentation fermentation that results in lactic acid as the sole product.

homolateral *a.* on, or *pert.* the same side.

homolecithal *a. appl.* eggs having little, evenly distributed yolk.

homolog *alt.* spelling of homologue.

homologous *a.* (1) *appl.* structures or other attributes in different species that resemble each other because of origin by common descent; (2) *appl.* chromosomes in a diploid organism which contain the same sequence of genes but are derived from different parents, and which pair with each other at meiosis; (3) *appl.* genes determining the same character; (4) *appl.* DNA or protein sequences that have some degree of similarity to each other because they have been derived from a common ancestral sequence by divergent evolution; (5) *appl.* structures having the same phylogenetic origin but not necessarily the same final structure or function, e.g. wings and legs in insects; (6) (*immunol.*) allogeneic *q.v.*

homologous end-joining a repair process for double-strand breaks in DNA, in which DNA recombination mechanisms enable the repair of the damaged helix by refer-

ence to a homologous portion of a sister chromatid or homologous chromosome.

homologous recombination (1) recombination between two DNAs of identical or very similar sequence, as in the recombination that occurs between homologous chromosomes at meiosis; (2) technique of targeted gene disruption in which a chromosomal gene is disrupted by the introduction into the cell of a mutant copy of the gene, which then undergoes recombination with the chromosomal gene, replacing it with the mutant copy.

homologue *n.* (1) any structure of similar evolutionary and developmental origin to another structure, but serving different functions; (2) one member of a pair of homologous chromosomes; (3) one member of a pair or set of homologous DNA or protein sequences.

homology *n.* resemblance by virtue of common descent. *adj.* **homologous**.

homology modelling a technique for predicting the three-dimensional structure of a protein on the basis of the similarity of its sequence to that of a protein of known structure. *alt.* comparative modelling.

homomallous *a.* curving uniformly to one side, *appl.* leaves.

homomixis *n.* the union of nuclei from the same thallus, as in homothallism.

homomorphic *a.* of similar size and structure.

homomorphism *n.* (1) condition of having perfect flowers of only one type; (2) similarity of larva and adult. *a.* **homomorphic**.

homomorphosis *n.* having a newly regenerated part like the one removed.

homomultimer *n.* a protein consisting of two or more identical subunits.

homonomous *a.* (1) *appl.* segmentation into similar segments; (2) following the same stages or processes, as of development or growth.

homonym *n.* a name which has been given to two different species. When an instance is discovered, the second named species must be renamed.

homo-oligomeric *a. appl.* proteins composed of several identical subunits.

homopetalous *a.* having all the petals alike.

homophilic *a.* binding like-to-like, *appl.* cell adhesion molecules that bind to identical molecules on other cells. *cf.* heterophilic.

homophyllous *a.* bearing leaves all of one kind.

homoplasmy *n.* the presence of organelles (e.g. mitochondria) of a single genetic type in the cytoplasm. *a.* **homoplasmic**. *cf.* heteroplasmy.

homoplast *n.* organism or organ formed from similar cells, as a coenobium.

homoplastic *a.* (1) similar in shape and structure but not origin; (2) *appl.* graft made into another individual of the same species.

homoplasy *n.* resemblance in form or structure between different organs or organisms due to evolution along similar lines rather than common descent. *a.* **homoplasious**. *alt.* **homoplasty**, convergent evolution.

homopolysaccharide *n.* polysaccharide made of only one type of monosaccharide subunit.

Homoptera *n.* group of insects that includes the plant bugs, aphids, cicadas and scale insects.

homopterous *a.* having wings alike.

homosequential *a.* *appl.* species of Diptera with polytene chromosomes that have exactly the same banding pattern.

homoserine *n.* an amino acid, not found in protein, involved in the biosynthesis of methionine and threonine.

homosporous *a.* *appl.* plants producing only one type of spore by meiosis, e.g. most mosses and ferns. *n.* **homospory**.

homostyly *n.* the condition that all flowers of the same species have styles of the same length. *a.* **homostylous**. *cf.* heterostyly.

homotaxis, homotaxy *n.* similar assemblage or succession of species or types in different regions or strata, not necessarily contemporaneous. *a.* **homotaxial**.

homothallic *a.* (1) *appl.* cells, thalli or mycelia of algae or fungi that can undergo sexual reproduction with a genetically similar strain, or with a branch of the same mycelium or thallus; (2) *appl.* strains of the yeast *Saccharomyces cerevisiae* in which switches of mating type take place in some individuals (and in which, therefore, conjugation can take place between members of the same strain). *see also* mating type. *n.* homothallism.

homotropous *a.* (1) turned in the same direction; (2) *appl.* ovules having micropyle and chalaza at opposite ends.

homotypic *a.* binding of like to like, e.g. of cell types, molecules.

homotypy *n.* (1) the equality of structures on both sides of the main axis of body; (2) serial homology, as of successive segments of some animals; (3) reversed symmetry.

homoxylous *a.* *appl.* wood without xylem vessels and consisting of tracheids.

homozygote *n.* a homozygous organism or cell.

homozygous *a.* *appl.* diploid organism, cell or nucleus that carries two identical alleles at a given gene locus. *n.* **homozygosity**. *cf.* heterozygous.

homunculus *n.* the miniature human foetus supposed to be present in sperm, according to proponents of 18th century preformation theory.

honest behaviour behaviour that conveys the individual's real intentions to another individual.

honey bee generally refers to *Apis mellifera*, the hive bee. *see also* bees.

honeydew *n.* (1) sugary exudate on leaves of many plants; (2) sweet liquid secreted by aphids.

honey guides nectar guides *q.v.*

honey-stomach in some insects, an expansion of the oesophagus in the anterior portion of the abdomen, serving to store ingested liquid which is regurgitated as required.

Hoogsteen base pairing non-standard base pairing that can be made between purine and pyrimidine bases.

hookworms *n.plu.* parasitic nematode worms that cause severe disease in humans, and including *Ancylostoma duodenale* and *Necator americanus*. Common and widespread in tropical areas, the larvae enter the body through the skin, and the adult worm lives in the intestine, abrading the intestinal walls and eventually causing severe anaemia and general debilitation.

hopanoids *n.plu.* compounds similar to sterols, found in the cell membranes of some bacteria.

hordaceous *a. pert.* or resembling barley.

hordein *n.* storage protein in barley grains.

horizon *n.* (1) soil layer of more-or-less well-defined character; (2) a layer of deposit characterized by definite fossil species and formed at a definite time.

horizontal cell type of nerve cell in retina, forming a layer with bipolar and amacrine cells and making lateral connections.

horizontal gene transfer the acquisition of genes by one species from another species. *alt.* lateral gene transfer.

hormogonium *n.* cyanobacterial filament between two heterocysts, which propagates a new organism when it breaks away. *plu.* **hormogonia**.

hormone *n.* a substance that is produced by one tissue and transported to another tissue where it induces a specific physiological response.

horn *n.* (1) the hollow projections on the head of many ruminant animals, consisting of layers of keratinized epidermis laid down on a bony base; (2) any projection resembling a horn; (3) anterior part of each uterus when posterior parts are united; (4) a tuft of ear feathers in owls; (5) a spine in fishes; (6) a tentacle in snails; (7) cornu *q.v.*; (8) keratin *q.v.*

horn corals Rugosa *q.v. cf.* horny corals.

hornworts *n.plu.* common name for members of the plant division Anthocerophyta *q.v.*

horny corals another name for the gorgonians *q.v. cf.* horn corals.

horological *a. appl.* flowers opening and closing at a particular time of day and night.

horotelic *a.* evolving at a standard rate.

horsehair worms Nematomorpha *q.v.*

horseradish peroxidase (HRP) enzyme found in the roots of horseradish and other plants, which reacts with certain substrates to leave a deposit of black granules. It is used esp. as a histochemical stain to trace axons of nerve cells.

horseshoe crabs common name for the Xiphosura, also called king crabs, a group of aquatic arthropods with affinities with the arachnids rather than the crustaceans, and often placed in the separate class Merostomata. They have a heavily chitinized body with the cephalothorax covered by a horseshoe-shaped carapace.

horsetails common name for the Sphenophyta *q.v.*

host *n.* (1) any organism in which another organism spends part or all of its life, and from which it derives nourishment or gets protection; (2) the recipient of grafted or transplanted tissue.

host-induced modification of enveloped viruses, the incorporation of host cell membrane material into the envelope, causing differences in the physical properties of virions propagated in different types of cell.

host range the range of different species, or cell types, that a pathogen can infect.

hotspot *n.* region of a gene or chromosome at which mutation or recombination is markedly increased.

housekeeping genes genes that are expressed in most cell types and which are concerned with basic metabolic activities common to all cells.

Hox genes ancient family of homeobox-containing genes present in all Metazoa and which are involved in specifying a cell's position and identity along the antero-posterior axis during development.

HPFH hereditary persistence of foetal haemoglobin *q.v.*

HPLC high-performance liquid chromatography *q.v.*

HPRT hypoxanthine guanine phosphoribosyl transferase *q.v.*

H$^+$ pump proton pump *q.v.*

HPV human papilloma virus.

HRP horseradish peroxidase *q.v.*

HSC haematopoietic stem cell *q.v.*

Hsp general abbreviation for heat-shock proteins.

H strand heavy strand of DNA, esp. *pert.* mammalian mitochondrial DNA which can be separated into H and L (light) strands on the basis of their density.

H substance complex carbohydrate antigen on red blood cells, the unmodified form of the basic ABO blood group antigen, found in persons of blood group O.

5-HT 5-hydroxytryptamine (serotonin) *q.v.*

HTH helix-turn-helix *q.v.*

HTLV-I human T-cell leukaemia virus, the cause of adult T-cell leukaemia, a rare cancer.

HTLV-II human T lymphotropic virus II, a retrovirus isolated from humans but causing no known disease.

hue *n.* colour, one of the three basic dimensions of perception of visible light by humans. *see also* brightness, saturation.

human chorionic gonadotropin (hCG) protein hormone produced by the developing conceptus and placenta after implantation and which is involved in maintenance of pregnancy.

human embryonic stem cell (hES cell, hESC) cultured embryonic stem cell derived from the inner cell mass of a human blastocyst. These cells have the potential for differentiating into any of the cell types of the human body.

Human Genome Project international publicly funded project that mapped and sequenced the complete human genome.

human immunodeficiency virus (HIV) a retrovirus transmitted maternally or by transfer of body fluids (sexually or by transfusion of infected blood), which infects CD4 T lymphocytes, leading to their eventual depletion and a severe immunodeficiency – acquired immune deficiency syndrome (AIDS). Two types of the virus have been found, HIV-1 and HIV-2.

humanized antibodies antibodies constructed by genetic engineering in which a desired antigen-binding site from a mouse antibody is inserted into a human antibody.

human leukocyte antigen HLA *q.v.*

humeral *a. pert.* shoulder region.

humerus *n.* the bone of the upper arm, or upper part of vertebrate forelimb.

humic *a. pert.* or derived from humus.

humic acid fraction that precipitates from a solution of humus in weak alkali on addition of acid.

humicolous *a.* living in the soil. *n.* **humicole.**

humification *n.* the production of humus in the soil by the action of microorganisms on plant and animal residues.

humin *n.* black insoluble residue left when humus is dissolved in dilute alkali.

humour, humor *n.* any body fluid, nowadays chiefly used in connection with the fluids of the eye. *see* aqueous humor, vitreous humour.

humoral *a.* (1) *appl.* immunity mediated by antibodies; (2) *appl.* antibodies circulating in blood and lymph.

humulone *n.* a bitter compound obtained from hops.

humus *n.* black organic material of complex composition which is the end-product of the microbial breakdown of plant and animal residues in the soil.

Huntington's disease autosomal dominant genetic disease characterized by the onset of mental and physical deterioration in middle age. It is caused by amplification of trinucleotide sequences within the affected gene. *alt.* **Huntington's chorea.**

HU protein a histone-like protein found complexed with DNA in bacteria.

Huxley's layer the middle layer of polyhedral cells in the inner root sheath of hair follicle.

HVR hypervariable region of antibody molecule.

hyaline *a.* (1) transparent or translucent; (3) free from inclusions; (4) *appl.* cartilage of smooth glassy appearance, lacking obvious fibres.

hyaline layer outer layer of matrix covering surface of sea urchin blastula. It is secreted by the cortical granules of the egg on fertilization.

hyalocyte *n.* cell secreting the vitreous humor of eye.

hyaloid *a.* transparent or translucent.

hyaloid artery central artery of retina running through hyaloid canal to back of lens, in foetal eye.

hyaloid canal canal running through vitreous body of eye, from optic nerve to back of lens.

hyaloid fossa anterior concavity of vitreous body in the eye, receptacle of lens.

hyaloid membrane delicate membrane enveloping vitreous body of the eye.

hyalopterous *a.* having transparent wings.

hyalospore *n.* transparent unicellular spore in some fungi.

hyaluronans *n.plu.* viscous high-molecular-weight polymers of *N*-acetylglucosamine and glucuronic acid, abundant in connective and other tissues. They act as lubricating agents in synovial fluid and form the cementing substance between animal cells. *alt.* **hyaluronate, hyaluronic acid.**

hyaluronidase *n.* enzyme that degrades hyaluronans, produced e.g. by various pathogenic bacteria and aiding them to invade tissues. EC 3.2.1.36, *r.n.* hyalurono-glucuronidase.

hybrid *n.* (1) progeny of a cross between parents of different genotype; (2) heterozygote *q.v.*; (3) any macromolecule (esp. DNA) composed of two or more portions of different origins. *v.* **hybridize**. *a.* **hybrid**.

hybrid-arrested translation technique for identifying the cDNA corresponding to an mRNA by its ability to pair with the mRNA *in vitro* to inhibit translation.

hybrid cell cell formed by fusion of cells from two different species in which the chromosomes are contained in a single large nucleus. *cf.* heterokaryon.

hybrid cline the serial arrangement of characters or forms produced by crossing species.

hybrid DNA (1) heteroduplex DNA *q.v.*; (2) DNA molecule composed of segments of different origin, as in recombinant DNA. *alt.* chimeric DNA.

hybrid dysgenesis the production of sterile progeny, showing chromosomal abnormalities and mutations, on crossing certain strains of the fruit fly *Drosophila melanogaster*. It may involve either the I-R system or the P-M system. Dysgenesis is seen in crosses of I males with R females and in crosses of P males with M females but not vice versa. *see* P factors, M cytotype.

hybridization *n.* (1) formation of a hybrid *q.v.*; (2) cross-fertilization; (3) DNA hybridization *q.v.*

hybridoma *n.* a hybrid cell line producing monoclonal antibodies. It is formed by fusion of a single antibody-producing B cell from the spleen and a myeloma cell. The resulting cell can both multiply indefinitely in culture and produce antibodies and is used to produce monoclonal antibodies for research and medical diagnostic procedures.

hybrid sterility sterility in an individual arising from the fact that it is a hybrid.

hybrid swarms populations consisting of descendants of species hybrids, as at borders between geographical areas populated by these species.

hybrid vigour the phenomenon often seen in crosses between two pure-bred lines of plants, that the hybrid is more vigorous than either of its parents, presumably owing to increased heterozygosity. *see also* overdominance.

hybrid zone geographical area in which two populations once separated by a geographical barrier hybridize after the barrier has broken down.

hydathode *n.* epidermal structure in plants specialized for secretion or exudation of water.

hydatid *n.* (1) any vesicle or sac filled with clear watery fluid; (2) sac containing encysted stages of larval tapeworms; (3) vestige of Müllerian duct constituting appendix of testis.

hydatiform *a.* resembling a hydatid.

hydatiform mole cyst-like growth arising in uterus from implantation of abnormal embryo.

Hydra small freshwater hydrozoan that has been used as an experimental animal in developmental studies.

hydranth *n.* an individual specialized for feeding in a hydrozoan colony.

hydrarch succession hydrosere *q.v.*

hydratase *n.* enzyme catalysing the hydration of a compound by acceptance of a molecule of water, and the removal of such added water. EC sub-subgroup 4.2.1. *r.n.* hydro-lyase. *cf.* hydrolase.

hydric *a.* having an abundant supply of moisture.

hydride ion *n.* a hydrogen atom with an additional electron.

hydrobiology *n.* study of aquatic plants and animals and their environment.

hydrobiont *n.* organism living mainly in water.

hydrocarbon *n.* molecule composed of hydrogen and carbon only.

hydrocarpic *a. appl.* aquatic plants having flowers that are fertilized out of the water but submerged for development of fruit.

hydrocaulus *n.* the 'stem' and 'branches' of a colonial hydroid.

Hydrocharitales *n.* order of aquatic herbaceous monocots comprising the family Hydrocharitaceae (frog's-bit).

hydrochoric *a.* (1) dispersed by water; (2) dependent on water for dissemination. *n.* **hydrochory**.

hydrocladia *n.plu.* the branches of certain hydrozoan colonies.

hydrocoel *n.* the water vascular system in echinoderms.

hydrocoles *n.plu.* animals living in water or a wet environment.

hydrocortisone *n.* glucocorticosteroid hormone produced by the cortex of the adrenal gland and very similar to cortisone in structure and properties. It has marked effects on carbohydrate metabolism and is an immunosuppressant. *alt.* cortisol.

hydrofuge *a.* water-repelling.

hydrogen (H) *n.* the lightest of all the chemical elements. In the free state it is a colourless odourless flammable gas (H_2). It is a constituent of all organic molecules and one of the essential elements for living organisms. *see also* deuterium, pH, tritium.

hydrogenase *n.* any of several enzymes that can use molecular hydrogen (H_2) for the reduction of a substance, present in e.g. the hydrogen-oxidizing bacteria.

hydrogenation *n.* addition of a hydrogen atom to a molecule, which is a reduction reaction. *cf.* dehydrogenation.

hydrogen bacteria bacteria that can use the oxidation of molecular hydrogen (H_2) as their source of energy and oxygen as the electron receptor. They are mostly facultative chemolithotrophs. *see also* hydrogen-oxidizing bacteria.

hydrogen bond the attraction between an electronegative atom with a lone pair of electrons and a hydrogen atom that is covalently bonded to another electronegative atom, e.g. –O.....H–N–. It is the strongest of the weak non-covalent intermolecular attractions and is of great importance in biology as it is one of the main forces governing e.g. the folding of a protein chain into its final functional three-dimensional structure, and the interactions of proteins with each other and with small molecules such as enzyme substrates. Hydrogen bonds also hold together the two DNA strands of a DNA molecule. Hydrogen bonding between water molecules is responsible for the high melting and boiling points of water (compared with those of e.g. methane) and its high surface tension.

hydrogen ion a proton, H^+.

hydrogen ion pump proton pump *q.v.*

Hydrogenobacteria *see* Aquifex–Hydrogenobacter group.

hydrogenosome *n.* organelle containing hydrogenases, found in anaerobic protozoa.

hydrogen-oxidizing bacteria diverse group of aerobic and anaerobic bacteria that oxidize molecular hydrogen (H_2) as their source of energy, using a variety of electron acceptors, depending on the species. Those aerobic species that use oxygen as the electron acceptor are generally called hydrogen bacteria.

hydrogen peroxide H_2O_2, potentially toxic product produced as a by-product of the reduction of oxygen in aerobic respiration. It is degraded by catalase to water and oxygen.

hydrogen sulphide H_2S, compound used as an energy source by some colourless sulphur bacteria, which oxidize it to elemental sulphur and then to sulphate.

hydroid *n.* (1) (*bot.*) empty water-conducting cell, joined with others to form a strand of water-conducting tissue in the stems of many mosses; (2) (*zool.*) one of the forms of individuals in the Hydrozoa, a class of solitary and colonial coelenterates, having a hollow cylindrical body closed at one end and with a mouth at the other surrounded by tentacles. *alt.* polyp.

hydrolase *n.* any enzyme that catalyses a hydrolysis. EC group 3.

hydrological cycle water cycle *q.v.*

hydro-lyase hydratase *q.v.*

hydrolysis *n.* the addition of the hydrogen and hydroxyl ions of water to a molecule, with its consequent splitting into two or more simpler molecules.

hydrolytic *a. pert.* or causing hydrolysis.

hydrome *n.* any tissue that conducts water.

hydromesophyte *n.* aquatic plant of temperate climates.

hydromorphic *a. appl.* soils containing excess water.

hydronasty *n.* plant movement induced by changes in atmospheric humidity.

hydronium ion (H_3O^+) the ion created by addition of a proton (H^+) to a water molecule, the usual fate of protons in solution.

hydropathy plot analysis of a protein sequence to determine stretches of hydrophobic amino acids, which, in a membrane protein, may indicate transmembrane regions.

hydrophilic *a.* water-attracting, *appl.* charged or polar chemical group or molecule that readily forms hydrogen bonds with water, thus tending to dissolve in water. *cf.* hydrophobic.

hydrophilous *a.* pollinated by the agency of water.

hydrophily *n.* pollination by water.

hydrophobia *n.* the aversion to water that is a symptom of rabies.

hydrophobic *a.* water-repelling or repelled by water, *appl.* non-polar chemical group or molecule that cannot form hydrogen bonds with water. Hydrophobic molecules tend to aggregate in water, excluding water from between them.

hydrophoric *a.* carrying water, *appl.* canal: the stone canal in echinoderms.

hydrophyllium *n.* one of leaf-like transparent bodies arising above and partly covering the sporosacs in a siphonophore.

hydrophyte *n.* (1) aquatic plant living on or in the water; (2) aquatic perennial herbaceous plant in which the perennating parts lie in water.

hydrophyton *n.* a complete hydrozoan colony.

hydroplanula *n.* stages between planula and actinula in larval stages of coelenterates.

hydropolyp *n.* a polyp of a hydrozoan colony.

hydroponics *n.* cultivation of plants without soil in nutrient-rich water, which is usually irrigated over some inert medium such as sand.

hydropote *n.* a cell or cell group, in some submerged leaves, easily permeable by water and salts.

hydropyle *n.* a specialized area in cuticular membrane of some insect embryos, for passage of water.

hydrorhiza *n.* branching root-like foot of a hydroid colony, which attaches it to the substratum.

hydrosere *n.* a plant succession originating in a wet environment.

hydrosinus *n.* an extension of the mouth cavity in some cyclostomes.

hydrosoma, hydrosome *n.* the conspicuously hydra-like stage in a coelenterate life history.

hydrosphere *n.* the portion of the planet which is water, e.g. the oceans, rivers, lakes, streams, and including soil water.

hydrospire *n.* long pouches running at the side of the ambulacral grooves and acting as respiratory structures in certain echinoderms.

hydrospore *n.* a zoospore when moving in water.

hydrostatic *a. appl.* organs of flotation, as air sacs in aquatic larvae of insects.

hydrostome *n.* the mouth of a hydroid polyp.

hydrotaxis *n.* movement or locomotion in response to the stimulus of water.

hydrotheca *n.* cup-like extension of perisarc around individual polyps in some colonial hydrozoans, into which the polyp may withdraw.

hydrothermal vent community community of organisms living around volcanic vents in the sea floor (hydrothermal vents) at great depths. Primary producers are chiefly chemoautotrophic sulphide-oxidizing bacteria that use the energy of sulphide oxidation to fix CO_2. They include free-living species (e.g. *Beggiatoa* spp.) and intracellular bacterial symbionts living in giant vestimentiferan tube worms.

hydrotropic *a. appl.* curvature of a plant organ towards a greater degree of moisture.

hydroxyapatite *n.* hydrated calcium phosphate $(Ca_{10}(PO_4)_6(OH)_2)$, major constituent of inorganic phase of bone. Also used as a material to which double-stranded DNA will bind in various separation techniques.

3-hydroxybutyrate a ketone body, formed by reduction of acetoacetate, which can act as a substrate for cellular respiration.

hydroxycobalamin(e) vitamin B_{12b}. *see* cobalamin.

hydroxyl *n.* chemical group (–OH) present in all alcohols, consisting of a hydrogen atom covalently linked to an oxygen atom.

hydroxylapatite hydroxyapatite *q.v.*

hydroxylase monooxygenase *q.v.*

hydroxylysine *n.* hydroxylated derivative of the amino acid lysine, modified after incorporation into a polypeptide chain. It is found in collagen.

hydroxyproline (Hyp) *n.* hydroxylated derivative of the amino acid proline,

5-hydroxytryptamine (5-HT)

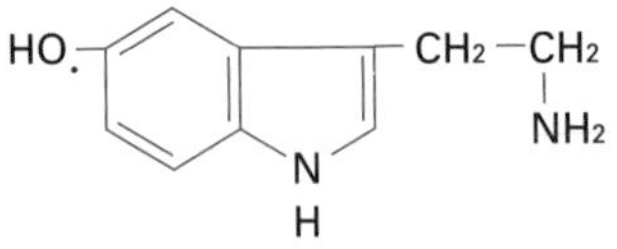

HO— (indole ring structure) —CH₂—CH₂—NH₂ with N—H

$$HO-\overset{\displaystyle}{\underset{\underset{H}{N}}{\bigcirc\!\!\!\bigcirc}}-CH_2-CH_2-NH_2$$

Fig. 23 5-Hydroxytryptamine.

modified after incorporation into a polypeptide chain. It is found in collagen.

5-hydroxytryptamine (5-HT) amine neurotransmitter in central nervous system. It is implicated in regulation of wakefulness and pain sensation. Also produced by platelets and other cell types and causes constriction of blood vessels by stimulating contraction of smooth muscle. *alt.* serotonin. *see* Fig. 23.

5-hydroxytryptaminergic serotonergic *q.v.*

Hydrozoa, hydrozoans *n., n.plu.* class of coelenterates with two body forms, hydroid (polyp) and medusa, generally occurring as different stages of the life cycle. They include solitary forms such as *Hydra*, branching colonial forms, and the siphonophores such as the Portuguese Man o' War which are colonies of several different types of modified polyps and medusae.

hygric *a.* (1) humid; (2) tolerating, or adapted to, humid conditions.

hygrochasy *n.* dehiscence of fruits when induced by moisture.

hygrokinesis *n.* movement induced by a change in humidity.

hygromesophyte *n.* plant of temperate climates that lives in water but is not aquatic.

hygromorphic *a.* structurally adapted to a moist habitat.

hygropetric *a. appl.* fauna of submerged rocks.

hygrophanous *a.* as if saturated with water.

hygrophilic, hygrophilous *a.* inhabiting moist or marshy places.

hygrophyte *n.* plant that thrives in plentiful moisture, but is not aquatic.

hygroreceptor *n.* a specialized cell or structure sensitive to humidity.

hygroscopic *a.* (1) sensitive to moisture; (2) absorbing water.

hygrotaxis *n.* movement in response to moisture or humidity.

hygrotropism *n.* plant growth movement in response to moisture or humidity.

hymen *n.* thin fold of mucous membrane at mouth of vagina.

hymeniferous *a.* having a hymenium.

hymenium *n.* in ascomycete and basidiomycete fungi, the distinct layer of spore-bearing structures, asci and basidia respectively, often interspersed with barren cells (paraphyses). *a.* **hymenial**. *plu.* **hymenia**.

Hymenomycetes *n.* group of basidiomycete fungi bearing their basidia in a well-defined layer (hymenium) which becomes exposed while the basidia are still immature. It comprises mushrooms and toadstools and bracket fungi.

Hymenoptera, hymenopterans *n., n.plu.* order of insects, including solitary and social species, comprising the ants, bees and wasps. They have two pairs of wings, and many have a pronounced waist between 2nd and 3rd abdominal segments. Males are haploid and females diploid, males developing from unfertilized ova. In colonial forms, a colony usually contains one reproductive female (the queen), sterile female workers, a few reproductive males, and (in ants) sterile soldiers.

hyobranchial *a. pert.* to the hyoid and branchial arches.

hyoepiglottic *a.* connecting hyoid and epiglottis.

hyoglossus *n.* an extrinsic muscle of the tongue, arising from greater cornu of hyoid bone.

hyoid *a.* (1) *pert.* or designating a bone or series of bones lying at the base of the tongue in mammals, developed from the hyoid arch of embryo; (2) in fishes, *pert.* the hyoid arch or 1st gill arch.

hyoid arch the 2nd branchial arch in vertebrate embryos, which develops into the first gill arch in fishes.

hyoid bone in humans, a horseshoe-shaped bone lying at base of tongue.

hyoideus *n.* nerve supplying mucosa of mouth and muscles of hyoid region.

hyomandibular cartilage the dorsal skeletal element of hyoid arch in fishes.

hyomental *a. pert.* hyoid and chin.

hyoplastron *n.* the 2nd lateral plate in shell of tortoises and turtles.

hyostylic *n. appl.* type of jaw suspension present in most modern bony fish in which the jaws are attached to the brain case mainly by the hyomandibula, and two out of three palatoquadrate (upper jaw) connections to the brain case are replaced by ligaments.

hyothyroid *a. pert.* hyoid bone and thyroid cartilage of larynx.

Hyp hydroxyproline *q.v.*

hypanthium *n.* in some flowers, a cup-shaped extension of the margin of the receptacle to which sepals, petals and stamens are attached.

hypanthodium *n.* inflorescence with a concave capitulum on whose walls the flowers are arranged.

hypantrum *n.* notch on vertebra of certain reptiles for articulation with a wedge-shaped extension (the hyposphene) on neural arch of neighbouring vertebra.

hypapophysis *n.* a ventral extension on vertebra.

hyparterial *a.* situated below an artery, *appl.* branches of bronchi below pulmonary artery.

hypaxial *a.* ventral or below vertebral column, *appl.* muscles.

hyperacute rejection rapid and untreatable rejection of transplanted tissue that is due to the presence in the host of pre-existing antibodies that react against antigens on the donor tissue blood vessels.

hypercholesterolaemia *n.* (1) raised levels of cholesterol in the blood; (2) the inherited condition familial hypercholesterolaemia, which in homozygotes results in a deficiency of LDL receptors, leading to increased blood cholesterol and deposition of cholesterol in nodules in tendons, premature atherosclerosis and childhood coronary artery disease.

hyperchromism *n.* increased absorbance (*q.v.*) seen e.g. when DNA separates into separate strands.

hypercoracoid *a. pert.* or designating upper bone at the base of pectoral fin in fishes.

hyperdactyly polydactyly *q.v.*

hyperdiploid *n.* cell or organism which, as a result of a translocation, has more than two copies of a particular chromosome segment.

hyperfeminization *n.* condition of a feminized male with female characteristics exaggerated, as in small size and weight.

hyperglycaemia *n.* excess glucose in blood.

hyperhaploid *n.* cell or organism containing supernumerary chromosomes.

hyper IgM syndrome immunodeficiency disease in which only IgM antibodies are made because of an inability to switch isotypes. It can be caused by several different genetic defects.

hyperimmune *n. appl.* individuals with large amounts of a given antibody in their blood, *appl.* antiserum obtained from such an individual.

hyperimmunization *n.* heightened state of immunity as a result of repeated immunization with the same antigen.

hyperkalaemia *n.* abnormally high concentration of potassium in the blood.

hyperkinetic *a.* over-active.

hypermasculinization *n.* condition of a masculinized female with male characteristics exaggerated, as in large proportions, appearance of male secondary sexual characters.

hypermetamorphosis *n.* kind of insect life history which includes two or more different kinds of larva.

hypermorph *n.* a mutant allele which produces a more exaggerated version of the effect of the wild-type gene.

hypermutation *n.* mutation occurring at a higher rate than the normal rate for that particular gene or species. *see also* somatic hypermutation.

hypernatraemia *n.* abnormally high concentration of sodium in the blood.

hyperosmotic *a. appl.* a solution of higher osmotic concentration than a given reference solution.

hyperparasite *n.* organism which is a parasite of, or in, another parasite.

hyperphagia *n.* increased food intake.

hyperpharyngeal *a.* dorsal to the pharynx.

hyperpituitarism *n.* overactivity of the pituitary gland, resulting in gigantism.

hyperplasia *n.* (1) excessive development due to an increase in the number of cells; (2) an abnormal increase in cell proliferation.

hyperploid *a.* (1) having extra chromosomes; (2) having too many copies of the gene in question.

hyperpnoea *n.* rapid breathing due to insufficient supply of oxygen.

hyperpolarization *n.* increase in the electrical potential difference across a membrane. With regard to living cells, it indicates the inside becoming more negative with respect to the outside. *cf.* depolarization.

hyperpolyploid *n.* polyploid cell or organism containing more than the normal number of chromosomes in each of its haploid sets.

hypersensitive *a.* (1) showing an exaggerated or otherwise unduly sensitive response to a stimulus; (2) (*immunol.*) showing an inappropriate, exaggerated or uncontrolled immune response to an antigen, as in allergic reactions and anaphylactic shock. *n.* hypersensitivity.

hypersensitive response in plant pathology, the rapid death of plant cells in response to infection with a fungal, bacterial or viral pathogen, a common defence mechanism in plants. It is generally also associated with other defence responses, e.g. accumulation of phytoalexins and lignification in neighbouring living cells. *alt.* **hypersensitive reaction, hypersensitivity response.**

hypersensitive site sites in chromatin where the DNA is susceptible to cleavage by DNase 1, and which are thought to represent sites where the chromatin is in a different state of packing from surrounding chromatin.

hypersensitivity *n.* an exaggerated or intense response to a particular stimulus or substance. *see also* immediate hypersensitivity, delayed hypersensitivity, type I, II, II, IV hypersensitivity reactions.

hyperstriatum ventralis pars caudalis in brain of birds, an integration centre for auditory and motor information in song control.

hypertelia, hypertely *n.* (1) excessive imitation of colour or pattern; (2) overdevelopment of canines of babirusa, an East Indian pig, the male of which has four large tusks.

hypertension *n.* raised blood pressure.

hyperthermia *n.* rise in body temperature above normal, which is used adaptively by some animals living in hot climates as a water-conserving mechanism.

hyperthermophile *n.* microorganism which requires very high temperatures (>80°C) for optimal growth. Hyperthermophiles are found in hot springs, geysers and deep-sea hydrothermal vents. Most known hyperthermophiles are members of the Archaea. *alt.* extreme thermophile.

hyperthyroidism *n.* overactivity of the thyroid gland, with excess production of thyroid hormone, resulting in increased metabolic rate, high blood pressure, protrusion of the eyeballs, rapid heart rate, thinness and emotional disturbances.

hypertonia *n.* excessive muscle tone.

hypertonic *a.* having a higher osmotic pressure than that of another solution. If the two solutions are separated by a semipermeable membrane water will flow into the hypertonic solution from the other. *cf.* hypotonic.

hypertriploid *a. appl.* cells with more than three sets of chromosomes.

hypertrophic *a.* (1) *appl.* waters grossly enriched with plant nutrients; (2) *appl.* a structure that arises from excessive growth.

hypertrophy *n.* excessive growth due to increase in the size of the cells. *cf.* hyperplasia.

hypervariable locus DNA sequence with an exceptionally high degree of polymorphism within a population.

hypervariable regions (HVR) portions of the variable regions of immunoglobulin and T-cell receptor chains that are particularly variable in amino acid sequence. They contribute to the antigen-binding site. They include the complementarity-determining regions. *cf.* framework regions.

hypha *n.* tubular filament which is the basic growth form of the vegetative phase of a fungus. Hyphal extension and branching produces the fungal mycelium. Hyphae may be continuous tubes of multinucleate protoplasm or may be partially or completely subdivided along their length by transverse partitions (septa) into uninucleate or binucleate compartments. Fungal hyphae are covered by a rigid cell wall, which in most cases contains chitin as well as or instead of cellulose. Similar filamentous vegetative structures in some algae are known as hyphae, as also are the acellular filaments of the prokaryotic actinomycetes. *plu.* **hyphae.** *a.* **hyphal.**

Hyphochytriomycota, hyphochytrids *n.*, *n.plu.* phylum of freshwater protists, now generally classified in the kingdoms Stramenopila or Chromista, commonly known as water moulds and formerly classified as fungi. They are parasitic on algae or fungi or saprobic on plant and insect debris, growing as fine threads and producing motile zoospores with one anterior tinsel flagellum. *alt.* **Hyphochytriomycetes.**

hyphomycetes *n.plu.* fungi that bear conidia free on the mycelium.

hyphopodium *n.* hyphal branch with enlarged terminal cell or haustorium for attaching the hypha to its host, as in some ascomycetes.

hypnody *n.* the long resting period of certain larvae.

hypnogenic *a.* sleep-inducing.

hypo- prefix from the Gk *hypo*, under. In anatomical terms often denoting situated under. In physiological and biochemical terms denoting a decrease in.

hypoachene *n.* achene developed from an inferior ovary.

hypobasal *n.* the lower segment of developing ovule, which ultimately gives rise to the root.

hypobasidium *n.* basal part of cell of septate basidium, in which nuclei unite and which gives rise to the epibasidium from which the basidiospores are budded off.

hypobenthos *n.* fauna of the sea bottom below 1000 m.

hypoblast *n.* (1) cells lining the blastocoel cavity in mammalian blastocyst, which give rise to the yolk sac endoderm, *alt.* primitive endoderm; (2) lower layer of cells in developing chick blastodisc.

hypobranchial *a.* (1) *pert.* lower or 4th segment of gill arch; (2) *appl.* space under gills in decapod crustaceans.

hypocalcaemia *n.* abnormally low level of calcium in the blood.

hypocalcaemic, hypocalcemic *n.* reducing the level of calcium in the blood.

hypocarp *n.* fleshy modified stalk of some fruits, as in cashew-apple.

hypocarpogenous *a.* having both flowers and fruit borne underground.

hypocentrum *n.* transverse cartilage that develops below nerve cord and becomes part of vertebral centrum. *plu.* **hypocentra.**

hypocercal *a.* having notochord terminating in lower lobe of tail fin.

hypocerebral *a. appl.* ganglion of stomatogastric system in insects, linked to frontal and ventral ganglia, also to corpora cardiaca.

hypochile, hypochilium *n.* in orchid flowers, the inner or basal part of lip when in two distinct parts.

hypochondrium *n.* abdominal region lateral to epigastric and above lumbar. *a.* **hypochondriac.**

hypochordal *a.* below the notochord.

hypochromic *a.* paler than usual.

hypochromicity *n.* the decrease in optical density of a duplex DNA in comparison to the value expected from the optical density of a mixture of its constituent nucleotides in free form, caused by interactions between the stacked bases in the duplex. *alt.* **hypochromism.**

hypocone *n.* (1) posterior internal cusp of upper molar teeth; (2) the part posterior to girdle in dinoflagellates.

hypoconid *n.* posterior cusp of lower molar on the cheek side.

hypoconule *n.* 5th or distal cusp of upper molar.

hypoconulid *n.* posterior middle cusp of lower molar teeth.

hypocoracoid *a. pert.* lower bone at base of pectoral fin in fishes.

hypocotyl *n.* that portion of stem below cotyledons in plant embryo, which eventually bears the roots.

hypodermis *n.* (1) in leaves, a layer of cells immediately underlying the epidermis; (2) layer of cells, often a syncytium, underlying the cuticle in nematode worms. *a.* **hypodermal.**

hypodiploid *a. appl.* cells with less than a complete diploid set of chromosomes.

hypogastric *a. pert.* lower abdomen.

hypogastrium *n.* lower central region of abdomen.

hypogeal, hypogean *a.* (1) living or growing underground; (2) *appl.* germination when cotyledons remain underground.

hypogenesis *n.* development without occurrence of alternation of generations.

hypogenous *a.* growing on the under surface of anything.

hypoglossal nerve 12th cranial nerve, controls muscles of tongue and floor of mouth. In anamniotes it is a spinal nerve.

hypoglottis *n.* the under part of the tongue.

hypoglycaemia *n.* abnormally low levels of glucose in the blood.

hypoglycaemic, hypoglycemic *a.* (1) *appl.* agents that tend to lower blood glucose level, such as insulin; (2) *pert.* hypoglycaemia.

hypognathous *a.* having the lower jaw slightly longer than the upper, with mouthparts on the underside, *appl.* insects.

hypogynium *n.* structure supporting ovary in flowers of sedges.

hypogynous *a. appl.* flowers having petals, sepals and stamens attached to the receptacle below the ovary. *n.* **hypogyny**.

hypohaploid *n.* cell or organism with one or several chromosomes missing from its haploid complement.

hypohyal *n.* a ventral skeletal element of the hyoid arch in fishes.

hypokalaemia *n.* abnormally low concentration of potassium in the blood.

hypolimnion *n.* the water between the thermocline and the bottom of a lake.

hypolithic *a.* found or living under stones.

hypomorph *n.* mutant allele that behaves in a similar way to the wild-type gene but has a weaker effect.

hyponasty *n.* the state of growth in a flattened structure when the underside grows more vigorously than the upper. *a.* **hyponastic**.

hyponatraemia *n.* abnormally low concentration of sodium in the blood.

hyponeural *a. appl.* system of transverse and radial nerves in echinoderms.

hyponeuston *n.* organisms swimming or floating immediately under the water surface.

hyponychium *n.* epidermal layer on which nail rests. *a.* **hyponychial**.

hyponym *n.* (1) generic name not founded on a type species; (2) a provisional name for a specimen.

hypo-osmotic *a. appl.* a solution of lower osmotic concentration than a given reference solution.

hypoparathyroidism *n.* condition due to decreased production or activity of parathyroid hormone. It is characterized by a fall in blood calcium levels which leads to a wide variety of symptoms.

hypopetalous *a.* having corolla inserted below, and not adjacent to, gynoecium.

hypopharyngeal *a. pert.* or situated below or on lower surface of pharynx.

hypopharynx *n.* projection from floor of mouth in dipteran insects, forming part of the food canal.

hypophragm *n.* lid closing opening of shell in some gastropod molluscs.

hypophyllous *a.* located or growing under a leaf.

hypophyseal, hypophysial *a. pert.* the hypophysis.

hypophysectomy *n.* excision or removal of the pituitary gland.

hypophysis pituitary *q.v.*

hypopigmentation *n.* lack of pigmentation.

hypopituitarism *n.* deficiency of pituitary hormones, resulting in a type of infantilism.

hypoplasia *n.* (1) developmental deficiency; (2) deficient growth.

hypoplastron *n.* the 3rd lateral bony plate in shell of turtles and tortoises.

hypoploid *a.* (1) having too few copies of the gene in question; (2) lacking one or more of the chromosomes of the normal haploid set.

hypopolyploid *n.* polyploid cell or organism lacking one or more chromosomes.

hypoptilum *n.* a small tuft of down near base of a feather.

hypopyge, hypopygium *n.* clasping organ of male dipterans.

hyporachis *n.* the stem of aftershaft (hypoptilum) of a feather.

hyporheic zone zone around a river, esp. those with gravel beds, in which river water and its microflora and fauna extends as groundwater throughout the surrounding land.

hyposkeletal *a.* lying beneath or internal to the endoskeleton.

hyposomite *n.* ventral part of body segment, as in certain cephalochordates such as amphioxus.

hyposperm *n.* the lower region of ovule or seed, below the level at which the integument or testa is free from the nucellus.

hyposphene *n.* wedge-shaped process on neural arch of vertebra of certain reptiles, which fits into a notch on next vertebra, in certain reptiles.

hypostasis *n.* case where expression of one gene is suppressed by another, non-allelic, gene. *a.* **hypostatic.**

hypostatic *a. appl.* a gene whose expression is prevented by another non-allelic gene.

hypostomatous *a.* (1) *appl.* leaf, having stomata on underside; (2) having mouth placed on lower or ventral side.

hypostome *n.* (1) conical projection containing mouth in hydrozoans; (2) the fold bounding posterior margin of oral aperture in crustaceans; (3) anterior and ventral part of insect head; (4) lower mouthpart in ticks, used to anchor the animal to the skin while it feeds.

hypostracum *n.* inner primary layer or endocuticle of exoskeleton in ticks and mites.

hypothalamus *n.* region of brain located below thalamus and forming greater part of floor of 3rd ventricle. It secretes various peptide hormones, including releasing factors for pituitary hormones. Involved in sexual development and the control of motivated behaviour such as eating, drinking and sex.

hypothallus *n.* (1) thin layer under sporangia in slime moulds; (2) undifferentiated hyphal growth or marginal outgrowth in lichens.

hypotheca *n.* younger or inner half of frustule in diatoms.

hypothecium *n.* the dense layer of hyphal threads below the hymenium in certain fungi.

hypothenar *a. pert.* the prominent part of palm of hand below base of little finger.

hypothermia *n.* a drop in body temperature below normal limits, which leads to death if an external heat source is not applied.

hypothesis *n.* a proposed explanation of a phenomenon or of a scientific problem that must be tested by experiment. *see also* null hypothesis. *cf.* theory.

hypothyroidism *n.* condition due to underactivity of thyroid gland and deficiency of thyroid hormones.

hypotonic *a.* having a lower osmotic pressure than that of another solution. If the two solutions are separated by a semipermeable membrane water will flow from the hypotonic solution to the other. *cf.* hypertonic.

hypotrematic *a. appl.* the lower lateral bar of branchial basket of cyclostomes.

hypotrichous *a.* having cilia mainly on the undersurface.

hypotrochanteric *a.* running below the trochanter.

hypotrophy *n.* (1) condition where wood grows more thickly on underside of a horizontal branch; (2) the condition where stipules or buds form on the underside.

hypotympanic *a.* situated below the tympanum, *pert.* quadrate bone.

Hypoviridae *n.* family of double-stranded RNA viruses that infect fungi.

hypovirulence factor double-stranded RNA virus found in isolates of plant pathogenic fungi with reduced virulence.

hypoxanthine *n.* 6-oxypurine, the purine base in the ribonucleoside inosine, found chiefly in tRNA. It is similar to adenine, but with the amino group replaced by a hydroxyl group. Also found as a breakdown product of purines.

hypoxanthine guanine phosphoribosyltransferase (HGPRT) enzyme catalysing the formation of inosine monophosphate or guanine monophosphate in the minor pathway of nucleic acid biosynthesis. The HGPRT gene is often used as a genetic marker in somatic cell genetics to enable selection of cells in HAT medium. *alt.* HPRT.

hypoxia *n.* transient low levels of oxygen. *a.* **hypoxic.**

hypselodont hypsodont *q.v.*

hypsilophodont *a.* having high-crowned teeth with transverse ridges on the cheekteeth grinding surfaces.

hypsodont *a. appl.* molar and premolar teeth with high crowns and short roots, as those of grazing mammals such as horses and cows.

hypural *a. pert.* bony structure formed by fused haemal spines of last few vertebrae, which supports the tail in certain fishes.

hyracoids *n.plu.* group of placental mammals including the hyrax, which have a rodent-like body and skull, but digits over a pad and bearing nails like elephants.

hysteranthous *a.* coming into leaf after flowering.

hysteresis *n.* (1) lag in one of two associated processes or phenomena; (2) lag in adjustment of external form to internal stresses.

hysterochroic *a.* gradually discolouring from base to tip, *appl.* ageing fruit bodies.

hysterotely *n.* the retention or manifestation of larval characteristics in pupa or imago, or of pupal characters in imago.

hysterothecium *n.* elongated apothecium with slits opening in moist conditions and closing in drought, as in certain fungi and lichens.

I

I (1) inosine *q.v.*; (2) isoleucine *q.v.*; (3) iodine *q.v.*; (4) Morisita's index of dispersion *q.v.*

IAA indole-2-acetic acid *q.v.*

i⁶Ade ⁶N-isopentenyladenine *q.v.*

IAMS International Association of Microbiological Societies.

IAN indole acetonitrile, a naturally occurring auxin.

IAP islet-activating protein. *see* pertussis toxin.

IAPT International Association of Plant Taxonomy.

I-band, I-disc the light band at either end of a sarcomere of striated muscle, representing a region of actin filaments only. The I-band is bounded at its outer edge by a dark Z-disc, the membrane to which the actin filaments are anchored at one end. *see* Fig. 28 (p. 425).

ICAM intercellular adhesion molecule *q.v.*

ICBN International Code of Botanical Nomenclature.

iccosome *n.* small fragment of membrane coated with immune complexes that fragments off follicular dendritic cells.

Ice Age Pleistocene *q.v.*

I-cell disease disease involving lysosomes, in which hydrolytic enzymes are secreted into the extracellular fluid instead of being packaged in lysosomes.

I-cells interstitial cells in coelenterates.

ichneumon wasps the Ichneumonidae, a family of hymenopteran insects that are parasitoids, laying their eggs in the larvae of other insects, esp. butterflies and moths. *alt.* **ichneumon flies.**

ichthyodont *n.* fossil tooth of fish.

ichthyofauna *n.* the fishes of a particular region, area or habitat.

ichthyoid *a. pert.*, characteristic of, or resembling, a fish. *alt.* **ichthyic.**

ichthyolite *n.* fossil fish or part of one.

ichthyology *n.* the study of fishes.

ichthyopterygium *n.* vertebrate limb when it is in the form of a fin.

Ichthyosauria, ichthyosaurs *n., n.plu.* group of Mesozoic aquatic reptiles with spindle-shaped body with fins and fin-like limbs.

ichthyostegalians, ichthyostegids *n.plu.* primitive extinct amphibians that lived from the Devonian to the Carboniferous periods, having many fish-like characteristics and sometimes considered to be primitive labyrinthodonts.

ICM inner cell mass *q.v.*

ICNB International Committee on Nomenclature of Bacteria.

ICNV International Committee on Nomenclature of Viruses.

iconic memory very brief memory that stores the impression of a scene.

iconotype *n.* a representation, drawing or photograph of a type specimen.

icosandrous *a. appl.* flowers, having 20 or more stamens.

ICSB International Committee on Systematic Bacteriology.

ICSH interstitial cell-stimulating hormone. *see* luteinizing hormone.

ICSU International Council of Scientific Unions.

ICTV International Committee on Taxonomy of Viruses.

ICZN International Commission on Zoological Nomenclature.

ID₅₀ dose of virus or other infectious agent at which 50% of the test units (e.g. animals, tissue cultures) become infected.

IDDM insulin-dependent diabetes mellitus, *see* type I diabetes.

ideal angle in phyllotaxis, the angle between successive leaf insertions on a stem when no leaf would be exactly above any lower leaf, 137°30′28″.

ideomotor *a. pert.* involuntary movement in response to a mental image.

idiobiology *n.* the study of individual organisms.

idiomorphs *n.plu.* alternative forms (alleles) of a genetic locus that lack significant sequence homology, e.g. the *MAT-1* and *MAT-2* alleles of the ascomycete mating-type locus.

idiopathic *a. appl.* diseases or conditions arising spontaneously and of unknown cause, not preceded or occasioned by another condition.

idiophase *n.* phase in secondary metabolism in which secondary metabolites are produced. *cf.* trophophase.

idiosoma *n.* the prosoma and opisthosoma, the body of ticks and mites.

idiotope *n.* antigenic determinant formed by part of the variable region of an antibody, the set of such idiotopes carried by a single antibody molecule being its idiotype.

idiotrophic *a.* capable of selecting food.

idiotype *n.* (1) the unique antigenic determinants formed by the variable region of an antibody molecule; (2) the total hereditary determinants of an individual.

IEP isoelectric point *q.v.*

IF intermediate filament *q.v.*

IF-1, IF-2, IF-3 initiation factors (*q.v.*) in bacterial protein synthesis.

IFN interferon (*q.v.*), such as IFN-α, IFN-β and IFN-γ.

Ig immunoglobulin *q.v.*

Igα, Igβ proteins that are part of the complete antigen receptor complex on B lymphocytes and which are involved in signalling from the receptor when antigen binds to the immunoglobulin part of the complex.

IgA immunoglobulin A *q.v.*

IgD immunoglobulin D *q.v.*

IgE immunoglobulin E *q.v.*

IGF-I somatomedin C *q.v.*

IGF-II insulin-like growth factor II *q.v.*

IgG immunoglobulin G *q.v.*

IgM immunoglobulin M *q.v.*

IκB protein that is complexed with the transcription factor NFκB, keeping it inactive in the cytoplasm.

IκB kinase (IKK) protein serine/threonine kinase that is involved in the intracellular signalling pathway that leads from a variety of receptors to the activation of the transcription factor NFκB.

IL-1, IL-2 etc. interleukins *q.v.*

Ile isoleucine *q.v.*

ileac, ileal *a. pert.* the ileum.

ileocaecal *a. pert.* ileum and caecum.

ileocolic *a. pert.* ileum and colon.

ileum *n.* (1) lower part of small intestine; (2) anterior end of hind-gut in insects.

iliac *a. pert.* or in region of ilium.

iliacus *n.* muscle stretching from upper part of iliac fossa to side of tendon of psoas major.

ilicium *n.* dorsal spine with modified tip for luring prey of angler fish (Lophiidae).

iliocaudal *a.* connecting ilium and tail, *appl.* muscle.

iliocostal *a.* in region of ilium and ribs, *appl.* muscles.

iliofemoral *a. pert.* ilium and femur, *appl.* ligament.

iliohypogastric *a. pert.* ilium and lower anterior part of abdomen, *appl.* a nerve.

ilio-inguinal *a.* in the region of ilium and groin, *appl.* a nerve.

ilio-ischadic *a. pert.* opening between ilium and ischium when these are fused at both ends.

iliolumbar *a.* in region of ilium and loins.

iliopectineal *a. appl.* an eminence marking the point of union of ilium and pubis.

iliopsoas *n.* the iliacus and psoas major when considered as one muscle.

iliotibial *a. appl.* tract of muscle at lower end of thigh.

iliotrochanteric *a.* uniting ilium and trochanter of femus, *appl.* ligament.

ilium *n.* dorsal bone in each half of pelvic girdle.

Illicidales *n.* order of woody dicots comprising shrubs, climbers and small trees, and including the two families Iliaceae (star anise) and Schisandraceae.

illuvial *a. appl.* layer of deposition and accumulation below the alluvial layer in soils. *alt.* B-horizon.

imaginal *a. pert.* an imago.

imaginal discs small sacs of undifferentiated epithelium in body of many insect larvae, which on metamorphosis produce the adult epidermal structures appropriate to each segment. Each disc specifies a single structure, e.g. leg disc, antennal disc, genital disc.

imago *n.* last or adult stage of insect metamorphosis, the perfect insect.

imbibition *n.* passive uptake of water, esp. by substances such as cellulose and starch, as in uptake of water by seeds before germination.

imbricate *a.* overlapping, as of scales.

imbricational *a.* overlapping, *appl.* layers of enamel deposited on sides of teeth during growth.

imbrication lines parallel growth lines of dentine.

Imd pathway the 'immunodeficiency pathway' in insects, which is involved in the recognition of Gram-negative bacteria in innate immunity.

imitative *a. appl.* e.g. form, habit, colouring, assumed for protection or aggression when one organism imitates another.

immaculate *a.* without spots or marks of a different colour.

immarginate *a.* without a distinct margin.

immature *a.* (1) still not fully developed; (2) not yet adult. *cf.* mature.

immediate early genes genes that are rapidly and transiently induced in the response of eukaryotic cells to agents that cause cell proliferation.

immediate hypersensitivity antibody-mediated hypersensitivity reactions that occur within minutes of exposure to antigen. *cf.* delayed hypersensitivity.

immigrant species species that migrate into an ecosystem or are introduced accidentally or deliberately by humans.

immobilization *n.* the locking up of elements essential for plant nutrition in organic matter in soil by soil microorganisms so that the elements are not available for plant growth.

immobilized enzyme in biotechnology, purified enzymes, or whole cells containing the required enzyme, which are immobilized by attachment to an inert solid matrix to increase the efficiency of enzyme use in industrial processes.

immortalized *a. appl.* mammalian cells that have become able to continue cell division indefinitely, such as the cells of a permanent cell line which has been derived on prolonged culture from a primary tissue culture. *n.* **immortalization**.

immune *a. appl.* an individual who has produced protective antibodies and/or activated T lymphocytes and memory B and T cells against a pathogenic microorganism and is therefore resistant to reinfection by that pathogen.

immune clearance the clearance of an antigen from the blood by formation of antibody–antigen complexes which are then scavenged by phagocytes.

immune complex complex of antibody and antigen. Deposition in tissues can cause tissue damage and disease due to activation of the complement system by the complex.

immune disease any disease involving immunological hypersensitivity reactions or autoimmune reactions.

immune memory immunological memory *q.v.*

immune response generally refers to an adaptive immune response but may also include the responses of innate immunity. Adaptive immune responses consist of two components. The production of antibodies circulating in the blood is known as the humoral immune response, and the production of cytotoxic and helper T cells as the cell-mediated or cellular immune response. *see also* adaptive immune response, allergy, antibody, antigen, B lymphocyte, clonal selection, immune system, immunity, innate immunity, plasma cell, primary immune response, secondary immune response, T lymphocyte.

immune surveillance the idea that the immune system recognizes and destroys most incipient cancer cells before they can develop.

immune system the cells and tissues in vertebrates which enable them to mount an innate immune response and an adaptive immune response to invading microorganisms, parasites and other foreign substances. The immune system protects the body from infection and establishes long-lasting specific immunity to reinfection. It is also responsible for the recognition and rejection of foreign cells in tissue and organ transplantation. Many invertebrates and plants have some form of innate immune system but only vertebrates have adaptive immunity. *see also* adaptive immune response, allergy, antibody, antigen, histocompatibility, immune response, immunity, innate immunity, leukocytes, lymphocyte, lymphoid, primary lymphoid organs, secondary lymphoid tissues.

immune tolerance *see* tolerance.

immunity *n.* (1) the ability to resist disease, usually referring to infectious disease. Innate immunity against infection is due to relatively nonspecific cellular and molecular systems that protect the organism against bacteria or viruses and is present in most animals and in plants. Lifelong immunity to a particular disease is a result of an adaptive immune response and may be acquired naturally by previous infection or is induced by vaccination with suitably treated microorganisms or their products. It is only found in vertebrates, *see also* adaptive immune response; (2) in bacteria, the inability of a bacterium carrying a prophage or a plasmid to be infected with another phage or plasmid of the same type.

immunization *n.* the administration of an antigen (e.g. by injection) that results in a specific immune response against the antigen. *alt.* inoculation or vaccination, when the antigen is derived from a pathogen and confers immunity against a disease. *v.* **immunize.**

immunoaffinity column chromatography type of affinity chromatography in which antibody bound to the column is used to purify its corresponding antigen from a mixture, or vice versa.

immunoassay *n.* any quantitative assay of a substance using its binding to specific antibody as the measuring technique.

immunoblot Western blot *q.v.*

immunocompetence *n.* the capacity to respond to antigen stimulation, the capacity to mount an immune response. *a.* **immunocompetent.**

immunocytochemistry immunohistochemistry *q.v.*

immunodeficiency *n.* any deficiency in the ability to mount an effective immune response. It may be due to various causes, such as the destruction of a class of helper T lymphocytes in AIDS, the non-production of immunoglobulin due to defects in the immunoglobulin genes, or the non-development of lymphocytes due to various genetic defects. *a.* **immunodeficient.** *see also* severe combined immunodeficiency.

immunodiagnostics *n.* the use of antibodies to detect and identify e.g. viruses, bacteria, hormones, drugs.

immunodiffusion *n.* detection of antigen or antibody by formation of a line of antibody–antigen precipitate where diffusing antigen and antibody meet in an agar gel.

immunoelectron microscopy electron microscopy where specimens have been stained with electron-dense material (e.g. gold particles) attached to a specific antibody in order to highlight specific structures.

immunoelectrophoresis *n.* technique for analysing antigens in a complex mixture in which the mixture is separated by electrophoresis before probing with labelled antibodies.

immunofluorescence microscopy microscopical technique in which cells are stained with antibodies labelled with fluorescent dyes in order to highlight specific structures within the cell.

immunofluorescence test any diagnostic test that uses antibodies labelled with fluorescent dyes to detect the desired antigen.

immunogen *a.* any substance that generates an immune response. *see also* antigen.

immunogenetics *n.* genetic analysis of the immune system.

immunogenic *a.* causing an immune response.

immunogenicity *n.* the ability of a substance to provoke an immune response when administered alone. *appl.* antigens.

immunoglobulin (Ig) *n.* highly variable protein synthesized by the B cells of the immune system and the plasma cells derived from them. Immunoglobulins occur in two forms: as membrane-bound immunoglobulin on the surface of B cells, where they act as the antigen receptors, and as antibodies secreted by plasma cells after the differentiation of B cells into plasma cells during an immune response. An immunoglobulin molecule consists of two identical light chains and two identical larger heavy chains. Light and heavy chains have a variable N-terminal region (V region), which varies in sequence among immunoglobulins with different antigen specificities, and a constant C-terminal region (C region) which is characteristic of the particular class or isotype to which the antibody belongs. *see also* antibody, B-cell receptor, C region, immunoglobulin A, immunoglobulin D, immunoglobulin E, immunoglobulin G, immunoglobulin genes, immunoglobulin M, isotypes, V region.

immunoglobulin A (IgA) class of immunoglobulin containing heavy chains of isotype α. It occurs as either a monomer or a dimer. The dimer is the main antibody in gut, and in saliva, tears, milk and colostrum.

immunoglobulin D (IgD) class of immunoglobulin containing heavy chains of isotype δ. It appears as surface-bound immunoglobulin on B cells subsequent to and in conjunction with IgM. Only very small amounts are secreted as antibody and its function is unknown.

immunoglobulin E (IgE) class of immunoglobulin containing heavy chains of isotype ε. It is involved in allergic inflammatory reactions and responses to intestinal parasites. It is bound by receptors on mast cells and some other leukocytes, where subsequent binding by specific antigens triggers release of cell granule contents.

immunoglobulin G (IgG) class of immunoglobulin containing heavy chains of isotype γ. It is produced towards the end of a primary immune response and in a secondary response and is the main class of antibody circulating in blood and lymph.

immunoglobulin genes genes encoding the polypeptide chains of immunoglobulins. There are three gene loci: the heavy-chain locus, the κ light-chain locus, and the λ light-chain locus. Each locus consists of many similar but not identical

'gene segments' encoding parts of the immunoglobulin chain. DNA rearrangements during lymphocyte development bring together one of each type of gene segment to form the sequence specifying a complete heavy- or light-chain gene. *see* C gene, D gene segment, J gene segment, V gene segment.

immunoglobulin-like domain protein domain related in structure to the characteristic protein domains of immunoglobulins.

immunoglobulin M (IgM) class of immunoglobulin containing heavy chains of isotype μ. It is the first type of immunoglobulin synthesized by developing B cells and is the first class of antibody secreted in a primary immune response. IgM antibody is a pentamer of five IgM molecules held together by a joining polypeptide (J chain).

immunoglobulin receptor the antigen receptor carried on the surface of the B cells of the immune system, which is a transmembrane form of immunoglobulin. *alt.* B-cell receptor. *see also* Fc receptor, poly-Ig receptor.

immunoglobulin superfamily large superfamily of proteins that contains the immunoglobulins, T-cell receptors, and many other cell-surface proteins, such as cell adhesion molecules, that have immunoglobulin-like domains.

immunohistochemistry *n.* techniques for locating particular proteins in cells which use specific antibodies labelled with an enzyme that will produce a coloured reaction product.

immunological memory the ability of the adaptive immune system to mount a larger and more rapid response (the secondary immune response) to an antigen already encountered, and which is mediated by long-lived memory cells (both T cells and B cells) activated in the initial immune response (the primary immune response).

immunological tolerance *see* tolerance.

immunology *n.* the study of the immune system.

immunophilin *n.* protein that binds to immunosuppressive drugs such as cyclosporin A and tacrolimus and is involved in their immunosuppressive effect.

immunoprecipitation *n.* the detection of a protein in a mixture by binding to its specific antibody, leading to precipitation of the protein–antibody complex.

immunoreceptor tyrosine-based activation motif (ITAM) short amino acid sequence containing two tyrosine residues, present in the cytoplasmic portions of many tyrosine-kinase-associated receptors on immune system cells, including the signalling chains associated with B-cell receptors and T-cell receptors. Phosphorylation of the tyrosines as a consequence of ligand binding to the receptor initiates an activating intracellular signalling pathway. *cf.* immunoreceptor tyrosine-based inhibitory motif.

immunoreceptor tyrosine-based inhibitory motif (ITIM) short amino acid sequence containing two tyrosine residues, present in the cytoplasmic portions of some tyrosine-kinase-associated receptors on immune system cells. Phosphorylation of the tyrosines as a consequence of ligand binding to the receptor initiates an intracellular signalling pathway that inhibits cellular activity.

immunosuppression *n.* suppression, e.g. by radiation and by certain drugs, of the ability to mount an immune response.

immunotoxin *n.* (1) an antitoxin that confers immunity against a disease; (2) an antibody conjugated with a toxin or drug that is specifically targeted to destroy or react with certain cells.

IMP inosine monophosphate *q.v.*

imparidigitate *a.* having an unequal number of digits.

imparipinnate *a.* pinnate with an odd terminal leaflet.

impedicellate *a.* having a very short or no pedicel (stalk).

imperfect *a. appl.* flowers lacking either stamens or carpels.

imperfect fungi Deuteromycetes *q.v.*

imperfect stage in ascomycete fungi, the asexual reproductive phase in their life history in which they bear conidia.

imperforate *a.* not pierced, *appl.* shells of foraminiferans that lack fine pores.

impermeable *a. appl.* barrier, e.g. a membrane, that will not let a given substance pass through it. *cf.* permeable.

impetigo *n.* skin infection caused by strains of staphylococci or streptococci.

implantation *n.* the embedding of the fertilized ovum in lining of uterus.

importance value a measure of the role of a species within a community, obtained by adding together its relative density, its relative dominance and the relative frequency at which it is found.

impregnation *n.* transfer of sperm from male to body of female.

imprinting *n.* (1) a form of learning in very young animals in which a particular stimulus normally provided by a parent becomes permanently associated with a particular response. Examples are when a young bird learns to follow a large moving object, usually the parent bird, or when young mammals follow their mother in response to the smell of her milk; (2) genomic imprinting *q.v.*; (3) sexual imprinting *q.v.*

impulse *see* action potential.

inactivation *n.* in neurophysiology, decline of sodium current component of action potential.

inantherate *a.* lacking anthers.

inappendiculate *a.* lacking appendages.

Inarticulata *n.* class of brachiopods in which the shells are joined by muscles only and not by a hinged joint. *cf.* Articulata.

inarticulate *a.* (1) not segmented; (2) not jointed.

inborn error of metabolism any of a diverse group of inherited conditions caused by a genetically determined deficiency of a particular enzyme or production of an impaired protein. *alt.* enzyme deficiency disease.

inbreeding *n.* (1) matings between related individuals, *alt.* consanguineous matings; (2) successive crossing between very closely related individuals, in e.g. laboratory animals, plants. This leads to the establishment of pure-breeding strains or varieties in which individuals are homozygous at a large proportion of their loci, or at selected loci.

inbreeding coefficient (F) (1) of an individual, the probability that the pair of alleles carried by the male and female gametes that produced it are identical by descent from a common ancestor, as a result of inbreeding; (2) a measure of the reduction of heterozygosity as a result of inbreeding, given by $F_S = (H_I - H_S)/H_S$, where H_I and H_S are heterozygosity among an inbred and outbred group of individuals of the same population respectively.

inbreeding depression loss of vigour following inbreeding, due to the expression of numbers of deleterious genes in the homozygous state.

incidence *n.* (1) occurrence of an event; (2) rate at which something occurs. *cf.* prevalence.

incipient lethal level concentration of a toxin at which 50% of the population of test organisms can live for an indefinite time.

incipient species populations that are diverging towards the point of becoming separate species, but which can still interbreed although they are prevented from doing so by some geographical barrier.

incised *a.* with deeply notched margin.

incisiform *a.* shaped like incisors.

incisive *a. pert.* or in region of incisors.

incisors *n.plu.* the cutting front (premaxillary) teeth of mammals.

included *a.* (*bot.*) having stamens and pistils not protruding beyond the corolla.

included niche the case where the niche of a species occurs completely within the niche space of another species.

inclusion bodies intracellular particles such as crystals formed of e.g. viruses or proteins.

inclusive fitness sum of an individual's own fitness plus all its influence on fitness in its relatives other than direct descendants.

incompatibility *n.* (1) genetically determined inability to cross-fertilize (in plants) or mate (in animals) successfully; (2) the rejection of transplanted tissue of a different tissue type; (3) of plasmids, *see* plasmid incompatibility.

incompatible *a.* (1) *see* incompatibility; (2) in plant pathology, *appl.* an interaction between host and pathogen that does not result in disease.

incomplete *a. appl.* flowers lacking sepals or petals or stamens or carpels.

incomplete dominance co-dominance *q.v.*

incomplete metamorphosis insect metamorphosis in which young are hatched in general adult form (but without wings and without mature sexual organs), and develop without a quiescent (pupal) stage. *cf.* complete metamorphosis.

incomplete penetrance the lack of expression of a genetic trait in some individuals possessing the genotype associated with the character.

incongruent *a.* not suitable or fitting, *appl.* surfaces of joints that do not fit properly.

incoordination *n.* (1) want of coordination; (2) irregularity of movement due to loss of muscle control.

incrassate *a.* thickened, becoming thicker.

incubation *n.* (1) the hatching of eggs by means of heat, natural or artificial; (2) the growth of a culture of microorganisms by keeping it for some time at an optimum temperature.

incubation period period between infection and the appearance of symptoms induced by pathogenic bacteria, viruses or other parasites, during which the pathogen is multiplying.

incubatorium *n.* temporary pouch surrounding mammary area, in which the egg of the echidna, the spiny anteater, is hatched.

incudal *a. pert.* the incus, one of the small bones of the middle ear.

incumbent *a.* (1) lying upon; (2) bent downwards to lie along a base.

incurrent *a.* (1) leading into; (2) afferent; (3) *appl.* ectoderm-lined canals which admit water in sponges, *alt.* ostia; (4) *appl.* inhalant siphons of molluscs; (5) *appl.* ostia in insect heart which admit blood.

incurvate *a.* curved upwards or bent back.

incus *n.* the middle of the chain of three small bones in the middle ear that transmit sound from the ear drum to the inner ear in mammals. It is shaped like an anvil.

indeciduate *a.* (1) not withering away, of sepals on fruit; (2) with the maternal part of the placenta not coming away at birth.

indeciduous *n.* (1) persistent; (2) not falling off at maturity; (3) everlasting; (4) evergreen.

indefinite *n.* not limited in e.g. size, number.

indehiscent *a. appl.* fruits, spores or sporangia which do not open to release the seeds or spores, the whole structure being shed.

indel abbreviation used in molecular genetics meaning insertion or deletion.

independent assortment Mendel's second law of inheritance, which describes the fact that the segregation of alleles at one gene occurs independently of the segregation of alleles at another. This law has had to be modified by the subsequent discovery of linkage, i.e. that alleles carried close together on the same chromosome tend to be inherited together.

indeterminate *a.* (1) indefinite; (2) undefined; (3) not classified.

indeterminate growth (1) growth from an apical meristem which forms an unrestricted number of lateral organs such as stems, branches or shoots indefinitely, not stopped by the development of a terminal bud; (2) indefinite prolongation and subdivision of an axis.

indeterminate inflorescence an inflorescence that grows by indeterminate branching because unlimited by the development of a terminal bud.

index *n.* (1) the forefinger or digit next to the thumb; (2) a number or formula expressing the ratio of one quantity to another.

index case any case of an infectious disease that exhibits new syndromes, types of pathogen or characteristics that indicate high potential for a new epidemic.

index fossil a fossil which typically occurs in a particular zone of a rock stratum and after which the zone is known.

index species an organism that lives only within a narrow range of environmental conditions and whose presence therefore indicates places where those conditions exist.

Indian region part of the Palaeotropical floristic kingdom comprising the Indian subcontinent, Sri Lanka, the Laccadives and Maldives.

Indian subregion subdivision of the Oriental zoogeographical region, comprising all the Indian subcontinent south of the Himalayas except for the southern tip.

indicator species (1) species characteristic of climate, soil and other conditions in a particular region or habitat; (2) dominant

species in a biotype; (3) species whose disappearance or disturbance gives early warning of the degradation of an ecosystem.

indifference curves iso-utility curves *q.v.*

indifferent species species that is not found in any particular community.

indigenous *a.* (1) belonging to the locality; (2) not imported; (3) native.

indirect competition limitation of population size or fertility of two or more organisms because of competition for the same limited resource.

indirect immunofluorescence type of immunofluorescence test in which the fluorescent label is attached to a secondary anti-immunoglobulin antibody, which is used to label the primary antigen–antibody complex.

indirect transmission transmission of infection from one person to another via an intermediate living or inanimate object.

individual distance the distance around a bird in a flock, which it defends while feeding or nesting.

individualism *n.* symbiosis in which the two parties together form what appears to be a single organism.

individuation *n.* (1) formation of interdependent functional units, as in a colonial organism; (2) process of developing into an individual.

Indo-Chinese subregion subdivision of the Oriental zoogeographical region, comprising southern China and South-East Asia except for the Malaysian peninsula.

indole-2-acetic acid (IAA) a naturally occurring auxin *q.v.*

indole acetonitrile (IAN) a naturally occurring auxin *q.v.*

indoleamines *n.plu.* the monoamine neurotransmitters, and hormones such as serotonin and melatonin, which contain an indole group.

Indo-Malayan subregion subdivision of the Oriental zoogeographical region, comprising the Malaysian peninsula, the Indonesian islands west of Wallace's line, and the Philippines.

Indo-Malaysian subkingdom subdivision of the Palaeotropical floristic kingdom, comprising the Indian subcontinent and the Himalayas, southernmost China, South-East Asia, the Malaysian archipelago,

Indonesia, Papua New Guinea and the Philippines. It is divided into the following regions: Continental South-East Asiatic, Indian, Malaysian.

induced *a. appl.* genes that are transcribed, or proteins that are synthesized, only in response to a specific stimulus. *cf.* constitutive. *see also* inducer, induction.

induced fit a model of enzyme–substrate interaction in which the substrate induces a conformational change in the active site of an enzyme on binding which enables the substrate to fit into the active site. *cf.* lock-and-key model.

induced mutations those occurring as a result of deliberate treatment with a mutagen. *cf.* spontaneous mutations.

induced pluripotent stem cell (iPS cell, iPSC) stem cell derived from an adult mammalian somatic cell that has been 'reprogrammed' to pluripotency by the introduction and expression of a small number of genes that are active in the pluripotent cells of the embryo. iPS cells resemble embryonic stem cells in the diversity of cell types they can give rise to.

inducer *n.* (1) any compound that specifically causes the synthesis of an enzyme, *cf.* repressor; (2) in bacteria, any small molecule that specifically causes the expression of genes specifying the enzymes required to metabolize it; (3) any chemical or physical stimulus that causes the expression of a specific gene; (4) in embryology, any substance produced by cells that influences neighbouring cells or tissues.

inducible *a.* (1) *appl.* proteins whose synthesis is stimulated in the presence of a specific agent (the inducer) which may be a chemical compound or a physical stimulus such as heat or light, *cf.* constitutive; (2) *appl.* genes whose transcription is similarly specifically stimulated.

induction *n.* (1) act or process of causing to occur; (2) (*dev.*) process in development whereby a cell or tissue directs neighbouring cells or tissues to develop in a particular way; (3) (*genet.*) the specific synthesis of proteins in response to some stimulus, involving the specific activation and transcription of the genes encoding the required proteins; (4) (*neurobiol.*)

lowering by one reflex of the threshold of another, spinal induction; (5) (*virol.*) of viruses, the production of infectious particles from a cell carrying a provirus or prophage.

inductive stimulus an external stimulus that influences the growth or behaviour of an organism; an internal stimulus causing the phenomenon of induction in development.

inductor organizer *q.v.*

indumentum *n.* (1) plumage of birds; (2) hairy covering in plants or animals.

indurated *a.* becoming firmer or harder.

indusial *a.* (1) containing insect larval cases, as certain limestones; (2) *pert.* an indusium.

indusiate *a.* having an indusium.

indusium *n.* (1) (*bot.*) an outgrowth of epidermis in plants, covering and protecting a sorus, as in ferns; (2) (*mycol.*) outgrowth hanging from top of stipe in some fungi; (3) (*zool.*) case of a larval insect. *plu.* **indusia.**

industrial melanics dark-coloured forms of otherwise light-coloured moths and other insects that have increased in industrial areas since the Industrial Revolution, selected, it is thought, by the need for camouflage against soot-blackened walls and trees.

induviae *n.plu.* scale leaves.

inequilateral *a.* (1) having two sides unequal; (2) having unequal portions on either side of a line drawn from umbo to gape of a bivalve shell.

inequivalve *a.* having the valves of shell unequal, *appl.* molluscs.

inerm(ous) *a.* (1) without means of defence and offence; (2) without spines.

inert *a.* (1) physiologically inactive; (2) *appl.* regions of heterochromatin, in which genes are not active.

inertia *n.* the ability of a living system to resist being disturbed.

infection *n.* invasion of the tissues of the body by bacteria, viruses, fungi, protozoans, and other internal parasites. *cf.* infestation.

infection thread structure formed by invagination of root hair cell through which the nitrogen-fixing rhizobia enter host tissue.

infectious *a.* (1) capable of causing an invasion; (2) capable of being transmitted from one organism to another.

infectivity *n.* a quantitative measure of the ability of a virus, bacteria or other parasite to cause an infection, measured by various assays.

inferior *a.* below; (1) growing or arising below another organ or structure; (2) *appl.* plant ovary having perianth inserted around the top; (3) *appl.* sepals, petals or stamens attached to the top of the ovary.

inferior colliculi paired structure on dorsal surface of midbrain, which receives auditory information from auditory nerve. *cf.* superior colliculi.

inferobranchiate *a.* with gills under margin of mantle, as certain molluscs.

inferolateral *a.* below and at, or towards, the side.

inferomedian *a.* below and about the middle.

inferoposterior *a.* below and behind.

infertile *a.* (1) not fertile as a result of the non-production of gametes or a failure of fertilization. *n.* **infertility**; (2) nonreproductive.

infestation *n.* invasion of the body by ectoparasites such as ticks or mites, which remain on the surface and do not enter tissues.

inflammation *n.* painful swelling and redness of tissues caused as a result of infection or tissue damage.

inflammasomes multiprotein complexes that recruit and activate the proteases caspase-1 and/or caspase-5, whose actions promote inflammation.

inflammatory cells cells that invade tissues and help to cause inflammation. Examples are macrophages, mast cells, eosinophils and neutrophils.

inflammatory mediator general term for any molecule involved in causing inflammation. Examples are histamine, chemokines (which attract white blood cells into inflamed tissues), leukotrienes, prostaglandins and tumor necrosis factor-alpha (TNF-α).

inflammatory reactions the cellular and molecular reactions that cause inflammation. They are due chiefly to the activities of macrophages and white blood cells that enter inflamed tissues.

inflammatory T cell Th1 cell *q.v.*

inflected, inflexed *a.* curved or abruptly bent inwards or towards the axis.

inflorescence *n.* (1) flower-head in flowering plants; (2) in mosses and liverworts, the area bearing the antheridia and archegonia.

inflorescence meristem apical shoot meristem from which an inflorescence develops.

influents *n.plu.* the animals present in a plant community, or those primarily dependent and acting upon the dominant plant species.

influenza virus *n.* respiratory disease RNA virus of the family Orthomyxoviridae, which causes the respiratory disease influenza. The virus undergoes mutation over short time periods, causing seasonal outbreaks. Epidemics and pandemics are caused by the emergence of new influenza strains with major genetic changes, to which most people are not immune. *see* genetic drift, genetic shift.

informational *a. appl.* molecules: DNA and RNA, the carriers of genetic information.

infra- (1) prefix derived from Gk *infra*, below; (2) in classification, denotes a group just below the status of a subgroup of the taxon following it, as in infraclass, the group below the subclass.

infra-axillary *a.* branching off just below the axil.

infrabranchial *a.* below the gills.

infrabuccal *a.* (1) below the cheeks; (2) below the buccal mass in molluscs.

infracentral *a.* below a vertebral centrum.

infraclass *n.* taxonomic grouping between subclass and order, e.g. Eutheria (mammals lacking pouches) and Metatheria (marsupials which bring forth live young) in the subclass Theria.

infraclavicle *n.* membrane bone occurring in pectoral girdle of some fishes.

infraclavicular *a.* below the clavicle.

infracortical *a.* below the cortex.

infracostal *a.* beneath the ribs, *appl.* muscles.

infradentary *a.* below the dentary bone.

infradian *a. appl.* biological rhythm with a period of less than 24 hours.

infrahyoid *a.* beneath the hyoid bone, *appl.* muscles.

infralabial *a.* beneath the lower lip.

infralittoral *a.* (1) *appl.* depth zone of lake which is permanently covered with rooted or floating macroscopic vegetation; (2) upper subdivision of the marine sublittoral zone.

inframarginal *a.* under the margin, or marginal structure.

inframaxillary *a.* beneath maxilla, *appl.* nerves.

infraneuston *n.* small aquatic animals living on the underside of the surface film of water, such as some mosquito larvae.

infraorbital *a.* beneath the orbit of the eye.

infrapatellar *a. appl.* pad of fat beneath patella.

infrapopulation *n.* all the organisms of a single species of parasite within a single host at a particular time.

infrared radiation (IR) electromagnetic radiation with wavelengths from *ca.* 800 μm to *ca.* 1 mm.

infrascapular *a.* beneath the shoulder blade.

infraspecific *a.* (1) occurring within a species, *appl.* e.g. variation, competition; (2) *pert.* a subdivision of a species, such as a subspecies or variety.

infrastapedial *a.* beneath the stapes of ear.

infrasternal *a.* below the breast bone.

infratemporal *a.* beneath the temporal bone.

infructescence anthocarp *q.v.*

infundibula *n.plu.* (1) passages surrounded by air cells in the lung; (2) *plu.* of infundibulum.

infundibulum *n.* (1) funnel-shaped organ or structure; (2) the siphon of a cephalopod; (3) outpushing of floor of brain that develops into the stalk of the pituitary gland; (4) conus arteriosus *q.v.*

infusoria, infusorians *n., n.plu.* term originating in the 19th century for microscopic animals such as protozoans and rotifers. Until quite recently used to denote the ciliate protozoans, but now no longer in general scientific use.

infusoriform *a. appl.* ciliated freeswimming larvae of Mesozoa.

ingesta *n.* sum total of material taken in by ingestion.

ingestion *n.* mode of nutrition in which solid food is swallowed or taken into the gut or food cavity. *a.* **ingestive**. *v.* **ingest**.

Ingoldian fungi certain aquatic hyphomycetes found on submerged leaves and twigs in well-aerated water.

ingression *n.* the penetration of superficial cells individually into the interior of the embryo.

ingroup *n.* in phylogenetic studies, the taxon under study. This is usually a group of closely related species considered to be monophyletic. *cf.* outgroup.

inguinal *a. pert.* or in region of the groin.

inhalant, inhalent *a.* adapted for breathing in or drawing in water, such as terminal pores of sponges and siphons of molluscs.

inheritance *see* heredity, Mendelian inheritance, non-Mendelian inheritance, polygenic inheritance, simple Mendelian trait.

inhibin A, inhibin B polypeptide hormones that inhibit the secretion of folliclestimulating hormone. They are produced by the gonads.

inhibition *n.* (1) prevention or checking of an action, process or biochemical reaction; (2) action of one neuron on another tending to prevent it from generating an impulse.

inhibitor *n.* (1) any agent which checks or prevents an action or process; (2) any substance which reversibly or irreversibly prevents the normal action of an enzyme without destroying the enzyme. Competitive inhibitors act by binding to the active site and preventing binding of substrate, non-competitive inhibitors act by binding to other parts of the enzyme.

inhibitory *a.* (1) tending to prevent a reaction or action, *appl.* e.g. stimuli, cells, compounds; (2) *appl.* neurons whose activity prevents adjacent neurons from firing; (3) *appl.* neurotransmitters whose action tends to prevent neurons from firing; (4) *appl.* synapses at which transmission of a signal tends to prevent firing in the postsynaptic neuron. *cf.* excitatory.

inhibitory postsynaptic potential (IPSP) hyperpolarizing potential produced in a postsynaptic neuron by neurotransmitter action. It tends to prevent an action potential being generated.

inion *n.* external protuberance of occipital bone.

initial *n.* self-renewing undifferentiated cell in a plant meristem. It produces both further initials, which remain in the meristem, and sister cells that go on to leave the meristem and differentiate.

initiation *n.* in animal behaviour, the response to a stimulus with behaviour that was not present before.

initiation codon the first codon of the protein-coding region in mRNA and the point at which translation starts. It almost invariably specifies methionine (in eukaryotes) or formylmethionine (in bacteria and mitochondria). *alt.* initiator codon.

initiation complex (1) transcription complex *q.v.*; (2) complex of small ribosomal subunit, aminoacyl-initiator tRNA and messenger RNA that is assembled before translation commences.

initiation factor any of the accessory proteins needed for the initiation of translation and which are involved in mRNA binding to the ribosome, initiator tRNA binding, and ribosome subunit association and dissociation. IF-2 and eIF-2 are guanine-nucleotide binding proteins. Designated IF-1, IF-2, IF-3 in bacteria, eIF-1, eIF-2, . . . eIF-6 in eukaryotic cells.

initiator codon initiation codon *q.v.*

initiator tRNA a specific tRNA that picks up methionine (in eukaryotes) or formylmethionine (in bacteria or mitochondria) and binds to the free small subunit of the ribosome as one of the first steps in the formation of the initiation complex in protein synthesis.

ink sac in certain cephalopods such as *Sepia*, a pear-shaped body in wall of the mantle cavity which contains the ink gland, secreting a black substance, ink or sepia, which is ejected as a means of defence.

inland wetland land that is covered for all or part of the year with fresh water, such as swamps, bogs, marshes, fens.

innate *a.* inborn, *appl.* behaviour which does not need to be learned.

innate immunity protection against infection as a result of the activation of fixed, relatively nonspecific defence mechanisms. *cf.* adaptive immunity.

innate releasing mechanism (IRM) an instinctive mechanism in an animal which is activated to produce a response by some external stimulus.

inner cell mass (ICM) mass of cells in the mammalian blastocyst that will give rise to an embryo.

inner ear bony cavity in skull behind middle ear. It contains the membranous labyrinth, the cochlea, and the utricle and saccule connecting them. The labyrinth is concerned with balance and sensing position and consists of three fluid-filled semicircular canals at right angles to each other. The spirally coiled cochlea is concerned with sensing pitch of sounds.

inner mitochondrial membrane *see* mitochondrion.

innervate *v.* to supply with nerves.

innervation *n.* the nerve supply to an organ or part.

innervation ratio the number of muscle fibres innervated by a single motor axon.

innominate *a.* nameless.

innominate artery artery that gives rise to arteries in head and forelimb.

innominate bone the hip bone or lateral half of the pelvic girdle.

innominate vein vein that joins up veins from head and forelimb.

inoculation *n.* (1) administration of a vaccine in order to induce protective immunity; (2) introduction of bacteria or other microorganisms, or plant and animal cells, into nutrient medium to start a new culture; (3) introduction of pathogen into a host.

inoculum *n.* cells, bacteria or spores used for starting off a culture.

inoperculate *a.* without an operculum or lid; (1) *appl.* spore capsules; (2) *appl.* fish lacking a gill cover.

inordinate *a.* not in any regular arrangement, *appl.* spores in an ascus.

inorganic *a. appl.* material or molecules which do not contain carbon.

inorganic chemistry the chemistry of substances other than those containing carbon.

inosine (I) *n.* nucleoside found in tRNA, formed from adenosine by replacement of the amino group with an oxygen, often present in the first anticodon position where it can pair with U, C or A. The base is hypoxanthine.

inosine monophosphate (inosinic acid, inosinate) (IMP) *n.* a nucleotide, biosynthetic precursor to adenine monophosphate and guanosine monophosphate.

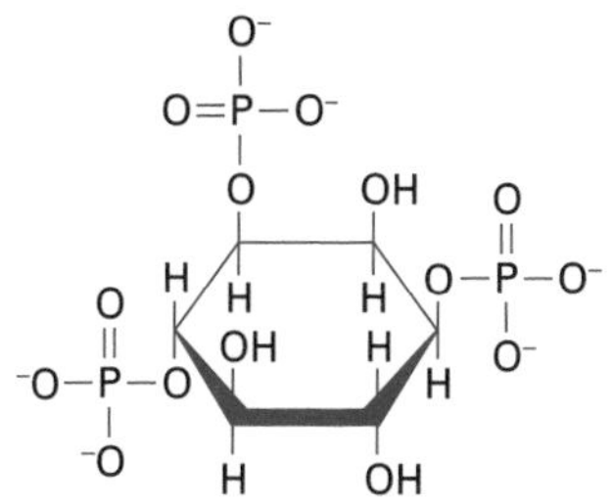

Fig. 24 Inositol 1,4,5-trisphosphate.

inositol *n.* a cyclic hexahydric alcohol, occurring in various forms, of which *myo*-inositol is the most important. It is a constituent of phospholipids (e.g. phosphatidylinositol) and also of phytic acid and phytin in plants.

inositol phospholipid pathway intracellular signalling pathway that starts with the generation of inositol trisphosphate and diacylglycerol by phospholipase C from inositol-containing phospholipids in the plasma membrane. *alt.* phosphoinositide pathway.

inositol trisphosphate (InsP₃, IP₃) inositol 1,4,5-trisphosphate, an intracellular signalling molecule produced by the action of phospholipase C on membrane phosphatidylinositol phosphate in response to stimulation of certain cell-surface receptors. Its chief effect is to stimulate the release of Ca^{2+} from intracellular stores such as the endoplasmic reticulum. Raising the level of free Ca^{2+} in the cytosol activates intracellular signalling pathways, leading to the appropriate cellular response. *see* Fig. 24.

inquilinism *n.* type of symbiosis in which an animal lives in the nest of another, is tolerated by the host and shares its food.

Insecta, insects *n., n.plu.* very large class of arthropods, found as fossils from the Devonian onwards, containing some three-quarters of all known extant species of animal. The insects include flies, bees and wasps, ants, butterflies and moths, beetles, dragonflies, grasshoppers and crickets, and many other orders. The segmented body is divided into distinct head, thorax and abdomen. The head bears one pair of antennae and paired mouthparts, and the thorax bears three pairs of walking legs and usually one or two pairs of wings. Other types of appendage may be present on the abdomen. The life history usually includes metamorphosis. *alt.* Hexapoda. *see* Appendix 3 for orders.

insecticide *n.* chemical that kills insects.

Insectivora, insectivores *n., n.plu.* large order of primitive insect-eating and omnivorous placental mammals known from Cretaceous times to the present, including hedgehogs, moles and shrews.

insectivorous *a.* insect-eating, *appl.* certain animals and carnivorous plants. *alt.* entomophagous. *n.* **insectivore**.

insemination *n.* (1) introduction of semen or sperm into female genital tract; (2) transfer of a fertilized ovum from one female to another.

inserted *a.* (1) attached; (2) united by natural growth, as petals to ovary.

insertion *n.* (1) point of attachment of organs, as of muscles, leaves; (2) point on which force of a muscle is applied; (3) mutation in which a segment of chromosome or a short sequence of bases is inserted in a gene.

insertional mutagenesis the production of a mutation as a result of the insertion of a piece of DNA into a gene by *in vitro* gene manipulation.

insertion sequence (IS) a simple type of transposon found in bacteria, consisting of around 800–1500 bp and carrying only the genetic functions for its own transposition. Insertion sequences are also found at the ends of some other transposons.

insessorial *a.* adapted for perching.

insight learning learning involving reasoning, as in humans and to a lesser extent in some animals.

insistent *a. appl.* hind-toe of certain birds, whose tip only reaches the ground.

in situ in the original place.

in situ **hybridization** technique for locating the site of a specific DNA sequence on a chromosome, by treating mitotic cells with a radioactively or fluorescently labelled nucleic acid probe complementary for that sequence. The probe binds to the DNA sequence and its position can be detected. The technique is also applied to detect and locate the synthesis of specific mRNAs in tissues. *see also* FISH.

insolation *n.* exposure to the sun's rays.

InsP₃ inositol trisphosphate *q.v.*

inspiration *n.* the act of drawing air or water into the respiratory organs.

instaminate *a.* lacking stamens.

instar *n.* larva of insect or other arthropod between moults.

instinct *n.* behaviour that occurs as an inevitable stereotyped response to an appropriate stimulus, sometimes equivalent to species-specific behaviour. *a.* **instinctive**.

instipulate *a.* lacking stipules.

instructive *a. appl.* developmental signal that changes the course of development of a tissue, *cf.* permissive.

instrumental conditioning, instrumental learning type of associative learning in which the animal initially succeeds as a result of trial and error, is rewarded, and eventually learns to perform the task correctly at the first attempt, the correct response being 'instrumental' in eliciting the reward. *alt.* operant conditioning.

insula *n.* (1) triangular eminence lying deep in lateral fissure of temporal lobe of brain; (2) islet of Langerhans *q.v.*; (3) blood island *q.v.*

insulator elements DNA sequences that prevent the spread of heterochromatin into adjacent chromatin. They are also able to prevent an enhancer associated with one gene from influencing the transcription of an adjacent gene. *alt.* boundary elements.

insulin *n.* polypeptide hormone produced by β-cells of the islets of Langerhans in the pancreas. It decreases the amount of glucose in the blood by promoting glucose uptake by cells and increasing the capacity of the liver to synthesize glycogen. Its action is antagonistic to glucagon, adrenal glucocorticoids and adrenaline, and its deficiency or reduced activity produces a raised blood sugar level and diabetes.

insulin-dependent diabetes mellitus (IDDM) type I diabetes *q.v.*

insulin-like growth factor I (IGF-I) polypeptide growth factor whose production is stimulated by growth hormone. Involved in post-embryonic and post-natal growth.

insulin-like growth factor II (IGF-II) polypeptide growth factor active in promoting growth in the embryo.

intectate *a.* without a tectum.

integral *a. appl.* membrane proteins that are firmly embedded in the membrane and difficult to remove. They are either transmembrane proteins or covalently linked to the membrane via lipids.

integrase *n.* enzyme that catalyses the integration of one DNA molecule into another by recombination.

integrate *n.* a piece of DNA inserted into another DNA molecule.

integrated pest management combined use of biological, chemical and cultivation methods to keep pests at an acceptable level.

integration *n.* (1) in the nervous system, the coordination of inputs from many neurons in a single neuron which combines them to produce a new signal for onward transmission; (2) the insertion by recombination of phage, plasmid, viral or other double-stranded DNA into another DNA molecule (e.g. bacterial chromosome or eukaryotic cell chromosome) to form a new continuous DNA molecule.

integrifolious *a.* with entire leaves.

integrin *n.* any of a large family of dimeric transmembrane glycoproteins that act as cell adhesion molecules on animal cells. Cells adhere to other cells and to extracellular matrix by binding of integrins to receptor proteins on these surfaces.

integument *n.* (1) a covering, investing, or coating structure or layer; (2) coat of ovule. *a.* **integumentary**.

intein *n.* an internal part of a protein sequence that is excised after translation to form an independent protein, while the two remaining portions (the exteins) are spliced together to make another protein. Inteins have been found in proteins from bacteria, archaea and eukaryotes. *alt.* protein intron.

intelligence *n.* a property of the mind that includes abilities such as the capacity for learning, abstract thought, understanding, reasoning, learning from experience, planning and problem solving.

intelligent design the idea that complex biological structures and some other aspects of the natural world must be designed by some supernatural intelligence, and are not the result of an undirected process such as natural selection or of physical laws. It differs from Creationism (*q.v.*) in not relying on the biblical account of creation. Intelligent design is dismissed as pseudoscience by the vast majority of scientists.

intentional behaviour behaviour that involves a mental representation of a goal that guides the behaviour.

intention movement preparatory motions an animal goes through before a complete behavioural response, e.g. the snarl before the bite.

inter- prefix derived from L. *inter*, between.

interaction domain protein domain that recognizes particular motifs and other domains on proteins and is involved in specific interactions with other proteins.

interactome *n.* collectively, the interactions that can be made between all the proteins expressed by a given genome.

interalveolar *a.* among alveoli, *appl.* cell islets.

interambulacral *a. appl.* area of echinoderm test between two ambulacral areas.

interatrial *a. appl.* groove and partition separating the two atria of the heart.

interatrial septum thin transparent sheet of nervous tissue from heart, containing neurons receiving input from the vagus nerve, and innervating the muscles of the atrium.

interauricular *a.* between auricles of heart.

interaxillary *a.* placed between the axils.

interband *n.* in polytene chromosomes, a lightly staining region of variable size alternating with densely staining regions (bands) along the length of the chromosome, each chromosome having a characteristic and invariant banding pattern.

interbrachial *a.* between arms or rays (e.g. of echinoderms).

interbranchial *a. appl.* partitions between successive gill slits.

interbreed *v.* to cross different varieties, species or genera of plants and animals.

interbreeding *a. appl.* a population whose members can breed successfully with each other.

intercalary *a.* (1) inserted between others; (2) *appl.* growth elsewhere than at growing point; (3) (*bot.*) *appl.* meristematic

layers between masses of permanent tissues; (4) (*zool.*) *appl.* veins between main veins of insect wings; (5) *appl.* discs: transverse wavy bands formed by boundaries of sarcomeres in heart muscle. *alt.* **intercalated**.

intercalation *n.* insertion between two other structures, as of certain mutagens such as acridine dyes, which insert between the base pairs in DNA.

intercarpal *a.* among or between carpal bones, *appl.* joints.

intercarpellary *a.* between the carpels.

intercellular *a.* among or between cells, *appl.* spaces, material.

intercellular adhesion molecules (ICAMs) family of cell-surface molecules of the immunoglobulin superfamily that bind to leukocyte integrins.

intercentral *a.* between two centra.

intercentrum *n.* a 2nd central ring in some types of vertebrae.

interchondral *a. appl.* articulations and ligaments between costal cartilages.

interchromosomal *a.* (1) between chromosomes, *appl.* recombination; (2) *appl.* fibrils that play a part in the beginning of cell-wall formation in plants.

interclavicle *n.* bone between clavicles of shoulder girdle in most reptiles.

interclavicular *a.* between the clavicles.

intercostal *a.* (1) between the ribs, *appl.* arteries, veins, glands, muscles; (2) between the veins of a leaf.

intercostobranchial *a. appl.* lateral branch of 2nd intercostal nerve which supplies upper arm.

intercrine chemokine *q.v.*

intercropping *n.* growing two types of crop on the same land, usually where one benefits the other, as legumes and cereals.

intercross *n.* the crossing of heterozygotes of the F_1 generation amongst themselves.

interdeferential *a.* between the vasa deferentia.

interdemic selection situation when entire breeding groups (demes) are the basic unit of natural selection.

interdigitating reticular cell dendritic cell *q.v.*

interfascicular *a.* (1) *appl.* parenchyma separating vascular bundles from each other in stems of some conifers and dicotyledons; (2) situated between the vascular bundles, *appl.* cambium.

interfemoral *a.* between the thighs.

interference *n.* (1) in virology, the case where the presence of one type of virus in a cell prevents concurrent infection with, or multiplication of, another virus; (2) in genetic recombination, the effect that the presence of one crossover reduces the probability of another occurring in its vicinity.

interference colours colours produced by optical interference between reflections from different layers of a surface.

interferons *n.plu.* family of small proteins produced by vertebrate cells in response to viral infection and as cytokines during an immune response. Interferons α and β (IFN-α and IFN-β) are induced by viral infection and have antiviral effects as a result of their inhibition of translation and stimulation of the breakdown of viral nucleic acid within the infected cell. Interferon γ (IFN-γ) is secreted by activated T lymphocytes during an immune response and is an important growth factor for lymphocytes and other immune system cells.

interfertile *a.* able to interbreed.

interganglionic *a.* connecting two ganglia, as nerve cords.

intergemmal *a.* between taste buds, *appl.* nerve fibres.

intergeneric *a.* between genera, *appl.* hybridization.

intergenic *a.* situated between genes.

interglacial *a. appl.* or *pert.* the ages between glacial ages, particularly of the Pleistocene.

interlamellar *a.* (1) *appl.* vertical bars of tissue joining gill lamellae of molluscs; (2) *appl.* compartments of lung book in scorpions and spiders; (3) *appl.* spaces between lamellae or gills of agarics.

interleukin *n.* any of a diverse group of cytokines produced by activated macrophages and lymphocytes during an immune response. They act on lymphocytes and other leukocytes to stimulate their proliferation and differentiation. Abbreviated IL-1, IL-2, IL-3, etc.

interleukin-2 (IL-2) one of the first cytokines produced by activated T cells, and

which acts in an autocrine manner to stimulate the proliferation and differentiation of the T cells that produce it.

interlittoral *a.* shallow marine zone to a depth of around 20 m.

interlobar *a.* (1) between lobes; (2) *appl.* sulci and fissures dividing cerebral hemispheres into lobes.

intermalar *a.* situated between the cheek bones.

intermaxilla *n.* (1) bone between maxillae; (2) premaxilla *q.v.*

intermediary *a. appl.* nerve cells receiving impulses from afferent cells and transmitting them to efferent cells.

intermediary metabolism metabolic pathways by which the basic molecular building blocks in a cell (e.g. monosaccharides, amino acids, nucleotides) are interconverted and incorporated into larger molecules.

intermediate *n.* in a chemical reaction, a chemical species that forms transiently along the reaction path from starting material to product.

intermediate compartment *cis* Golgi network *q.v.*

intermediate filament (IF) one of the three main components of the eukaryotic cell cytoskeleton. A tough and durable type of filament that forms a network in eukaryotic cells subject to stress. Examples are the keratin filaments of skin cells, the neurofilaments of nerve cell axons, the vimentin filaments of connective tissue cells, and the lamin filaments of the nuclear lamina. Intermediate filaments are constructed of subunits of various types of intermediate filament proteins.

intermediate host organism in which a parasite lives for part of its life cycle but in which it does not become sexually mature.

intermediate junction type of adhesive junction between epithelial cells which runs in a band around the circumference of the upper part of the cell, just below the tight junction. *alt.* belt desmosome, adhaerens junction, zonula adhaerens.

intermediate memory type of memory that lasts longer than short-term memory but not as long as long-term memory.

intermedin melanocyte-stimulating hormone *q.v.*

intermedium *n.* a small bone of carpus and tarsus.

intermembrane space space between the two membranes surrounding organelles such as mitochondria and chloroplasts, and between the inner and outer membranes of bacterial envelopes.

intermolecular *a.* between molecules, *appl.* e.g. hydrogen bonds, distances.

intermuscular *a.* between or among muscle fibres.

internal *a.* (1) located on inner side; (2) nearer middle axis; (3) located or produced within.

internal fertilization fertilization of eggs by sperm that takes place inside the body of the female, as e.g. in insects, mammals and birds. *cf.* external fertilization.

internal phloem primary phloem found internal to primary xylem.

internal respiration the biochemical, intracellular reactions of respiration.

internarial *a.* between the nostrils, *appl.* septum.

International Unit (IU) (1) unit used to measure vitamin content of foods; (2) unit of enzyme activity, 1 IU, defined as the amount of enzyme that will catalyse the conversion of 1 micromole of substrate in 1 minute.

interneuron *n.* small neuron in grey matter of central nervous system, interposed between afferent and efferent neurons in spinal cord, e.g. between sensory and motor neurons in a reflex arc, and forming connections between neural pathways generally.

internode *n.* the part of a plant stem between two leaf origins (nodes) or two joints.

internuncial *a.* intercommunicating, as paths of transmission or nerve fibres.

interoceptor *n.* sense organ or receptor that detects stimuli originating inside the body. *cf.* exteroceptor.

interocular *a.* between the eyes.

interoptic *a.* between the optic lobes of brain.

interorbital *a.* between the orbits of the eye, *appl.* sinus.

interosseous *a.* between bones, *appl.* arteries, ligaments, muscles, membranes, nerves. ·

interparietal *n.* in many vertebrates, a bone arising between parietals and supraoccipital.

interpeduncular *a.* *appl.* fossa between cerebral peduncles, and to ganglion.

interphase *n.* (1) the period between one mitosis and the next in a eukaryotic cell. It is the period during the cell cycle during which growth occurs, the chromosomes are in the uncondensed state, transcription occurs, and during which the DNA is replicated. *see* Fig. 8 (p. 106); (2) the period sometimes occurring between the 1st and 2nd meiotic division.

interpositional growth of cells, by interposition between neighbouring cells without loss of contact.

interradius *n.* a radius between the four primary radii of a radially symmetrical animal.

interrenal *a.* between the kidneys, *appl.* veins.

interrenal body a gland situated between kidneys of elasmobranch fishes, representing the adrenal cortex of higher vertebrates.

interrupted *a.* (1) with continuity broken; (2) irregular; (3) asymmetrical.

interruptedly pinnate pinnate with pairs of small leaflets occurring between larger ones, *appl.* leaves.

interscapular *a.* (1) between the shoulder blades, *appl.* e.g. feathers; (2) *appl.* brown fat: the so-called hibernating gland, found between the shoulder blades of some rodents.

interscutal *a.* between scuta or scutes.

intersegmental *a.* (1) between segments; (2) between spinal segments, *appl.* axons, septa.

interseminal *a.* between seeds or ovules, *appl.* scales in certain gymnosperms.

interseptal *a.* *appl.* spaces between septa or partitions.

intersex *n.* (1) organism with characteristics intermediate between a typical male and a typical female of its species; (2) organism first developing as a male or female, then as an individual of the opposite sex.

interspecific *a.* between distinct species, *appl.* crosses, as mule, hinny, cattalo, tigon.

interspecific competition competition between the members of different species for the same resource. *cf.* intraspecific (2).

interspersed repeated DNA repeated DNA sequences occurring as individual copies interspersed throughout the genome, rather than as blocks of tandemly repeated sequences.

interspinal *a.* occurring between spinal processes or between spines, *appl.* bones, ligaments, muscles.

intersterility *n.* incapacity to interbreed.

intersternal *a.* (1) between the sterna; (2) *appl.* ligaments connecting manubrium and body of sternum.

interstitial *a.* occurring in interstices or spaces; (1) *appl.* flora and fauna living between sand grains or soil particles; (2) *appl.* cells, *see* interstitial cells; (3) *appl.* fluid, *see* extracellular fluid.

interstitial cells (1) connective tissue cells in the vertebrate testis. They lie between the tubules and produce testosterone; (2) small cells lying in the interstices between other cells in coelenterates and which give rise to various cell types. *alt.* I-cells.

interstitial cell-stimulating hormone (ICSH) luteinizing hormone *q.v.*

intertemporal *n.* paired membrane bone, part of sphenoid complex.

intertentacular organ structure in certain bryozoans through which eggs are released.

intertidal *a. appl.* shore organisms living between high- and low-water marks. *see* Fig. 34 (p. 605).

intertrochanteric *a.* between trochanters, *appl.* crest, line.

intertrochlear *a. appl.* an ulnar ridge fitting into a groove of the humerus.

intertubular *a.* (1) between tubules, esp. kidney tubules, *appl.* capillaries; (2) between seminiferous tubules.

intervarietal *a. appl.* crosses between two distinct varieties of a species.

intervening sequence intron *q.v.*

interventricular *a.* (1) between the ventricles of heart or brain; (2) *appl.* foramen: foramen of Monro, the passage between the third and lateral ventricles in brain.

intervertebral *a.* occurring between the vertebrae, *appl.* e.g. discs, fibrocartilages, veins.

intervillous *a.* (1) occurring between villi; (2) *appl.* spaces in placenta filled with maternal blood.

interxylary *a.* between xylem strands, *appl.* phloem.

intestine *n.* that part of the alimentary canal from the pyloric end of the stomach to the anus. In humans it comprises the duodenum, jejunum, ileum, caecum, colon and rectum in that order. *a.* **intestinal**.

intima *n.* layer of tissue forming the innermost lining of a part or organ. *alt.* tunica intima.

intine *n.* the cellulosic inner layer of wall of pollen grain.

intolerant *a.* incapable of living in a particular set of conditions.

intra- prefix derived from L. *intra*, within, and signifying within, inside. *cf.* inter-.

intrabulbar *a.* within a taste bud.

intracapsular *a.* contained within a capsule.

intracardiac endocardiac *q.v.*

intracellular *a.* within a cell.

intracellular motility movement that is generated within a cell, e.g. cytoplasmic streaming, muscle contraction, intracellular transport.

intracellular signalling pathways the biochemical pathways that carry signals from cell-surface receptors to their intracellular destination. They are made up of sets of interacting proteins and some also involve small molecules such as cyclic AMP and inositol trisphosphate.

intracellular transport the directed transport and movement of organelles within the cell. Material is transported between certain organelles, and to the plasma membrane for secretion, in the form of small membrane-bounded vesicles. Large particles are also moved through the cytoplasm.

intrachromosomal *a. appl.* duplication or rearrangement occurring within a chromosome.

intracistronic complementation the ability of two mutations in one gene (a cistron) to produce a normal phenotype (to complement each other) when present in the same cell.

intraclonal *a.* within a clone, *appl.* differentiation.

intracortical *a.* within the cortex, *appl.* nerve tracts linking different parts of the cerebral cortex.

intrademic selection selection within a local breeding group.

intradermal *a.* within the dermis of skin.

intraepithelial *a.* occurring within epithelium, *appl.* glands.

intrafascicular *a.* within a vascular bundle.

intrafoliaceous *a. appl.* stipules encircling stem and forming a sheath.

intrafusal *a.* (1) *appl.* muscle fibres within muscle spindle; (2) *appl.* nerve endings in tendons.

intragemmal *a.* within a taste bud.

intrageneric *a.* among members of the same genus.

intragenic *a.* within a gene, *appl.* e.g. mutation, recombination, spacer.

intragenic suppression *see* back mutation.

intrajugular *n.* projection in middle of jugular notch of occipital bone.

intralamellar *a.* within a lamella, *appl.* trama of gill-bearing fungi.

intramembrane particles protein particles seen studding the interior faces of freeze-fractured membranes.

intramembranous *a.* within a membrane, *appl.* bone development.

intramolecular *a.* occurring or existing within a molecule.

intranarial *a.* inside the nostrils.

intranuclear *n.* within the nucleus.

intraparietal *a.* (1) enclosed within an organ; (2) within the parietal lobe of the brain.

intrapleural *a.* within the thoracic cavity.

intrasexual *a. appl.* selection amongst competing individuals of the same sex.

intraspecific *a.* (1) within a species, *appl.* e.g. variation; (2) *appl.* competition between the members of the same species for the same resource, e.g. mates, food, territory.

intrastelar *a.* within the central vascular tissue (stele) of a stem or root.

intrathecal *a.* within the meninges of the spinal cord.

intrauterine *a.* within the uterus.

intravaginal *a.* (1) within vagina; (2) contained within a sheath, as grass leaves or branches.

intravascular *a.* within blood vessels.

intraventricular *a.* within a ventricle, *appl.* caudate nucleus of corpus striatum, seen within a ventricle of brain.

intravesical *a.* within the bladder.

intravitelline *a.* within the yolk of an egg or ovum.

intrazonal *a.* (1) within a zone; (2) *appl.* locally limited soils, differing from prevalent or normal soils of the region or zone.

intrinsic *a.* (1) inherent; (2) inward; (3) *appl.* inner muscles of a part or organ; (4) *appl.* rate of natural increase in a population having a balanced age distribution; (5) *appl.* sensation of brightness due to the differential response of retinal cells to different wavelengths; (6) *appl.* membrane proteins: proteins which span the whole membrane. *cf.* extrinsic.

intrinsic factor glycoprotein secreted by stomach, binding cobalamin (vitamin B_{12}) in the intestine and carrying it into the blood. Its deficiency causes pernicious anaemia due to impaired cobalamin absorption. *alt.* gastric intrinsic factor.

intrinsic isolating mechanism any genetic mechanism preventing interbreeding.

intrinsic pathway series of reactions in blood leading to clot formation, triggered by contact with an abnormal surface such as the walls of a glass vessel, the final stages being the same as in the extrinsic pathway.

intrinsic rate of increase the fraction by which a population is growing at each instant of time, symbolized by r.

introduced *a. appl.* plants and animals not native to the country and thought to have been brought in by humans.

introgression *n.* the gradual diffusion of genes from the gene pool of one species into another when there is some hybridization between the two species as a result of incomplete genetic isolation.

intromission *n.* insertion of the erect penis into the vagina during copulatory behaviour.

intromittent *a.* adapted for insertion, *appl.* male copulatory organs.

intron *n.* a non-coding nucleotide sequence, one or more of which interrupt the coding sequence of most eukaryotic genes. Many genes have substantial numbers of long intron sequences that take up most of the total length of the gene. Introns are transcribed into RNA as part of the primary RNA transcript, and subsequently removed by RNA splicing to produce a functional mRNA or other RNA. They are extremely rare in prokaryotic DNA. Some introns are self-splicing, i.e. they can remove themselves from RNA without the aid of any protein enzymes. Self-splicing introns consist of two main classes, group I and group II, in which the detailed biochemistry of the splicing reaction is different. *alt.* intervening sequence. *cf.* exon. *see also* RNA splicing.

intron–exon boundary the boundary between an intron and its adjacent exon in a gene or primary RNA transcript, each intron having a 3′ and a 5′ boundary. The boundaries are the sites of RNA splicing.

introrse *a.* (1) turned inwards or towards the axis; (2) of anthers, opening towards the centre of the flower.

introvert *a.* that which can be drawn inwards, e.g. anterior region of some polyps or of certain annelid worms.

intumescence *n.* (1) the process of swelling up; (2) a swollen or tumid condition.

intussusception *n.* growth in surface extent or volume by intercalation of new material among that already present. *cf.* accretion, apposition.

inulase inulinase *q.v.*

inulin *n.* linear polysaccharide made up of fructose units, a storage carbohydrate in the roots, rhizomes and tubers of many Compositae.

inulinase *n.* enzyme hydrolysing inulin to fructose. EC 3.2.1.7. *alt.* inulase.

invagination *n.* (1) the sinking in of a wall of a hollow structure such as a vessel or a blastula, which draws the exterior layer into the interior, forming a cavity; (2) involution or turning inside out of a tube; (3) drawing into a sheath. *v.* **invaginate**.

invariant chain polypeptide that binds to peptide-binding site in a MHC class II molecule in the endoplasmic reticulum, preventing premature binding of antigen peptides.

inversion *n.* (1) a turning of a part inward, or inside out, or upside down; (2) (*genet.*) chromosomal rearrangement in which a sequence of genes appears in the reverse of its normal order on the chromosome, due to the repositioning of a segment of chromosome; (3) (*biochem.*) hydrolysis of sucrose to glucose and fructose.

invertase *n.* enzyme found in plants, fungi and bacteria that hydrolyses terminal non-

reducing β-D-fructofuranoside residues in β-D-fructofuranosides (e.g. sucrose). EC 3.2.1.26, *r.n.* β-D-fructofuranosidase.

invertebrates *n.plu.* general term for all animals without backbones, i.e. all groups except the vertebrates.

inverted repeats in nucleic acids, two adjacent or nearby copies of an identical sequence, one being in reverse orientation.

inverted U-shaped curve a dose-response curve in the form of an inverted U. It indicates a response in which increasing doses of drug elicit increasing responses at first, but response declines at higher doses.

invert sugar sucrose *q.v.*

investing bone membrane bone *q.v.*

investment *n.* the outer covering of a part, organ, animal or plant.

in vitro appl. biological processes and reactions occurring in (i) cells or tissues grown in culture or (ii) cell extracts or synthetic mixtures of cell components.

in vitro **fertilization (IVF)** the fertilization of an ovum outside the mother's body, generally followed by replacement of the fertilized egg into the mother or into a pseudopregnant 'foster mother' where it develops normally.

in vitro **translation or transcription** translation or transcription carried out in isolated extracts of cells, or defined systems consisting of the appropriate enzymes and accessory proteins.

in vivo occurring in a living organism.

involucrate *a.* bearing involucres.

involucre *n.* (1) circle of bracts at the base of a compact flowerhead; (2) leaves surrounding the groups of antheridia and archegonia in mosses and liverworts. *a.* **involucral**.

involucrum *n.* (1) the notum of metathorax in Orthoptera; (2) layer of bone formed around dead bone in some diseased conditions.

involuntary *a.* not under the control of the will, *appl.* e.g. movements.

involuntary muscle smooth muscle *q.v.*

involute *a.* (1) having edges rolled inwards at each side, *appl.* leaves; (2) tightly coiled, *appl.* shells.

involution *n.* (1) reduction of enlarged, modified or deformed conditions to normal; (2) decrease in size, or other structural or functional changes, as in old age; (3) a rolling inwards, as of edges of leaves; (4) movement of exterior cells to interior, as in gastrulation; (5) *a.* resting, *appl.* spores or stage in life cycle.

iodine (I) *n.* halogen element which forms a black shiny volatile solid in the elemental state (I_2), and is present in small amounts in seawater and concentrated in seaweed. It is an essential micronutrient for humans and other vertebrates as it is a constituent of some thyroid hormones. The radioisotope ^{131}I (half-life 8.6 days) is used as a tracer to diagnose and treat thyroid gland disorders.

iodophilic *a.* (1) staining darkly with iodine solution; (2) *appl.* bacteria that stain blue with iodine.

iodopsin *n.* photosensitive protein pigment with rhodopsin-type protein component, in the cones of the retina. *alt.* visual violet.

iodothyroglobulin *n.* compound of iodine and thyroglobulin, extractable from the thyroid gland.

ion *n.* atom or molecule that has acquired an electrical charge by gaining or losing one or more electrons.

ion channel membrane protein that forms an aqueous pore in the membrane, allowing the passive flow of ions across the membrane. Ion channels may be permanently open or may be gated, i.e. open only in response to a specific stimulus. Different ion channels convey different ions, being divided broadly into those conveying anions and those that carry cations. *see* Fig. 26 (p. 392).

ion exchange the adsorption of ions of one type onto a resin, or clay particles in soil, in exchange for others which are lost into solution.

ion-exchange chromatography separation of molecules such as proteins on the basis of their net charge by differential binding to a column of carboxylated polymer, positively charged molecules binding to the column.

ionic bond, ion pair electrostatic bond *q.v.*

ionic coupling the direct passive flow of ions from the cytoplasm of one animal cell to an adjacent cell through gap junctions.

ionic detergent detergent in which the polar end of the molecule is charged (e.g. sodium dodecyl sulphate). It is a strong detergent that can denature (unfold) proteins.

ionizing radiation short wavelength high-energy radiation such as gamma-rays and fast-moving particles such as alpha- and beta-particles emitted by radioisotopes. They cause the formation of ions in tissues, thus contributing to DNA and tissue damage.

ionophore *n.* small hydrophobic molecule that dissolves in lipid bilayers and increases the permeability of the membrane to ions.

ionophoresis *n.* movement of ions under the influence of an electric current. *a.* **ionophoretic**.

ionotropic *a. appl.* neurotransmitter receptor that contains an ion channel which is opened when the neurotransmitter binds to the receptor. *cf.* metabotropic.

ion pump protein that actively transports an ion across a biological membrane against a concentration gradient.

IP$_3$ inositol trisphosphate *q.v.*

iPS cell, iPSC induced pluripotent stem cell *q.v.*

ipsilateral *a. pert.* or situated on the same side. *cf.* contralateral.

IPSP inhibitory postsynaptic potential *q.v.*

IPTG isopropylthiogalactoside *q.v.*

IR infrared radiation *q.v.*

IRE iron-response element *q.v.*

iridal, iridial *a. pert.* the iris of the eye.

Iridales *n.* order of herbaceous monocots and including the families Corsiaceae, Iridaceae (iris) and others.

iridocytes, iridophores *n.plu.* (1) guanine-containing granules, bodies or plates of which the reflecting, silvery or iridescent tissue of skin of fish and reptiles is composed; (2) iridescent cells in integument of some cephalopods.

Iridoviridae *n.* family of enveloped, double-stranded DNA viruses that infect insects (e.g. insect iridescent viruses) and vertebrates.

iris *n.* thin circular contractile disc with a central aperture, the pupil, lying in front of the lens in the vertebrate eye.

iris cells pigment cells surrounding cone and retinula of an ommatidium of a compound eye.

IRM innate releasing mechanism *q.v.*

iron (Fe) *n.* metallic element that is an essential micronutrient for living organisms. Fe atoms are part of prosthetic groups, such as haem, in many enzymes and other proteins. In humans, a major requirement for iron is for haemoglobin synthesis.

iron–copper centre active site in the respiratory enzyme cytochrome oxidase at which molecular oxygen is held by an iron and a copper atom while being converted to water.

iron-oxidizing bacteria diverse group of aerobic bacteria and archaea that can use ferrous iron (Fe^{2+}) as their sole source of energy, oxidizing it to ferric iron (Fe^{3+}). Most also oxidize sulphur and are obligately acidophilic.

iron-response element (IRE) control sequence in DNA that binds an iron-binding gene regulatory protein, and by which a gene is regulated by iron.

iron–sulphur (FeS) protein any of a group of proteins containing a prosthetic group of non-haem iron complexed with sulphur atoms (iron–sulphur centres). They are components of photosynthetic electron-transport chains. There are various types of centres: FeS (one Fe complexed with four S atoms), Fe_2S_2 and Fe_4S_4. *alt.* non-haem iron protein.

irreciprocal *a.* (1) not reversible; (2) one-way.

irregular *a. appl.* flowers showing bilateral rather than radial symmetry. *alt.* zygomorphic.

irritability *n.* capacity to receive external stimuli and respond to them.

irrorate *a.* (1) covered as if by minute droplets; (2) with minute colour markings, as the wings of some butterflies.

IS1, IS2, etc. insertion sequence *q.v.*

isandrous *a.* having similar stamens, their number equalling the number of sections of the corolla.

isanthous *a.* having uniform or regular flowers.

ischaemia *n.* tissue damage and localized death due to lack of oxygen. *a.* **ischaemic.**

ischiadic *a. pert.* or in region of hip.

ischiatic sciatic *q.v.*

ischiopodite *n.* proximal joint of walking legs of certain crustaceans, or of maxillipeds.

ischiopubic *a. appl.* gap between ischium and pubis.

ischiopubis *n.* a fused ischium and pubis.

ischiorectal *a. pert.* ischium and rectum, *appl.* fossa and muscles.

ischium *n.* the ventral and posterior bone of each half of pelvic girdle of vertebrates except fishes, often fused to pubis.

iscom immune stimulatory complex: an adjuvant composed of antigen complexed with lipids.

IS element insertion sequence *q.v.*

isidia *n.plu.* coral-like soredia on surface of some lichens. *sing.* **isidium.**

island biogeography the study of the flora and fauna of islands with a view to understanding the nature and evolution of biodiversity in an isolated environment.

islets of Langerhans groups of cells scattered throughout the pancreas that secrete glucagon (from α or A cells) or insulin (from β or B cells).

isoaccepting *a. appl.* tRNAs which pick up the same amino acid and are charged by the same aminoacyl-tRNA synthetase.

isoagglutination *n.* (1) agglutination of erythrocytes by blood from another member of the same species or of the same blood group; (2) agglutination of sperm by agglutinins of ova of the same species.

isoagglutinin *n.* agglutinin of ova which reacts with sperm of the same species.

isoalleles *n.plu.* alleles that are identical in their gross phenotypic effects but which can be distinguished at the DNA or protein product level.

isoantigen alloantigen *q.v.*

isobilateral *a. appl.* a form of bilateral symmetry where a structure is divisible in two planes at right angles.

isocarpous *a.* having carpels and perianth divisions equal in number.

isocercal *a.* with vertebral column ending in median line of caudal fin.

isochromatic, isochromous *a.* (1) uniformly coloured; (2) equally coloured.

isochromosome *n.* abnormal chromosome with two genetically identical arms.

isochronous *a.* (1) having equal duration; (2) occurring at the same rate.

isocitrate *see* isocitric acid.

isocitrate dehydrogenase either of two enzymes. One (EC 1.1.1.41, NAD-requiring) is located in mitochondria and catalyses formation of α-ketoglutarate from isocitrate, a regulatory step in the tricarboxylic acid cycle. The other (EC 1.1.1.42, NADP-requiring) is located in the cytoplasm.

isocitric acid, isocitrate six-carbon intermediate of the tricarboxylic acid cycle, formed from citrate via *cis*-aconitate in reactions catalysed by the enzyme aconitase.

isocortex *n.* the six outermost cell layers of the cerebral cortex.

isocytic *a.* with all cells equal.

isodactylous *a.* with all digits of equal size.

isodemic *a.* (1) with, or *pert.*, populations composed of an equal number of individuals; (2) *appl.* lines on a map which pass through points representing equal population density.

isodiametric *a.* (1) having equal diameters, *appl.* cells or other structures; (2) *appl.* rounded or polyhedral cells.

isodont homodont *q.v.*

isodynamic *a.* (1) of equal strength; (2) providing the same amount of energy, *appl.* foods.

isoelectric focusing the separation of proteins on the basis of charge, by their differential migration on electrophoresis in a pH gradient, proteins of different charge each ceasing their migration at their isoelectric point.

isoelectric point (IEP) the pH at which an amphoteric molecule, such as a protein, carries no net charge, being a definite value for each protein.

isoenzyme *n.* any one of a group of enzyme variants of slightly different sequence, encoded by different genes, that catalyse the same reaction in the same pathway but differ in properties such as optimum pH or rate of reaction. They can be distinguished by isoelectric point or electrophoretic mobility. Different tissues often express different isoenzymes. *alt.* **isozyme**.

Isoetales *n.* order of Lycopsida having linear leaves, and a 'corm' with complex secondary thickening, and including the quillworts.

isoforms *n.plu.* (1) the different forms of a polymorphic protein; (2) two or more proteins or RNAs that are produced from the same gene by differential transcription and/or differential RNA splicing.

isogametangiogamy *n.* the union of similar gametangia.

isogamete *n.* one of a pair of gametes that are morphologically similar.

isogamy *n.* the fusion of gametes that are morphologically similar, i.e. of equal size and similar structure. *a.* **isogamous**.

isogeneic syngeneic *q.v.*

isogenes *n.plu.* lines on a (geographic) map which connect points where the same gene frequency is found.

isogenetic *a.* (1) arising from the same or a similar origin; (2) of the same genotype.

isogenic homozygous *q.v.*

isogenomic *a.* containing similar sets of chromosomes, *appl.* nuclei.

isogenous *a.* of the same origin.

isognathous *a.* having both jaws alike.

isogonal *a.* forming equal angles, *appl.* branching.

isograft *n.* tissue graft taken from another individual of the same genotype as the recipient.

isohaline *a. appl.* water having the same level of salinity.

isokont isomastigote *q.v.*

isolate *n.* (1) a breeding group limited by isolation; (2) the first pure culture of a microorganism derived e.g. from soil or tissues.

isolateral *a.* having equal sides, *appl.* leaves with palisade tissue on both sides.

isolating mechanisms mechanisms that prevent breeding between two populations of the same species, and eventually lead to speciation. The most important is geographical isolation, but there may also be genetic, anatomical and behavioural barriers to successful interbreeding.

isolation *n.* prevention of mating between breeding groups owing to spatial, topographical, ecological, morphological, physiological, genetic, behavioural or other factors.

isolecithal *a. appl.* eggs with yolk distributed nearly equally throughout.

isoleucine (Ile, I) *n.* amino acid with a nonpolar side chain, stereoisomer of leucine, constituent of proteins, and essential in diet of humans and other animals.

isomastigote *a.* having flagella or cilia of the same length.

isomer *n.* any one of a set of molecules having the same kind and number of atoms but differing in arrangement of the atoms and in physical and (sometimes) chemical properties. *a.* **isomeric**. *n.* **isomerism**. *see also* anomer, enantiomer, epimer, optical isomer, racemate, tautomerism, stereoisomer.

isomerases *n.plu.* enzymes that catalyse rearrangements of atoms within molecules. Examples are epimerases, mutases, racemases, tautomerases. EC class 5.

isomere *n.* a homologous structure or part.

isomeric *a. appl.* an isomer of a molecule.

isomerism *n.* the existence of isomers of a molecule.

isomerization *n.* the conversion of one isomer into another.

isomerous *a.* (1) having equal numbers of different parts; (2) *appl.* flowers with equal numbers of parts in each whorl.

isometric *a.* (1) of equal measure or growth rate; (2) *appl.* contraction of muscle under tension without change in length.

isometry *n.* growth of a part at the same rate as the standard or as the whole.

isomorphic, isomorphous *a.* (1) superficially alike; (2) *appl.* alternation of haploid and diploid phases in morphologically similar generations.

isomorphism *n.* apparent similarity of individuals of different race or species.

isonym *n.* a new name, e.g. of species, based upon oldest name or basinym.

iso-osmotic *a. appl.* two solutions of the same osmotic concentration. *see* isotonic.

⁶N-isopentenyladenine (i⁶Ade) *n.* naturally occurring cytokinin in plants and a constituent of some transfer RNAs.

isopetalous *a.* having similar petals.

isophagous *a.* feeding on one or allied species, *appl.* fungi.

isophane *n.* a line connecting all places within a region at which a biological phenomenon, e.g. flowering of a plant, occurs at the same time.

isophene *n.* a contour line delimiting an area corresponding to a given frequency of a variant form.

isophyllous *a.* having uniform foliage leaves, on the same plant.

isophytoid *n.* an 'individual' of a compound plant not differentiated from the rest.

isoplankt *n.* a line representing on a map the distribution of equal amounts of plankton, or of particular species.

isoploid *a.* with an even number of chromosome sets in somatic cells.

Isopoda, isopods *n., n.plu.* group of marine, freshwater and terrestrial malacostracan crustaceans, including the woodlice and water slaters, having a dorsoventrally flattened body and no carapace.

isopodous *a.* having the legs alike and equal.

isopogonous *a.* having the two sides of feather equal and similar.

isopolyploid *a. appl.* polyploid with an even number of chromosome sets, e.g. tetraploid, hexaploid.

isoprene *n.* five-carbon unsaturated aliphatic hydrocarbon that is a subunit of many lipids.

isoprenoid *n.* any of a large and varied group of lipids built up of five-carbon isoprene units. They include carotenoids, terpenes, natural rubber and the side chains of e.g. chlorophyll and vitamin K.

isopropylthiogalactoside (IPTG) a non-metabolized inducer of the *lac* operon in *E. coli.*

Isoptera *n.* order of social insects comprising the termites and their relatives, which live in large organized colonies containing reproductive forms (the queen and the king) and non-reproductive wingless soldiers and workers, all offspring of the king and queen. Termites have a gut flora that enables them to digest wood, and they can be serious pests, devouring wooden buildings, trees and paper.

isosporous homosporous *q.v.*

isostemonous *a.* having stamens equal in number to that of sepals or petals.

isotelic *a.* (1) exhibiting, or tending to produce, the same effect; (2) homoplastic *q.v.*

isotomy *n.* forking repeatedly in a regular manner.

isotonic *a.* (1) of equal tension; (2) of equal osmotic pressure, *appl.* solutions; (3) *appl.* contraction of muscle with change in length. *cf.* isometric.

isotonicity *n.* normal tension under pressure or stimulus.

isotope *n.* a form of a chemical element having the same atomic number (number

of protons) and identical chemical properties as another, but differing in atomic mass
as a result of a different number of neu-
trons in the atomic nucleus. *a.* **isotopic**.

isotopic fractionation the differential
incorporation of particular isotopes of
e.g. carbon (^{12}C), sulphur (^{32}S), or other
elements into biological material. *see also*
carbon isotope ratio.

isotropic, isotropous *a.* (1) singly
refracting in polarized light, *appl.* the
light stripes of striated muscle under the
microscope; (2) symmetrical around longitudinal axis; (3) not influenced in any one
direction more than another, *appl.* growth
rate; (4) without predetermined axes, as
some ova. *n.* **isotropy**.

isotypes *n.plu.* the different heavy-chain
types (α, δ, ε, γ, μ) and light-chain
types (κ, λ) of immunoglobulins. They
are the result of sequence differences
in the constant regions of the heavy and
light chains. The isotype of the heavy
chain determines the functional class of
the antibody. *see* immunoglobulin A,
immunoglobulin D, immunoglobulin E,
immunoglobulin G, immunoglobulin M.

isotype switching class switching *q.v.*

iso-utility curves in the study of animal
behaviour and ecology, curves joining
all points of equal utility or benefit, and
which show how different behaviours (e.g.
feeding behaviours) can result in the
same quantitative benefit. *alt.* indifference
curves.

isoxanthopterin *n.* colourless pterin in
the wings of cabbage butterflies and in
eyes and bodies of other insects. *alt.*
leucopterin B.

isozoic *a.* inhabited by similar animals.

isozyme isoenzyme *q.v.*

isthmus *n.* narrow structure connecting
two larger parts.

ITAM immunoreceptor tyrosine-based activation motif *q.v.*

iter *n.* a passage or canal, as those of middle
ear and in brain.

iteration *n.* repetition, as of similar trends in
successive branches of a taxonomic group.

iteroparity *n.* production of offspring by
an organism in successive groups. *a.*
iteroparous. *cf.* semelparity.

ITIM immunoreceptor tyrosine-based inhibitory motif *q.v.*

ITM intermediate memory *q.v.*

IU International Unit *q.v.*

IVF *in vitro* fertilization *q.v.*

ivory *n.* dentine of teeth, usually that of
tusks, as of elephant and narwhal.

IVS intervening sequence. *see* intron.

J

J joule *q.v.*

Jaccard index, Jaccard coefficient of similarity measure of the similarity (*see* coefficient of community) between two plant communities based on their species composition. It can be calculated from the formula $J = c/(a + b - c)$ where $a =$ no. of species at site A, $b =$ no. of species at site B, and $c =$ no. of species common to each site. J will lie between 0 and 1. The greater the value of J, the more similar the sites are in their species composition. *cf.* Gleason's index, Kulezinski index, Morisita's similarity index, Simpson's index of floristic resemblance, Sørensen similarity index.

jacket cell cell in antheridium that gives rise to the wall.

Jacobson's cartilage vomeronasal cartilage supporting Jacobson's organ.

Jacobson's organ a diverticulum of the olfactory organ in many vertebrates, often developing into an epithelium-lined sac opening into the mouth.

jactitation *n.* scattering of seeds by a censer mechanism.

jaculatory *a.* (1) darting out; (2) capable of being emitted.

jaculatory duct portion of vas deferencs which can be protruded, in many animals.

jaculiferous *a.* bearing dart-like spines.

Janus kinases (JAKs) family of cytoplasmic protein tyrosine kinases associated with the receptors for many cytokines, and which activate the STAT family of transcription factors.

Java man fossil hominid found in Java and originally called *Pithecanthropus erectus*, now called *Homo erectus*, and dating from mid-Pleistocene.

jaw-foot maxilliped *q.v.*

jaws *n.plu.* (1) in vertebrates, a skeletal structure, the upper and lower jaws, forming part of the mouth, bearing the teeth or horny tooth-plates, the upper jaw articulating with the braincase, the lower jaw movable, articulating with the upper jaw to open and shut the mouth; (2) structures of similar function and mechanism in invertebrates.

J chain joining chain, a short polypeptide holding individual immunoglobulin molecules together in IgM pentamers and in oligomers of IgA.

jecoral *a.* of or *pert.* the liver.

jejunum *n.* part of small intestine between duodenum and ileum in mammals. It has large villi and is the main absorptive region.

jellyfish common name for the Scyphozoa *q.v.*

jelly fungi common name for a group of basidiomycete fungi whose fruiting bodies are typically of a jelly-like consistency, e.g. the funnel-shaped *Tremella* and the ear-shaped *Auricularia*.

jelly of Wharton gelatinous connective tissue surrounding the vessels of the umbilical cord.

Jerne plaque assay assay for B cells that produce antibodies specific for a cell-surface antigen, based on complement-mediated lysis.

J gene segment any one of the 'joining' gene segments found in all immunoglobulin and T-cell receptor loci. They are short DNA sequences one of which is spliced to a V gene segment (or to a V + D sequence) to produce a DNA sequence coding for the variable region of the polypeptide chains.

alt. **J segment, J region**, which may also designate the corresponding region of the polypeptide chain. *see also* D gene segment, gene rearrangement, variable region, V gene segment.

Johnston's organ sensory organ in 2nd segment of insect antennae, concerned with balance or with sensing sound or mechanical vibrations.

joining chain J chain *q.v.*

joining gene segment J gene segment *q.v.*

joint *n.* the articulation between two neighbouring parts, such as the knee joint between the tibia and femur of a leg.

jordanon *n.* a true-breeding unit below the species level, with little variability, such as a race, subspecies or variety. *alt.* microspecies.

joule (J) *n.* the derived unit of energy or work in the SI system, which is used instead of the calorie. One joule of work is done whenever a force of 1 N (equivalent to 1 kg m s^{-2}) is sustained over 1 m in that force's direction. Therefore, 1 J is equivalent to 1 kg m^2 s^{-2}. 1 J = 0.23892 calories.

J region region of an immunoglobulin or T-cell receptor polypeptide chain that is encoded by a J gene segment.

J-shaped growth curve type of growth curve exhibited by populations that grow exponentially and then crash to very low numbers, typical of organisms with many generations per year, e.g. planktonic algae. *cf.* S-shaped growth curve.

Juan Fernandez region region of the Neotropical floristic kingdom comprising the Juan Fernandez islands off the west coast of Chile.

jubate *a.* with a mane-like growth.

jugal *n.* a component of the cheekbone in vertebrate skull.

jugate *a.* having pairs of leaflets.

Juglandales *n.* order of dicot trees, often aromatic, with pinnate leaves and comprising two families Juglandaceae (walnut) and Rhoiptelaceae.

jugular *a.* (1) *pert.* neck or throat; (2) *appl.* vein, the main veins carrying blood from the head in vertebrates; (3) *appl.* ventral fins beneath and in front of pectoral fins.

jugum *n.* (1) (*bot.*) a pair of opposite leaflets or leaves; (2) (*zool.*) small lobe on posterior of forewing of some moths; (3) ridge or depression connecting two structures.

jumping genes colloquial term for transposable genetic elements. *see also* controlling element, insertion sequence, transposition, transposon.

Juncales *n.* order of herbaceous monocots with long, narrow, channelled or grasslike leaves, and comprising the families Juncaceae (rush) and Thurniaceae.

junction *see* cell junction, gap junction, septate junction.

junctional complex in epithelial cells, the region of attachment between neighbouring cells. *see* desmosome, hemidesmosome, tight junction.

junctional diversity the diversity in immunoglobulins and T-cell receptors that is due to variation created at the junctions between gene segments during the gene rearrangement process.

junk DNA colloquial term for any DNA in the genome that appears to have no coding, regulatory or structural function. It is used in particular for the large amount of DNA in the genomes of plants and animals which is composed of repetitive noncoding sequences.

Jurassic *n.* geological period lasting from *ca.* 200 million years ago to 146 million years ago, after the Triassic and before the Cretaceous. The vegetation included conifers, cycads, ferns, ginkgos and tree ferns. The dominant land animals were dinosaurs, esp. the giant sauropods (e.g. *Brachiosaurus*) and theropods. The first birds appeared in the late Jurassic.

juvenal, juvenile *a.* youthful, *appl.* plumage replacing nestling down of 1st plumage.

juvenile *n.* young bird or other animal, before it has acquired full adult plumage or form.

juvenile hormone insect hormone that maintains larval state and prevents premature metamorphosis into an adult form.

juvenile-onset diabetes type 1 diabetes mellitus *q.v.*

juxta- prefix derived from Latin *juxta*, close to.

juxta-articular *a.* near a joint or articulation.

juxtaglomerular cells cells surrounding the arteriole feeding the network of blood capillaries around the dilated end of a kidney tubule, and which secrete renin.

juxtamedullary *a.* near medulla, *appl.* inner portion of zona reticularis of adrenal glands.

juxtanuclear *a.* beside the nucleus, *appl.* bodies: basophil deposits in cytoplasm of vitamin D-deficient parathyroid cells.

K

κ *see* kappa.

K (1) symbol for the chemical element potassium *q.v.*; (2) lysine *q.v.*

K (1) symbol for the carrying capacity (*q.v.*) of the environment; (2) equilibrium constant *q.v.*

*K*ₐ association constant *q.v.*

*K*𝒹 dissociation constant *q.v.*

*K*ₘ Michaelis constant. *see* Michaelis–Menten kinetics.

kainate *n.* structural analogue of glutamate, used to define a class of glutamate receptors in central nervous system.

kairomone *n.* chemical messenger or pheromone emitted by one species which has an effect on a member of another species, sometimes to the detriment of the transmitter. An example would be a chemical that attracts a male to a female and also attracts a predator.

kallidin lysylbradykinin *q.v.*

kallikrein *n.* plasma kallikrein (EC 3.4.21.34) is a serine peptidase that cleaves kininogen to produce kinin (bradykinin). Tissue kallikrein (EC 3.2.21.35) cleaves kininogen to produce lysylbradykinin. *alt.* kininogenase.

kanamycin kinase enzyme conferring resistance to the antibiotics kanamycin, neomycin and related compounds. It phosphorylates them, rendering them inactive. EC 2.7.1.95. *alt.* neomycin phosphotransferase.

kanamycins *n.plu.* antibiotics produced by *Streptomyces* spp. that interfere with bacterial protein synthesis.

kappa (κ) *n.* one of the two types of light chain found in mammalian antibodies.

Kartagener's syndrome recessive genetic defect in humans leading to non-functional cilia on e.g. respiratory epithelium. Situs inversus also sometimes occurs.

karyaster *n.* star-shaped group of chromosomes.

karyo- prefix derived from Gk *karyon*, nut (= nucleus), signifying *pert.* nucleus or chromosomes.

karyoclasis *n.* breaking down of cell nucleus. *a.* **karyoclastic.**

karyogamy *n.* fusion of the nuclei of two gametes after cytoplasmic fusion.

karyokinesis *n.* nuclear division. *see* mitosis.

karyology *n.* study of the nucleus and the chromosomes.

karyolysis *n.* disintegration of cell nucleus. *a.* **karyolytic.**

karyomixis *n.* mingling or union of nuclear material of two gametes.

karyon *n.* the cell nucleus. *plu.* **karya.**

karyopherin nuclear transport receptor *q.v.*

karyoplast *n.* an isolated nucleus, with a small amount of cytoplasm adhering.

karyorhexis, karyoschisis *n.* fragmentation of cell nucleus.

karyosome *n.* structure in nucleus of some protozoa, analogous to nucleolus.

karyota *n.plu.* nucleated cells.

karyotype *n.* photographic representation of the chromosome complement of a cell, with individual mitotic chromosomes arranged in pairs in order of size.

Kaspar Hauser experiment experiment in which an animal is reared in complete isolation from members of its own species.

kasugamycin *n.* antibiotic that blocks initiation of translation in bacteria.

katagenesis *n.* retrogressive evolution.

katakinetic *a.* *appl.* processes leading to discharge of energy.

katal *n.* derived SI unit of enzyme activity, defined as the amount of enzyme that will catalyse the conversion of 1 mole of substrate in 1 second. 1 nanokatal = 0.06 IU.

kataphoric *a. appl.* passive action.

kataplexy cataplexis *q.v.*

katharobic *a.* living in clean waters, *appl.* protists.

katharometer *n.* thermal conductivity detector device for measuring the composition of sample vapours emerging from gas–liquid chromatography.

kb, kbp kilobase or kilobase pairs = 1000 bases or base pairs of DNA.

kcal kilocalorie *q.v.*

K cell *see* killer cells.

K⁺ channel potassium channel *q.v.*

kD, kDa, kdal kilodalton *q.v.*

KDEL C-terminal amino-acid sequence that directs the retention of proteins in the endoplasmic reticulum.

keel *n.* (1) narrow ridge; (2) ridge on sternum of flighted birds and bats to which wing muscles are attached; (3) boat-shaped structure formed by two anterior petals in flowers of the pea family.

kelp *n.* common name for seaweeds of the Laminariales, marine multicellular brown algae with a large broad-bladed thallus attached to the substratum by a tough stalk and holdfast.

kenenchyma *n.* tissue devoid of its living contents, e.g. cork.

Kennedy's disease X-linked spinal and bulbar muscular atrophy, a genetic disease characterized by progressive muscle weakness and atrophy. It is caused by amplification of trinucleotide repeats in the gene for the androgen receptor.

kenosis *n.* (1) process of voiding, or condition of having voided; (2) exhaustion; (3) inanition.

kentragon *n.* larval stage following the cypris stage of parasitic cirripeds (barnacles), which penetrates into the body of host.

keraphyllous *a. appl.* layer of a hoof between horny and sensitive parts.

kerasin *n.* cerebroside from brain yielding on hydrolysis a fatty acid (lignoceric acid), galactose and sphingosine.

keratan *n.* polysaccharide of galactose and *N*-acetylglucosamine which is part of a proteoglycan from cartilage and other connective tissues, where it occurs as keratan sulphate.

keratin *n.* fibrous protein rich in cysteine, constituent of intermediate filaments (kera-

tin filaments). It is the chief material in horn, hair, nails and the upper flaky layer of skin.

keratinization *n.* intracellular deposition of keratin to form an inert horny material, e.g. nails, claws, horns, outer layers of skin. *alt.* cornification.

keratinizing *a.* becoming horny due to intracellular deposition of keratin, *appl.* cells of epidermis (skin cells, hair shaft cells, cells of fingernails and toenails, epidermal cells at base of horns). *alt.* cornifying.

keratinocyte *n.* epidermal skin cell that synthesizes keratin.

keratinolytic *a.* breaking down keratin, *appl.* enzymes.

keratinophilic *a.* growing on horn or other keratinized substrate, *appl.* certain fungi.

keratinous *a. pert.*, containing, or formed by, keratin.

keratocyte *n.* epithelial cell, containing keratin filaments, of the epidermis of fish and frogs. They can migrate rapidly to close wounds.

keratogenous *a.* horn-producing.

keratohyalin *n.* substance formed in the middle layer of vertebrate epidermis (skin), preceding full keratinization.

keratoid *a.* resembling horns.

keratose *a.* having horny fibres in skeleton, as certain sponges.

kernel *n.* the inner part of a seed containing the embryo.

keroid keratoid *q.v.*

ketogenesis *n.* the production of ketone bodies, which occurs especially during starvation or fasting.

ketogenic *a. appl.* amino acids that give rise to ketone bodies during their oxidation to carbon dioxide and water, e.g. leucine, phenylalanine and tyrosine.

ketogenic hormone lipotropin *q.v.*

α-ketoglutaric acid, α-ketoglutarate five-carbon carboxylic acid intermediate of the tricarboxylic acid cycle, which is decarboxylated to yield succinyl-CoA, a reaction catalysed by the α-ketoglutarate dehydrogenase complex. It can also be aminated to form glutamate.

ketone body any of a group of compounds such as acetoacetate, D-3-hydroxybutyrate and acetone. They are formed in the liver from acetyl-CoA produced during fatty acid oxidation, especially during fasting or in conditions such as diabetes. They can be used by the brain as an alternative fuel to glucose.

ketose *n.* any monosaccharide containing a keto (C=O) group. *cf.* aldose.

ketosis *n.* change in the energy source for the brain from glucose to ketone bodies which occurs e.g. during starvation.

key *n.* (1) means of identifying objects or organisms by a series of questions with alternative answers, each answer leading on to another question or a positive identification; (2) winged nutlet hanging in clusters, as in ash. *alt.* samara.

keystone species species that have a key role in an ecosystem, affecting many other species, and whose removal leads to a series of extinctions within the system.

kidney *n.* one of a pair of organs in vertebrates concerned with excretion of nitrogenous waste as urine and regulation of water balance.

kidney duct tube through which urine is excreted to the exterior in fishes and amphibians.

kidney tubule one of numerous convoluted fine tubules in cortical region of kidney that carry urine from glomeruli to the collecting ducts. In mammals and birds, water is reabsorbed from the urine during passage through the tubules.

killed vaccine vaccine made of whole but inactivated viruses or bacteria.

killer cell *see* cytotoxic T cell, natural killer cells.

killer cell immunoglobulin-like receptors (KIRs) one of the two main families of cell-surface receptors expressed by the natural killer cells (NK cells) of the immune system and which regulate NK cells' activity. They are members of the immunoglobulin superfamily and include receptors that activate and receptors that inhibit the killing activity of an NK cell.

killer cell lectin-like receptors (KLRs) one of the two main families of cell-surface receptors expressed by the natural killer cells (NK cells) of the immune system and which regulate NK cells' activity. They are homologous to C-type lectins and include receptors that activate and receptors that inhibit the killing activity of an NK cell.

kilobase (kb), kilobase pair (kbp) unit of length used for nucleic acids and poly-nucleotides, corresponding to 1000 bases or base pairs.

kilocalorie (kcal) *n.* 1000 calories.

kilodalton (kD, kDa, kdal) unit of mass equal to 1000 daltons, or 1000 units of molecular mass, also sometimes abbreviated to K (e.g. a 30K protein), used chiefly for proteins.

kilojoule (kJ) *n.* 1000 joules.

kinaesthesis *n.* perception of movement due to stimulation of muscles, tendons and joints.

kinaesthetic *a. pert.* sense of movement or muscular effort.

kinase *n.* any enzyme that transfers a phosphoryl (phosphate) group from ATP or other high-energy phosphates to an acceptor molecule. EC 2.7.1–4. *alt.* phosphokinase. *see also* protein kinase.

kindling *n.* long-term electrophysiological and morphological changes produced in neuronal tissue following repeated electrical stimulation, and which can e.g. result in a seizure after a mild stimulus many months afterwards.

kindred *n.* group of people related by marriage or ancestry.

kinesin *n.* protein with ATPase/GTPase activity that can move along microtubules and thus can act as a molecular motor for various types of movement within cells. *see* axonal transport.

kinesis *n.* (1) random movement; (2) an orientation movement in which the organism swims at random until it reaches a better environment, the movement depending on the intensity, not the direction, of the stimulus.

kinesth- *see* kinaesth-.

kinete *n.* motile zygote stage in the life cycle of piroplasms.

kinethmoid *n.* small bone in fish skull intermediate between premaxilla and cranium, involved in protrusion of premaxillae.

kinetic *a.* (1) *pert.* movement; (2) active; (3) *appl.* energy employed in producing or changing motion.

kinetics *n.* the sequence of events involved in a movement or activity and the rates at which they occur. *see also* enzyme kinetics.

kinetid *n.* unit composed of one or more kinetosomes, and their cilia, and the ribbons of microtubules and striped fibres (kinetodesmata) that surround them, in ciliate protozoa.

kinetin *n.* adenine derivative with cytokinin activity, probably not occurring naturally.

kinetium *n.* row of kinetosomes, in ciliate protozoa. *plu.* **kinetia.**

kinetoblast *n.* ciliated membrane of aquatic larvae, used for locomotion.

kinetochore *n.* densely staining fibrillar region at the centromere of a chromosome, to which spindle microtubules attach during meiosis or mitosis. It is composed of specialized proteins that bind specifically to the centrosomal DNA and associated proteins.

kinetochore fibres, kinetochore microtubules microtubules of the mitotic or meiotic spindle which are attached to kinetochores of chromosomes at one end and to a pole of the spindle at the other.

kinetodesma *n.* striped fibre that is part of a kinetid in ciliate protozoans. *plu.* **kinetodesmata.**

kinetoplast *n.* large mitochondrion associated with the base of the flagellum in some flagellate protozoans.

kinetoplastids *n.plu.* flagellate protozoans that contain a kinetoplast. This group includes the parasitic trypanosomes.

kinetosome *n.* (1) basal body of cilium in ciliate and flagellate protozoans; (2) one of a group of granules occupying the polar plate region in sporogenesis in mosses.

kinetospore *n.* a motile spore.

king *n.* in social hymenopterans or termites, a male reproductive individual.

king crabs horseshoe crabs *q.v.*

kingdom *n.* (1) in taxonomy, the name given to a primary division of living organisms. Until quite recently, five kingdoms were generally recognized: Prokaryotae (bacteria and other prokaryotes), Protoctista (or Protista) (simple eukaryotic organisms such as the protozoa and algae), Fungi, Plantae (multicellular green plants other than algae) and Animalia or Metazoa (multicellular animals). More recently, the living world has been divided on the evidence of DNA sequence data into three primary superkingdoms or domains: Bacteria (*q.v.*), Archaea (*q.v.*) and Eukarya (*q.v.*). The

Bacteria and Archaea correspond to the Prokaryotae, while the Eukarya comprises all the remaining kingdoms. In addition, the taxonomy of the unicellular eukaryotes in particular is under almost constant redefinition, as molecular evidence for their relationships and evolutionary history accumulates. Kingdoms are divided into phyla for animals and unicellular eukaryotes, and divisions for plants. *see* Appendices 1–6; (2) biogeographical kingdom *q.v.*

kinin *n.* (1) bradykinin, lysylbradykinin (kallidin) *q.v.*; (2) cytokinin *q.v.*

kininogen *n.* precursor protein from which a kinin is produced by post-translational cleavage by kallikrein. Also acts as a cofactor in the blood clotting pathway, helping to activate Factor XII.

kinoplasmasomes *n.plu.* phragmoplast fibres seen at periphery of cell plate in plant cell division.

Kinorhyncha, kinorhynchs *n.*, *n.plu.* phylum of marine microscopic pseudocoelomate invertebrate animals having bodies of jointed spiny segments and spiny heads.

kin selection the selection of genes in a population due to individuals favouring or disfavouring the survival of relatives, other than offspring, who possess the same genes by common descent.

kinship *n.* possession of a common ancestor in the not too distant past.

kinship, coefficient of coefficient of kinship *q.v.*

KIR killer cell immunoglobulin-like receptor *q.v.*

kirromycin *n.* antibiotic that inhibits elongation of the polypeptide chain during translation in bacteria by inhibiting the elongation factor EF-Tu.

kJ kilojoule *q.v.*

Kjeldahl analysis technique widely used to determine the total nitrogen content of a tissue.

K⁺ leak channel ion channel in plasma membrane of animal cells that allows potassium ions (K^+) to leak passively out of the cell down their concentration gradient. Together with the Na^+-K^+ ATPase, it is involved in generating and maintaining plasma membrane electrical potential.

Klenow fragment the fragment of DNA polymerase I from *Escherichia coli* that contains both the polymerase and $3' \rightarrow 5'$ exonuclease activity but lacks the $5' \rightarrow 3'$ exonuclease activity.

kleptobiosis *see* cleptobiosis.

kleptoparasitism *n.* type of parasitism in which the female searches out the prey or stored food of another female, usually of a different species, and takes it for her own offspring. *alt.* cleptoparasitism.

Klinefelter's syndrome syndrome occurring in men having genetic constitution XXY, resulting in underdevelopment of male sexual organs and sterility.

klinokinesis *n.* movement in which an organism continues to move in a straight line until it meets an unfavourable environment, when it turns, resulting in its remaining in a favourable environment, the frequency of turning depending on the intensity of the environmental stimulus.

klinotaxis *n.* taxis in which an organism orients itself in relation to a stimulus by moving its head or whole body from side to side symmetrically in moving towards the stimulus, and so compares the intensity of the stimulus on either side.

KLRs killer cell lectin-like receptors *q.v.*

knee *n.* (1) joint between tibia and femur; (2) root that emerges above water or ground in certain trees living in swampy land; (3) joint in stem of some grasses.

knee-jerk reflex reflex composed of a simple neural circuit of sensory neuron (stretch receptor), spinal interneuron, and motor neuron, that makes the lower leg swing up when tapped just below the knee.

knob *n.* (1) in cytogenetics, a darkly staining chromomere that identifies a particular chromosome; (2) in parasitology, cells infected with certain strains of the malaria parasite *Plasmodium falciparum*, which produce knobs that attach to endothelial cells and block cerebral vessels, giving rise to cerebral malaria.

knock-in *see* gene knock-in.

knock-out *see* gene knock-out.

knot *n.* in wood, the base of a branch surrounded by concentric layers of new wood and hardened under pressure.

Koch's postulates the criteria that need to be satisfied to prove that a particular microorganism is the cause of a disease: (i) that the microbe be found in the body in all cases of the disease, (ii) that it be isolated from a disease case and grown in a series of pure cultures *in vitro*, and (iii) that it reproduce the disease on the inoculation of a pure culture into a susceptible animal.

koilin lining the grinding surface inside a bird's gizzard.

Korarchaeota *n.* a taxonomic grouping within the Archaea that includes so-far uncultured microbes, known only from their DNA, and which are inhabitants of hot terrestrial springs. *cf.* Crenarchaeota, Euryarchaeota.

Kornberg cycle glyoxylate cycle *q.v.*

Kornberg enzyme DNA polymerase involved in DNA repair.

Kranz anatomy type of leaf anatomy in plants with C4 photosynthesis, in which photosynthetic outer mesophyll cells are arranged in a 'wreath' around inner bundle sheath cells. This enables the intercellular transport of the C4 acids, which are the first products of photosynthesis, from the mesophyll cells to the bundle sheath cells in which they are decarboxylated to provide CO_2 for the Calvin cycle. C3 metabolites are transported the other way to act as a substrate for the initial CO_2 fixation.

krasnozem *n.* deep friable red loamy soil found in the subtropics and developed from base-rich parent materials.

Krebs cycle tricarboxylic acid cycle *q.v.*

Krebs–Henseleit cycle urea cycle *q.v.*

krill *n.* planktonic crustaceans, which are abundant in the oceans and form the principal food of the filter-feeding whales.

kringle *n.* a small protein domain found in a number of different proteins.

K-selected species species selected for its superiority in a stable environment. Members of such species typically have a slow development, relatively large size, and produce only a small number of offspring at a time.

K selection selection favouring superiority in stable, predictable environments in which rapid population growth is unimportant. *cf.* r selection.

K–T boundary the boundary between the Cretaceous (*Kreidezeit* in German) and the Tertiary, at *ca.* 65 million years, which was the time of a mass extinction that included the dinosaurs.

Kulezinski index index of similarity (*see* coefficient of community) between two plant communities based upon their species composition. It is calculated by dividing the number of species in common by the sum of the total number of species in each community. *cf.* Gleason's index, Jaccard index, Morisita's similarity index, Simpson's index of floristic resemblance, Sørensen similarity index.

Kupffer cells star-shaped phagocytic cells (macrophages) found in liver sinusoids and which ingest defunct red blood cells.

kurtosis *n.* statistical term that describes the degree of 'peakedness' of a frequency distribution, i.e. the sharpness of the peak around the mean, especially when compared with a normal distribution curve.

kuru *n.* transmissible spongiform encephalopathy that used to affect people in Papua New Guinea and was caused by the consumption of infected human brains. *see* prion.

kwashiorkor *n.* deficiency disease caused by an insufficiency of protein.

kynurenine *n.* metabolic product of tryptophan, and a precursor of some ommatochromes and other pigments in insects.

L

λ wavelength (of light). *see also* lambda.

L leucine *q.v.*

L-, D- prefixes denoting particular molecular configurations, defined according to convention, of certain chiral molecules esp. monosaccharides and amino acids, the L configuration being a mirror image of the D. In living cells such molecules usually occur in one or other of these configurations but not both (e.g. glucose as D-glucose, amino acids always in the L form in proteins).

labella *n.* pair of grooved lobes at end of labium in some dipteran insects, for mopping up liquid food.

labellate *a.* furnished with labella or small lips.

labelled *a. appl.* a molecule made detectable and traceable by incorporation of a radioactive element or by linkage to a detectable chemical tag – the label.

labia *n.plu.* (1) lips; (2) lip-like structures; (3) *plu.* of labium.

labial *a. pert.* labium.

labial palp lobe-like structure near mouth of certain insects.

labia majora outer lips of vulva.

labia minora inner lips of vulva.

labiate *a.* (1) lip-like; (2) possessing lips or thickened margins; (3) *n.* any member of the dicot family Labiatae, which includes the mints and balsams. They are characterized by typically square stems, opposite decussate leaves and flowers with a corolla divided into two lips.

labidophorous *a.* possessing pincer-like organs.

labiella *n.* mouthpart of millipedes.

labile *a.* (1) readily undergoing change; (2) unstable; (3) *appl.* genes that have a tendency to mutate.

lability *n.* in evolutionary theory, the ease and speed with which particular categories of traits evolve.

labiodental *a. pert.* lip and teeth, *appl.* the surface of tooth nearest to lip.

labium *n.* (1) a lip or lip-shaped structure; (2) in insects the fused 2nd maxillae, forming the lower mouthpart; (3) inner margin of mouth of gastropod shell. *plu.* **labia**.

Laboulbeniomycetes *n.* group of highly specialized ascomycete fungi, parasitic on the outer surface of insects and arachnids. They have an ascogonium with a trichogyne and fertilization by spermatia.

labrum *n.* (1) anterior mouthpart of some arthropods; (2) outer margin of gastropod shell. *a.* **labral**.

labyrinth *n.* (1) the complex convoluted membranous and bony structures of the inner ear. *see* ear; (2) much folded accessory respiratory organ above gills in some fishes; (3) any of various other convoluted structures.

labyrinthodont *a.* having teeth with a complicated arrangement of dentine.

Labyrinthodontia, labyrinthodonts *n.,* *n.plu.* subclass of extinct early amphibians, of the late Paleozoic and Mesozoic, with labyrinthodont teeth. They included temnospondyls, anthracosaurs and ichthyostegalians.

Labyrinthulomycota *n.* the net slime moulds or slime nets, a phylum of colonial non-photosynthetic eukaryotic microorganisms that form colonies of cells that move and grow within a slime track that they secrete. Found mainly in estuarine and nearshore habitats associated with leaves, algae and organic debris. Two families are recognized: Labyrinthulaceae

353

and Thraustochytriaceae. They have been classified variously as fungi or protists and are now generally included in the Stramenopila. *cf.* Mycetozoa.

lac *n.* (1) resinous secretion of lac glands of certain insects. Some types are used to make shellac; (2) designation for the lactose operon of bacteria, which encodes enzymes involved in the uptake and metabolism of lactose to glucose.

laccate *a.* appearing as if varnished.

lacertiform *a.* shaped like a lizard.

Lacertilia, lacertids *n.*, *n.plu.* suborder of reptiles containing the lizards (e.g. geckos, iguanas, agamas, skinks, bearded lizards, monitors). Most species are four-legged, some running on their hind-legs, but some (e.g. slow-worms) are legless. They include insectivorous, herbivorous and carnivorous species, and are adapted to a wide range of habitats, including very dry regions.

lacewings *see* Neuroptera.

lacinia *n.* (1) (*bot.*) segment of finely cut leaf or petal; (2) slender projection from a thallus; (3) (*zool.*) extension of posterior portion of proglottis; (4) inner branch of maxilla in insects.

laciniate *a.* (1) irregularly cut, as of petals; (2) fringed.

laciniform *a.* fringe-like.

laciniolate *a.* minutely incised or fringed.

lacinula *n.* (1) a small lacinia; (2) the inflexed sharp point of a petal. *plu.* **lacinulae.**

lacinulate *a.* having lacinulae.

lacrimal *a.* (1) secreting or *pert.* tears; (2) *pert.* or situated near the tear gland, *appl.* e.g. artery, duct, nerve.

lacrimal bone bone in skull near the tear gland.

lacrimal gland gland in the corner of the eye which secretes tears. *alt.* tear gland.

lacrimiform *a.* tear-shaped, *appl.* spores.

lacrimonasal *a.* *pert.* lacrimal and nasal bones and duct.

lacrimose *a.* bearing tear-shaped append-ages, *appl.* gills of certain fungi.

lacrioid *a.* tear-shaped.

lactalbumin *n.* an albumin protein present in milk.

β-lactam antibiotic any of a large group of antibiotics, including the penicillins

and cephalosporins, that contain a β-lactam group.

β-lactamase *n.* enzyme secreted by penicillin-resistant bacteria, which destroys penicillin's antibiotic activity, hydrolys-ing the β-lactam ring. EC 3.5.2.6. *alt.* penicillinase.

lactase *n.* enzyme that hydrolyses terminal non-reducing β-D-galactose residues in β-D-galactosides (e.g. lactose) to glucose and galactose. EC 3.2.1.23, *r.n.* β-galactosidase.

lactate *see* lactic acid.

lactate dehydrogenase enzyme catalys-ing reduction of pyruvate by NADH to lactic acid in animal tissues and some bacteria. EC 1.1.1.27.

lactation *n.* (1) secretion of milk in mam-mary glands; (2) period during which milk is secreted.

lacteals *n.plu.* (1) lymphatic vessels of the small intestine; (2) (*bot.*) ducts that carry latex.

lacteous *a.* milky in appearance or texture.

lactescent *a.* producing milk or latex.

lactic *a.* *pert.* milk.

lactic acid, lactate three-carbon carboxylic acid, usually formed by the reduction of pyruvate. It is formed in animal cells, esp. muscle, when insufficient oxygen is supplied for the full oxidation of sugars. It is also produced during fermen-tation of sugars by certain bacteria, esp. lactobacilli.

lactic acid bacteria lactobacilli *q.v.*

lactifer *n.* any latex-containing cell, series of cells or duct.

lactiferous *a.* (1) forming or carrying milk; (2) carrying latex.

lactific *a.* milk-producing.

lactobacillus *n.* member of the bacterial genus *Lactobacillus*, Gram-positive bac-teria characteristically producing lactic acid as an end-product of anaerobic res-piration, and responsible for souring milk. *plu.* **lactobacilli.**

lactoferrin *n.* an iron-binding protein in animals.

lactogenesis *n.* initiation of milk secretion.

lactogenic *a.* (1) *pert.* or stimulating secre-tion of milk; (2) *appl.* hormone: prolactin *q.v.*; (3) *appl.* interval between parturition and ovulation, or between parturition and menstruation.

lactoglobulin *n.* milk protein soluble in ammonium sulphate but insoluble in water alone.

lactoperoxidase *n.* peroxidase (*q.v.*), an antimicrobial enzyme in milk and saliva.

lactoprotein *n.* any of the proteins in milk.

lactose *n.* disaccharide composed of glucose and galactose, abundant in milk.

lactose intolerance inability to digest lactose, leading to abdominal symptoms when large amounts are ingested, e.g. in milk. It is shown by many adults as a result of decreased production of the enzyme lactase after weaning. It may also be a genetically determined condition, present from birth, which is due to a deficiency in the gene for lactase.

lactosis lactation *q.v.*

lacuna *n.* a space or cavity. *plu.* **lacunae**.

lacunar *a.* having, resembling, or *pert.* lacunae.

lacunate *a.* possessing or forming lacunae.

lacunose *a.* having many cavities.

lacunosorugose *a.* having deep furrows or pits, as some seeds and fruits.

lacunula *n.* minute cavity or air space.

lacustrine *a. pert.*, or living in or beside, lakes.

lacZ the bacterial gene for β-galactosidase, which is often used as a marker in a fusion gene to detect expression of an attached promoter in e.g. particular cell types, as expression of β-galactosidase can be detected histochemically.

laeotropic *a.* inclined, turned or coiled to the left.

laevigate levigate *q.v.*

laevo-rotatory *a. appl.* optically active molecules that rotate a beam of plane polarized light in an anticlockwise direction. The prefix (–) or *l* is often used for such compounds.

lageniform *a.* shaped like a flask.

lagging strand, lagging chain in DNA replication *in vivo*, the new DNA strand that is synthesized discontinuously at the replication fork. *cf.* leading strand.

Lagomorpha, lagomorphs *n., n.plu.* the rabbits, hares and pikas. An order of herbivorous mammals known from the Eocene, with skulls and dentition similar to rodents, but with a second pair of incisors, and with hindlimbs modified for leaping.

lagopodous *a.* having hairy or feathered feet.

lag phase the first phase of growth of a bacterial culture, in which there is no appreciable increase in cell numbers.

LAI leaf area index *q.v.*

LAK lymphokine-activated killer cells *q.v.*

laking haemolysis *q.v.*

Lamarckism *n.* a theory of evolution chiefly formulated by the French scientist J.B. de Lamarck in the 18th century, which embodied the principle, now known to be mistaken, that somatic characteristics acquired by an organism during its lifetime can be inherited.

lambda (λ) *n.* (1) DNA bacteriophage that infects *Escherichia coli* and whose genetic structure and function have been minutely dissected. Lambda is a temperate bacteriophage that may persist as a prophage integrated into the bacterial DNA or multiply within the bacterial cell, eventually destroying it. It is used as a vector in recombinant DNA work; (2) (*anat.*) the junction of the lambdoid and sagittal sutures of skull; (3) (*immunol.*) one of the two types of light chain found in immunoglobulins.

lambda particles (1) cytoplasmic inclusions in the ciliate *Paramecium*; (2) particles of phage lambda.

lambdoid *a.* (1) *appl.* phages genetically and morphologically resembling phage lambda; (2) lambda-shaped, *appl.* cranial suture joining occipital and parietal bones.

lamella *n.* (1) any thin or plate-like structure; (2) a gill of a mushroom or toadstool; (3) a layer of cells. *plu.* **lamellae**.

lamellar, lamellate *a.* composed of thin plates.

lamellasome *n.* layered membranous structure in cyanobacteria which bears the biochemical apparatus of photosynthesis.

lamellated corpuscles Pacinian bodies *q.v.*

lamellibranch(iate) *a.* (1) having plate-like gills on each side; (2) with bilaterally compressed symmetrical body, like a bivalve.

lamellibranchs *n.plu.* the Lamellibranchia, a large subclass of bivalve molluscs, including e.g. clams, cockles, mussels.

lamellicorn *a.* having segments of antennae expanded into flattened plates.

lamelliferous *a*. having small plates or scales.

lamelliform, lamelloid *a*. plate-like.

lamellipodium *n*. thin sheet-like pseudopodial extension temporarily put forward by animal cells such as fibroblasts or lymphocytes when moving over a surface. *plu*. **lamellipodia**.

lamellirostral *a*. having inner edges of bill bearing lamella-like ridges.

lamellose *a*. (1) containing lamellae; (2) having a lamellar structure.

Lamiales *n*. order of dicot herbs, shrubs and trees including Verbenaceae (verbena), Lamiaceae (e.g. mint) and others.

lamin *n*. any of a small family of proteins that form intermediate filaments in the nuclear lamina.

lamina *n*. (1) thin layer, plate or scale; (2) layer of cell bodies in cerebral cortex and other brain structures; (3) blade of leaf or petal; (4) flattened part of a thallus.

lamina basalis Bruch's membrane *q.v.*

lamina cribrosa region of sclera at site of attachment of optic nerve and with perforations for axons of retinal ganglion cells.

lamina fusca inner layer of sclera, adjoining lamina suprachoroidea.

lamina propria layer of loose connective tissue in the mucosa of gut and other tubular tracts, which houses the bases of glands and contains blood and lymph vessels and mucosa-associated lymphoid tissues. It lies immediately under the epithelium and basement membrane.

laminar *a*. (1) consisting of plates or thin layers; (2) (*bot*.) *appl*. placentation of ovule, attachment over the surface of carpel.

laminarian *a*. *appl*. zone between low tide line to about 30 m depth, i.e. the zone typically inhabited by *Laminaria* seaweeds.

laminarin *n*. any of various carbohydrates which are the main food reserves in brown algae and are stored in solution. They consist mainly of glucose units but some contain mannitol.

lamina suprachoroidea delicate tissue layer between choroid and sclera of eye.

laminated *a*. composed of thin plates, *appl*. plant cuticle.

lamina terminalis thin layer of grey matter forming anterior boundary of 3rd ventricle of brain.

lamination *n*. (1) the formation of thin plates or layers; (2) arrangement in layers, as of the nerve cell bodies of cerebral cortex.

lamina vasculosa outer layer of choroid in retina beneath suprachoroid membrane.

laminiform *a*. (1) like a thin layer or layers; (2) like a leaf blade; (3) laminar *q.v.*

laminin *n*. glycoprotein of the extracellular matrix.

laminiplantar *a*. having scales of metatarsus meeting behind in a smooth ridge.

lampbrush chromosome type of bivalent chromosome formed in many vertebrate oocyte nuclei during extended meiosis. The chromosomes are in an extended state, with loops of chromatin folded out from each chromatid at the chromomeres to give an appearance rather like a bottlebrush (or the brush that was used to clean oil lamps). The chromatin loops are being transcribed, as can be seen in the electron microscope.

lampreys *n.plu*. common name for primitive fish-like freshwater and marine chordates of the order Petromyzoniformes, in the class Agnatha, which as adults have sucking and rasping mouthparts.

lamp shells common name for the Brachiopoda *q.v.*

lanate *a*. covered with short woolly hairs.

lancelets common name for the Cephalocordata *q.v.*

lanceolate *a*. slightly broad or tapering at base, and tapering to a point at tip, *appl*. leaves.

lance-oval *a*. having a shape intermediate between lanceolate and ovate, *appl*. leaves.

lancet *n*. one of the paired parts ventral to stylet of sting in bees and wasps.

landscape genetics the combination of information on geographical landscape features with analysis of molecular markers in order to understand how environmental factors affect the dispersal of individuals and the size and density of populations.

Langerhans' cell immature dendritic cell found in the skin and parts of gastrointestinal tract. It is involved in the uptake of antigen, which it then transports to secondary lymphoid organs to initiate an immune response.

Langerhans, islets of *see* islets of Langerhans.

laniary *a.* adapted for tearing, *appl.* canine tooth.

laniferous, lanigerous *a.* wool-bearing or fleecy.

lantern *n.* (1) Aristotle's lantern *q.v.*; (2) a light-emitting organ, as of lantern fishes.

lanuginose, lanuginous *a.* covered in down.

lanugo *n.* the downy covering on a foetus which begins to be shed before birth.

lapidicolous *a. appl.* animals that live under stones.

lappaceous *a.* (1) like a burr; (2) prickly.

lappet *n.* (1) any of various hanging, lobe-like structures; (2) wattle of a bird.

LAR leaf area ratio *q.v.*

large intestine the caecum, colon and appendix in some vertebrates, sometimes used for the colon only.

lariat *n.* tailed circle composed of intron RNA that is formed when an intron is spliced out of primary transcript RNA.

larmier dacryocyst *q.v.*

larva *n.* independently living, post-embryonic stage of an animal that is markedly different in form from the adult and which undergoes metamorphosis into the adult form, e.g. caterpillar, grub, tadpole. *plu.* **larvae.** *a.* **larval.**

Larvacea *n.* class of tunicates (urochordates) which retain the larval 'tadpole' form throughout their lives.

larvicide *n.* an agent that kills insect larvae.

larviform *a.* shaped like a larva.

larviparous *a.* giving birth to offspring at the larval stage.

larvivorous *a.* larva-eating.

larvule *n.* a young larva.

laryngeal *a. pert.* or near the larynx, *appl.* e.g. artery, vein, nerve.

laryngeal prominence in primates, a subcutaneous projection of the thyroid cartilage in front of the throat, causing a ridge on the ventral surface of the neck, and more pronounced in males. *alt.* Adam's apple.

larynges *plu.* of larynx.

laryngopharynx *n.* part of pharynx between soft palate and oesophagus.

laryngotracheal *a.* (1) *pert.* larynx and trachea; (2) *appl.* chamber into which lungs open in amphibians.

larynx *n.* in mammals, the organ in throat that produces sound, the voice box. It is the upper end of the windpipe stiffened by cartilage and with two membranes (vocal chords) each extending half-way across the windpipe leaving a narrow slit between them. Sound is produced when air is driven through the slit, setting the vocal chords vibrating. *a.* **laryngeal.** *plu.* **larynges.**

laser capture microdissection technique for isolating a very small area, even a single cell, from a tissue. The tissue is coated with a thin plastic film and a laser is then used to fuse the required area to the plastic, cut round it and remove it.

lash flagellum flagellum in which the main filament ends in a thinner portion, the lash.

lasso *n.* a contractile filamentous noose formed by certain soil fungi and used to trap nematodes.

lasso cell colloblast *q.v.*

last universal common ancestor (LUCA) the hypothetical microorganism from which all present-day organisms are descended.

late *a.* (1) *appl.* period from the start of virus nucleic acid replication within a cell to the release of infectious bacteriophage or virus; (2) *appl.* bacteriophage or viral genes expressed at later stages of infection.

latebricole *a.* living in holes.

latency *n.* property of enzymes in a cell extract which fail to show maximum activity unless treated with detergents. Such enzymes are membrane-bound enzymes.

latent *a.* (1) lying dormant but capable of development under certain circumstances, *appl.* buds, resting stages; (2) *appl.* characteristics that will become apparent under certain conditions; (3) *appl.* virus infection in which the virus remains quiescent for long periods of time, symptoms appearing as the virus resumes multiplication.

latent bodies the resting stage of certain flagellate blood parasites.

latent period reaction time *q.v.*

laterad *a.* (1) towards the side; (2) away from the axis.

lateral *a. pert.*, or situated at, the side.

lateral bud bud arising in the axil of a leaf at the node of a stem, and which will develop into a side shoot.

lateral diffusion usually refers to the ability of membrane proteins and membrane lipids to diffuse in the plane of the lipid bilayer.

lateral ganglia ganglia of the autonomic nervous system. In mammals they lie in two chains alongside the aorta and are linked to each other by nerve fibres. *cf.* collateral ganglia.

lateral gene transfer horizontal gene transfer *q.v.*

lateral geniculate nucleus organized region of cell bodies in each cerebral hemisphere, at which optic nerve fibres terminate, and from which impulses are relayed to the visual area of the cerebral cortex.

lateralia *n.plu.* the lateral plates of barnacles.

lateral inhibition (1) mechanism in compound eyes which enhances contrast at boundaries between lighter and darker parts of the visual field; (2) (*dev. biol.*) developmental mechanism ensuring that repeating structures such as insect sensillae or the individual units (ommatidia) of a compound eye are evenly spaced. The developing cell or structure inhibits the differentiation of similar elements within a given distance away.

lateralis organ neuromast *q.v.*

lateral line longitudinal line on each side of the body in fishes. It marks the position of cutaneous sensory cells of the acoustico-lateralis system concerned with the perception of movement and sound waves in water, the cells on the lateral line being known collectively as the lateral line system.

lateral line organ neuromast *q.v.*

lateral meristems dividing tissues in plants that are concerned with the production of secondary tissues rather than with the primary apical growth of the plant. They include the vascular cambium (producing the vascular bundles) and the cork cambium (producing the outer layer of cork or bark).

lateral plate mesoderm mesoderm that gives rise to the splanchnic and somatic mesoderm in vertebrates, from which internal organs and blood system will develop. *cf.* somite.

lateral roots roots which branch off the primary root, and which themselves may give rise to further lateral roots.

lateral ventricle large fluid-filled cavity in centre of cerebral hemisphere of brain.

laterigrade *a.* walking sideways, like a crab.

laterinerved *a.* with lateral veins, *appl.* leaves.

laterite *a. appl.* tropical red soils containing alumina and iron oxides and little silica owing to leaching under hot moist conditions.

laterobronchi *n.plu.* secondary bronchi arising from the mesobronchus in birds.

laterosensory *a. appl.* lateral-line system in fishes. *see* lateral line.

laterosphenoid *n.* bone on the mid-line of the reptilian skull, behind the orbit and above the palate.

late wood wood formed in the later part of the growing season. It shows as a dense ring of small cells in the annual ring. *alt.* summer wood.

latex *n.* thick milky or clear juice or emulsion of diverse composition present in plants such as rubber trees, spurges, and in certain agaric fungi.

lathyrism *n.* disease of animals characterized by fragile collagen. It is caused by eating seeds of certain *Lathyrus* species that contain β-aminopropionitrile, which inhibits essential post-translational modification of collagen.

laticifer lactifer *q.v.*

laticiferous *a.* conveying latex, *appl.* cells, tissues.

latifoliate *a.* with broad leaves.

Latin name *see* binomial nomenclature.

latiplantar *a.* having hinder tarsal surface rounded.

latirostral *a.* broad-beaked.

latiseptate *a.* having a broad septum in the siliqua.

latitudinal furrow a cleavage furrow running round the dividing fertilized egg above and parallel to the equatorial furrow.

latosol *n.* leached red or yellow tropical soil.

Laurales *n.* order of dicot trees, shrubs and climbers, with aromatic ethereal oils in the cells, and including the families Calycanthaceae (calycanthus), Lauraceae (laurels) and others.

Laurasia *n.* former northern land mass or supercontinent that consisted of present-day North America, Europe and Northern Asia before they were separated by continental drift. It broke off from the massive supercontinent Pangaea mainly in the Jurassic. *see also* Gondwana.

laurilignosa *n*. type of subtropical forest and bush composed of laurel.

laurinoxylon *n*. fossil wood.

law of equal segregation *see* segregation of alleles.

law of independent assortment *see* independent assortment.

law of the minimum rule stating that the factor for which an organism or species has the narrowest range of tolerance or adaptability limits its existence.

law of tolerance (*ecol.*) rule stating that the distribution of a species will be limited by its range of tolerance for local environmental factors. *alt.* Shelford's law of tolerance.

lax *a*. loosely clustered, *appl.* panicle of flowers.

layer *n*. horizontal stratum in a plant community. The following are usually distinguished: the tree layer or canopy, the shrub layer, comprising the shrubby understorey, the herb layer of grass and herbaceous plants, and the ground (moss) layer, comprising the ground surface and lichens and mosses.

LC lethal concentration *q.v.*

LC$_{50}$ concentration of any toxic chemical that kills 50% of the organisms in a test population per unit time.

LCA family of lectins isolated from lentils, *Lens culinaris*.

L chain light chain *q.v.*

LCM lymphocytic choriomeningitis virus *q.v.*

LCR locus control region *q.v.*

LD lethal dose *q.v.*

LD$_{50}$ a measure of infectivity for viruses or of toxicity of chemicals, the dose at which 50% of test animals die.

LDL low-density lipoprotein *q.v.*

LDL receptor receptor on non-hepatic cells which is specific for low-density lipoprotein (LDL). It is important in uptake of cholesterol and its clearance from the circulation.

leaching *n*. (1) process by which chemicals in the upper layers of the soil are dissolved and carried down into lower layers; (2) microbial leaching, extraction of metals from their ores by microbial action.

lead (Pb) *n*. metallic element, a 'heavy metal' which is toxic to many organisms and is an environmental pollutant.

leader *n*. (1) topmost growing shoot or main branch of tree; (2) leader sequence *q.v.*

leader peptide signal peptide *q.v.*

leader sequence (1) region in mRNA (and DNA) preceding the coding region, which is transcribed but is not translated, *alt.* untranslated leader, 5′ UTR; (2) translated leader *q.v.*; (3) signal sequence *q.v.*

leading strand, leading chain in DNA replication *in vivo*, the new DNA strand that is synthesized continuously at the replication fork. *cf.* lagging strand.

leaf *n*. an expanded, flattened or needle-like outgrowth from plant stem, usually green and the main photosynthetic organ of most plants. *plu.* **leaves**.

leaf area index (LAI) of a given area of vegetation, the total area of photosynthetic leaf surface divided by the area of soil covered.

leaf area ratio (LAR) the ratio of the photosynthetic surface area of a leaf to its dry weight.

leaf blade the thin flat part of a leaf.

leaf buttress lateral prominence on shoot axis caused by an underlying leaf primordium.

leaf cushions prominent persistent leaf bases on stems of some trees, e.g. palms, and furnishing diagnostic characters in some fossil plants.

leaf divergence the fixed proportion of the circumference of the stem by which each leaf is separated from the next.

leaf gap region of ground tissue interrupting the pattern of the stele, resulting from the divergence of vascular tissue (the leaf trace) away from stele to leaf.

leaf hair trichome *q.v.*

leaf insects common name for some members of the Phasmida whose bodies mimic leaves in form.

leaflet *n*. (1) a small leaf; (2) individual unit of a compound leaf; (3) (*mol. biol.*) one layer of a lipid bilayer.

leaf mine a blotch on a leaf indicating where the larva of certain insect species (e.g. some moths) has eaten the tissue between the upper and lower leaf epidermis.

leaf mosaic the arrangement of leaves on a plant that results in minimum overlap and maximum exposure to sunlight.

leaf scar the trace, usually covered with a corky layer, left on stem after leaf has fallen.

leaf sheath extension of leaf base sheathing the stem, as in grasses.

leaf stalk petiole *q.v.*

leaf trace vascular tissue extending from stele of stem into base of leaf.

leaky *a. appl.* mutations with some residual function. *see also* hypomorph.

learning *n.* any process in an animal in which its behaviour becomes consistently modified as a result of experience, and including conditioning, habituation and imprinting. *see also* long-term potentiation.

learning set in animal behaviour, the apparent learning of a general rule for solving a set of problems.

leberidocyte *n.* cell containing glycogen, present in blood of arachnids at moulting.

lechriodont *a.* with vomerine and pterygoid teeth in a row nearly transverse.

lecithin phosphatidylcholine *q.v.*

lecithoprotein *n.* lipoprotein in which lecithin (phosphatidylcholine) is the lipid component.

lecithotrophic *a.* feeding on stored yolk, as in some sea urchin larvae.

lecithovitellin *n.* lipoprotein composed of lecithin (phosphatidylcholine) and the yolk protein vitellin.

lectin *n.* any of a group of plant proteins that can agglutinate animal cells *in vitro* by binding to specific sugar residues in membrane glycoproteins. Some lectins also have mitogenic activity. Their natural role in plants is likely to be in cell–cell adhesion, cell recognition and defence against infection. *see also* C-type lectins.

lectin-mediated pathway of complement activation *see* complement activation.

lectotype *n.* specimen chosen from syntypes to designate type species.

leech common name for a member of the Hirudinea *q.v.*

left–right asymmetry the normal asymmetrical arrangement of the internal organs of the human body.

leghaemoglobin *n.* red oxygen-binding protein pigment, resembling haemoglobin, found in the nitrogen-synthesizing root nodules of leguminous plants.

Legionnaire's disease respiratory disease caused by the bacterium *Legionella pneumophila. alt.* **legionellosis**.

legume *n.* (1) a pod, a type of fruit derived from a single carpel and which splits down both sides at maturity, characteristic of the pea family; (2) any member of the Leguminosae, e.g. peas, beans, clovers, vetches, gorse, broom.

legumin *n.* protein present in seeds of leguminous plants.

Leguminosae *n.* large family of dicotyledonous plants, commonly called legumes or leguminous plants, and including trees, shrubs, herbs and climbers. They have typical sweet-pea shaped flowers and fruit in the form of pods. They include peas, beans, clovers and vetches. *see also* Fabales.

leguminous *a.* (1) *pert.* Leguminosae; (2) *pert.*, or consisting of, peas, beans or other legumes.

leimocolous *a.* inhabiting damp meadows.

leiosporous *a.* with smooth spores.

leiotrichous *a.* having straight hair.

Leishmania genus of parasitic protozoa, infecting humans and other mammals, with sandflies as the intermediate host and vector. *L. donovani* causes the chronic and often fatal tropical disease visceral leishmaniasis, *L. tropica* causes cutaneous leishmaniasis or tropical sore.

leishmanial *a. appl.* short stout forms of trypanosome lacking a free flagellum.

Leitneriales *n.* order of resinous dicot shrubs comprising the family Leitneriaceae with the single genus *Leitneria*.

lek *n.* special arena removed from nesting and feeding grounds, used for communal courtship display (lekking) preceding mating in some birds (e.g. ruffs and black grouse). The term is sometimes applied to similar areas used by other animals for communal displays.

lekking *n.* highly ritualized sexual display by birds such as black grouse, which takes place on a particular display ground, the lek, and which precedes mating.

lemma *n.* lower of the two bracts enclosing a floret (individual flower) in grasses.

lemniscus *n.* band of white matter in midbrain and medulla oblongata.

lens *n.* (1) transparent structure in the eye through which light is focused onto the retina, the crystalline lens of the vertebrate eye being formed from prism-shaped, refractile dead cells filled with the protein crystallin; (2) modified portion of the cornea in front of each element of a compound eye; (3) modified cells of luminescent organ in certain fishes.

lens fibres elongated, lifeless, crystallin-filled cells making up the lens of the eye.

lens placode local thickening of the ectoderm opposite the optic vesicle, which invaginates to form the lens pit, which then closes to become the lens vesicle, developing into the lens. *alt.* lens rudiment.

lentic *a.* (1) *appl.* standing water; (2) *appl.* organisms living in swamp, pond, lake or any other standing water.

lenticel *n.* pore in periderm of trees and shrubs, allowing the passage of air to internal tissues.

lenticula *n.* (1) a spore case in certain fungi; (2) lenticel *q.v.*; (3) lentigo or freckle.

lenticular *a.* (1) shaped like a double convex lens; (2) *pert.* lenticels.

lenticulate *a.* (1) meeting in a sharp point; (2) depressed, circular and often ribbed.

lentiform *a.* lentil-shaped.

lentiform glands lymphoid glands situated between pyloric glands.

lentiform nucleus the putamen and globus pallidus. *see* basal ganglia.

lentigerous *a.* having a lens.

lentiginose, lentiginous *a.* freckled, speckled, or bearing many small dots.

lentiviruses *n.plu.* subfamily of non-oncogenic slow-acting retroviruses, which cause chronic infections that only become manifest years after infection. They include HIV (human immunodeficiency virus).

lepidic *a.* (1) consisting of scales; (2) *pert.* scales.

lepidodendroid *a.* having scale-like leaf scars.

Lepidodendron a genus of fossil tree fern (scale trees) with small leaves producing scale-like leaf scars.

lepidoid *a.* resembling a scale or scales.

lepidomorium *n.* small scale or unit of composite scale, with bony base and conical crown of dentine, containing a pulp cavity and sometimes covered with enamel. *a.* **lepidomorial**.

lepidophyte *n.* fossil fern.

Lepidoptera, lepidopterans *n.*, *n.plu.* order of insects commonly known as moths

and butterflies. Their bodies and wings are covered by small scales, often brightly and variously coloured, forming characteristic patterns. They undergo complete metamorphosis, the larval (caterpillar) stage giving rise to a pupa in which metamorphosis occurs, with development of adult structures such as the two pairs of wings, the legs and compound eyes. Adult Lepidoptera feed largely on nectar, through a hollow proboscis. *a.* **lepidopterous**.

Lepidosauria, lepidosaurs *n.*, *n.plu.* subclass of reptiles comprising the lizards, snakes and amphisbaenians, and the tuatara, with a diapsid skull, and with limbs and limb girdles unspecialized, reduced or absent.

lepidosis *n.* character and arrangement of scales on an animal.

lepidote *a.* covered with minute scales.

lepidotrichia *n.plu.* bony unjointed rays at the edge of fins in teleost fishes.

lepospondyls *n.plu.* a group of extinct amphibians having hour-gluss-shaped vertebrae. *adj.* **lepospondylous**.

leprosy *n.* chronic disease caused by the bacterium *Mycobacterium leprae*, leading to extensive tissue damage, deformity and disability.

leptin *n.* protein hormone secreted by fat cells and which acts at leptin receptors in the brain to lessen hunger and discourage eating. It is thought to be a means of monitoring the amount of body fat. When the amount of fat is low, leptin is not produced.

lepto- prefix derived from Gk *leptos*, slender.

leptocaul *a.* having a slender primary stem.

leptocentric *a.* *appl.* concentric vascular bundle with phloem at centre.

leptocephaloid *a.* resembling or having the shape of eel larvae.

leptocephalus *n.* translucent larva of certain eels, before the elver stage.

leptocercal *a.* with a long slender tapering tail, *appl.* some fishes.

leptodactylous *a.* having slender fingers.

leptodermatous *a.* thin-skinned.

leptoid *n.* living food-conducting cell joined with others to form a simple conducting tissue in stems of some mosses.

leptoma *n.* thin area in the wall of a gymnosperm pollen grain, through which the pollen tube emerges.

leptome *n.* (1) sieve elements and parenchyma of phloem; (2) similar conducting elements in bryophytes.

leptomonad *a. appl.* long slender form of trypanosome with a free flagellum.

leptonema *n.* fine thread-like chromosome that appears at leptotene stage of meiosis.

leptophyllous *a.* (1) with slender leaves; (2) having a small leaf area, under 25 mm³.

leptosome *a.* tall and slender.

leptospirosis *n.* disease commonly characterized by jaundice and nephritis, caused by pathogenic species of the spirochaete bacterium *Leptospira.* Humans usually contract the disease from infected pets or from wild rodents. *alt.* Weil's disease.

leptosporangiate *a. appl.* ferns in which the sporangia develop from a single initial cell, which first produces a stalk and then a capsule. *cf.* eusporangiate.

leptosporangium *n.* stalked sporangium, in ferns. *plu.* **leptosporangia**.

leptotene *n.* early stage in the prophase of meiosis in which the chromatin is beginning to become compacted and the chromosomes show up under the microscope as fine threads.

leptotrombicula *n.* larval form of a trombicula, a mite transmitting scrub typhus (tsutsugamushi disease), a rickettsial disease.

leptoxylem *n.* rudimentary wood tissue.

leptus *n.* the six-legged larva of mites.

Lesch–Nyhan syndrome inherited disease characterized by an almost complete deficiency of the enzyme hypoxanthine-guanine ribosyltransferase and by symptoms of self-mutilation, mental deficiency and spasticity.

lesion *n.* area of tissue destruction.

lesser apes the gibbons and their extinct relatives, back to the divergence of this lineage from that of the great apes (*q.v.*).

lesser omentum a fold of peritoneum which connects the stomach and liver and supports the hepatic vessels.

lestobiosis *n.* the relation in which colonies of small species of insect nest in the walls of the nests of larger species and enter their chambers to prey on the brood or rob food stores.

lethal *a.* (1) causing death; (2) of a parasite, fatal or deadly in relation to a particular host; (3) *appl.* mutations or alleles which when present in an embryo cause its death at an early stage.

lethal concentration (LC) where death is the criterion of toxicity, the results of toxicity tests are expressed as a number (LC$_{50}$, LC$_{70}$) which indicates the percentage of test organisms killed at a particular concentration over a given exposure time, e.g. the 48-hour LC$_{70}$ is the concentration of a toxic material that kills 70% of the test organisms in 48 hours.

lethal dose (LD) dose of a toxic chemical or of a pathogen that kills all the animals in a test sample within a certain time. *cf.* median lethal dose.

lethality *n.* (1) the capacity to cause death; (2) ratio of fatal cases to the total number of cases affected by a disease or other harmful agent.

lethal synthesis the synthesis *in vivo* of a metabolic poison from a substance that is not itself toxic.

leuc- *see also* leuk-

leucine (Leu, L) *n.* α-amino isocaproic acid, an amino acid with a non-polar hydrocarbon side chain, a constituent of protein and essential in human and animal diet.

leucine-rich repeat (LRR) protein motif containing leucine residues, present in the extracellular domains of many plant transmembrane serine/threonine kinases, and in the extracellular portions of the Toll-like receptors of animals.

leucine zipper DNA-binding protein structure in which two α-helices containing leucine are held together by formation of a coiled-coil in one part, while the two separate arms grip the DNA. It is found in many proteins involved in gene regulation.

leucism *n.* presence of white plumage or fur in animals with pigmented eyes and skin. *cf.* albinism.

leuco- prefix derived from Gk *leukos*, white.

leucoanthocyanidins *n.plu.* a group of colourless flavonoids.

leucocarpous *a.* with white fruit.

leucocidin leukocidin *q.v.*

leucocyte *alt.* spelling of leukocyte.

leucoplast(id) *n.* colourless plastid, such as the starch-containing amyloplast, in cells of plant epidermis and internal tissues.

leucopterin *n.* white pterin pigment of cabbage white butterflies (*Pieris*) and other Lepidoptera and wasps, and which can be reduced to xanthopterin. *see also* isoxanthopterin.

leucosin *n.* storage polysaccharide forming whitish granules in some yellow-brown algae.

leukaemia *n.* malignant disorder of white blood cells in which precursors proliferate and fail to differentiate.

leukaemogenesis *n.* the generation of a leukaemia.

leukemia *alt.* spelling of leukaemia.

leukin *n.* basic polypeptide extracted from leukocytes and active against Gram-positive bacteria.

leuko- *see also* leuco-

leukoblast myeloblast *q.v.*

leukocidin *n.* protein toxin that lyses white blood cells, produced by bacteria such as *Staphylococcus aureus* and *Streptococcus pyogenes*.

leukocytes *n.plu.* the colourless cells of the blood, commonly known as white blood cells. They consist of the basophils, eosinophils, neutrophils, lymphocytes and monocytes. All derive from a common progenitor in bone marrow and are involved in protecting the body from infection. Basophils, eosinophils and neutrophils are known collectively as granulocytes or polymorphonuclear leukocytes. Monocytes become macrophages when they enter tissues.

leukocytolysis *n.* breakdown or disintegration of white blood cells. *alt.* leukolysis.

leukocytosis leukosis *q.v.*

leukopenia *n.* reduction in the number of circulating white blood cells, characteristic of many diseases.

leukopoiesis *n.* generation of white blood cells.

leukosis *n.* an increase in the numbers of circulating white blood cells. *alt.* leukocytosis.

leukotrienes *n.plu.* class of compounds derived from arachidonic acid and released from mast cells in local inflammatory reactions, formerly known as slowreacting substance of anaphylaxis (SRS-A). They cause smooth muscle contraction (leukotriene E4) and attract polymor-phonuclear leukocytes and eosinophils to sites of injury or infection (leukotriene B4).

levan *n.* any polysaccharide made up of fructose units.

levator *n.* a muscle serving to raise an organ or part.

levigate *a.* made smooth, polished.

ley *n.* temporary agricultural grassland, which is sown and used as a crop.

Leydig cells in interstitial tissue of testis that secrete testosterone.

Leydig's organs minute organs on antennae of arthropods, possibly chemoreceptors.

L-form *n.* stage of certain mycoplasmas in which they will pass through the normal bacterial filter and which consists of specialized reproductive bodies, produced in extreme conditions.

LH luteinizing hormone *q.v.*

LHC light-harvesting complex *q.v.*

LHCP light-harvesting chlorophyll-protein. *see* photosynthetic unit.

LHRH luteinizing hormone-releasing hormone *q.v.*

liana *n.* any woody climbing plant of tropical and semitropical forests.

Lias *n.* marine and estuarine deposits of the Jurassic period, containing remains of fossil cycads, insects, ammonites and saurians. *a.* **Liassic**.

libriform *a. appl.* woody fibres with thick walls and simple pits.

Librium trade name for one of the commonly used benzodiazepines *q.v.*

lice *plu.* of louse.

lichen *n.* a composite organism formed from the symbiotic association of certain basidiomycete or ascomycete fungi and a green alga or a cyanobacterium. This forms a simple thallus, found e.g. encrusting rocks and tree trunks.

lichen acids diverse secondary metabolites produced by lichen fungi and which contribute to weathering of rocks on which the lichens grow.

lichenase *n.* enzyme that breaks down lichenin to glucose, and so digests lichens. It is found in the gut of reindeer and caribou and some gastropods. EC 3.2.1.6, *r.n.* endo-1,3(4)-β-D-glucanase.

lichenicole, lichenicolous *a.* living or growing on lichens.

lichenin *n.* glucose polysaccharide present in the walls of lichen fungi, and also in some seeds, esp. oats, and which is hydrolysed by the enzyme lichenase to glucose.

lichenization *n.* (1) production of a lichen by alga and fungus; (2) spreading or coating of lichens over a substrate; (3) effect of lichens on their substrates.

lichenoid *a.* resembling a lichen.

lichenology *n.* the study of lichens.

lichenometry *n.* estimation of lichen growth to measure minimum time elapsed since lichens first developed on the substrate in question.

lichen starch lichenin *q.v.*

Lieberkühn's crypts crypts of Lieberkühn *q.v.*

Liebig's law (1) the nutrient least plentiful in proportion to the requirements of plants limits their growth; (2) law of the minimum *q.v.*

lienal *a. pert.* spleen.

lienculus *n.* an accessory spleen.

lienogastric *a. pert.* spleen and stomach, *appl.* artery supplying parts of spleen and parts of stomach and pancreas.

life *n.* living organisms can be distinguished from other complex physicochemical systems by their storage and transmission of molecular information in the form of nucleic acids, their possession of enzyme catalysts, their energy relations with the environment and their internal energy conversion processes (e.g. photosynthesis, respiration and other enzyme-catalysed metabolic activities), their ability to grow and reproduce, and their ability to respond to stimuli (irritability). Viruses, which satisfy only some of these criteria, are also generally considered as part of the living world. *see also* origin of life.

life cycle the various stages an individual organism passes through from its origin to maturity and reproduction.

life-cycle assessment type of analysis used in environmental cost–benefit analysis, which can be implemented in various ways. It essentially calculates the environmental impact of a product or process throughout its lifetime from creation to waste, taking into account also any hazards and improvements.

life expectancy *see* average life expectancy.

life form the typical adult form of a species.

life tables demographic data required to calculate e.g. the intrinsic rate of increase of a population. They comprise the survivorship schedule, which gives the number of individuals surviving to each particular age, and the fertility schedule, which gives the average number of female offspring that will be produced by a single female at each particular age. From these the net reproductive rate, R_0, which is the average number of female offspring produced by each female during her lifetime, can be calculated. The intrinsic rate of increase, r, of the population can be computed from survivorship and fertility schedules using the Euler–Lotka equation.

ligament *n.* (1) strong fibrous band of tissue connecting two or more movable bones or cartilages; (2) band of elastic tissue forming the hinge of a bivalve shell.

ligand *n.* any molecule that binds specifically to another molecule. Examples are a hormone binding to its receptor, an inhibitor binding to an enzyme, oxygen binding to haemoglobin, an antigen binding to an antibody.

ligand-gated *appl.* ion channels in cell membranes which are stimulated to open or close by the binding of a particular molecule (the ligand).

ligase *n.* any of a class of enzymes that catalyse the joining together of two molecules coupled with the breakdown of a pyrophosphate bond in ATP or a similar nucleoside triphosphate. They include the synthetases, carboxylases, and the DNA and RNA ligases. EC 6.

ligation *n.* of DNA, the joining of two molecules of DNA end to end by the enzyme DNA ligase to form a continuous DNA. The enzyme catalyses the formation of a phosphodiester bond in the sugar–phosphate backbone.

light chain (L chain) (1) the smaller of the two types of polypeptide chain in an immunoglobulin molecule. In each molecule there are two identical light chains; (2) in myosin, any of four polypeptide chains attached to the globular heads of the molecule.

light-harvesting complex (LHC) complex of chlorophyll (and other pigments) and protein that collects light energy and passes it to the photosynthetic reaction centre. In green plants and green algae there are two light-harvesting complexes, one associated with photosystem I and one with photosystem II.

light microscopy type of microscopy that uses visible light and optical lenses to create the magnified image, the instrument used being known as a light microscope or optical microscope. *alt.* optical microscopy.

light reaction in photosynthesis, reactions occurring in the thylakoid membranes of chloroplasts, in which light energy drives the synthesis of NADP and ATP. *cf.* dark reactions.

ligneous *a.* woody, or resembling wood in structure.

lignescent *a.* developing the character of woody tissue.

lignicole, lignicolous *a.* growing or living on or in wood.

lignification *n.* (1) wood formation; (2) the thickening of plant cell walls by deposition of lignin, which occurs in both primary and secondary walls.

lignin *n.* hard material found in walls of cells of xylem and sclerenchyma fibres in plants and which stiffens the cell wall. It is a very variable cross-linked polymer of phenylpropane units such as coniferyl alcohol (guaiacyl), sinapyl alcohol (syringyl) or hydroxycinnamyl alcohol.

lignivorous *a.* eating wood, *appl.* various insects.

lignocellulose *n.* lignin and cellulose combined, a constituent of woody tissue.

lignolytic *a.* lignin-degrading.

lignosa *n.* vegetation made up of woody plants.

ligula *a.* band of white nerve fibres in dorsal wall of 4th ventricle of brain.

ligula, ligule *n.* (1) thin flattened outgrowth at junction of leaf blade and leaf sheath or petiole; (2) small scale on upper surface of leaf base in some club mosses and quillworts; (3) a strap-shaped corolla.

ligular *a.* tongue-shaped.

ligulate *a.* (1) having or *pert.* ligules; (2) having a strap-shaped corolla, like a ray floret of flowers of the Compositae; (3) *appl.* flowerhead of strap-shaped florets.

liguliferous *a.* having ligulate flowers only.

Ligustrales Oleales *q.v.*

Liliales *n.* order of monocot plants, growing from rhizomes or bulbs, mostly herbaceous, and including the families Liliaceae (lily), Agavaceae (agave), Alliaceae (onion), Amaryllidaceae (daffodil), Dioscoreaceae (yam) and others. *alt.* **Liliiflorae**.

Liliopsida *n.* in some plant classifications the name for the class containing the monocotyledons.

limacel(le) *n.* concealed vestigial shell of slugs.

limaciform *a.* (1) slug-shaped; (2) like a slug.

limacine *a. pert.* slugs.

limb *n.* (1) arm, leg or wing in vertebrates; (2) expanded part of calyx or corolla, the base of which is tubular; (3) a branch of a tree.

limbate *a.* (1) with a border; (2) bordered and having a differently coloured edge.

limb bud small protuberance on side of vertebrate embryo from which an arm, leg or wing develops.

limbic *a.* bordering.

limbic system the brain regions involved in emotions, learning and memory, consisting of (in each hemisphere) olfactory bulb, cingulate cortex, thalamus, hippocampus, fornix, mamillary body and amygdala.

limbus *n.* (1) any border if distinctly marked off by colour or structure; (2) transitional zone between cornea and sclera.

limicolous *a.* living in mud.

liminal *a. pert.* a threshold, *appl.* minimal stimulus or quantitative difference in stimulus that is perceptible. *cf.* subliminal.

limit dextrins branched oligosaccharide fragments formed as one of the final products of starch hydrolysis by α-amylase.

limiting factor any single factor (e.g. nutrient, temperature, space) that limits e.g. a biochemical process, the growth of an organism, or its abundance or distribution.

limiting membrane layer of connective tissue behind the vertebrate retina.

limivorous *a.* mud-eating, *appl.* certain aquatic animals.

limnetic *a.* (1) living in, or *pert.*, marshes or lakes; (2) living in open water; (3) *appl.* zone of deep water between surface and compensation depth (depth at which photosynthesis cannot be supported owing to insufficient light).

limnium *n.* a lake community.

limnobiology *n.* the study of life in standing waters, i.e. ponds, marshes, lakes.

limnobios *n.* freshwater plants and animals collectively.

limnobiotic *a.* living in freshwater marshes.

limnology *n.* the study of the biological and other aspects of standing waters.

limnophilous *a.* living in freshwater marshes.

limnophyte *n.* a pond plant.

limnoplankton *n.* the floating microscopic life in freshwater lakes, ponds and marshes.

Limulus a genus of horseshoe crab *q.v.*

***Limulus* assay** highly sensitive assay for bacterial endotoxin, used in the preparation of pharmaceuticals, which employs extracts of amoebocytes of the horseshoe crab, *Limulus*. Endotoxin forms a precipitate with the extract.

lincomycin *n.* antibiotic produced by actinomycetes and which inhibits bacterial protein synthesis.

LINE long interspersed element, intermediate-repeat retrotransposon-like DNA sequence found in mammalian genomes, exemplified by the mammalian L1 elements. LINEs are 6–7 kb long and are present in many thousands of copies (~600,000 copies in the human genome). *see also* SINE.

linea *n.* a line-like structure or mark.

lineage *n.* organisms allied by common descent. *see also* cell lineage.

lineage group group of species allied by descent from a common ancestor.

linear *a. appl.* leaves, the long narrow leaves which are characteristic of monocotyledons.

linear-ensate *a.* between linear and sword-shaped, *appl.* leaves.

linear growth type of growth in which the amount increases by the same amount over each set period (e.g. a year). *cf.* exponential growth.

linear-lanceolate *a.* between linear and lanceolate in shape, *appl.* leaves.

linear-oblong *a.* between linear and oblong in shape, *appl.* leaves.

lineolate *a.* marked by fine lines or striae.

line transect the recording of types and numbers of plants along a measured line.

Lineweaver–Burk plot a way of plotting the course of an enzymatic reaction in which reciprocals of the velocity of the reaction and of the substrate concentration are plotted against each other to give a straight line.

lingua *n.* a tongue or tongue-like structure.

lingual *a. pert.* tongue, *appl.* e.g. artery, nerve, vein.

linguiform *a.* tongue-shaped.

lingula *n.* small tongue-like projection of bone or other tissue.

lingulate *a.* shaped like a short broad tongue. *cf.* ligulate.

linin *n.* (1) protein of flax seed; (2) a bitter purgative substance obtained from purging flax.

linkage *n.* (*genet.*) the case when a set of alleles of different genes from one of the parents are inherited together, in opposition to Mendel's law of independent assortment. It is usually due to the genes being close together on the same chromosome and thus no recombination occurring between them.

linkage disequilibrium condition in which certain alleles at two linked loci are non-randomly associated with each other. This is either because of very close physical proximity which virtually precludes recombination between the two loci, or because the allele combination is under some form of selective pressure.

linkage group the genes carried on any one chromosome.

linkage map genetic map *q.v.*

linker *n.* short stretch of synthetic DNA, usually containing a restriction site, which may be used to connect two different DNA molecules to form a recombinant DNA. It allows the DNA of interest to be easily recovered by treatment with the appropriate restriction enzyme after cloning.

linking number number of turns of one strand of a closed circular DNA about the other, $L = T$ (degree of twisting of the double helix) + W (extent of supertwisting).

Linnean *a. pert.* or designating the system of binomial nomenclature and classification established by the 18th century Swedish biologist Carl von Linné or Linnaeus.

linneon *n.* a taxonomic species distinguished on purely morphological grounds, esp. one of the large species described by Linnaeus or other early naturalists.

linoleic acid common 24-carbon unsaturated fatty acid, essential for growth in mammals and necessary in the diet.

linolenic acid common 20-carbon unsaturated fatty acid, necessary for growth in mammals but not essential in the diet as it can be synthesized from linoleic acid.

lipase *n.* (1) any of a group of widely distributed enzymes (produced esp. by pancreas) hydrolysing triacylglycerols to diacylglycerols plus a fatty acid anion. EC 3.1.1.3, *r.n.* triacylglycerol lipase, and EC 3.1.1.34, *r.n.* lipoprotein lipase; (2) sometimes used for any enzyme that breaks down fats.

lipid *n.* any of a diverse class of compounds found in all living cells, insoluble in water but soluble in organic solvents such as ether, acetone and chloroform. Lipids include fats, oils, triacylglycerols (triglycerides), fatty acids, glycolipids, phospholipids and steroids, some lipids being essential components of biological membranes, others acting as energy stores and fuel molecules for cells.

lipid anchor any of various types of lipid that are covalently linked post-translationally to some proteins and serve to attach them to the cytoplasmic side of the plasma membrane.

lipidation *n.* the covalent attachment of a fatty acid to a protein.

lipid bilayer double layer of molecules formed by phospholipids in an aqueous environment. Each molecule is oriented with the hydrophilic group on the outside and the hydrophobic group to the interior of the layer. This is the basic structure of biological membranes in most organisms. *alt.* bilayer, bimolecular sheet. *cf.* lipid monolayer. *see* Fig. 26 (p. 392).

lipid bodies lipid storage structures found in oil-rich plant seeds, composed of a large droplet of triacylglycerol surrounded by a single-layered membrane. *alt.* oil bodies.

lipid droplet large droplet of (usually) triacylglycerols found in cells specialized for fat storage.

lipid-exchange protein lipid-transfer protein *q.v.*

lipid monolayer (1) type of membrane structure present in the Archaea, in which diglycerol tetraethers span the membrane as single molecules with a hydrophilic glycerol moiety on each side of the membrane; (2) membrane formed from a single layer of phospholipid or fatty acid, found e.g. around fat droplets in cells.

lipid raft an area of plasma membrane with a lipid composition distinct from that of the surrounding membrane, enriched in sphingolipids, cholesterol and membrane proteins.

lipid-transfer protein, lipid-transport protein cytosolic protein that carries membrane lipids from their site of synthesis at the endoplasmic reticulum to chloroplast and mitochondrial membranes.

lipoamide *n.* enzyme cofactor derived from lipoic acid, and which is an acyl group carrier esp. important in carbohydrate metabolism.

lipoate *see* lipoic acid.

lipochroic *a.* with pigmented oil droplets.

lipochrome *n.* a fat-soluble pigment.

lipocyte *n.* cell specialized for lipid production and storage. *alt.* (in animals) adipocyte, fat cell.

lipofection *n.* transfer of material into a cell by enclosing it in liposomes, which fuse with the cell membrane.

lipofuscin granules residual orange fluorescent bodies seen in ageing cells, derived from lysosomes. *alt.* age pigments.

lipogenesis *n.* synthesis of fatty acids and other lipids.

lipogenous *a.* fat-producing.

lipoglycan *n.* long-chain heteropolysaccharide covalently linked to a membrane lipid. Found in the cell membrane of many mycoplasmas.

lipoic acid, lipoate *n.* 1,2-dithiolane-3-valeric acid, a compound composed of a fatty acid (valeric acid) and disulphide. It is required for carbohydrate metabolism as a precursor for the enzyme cofactor lipoamide.

lipoid (1) *a.* resembling a fatty substance; (2) *n.* a substance that is not a true lipid but resembles one in certain properties and is extracted in organic solvents. Examples are sterols and steroids.

lipolysis *n*. (1) enzymatic breakdown of fats, as during digestion; (2) breakdown of triacylglycerols in adipose cells during mobilization of food reserves.

lipolytic *a*. (1) *pert*. lipolysis; (2) *appl*. enzymes capable of breaking down fat; (3) *appl*. hormone: lipotropin *q.v.*

lipopalingenesis *n*. loss of a developmental stage or stages during evolution.

lipopexia *n*. deposition and storage of fats in tissues.

lipopolysaccharide (LPS) *n*. usually refers to the complex compound of lipid and polysaccharide that is the main constituent of the outer membrane of Gram-negative bacteria. It produces fever and other symptoms in humans and other mammals, and by stimulating the systemic overproduction of the cytokine TNF-α it contributes to the septic shock that sometimes follows a blood infection. *alt.* endotoxin.

lipopolysaccharide layer outer membrane of Gram-negative bacteria.

lipoprotein *n*. complex of lipid and protein.

liposome *n*. (1) artificially constructed sphere of lipid bilayer enclosing an aqueous compartment. Liposomes are used in experimental biology to study properties of biological membranes and clinically as a possible means of delivering drugs to cells more efficiently, *alt.* lipid vesicle; (2) fatty droplet in cytoplasm, esp. of an egg cell.

lipoteichoic acid any of a class of glycerol-containing teichoic acids covalently bound to membrane lipids in the cell walls of Gram-positive bacteria.

lipotropic *a*. concerned with the mobilization of storage lipids and their breakdown into fatty acids and triacylglycerols.

lipotropin (LPH) *n*. in mammals, either of two peptide hormones produced by the anterior pituitary gland which stimulate lipolysis.

lipovitellin *n*. lipoprotein in amphibian egg yolk.

lipoxenous *a*. leaving the host before development is complete, *appl.* parasites.

lipoxidase lipoxygenase *q.v.*

lipoxin *n*. member of a class of molecules produced from arachidonic acid in cells in response to injury or during inflammation. They can cause contraction of smooth muscle, vasodilation, chemotaxis, hyperfiltration in the kidney, and inhibition of natural killer cell activity.

lipoxygenases *n.plu.* enzymes that catalyse the addition of a molecule of oxygen to the double bonds of certain unsaturated fatty acids and their derivatives. They are involved *inter alia* in the synthesis of some eicosanoid chemical mediators (e.g. leukotrienes and lipoxins).

liquor folliculi fluid surrounding the ovum in a Graafian follicle.

lirella *n*. type of long apothecium in some lichens.

Lissamphibia *n*. in some classifications a subclass of amphibians containing all extant species and divided into three orders: Salientia (Anura), Urodela and Apoda.

lissencephalous *a*. having few or no convolutions on the surface of the brain.

listeriosis *n*. disease caused by the bacterium *Listeria monocytogenes*, which is an opportunistic pathogen and can cause disease if ingested in contaminated food.

lithite *n*. (1) calcareous secretion in ear; (2) statolith *q.v.*

lithocarp *n*. a fossil fruit.

lithocyst *n*. (1) minute sac or groove containing statoliths, found in many invertebrates; (2) enlarged cells of plant epidermis, in which cystoliths are formed.

lithocyte *n*. large cell in hydrozoan statocyst (gravity detector) containing a concretion of calcium salts, whose movement under gravity is detected by an adjacent sensory cell.

lithodomous *a*. living in holes or clefts in rock.

lithogenous *a*. rock-forming or rock-building, as the reef-building corals.

lithophagous *a*. (1) stone-eating, as some birds; (2) rock-burrowing, as some molluscs and sea urchins.

lithophyll *n*. fossil leaf or leaf impression.

lithophyte *n*. plant growing on rocky ground.

lithosere *n*. plant succession originating on rock surfaces.

lithosol *n*. soil that develops at high altitudes on resistant parent materials that withstand weathering. It is a humus-rich, shallow, stony soil.

lithosphere *n.* (1) the Earth's crust; (2) the non-living, non-organic part of the environment, such as rocks or the mineral fraction of soil.

lithotomous *a.* stone-boring, as certain molluscs.

lithotroph chemolithotroph *q.v. a.* **lithotrophic.**

litter *n.* (1) (*bot.*) partly decomposed plant residues on the surface of soil; (2) (*zool.*) offspring produced at a single multiple birth.

littoral *a.* (1) growing or living near the seashore; (2) *appl.* zone between high- and low-water marks. *see* Fig. 34 (p. 605); (3) *appl.* zone of shallow water and bottom above compensation depth (depth at which photosynthesis cannot be supported) in lakes.

Littré's glands mucus-secreting glands of the urethra.

lituate *a.* forked, with prongs curving outwards.

liver *n.* (1) in vertebrates, a glandular organ closely associated with gut, developing mainly from embryonic gut epithelium. It secretes bile and is a key organ in the metabolism of foodstuffs after digestion and storage of carbohydrate as glycogen. The specialized epithelial cells of the liver are known as hepatocytes; (2) in some invertebrates, a glandular digestive organ.

liverworts *n.plu.* common name for members of the plant division Hepatophyta *q.v.*

live vaccine vaccine made from active but non-pathogenic viruses of the same or similar type as that which it is intended to protect against. Such viruses can multiply to a limited extent in the vaccinated host and stimulate protective immunity, but do not cause any severe clinical symptoms.

living fossil extant species of ancient lineage that has remained morphologically unchanged for a very long time, whose only close relatives are fossils and which in some cases was itself thought to be extinct. Examples are the coelacanth, the ginkgo and the metasequoia.

LMM light meromyosin. *see* meromyosin.

loam *n.* rich friable soil consisting of a fairly equal mixture of sand and silt and a smaller proportion of clay.

lobar *a. pert.* a lobe.

lobate *a.* divided into lobes.

lobed *a. appl.* leaves having margin cut up into rounded divisions that reach less than halfway to midrib.

lobe-finned fishes common name for lungfishes and crossopterygians.

lobopodia *n.plu.* blunt-ended pseudopodia of some protozoans.

lobose lobate *q.v.*

lobular *a.* like or *pert.* small lobes.

lobulate *a.* divided into small lobes.

lobule *n.* a small lobe.

local-circuit neurons neurons in the brain that form part of a localized network or circuit within a particular region. *cf.* projection neurons.

local faciation lociation *q.v.*

localization of function in brain, the situation that different areas of the brain are concerned with different tasks, e.g. the visual area in the occipital lobe with processing incoming signals from the eye, the motor areas with generating outgoing signals to muscles, and so on.

localization of sensation identification on surface of body of exact spot affected.

local potential post-synaptic potential *q.v.*

local sequence alignment alignment of two sequences for only small regions along their length.

locellate loculate *q.v.*

locellus *n.* a small compartment of plant ovary.

loci *plu.* of locus.

lociation *n.* local differences in abundance or proportion of dominant species. *alt.* local faciation.

lock-and-key model theory first proposed at end of 19th century, that the active site on an enzyme is an exact three-dimensional fit for the substrate, also applied to other specific interactions such as antigen–antibody binding. Although in essence correct, *see also* induced fit.

locomotion *n.* movement of a cell or organism from one place to another.

locular, loculate *a.* (1) *pert.* locules; (2) containing or composed of locules.

locule *n.* small chamber or cavity, e.g. chamber containing ovules in ovary of flower or cavity containing asci in fungal stroma.

loculi *plu.* of loculus.

Loculoascomycetes *n.* taxon of ascomycete fungi parasitic on plants and insects, and which bear their asci in cavities in a loose hyphal stroma. They include the sooty moulds, and agents of various scab (e.g. *Venturia*, apple scab), leaf spot (e.g. *Mycosphaerella* on strawberries) and anthracnose diseases of plants. Some form lichens.

loculose *a.* (1) having several locules; (2) partitioned into small cavities.

loculus locule *q.v.*

locus *n.* (1) gene locus *q.v.*; (2) location of a stimulus. *plu.* **loci**.

locus coeruleus pair of minute bodies of grey matter lying just beneath the floor of the midbrain. They contain the cell bodies of noradrenergic neurons.

locus control region (LCR) region of DNA that controls the expression of a distant gene or gene cluster, most probably by controlling the state of chromatin.

locust *n.* common name for many members of the Orthoptera *q.v.*

locusta *n.* spikelet, of grasses.

lodicule *n.* small body at the base of each carpel in grass florets.

lod score log odds ratio, a measure of the likelihood that two genes are linked. It is the $\log_{10}$ of the ratio of the odds on linkage between two loci at a given recombination fraction.

loess *n.* a clay soil formed from wind-blown particles which is very fertile when mixed with humus.

logarithmic phase the rapid stage in growth of a bacterial culture when increase follows a geometric progression.

logistic curve an S-shaped curve initially rising slowly, then steeply, and finally flattening out, and which is characteristic of the growth and stabilization of a population.

log odds *see* lod score.

loma *n.* thin membranous flap forming a fringe round an opening.

lomasomes *n.plu.* invaginations of cell membranes found in certain algae, vascular plants and fungi.

loment, lomentum *n.* a pod constricted between seeds. *plu.* **lomenta**.

lomentaceous *a. pert.* resembling or having lomenta (pods).

long bones the long bones of the limbs: femur, humerus, radius, ulna.

long-day plants plants that will only flower if the daily period of light is longer than some critical length. The critical factor is in fact the period of continuous darkness they are exposed to. *cf.* day-neutral plants, short-day plants.

long-germ development type of insect embryonic development in which the whole of the embryo is specified in the germ band at much the same time. *cf.* short-germ development.

longicorn *a.* having long antennae, *appl.* certain beetles.

long interspersed element *see* LINE *q.v.*

longipennate *a.* having long wings or long feathers.

longirostral, longirostrate *a.* having a long beak.

longisection *n.* section along or parallel to a longitudinal axis.

long period interspersion type of sequence arrangement within some eukaryotic genomes in which rather long moderately repetitive DNA sequences alternate with long sequences of nonrepetitive DNA.

long-term depression (LTD) phenomenon demonstrated in some parts of brain, in which stimulation of the area by bursts of electrical stimuli of moderately low frequency results in a decreased magnitude of response on subsequent normal stimulation, an effect that lasts for many hours.

long terminal repeat (LTR) a nucleotide sequence repeated at each end of an integrated retroviral provirus. It is generated from the ends of the virus during the insertion process and contains control sites for initiation of transcription.

long-term memory (LTM) the process whereby information is stored for weeks, months or years in the brain and the behaviour of recalling that information. *cf.* short-term memory.

long-term potentiation (LTP) phenomenon demonstrated in hippocampus, and some other parts of brain, which has been proposed by some as a model for studying synaptic changes that underlie learning and memory. Stimulation of the area by bursts of electrical stimuli of moderately high frequency results in an increased magnitude of response on subsequent normal stimulation, an effect that lasts for many hours.

loop of Henle a loop of thin-walled mammalian kidney tubule that passes from cortex into medulla and back into cortex. Movements of water, sodium and urea through the walls lead to a reabsorption of water and production of concentrated urine.

loph *n.* crest connecting cones in teeth and forming a ridge.

lophium *n.* a community living on a ridge.

lophobranchiate *a.* with tufted gills.

lophocercal *a.* having a ray-less caudal fin in the form of a ridge around the end of the vertebral column.

lophodont *a.* having transverse ridges on the cheek-teeth grinding surface.

Lophophorata *see* lophophore, Lophotrochozoa.

lophophorate *a.* having a lophophore.

lophophore *n.* horseshoe-shaped crown of tentacles surrounding the mouth, characteristic of animals of the phyla Brachiopoda, Ectoprocta (Bryozoa) and Phoronida.

lophoselenodont *a.* having cheek teeth ridged with crescentic cuspid ridges on grinding surface.

lophotrichous *a.* (1) having long whip-like flagella; (2) with a tuft of flagella at one pole.

Lophotrochozoa, lophotrochozoans *n.*, *n.plu.* one of the two major divisions of the protostomes (the other being the Ecdysozoa), which is subdivided into the worm-like Trochozoa (phyla Nemertini, Mollusca, Sipuncula, Echiura, Pogonophora and Annelida), and the Lophophorata (phyla Brachiopoda, Phoronida, Entoprocta and Bryozoa), which possess a feeding apparatus of hollow tentacles called a lophophore.

loral *a. pert.* or situated at the lore.

lorate *a.* strap-shaped.

lordosis *n.* posture adopted by a sexually receptive female rat.

lore *n.* space between bill and eyes in birds.

Lorenzini's ampulla ampulla of Lorenzini *q.v.*

lorica *n.* a protective external case.

loricate *a.* covered with a protective coat of scales.

Loricifera, loriciferans *n.*, *n.plu.* small phylum of tiny marine multicellular pseudocoelomate animals living in sediments. They have spiny heads and abdomens covered with a case of spiny plates called a lorica.

lorulum *n.* the small strap-shaped and branched thallus of some lichens.

loss-of-function *appl.* mutations that cause no functional gene product to be produced.

loss-of-heterozygosity (LOH) situation where one homologous chromosome or part of a homologous chromosome is lost or mutated in a somatic cell. When the loss involves a normal copy of a tumour suppressor gene in cells that already have one mutant copy of that gene, the cell can become cancerous. This is the basis of the greatly increased risk of cancer in people who inherit one mutant copy of such a gene.

lotic *a.* (1) *appl.* or *pert.* running water; (2) living in a brook or river.

louse *n.* common name for various small insects of the orders Psocoptera (book lice) and Anoplura (Siphunculata, the sucking lice, and Mallophaga, the biting lice), and for crustaceans of the orders Isopoda (woodlice) and Branchiura. Some are ectoparasites of humans and other animals and can transmit disease, e.g. typhus fever transmitted by the common head or body louse. *plu.* lice.

low-density lipoprotein (LDL) group of plasma lipoproteins rich in cholesterol esters, synthesized from very low-density lipoprotein (VLDL), and which transport cholesterol to peripheral tissues and regulate *de novo* cholesterol synthesis. *see also* LDL receptor.

lower shore zone of seashore that extends from the lowest low-water level to the average low-water level and which is therefore only uncovered occasionally and for short periods.

lowland *n.* land up to *ca.* 700 m altitude. What is considered a lowland will vary depending on the topography of the area.

loxodont *a.* having molar teeth with shallow grooves between the ridges.

LP protein with lectin-like activity isolated from *Limulus polyphemus.*

LPH lipotropin *q.v.*

LPS lipopolysaccharide *q.v.*

LRR leucine-rich repeat *q.v.*

LSD lysergic acid diethylamide *q.v.*

L strand light strand of DNA, esp. *pert.* mammalian mitochondrial DNA which can be separated into H (heavy) and L strands on the basis of their density.

LT lymphotoxin *q.v.*

LTD long-term depression *q.v.*

LTH prolactin *q.v.*

LTM long-term memory *q.v.*

LTP long-term potentiation *q.v.*

LTR long terminal repeat *q.v.*

LUCA last universal common ancestor *q.v.*

luciferase *n.* enzyme present in all luminescent organisms which reacts with the activated substrate luciferin. The reaction results in light emission as luciferin is oxidized and returns to the ground state.

luciferin *n.* light-emitting compound whose enzymatic oxidation is responsible for bioluminescence in organisms such as deep-sea fishes, coelenterates, fire-flies and glow-worms.

lumbar *a. pert.* or near the region of the loins, i.e. the lower back in humans.

lumbarization *n.* fusion of lumbar and sacral vertebrae.

lumbocostal *a. pert.* loins and ribs, *appl.* arch, ligament.

lumbosacral *a. pert.* region of loins and termination of vertebral column (sacrum).

lumbrical *a.* (1) like a worm in appearance; (2) *appl.* four small muscles in palm of hand and sole of foot.

lumbricid *a. appl.* worms of the genus *Lumbricus* and close relatives, i.e. earthworms.

lumbriciform *a.* like a worm in appearance.

lumbricoid *a.* resembling an earthworm.

lumen *n.* internal space of any tubular or sac-like organ or subcellular organelle.

luminal *a.* within or *pert.* a lumen.

luminescence *see* bioluminescence *q.v.*

luminescent *a.* emitting light.

luminescent organs light-emitting organs, present in various animals.

lunar, lunate *a.* somewhat crescent-shaped.

lunar bone, lunatum *n.* a wrist bone, the middle of the three carpals nearest the arm.

lunar rhythms physiological or behavioural patterns influenced by the lunar cycle, which occur in both marine and terrestrial organisms. *see also* tidal rhythms.

lunette *n.* transparent lower eyelid in snakes.

lung *n.* organ specialized for the respiratory uptake of oxygen directly from air and release of carbon dioxide to the air. In vertebrates, lungs are present in air-breathing fishes (the lungfishes, Dipnoi), and tetrapods. In mammals the lungs are paired masses of spongy tissue made up of finely divided airways lined with moist epithelium. These airways extend from the bronchi and end in small sacs, the alveoli, providing a large surface area for gas exchange between air and bloodstream. The lung of terrestrial molluscs is a cavity under the mantle lined with vascular tissue.

lung book one type of respiratory organ in spiders and some other arachnids, formed of thin hollow 'leaves' filled with blood, at which gas exchange takes place.

lungfishes *n.plu.* lobe-finned bony air-breathing fish of the subclass Dipnoi, represented by only four extant genera.

lunula, lunule *n.* (1) small crescent-shaped marking; (2) white opaque crescent on a nail near its root.

lunulet *n.* a small bundle.

lupulin *n.* a yellow resinous powder on the flower scales of hops, containing humulone and lupulone.

lupuline, lupulinous *a.* resembling a group of hop flowers.

lupulon(e) *n.* a bitter antibiotic compound obtained from lupulin, effective against fungi and various bacteria.

Lusitanian *a. appl.* certain plants and animals that occur both in the Iberian Peninsula and in coastal regions of the far west of the British Isles, e.g. in western Ireland and Cornwall, and which are thought to be relics of an interglacial period.

luteal *a.* (1) *pert.* or like the cells of the corpus luteum; (2) *appl.* hormones: progesterone, relaxin.

lutein *n.* yellow carotenoid pigment present in the leaves and flowers of many plants, in green and brown algae, in egg yolk and in the corpus luteum.

lutein cells yellow cells found in the corpus luteum of the ovary and formed either from the follicle cells or the cells of the theca interna.

luteinization *n.* formation of the corpus luteum.

luteinizing hormone (LH) glycoprotein hormone secreted by the anterior pituitary. It acts on the gonads to stimulate testosterone production and interstitial cell formation in the testes, and oocyte maturation, ovulation and formation of corpus luteum in the ovary.

luteinizing hormone-releasing hormone (LHRH) decapeptide hormone produced by the hypothalamus which causes secretion of luteinizing hormone by cells of anterior pituitary. A similar peptide acts as a neurotransmitter in the frog.

luteolysis *n.* breakdown of the corpus luteum that occurs during the cycle of ovulation in mammals.

Luteoviridae *n.* family of plant viruses containing isometric single-stranded positive-strand RNA viruses that typically cause yellowing of the leaves, e.g. barley yellow dwarf virus.

luxury genes genes that are expressed only in a particular cell type and specify proteins produced only by that cell type (usually in large amounts). Examples are ovalbumin in the chick oviduct, globin in red blood cell precursors.

lyase *n.* any of a large class of enzymes catalysing the cleavage of C–C, C–O, C–S and C–N bonds by means other than hydrolysis or oxidation. They have two substrates in one reaction direction, and one in the other, in this direction eliminating a molecule (e.g. carbon dioxide, water) and creating a double bond. They include the decarboxylases, aldolases, hydratases, dehydratases and synthases. EC group 4.

lychnidiate *a.* luminous.

lycopene *n.* the red carotenoid pigment of tomatoes and other fruits.

Lycophyta, lycophytes *n., n.plu.*, one of the four major divisions of extant seedless vascular plants, with 10–15 living genera comprising club mosses, *Selaginella* and the aquatic quillworts. They are characterized by a sporophyte with roots, stems and small scale-like leaves arranged spirally on the stem, with sporangia solitary and borne on or associated with a sporophyll. Fossil lycophytes are found from the Devonian onwards and include extinct forms such as the woody tree-like lepidodendroids (scale trees), which are the dominant plants of the Carboniferous coal measures.

Lycopodophyta, Lycopodiales, lycopods, Lycopsida, lycopsids *see* Lycophyta.

lygophil *a.* preferring shade or darkness.

Lyme disease tick-borne disease, caused by the spirochaete *Borrelia burgdorferi*, initially characterized by influenza-like symptoms, but progressing if untreated to a chronic disease with neurological symptoms.

lymph *n.* fluid bathing all tissue spaces and which drains into afferent lymphatic vessels and is carried to lymph nodes and mucosal-associated lymphoid tissues. It is returned to the blood via afferent lymphatic vessels and the thoracic duct, which drains into the left subclavian vein.

lymphatic (1) *n.* vessel carrying lymph; (2) *a. pert.* or conveying lymph.

lymphatic system network of fine vessels extending throughout the body in vertebrates, connected at points to the blood circulatory system and at other points to secondary lymphoid tissues. The lymphatic vessels originate in connective tissues and drain fluid from the intercellular spaces. When contained in lymphatic vessels this fluid is called lymph. Lymph also contains white blood cells.

lymph gland lymph node *q.v.*

lymph heart contractile expansion of lymph vessel where it opens into a vein, in many vertebrates.

lymph node small secondary lymphoid organ present at intervals along lymphatics. Antigens entering the lymphatic system are trapped in lymph nodes and there stimulate the proliferation and differentiation of T and B lymphocytes. Lymph nodes are one of the sites of antibody production.

lymphoblast *n.* lymphocyte dividing after stimulation with antigen.

lymphocyte *n.* type of mononuclear white blood cell present in large numbers in lymphoid tissues and circulating in blood and lymph, and involved in protecting the body against infection. B lymphocytes (B cells) and T lymphocytes (T cells) are small lymphocytes responsible for antigen-specific adaptive immune responses. Natural killer cells (NK cells) are large granular lymphocytes with a role in innate immunity in killing virus-infected cells and producing cytokines that influence the subsequent course of the adaptive immune response, and their killing activity is not antigen-specific.

lymphocytic choriomeningitis virus (LCM) virus of the Arenaviridae, infects the membranes of the brain provoking a cellular immmune response.

lymphogenic *a.* produced in lymphoid tissue.

lymphogenous *a.* lymph-forming.

lymphoid *a. appl.* tissue or organ composed largely of lymphocytes. Primary lymphoid organs are the bone marrow and thymus, in which lymphocytes are produced. These then migrate to the secondary lymphoid tissues or organs (e.g. lymph nodes, spleen, and mucosal-associated lymphoid tissues) where they encounter foreign antigens, undergo final differentiation and participate in immune reactions.

lymphoid follicle aggregations of activated proliferating B cells that arise in secondary lymphoid tissues during an adaptive immune response.

lymphokine *n.* any cytokine produced by activated lymphocytes.

lymphokine-activated killer cells (LAK) cytotoxic cells produced by the activation of T cells by the interleukin IL-2.

lymphoma *n.* a cancer of lymphoid tissues.

lymphopoiesis *n.* production and differentiation of lymphoid cells.

lymphotoxin (LT) *n.* cytokine secreted by activated Th1 lymphocytes, and which is known to be toxic to some cell types and to activate endothelial cells to enhance inflammation.

Lyonnet's glands paired accessory silk glands in lepidopterous larvae.

lyophil *a. appl.* solutes which after evaporation to dryness go readily into solution again on addition of liquid.

lyosphere *n.* thin film of water surrounding a colloidal particle.

lyra *n.* (1) triangular layer of cells joining lateral parts of fornix of brain, marked with fibres in the form of a lyre; (2) a lyre-shaped pattern, as on some bones; (3) series of chitinous rods forming the stridulating organ in certain spiders.

lyrate *a.* lyre-shaped, *appl.* leaves.

lyriform *a.* lyre-shaped.

Lys lysine *q.v.*

lysate *a.* a suspension of cells that have been broken up (lysed).

lyse *v.* to dissolve or break open a cell so that its contents are released.

Lysenkoism *n.* a doctrine promoted by the Soviet agriculturalist Trofim Lysenko in the 1930s and 1940s, which was based on the idea that acquired characteristics could be inherited. This mistaken theory became for some time the only officially permitted theory of genetics in the USSR and led to the suppression of the teaching of Mendelian genetics and of work based on modern genetic concepts, and the persecution of geneticists who opposed Lysenkoism.

lysergic acid diethylamide (LSD) hallucinogenic drug whose effects sometimes mimic the symptoms of some psychotic conditions, esp. schizophrenia. It acts at serotonin receptors.

lysigenic, lysigenous *a. appl.* formation of tissue cavities caused by degeneration of cell walls in centre of mass.

lysin *n.* any substance that can promote the lysis or dissolution of cells.

lysine (Lys, K) *n.* diaminocaproic acid, an amino acid with a basic side chain, constituent of protein and essential in the human diet.

lysis *n.* the breaking down or dissolution of cells.

lysogen *n.* bacterium in which lysogeny has been established.

lysogenesis *n.* the action of lysins.

lysogenic *a.* (1) *pert.* or involved in lysogeny; (2) *appl.* bacterium carrying a prophage integrated into its DNA; (3) *appl.* bacteriophage capable of causing lysogeny, *alt.* temperate.

lysogenic immunity ability of a prophage to prevent another phage of the same type from becoming established in the same bacterium.

lysogenization *n.* the production of lysogeny by a phage. *v.* **lysogenize**.

lysogeny *n.* (1) in bacteria, the condition where a bacterium infected with a phage carries the phage DNA integrated into the bacterial chromosome as a prophage in which most phage functions are repressed; (2) (*bot.*) formation of cavities within tissues by breakdown of cell walls in centre of mass.

lysophospholipid *n.* glycerol-based phospholipid in which only two of the three possible positions on glycerol are esterified to acyl groups.

lysosome *n.* membrane-bounded organelle in eukaryotic cells, esp. animal cells, which has an acid internal environment (pH~5) when fully mature. Lysosomes are derived from vesicles that bud from the Golgi apparatus and are part of the cell's endomembrane system. They fuse with endosomes and phagosomes carrying material into the cell, and contain many different hydrolytic enzymes that digest damaged cellular components and material taken into the cell by endocytosis. The end-products of digestion are transported out of the lysosome for use in the cytosol. In plant cells, the digestive function of the lysosome is taken by the vacuole. *a.*

lysosomal. *see* Fig. 7 (p. 105). *see also* endosome, phagolysosome.

lysozyme *n.* enzyme found esp. in animal secretions such as tears, in white of egg, and in some microorganisms. It splits the glycosidic bond between certain residues in mucopolysaccharides and mucopeptides of bacterial cell walls, resulting in bacteriolysis, and thus acts as an antibacterial agent. EC 3.2.1.17.

lysylbradykinin *n.* polypeptide which causes dilation of blood vessels and contraction of smooth muscle. Derived by post-translational cleavage from kininogen by tissue kallikrein. *alt.* kallidin, kinin.

lytic *a.* (1) *pert.* lysis; (2) *pert.* a lysin; (3) *appl.* virus infection of a cell (or phage infection of a bacterium) in which virus multiplication within the cell eventually causes the cell to break open (lyse) with release of new virus.

lytic complex the complex of complement components (C5, 6, 7, 8, 9) which assembles on the surface of a cell and forms a pore through the cell membrane, leading to cell lysis.

lytta *n.* (1) a worm-like structure of muscle, fatty and connective tissue or cartilage, under the tongue of carnivores such as dog; (2) cantharis, a blister beetle.

M

M methionine *q.v.*

M_r relative molecular mass *q.v.*

Macaronesian region part of the Boreal floristic kingdom comprising the islands off the west coast of Africa, e.g. Madeira, the Canary Islands and the Cape Verde Islands.

macerate (1) *n.* softened tissue in which cells have been separated; (2) *v.* to soften or wear away by digestion or other means.

machair *n.* herb-rich calcareous grassland on shell-sand on the western Scottish coast.

machairodont *a.* sabre-toothed.

mAChR muscarinic acetylcholine receptor *q.v.*

macrandrous *a.* having large male plants or male elements.

macraner *n.* male ant of unusually large size.

macrergate *n.* worker ant of unusually large size.

macro- prefix derived from Gk *makros*, large.

macroalga *n.* multicellular alga, e.g. a seaweed. *cf.* microalga.

macroarray *n.* a protein or DNA array with the components spotted onto a nylon membrane or similar large-sized format.

macroarthropod *n.* medium- or large-sized arthropod whose size is measured in millimetres or centimetres rather than microscopic units.

macrobiota *n.* organisms of a size larger than *ca.* 1 cm in a habitat or ecosystem (esp. soil). *cf.* microbiota, mesobiota.

macrobiotic *a.* long-lived.

macroblast *a.* a large cell.

macrocarpous *a.* having large fruit.

macrocephalous *a.* with a large head.

macrochaeta *n.* a large bristle, as on body of some insects. *plu.* **macrochaetae**.

macrochromosomes *n.plu.* the relatively large chromosomes in a nucleus.

macroclimate *n.* the climate over a relatively large area, generally synonymous with climate in the usual sense.

macroconidium *n.* a large asexual spore or conidium.

macroconjugant *n.* the larger individual of a conjugating pair.

macrocyclic *a. appl.* rust fungi that produce basidiospores, teleutospores and at least one other type of spore (aeciospores and/or uredospores) during their life cycle.

macrocyst *n.* (1) large reproductive cell of certain fungi; (2) large cyst or case for spores.

macrocystidium *n.* long bladder-like sterile cell in the hymenium of some gasteromycete fungi.

macrodactylous *a.* with long digits.

macrodont *a.* with large teeth.

macroecology *n.* an approach to ecology that aims to identify general principles or large-scale statistical patterns of variation in ecological systems, using data on variation in large numbers of 'units' (either members of the same species, or species as a whole), in order to understand the abundance and distribution of species at large spatial and time scales, using first principles of evolution, ecology and historical contingency.

macroelement macronutrient *q.v.*

macroevolution *n.* evolutionary processes extending through geological time, leading to the evolution of markedly different genera and higher taxa. *cf.* microevolution.

macrofauna *n.* animals whose size is measured in centimetres rather than microscopic units.

macrofibril *n.* cellulose fibril made of bundles of microfibrils, in plant cell wall. Macrofibrils are scattered throughout the primary wall and arranged in a more ordered fashion in secondary walls.

macrogamete *n.* the larger of two gametes in a heterogametic organism, usually considered equivalent to the ovum or egg.

macrogametocyte *n.* the mother cell that produces a macrogamete, esp. in protists. *alt.* megagametocyte.

macroglia *n.* general term for certain neuroglial cells which may apply to astrocytes and oligodendroglia, or also to ependyma.

macroglobulinaemia *n.* an excess of IgM pentamers (macroglobulin) in the blood.

macroglossate *a.* having a large tongue.

macrognathic *a.* having especially well developed jaws.

macrogyne *n.* female ant of unusually large size.

macroinvertebrate *n.* any invertebrate or invertebrate larva whose size is measured in millimetres or centimetres rather than microscopic units. Such species are one of the main groups of organisms sampled in surveys of water quality.

macrolecithal megalolecithal *q.v.*

macromere *n.* the larger of the cells produced as a result of unequal cleavage in the fertilized eggs of some animals. *cf.* micromere.

macromerozoite *n.* one of many divisions produced by the macroschizont stage of sporozoan parasites.

macromesentery *n.* radial partition (mesentery) of internal cavity in sea anemones and their relatives (Anthozoa) which extends all the way from body wall to pharynx.

macromolecule *n.* general term for protein, nucleic acid or polysaccharide or other very large polymeric organic molecule. *alt.* **macromolecular.**

macromutation *n.* (1) simultaneous change in several different characters; (2) hypothetical step in evolution involving a single mutational change of large effect.

macronotal *a.* with a large notum, as queen ant.

macront *n.* the larger of two sets of cells formed after schizogony in some sporozoans, the macront giving rise to the macrogametes.

macronucleocyte *n.* a white cell of blood in insects, having a large nucleus.

macronucleus *n.* the larger of the two types of nucleus found in ciliate protozoans. It corresponds in function to the nucleus of a eukaryotic somatic cell. It disappears during sexual reproduction and is renewed from the micronucleus. It contains much of the genome in multiple and fragmented form.

macronutrient *n.* chemical element required in large amounts by living organisms for proper growth and development, being a major constituent of living matter. Examples are carbon, nitrogen, oxygen, phosphorus, calcium.

macroparasite *n.* a parasite that is visible to the naked eye, e.g. arthropods.

macrophage *n.* large phagocytic cell found in many types of tissue. Macrophages derive from monocytes, a type of white blood cell, that have entered tissues. Macrophages ingest and destroy invading microorganisms and also scavenge dead and damaged cells and cellular debris. In the presence of infection they also produce cytokines that influence the adaptive immune response.

macrophage colony-stimulating factor (M-CSF) cytokine that stimulates the growth of monocytes and macrophages. It is produced e.g. by bone marrow stromal cells and macrophages. The receptor for M-CSF is encoded by the proto-oncogene *fms*.

macrophage-like cells phagocytic cells of the fixed reticuloendothelial system in lymphoid tissue, lungs, liver and kidneys.

macrophagous *a.* feeding on relatively large masses of food. *cf.* microphagous.

macrophanerophytes *n.plu.* trees.

macrophyll, macrophyllous megaphyll, megaphyllous *q.v.*

macrophyte *n.* large aquatic plant (e.g. water lily, water crowfoot) as opposed to the phytoplankton and small plants like duckweed.

macropinocytosis *n.* the non-specific uptake of fluid and small particles such as bacteria by endocytosis, with the formation of large intracellular vesicles known as macropinosomes.

macroplankton *n.* the larger organisms drifting with the surrounding water, e.g. jellyfish, sargassum weed.

macropodous *a.* (1) having a long stalk, *appl.* leaf or leaflet; (2) having hypocotyl large in relation to rest of embryo; (3) long-footed.

macropterous *a.* with unusually large fins or wings.

macroschizogony *n.* (1) multiplication of macroschizonts; (2) schizogony giving rise to macromerozoites.

macroschizont *n.* stage in life cycle of some sporozoan blood parasites, developing from the sporozoite and giving rise to macromerozoites.

macrosclere megasclere *q.v.*

macrosclereids *n.plu.* relatively large columnar sclereids, as in coats of some leguminous seeds.

macroscopic *a.* visible to the naked eye.

macrosepalous *a.* with especially large sepals.

macroseptum macromesentery *q.v.*

macrosiphon *n.* large internal siphon of certain cephalopods.

macrosmatic *a.* with a well developed sense of smell.

macrospecies *n.* a large polymorphic species, usually with several to many subdivisions.

macrosplanchnic *a.* large-bodied and short-legged.

macrosporangium megasporangium *q.v.*

macrospore *n.* (1) large anisospore or gamete of some protozoans; (2) megaspore *q.v.*

macrosporogenesis *n.* production of macrospores.

macrosporophore *n.* a leafy megasporophyll, i.e. a structure bearing the female sex organs in plants.

macrosporophyll megasporophyll *q.v.*

macrosporozoite *n.* a large sporozoite in which fusion takes place between the products of cell division.

macrostomatous *a.* with a very large mouth.

macrostylous *a.* with long styles.

macrosymbiont *n.* the larger of two organisms living in symbiosis.

macrotous *a.* with large ears.

macrotrichia *n.plu.* the larger bristles on the wings or body of insects.

macrozoospore *n.* a large motile spore.

macruric *a.* long-tailed.

macula *n.* (1) a spot or patch of colour; (2) a small pit or depression. *plu.* **maculae**. *a.* **macular**.

macula cribrosa area on wall of vestibule of ear, perforated for passage of auditory nerve fibres.

macula lutea oval yellow area at fovea of retina in eyes of some mammals, also called the yellow spot.

maculate *a.* spotted.

maculation *n.* the arrangement of spots on an animal.

Madagascan region part of the Palaeotropical floristic kingdom comprising the island of Madagascar and its neighbouring islands.

mad cow disease bovine spongiform encephalopathy. *see* transmissible spongiform encephalopathies.

madescent *a.* (1) becoming moist; (2) slightly moist.

madrepore *n.* a branching, stony, reef-building coral of the order Scleractinia.

madreporite *n.* in echinoderms, a perforated plate at the end of the stone canal of the water vascular system.

MADS box sequence characterizing a class of gene regulatory proteins most abundant in plants, where they are involved in flower development.

Magendie's foramen central aperture in roof of 4th ventricle of brain, connecting with the subarachnoid space.

maggot *n.* a worm-like insect larva, without appendages or distinct head.

magnesium (Mg) *n.* metallic element. It is an essential nutrient for living organisms.

magnetic resonance imaging (MRI) non-invasive brain imaging technique based on the molecular effects of an applied magnetic field, which can detect small structural changes in the brain. *see also* functional magnetic resonance imaging.

magnetosomes *n.plu.* particles of the iron mineral magnetite Fe_3O_4 present in some bacteria, and which enable them to orient themselves in a magnetic field.

magnetotaxis *n.* directed movement within a magnetic field, shown by some bacteria which contain small particles of magnetite within their cells. *a.* **magnetotactic**.

magnetotropism *n.* movement or growth of an organism in response to lines of magnetic force.

magnocellular layer a layer in the lateral geniculate nucleus of primate brain, comprising layers 1 and 2 and composed of large cell bodies.

Magnoliales *n.* order of dicot trees and shrubs including the families Annonaceae (custard apple), Magnoliaceae (magnolia), Myristicaceae (nutmeg), and others.

Magnoliophyta *n.* the name for the angiosperms in some plant classifications.

Magnoliopsida *n.* the name for the class comprising the dicotyledons in some classifications.

maintenance behaviour animal behaviour involved in carrying out day-to-day activities such as searching for food, mating, reproduction or avoidance of extreme environments.

maintenance energy the energy required to simply maintain cellular structure and integrity.

maintenance ration food required to maintain an animal when the production term (P) in the energy budget is zero.

maize *n. Zea mays*, a member of the Gramineae, originating and first domesticated in the Americas and now widely grown as a crop plant for human and animal feed. *alt.* (US) corn.

major element macronutrient *q.v.*

major gene a gene having a pronounced phenotypic effect, as distinguished from a modifying gene.

major groove the wider of the two helical grooves in a double-helical DNA molecule.

major histocompatibility antigen, major histocompatibility molecule MHC molecule *q.v.*

major histocompatibility complex (MHC) large cluster of genes encoding the polymorphic major histocompatibility molecules (MHC molecules *q.v.*) and some other surface proteins of immune system cells, some of the complement proteins, and the TAP proteins, among others. In humans it is known as the HLA complex, in mice the H-2 complex. *see also* minor histocompatibility antigen.

major locus genetic locus that has a major role in determining a phenotypic trait.

major worker member of the largest worker subcaste. In ants this is equivalent to a soldier.

mala *n.* cheek, in vertebrates, and corresponding region of head in invertebrates.

malacoid *a.* soft-textured.

malacology *n.* the study of molluscs.

malacophilous *a.* pollinated by the agency of gastropod molluscs, generally snails and slugs.

malacophyllous *a.* with soft and fleshy leaves.

malacopterous *a.* with soft fins.

malacospermous *a.* having seeds covered by a soft coat.

Malacostraca, malacostracans *n.*, *n.plu.* a subclass of crustaceans containing e.g. the crabs, lobsters, crayfish, shrimps and woodlice, all of which have some form of hard carapace covering part of the body.

malacostracous *a.* soft-shelled.

Malagasy subregion subdivision of the Ethiopian zoogeographical region, comprising the island of Madagascar and neighbouring islands.

malar *a.* (1) *pert.* or in region of cheek; (2) *appl.* bone: the jugal bone or cheekbone.

malaria *n.* sometimes fatal disease characterized by recurrent fevers, caused by parasitic protozoa of the genus *Plasmodium*, of which *P. vivax* causes the most severe form, falciparum malaria. The intermediate hosts and vectors of *Plasmodium* species are anopheline mosquitoes.

malate *see* malic acid.

malate dehydrogenase enzyme that catalyses the reversible conversion of pyruvate or oxaloacetate to malate. Enzymes (EC 1.1.1.37–39) involved in the tricarboxylic acid cycle use NAD as an electron-accepting cofactor, those (EC 1.1.1.40 and 1.1.1.82) involved in β-oxidation reactions and in the C4 pathway of carbon dioxide fixation in tropical plants use NADP. The latter, NADP-dependent malate dehydrogenases, are also known as malic enzyme.

Malayan *a. appl.* and *pert.* the zoogeographical subregion including Malaysia, Indonesia west of Wallace's line and the Philippines.

Malaysian region part of the Palaeotropical floristic kingdom comprising the southern part of the Malaysian peninsula, the Philippines, the islands of Indonesia and Papua New Guinea.

MALDI-TOF mass spectrometry *see* mass spectrometry.

male *n.* individual whose sex organs contain only male gametes, symbol ♂.

male haploidy arrhenotoky *q.v.*

male pronucleus the nucleus the sperm contributes to the zygote.

male-sterile *appl.* mutants of normally hermaphrodite plants which do not produce viable pollen. They are of importance in plant breeding as they can be used to block self-fertilization and force cross-fertilization, thus avoiding the need for laborious emasculation of the plants used as the female parent. *see* cytoplasmic male sterility.

malic acid, malate four-carbon acid of the tricarboxylic acid cycle, where it is oxidized to oxaloacetate by malate dehydrogenase (EC.1.1.1.37). It is also part of other metabolic pathways where it may be oxidized to pyruvate by malate dehydrogenases EC.1.1.1.38–40. It gives the tart taste to fruits such as apples.

malic enzyme *see* malate dehydrogenase.

malignant *a. appl.* neoplasms that have become invasive and can spread to other sites in the body. *alt.* cancerous.

malleate *a.* hammer-shaped.

mallee scrub vegetation consisting of low bushes of *Eucalyptus* species, typical of dry subtropical regions of South-east and South-west Australia.

malleoincudal *a. pert.* malleus and incus of middle ear.

malleolar *n.* the vestigial fibula of ruminants.

malleolus *n.* (1) prolongation of lower end of tibia or fibula; (2) a club-shaped or hammer-shaped projection.

malleus *n.* the small hammer-shaped bone connecting the eardrum and incus in middle ear of mammals.

Mallophaga *n.* order of insects known as bird lice or biting lice. They are ectoparasitic on birds, and have biting mouthparts, secondarily no wings and slight metamorphosis.

malloplacenta *n.* non-deciduate placenta with villi evenly distributed, as in cetaceans and some ungulates.

malonic acid *n.* organic acid that acts as a metabolic poison, blocking cellular respiration by reversibly inhibiting succinate dehydrogenase. Occurs as an end-product of metabolism in some plants.

Malpighian body (1) nodular mass of lymphoid tissue ensheathing the smaller arteries in the spleen; (2) in vertebrate kidney, the glomerulus of convoluted capillaries and the Bowman's capsule that encloses it.

Malpighian corpuscle *see* Malpighian body.

Malpighian layer in vertebrate skin, the innermost layer of the epidermis, containing dividing epidermal cells and cells containing melanin. *alt.* rete mucosum, basal layer. *see* Fig. 35 (p. 622).

Malpighian tubule fine, thin-walled excretory tubule, present in large numbers, leading into posterior part of gut in insects.

MALT mucosal-associated lymphoid tissue *q.v.*

maltase *n.* enzyme that hydrolyses maltose to glucose. *see* glucosidases.

maltodextrins *n.plu.* polysaccharides formed during incomplete hydrolysis of starch.

maltose *n.* disaccharide of glucose, produced by hydrolysis of starch with amylase. It does not occur widely in the free state, but is produced by germinating barley. Hydrolysed by maltase to glucose.

malt sugar maltose *q.v.*

Malvales *n.* order of dicot trees, shrubs and herbs, often mucilaginous, and including the families Malvaceae (mallow, cotton), Sterculiaceae (cocoa), Tiliaceae (linden), Bombacaceae (baobab, silk cotton) and others.

malvidin *n.* a mauvish anthocyanin pigment.

mamilla *n.* (1) nipple; (2) nipple-shaped structure. *plu.* **mamillae.**

mamillary bodies two white bodies enclosing grey matter in hypothalamus in the brain, situated beneath floor of 3rd ventricle. *alt.* corpora mamillaria.

mamillary process or tubercle metapophysis *q.v.*

mamillate *a.* covered with small protuberances.

mamillothalamic tract bundle of nerve fibres running from the corpora mamillaria to the thalamus. *alt.* bundles of Vicq d-Azyr.

mamma *n.* mammary gland. *plu.* **mammae.**

Mammalia, mammals *n., n.plu.* class of homoiothermic tetrapod vertebrates, known from the late Triassic to present, in which the female produces milk from mammary glands and suckles her young. Except in the egg-laying monotremes, the young develop inside the mother in the uterus and are born at a more or less mature stage. Other distinguishing characteristics of mammals are a four-chambered heart and hair on the body. The three main groups of mammals are the monotremes, the marsupials and the eutherians. *see* Appendix 3 for mammalian orders.

mammal-like reptiles common name for extinct reptiles of the subclass Synapsida, living from the Carboniferous to the Triassic, with synapsid skulls. They included the orders Pelycosauria and Therapsida, of which the therapsids were direct ancestors of the mammals.

mammalology *n.* the study of mammals.

mammary *a. pert.* the breast, *appl.* e.g. arteries, veins, tubules.

mammary gland gland secreting milk in female mammals. *alt.* mamma.

mammiferous *a.* (1) developing mammae; (2) secreting milk.

mammiform *a.* shaped like a breast, *appl.* cap of some fungi.

mammilla mamilla *q.v.*

manca *n.* larval (juvenile) stage of some isopods.

manchette armilla *q.v.*

Manchurian subregion subdivision of the Palaearctic zoogeographical region, comprising central and northern China, and Japan except for the northernmost island.

mandible *n.* (1) in vertebrates, the lower jaw, comprised either of a single bone or several; (2) in arthropods, a set of paired mouthparts usually used for biting.

mandibular *a. pert.* the lower jaw in vertebrates, or arthropod mandible, *appl.* canal, foramen, nerve, notch. *alt.* submaxillary.

mandibular arch the 1st branchial arch in vertebrate embryos, which develops in various ways in different vertebrates.

mandibulate *a.* (1) having a lower jaw; (2) having functional jaws; (3) having mandibles, *appl.* arthropods.

mandibuliform *a.* resembling or used as a mandible, *appl.* certain insect maxillae.

mandibulohyoid *a.* in the region of the mandible and hyoid.

manducation *n.* chewing or mastication.

manganese (Mn) *n.* chemical element that is an essential micronutrient for plants. It is a prosthetic group in certain enzymes.

manicate *a.* covered with entangled hairs or matted scales.

manna *n.* the hardened exudate of bark of certain trees such as the European ash, *Fraxinus ornus*, and similar substances in other plants such as tamarisk, where its production is caused by infestation with scale insects.

mannan *n.* any polysaccharide composed predominantly of mannose. *see also* glucomannans.

mannitol *n.* sweet-tasting polyhydroxyalcohol derivative of mannose or fructose found in many plants and some algae. Used commercially as a sweetening agent.

mannose *n.* six-carbon aldose sugar found in glycoproteins and in many polysaccharides, esp. those of plant cell walls.

mannose-binding lectin (MBL) blood protein that binds to bacteria and can initiate activation of the complement system by the lectin-mediated pathway. *alt.* mannose-binding protein.

mannose-6-phosphate receptor receptor protein in the membranes of the *trans* Golgi apparatus that recognizes proteins tagged with mannose 6-phosphate and directs them into lysosomes.

manoxylic *a.* having soft wood containing much parenchyma, as in cycads.

mantid *n.* common name for many members of the Dictyoptera *q.v.*

mantle *n.* (1) (*bot.*) external sheath of fungal mycelium covering the plant rootlets in a mycorrhiza; (*zool.*) (2) fold of soft tissue underlying shell in molluscs, barnacles and brachiopods, and which usually encloses a space, the mantle cavity, between it and the body proper; (3) body wall of ascidians; (4) feathers of bird between neck and back.

mantle cell a cell of the coat or outer covering of a sporangium.

mantle layer layer of embryonic tissue in spinal cord that represents the future columns of grey matter.

mantle lobes dorsal and ventral flaps of mantle in bivalve molluscs.

mantle zone (1) the outer layer of cells of a germinal centre *q.v.*; (2) mantle layer *q.v.*

manual *n.* quill feather borne on the 'hand' region of wing. *alt.* primary feather.

manubrium *n.* tube bearing the mouth hanging down from the undersurface of a medusa.

manus *n.* hand, or part of forelimb corresponding to it, as present in amphibians, reptiles and mammals. *plu.* **manus**.

manyplies omasum *q.v.*

MAP microtubule-associated protein *q.v.*

map distance the relative distance apart of two gene loci on the same chromosome measured by the extent of recombination between them, expressed as the percentage of recombinants in the total progeny, 1 map unit = 1% recombination.

MAP kinase (MAPK) mitogen-associated protein kinase. A serine/threonine protein kinase whose activity is stimulated by the action of many growth factors and other cytokines. It lies at the end of an intracellular signalling pathway and phosphorylates and activates transcription factors, resulting in changes in gene expression.

MAP kinase cascade a series of three protein kinases, acting sequentially to activate the next in the series, culminating in the activation of MAP kinase. It is a part of many intracellular signalling pathways.

MAP kinase kinase (MAPKK) protein serine/threonine kinase that phosphorylates and activates MAP kinase. It is activated by phosphorylation by MAP kinase kinase kinase.

MAP kinase kinase kinase (MAPKKK) the first protein kinase in the MAP kinase cascade. It is activated by a small G protein such as Ras and phosphorylates and activates MAP kinase kinase.

MAPKK MAP kinase kinase *q.v.*

MAPKKK MAP kinase kinase kinase *q.v.*

mapping function formula for using frequencies of recombinants to calculate map distances corrected for the occurrence of multiple crossovers.

map unit *see* map distance. *alt.* centimorgan.

maquis *n.* vegetation composed of low-growing xerophilous shrubs, found in the Mediterranean area.

marasmus *n.* deficiency disease caused by general undernutrition, generally found in infants.

marble gall type of gall found on oak and caused by larvae of a species of wasp.

marcescent *a.* withering but not falling off, *appl.* calyx or corolla persisting after fertilization.

marcid *a.* withered, shrivelled.

marginal *a.* (1) *pert.*, at or near the margin, edge or border; (2) *appl.* venation of leaf or insect wing; (3) *appl.* a convolution of frontal lobe of brain; (4) *appl.* plates around edge of carapace of turtles and tortoises; (5) (*bot.*) *appl.* placentation of ovules, the attachment of ovules to margin of carpel.

marginalia *n.plu.* sponge spicules which project beyond the body surface.

marginal meristem meristematic tissue that arises on either side of axis of developing leaf and from which growth occurs.

marginal veil a secondary growth around edge of cap in boletes and agaric fungi.

marginal zone the equatorial zone of mesoderm and endoderm of the amphibian blastula, which moves inside through the blastopore at gastrulation.

marginate *a.* having a distinct margin in structure or colouring.

marginella *n.* ring formed by part of cutis proliferating beyond margin of gills, in certain fungi with an exposed hymenium.

marginiform *a.* like a margin or border in appearance.

marginirostral *a.* forming the edges of a bird's bill.

marijuana, marihuana *n.* (1) the drug cannabis *q.v.*; (2) sometimes used for the whole plant.

marita *n.* sexually mature stage in trematode life history. *a.* **marital**.

marker *n.* (1) an identifying feature; (2) a gene or other DNA of known location and effect which is used e.g. to track the inheritance of other genes whose exact location is not yet known.

marmorate *a.* of marbled appearance.

marrow bone marrow *q.v.*

marsh *n.* plant community developing on wet but not peaty soil.

Marsupialia, marsupials *n.*, *n.plu.* the only order in the Metatheria, a group of mammals found only in Australia and the Americas (mainly South America), in which the placenta does not develop or is not as efficient as that of eutherian mammals, so that the young are born in a very immature state. They then migrate to a pouch (marsupium) where they are suckled until relatively mature. Marsupials include opossums and shrew opossums (South America), and in Australia, kangaroos, wallabies, koalas, possums and many other species adapted to fill ecological niches that are filled by eutherian mammals elsewhere. *alt.* metatherians.

marsupium *n.* the pouch in which marsupials suckle their young.

mask *n.* a hinged prehensile structure, corresponding to adult labium, peculiar to dragonfly nymph.

masked *a.* (1) *appl.* virus whose presence in cell is difficult to detect because its multiplication is prevented by superinfection with another virus; (2) personate *q.v.*

masseter *n.* muscle that raises lower jaw and assists in chewing.

mass extinction events various episodes in geological history in which numerous groups of organisms disappear from the fossil record over a relatively short time, e.g. at the end of the Permian, when *ca.* 96% of marine species went extinct, and at the end of the Cretaceous, when the dinosaurs and other groups became extinct. The causes of each mass extinction event are still a matter of debate.

mass flow a theory of the way in which materials are translocated through phloem. It proposes that the cause of movement is the difference in the hydrostatic pressure at each end of a sieve tube, resulting in flow of contents along the tube.

massively parallel DNA sequencing very rapid automated DNA sequencing methods carried out in specialized instruments in which the nucleotide sequences of hundreds of thousands of overlapping short (100–200 nucleotides) fragments of DNA are determined in parallel and the complete DNA sequence reconstructed from this data. *see* pyrosequencing.

mass provisioning the act of storing all of the food required for the development of a larva at the time the egg is laid.

mass spectrometry (MS) a technology used in physics and chemistry to determine the precise masses of ionized molecules and identify and quantify them. It is now used in proteomics to rapidly characterize and identify proteins from the mass spectra of their ionized peptide fragments. A mass spectrometer consists of a means of producing the ions (e.g. by electrospray ionization (ESI) or matrix-assisted laser desorption/ionization (MALDI)), a mass analyser that measures the mass-to-charge-ratio (m/z) of the ionized analytes and a detector that registers the number of ions at each m/z value. Various types of mass spectrometry are used in proteomics, e.g. MALDI mass spectrometry, MALDI-time-of-flight (TOF) mass spectrometry, and MS/MS tandem mass spectrometry in which selected peptides are further fragmented for another round of analysis.

massula *n.* (1) mass of microspores in a sporangium of some pteridophytes; (2) mass of pollen grains in orchids.

mast *n.* the fruit of beech and some related trees.

mastax *n.* pharynx of rotifers, containing jaw-like structures used to grind small food.

mast cell inflammatory cell derived from bone marrow, often amoeboid, with large nucleus and very granular cytoplasm, and bearing receptors for IgE antibodies on its surface. Mast cells are found in connective and fatty tissue and generally carry IgE bound to the receptors. When specific antigen binds to the IgE, the mast cell secretes granules containing histamine, heparin and other inflammatory substances. Mast cells are involved in allergic reactions mediated by IgE antibodies.

mastication *n.* process of chewing food until reduced to small pieces or a pulp. *v.* **masticate**.

masticatory stomach gastric mill *q.v.*

mastigium *n.* defensive posterior lash of some larvae.

mastigobranchia *n.* an outgrowth from basal joint of crustacean limb and extending upwards in a thin sheet between gills.

Mastigomycota *n.* in some classifications, a major division of the Fungi including the classes Chytridiomycetes, Hypho-chytridiomycetes, Plasmodiophoromycetes and Oomycetes, i.e. simple, generally aquatic, often microscopic fungi with motile flagellate zoospores and/or gametes.

mastigoneme *n.* filaments projecting laterally from the flagellar sheath in the 'tinsel' flagella of some flagellates.

Mastigophora *n.* name for the flagellates when considered as protozoa. *see* Zoomastigina.

mastoid *a.* nipple-shaped, *appl.* a projection from the temporal bone of mammalian skull, and to foramen, fossa and notch associated with it.

mastoideosquamous *a. pert.* mastoid and squamous parts of temporal bone.

masto-occipital *a. pert.* occipital bone and mastoid process of temporal bone.

mastoparietal *a. pert.* parietal bone and mastoid process of temporal.

MAT mating-type locus (*q.v.*) of yeasts.

maternal *a. pert.* or originating in the mother.

maternal developmental determinants RNAs and proteins laid down in the ovum in the mother and which direct the development of the zygote after fertilization. *alt.* maternal factors.

maternal-effect genes genes that must be functional in the mother to produce a normally developing embryo, regardless of the genetic constitution of the embryo itself. In general they specify RNAs and proteins that are laid down in the egg in the mother and which direct the earliest stages of development, which are completed before the embryo's own genes (the zygotic genes) become active. Maternal-effect mutations are mutations that render such genes inactive in the mother, resulting in an inviable or abnormal embryo, even when the embryo itself is heterozygous for the mutant gene.

maternal factors maternal developmental determinants *q.v.*

maternal inheritance (1) the inheritance of genes carried by mitochondria, chloroplasts, and any other cytoplasmic genes. This occurs through the maternal line only, as only the egg contributes an appreciable amount of cytoplasm to the zygote; (2) preferential survival in a cross of genetic markers provided by the mother. *cf.* maternal-effect genes.

mating factor protein secreted by cells of different mating types in yeasts and other unicellular organisms. In *Saccharomyces cerevisiae* they are called **a**-factor and α-factor. They attract cells of the opposite mating type and induce conjugation. *alt.* **mating pheromone**.

mating type genetically determined property of microorganisms (e.g. bacteria, protozoa, fungi and algae) which determines their ability to conjugate with other individuals of the same species and reproduce sexually. In most microorganisms, cells that can conjugate are said to be of opposite mating types. In some, e.g. ciliates, individuals that can conjugate are said to belong to the same mating type.

mating-type locus (MAT) genetic locus that determines mating type in budding yeast, either **a** or α. In some strains rearrangements of DNA sequence at the locus occur, resulting in a switch in mating type. *see also* cassette model.

matriclinal *a.* with inherited characteristics more maternal than paternal.

matrifocal *a. pert.* a society in which most of the activities and personal relationships are centred on the mothers.

matrilineal *a.* passed from a mother to her offspring.

matrilinear inheritance maternal inheritance *q.v.*

matrix *n.* (1) medium in which a substance or object is embedded; (2) extracellular matrix *q.v.*; (3) in mitochondria, the inner region enclosed by the inner mitochondrial membrane; (4) nuclear matrix *q.v.*; (5) part beneath body and root of nail; (6) object upon which a lichen or fungus grows; (7) substance in which a fossil is embedded; (8) (*bioinf.*) *see* substitution matrix.

matrix-assisted laser desorption ionization-time-of-flight (MALDI-TOF) *see* mass spectrometry.

matrix-attachment regions DNA sequences in chromatin that have been proposed to attach chromosomes to the nuclear matrix. *alt.* scaffold-attachment regions.

matromorphic *a.* resembling the female parent in morphological characters.

mattorral chaparral *q.v.*

mattula *n.* fibrous network covering petiole bases of palms.

maturase RNA maturase *q.v.*

maturation *n.* (1) ripening; (2) the process of becoming mature and fully functional; (3) the phase in virus life cycle when complete virus particles are being assembled in the host cell; (4) automatic development of a behaviour pattern which becomes progressively more complex as the animal matures and which does not involve learning; (5) *a. appl.* maturing processes.

maturation divisions nuclear and cell divisions by which gametes are produced from primary gametocytes, during which meiosis occurs.

maturation-promoting factor (MPF) cytoplasmic protein factor first identified in mature *Xenopus* eggs which can induce the onset of meiosis in immature eggs. It is also present in the cytoplasm of mammalian and other eukaryotic cells, where it induces mitosis in the normal cell cycle. It is a complex of a cyclin and a protein kinase. *alt.* M-phase-promoting factor.

mature *a.* (1) fully developed and capable of reproduction; (2) fully differentiated and functional, *appl.* cells.

Mauthner cell one of a pair of large neurons in medulla of brain of teleost fish that control tail movement.

Mauthner's cells cells forming a layer between myelin sheath and axon membrane.

Maxam–Gilbert sequencing one of the first methods of rapid DNA sequencing, invented in the 1970s.

maxilla *n.* (1) the upper jaw; (2) paired appendage on head or cephalothorax of most arthropods, posterior to the mandible, present in one or two pairs and modified in various ways in different groups. *plu.* **maxillae**.

maxillary *a. pert.* or in region of upper jaw in vertebrates, or of arthropod maxilla.

maxillary glands paired excretory organs opening at base of maxilla in Crustacea.

maxilliferous *a.* bearing maxillae.

maxilliform *a.* like a maxilla.

maxilliped *n.* any arthropod limb in the mouth region which is modified to assist in feeding. *alt.* gnathopod.

maxillodental *a. pert.* jaws and teeth.

maxillojugal *a. pert.* jaws and cheek bone.

maxillolabial *a. pert.* maxilla and labium, *appl.* dart in ticks.

maxillomandibular *a. appl.* arch forming jaws of primitive fishes.

maxillopalatine *a. pert.* jaw and palatal bones.

maxillopharyngeal *a. pert.* lower jaw and pharynx.

maxillopremaxillary *a.* (1) *pert.* whole of upper jaw; (2) *appl.* jaw when maxilla and premaxilla are fused.

maxilloturbinal *n.* bone arising from the lateral wall of nasal cavity, which supports sensory epithelium.

maxillule *n.* 1st maxilla in crustaceans, where there is more than one pair. *alt.* **maxillula**.

maxim *n.* an ant of the large worker or soldier caste. *cf.* minor worker.

maximum allowable concentration of pollutants, the concentration deemed in regulations to be safe to healthy adults in the workplace, assuming they are not exposed to the pollutant outside working hours.

maximum likelihood in phylogenetic analysis, a method of phylogenetic tree building that calculates the probability of a given tree out of all possible trees, given the particular model of evolution used to build the tree. The required tree is that with the greatest probability.

maximum parsimony in phylogenetic analysis, a method of phylogenetic tree building that minimizes the number of mutations required to connect the present-day sequences being used to build the tree.

maximum permissible body burden the concentration of a radioisotope that will not deliver more than the maximum permissible dose to any organ, if inhaled or ingested at a normal rate.

maximum permissible dose the dose of ionizing radiation, accumulated over a given time, that is considered not to result in any harmful effects to the individual over their lifetime or to cause genetic damage that might affect their descendants.

maximum sustainable yield the maximum crop or yield that can be harvested from a plant or animal population each year without harming it.

maxithermy *n.* the maintenance of body temperature at a maximum for as long as possible. *a.* **maxithermic**, *appl.* animals that can do this.

mayfly *n.* common name for a member of the Ephemeroptera *q.v.*

Mb, Mbp megabase or megabase pair, a million (10^6) bases or base pairs, in a nucleic acid.

M band dark transverse band seen across middle of skeletal muscle sarcomere in longitudinal section. It corresponds to the bare central portions of the myosin filaments.

MBL mannose-binding lectin *q.v.*

mC methylcytosine. *see also* DNA methylation, 5-methylcytosine.

M cells specialized epithelial cells that collect antigen and pass it into mucosal-associated lymphoid tissues.

MCP methyl-accepting chemotaxis protein *q.v.*

M-CSF macrophage colony-stimulating factor *q.v.*

M cytotype a maternal cytotype of *Drosophila melanogaster* which when crossed with males carrying P factors produces dysgenic progeny as the result of activation of P element transposition when exposed to the M cytotype. *see* hybrid dysgenesis.

meadow *n.* permanent grassland, esp. one that is mown for hay and not grazed in summer. *cf.* ley, pasture, water meadow.

mealworm *n.* larva of the beetle *Tenebrio*, which lives in grain stores.

mealy *a.* covered with a powder resembling flour or coarse ground cereal.

mealy bug scale insect *q.v.*

mean *n.* an average. Two types of mean are recognized, the arithmetic mean and the geometric mean. For a set of *n* data points *a*, *b*, *c*, . . . , the arithmetic mean equals $(a + b + c + \dots)/n$, whereas the geometric mean (applicable only to positive numbers) equals $(a \times b \times c \times \dots)^{1/n}$. For example, the arithmetic mean of 3, 4, 5 equals $(3 + 4 + 5)/3 = 4$, whereas the geometric mean equals $(3 \times 4 \times 5)^{1/3} = 3.915$. The terms average and mean are generally taken to be synonymous with the arithmetic mean.

meatus *n.* passage or channel, such as auditory meatus in the ear, nasal meatus in the nose.

mechanically gated ion channel ion channel which opens in response to a mechanical stress such as stretching, contact or pressure on the cell.

mechanical tissue supporting tissue *q.v.*

mechanoreceptor *n.* specialized sensory structure sensitive to mechanical stimuli such as extension, contact, pressure or gravity.

mechanosensory *a.* sensitive to mechanical stimuli. *see* mechanoreceptor.

Meckel's cartilage in elasmobranch fish, the skeletal element forming lower jaw, and which in other vertebrates forms the axis around which the bones of lower jaw are arranged.

meconidium *n.* the sessile or stalked medusa lying outside the capsule of the reproductive structure of certain hydrozoans.

meconium *n.* (1) waste products of pupa or other embryonic form; (2) contents of intestine of newborn mammal.

Mecoptera *n.* order of slender carnivorous insects with complete metamorphosis, commonly called scorpion flies, having biting mouthparts, long slender legs, and membranous wings lying along the body in repose.

media *n.* (1) the longitudinal vein running through the central region of most insect wings; (2) *plu.* of medium.

mediad *a.* towards but not quite on the midline or axis.

medial (1) *a.* situated in the middle; (2) *n.* the main middle vein of insect wing.

medial geniculate nucleus brain nucleus lying in the auditory pathway. It receives outputs from the inferior colliculus, and sends outputs to the auditory cortex.

median *a.* (1) lying or running in axial plane; (2) intermediate; (3) middle; (4) *n.* the middle value when a set of values is arranged in order of magnitude.

median eminence part of hypothalamus situated immediately above the pituitary gland. It synthesizes and secretes peptide-releasing hormones that are transported to the anterior pituitary via the hypophyseal portal blood system.

median lethal dose (LD$_{50}$) dose of a toxic chemical or of a pathogen that kills 50% of the animals in a test sample within a certain time.

median nerve nerve arising from median and lateral cord of brachial plexus, with branches in forearm.

mediastinal *a. pert.* or in region of mediastinum.

mediastinum *n.* the part of thoracic cavity on the midline between the right and left pleura.

mediator *n.* (1) a nerve cell connecting or intermediate between a receptor and effector; (2) any molecule which is involved in and influences a biochemical pathway or a physiological process, *alt.* chemical mediator. *see also* inflammatory mediator.

media worker in ants with three or more worker subcastes, an individual belonging to the medium-sized subcaste(s).

mediocentric *a. appl.* a chromosome having a centrosome in the centre, dividing it into two more-or-less equal arms.

mediocubital *n.* a cross vein between posterior media and cubitus of insect wing.

mediodorsal *a.* in the dorsal middle line.

mediolateral *a. appl.* axis, from the median plane outward to both sides.

mediolecithal *a. appl.* eggs having a moderate amount of yolk.

mediopalatine *n.* between palatal bones, *appl.* a cranial bone of some birds.

mediopectoral *a. appl.* middle part of sternum (breast bone).

mediostapedial *a. appl.* that part of columella of the ear external to stapes.

mediotarsal *a.* between tarsal bones.

medioventral *a.* in the ventral middle line.

mediterranean *appl.* climate characterized by hot dry summers and mild wet winters, as found around the Mediterranean Sea.

Mediterranean region part of the Boreal floristic kingdom comprising Southern Europe and North Africa, around the Mediterranean Sea.

Mediterranean subregion subdivision of the Palaearctic zoogeographical region, comprising Europe south of the Alps and Pyrenees, and North Africa, including the Sahara.

medithorax *n.* middle part of the thorax.

medium *n.* nutritive material in which microorganisms, cells and tissues are grown in the laboratory. *plu.* **media**.

medulla *n.* (1) central part of an organ or tissue, interior to the cortex; (2) marrow of

bones; (3) medulla oblongata *q.v.*; (4) pith or central region of plant stem; (5) loose hyphae in a tangled fungal structure such as rhizomorphs or fruit bodies.

medulla oblongata bulbous upward prolongation of the spinal cord, the lowest part of the brain, controlling heartbeat, respiration and blood pressure.

medullary *a. pert.* or in region of medulla.

medullary canal central cylindrical hollow of a long bone which contains the marrow.

medullary phloem internal phloem in a vascular bundle with two layers of phloem, as in cucurbits.

medullary ray (1) (*bot.*) rays of parenchyma tissue extending from pith to outer edge of vascular tissue in plant stem; (2) (*zool.*) bundles of straight urine-carrying tubules in medulla of kidney.

medullary sheath (1) the fatty sheath (myelin) surrounding the axis of some nerve fibres. *alt.* myelin sheath; (2) (*bot.*) ring of protoxylem around the pith in certain stems.

medullated *a.* (1) *appl.* nerve fibres having a myelin sheath; (2) (*bot.*) *appl.* plant stems having pith.

medulliblasts *n.plu.* cells of embryonic nervous tissue, arising in the medulla or central portion, and which produce neuroblasts (embryonic neurons) and spongioblasts (embryonic glial cells).

medullispinal *a. pert.* the medulla or central portion of the spinal cord.

medusa *n.* one of the forms of individuals of coelenterates of the classes Hydrozoa (hydroids) and Scyphozoa (jellyfish). It is bell-shaped, with a tube hanging down in the centre ending in a mouth, and tentacles around the edge of the bell, and is the form commonly called a jellyfish. It forms the free-swimming sexual reproductive stage of most hydrozoans, and is large and conspicuous in jellyfish. *plu.* **medusae**.

medusoid *a.* resembling or developing into a medusa.

mega- prefix derived from Gk *megas*, large.

megacephalic *a.* (1) with abnormally large head; (2) having a cranial capacity, in humans, of over 1450 cm³.

Megachiroptera, megachiropterans *n., n.plu.* the fruit-eating bats (flying foxes), of the mammalian order Chiroptera. *cf.* Microchiroptera.

megachromosomes *n.plu.* large chromosomes forming an outer set in certain sessile ciliate protozoans. *cf.* microchromosomes.

megacins *n.plu.* a class of bacteriocins produced by strains of *Bacillus megatherium*.

megagamete *n.* in sporozoans, a cell that develops from a megagametocyte and is regarded as equivalent to an ovum.

megagametocyte *n.* in sporozoans, a cell developing from a merozoite and giving rise to a megagamete.

megagametogenesis *n.* the development of a megaspore into the megagametophyte.

megagametophyte *n.* in heterosporous plants, the female gametophyte, which develops from a megaspore.

megakaryocyte *n.* giant amoeboid cell in bone marrow with a single lobed nucleus, and which gives rise to blood platelets.

megalaesthetes *n.plu.* sensory organs, sometimes in the form of eyes, in chitons (Amphineura).

megalecithal megalolecithal *q.v.*

megaloblast *n.* a large erythroblast precursor cell.

megalolecithal *a.* containing large amounts of yolk.

megalops *n.* a larval stage of certain crustaceans such as crabs, which has large stalked eyes and a crab-like cephalothorax. *a.* **megalopic**.

megalospheric *a. appl.* to many-chambered shells of foraminifera which have a large initial chamber.

megamere macromere *q.v.*

megameric *a.* with relatively large parts.

meganephridia *n.plu.* large nephridia, occurring as one pair per segment.

meganucleus macronucleus *q.v.*

megaphanerophyte *n.* a tree exceeding 30 m in height.

megaphyll *n.* (1) type of leaf present in most vascular plants, being relatively large and having a network of veins (vascular tissue), and which is associated with the presence of leaf gaps in the stele of the stem; (2) a large leaf, esp. as produced by ferns.

megaphyllous *a.* having relatively large leaves.

megaplankton macroplankton *q.v.*

megasclere *n.* skeletal spicule of general supporting framework of a sponge.

megasorus *n.* sorus containing megasporangia.

megasporangiate *a.* composed of or producing megasporangia, *appl.* cones.

megasporangium *n.* (1) a megaspore-producing sporangium; (2) nucellus of ovule, sometimes used incorrectly for whole ovule. *plu.* **megasporangia**.

megaspore *n.* (1) in heterosporous plants, the spore that gives rise to the female gametophyte, and which is formed in a megasporangium; (2) the larger spore, in any organism that produces two types of spore.

megaspore mother cell the diploid cell in the megasporangium of a heterosporous plant, in which meiosis will occur to produce one or more megaspores.

megasporocyte *n.* the embryo sac mother cell in a plant ovary, a diploid cell that undergoes meiosis, producing four haploid megaspores.

megasporophyll *n.* leaf or leaf-like structure on which a megasporangium develops. In flowering plants it is called a carpel.

megatherm *n.* (1) a tropical plant; (2) a plant requiring moist heat and thriving in temperatures between 20 and 35°C.

megazooid *n.* the larger zooid resulting from binary or other fission.

megazoospore *n.* (1) a large zoospore, as in the reproduction of certain radiolarian protozoans; (2) a zoogonidium of certain algae.

megistotherm *n.* a plant that thrives at more-or-less uniformly high temperature.

meio- prefix derived from Gk *meion*, less.

meiocyte *n.* a cell destined to undergo meiosis and produce gametes.

meiofauna mesofauna *q.v.*

meiolecithal *a.* having little yolk.

meiophase *n.* in an organism's life cycle the stage at which meiosis occurs, with reduction of chromosome number from diploid to haploid.

meiophylly *n.* the suppression of one or more leaves in a whorl.

meiosis *n.* a type of nuclear division which results in daughter nuclei each containing half the number of chromosomes of the parent, i.e. chromosome number is reduced from diploid to haploid. Meiosis is preceded by chromosome replication and comprises two distinct nuclear divisions, the 1st and 2nd meiotic divisions, which may be separated by cell division. The reduction in chromosome number takes place during the 1st division, when paired homologous chromosomes separate and are segregated into different nuclei. The second division resembles mitosis, when the two chromatids of each replicated chromosome separate and are segregated into different nuclei. The 1st division of meiosis is conventionally divided into the following stages: leptotene, zygotene, pachytene, diplotene, diakinesis, metaphase I, anaphase I, telophase I. *see* Fig. 25. *cf.* mitosis. *see also* reduction division.

meiosporangium *n*. thick-walled diploid sporangium producing haploid zoospores by meiosis.

meiospore *n*. (1) spore produced by meiosis; (2) uninucleate haploid spore formed in a meiosporangium.

meiosporic *a. appl.* a fungus: its sexual form.

meiosporocyte megaspore mother cell *q.v.*

meiostemonous *a*. having fewer stamens than sepals and petals.

meiotaxy *n*. suppression of a whole whorl of leaves or floral parts.

meiotherm *n*. a plant that thrives in a cool temperate environment.

meiotic *a. pert.* or produced by meiosis.

meiotic apogamy generative apogamy *q.v.*

meiotic drive any mechanism that operates during meiosis in heterozygotes to produce a disproportionate representation of

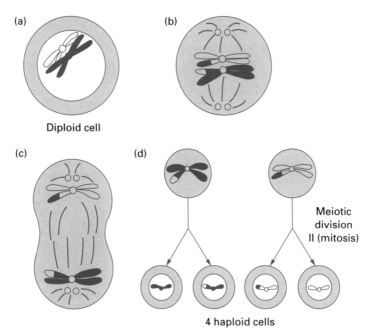

Fig. 25 Meiosis. (a) Diploid nucleus, prophase I; (b) metaphase I; (c) nuclear division I; (d) meiotic division II (equivalent to a mitotic division), producing four haploid nuclei from the original diploid nucleus. For clarity, only one pair of homologous chromosomes is shown throughout.

one member of a chromosome pair in the gametes.

meiotic spindle *see* spindle.

Meissner's corpuscles sensory nerve endings associated with sense of pain in skin of fingers, lips, etc. *see* Fig. 35 (p. 622).

Meissner's plexus a plexus of nerve fibres in submucous coat of small intestine wall.

Melanconia *n.* form class of deuteromycete fungi that reproduce by conidia borne in acervuli. It includes many plant pathogenic fungi causing anthracnose diseases (e.g. *Colletotrichum*, *Marssonina*). *alt.* **Melanconiales**.

Melanesian and Micronesion region part of the Polynesian floristic kingdom comprising the island groups of Melanesia and Micronesia.

melanin *n.* any of a range of black or brown pigments produced from tyrosine by the enzyme tyrosinase and giving colour to animal skin and hair. Also present in some plants.

melanism *n.* (1) excessive development of black pigment; (2) industrial melanism *q.v.*

melanoblast *n.* immature melanocyte (*q.v.*).

melanocyte *n.* pigment cell, in which the dark pigment melanin is synthesized. In the skin, melanocytes are found in the Malpighian layer.

melanocyte-stimulating hormone (MSH) peptide hormone produced by the pars intermedia of the adenohypophysis (pituitary gland) from the precursor pro-opiocorticotropin. It causes dispersal of melanin granules in chromatophores, resulting in a generalized darkening of the skin. *alt.* intermedin.

melanogen *n.* a colourless compound formed by reduction of the red oxidation product of tyrosine, and which is oxidized to melanin.

melanogenesis *n.* formation of melanin.

melanoids *n.plu.* dark-brown or black pigments related to melanin.

melanoma *n.* a dark pigmented mole on skin. Some forms, e.g. malignant melanoma, are highly invasive cancers.

melanophore *n.* (1) melanocyte *q.v.*; (2) chromatophore *q.v.*

melanosome *n.* (1) pigment granule synthesizing melanin and rich in the enzyme tyrosinase; (2) dark pigment mass associated with ocellus, as in certain dinoflagellates.

melanosporous *a.* with dark-coloured spores.

melanotic *a. appl.* animals and plants which are much darker than the usual colour, as a result of unusual or excessive production of melanin.

melatonin *n.* *N*-acetyl-5-methoxytryptamine, an indoleamine hormone released from the pineal gland. In most vertebrates it is released almost exclusively at night. In humans it has been implicated in regulating circadian rhythms such as sleep rhythms. In animals with chromatophores in the skin, it causes the aggregation of melanin granules, causing lightening of skin colour.

meliphagous *a.* honey-eating.

melittophile *n.* an organism that must spend at least part of its life cycle with bee colonies.

melliferous *a.* honey-producing.

mellisugent *a.* honey-sucking.

mellivorous *a.* feeding on honey.

melting *n.* separation of the strands of the DNA double helix by heating, or in acid or alkaline conditions.

melting temperature (T_m) temperature at which half the macromolecules in a solution are denatured (i.e. protein tertiary structure lost or DNA dissociated into single strands).

member *n.* (1) a limb or organ; (2) a well-defined part or organ of a plant.

membranaceous *a.* having the consistency or structure of a membrane.

membranal *a. pert.*, or within, membranes.

membrana propria basement membrane *q.v.*

membrane *n.* (1) a thin layer of organic material, skin or other tissue covering a part of an animal or plant, or separating different layers of tissue; (2) of cells, the organized layer of molecules forming the boundary of the cell (the cell membrane or plasma membrane) and of intracellular organelles, and which acts as a selective permeability barrier to substances entering and leaving the cell or organelle. Cellular membranes are composed of two layers of phospholipids (the lipid bilayer) in which proteins are embedded. The lipid bilayer is impermeable to ions and to most water-soluble molecules, which are transported across membranes by means of proteins embedded in the membrane. *see* Fig. 26. *see also* fluid mosaic model; (3) thin sheet of nylon or nitrocellulose material that acts as a support to which DNA or other macromolecules can be stably bound for further analysis.

membrane anchor type of linkage involving a lipid molecule that attaches some proteins to one face of a cellular membrane.

membrane attack complex *see* terminal complement components.

membrane bone dermal bone *q.v.*

membrane domain an area of a cell membrane distinguished by a particular protein and/or lipid composition, and which is isolated from other membrane domains by morphological features such as tight junctions.

membrane filter sheet of e.g. cellulose acetate, cellulose or polycarbonate with tiny holes too small to let bacteria pass through. Used to sterilize heat-sensitive liquids or to retain microorganisms for study.

membrane lipid *see* cholesterol, glycolipid, phospholipid, sphingolipid.

membranelle *n.* undulating structure formed by fusion of clumps of cilia in some ciliate protozoans and some rotifers. *alt.* **membranella**.

membrane potential electrical potential difference present across the plasma membranes of all living cells, so that the cytoplasmic side of the membrane is negative with respect to the exterior. It is generated by the movements of ions in and out of the cell. Rapid large changes in membrane potential convey signals in electrically excitable cells such as muscle and nerve.

membrane protein any protein that forms part of a cellular membrane. Membrane proteins may be attached to the external or the cytoplasmic faces of the membrane or span the membrane completely (transmembrane proteins). Membrane proteins have many functions, in particular the transport of materials across the membrane and as cell-surface receptors for extracellular signals. *see* Fig. 26. *see also* membrane transport, membrane transport protein.

membrane skeleton highly regular cytoskeleton underlying the red blood cell plasma membrane and giving shape to the cell.

membrane traffic the flow of membrane between different endomembrane compartments within eukaryotic cells (e.g.

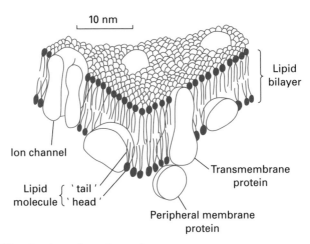

Fig. 26 The structure of a cell membrane.

between endoplasmic reticulum, Golgi apparatus and plasma membrane). Membrane is conveyed from one compartment to another in the form of small vesicles that pinch off one set of membranes and fuse with another.

membrane transport the protein-mediated passage of substances across cellular membranes. *see* active transport, coupled transport, facilitated diffusion, membrane transport proteins, passive transport, protein translocation.

membrane transport protein protein that carries a specific substance across a cellular membrane which is otherwise impermeable to it. *see* antiport, carrier protein, ion channel, symport.

membrane vesicle small intracellular sac of membrane.

membraniferous *a.* enveloped in or bearing a membrane.

membranoid *a.* resembling a membrane.

membranous *a.* (1) formed of membrane; (2) thin, dry and delicate, often transparent.

membranous cranium during embryonic development, a covering derived from mesoderm that encloses the brain.

membranous labyrinth the cochlea, semicircular canals, and associated structures of the inner ear, separated from the surrounding bone by perilymph.

membranula membranelle *q.v.*

membranule *n.* a small opaque area in anal area of wing of some dragonflies.

memory cells long-lived antigen-specific B and T lymphocytes which are produced in a primary immune response and are the basis of immunity against re-infection. They are stimulated into activity on subsequent encounters with the same antigen, when they rapidly differentiate into effector cells. *cf.* plasma cell. *see* secondary immune response.

MeNA α-napththylacetic acid methyl ester, a volatile compound used as an artificial auxin.

menacme *n.* interval between first and last menstruation, i.e. life between menarche and menopause.

menadione *n.* a vitamin K_2 analogue (2-methyl-1,4-naphthoquinone) sometimes known as vitamin K_3. *see* vitamin K.

menarche *n.* (1) first menstruation; (2) age at first menstruation.

Mendelian genetics (1) the rules of inheritance first formulated by Gregor Mendel in the middle of the 19th century. *see* Mendel's laws; (2) Mendelian inheritance *q.v.*

Mendelian inheritance inheritance of genes or characters that occurs according to Mendel's laws. *cf.* maternal inheritance, non-Mendelian inheritance.

Mendel's laws laws first proposed by Gregor Mendel in the 19th century that describe some basic rules of inheritance in sexually reproducing organisms. In breeding experiments with the garden pea, Mendel observed that a trait such as flower colour can occur in two alternative forms (alleles), which are passed on unchanged to the offspring, although the effects of one type of allele (recessive) can be masked to the observer by the presence of a dominant partner. This contrasted with the concept of blending inheritance in favour at the time. He realized that alleles must always occur in pairs and that only one allele of a pair is passed on by each parent to an offspring. Mendel's first law concerns the equal segregation of alleles, the second concerns the independent assortment of genes. *see* independent assortment, segregation of alleles.

Mendosicutes Archaea *q.v.*

meningeal *a. pert.* or in region of meninges.

meninges *n.plu.* the three membranes enclosing the brain and spinal cord. From outside to inside, they are the dura mater, arachnoid and pia mater. *sing.* **meninx.**

meningitis *n.* inflammation of the meninges.

meningococcus *n.* the bacterium *Neisseria meningitidis*, the causal agent of one type of bacterial meningitis.

meningocyte *n.* a phagocytic cell of the subarachnoid space (the space between arachnoid and pia mater).

meningosis *n.* attachment by means of membranes.

meningospinal *a. pert.* membranes surrounding the spinal cord.

meningovascular *a. pert.* meningeal blood vessels.

meninx *sing.* of meninges.

meniscus *n.* (1) a fibrocartilage sandwiched between articulating faces of joints subject to violent concussion; (2) intervertebral disc; (3) the end of a touch receptor, being the terminal extension of axon and tactile corpuscle.

menognathous *a.* with biting jaws, *appl.* insects.

menopause *n.* cessation of ovulation and menstruation in women.

menorhynchous *a.* with persistent sucking mouthparts, *appl.* insects.

menotaxis *n.* (1) compensatory movements to maintain a given direction of body axis in relation to sensory stimuli, esp. light, but not necessarily moving towards or away from it; (2) maintenance of visual axis during locomotion.

mensa *n.* grinding surface of a tooth.

menses *n.plu.* menstrual discharge.

menstrual *a.* (1) *pert.* menstruation, *see* menstrual cycle; (2) monthly; (3) lasting for a month, as of flower.

menstrual cycle monthly cycle of ovulation and menstruation in human females and some other primates, in which the lining of the uterus thickens in preparation to receive the fertilized ovum. If fertilization and conception does not occur the uterine lining is shed along with the unfertilized ovum in a short period of menstrual bleeding.

menstruation *n.* discharge of unfertilized ovum plus layer of uterine wall that occurs periodically in humans and some other higher mammals.

mental *a.* (1) *pert.* the mind; (2) *pert.* or in the region of the chin, *appl.* nerve, spines, tubercle, muscle.

mentigerous *a.* supporting or bearing the chin.

mentum *n.* the chin, or a similar region in some invertebrates.

MEPP miniature end-plate potential *q.v.*

mercury (Hg) *n.* metallic element forming a heavy silvery liquid in elemental form. Mercury pollution from industrial uses is an important environmental contaminant as mercury and its organic compounds are persistent and highly toxic.

mericarp *n.* a one-seeded unit that breaks off a composite fruit at maturity.

mericlinal chimaera chimaeric plant in which the cells of the different genotype are confined to only a sector of a layer of the meristem and thus only give rise to a sector of the shoot. *cf.* periclinal chimaera.

meridional *a.* running from pole to pole of a structure, as along a meridian.

meridional furrow longitudinal furrow extending from pole to pole of a fertilized egg undergoing cleavage.

meridiungulates *n.plu.* extinct group of South American hoofed mammals, present during the Tertiary, which included the camel-like litopterns and the notoungulates.

merisis *n.* increase in size owing to cell division.

merismatic *a.* dividing or separating into cells or segments.

merispore *n.* a segment or spore of a multicellular spore body.

meristele *n.* the branch of a stele supplying a leaf. *alt.* leaf trace.

meristem *n.* plant tissue capable of undergoing mitosis and so giving rise to new cells and tissues. It is located at growing tips of shoot and root (apical meristems), in the cambium and cork cambium encircling some plant stems (lateral meristems), and in leaves and fruits. *a.* **meristematic**, **meristemic**.

meristematic ring in developing plant embryo, a tube of meristematic tissue between cortex and pith, subtending the apical meristems and giving rise to the vascular tissues.

meristic *a.* (1) segmented; (2) divided off into parts.

meristic variation changes in numbers of parts or segments, and in the geometrical relations of parts.

meristo- word element from Gk *meristos*, divided.

meristogenetic, meristogenous *a.* (1) developing from a meristem; (2) developing from a single hyphal cell or group of contiguous cells.

Merkel cell, Merkel's discs oval-shaped sensory receptor endings sensitive to touch, in skin and in submucosa of mouth.

mermaid's purse the horny egg case of elasmobranch fishes.

mero- word element from Gk *meros*, part, or from Gk *mēros*, thigh.

meroandry *n.* the condition of having a reduced number of testes.

meroblast *n.* intermediate stage between schizont and merozoite in some sporozoan parasites.

meroblastic *a. appl.* eggs that undergo only partial cleavage during development, as a result of the presence of large amounts of yolk. The embryo develops from a small cap of cells on top of the yolk, as in the eggs of birds and reptiles.

merocerite *n.* 4th segment of crustacean antenna.

merocrine *a. appl.* glands whose secretion accumulates below the free surface of the cells through which it is released without destroying the cells, e.g. goblet cells and sweat glands.

merocytes *n.plu.* nuclei formed by repeated division of supernumerary sperm nuclei, as in fertilized egg of selachians, reptiles and birds.

merodiploid *n.* cell that is partially diploid, e.g. a bacterium that is partially diploid as a result of the introduction of part of a genome from another bacterium of the same species.

merogamete *n.* individual protozoan smaller than normal cells of the species and functioning as a gamete.

merogenesis *n.* (1) formation of parts; (2) segmentation.

merognathite *n.* 4th segment of crustacean mouthpart.

merogony *n.* development of normal young of small size from part of an egg in which there was no female pronucleus.

meroistic *a. appl.* ovariole containing nutritive or nurse cells.

meromixis *n.* *(microbiol.)* type of genetic exchange found in bacteria, where transfer of genetic material is in one direction only. *adj.* **meromictic.**

meromixis *n.* type of genetic exchange found in bacteria, where transfer of genetic material is in one direction only.

meromorphosis *n.* regeneration of a part with the new part being less than that lost.

meromyarian *a.* with only a few longitudinal rows of muscle cells, as in some nematodes.

meromyosin *n.* either of two fragments formed by cleavage of myosin by trypsin, light meromyosin (LMM) containing the 'tail' of the molecule, and heavy meromyosin (HMM) the globular 'heads' of the molecule with ATPase activity.

meron *n.* (1) posterior portion of coxa of insects; (2) sclerite between coxa of middle and hind leg in dipterans.

meront *n.* (1) any unicell formed by cleavage or schizogony; (2) uninucleate schizont stage in some sporozoan parasites, succeeding the planont stage.

meroplankton *n.* (1) temporary plankton, consisting of eggs and larvae; (2) seasonal plankton.

meropodite *n.* (1) femur in spiders; (2) 4th segment of thoracic appendage in crustaceans.

merosome *n.* body segment or somite.

merospermy *n.* condition where the nucleus of a sperm does not fuse with that of the egg, and development of a new individual is parthenogenetic.

merosporangium *n.* outgrowth from the apex of a sporangiophore, producing a row of spores.

merosthenic *a.* with unusually developed hindlimbs.

Merostomata *n.* a class of aquatic arthropods, the horseshoe crabs, which breathe by gills and have chelicerae (claws) and walking legs on the prosoma, and an opisthosoma with some segments lacking appendages.

merotomy *n.* segmentation or division into parts.

merozoite schizozoite *q.v.*

merozygote *n.* a zygote containing only part of the genome of one of the two cells or gametes from which it is formed.

mesadenia *n.plu.* mesodermal accessory genital glands in insects.

mesal *a.* medial (*q.v.*) or mesial (*q.v.*).

mesarch *a.* (1) *appl.* xylem having metaxylem developing in all directions from the protoxylem, characteristic of ferns; (2) having the protoxylem surrounded by metaxylem; (3) beginning in a mesic environment, *appl.* seres.

mesaxonic *a.* with the line dividing the foot passing up the middle digit, as in perissodactyls.

mescaline *n.* hallucinogenic alkaloid obtained from the mescal cactus, *Lophophora williamsii.*

mesectoderm *n.* cells in embryo that will give rise to ectoderm and mesoderm.

mesencephalic *a.* *pert.* or involving mesencephalon.

mesencephalon midbrain *q.v.*

mesenchymal stem cell (MSC) type of multipotent stem cell found in bone marrow stromal tissue, adult muscle, dental pulp and many other tissues, which is able to give rise to osteoblasts, myocytes, chondrocytes, adipocytes and endothelial cells, among others. Mesenchymal stem cells also have immunosuppressive properties and are being tested as a treatment for graft-versus-host disease and in tissue regeneration. *alt.* multipotent marrow stromal cell.

mesenchyme *n.* undifferentiated mesoderm-derived cells of embryo which will differentiate into muscle and connective tissue. *alt.* **mesenchymal**.

mesendoderm *n.* cells in embryo that will give rise to both endoderm and mesoderm.

mesenterial *a.* *pert.* a mesentery, *appl.* filaments of anthozoans.

mesenteric *a.* (1) *pert.* or associated with a mesentery, *appl.* e.g. arteries, glands, nerves, veins; (2) *appl.* lymph nodes associated with the small intestine.

mesenteriole *n.* a fold of peritoneum derived from mesentery and retaining the vermiform process or appendix in position.

mesenterium mesentery *q.v.*

mesenteron *n.* (1) the mid-gut in embryo, the portion of the alimentary canal lined with endoderm and derived from the archenteron; (2) in corals and sea anemones, the main digestive cavity as distinct from spaces between mesenteries.

mesentery *n.* (1) a fold of the peritoneum serving to hold viscera in position; (2) a muscular partition extending inwards from body wall in coelenterates.

mesepimeron *n.* portion of lateral exoskeletal plate of insect mesothorax.

mesethmoid *n.* bone of skull separating the nasal cavities.

mesiad *a.* towards or near the middle plane.

mesial, mesian *a.* in the middle vertical or longitudinal plane.

mesic *a.* conditioned by a temperate moist climate.

meso-, mesi-, mes- word elements from Gk *mesos*, middle, signifying situated in the middle, intermediate, neither to one end or the other of a range of conditions.

mesobenthos *n.* animal and plant life of the sea bottom at depths between 200 and 1000 m.

mesobiota *n.* the population of organisms in an ecosystem or habitat (esp. soil) which range in size from *ca.* 200 μm to 1 cm, i.e. larger than bacteria and unicellular algae, and smaller than the large soil organisms such as earthworms. It includes organisms such as mites and nematode worms. *cf.* macrobiota, microbiota.

mesoblastic *a.* *pert.* or developing from the mesoderm of an embryo.

mesobranchial *a.* *pert.* middle gill region.

mesobronchus *n.* in birds, the main trunk of a bronchus giving rise to secondary bronchi.

mesocaecum *n.* the mesentery connected with the caecum.

mesocardium *n.* (1) an embryonic mesentery binding the heart to the wall of the pericardium; (2) part of pericardium enclosing veins or aorta.

mesocarp *n.* the middle layer of the pericarp of a fruit, e.g. the flesh of fruits such as plums and cherries.

mesocentrous *a.* ossifying from a median centre.

mesocephalic *a.* having a cranial capacity (in humans) of between 1350 and 1450 cm^3.

mesocercaria *n.* trematode larval stage between cercaria and metacercaria.

mesochilium *n.* the middle portion of labellum of orchid flower.

mesocoel *n.* (1) middle portion of coelomic cavity; (2) the second of three main parts of coelom in molluscs; (3) the cavity of the mesencephalon, *alt.* aqueduct of Sylvius.

mesocole *n.* animal living in conditions not very wet or very dry.

mesocolic *a.* *pert.* mesocolon, *appl.* lymph glands.

mesocolon *n.* fold of the peritoneum attaching colon to dorsal wall of abdomen.

mesocoracoid *a.* situated between hypo- and hypercoracoid, *appl.* central part of coracoid arch of certain fishes.

mesocotyl *n.* internode between scutellum and coleoptile in grass seeds.

mesocycle *n.* layer of tissue between phloem and xylem in plant stems with a single stele.

mesodaeum *n.* endodermal part of embryonic digestive tract, between stomodaeum and proctodaeum.

mesoderm *n.* one of the three germ layers in triploblastic animals. It forms the layer of cells lying between ectoderm and endoderm in the gastrula, and gives rise to muscle and connective tissues. In vertebrates it also gives rise to the blood, the vascular system and heart, much of the kidney, and the dermis of the skin.

mesodermal *a. pert.*, derived from, or developing from mesoderm.

mesodont *a. appl.* stag beetles having medium-sized mandibles.

mesofauna *n.* animals of size from 200 µm to 1 cm. *alt.* meiofauna.

mesogamy *n.* entry of the pollen tube into an ovule through the funicle or the integument.

mesogaster *n.* the fold of peritoneum supporting the stomach.

mesogastric *a. pert.* mesogaster, mesogastrium, or middle gastric region.

mesogastrium *n.* (1) fold of peritoneum connecting stomach with dorsal abdominal wall in embryo; (2) middle abdominal region.

mesogenous *a.* produced at or from the middle.

mesoglia oligodendroglia *q.v.*

mesogloea *n.* gelatinous, non-cellular layer between the inner and outer body wall in sponges and coelenterates.

mesognathion *n.* the lateral segment of premaxilla, bearing lateral incisor.

mesohaline *a. appl.* brackish water of salinity between 5 and 15 parts per thousand.

mesohepar *n.* fold of peritoneum supporting the liver.

mesohydrophytic *a.* growing in temperate regions but requiring much moisture.

mesokaryote *a. appl.* the nucleus of dinoflagellates, in which the chromosomes are permanently condensed and which does not undergo a conventional mitosis.

mesolecithal *a.* having a moderate yolk content.

mesolimbocortical system system of dopaminergic neurons that projects to the limbic system (amygdala, hippocampus, septum).

mesome *n.* the axis or main stem of a plant regarded as a morphological unit.

mesomere *n.* (1) the middle zone of coelomic pouches in embryo; (2) somite *q.v.*

mesometrium *n.* the fold of peritoneum supporting the uterus and connecting tubes.

mesomicrotherm *n.* plant living in a temperate climate which can resist low winter temperatures.

mesomitosis *n.* mitosis that is carried out within an intact nuclear envelope.

mesomorphic *a.* of normal or average structure, form or size, or intermediate between extremes. *n.* **mesomorph**.

meson *n.* the central plane of an object or a region of it.

mesonephric *a. pert.* mesonephros, *appl.* tubules, *appl.* duct.

mesonephridium *n.* nephridium or excretory organ of some invertebrates, derived from mesoderm.

mesonephros *n.* the middle of the three pairs of renal organs of vertebrate embryos, which persist as the adult kidneys of anamniotes.

mesonotum *n.* dorsal part of insect metathorax.

mesoparasite *n.* type of parasite intermediate between ecto- and endoparasites.

mesopelagic *a. pert.* or inhabiting the ocean at depths between 200 and 1000 m.

mesoperidium *n.* middle layer, between endoperidium and exoperidium, of the peridium of some fungi.

mesopetalum *n.* labellum of orchid.

mesophanerophyte *n.* a tree from 8 to 30 m in height.

mesophil(ic) *a.* thriving at moderate temperatures, between 20 and 45°C when *appl.* bacteria. *n.* **mesophile**.

mesophilous *a. appl.* plants associated with neutral soils.

mesophloem *n.* middle or green bark.

mesophragma *n.* piece of chitinous exoskeleton descending into interior of insect body with postscutellum for base.

mesophyll *n.* (1) the internal parenchyma tissue of a leaf, which is usually photosynthetic; (2) a leaf of moderate size.

mesophyllous *a.* having leaves of moderate size.

mesophyte *n.* a plant thriving in a temperate climate with a moderate amount of moisture.

mesoplankton *n*. (1) plankton at depths of 200 m downwards; (2) drifting organisms of medium size.

mesoplastron *n*. bony plate of shell of some turtles, between 2nd and 3rd lateral plates.

mesopleurite, mesopleuron *n*. lateral sclerite of mesothorax, as in dipterans.

mesopodial *a*. having a supporting structure, such as a stipe, in a central position.

mesopodium *n*. (1) leaf stalk or petiole region of leaf; (2) middle part of molluscan foot; (3) the metacarpus or metatarsus.

mesopterygium *n*. the middle of three basal pectoral fin cartilages in recent elasmobranchs.

mesopterygoid *n*. the middle of three pterygoid bone elements in teleost fishes.

mesoptile *n*. the second type of down feather that develops in nestling birds. *cf*. protoptile.

mesorchium *n*. fold of peritoneum supporting testis.

mesorectum *n*. fold of peritoneum supporting rectum.

mesorhinal *a*. between nostrils.

mesosalpinx *n*. the broad ligament enclosing the uterus.

mesosaprobic *a*. *appl*. aquatic habitats having a decreased quantity of oxygen and substantial organic decomposition.

α-mesosaprobic category in the saprobic classification of river organisms consisting of those, e.g. the water-louse (*Asellus*), that can live in polluted water in which decomposition is partly aerobic and partly anaerobic. *cf*. β-mesosaprobic, mesosaprobic, oligosaprobic, polysaprobic.

β-mesosaprobic category in the saprobic classification of river organisms comprising those that can live in water mildly polluted with organic pollutants, in which organic decomposition is mainly aerobic, e.g. the three-spined stickleback (*Gasterosteus aculeata*) and Canadian pondweed (*Elodea canadensis*). *cf*. α-mesosaprobic, mesosaprobic, oligosaprobic, polysaprobic.

mesosaur *n*. member of the order Mesosauria, an order of lower Permian anapsid reptiles which were fish-eating and slender with long jaws.

mesoscapula *n*. the spine on the scapula when considered a distinct unit.

mesoscutellum *n*. scutellum of insect mesothorax.

mesoscutum *n*. scutum of insect mesothorax.

mesosoma *n*. (1) anterior, broader part of abdomen in scorpions; (2) middle portion of body of certain invertebrates, esp. when their original segmentation is obscured.

mesosome *n*. an invagination of the bacterial cell membrane.

mesosperm *n*. the integument surrounding nucellus of ovule.

mesosporium *n*. the middle of three layers in coat of some spores.

mesosternum *n*. (1) middle part of sternum (breast bone) of vertebrates; (2) sternum of mesothorax in insects.

mesostriatal system dopaminergic neurons originating in the substantia nigra and neighbouring tegmental areas and projecting to the striatum (caudate nucleus, globus pallidus, putamen) and nucleus accumbens and olfactory tubercle. Its degeneration is implicated in Parkinson's disease.

mesostylous *a*. having styles of intermediate length, *appl*. heterostylous flowers.

mesotarsus *n*. tarsus of middle limb in insects. *a*. **mesotarsal**.

mesotheic *a*. neither highly susceptible nor completely resistant to parasites or infection.

mesothelial *a*. *pert*. mesothelium.

mesothelioma *n*. rare type of tumour, derived from mesothelial cells of peritoneum, pleura or pericardium. Pleural mesotheliomas have been found at a greater than normal frequency in workers exposed to asbestos.

mesothelium *n*. (1) epithelium-like layer of mesoderm-derived flattened squamous cells lining serous cavities (e.g. pericardial and pleural cavities); (2) mesoderm bounding embryonic coelom and giving rise to muscular and connective tissue.

mesotherm *n*. plant thriving in moderate heat, within the range 12–19°C, as in a warm temperate climate.

mesothoracic *a*. *pert*. or in region of mesothorax.

mesothorax *n*. the middle segment of thoracic region of insects. *cf*. prothorax, metathorax.

mesotroch *n*. type of larva of annulates which has a circlet of cilia round the middle of body.

mesotrophic *a.* (1) having partly autotrophic and partly saprobic nutrition; (2) obtaining nourishment partly from an outside source; (3) partly parasitic; (4) providing a moderate amount of nutrition, *appl.* environment.

mesotropic *a.* turning or directed toward the middle or the median plane.

mesovarium *n.* fold of peritoneum supporting the ovary.

mesoventral *a.* in middle part of ventral region.

mesoxerophilous *a. appl.* plant of temperate climates showing adaptation to dry conditions and thriving in such conditions. *n.* **mesoxerophyte**.

Mesozoa, mesozoans *n.*, *n.plu.* phylum of extremely simple, small multicellular marine invertebrates, which live as parasites in the kidneys of marine invertebrates. The adult has a body composed of two layers of cells, no muscular or nervous system, and lacking a body cavity and all organs except a gonad. Infusoriform and vermiform larvae are produced. The Mesozoa include the dicyemids, heterocyemids and orthonectids.

Mesozoic *n.* geological era lasting from *ca.* 251 million years ago to 65 million years ago and comprising the Triassic, Jurassic and Cretaceous periods.

message in molecular biology, messenger RNA *q.v.*

messenger RNA (mRNA) RNA found in all cells which acts as a template for protein synthesis. Each different mRNA is a complementary copy of a strand of DNA of a single protein-coding gene (or, in bacteria, often a set of adjacent genes) and is produced by transcription using the DNA as a template. In eukaryotes, mRNA is the product of extensive processing of the primary RNA transcript in the nucleus after transcription. *alt.* message. *see also* RNA, RNA processing, splicing.

mestome *n.* inner sheath of thick-walled cells of bundle sheath.

Met methionine *q.v.*

meta- (1) prefix derived from Gk *meta*, after, signifying posterior, as in metathorax, the 3rd and last thoracic segment in insects; (2) prefix derived from Gk *meta*, change of, as in metamorphosis, a change in form.

meta-analysis *n.* analysis of a particular topic that makes use of existing published data, often from many different studies.

metabasidium *n.* cell in which meiosis occurs in basidium of basidiomycete fungi.

metabiosis *n.* the beneficial exchange of factors (e.g. nutrients, vitamins) between species.

Metabola pterygotes *q.v.*

metabolic *a. pert.* metabolism.

metabolic activation conversion of foreign compounds into chemically reactive (often toxic or carcinogenic) forms in the body, esp. by enzymes in the liver.

metabolic compartment in eukaryotic cells, any membrane-bounded organelle or space, such as mitochondrion, chloroplast or cytosol, in which particular metabolic processes are segregated.

metabolic cooperation (1) syntrophy *q.v.*; (2) complementary metabolic transformations, such as those of nitrosifying and nitrifying bacteria, which together can oxidize NH_3 to NO_3^-.

metabolic pathway a series of enzyme-catalysed biochemical reactions in a cell in which the product of one reaction becomes the substrate for the next. Metabolic pathways are used to convert one compound into another, to build up large macromolecules from smaller units, or to break down compounds to release usable energy.

metabolic profiling the analysis of cell extracts to determine the presence of a large number of metabolites and analytes at the same time.

metabolic rate a measure of the rate of metabolic activity in a living organism. It is the rate at which an organism uses energy to sustain essential life processes such as respiration, growth, reproduction and, in animals, processes such as blood circulation, muscle tone and activity. It can be determined in numerous ways: (i) as the total heat produced over a given period, (ii) as oxygen consumption (and sometimes also carbon dioxide production) over a given period, which although easier to measure, only gives the contribution of aerobic metabolism, (iii) as the energy content of the food eaten over a given period, and (iv) by the fate of isotopically labelled water. *see* average daily metabolic rate, basal metabolic rate, energy budget, field metabolic rate, respiratory quotient, resting metabolic rate, standard metabolic rate.

metabolic scope the range of metabolic rate shown by an animal, which is the difference between the resting metabolic rate and the maximum rate of energy expenditure of which the animal is capable at maximum activity.

metabolic water water produced by oxidative processes (e.g. respiration) within the body.

metabolism *n.* the integrated network of biochemical reactions that supports life in a living organism. *see also* anabolism, catabolism, metabolic pathway.

metabolite *n.* any chemical compound involved in, or a product of, metabolism. *see also* secondary metabolites.

metabolome *n.* the total complement of small-molecule metabolites in a cell in any given physiological state.

metabolomics *n.* the study of the metabolic networks of a cell as a whole.

metaboly *n.* change of shape resulting in movement, as in euglenoids and other flagellates.

metabotropic *a. appl.* receptor for a neurotransmitter which does not contain an ion channel, but which acts via G proteins and/or intracellular second messengers.

metabranchial *a. pert.* or in region of posterior gills.

metacarpal *n.* one of the bones joining wrist and fingers.

metacarpus *n.* the skeletal part of hand between wrist and fingers, typically consisting of five cylindrical bones, the metacarpals. *a.* **metacarpal**.

metacentric *a. appl.* chromosomes with the centromere halfway along and which appear V-shaped when segregating during mitosis or meiosis. *n.* a metacentric chromosome.

metacercaria *n.* the stage in the life cycle of endoparasitic flukes that develops from a cercaria, in which the cercaria loses its tail. In schistosomes this occurs after infection of the vertebrate host by cercariae. In other species metacercariae infect the vertebrate host. The adult parasite develops directly from a metacercaria.

metacestode *n.* larval tapeworm found in the intermediate host.

metachroic *a.* changing colour, as older tissue in fungi.

metachromasy *n.* change in colour. *a.* **metachromatic**.

metachromatin *see* volutin granules.

metachromy *n.* change in colour, as in flowers as they age.

metachronal *a.* one acting after another, *appl.* e.g. rhythm of movement of cilia, the legs of centipedes and millipedes.

metachrosis *n.* ability to change colour by expansion or contraction of pigment cells.

metacoel *n.* (1) posterior part of the coelom, as in molluscs; (2) anterior extension of 4th ventricle in brain.

metacommunication *n.* communication about the meaning of other acts of communication, such as the posture adopted by a dominant male, which signals to other males its status and likely behaviour if attacked.

metacone *n.* posterior external cusp of upper molar tooth.

metaconid *n.* posterior internal cusp of lower molar tooth.

metaconule *n.* posterior secondary cusp of upper molar tooth.

metacoracoid *n.* posterior part of coracoid.

metacromion *n.* posterior branch of acromion of scapular spine.

metacyclic *a. appl.* infective short broad forms of trypanosome that develop in the insect vector salivary gland and which are passed on to the next host.

metadiscoidal *a. appl.* placenta in which villi are at first scattered and later restricted to a disc, as in primates.

metadromous *a.* with primary veins of segment of leaf arising from upper side of midrib.

metafemale *n.* a female *Drosophila* with a normal diploid set of autosomes but having three X chromosomes. *alt.* superfemale.

metagastric *a. pert.* posterior gastric region.

metagenesis *n.* alternation of generations, esp. of asexual and sexual generations, esp. in animals.

metagnathous *a.* (1) having mouthparts for biting in the larval stage and sucking in the adult, as in certain insects; (2)

having the points of the beak crossed, as in crossbills.

metagyny protandry *q.v.*

metallic *a.* iridescent, *appl.* colours due to interference by fine striations or thin plates on surface, as in insects.

metalloenzyme *n.* any enzyme containing metal ions.

metalloprotein *n.* any protein containing metal ions.

metallothionein *n.* protein involved in cell detoxification mechanisms for various heavy metals.

metaloph *n.* the posterior crest of a molar tooth, uniting metacone, metaconule and hypocone.

metamale *n.* a male *Drosophila* with one X chromosome and three sets of autosomes. *alt.* supermale.

metamere *n.* a body segment or somite.

metameric *a.* having the body divided into a number of segments more-or-less alike. *n.* **metamerism**.

metamerized *a.* segmented.

metamitosis *n.* mitosis in which nuclear membrane breaks down.

metamorphosis *n.* (1) transformation of one structure into another, most commonly referring to the radical change in form and structure undergone by some animals between embryo and adult stage, as in insects and amphibians. The embryo develops into a free-living larva which is unlike the adult in form. This then undergoes metamorphosis to produce the adult. In insects, incomplete metamorphosis occurs in e.g. locusts and grasshoppers, where the larval form is relatively similar to the adult and changes gradually towards the adult form at each moult, and in which there is no non-feeding pupal stage. Complete metamorphosis occurs in e.g. butterflies and flies, in which the larval caterpillar or maggot stage is quite unlike the adult in form and internal structure, and undergoes a radical remodelling during a non-feeding pupal stage; (2) homeosis *q.v.*; (3) interference with normal symmetry in flowers.

metamorphy homeosis *q.v.*

metanauplius *n.* larval stage of crustaceans, succeeding the nauplius stage.

metandry *n.* meroandry with retention of posterior pair of testes only.

metanephric *a. pert.* or in region of the kidney.

metanephridium *n.* nephridial tubule with opening into the coelom.

metanephros *n.* organ arising behind mesonephros in vertebrate embryos, and replacing it as the functional kidney in amniotes.

metanotum *n.* notum of insect metathorax.

metanucleus *n.* egg nucleolus after extrusion from germinal vesicle.

metaphase *n.* stage in mitosis or meiosis when the chromosomes have become aligned on the equator of the cell with the centromeres lying along the spindle equator. In metaphase I (1st meiotic division) the aligned structures are paired duplicated homologous chromosomes, in metaphase II (2nd meiotic division) they are individual duplicated chromosomes.

metaphase plate imaginary plane halfway between opposite poles of the spindle and at right angles to the spindle axis and on which the centromeres of the chromosomes lie at metaphase of mitosis or meiosis. *alt.* equatorial plate.

metaphase spread a squashed preparation of the stained chromosomes from a cell in metaphase in which individual chromosomes are separated and can be distinguished.

metaphloem *n.* primary phloem formed after the protophloem.

metaphysis *n.* vascular part of the shaft of a limb bone that adjoins the epiphyseal cartilage at each end of the bone.

Metaphyta, metaphytes *n., n.plu.* multicellular plants.

metaplasia transdifferentiation *q.v.*

metaplastic *a. pert.* metaplasia, transdifferentiated.

metapleural *a.* posteriorly and laterally situated.

metapneustic *a. appl.* insect larvae with only the terminal pair of spiracles open.

metapodeon *n.* that part of insect abdomen behind the 'waist' or podeon.

metapodium *n.* (1) portion of foot between tarsus and digits; (2) in four-footed animals, either the metacarpus or metatarsus.

metapophysis *n.* prolongation of a vertebral articular projection esp. in lumbar region, developed in some vertebrates.

metapostscutellum *n.* postscutellum of insect metathorax.

metaprescutellum *n.* prescutellum of insect metathorax.

metaproteomics *n.* the large-scale study of the totality of proteins expressed by complex microbial communities in the environment.

metapterygium *n.* the posterior basal fin cartilage, pectoral or pelvic, of recent elasmobranchs.

metapterygoid *n.* posterior of three pterygoid elements in some lower vertebrates.

metarteriole *n.* a small branch of an arteriole between arteriole and capillaries.

metascolex *n.* organ formed by enlargement of the neck area directly behind scolex in some cestodes.

metascutellum *n.* scutellum of insect metathorax.

metascutum *n.* scutum of insect metathorax.

metasoma *n.* the posterior region of opisthosoma of arachnids and some crustaceans.

metasomatic *a. pert.* or situated in metasoma.

metastasis *n.* (1) a change in state, position, form or function; (2) the migration of cancer cells to colonize tissues and organs other than those in which they originated; (3) a secondary tumour caused by migration of cancer cells to another tissue. *plu.* **metastases**. *a.* **metastatic**. *v.* **metastasize**.

metasternum *n.* (1) the sternum of insect metathorax; (2) posterior part of sternum of higher vertebrates.

metasthenic *a.* with well-developed posterior part of body.

metastoma *n.* (1) the two-lobed lip of crustaceans; (2) hypopharynx of myriapods.

metastomial *a.* behind the mouth region, *appl.* segment posterior to peristomium or buccal segment in annelids.

metastructure *n.* structure as distinguished at the ultramicroscopic level.

metatarsal *n.* one of the bones joining ankle and toes.

metatarsus *n.* (1) skeleton of vertebrate foot between tarsus and toes, comprising the metatarsals; (2) 1st segment of insect tarsus. *a.* **metatarsal**.

Metatheria, metatherians marsupials *q.v.*

metathorax *n.* posterior (3rd) segment of insect thorax.

metatracheal *a. appl.* wood in which xylem parenchyma is located independently of the vessels and scattered throughout the annual ring.

metatranscriptomics *n.* the large-scale study of the transcriptomes (expressed genes) of a group of interacting organisms or species, usually in the context of complex microbial communities in the environment.

metatroch *n.* in a trochophore, a circular band of cilia behind the mouth.

metatype *n.* a topotype of the same species as the holotype or lectotype.

Metaviridae *n.* family of double-stranded DNA viruses with a reverse transcription step in their replication, infecting fungi (yeast Ty3 virus) and invertebrates (*Drosophila* gypsy virus).

metaxenia *n.* the case where there is a difference in the influence of pollen from different parents on the development of a fruit.

metaxylem *n.* primary xylem developing after the protoxylem and before the secondary xylem, if present, and which is distinguished by wider vessels and tracheids.

Metazoa, metazoans *n., n.plu.* multicellular animals, sometimes more strictly applied only to those multicellular animals with cells organized into tissues and possessing nervous tissue.

metecdysis *n.* in arthropods, period after moult when the new cuticle hardens.

metencephalon *n.* (1) that part of the developing hindbrain that will form the cerebellum, pons and intermediate part of 4th ventricle.

metepimeron *n.* epimeron of insect metathorax.

metepisternum *n.* episternum of insect metathorax.

metestrus *alt.* spelling of metoestrus.

methaemoglobin (MetHb) haemoglobin with the haem iron in the ferric state and unable to bind oxygen. It is found in small amounts in the blood and is produced by the action of oxidizing agents such as nitrite and chlorate poisons. *alt.* ferrihaemoglobin.

methane *n.* CH_4, a gas derived from the anaerobic breakdown of organic matter.

It is produced by rotting vegetation and in the digestive tracts of ruminants. It is an important greenhouse gas. *alt.* marsh gas, natural gas.

methanogen *n.* microorganism that can generate the gas methane (CH_4). Methanogens are archaea and can produce methane by the reduction of CO_2, formate, acetate, carbon monoxide or methyl substrates under anaerobic conditions, using electrons derived from hydrogen or from the substrates themselves. They are found in anoxic environments, such as marine and freshwater muds, the digestive tracts of animals and in sewage treatment plants.

methanogenesis *n.* the generation of the gas methane (CH_4) by living organisms, the methanogens.

methanogenic *a.* generating methane.

methanol *n.* methyl alcohol.

methanotrophs *n.plu.* diverse group of aerobic bacteria that can utilize methane as an electron donor for energy generation and as a sole carbon source.

MetHb methaemoglobin *q.v.*

methicillin *n.* a β-lactamase-resistant semi-synthetic penicillin. *see also* MRSA.

methionine (Met, M) *n.* amino acid with a non-polar sulphur-containing side chain, constituent of proteins, essential in human diet, and which provides sulphur and methyl groups for metabolic reactions. It is the first amino acid to be inserted in all eukaryotic polypeptide chains, formylmethionine being used in bacteria and mitochondria.

methotrexate amethopterin *q.v.*

methyl-accepting chemotaxis protein (MCP) receptor protein in bacteria that detects an external stimulus that generates chemotaxis.

methylase methyltransferase *q.v.*

methylation *n.* addition of a methyl group ($-CH_3$) to a molecule. *see also* DNA methylation, histone methylation.

5-methylcytosine (mC) *n.* modified base formed by enzymatic methylation of cytosine *in situ* in DNA. Methylcytosine can cause mutations by spontaneous deamination to thymine, resulting in substitution of adenine for cytosine at subsequent DNA replication.

methyl group the chemical group $-CH_3$.

methylmercury *n.* soluble and highly toxic compound of mercury formed in the environment by microbial methylation of mercury.

methylotrophs *n.plu.* diverse group of bacteria that can utilize one-carbon compounds, such as methane and methanol, as sole carbon source.

methyl red test clinical diagnostic test to detect bacteria that are mixed-acid fermenters. Used to differentiate enteric bacteria.

methyltransferase *n.* any of a group of enzymes that transfer methyl groups ($-CH_3$) from one molecule to another, e.g. DNA methyltransferase (EC 2.1.1.37) which methylates cytosine residues in DNA using *S*-adenosylmethionine as the methyl donor. *alt.* methylase, transmethylase.

metochy *n.* relationship between a neutral guest insect and its host.

metoecious *a. appl.* parasites that are not host-specific.

metoestrus *n.* the luteal phase, the period when activity subsides after oestrus.

metope *n.* the middle frontal portion of a crustacean.

metopic *a.* (1) *pert.* forehead; (2) *appl.* frontal suture of skull.

metoxenous metoecious *q.v.*

metra uterus *q.v.*

mevalonic acid six-carbon organic acid, an intermediate in cholesterol biosynthesis.

Mexican subregion subdivision of the Neotropical zoogeographical region, comprising Central America.

Mg symbol for the chemical element magnesium *q.v.*

MHC major histocompatibility complex *q.v.*

MHC antigens major histocompatibility antigens. *see* MHC molecules.

MHC class I molecules one of the two classes of polymorphic MHC molecule. MHC class I molecules are present on most cells of the body. In immune responses, they bind peptides generated via the cytosolic protein degradation pathway (e.g. peptides derived from viruses infecting cells) and present these peptide antigens to CD8 T cells. In humans the three types of MHC class I molecule are HLA-A, HLA-B and HLA-C.

MHC class II molecules one of the two classes of polymorphic MHC molecule.

MHC class II molecules are present chiefly on the professional antigen-presenting cells of the immune system – macrophages, B cells and dendritic cells. In immune responses, they bind peptides generated via the endocytic pathway from extracellular material or from ingested microorganisms, and present these peptide antigens to CD4 T cells. In humans the three main types of MHC class II molecule are HLA-DP, HLA-DQ and HLA-DR.

MHC molecules highly polymorphic cell-surface glycoproteins encoded by the major histocompatibility complex in vertebrates. They are involved in recognition of foreign antigens in immune responses and are themselves the antigens reacted against in rejection of transplanted organs. They consist of two main classes: MHC class I molecules (*q.v.*), which are present on most cells of the body, and MHC class II molecules (*q.v.*), which are normally restricted to certain cells of the immune system. MHC molecules bind peptide antigens and present them to the T cells of the immune system. They are also the determinants of the clinical tissue type of an individual. MHC molecules are highly polymorphic, with hundreds of different genetic variants of some types being present in the population. Thus the MHC molecules of one individual are rarely all identical to those of another unrelated individual. Transplantation of organs not matched for MHC type with the recipient leads to transplant rejection unless normal immune function is suppressed by drugs. MHC molecules are known as HLA molecules in humans and H-2 molecules in mice. *alt.* MHC antigens, H-2 antigens (in mouse), histocompatibility antigens, HLA antigens (in humans), transplantation antigens. *see also* antigen presentation, antigen processing, MHC restriction.

MHC restriction (1) the fact that a T-cell receptor recognizes a combination of a particular peptide antigen and a particular MHC molecule rather than the peptide antigen itself. T cells with receptors that can recognize peptide antigens presented by the body's own MHC molecules, and which are thus of use in recognizing and reacting against foreign peptides presented by these MHC molecules during an infection, are selected during their development in the thymus; (2) T cells with different functions recognize foreign antigen in conjunction with different classes of MHC molecules, and are thus restricted to interacting with certain types of cell only. Helper T cells (CD4 T cells) recognize antigen only in conjunction with MHC class II molecules, which are expressed mainly by professional antigen-presenting cells, while cytotoxic T cells (CD8 T cells) recognize antigen only in conjunction with MHC class I molecules, which are expressed by most cell types.

MIC minimum inhibitory concentration *q.v.*

micelle *n.* hollow spherical structure formed by aggregates of amphipathic molecules such as phospholipids, glycolipids and detergents in water, with hydrophilic groups on the outside and polar groups inside.

Michaelis–Menten kinetics set of equations that describe the properties of many enzyme-catalysed reactions, from which two characteristic parameters of such reactions are derived: K_m (the Michaelis constant) and V_{max}. K_m = substrate concentration at which the reaction rate is half its maximal value, V_{max} = maximal rate of an enzyme-catalysed reaction under steady-state conditions.

micraesthetes *n.plu.* the smaller sensory organs of chitons (Amphineura).

micrander *n.* a dwarf male, as in certain green algae.

micraner *n.* a dwarf male ant.

micrergate *n.* a dwarf worker ant.

micro- prefix derived from Gk *mikros*, small.

microaerophile *n.* an aerobic organism that can grow only at reduced oxygen tension relative to that of air. *a.* **microaerophilic**.

microalga *n.* unicellular alga visible only under a microscope. *cf.* macroalga.

microarray *n.* a DNA or protein array in a small-sized format such as a microscope slide, containing many hundreds to thousands of individual spots.

microarthropod *n.* arthropod of microscopic size. *cf.* macroarthropod.

microautoradiography *n.* technique that uses radioactively labelled probes or radioactively labelled substrates and autoradiography to study single cells of microorganisms.

microballistics biolistics *q.v.*

microbe microorganism *q.v.*

microbial *a. pert.*, composed of, or caused by microorganisms.

microbial leaching *see* leaching.

microbial mat layered community of microorganisms forming a mat about 2 cm thick, found esp. in hot springs and shallow marine environments. The top layer is usually cyanobacteria, with anoxygenic phototrophic bacteria in the lower layers and chemoorganotrophic bacteria in the light-limited bottom layer.

microbial plastics *see* biopolymers.

microbiology *n.* the study of microorganisms.

microbiota *n.* organisms of microscopic size in any ecosystem or habitat (esp. soil). It includes e.g. bacteria, unicellular algae, protozoa, slime moulds and fungal mycelium and spores. *cf.* macrobiota, mesobiota.

microbivore *n.* animal that feeds on microorganisms. *a.* **microbivorous**.

microbody *n.* any of a diverse class of small spherical organelles, including glyoxysomes and peroxisomes.

microcell *n.* cell in which all chromosomes except one or two have been eliminated. Used in making somatic cell hybrids for gene mapping.

microcephalic *a.* with an abnormally small head, having cranial capacity (in humans) of less than 1350 cm³.

microchaeta *n.* small bristle, as on body of some insects.

Microchiroptera, microchiropterans *n., n.plu.* small mainly insectivorous bats of the mammalian order Chiroptera. *cf.* Megachiroptera.

microchromosomes *n.plu.* chromosomes considerably smaller than the other chromosomes of the same type in the nucleus, and usually centrally placed in the metaphase plate during mitosis. *alt.* M chromosomes.

microclimate *n.* the climate within a very small area or in a particular habitat.

micrococcus *n.* bacterium of the family Micrococcaceae, aerobic or facultatively anaerobic Gram-positive cocci found in soil and water. *plu.* **micrococci**. *a.* **micrococcal**.

microcolony *n.* small colony of bacteria encased in an adhesive polysaccharide covering, which develops when bacteria are growing on a surface.

microconidium *n.* minute conidium, produced by some ascomycete fungi, which can behave as a male sex cell or germinate to give rise to a mycelium. *plu.* **microconidia**. *alt.* aleuriospore.

microconjugant *n.* motile free-swimming conjugant or gamete that attaches itself to a macroconjugant and fertilizes it.

microcosm *n.* a world in miniature, a community that is a miniature version of a larger whole.

microcyclic *a.* (1) *appl.* rust fungi that produce only teleutospores and basidiospores during the life cycle; (2) *appl.* short and simple life cycles; (3) having a haplophase or gametophyte stage only.

microcyst *n.* a spherical thick-walled resting spore formed by some bacteria.

microcyte *n.* red blood cell about half the size of normal, numerous in certain diseases.

microdeletion *n.* mutation consisting of the deletion of a single base pair.

microdissection *n.* technique for manipulating and operating upon live microorganisms and individual cells using a microscope to view the specimen being manipulated.

microdont *a.* with comparatively small teeth.

microelectrode *n.* very fine electrode used for recording from single cells.

microelements trace elements *q.v.*

microendemic *a.* restricted to a very small area.

microenvironment microhabitat *q.v.*

microevolution *n.* (1) evolutionary change consisting in alterations in gene frequencies in the population, or in chromosome structure or chromosome numbers due to mutation and recombination, and which can be noticed over a relatively short time, e.g. the acquisition of resistance to a pesticide amongst insects; (2) the relatively small evolutionary changes that differentiate the members of geographical races, subspecies or sibling species. *cf.* macroevolution.

microfauna *n*. animals less than 200 μm long, only visible under the microscope. Protozoa are generally included in the microfauna.

microfibril *n*. microscopic fibre composed of protein (e.g. keratin fibres of hair) or polysaccharide (e.g. cellulose microfibrils in plant cell walls).

microfilament *n*. one of the three main elements of the eukaryotic cell cytoskeleton. Microfilaments are protein threads composed of globular actin subunits. In association with myosin they form contractile structures including the contractile apparatus of muscle cells. *alt*. actin filament.

microfilaria *n*. embryonic form of certain parasitic threadworms. *plu*. **microfilariae**.

microflora *n*. (1) the microorganisms (bacteria, unicellular fungi and algae) living in or on an organism, or in a particular habitat or ecosystem; (2) (*bot*.) the dwarf flora of high mountains.

microfluidics technology in which chemical reactions, assays or sample preparation processes are carried out in automated fashion in minute reaction chambers and channels in preformed plastic devices using very small amounts of reactants.

microfossil *n*. any microscopic fossil, such as those of prokaryotic and eukaryotic microorganisms, spores, pollen, and microscopic animals and plants.

microfungi *n.plu*. the yeasts and moulds, as opposed to mushrooms and toadstools.

microgamete *n*. the smaller of two conjugant gametes, regarded as the male.

microgametoblast *n*. intermediate stage between microgametocyte and microgamete in certain sporozoans.

microgametocyte *n*. cell developed from merozoite in certain protozoans, giving rise to microgametes.

microgametogenesis *n*. development of microgametes or spermatozoa.

microgametophyte *n*. in heterosporous plants, the male gametophyte, which develops from a microspore.

microglia *n.plu*. type of neuroglia composed of small migratory phagocytic cells of mesodermal origin and more common in grey than in white matter. *a*. **microglial**.

β$_2$-microglobulin *n*. small protein composed of a single immunoglobulin-like domain and which is one of the two polypeptide chains in MHC class I molecules. It is encoded by a gene outside the MHC.

micrograph *n*. photograph of an image obtained by microscopy.

microgyne *n*. dwarf female ant.

microhabitat *n*. (1) the immediate environment of an organism, esp. a small organism; (2) a small place in the general habitat distinguished by its own set of environmental conditions.

microheterogeneity *n*. small variations in nucleotide sequence seen e.g. in individual repeating units of a tandemly repeated gene cluster.

microinjection *n*. the introduction of substances into a single cell by injection with special instruments (e.g. micropipette).

microinsertion *n*. mutation caused by insertion of a single base pair.

microlecithal *a*. *appl*. ova containing little yolk.

micromanipulator *n*. a device that allows the precise manipulation of fine instruments such as microelectrodes or microdissection tools over very small distances.

micromere *n*. the smaller of the cells produced as a result of unequal cleavage in the fertilized eggs of some animals. *cf*. macromere.

micromerozoite *n*. cell derived from microschizont and developing into gametocyte in certain sporozoan blood parasites.

micromesentery *n*. an incomplete radial partition (mesentery) of internal cavity in sea anemones and their relatives (Anthozoa) which does not extend all the way from body wall to pharynx.

micrometre (μm) *n*. unit of microscopic measurement, being 10^{-6} m or one-thousandth of a millimetre. Formerly called a micron. *alt*. (US) **micrometer**.

micromutation point mutation *q.v*.

micron micrometre *q.v*.

micronemic, micronemous *a*. *pert*. or having small hyphae.

micronephridia *n*. small nephridia.

micront *n*. a small cell formed by schizogony, itself giving rise to microgametes.

micronucleus *n*. the smaller of the two types of nucleus found in ciliate protozoans. It is a diploid nucleus which undergoes meiosis and is involved in sexual reproduction but which does not produce RNA. *cf.* macronucleus.

micronutrient *n.* chemical element or organic compound required in small amounts by living organisms for proper growth and development, e.g. a trace element (such as zinc, iron, copper) or a vitamin. For plants and some microorganisms, the term micronutrient refers to the trace elements only. *see also* growth factor.

microorganism *n.* general term for bacteria, viruses, unicellular algae and protozoans, and microscopic fungi (yeasts and moulds), which are all of microscopic or ultramicroscopic size. *alt.* microbe.

microparasite *n.* any parasite of microscopic size.

microphage neutrophil *q.v.*

microphagic, microphagous *a.* (1) feeding on minute organisms or particles; (2) feeding on small prey. *n.* **microphagy.**

microphanerophyte *n.* small tree or shrub from 2 to 8 m in height.

microphil(ic) *a.* tolerating only a narrow range of temperature, *appl.* certain bacteria.

microphyll *n.* (1) simple leaf containing only a single strand of vascular tissue, the leaf type present in horsetails, club mosses and the Psilophyta; (2) a small leaf, esp. in ferns and their relatives.

microphyllous *a.* having small leaves.

micropipette *n.* very fine glass pipette used to impale a single cell for electrical recording or introduction of DNA.

microplankton *n.* small organisms drifting with the surrounding water, somewhat larger than those of the nanoplankton.

Micropodiformes Apodiformes *q.v.*

micropodous *a.* with rudimentary or small foot or feet.

micropropagation *n.* propagation of plants through tissue culture.

micropterous *a.* (1) having small hindwings invisible until tegmina are expanded, as in some insects; (2) having small fins.

micropyle *n.* (*bot.*) (1) small pore or channel in coat at apex of ovule through which the pollen tube enters; (2) corresponding aperture in testa of seed between hilum and point of radicle; (*zool.*) (3) aperture in

the membrane of some animal eggs through which sperm enter; (4) pore in the coat of sponges through which gemmules escape.

micropyle apparatus raised processes or porches, sometimes of elaborate structure, developed round the micropyle of some insect eggs.

microRNAs (miRNAs) large family of small regulatory ribonucleic acids 21–25 nucleotides long in animals and plants, which repress gene expression through interactions with specific messenger RNAs (mRNAs), leading either to repression of translation or actual destruction of the mRNA.

microsatellite DNA *n.* repetitive DNA based on a dinucleotide sequence repeated in series many times. The numbers of repeats at each microsatellite locus are very variable between individuals, and microsatellite DNA can be used as a marker in genetic mapping and for DNA fingerprinting.

microsaurs *n.plu.* extinct order of lepospondyl amphibians having an elongated or reptilian shape and well-developed limbs each with four or fewer fingers.

microschizogony *n.* schizogony (fission into several new cells) resulting in small merozoites, as of some protozoans.

microschizont *n.* a cell functioning as a male gamete, produced by schizogony, in some protozoans.

microsclere *n.* one of numerous small spicules scattered in tissues of sponges.

microsclerotium *n.* a microscopic sclerotium.

microscope *n.* instrument that gives an enlarged image of the object under study. *see* compound microscope, electron microscope, scanning electron microscope, scanning-tunnelling electron microscope, transmission electron microscope, tunnelling electron microscope, ultraviolet microscope. *see also* microscopy.

microscopy *n.* the study of objects using a microscope. *see* bright-field microscopy, dark-field microscopy, fluorescence microscopy, immunoelectron microscopy, immunofluorescence microscopy, light microscopy, phase-contrast microscopy.

microseptum micromesentery *q.v.*

microsere *n.* a successional series of plant communities in a microhabitat.

microsmatic *a.* with a feebly developed sense of smell.

microsomal *a.* (1) *appl.* enzymes associated with the endoplasmic reticulum and which spin down in the microsomal faction on high-speed centrifugation of a cell homogenate; (2) *appl.* enzymes associated with peroxisomes. *see* microsomes.

microsomes *n.plu.* (1) the smallest sized particles spun down from cell homogenates by centrifugation at high speed (60,000*g*). They are chiefly ribosomes and small membrane-bounded vesicles, including those formed from fragmented endoplasmic reticulum membrane; (2) name formerly used by cytologists for any small granules in the cytoplasm, now usually restricted to peroxisomes; (3) aggregations of ribosomes.

microsomia *n.* dwarfism.

microsorus *n.* a sorus containing microsporangia.

microspecies *n.* a taxonomic unit below the species level, such as a race, subspecies or variety.

microsphere *n.* structure formed by heating polypeptides, which can absorb various organic molecules from aqueous solution.

microspheric *a. appl.* foraminiferans when the initial chamber of the shell is small.

microspike *n.* hair-like cytoplasmic extension put out by animal cells in tissue culture as they settle on to a surface or move over it.

microsplanchnic *a.* small-bodied and long-legged.

microsporangiate *a.* composed of or producing microsporangia, *appl.* cones.

microsporangium *n.* (1) sporangium containing a number of microspores; (2) pollen sac in spermatophytes. *plu.* **microsporangia**.

microspore *n.* (1) in heterosporous plants, the spore that gives rise to the microgametophyte (the male gametophyte), and which is formed in a microsporangium; (2) in any organism that produces two types of spore, the smaller spore.

Microsporida, microsporidians *n.*, *n.plu.* group of obligately parasitic eukaryotic microorganisms, formerly classified as protozoa, that have very small genomes and lack mitochondria. One known human parasite, *Encephalitozoon cuniculi. alt.* Archezoa.

microsporocyte *n.* microspore mother cell, which undergoes meiosis to produce haploid microspores.

microsporogenesis *n.* formation of microspores within the microsporangia, pollen sacs, of the anther.

microsporophore microsporangium *q.v.*

microsporophyll *n.* small leaf or leaf-like structure on which a microsporangium develops.

microsporozoite *n.* a smaller endogenous sporozoite of sporozoan parasites.

microstome *n.* a small opening or orifice.

microstrobilus *n.* a male cone composed of microsporophylls, as in gymnosperms.

microstylous *a.* having short styles, *appl.* heterostylous flowers.

microsymbiont *n.* the smaller of two symbiotic organisms.

microteliospore *n.* a spore produced in a microtelium.

microtelium *n.* sorus of microcyclic rust fungi.

microtherm *n.* a plant of the cold temperate zone, which can grow below 12°C.

microthorax *n.* a term applied to the cervix or neck in insects when it is thought to represent a reduced prothorax.

microtitre plate plastic slab containing rows of small wells in which the samples to be tested are placed.

microtome *n.* machine with a sharp metal blade for slicing tissue into sections for microscopy.

microtomy *n.* the cutting of thin sections of tissues or other material, for examination by microscopy.

microtrichia *n.plu.* small unjointed hairs in insect wings.

microtubule *n.* one of the three main structural components of the eukaryotic cell cytoskeleton. Microtubules are fine hollow protein tubes composed of molecules of the globular protein tubulin. As well as being involved in the intracellular transport of materials and movement of organelles, they form the mitotic and meiotic spindles. Microtubules also make up the cores of eukaryotic flagella and cilia and are responsible for their motility. Motor proteins associated with micro-tubules include dynein and kinesin.

microtubule-associated protein (MAP) any of various proteins found in close association with the microtubules of the cytoskeleton.

microtubule-organizing centre (MTOC) the structure within a eukaryotic cell which initiates microtubule formation, e.g. the centrosome of animal cells.

microvillus *n.* thin finger-like projection from epithelial cell surface, found esp. on intestinal and renal epithelium, where the microvilli collectively form a brush border. Individual microvilli are stiffened by an internal core of actin filaments. *plu.* **microvilli.**

microzooid *n.* free-swimming ciliated cell budded off by certain protozoans.

microzoospore *n.* a small motile spore.

mictic *a.* capable of reproducing by apomixis.

micton *n.* species resulting from interspe-cific hybridization and whose individuals are interfertile.

micturition *n.* urination.

mid-blastula transition the point in amphibian embryonic development at which the zygote's own genes become active.

midbody the remaining narrow strand of cytoplasm that joins two animal cells towards the end of cell division, before cleavage is completed. It contains the remains of the spindle.

midbrain *n.* that portion of the brain com-prising the corpora quadrigemina, cerebral peduncles and the aqueduct of Sylvius.

middle canal one of the three parallel canals in the cochlea of inner ear.

middle commissure grey matter connect-ing thalami across the 3rd ventricle.

middle ear cavity bounded by ear drum on one side, the bones of the skull and the bony wall of the inner ear on the other. It is spanned by a bridge formed from three small bones, the auditory ossicles, through which sound vibrations are conducted from eardrum to the oval window of inner ear.

middle lamella layer of material, largely pectin, that is present between two adjoin-ing plant cell walls.

middle shore zone of seashore between the average low-tide level and average high-tide level, which is usually the most extensive zone and is covered by the sea twice a day.

mid-grass community grassland contain-ing grasses of medium height, over 60 cm but under 2 m.

mid-gut mesenteron *q.v.*

midparent value for any given character (e.g. wing length in *Drosophila*), the aver-age of the values for the two parents.

midpiece *n.* the middle portion of a sperm, consisting of a portion of the flagellum sheathed by mitochondria.

midrib *n.* the large central vein of a leaf.

migration *n.* (1) seasonal journey to a dif-ferent location made by many animal species, usually in response to changes in seasonal climate and food supply, often travelling very long distances along prede-termined routes; (2) the movement of plants into new areas.

mildew *n.* fungal disease of plants mani-fested by a downy or powdery fungal coating on leaf or other affected parts. Mildews are caused by fungi from several different groups: powdery mildews are caused by ascomycetes, downy mildews by phycomycetes.

miliary *a.* (1) of granular appearance; (2) consisting of small and numerous grain-like parts.

milk *n.* liquid secreted from mammary glands, used by mammals to feed their young. It is rich in fat, protein and sugar (mainly lactose).

milk sugar lactose *q.v.*

milk teeth first dentition of mammals, shed after or before birth. *alt.* deciduous teeth.

mill chamber part of alimentary canal in some crustaceans, in which food is broken down by movement of chitinous plates.

milleporine *n. appl.* stony corals of the order Milleporina, which have colonies of two kinds of polyp living in pits on the surface of a massive calcareous skeleton (a corallium) and a brief medusoid stage.

millet *n.* small-grained plants of the mono-cot family Gramineae cultivated for their seed and including *Sorghum vulgare*, *Setaria italica*, *Pennisetum typhoideum* and *Panicum* species.

millimicron *n.* former term for nanometre, being one-thousandth of a micron (micrometre).

millipede *n.* common name for a member of the Diplopoda (*q.v.*).

milt *n.* testis or sperm of fishes.

mimesis *n.* (1) mimicry *q.v.*; (2) the effect of the actions of one animal of a group on the activity of the others.

mimetic *a. pert.* or exhibiting mimicry.

mimicry *n.* (1) the resemblance of one animal to an animal of a different species so that a third animal is deceived into confusing them; (2) the resemblance of an animal or plant to an inanimate object, or of an animal or part of an animal to a plant or part of a plant, usually for the purposes of camouflage. *a.* **mimetic.** *see also* Batesian mimicry, Müllerian mimicry.

mine *see* leaf mine.

mineralization *n.* breakdown of organic matter into inorganic compounds, carried out chiefly by decomposer microorganisms. For carbon it occurs chiefly during respiration, when carbon dioxide is returned to the environment.

mineralocorticoid *n.* steroid hormone secreted by the adrenal cortex which is involved in regulation of water and electrolyte balance in the body. An example is aldosterone.

mineralocorticoid receptor (MR) intracellular receptor protein for mineralocorticoid steroid hormones. When complexed with the hormone it acts as a transcription factor, regulating the expression of particular genes.

miniature end-plate potential (MEPP) electrical potential produced in muscle cell membrane by release of very small amounts of acetylcholine from an unstimulated nerve fibre.

minicell *n.* abnormally small bacterial cell produced by abnormal cell division.

minima minor worker *q.v.*

minimal medium culture medium containing a basic set of nutrients only, on which normal wild-type organisms can grow, but which cannot support the growth of metabolic mutants.

minimum inhibitory concentration (MIC) the smallest amount of an antimicrobial agent that is required to inhibit the growth of the test organism.

minimum, law of the law of the minimum *q.v.*

minimum lethal dose (MLD) minimum dose of any agent sufficient to cause 100% mortality in the test population.

minimus *n.* 5th digit of hand or foot.

minisatellite DNA *n.* DNA locus about 1–5 kb long, composed of tandem repeats of a short DNA sequence (15–100 nucleotides). The size of any given minisatellite locus is very variable within the population, and these loci can be used as markers for genetic mapping and DNA fingerprinting. *alt.* variable number tandem repeats (VNTRs).

minor elements trace elements *q.v.*

minor gene a gene which has a small effect individually but contributes to a multifactorial phenotypic trait.

minor groove the narrower of the two helical grooves in a double-helical DNA molecule.

minor histocompatibility antigen any of the many antigens encoded outside the major histocompatibility complex that also provoke rejection of transplanted tissues. They represent various proteins that can differ antigenically between genetically different members of the same species.

minor worker a member of the smallest worker subcaste, esp. in ants.

Miocene *n.* a geological epoch of the Tertiary sub-era, the earliest epoch of the Neogene period, between the Oligocene and Pliocene, lasting from *ca.* 23 million years ago to 5 million years ago. Continents had separated and were drifting towards their present positions. Grasslands extended in the generally cooler and drier climate, and mammals were highly diverse, with large numbers of grazing herbivores, including many extinct groups, such as the chalcitheres. Animals included mastodons, rhinos, deer, giraffes, horses, feline and canine carnivores, and anthropoid apes.

miracidium *n.* ciliated larval stage of gut, liver and blood flukes, which hatches out of the egg and infects the snail host.

mire *n.* bog or fen, usually referring to a peatland but also used to describe a fen developing on mineral soils. *see* blanket mire, raised mire, topogenous mire.

miRNAs microRNAs *q.v.*

MIS Müllerian inhibiting substance *q.v.*

mismatch repair type of DNA repair in which an incorrect mismatched nucleotide, or short sequence of mismatched nucleotides, is excised from DNA and the correct nucleotides inserted using the intact DNA strand as template.

mispairing *n.* the condition of having a base in one strand of DNA which does not pair correctly with its opposite number in the other strand of the double helix.

missense mutation mutation in which one base pair is altered in DNA and causes an amino acid change in the protein product of the gene.

missense suppressor mutant gene specifying a mutant tRNA with an altered anticodon that overcomes the effects of a missense mutation in another gene. It achieves this by inserting an acceptable amino acid during translation of the mutant gene.

Mississippian *a.* Lower Carboniferous in North America.

mitDNA mitochondrial DNA *q.v.*

mite *n.* common name for many members of the Acarina *q.v.*

miter mitra *q.v.*

mitochondria *plu.* of mitochondrion.

mitochondrial ATPase *see* ATP synthases.

mitochondrial cloud Balbiani body *q.v.*

mitochondrial DNA (mtDNA) the small DNA genome contained in a mitochondrion.

mitochondrial sheath an envelope containing mitochondria that forms a sheath around the base of sperm flagellum.

mitochondrion *n.* organelle that is the site of aerobic respiration and ATP generation in eukaryotic cells. It is surrounded by two membranes, the inner one much convoluted and carrying respiratory electron-transport chains and ATP synthases. The mitochondrial matrix enclosed by the inner membrane contains the enzymes of the tricarboxylic acid cycle. Mitochondria contain a small circular DNA that specifies tRNAs, rRNAs and some mitochondrial proteins. *plu.* **mitochondria.** *see* Fig. 7 (p. 105). *see also* chemiosmotic theory, endosymbiont hypothesis, oxidative phosphorylation, respiration.

mitochondriopathy *n.* disease caused by a defect in mitochondrial function.

mitogen *n.* any substance that causes mitosis and cell division.

mitogen-activated protein kinase MAP kinase *q.v. see also* MAP kinase cascade, MAP kinase kinase, MAP kinase kinase kinase.

mitogenic *a.* inducing mitosis and cell division.

mitomycin C antibiotic produced by *Streptomyces caespitosus* which inhibits nuclear division, DNA and protein synthesis in mammalian cells, and is used clinically as an anti-tumour agent.

mitosis *n.* the typical process of nuclear division in eukaryotic cells. It is preceded by chromosome replication. During mitosis, the two chromatids of each duplicated chromosome separate and are segregated into different daughter nuclei. Mitosis thus results in two identical daughter nuclei of identical genetic constitution to the parent cell. It is conventionally divided into the following stages: prophase, prometaphase, metaphase, anaphase and telophase. Mitosis is usually followed immediately by cell division. *see* Fig. 27 (p. 412). *cf.* meiosis.

mitosporangium *n.* thin-walled diploid zoosporangia that produces diploid zoospores.

mitospore *n.* uninucleate diploid zoospore produced by mitosis.

mitosporic *a. appl.* a fungus: its asexual form.

mitotic *a. pert.* or produced by mitosis.

mitotic chromosome chromosome as it appears at mitosis, as a compact, thickened, darkly staining rod.

mitotic index the number of cells per 1000 undergoing mitosis at the time of counting.

mitotic non-disjunction *see* non-disjunction.

mitotic recombination genetic recombination occurring between homologous chromosomes during mitosis in a somatic cell. If the cell is heterozygous for the genetic loci exchanged, a daughter cell of a different phenotype from that of the rest of the tissue may be produced.

mitotic spindle

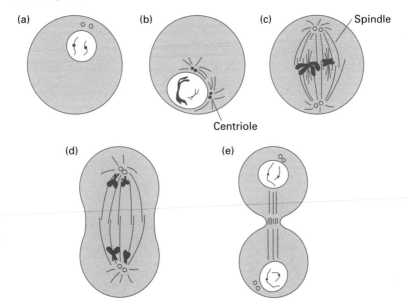

Fig. 27 Mitosis. (a) Diploid nucleus; (b) prophase; (c) metaphase; (d) anaphase; (e) telophase and cytokinesis. For clarity only two chromosomes are shown.

mitotic spindle *see* spindle.

mitra *n.* helmet-shaped piece of calyx or corolla.

mitral cell pyramidal neuron with thick basal dentrites, found in molecular layer of olfactory bulb of brain.

mitral valve valve of the left auriculoventral orifice of the heart. *alt.* bicuspid valve.

mitriform *a.* shaped like a mitre.

mixed-acid fermentation a basic fermentation pattern in enteric bacteria in which acetic, lactic and succinic acids are produced in significant amounts, along with other products such as ethanol and CO_2, but no butanediol. *cf.* 2,3-butanediol fermentation.

mixed forest, mixed woodland forest in which at least 20% of the trees are of species other than the dominant species.

mixed function oxygenase, mixed function oxidase monooxygenase *q.v.*

mixed lymphocyte response (MLR) the reaction seen when lymphocytes from two genetically distinct individuals are cultured together. Usually, the lymphocytes from one source are treated to prevent them undergoing cell division. The T lymphocytes from the untreated sample mount a reaction against the 'foreign' MHC molecules on the inactivated lymphocytes and proliferate.

mixed spinal nerves spinal nerves after union of the dorsal (afferent) and ventral (efferent) roots. They contain both motor and sensory nerve fibres.

mixed tissue tissue containing different cell types, all originating from the same group of founder cells.

mixipterygium *n.* clasper of male elasmobranch fishes, the medial lobe of the pelvic fin. *alt.* **mixopterygium**.

mixis *n.* sexual reproduction, esp. the fusion of gametes. *a.* **mictic**.

mixoploidy *n.* condition of having cells or tissues with different chromosome numbers in the same individual, as in a chimaera or a mosaic.

mixotroph *n.* organism that can use an inorganic compound as an energy source and an organic compound as a carbon source. *a.* **mixotrophic.**

MLD minimum lethal dose *q.v.*

MLV murine leukaemia virus, an oncogenic retrovirus.

MLR mixed lymphocyte response *q.v.*

MMTV mouse mammary tumour virus, an oncogenic retrovirus.

Mn symbol for the chemical element manganese *q.v.*

mnemonic *a. pert.* memory.

mnemotaxis *n.* movement directed by memory, e.g. returning to feeding place or home.

Mo symbol for the chemical element molybdenum *q.v.*

mobile genetic elements transposable genetic elements *q.v.*

mobile ion carrier a type of ionophore that makes a membrane permeable to ions by diffusing across the membrane carrying the ion.

modality *n.* the qualitative nature of a sense or stimulus, e.g. taste, smell, hearing and sight are different modalities of sensory experience.

modal number the most frequently occurring chromosome number in a taxonomic group.

mode *n.* in a distribution of values, the most frequently occurring value.

model organism name given to the relatively few organisms, such as the fruit fly *Drosophila melanogaster*, the budding yeast *Saccharomyces cerevisiae*, and the mouse, that have been studied experimentally over many years and for which a large body of knowledge about their genetics, development and physiology has accumulated.

moderate *a. appl.* animal viruses that can form a stable association with an infected cell, not destroying the cell, producing few if any virus particles, and being passed on from a cell to its progeny. *cf.* virulent.

moderator *n.* band of muscle checking excessive distention of right ventricle, as in the heart of some mammals.

modern humans the extant species *Homo sapiens*, to which all living humans belong, which originated around 200,000 years ago. *alt.* anatomically modern humans.

modification *n.* (1) in bacteria, the selective methylation of DNA (at cytosine or adenine residues) which protects the bacterium's DNA against degradation by its own restriction enzymes; (2) phenotypic change due to environment or use.

modifier *a.* (1) *appl.* a mutation that modifies the phenotypic effect of another mutation; (2) *appl.* gene that modifies the effect of a gene at a different locus; (3) *appl.* any factor that modifies the effect of another factor.

modulation *n.* (1) an alteration in a cell, produced by environmental stimuli, which does not change its essential character; (2) variation in the strength of a signal delivered to a cell, or its response to it, caused by some second stimulus – the modulator.

modulator *n.* (1) a band of the visible light spectrum, localized in the red–yellow, green and blue regions, which evokes colour sensation; (2) the component in a system which is responsible for maintaining a steady state by controlling specific reactions; (3) the agency that selects the appropriate route of transmission between receptor and effector; (4) any stimulus that varies the response of a cell or organ to a preceding stimulus.

molality *n.* a way of expressing the concentration of a solute in solution: a molal solution contains 1 mole of solute per kg of solvent.

molar *a.* (1) adapted for grinding, *appl.* teeth; (2) containing one gram-molecule or mole per litre, *appl.* solutions.

molarity *n.* a way of expressing the concentration of a solution of a solute: a molar solution (1 M) contains 1 mole of solute per litre of solution, a 5 M solution contains 5 moles of solute per litre.

mold *alt.* spelling of mould *q.v.*

mole (mol) the SI unit for the amount of a substance that contains as many elementary units (e.g. atoms, molecules, ions as appropriate) as there are atoms in 0.012 kilograms of ^{12}C (6×10^{23}). For example, a mole is M grams of a substance where M is its numerical molecular mass (for a molecule) or atomic mass (for an element). Thus 1 mole of glucose = 180 grams of glucose, 1 mole of carbon = 12 grams of carbon.

molecular *a. pert.*, or composed of, molecules.

molecular biology study of biological phenomena at the molecular level.

molecular chaperone chaperone *q.v.*

molecular clock a way of measuring the time elapsed since the divergence of two present-day lineages from their common ancestor by comparing corresponding DNA sequences from two extant species and counting the differences that have accumulated between them. Over long periods of time the rate of certain types of unselected nucleotide change appears to be directly proportional to time elapsed, if measurements are restricted to appropriate genes and closely related lineages. Such clocks may be calibrated in real time by comparison of sequences from species whose time of divergence is well established from the fossil record. *alt.* evolutionary clock.

molecular cloning DNA cloning *q.v.*

molecular evolution the changes that occur in DNA and in proteins as a result of mutation over long periods of time, which may alter function, eventually giving rise to novel genes and proteins.

molecular genetics study of the molecular structure of DNA, the structure and arrangement of genes, and the biochemical basis of gene expression and its regulation.

molecular layer external layer of cortex and cerebellum. *alt.* plexiform layer.

molecular mass sum of the atomic masses of all the atoms in a molecule. It is generally expressed in daltons or kilodaltons (1 kDa = 1000 daltons), where 1 dalton = 1.000 on the atomic mass scale, a unit of mass very nearly equal to the mass of a hydrogen atom. The term molecular weight, although not strictly equivalent, is often used as a synonym. *see also* relative molecular mass.

molecular mimicry the induction of antibodies and T cells against infectious agents that also cross-react with the body's own antigens.

molecular motor motor protein *q.v.*

molecular phylogeny the tracing of evolutionary relationships by the compar-

ison of DNA and protein sequences from different organisms.

molecular taxonomy the use of molecular characteristics (e.g. DNA and protein sequences, the presence of particular chemical compounds) to construct classifications of organisms.

molecular weight *see* molecular mass, relative molecular mass.

molecule *n.* group of atoms held together by covalent bonds.

Mollicutes *n.* in some classifications a class of the Bacteria that includes the rickettsiae, chlamydiae and mycoplasmas, which all lack cell walls and are intracellular parasites.

Moll's glands modified sweat glands between follicles of eyelashes. *alt.* ciliary glands.

Mollusca, molluscs *n., n.plu.* large and diverse phylum of soft-bodied, usually unsegmented, coelomate animals, many of which live enclosed in a hard shell. They include the classes Gastropoda (e.g. winkles, whelks, slugs, snails, sea slugs), Bivalvia (e.g. clams, cockles), and other smaller classes of shells, and the Cephalopoda (nautilus, squids and octopuses). The coelom is small and the main body cavity is a blood-filled haemocoel. Molluscs have a heart and blood system, and well-developed sense organs and nervous system, esp. in the Cephalopoda.

molluscicide *n.* chemical that kills molluscs, e.g. snails.

molluscoid *a. pert.* or resembling a mollusc.

mollusk *alt.* (US) spelling of mollusc *q.v.*

molt *alt.* spelling of moult *q.v.*

molten globule the stage in the folding of a globular protein when the protein has adopted its general conformation and main elements of secondary structure but is still more open and less ordered than the final native state.

moltinism *n.* the condition in which different strains of a species undergo a different number of larval moults.

molybdenum (Mo) *n.* chemical element which is an essential micronutrient for plants. It is a prosthetic group in certain enzymes including bacterial nitrogenases.

monacanthid *a.* with one row of ambulacral spines, as some starfishes.

monactinal *a.* with a single ray, *appl.* spicules of sponges.

monactinellid *a. appl.* certain sponges that bear spicules with only a single ray.

monad *n.* (1) single-celled organism or flagellate cell; (2) a single cell, instead of a tetrad, arising from meiosis.

monadelphous *a.* having stamens united into one bundle by union of filaments.

monamniotic *a.* having one amnion, *appl.* uniovular (identical) twins.

monandrous *a.* (1) having only one stamen; (2) having only one antheridium; (3) having only one male mate.

monanthous *a.* having only one flower.

monarch *a.* with only one protoxylem strand, or only one vascular bundle, *appl.* plant stem.

monaural *a. pert.* hearing through one ear only.

monaxial *a.* (1) having one line of axis; (2) having inflorescence developed on the primary axis.

monaxonic *a.* with one axon, *appl.* nerve cell.

monecious *alt.* spelling of monoecious.

monembryonic *a.* producing one embryo at a time.

Monera old name for the prokaryotes *q.v.*

monestrous *alt.* spelling of monoestrus.

Monilia *n.* large form class of deuteromycete fungi that reproduce by oidia or by budding, or by conidia not borne in pycnidia or acervuli. It includes *Penicillium* and *Aspergillum*, the false yeasts (e.g. *Cryptococcus*), and fungi that cause skin diseases in humans and animals (e.g. ringworm, *Microsporum*) and the serious human fungal pathogens *Blastomyces* and *Histoplasma*.

monilicorn *a.* having antennae with appearance like a chain of beads.

moniliform *a.* (1) arranged like a chain of beads, *appl.* spores, hyphae, antennae; (2) constricted at regular intervals, so appearing like a chain of beads.

monimostylic *a.* having quadrate united to squamosal, and sometimes to other bones, as in some reptiles.

monkeys *see* Primates.

mono- prefix derived from Gk *monos*, single, signifying one, having one of, borne singly.

monoallelic *a. appl.* a polyploid in which all the alleles at a locus are the same.

monoamine neurotransmitters dopamine, adrenaline, noradrenaline (*all q.v.*).

monoamine oxidase enzyme that inactivates the catecholamine neurotransmitters dopamine, adrenaline and noradrenaline by oxidative removal of the amino group. *alt.* **monoaminooxidase**.

monoblast *n.* cell that develops into a monocyte.

monoblastic *a. appl.* embryos with a single undifferentiated germ layer.

monocardian *a.* having one auricle and ventricle.

monocarpellary *a.* containing or consisting of a single carpel.

monocarpic *a. appl.* plants that die after bearing fruit once. *n.* **monocarp**.

monocarpous *a.* having only one carpel to a gynoecium.

monocaryon monokaryon *q.v.*

monocentric *a.* (1) having a single centromere, *appl.* a chromosome; (2) having a single centre of growth and reproduction, *appl.* fungal thallus. *cf.* polycentric.

monocephalous *a.* with one capitulum only, *appl.* flowers.

monocercous uniflagellate *q.v.*

monocerous *a.* having one horn.

monochasium *n.* a branched flowerhead with main axes producing one branch each.

monochlamydeous *a. appl.* flowers having a calyx but no corolla, or having only one whorl of perianth segments.

monochorionic *a.* having only one chorion, *appl.* uniovular twins.

monochromatic *a.* (1) having only one colour; (2) colour blind, seeing variations in brightness but no colours.

monochronic *a.* happening or originating only once.

monocistronic *a. appl.* mRNA coding for only one polypeptide chain, i.e. some bacterial and all eukaryotic mRNAs.

monoclinous *a.* (1) hermaphrodite; (2) having stamens and pistil in each flower; (3) having antheridium and oogonium originating from the same hypha.

monoclonal antibody antibody produced by a single clone of B cells and thus consisting of a population of identical antibody molecules all specific for the same antigenic determinant.

monocolpate *a. appl.* pollen grains with one groove, through which the pollen tube emerges.

monocondylar *a.* having a single occipital condyle (point of articulation of skull with spinal column), as in birds and reptiles.

monocots monocotyledons *q.v.*

Monocotyledon(e)ae, Monocotyledones, monocotyledons *n., n.plu., n.plu.* class of angiosperms having an embryo with only one cotyledon, parts of the flower usually in threes, leaves with parallel veins, and vascular bundles scattered throughout the stem. They include familiar bulbs such as daffodils, snowdrops and lilies, and the cereals and grasses such as maize, wheat and rice. *see* Appendix 1.

monocotyledonous *a.* (1) *pert.* monocotyledons; (2) *appl.* embryo with only one cotyledon.

monocratic *a.* with the four spores of a meiotic tetrad being of the same sex.

monocular *a. pert.* one eye only.

monocule *n.* a one-eyed animal, as certain insects and crustaceans.

monoculture *n.* large area covered by a single species (or for crops, a single variety) of plant, esp. if grown year after year.

monocyclic *a.* (1) having a single cycle; (*bot.*) (2) *appl.* annual herbaceous plants; (3) *appl.* flowers with a single whorl.

monocystic *a.* with one stage of encystation.

monocyte *n.* large phagocytic white blood cell with a single oval or horseshoe-shaped nucleus. It enters tissues, where it becomes a macrophage. *alt.* mononuclear leukocyte.

monodactylous *a.* with a single digit, or a single claw.

Monodelphia Eutheria. *see* eutherians.

monodelphic *a.* (1) having uteri more or less united, as in placental mammals; (2) having a single uterus, as *appl.* certain nematodes.

monodelphous monadelphous *q.v.*

monodesmic *a.* (1) (*zool.*) *appl.* scales formed of fused small bony scales with a continuous covering of dentine; (2) (*bot.*) having a single vascular bundle.

monodont *a.* having a single persistent tooth, as in male narwhal.

monoecious *a.* (1) having male and female flowers on the same plant; (2) with male and female sex organs on same gametophyte; (3) having microsporangia and megasporangia on same sporophyte.

monoestrous *a.* having only one period of oestrus in a sexual season. *cf.* polyoestrous.

monofactorial unifactorial *q.v.*

monogalactosyl diacylglycerol glycolipid found in plant cell membranes.

monogamous *a.* consorting with one mate only, usually for the whole of the animal's lifetime. *n.* **monogamy.**

monoganglionic *a.* having a single ganglion.

monogastric *a.* with one gastric cavity.

Monogenea, monogeneans *n., n.plu.* class of parasitic flatworms consisting of the skin and gill flukes, ectoparasites mainly of fish and amphibians. They have a flattened leaf-shaped body and a simple life cycle on one host.

monogenetic *a.* (1) *appl.* parasites completing their life cycle in a single host; (2) *appl.* origin of a new form at a single place or period. *n.* **monogenesis.**

monogenic *a.* (1) controlled by a single gene; (2) producing offspring all of the same sex.

monogenomic *a.* having a single set of chromosomes.

monogenous *a.* asexual, as *appl.* reproduction.

monogeny *n.* production of offspring consisting of one sex, either male or female. *a.* **monogenic.**

monogoneutic *a.* breeding once a year.

monogony *n.* asexual reproduction, including schizogony and budding.

monogynoecial *a.* developing from one pistil.

monogynous *a.* (1) having one pistil; (2) having one carpel to a gynoecium; (3) consorting with only one female.

monogyny *n.* (1) the tendency of a male to mate with only one female; (2) in social insects, the existence of a single functional queen in the colony.

monohybrid (1) *n.* hybrid offspring of parents that differed in one character; (2) *a.* heterozygous at a single gene locus.

monohybrid cross the intercrossing of the identical heterozygous F1 individuals (monohybrids) resulting from the cross of pure-breeding parents differing in only one character.

monokaryon *n.* cells of a hypha containing one nucleus.

monokont *a.* having a single flagellum.

monolayer *n.* a single homogeneous layer of units, as of molecules or cells.

monolepsis *n.* transmission of characteristics from only one parent to progeny. *a.* **monoleptic**.

monomastigote *a.* having one flagellum, as in certain protists.

monomeniscous *a.* having an eye with one lens.

monomer *n.* (1) molecule which is the repeating unit of a polymer, e.g. amino acids are the monomers in proteins; (2) molecule consisting of a single unit, e.g. a protein consisting of one polypeptide chain.

monomeric *a.* (1) *pert.* one segment; (2) derived from one part; (3) *appl.* protein molecule composed of a single polypeptide chain.

monomerosomatous *a.* having body segments all fused together, as in certain arthropods.

monomerous *a.* consisting of one part only, *appl.* flowers.

monometrosis *n.* colony foundation by one female, as by the queen in some social hymenopterans.

monomolecular layer monolayer *q.v.*

monomorphic *a.* (1) *appl.* species in which all individuals look alike; (2) developing with no or very slight change from stage to stage, as certain protozoans and insects; (3) producing spores of one kind only. *cf.* dimorphic, polymorphic.

monomorphic locus a gene locus for which the most common homozygote has a frequency of more than 90% in the population.

monomorphism *n.* in entomology, the existence of only a single worker subcaste within an insect species or colony.

mononemic *a.* consisting of one strand.

mononeuronic *a.* (1) with one nerve; (2) *appl.* chromatophores with single type of innervation.

mononuclear, mononucleate *a.* with a single nucleus.

mononuclear cells of the blood, the monocytes and lymphocytes.

mononychous *a.* having a single or uncleft claw.

mononym *n.* (1) designation consisting of one term only; (2) name of a monotypic genus.

monooxygenase *n.* any of a class of oxidoreductase enzymes which catalyse oxidation reactions in which one atom of oxygen is incorporated into the product and one is reduced to form water. *alt.* mixed-function oxygenase, mixed-function oxidase.

monopectinate *a.* having one margin furnished with teeth like a comb.

monopetalous *a.* (1) having one petal only; (2) having petals united all round.

monophagous *a.* subsisting on one kind of food, *appl.* insects feeding on plants of one genus only, or insects restricted to one species or variety of food plant.

monophasic *a.* *appl.* the shortened life-cycle of some trypanosomes, lacking the active stage.

monophenol monooxygenase polyphenol oxidase *q.v.*

monophyletic *a.* derived from a common ancestor, *appl.* taxa derived from and including a single founder species.

monophyllous *a.* (1) having one leaf only; (2) having calyx in one piece.

monophyodont *a.* having only one set of teeth, the milk teeth being absorbed in the foetus or being absent altogether.

monoplacid *a.* with one plate only.

Monoplacophora *n.* a mainly extinct class of molluscs with a shell like a limpet, the living forms (e.g. *Neopilina*) being known only from the deep-sea bed.

monoplanetic *a.* with one stage of motility in life history, *appl.* certain fungi.

monoplastic *a.* (1) persisting in one form; (2) having one chloroplast, *appl.* cells.

monoploid *a.* (1) having one set of chromosomes, true haploid; (2) in a polyploid series, having the basic haploid chromosome number.

monopodal *a.* (1) having one supporting structure; (2) having one pseudopodium.

monopodial *a.* branching from one primary axis with the youngest branches arising at the apex.

monopodium *n.* a single main or primary axis from which all main lateral branches develop.

monopolar *a. appl.* neuron with a single branch, an axon, that divides into two and extends in two directions.

monopyrenic, monopyrenous *a.* single-stoned, as of fruit.

monorchic *a.* having one testis.

monorefringent isotropic *q.v.*

monorhinal *a.* (1) having only one nostril; (2) *pert.* one nostril.

monosaccate *a. appl.* pollen grains with one air bladder.

monosaccharide *n.* any of a class of simple carbohydrates, all being reducing sugars, with the general formula $(CH_2O)_n$, where n is greater than 3. The simplest monosaccharides are the trioses (e.g. glyceraldehyde, $n = 3$). Others include the pentoses (e.g. ribose, $n = 4$), and the hexoses (e.g. glucose and fructose, $n = 6$).

monosepalous *a.* having one sepal.

monosiphonic, monosiphonous *a. appl.* algae having a single central tube in filament.

monosomic *a.* (1) *appl.* an unpaired X chromosome or a chromosome lacking its homologous partner; (2) *appl.* diploid cells or organisms in which one partner of a pair of homologous or sex chromosomes has been lost, e.g. XO cells are monosomic for X.

monosomy *n.* the absence of a single chromosome from the diploid set.

monospecific *a.* (1) having only one species, *appl.* genus, family or other taxonomic group; (2) *appl.* antibody reacting with only one antigenic determinant.

monospermic, monospermous *a.* (1) one-seeded; (2) fertilized by entrance of only one sperm into ovum.

monospermy *n.* fertilization normally achieved by penetration of one sperm into the ovum.

monospondylic *a. appl.* vertebrae with only one vertebral ring or central hole.

monospore *n.* a simple or undivided spore.

monosporic *a.* originating from a single spore.

monosporous *a.* having only a single spore.

monostachyous *a.* having only one spike.

monostele protostele *q.v.*

monostichous *a.* (1) arranged in a single row; (2) along one side of an axis.

monostigmatous *a.* with one stigma only.

monostromatic *a.* having a single-layered thallus, *appl.* algae.

monostylous *a.* having one style only.

monosulcate *a. appl.* pollen grains with a single groove on the surface away from that through which pollen tube emerges.

monosy *n.* separation of parts normally fused.

monosymmetrical zygomorphic *q.v.*

monotaxic *a.* belonging to the same taxonomic group.

monothalamous *a.* (1) with a single chamber or locule; (2) *appl.* fruits formed from single flowers.

monothecal *a.* single-chambered.

monothetic *a. appl.* classification based on only one or a few characteristics, such as a classification of plants based on the number of stamens. *cf.* polythetic.

monotokous *a.* uniparous, having one offspring at birth. *alt.* **monotocous.**

Monotremata, monotremes *n.*, *n.plu.* order of primitive mammals that lay eggs, have mammary glands without nipples, and no external ears. The only extant species are the duck-billed platypus of Australia (*Ornithorhynchus*) and the spiny anteaters (*Tachyglossus* and *Zaglossus*), which are found in Australia and New Guinea.

monotrichous *a.* having a single polar flagellum.

monotrochous *a.* having a trochanter in a single piece, as in most stinging Hymenoptera.

monotrophic *a.* subsisting on one kind of food.

monotropic *a.* (1) turning in one direction only; (2) visiting only one kind of flower, *appl.* insects.

monotropoid *a. appl.* mycorrhizas formed on members of the Monotropaceae, plants lacking chlorophyll and which are dependent on the mycorrhiza for their carbon and energy source. An extensive root-ball of fungal and root tissue is formed, which also forms connections with the ectomycorrhizal roots of nearby green plants.

monotype *n.* single type which constitutes species or genus.

monotypic *a.* (1) *appl.* genera having only one species; (2) *appl.* species having no subspecies.

monovalent univalent *q.v.*

monovoltine univoltine *q.v.*

monovular uniovular *q.v.*

monoxenic, monoxenous *A.* inhabiting one host only, *appl.* parasites.

monoxylic *a.* having wood formed as a continuous ring.

monozoic *a.* producing one sporozoite only.

monozygotic *a.* originating from a single fertilized ovum (zygote), as identical twins (MZ twins). *cf.* dizygotic.

Monro, foramen of, Monro's foramen foramen of Monro *q.v.*

monsoon rain forest type of rain forest that develops in tropical and subtropical regions with a high annual rainfall but marked dry and rainy seasons (monsoon rainfall). It consists of deciduous trees and shrubs that lose their leaves in the dry season. *alt.* monsoon forest.

mons pubis, mons Veneris prominence of fatty subcutaneous tissue in front of pubic bone.

montane *a. pert.* mountains.

monticolous *a.* inhabiting mountainous regions.

monticulus *n.* largest part of the superior vermis of cerebellum.

moor *n.* open area of upland acid peat, with vegetation of heathers, sedges and certain grasses (e.g. *Molinia, Caerulea*).

mor *n.* acid humus of cold wet soils which inhibits action of soil organisms and may form peat. *cf.* mull.

morbidity *n.* illness.

morbilliviruses *n.* group of RNA viruses of the paramyxovirus family, related to canine distemper virus.

mores *n.plu.* groups of organisms preferring the same habitat, having the same reproductive season, and agreeing in their general reactions to the physical environment.

morgan *n.* a unit of distance on a genetic map, in which the mean number of recombinations is 1, named after the geneticist Thomas Hunt Morgan. *see also* centimorgan.

moriform *a.* (1) shaped like a mulberry; (2) formed in a cluster resembling an aggregate fruit.

Morisita's index of dispersion (I) a number that represents the spatial distribution (dispersion) of individuals within a population.

Morisita's similarity index (C$_v$) an index of the similarity (*see* coefficient of community) between two communities that is weighted by the population sizes of species present in the two communities. *cf.* Gleason's index, Jaccard index, Kulezinski index, Simpson's index of floristic similarity, Sørensen similarity index.

morph *n.* one of the forms present in a polymorphic population.

morphactins *n.plu.* group of substances derived from fluorine-9-carboxylic acids, which affect plant growth and development.

morphallaxis *n.* (1) type of regeneration that involves repatterning of existing tissue without new growth; (2) gradual growth or development into a particular form. *a.* **morphallactic.**

morphine *n.* the chief alkaloid of opium, used clinically to relieve pain, but produces dependency on long-term use.

morphogen *n.* any substance or other agent (e.g. gravity) which specifically influences pattern formation and morphogenesis rather than simply stimulating non-specific features of development such as growth.

morphogenesis *n.* (1) the development of shape and structure; (2) origin and development of organs or parts of organisms. *alt.* **morphogeny.**

morphogenetic *a.* (1) *pert.* or inducing morphogenesis; (2) *appl.* hormones: hormones that influence growth, development and/or metamorphosis of organisms. Examples are thyroxine, ecdysone and juvenile hormone.

morphogenetic field equivalence group *q.v.*

morphogenetic furrow a furrow that moves over the eye imaginal disc in certain larval insects. Cells through which the furrow has passed begin to differentiate into prospective ommatidia of the adult compound eye.

morpholino-RNA a stable synthetic RNA analogue that can be used instead of natural RNA in antisense experiments.

morphologic, morphological *a. pert.* the form and structure of an organism.

morphological index ratio expressing the relative sizes of trunk and limbs.

morphological species *see* morphospecies.

morphology *n.* (1) the form and structure of an organism; (2) the study of form and structure.

morphoplankton *n.* plankton organisms rendered buoyant by small size, body shape, or structures containing oily globules, mucilage or gas.

morphosis *n.* (1) the manner of development of part of an organism; (2) the formation of tissues. *a.* **morphotic**.

morphospecies *n.* group of individuals which are considered to belong to the same species on grounds of morphology alone.

morphotype *n.* type specimen of one of the forms of a polymorphic species.

mortality *n.* death.

morula *n.* solid globular mass of cells, the product of the first rounds of cell division of a fertilized egg, the stage preceding the blastula or blastocyst.

mosaic *n.* (1) disease of plants characterized by mottling of leaves and caused by various viruses, e.g. tobacco mosaic, cucumber mosaic; (2) organism whose body cells are a mixture of two or more different genotypes. Examples are human and other mammalian females, who have one of their X chromosomes inactivated at random early in development. Their adult tissues therefore usually contain a mixture of cells containing different active X chromosomes.

mosaic eggs eggs whose development appears to be directed largely by cytoplasmic determinants localized to different parts of the egg. The blastomeres formed by early cleavage develop to a large extent independently of each other, and are unable to substitute for each other if one is removed.

mosaicism *n.* the condition of being a mosaic *q.v.*

moschate *a.* having or resembling the odour of musk.

moss *n.* common name for a member of the Bryophyta, a division of non-vascular, spore-bearing green plants. The mosses are divided into three classes: the 'true' mosses, the sphagnum mosses, and the granite mosses. Several plants commonly called mosses belong to other groups: reindeer moss is a lichen, club mosses and Spanish moss are vascular plants, sea moss and Irish moss are algae.

moss animals common name for the Bryozoa *q.v.*

moss layer the lowest horizontal ecological stratum of a plant community, comprising the ground surface and its immediate plant cover such as mosses and lichens. *alt.* ground layer.

mossy fibres nerve fibres branching profusely around cells of the internal layer of the cerebellar cortex.

moth *n.* common name for a member of the Lepidoptera having the antennae tapering to a point and not clubbed.

mother cell cell which gives rise to other cells by division.

motif *n.* (1) in DNA or RNA, a distinctive sequence or pattern of nucleotides that is characteristic of a particular class of genes or regulatory elements; (2) in proteins, a distinctive sequence or pattern of amino acids, or an element of structure, that is characteristic of a particular class of proteins, or is shared with different proteins.

motility *n.* movement, or the capacity for movement.

motile *a.* capable of spontaneous movement.

motivation *n.* internal factors controlling behaviour in an animal, which lead to its achieving a goal or satisfying a need.

motivational state the combined effect of the physiological state of an animal and its perception of stimuli from the environment, which determines behaviour.

motoneuron motor neuron *q.v.*

motor *a. pert.* or connected with movement, *appl.* e.g. nerves.

motor areas areas of the brain where muscular activity is initiated and coordinated.

motor cortex area of cerebral cortex in brain concerned with initiation of voluntary movement.

motor end organ, motor end plate the structure formed where the axon of a motor neuron terminates on a skeletal muscle fibre.

motor neuron nerve cell that carries impulses away from the central nervous system (brain and spinal cord) to an effector muscle. *alt.* motoneuron. *see also* alpha motor neuron, gamma efferent motor neuron.

motor protein protein such as myosin, dynein and kinesin, which can move along microfilaments or microtubules. By their actions, motor proteins cause muscle contraction, transport of vesicles within the cell, cytoplasmic streaming, and the movement of eukaryotic cilia and flagella. They have an intrinsic ATPase activity that powers their movement.

motor unit a motor neuron and its associated muscle fibres.

mould *n*. common name for fungi that grow as a fluffy mycelium over a substrate and do not produce large macroscopic fruiting bodies.

moult *n*. the periodic shedding of outer covering, whether of feathers, hair, skin or cuticle. In crustaceans and arthropods, it is necessary during larval growth as the exoskeleton, once hardened, does not allow further internal growth.

moulting glands ecdysial glands *q.v.*

moulting hormone ecdysone *q.v.*

moultinism moltinism *q.v.*

mouthpart *a*. appendage around mouth of arthropods.

m.p. melting point.

MPF maturation-promoting factor *q.v.*

M phase the period during the cell cycle when mitosis and cell division occurs. The term is sometimes used to denote the phase of mitosis only. *see* Fig. 8 (p. 106).

M-phase-promoting factor maturation-promoting factor *q.v.*

MR mineralocorticoid receptor *q.v.*

MRI magnetic resonance imaging *q.v.*

mRNA messenger RNA *q.v.*

MRSA methicillin-resistant *Staphylococcus aureus*.

MS mass spectrometry *q.v.*

MS/MS *see* mass spectrometry.

MSC mesenchymal stem cell *q.v.*

MSH melanocyte-stimulating hormone *q.v.*

mtDNA mitochondrial DNA *q.v.*

MTOC microtubule-organizing centre *q.v.*

Mu DNA bacteriophage of *E. coli* which can insert into the bacterial genome in a manner analogous to transposons.

mucid *a*. mouldy or slimy.

muciferous body protrusible organelle of euglenoids and dinoflagellates.

mucific *a*. mucus-secreting.

mucigel *n*. gelatinous material on the surface of roots in soil, comprising a mixture of plant mucilages, bacterial capsules and slime layers and colloidal soil particles.

mucilage *n*. general term for complex substances composed of various types of polysaccharides, becoming viscous and slimy when wet, widely occurring in plants, and secreted by plant roots and by bacteria (the capsule or slime layer).

mucilaginous *a*. *pert.*, containing, or composed of mucilage.

mucin *n*. general term for glycoproteins found in secretions such as saliva and mucus. *alt.* mucoid.

muciparous *a*. mucus-secreting.

mucivorous *a*. feeding on plant juices, *appl.* insects.

mucocellulose *n*. cellulose mixed with various glycoproteins as in some seeds and fruit.

mucocutaneous *a*. *appl.* skin and mucous membranes.

mucoid mucin *q.v.*

mucoid, mucoidal *a*. (1) *appl.*, *pert.* or resembling mucus, *alt.* mucous; (2) *appl.* method of feeding used by some molluscs, which extrude mucus from the mouth and then re-ingest it with food particles attached to it; (3) *appl.* colonies of encapsulated bacteria, which have a viscous, sticky appearance and texture due to the large amount of carbohydrate capsule.

mucolipidosis *n*. hereditary disease resulting from secretion of lysosomal enzymes instead of their direction into lysosomes.

mucolytic *a*. breaking down mucus or mucilage.

mucopeptide peptidoglycan *q.v.*

mucopolysaccharide glycosaminoglycan *q.v.*

mucopolysaccharidosis *n*. disease characterized by massive accumulation of mucopolysaccharides (glycosaminoglycans) in lysosomes in cells.

mucoprotein *n*. (1) glycoprotein, esp. those found in mucous secretions; (2) proteoglycan *q.v.*

Mucorales *n*. order of zygomycete fungi with a well-developed mycelium and non-motile spores contained in a stalked sporangium, most living as saprophytes on dung or decaying plant and animal matter. It includes the bread moulds (e.g. *Mucor*, *Rhizopus*) and the dung fungus *Pilobolus*.

mucosa *n.* the wall of tubular structures such as gut, respiratory tract, urinary and genital tracts, which consists of several distinct layers of tissue: the epithelium, containing mucus-secreting cells and other glands, which is separated from the connective tissue lamina propria by a basement membrane, and the whole surrounded by a layer of smooth muscle, the tunica muscularis. *alt.* mucous membrane, tunica mucosa.

mucosa-associated lymphoid tissue (MALT) secondary lymphoid tissue found within the epithelium and connective tissue of the walls (mucosa) of gastrointestinal, respiratory and urinogenital tracts.

mucoserous *A.* secreting mucous and fluid.

mucous *A.* secreting, containing or *pert.* mucus.

mucous membrane any epithelial layer secreting mucus, e.g. the linings of the nasal pasages, reproductive tract, gut. The term is sometimes applied to the whole mucosa.

mucro *n.* (1) sharp point at termination of an organ or other structure; (2) small awn; (3) pointed keel or sterile third carpel, as in pine.

mucronate, mucroniferous *A.* abruptly terminated by a sharp spine.

mucronulate *A.* tipped with a small mucro.

mucus *n.* (1) slimy material rich in glycoproteins, secreted by goblet cells of mucous membranes or by mucous cells of a gland; (2) similar slimy secretion produced on the external body surface of many animals.

mull *n.* humus of well-aerated moist soils, formed by the action of soil organisms on plant debris and favouring plant growth.

Müller cells glial cells of retina.

Müller's fibres neuroglia forming a framework supporting the nervous tissue of the retina.

Müllerian ducts pair of ducts developing in the early vertebrate embryo, running alongside the Wolffian ducts, that will give rise to the oviducts in females.

Müllerian-inhibiting substance (MIS) hormone secreted by the developing testis in a male mammalian embryo which prevents further development of the Müllerian ducts and tisssues which would otherwise become the uterus, cervix and upper vagina. *alt.* anti-Müllerian duct hormone, Müllerian regression hormone.

Müllerian mimicry the resemblance of two animals to their mutual advantage, for example the yellow and black stripes of wasps and of the unpleasant-tasting cinnabar moth caterpillars. This leads a predator that has encountered one subsequently also avoiding the other. *cf.* Batesian mimicry.

multi- prefix derived from L. *multus*, many.

multiarticulate *a.* many-jointed.

multiaxial *a.* (1) having or *pert.* several axes; (2) allowing movement in many planes.

multicamerate *a.* (1) with many chambers; (2) multiloculate.

multicarinate *a.* with many ridges or keels.

multicarpellary *a.* with compound gynoecium consisting of many carpets.

multicauline *a.* with many stems.

multicellular *a.* (1) many-celled, *appl.* eukaryotic organisms (e.g. plants and animals) consisting of large numbers of cells specialized for different functions and organized into a cooperative structure; (2) consisting of more than one cell. *n.* **multicellularity**.

multicentral *a.* with more than one centre of growth or development.

multiciliate *a.* with some or many cilia.

multicipital *a.* with many heads or branches arising from one point.

multicomponent viruses viruses in which different parts of the viral genome are packaged into separate virus particles.

multicopy plasmid plasmid present in a bacterial cell in several copies, the number being characteristic for different plasmids.

multicostate *a.* (1) with many ribs or veins, *appl.* leaves; (2) with many ridges.

multi-CSF interleukin-3, a glycoprotein cytokine that stimulates the growth *in vitro* of white blood cell precursors of all types.

multicuspid(ate) *a.* with several cusps, *appl.* molar teeth.

multidentate *a.* with many teeth or indentations.

multideterminant *a. appl.* antigen carrying more than one antigenic determinant.

multidigitate *a.* many-fingered.

multidomain proteins proteins composed of a single polypeptide with more than one distinct protein domain.

multidrug resistance simultaneous resistance to a variety of unrelated anti-cancer drugs shown by many tumour cells. It is due to the presence of transport proteins of the ABC superfamily that transport the drugs out of the cell.

multienzyme complex a complex of different enzyme molecules, usually all catalysing different steps in a particular pathway. Examples are the cellulosesynthesizing enzyme complex that is formed in the Golgi apparatus and delivered to the cell wall in plant cells, and pyruvate dehydrogenase, which is a complex of 44 molecules with three distinct enzymatic activities.

multifactorial *a. appl.* phenotypic traits. *see* polygenic.

multifactorial inheritance inheritance of phenotypic characters determined by the action of several independent genes.

multifarious polystichous *q.v.*

multifascicular *a.* containing or *pert.* many small bundles or fascicles.

multifid *a.* having many clefts or divisions.

multiflagellate *a.* having several or many flagella.

multiflorous *a.* having many flowers.

multifoliate, multifoliolate *a.* having many leaves or many small leaves respectively.

multiform *a.* (1) occurring in, or containing, different forms; (2) *appl.* layer: the inner cell layer of cerebral cortex.

multifunctional proteins proteins containing various enzymatic or other activities within a single molecule.

multigene family gene family *q.v.*

multigyrate *a.* intricately folded.

multijugate *a.* having many pairs of leaflets.

multilacunar *a.* (1) having many lacunae; (2) having a number of leaf gaps, *appl.* nodes.

multilaminate *a.* composed of several layers.

multilineage *a. appl.* cells capable of differentiating into a variety of different cell types.

multilobate, multilobulate *a.* with many lobes or lobules respectively.

multilocular, multiloculate *a.* many-chambered, *appl.* e.g. ovary (of flower), shells.

multimer *n.* (1) protein molecule made up of more than one polypeptide chain (protein subunit); (2) protein complex made up of several different protein molecules. *a.* **multimeric.**

multinervate *a.* with many nervures (of wing or leaf) or nerves.

multinodal, multinodate *a.* with many nodes.

multinomial *a. appl.* a name or designation composed of several terms.

multinucleate *a.* with several or many nuclei.

multinucleolate *a.* with several or many nucleoli.

multiovulate *a.* with several or many ovules.

multiparous *a.* (1) bearing several, or more than one, offspring at a birth; (2) (*bot.*) developing several or many lateral axes.

multipass *n. appl.* transmembrane proteins in which the protein chain threads back and forth several times across the lipid bilayer. *alt.* serpentine.

multipennate *a. appl.* muscle containing a number of extensions of its tendon of insertion.

multiperforate *a.* having more than one perforation.

multipinnate *a.* (1) having many leaflets; (2) with each leaflet or division divided pinnately, and then pinnately subdivided, *appl.* leaves.

multiple adaptive peaks case where there is more than one combination of allele frequencies (with respect to a given set of genes) that confer equal or similar fitness.

multiple alignment an alignment of more than two nucleotide or amino-acid sequences. *see* alignment.

multiple alleles the existence in the population of more than two alleles for a given gene locus. *alt.* genetic polymorphism.

multiple corolla a corolla with two or more whorls of petals.

multiple drug-resistant bacteria bacteria carrying resistance genes to a number of commonly used antibiotics.

multiple fission (1) repeated division; (2) division into a large number of parts or spores.

multiple fruit fruit developing from the gynoecia (carpels) of more than one flower, e.g. pineapple.

multiple myeloma a cancer of the bone, derived from plasma cells.

multiple sclerosis autoimmune disease caused by an immune response against the myelin sheath of nerve cells, which causes sclerotic plaques of demyelinated tissue in the white matter of the central nervous system.

multiplicate *a.* (1) consisting of many; (2) having many folds.

multipolar *a.* (1) *appl.* nerve cells with more than two main cellular processes (i.e. more than one main dendron and axon); (2) *appl.* mitosis in which more than two poles are formed, normal in certain sporozoan protozoans, but otherwise pathological.

multiporous *a.* having many pores.

multipotent *a.* capable of giving rise to a particular and limited set of different cell types, *appl.* adult stem cells such as mesenchymal stem cells and haematopoietic stem cells. *cf.* pluripotent.

multipotent marrow stromal cell mesenchymal stem cell *q.v.*

multiradiate *a.* many rayed, *appl.* spicules of certain sponges.

multiradicate *a.* with many roots or rootlets.

multiramose *a.* much branched.

multiseptate *a.* having numerous partitions.

multiserial, multiseriate *a.* (1) arranged in many rows; (2) *appl.* xylem rows more than one cell wide; (3) *appl.* spores in rows in ascus.

multispecificity *n.* capacity of certain antibody molecules to bind to a number of different antigens. *alt.* polyspecificity.

multistaminate *a.* having many stamens.

multisulcate *a.* much furrowed.

multitentaculate *a.* having numerous tentacles.

multituberculate *a.* having several or many small humps or protuberances.

multituberculates *n.plu.* class of extinct herbivorous mammals existing in the Jurassic to Eocene, with affinities to present-day monotremes.

multivalent *a.* (1) *appl.* antibodies with more than one antigen-binding site; (2) *appl.* antigens with more than one antigenic determinant; (3) *n.* structure formed by the association of more than two chromosomes during meiosis in polyploids.

multivalve *a.* *appl.* shell composed of more than two pieces or parts.

multivesicular body endosome at an early stage after endocytosis, when it contains large numbers of pinched-off membrane-bounded vesicles.

multivoltine *a.* having more than one brood in a year, *appl.* some birds.

multocular *a.* many-eyed.

multungulate *a.* with the hoof in more than two parts.

mune mores *q.v.*

mural *a.* (1) constituting or *pert.* a wall; (2) growing on a wall.

muralium *n.* structure formed by layers one cell thick, such as the internal structure of the liver.

muramic acid monosaccharide found in the peptidoglycans of cell walls of members of the Bacteria. It is an *N*-acetylaminoacid consisting of *N*-acetylglucosamine condensed with lactic acid at carbon 3 of the sugar. *alt.* N-acetylmuramic acid.

murein *n.* any of the peptidoglycans found in the cell walls of the Bacteria.

muricate *a.* (1) formed with sharp points; (2) covered with short sharp outgrowths; (3) studded with oxalic acid crystals.

muriform *a.* (1) (*bot.*) like a brick wall, *appl.* a parenchyma tissue arranged with cells in overlapping rows, in medullary ray of dicotyledons and in cork; (2) (*zool.*) shaped like a morula, *appl.* coelomocytes.

muscarine *n.* a ptomaine base, found in the fly agaric toadstool, *Amanita muscaria*, and other plants.

muscarinic *a.* *appl.* acetylcholine receptors stimulated by muscarine and similar drugs. *cf.* nicotinic.

Musci mosses *q.v.*

muscicoline *a.* living or growing among or on mosses.

muscimol *n.* hallucinogenic plant alkaloid, binds to GABA receptors in brain.

muscle *n.* contractile tissue in animals that is involved in movement of the organism and which also forms part of many internal organs. Muscle cells contain assemblages of protein microfibrils which contract simultaneously, usually in response to a nervous or chemical stimulus. There are three main types of muscle in vertebrates: (i) striated or striped muscle, which forms the muscles attached to the skeleton, (ii) the smooth muscle associated with many organs and which forms the contractile layer surrounding arteries and gut, and (iii) the cardiac muscle of the heart. *see* cardiac muscle, smooth muscle, striated muscle.

muscle cell any of various types of specialized contractile cell making up the muscular tissues of the body, the four main types in mammals being skeletal muscle cells, cardiac muscle cells, smooth muscle cells and myoepithelial cells. All contract by means of a contractile apparatus formed of the proteins actin and myosin. Skeletal muscle cells, also called muscle fibres, are often very large, roughly spindle shaped, multinucleate, and develop from the fusion of several immature cells. They contain orderly arrays of the proteins actin and myosin giving the cells a striated appearance. Heart (cardiac) muscle cells are smaller and uninucleate and also striated. Smooth muscle cells (found e.g. in the wall of the gut and blood vessels) are small spindle-shaped cells, unstriated and uninucleate. Myoepithelial cells are contractile epithelial cells that form e.g. the dilator muscle of the iris and the contractile tissue of some glands. *see* Fig. 28.

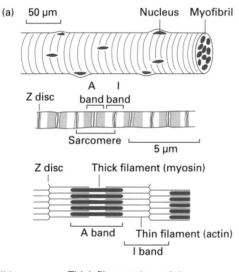

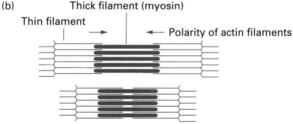

Fig. 28 (a) The structure of a skeletal muscle cell and (b) the mechanism of contraction.

muscle contraction *see* sliding filament mechanism.

muscle fibre *see* muscle cell.

muscle segment myomere *q.v.*

muscle spindle sensory receptor in muscle which monitors degree of stretch. It consists of a spindle-shaped connective tissue sheath containing small modified muscle fibres, with nerve fibres entering each spindle and forming spirals or arborizations around individual muscle fibres.

muscoid *a.* (1) moss-like; (2) mossy.

muscology *n.* study of mosses.

Muscopsida mosses *q.v.*

muscular *a. pert.* or consisting of muscle.

muscular dystrophy *see* Becker muscular dystrophy, Duchenne muscular dystrophy.

musculature *n.* the system or arrangement of muscles as a whole.

musculocutaneous *a. pert.* muscles and skin, *appl.* veins and nerves of limbs which supply muscles and skin.

musculo-epithelial cells cells in the gastrodermis of coelenterates, with a cell body and long contractile processes.

musculophrenic *a.* (1) supplying diaphragm and body wall muscles; (2) *appl.* artery: a branch of the internal mammary artery.

musculospiral *a.* (1) *appl.* radial nerve which passes spirally down humerus; (2) *appl.* spiral arrangement of muscle fibres.

musculotendinous *a. pert.* muscle and tendon.

mushroom *n.* common name for edible basidiomycete fungi, esp. of the genus *Agaricus*.

mushroom bodies corpora pedunculata *q.v.*

mushroom gland large seminal vesicles of certain insects.

mustelids *n.plu.* members of the family Mustelidae: weasels, stoats, badgers, otters, polecats, martens.

mutagen *n.* any agent that can cause a mutation. *a.* **mutagenic**.

mutagenesis *n.* the production of mutations by agents such as X-rays or chemicals. *see also* cassette mutagenesis, site-directed mutagenesis.

mutagenic *a.* capable of causing a mutation, *appl.* radiation, chemicals or other agents. *n.* **mutagenicity**.

mutagenize *v.* to treat with a mutagen.

mutant *n.* organism that differs from its parent or immediate precursor cell in some physical or biochemical characteristic(s) because of an alteration (mutation) in the genetic material. *a.* **mutant**.

mutase *n.* any of a group of enzymes that catalyse the intramolecular shift of a chemical group and which are classified amongst the isomerases in EC class 5.

mutate *v.* to undergo mutation.

mutation *n.* a change in the amount or chemical structure of DNA resulting in a change in the characteristics of an organism or an individual cell. The changes are due to alterations in, or non-production of, proteins (or RNAs) specified by the altered DNA. Mutations occurring in body cells of multicellular organisms are called somatic mutations and are only passed on to the immediate descendants of those cells. Mutations occurring in germline cells can be inherited by the offspring. Mutations can occur spontaneously as a result of errors in normal cell processes, e.g. DNA replication, or can be induced by certain chemicals or radiation. Alterations in DNA that do not cause any phenotypic change are also sometimes called mutations (silent mutations or neutral mutations). *see also* back mutation, base substitution, deletion, frameshift, insertion, neutral mutation, point mutation, translocation, transposition, transversion, revertant, silent mutation, wild type.

mutational load the reduction in population fitness due to the accumulation of deleterious mutations.

mutation breeding the use of induced mutations to provide additional variation in plant breeding programmes.

mutation event the actual occurrence of a mutation.

mutation frequency the proportion of individuals within a population that carry a particular mutation. *cf.* mutation rate.

mutation pressure changes in gene frequencies in a population that are brought about by mutational change alone.

mutation rate the number of mutations occurring in a population over some specified unit of time. The units used as the time denominator are commonly the total number of cell divisions or the total generation spans of the cells or organisms. The spontaneous mutation rate seems to vary between different organisms and between different genes. *cf.* mutation frequency.

mutator *a.* (1) *appl.* genes that increase the general mutation rate; (2) *appl.* phage that causes mutations in host DNA when it inserts into the genome.

mutualism *n.* a special case of symbiosis in which both partners benefit from the association. *a.* **mutualistic**.

mutuality *n.* evolutionary strategy in regard to animal communication where both signaller and receiver benefit from the interaction.

MW molecular weight. *see* molecular mass.

myarian, myaric *a. pert.* musculature.

myasthenia gravis autoimmune disease in which antibodies against skeletal muscle acetylcholine receptors are produced.

mycangium *n.* specialized pouch in which fungus is carried within the cuticle of bark and ambrosia beetles. *plu.* **mycangia**.

mycelia *plu.* of mycelium.

mycelial *a. pert.* or composed of mycelium.

mycelial cord thick strand of hyphae formed by some fungi. *alt.* rhizomorph.

Mycelia Sterilia a diverse group of fungi without any currently known conidial (asexual) or sexual reproductive stages.

mycelioid *a.* (1) resembling a mycelium; (2) growing in the form of a mycelium.

mycelium *n.* network of hyphae forming the characteristic vegetative phase of many fungi, often visible as a fluffy mass or mat of threads. *plu.* **mycelia**.

Mycetae Fungi *q.v.*

mycetismus *n.* mushroom poisoning.

mycetocyte *n.* one of follicle cells at posterior pole of oocyte through which the egg of aphids and other bugs may be infected by symbionts.

mycetogenetic *a.* produced by a fungus.

mycetoid *a.* like a fungus.

mycetome *n.* specialized organ in some blood-feeding insects that carries symbiotic microorganisms that provide nutrients to the insects.

mycetophage *n.* an organism that eats fungi.

Mycetozoa *n.* the slime moulds (*q.v.*), which include the acellular plasmodial slime moulds (myxogastrids), the cellular slime moulds (dictyostelids) and the protostelid slime moulds. *alt.* Myxomycota.

mycina *n.* a spherical stalked apothecium of certain lichens.

myco-, myce-, mycet- prefixes derived from Gk *mykes*, fungus.

mycobacteria *n.plu.* bacteria of the family Mycobacteriaceae, Gram-positive non-motile rods, some species found in soil, others pathogenic for humans and animals, e.g. *Mycobacterium tuberculosis*, which causes tuberculosis, and *M. leprae*, the causal agent of leprosy. *sing.* **mycobacterium**.

mycobiont *n.* the fungal component of a lichen.

mycobiota *n.* the fungi of an area or region.

mycoderm *n.* a superficial film of bacteria or yeast that develops during alcoholic fermentation.

mycoecotype *n.* the habitat type of mycorrhizal or parasitic fungi.

mycoflora *n.* all fungi growing in a specified area or region, or within an organism.

mycogenetics *n.* the genetics of fungi.

mycoherbicide *n.* a fungus used as a biological control agent for a weed.

mycoid *a.* like a fungus.

mycology *n.* the study of fungi.

mycolysis *n.* the lysis or disintegration of fungi, as by bacteria.

mycoparasite *n.* a fungus that parasitizes another fungus.

mycophage *n.* a virus infecting fungi.

mycophagy *n.* feeding on fungi.

mycophthorous *a.* fungus-destroying, *appl.* or *pert.* fungi parasitizing other fungi.

Mycophycophyta lichens *q.v.*

mycoplasma *n.* any of a group of almost sub-microscopic Bacteria of very simple internal structure. They lack the typical rigid bacterial cell wall, occur in a variety of morphological forms, are obligate intracellular parasites and are responsible for several animal diseases. Formerly called the pleuropneumonia-like organisms (PPLOs). *plu.* **mycoplasmas**.

mycoplasma-like organism (MLO) any of a group of almost sub-microscopic plant-pathogenic, motile Bacteria lacking cell walls, which are obligate intracellular parasites and the cause of some diseases that cause leaf yellowing and stunting in plants. They are similar to spiroplasmas but are not helical.

mycorrhiza *n.* symbiotic association between plant roots and certain fungi, in which a sheath of fungal tissue (the mantle) encloses the smallest rootlets. There are two main types. In ectomycorrhizas the fungal hyphae penetrate between the cells of the epidermis and cortex. In endomycorrhizas they penetrate the plant cells themselves and an external fungal sheath is often lacking. Mycorrhizas are essential for optimum growth and development in many trees, shrubs and herbaceous plants. *plu.* **mycorrhizae, mycorrhizas.** *a.* **mycorrhizal.**

mycorrhizoma *n.* association of fungi and a rhizome.

mycosis *n.* animal disease caused by a fungus. *plu.* **mycoses.**

mycosis fungoides T-cell lymphoma of the skin.

mycostasis *n.* (1) inhibition of germination of fungal spores in the soil in the absence of nutrients; (2) inhibition of fungal growth generally. *a.* **mycostatic.** *alt.* fungistasis.

mycosterols *n.plu.* sterols originally found in fungi but now known to be much more widely distributed, e.g. ergosterol, fucosterol.

mycotic *a.* caused by fungi.

mycotoxin *n.* any toxin produced by a fungus.

mycotrophic *a. appl.* plants living symbiotically with fungi.

mycteric *a. pert.* nasal cavities.

mydriasis *n.* dilation of pupil of the eye.

myelencephalon *n.* the posterior part of the developing hindbrain, which will form the medulla oblongata and lower part of 4th ventricle.

myelic *a. pert.* medulla of spinal cord.

myelin, myelin sheath *n.* fatty material forming an insulating sheath around some nerve fibres. It is formed of Schwann cell bodies wrapped round and round the axon to form concentric layers of cell membrane.

myelinated *a.* having a myelin sheath, *appl.* nerve fibres.

myelination *n.* formation of a myelin sheath.

myelo- prefix derived from Gk *myelos*, marrow.

myelocoel *n.* the canal of the spinal cord.

myelocyte *n.* bone marrow cell.

myeloid *a.* (1) *pert.* bone marrow; (2) *appl.* tissue in which haemopoiesis occurs in vertebrates, such as bone marrow, and liver and spleen in embryos; (3) *appl.* cells: monocytes, macrophages, mast cells, and leukocytes other than lymphocytes; (4) (*neurobiol.*) resembling myelin.

myeloid lineage bone-marrow derived blood cells with the exception of lymphocytes.

myeloid stem cell stem cell in bone marrow that gives rise to all blood cells except lymphocytes.

myeloma *n.* a cancer of plasma cells (B cells) which arises in the bone marrow.

myelomere *n.* segment of the spinal cord.

myeloperoxidase *n.* peroxidase (*q.v.*), one of the main enzymes involved in producing toxic antimicrobial agents in phagocytes. It forms hypochlorous acid (HOCl) from chloride ions and hydrogen peroxide. The HOCl reacts with a second molecule of hydrogen peroxide, forming toxic singlet oxygen ($^1O_2$).

myelopoiesis *a.* (1) formation and development of white blood cells; (2) formation of bone marrow.

myenteric plexus layer of ganglia between the circular and longitudinal layers of muscular coat of small intestine. *alt.* Auerbach's plexus.

myenteron *n.* the muscular coat of the intestine.

myiasis *n.* invasion of living tissue by the larvae of certain flies.

mylohyoid *a.* in region of hyoid and posterior part of mandible.

myo- prefix derived from Gk *mys*, muscle.

myoblast *n.* undifferentiated mesenchymal cell present in embryo and adult (as muscle satellite cell), several of which fuse together to form a multinucleate skeletal muscle cell.

myocardium *n.* the muscular wall of the heart. *a.* **myocardial.**

myocoel *n.* part of the coleom enclosed in a myotome.

myocomma *n.* a ligamentous connection between successive myomeres. *alt.* myoseptum.

myocyte muscle cell *q.v.*

myodynamic *a. pert.* muscular force or contraction.

myoelastic *a. appl.* tissue composed of unstriated muscle fibres (smooth muscle) and elastic connective tissue fibres.

myoepicardial *a. appl.* a covering consisting of the mesocardium walls, destined to form the muscular and epicardial walls of the heart.

myoepithelial cell (1) contractile epithelial cell; (2) musculo-epithelial cell *q.v.*

myofibril *n.* individual contractile thread in the cytoplasm of a muscle cell, each myofibril being composed of many actin and myosin filaments. *see* Fig. 28 (p. 425).

myogenesis *n.* the differentiation and development of muscle.

myogenic *a.* (1) having origin in muscle cells; (2) *appl.* contractions arising in muscle cells spontaneously and independently of nervous stimulation, e.g. those that cause the heart to beat.

myoglobin *n.* oxygen-carrying globular haem protein consisting of a single polypeptide chain, involved in oxygen storage and transport in vertebrate muscle.

myohaemoglobin myoglobin *q.v.*

myoid (1) *a.* resembling or composed of muscle fibres, *appl.* striated cells of thymus; (2) *n.* contractile proximal part of filament of rods and cones of retina.

myology *n.* the study of muscles.

myomere *n.* segment of muscle separated from the next by a thin sheet of connective tissue.

myometrial *a. pert.* myometrium.

myometrium *n.* the muscular uterine wall.

myonema, myoneme *n.* minute contractile thread in protists.

myoneural neuromuscular *q.v.*

myopathy *n.* any disease or condition leading to degeneration and loss of function of muscles.

myopia *n.* short-sightedness.

myoplasm *n.* contractile portion of a muscle cell or fibre.

myoseptum myocomma *q.v.*

myosin *n.* ubiquitous motor protein of eukaryotic cells, which interacts with actin to form a contractile complex. There are various types. The myosin II that comprises the thick filaments of striated muscle has a long α-helical 'tail' and two globular 'heads' containing ATPase activity, which form cross-bridges with the actin thin filaments. A cycle of ATPase hydrolysis powers the movement of the myosin head and contraction of the muscle. Other types of myosin (e.g. myosin I with a single ATPase head and a short tail), are found in non-muscle cells. *see also* meromyosin.

myosin heavy chain one of the two paired chains that form the tail and globular heads of a myosin II molecule.

myosin light chain one of the four small paired polypeptide chains that are non-covalently associated with the heads of a myosin II molecule.

myosis *n.* contraction of pupil of eye.

myositis *n.* inflammation of skeletal muscle.

myotendinal, myotendinous musculotendinous *q.v.*

myotendinous junction a type of focal adhesion at which a muscle cell attaches to a tendon.

myotic *a.* causing or *pert.* myosis or pupillary contraction.

myotome *n.* (1) one of a series of hollow cubes of mesenchyme that develops into muscle tissue in early vertebrate embryo; (2) a muscular segment of primitive vertebrates and segmented invertebrates.

myotonia *n.* muscular tension or tonicity.

myotube *n.* a stage in development of a skeletal muscle cell, formed from fusion of individual immature myoblasts into an elongated syncytium. It differentiates into a muscle fibre.

Myriapoda, myriapods *n., n.plu.* centipedes and millipedes and their relatives, terrestrial arthropods characterized by possession of a distinct head with a pair of antennae followed by numerous similar segments each bearing legs.

Myricales *n.* order of dicot trees and shrubs comprising the family Myriaceae (sweet gale).

myriophylloid *a.* having a much-divided thallus, *appl.* certain algae.

myriosporous *a.* having many spores.

myristic acid a 14-carbon saturated fatty acid.

myristoylation *n.* the covalent addition of myristic acid to a protein, through which the protein is attached to the cell membrane.

myrme- prefix derived from Gk *myrmex*, an ant.

myrmecioid complex one of the two major taxonomic subgroups of ants, exemplified by the subfamily Myrmeciinae.

myrmecochore *n.* oily seed modified to attract and be spread by ants.

myrmecole *n.* an organism occupying ants' nests.

myrmecology *n.* the study of ants.

myrmecophobic *a.* repelling ants, *appl.* plants with special glands or hairs that check ants.

myrmecophagous *a.* ant-eating.

myrmecophil(e) *n.* (1) guest insect in an ants' nest; (2) organism that must spend some part of its life cycle with ant colonies.

myrmecophilous *a.* (1) *appl.* flowers, pollinated by the agency of ants; (2) *appl.* fungi, serving as food for ants; (3) *appl.* spiders, living with, preying on or mimicking ants.

myrmecophyte *n.* a plant pollinated by ants, or one that benefits from ant inhabitants and has special adaptations for housing them.

myrosin sinigrin *q.v.*

myrrh *n.* fragrant resin obtained from plants of the genus *Commiophora*.

Myrtales *n.* order of dicots, mostly shrubs and trees, including Lythraceae (loosestrife), Myrtaceae (myrtle), Onagraceae (evening primrose), Punicaceae (pomegranate), Rhizophoraceae (mangrove) and others.

Myrtiflorae Myrtales *q.v.*

mystacial *a. appl.* a pad of thickened skin on either side of snout of some mammals, and the tactile hairs and vibrissae (whiskers) borne on it.

Mystacocarida *n.* class of small crustaceans, similar to copepods, living in marine sands and having no carapace, no clear divisions of the body and simple appendages.

mystax *n.* (1) group of hairs above mouth of certain insects; (2) the tactile hairs or whiskers on an animal's snout.

Mysticeti *n.* suborder of placental mammals, the baleen or whalebone whales, including the blue whale, the right whales, and rorquals. *alt.* Balaenoidea.

myxamoeba *n.* an amoeboid cell produced from a germinating spore in slime moulds.

myxinoids *n.plu.* the order of cyclostomes consisting of hagfish.

myxo- prefix derived from Gk *myxa*, slime.

Myxobacteria *n.* group of Bacteria that includes the myxobacteria, and the gliding bacteria that do not form fruiting bodies.

myxobacteria *n.plu.* flexible rod-shaped bacteria with a gliding movement. Individual cells aggregate into multicellular 'fruiting bodies' containing resting spores called myxospores. *sing.* **myxobacterium**.

myxogastrid slime moulds plasmodial slime moulds *q.v.*

Myxomycota *n., n.plu.* in some classifications a division of protists containing the slime moulds. *alt.* Mycetozoa.

myxopodium *n.* a slimy pseudopodium.

myxosporangium *n.* fruit body of slime moulds.

myxospore *n.* (1) spore of a slime mould or of myxobacteria; (2) spore separated off by slimy disintegration of the hypha.

Myxozoa *n.* phylum of eukaryotic parasitic microorganisms, thought to be related to the cnidarians, with a multinucleate spore-producing stage, which have a life cycle involving invertebrate and vertebrate (mainly fish) hosts. *Myxobolus* species are the cause of economically serious disease in cyprinid fish.

Myzostomaria *n.* class of annelids, ectoparasitic on echinoderms, and almost circular in shape.

MZ twins *see* monozygotic.

N

N (1) symbol for the chemical element nitrogen *q.v.*; (2) asparagine *q.v.*; (3) denotes a position in a consensus DNA sequence at which there is no apparent preference for any particular nucleotide.

***N*ₑ** effective population size *q.v.*

Na symbol for the chemical element sodium *q.v.*

NA (1) neuraminidase *q.v.*; (2) noradrenaline *q.v.*

NAA α-naphthaleneacetic acid, a synthetic auxin, used to induce roots in cuttings and prevent premature fruit drop in commercial crops.

Na⁺-Ca²⁺ exchanger *see* sodium–calcium exchanger.

Na⁺ channel *see* sodium channel.

nAChR nicotinic acetylcholine receptor *q.v.*

nacre *n.* mother of pearl, iridescent inner layer of many mollusc shells, and the substance of pearls.

nacreous *a.* yielding or resembling nacre.

nacrine *a.* (1) mother-of-pearl colour; (2) *pert.* nacre.

NAD nicotinamide adenine dinucleotide *q.v.*

NADH reduced form of nicotinamide adenine dinucleotide.

NADH dehydrogenase respiratory chain enzyme, an iron-sulphur flavoprotein with a FMN prosthetic group, which transfers electrons from NADH to ubiquinone. EC 1.6.99.3. *alt.* **NADH-Q reductase**.

NADP nicotinamide adenine dinucleotide phosphate *q.v.*

NADPH reduced form of nicotinamide adenine dinucleotide phosphate.

NADPH oxidase complex enzyme complex in phagosomes that produces toxic reactive oxygen species (e.g. superoxide, hypochlorite, hydrogen peroxide, hydroxyl radicals and nitric oxide) that kill bacteria engulfed by the phagocyte.

naiad *n.* aquatic nymph stage of certain insects such as dragonflies and mayflies.

naidid *a. appl.* freshwater worms of the genus *Nais*, which are often found in increased numbers in water subject to organic pollution.

naive *a. appl.* mature lymphocytes that have not yet encountered their specific antigen.

Najadales *n.* order of aquatic and semi-aquatic monocots including Potamogeton-aceae (pondweed), Zosteraceae (eel-grass) and others.

Na⁺-K⁺ ATPase transport protein in animal cell plasma membranes. It actively transports Na⁺ out of and K⁺ into animal cells using energy derived from ATP hydrolysis. The ion gradients set up across the plasma membrane by its action are largely responsible for the membrane potential, control of cell volume, and some active transport of sugars and amino acids into cells. *alt.* Na⁺-K⁺ pump, sodium pump, sodium–potassium pump.

naked virus non-enveloped virus.

nalidixic acid antibacterial antibiotic that interferes with DNA gyrase activity.

naloxone *n.* compound that blocks opiate receptors.

NAM *N*-acetylmuramic acid *q.v.*

NANA *N*-acetylneuraminic acid *q.v.*

nanander *n.* a dwarf male, in plants. *a.* **nanandrous**.

nanism *n.* dwarfism.

nano- prefix derived from Gk *nanos*, dwarf, signifying small, or smallest.

nanoid *a.* dwarfish.

nanometre (nm) *n.* unit equivalent to 10^{-9} m, a thousandth of a micrometre or 10 Ångström units. 1 nanometre was formerly called a millimicron.

nanophanerophyte *n.* shrub under 2 m tall.

nanophyllous *a.* having minute leaves.

nanoplankton *n.* microscopic floating plant and animal organisms.

nanotechnology *n.* the science of manipulating materials on an atomic or molecular scale, esp. in the design of molecular machines.

nanous *a.* smaller than normal.

naphthaquinone *n.* derivative of quinone from which vitamin K is synthesized.

napiform *a.* turnip-shaped, *appl.* roots.

NAR net assimilation rate *q.v.*

narcolepsy *n.* rare disorder in which the patient is afflicted with frequent, uncontrollable attacks of sleep lasting 5 to 30 minutes.

narcosis *n.* state of unconsciousness or stupor produced by a drug.

narcotic *a. appl.* drugs that can produce a state of unconsciousness, sleep or numbness.

nares *n.plu.* nostrils *q.v. sing.* **naris**.

nares, anterior openings of olfactory organ to exterior. *alt.* nostrils.

nares, posterior openings of olfactory organ into pharynx or throat.

narial *a.* (1) *pert.* the nostrils; (2) *appl.* septum, the partition between nostrils.

naricorn *n.* the terminating horny part of nostril of certain birds such as albatross.

nariform *a.* shaped like nostrils.

naris *sing.* of nares.

narrow heritability *see* heritability.

narrow-spectrum antibiotic antibiotic that acts on only a limited range of microorganisms.

nasal (1) *a. pert.* the nose; (2) *n.* a nasal scale, plate or bone.

nascent DNA, nascent RNA newly synthesized DNA or RNA.

nasion *n.* middle point of nasofrontal suture.

nasoantral *a. pert.* nose and cavity of upper jaw.

nasobuccal *a.* (1) *pert.* nose and cheek; (2) *appl.* nose and mouth cavity.

nasociliary *a. appl.* branch of ophthalmic nerve, with internal and external nasal branches, and giving off the long ciliary and other nerves.

nasofrontal *a.* (1) *appl.* part of superior ophthalmic vein which communicates with the angular vein; (2) *appl.* suture between nasal and frontal bones.

nasolabial *n. pert.* nose and lip.

nasolacrimal *a. appl.* canal from lacrimal sac to inferior meatus of nose through which the tear duct passes.

nasomaxillary *a. pert.* nose and upper jaw.

naso-optic *a. appl.* embryonic groove between nasal and maxillary process.

nasopalatine *a.* (1) *pert.* nose and palate; (2) *pert.* canal communicating with vomeronasal organs.

nasopharyngeal *a. pert.* nose and pharynx, or nasopharynx.

nasopharynx *n.* that part of pharynx or throat continuous with the posterior openings of the nasal passsages.

nasoturbinal *a. appl.* outgrowths from lateral wall of nasal cavity increasing the area of sensory surface.

nastic movement plant movement caused by a diffuse non-directional stimulus. It is usually a growth movement but may be caused by a change in turgidity, as in the sensitive plant (*Mimosa* sp.) that droops on contact.

nasty nastic movement *q.v.*

nasus *n.* snout-like organ of soldiers of some species of termite, used to eject poisonous or sticky fluid at invaders.

nasute *a.* possessing a nasus.

nasute, nasutus *n.* a type of soldier termite.

natal *a. pert.* birth.

natality *n.* birth rate.

natant *a.* floating on surface of water.

natatorial *a.* formed or adapted for swimming.

natatory *a.* (1) swimming habitually; (2) *pert.* swimming.

nates *n.plu.* buttocks.

National Vegetation Classification (NVC) United Kingdom national survey of plant communities and vegetation types, which started in 1975, and which has devised a standard nomenclature for the types of plant community found in Britain.

native *a.* (1) *appl.* animals and plants which originate in district or area in which they live; (2) fully folded, entire and undenatured, *appl.* protein.

native species indigenous species that is normally found as part of a particular ecosystem.

natriferic *a.* transporting sodium.

natriuresis *n.* excessive loss of sodium in urine, disturbing electrolyte balance in body.

natriuretic peptide *see* atrial natriuretic peptide.

natural classification a taxonomic classification that groups organisms or objects together on the basis of the sum total of all their characteristics, and tries to indicate evolutionary relationships. *cf.* artificial classification.

natural decrease the rate of decline of a population, calculated by subtracting the number of births from the number of deaths in a given period.

natural gas methane *q.v.*

natural increase the rate of growth of a population, calculated by subtracting the number of deaths from the number of births in a given period.

naturalized *a. appl.* alien species that have become successfully established.

natural killer cells (NK cells) large granular lymphocytes, which do not possess specific antigen receptors but which carry a wide range of invariant receptors by which they can recognize and kill cells infected with some viruses. They are part of the innate immune system.

natural selection the process by which evolutionary change is chiefly driven according to Darwin's theory of evolution. Environmental factors such as climate, disease, competition from other organisms, and availability of certain types of food will lead to the preferential survival and reproduction of those members of a population genetically best fitted to deal with them. Continued selection will therefore lead to certain genes becoming more common in subsequent generations. Such selection, operating over very long periods of time, is believed to be able to give rise to the considerable differences now seen between different organisms. *see also* neutral theory of evolution.

naupliiform *a.* superficially resembling a nauplius, *appl.* larvae of certain hymenopterans.

nauplius *n.* earliest larval stage of many crustaceans, with three pairs of appendages. *plu.* **nauplii**.

nautiliform *a.* shaped like a nautilus shell.

nautiloid *n.* member of the subclass Nautiloidea of the cephalopod molluscs, typified by the pearly nautilus (*Nautilus*). Nautiloids have a spiral, many-chambered shell from which the head and tentacles emerge.

navicular, naviculate *a.* boat-shaped.

naviculare *n.* (1) a boat-shaped bone of the mammalian carpus; (2) tarsal bone between talus and cuneiform bones.

ncRNA non-coding RNA *q.v.*

NDV Newcastle disease virus, a paramyxovirus.

NE norepinephrine. *see* noradrenaline.

neala *n.* fan-like posterior lobe of hindwing of some insects.

neallotype *n.* type specimen of the opposite sex to that of the specimen previously chosen for designation of a new species.

nealogy *n.* the study of young animals.

Neanderthal *n.* species of archaic human, *Homo neanderthalensis*, who lived in the Old World during the Pleistocene. Now generally considered a separate species from *Homo sapiens*.

Nearctic region zoogeographical region in Wallace's classification, consisting of Greenland, and North America down to northern Mexico. It is divided into the Alleghany, Californian, Canadian and Rocky Mountain subregions.

necrocytosis *n.* cell death.

necrogenic *a.* promoting necrosis in the host, *appl.* certain parasitic fungi.

necrogenous *a.* living or developing in dead bodies.

necrophagous, necrophilous *a.* feeding on dead bodies.

necrophoresis *n.* transport of dead members of the colony away from the nest.

necrophoric *a.* carrying away dead bodies, *appl.* certain insects.

necrosis *n.* death of cells or tissues as a result of external trauma such as physical damage or lack of oxygen. It results in cell lysis and damage to surrounding tissues. *a.* **necrotic.** *cf.* apoptosis.

necrotizing *a.* undergoing necrosis, *appl.* tissue; causing necrosis.

necrotoxin *n.* a toxin which causes necrosis of tissue.

necrotroph *n.* fungus living off dead host plant tissue.

nectar *n.* (1) sweet-tasting liquid secreted by the nectaries of flowers and certain leaves to attract insects, and some birds, for pollination; (2) in some fungi, a liquid exuding from fruiting body and containing spores.

nectar gland *n.* (1) gland secreting nectar in flowers, and in some leaves; (2) gland secreting the sweet honeydew in aphids. *alt.* **nectary.** *see* Fig. 18 (p. 239).

nectar guides markings on petals of flowers that guide insects to the nectar, thus making cross-fertilization more likely.

nectariferous *a.* producing or carrying nectar.

nectarivorous, nectivorous *a.* nectar-eating.

nectary nectar gland *q.v.*

nectocalyx nectophore *q.v.*

necton nekton *q.v.*

nectophore *n.* in a siphonophore, a medusoid individual modified for swimming, clusters of which propel the colony through the water. *alt.* nectocalyx.

nectopod *n.* an appendage modified for swimming.

NEFA non-essential fatty acids. Those fatty acids that can be synthesized *de novo* and therefore do not need to be supplied in the diet.

negative assortative mating preferential mating between partners who are phenotypically unlike.

negative control type of control of gene expression in which the regulatory protein (repressor) switches off gene transcription. In the absence of the repressor, the gene is expressed. *cf.* positive control.

negative feedback type of control mechanism in which the end-product of a particular pathway or process inhibits the first step in the pathway or steps in other pathways that contribute to the overall process.

negative reinforcement a stimulus or series of stimuli which are unpleasant to an animal and so diminish its response to the stimulus or cause avoidance reactions.

negative selection selection process that occurs for developing T lymphocytes in the thymus in which potentially self-reactive T cells are eliminated.

negative staining technique in electron microscopy in which the specimen is surrounded by a heavy-metal stain which provides a 'negative impression' of the object's outline and surface features. Similar techniques in which the background is stained to show up an unstained specimen are used in light microscopy.

negative-strand RNA viruses viruses with a single-stranded RNA genome that is complementary to the messenger RNA, and from which mRNA is made by a viral transcriptase.

negative supercoiling twisting of a circular DNA molecule in a direction opposite to that of the right-handed double helix.

negative tropism tendency to move or grow away from the source of the stimulus, e.g. plant shoots show negative geotropism.

Negri bodies virus inclusions in nerve cells infected with rabies.

neighbour joining a method of building a phylogenetic tree in which the two most similar sequences are connected together through a node, the two next through another node and so on and then the most similar pairs are connected through a further node, and so on. It produces an additive, non-ultrametric, unrooted tree.

nekton *n.* the organisms swimming actively in water.

Nelumbonales *n.* order of large aquatic herbaceous dicots and including the single family Nelumbonaceae (Indian lotus).

nematoblast *n.* the cell from which a nematocyst develops.

nematocalyx *n.* in some colonial hydrozoans, a small polyp that has no mouth but engulfs organisms by pseudopodia.

nematoceran, nematocerous *a.* possessing thread-like antennae.

nematocide *n.* chemical that kills nematode worms.

nematocyst *n.* stinging cell of sea anemones, jellyfishes and other coelenterates, containing a long coiled thread which is discharged on contact and pierces prey. Sometimes refers only to the contents of the cell, the cell itself being termed a nematoblast or cnidoblast.

Nematoda, nematodes *n.*, *n.plu.* roundworms. Slender, pseudocoelomate, unsegmented worms circular in crosssection. Some (eelworms) are serious parasites of plants, others are parasitic in animals and some are free-living in soil and marine muds. Parasitic nematodes causing severe diseases in humans include the hookworms *Ancyclostoma* and *Necator*, *Trichinella* (causing trichinellosis) and *Wucheria* (the cause of elephantiasis). The soil nematode *Caenorhabditis elegans* is an important experimental organism in genetic and developmental research.

nematogen *n.* one of the reproductive phases of dicyemids.

nematoid *a.* thread-like or filamentous.

nematology *n.* the study of nematodes.

Nematomorpha, nematomorphs *n.*, *n.plu.* phylum of pseudocoelomate worms, sometimes known as horsehair worms, that are free-living in soil or fresh water as adults, and parasitic in arthropods when young. *alt.* threadworms.

nematophore nematocalyx *q.v.*

nematosphere *n.* the enlarged end of a tentacle in some sea anemones.

nematozooid *n.* a defensive zooid in hydrozoans.

Nemertina, nemertines *n.*, *n.plu.* phylum of long, slender, acoelomate, marine worms, flattened dorsoventrally, e.g. the bootlace worm (*Lineus*). Most live on shores around the low tide line. They have a mouth and anus, a simple blood system, and a typical muscular proboscis, which is extended to catch prey. *alt.* **Nemertea, nemerteans,** proboscis worms, ribbon worms.

nemoral *a.* living at the edges of woodlands, or in open woodland.

nemorose, nemoricole *a.* inhabiting open woodland places.

N-end rule the fact that certain amino-acid residues tend to promote the degradation of a protein via proteasomes if present at the N-terminus.

neo- prefix derived from Gk *neos*, young, signifying young or new.

neoblast *n.* one of the undifferentiated cells forming primordium of regeneration tissue after wounding.

neocarpy *n.* production of fruit by an otherwise immature plant. *alt.* **neocarpic**.

neocentromere *n.* new centromere that forms on a chromosome at a position other than the normal centromere position.

neocerebellum *n.* region of cerebellum that receives nerve fibres predominantly from the pons.

neocortex *n.* the evolutionarily most recent part of the mammalian brain. It comprises the cerebral cortex, excluding the hippocampus, limbic system and olfactory bulb.

neo-Darwinism the modern version of the Darwinian theory of evolution by natural selection, incorporating the principles of genetics and still placing emphasis on natural selection as a main driving force of evolution.

neoencephalon *n.* the telencephalon, or latest evolved anterior portion of brain.

Neogene the most recent period in the Tertiary, lasting from *ca.* 23 million years ago to the present day and comprising the Miocene, Pliocene, Pleistocene and Holocene epochs.

neogenesis *n.* (1) new tissue formation; (2) regeneration *q.v.*

Neolaurentian *a. pert.* or *appl.* early Proterozoic era.

Neolithic *a. appl.* or *pert.* the New or polished Stone Age, characterized by the use of polished stone tools and weapons and the appearance of settled cultivation.

neomorph *n.* (1) a structural variation from the type; (2) mutant allele that produces changes in developmental processes, resulting in the appearance of a new character.

neomorphosis *n.* regeneration when the new part is unlike anything in the body.

neomycin *n.* aminoglycoside antibiotic synthesized by *Streptomyces fradiae*, used clinically against Gram-negative bacteria. It inhibits bacterial protein synthesis.

neonate *n.* newborn animal. *a.* **neonatal**.

neontology *n.* the study of existing organic life. In the study of evolutionary biology neontologists are those who study evolution by comparisons between living organisms, whereas palaeontologists study evolution through the fossil record.

neonychium *n.* (1) soft pad enclosing each claw of embryos of clawed vertebrates, to prevent tearing of foetal membranes; (2) horny claw pad present in birds before hatching.

neopallium neocortex *q.v.*

neoplasia *n.* cell proliferation, often uncontrolled, producing additional tissue, often used with reference to the malignant proliferation of cancer cells.

neoplasm *n.* an abnormal growth of cells, sometimes self-limiting as in benign tumours, sometimes malignant.

neoplastic *a. appl.* cells or tissue arising as a result of uncontrolled growth. *alt.* (in some cases) malignant, tumorous.

neoplastic transformation the changes that take place in a cell as it becomes cancerous.

neoptile down feather *q.v.*

Neornithes *n.* subclass of birds (Aves) including all extant modern birds.

neostigmine *n.* a plant alkaloid, an inhibitor of the enzyme acetylcholinesterase. *alt.* prostigmine.

neoteny *n.* retention of larval characters beyond the normal period and into the sexually mature adult, as in some amphibians. *a.* **neotenic, neotenous**.

neothalamus *n.* that part of the thalamus consisting of nuclei with connections to association areas in the cerebral cortex.

Neotropical kingdom floristic kingdom consisting of southern Mexico, Central and South America, and the Caribbean. It is divided into the Amazon, Andean, Caribbean, Juan Fernandez, Pampas, South Brazilian and Venezuelan regions.

Neotropical region zoogeographical region in Wallace's classification consisting of southern Mexico, Central and South America, and the Caribbean. It is divided into the Antillean, Brazilian, Chilean and Mexican subregions.

neotype *n.* (1) a new type; (2) a new type specimen from the original locality.

neoxanthin *n.* xanthophyll carotenoid pigment found in some protists, esp. green algae and euglenoids.

Nepenthales *n.* order of herbaceous carnivorous dicots with leaves adapted for trapping small animals, and comprising the families Droseraceae (sundew) and Nepenthaceae (pitcher plants).

nephric *a. pert.* kidney.

nephridia *plu.* of nephridium.

nephridial *a.* (1) *pert.* kidney, usually *appl.* the small excretory tubules; (2) *pert.* excretory organ or nephridium of invertebrates.

nephridioblast *n.* ectodermal cell that gives rise to a nephridium.

nephridiopore *n.* external opening of excretory organs (nephridia) in invertebrates.

nephridiostome *n.* ciliated opening of a nephridium into the coelom.

nephridium *n.* (1) excretory organ having function of kidney in invertebrates; (2) embryonic kidney tubule in vertebrates. *plu.* **nephridia**.

nephroblast *n.* embryonic cell that gives rise ultimately to nephridia.

nephrocoel *n.* the cavity of a nephrotome.

nephrocyte *n.* any of various cells in sponges, ascidians and insects, that store and discharge waste products.

nephrodinic *a.* having one duct serving for both excretory and genital purposes.

nephrogenic *a.* (1) *pert.* development of kidney; (2) *appl.* cord or column of fused mesodermal cells giving rise to tubules of mesonephros.

nephrogenous *a.* produced by the kidney.

nephroid reniform *q.v.*

nephrolytic *a. pert.* or designating enzymatic action destructive to kidneys.

nephromere nephrotome *q.v.*

nephromixium *n.* compound excretory organ comprising flame cells and coelomic funnel and acting both as an excretory organ and genital duct.

nephron *n.* the individual structural and functional unit of a vertebrate kidney, comprising glomerulus, Bowman's capsule and a convoluted tubule.

nephrostome *n.* opening of nephridial tubule into body cavity.

nephrotome *n.* that part of a somite developing into an embryonic excretory organ. *alt.* nephromere.

neritic *a. pert.*, or inhabiting, coastal waters only, as distinct from oceanic.

neritopelagic *a. pert.* or inhabiting the sea above the continental shelf.

Nernst equation the force tending to drive an ion across a membrane is made up of two components: one due to the electrical membrane potential and one due to the concentration gradient. At equilibrium at a temperature of 37°C, for an ion with a single positive charge, the two forces are balanced in the Nernst equation $V = 62 \log_{10}(C_o/C_i)$, where V = membrane potential in millivolts and C_o and C_i are the concentrations of the ion outside and inside the membrane, respectively.

nervate *a.* having nerves or veins, *appl.* leaves, insect wings.

nervation, nervature *n.* venation of leaves or insect wings.

nerve *n.* (1) bundle of many nerve fibres (axons) of individual neurons, connecting the central nervous system with other parts of the body. Discrete bundles of nerve fibres within the nerve are each enclosed in a sheath of connective tissue (perineurium), and the whole nerve is covered in an epineurium of fibrous connective tissue; (2) (*bot. & zool.*) vein of leaf or insect wing, leaf veins being strands of vascular tissue, the veins of insect wings being extensions of the tracheal system.

nerve canal a canal for passage of nerve to pulp of a tooth.

nerve cell neuron *q.v.*

nerve centre group of nerve cells associated with a particular function.

nerve cord in invertebrates, a bundle of nerve fibres, or chain of ganglia and interconnecting nerve fibres, running the length of the body.

nerve ending the terminal portion of axon of a neuron, or receptor portion of a sensory neuron, modified in various ways.

nerve fibre axon (*q.v.*) of nerve cell, many individual nerve fibres making up a nerve.

nerve growth factor (NGF) protein of the neurotrophin family of neurotrophic factors which is produced by the target tissues of sympathetic neurons, such as smooth muscle, and of some sensory neurons, and which is needed for the survival of the innervating neurons. It also stimulates the outgrowth of neurites from developing neurons of these types.

nerve impulse *see* action potential.

nerve net simple network of nerve cells in body wall of coelenterates, and some other invertebrates, connecting sensory cells and muscular elements.

nervicolous *a.* living or growing in the veins of a leaf.

nerviduct *n.* passage for nerves in cartilage or bone.

nervous *a.* (1) *pert.* nerves; (2) *appl.* tissue composed of neurons.

nervous system highly organized system of electrically active cells (nerve cells or neurons), which generate and convey signals in the form of electrical impulses. A nervous system is present in all multicellular animals except sponges, and is most highly developed in vertebrates. It receives and coordinates inputs from the environment and from the body through sensory receptors, and conveys executive commands to muscles and glands, enabling the animal to sense and respond rapidly to external and internal stimuli. In all but the most primitive nervous systems, the nerve cells are organized into nerves and aggregates of nerve cell bodies (ganglia). The vertebrate nervous system consists of a brain and spinal cord, which constitute the central nervous system, and a peripheral nervous system, consisting of sensory cells and the peripheral nerves and their branches, which convey signals to and from the central nervous system. *see also* brain, central nervous system, nerve, nerve net, neuromuscular junction, neuron, peripheral nervous system, synapse.

nervule *n.* tiny branch of a vein in an insect wing.

nervure *n.* (1) (*zool.*) one of the rib-like structures which support the membranous wings of insects. They are branches of the tracheal system; (2) (*bot.*) a leaf vein.

nervus lateralis a branch of the vagus nerve in fishes, connecting sensory lateral line with brain.

nervus terminalis preoptic nerve *q.v.*

nessoptile neoptile *q.v.*

nest epiphyte an epiphyte which builds up a store of humus around itself for growth.

nest provisioning returning regularly to nests to bring food to developing offspring, as in some solitary wasps.

net assimilation rate (NAR) the increase in dry weight of an individual plant per unit time, with reference to the total area involved in assimilation.

net efficiency a measure of the efficiency of an organism in converting its assimilated food to protoplasm.

net photosynthesis photosynthesis measured as the net uptake of carbon dioxide into the leaf, equal to gross photosynthesis less carbon dioxide released during respiration.

net plasmodium the kind of plasmodium found in some slime moulds, where the cells are connected by cytoplasmic strands, forming a net.

net production the amount of food in an ecosystem available for the primary consumers, being the gross primary production minus the amount of biomass used in respiration by primary producers.

net reproductive rate average number of offspring a female produces during her lifetime, symbolized by R_0.

net slime mould *see* Labyrinthulomycota.

netted-veined with veins in the form of a fine network, *appl.* leaves, insect wings.

neural *a. pert.* or closely connected with nerves, or nervous system, or nervous tissue.

neural arc the afferent and efferent neuronal connections running between sensory receptor and the effector.

neural arch arch on dorsal surface of vertebra for passage of the spinal cord.

neural canal canal through backbone formed by neural arches of individual vertebrae, through which runs the spinal cord.

neural crest ridge of ectoderm that forms above the neural tube during early embryogenesis in vertebrates. The cells of the neural crest migrate to form a variety of structures including the melanocytes of the dermis, the dorsal root ganglia of the sensory nervous system and the autonomic nervous system, the adrenal medulla and some skeletal elements in the face.

neural folds the edges of the neural plate which rear up and join to form the neural tube.

neural induction in vertebrates, the induction of the neural tube from ectoderm by the underlying mesoderm.

neural lobe of pituitary body, *see* neurohypophysis.

neural map the point-to-point mapping of sensory information, e.g. visual information, from the sensory organ onto the areas of the brain that process and interpret it.

neural network (1) a system of interconnected neurons, proposed as the basis of higher brain functions such as perception, memory and learning. A perception or a memory, for example, is thought to result from the pattern of activity within the network, which can be varied by varying the strength and number of the connections between the elements; (2) artificial neural network *q.v.*

neural plate (1) band of thickened ectoderm down midline of back of early chordate embryo from which the neural tube develops; (2) a lateral member of neural arch; (3) one of a median row of bony plates in carapace of turtle.

neural retina the photoreceptor and nerve cell layers of the retina.

neural shields horny shields above neural plates of turtles.

neural spine dorsal projection on a vertebra.

neural stalk infundibulum of neurohypophysis.

neural stem cell (NSC) pluripotent cell in animal embryos capable of giving rise to neurons and glial cells. Neural stem cells persist in certain areas of the adult mammalian brain.

neural tube tube of ectoderm formed down the back of early vertebrate embryo. It will develop into the brain and spinal cord.

neuraminic acid *see* N-acetylneuraminic acid.

neuraminidase *n.* enzyme that removes terminal *N*-acetylneuraminic acid (sialic acid) from carbohydrate side chains of glycoproteins. Sometimes known as receptor-destroying enzyme because by splitting neuraminic acid from the cell-surface glycoproteins that act as receptors for some viruses it destroys their receptor properties. Some viruses, e.g. myxoviruses and paramyxoviruses, carry neuraminidases on their surface. *alt.* sialidase.

neurapophysis *n.* one of the two plates growing from centrum of vertebra and meeting over the spinal cord to form the neural spine.

neuraxis *n.* (1) the cerebrospinal axis; (2) axon *q.v. alt.* neuraxon.

neurectoderm *n.* the ectodermal cells destined to form the earliest rudiment of the nervous system, as distinct from epidermal ectoderm.

neuregulins *n.plu.* proteins released from axon terminals of developing motor neurons and which stimulate synthesis of acetylcholine receptors on the muscle cell membrane. *alt.* ARIAs.

neurenteric *a. pert.* cavity of neural tube and enteric cavity, *appl.* canal temporarily connecting posterior end of neural tube with posterior end of archenteron.

neuric neural *q.v.*

neurilemma, neurolemma *n.* thin sheath investing the myelin sheath of a nerve fibre. *a.* **neurilemmal, neurolemmal.**

neurine *n.* a ptomaine with a fishy smell obtained mainly from brain, bile and egg yolk, and which is formed from choline in putrefying meat.

neurite *n.* general term for an extension from a nerve cell body, especially in developing neurons when axons and dendrites cannot yet be distinguished.

neuroanatomy *n.* the study of the structure and anatomy of the brain and nervous system.

neurobiology *n.* the study of the morphology, physiology, biochemistry and development of the brain and nervous system, and the biochemical and cell biological basis of brain function. It is not usually considered to include psychology and cognitive psychology.

neuroblast *n.* cell from which a neuron is formed.

neuroblastoma *n.* tumour originating from immature cells of the nervous system.

neurocele neurocoel *q.v.*

neurocentral *a. appl.* two vertebral synchondroses persisting during the first few years of human life.

neurochord *n.* (1) a giant nerve fibre; (2) a primitive tubular nerve cord.

neurocoel *n.* the central cavity of the central nervous system.

neurocranium *n.* the cartilaginous or bony case containing the brain and capsules

of special sense organs (e.g. eyes and ears).

neurocrine *a. pert.* secretory function of nervous tissue or cells. *alt.* neurosecretory.

neurodegenerative *a. appl.* diseases and conditions of the brain or other parts of the nervous system which result from the progressive death of neurons and loss of function, e.g. Parkinson's disease and Alzheimer's disease in humans, scrapie in sheep and bovine spongiform encephalopathy in cattle.

neuroectoderm *n.* that portion of the ectoderm giving rise to the nervous system.

neuroendocrine neurosecretory *q.v.*

neuroendocrine system the hypothalamus and pituitary, secretion of hormones from the pituitary being regulated by hormones secreted by the neurons of the hypothalamus.

neuroepithelium *n.* a superficial layer of cells where it is specialized for sensory reception.

neurofibrillary tangles abnormal aggregation of paired helical protein filaments within neurons, destroying their function, which occurs in brains of patients with Alzheimer's disease.

neurofibrils *n.plu.* very fine protein fibres (neurofilaments, microtubules and actin filaments) running longitudinally in axons and dendrites of nerve cells and forming a complex meshwork in the cell body.

neurofibromatosis *n.* disease with tumours of cutaneous nerves and café-au-lait spots on the skin.

neurofilaments *n.plu.* longitudinal intermediate filaments providing internal support for the axon of a neuron.

neurogenesis *n.* (1) formation of the nervous system during embryogenesis; (2) development of nerves.

neurogenic *a.* (1) induced by nervous stimulation, e.g. muscular contraction and secretion from some glands; (2) giving rise to nervous tissue or nervous system.

neuroglia *n.plu.* the cells of the central nervous system other than neurons. They include astrocytes, oligodendrocytes, microglia, and ependymal cells. *alt.* glia. *a.* **neuroglial.**

neurohaemal *a.* (1) *appl.* nerve endings in close relationship with blood vessels and discharging neurosecretory material into blood; (2) *appl.* organ: organ such as corpora cardiaca in insects in which secretion from numbers of neurosecretory cells is stored and released into the blood.

neurohormone *n.* any hormone produced by neurosecretory cells, usually in the brain. Neurohormonal activity is distinguished from that of classical neurotransmitters as it can have effects on cells distant from the source of the hormone.

neurohumoral *a. pert.* hormones released into the general circulation from the brain.

neurohypophysis *n.* the posterior part of the pituitary gland, containing secretory nerve endings from the hypothalamus and producing vasopressin and oxytocin, among other hormones. It comprises the pars nervosa or neural lobe, and the neural stalk (infundibulum).

neurolemma neurilemma *q.v.*

neuroleptic *a.* anti-psychotic, *appl.* drugs used to treat psychotic disorders.

neuroleukin *n.* glucose-6-phosphate isomerase, promotes growth and survival of cultured spinal and sensory neurons.

neurology *n.* the study of the morphology, physiology and pathology of the nervous system, in present-day usage sometimes implying a more clinical orientation than e.g. neurobiology or neuroscience.

neurolymph cerebrospinal fluid *q.v.*

neurolysis *n.* the lysis or disintegration of nerve tissue.

neuromast *n.* a group of hair cells (sensory cells) comprising a sensory unit of the acoustico-lateralis system in fishes and some amphibians.

neuromere *n.* (1) segment of vertebrate spine between the attachment points of successive pairs of spinal nerves; (2) segmental ganglion of annelids and arthropods.

neuromodulation *n.* proposed mode of action of some chemical transmitters in brain, which affect the activity of a neuronal pathway by influencing the efficiency of synaptic transmission.

neuromuscular *a. pert.* or involving both nerves and muscles.

neuromuscular junction a specialized type of chemical synapse where an axon terminal of a motor neuron contacts a muscle cell.

neuronal *a. pert.* neurons or nerve cells.

neuronal transplantation the transplantation of foetal neurons into the brains of adults to try and restore functions lost by disease or damage.

neuron *n.* nerve cell, basic unit of the nervous system, specialized for the generation and conveyance of electrical impulses. A typical nerve cell consists of a rounded cell body, which contains the nucleus and other organelles, from which cytoplasmic extensions project. These are the dendrites, which receive signals from other neurons, and the axon, which conducts impulses outward from the cell body. *alt.* neuron.

neuropeptide *n.* any of many small peptides produced by the nervous system, some of which act as neurotransmitters, others as neuromodulatory hormones.

neuropil *n.* network of axons, dendrites and synapses in neural tissue.

neuroplasm *n.* cytoplasm of neuron, not including the fibrillar components.

neuropodium *n.* ventral lobe of polychaete parapodium.

neuropore *n.* the opening of neural tube or neurocoel to the exterior.

Neuroptera *n.* order of insects with complete metamorphosis, including alder flies, lacewings and ant lions, having long antennae, biting mouthparts and two pairs of membranous wings held roof-like over the abdomen in repose.

neurose *a.* having numerous veins, *appl.* leaves and insect wings.

neurosecretion *n.* the release of neurotransmitters, neurohormones and neuromodulatory compounds from nerve cells.

neurosecretory *a. appl.* nerve cells that secrete substances that travel via the blood to their targets.

neurosensory *a. appl.* the epithelial sensory cells of coelenterates.

neuroskeleton *n.* the cytoskeleton of a neuron.

neurotendinous *a.* (1) *pert.* nerves and tendons; (2) *appl.* nerve endings in tendons.

neurotome neuromere *q.v.*

neurotoxin *n.* any poison that acts specifically on the nervous system. *a.* **neurotoxic**.

neurotransmitter *n.* chemical liberated by the axon terminal of a neuron in response to an electrical impulse in the neuron and which transmits the neuronal signal across a synapse to another neuron or a muscle fibre. Neurotransmitters are released by the pre-synaptic neuron and interact with receptors on the dendrites (or other parts) of the post-synaptic neuron. Some common neurotransmitters are acetylcholine, noradrenaline (norepinephrine), dopamine and 5-hydroxytryptamine (serotonin).

neurotrophic *a.* required for the growth and survival of nerve cells, *appl.* factors: protein growth factors that act specifically on nerve cells.

neurotrophins *n.plu.* family of neurotrophic factors which includes brain-derived neurotrophic factor (BDNF), nerve growth factor, neurotrophin-3 and neurotrophin-4.

neurotropic *a. appl.* viruses and bacteria that infect nerve cells or toxins that act on nerve cells.

neurotubule *n.* microtubule of neuron, numbers of which run longitudinally along axon.

neurovascular *a. appl.* nerves and blood vessels.

neurula *n.* early chordate embryo at the stage of formation of the neural tube. It develops from the gastrula.

neurulation *n.* formation of the neural tube in a chordate embryo.

neuston *n.* organisms floating or swimming in surface water, or inhabiting surface film.

neuter *n.* (1) a non-fertile female of social insects; (2) a castrated animal; *a.* (3) sexless, neither male nor female; (4) having neither functional stamens nor pistils.

neutral *a.* (1) neither acid nor alkaline, pH 7.0; (2) achromatic, as white, grey and black; (3) day-neutral *q.v.*; (4) *appl.* changes in nucleotide or amino acid sequence that have no effect on function of the gene product and therefore no effect on phenotype.

neutral allele allele formed as the result of a neutral mutation.

neutral fat triacylglycerol *q.v.*

neutralization *n.* inactivation of a virus, bacterium or toxin by formation of a complex with specific antibody.

neutral mutation a mutation that confers no selective advantage or disadvantage on the individual.

neutral polymorphism a genetic polymorphism within a population in which the relative frequencies of the different forms are the result of chance and the action of intrinsic genetic mechanisms and are not being maintained by selection.

neutral theory of evolution the hypothesis, first proposed in the 1960s, that evolutionary change at the molecular level can be due to genetic changes (neutral mutations) that are not subject to natural selection. This idea is now accepted to some degree or other by most evolutionary biologists and considered as complementary to the theory of natural selection.

neutrophil *n.* phagocytic white blood cell with granules that stain with neutral stains. It enters infected tissues in large numbers and engulfs and kills microorganisms. *alt.* polymorphonuclear leukocyte.

neutrophile *a. appl.* bacteria with optimum pH for growth of between 6 and 8.

neutrophilic *a.* staining only with neutral stains.

New Caledonian region part of the Palaeotropical floristic kingdom, consisting of the islands of New Caledonia in the south-western Pacific.

newt *n.* common name for the genera *Triturus*, *Taricha* and *Notophthalamus* (in the family Salamandridae) of tailed amphibians (urodeles). They return to the water to breed and lay their jelly-coated eggs singly.

New Zealand region floristic region consisting of New Zealand and its offshore islands, a subdivision of the Antarctic kingdom.

New Zealand subregion subdivision of the Australian zoogeographical region, comprising New Zealand and its offshore islands.

nexin *n.* protein linking adjacent microtubules in cilia and flagella. *see also* protease nexin.

nexus *n.* region of fusion of plasma membrane between two electrically excitable cells.

NFκB gene regulatory protein activated as a result of a variety of extracellular signals associated with infection and stress, and in development.

NGF nerve growth factor *q.v.*

niacin *n.* nicotinic acid, a member of the vitamin B complex, vitamin B_7, found in all living cells as the nicotinamide moiety of the enzyme cofactors NAD and NADP. It can be used to treat pellagra in humans. Blood, liver, legumes and yeast are particularly rich sources. *alt.* pellagra-preventive factor.

niche ecological niche *q.v.*

niche construction the activities of organisms that bring about changes in their environment that form their ecological niche (*q.v.*).

niche diversification alpha diversity *q.v.*

niche overlap the situation where two or more species use the same resources or the same habitat within a community and thus share the same ecological niche, leading to competition between them. Two species with identical niche requirements cannot coexist in the same community, but niche overlap may be seen where there are small and difficult-to-distinguish ecological differences between species.

nick translation limited DNA synthesis initiated at a single-strand break (nick) and which displaces the homologous DNA strand from the template. It is carried out by bacterial DNA polymerase I and has a repair function *in vivo*. It is widely used *in vitro* to introduce radioactively labelled nucleotides into DNA.

nicotinamide *n.* amide derived from nicotinic acid (niacin). It is a constituent of the cofactors NAD and NADP and may be used like niacin to treat pellagra in humans.

nicotinamide adenine dinucleotide (NAD⁺) important cofactor composed of nicotinamide, adenine, two riboses and two phosphate groups. It is found in all living cells where it acts as a hydrogen (electron) acceptor in many biochemical reactions and is reduced to NADH. In this form it is a hydrogen (electron) donor, donating electrons esp. to the respiratory chain. *see* Fig. 29.

nicotinamide adenine dinucleotide phosphate (NADP⁺) important cofactor composed of NAD with an extra phosphate group attached. It is found in all

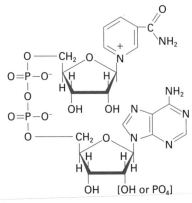

Fig. 29 Nicotinamide adenine dinucleotide (NAD⁺) and nicotinamide adenine dinucleotide phosphate (NADP⁺). The form illustrated is the oxidized form. In the reduced forms, NADH and NADPH, a hydrogen atom plus an electron is added to the aromatic ring at the top of the figure.

living cells where it acts as a hydrogen (electron) acceptor, esp. in biosynthetic pathways. It is reduced to NADPH. In this form it is a donor of hydrogen (electrons). In plants, NADPH is one of the primary products of the light reactions of photosynthesis and is used as the electron donor in the Calvin cycle of carbon fixation. *see* Fig. 29.

nicotine *n.* alkaloid obtained from the tobacco plant *Nicotiana tabacum*, toxic to many animals because it mimics the action of the neurotransmitter acetylcholine at neuromuscular junctions.

nicotinic *a.* (1) *pert.* nicotine, *appl.* acetylcholine receptors stimulated by nicotine; (2) resembling nicotine in its effects, *appl.* various substances that also act at the class of acetylcholine receptors sensitive to nicotine.

nicotinic acid niacin *q.v.*

nictitating membrane the third eyelid, a membrane that can be passed over the eye and helps to keep it clean in reptiles, birds and some mammals.

nidation implantation *q.v.*

nidicolous *a.* living in the nest for a time after hatching.

nidification *n.* nest building and the behaviour associated with it.

nidifugous *a.* leaving the nest soon after hatching.

nidogen *n.* extracellular matrix glycoprotein found in most basal laminae. It interacts with laminin. *alt.* entactin.

nidus *n.* (1) nest; (2) a nest-like hollow; (3) a cavity for development of spores.

Nieuwkoop centre early organizer region in *Xenopus* blastula, which establishes the Spemann organizer.

nigrescent *a.* (1) blackish; (2) turning black.

nigropunctate *a.* black-spotted.

ninhydrin *n.* reagent that gives an intense blue colour with amino acids (yellow with proline).

Nissl bodies large granular particles in the cytoplasm of neurons, staining with basic dyes. They are composed of rough endoplasmic reticulum and polyribosomes. *alt.* **Nissl granules**.

Nissl stain dye that stains RNA and which is used to outline cell bodies in nervous system tissue.

nitrate bacteria bacteria in the soil that convert nitrite to nitrate. *see* nitrifier.

nitrate reductase general name for enzymes that catalyse the conversion of nitrate to nitrite. EC 1.6.6.1, 1.6.6.2, 1.6.6.3, 1.9.6.1 and 1.7.99.4.

nitrate reduction test clinical diagnostic test for ability of bacteria to reduce nitrate to nitrite, to aid identification of enteric bacteria, which are usually positive in this test.

nitric oxide (NO) gaseous local signalling molecule in animals and plants that is synthesized by the deamination of arginine by the enzyme nitric oxide synthase. It is produced e.g. in endothelial cells of blood vessels in response to acetylcholine, causing smooth muscle relaxation and blood vessel dilation. In many target cells it activates guanylate cyclase, resulting in the production of cyclic GMP. It is also produced by activated macrophages and neutrophils in response to infection and by many types of nerve cells. In plants NO is involved in responses to injury or infection.

nitrification *n.* the oxidation of ammonium ion to nitrite, and the oxidation of nitrite to nitrate, carried out chiefly by a few groups of soil bacteria (nitrifiers), mainly genera *Nitrosomonas* and *Nitrobacter* and also by a few species of fungi.

nitrifier *n.* any of a group of autotrophic aerobic soil bacteria that can either oxidize ammonia to nitrite (e.g. *Nitrosomonas*) or nitrite to nitrate (e.g. *Nitrobacter*). *alt.* **nitrifying bacteria**.

nitrite bacteria bacteria in the soil that convert ammonium to nitrite. *see* nitrifier.

nitrite-oxidizing bacteria nitrifying bacteria such as *Nitrobacter*, which oxidize nitrite to nitrate in the soil.

nitrite reductase enzyme that catalyses the conversion of nitrite to ammonium hydroxide. EC 1.7.1.4.

nitrocellulose a material to which nucleic acids and proteins adhere and which is used in sheet form (nitrocellulose membrane) as a support for these molecules in many applications in molecular biology. *see* Northern blotting, Southern blotting, Western blot.

nitrocobalamine *n.* vitamin $B_{12}c$. *see* cobalamin.

nitrofuran *n.* antibiotic that interferes with bacterial protein synthesis.

nitrogen (N) *n.* gaseous element, which in the free state is a colourless odourless unreactive gas (N_2), which makes up 78% of the Earth's atmosphere by volume. An essential macronutrient for living organisms as it is a component of organic molecules such as proteins and nucleic acids. *see* nitrogen cycle, nitrogen fixation.

nitrogen-15 (^{15}N) naturally occurring stable isotope of nitrogen, accounting for about 0.4% of atmospheric N_2.

nitrogenase *n.* enzyme complex in nitrogen-fixing microorganisms which catalyses the reduction of elemental nitrogen (N_2) to ammonia. The most widespread form contains molybdenum. Other nitrogenases containing vanadium and iron have been found.

nitrogen assimilation in plants, the uptake of nitrogen from the soil in the form of ammonia, nitrites and nitrates.

nitrogen balance equilibrium state of body in which nitrogen intake and excretion are equal. *alt.* nitrogen equilibrium.

nitrogen cycle the sum total of processes by which nitrogen circulates between the atmosphere and the biosphere or any subsidiary cycles within this overall process. Atmospheric elemental nitrogen (N_2) is converted by a few groups of soil and aquatic bacteria into inorganic nitrogenous compounds, primarily ammonia, by the process of nitrogen fixation. These inorganic compounds are incorporated into plants and bacteria and thence into animals, with the synthesis of complex nitrogen-containing organic molecules in their tissues. Organic nitrogen-containing compounds are subsequently broken down by bacteria and fungi (ammonification and nitrification) to generate inorganic nitrogen compounds – ammonia, nitrites and nitrates – which may be used by plants as nutrients, or may be converted to elemental nitrogen or nitrous oxide by certain bacteria (denitrification), thus releasing nitrogen to the atmosphere. The cycle also incorporates non-biological exchanges of nitrogen between atmosphere and biosphere as in the precipitation of inorganic nitrogen compounds in rainwater and the fixation of atmospheric nitrogen by lightning.

nitrogen equilibrium nitrogen balance *q.v.*

nitrogen fixation the process whereby atmospheric elemental nitrogen (dinitrogen, N_2) is reduced to ammonia (NH_3), and which is carried out in the living world only by some free-living bacteria and cyanobacteria and by a few groups of bacteria in symbiotic association with plants (the *Rhizobium*–legume association and the actinomycete–non-legume associations). The reaction is catalysed by the enzyme nitrogenase. Biological nitrogen fixation is the chief process by which atmospheric nitrogen enters the biosphere and becomes available as a nutrient to other organisms, although industrial nitrogen fixation to produce nitrogen fertilizers is now also of considerable significance. A smaller amount of atmospheric nitrogen is also fixed by conversion into nitrogen oxides by the action of lightning. *see also* nitrogen cycle.

nitrogenous *a. pert.* or containing nitrogen.

nitrogenous bases in biology, refers to the bases adenine, guanine, cytosine, thymine and uracil, which are present in nucleic acids.

nitrophilous *a.* nitrogen-loving, *appl.* plants.

nitrosamine *n.* potential carcinogen produced by reaction of dietary nitrates with secondary amines at low pH.

nitrosifying bacteria ammonia-oxidizing bacteria *q.v.*

Nitrospira *n.* a phylum of the Bacteria distinguished on DNA sequence criteria.

NK cells natural killer cells *q.v.*

N-linked glycosylation *see* glycosylation.

NLRs NOD-like receptors *q.v.*

NLS nuclear localization sequence *q.v.*

NMDA *N*-methyl-D-aspartate, structural analogue of glutamate.

NMDA receptor one type of receptor for the neurotransmitter glutamate in the brain, so-called from its specific interaction with *N*-methyl-D-aspartate. NMDA receptors are ligand-gated ion channels.

NMR spectroscopy nuclear magnetic resonance spectroscopy *q.v.*

N-nucleotides, N-regions short sequences of nucleotides, not encoded in the genome, which are added at the junctions between V, J and D gene segments during the rearrangement of immunoglobulin and T-cell receptor genes. They contribute to antibody and T-cell receptor diversity.

NO nitric oxide *q.v.*

nociception *n.* sensing of painful or injurious stimuli.

nociceptive *a.* (1) *appl.* stimuli which tend to injure tissue or produce pain; (2) *appl.* reflexes which protect from injury.

nociceptor *n.* receptor sensitive to injurious or painful stimuli.

nocodazole *n.* drug that prevents assembly of microtubules.

noctilucent *a.* emitting light. *see* bioluminescence, phosphorescence.

nocturnal *a.* (1) seeking food and moving about only at night; (2) occurring only at night.

NOD family of intracellular proteins in mammals that recognize components of bacterial cell walls and are involved in signalling the presence of infection and stimulating the production of antimicrobial defenses.

nodal *a. pert.* a node or nodes.

Nodaviridae *n.* family of single-stranded, positive-strand RNA viruses of insects.

node *n.* (1) (*bot.*) knob or joint of a plant stem at which leaves arise; (2) (*zool.*) aggregation of specialized cardiac muscle cells in heart, e.g. atrioventricular and sinuatrial nodes; (3) nodes of Ranvier *q.v.*; (4) branch point on a phylogenetic tree, representing an ancestral sequence or taxon that then diverges to form two new taxa.

nodes of Ranvier constrictions of the myelin sheath occurring at intervals along myelinated nerve fibres, and at which the axon membrane is exposed.

NOD-like receptors (NLRs) evolutionarily ancient family of intracellular proteins present in both plants and animals that recognize components of bacterial cell walls and are involved in signalling the presence of infection and stimulating the production of antimicrobial defenses.

nodose *a.* having knots or swellings.

nodular *a. pert.*, or like a nodule or knot.

nodulated *a.* bearing nodules, in plants *appl.* esp. to roots bearing nodules containing nitrogen-fixing bacteria.

nodulation *n.* formation of nitrogen-fixing root nodules on plant roots.

nodule *n. see* root nodule.

noduliferous *a.* bearing nodules.

nodulins *n.plu.* nodule-specific proteins of host origin produced by legumes infected with nitrogen-fixing rhizobia, and which are involved in establishing symbiotic nitrogen fixation.

Nomarski differential-interference-contrast microscopy a type of optical microscopy that produces a high-contrast image of unstained living cells and tissue.

nomen nudum a name not valid because when it was originally published the organism to which it referred was not adequately described, defined or sketched.

nomenspecies *n.* a group of individuals bearing a binomial name, whatever its status in other respects.

non-adaptive *a. appl.* traits that tend to decrease an organism's genetic fitness.

non-allelic *a.* (1) not affecting the same gene, *appl.* mutations in different genes; (2) *appl.* different members of a gene family that encode versions of the same protein.

non-associative learning learning in response to a single stimulus. *see* dis-habituation, habituation, sensitization. *cf.* associative learning.

non-autonomous *appl.* situation in which the cell that produces a particular protein (or carries a particular mutation) causes changes in other cells that do not produce it (or carry the mutation).

non-coding *a. appl.* a DNA or RNA sequence that does not encode a protein or part of a protein. *see also* microRNAs.

non-coding RNAs (ncRNAs) RNAs that do not encode protein. They include structural RNAs such as rRNA and tRNA, the RNAs in spliceosomes and other small structural RNAs, and also a large group of small regulatory RNAs of between ~21–30 nucleotides long that are produced by the specific processing of longer transcripts. These act mainly as gene regulatory RNAs and include microRNA, Piwi-interacting RNA, small interfering RNA (all *q.v.*).

non-competitive inhibitor *see* inhibitor.

non-conjugative *a. appl.* plasmids that cannot direct their own transfer by inducing conjugation between two bacteria.

non-conjunction *n.* the failure of homologous chromosomes to pair at meiosis.

non-covalent bond general term for a variety of attractive forces between atoms and molecules that do not involve the sharing of electrons. Such bonds are weaker than covalent bonds and can be disrupted by environmental changes in ionic strength or temperature. *see* electrostatic bond, hydrogen bond, van der Waals forces. *cf.* covalent bond.

non-cyclic photophosphorylation the type of photosynthesis in which both ATP and NADPH are generated directly, as in green plants and cyanobacteria.

non-declarative memory memory of learned information that is shown by performance rather than by conscious recollection. *alt.* procedural memory.

non-degradable *a. appl.* material which cannot be broken down by the natural processes of decomposition by microorganisms, and which therefore persists in the environment.

non-disjunction *n.* failure of a pair of chromatids to separate and go to opposite poles at mitosis or meiosis, which results in aneuploidy in the daughter cells.

non-essential amino acids amino acids which can be synthesized in the body and are not required in the diet: for humans these are Ala, Asn, Asp, Cys, Glu, Gln, Gly, Pro, Ser, Tyr.

non-haem iron protein iron–sulphur protein *q.v.*

non-histone protein any of numerous types of protein, other than histones, associated with eukaryotic chromosomes.

non-homologous end-joining a repair process for double-strand breaks in DNA, in which the broken ends are juxtaposed and ligated.

non-ionic detergent detergent in which the polar group at the end of the molecule is not charged. Triton X-100 is an example. Such detergents are relatively mild and can be used to solubilize membrane proteins without loss of all function. *cf.* ionic detergent.

non-medullated fibres grey or yellowish-grey nerve fibres, lacking myelin sheaths. They make up most of the sympathetic system and some of the central nervous system.

non-Mendelian *a. appl.* genes or characters which are not inherited according to Mendel's laws, e.g. mitochondrial or chloroplast genes. *n.* **non-Mendelian inheritance**.

non-pathogenic *a. appl.* or *pert.* a microorganism that does not cause disease.

non-permissive *a.* (1) (*genet.*) *appl.* conditions under which an organism or cell carrying a conditional lethal mutation displays the mutant phenotype and dies or becomes severely defective; (2) (*virol.*) *appl.* conditions in which virus infection of a cell does not occur, and which may be due to environmental or genetic factors (i.e. species differences in susceptibility).

non-polar *a. appl.* molecules or covalent bonds in which the bonding electrons are attracted equally to both atoms in the bond, and thus there is no accumulation of positive or negative charge. Non-polar molecules are generally insoluble in water. *cf.* polar.

non-porous wood secondary xylem with no vessels.

non-reciprocal recombination a genetic recombination event such as the integration of one DNA molecule into another, in which there is not an exact exchange of parts of the two DNAs undergoing recombination. *cf.* reciprocal recombination.

non-repetitive DNA single-copy DNA *q.v.*

non-retroviral retrotransposon a type of transposable element that is thought to be derived from a cellular RNA. It encodes a reverse transcriptase and moves via an RNA intermediate. An example is the L1 element of the human genome.

nonsense codon termination codon *q.v.*

nonsense mutation mutation which generates one of the nonsense (termination) codons UAA, UAG or UGA, resulting in premature termination of polypeptide synthesis during translation.

nonsense suppressor mutation which generates a tRNA that inserts an amino acid at a nonsense codon.

non-shivering thermogenesis (NST) the generation of large amounts of heat by metabolism, without shivering, of which some mammals are capable. *see* brown fat.

non-spiking *a. appl.* neurons that do not produce impulses.

non-storied *a. appl.* cambium in which initials are not arranged in horizontal series on tangential surfaces.

non-striated muscle smooth muscle *q.v.*

non-sulphur purple bacteria group of photosynthetic heterotrophic bacteria containing purple pigments.

non-synonymous *a. appl.* mutations that result in the substitution of one amino acid for another in a protein sequence. *cf.* synonymous.

non-transmissible *a. appl.* diseases not caused by infection with microorganisms, e.g. cancer, diabetes, cardiovascular disease, and which therefore cannot be transmitted from one individual to another.

non-viable incapable of surviving or developing.

NOR nucleolar organizer (region) *q.v.*

noradrenaline (NA) *n.* catecholamine closely related to adrenaline and with similar effects, secreted with it in small amounts by the adrenal medulla. Also acts as a neurotransmitter in sympathetic and enteric nervous systems, being secreted by sympathetic nerve endings in internal organs such as the gut, heart and spleen, and in various pathways in the central nervous system. *alt.* (US) norepinephrine. *see* Fig. 3 (p. 15).

noradrenergic *a. appl.* nerve fibres, of the sympathetic system, that liberate noradrenaline from their terminals.

norepinephrine (NE) preferred name for noradrenaline in North America.

norma *n.* view of the skull as a whole from certain points: basal, vertical, frontal, occipital and lateral.

normal distribution statistical term for a distribution in which most values fall near a central point, producing a bell-shaped curve. *alt.* Gaussian distribution.

normalizing selection stabilizing selection *q.v.*

normoblast *n.* immature nucleated mammalian red blood cell, derived from an erythroblast and which develops into an erythrocyte.

normoxic *a. appl.* conditions of normal oxygen concentration.

North African Desert region part of the Palaeotropical floristic kingdom, stretching from the north-western coast of Africa across the Sahara to Egypt and the Arabian peninsula as far as the western edge of the Himalayas.

North and East Australian region part of the Australian floristic kingdom, consisting of Tasmania and a wide temperate and tropical belt along the eastern and north-eastern side of Australia.

Northeast African Highland and Steppe region part of the Palaeotropical floristic kingdom, consisting of the Ethiopian highlands in North-eastern Africa and the land to their east as far as the eastern African coast, the bottom tip of the Red Sea and the tip of the Arabian peninsula.

Northern blotting technique in which RNAs separated by gel electrophoresis are transferred by blotting onto a suitable medium (e.g. a nitrocellulose membrane) for subsequent hybridization with radioactive probes for identification and isolation of RNAs of interest. Named by analogy with the Southern blotting technique. *cf.* Western blot.

nosocomial *a.* hospital-acquired, *appl.* infections.

nosogenic pathogenic *q.v.*

nosology *n.* (1) branch of medicine dealing with the classification of diseases; (2) pathology *q.v.*

nostrils *n.plu.* exterior openings of the nose. *alt.* nares.

notal *a.* (1) dorsal; (2) *pert.* the back; (3) *pert.* notum.

notate *a.* marked with lines or spots.

Notch signalling pathway important intercellular signalling pathway in animal development, used in many different contexts during embryogenesis for determining cell fate and essential for embryonic survival. It is initiated by binding of the cell-surface receptor protein Notch by a cell-surface ligand on a neighbouring cell. The cytoplasmic tail of Notch is cleaved off and migrates to the nucleus where it forms a complex with a CSL protein and activates transcription of genes containing a Notch-responsive control sequence.

nothocline hybrid cline *q.v.*

Nothosauria, nothosaurs *n., n.plu.* an extinct order of streamlined fish-eating marine reptiles.

notocephalon *n.* dorsal shield of legbearing segments in some ticks and mites.

notochord *n.* slender rod of cells of mesodermal origin running along the back in the early chordate embryo and which directs formation of the neural tube. It persists in primitive chordates but in vertebrates is replaced by the spinal column.

notochordal *a. pert.* or enveloping notochord, *appl.* e.g. sheath.

notogaster *n.* posterior dorsal shield of certain mites and ticks.

notogenesis *n.* the development of the notochord.

notonectal *a.* swimming with back downwards.

notopleural suture in dipterans, a lateral suture separating the mesonotum from the pleuron.

notopodium *n.* dorsal lobe of polychaete parapodium.

nototribe *a. appl.* flowers whose anthers and stigma touch back of insect as it enters corolla, a device for securing cross-fertilization.

notum *n.* dorsal portion of an insect segment.

novobiocin *n.* antibiotic synthesized by the actinomycete *Streptomyces niveus*. Interferes with DNA gyrase.

NSC neural stem cell *q.v.*

NPC nuclear pore complex *q.v.*

NST non-shivering thermogenesis *q.v.*

N-terminal, N terminus *see* amino terminus.

NTPs nucleoside triphosphates *q.v.*

nucellus *n.* in seed plants, the megasporangial tissue which persists around the megaspore (embryo sac) and the megagametophyte that develops within it, and eventually forms the inner layer of the ovule wall.

nuchal *a. pert.* nape of neck.

nuciferous *a.* nut-bearing.

nucivorous *a.* nut-eating.

nuclear *a. pert.* a nucleus.

nuclear disc star-like structure formed by chromosomes at equator of spindle during mitosis.

nuclear envelope the double membrane surrounding the eukaryotic nucleus. *see also* nuclear lamina.

nuclear export receptor, nuclear export signal *see* nuclear transport receptors.

nuclear gene, nuclear genome those genes present on the chromosomes in the nucleus, as opposed to mitochondrial or chloroplast genes.

nuclear hormone receptor *see* nuclear receptor.

nuclear import receptor *see* nuclear transport receptors.

nuclear lamina fibrous protein layer underlying the inner nuclear membrane.

nuclear layer (1) internal layer of cerebral cortex; (2) inner nuclear layer of retina, between inner and outer plexiform layers, and outer nuclear layer, between outer plexiform layer and basement membrane.

nuclear localization sequence (NLS) sequence of amino acids or a surface feature (signal patch) that occurs in proteins destined to enter the nucleus, and which aids their transport through nuclear pores and their maintenance in the nucleus. *alt.* nuclear localization signal, nuclear location sequence, nuclear recognition sequence.

nuclear magnetic resonance spectroscopy (NMR) method for determining the three-dimensional structure of small proteins (<20,000 molecular mass) and other small molecules in solution.

nuclear matrix the insoluble proteinaceous material left in the nucleus after certain biochemical extraction procedures. It has been proposed to form a framework on which the chromosomes are organized.

nuclear membrane the double membrane surrounding the eukaryotic nucleus, or one of these membranes.

nuclear plate equatorial plate *q.v.*

nuclear pore structure formed where inner and outer nuclear membranes join and which connects cytoplasm and nucleoplasm. They are present in large numbers in the nuclear envelope and form channels through which macromolecules pass from nucleus to cytoplasm or vice versa.

nuclear pore complex (NPC) large and elaborate structure composed of more than 50 different proteins that surrounds a nuclear pore and is involved in the selective transport of macromolecules through the pore.

nuclear receptor any of a large superfamily of intracellular receptor proteins for compounds such as retinoids and steroid and thyroid hormones. The receptors are gene regulatory proteins which become active only after binding their ligand. Receptors are either located in the cytosol, and enter the nucleus after binding ligand, or are already bound to DNA in the nucleus. The ligand–receptor complex acts at specific control sequences to activate target genes.

nuclear recognition sequence nuclear localization sequence *q.v.*

nuclear scaffold *see* nuclear matrix *q.v.*

nuclear spindle organized system of microtubules which spans the centre of a cell undergoing mitosis or meiosis. It forms a framework that attaches chromosomes and guides their movement and segregation into the two daughter nuclei. The spindle develops from material surrounding the centrioles in animal cells and a similar region in plant cells.

nuclear transplantation the transfer of an intact nucleus from one cell to another which has had its own nucleus removed. The technique is used to study the expression of genes in different cytoplasmic environments, and in cloning. *see also* somatic cell nuclear transfer.

nuclear transport receptors proteins that selectively guide and transport proteins and RNAs into and out of the nucleus. Both import receptors and export receptors bind to recognition sequences on the macromolecule or complex to be transported and to nucleoporins of the nuclear pore complex. For RNA export, the receptors bind to recognition signals on proteins complexed with the RNA. *alt.* karyopherins.

nuclease *n.* any of a class of enzymes that degrade nucleic acids into shorter oligonucleotides or single nucleotide subunits by hydrolysing sugar–phosphate bonds in the nucleic acid backbone. *see* deoxyribonuclease, endonuclease, exonuclease, ribonuclease.

nucleate *a.* containing a nucleus.

nucleation *n.* initiation of the formation of a structure, such as a microtubule, by a pre-existing 'seed' structure, onto which components start assembling.

nuclei *plu.* of nucleus.

nucleic acid *n.* deoxyribonucleic acid (DNA) or ribonucleic acid (RNA). Nucleic acids are very large linear molecules containing C, H, O, N and P, which on partial hydrolysis yield nucleotides and nucleosides, and which are composed of one (RNA) or two (DNA) polynucleotide chains. They are essential components of all living cells, where they are the carriers of genetic information (DNA and mRNA), components of ribosomes (rRNA), and involved in deciphering the genetic code (tRNA). *see also* DNA (deoxyribonucleic acid), genetic code, messenger RNA, nucleotide, polynucleotide, ribosomal RNA, RNA (ribonucleic acid), transfer RNA.

nucleic acid hybridization *see* DNA hybridization, *in situ* hybridization.

nucleocapsid *n.* the nucleic acid together with the protein coat of an enveloped virus.

nucleoid (1) *a.* resembling a nucleus; *n.* (2) the 'chromosome' in prokaryotes. It is not enclosed in a membrane; (3) a nucleus-like body occurring in some red blood cells; (4) the dense region seen in some viruses which represents the nucleic acid.

nucleolar *a. pert.* a nucleolus.

nucleolar organizer (NOR) region of a chromosome associated with the nucleolus, corresponding to a cluster of rRNA genes.

nucleolus *n.* region of the nucleus where rRNA is synthesized. It consists of a fibrillar core surrounded by a granular region, and is the site of assembly of ribosomal subunits and other ribonucleoprotein complexes. *plu.* **nucleoli**. *see* Fig. 7 (p. 105).

nucleolysis *n.* disintegration of a cell nucleus.

nucleoplasm *n.* the ground substance of the nucleus (excluding the nucleolus) internal to the nuclear membrane and excluding the chromatin. *alt.* nuclear sap.

nucleoplasmin *n.* a molecular chaperone involved in the assembly of nucleosomes in eukaryotic cells.

nucleoporin *n.* protein of the nuclear pore complex to which nuclear transport receptors bind.

nucleoprotein *n.* any complex of protein and nucleic acid.

nucleosidase *n.* (1) any of a class of enzymes that split nucleosides (and nucleotides) into a nitrogenous base and a pentose (or pentose phosphate). EC 3.2.2, *alt.* nucleotide phosphorylase; (2) *r.n.* for enzyme splitting N-ribosyl-purine into purine and ribose, EC 3.2.2.1.

nucleoside *n.* any of a group of compounds consisting of a purine or pyrimidine base (commonly adenine, cytosine, guanine, thymine, uracil) linked to the sugar ribose or deoxyribose, and including adenosine, cytidine, guanidine, thymidine, uridine.

nucleoside triphosphates (NTPs) collective name for the four nucleotides required for RNA synthesis.

nucleoside phosphorylase nucleosidase *q.v.*

nucleosome *n.* repeating structural unit in eukaryotic chromatin in which DNA is wound round a protein core composed of two each of the four histones H2A, H2B, H3 and H4. Nucleosomes are linked by a stretch of DNA associated with histone H1.

nucleosome-remodelling complex ATPase-containing protein complex that acts to change the structure of a nucleosome temporarily such that DNA becomes less tightly bound to the histone core. This can lead to increased accessibility of the DNA to other proteins, to repositioning of the histone core on the DNA, to the transfer of the histone core from one DNA molecule to another, or to the exchange of histones in the nucleosome.

nucleotidase *n.* any of several enzymes that hydrolyse nucleotides into nucleosides and orthophosphate. EC 3.1.3.5–7.

nucleotide *n.* phosphate ester of a nucleoside, consisting of a purine or pyrimidine base linked to a ribose or deoxyribose phosphate (up to three phosphate groups linked in series). The purine nucleotides chiefly have adenine or guanine as the base, the pyrimidine nucleotides cytosine, thymine or uracil. Nucleotides are the basic chemical subunits of DNA and RNA. Nucleotides containing deoxyribose are called deoxyribonucleotides, those containing ribose are called ribonucleotides. Nucleotides containing one phosphate group are known as nucleoside phosphates, those containing two phosphate groups as nucleoside diphosphates, and those containing three phosphate groups as nucleoside triphosphates, such as adenosine triphosphate (ATP).

nucleotide excision repair type of DNA repair used to repair damage due to formation of pyrimidine dimers or bulky adducts in one of the strands. The damaged portion of the DNA strand is excised and resynthesized using the undamaged strand as template.

nucleotide sequence order of nucleotide subunits in a nucleic acid. In mRNA and protein-coding DNA the nucleotide sequence determines the amino acid sequence of the protein specified. *alt.* base sequence, DNA sequence.

nucleotide sequencing *see* DNA sequencing.

nucleotidyltransferase *n.* any of a class of enzymes that transfer a nucleotide (in this case a nucleoside monophosphate) from one compound to another, and which include e.g. adenylyltransferases, uridylyltransferases and the DNA and RNA polymerases. EC 2.7.7.

nucleus *n.* (1) large dense organelle bounded by a double membrane, present in eukaryotic but not prokaryotic cells, and which contains the chromosomes. Transcription and RNA processing take place in the nucleus. Its function is essential to the survival of the cell. *see* Fig. 7 (p. 105); (2) the centre of any structure, around which it grows; (3) (*neurobiol.*) any of various masses of grey matter (nerve cell bodies) in the central nervous system. *plu.* **nuclei**.

nucleus ambiguus cells in medulla oblongata from which originate the motor fibres of glossopharyngeal and vagus, and of cerebral part of spinal accessory nerves.

nucleus pulposus the soft core of an intervertebral disc, remnant of notochord.

nucleus robustus archistrialis in brain of birds, a main centre for motor organization in the forebrain.

nuculanium *n.* berry formed from a superior ovary.

nucule *n.* nutlet.

nudibranch *a.* lacking a protective cover over the gills, *appl.* a group of shell-less molluscs (Nudibranchia), the sea slugs.

nudibranchiate *a.* having gills not covered by a protective shell or membrane, in a branchial chamber, *appl.* certain molluscs.

nudicaudate *a.* having a tail not covered in hair or fur.

nudicaulous *a. appl.* or having a stem without leaves.

nudiflorous *a.* having flowers without glands or hairs.

Nuhn, glands of anterior lingual glands.

null allele mutant allele that results in an absence of functional gene product, the mutation itself being termed a null or amorphic mutation. Null alleles are usually recessive.

null hypothesis in planning a scientific experiment, the hypothesis that would give a certain set of experimental results in the conditions of the experiment. If the observed results depart significantly from these expected results the null hypothesis is unlikely to be true.

nullipennate *a.* without flight feathers.

nullisomic *a. appl.* an organism or cell which has both members of a pair of chromosomes missing. *n.* **nullisomic**.

nullizygous *a.* homozygous for a null allele or mutation.

null mutation mutation that causes a complete loss of the normal function of the gene.

numerical *a. appl.* hybrid of parents with different chromosome numbers.

numerical abundance the number of individuals of a species present in a given area.

numerical chromosome mutation a mutation that involves a change in the number of chromosomes in a cell.

numerical taxonomy classification of organisms by a quantitative assessment of their phenotypic similarities and differences, not necessarily leading to a phylogenetically based classification.

nummulation *n.* the tendency of red blood cells to adhere together like piles of coins (rouleaux).

nummulite *n.* coin-shaped fossil foraminiferan.

nummulitic *a.* containing nummulites.

nunatak *a.* an area, on a mountain or plateau, which has escaped past environmental changes, such as glaciation, and in which plants and animals of earlier floras and faunas have survived.

nuptial flight flight taken by queen bee when fertilization takes place.

nurse cells (1) single cells or layers of cells attached to or surrounding an oocyte and which provide nutrients to it and synthesize specific proteins and mRNAs which pass into the oocyte and remain inactive until after the egg is fertilized and development begins; (2) Sertoli cells in testis; (3) certain other cells that provide nourishment and sometimes storage material to other cells.

nurture *n.* the sum total of environmental influences on a developing individual.

nut *n.* a dry, indehiscent, one- or two-seeded, one-chambered fruit with a hard woody pericarp (the shell), such as an acorn.

nutant *a.* (1) bent downwards; (2) drooping.

nutation *n.* rotational curvature of the growing tip of a plant; slow rotating movement by pseudopodia.

nutlet *n.* (1) a small nut; (2) an individual achene of a fruit such as beechmast.

nutraceutical *n.* a foodstuff that provides health benefits.

nutrient *n.* any substance used or required by an organism as food. *see also* macronutrient, micronutrient.

nutrient cycles the exchanges of elements between the living and non-living components of an ecosystem.

nutrition *n.* the process by which an organism obtains from its environment the energy and the chemical elements and compounds it needs for survival and growth. *see* autotroph, chemotroph, heterotroph, phototroph.

nutritive *a.* concerned with nutrition.

NVC National Vegetation Classification *q.v.*

nyctanthous *a.* flowering by night.

nyctinasty *n.* 'sleep' movements in plants, involving a change in the position of leaves, petals, etc., as they close at night or in dull weather in response to a change in the level of light and/or temperature. *a.* **nyctinastic.** *alt.* **nyctitropism.**

nyctipelagic *a.* rising to the surface of the sea only at night.

nyctitropism nyctinasty *q.v.*

nyctoperiod *n.* daily period of exposure to darkness.

nymph *n.* a juvenile form without wings or with incomplete wings in insects with incomplete metamorphosis, i.e. when the change at each moult is small and the larvae are relatively similar to the adult form. *a.* **nymphal.**

Nymphales *n.* order of herbaceous aquatic dicots including families Ceratophyllaceae (hornwort) and Nymphaceae (water lily).

nymphochrysalis *n.* pupa-like resting stage between larval and nymphal form in certain mites.

nymphosis *n.* the process of changing into a nymph or pupa.

nystagmus *n.* condition when the eye makes large involuntary oscillatory movements, as in dizziness.

nystatin *n.* an antibiotic with antifungal activity.

O

O symbol for the chemical element oxygen *q.v.*

oak-apple *n.* type of hard spherical gall found on stems and leaves of oak and caused by larvae of a species of gall wasp.

oar feathers the wing feathers used in flight.

ob- prefix derived from L. *ob*, against, signifying the other way round, obversely, esp. when prefixed to names of leaf shapes.

obcompressed *a.* flattened in a vertical direction.

obconic *a.* shaped like a cone, but attached at its apex.

obcordate, obcordiform *a.* inversely heart-shaped, *appl.* leaves which have stalk attached to apex of heart.

obcurrent *a.* converging, and attaching at point of contact.

obdeltoid *a.* more-or-less triangular with point of attachment at apex of triangle.

obdiplostemonous *a.* with stamens in two whorls, the inner opposite the sepals and the outer opposite the petals.

obelion *n.* the point on skull between parietal foramina, on sagittal suture.

obesity *n.* an excessive increase in body weight due to the deposition of fat tissue. Clinically defined for adults as a body mass index (weight (kg)/height (m)^2) of more than 30.

obimbricate *a.* with regularly overlapping scales, with the overlapped ends downwards.

oblanceolate *a.* inversely lanceolate, of leaves.

obligate *a.* (1) obligatory; (2) limited to one mode of life or action.

obligate parasite an organism that can only live as a parasite.

oblique *a. appl.* leaves, asymmetrical.

obliquely striated muscle type of muscle found in nematodes, molluscs and annelid worms, in which the arrangement of the myofibrils leads to a diagonal striation, and which is capable of greater extension and contraction than vertebrate striated muscle.

obliquus *n.* an oblique muscle, as of ear, eye, head, abdomen.

obliterate *a.* (1) indistinct, *appl.* markings on insects; (2) suppressed.

obliterative coloration type of coloration in which parts of an organism normally exposed to the brightest light are shaded more darkly, ensuring that it blends with its background more effectively.

obovate *a.* inversely egg-shaped, *appl.* leaf with narrow end attached to stalk.

obpyriform *a.* inversely pear-shaped.

obsolescence *n.* (1) the gradual reduction and eventual disappearance of a species; (2) gradual cessation of a physiological process, or of a structure becoming disused, over evolutionary time. *a.* **obsolescent**.

obsolete *a.* (1) wearing out or disappearing; (2) *appl.* any character that is becoming less and less distinct in succeeding generations; (3) (*bot.*) *appl.* calyx united with ovary or reduced to a rim.

obsubulate *a.* (1) reversely awl-shaped; (2) narrow and tapering from tip to base.

obtect *a. appl.* pupa with wings and legs held close to body.

obturator *n.* any of various structures which close off a cavity.

obturator foramen oval foramen between ischium and pubis.

obtuse *a.* with blunt or rounded end, *appl.* leaves.

obtusilingual *a.* with a blunt short tongue.

obumbrate *a.* with some structure overhanging the parts so as partially to conceal them.

obverse *a.* with base narrower than apex.

obvolute *a.* overlapping.

obvolvent *a.* bent downwards and inwards, *appl.* wings and elytra of some insects.

Occam's or Ockham's razor the principle that where several hypotheses are possible, the simplest is chosen, first proposed by William of Ockham, a medieval scholastic philosopher.

occasional species one which is found from time to time in a community but is not a regular member of it.

occipital *a. pert.* to the back of the head or skull.

occipital condyles two knob-like protrusions on back of skull of amphibians and mammals which articulate with the atlas of backbone.

occipital foramen (1) posterior opening of head in insects; (2) the foramen magnum of skull in vertebrates.

occipitalia *n.plu.* the parts of the cartilaginous brain case forming the back of head. *alt.* occipital bones.

occipital lobe of brain, the hind part of cerebral hemisphere.

occipito-atlantal *a.* (1) *appl.* membrane closing gap between skull and neural arch of atlas in amphibians; (2) *appl.* dorsal (posterior) and ventral (anterior) membranes between margin of foramen magnum and atlas in mammals.

occipito-axial *a. appl.* ligament or tectorial membrane connecting occipital bone with axis.

occipitofrontal *a.* (1) *appl.* longitudinal arc of skull; (2) *appl.* tract of fibres running from frontal to occipital lobes of cerebral hemispheres.

occiput *n.* (1) back of head or skull; (2) dorsolateral region of insect head.

occluding junction intercellular contact that seals cells together in an epithelial sheet, preventing molecules from leaking from one side of the epithelium to the other, such as the tight junctions of vertebrates and the septate junctions of invertebrates.

occlusal *a.* (1) contacting the opposing surface; (2) *appl.* teeth which touch those of the other jaw when jaws are closed.

occlusion *n.* (1) closure or blocking of a duct, vessel or tubule; (2) overlapping activation of several motor neurons by simultaneous stimulation of several afferent nerves.

occlusor (1) *n.* a closing muscle; (2) *a. appl.* muscles of operculum or movable lid.

oceanic *a.* inhabiting the open sea, where the sea is deeper than 200 m.

oceanodromous *a.* migrating only within the ocean, *appl.* fishes. *cf.* potamodromous.

ocellate *a.* like an eye or eyes, *appl.* markings.

ocellated *a.* having ocelli, or eye-like spots or markings.

ocellus *n.* (1) (*zool.*) simple eye or eye-spot found in many invertebrates; (2) dorsal eye in insects; (3) eye-like marking; (4) (*bot.*) large cell in leaf epidermis specialized for reception of light. *plu.* **ocelli**. *a.* **ocellar**.

ochraceous, ochreous *a.* ochre-coloured.

ochre the termination (stop) codon UAA.

ochrea ocrea *q.v.*

ochroleucous *a.* yellowish-white, buff-coloured.

ochrophore *n.* cell containing yellow pigment.

ochrosporous *a.* having ochre-coloured spores.

Ockham's razor Occam's razor *q.v.*

ocrea *n.* (1) tubular sheath-like expansion at base of petiole; (2) a sheath; (3) partial covering of stipe of some toadstools, formed by fragments of the disintegrated veil.

ocreaceous *a.* like an ocrea, *appl.* various structures in plants and animals.

ocreate *a.* (1) having an ocrea; (2) booted; (3) sheathed; (4) intrafoliaceous *q.v.*

octa- prefix derived from Gk *octa*, eight, signifying having eight of, arranged in eights, etc.

octactine *n.* sponge spicule with eight rays.

octad *n.* group of eight cells originating by division of a single cell.

octagynous *a.* (1) having eight pistils or styles; (2) having eight carpels to a gynoecium.

octamerous *a. appl.* organs or parts of organs when arranged in eights.

octandrous *a.* having eight stamens.

octant *n.* one or all of the eight cells formed by division of the fertilized ovum in plants and animals.

Octapoda *n.* order of Cephalopoda whose members have eight tentacles and no shell (e.g. octopus).

octarch *a.* (1) *appl.* stems with steles having eight alternating groups of phloem and xylem; (2) with eight vascular bundles.

octopamine *n.* a catecholamine-like neurotransmitter.

octo- prefix derived from Gk *octa*, eight, signifying having eight of, arranged in eights, etc.

Octocorallia *n.* the soft corals, a subclass of colonial cnidaria with 8-fold radial symmetry, comprising the alcyonacean and other soft corals, the organ-pipe corals, the sea fans and sea whips (gorgonians), the sea pens and sea pansies, and the blue coral.

octopetalous *a.* having eight petals.

octoploid *a.* having eight haploid chromosome sets in somatic cells.

octopod *a.* having eight tentacles, arms or feet.

octoradiate *a.* having eight rays or arms.

octosepalous *a.* having eight sepals.

octosporous *a.* having eight spores.

octostichous *a.* (1) arranged in eight rows; (2) having leaves arranged in eights.

octozoic *a. appl.* a spore, of gregarines, containing eight sporozoites.

ocular *a. pert.*, or perceived by, the eye.

ocular dominance columns slabs of cells in visual cortex, perpendicular to the surface, which respond to stimuli from one or the other eye.

oculate *a.* having eyes or eye-like spots.

oculiferous, oculigerous *a.* having eyes.

oculofrontal *a. pert.* region of forehead and eye.

oculomotor *a.* causing movements of the eyeball, *appl.* 3rd cranial nerve, which controls four of the six small muscles moving the eyeball.

oculonasal *a. pert.* eye and nose.

oculus *n.* (1) the eye; (2) a leaf bud in a tuber.

OD optical density. *see* absorbance.

Oddi's sphincter muscle fibres surrounding duodenal end of common bile duct.

odd-pinnate pinnate with one terminal leaflet.

odd-toed ungulates perissodactyls *q.v.*

Odonata *n.* order of insects that includes the dragonflies and damselflies. They are winged carnivorous insects with brilliant metallic colouring, whose eggs are laid in water and which develop through an aquatic nymph (larval) stage that has gills.

odontoblast, odontocyte *n.* (1) one of the columnar cells on outside of dental pulp that secretes dentine of tooth; (2) one of the cells giving rise to teeth of molluscan radula.

Odontoceti *n.* the toothed whales, a suborder of the Cetacea that includes the sperm whale, killer whale, narwhal, porpoises and dolphins. They are all predatory, feeding on fish and other marine animals.

odontoclast *n.* one of the large multinucleate cells that absorb roots of milk teeth and destroy dentine.

odontogeny *n.* the origin and development of teeth.

odontoid *a.* (1) tooth-like; (2) *pert.* the odontoid process.

odontoid process tooth-like peg around which the atlas or first cervical vertebra rotates.

odontology *n.* dental anatomy, histology, physiology and pathology.

odontophore *n.* the tooth-bearing organ in molluscs.

odontorhynchous *a.* having inner edge of bill bearing plate-like ridges.

odontosis *n.* (1) dentition *q.v.*; (2) tooth development.

odontostomatous *a.* having toothbearing jaws.

odorant *n.* any substance that stimulates the sensory receptors in the nose – the odorant receptors – and thus has a smell.

odorimetry *n.* measurement of the strength of the sense of smell, using substances of known ability to stimulate olfaction.

odoriphore osmophore *q.v.*

oecium *n.* calcareous or chitinous covering of a hydrozoan colony.

oedema *n.* swelling of a tissue through increase in tissue fluid. *alt.* edema.

oenocyte *n.* one of the large cells from clusters which surround trachea and fatbody of insects and undergo changes related to the moulting cycle.

oesophageal *a. pert.* or near oesophagus.

oesophageal bulbs two swellings on the oesophagus in nematodes, the posterior of which, the pharynx, exhibits rhythmical pumping movements.

oesophageal glands or pouches in earthworms, outgrowths of the oesophagus which secrete calcium carbonate.

oesophagus *n.* that part of alimentary canal between pharynx and stomach or part equivalent thereto. *alt.* esophagus, gullet.

oestradiol *n.* a major oestrogen in ovarian follicular fluid and also produced by the placenta. It is responsible for the development and maintenance of secondary female sexual characteristics, the maturation and cyclic function of accessory sexual organs and the development of the duct system in mammary glands.

oestrin oestrone *q.v.*

oestriol *n.* oestrogen present in urine of pregnant women.

oestrogenic *a.* (1) inducing oestrus, *appl.* hormones; (2) *pert.* oestrogen.

oestrogens *n.plu.* vertebrate steroid hormones, the principal female sex hormones, synthesized chiefly by the ovary and placenta in females and responsible for the development and maintenance of secondary female sexual characteristics and the growth and function of female reproductive organs. Similar compounds have been found in plants.

oestrone *n.* derivative of oestradiol with similar activity, excreted in urine of some pregnant mammals.

oestrous *a. pert.* oestrus.

oestrus *n.* period of sexual heat and fertility in a female mammal when she is receptive to the male.

oestrus cycle reproductive cycle in female mammals in the absence of pregnancy. It consists of oestrus, when ovarian follicles mature and ovulation takes place, metoestrus (luteal phase) and pro-oestrus.

officinal *a.* used medicinally, *appl.* plants.

oidia *plu.* of oidium. *see* oidiospore.

oidiophore *n.* specialized hypha that cuts off oidia from its tip.

oidiospore, oidium *n.* fungal spore formed by transverse segmentation of a hypha. *plu.* oidia.

oil bodies, oil storage bodies lipid bodies *q.v.*

oil gland (1) any gland secreting oils; (2) in birds, a gland in the skin that secretes oil used in preening the feathers.

oils *n.plu.* glycerides and esters of fatty acids which are liquid at 20°C, the fatty acids being in general less saturated than in fats. *cf.* fat.

Okazaki fragments lengths of DNA of around 1000 nucleotides that are formed during DNA replication by the discontinuous synthesis of the lagging DNA strand on the 5′ to 3′ template strand. They are joined by DNA ligase into a continuous DNA strand.

old-growth forest uncut virgin forest containing trees that are hundreds and sometimes thousands of years old, as in the forests of Douglas fir, western hemlock, giant sequoia and redwoods in the western United States.

oleaginous *a.* containing or producing oil.

Oleales *n.* order of dicot trees, shrubs and climbers comprising the family Oleaceae (olive, privet).

olecranon *n.* large bony process at upper end of ulna.

oleic acid common unsaturated 12-carbon fatty acid.

oleiferous *a.* producing oils.

olein *n.* a fat, containing oleic acid, liquid at ordinary temperatures, found in animal and plant tissues.

oleoplast, oleosome elaioplast, elaiosome *q.v.*

olfaction *n.* (1) the sense of smell; (2) the process of smelling.

olfactory *a. pert.* sense of smell, *appl.* stimuli, organs, tract of nerve fibres.

olfactory bulb, olfactory lobe lobe projecting from anterior lower margin of a cerebral hemisphere in the vertebrate brain. It contains the terminations of the olfactory nerves and is concerned with the sense of smell.

olfactory epithelium epithelial lining of nasal cavity that contains the olfactory receptor neurons that sense odours.

olfactory lobe olfactory bulb *q.v.*

olfactory nerves 1st cranial nerves in vertebrates, sensory nerves running from olfactory organs to olfactory bulb in forebrain.

olfactory neuron ciliated chemoreceptor cell in nasal epithelium that senses a specific set of odours by means of a G-protein-coupled receptor and transmits signals to the olfactory nerve. *alt.* olfactory receptor neuron.

olfactory pit (1) olfactory organ in the form of a small pit or hollow, in certain invertebrates; (2) embryonic ectodermal sac which gives rise to nasal cavity, olfactory epithelium and vomeronasal organ in vertebrates.

olfactory receptor one of a large number of G-protein-coupled receptors that sense odours and are present on the surface of olfactory neurons, with each neuron having only one type of receptor.

olfactory spindle sensory cell structure associated with olfactory nerve in antennule of decapod crustaceans. *alt.* lobus osphradicus.

olig- prefix derived from Gk *oligos*, few, signifying having few, having little of, etc.

oligacanthous *a.* bearing few spines.

oligandrous *a.* having few stamens.

oligarch *a.* having few vascular bundles or elements.

oligocarpous *a.* having few carpels.

Oligocene *n.* geological epoch in the Tertiary sub-era, the last epoch in the Paleogene period, between the Eocene and Miocene, lasting from *ca.* 34 million years ago to 23 million years ago. Ancestors of modern orders of horses, deer, cats, dogs, primates and birds are present as well as extinct mammalian groups. Open grasslands develop. First elephants appear. Marsupials diversify in Australia.

Oligochaeta, oligochaetes *n., n.plu.* class of annelid worms characterized by possession of a few bristles (setae or chaetae) on each segment and no parapodia, and which includes the earthworms.

oligoclonal *a.* in immunology, *appl.* a response involving only a few clones of lymphocytes.

oligodendrocyte *n.* type of small neuroglial cell predominant in white matter of the central nervous system, forming myelin sheaths around axons.

oligodendroglia the oligodendrocytes of the brain, collectively.

oligogenic *a. appl.* characters controlled by a few genes responsible for major heritable changes.

oligoglia oligodendroglia *q.v.*

oligogyny *n.* the occurrence of two to several functional queens in a single colony of social insects.

oligohaline *a. appl.* brackish water with salinity of 0.5–5 parts per thousand.

oligolecithal *a. appl.* eggs containing little yolk.

oligolectic *a.* selecting only a few, *appl.* insects visiting only a few different food plants or flowers.

oligolobate *a.* divided into only a small number of lobes.

oligomer *n.* a molecule composed of only a few monomer units.

oligomerous *a.* having one or more whorls with fewer members than the rest.

oligomycin *n.* antibiotic that inhibits ATP synthesis in mitochondria by interacting with one of the proteins in mitochondrial ATP synthase.

oligonucleotide *n.* short chain of nucleotides. *see also* polynucleotide.

oligopeptide *n.* small polypeptide.

oligophagous *a.* restricted to a single order, family or genus of food plants, *appl.* insects.

oligophyletic *a.* derived from a few different lines of descent.

oligopneustic *a. appl.* insect respiratory system in which only a few spiracles are functional.

oligopod *a.* (1) having few legs or feet; (2) having thoracic legs fully developed.

oligorhizous *a.* having few roots, *appl.* certain marsh plants.

oligosaccharide *n.* a molecule composed of only a few (4–20) monosaccharide units.

oligosaprobic *a.* (1) *appl.* category in the saprobic classification of river organisms consisting of those that can only live in water unpolluted by organic pollutants, e.g. brown trout (*Salmo trutta*) and stonefly nymphs (plecopterans). *cf.* α-mesosaprobic, β-mesosaprobic, polysaprobic; (2) *appl.* aquatic environment with a high dissolved oxygen content and little organic decomposition.

oligospermous *a.* having few seeds.

oligosporous *a.* producing or having few spores.

oligostemonous *a.* having few stamens.

oligothermic *a.* tolerating relatively low temperatures.

oligotokous *a.* bearing few young. *alt.* oligotocous.

oligotrophic *a.* (1) providing or *pert.* inadequate nutrition; (2) *appl.* waters relatively low in nutrients, e.g. the open oceans compared with the continental shelves, and e.g. some lakes whose waters are low in dissolved minerals and which cannot support much plant life; (3) *appl.* microorganism that thrives and predominates in a nutrient-poor environment. *n.* **oligotroph**.

oligotrophophyte *n.* plant that will grow on poor soil.

oligotrophy *n.* the ability to live in a nutrient-poor environment, *appl.* e.g. to some soil actinomycetes.

oligotropic *a.* visiting only a few allied species of flowers, *appl.* insects.

oligoxenous *a. appl.* parasites adapted for life in only a few species of hosts.

O-**linked glycosylation** *see* glycosylation.

olivary nucleus, olive one of several nuclei in brain situated in the medulla just below the pons.

omasum *n.* in ruminants, the third chamber of the stomach, through which food must pass from rumen to abomasum, and whose structure prevents mixing of rumen and abomasum contents.

ombrogenous *a. appl.* wet habitats arising from precipitation rather than from water in the ground.

ombrophile *a.* adapted to living in a rainy place, *appl.* plants, leaves.

ombrophobe *n.* plant that does not thrive under conditions of heavy rainfall.

ombrophyte *n.* plant adapted to rainy conditions.

omega peptidase hydrolytic enzyme that catalyses the removal of terminal residues of a peptide that are substituted, cyclized or linked by isopeptide bonds (peptide linkages other than those of α-carboxyl to α-amino groups). EC 3.4.19.

omentum *n.* fold of peritoneum either free or acting as connecting link between viscera. *a.* **omental**.

ommateum *n.* a compound eye.

ommatidium *n.* individual facet of insect or crustacean compound eye, comprising a hexagonal tube with a lens of transparent cuticle at the external face, from which a cone of transparent jelly (the crystalline cone) projects backwards. The sides of the ommatidium are composed of pigmented cells. At the base is a cup-shaped retinula of sensory photoreceptor cells containing light-sensitive pigment. *plu.* **ommatidia**. *a.* **ommatidial**.

ommatochromes, ommochromes *n.plu.* yellow, red and brown pigments of arthropods, found in eye and body.

ommatophore *n.* a movable stalk bearing an eye, in arthropods.

omni- prefix derived from L. *omnis*, all.

Omnibacteria *n.* name sometimes given to a large grouping of aerobic or facultatively anaerobic heterotrophic Gram-negative bacteria that includes the enterobacteria, vibrios, aeromonads, and the stalked, budding and sheathed bacteria.

omnicolous *a.* able to grow on a variety of substrates, *appl.* lichens.

omnivore *n.* animal that eats both plant and animal food. *a.* **omnivorous**.

omohyoid *a. pert.* shoulder and hyoid, *appl.* muscle.

omphalic umbilical *q.v.*

omphalodisc *n.* apothecium with a small central protuberance, as in certain lichens.

omphalogenesis *n.* development of the umbilical vesicle and cord.

omphaloid *a.* (1) like a navel; (2) umbilicate *q.v.*

omphalomesenteric *a. pert.* umbilicus and mesentery, *appl.* arteries, veins, ducts.

onchocerciasis *n.* tropical disease that causes impaired vision and eventual blindness. It is caused by infection with the parasitic roundworm *Onchocerca volvulus*, which is transmitted to humans by blackflies. *alt.* river blindness.

onchosphere oncosphere *q.v.*

oncofoetal antigens antigens found on the surface of cancer cells and also on embryonic cells but not on normal adult cells.

oncogene *n.* a gene carried by a tumour virus or a cancer cell, which is solely or partly responsible for tumorigenesis.

Oncogenes are altered versions of normal cellular genes involved in the control of cell division or differentiation. The normal counterpart to an oncogene is sometimes known as a proto-oncogene, but the term oncogene is often used loosely to denote any gene, altered or not, which is a potential oncogene. Viral oncogenes are symbolized as v-*onc*, where *onc* may be a wide range of genes. The normal counterpart is often symbolized as c-*onc*. *see also* tumour suppressor genes.

oncogenesis *n.* the generation and development of a tumour.

oncogenic *a.* capable of causing a tumour.

oncolytic *a.* capable of destroying cancer cells.

oncomiracidium *n.* free-swimming ciliated larval form of skin and gill flukes (monogenean flatworms). *plu.* **oncomiracidia**.

oncomouse *n.* strain of transgenic mice that carries an activated oncogene, and which is used experimentally to study oncogenesis.

oncoprotein *n.* protein encoded by an oncogene, and which can cause transformation if introduced into a cell.

oncornavirus acronym for oncogenic RNA virus. *see* oncoviruses, RNA tumour viruses.

oncosphere *n.* spherical, hooked larva that hatches from tapeworm egg and develops into a cysticercoid.

oncoviruses *n.plu.* a subfamily of retroviruses consisting of the RNA tumour viruses. Divided into types C, B and D. Formerly known as oncornaviruses.

one gene–one enzyme hypothesis the original form of the idea that each gene specifies one polypeptide chain, developed by Beadle and Tatum in the 1930s and 1940s from the study of biochemical mutants.

ontogenesis ontogeny *q.v.*

ontogeny *n.* the development of an individual. *alt.* **ontogenesis**. *a.* **ontogenetic**. *cf.* phylogeny.

ontology *n.* a controlled vocabulary used to describe concepts and their interrelationships within a particular field of knowledge, e.g. protein functions.

onychium *n.* (1) the layer below the nail; (2) pulvillus *q.v.*; (2) a special false

articulation to bear claws at end of tarsus in some spiders.

onychogenic *a.* capable of producing a nail or nail-like substance, *appl.* material in nail matrix, and cells forming fibrous substance and cuticula of hairs.

Onychophora, onychophorans *n., n.plu.* class of primitive worm-like terrestrial arthropods with a soft flexible cuticle, which live in damp habitats in warm climates. They are sometimes treated as a separate phylum. Onycophorans possess a pair of antennae and a pair of stiff jaw appendages, followed by a number of pairs of unjointed walking legs. There is only one extant genus, *Peripatus*. *alt.* velvet worms.

ooangium archegonium *q.v.*

ooapogamy *n.* parthenogenetic development of an unfertilized ovum.

ooblast *n.* in some red algae, a tubular outgrowth from the carpogonium through which the fertilized nucleus or its derivatives pass into the auxiliary cell.

oocyst *n.* cyst formed around two conjugating gametes in sporozoan protozoans.

oocyte *n.* female germ cell in which meiosis occurs, in animals. Oocytes undergoing the 1st meiotic division are often termed primary oocytes, after which they become secondary oocytes, which undergo the 2nd meiotic division to become mature eggs.

oocyte injection technique for studying gene expression by injecting purified DNA or mRNA into the nucleus or cytoplasm of (usually) *Xenopus* oocytes.

ooecium ovicell *q.v.*

oogamete *n.* female gamete, esp. a large non-motile one containing food material for the zygote.

oogamous *n.* reproducing by oogamy.

oogamy *n.* the union of unlike gametes, usually a large non-motile female gamete and a small motile male gamete.

oogenesis *n.* formation, development and maturation of the female gamete or ovum.

oogloea *n.* egg cement.

oogonial *a. pert.* an oogonium.

oogonium *n.* (1) diploid precursor to female germ cells. In animals, it becomes an oocyte which undergoes meiosis to produce the mature ovum; (2) female reproductive organ in fungi and algae. *plu.* **oogonia**.

ooid *a.* egg-shaped or oval.

ookinesis *n.* the mitotic stages of nuclear division in the maturation and fertilization of eggs.

oology *n.* the study of birds' eggs.

Oomycota, oomycetes *n., n.plu.* phylum of simple non-photosynthetic, saprobic or parasitic, unicellular or filamentous protists, now classified in the Stramenopila or the Chromista, formerly classifed as fungi. Unlike most fungi their cell walls contain cellulose. Sexual reproduction is oogamous and they reproduce asexually by motile zoospores. They include the water moulds (e.g. *Saprolegnia*), and the causative organisms of several important plant diseases, e.g. downy mildew of grapes (*Plasmopora*) and potato blight (*Phytophthora infestans*).

oophoridium *n.* the megasporangium in certain plants.

oophyte *n.* the gametophyte in certain lower plants.

ooplasm *n.* the cytoplasm of an egg.

oopod *n.* component part of sting or ovipositor.

oosphere *n.* a female gamete or egg, esp. as produced in an oogonium by algae and oomycetes.

oosporangium oogonium *q.v.*

oospore *n.* thick-walled zygote that arises from a fertilized oosphere in oomycete fungi, algae and protozoa.

oostegite *n.* plate-like structure on basal portion of thoracic limb in crustaceans, which helps to form a receptacle for the egg and acts as a brood pouch.

oostegopod *n.* thoracic appendage bearing an oostegite (egg receptacle) in crustaceans.

ootheca *n.* egg case in certain insects.

ootocoid *a.* giving birth to young at a very early stage and then carrying them in a pouch, as marsupials.

ootokous *a.* egg-laying. *alt.* **ootocous**.

ootype *n.* structure in the female reproductive system of flatworms in which the fertilized zygote and yolk cells are enclosed in a capsule to form the egg.

ooze *n.* (1) a deposit containing skeletal parts of minute organisms and covering large areas of the ocean floor; (2) soft mud.

oozoid *n.* any individual developed from an egg.

oozoite *n.* the asexual parent, in ascidians.

OP osmotic pressure *q.v.*

opal the termination (stop) codon UGA.

Opalinata, opalinids *n., n.plu.* group of multiflagellate protists of the phylum Sarcomastigophorea, which are parasitic in the guts of amphibians, reptiles and fish. They are characterized by the falx, a structure made up rows of kinetosomes, each supporting a flagellum.

opal mutation mutation generating an opal termination codon UGA and resulting in premature termination of synthesis of the protein product of the mutated gene.

opal suppressor a mutant gene producing a tRNA that overcomes the effects of an opal mutation by inserting an amino acid at UGA.

open *a.* (1) *appl.* arrangement of floral parts where perianth segments do not meet at the edges, as in Cruciferae; (2) *appl.* plant community that does not completely cover the ground but leaves bare areas that can subsequently be colonized by other plants.

open forest area covered with trees sufficiently widely spaced not to form a closed canopy. *cf.* closed forest.

open reading frame (ORF) a stretch of DNA that contains a signal for the start of translation, followed in the correct register by a sufficient length of amino acid-encoding nucleotide triplets to form a protein, followed by a signal for termination of translation. It denotes a stretch of DNA that has been identified by sequence alone as a possible-protein coding gene, but whose encoded protein and its function is unknown.

operant behaviour spontaneous animal behaviour that occurs without any apparent stimulus.

operant conditioning type of procedure for studying animal behaviour in which rewards and punishments are used to select, strengthen or weaken behaviour patterns.

operational taxonomic unit (OTU) (1) any group of living organisms, e.g. genus or species, designated by numerical taxonomy; (2) any object that can be used as the unit to be ranked in a phylogenetic tree, e.g. a species, protein sequence or DNA sequence.

operator *n.* control region of DNA present in many bacterial operons. It contains the nucleotide sequence to which the repressor (or apoinducer) binds, thus preventing (or allowing) transcription of the operon.

opercula *plu.* of operculum.

opercular (1) *n.* posterior bone of operculum in fishes; *a.* (2) *pert.* operculum; (3) *appl.* fold of skin covering gills in tadpoles.

operculate *a.* (1) having a lid, *appl.* e.g. spore capsules; (2) having a covering (operculum) over the gills, as most fishes.

operculiform *a.* lid-like.

operculigenous *a.* producing or forming a lid.

operculum *n.* (1) lid, or covering flap, as on spore capsules, eggs of some invertebrates; (2) gill cover in fishes; (3) flap covering the nostrils and ears in some birds; (4) movable plates in shell of barnacle; (5) lid-like structure closing mouth of shell in some gastropod molluscs; (6) small bone in middle ear of amphibians, lying on the oval window. *plu.* **opercula**.

operon *n.* a type of gene organization in bacteria, in which the genes coding for the enzymes of a metabolic pathway are clustered together in the DNA and transcribed together into a single mRNA. This mRNA is then translated to give the individual proteins. The expression of all the genes in an operon is controlled by a single promoter.

ophidians *n.plu.* snakes.

ophiocephalous *a.* (1) snake-headed; (2) *appl.* small pedicellariae of some echinoderms having broad jaws with toothed edges.

ophiopluteus *n.* the pluteus larva of brittle stars.

ophiurans Ophiuroidea *q.v.*

ophiuroid *a.* (1) *pert.* or resembling a brittle star; (2) *appl.* cells: multiradiate or spiculate sclereids.

Ophiuroidea *n.* class of echinoderms commonly known as brittle stars, having a star-shaped body with the arms clearly marked off from the central disc.

ophthalmic *a.* (1) *pert.* eye; (2) *appl.* nerve: division of 5th cranial nerve conveying sensation from eye orbit, forehead and front of scalp; (3) *appl.* an artery arising from internal carotid; (4) *appl.* inferior and superior veins of eye orbit.

ophthalmic rete network of arteries and veins in the head of birds, which acts as a heat exchanger to allow the brain to be kept at a temperature below that of the arterial blood leaving the heart.

ophthalmogyric *a.* bringing about movement of the eye.

ophthalmophore ommatophore *q.v.*

ophthalmopod *n.* eye-stalk, as of decapod crustaceans.

opiate *n.* any compound mimicking the effects of opium at its receptors.

opiate receptors class of chemoreceptors in brain and gut that bind morphine and other opiates, and whose natural substrates are the peptide enkephalins and endorphin.

Opiliones *see* harvestman.

opioid *a.* having opiate-like activity, *appl.* peptides such as the enkephalins and endorphin.

opioid receptors cell-surface receptor proteins for opiates and for the opioid peptides such as enkephalins and endorphins. There are three types: μ, κ and δ.

opisth- prefix derived from Gk *opisthe*, behind.

opisthial *a.* posterior, *appl.* pore or stomatal margin.

opisthobranch *n.* member of the mollusc subclass Opisthobranchia (e.g. sea slugs, sea hares), which are marine, and in which the shell is much reduced or absent.

opisthocoelous *a.* having the centrum concave behind, *appl.* vertebrae.

opisthodetic *a.* lying posterior to beak or umbo, *appl.* ligaments in some bivalve shells.

opisthogenesis *n.* development of segments or markings proceeding from the posterior end of the body.

opisthoglossal *a.* having tongue fixed in front, free behind.

opisthognathous *a.* (1) having retreating jaws; (2) with mouthparts directed backwards, *appl.* head of insects.

opisthogoneate *a.* (1) having the genital aperture at hind end of body, *appl.* arthropods: insects and centipedes.

opisthohaptor *n.* posterior sucker or disc in trematodes.

opisthokont *a.* with flagellum or flagella at posterior end.

Opisthokonta *n.* proposed major taxonomic grouping, based on molecular phylogenetics, that includes animals, true fungi and some unicellular eukaryotes including the choanoflagellates, myxozoans and microsporidians.

opisthomere *n*. terminal plate on abdomen of female earwig.

opisthonephros *n*. renal organ of embryo, consisting of meso- and metanephric series of tubules.

opisthosoma *n*. posterior section of the body in arachnids and some other invertebrates.

opisthure *n*. the projecting tip of vertebral column.

opium *n*. addictive alkaloid drug obtained from the opium poppy *Papaver somniferum*, consisting of the dried milky juice from the slit poppy capsules. It acts as a stimulant, narcotic and hallucinogen. It was formerly widely used to ease pain but is now replaced for that purpose by its derivative alkaloids such as morphine.

opophilous *a*. feeding on sap.

opossum *n*. small arboreal marsupial from the Americas (most species are from South America, one species from North America). Although often called a 'possum' it should not be confused with the Australasian possums, which are only distantly related.

opponens *a*. *appl*. muscles which cause the fingers or toes to approach one another.

opponent-process hypothesis hypothesis of colour vision introduced by Hering in the late 19th century. It proposed that there are three pairs of opposed colours, and that colour vision involves three physiological processes with opposed positive and negative values. *cf*. trichromatic theory.

opportunistic *a*. (1) *appl*. microorganisms that are normally non-pathogenic but that can cause disease in immunosuppressed or otherwise debilitated individuals; (2) *appl*. species specialized to exploit newly opened habitats.

opposable *a*. *appl*. the thumb of primates, which can be brought together with fingers in a grasping action, enabling objects to be held.

opposite *a*. *appl*. leaves or other organs that form a pair opposite each other on the stem.

opsiblastic *a*. with delayed cleavage, *appl*. eggs having a dormant period before hatching. *cf*. tachyblastic.

opsigenes *n.plu*. structures formed or becoming functional long after birth.

opsin *n*. protein component of the various vertebrate visual pigments, differing slightly from pigment to pigment, these differences determining the spectral sensitivity of the pigment.

opsonic *a*. *pert*. or affected by opsonins.

opsonin *n*. a protein (antibody or complement) whose binding to the surface of a virus particle or bacterium facilitates its engulfment by a phagocytic cell.

opsonization *n*. process whereby foreign particles become coated with antibody or complement, making them more readily ingested by phagocytic cells.

optic *a*. *pert*. vision or the eye.

optical *a*. (1) *pert*. vision; (2) of or concerned with light, e.g. optical microscope.

optical density (OD) absorbance *q.v.*

optical isomer either of two optically active enantiomeric (mirror image) forms of a molecule, crystal or compound, one of which rotates a beam of plane-polarized light to the right (dextrorotatory, *d* or +) and the other to the left (laevorotatory, *l* or −).

optical microscopy light microscopy *q.v.*

optical section an image of a section through an object at a given depth, produced by a laser beam scanning across the object at a given depth, excluding light from other parts of the object, and thus producing a sharp image of the scanned plane.

optical tweezers an intense focused laser beam at a wavelength that refracts through a transparent object of interest, such as a subcellular organelle, keeping the object positioned at the focus of the laser. By moving the laser, the object can be moved around. *alt*. **optical forceps, optical trap**.

optic axis line between central points of anterior and posterior curvature or poles of eyeball.

optic bulb expansion of the embryonic optic vesicle, later invaginated to form the optic cup from which retina develops.

optic chiasm(a) X-shaped structure below frontal lobes of brain in which the optic nerves from right and left eyes meet. Nerve fibres from the inner half of each retina cross over to form two optic tracts composed of fibres from the right and left halves respectively of both retinas. The right tract carries all signals from the left visual field and the left tract those from the right visual field.

optic lobes part of brain concerned with the processing of visual signals, consisting of a large part of the occipital lobes. *alt.* visual areas, visual cortex.

optic nerves 2nd cranial nerves in vertebrates, concerned with vision. The optic nerves run from the retinas of the eyes to the optic chiasm and thence to the lateral geniculate nuclei.

opticociliary *a. pert.* optic and ciliary nerves.

opticon *n.* inner zone of optic lobe in insects.

opticopupillary *a. pert.* optic nerve and pupil.

optic radiation neural pathways from lateral geniculate nucleus to visual cortex.

optic tectum the brain area to which the optic nerve connects in amphibians and birds.

optimal *a.* (1) the most efficient, the most cost-effective; (2) *appl.* various animal behaviours such as optimal foraging or optimal reproductive strategy. It may be used in a long-term sense to indicate a behaviour that results in an animal leaving the largest possible number of viable offspring, or in a short-term sense to indicate behaviour that optimizes the energy collected in a certain time, as in optimal foraging, or minimizes the metabolic energy expended in achieving a goal.

optimal yield the highest rate of increase that a population can sustain in a given environment.

optimum *n.* the condition or set of conditions that gives the best result.

optocoel *n.* the cavity in the optic lobes of brain.

optokinetic *a. pert.* movement of the eyes.

optomotor *a. appl.* reflex of turning head or body in response to stimulus of moving stripes.

Opuntiales Cactales *q.v.*

ora *n.plu.* mouths. *sing.* **os.**

orad *a.* towards the mouth or mouth region.

oral *a.* (1) *pert.* or belonging to the mouth; (2) on the same side as the mouth (of radially symmetrical animals such as echinoderms).

oral disc in anthozoan polyps, circular flattened area surrounded by tentacles with the mouth at the centre.

oral siphon in urochordates, a narrowing of the body near the mouth.

oral valve in crinoids, one of five low triangular flaps separating the ambulacral grooves.

ora serrata the wavy margin of retina, where the nervous tissue ends.

orbicular *a.* (1) (*bot.*) *appl.* leaves, round or shield-shaped with leaf-stalk attached in the centre; (2) (*zool.*) *appl.* eye muscles.

orbicularis *n.* muscle whose fibres surround an opening.

orbiculate *a.* nearly circular in outline.

orbit *n.* (1) bony cavity in which eye is situated; (2) skin around eye in bird; (3) conspicuous zone around compound eye in insects; (4) hollow in arthropod cephalothorax from which eye-stalk arises.

orbital *a. pert.* the orbit.

orbitofrontal cortex region of cortex in the human brain involved in cognitive processes such as decision making. It is located within the frontal lobes, just above the eyes.

orbitomalar *a. pert.* orbit of eye and malar bone.

orbitonasal *a. pert.* orbit of eye and nasal portions of adjoining bones.

orbitosphenoid *a.* (1) *pert.* paired cranial elements lying between presphenoid and frontal; (2) *appl.* bone with foramen for optic nerve.

orb web the most familiar type of spider's web, with a spiral centre supported on radiating threads anchored to a roughly triangular frame.

ORC origin recognition complex *q.v.*

Orchidales *n.* order of monocot herbs, with tubers or stems swollen into pseudobulbs, often epiphytic or sometimes saprophytic, often with showy flowers, and comprising the family Orchidaceae (orchids).

orchid mycorrhiza mycorrhiza formed by basidiomycete fungi on the embryos of orchids (family Orchidaceae), which are necessary for the embryo's successful development.

orchitic *a.* testicular *q.v.*

orculaeform *a.* cask-shaped, *appl.* spores of certain lichens.

order *n.* taxonomic group of related organisms ranking between family and class.

ordinate *a.* having markings in rows.

ordinatopunctate *a. appl.* rows of dots.

Ordovician *n.* geological period lasting from *ca.* 488 million years ago to 441 million years ago, following the Cambrian and preceding the Silurian. At this time, most land was part of the southern supercontinent Gondwana. Marine life included graptolites, trilobites, brachiopods, conodonts, primitive fish, cephalopods, crinoids and gastropods. Plants may have invaded the land for the first time.

ORF open reading frame *q.v.*

organ *n.* any part or structure of an organism adapted for a special function or functions, e.g. heart, stomach, kidney.

organelle *n.* structure within a eukaryotic cell in which certain functions and processes are localized.

organ of Corti structure running the length of the mammalian cochlea, located on the basilar membrane and concerned with sound perception. It consists of rows of sensory hair cells supported on a double row of arching rods (Corti's rods) and which are sensitive to vibrations in the surrounding fluid set up by sound waves transmitted from the outer ear. It is covered by a membrane, the tectorial or Corti's membrane.

organ of Golgi cylindrical sensory receptor at junction of tendons and muscles.

organic *a.* (1) *pert.*, derived from, or showing the properties of, a living organism; (2) containing carbon, *appl.* molecules.

organic chemistry the chemistry of carbon-containing compounds.

organic farming farming without the use of artificial fertilizers and pesticides.

organicism *n.* the integration of an organism as a unit.

organific *a.* making an organized structure.

organism *n.* any living thing.

organismal, organismic *a.* (1) *pert.* an organism as a whole; (2) *appl.* or *pert.* factors or processes involved in maintaining the life of an organism.

organized *a.* (1) exhibiting characteristics of, or behaving like, an organism; (2) *appl.* growth of cells in tissue culture that resembles their normal organization in tissue.

organizer *n.* a part of an embryo which directs the development of other parts. It commonly refers esp. to the mesoderm of the dorsal lip of the blastopore in amphibians (Spemann's organizer) and similar regions in other vertebrates, which influence much subsequent development and are known as primary organizers. *alt.* **organizing region.**

organogenesis *n.* the formation and development of organs.

organogenic *a.* (1) due to the activity of an organ; (2) *pert.* organogenesis.

organoid *a.* having a definite organized structure, *appl.* certain plant galls.

organoleptic *a. appl.* a stimulus capable of affecting the sensory organs.

organology *n.* the study of organs of plants and animals.

organophyly *n.* the phylogeny of organs.

organoplastic *a.* capable of forming, or producing, an organ.

organotroph heterotroph *q.v.*

organotrophic *a.* (1) *pert.* formation and nourishment of organs; (2) *appl.* organisms that require organic compounds as a carbon source. *alt.* heterotrophic.

organotypic *a. appl.* organized growth of cultured cells in a form resembling a tissue.

orgasm *n.* immoderate excitement, esp. sexual.

ori origin of replication in DNA.

Oriental region zoogeographical region in Wallace's classification, consisting of the Indian subcontinent south of the Himalayas, South-east Asia, and Indonesia apart from Sulawesi and New Guinea. It is separated from the Australian region by Wallace's line. It is divided into the Ceylonese, Indian, Indo-Chinese, and Indo-Malaysian subregions.

orientation *n.* alteration in position shown by organs or organisms under a stimulus.

orientation column feature detection column *q.v.*

orientation response physiological response of an organism to a sudden change (e.g. a sound or a novel sight) in its environment.

orifice *n.* mouth or aperture.

origin *see* origin of replication.

original antigenic sin tendency to make antibody responses to epitopes shared between the originally encountered strain of a virus and subsequent related viruses, while not responding to different epitopes on the new strain.

origin of life modern scientific theories on the origin of life propose an origin on the early Earth, some 4000 million years ago, from simple carbon-containing compounds formed in the atmospheric conditions of the time. The evolution of a self-replicating molecule would have been essential at some point. What that self-replicating molecule would have been is still a matter for debate. *see also* ribozyme, RNA world.

origin of replication (ori) site in a DNA molecule at which DNA replication can begin.

origin recognition complex (ORC) large multisubunit protein that is bound to a site in an origin of replication in DNA, marking it for subsequent recognition by the biochemical machinery required to replicate the DNA (e.g. helicase, DNA polymerase).

ornis avifauna *q.v.*

ornithic *a. pert.* birds.

ornithine *n.* diaminovaleric acid, an amino acid involved in the urea cycle, and in birds excreted with one of its derivatives, ornithuric acid.

ornithine cycle urea cycle *q.v.*

Ornithischia, ornithischians *n., n.plu.* order of Mesozoic dinosaurs, commonly called the bird-hipped dinosaurs, having a pelvis resembling that of a modern bird. They were all herbivorous and included both bipedal (ornithopods) and quadrupedal (ceratopians, stegosaurs and ankylosaurs) members. Modern birds are not, however, thought to have descended from this group of dinosaurs.

ornithology *n.* the study of birds.

ornithophilous *a. appl.* flowers pollinated through the agency of birds. *n.* **ornithophily**.

ornithuric acid dibenzoylornithine, a constituent of bird droppings.

oroanal *a.* (1) serving as mouth and anus; (2) connecting mouth and anus.

orobranchial *a. pert.* mouth and gills, *appl.* epithelium.

oronasal *a. pert.* or designating groove connecting mouth and nose.

oropharynx *n.* the cavity of mouth and pharynx.

orotate *n.* pyrimidine base which is a precursor of the pyrimidine nucleotide orotidylate.

orotidylic acid, orotidylate pyrimidine nucleotide which is decarboxylated to form uridine monophosphate (UMP).

orphan virus a virus not known to cause a disease.

orphan receptor a receptor whose ligand is as yet unknown.

orphon *n.* an isolated gene that is related in sequence to members of a gene cluster elsewhere in the genome.

ortet *n.* the ancestor of a clone.

orthal *a.* straight up and down, *appl.* jaw movement.

orthaxial *a.* with a straight axis, or vertebral axis, *appl.* tail fin.

ortho- prefix derived from Gk *orthos*, straight.

orthoblastic *a.* with a straight germ band.

orthochromatic *a.* (1) of the same colour as the stain; (2) staining positively.

orthocladous *a.* straight-branched.

orthodentine *n.* dentine pierced by numerous more-or-less parallel dentinal tubules.

orthodromic *a.* moving in the normal direction, *appl.* conduction of nerve impulse.

orthoenteric *a.* having alimentary canal along ventral body surface, *appl.* certain ascidians.

orthogenesis *n.* evolution along some apparently predetermined line independent of natural selection or other external forces.

orthognathous *a.* (1) having straight jaws; (2) having axis of head at right angles to body, as some insects.

orthograde *a.* walking with the body in a vertical position.

orthokinesis *n.* movement in which the organism changes its speed when it meets an unfavourable environment, the speed depending on the intensity of the stimulus.

orthologous *a. appl.* equivalent genes in different species that are homologous because they have both evolved in a direct line from a common ancestral gene (e.g. the α-globin gene from humans is orthologous with the α-gene of horses, whereas the β-globin gene in both humans or horses is not, having arisen by duplication of an ancestral gene). *n.* **orthologue.** *cf.* paralogous.

Orthomyxoviridae, orthomyxoviruses *n., n.plu.* family of large, enveloped viruses with single-stranded negative-strand segmented RNA genomes, and including influenza.

Orthonectidea *n.* class of the invertebrate phylum Mesozoa that includes the orthonectids, parasites of the kidneys of flatworms, nemertines, polychaetes, bivalve molluscs and echinoderms. They have a free-swimming sexual generation.

orthophosphate (Pi) HPO_4^{2-} ($+ H_2PO_4^-$).

orthoploid *a.* (1) with even chromosome number; (2) polyploid with complete and balanced genomes.

Orthoptera, orthopterans *n., n.plu.* order of insects that includes crickets, locusts and grasshoppers (in some classifications also cockroaches and mantises). They have long antennae, biting mouthparts, long narrow tough forewings, and broad membranous hindwings, usually also having enlarged hindlegs for jumping, and stridulating organs, which produce the characteristic sounds of these insects.

orthoradial *a. appl.* cleavage where divisions are disposed symmetrically around egg axis.

orthoselection *n.* natural selection acting continuously in the same direction over a long period.

orthosomatic *a.* having a straight body, *appl.* certain larval insects.

orthospermous *a.* with straight seeds.

orthospiral *a. appl.* coiling of parallel sister chromatids interlocked at each twist. *alt.* plectonemic.

orthostichous *a.* arranged in a vertical row, *appl.* leaves.

orthostichy *n.* the vertical row of leaves formed by the leaves immediately above each other in a spiral arrangement.

orthotopic *a.* in the proper place, *appl.* transplantation. *cf.* heterotopic.

orthotopy *n.* (1) natural placement; (2) existence of an organism in its natural habitat.

orthotropism *n.* (1) growth in a straight line, as of tap roots; (2) condition of tending to be oriented in the line of action of a stimulus. *a.* **orthotropic.**

orthotropous *a. appl.* ovules having chalaza, hilum and micropyle in a straight line so that ovule is not inverted.

orthotype *n.* genotype originally designated.

os *n.* (1) mouth. *plu.* **ora;** (2) bone. *plu.* **ossa.**

oscitate *v.* to yawn, to gape.

oscula *plu.* of osculum *q.v.*

osculate *a.* to have characteristics intermediate between two groups.

osculum *n.* large pore in body wall of sponge through which water flows out from the body cavity. *plu.* **oscula.** *a.* **oscular.**

osmatic *a.* having a sense of smell.

osmeterium *n.* organ borne on the 1st thoracic segment of larva of some butterflies, emitting a smell.

osmiophilic *a.* staining readily with osmium stains.

osmoconformer *n.* organism which does not regulate the osmotic concentration of its internal fluids, which therefore vary with the osmotic concentration of the external environment. Examples are some estuarine invertebrates.

osmolality *n.* a quantitative measure of the osmotic concentration of a solution, usually expressed in osmoles (1 osmole = 1 molal, *see* molality). *cf.* osmolarity.

osmolarity *n.* the capacity of a solution to induce osmosis. The osmolarity of the interior of a cell is due mainly to the high concentration of small charged organic molecules and their counterions. *alt.* tonicity. *cf.* osmolality.

osmomorphosis *n.* change in shape or structure due to changes in osmotic pressure, such as those due to changes in salinity of the surrounding water.

osmophile *n.* microorganism able to live in an environment high in sugar.

osmophore *n.* the group of atoms responsible for the odour of a compound, and which binds to the receptors on olfactory neurons.

osmoreceptor *n.* cell stimulated by changes in osmotic pressure, e.g. the cells reacting to osmotic changes in the blood.

osmoregulation *n.* in animals, regulation of the osmotic pressure of body fluids by controlling the amount of water and/or salts in the body.

osmoregulator *n.* organism which actively regulates the osmotic concentration of its internal fluids.

osmosis *n.* diffusion of a solvent, usually water, through a semipermeable membrane from a dilute to a concentrated solution, or from the pure solvent to a solution.

osmotaxis *n.* a movement in reponse to changes in osmotic pressure.

osmotic *a. pert.* osmosis.

osmotic activity the ability of a solution to cause osmosis. *see* osmotic pressure.

osmotic balance the adjustments made by a living cell to intercellular and extra-cellular concentrations of solutes and ions to prevent the usual tendency of water to flow into the cell from the environment by osmosis.

osmotic pressure (OP) a measure of the osmotic activity of a solution, defined as the minimum pressure that must be exerted to prevent the passage of pure solvent into the solution when the two are separated by a semipermeable membrane. Osmotic pressure is proportional to solute concentration.

osmotroph *n.* any heterotrophic organism (e.g. fungi and bacteria) that absorbs organic substances in solution. *a.* **osmotrophic**. *cf.* phagotroph.

osphradium *n.* chemical sense organ associated with visceral ganglia in many molluscs.

osphresis *n.* the sense of smell.

ossa *n.plu.* bones, *plu.* of os.

ossa sutura or triquetra sutural bones *q.v.*

osseous *a.* composed of or resembling bone.

osseous labyrinth the vestibule, semi-circular canals and cochlea of inner ear embedded in the temporal bone.

ossicle *n.* (1) any small bone or other calcified hard structure such as a plate of exoskeleton in echinoderms; (2) any of the three small bones of the middle ear.

ossicone *n.* the bony core of horn of ruminants (e.g. sheep, cows).

ossicular *a. pert.* ossicles.

ossiculate *a.* having ossicles.

ossification *n.* formation of bone, replacement of cartilage by bone.

ossify *v.* to change to bone.

Ostariophysi *n.* superorder of teleost fishes that includes the carps, minnows, catfishes and loaches, characterized by the presence of Weberian ossicles.

Osteichthyes *n.* the bony fishes, a class comprising all fish except for the Agnatha and the Selachii, and which have a bony skeleton, usually an air bladder (swim-bladder) or lung, and a cover (operculum) over gill openings. *see* Appendix 3.

osteoblast *n.* bone-forming cell that secretes the bone matrix. *alt.* bone cell.

osteochondral *a. pert.* bone and cartilage.

osteochondrous *a.* consisting of both bone and cartilage.

osteoclasis *n.* destruction of bone by osteoclasts.

osteoclast *n.* large multinucleate cell, related to macrophages, which destroys bone or other matrix, whether calcified or cartilaginous, during bone formation and remodelling. *alt.* giant cell.

osteocranium *n.* the bony cranium as distinguished from the cartilaginous cranium or chondrocranium.

osteocyte *n.* non-secreting, non-dividing cell derived from an osteoblast, found in calcified bone. *alt.* bone cell.

osteodentine *n.* a type of dentine that is very similar to bone in structure.

osteodermis *n.* (1) dermis that is more-or-less ossified; (2) a bony dermal plate.

osteogen *n.* tissue which forms and alters bone.

osteogenesis *n.* formation and growth of bones.

osteogenic *a. pert.* or causing formation of bone. *alt.* **osteogenetic**.

osteoid (1) *n.* the uncalcified bone matrix, collagenous material secreted by bone cells (osteoblasts) and which forms the basis of bone; (2) *a.* bone-like.

osteolepid *a.* having a skin armoured with bony scales, as the Crossopterygii.

osteology *n.* the study of the structure, nature and development of bones.

osteolysis *n.* breakdown and dissolution of bone.

osteomere *n.* a segment of vertebrate skeleton.

osteon bone type of bone composed of overlaping cylinders (osteons) each comprised of concentric layers (lamellae), the collagen fibres of each layer running at right angles to the next, surrounding a central canal (Haversian canal).

osteonectin *n.* protein specific to bone and which is involved in the growth of hydroxyapatite crystals during calcification.

osteopetrosis *n.* condition where bone becomes excessively thick and dense.

osteoplastic *a.* (1) producing bone; (2) *appl.* cells: osteoblasts.

osteoporosis *n.* excessive erosion of the bone matrix leading to bone weakening.

osteosarcoma *n.* tumour of bone.

osteoscute *n.* a bony external shield or plate, as in armadillos.

ostia *plu.* of ostium.

ostiole *n.* a small aperture or opening, as in the wall of sponges. *alt.* ostium. *a.* **ostiolar**.

ostium *n.* (1) any mouth-like opening; (2) opening in arthropod heart through which blood enters the pericardial cavity; (3) opening from exterior into body cavity of sponges through lateral wall, through which water is drawn in. *plu.* **ostia**. *a.* **ostial**.

Ostracoda, ostracods *n.*, *n.plu.* group of small aquatic crustaceans having a bivalved carapace enclosing head and body, and reduced trunk and abdominal limbs.

ostracoderms *n.plu.* extinct Palaeozoic jawless fishes (Agnatha), which were armoured with an exoskeleton of dermal bone.

otic *a. pert.* ear and region of ear.

otitis media infection of the middle ear.

otoconia, otoconites *n.plu.* minute grains of calcium carbonate found in membranous labyrinth of inner ear.

otocyst statocyst *q.v.*

otolith *n.* calcareous particle found in fluid of semicircular canals, utricle and saccule of inner ear. The movement of otoliths under gravity in response to changes in position of the head stimulates sensory cells, giving the animal its position with respect to gravity and allowing it to balance. *alt.* statolith.

otolith organ structure in ear of fish comprising an otolith and associated hair cells (sensory cells).

oto-occipital *n.* bone formed by fusion of opisthotic with exoccipital.

ototoxicity *n.* toxicity of some drugs for structures in the ear.

OTU operational taxonomic unit *q.v.*

ouabain *n.* G-strophanthidin, a plant glycoside which is a specific inhibitor of the Na$^+$-K$^+$ ATPase in eukaryotic cell membranes, and whose digitalis-like effects on the heart are mediated by inhibition of Na$^+$-K$^+$ ATPase.

Ouchterlony double diffusion technique for measuring the amount of antigen in a sample and/or the degree of identity of two antigens, based on diffusion of antigen and antibody from wells in an agar slab, a line of precipitation being formed in the gel where they meet if the antigen reacts with the antibody.

outbreeding *n.* the mating of individuals who are not closely related. *a.* **outbred**. *alt.* **outcrossing**. *cf.* cross-fertilization, inbreeding.

outer membrane of Gram-negative bacteria, a second membrane outside the peptidoglycan cell wall, composed largely of lipopolysaccharide. Unlike the plasma membrane it is permeable to water and other small molecules. *alt.* lipopolysaccharide layer. *see also* chloroplast, mitochondrion.

outer phalangeal cells Deiter's cells *q.v.*

outgroup, outliers a taxon or group of taxa that are not monophyletic with the taxa of interest being used to construct a phylogenetic tree or taxonomic cladogram. They are included in the tree to indicate the position of the root of the main part of the tree.

ova *plu.* of ovum.

ovalbumin *n.* a glycoprotein, chief protein constituent of egg white.

oval window the upper of two membrane-covered openings in the bony wall between the tympanic cavity (middle ear) and vestibule of inner ear. *alt.* fenestra ovalis.

ovarian *a. pert.* an ovary.

ovarian follicle *see* Graafian follicle.

ovariole *n.* egg tube of insect ovary.

ovariotestis *n.* reproductive organ when both male and female elements are formed, as in case of sex reversal.

ovarium ovary *q.v.*

ovary *n.* in plants and animals the reproductive organ in which female gametes or
egg cells are produced. In flowering plants
it comprises the enlarged portion of the
carpel(s) containing the ovules and after
fertilization develops into the fruit containing the seeds. *see* Fig. 18 (p. 239).

ovate *a.* egg-shaped in outline and attached
by the broader end, *appl.* leaves.

ovate-acuminate *a.* having an ovate blade
with a very sharp point.

ovate-ellipsoidal *a.* ovate, approaching
ellipsoidal, *appl.* leaves.

ovate-lanceolate *a.* having leaf-blade
intermediate between ovate and lanceolate.

ovate-oblong *a.* having leaf-blade oblong
with one end narrower.

overdominance *n.* the condition where
a heterozygote phenotype lies outside
the range of either of the homozygote
phenotypes.

overflow *a. appl.* behaviour in which an inappropriate response occurs to a certain
stimulus in order to satisfy certain drives,
such as a dog displaying maternal care to
a bone.

overlapping genes an uncommon genetic
arrangement where part of the protein-
coding DNA sequence of one gene forms
part or all of the protein-coding sequence
of another, usually not in the same reading frame.

overshoot *n.* (1) situation in which the
population size of a species temporarily
exceeds the carrying capacity of its habitat, leading to a sharp reduction in the
population; (2) of a nerve impulse, the
brief reversal of the membrane potential
during an action potential.

ovicapsule *n.* egg case or ootheca.

ovicell *n.* a specialized chamber in which
embryos develop in certain bryozoans.

oviducal *a. pert.* oviduct.

oviduct *n.* tube that carries eggs from ovary
to exterior. *a.* **oviducal**.

oviferous, ovigerous *a.* serving to carry
eggs.

oviform *a.* oval.

oviger *n.* egg-carrying leg of certain
arachnids.

oviparous *a.* egg-laying. *n.* **oviparity**.

oviposition *n.* laying of eggs on a surface,
by insects and fishes.

ovipositor *n.* (1) specialized structure in
insects for depositing eggs; (2) tubular
extension of genital orifice in fishes.

ovisac *n.* (1) egg case or receptacle;
(2) brood pouch *q.v.*

ovocyst, ovocyte, ovogenesis oocyst,
oocyte, oogenesis *q.v.*

ovoid *a.* egg-shaped.

ovomucoid *n.* a glycoprotein in white of egg.

ovo-testis *n.* (1) the reproductive organ
of naturally hermaphrodite animals, which
produces both eggs and sperm; (2) organ
composed of ovarian and testicular tissue
found in some cases of human pseudohermaphroditism.

ovovitellin *see* vitellin.

ovovitelline duct duct lined with yolk
glands that produce yolk cells.

ovoviviparous *a. pert.* organisms that
produce an egg with a persistent outer
covering, but which hatches within the
maternal body. *n.* **ovoviviparity**.

ovular *a.* like or *pert.* an ovule.

ovulate (1) *a.* producing ovules, *appl.*
cones; (2) *v.* to release an egg or eggs
from the ovary. *n.* **ovulation**.

ovulatory *a. pert.* ovulation.

ovule *n.* in seed plants, the structure
consisting of the megagametophyte and
megaspore. It is surrounded by the nucellus and enclosed in an integument, and
develops into a seed after fertilization. *see*
Fig. 18 (p. 239).

ovuliferous *a.* (1) bearing or containing
ovules; (2) *appl.* the scales bearing one or
more ovules that develop on the bract
scales of cones.

ovum *n.* a female gamete *alt.* egg, egg cell.
plu. **ova**. *see also* oocyte.

oxalic acid, oxalate simple organic carboxylic acid $(COOH)_2$, excreted as calcium
oxalate by many fungi, forming crystals.
Oxalates also occur as by-products in various plant tissues and in urine and are also
found in the mantles of certain bivalves.

oxaloacetic acid, oxaloacetate four-
carbon carboxylic acid, a component of the
tricarboxylic acid cycle, accepting acetyl
groups from acetyl CoA to form citric
acid.

oxalosuccinic acid, oxalosuccinate six-carbon carboxylic acid, component of the tricarboxylic acid cycle, decarboxylated to α-ketoglutarate.

oxidase *n*. any enzyme that catalyses oxidation/reduction reactions using molecular oxygen as acceptor. *cf*. reductase, dehydrogenase.

oxidation *n*. the addition of oxygen, loss of hydrogen or loss of electrons from a compound, atom or ion. *cf*. reduction.

β-oxidation pathway metabolic pathway by which fatty acids are degraded in the mitochondria to yield acetyl-CoA.

oxidation–reduction enzymes oxidoreductases *q.v.*

oxidation–reduction potential redox potential *q.v.*

oxidative phosphorylation the formation of ATP from ADP and orthophosphate as the result of aerobic respiration. *cf*. photophosphorylation. *see also* chemiosmosis.

oxidative stress stress placed on cells, mainly through damage to proteins and DNA, by the production of large amounts of free oxygen radicals in situations such as responses to infection.

oxidoreductase *n*. any of a class of enzymes catalysing the oxidation of one compound with the reduction of another, and including the dehydrogenases, catalases, oxidases, peroxidases and reductases.

oxoglutarate *n*. currently the chemically correct name for ketoglutarate, which is, however, still widely used.

oxy- when prefixing the name of an oxygen-carrying blood pigment such as haemoglobin denotes the oxygenated form.

oxyaster *n*. stellate sponge spicule with pointed rays.

oxybiotic *a*. living in the presence of oxygen.

oxychlorocruorin *n*. chlorocruorin with oxygen bound, as in the aerated blood of certain polychaete worms.

oxydactylous *a*. having slender tapering digits.

oxyerythrocruorin *n*. erythrocruorin with oxygen bound, as in the aerated blood of many annelids and molluscs.

oxygen (O) *n*. gaseous element, in the free state a colourless odourless gas (O_2), but can also form ozone (O_3). As O_2 it forms 21% of the atmosphere by volume. It is highly reactive, and is the most abundant element on Earth, occurring also in Earth's crust and in many types of organic molecules. An essential element for living organisms. Many organisms also require molecular oxygen (O_2) to carry out respiration. *see also* aerobic respiration, oxygen cycle, ozone.

oxygenase *n*. enzyme that incorporates oxygen from molecular oxygen (O_2) into organic compounds. *see* dioxygenase, monooxygenase.

oxygen cycle the circulation of oxygen through the biosphere and the atmosphere. Oxygen is released by green plants and algae and by some bacteria, e.g. cyanobacteria, during photosynthesis, and is taken up by aerobic organisms, including plants, for respiration.

oxygen debt deficit in stored chemical energy which builds up when a normally aerobic tissue such as muscle is working with an inadequate oxygen supply. Once oxygen is restored, the tissue then consumes oxygen above the normal rate for some time until energy supplies are restored by respiration.

oxygen dissociation curve (1) graph of percentage saturation of haemoglobin with oxygen against concentration of oxgen, which gives information about the dissociation of oxyhaemoglobin under different environmental conditions or in different animals; (2) any graph showing the dissociation of oxygen from a substance.

oxygenic photosynthesis type of photosynthesis in which oxygen is produced, and which is the type of photosynthesis carried out by plants, algae and cyanobacteria.

oxygenotaxis oxytaxis *q.v.*

oxygenotropism oxytropism *q.v.*

oxygen quotient (Q_{O2}) the volume of oxygen (in microlitres gas at normal temperature and pressure) taken in per hour per milligram dry weight.

oxygen sag curve curve obtained when the dissolved oxygen in a water-course receiving a source of organic pollution is plotted against distance downstream of the discharge. It shows a sharp decrease in dissolved oxygen immediately downstream of the discharge, followed by a gradual increase as one goes farther downstream.

oxygnathous *a.* with more-or-less sharp jaws.

oxyhaemerythrin *n.* haemerythrin with oxygen bound, as in the blood of some annelids.

oxyhaemocyanin *n.* haemocyanin with oxygen bound, as in the aerated blood of many molluscs and arthropods.

oxyhaemoglobin *n.* haemoglobin with oxygen bound. It is formed when the concentration of oxygen is high, as in the lungs, and releases oxygen when oxygen concentration is low, as in the tissues.

oxymyoglobin *n.* myoglobin with oxygen bound. It is formed when oxygen concentration is high and releases oxygen when oxygen concentration is low.

oxyntic *a.* acid-secreting, *appl.* cells in the gastric gland of the stomach which secrete hydrochloric acid (HCl).

oxyphil(ic) *a.* having a strong affinity for acidic stains. *alt.* acidophil, acidophilic.

Oxyphotobacteria *n.* class of photosynthetic Bacteria consisting of the cyano-bacteria, i.e. those bacteria that produce oxygen as a by-product of photosynthesis.

oxytaxis *n.* a taxis in response to the stimulus of oxygen. *a.* **oxytactic**.

oxytetracycline *n.* broad-spectrum tetracycline antibiotic, which inhibits bacterial protein synthesis. *alt.* terramycin.

oxytocic *a.* accelerating parturition.

oxytocin *n.* peptide hormone secreted by the neurohypophysis (the posterior lobe of the pituitary), and which in mammals induces contraction of smooth muscle, especially uterine muscle.

oxytocinergic *a. appl.* neurons secreting oxytocin.

oxytropism *n.* tendency of organisms or organs to be attracted by oxygen.

ozone *n.* O_3, gas formed from oxygen (O_2) under the action of short-wavelength ultraviolet radiation in the stratosphere, where it forms the ozone layer. This absorbs considerable solar ultraviolet radiation and shields the Earth's surface from its harmful effects. Ozone is also formed as a pollutant in the lower atmosphere from e.g. nitrogen oxides. It is damaging to herbaceous plants at levels greater than 100 parts per billion.

P

P (1) symbol for the chemical element phosphorus *q.v.*; (2) proline *q.v.*

P$_f$, P$_{fr}$ the plant pigment phytochrome in its activated form, produced by the action of red light on P$_r$.

P$_i$ inorganic phosphate or orthophosphate.

P$_r$ the inactive form of phytochrome.

P$_1$ parents, **P$_2$**, grandparents, in genetic crosses.

Δp protonmotive force *q.v.* *see* chemiosmotic theory.

p53 tumor suppressor gene that is activated by damage to DNA and encodes a gene-regulatory protein (p53) that is involved in blocking further progression through the cell cycle or in inducing apoptosis of the damaged cell.

P430 bound ferredoxin, a component of photosystem I.

P680 reaction centre of photosystem II.

P700 reaction centre of photosystem I.

P730 the plant pigment phytochrome in its activated form.

PABA *p*-aminobenzoic acid *q.v.*

Pacchionian bodies discrete masses of subarachnoid tissue covered by arachnoid membrane and pressing into dura mater.

pacemaker *n.* (1) cell or region of organ determining rate of activity in other cells or organs; (2) in heart, the sinuatrial node, which initiates and maintains the normal heartbeat; (3) *appl.* neurons that set up and maintain rhythmic activity without further external stimuli.

pachy- prefix derived from Gk *pachys*, thick.

pachycarpous *a.* with a thick pericarp, *appl.* fruit.

pachycaulous *a.* with a thick or massive primary stem and root.

pachydermatous *a.* with thick skin or covering.

pachyderms *n.plu.* group of large non-ruminant hoofed mammals with thick tough hides, e.g. elephant, rhinoceros.

pachyphyllous *a.* thick-leaved.

pachytene *n. appl.* stage in prophase of 1st division of meiosis in which homologous chromosomes are associated as bivalents.

Pacific North American region part of the Holoarctic (Boreal) floristic kingdom, consisting of North America west of the Rocky Mountains from southern Alaska south to the Mexican border.

Pacinian bodies, Pacinian corpuscles sensory receptors in joints and skin consisting of a connective tissue capsule with core of cells innervated by sensory nerve endings, and sensitive to pressure. *see* Fig. 35 (p. 622).

packing motifs distinct arrangements of protein secondary structure elements that are defined by the number and types of elements present and the angles between them, and which can be used to classify tertiary protein structures.

packing ratio of DNA, the ratio of the length of the extended DNA molecule to the length of the fibre into which it is folded.

Paeoniales *n.* order of herbaceous dicots, some shrubs, with large deeply cut leaves and large showy flowers. Comprises the family Paeoniaceae (peony).

PAGE polyacrylamide gel electrophoresis *q.v.*

PAH polycyclic aromatic hydrocarbon *q.v.*

pain *n.* unpleasant sensory and emotional experience associated with actual or potential tissue damage.

pain fibres nerves that carry pain signals.

paired bodies small bodies lying close to sympathetic nerve chain in elasmobranch fishes, representing the adrenal medulla.

paired fins pelvic and pectoral fins of fishes.

pairing *n.* the synapsis of homologous chromosomes during the zygotene stage of meiosis, when they come to lie side by side.

pair-rule genes genes acting early in *Drosophila* embryogenesis which are involved in delimiting the parasegments.

pairwise alignment an alignment of two protein or nucleotide sequences. *cf.* multiple alignment.

Palaearctic region zoogeographical region in Wallace's classification, consisting of Europe, North Africa, Western Asia, Siberia, northern China and Japan. It is divided into the European, Manchurian, Mediterranean, and Siberian subregions.

palaeo- prefix derived from Gk *palaios* ancient. *alt.* paleo-.

palaeobiology *n.* the study of the biology of extinct plants, animals and microorganisms.

palaeobotany *n.* study of fossil plants and plant impressions.

Palaeocene Paleocene *q.v.*

palaeocerebellum *n.* the phylogenetically older region of cerebellum, receiving spinal and vestibular afferent nerve fibres. *cf.* neocerebellum.

palaeocortex *n.* evolutionarily ancient part of brain including the limbic system and olfactory bulb. *cf.* archicortex.

palaeocranium *n.* type of skull or stage in development extending back no farther than the vagus nerve.

palaeodendrology *n.* the study of fossil trees and tree impressions.

palaeoecology *n.* the study of the relationship between past organisms and the environment in which they lived.

palaeoencephalon *n.* the primitive vertebrate brain.

palaeogenetic *a. appl.* atavistic features which are fully developed in adult although normally only characteristic of embryonic stage.

palaeogenetics *n.* the application of the principles of genetics to interpretation of the fossil record and the evolution of now extinct species.

Palaeognathae *n.* the ratites, flightless birds of the subclass Neornithes such as the kiwis, cassowaries and ostrich, which are secondarily flightless.

Palaeolithic *a. appl.* or *pert.* the Old Stone Age, characterized by a hunter-gatherer economy and chipped stone tools.

Palaeoniscoidei, Palaeonisciformes, palaeoniscids *n., n.plu., n.plu.* group of actinopterygian fishes, most now extinct, existing from the Devonian to the present day and including the bichars of the Nile. They are carnivorous with large sharp teeth and usually a very heterocercal tail.

palaeontology *n.* the study of past life on Earth from fossils and fossil impressions.

palaeopallium *n.* the olfactory region of the cerebral hemispheres, comprising olfactory bulbs and tubercles, pyriform lobes, hippocampus and fornix.

palaeosere *n.* the development of vegetation throughout the Paleozoic.

palaeospecies *n.* a group of extinct organisms that are assumed to have been able to interbreed and so are placed in the same species.

Palaeotropical kingdom one of the five generally recognized floristic kingdoms, consisting of Africa except for the northern part around the Mediterranean Sea and the tip of Southern Africa (but including Madagascar), Arabia and Asia, including the Indian subcontinent and the islands of the Indian Ocean south from the Himalayas to New Guinea. It is divided into the African, Indo-Malaysian and Polynesian subkingdoms and the following regions: Ascension and St Helenan, Continental Southeast Asian, East African Steppe, Hawaiian, Indian, Madagascan, Malaysian, Melanesian and Micronesian, New Caledonian, North African Desert, Northeast African Highland and Steppe, Polynesian, Sudanese Park Steppe, West African Rainforest, South African Steppe. *all q.v.*

Palaeozoic Paleozoic *q.v.*

palaeozoology *n.* the study of the biology of extinct animals from fossils and fossil impressions.

palama *n.* the foot-webbing of aquatic birds.

palatal, palatine *a. pert.* palate.

palate *n.* (1) roof of mouth in vertebrates; (2) roof of pharynx in insects; (3) (*bot.*) projection of lower lip of snapdragon-type corolla.

palatine *n.* a bone or cartilage of the hard palate in vertebrates.

palatoglossal *a.* (1) *pert.* palate and tongue; (2) *appl.* a muscle: glossopalatinus *q.v.*

palatonasal *a. appl.* palate and nose.

palatopharyngeal *a. pert.* palate and pharynx.

palatopterygoid *a. pert.* palate and pterygoid.

palatoquadrate *a.* connecting palatine and quadrate bones or cartilages, *appl.* dorsal cartilage of mandibular arch.

palatoquadrate cartilage paired skeletal element forming upper jaw in elasmobranch fishes.

palea *n.* the upper of the two bracts enclosing a individual flower in grasses. *plu.* **paleae.**

paleo- *see also* palaeo-.

Paleocene *n.* the earliest epoch of the Tertiary geological sub-era and the Paleogene period, before the Eocene and lasting from *ca.* 65 to 55 million years ago. It is the first epoch after the mass extinction at the end of the Cretaceous period and saw the diversification of primitive mammals, such as the now extinct multituberculates, after the extinction of the dinosaurs.

Paleogene *n.* geological period lasting from *ca.* 65 million years ago to 23 million years ago, the earliest part of the Tertiary sub-era. It comprises the Paleocene, Eocene and Oligocene epochs and saw the rise of the mammals and their evolution into the ancestors of modern forms.

Paleozoic *n.* geological era lasting from *ca.* 542 million years ago to 251 million years ago and comprising the Cambrian, Ordovician, Silurian, Devonian, Carboniferous and Permian periods.

palette *n.* the modified cupule-bearing tarsus of anterior leg in male beetles.

paliform *a.* like an upright stake.

palinal *a.* from behind forwards, *appl.* e.g. jaw movements of elephants.

palindrome *n.* in DNA, a base sequence that reads exactly the same from left to right as from right to left. *a.* **palindromic.**

palingenesis *n.* (1) abrupt metamorphosis; (2) the recapitulation of ancestral stages in the development of their descendants. *alt.* recapitulation.

palingenetic *a.* (1) of remote or ancient origin; (2) *pert.* palingenesis.

palisade tissue the layer or layers of photosynthetic cells beneath the epidermis of many foliage leaves.

pallaesthesia *n.* vibratory sensation, as felt e.g. in bones.

pallet *n.* a shelly plate on a bivalve siphon.

palliate *n.* having a mantle or similar structure.

pallidium globus pallidus *q.v.*

palliopedal *a. pert.* molluscan mantle and foot.

pallioperitoneal *a. appl.* complex of organs in some molluscs, including heart, renal organs, gonads and ctenidia.

pallium *n.* (1) the mantle of molluscs and brachiopods; (2) cerebral cortex *q.v.*

Palmaceae, palms *n., n.plu.* tropical and subtropical family of monocots typically with large leathery fan-like leaves. They include climbing and tree species. The fruit is a berry or drupe. Cultivated species include the date palm (*Phoenix dactylifera*), oil palms (*Elaeis* spp.) and the coconut palm (*Cocos nucifera*).

palmaesthesia *n.* sensing sounds by the vibrations they make in the bones.

palmar *a. pert.* palm of hand.

palmate *a.* (*bot.*) (1) *appl.* leaves divided into lobes arising from a common centre; (2) *appl.* tuber shaped like a hand, as in some orchids; (3) (*zool.*) having anterior toes webbed, as in most aquatic birds.

palmatifid *a. appl.* leaves divided into lobes to about the middle, at acute angles to each other.

palmatilobate *a.* palmate with rounded lobes and divisions halfway to base, of leaves.

palmatipartite *a.* palmate with divisions more than halfway to base, of leaves.

palmatisect *a.* palmate with divisions nearly to base, of leaves.

palmella *n.* sedentary stage of certain algae, the cells dividing within a jelly-like mass and producing motile gametes. *a.* **palmelloid.**

palmigrade *a.* walking on the sole of the foot.

palmiped (1) *a.* web-footed; (2) *n.* a web-footed bird.

palmitic acid common 16-carbon saturated fatty acid.

palmitin *n.* a fat present in adipose tissue, milk and palm oil.

palmitoylation *n.* the covalent addition of the fatty acid palmitic acid to a protein, a linkage by which some proteins are attached to cell membranes.

palmoid *a. appl.* palms and palm-like forms.

palmula *n.* terminal lobe or process between paired claws of insect feet.

palp *n.* (1) labial feeler of an insect; (2) other similarly situated appendages in other invertebrates. *a.* **palpal**.

palpacle *n.* the tentacle of a dactylozooid individual of a siphonophore.

palpal *a. pert.* a palp.

palpate (1) *a.* having palps; (2) *v.* to examine by touch.

palpebral *a. pert.* eyelids.

palpifer, palpiger *n.* in insects, a lobe of maxilla or other mouthpart which bears a palp.

palpiform *n.* resembling a palp.

Palpigrada *n.* order of very small arachnids having a jointed flagellum on the last segment of the opisthosoma.

palpimacula *n.* sensory area on labial palps of certain insects.

palpocil *n.* stiff sensory filament of tactile sensory cells of some coelenterates.

palpon *n.* in colonial hydrozoans, an individual hydroid modified for catching prey and for defence, being long and slender, usually with tentacles and without a mouth.

palpule, palpulus *n.* a small palp or feeler.

paludal *a.* (1) marshy; (2) *pert.*, or growing in, marshes or swamps.

paludicole *a.* living in marshes.

palus *n.* a small stake-like structure. *plu.* **pali**.

palustral, palustrine *a.* growing in marshes or swamps.

palynology *n.* (1) the study of pollen and its distribution, *alt.* pollen analysis; (2) the study of spores.

palynomorph *n.* general term for microfossils with organic walls, including pollen and fossil spores.

PAM percent accepted mutations *q.v.*

PAMP pathogen-associated molecular pattern *q.v.*

Pampas region part of the Neotropical floristic kingdom, consisting of the pampas of Argentina.

pampiniform *a.* (1) like a tendril; (2) *appl.* convoluted vein plexus of the spermatic cord which acts as a countercurrent heat exchanger to cool the testes; (3) *appl.* body: a small collection of tubules anterior to ovary, the remnant in adult of embryonic mesonephros.

pamprodactylous *a.* with all the toes pointing forward.

PAN *see* peroxyacyl nitrates.

pancolpate *a.* of pollen grains, having many furrows.

pancreas *n.* glandular organ associated with the gut in most vertebrates, and secreting the hormones insulin and glucagon from endocrine glands and digestive enzymes from exocrine glands.

pancreastatin *n.* peptide produced by pancreas which inhibits glucose-induced release of insulin from pancreas.

pancreatic *a. pert.* pancreas.

pancreatic juice secretion containing digestive enzymes and enzyme precursors secreted into the gut from the pancreas. It contains trypsinogen, chymotrypsinogen, procarboxypeptidases, lipase, α-amylase, maltase and ribonuclease.

pancreaticoduodenal *a. pert.* pancreas and duodenum, *appl.* arteries, veins.

pancreatic ribonuclease an endoribonuclease (originally isolated from mammalian pancreas) that specifically cleaves RNA on the 3′ side of C or U. EC 3.1.27.5.

pancreatic trypsin inhibitor protein produced in the pancreas which binds specifically to the enzyme trypsin, inhibiting it. *alt.* antitrypsin.

pancreozymin cholecystokinin *q.v.*

Pandanales *n.* order of monocots, mainly sea coast or marsh plants with tall stems supported by aerial roots, and leaves running in spirals, comprising the family Pandanaceae (screw pine).

pandemic (1) *a.* very widely distributed; (2) *a. appl.* outbreak of a disease that affects large numbers of people at the same time and spreads all over the world, e.g. the influenza pandemic of 1918. *cf.* epidemic, endemic.

panduriform *a.* fiddle-shaped, *appl.* leaves.

Paneth cells cells at base of crypts of Lieberkühn in small intestine, which secrete cryptdins and other antimicrobial peptides.

Pangaea *n.* a supercontinent present from the Permian to the Jurassic which was made up of all the present continents fitted together before their separation by continental drift and which stretched from pole to pole. *see also* Gondwana, Laurasia. *alt.* **Pangea.**

pangamic *a. appl.* indiscriminate or random mating. *n.* **pangamy.**

pangenesis *n.* a now discarded theory that hereditary characteristics were carried and transmitted by gemmules from individual body cells.

panicle *n.* a branched flowerhead, strictly one in which the branches are alternate and side branches also branch alternately.

panicoid *a. appl.* millets of the genus *Panicum.*

paniculate *a.* having flowers arranged in panicles.

panmictic *a.* characterized by, or resulting from, random matings.

panmixia *n.* indiscriminate interbreeding. *alt.* **panmixis.**

panniculus carnosus a thin layer of muscle fibres in dermis which is involved in moving or twitching the skin.

panning *n.* method of separating cell subpopulations by means of monoclonal antibodies bound to culture dishes.

pannose *a.* resembling cloth in texture.

panoistic *a. appl.* ovariole in which nutritive cells are absent, the egg yolk being synthesized by the epithelial cells of the follicle.

panphotometric *a. appl.* leaves oriented to avoid maximum direct sunlight.

panphytotic *n.* a pandemic affecting plants.

panspermia *n.* a theory popular in the 19th century and which enjoys periodic revivals, that life did not originate on Earth but arrived in the form of bacterial spores or viruses from an extraterrestrial source.

pansporoblast *n.* a cell complex of certain sporozoan protozoans, the Neosporidia, producing sporoblasts and spores.

panthalassic *a.* living in both coastal and oceanic waters.

panting centre region of hypothalamus whose stimulation causes the rate of respiration to quicken.

pantodonts *n.plu.* group of extinct North American and Asian herbivorous placental mammals existing from the Paleocene to the Oligocene.

pantonematic *a. appl.* flagella with longitudinal rows of fine hairs along their axis.

pantostomatic *a.* capable of ingesting food at any part of the surface, as amoebae and similar organisms.

pantothenic acid vitamin B_3 or B_5, a precursor of coenzyme A. It is necessary for growth in various animals, and is the rat anti-grey hair and chick antidermatitis factor.

pantotheres *n.plu.* order of extinct trituberculate mammals of the Jurassic, possible ancestors of living therians, having molar teeth showing the basic pattern found in living forms.

pantropic *a.* (1) invading many different tissues, *appl.* viruses; (2) turning in any direction.

pantropical *a.* distributed throughout the tropics, *appl.* species.

papain *n.* endopeptidase found in the fruit juice and leaves of pawpaw (papaya), *Carica papaya*, and used commercially as a meat tenderizer. EC 3.4.22.2.

Papaverales *n.* order of mainly herbaceous dicots, some shrubs and small trees, and including the families Fumariaceae (fumitory), Hypecoaceae and Papaveraceae (poppy).

paper chromatography chromatography in which paper is used as the support medium on which the substances to be separated migrate differentially in the solvent.

papilionaceous *a.* like a butterfly, *appl.* flowers like sweet pea, gorse and broom, with a corolla of five petals, one enlarged upright standard or vexillum, two anterior united to form a keel or carina, and two lateral, the wings or alae.

papilla *n.* (1) small projection or protuberance, as clusters of taste buds on tongue; (2) conical structure, as nipple. *plu.* **papillae.**

papillary *a.* (1) *pert.* or with a papilla; (2) *appl.* a process of caudal lobe of liver; (3) *appl.* muscle between walls of ventricles of heart and chordae tendineae; (4) *appl.* a layer of the dermis.

papillate *a.* (1) covered with papillae; (2) like a papilla.

papilliform *a.* like a papilla in shape.

papilloma *n.* wart, or similar small benign tumour.

Papillomaviridae *n.* family of small, non-enveloped, double-stranded DNA viruses that includes the papillomaviruses that cause common warts. Some papillomaviruses are a cause of cervical cancer.

pappiferous *a.* bearing a pappus.

pappose, pappous *a.* (1) having limb of calyx developed as a tuft of hairs or bristles (pappus); (2) downy, or covered with feathery hairs.

pappus *n.* a circle or tuft of bristles, hairs or feathery processes in place of a calyx on florets of flowers of the Compositae. It persists on the seed, aiding its dispersal by wind.

papulae *n.plu.* hollow contractile pustules on epidermis of some echinoderms, such as starfish, having a respiratory function.

papyraceous, papyritic *a.* like paper in texture.

PAR photosynthetically active radiation *q.v.*

para- prefix derived from Gk *para*, beside, signifying situated near, or surrounding.

para-aminobenzoic acid *see* *p*-aminobenzoic acid.

para-aortic *a. appl.* chromaffin bodies situated alongside the abdominal aorta.

Parabasala, Parabasalia, parabasalids *n., n., n.plu.* a taxonomic grouping of non-photosynthetic flagellate unicellular eukaryotes that lack mitochondria and contain prominent Golgi-body-like parabasal bodies surrounding the nucleus. Examples are *Trichomonas* (*T. vaginalis* is a human pathogen) and *Trichonympha*.

parabiosis *n.* (1) the condition of being conjoined either from birth, as Siamese twins, or experimentally as laboratory animals; (2) use of the same nest by different species of ant, which, however, keep their broods separate. *a.* **parabiotic**.

parablastic *a. appl.* large nuclei of eggs laden with yolk granules.

parabranchia *n.* feathery chemoreceptor organ in molluscs, so-called from its superficial resemblance to gills.

parabranchial cavity space between gill opening and gill.

parabronchus *n.* one of numerous small tubules in lungs of birds, connecting dor-

sobronchi and ventrobronchi, and in which respiratory gas exchange takes place.

Paracanthopterygii *n.* advanced group of teleost fishes, existing from the Eocene to the present day and including the cod family.

paracardial *a.* surrounding the neck of the stomach.

paracellular transport transport of water and solutes through the intercellular junctions of epithelial cells.

paracentral *a.* (1) situated near the centre, *appl.* lobule, gyrus, fissure; (2) *appl.* retinal area surounding fovea centralis.

paracentric *a.* (1) on same side of centromere of a chromosome; (2) *appl.* chromosomal rearrangements involving only one chromosome arm; (3) *appl.* chromosomal inversions which do not include the centromere.

parachordal *a.* on either side of the notochord.

parachrosis *n.* (1) process or condition of changing colour; (2) discoloration; (3) fading.

parachute *n.* (1) special structure of a seed, such as aril, caruncle, pappus or wing, which aids dispersal by wind; (2) fold of skin used for gliding as in flying squirrels.

paracme *n.* (1) the evolutionary decline of a taxon after reaching the highest point of development; (2) the declining or senescent period in the life history of an individual.

paracoel *n.* lateral ventricle or cavity of cerebral hemisphere.

paracondyloid *a. appl.* projection of occipital bones occurring beside condyles in some mammals.

paracone *n.* the front outside cusp of upper molar tooth.

paraconid *n.* the front inside cusp of lower molar tooth.

paracorolla *n.* a corolla appendage such as a corona.

paracortex *n.* the region of a lymph node between the medulla and the lymphoid follicles, where most T cells are found. *alt.* T-cell zone.

paracrine *a. appl.* cytokines acting at short range on neighbouring cells.

paracymbium *n.* accessory part of the cymbium, between tarsus and tibia in some spiders.

paracyte *n.* modified cell extruded from embryonic tissue into yolk as in some insects.

paracytic *a. appl.* stomata of a type in which subsidiary cells lie alongside the stoma parallel to the long axis of guard cells, formerly called rubiaceous.

paracytoids *n.plu.* minute pieces of chromatin extruded from nuclei of embryonic cells and shed into the blood, as in certain insects.

parademe *n.* an apodeme arising from the edge of a sclerite.

paraderm *n.* the delicate limiting membrane of a pronymph.

paradermal *a. appl.* section cut parallel to the surface of a flat organ such as a leaf. *cf.* tangential.

paradesmus *n.* secondary connection between centrioles outside nucleus in mitosis of some flagellate protozoans.

paradidymis *n.* body of convoluted tubules anterior to the lower part of spermatic cord, representing posterior part of embryonic mesonephros.

paraesophageal paraoesophageal *q.v.*

parafacialia *n.plu.* narrow parts of head capsule between frontal suture and eyes, as in certain Diptera.

paraflagellum *a.* a subsidiary flagellum.

paraflocculus *n.* cerebellar lobule lateral to flocculus.

parafrons *n.* area between eyes and frontal suture in certain insects.

parafrontals *n.plu.* the continuation of genae between eyes and frontal suture in insects.

paraganglia *n.plu.* scattered clusters of cells secreting adrenaline alongside aorta.

paragaster *n.* central cavity lined with choanocytes, into which ostia open, in sponges.

paragastric *a.* (1) *appl.* passages or cavities in branches of sponges; (2) *appl.* paired blind canals from infundibulum to oral cone of ctenophores.

paragenesis *n.* (1) condition in which an interspecific hybrid is fertile with the parental species but not with other similar hybrids; (2) a subsidiary mode of reproduction.

paragenetic *a. appl.* a mutation affecting the expression rather than the structure of a gene.

paraglenal hypercoracoid *q.v.*

paraglobulin *n.* one of the globulins present in blood serum.

paraglossa *n.* (1) appendage on either side of labium of insects, *alt.* labella; (2) a paired cartilage of the chondrocranium.

paraglossum *n.* median cartilaginous or bony prolongation of copula supporting the tongue, as in birds.

paraglycogen *n.* carbohydrate food reserve in protozoans, resembling glycogen.

paragnatha *n.plu.* (1) paired, delicate, unjointed processes of maxilla of certain arthropods; (2) buccal denticles of certain polychaete worms.

paragnathous *a.* with mandibles of equal length, *appl.* birds.

paragula *n.* region beside gula on insect head.

paraheliode *n.* parasol-like arrangement of spines in certain cacti.

paraheliotropism paraphototropism *q.v.*

parahormone *n.* substance that acts like a hormone but is a product of the ordinary metabolism of cells.

paralectotype *n.* specimen of a series used to designate a species, which is later designated as a paratype.

paralimnic *a. pert.* or inhabiting the lake shore.

parallel β-sheet type of β-sheet secondary structure in proteins in which adjacent strands run in the same directions. It is formed by the bringing together of non-contiguous regions of the protein chain.

parallel descent, parallel evolution (1) evolution in a similar direction in different groups; (2) the independent acquisition of similar traits in two related species.

parallel distributed processing one hypothesis of brain function in which various aspects of information from e.g. the visual field (such as colour, movement, form) are encoded in many separate neuronal channels at source and throughout much of their analysis in the cortex. The final perception is due to activation of connections between channels and the simultaneous activity of neurons in several different areas.

parallel DNA sequencing massively parallel DNA sequencing *q.v.*

parallel flow the flow of two fluids in the same direction.

parallelinervate, parallelodrome *a. appl.* leaves with veins parallel.

parallelotropism orthotropism *q.v.*

parallel venation leaf venation in which veins run parallel longitudinally along the leaf, as in monocotyledons.

paralogous *a.* (1) *appl.* similarities in anatomy that are not related to common descent or similar function; (2) *appl.* two homologous genes in the same or different genomes that are similar because they derive from a gene duplication (e.g. α- and β-globins). *n.* **paralogue, paralogy.** *cf.* orthologous.

paramastigote *a.* having one long principal flagellum and a short accessory one, as certain flagellate protozoans.

paramastoid *a.* (1) beside the mastoid; (2) *appl.* two paraoccipital processes of exoccipitals; (3) *appl.* a process projecting from the jugular process.

paramesonephric ducts Müllerian ducts *q.v.*

parametrium *n.* fibrous tissue partly surrounding uterus.

paramitosis *n.* nuclear division, as in some protozoans, in which the chromosomes are not regularly arranged on equator of spindle and tend to cohere at one end when separating.

paramorph *n.* (1) any variant form or variety; (2) a form induced by environmental factors without underlying genetic change.

paramutation *n.* the condition when one allele (the paramutagenic allele) influences the expression of another (the paramutable allele) at the same locus when they are combined in a heterozygote.

paramutualism facultative symbiosis *q.v.*

paramylon *n.* substance allied to starch, present in certain algae and flagellates. *alt.* **paramylum.**

paramyosin *n.* protein present in filaments of unstriated muscle, as of molluscs.

Paramyxoviridae, paramyxoviruses *n.*, *n.plu.* family of large, enveloped, single-stranded, negative-strand RNA viruses including measles, mumps and Newcastle disease virus of swine.

paranasal *a. appl.* air sinuses in bones of upper jaw and face.

paranema *n.* sterile filament in reproductive organs of algae, mosses and ferns.

paranemic *a.* (1) *appl.* DNA structures in which the two strands are not intertwined in a double helix, as may occur in limited regions of heteroduplex DNA during recombination; (2) *appl.* a double spiral structure in which the two strands do not interlock at each turn but lie side by side. *alt.* anorthospiral.

paranephric *a.* beside the kidney, *appl.* a fatty body behind the connective tissue ensheathing the kidney.

paranephrocyte arthrocyte *q.v.*

paranotum *n.* one of the lateral expansions of the arthropod notum or tergum, thought to be the structures that have evolved into wings in some insects. *a.* **paranotal.**

Paranthropus genus of fossil hominins from Southern Africa comprising the 'robust' australopithecines.

paraoesophageal *a. appl.* nerve fibres connecting 'brain' or cerebral ganglion with suboesophageal ganglion in some invertebrates.

parapatric *a. appl.* distribution of species or other taxa that meet in a very narrow zone of overlap. *n.* **parapatry.**

parapet *n.* circular fold in body wall below margin of disc in sea anemones.

paraphasia *n.* language disorder in which a word is substituted by a sound, an incorrect word or an unintended word.

paraphototropism *n.* tendency of plants to turn edges of leaves towards intense illumination, thus protecting the surfaces.

paraphyletic *a. appl.* a group, such as the reptiles, which has evolved from and includes a single ancestral species (known or hypothetical), but which does not contain all the descendants of that ancestor. *n.* **paraphyly.**

paraphyll(ium) *n.* one of the branching chlorophyll-containing outgrowths arising between leaves or from the leaf bases, in mosses.

paraphysis *n.* in fungi, sterile elongated cell interspersed amongst asci or basidia in hymenium of some ascomycetes and basidiomycetes. *plu.* **paraphyses.**

parapineal *a. appl.* the eye-like parietal organ of brain of cyclostomes and some reptiles, the pineal body in other vertebrates.

parapleuron *n.* anterior lateral plate of exoskeleton at base of meta- and meso-thoracic segments, in insects.

parapodium *n.* (1) paired lateral flap of tissue on segment of polychaete worms, bearing numerous chaetae and used for locomotion; (2) lateral undulating extension of foot in some molluscs, used for propulsion. *plu.* **parapodia**. *a.* **parapodial**. *alt.* **parapod**.

parapolar *a.* beside the pole.

parapophysis *n.* transverse projection arising from centrum of vertebra.

parapostgenal *a. appl.* thickened portion of occiput in insects.

paraprostate *a.* anterior bulbo-urethral glands.

parapsid *a.* having skull with one dorsal temporal fenestra on each side.

paraquadrate squamosal *q.v.*

pararectal *a.* beside rectum.

parasegment *n.* a developmental unit in *Drosophila* and other insect embryos, delimited before visible segmentation, and which consists of the posterior compartment of one prospective segment and the anterior compartment of the next.

parasematic *a. appl.* markings, structures or behaviour tending to mislead or deflect attack by an enemy. *cf.* aposematic, episematic.

paraseme *n.* misleading appearance of markings, as an ocellus near the tail of fishes.

paraseptal *a. appl.* cartilage that almost encloses the vomeronasal organ.

parasexual *a. appl.* or *pert.* genetic recombination occurring other than during meiosis.

parasexual cycle a cycle of plasmogamy, karyogamy and haploidization in some fungi that superficially resembles true sexual reproduction but which may take place at any time in the life cycle.

parasite *n.* organism that for all or some part of its life derives its food from a living organism of another species (the host). It usually lives in or on the body or cells of the host, which is usually harmed to some extent by the association. *a.* **parasitic**.

parasite chain a food chain passing from large to small organisms.

parasitic castration (1) castration caused by the presence of a parasite, as in male crabs infested by the barnacle *Sacculina*; (2) sterility in various other plants or animals caused by a parasite attacking sex organs.

parasitic male dwarf male which is parasitic on its female and has a reduced body in all but the sex organs, as in some deep-sea fish.

parasitism *n.* special case of symbiosis in which one partner (the parasite) receives advantage to the detriment of the other (the host).

parasitocoenosis *n.* the whole complex of parasites living in any one host.

parasitoid *n.* organism alternately parasitic and free-living and whose parasitism ultimately kills its host. Examples are certain insects (e.g. ichneumon wasps) in which the adults are free-living but lay their eggs in the bodies of other insect larvae, in which the larvae develop, consuming host tissues and killing the host on hatching and emergence.

parasitology *n.* the study of parasites, esp. animal parasites.

parasocial *a. appl.* social group in which some features of complete eusociality are absent.

parasol paraheliode *q.v.*

parasphenoid *n.* skull bone forming floor of cranium in certain vertebrates.

parasporal *a. appl.* protein bodies formed within the cell during sporulation in some bacteria.

parasporangium *n.* sporangium containing paraspores.

paraspore *n.* a spore formed from a cortical somatic cell, as in some algae.

parastamen, parastemon staminode *q.v.*

parasternum *n.* the sum total of abdominal ribs in some reptiles.

parastichy *n.* a descending curved row formed by leaf primordia at the growing apex of a shoot.

parasymbiosis *n.* the living together of organisms without mutual harm or benefit.

parasympathetic *a. appl.* components of the parasympathetic nervous system, as parasympathetic ganglion.

parasympathetic nervous system part of the autonomic nervous system that controls involuntary muscular movement in blood vessels and gut, and glandular secretions from the eye, salivary glands, bladder, rectum and genital organs. Also includes the vagus nerve supplying the viscera as a whole. Parasympathetic nerve fibres are contained within the last five cranial nerves and the last three spinal nerves and terminate at parasympathetic ganglia near or in the organ they supply. The actions of the parasympathetic system are broadly antagonistic to those of the sympathetic system, lowering blood pressure, slowing heartbeat and stimulating the process of digestion. The chief neurotransmitter in the parasympathetic system is acetylcholine.

parately *n.* evolution from material unrelated to that of type, but resulting in superficial resemblance.

paratenic *a. appl.* to a host in which parasite development does not occur and which serves to bridge an ecological gap in the parasite's life cycle.

paraterminal *a. appl.* bodies constituting part of the anterior median wall of lateral ventricles of brain in amphibians and reptiles.

paratestis *n.* small reddish-yellow fatty body in some male newts and salamanders which produces hormones regulating the appearance of mating coloration.

parathecium *n.* (1) peripheral layer of apothecium, as in cup fungi; (2) peripheral layer of hyphae in lichens.

parathormone parathyroid hormone *q.v.*

parathyroid glands four small brownish-red endocrine glands near or within the thyroid in mammals, which secrete parathyroid hormone.

parathyroid hormone (PTH) protein hormone secreted by parathyroid glands which stimulates increase in bone resorption, resulting in increased calcium and phosphate in blood. It also stimulates increased resorption of calcium and magnesium and decreased resorption of phosphate in kidney tubules. It is essential for normal skeletal development.

paratoid *a. appl.* double row of poison glands extending along back of certain amphibians, as of salamanders.

paratomium *n.* side face of a bird's beak, between the cutting edge and the median longitudinal ridge.

paratomy *n.* asexual reproduction by the generation of chains of new individuals, followed by their separation, found in some annelids.

paratonic *a.* (1) stimulating or retarding; (2) *appl.* movements induced by external stimuli, e.g. tropisms or nastic movements.

paratracheal *a.* with xylem parenchyma cells around or close to vascular tissue.

paratrophic *a. appl.* method of nutrition of obligate parasites.

paratympanic *a.* (1) medial and dorsal to the tympanic cavity; (2) *appl.* a small organ with sensory epithelium innervated from geniculate ganglion in many birds.

paratype *n.* specimen described at same time as the one regarded as type specimen of a new genus or species.

para-urethral *a. appl.* mucus-secreting glands associated with urethra, Littré's glands.

paravertebral *a.* alongside the spinal column, *appl.* trunk of sympathetic nerve.

paravesical *a.* beside the bladder.

paraxial *a.* (1) alongside the axis; (2) *appl.* the mesoderm in vertebrate embryos that is situated immediately on either side of the axis alongside the notochord and neural tube, and which gives rise to the somites and the mesoderm of the branchial arches. *cf.* axial, splanchnic.

paraxon *n.* a lateral branch of the axon of a neuron.

paraxonic *a.* (1) *pert.* or having an axis outside the usual axis; (2) with axis of foot between 3rd and 4th digits, as in artiodactyls.

Parazoa *n.* term sometimes used for those multicellular animals, such as sponges, having a loose organization of cells and not forming distinct tissues or organs.

parencephalon cerebral hemisphere *q.v.*

parenchyma *n.* (1) (*bot.*) soft plant tissue composed of thin-walled, relatively undifferentiated cells, which may vary in structure and function; (2) (*zool.*) solid layer of tissue between muscle layer and gut in platyhelminths, composed of many different types of cell; (3) ground tissue of an organ.

parenchymatous *a. pert.* or found in parenchyma.

parenchymula *n.* flagellate sponge larva with internal cavity filled with gelatinous material.

parental generation the parent individuals in a genetic cross, designated P_1.

parental genomic imprinting *see* genomic imprinting.

parental investment any behaviour towards offspring that increases the chances of the offspring's survival at the cost of the parent's ability to invest in other offspring.

parenthosome *n.* structure formed from endoplasmic reticulum and covering the pore in the dolipore septa of certain fungi.

parethmoid ectethmoid *q.v.*

parhomology *n.* apparent similarity of structure.

paries *n.* (1) the central division of a compartment of barnacles; (2) the wall of a hollow structure, as of tympanum, or of honeycomb. *plu.* **parietes**.

parietal *a.* (1) *pert.*, next to, or forming part of the wall of a structure; (2) *appl.* placentation of plant ovary: ovules borne in rows on carpel wall or extensions of it.

parietal bone a paired bone of roof of skull.

parietal cells oxyntic cells *q.v.*

parietal lobe of brain, dorsal lobe of cerebral cortex, lying behind central sulcus.

parietal organ the pineal body in some lower vertebrates, where it has a photoreceptor function.

parietal vesicle dilated distal part of pineal stalk.

parietes *plu.* of paries.

parietobasilar *a. appl.* muscles between pedal disc and lower part of body wall in sea anemones.

parietofrontal *a. appl.* a skull bone in place of parietals and frontals, as in lungfishes.

parietomastoid *a.* connecting mastoid and parietal bone, *appl.* a suture.

parieto-occipital sulcus fissure between parietal and occipital lobes of cerebral hemisphere.

parietotemporal *a.* (1) *pert.* parietal and temporal regions; (2) *appl.* a branch of the middle cerebral artery.

paripinnate *a.* pinnate without a terminal leaflet.

parity *n.* the number of times a female has given birth, regardless of the number of offspring produced at any one birth.

parivincular *a.* bivalve hinge ligament attached along whole edge of shell.

parkinsonism *n.* the symptoms of muscular rigidity, weakness and tremor at rest characteristic of Parkinson's disease, caused by disorders in the dopaminergic pathways in brain.

parolfactory *a. appl.* an area and sulcus adjoining the small triangular space in olfactory lobe.

paronychia *n.* bristles on pulvillus of insect foot.

parosteosis *n.* bone formation in tracts normally composed of fibrous tissue.

parotic *n.* bony projection formed by fusion of exoccipital and opisthotic elements of skull in some reptiles and fishes.

parotid glands (1) paired salivary glands opening into the mouth, in some mammals; (2) in some amphibians, large swellings on side of head, formed of aggregated cutaneous glands, sometimes poisonous.

parovarium *n.* small collection of tubules anterior to ovary, the remnants in adult of embryonic mesonephros.

paroxysmal nocturnal haemoglobinaemia disease due to deficiency in a complement regulatory protein, so that complement triggers episodes of haemolysis.

pars *n.* a part of an organ.

pars distalis a part of the adenohypophysis *q.v.*

parsimony principle in molecular taxonomy and phylogeny, the principle that organisms from closely related species (i.e. species that diverged more recently) will have fewer nucleotide differences in their DNA than those that diverged longer ago. *see* maximum parsimony.

pars intercerebralis in an insect's forebrain, the region containing neurosecretory cells.

pars intermedia part of the glandular tissue of the pituitary, producing lipotropin and melanocyte-stimulating hormone.

pars nervosa the neural lobe of the pituitary, containing neurosecretory neurons projecting from the hypothalamus and secreting e.g. oxytocin and vasopressin.

pars tuberalis part of the adenohypophysis *q.v.*

parthenita *n.* unisexual stage of trematodes in intermediate host.

parthenocarpic *a. appl.* fruit lacking seeds, which has developed without fertilization.

parthenogenesis *n.* reproduction from a female gamete without fertilization by a male gamete.

parthenogenetic *a.* (1) *appl.* organisms produced by parthenogenesis; (2) *appl.* agents that can induce an unfertilized ovum to develop further.

parthenogonidia *n.plu.* zooids of a protozoan colony, with the function of asexual reproduction.

parthenote *n.* a parthenogenetically produced haploid organism.

partial diploid bacterium or other cell carrying two copies of some but not all its genes.

partial dominance co-dominance *q.v.*

partial parasite semiparasite *q.v.*

partial veil inner veil of certain basidiomycete fungi, growing from stipe towards edge of cap and becoming separated to form the cortina.

particulate inheritance refers to the fact that hereditary characteristics are transmitted by discrete entities (genes) which themselves remain unchanged from generation to generation. *cf.* blending inheritance.

partite *a.* divided nearly to the base.

partition chromatography separation technique in which the materials to be separated are selectively partitioned between two solvents.

Partitiviridae *n.* family of double-stranded RNA viruses infecting plants, e.g. white clover cryptic virus.

parturition *n.* the act of giving birth.

parumbilical *a. appl.* small veins from anterior abdominal wall to portal and iliac veins.

parvicellular *a.* consisting of small cells.

parvocellular layer layer of lateral geniculate nucleus of primate brain, consisting of layers 3, 4, 5 and 6, and composed of small neurons.

Parvoviridae, parvoviruses *n.*, *n.plu.* family of small, non-enveloped, single-stranded DNA viruses that includes Aleutian disease of mink, adeno-associated viruses and human parvoviruses.

PAS periodic acid-Schiff reagent *q.v.*

pascual *a. pert.* pastures, *appl.* flora.

pasculomorphosis *n.* changes in the structural features of plants as a result of grazing.

passage cells thin-walled exodermal or endodermal cells of root which permit passage of solutes, and are usually associated with cells with thick walls, as companion cells of phloem.

Passeriformes, passerines *n.*, *n.plu.* large order of birds, which includes small and medium-sized perching birds and songbirds such as crows, tits, warblers, thrushes and finches.

Passiflorales *n.* order of dicot shrubs, herbs, climbers and small trees and including the families Caricaceae (pawpaw) and Passifloraceae (passion flower).

passive immunity short-lived immunity acquired by transfer of preformed antibodies, as from mother to foetus across placenta and in mother's milk to infant.

passive transport unaided diffusion of small uncharged molecules or protein-mediated transport (facilitated diffusion) of ions and other charged molecules across a biological membrane in the direction of concentration gradient and electrochemical gradient, and which does not require energy.

Pasteur effect (1) the observation of Louis Pasteur (originally in yeast) that glycolysis is inhibited by aerobic conditions (i.e. by respiration); (2) sometimes incorrectly said to be the ability of a normally anaerobic organism to oxidize sugar completely to carbon dioxide and water in the presence of oxygen, which is a later interpretation of the effect.

pasteurization *n.* method of partial sterilization, used for milk, wine and other beverages, by heating at 62°C for 30 minutes or at 72°C for 15 seconds, followed by rapid cooling.

pasture *n.* grassland used for grazing.

patabiont *n.* animal that spends all its life in the litter on the forest floor.

patacole *n.* animal that lives temporarily in the litter on the forest floor.

patagial *a.* of or *pert.* a patagium, or flying membrane.

patagiate *a.* having a patagium.

patagium *n.* (1) the membranous expansion between fore- and hindlimbs of bats; (2) the extension of skin between fore- and hind-limbs of flying lemurs and flying squirrels.

Patagonian region part of the Antarctic floristic kingdom, consisting of the southern part of South America and the Falkland islands (Malvinas).

patch clamp technique in neurophysiology in which a tiny portion of cell membrane is isolated from the rest of the cell membrane by sealing to the tip of a microelectrode, allowing recording from single ion channels.

patching *n.* clustering of membrane proteins seen on eukaryotic cells when treated with lectins or specific antibody, caused by lateral movement of membrane protein in the fluid lipid bilayer as a result of crosslinking by lectins or antibodies.

patella *n.* (1) the knee cap or elbow cap; (2) segment between femur and tibia in certain arachnids; (3) 4th segment of spider's leg; (4) a genus of limpet; (5) a rounded apothecium of lichens.

patellar *a. pert.* a patella.

patellaroid, patelliform *a.* (1) pan-shaped; (2) like a bordered disc.

patent *a.* open or spreading widely.

pateriform *a.* saucer-shaped.

paternal *a. pert.* or originating in the father.

pathetic *a. appl.* nerve: 6th cranial nerve, controls a muscle of eye.

pathogen *n.* any disease-causing organism.

pathogen-associated molecular patterns (PAMPs) the molecules typically associated with groups of pathogens, e.g. the lipopolysaccharide of the outer membrane of Gram-negative bacteria, and which are recognized by receptors on cells of the innate immune system, thus alerting the body to the presence of infection.

pathogenesis *n.* the origin or cause of the pathological symptoms of a disease.

pathogenic *n.* causing disease, *appl.* a parasite (esp. a microorganism) in relation to a particular host.

pathogenicity *n.* the ability of a parasite to damage the host and to cause disease.

pathogenicity island in genomes of pathogenic bacteria, a cluster of genes that determine the ability of the bacterium to cause disease.

pathology *n.* (1) science dealing with disease or dysfunction; (2) the characteristic symptoms and signs of a disease.

patina *n.* circle of plates around calyx of crinoids.

patoxene *n.* animal that occurs accidentally in the litter on the forest floor.

patriclinal, patriclinous *a.* with hereditary characteristics more paternal than maternal. *alt.* **patroclinal, patroclinous.**

patrilineal *a.* (1) *pert.* paternal line; (2) passed from male parent.

patristic *a.* in plants, *pert.* similarity due to common ancestry.

pattern formation the generation of spatial patterns of differentiated cells to form tissues, and eventually organs and morphological structures, within a multicellular embryo.

pattern recognition receptors (PRR) soluble molecules or receptors on phagocytes and other cells of the innate immune system that recognize molecular patterns characteristic of broad classes of pathogenic microorganism (pathogen-associated molecular patterns). Examples are the Toll-like receptors and mannose-binding lectin.

patulent, patulose, patulous *a.* spreading, open or expanding.

paturon *n.* basal segment of spider's chelicera, used for crushing prey.

paucilocular *a.* containing, or composed of, few small cavities or loculi.

paucispiral *a.* with few coils or whorls.

paunch rumen *q.v.*

Pauropoda *n.* class of arthropods allied to the myriapods, which have 12 body segments, 9 of which carry appendages.

pavement epithelium epithelium formed of cuboid cells, or a simple squamous epithelium.

Pavlovian conditioning classical conditioning *q.v.*

paxilla *n.* a thick plate, supporting calcareous pillars, the summit of each covered by a group of small spines, in certain starfish. *plu.* **paxillae.** *alt.* **paxillus.**

paxillate *a.* having paxillae.

paxilliform *a.* shaped like a paxilla.

PBLs peripheral blood leukocytes (*q.v.*), sometimes used to mean only peripheral blood lymphocytes.

PBS phosphate-buffered saline.

PCB polychlorinated biphenyl *q.v.*

PCNA DNA-replication protein that forms a 'sliding clamp' that encircles the DNA and keeps DNA polymerase on the DNA template.

PCP planar cell polarity *q.v.*

PCR polymerase chain reaction *q.v.*

PCR cycle photosynthetic carbon reduction cycle. *see* Calvin cycle.

PD$_{50}$ a measure of activity for certain viruses, the dose at which 50% of test animals show paralysis.

PDB Protein Data Bank *q.v.*

PDE phosphodiesterase *q.v.*

PDGF platelet-derived growth factor *q.v.*

PDI protein disulphide isomerase *q.v.*

p-distance the genetic distance between two sequences calculated as the proportion of nucleotide or amino-acid differences between them. Defined as D/N, where N is the total number of positions in the pairwise alignment ignoring gaps and D is the total number of differences.

PDK phosphatidylinositol-dependent protein kinase.

pearl *n.* the abnormal growth formed around a minute grain of sand or other foreign matter which gets inside the shell of certain bivalve molluscs, and which consists of many layers of nacre.

peat *n.* dark-brown organic deposit consisting of partly decomposed plant material, associated with wetland areas. Peat accumulates when the rate of production of vegetation exceeds the rate at which it is decomposed by microorganisms. Their action is inhibited primarily by a lack of oxygen, but other factors such as a lack of nutrients, low temperatures or low pH can contribute. *see* blanket peat, fen peat.

peat mosses *see* sphagnum mosses.

pecking order a social hierarchy, esp. in birds, ranging from the most dominant and aggressive animal down to the most submissive. *alt.* **peck order.**

pecorans *n.plu.* giraffes, deer and cattle.

pectate *see* pectic acid.

pecten *n.* (1) any comb-like structure; (2) a process of inner retinal surface in reptiles, expanded into a folded quadrangular plate in birds; (3) ridge of the superior branch of the pubic bone; (4) part of anal canal between internal sphincter and anal valves; (5) part of the stridulating organ of certain spiders; (6) sensory abdominal appendage of scorpions; (7) genus of scallops; (8)

comb-like assemblage of sterigmata. *plu.* **pectines.**

pectic *a. pert.* pectin.

pectic acid a polygalacturonan which is part of pectin.

pectic enzymes enzymes that hydrolyse the components of pectin, and including polygalacturonase, pectinesterase, pectin lyase and pectate lyase.

pectin *n.* a group of highly variable complex polysaccharides found in middle lamella of plant primary cell walls, rich in galacturonic acid, and forming a gel when isolated, and also containing arabinose, galactose and rhamnose residues.

pectinal *a. pert.* a pecten.

pectinase *n.* (1) mixture of glycosidases that hydrolyse the various components of pectin; (2) sometimes refers to polygalacturonase (EC 3.2.1.15). *see* polygalacturonan.

pectinate *a.* shaped like a comb.

pectineal *a.* (1) comb-like; (2) *appl.* a ridge on femur and its attached muscle, the pectineus muscle.

pectinellae *n.plu.* transverse comb-like membranelles constituting adoral ciliary spiral of some ciliate protozoans.

pectines *plu.* of pecten.

pectinesterase *n.* enzyme catalysing the removal of methyl groups from the methylated galacturonans occurring in pectin, producing methanol and pectic acid. EC 3.1.1.11.

pectineus *n.* a flat muscle between pecten of pubis and the upper medial part of femur.

pectinibranch *a.* having one margin only furnished with teeth like a comb, *appl.* gills in certain molluscs.

pectiniform pectinate *q.v.*

pectinogen protopectin *q.v.*

pectinolytic *a.* able to break down pectin, *appl.* bacteria.

pectocellulose *n.* pectin mixed with cellulose as in fleshy roots and fruits.

pectolytic *a.* breaking down pectin, *appl.* enzymes.

pectoral *a. pert.* chest, in region of chest.

pectoral fin the fin on the side of body of fish.

pectoral girdle in vertebrates, a skeletal support in the shoulder region for attachment of fore-fins or forelimbs, made up of a hoop of cartilages or bones, usually the scapula, clavicle and coracoid.

pectoralis *n.* breast muscle connecting breastbone with humerus, much enlarged in birds where it is a main flight muscle.

pectoralis major, pectoralis minor outer and inner chest muscles connecting ventral chest wall with shoulder and humerus.

pectose protopectin *q.v.*

pectosinase protopectinase *q.v.*

ped *n.* relatively robust aggregate of soil particles, capable of withstanding several cycles of wetting and drying. Individual peds are clearly visible, being separated from one another by lines of weakness.

pedal *a.* (1) *pert.* foot or feet; (2) *appl.* disc: base of a sea anemone.

pedalfer *n.* any of a group of soils, in humid regions, usually characterized by the presence of aluminium and iron compounds and the absence of carbonates.

pedate *a.* with toe-like parts.

pedatipartite *a. appl.* a type of palmate leaf having three main divisions and the two outer divisions subdivided one or more times.

pedatisect *a.* in a pedate arrangement and with divisions nearly to midrib, *appl.* leaves.

pedicel *n.* small, short stalk of leaf, fruit or sporangium (*see* Fig. 18, p. 239), or foot-stalk of a fixed organism or of an organ.

pedicellariae *n.plu.* minute pincer-like structures studding the surface of some echinoderms (e.g. starfish) which grab, kill and discard small animals that touch them. *sing.* **pedicellaria.**

pedicellate *a.* (1) supported by a small stalk; (2) *appl.* Hymenoptera whose thorax and abdomen are joined by a short stalk.

pedicellus pedicel *q.v.*

pedicle *n.* short stem or stalk.

pediculates *n.plu.* order of very specialized marine bony fishes, including anglerfish and batfishes, which have anterior portion of dorsal fin modified as a lure.

pediferous *a.* (1) having feet; (2) having a foot-stalk.

pedigree *n.* in genetics, a diagram showing the ancestral history of a group of related individuals.

pedipalp *n.* in spiders and other arachnids, a small paired leg-like appendage, immediately anterior to the walking legs. In male spiders it is used in mating to transfer sperm into female, sometimes modified as a pincer or claw.

pedocal *n.* any of a group of soils, of arid and semiarid regions, characterized by presence of calcium carbonate (lime).

pedogamy *a.* type of autogamy in protozoans where gametes are formed after multiple division of nucleus.

pedogenesis *n.* (1) reproduction in young or larval stages, as in the axolotl; (2) spore production in immature fungi; (3) formation of soil.

pedogenic *a. pert.* the formation of soil.

pedology *n.* soil science.

pedomorphosis *n.* retention of juvenile traits in adults. *a.* **pedomorphic.**

pedonic *a. appl.* organisms of freshwater lake bottoms.

peduncle *n.* (1) (*bot.*) the stalk of a flowerhead; (2) (*zool.*) the stalk of sedentary protozoans, crinoids, brachiopods and barnacles; (3) link between thorax and abdomen in arthropods; (4) (*neurobiol.*) band of nerve fibres joining different parts of brain.

pedunculate *a.* growing on or having a peduncle.

PEG polyethylene glycol *q.v.*

Pekin(g) man a fossil hominid found near Beijing in the 1920s and at first called *Sinanthropus pekinensis*, then *Pithecanthropus pekinensis*, and now classified as *Homo erectus*.

pelage *n.* the hairy, furry or woolly coat of mammals.

pelagic *a.* living in the sea or ocean at middle or surface levels.

pelargonidin *n.* red flavonoid pigment present in many flowers.

pelasgic *a.* moving from place to place.

Pelecaniformes *n.* order of large aquatic, fish-eating birds with all four toes webbed, including the pelicans, cormorants, boobies and gannets.

Pelecypoda Bivalvia *q.v.*

P elements family of DNA-only transposons found in *D. melanogaster*, involved in hybrid dysgenesis and also used experimentally to introduce genes into the germline to produce transgenic insects.

pellagra *n.* deficiency disease caused by insufficient tryptophan and nicotinate (niacin) in the diet, can be treated with niacin or nicotinamide.

pellagra-preventive factor nicotinamide or niacin *q.v.*

pellicle *n.* (1) thin flexible outer layer, as formed e.g. by the plasma membrane and its underlying protein strips in *Euglena*; (2) any delicate surface or skin-like growth; (3) the skin-like aggregation of bacteria or yeasts on the surface of liquid media.

pelliculate *a.* having a pellicle on outer surface.

pellions *n.plu.* ring of plates supporting suckers of echinoids.

pelma *n.* the sole of the foot.

pelophilous *a.* growing on clay.

peloria, pelory *n.* (1) condition of abnormal regularity; (2) modification of structure from irregularity to regularity. *a.* **peloric.**

pelotons *n.plu.* coils of hyphae of the mycorrhizal fungus in meristematic tissue of orchid embryos.

pelta *n.* shield-shaped apothecium of certain lichens.

peltate *a.* (1) shield-shaped; (2) having the stalk inserted at or near the middle of the under surface, not near the edge.

peltinervate *a.* having veins radiating from near the centre, as of a peltate leaf.

pelvic *a. pert.* or situated at or near pelvis.

pelvic fins paired fins on underside of body in fish, representing the hindlimbs of land vertebrates.

pelvic girdle in vertebrates, a skeletal support in the hip region for attachment of hind-fins or hindlimbs, made of a hoop of cartilages or bones, in tetrapods usually ilium, ischiopubis or ischium, and pubis.

pelvis *n.* (1) the cavity in vertebrates formed by pelvic girdle along with coccyx and sacrum; (2) of kidney, the expansion of ureter at its junction with kidney. *a.* **pelvic.**

pelvisternum *n.* the epipubis when separate from pubis.

pelycosaurs *n.plu.* extinct order of primitive mammal-like reptiles of the Carboniferous to Permian, having a primitive sprawling gait and including the sail-back lizards.

pen *n.* (1) the horny beak of certain cephalopods; (2) a primary wing feather; (3) a female swan. *see also* sea pens.

pendent *a.* hanging down.

pendulous *a.* (1) bending downwards from point of origin; (2) overhanging.

penetrance *n.* the percentage of individuals possessing a particular genotype who show the associated phenotype, i.e. in whom the trait is expressed. A trait shows complete penetrance if it is expressed in all persons who carry the corresponding genotype. It shows incomplete penetrance if it is not expressed at all in some individuals who carry the genotype.

penetration path path of sperm in ooplasm to the female pronucleus.

penicillate *a.* resembling a small paintbrush, *appl.* heads of conidia of fungi of the genus *Penicillium*.

penicillin *n.* any of various antibiotics, based on a β-lactam ring structure, produced by the mould *Penicillium notatum* and related species, and which inhibit bacterial cell-wall synthesis, leading to osmotic lysis.

penicillinase β-lactamase *q.v.*

penicillin-binding proteins the transpeptidases that cross-link peptide chains of the peptidoglycans of bacterial cell walls, and which are inhibited by penicillin.

penis *n.* the male copulatory organ.

penna *n.* a contour feather in birds, as distinguished from a plume or down feather.

pennate *a.* (1) divided in a feathery manner; (2) feathered; (3) having a wing; (4) in the shape of a wing; (5) *appl.* diatoms that are bilaterally symmetrical in valve view.

Pennsylvanian *a.* (1) *appl.* and *pert.* an epoch of the Carboniferous era, lasting from 318 to 299 million years ago; (2) *appl.* coal measures in North America.

pensile *a.* hanging.

penta- prefix derived from Gk *pente*, five, signifying having five of, arranged in fives, etc.

pentactinal *a.* five-rayed or five-branched.

pentadactyl *a.* having all four limbs normally terminating in five digits.

pentadactyl limb the limb with five digits characteristic of tetrapods.

pentadelphous *a.* applied to flowers having anthers consisting of five clusters of more-or-less united filaments.

pentafid *a.* in five divisions or lobes.

pentagynous *a.* (1) having five pistils or styles; (2) having five carpels to the gynoecium.

pentamer *n.* (1) an assembly of five subunits; (2) five contiguous nucleotides or amino acids in a nucleic acid or protein sequence.

pentamerous *a.* (1) composed of five parts; (2) *appl.* flowers with parts arranged in whorls of five or a multiple of five.

pentandrous *a.* having five stamens.

pentaploid *a.* having five times the normal haploid number of chromosomes.

pentapterous *a.* with five wings, as some fruits.

pentaradiate *a.* with a body built on a five-rayed plan.

pentarch *a.* (1) with five alternating groups of xylem and phloem; (2) with five vascular bundles.

pentasepalous *a.* with five sepals.

pentaspermous *a.* with five seeds.

pentastemonous *a.* with five stamens.

pentasternum *n.* sternite of 5th segment of prosoma or 3rd segment of podosoma in ticks and mites.

pentastichous *a.* arranged in five vertical rows, *appl.* leaves.

Pentastomida, pentastomids *n.*, *n.plu.* phylum of flat, soft-bodied, worm-like parasites, commonly known as tongue worms, that live in the nasal passages of vertebrates, mainly tropical and subtropical. They have three larval stages, the first taking place in the egg in the nasal passages of herbivorous mammals. *alt.* **Pentastoma, pentastomes.**

pentosan *n.* polysaccharide made of linked pentose units, such as xylans and arabinans.

pentose *n.* monosaccharide having the formula $(CH_2O)_5$, e.g. arabinose, xylose and ribose.

pentose phosphate pathway (1) the oxidative pentose phosphate pathway (PPP) consists of a group of reactions that generates biochemical reducing power in the form of NADPH from the partial oxidation of glucose-6-phosphate, producing carbon dioxide and ribose-5-phosphate. This undergoes conversion to triose phosphates, sometimes in a cycle, regenerating glucose-6-phosphate. It is the only source of NADPH in red blood cells, and is active in adipose cells rather than muscle. *alt.* hexose monophosphate shunt, phosphogluconate oxidative pathway; (2) the reductive pentose phosphate pathway (RPPP) is also present in plants and some

bacteria and is involved in carbon dioxide fixation into hexoses in photosynthesis. *see* Calvin cycle.

pentraxins *n.plu.* family of blood proteins that act in the acute-phase response.

peonidin *n.* red anthocyanin pigment found in plants.

PEP phosphoenolpyruvate *q.v.*

pepo *n.* an inferior many-seeded pulpy fruit, e.g. melons, cucumbers, squashes.

pepsin *n.* endopeptidase in gastric juice which hydrolyses proteins to peptides. It is formed from pepsinogen by cleavage of one peptide bond by hydrochloric acid in the stomach. EC 3.4.23.1, *r.n.* pepsin A.

pepsinogen *n.* inactive precursor of the enzyme pepsin. It is secreted into the stomach by cells of the gastric mucosa and activated by hydrochloric acid secreted by oxyntic cells.

pepstatin *n.* hexapeptide which is an inhibitor of the enzyme pepsin and other acid endopeptidases.

peptic *a.* relating to or promoting digestion.

peptic cells cells of the gastric gland that secrete prorennin and pepsinogen, precursors to the digestive enzymes rennin and pepsin.

peptidase *n.* enzyme that cleaves peptide bonds by hydrolysis. EC 3.4. Endopeptidases (*alt.* proteases, proteinases) cleave internal peptide bonds in proteins, exopeptidases trim residues individually or as di- or tripeptides from the ends of the protein chain, *see* aminopeptidase, carboxypeptidase, dipeptidase, omega peptidase.

peptide *n.* a chain of a relatively small number (up to 100) of amino acids linked by peptide bonds.

peptide bond covalent bond joining the α-amino group of one amino acid to the carboxyl group of another with the loss of a water molecule. It is the bond linking amino acids together in a protein chain. *see* Fig. 30 (p. 490).

peptide map *see* protein fingerprinting.

peptide nucleic acid (PNA) molecule consisting of a peptide backbone with purine and pyrimidine bases attached as side chains, which has been proposed as an information-carrying molecule that might have pre-dated RNA.

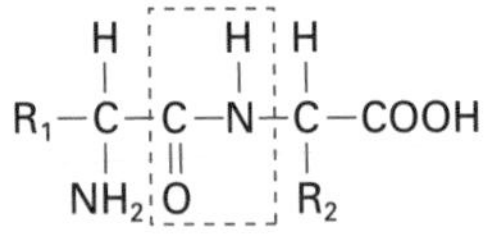

Fig. 30 Peptide bond.

peptide transporters membrane transport proteins (the TAP proteins) that carry peptides produced by breakdown of proteins in the cytosol into the endoplasmic reticulum.

peptidoglycans *n.plu.* class of macromolecules in which linear polysaccharide chains are extensively crosslinked by short peptides. They are characteristic components of the cell walls of bacteria.

peptidylprolyl isomerase (PPI) enzyme involved in the folding of newly synthesized proteins. EC 5.2.1.8. *alt.* cyclophilin, peptidylprolyl *cis-trans* isomerase.

peptidyl transferase enzymatic activity of the 23S rRNA of the large ribosome subunit which catalyses the formation of peptide bonds during protein synthesis on the ribosomes. It is an integral part of the large ribosomal subunit.

peptidyl-tRNA a tRNA attached to the whole polypeptide chain synthesized so far, formed on ribosome at each cycle of amino acid addition during protein synthesis.

peptone *n.* polypeptide product of hydrolysis of proteins by enzymes such as pepsin.

peptonephridia *n.plu.* the anterior nephridia, which function as digestive glands, in some oligochaete worms.

percent accepted mutation (PAM) in relation to the evolutionary comparison of amino-acid sequences, the number of amino acid changes per 100 amino acids. A unit of evolution used in the construction of some types of substitution matrix used in sequence alignment. *alt.* point accepted mutation.

percentage successive mortality apparent mortality *q.v.*

percent identity in bioinformatics, the percentage of identical residues at matching positions in two or more aligned sequences, used as a quantitative measure of their similarity.

percept *n.* a conscious mental image of a perceived object.

perception *n.* the mental interpretation of physical sensations produced by stimuli from the outside world.

percurrent *a.* extending throughout length, or from base to apex.

perdominant *n.* a species present in almost all the associations of a given type.

pereion *n.* the thorax of crustaceans.

pereiopod *n.* walking limb of crustaceans such as crayfish, crabs and lobsters, present as four pairs on cephalothorax.

perennating *a.* overwintering, *appl.* roots, buds.

perennation *n.* of a plant, survival from year to year.

perennial *n.* plant which persists for several years.

perennibranchiate *a.* having gills persisting throughout life, as certain amphibians.

perfect *a. appl.* flowers containing both stamens and carpels.

perfect stage *a.* in ascomycete fungi, *appl.* the sexual reproductive phase in their life history in which they produce asci.

perfoliate *a. appl.* a leaf with basal lobes united so that it entirely surrounds the stem.

perforate *a.* (1) having pores, as corals, foraminiferans; (2) *appl.* leaves having small translucent spots when held up against light, as Perforate St John's Wort.

perforation plate perforate septum or area of contact between cells or elements of xylem vessels.

perforin *n.* cytotoxic protein produced by cytotoxic T cells.

peri- prefix derived from Gk *peri*, around, and signifying surrounding, or situated around.

perianth *n.* (1) the outer whorl of floral leaves of a flower when not clearly divided into calyx or corolla; (2) collectively, the calyx and corolla; (2) the cover or sheath surrounding the archegonia in some mosses; (3) tubular sheath surrounding developing sporophyte in leafy liverworts.

periaqueductal grey area grey matter surrounding the aqueduct in the brain stem.

periaxial *a.* (1) surrounding an axis or an axon; (2) *appl.* space between the axon membrane and the myelin sheath in a medullated nerve fibre.

periblast *n*. (1) the outside layer, epiblast, or blastoderm of an insect embryo; (2) syncytium formed by fusion of small marginal blastomeres and not forming part of mammalian embryo.

periblastesis *n*. envelopment by surrounding tissue, as of lichen gonidia.

periblastic *a*. (1) *pert*. periblast; (2) superficial, as *appl*. segmentation of fertilized ovum.

periblem *n*. the meristem that produces the cortex.

peribranchial *a*. (1) around the gills; (2) *appl*. type of budding in ascidians; (3) *appl*. atrial cavity in ascidians and lancelet; (4) *appl*. circular spaces surrounding basal parts of papulae of starfish.

peribulbar *a*. (1) around the eyeball; (2) surrounding a taste bud, *appl*. nerve fibres.

pericambium *n*. layer of dividing cells around the stele of plant axis, also called the pericycle.

pericapillary *a*. *appl*. cells in contact with outer surface of walls of capillaries.

pericardiac, pericardial *a*. (1) *pert*. the pericardium; (2) surrounding the heart, *appl*. cavity, septum; (3) *appl*. paired excretory organs in lamellibranchs; (4) *appl*. cells: cords of nephrocytes in certain insects.

pericardium *n*. membrane surrounding the heart, delimiting a pericardial cavity which is sometimes itself called the pericardium.

pericarp *n*. the tissues of a fruit that develop from the ovary wall, comprising an outer skin, sometimes a fleshy mesocarp, and an inner endocarp.

pericaryon *alt*. spelling of perikaryon.

pericellular *a*. surrounding a cell.

pericentral *a*. around or near centre.

pericentric *a*. (1) *pert*. or involving the centromere of a chromosome; (2) *appl*. breaks in arms of a chromosome on either side of the centromere; (3) *appl*. inversions that include the centromere.

pericentriolar material fibrous matrix surrounding the centrioles in a centrosome in animal cells. It contains numerous rings of γ-tubulin, which form nucleating centres for microtubules of the spindle. *alt*. centrosome matrix.

perichaetial *a*. *pert*. a perichaetium.

perichaetine *a*. having a ring of chaetae encircling the body.

perichaetium *n*. one of the membranes or leaves enveloping archegonia or antheridia of bryophytes. *a*. **perichaetial**.

perichondral *a*. *appl*. ossification in cartilage beginning on the outside and spreading inwards.

perichondrium *n*. fibrous envelope of connective tissue surrounding cartilage. *a*. **perichondrial**.

perichondrostosis *n*. ossification in cartilage beginning outside and spreading inwards.

perichordal *a*. enveloping, or near, the notochord.

perichoroidal *a*. surrounding the choroid, *appl*. lymph space.

periclinal *a*. (1) *appl*. division of cells parallel to the surface of the structure in which they occur; (2) *appl*. system of cells parallel to surface of apex of growing point; (3) *appl*. two tissues in which one completely surrounds the other.

periclinal chimaera chimaeric plant in which the cells of different genotype comprise a whole layer of the meristem. *cf*. mericlinal chimaera.

periclinium *a*. the involucre of a composite flower.

pericranium *n*. fibrous membrane investing skull.

pericycle *n*. the external layer of stele (primary vascular tissue in stem), the layer between endodermis and conducting tissue.

pericyclic *a*. *appl*. fibre situated on the outer edge of the vascular region in plant axis and usually arising in the primary phloem.

pericyte *n*. connective tissue cell related to a smooth muscle cell, found in the walls of small blood vessels and essential for maintaining their integrity.

periderm *n*. a three-layered tissue that replaces the epidermis in most stems and roots having secondary growth. It is produced by the cork cambium, which on its outer side produces a layer of cork tissue containing suberin, which is non-living at maturity, and on its inner side produces the phelloderm, a living parenchyma tissue. Commonly known as the bark.

peridesm *n*. tissue surrounding a vascular bundle.

peridial *a*. *pert*. a peridium *q.v.*

perididymis *n*. the dense white connective tissue surrounding testis.

peridinin *n*. xanthophyll pigment found in algae of the Pyrrophyta.

peridiole *n*. one of the small 'eggs' of bird's nest fungi, having a hard waxy wall and containing basidiospores. It acts as a propagative unit, being splashed out of the peridium (the 'nest') by rain.

peridium *n*. the outer wall of a fruiting body, esp. in fungi.

peridural *a. appl*. perimeningeal space at a later stage in development.

perienteric *a*. surrounding the enteron, or gut.

periesophageal perioesophageal *q.v.*

perifibrillar *a*. surrounding a fibril, *appl*. the cytoplasm surrounding the neurofibrils in axon.

perifoliary *a*. surrounding a leaf margin.

periganglionic *a*. surrounding a ganglion.

perigastric *a*. surrounding the viscera, *appl*. abdominal cavity.

perigastrium *n*. the body cavity.

perigemmal *a*. surrounding a taste bud, *appl*. nerve fibres, spaces.

perigenous *a*. borne or growing on both sides of a leaf or other structure, or borne on all sides of a structure.

perigonium *n*. (1) floral envelope or perianth; (2) involucre around antheridium in mosses.

perigynium *n*. (1) membranous envelope or pouch of archegonium in liverworts; (2) involucre in mosses.

perigynous *a. appl*. flowers in which petals and stamens are attached to the extended margin of the receptacle. *n*. **perigyny**.

perihaemal *a. appl*. the canals of the blood vascular system in echinoderms, which enclose the haemal strands.

perikaryon *n*. the cell body containing the nucleus in nerve cells, and in other structures such as the syncytial tegument of flatworms in which many cell bodies in the deep layer of the skin are connected to an outer continuous layer of cytoplasm.

perikymata *n.plu*. ridges in enamel of teeth caused by incremental growth.

perilymph *n*. fluid surrounding membranous labyrinth of the ear separating it from the bony labyrinth, contained in the perilymphatic space.

perimeningeal *a. appl*. space between the layer of connective tissue lining canal of vertebral column and skull, and the meninges.

perimysium *n*. connective tissue binding muscle fibres into bundles and into muscles, and continuing into tendons.

perinaeum, perinaeal perineum, perineal *q.v.*

perineal *a. pert*. perineum, *appl*. e.g. artery, nerve, gland.

perinephrium *n*. the enveloping adipose and connective tissue around kidney.

perineum *n*. part of body surface delimited by scrotum or vulva in front, anus behind and side of thigh to the side.

perineural *a*. surrounding a nerve or nerve cord.

perineurium *n*. the fibrous connective tissue sheath around a bundle of nerve fibres within a nerve.

perineuronal *a*. surrounding a nerve cell or cells.

perinium *n*. outer coat of microspores of certain pteridophytes.

perinuclear space the space between the outer and inner nuclear membranes.

periocular *a*. surrounding the eyeball within the orbital cavity.

period *n*. in geological time, a subdivision of an era, e.g. the Jurassic is a period of the Mesozoic era.

periodic acid-Schiff reagent (PAS) a dye used for staining proteins rich in carbohydrate side chains.

periodicity *n*. (1) the fulfilment of functions at regular intervals or periods; (2) in a cyclic or rhythmic reaction or process, the interval between two peaks of activity; (3) of DNA, the number of base pairs per turn of the double helix.

periodontal *a*. covering or surrounding a tooth, *appl*. e.g. membrane, tissue of gums.

perioesophageal *a*. surrounding oesophagus, *appl*. a nerve ring.

periople *n*. thin outer layer of hoof of equines.

periopticon *n*. in insects, the region of the optic lobes nearest to the eye.

periorbital *a*. surrounding the eye socket.

periosteum *n*. fibrous membrane investing the surface of bones.

periostracum *n*. the external layer of most mollusc and brachiopod shells.

periotic *n*. bone enclosing parts of membranous labyrinth of inner ear.

peripatric *a*. speciation resulting from isolation of a small population on the edge of a larger ancestral population.

peripetalous *a*. surrounding petals or a petaloid structure.

peripharyngeal *a*. surrounding or encircling the pharynx, *appl*. cilia of ascidians and cephalochordates.

peripheral *a*. (1) distant from the centre; (2) near circumference; (3) *appl*. membrane proteins that are weakly attached to membrane and easily removed; (4) (*immunol.*) *pert*. or *appl*. tissues other than those of the primary lymphoid organs, i.e. all secondary lymphoid tissues.

peripheral blood leukocytes (PBLs) white blood cells circulating in the blood, i.e. excluding those resident in tissues. The abbreviation PBLs is also used for circulating lymphocytes only.

peripheral lymphoid organ, peripheral lymphoid tissues secondary lymphoid tissues *q.v.*

peripheral membrane protein protein attached to one face of the membrane by noncovalent interactions and that can be removed by treatments that leave the lipid bilayer intact.

peripheral nervous system the nervous system of vertebrates other than the brain and spinal cord. It consists of sensory receptors of trunk, limbs and internal organs, and nerves other than the cranial nerves.

peripheral tolerance tolerance to self-antigens acquired by the mature lymphocyte population as a result of self-reactive mature lymphocytes being rendered unresponsive on contact with their corresponding antigen.

peripherical *a*. *appl*. a plant embryo more-or-less completely surrounding the endosperm in seed.

periphloem *n*. phloem sheath or pericambium *q.v.*

periphloic *a*. having phloem outside a centric xylem in vascular bundle.

periphoranthium periclinium *q.v.*

periphorium *n*. fleshy structure supporting ovary and to which stamens and corolla are attached.

periphyllum *n*. scale at base of ovary in grasses, supposed to represent part of perianth.

periphysis *n*. in some fungi, short sterile hair fringing the inside of a pore or aperture in fruiting body.

periphyton *n*. plants and animals adhering to parts of rooted aquatic plants.

peripileic *a*. arising from around the margin of cap in agaric fungi.

periplasm *n*. (1) cytoplasm surrounding yolk in ova with a central yolk; (2) layer of protoplasm surrounding the oosphere in some oomycetes; (3) gel-filled space between the plasma membrane and the cell wall in bacteria, *alt*. periplasmic space. *a*. **periplasmic**.

periplasmodium *n*. mass of multinucleate protoplasm derived from tapetal cells and enclosing developing spore.

periplast *n*. (1) outer covering of cells of some algae of the Cryptophyta; (2) intercellular substance or stroma of tissues.

peripneustic *a*. having spiracles arranged all along sides of body, the usual condition in most insect larvae.

peripodial *a*. (1) surrounding an appendage; (2) *appl*. membrane covering wing bud in insects.

periportal *a*. (1) *pert*. transverse fissure of the liver; (2) *appl*. connective tissue partially separating lobules and forming part of the hepatobiliary capsule.

peripyle *n*. one of the apertures, additional to the astropyle, of the central capsule of some radiolarians.

perisarc *n*. chitinous outer casing of common tissue connecting individuals in some colonial hydrozoans.

periscleral *a*. *appl*. lymph space external to sclera of the eye.

perisome *n*. (1) a body wall; (2) the surface layer of echinoderms.

perisperm *n*. in some seeds, a storage tissue formed by proliferation of the nucellus rather than the endosperm.

perispiracular *a*. (1) surrounding a spiracle; (2) *appl*. glands with oily secretion in certain aquatic insect larvae.

perisporangium *n*. (1) membrane covering a sorus; (2) indusium of ferns.

perispore *n*. (1) a spore covering; (2) transient outer membrane enveloping a spore; (3) mother cell in algal spores.

perissodactyl *a.* with an uneven number of digits.

Perissodactyla, perissodactyls *n.*, *n.plu.* order comprising the odd-toed ungulate mammals, in which the weight is borne mainly on the 3rd toe, and which includes the horse, tapir and rhinoceros. *cf.* Artiodactyla.

peristalsis *n.* successive contractions of a muscular organ such as the gut, which moves gut contents along. Some organisms such as earthworms move by peristalsis of their body wall. *a.* **peristaltic**.

peristigmatic perispiracular *q.v.*

peristome, peristomium *n.* (1) (*bot.*) ring of teeth around mouth of spore capsule of mosses; (2) (*zool.*) region surrounding the mouth, in ciliate protozoans, starfish, annelid worms, insects, echinoderms, etc.

perisystole *n.* interval elapsing between diastole and systole of heart.

perithecium *n.* flask-shaped structure opening in a terminal hole (ostiole) and which contains the asci in some groups of ascomycete fungi, the flask fungi or Pyrenomycetes.

perithelium *n.* connective tissue associated with capillaries.

peritoneal *a.* *pert.* peritoneum, *appl.* e.g. cavity, membrane, fossa.

peritoneum *n.* membrane partly applied to abdominal walls, partly extending over the organs contained in abdominal cavity, delimiting the peritoneal cavity.

Peritrichia, peritrichans *n.*, *n.plu.* group of ciliate protozoans which are usually fixed permanently to the substratum and have few cilia, e.g. *Vorticella*.

peritrichous *a.* *appl.* flagella, distributed all over the cell surface.

peritrochium *n.* (1) a band of cilia; (2) a circularly ciliated larva.

peritrophic *a.* (1) (*zool.*) *appl.* a chinitous acellular sheet that folds around food in mid-gut of insects; (2) (*bot.*) *appl.* mycorrhiza with special fungal populations on root surfaces.

periurethral *a.* surrounding the urethra, *appl.* glands, homologues of prostate glands.

perivascular *a.* (1) surrounding a blood vessel, *appl.* lymph channels, fibres,

spaces; (2) (*bot.*) surrounding the vascular cylinder, *appl.* fibres.

perivascular feet terminal enlargements of processes of astrocytes in contact with minute blood vessels.

perivisceral *a.* surrounding the viscera, *appl.* body cavity.

perivitelline *a.* *appl.* space between ovum and zona pellucida.

perixylic *a.* having xylem outside centric phloem, *appl.* vascular bundles.

perizonium *n.* the membrane or siliceous wall enveloping the autospore or zygote in diatoms.

perlecan *n.* a heparan sulphate-containing proteoglycan of the basal lamina.

permafrost *n.* a layer of permanently frozen soil, usually sub-surface, which occurs where temperatures are low enough, and which is a feature of the tundra in polar regions. It is generally covered in warmer months with a thin partially melted layer, which in the tundra supports a vegetation of mosses, lichens, grasses and small herbaceous plants. The permafrost layer may be up to hundreds of metres thick.

permanent cartilage cartilage that remains unossified throughout life, as opposed to temporary cartilage, which is ossified into bone.

permanent hybrid heterozygote which breeds true because of the elimination of certain homozygous genotypes by lethal factors in the genotype.

permanent memory type of memory that appears to last the lifetime of an individual.

permanent teeth, permanent dentition set of teeth developed after milk or deciduous teeth, the second set of teeth, third set of some mammals. Some mammals do not develop a second set of teeth.

permanent tissue tissue consisting of cells which have completed their period of growth, are fully differentiated, and subsequently change little until they die.

permeable *a.* allowing a given substance to diffuse freely across it, *appl.* membranes. *n.* **permeability**.

permeants *n.plu.* animals which move freely from one community or habitat to another.

permease *n.* any of various membrane proteins in bacteria responsible for carrying solutes into the cell.

Permian *n.* geological period at the end of the Paleozoic, lasting from *ca.* 299 million years ago to 250 million years ago, between the Carboniferous and Triassic periods. Most of the land was gathered together in the vast supercontinent of Pangaea. Marine life included ammonites, brachiopods, corals, nautiloids and fishes. On land, gymnosperms became the dominant trees and the first conifers appeared towards the end of the period. Reptiles such as the pelycosaurs and the therapsids (the mammal-like reptiles) became the dominant vertebrates on land. The Permian ended in a mass extinction event in which *ca.* 95% of marine species, including the trilobites, are estimated to have gone extinct.

permissive *a.* (1) (*genet.*) *appl.* conditions under which an organism or cell carrying a conditional lethal mutation does not display the mutant phenotype; (2) (*virol.*) *appl.* conditions allowing normal infection and multiplication of a virus in a cell, and which may be environmental or genetic (i.e. species differences in susceptibility); (3) (*dev.*) *appl.* a developmental signal that merely allows a certain process to proceed. *cf.* instructive.

peronaeus peroneus *q.v.*

peronate *a.* (1) covered with woolly hairs; (2) surrounded by volva, *appl.* stalk; (3) powdery or mealy externally.

peroneal *a.* *pert.*, or lying near, the fibula, *appl.* artery, nerve, tubercle.

peroneotibial *a.* in region of fibula and tibia, *appl.* certain muscles.

peroneus *n.* any of several muscles arising from the fibula in the lower leg, as two lateral muscles, longus and brevis, and an anterior muscle, tertius.

peronium *n.* in some jellyfish, one of the cartilaginous processes ascending from the margin of the disc towards the centre.

peropod *a.* with rudimentary limbs.

peroral *a.* *appl.* membrane formed by the concrescence of rows of cilia around the cytopharynx of ciliate protozoans.

per os by mouth.

peroxidase *n.* haem-containing enzyme that reduces hydrogen peroxide (H_2O_2) and organic peroxides, using NADH or other compounds as a hydrogen donor, producing water and the oxidized donor. EC. 1.11.1.7. *alt.* lactoperoxidase, myeloperoxidase.

peroxins *n.plu.* proteins that are involved as docking proteins and receptors in the targeted import of proteins into peroxisomes.

peroxiredoxins *n.plu.* family of thiol-containing proteins that reduce hydrogen peroxide and organic peroxides, helping to protect cells against oxidative stress.

peroxisome *n.* a small organelle bounded by a single membrane and containing catalase and peroxidases. In liver cells is believed to be important in detoxification reactions (e.g. of ethanol). *see also* glyoxysome. *alt.* microbody.

peroxisome proliferator-activated receptor (PPAR) any of a family of gene regulatory nuclear receptors, whose ligands include cholesterol and fatty acids, and which regulate lipid storage and metabolism as well as fat-cell differentiation from fibroblast-like precursor cells.

peroxyacyl nitrates (PANs) a class of organic compounds with the general formula $RCO.O_2.NO_2$ which are formed in photochemical smogs. They are respiratory irritants and lachrymators in humans and are highly damaging to plants, which are sensitive to concentrations of around 0.05 ppm. Known by either of the abbreviations PAN or PANs, the first of which is also used for a specific peroxyacyl nitrate, namely peroxyacetyl nitrate ($CH_3CO.O_2.NO_2$).

perradius *n.* one of the four main radii of a radially symmetrical animal.

perseveration *n.* the persistence of a response after the original stimulus has ceased.

persistent *a.* (1) remaining attached until maturity, as corolla or perianth on a developing fruit; (2) *appl.* teeth with continuous growth; (3) *appl.* viral infection in which cell remains alive and continues to produce virus over a long period of time. Characteristic of infections with enveloped viruses. *cf.* latent (3), lytic.

person *n.* an individual or zooid in a colonial coral, hydrozoan or siphonophore.

Personatae Scrophulariales *q.v.*

personate *a. appl.* corolla of two lips, touching and with a projection of the lower closing the throat of the corolla, as in snapdragons.

perthophyte *n.* parasitic fungus that obtains nourishment from host tissues after having killed them by a poisonous secretion.

perthrotroph necrotroph *q.v.*

pertusate *a.* pierced at apex.

pertussis *n.* whooping cough, caused by the bacterium *Bordetella pertussis*.

pertussis toxin (PT) protein toxin from *Bordetella pertussis*. Used experimentally to study G protein activity, as it induces permanent activation of some G proteins.

perula *n.* a leaf bud scale.

pervalvar *a.* dividing a valve longitudinally.

pervious *a. appl.* nostrils with no septum between nasal cavities.

pes *n.* a foot, base or foot-like structure, as in certain parts of brain, branches of facial nerve. *plu.* **pedes**.

pessulus *n.* an internal dorsoventral rod at lower end of trachea in syrinx of birds.

pesticide *n.* general term for any chemical agent that kills animal pests.

PEST sequence peptide motif rich in proline, glutamate, serine and threonine that targets proteins for degradation in eukaryotic proteasomes.

PET positron emission tomography. *see* PET scan.

petal *n.* modified sterile leaf, often brightly coloured and with others forming the corolla, or inner series of perianth segments, of a flower. *see* Fig. 18 (p. 239).

petaliferous *a.* bearing petals.

petaliform *a.* petal-shaped, petal-like.

petalody *n.* the conversion of other parts of a flower into petals.

petaloid *a.* resembling a petal.

petaloideous *a. appl.* monocotyledons with a coloured perianth.

petalomania *n.* an unusual multiplication of petals.

petiolar *a. pert.* a petiole.

petiolate *a.* having a petiole.

petiole *n.* (1) (*bot.*) the stalk of a leaf; (2) (*zool.*) slender stalk connecting thorax and abdomen in certain insects such as wasps; (3) any short, slender stalk-like structure.

petiolule *n.* the stalk of a leaflet of a compound leaf.

petite mutants yeast strains isolated as slow-growing small colonies on solid medium, which lack mitochondrial function due to mutations in nuclear or mitochondrial genes.

Petri dish, Petri plate shallow circular glass or plastic dish used for growing microorganisms or cultured cells in a suitable medium. *alt.* plate.

petrification *n.* fossilization through saturation with mineral matter in solution, subsequently turned to solid.

petrohyoid *a. pert.* hyoid and petrous part of temporal bone of skull.

petromastoid *a. pert.* mastoid process and petrous portion of temporal bone of skull.

petro-occipital *a. pert.* occipital and petrous part of temporal bone of skull, *appl.* a fissure.

petrophyte *n.* a rock plant.

petrosal *a.* (1) made of compact bone; (2) *appl.* otic bones of fishes; (3) *appl.* a sphenoidal process, to a ganglion of glossopharyngeal nerve, and to nerves and sinus in region of petrous portion of temporal bone; (4) *appl.* bone: the periotic *q.v. alt.* **petrous**.

petrosphenoidal *a. pert.* sphenoid and petrous portion of temporal bone, *appl.* fissure.

petrosquamosal *a. pert.* squamosal and petrous portion of temporal bone, *appl.* sinus and suture.

petrotympanic *a.* (1) *pert.* tympanum and petrous portion of temporal bone; (2) *appl.* fissure: the fissure in the temporal bone of mammals which holds the Folian process of the malleus of middle ear.

petrous *a.* (1) very hard or stony; (2) *appl.* a pyramidal portion of the temporal bone behind sphenoid and occipital; (3) *appl.* a ganglion on its lower border. *alt.* **petrosal**.

PET scan non-invasive imaging technique using positron emission tomography to detect functioning of different brain areas in different tasks.

petunidin *n.* purple anthocyanin pigment found in plants.

Peyer's patches lymphoid tissue present in the connective tissue of the intestinal mucosa.

P face in freeze-fractured membranes, the face representing the hydrophobic interior of the cytoplasmic half of the lipid bilayer.

P factors P elements *q.v.*

Pflüger's cords columns of cells growing from germinal epithelium into stromatic tissue of embryo and which give rise to the gonads.

PFU plaque-forming unit *q.v.*

PG prostaglandin *q.v.*

PGA (1) phosphoglycerate *q.v.*; (2) pteroylglutamic acid, *see* folic acid; (3) prostaglandin A.

PGB, PGE, PGF classes of prostaglandins *q.v.*

P granules ribonucleoprotein particles located at the posterior pole of the nematode egg, where they are segregated into the germ-cell precursor cell.

pH a measure of the acidity of a solution, the negative $\log_{10}$ of the hydrogen ion concentration. The pH of a neutral solution is 7, that of acid solutions less than 7 and of alkaline solutions greater than 7.

PHA phytohaemagglutinin *q.v.*

phacoid *a.* lentil- or lens-shaped.

phaeic *a.* of a dusky colour.

phaeism *n.* (1) duskiness; (2) *appl.* colouring of butterflies due to incomplete melanism.

phaeochrome chromaffin *q.v.*

phaeochrous *a.* of a dusky colour.

phaeomelanin *n.* a brownish melanin.

phaeophyll *n.* the colouring matter of brown algae, a mixture of fucoxanthin, xanthophyll, chlorophyll and carotene.

Phaeophyta *n.* the brown algae or brown seaweeds. Mainly multicellular and marine, they can reach large sizes, as in the kelps, and some groups possess complex internal structure. Their green chlorophyll is masked by the brown pigment fucoxanthin so that they appear brownish. Carbohydrate reserves are in the form of laminarin.

phaeophytin pheophytin *q.v.*

phaeoplast *n.* pigmented plastid of brown algae.

phaeosome *n.* an optic organelle in some epidermal cells of annelid worms.

phaeospore *n.* a spore containing phaeoplasts.

phage bacteriophage *q.v.*

phage conversion phenomenon in which genes carried by a temperate phage change the phenotype of the host bacterium, e.g. alteration of O-chain polysaccharide in salmonellae, and the production of phage-specified toxin by *Corynebacterium diphtheriae*.

phage display library collection of phages carrying a diversity of genes encoding immunoglobulin variable regions, and which each display a different variable region on their surface.

phagemid *n.* a cloning plasmid containing the origin of replication of a filamentous phage, so that the cloned DNA can be packaged into phage capsids after replication.

phagocyte *n.* (1) cell specialized to carry out phagocytosis. In mammals the principal phagocytic cells are monocytes, neutrophils and macrophages; (2) in plants, a root cell with a lobed nucleus, capable of digesting endotrophic fungal filaments.

phagocytic *a. pert.* or effecting phagocytosis.

phagocytic vacuole phagosome *q.v.*

phagocytose *v.* to carry out phagocytosis.

phagocytosis *n.* uptake of large solid particles (including other cells) into a cell by the process of endocytosis. It is seen e.g. in amoeboid protozoans engulfing their prey and in cells such as macrophages, which ingest and destroy invading microorganisms and scavenge damaged and dying cells.

phagolysis *n.* the lysis or disintegration of phagocytes.

phagolysosome *n.* a cytoplasmic vesicle formed by the fusion of a phagosome and lysosome.

phagosome *n.* a large membrane-bounded vesicle in the cytoplasm of a phagocytic cell which contains material taken up by phagocytosis. *alt.* phagocytic vacuole.

phagotroph *n.* any heterotrophic organism (e.g. many animals and protozoans) that ingests nutrients as solid particles. *cf.* osmotroph.

phagozoite *n.* animal which feeds on disintegrating or dead tissue.

phalange phalanx *q.v.*

phalangeal *a. pert.* or resembling phalanges, like segmented fingers.

phalanges *n.plu.* the bones of the fingers and toes of vertebrates.

phalanx *n.* (1) one of several bones (phalanges) in fingers or toes; (2) bundle of stamens united by filaments; (3) a taxonomic group, never precisely defined, but usually used for a group resembling a subfamily.

phallic *a.* (1) *pert.* phallus; (2) *appl.* gland secreting substance for spermatophores, as in certain insects.

phalloidin *n.* alkaloid obtained from *Amanita phalloides* which binds to actin filaments and prevents cell movement.

phallus *n.* (1) the embryonic structure which becomes penis or clitoris; (2) the penis; (3) external genitalia of a male insect; (4) the fruiting body of the stinkhorn fungi.

phanerogams *n.plu.* (1) all seed-bearing plants; (2) formerly used for plants with conspicuous flowers. *a.* **phanerogamic**.

phanerophyte *n.* tree or shrub with aerial dormant buds.

phaneroplasmodium *n.* in slime moulds, a thick opaque plasmodium with veins having clearly defined endoplasm and ectoplasm.

Phanerozoic *n.* geological eon comprising the Paleozoic, Mesozoic and Cenozoic eras, lasting from around 542 million years ago to the present day.

phaoplankton *n.* surface plankton living at depths at which light penetrates.

pharate *a. appl.* instar (larval stage of some insects) within previous cuticle before moulting.

pharmacodynamics *n.* the study of the action of drugs, including all aspects of their behaviour in the body, i.e. transport to tissues, persistence in blood stream and tissues, as well as their immediate biochemical activity.

pharmacogenetics, pharmacogenomics *n.* the study of the genetic basis of deleterious responses or non-responses to therapeutic drugs.

pharmacological *a. appl.* or *pert.* medicinal drugs and their actions.

pharmacology *n.* the study of the action of medicinal drugs and other biologically active chemicals.

pharmacophore *n.* the part of a molecule causing the specific physiological effects of a drug.

pharotaxis *n.* the movement of an animal towards a definite place, the stimulus for which is acquired by conditioning or learning.

pharyngeal *a. pert.* pharynx.

pharyngeal pouches infoldings of endoderm between branchial arches in pharyngeal region of vertebrate embryos. *cf.* branchial clefts.

pharyngobranchial *n.* a dorsal skeletal element of gill arch.

pharyngopalatine *a. pert.* pharynx and palate, *appl.* arch and muscle.

pharyngotympanic *a. pert.* tube connecting pharynx and tympanic cavity, the auditory or Eustachian tube.

pharynx *n.* in humans and other vertebrates, the throat. In other animals the gullet or anterior part of the alimentary canal or enteron following the mouth.

phase-contrast microscopy type of light microscopy which enables living unstained cells and tissue to be studied. A special ring in the objective lens of the microscope enables the differential diffraction of light by different parts of the cell to be detected to give a high-contrast image.

phaseolin *n.* a globulin protein obtained from the seeds of the bean *Phaseolus*.

phase problem the problem that, in the X-ray diffraction data from a protein, only the amplitude of the wave is determined, but to compute a structure, the phase of the wave must also be known. The phase is often determined experimentally by incorporating one or more heavy atoms into the protein, whose positions can be determined independently.

phase variation in *Salmonella* species the change from expression of one gene for flagellin (a flagellar protein) to another non-allelic flagellin gene. This occurs at a regular frequency within a population (around once every 1000 cell divisions). It was originally detected by a change in antigenic specificity within the population.

phasmid *n.* (1) one of a pair of posterior sense organs in nematodes, possibly detecting chemical stimuli; (2) stick or leaf insect, a member of the insect order Phasmida.

Phasmida *n.* order of insects including the stick insects, which are long slender insects with long legs, and the leaf insects, which have flattened bodies with leaf-like flaps on their limbs. Both are well camouflaged in the bushes and trees in which they live.

PH domain pleckstrin homology domain *q.v.*

Phe phenylalanine *q.v.*

phellem(a) *n.* (1) cork; (2) cork and the non-suberized layers forming an external zone of periderm produced by the cork cambium.

phelloderm *n.* the secondary cortex of parenchyma cells filled with suberin that is formed by and on the inner side of the cork cambium. *alt.* secondary cortex.

phellogen cork cambium *q.v.*

phelloid *a.* cork-like. *n.* non-suberized cell layer in outer periderm.

phencyclidine *n.* drug originally developed as an anaesthetic but whose use was discontinued as it produced undesirable side-effects. It is now abused as a hallucinogenic drug.

phene *n.* a phenotypic character which is genetically determined.

phenetic *a. appl.* classification purely based on similarities in phenotypic characters, not necessarily reflecting relationships by evolutionary descent.

phenocontour *n.* (1) contour line on a map showing the distribution of a certain phenotype; (2) a line connecting all places within a region at which a biological phenomenon, e.g. flowering of a plant, occurs at the same time; (3) contour line delimiting an area corresponding to a given frequency of a variant form.

phenocopy *n.* a modification produced by environmental factors which simulates a genetically determined change.

phenocritical period for a particular gene, the time during development when its expression is required.

phenogram *n.* a tree-like diagram showing the conclusions of numerical taxonomy.

phenological *a.* (1) *pert.* phenology; (2) *appl.* isolation of species owing to differences in flowering or breeding season.

phenology *n.* recording and study of periodic biological events, such as flowering, breeding and migration, in relation to climate and other environmental factors.

phenome *n.* all the phenotypic characteristics of an organism.

phenomenology phenology *q.v.*

phenon *n.* group of organisms placed together by numerical taxonomy.

phenotype *n.* (1) the visible or otherwise measurable physical and biochemical characteristics of an organism, resulting from the interaction of genotype and environment; (2) a group of individuals exhibiting the same phenotypic characters.

phenotypic *a. pert.* phenotype. *cf.* genotypic.

phenotypic plasticity the range of variability shown by the phenotype in response to environmental fluctuations.

phenylalanine (Phe, F) *n.* amino acid with an aromatic side chain, constituent of protein, essential in human diet.

phenylketonuria *n.* inborn error of metabolism due to absence or deficiency of phenylalanine hydroxylase, leading to accumulation of phenylalanine in all body fluids, and to mental retardation and early death if untreated.

pheophytin *n.* blue-black pigments, components of the photosynthetic electron transport chain, derived from chlorophylls *a* and *b* by removal of the magnesium atom. *alt.* **phaeophytin**.

pheromone *n.* a chemical released in minute amounts by one organism which is detected and acts as a signal to another member of the same species. Examples are the volatile sexual attractants released by some female insects, which can attract males from a distance. Some pheromones act as alarm signals.

phialide *n.* a small bottle-shaped outgrowth of hypha in some fungi, from which spores (phialospores) are produced.

phialiform *a.* cup-shaped or saucer-shaped.

phialophore *n.* a hypha which bears a phialide.

phialopore *n.* the opening in a hollow daughter colony or gonidium of the protist *Volvox*.

phialospore *n*. a spore or conidium borne at tip of a phialide.

-phil, -philous suffixes derived from Gk *philein*, to love, denoting loving, or thriving in.

Philadelphia chromosome abnormal chromosome 22 (22q–) present in leukaemic cells of many patients with chronic myelogenous leukaemia. The end of the long arm of 22 has been exchanged with the end of the long arm of chromosome 9.

philopatry *n*. tendency of an organism to stay in or return to its home area.

philoprogenitive *a*. having many offspring.

philtrum *n*. the depression on upper lip beneath septum of nose.

phi torsion angle (φ) the angle of rotation between the N–C_α bond and the peptide bond in the backbone of a polypeptide chain.

phlebenterism *n*. the condition of having branches of the intestine extending into other organs, as arms and legs.

phlobaphaenes *n.plu.* phenolic compounds, derivatives of tannins, producing yellow, red or brown colours in fern ramenta, roots and sections of wood.

phloem *n*. the principal food-conducting tissue of vascular plants, extending throughout the plant body. It is composed of elongated conducting vessels, sieve tubes (in angiosperms) or sieve cells (in ferns and gymnosperms), both containing clusters of pores (sieve areas) in the walls, through which the protoplasts of adjacent cells communicate. Sugars and amino acids are the main nutrients transported via the phloem. Parenchymatous companion cells (or albuminous cells in gymnosperms) closely associated with the conducting elements are involved in the delivery to and uptake of material from the phloem. Phloem also contains supporting fibres (bast). *see also* vascular bundle, xylem.

phloem loading the active transport of sugars into the phloem at their site of synthesis.

phloem-mobile *appl.* ions that can be transported via the phloem, e.g. K^+, Cl^-, $H_2PO_4^{2-}$, but not Ca^{2+}, which is termed phloem-immobile.

phloem parenchyma thin-walled parenchyma associated with the sieve tubes of phloem.

phloem sheath pericycle *q.v.*

phloeodic *a*. having the appearance of bark.

phloic *a. pert.* phloem.

phlorizin *n*. plant glucoside from roots, used experimentally as an uncoupler of electron transport and ATP synthesis in chloroplasts.

phobotaxis *n*. an avoidance reaction.

phocids *n.plu.* members of the Phocidae: the seals.

Phoenicopteriformes *n*. order of birds in some classifications, including the flamingoes.

pholidosis *n*. the arrangement of scales, as on scaled animals.

Pholidota *n*. order of placental mammals known from the Pleistocene or possibly Oligocene, the only living member being the pangolin (scaly anteater), having no teeth, and the body covered with imbricated scales.

phonation *n*. production of sounds, e.g. by insects.

phonoreceptor *n*. a receptor for sound waves.

phoranth(ium) *n*. the receptacle of flower-heads of Compositae.

phorbol esters polycyclic alcohols derived from croton oil (e.g. 12-O-tetradecanol phorbol-13 acetate) which activate protein kinase C as a result of their resemblance to diacylglycerol.

phoresia, phoresy *n*. the carrying of one organism by another, without parasitism, as in certain insects, or as in the carriage of flower pollen by insects.

Phoronida, phoronids *n*., *n.plu.* small phylum (only 15 species) of marine worm-like coelomate animals that secrete chitinous tubes in which they live. The mouth is surrounded by a horseshoe-shaped crown of tentacles (a lophophore) which projects from the tube.

phoront *n*. encysted stage leading to formation of trophont in life cycle of some ciliate protozoans.

phosphagen creatine phosphate *q.v.*

phosphatase *n*. any of a large group of widely distributed enzymes catalysing the hydrolysis of organic phosphate esters.

Protein phosphatases catalyse the hydrolytic removal of phosphate groups from proteins. EC 3.1.3. *see also* phosphorylation.

phosphate *n.* the phosphate anion PO_4^{3-} or a salt of phosphoric acid (e.g. potassium phosphate, K_3PO_4). Phosphates are essential to the metabolism of living organisms because inorganic phosphate is required for the synthesis of the energy-storage molecule ATP. Plants and microorganisms take up phosphorus mainly in the form of phosphates, and various phosphates are used as fertilizers. Excess phosphate washed into streams and lakes contributes to eutrophication and the formation of algal blooms. *see also* phosphorus cycle, phosphoryl group.

phosphate bonds covalent bonds involving a phosphorus atom. *see* phosphoanhydride bond, phosphoester bond.

phosphatidalcholine, phosphatidalethanolamine, phosphatidalserine *see* plasmalogen *q.v.*

phosphatide phospholipid *q.v.*

phosphatidylcholine *n.* phosphoglyceride with a choline alcohol group, the principal phospholipid in most membranes of higher organisms. *alt.* lecithin.

phosphatidylethanolamine *n.* phosphoglyceride with ethanolamine as the alcohol group, a common phospholipid of cell membranes. *alt.* (formerly) cephalin.

phosphatidylinositol *n.* phosphoglyceride with inositol as the alcohol group, a common phospholipid of cell membranes.

phosphatidylinositol 4,5-bisphosphate (PIP2, PI(4,5)P2) membrane phospholipid that is broken down by phospholipase C to form diacylglycerol and inositol trisphosphate.

phosphatidylinositol 3-kinase (PI 3-kinase, PI3K) a kinase that is activated by some cell-surface receptors and takes part in intracellular signalling by phosphorylating inositol phospholipids at the C3 position on the inositol ring. It is involved in pathways leading to cell growth and to cell survival. EC 2.7.1.137.

phosphatidylinositol phosphates forms of the membrane phospholipid phosphatidylinositol that are phosphorylated on the inositol ring. Some, e.g. phosphatidylinositol 4,5-bisphosphate, are part of intracellular signalling pathways.

phosphatidylserine *n.* phosphoglyceride with serine as the alcohol group, a common phospholipid of cell membranes.

phosphene *n.* an impression of light on the retina as a result of a stimulus other than rays of light.

phosphoanhydride bond covalent chemical bonds –P–O–P– linking two phosphate groups in series, as in ATP.

phosphodiesterase (PDE) *n.* any of a group of enzymes that hydrolyse phosphodiester linkages. They include cyclic AMP phosphodiesterase, which converts cyclic AMP to adenosine monophosphate.

phosphodiester linkage the covalent bonds $–O–P(O_2)–O–$ that e.g. form the linkage between nucleotide residues in a polynucleotide chain. *see* Fig. 31.

phosphoenolpyruvate (PEP) *n.* three-carbon intermediate in the conversion of phosphoglycerate to pyruvate in glycolysis. Its conversion to pyruvate generates ATP.

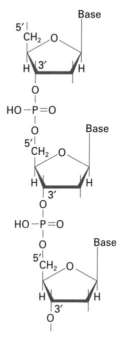

Fig. 31 Phosphodiester bonds in a polynucleotide.

It is also an important intermediary metabolite in biosynthesis of glycogen, neuraminic acid and phenylalanine.

phosphoester bond covalent bond that links a phosphate group to another molecule, e.g. $P(O_3)O–C–$.

phosphofructokinase *n.* (1) enzyme that phosphorylates fructose 6-phosphate to fructose 1,6-bisphosphate in glycolysis and other metabolic pathways (*r.n.* 6-phosphofructokinase, EC 2.7.1.11); (2) the enzyme using fructose 1-phosphate in the same reaction (*r.n.* 1-phosphofructokinase, EC 2.7.1.56). *alt.* fructokinase.

phosphoglucomutase *n.* widely distributed phosphotransferase which catalyses the conversion of glucose 1-phosphate to glucose 6-phosphate using glucose 1,6-bisphosphate (diphosphate) as donor. EC 5.4.2.2.

phosphogluconate oxidative pathway pentose phosphate pathway *q.v.*

phosphoglucose isomerase glucosephosphate isomerase *q.v.*

phosphoglyceric acid, phosphoglycerate (PGA) three-carbon monosaccharide, occurring in all cells as 2- or 3-phosphoglycerate, an important intermediate in photosynthesis, respiration and carbohydrate metabolism.

phosphoglycerides *n.* phospholipids based on the three-carbon alcohol glycerol phosphate. They consist of a glycerol backbone to which two fatty acid chains are attached, and a phosphorylated alcohol (e.g. choline, ethanolamine). They are constituents of cell membranes.

phosphoglyceromutase widely distributed phosphotransferase which catalyses the conversion of 2-phosphoglycerate to 3-phosphoglycerate using 2,3-bisphosphoglycerate as donor. EC 5.4.2.1.

phosphohexose isomerase glucosephosphate isomerase *q.v.*

phosphohistidine *n.* histidine residue phosphorylated by the action of a histidine kinase.

phosphoinositide *n.* phospholipid containing inositol.

phosphoinositide pathway (PI pathway) intracellular signalling pathway in which stimulation of a cell-surface receptor leads to the breakdown of phosphatidylinositol phosphates in the membrane by phospholipase C to form soluble inositol trisphosphate and membrane-bound diacylglycerol. Both these compounds have second messenger activity, stimulating the release of Ca^{2+} from intracellular stores and the activation of protein kinase C, respectively.

phosphokinase *n.* enzyme that catalyses the addition of phosphate groups to a molecule. *alt.* kinase.

phospholipase (PLC) *n.* any of a group of enzymes which catalyse the hydrolysis of membrane phospholipids at C–O bonds (phospholipases A, which cleave off one of the fatty acid moieties) or P–O bonds (phospholipases C and D). Phospholipases C (e.g. PLC-β and PLC-γ) generate the intracellular signalling molecules inositol trisphosphate and diacylglycerol from membrane lipids.

phospholipid *n.* any of a group of amphipathic lipids with either a glycerol or a sphingosine backbone, fatty acid side chains and a phosphorylated alcohol headgroup, which form the lipid bilayer in all biological membranes. They include the glycerolipids phosphatidylcholine, phosphatidylethanolamine, phosphatidylinositol and phosphatidylserine, and the sphingolipid sphingomyelin.

phospholipid exchange protein watersoluble protein that carries a phospholipid from one membrane to another within a cell. *alt.* phospholipid transfer protein.

phospholipid translocator enzyme that catalyses the flip-flop of phospholipids from one layer of a biological membrane to the other. *see* flippase, scramblase.

phosphoprotein *n.* protein carrying phosphoryl groups, which are added by protein kinases after the protein has been synthesized.

phosphoprotein phosphatase protein phosphatase *q.v.*

phosphorescence *n.* the luminescence of marine protozoans, copepods and the majority of deep-sea animals.

phosphoribosylpyrophosphate (PRPP) an activated form of ribose phosphate which is an important intermediate in biosynthesis of aromatic amino acids and purine and pyrimidine nucleotides.

phosphoroclastic reaction *a.* the reaction that produces molecular hydrogen (H_2) from pyruvate.

phosphorolysis *n.* cleavage of a chemical bond by orthophosphate (P_i), in biochemical reactions catalysed by phosphorylases. *cf.* hydrolysis.

phosphorus (P) *n.* non-metallic element that exists in nature only in the combined state, mainly as phosphates in minerals (e.g. apatite) and in organic matter. It is an essential nutrient for living organisms. The radioactive isotope phosphorus-32 (^{32}P) is used widely as a label in phosphorus-containing molecules. *see also* phosphate, phosphorus cycle.

phosphorus cycle the movement of phosphorus within the biosphere and between the biosphere and the inorganic environment. Phosphorus released to the soil in the form of phosphates through the weathering of minerals is taken up by plants and microorganisms and from them passes into animals. Phosphorus is released back to the soil in animal wastes and by decomposition of organic matter.

phosphorylase *n.* (1) recommended name for a group of enzymes that catalyse the progressive breakdown of glucose polysaccharides such as starch or glycogen by phosphorolysis, giving glucose 1-phosphate, EC 2.4.1.1; (2) any of a group of enzymes that catalyse the transfer of glucose residues from a glucose oligo- or polysaccharide to orthophosphate giving glucose 1-phosphate as a product, e.g. maltose phosphorylase (EC 2.4.1.8).

phosphorylase kinase protein kinase that phosphorylates the enzyme glycogen phosphorylase and activates it.

phosphorylation *n.* the covalent addition of a phosphate (phosphoryl) group to a molecule. The enzymatic phosphorylation of proteins at specific amino acids by protein kinases is a widespread means of rapidly altering a protein's activity in response to intracellular or extracellular signals. The phosphate donor for protein phosphorylation is usually ATP.

phosphorylation potential index of the energy status of a cell in terms of potential transferable phosphate groups, being calculated as the ratio of the concentration of ATP to the product of the concentrations of ADP and inorganic phosphate.

phosphoryl group $-PO_4^{2-}$, phosphate group when attached to another molecule.

phosphoserine *n.* phosphorylated derivative of the amino acid serine in proteins, modified after incorporation into the protein chain. Enzymatic phosphorylation and dephosphorylation of serine by serine/threonine protein kinases is a mechanism for regulating the activity of certain proteins.

phosphothreonine *n.* phosphorylated derivative of the amino acid threonine in proteins, modified by protein kinases after incorporation into the protein chain. Enzymatic phosphorylation and dephosphorylation of threonine by serine/threonine protein kinases is a mechanism for regulating the activity of certain proteins.

phosphotransferase system bacterial transport system for moving certain metabolites over the cell membrane in which the substance being transported is phosphorylated.

phosphotyrosine *n.* phosphorylated derivative of the amino acid tyrosine in proteins, modified after incorporation into the protein chain. Enzymatic phosphorylation and dephosphorylation of tyrosine residues by tyrosine protein kinases is a mechanism for regulating the activity of some proteins. Phosphorylation of tyrosine in the cytoplasmic tails of a class of receptor proteins is a means of signalling their activation by ligand. *see* receptor tyrosine kinases.

phosphovitin, phosvitin *n.* a phosphoprotein of amphibian egg yolk.

photic *a. appl.* zone of surface waters penetrated by sunlight.

photo- prefix derived from Gk *phos*, light, indicating response to, sensitivity to or causation by light.

photoassimilate *n.* the carbon-containing compounds produced as a result of photosynthesis.

photoautotroph *n.* organism using light as an energy source and carbon dioxide as the main source of carbon, such as green plants and some bacteria.

photobiodegradable *a. appl.* materials whose chemical structure is altered by light in such a way that makes them more susceptible to microbial degradation.

photobiont *n.* the photosynthetic component in a lichen. *alt.* phycobiont.

photobleaching *n.* loss of colour by photosensitive pigments such as rhodopsin on exposure to light.

photoceptor photoreceptor *q.v.*

photochemical *a. appl.* and *pert.* chemical changes brought about by light.

photochemical reaction centre reaction centre *q.v.*

photochromatic *a. appl.* interval between achromatic and chromatic thresholds.

photochromic effect change of colour brought about by light.

photochromogenesis *n.* pigment synthesis only in the presence of light.

photodinesis *n.* protoplasmic streaming induced by light.

photodynamics *n.* the study of the effects of light on plants.

photogen *n.* light-producing organ or substance.

photogene *n.* a gene whose expression is controlled by light.

photogenic *a.* (1) light-producing; (2) luminescent.

photogenin luciferase *q.v.*

photoheterotroph *n.* organism that uses light as a source of energy, but derives much of its carbon from organic compounds, such as the photosynthetic non-sulphur purple and green bacteria.

photoinhibition *n.* inhibition by light, e.g. of germination.

photokinesis *n.* random movement in response to light.

photolabile *a. appl.* substances such as retinal pigments which undergo a chemical change on exposure to light.

photolithotroph photoautotroph *q.v.*

photolyase *n.* enzyme that catalyses the repair of pyrimidine dimers in UV-irradiated DNA in the presence of light. EC 4.1.99.3, *r.n.* deoxyribopyrimidine photo-lyase. *alt.* photo-reactivating enzyme.

photolysis *n.* splitting a compound or molecule by the action of light, e.g. the splitting of water into hydrogen and oxygen.

photometer *n.* instrument that measures the turbidity of a solution by the amount of light it lets pass through.

photomorphogenesis *n.* any effect on plant growth produced by light.

photomotor reflex change in size of pupil of eye with sudden change in light intensity.

photon *n.* the particulate unit of light, carrying a quantum of energy. 1 mol photons (or 1 mol quanta) is the number of photons corresponding to Avogadro's number of particles (6.023×10^{23}) and is the number of photons required to convert 1 mol of a substance to another form with 100% efficiency if captured in a single step. Photon number incident on a surface normal to the beam in a given time is the photon flux (often called photon flux density, not recommended) and is measured in mol m^{-2} s^{-1}.

photonasty *n.* response of plants to diffuse light stimuli, or to variations in illumination.

photoorganotroph photoheterotroph *q.v.*

photopathy *n.* pronounced movement in relation to light, usually away from it as in negative phototaxis or negative phototropism.

photoperiod *n.* (1) duration of daily exposure to light; (2) the length of day favouring optimum functioning of an organism.

photoperiodicity, photoperiodism *n.* the response of an organism to the relative duration of day and night. Examples are the flowering of plants and mating of animals, which can be triggered by the lengthening or shortening of the days as the seasons change.

photophase *n.* the developmental stage of a plant when it shows definite requirements as to duration and intensity of light and temperature.

photophilous *A.* seeking, and thriving in, strong light.

photophobic *A.* not tolerating light, shunning light.

photophore *n.* (1) light-emitting organ which directs light ventrally in some deep-sea fish, crustaceans and cephalopods, and so camouflages the silhouette from ventral view; (2) any light-emitting organ.

photophosphorylation *n.* the formation of ATP using energy from light during photosynthesis. *cf.* oxidative phosphorylation.

photopia *n.* adaptation of the eye to light.

photopic system the system of cones in the retina of the eye.

photopigment *n.* any light-sensitive pigment.

photoplagiotropy *n.* tendency to take up a position transverse to the incident light.

photoprotein *n.* a protein sensitive to light.

photopsin *n.* protein component of the violet retinal cone pigment iodopsin.

photoreactivation *n.* reactivation or repair of inactivated or damaged protein or DNA by the stimulus of light, which initiates a light-dependent enzymatic reaction.

photoreceptor *n.* (1) sense organ responding to light (e.g. eye); (2) cell or part of a cell sensitive to light (e.g. rod or cone cells in the retina); (3) molecule sensitive and responding to light, such as rhodopsin in retina, chlorophyll in plants.

photoregulation *n.* the regulation of genes, or a physiological or developmental process, by light.

photorespiration *n.* type of 'wasteful' respiration occurring in green plants in the light. It is different from normal mitochondrial respiration, consuming oxygen and evolving carbon dioxide using chiefly glycolate derived from the primary photosynthate as substrate. It occurs to a much greater extent in C3 plants than in C4 plants.

photospheres *n.plu.* luminous organs of crustaceans.

photosynthate *n.* product(s) of photosynthesis.

photosynthesis *n.* (1) the use of the energy in sunlight to power biosynthesis in living organisms; (2) in green plants, algae and cyanobacteria, it is the synthesis of carbohydrate from carbon dioxide as a carbon source and water as a hydrogen donor with the release of oxygen as a waste product, using light energy captured by the pigments such as chlorophyll. The primary products are ATP and NADPH, formed in the presence of light (the light reaction). These are then used as energy source and electron donor, respectively, for carbohydrate synthesis (the dark reaction). In green plants and algae, photosynthesis takes place in chloroplasts; (3) a similar process in other bacteria, but using hydrogen donors other than water and pro-

ducing waste products other than oxygen. *a.* **photosynthetic.** *see also* bacteriochlorophylls, Calvin cycle, chlorophyll, cyclic photophosphorylation, photophosphorylation, photosystem I and II, reaction centre.

photosynthetically active radiation (PAR) radiation capable of driving the light reactions of photosynthesis, wavelength 380–710 nm.

photosynthetic carbon reduction cycle (PCR cycle), photosynthetic cycle Calvin cycle *q.v.*

photosynthetic efficiency the conversion factor relating the energy falling per unit area on a photosynthetic tissue and the energy value of the biochemical compounds produced.

photosynthetic electron-transport chain chain of electron-transporting proteins embedded in chloroplast thylakoid membranes, which carry electrons derived from the action of light on chlorophyll and pump protons across the thylakoid membrane, with the associated synthesis of ATP and NADPH. *cf.* respiratory chain.

photosynthetic quotient ratio between the volume of oxygen produced and the volume of carbon dioxide used in photosynthesis.

photosynthetic reaction centre *see* reaction centre.

photosynthetic unit functional unit in photosynthesis, composed of several hundred chlorophyll molecules, a reaction centre and accessory pigments.

photosynthetic zone of sea or lakes, the vertical zone in which photosynthesis can take place, between surface and compensation point.

photosystem I (PSI) and photosystem II (PSII) the multimolecular light-capturing and electron-transporting complexes of protein, chlorophyll and other pigments, found in thylakoid membranes of chloroplasts and involved in the light reaction of photosynthesis. Activation of photosystem II by sunlight leads to the splitting of water, the generation of ATP and a flow of electrons to photosystem I, whose activation by light then leads to the reduction of NADP to NADPH.

phototaxis *n.* movement in response to light. Positive phototaxis is movement towards a light source. Negative photo-taxis is movement away from a light source. *a.* **phototactic.**

phototransduction *n.* the reception and interpretation by a cell of a signal in the form of light.

phototroph *n.* organism using sunlight as a source of energy. *a.* **phototrophic.**

phototropin *n.* plant protein sensitive to blue light and involved in phototropism.

phototropism *n.* growth movement of plants in response to the stimulus of light.

phototropy *n.* reversible change in the colour of a substance while it is illuminated.

phragma *n.* a septum or partition. *plu.* **phragmata.**

Phragmobasidiomycetes, Phragmo-basidiomycetidae *n.* basidiomycete fungi which form basidiospores on a septate basidium. They include the rust and smut fungi.

phragmobasidium *n.* septate basidium forming four cells.

phragmocone *n.* in certain molluscs, the cone of a shell divided internally by a series of partitions perforated by a siphuncle.

phragmoplast *n.* the vesicles laid down across the middle of a plant cell undergoing cell division. They are the basis for the formation of the cell plate and new cell wall.

phragmosis *n.* the use of part of the body by reptiles or amphibians to close a burrow.

phragmosome microbody *q.v.*

phragmospore *n.* septate fungal spore.

phratry *n.* a loose term in classification, never generally adopted or precisely defined, but often used to mean a subtribe.

phreaticolous *a. appl.* organisms living in underground fresh water.

phreatophyte *n.* plant with very long roots that reach the water table.

phrenic *a.* in region of diaphragm, *appl.* e.g. artery, nerve.

phrenicocostal *a. appl.* a narrow slit or sinus between costal and diaphragmatic pleurae.

phrenicolienal *a. appl.* ligament forming part of peritoneum reflected over spleen and extending to diaphragm.

phrenicopericardiac *a. appl.* ligament extending from diaphragm to pericardium.

phyco- prefix derived from Gk *phykos,* seaweed, signifying to do with algae.

phycobilins *n.plu.* light-harvesting pigments in cyanobacteria and red algae: phycoery-thrin, phycocyanin, allophycocyanin.

phycobiliprotein *n.* protein pigment with a phycobilin chromophore.

phycobilisome *n.* one of a number of small particles present on photosynthetic lamellae of some red algae and cyano-bacteria, and which contain phycobilins.

phycobiont *n.* the algal partner in a sym-biosis, e.g. in lichens and in certain marine invertebrates.

phycochrome *n.* general term for a pigment found in algae.

phycochrysin *n.* golden-yellow pigment present in chromophores of the golden-brown algae (Chrysophyta).

phycocoenology *n.* study of algal communities.

phycocyanin *n.* a blue phycobiliprotein that gives cyanobacteria their colour and is also present in red algae and Cryptophyta. It acts as a light-harvesting pigment.

Phycodnaviridae *n.* family of nonenveloped isometric double-stranded DNA viruses that infect algae.

phycoerythrin *n.* a red phycobiliprotein that gives red algae their colour and is also present in some cyanobacteria and Cryptophyta. It acts as a light-harvesting protein pigment.

phycology *n.* the study of the algae.

Phycomycetes *n.* general name for organ-isms of the Chytridiomycota, Hyphochy-tridiomycota, Plasmodiophoromycota and Oomycota.

phycophaein *n.* brown pigment in brown algae, an oxidation product of fucosan.

Phycophyta *n.* in some classifications the name for the algae.

phycoplast *n.* system of microtubules that develops parallel to the plane of nuclear division in dividing cells of green algae of the Chlorophyta.

phycoxanthin *n.* yellow or brownish-yellow pigment present in some algae.

phyla *plu.* of phylum.

phylacobiosis *n.* mutual or unilateral pro-tective behaviour, as of certain ants. *a.* **phylacobiotic.**

Phylactolaemata *n.* in some classifications a class of freshwater Bryozoa.

phylembryo *n.* a developmental stage in brachiopods at completion of the embryonic shell.

phyletic *a.* (1) *pert.* a phylum or a major branch of an evolutionary lineage; (2) *appl.* a group of species related to each other by common descent; (3) *pert.* a line of direct descent.

phyletic evolution sequence of evolutionary changes leading to a sequence of species or forms arising through time in a single line of descent.

phyletic gradualism the idea that evolutionary change is built up in small steps, the change in any single generation being extremely small. Each stage must have a selective advantage, however marginal, eventually giving rise to new forms, organs and functions by a cumulative effect. The term phyletic gradualism is sometimes also used to refer to the view that evolution proceeds at a steady rate by such imperceptible small changes, and in this usage is often contrasted with the idea of punctuated equilibria and a non-uniform rate of evolution.

phyllade *n.* reduced scale-like leaf.

phyllary *n.* bract of the involucre in Compositae.

phyllidium bothridium *q.v.*

phyllo- prefix derived from Gk *phyllon*, leaf.

phyllobranchia *n.* a gill consisting of numbers of lamellae or thin plates.

phylloclade *n.* (1) a green flattened photosynthetic stem; (2) in cacti, a green, rounded stem functioning as a leaf; (3) an assimilative branch of a 'shrubby' thallus of lichens.

phyllocyst *n.* the rudimentary cavity of a hydrophyllium or protective medusoid.

phyllode *n.* winged leaf-stalk with flattened surfaces placed laterally to stem, functioning as a leaf.

phyllody *n.* metamorphosis of an organ into a foliage leaf.

phylloerythrin *n.* red pigment derived from chlorophyll and occurring in bile of herbivorous mammals.

phyllogen *n.* meristematic cells that give rise to leaf primordium.

phyllogenetic *a.* producing or developing leaves.

phylloid (1) *a.* leaf-like; (2) *n.* the leaf regarded as a flattened branch or telome.

phyllomania *n.* abnormal leaf production.

phyllome *n.* the leaf structures of a plant as a whole.

phyllomorphosis *n.* (1) metamorphosis of a plant organ into a foliage leaf; (2) variation in type of leaf at different seasons.

phyllophagous *a.* feeding on leaves.

phyllophore *n.* terminal bud or growing point of palms.

phyllophorous *a.* bearing or producing leaves.

phylloplane *n.* the leaf surface.

phyllopode *n.* sheathing leaf-base in quillworts (Isoetales).

phyllopodium *n.* (1) the axis of a leaf; (2) the plant stem when regarded as a pseudoaxis formed of fused leaf bases (as in some palms); (3) (*zool.*) a leaf-like swimming appendage in some crustaceans such as the branchiopods.

phylloquinone *see* vitamin K.

phyllorhiza *n.* a young leaf with a root.

phyllosiphonic *a.* with insertion of leaf trace disturbing axial stele tissue. *cf.* cladosiphonic.

phyllosoma *n.* larval stage of crawfish, being a broad thin schizopod larva.

phyllosperm *n.* a plant having seeds or spores borne on the leaves.

phyllosphere *n.* the leaf surfaces.

phyllospondylous *a.* *appl.* vertebrae consisting of a hypocentrum and neural arch, both contributing to hollow transverse processes.

phyllosporous *a.* having sporophylls like foliage leaves.

phyllotactic *a.* *pert.* phyllotaxis, *appl.* the fraction of circumference of stem between successive leaves, representing the angle of their divergence.

phyllotaxis, phyllotaxy *n.* the arrangement of leaves on an axis or stem, which for spirally arranged leaves may be expressed as the number of circuits of the stem that have to be made and the number of leaves that have to be passed to progress from the point of attachment of one leaf to that immediately above it (e.g.

1/2 for leaves positioned 180° apart). *see also* orthostichy, parastichy.

phyllozooid *n.* shield-shaped medusa with a protective function in a siphonophore colony.

phylogenesis phylogeny *q.v.*

phylogenetic *a. pert.* the evolutionary history and line of descent of a species or higher taxonomic group.

phylogenetic classification classification based on evolutionary relationships, in which all taxa aim to correspond to monophyletic lineages.

phylogenetics *n.* (1) the line of descent of a species or higher taxon; (2) approach to classification that attempts to reconstruct evolutionary genealogies and the historical course of speciation.

phylogenetic tree diagram like a family tree that represents evolutionary relationships between different taxa or between different proteins.

phylogeny *n.* the evolutionary history and line of descent of a species or higher taxonomic group. *cf.* ontogeny.

phylotypic *a.* typical of a phylum.

phylotypic stage embryonic stage at which all embryos of a given taxonomic group, e.g. the vertebrates, resemble each other most closely and show typical features of the group.

phylum *n.* in classification, a primary grouping consisting of animals constructed on a similar general plan, and thought to be evolutionarily related. In plants the similar category is called a division. Examples: Cnidaria (sea anemones, jellyfish, corals, etc.), Porifera (sponges), Platyhelminthes (flatworms, flukes and tapeworms), Mollusca (molluscs), Arthropoda (spiders, insects, crustaceans) and Chordata (includes the vertebrates). *plu.* **phyla.**

physa *n.* the rounded base of burrowing sea anemones.

physical containment in genetic engineering and in microbiology generally, the level of physical security and safety required in laboratory procedures, different levels being recommended for work involving microorganisms of differing degrees of pathogenicity.

physical mapping methods of gene mapping that rely on finding the actual positions of genes and other DNA sequences by means such as cytogenetics and restriction mapping rather than finding their relative positions by recombination-based genetic mapping methods.

physiogenic *a.* (1) caused by the activity of an organ or part; (2) caused by environmental factors.

physiographic succession plant succession influenced mainly by topography and local climate.

physiological races outwardly similar varieties within a species which differ in their physiology as a result of genetically determined factors. Examples are strains of plant pathogenic fungi that differ in their virulence towards different varieties of the host species.

physiology *n.* that part of biology dealing with the functions and activities of organisms, as opposed to their structure. *a.* **physiological.**

physoclistous *a.* having no channel connecting swim bladder and digestive tract, as in most teleosts.

physodes *n.plu.* (1) spindles of phlorglucin contained in plasmodium of certain Sarcodina; (2) fucosan vesicles.

physostigmine *n.* an alkaloid derived from the Calabar bean, an inhibitor of the enzyme acetylcholinesterase. *alt.* eserine.

phytal zone shallow lake bottom and its rooted vegetation.

phytanyl *n.* branched hydrocarbon present in the membrane lipids of Archaea.

phytic acid hexaphosphoinositol, the phosphate derivative of the sugar alcohol *myo*-inositol, found chiefly in seeds.

phytin *n.* magnesium calcium phytate, a phosphate storage substance in seeds, a salt of phytic acid.

phyto- prefix derived from Gk *phyton*, plant.

phytoactive *a.* stimulating plant growth.

phytoalexins *n.plu.* substances produced by plant cells in response to wounding or attack by parasitic fungi or bacteria. They are involved in resistance to infection and in limiting the damage caused by accidental wounding.

phytobiotic *a.* living within plants, *appl.* certain protozoans.

phytochemistry *n.* the chemistry of plants.

phytochoria *n.plu.* floristic realms, kingdoms or regions. *sing.* **phytochorion.**

phytochory *n.* dissemination of pathogens through the agency of plants.

phytochrome *n.* a light-sensitive protein pigment in plants. It exists in two forms: P_r, which is sensitive to red light, which converts it into P_{fr} which is sensitive to far-red light. Far-red light converts P_{fr} back to P_r. P_{fr} is active in stimulating developmental processes such as flowering in short-day plants and germination in some seeds, whereas P_r is inactive.

phytocoenology *n.* the study of plant communities.

phytocoenosis *n.* the assemblage of plants living in a particular locality.

phytoedaphon *n.* microscopic soil flora.

phytogenesis *n.* the evolution or development of plants.

phytogenetics *n.* plant genetics.

phytogenous *a.* of vegetable origin, produced by plants.

phytogeographical kingdoms *see* **floristic kingdoms.**

phytogeography *n.* study of the geographical distribution of plants.

phytoglycogen *n.* large highly branched polymer of glucose, found in some plants, and which is abundant in 'waxy' mutants of maize.

phytohaemagglutinin (PHA) *n.* a lectin, a protein isolated from the kidney bean *Phaseolus* and which acts as a mitogen on certain animal cells.

phytohormones *n.plu.* plant hormones, including auxins, gibberellins, cytokinins, abscisic acid. *see individual entries.*

phytoid *a.* plant-like.

phytokinin cytokinin *q.v.*

phytol *n.* product of hydrolysis of chlorophyll that is used in the synthesis of vitamins E and K. It is a long-chain alcohol forming the tail of the chlorophyll molecule.

phytolith *n.* minute mineral particle, e.g. hydrate of silica, when present in plant tissue. Found esp. in grasses.

phytomass *n.* plant biomass.

phytome *n.* plants considered as an ecological unit.

phytomorphosis *n.* changes in structural features of plants as a result of fungal and bacterial infection.

phytoparasite *n.* any parasitic plant.

phytopathogen *n.* any organism that causes disease in plants.

phytopathology *n.* the study of plant diseases.

phytophage *n.* animal that feeds on plants, usually refers to the smaller sap-sucking and leaf-eating insects rather than the larger herbivores. *a.* **phytophagous.**

phytophysiology *n.* plant physiology.

phytoplankton *n.* all photosynthetic plankton, including e.g. unicellular algae and cyanobacteria.

phytosis *n.* a disease caused by fungi.

phytosociology *n.* the study of all aspects of the ecology of plants and the influences on them.

phytosphingosine *see* sphingosine.

phytosterols *n.plu.* sterols such as sitosterol, originally isolated from plant material.

phytosuccivorous *a.* living on plant juices.

phytotelma *n.* a small aquatic environment held by a plant, such as the cup formed by the leaf bases of bromeliads, or a water-filled hole in a tree, which forms a small habitat. *plu.* **phytotelmata.**

phytotoxic *a.* toxic to plants.

phytotoxin *n.* any toxin originating in a plant.

phytotype *n.* representative type of plant.

PI phosphoinositide *q.v.*

pia mater delicate vascular membrane, the innermost of the three membranes surrounding the brain and spinal cord.

Piciformes *n.* order of birds including the woodpeckers.

Picornaviridae, picornaviruses *n., n.plu.* family of small, non-enveloped, single-stranded, positive-strand RNA viruses including polio virus, the rhinoviruses that cause common colds, and the aphthoviruses which include foot-and-mouth disease.

piezoelectric *a.* becoming electrically polarized when subjected to mechanical stress.

pigment *n.* colouring matter in plants and animals.

pigmentation *n.* disposition of colouring matter in an organ or organism.

pigment cell (1) chromatophore *q.v.*; (2) chromatocyte *q.v.*

PI 3-kinase phosphatidylinositol 3-kinase *q.v.*

pigmy *see* pygmy.

pileate *a.* having a pileus.

pileated *a.* crested, *appl.* birds.

pileocystidium *n.* sterile hair-like structure on the cap of certain basidiomycete fungi.

pileolus *n.* a small cap, as of many small toadstools.

pileorhiza *n.* (1) a root covering; (2) root cap *q.v.*

pileum *n.* the top of the head in birds. *plu.* **pilea**.

pileus *n.* the umbrella-shaped cap of mushrooms and toadstools. *plu.* **pilei**.

pili *plu.* of pilus.

pilidium *n.* (1) the characteristic helmet-shaped larva of nemertine worms; (2) a hemispherical apothecium of certain lichens.

pilifer *n.* part of labrum of some lepidopterans.

piliferous *a.* bearing or producing hair, *appl.* outermost layer of root which gives rise to root hairs.

pilomotor *a.* (1) causing hairs to move, *appl.* non-myelinated nerve fibres innervating muscles to hair follicles.

pilose *a.* downy or hairy.

pilus *n.* fine tube-like filamentous appendage produced by a 'male' bacterium which connects to the 'female' bacterium and through which DNA passes at conjugation. *plu.* **pili**. *alt.* sex pilus.

Pinaceae, pines *n., n.plu.* family of coniferous trees and shrubs that bear their needles in bunches.

pinacocytes *n.plu.* the flattened plate-like cells of outer epithelium of sponges.

pinacoderm *n.* the external layer of body wall of sponges.

pincushion gall reddish thread-like growth found on roses and caused by larvae of the gall wasp *Diplolepis rosae*.

pineal body in vertebrates, an outgrowth from the 1st cerebral vesicle, which has endocrine functions. It secretes vasotocin in mammals and melatonin. In lower vertebrates it is visible as the pineal or median eye. In higher vertebrates it is embedded in nervous tissue. In mammals it is involved in the regulation of certain biological rhythms by the release of melatonin. *alt.* **pineal gland**.

pineal region part of brain giving rise to pineal and parapineal organs.

pineal sac end vesicle of epiphysis, as in *Sphenodon*, the tuatara.

pineal stalk the connection between the pineal body and rest of brain.

pineal system the parietal organ and associated structures, such as pineal sac, pineal stalk and pineal nerves, parapineal organ and epiphysis.

pin-eyed having stigma at mouth of tubular corolla, with shorter stamens. *cf.* thrum-eyed.

pin feather a young feather, esp. one just emerging through the skin and still enclosed in sheath.

pinna *n.* (1) outer ear, thin cartilaginous structure covered with skin; (2) a bird's feather or wing; (3) fin or flipper; (4) (*bot.*) leaflet of pinnate leaf; (5) branch of pinnate thallus. *plu.* **pinnae**.

pinnate *a.* (1) divided in a feathery manner; (2) having lateral processes; (3) (*bot.*) of a compound leaf, having leaflets on each side of an axis or midrib.

pinnatifid *a. appl.* leaves lobed halfway to midrib.

pinnatilobate *a.* with leaves pinnately lobed.

pinnation *n.* the pinnate condition.

pinnatipartite *a.* with leaves lobed three-quarters of the way to the base.

pinnatiped *a.* with lobed toes, as certain birds.

pinnatisect *a.* with leaves lobed almost to base or midrib.

pinnatodentate *a.* pinnate with toothed lobes, of leaves.

pinnatopectinate *a.* pinnate with comb-like lobes.

pinniform *a.* feather-shaped or fin-shaped.

pinninervate *a.* with veins disposed like parts of a feather, *appl.* leaves.

Pinnipedia, pinnipeds *n., n.plu.* the seals, walruses and sealions.

pinnule *n.* (1) (*bot.*) a secondary branch of a compound pinnate leaf; (2) (*zool.*) projection from arms of sea lilies, two rows of which fringe each arm.

pinocytosis *n.* uptake of droplets of liquid into a eukaryotic cell by the process of endocytosis. *a.* **pinocytotic**.

pinocytotic vesicle intracellular membrane-bounded vesicle that is formed as a result of pinocytosis.

Pinophyta the gymnosperms *q.v.*

pinosome *n.* a vesicle containing material taken up by pinocytosis.

pinworm *see* Nematoda.

pioneer community the organisms that establish themselves on bare ground at the start of a primary succession.

pioneer species first species, usually mosses, lichens and microorganisms, that colonize a bare site as the first stage in a primary succession.

PIP₂ phosphatidylinositol bisphosphate *q.v.*

PI pathway phosphoinositide pathway *q.v.*

Piperales *n.* order of woody and herbaceous dicots comprising the families Piperaceae (pepper, the spice) and Saururaceae (lizard's tail).

piperidine *n.* alkaloid obtained from pepper, *Piper nigrum*.

piriform *a.* pear-shaped, *appl.* muscle of the buttocks, the musculus piriformis.

piroplasm(a) *n.* parasitic protozoan of the class Piroplasmea, which infect red blood cells. *Babesia* is the causal agent of red water fever of cattle. Piroplasms are characterized by an intracellular stage in which the parasite is not contained in a vacuole, but has its cell membrane in direct contact with the cytoplasm of the host cell.

piRNA Piwi-interacting RNAs *q.v.*

Pisces *n.* the fishes. *see* Appendix 3.

piscicolous *a.* living in fish.

pisciform *a.* like a fish.

piscine *a. pert.* fishes.

piscivorous *a.* fish-eating.

pisiform *a.* pea-shaped, *appl.* to a carpal bone.

pisohamate *a. appl.* a ligament connecting pisiform and hamate bones.

pisometacarpal *a. appl.* a ligament connecting pisiform bone with 5th metacarpal.

pistil *n.* (1) the carpels collectively when fused into a single structure; (2) each carpel with its stigma and style in flowers with carpels separate.

pistillate *a.* bearing pistils, *appl.* a flower bearing pistils but no stamens. *cf.* staminate.

pistillidium archegonium *q.v.*

pistillode *n.* a rudimentary or non-functional pistil.

pistillody *n.* the conversion of any organ of a flower into carpels.

pit *n.* minute, wall-free area in surface of plant cell. *see also* coated pit.

pitcher *n.* modification of a leaf or part of a leaf into a hollow organ to trap and digest insects, e.g. in the pitcher-plants *Nepenthes* and *Sarracenia*.

pit fields areas of small pits or depressions in the plant cell primary wall.

pith *n.* the central parenchymatous tissue present in the stems of some dicotyledons.

pithecanthropines *n.plu.* fossil hominids formerly placed in the genus *Pithecanthropus* and now considered to be the species *Homo erectus* and including Java man.

pit-pair the two corresponding pits in cell walls of adjacent plant cells and the plasma membrane separating them.

pituicyte *n.* glial cell in the neural lobe of pituitary body.

pituitary, pituitary body, pituitary gland in vertebrates, an endocrine gland attached to the undersurface of the brain below the hypothalamus by a short stalk. It secretes a number of important hormones such as adrenocorticotropin (ACTH), prolactin, the gonadotropins, thyroid-stimulating hormone, oxytocin and vasopressin. It consists of two parts, the glandular adenohypophysis and the neuroendocrine neurohypophysis. *alt.* hypophysis.

pivot joint joint in which movement is limited to rotation.

Piwi-interacting RNAs (piRNAs) type of small non-coding RNAs that act in a complex with Piwi proteins to silence gene expression. Their particular targets are thought to be retrotransposons and other genetic elements in germline cells.

pK_a a measure of the strength of an acid, given by $\log_{10}(1/K_a)$, where K_a is the acid dissociation constant.

PKA protein kinase A *q.v.*

PKB protein kinase B *q.v.*

PKC protein kinase C *q.v.*

PKG protein kinase G *q.v.*

placebo effect the effect where patients report alleviation of symptoms by a compound, even though the experimenter knows that the substance has no pharmacological effect.

placenta *n.* (1) (*bot.*) that part of the plant ovary where the ovules originate and remain attached until maturity, *see* Fig. 18 (p. 239); (2) (*zool.*) in mammals, a double-layered spongy, vascular tissue, formed from maternal and foetal tissue in wall of uterus, and in which the blood vessels of mother and foetus are in close proximity, allowing exchange of nutrients and respiratory gases. The placenta also produces various hormones involved in the maintenance of pregnancy, e.g. chorionic gonadotropin. Eutherian mammals (placental mammals) form various types of long-lived placenta involving both yolk sac and chorion. Marsupials form a short-lived placenta involving only the yolk sac.

placental *a.* (1) *pert.* a placenta; (2) secreted by placenta.

placental mammals mammals which develop a persistent placenta, the eutherians. *cf.* marsupials.

placentate *a.* having a placenta developed.

placentation *n.* (1) the manner in which ovules are attached to the plant ovary wall; (2) the type of placenta in mammals.

placode *n.* embryonic plate-like structure, e.g. lens placode, the structure from which the lens of the vertebrate eye develops.

Placodermi, placoderms *n., n.plu.* extinct class of early Devonian to early Carboniferous jawed primitive fish, with archaic jaw suspension, crushing dental plates and bony dermal plates on the head and thorax.

Placodontia, placodonts *n., n.plu.* extinct order of fully aquatic Triassic marine reptiles, having a short armoured body, some with a turtle-like carapace.

placoid *a.* plate-like.

placoid scale *see* denticle.

Placozoa *n.* phylum of extremely simple metazoans consisting of a single known species, *Trichoplax adhaerens*, a round flattened sac-like organism with a fluid-filled internal cavity, and cilia covering the body surface. It lacks differentiated tissues, organs, or any discernible head or tail or bilateral symmetry.

placula *n.* (1) an embryonic stage in urochordates, a flattened blastula with a small blastocoel; (2) developmental stage in the colonial protist *Volvox*.

plagioclimax *n.* stage in plant succession preceding the natural climax but which persists because of e.g. human intervention.

plagiopatagium *n.* (1) part of bat wing membrane posterior to arm; (2) part of the patagium between fore- and hindfeet in flying lemurs.

plagiosaurs *n.plu.* extinct order of advanced labyrinthodont amphibians, existing from Permian to Triassic times and having a very wide flat body and a body armour of interlocking plates.

plagiosere *n.* ecological succession deviating from its natural course as a result of continuous human intervention.

plagiotropism *n.* growth tending to incline a structure from the vertical plane to the oblique or horizontal as in lateral roots and branches. *a.* **plagiotropic**.

plague *n.* disease caused by the bacterium *Yersinia pestis*, transmitted by fleas (bubonic plague) or by breathing in infected droplets (pneumonic plague).

plagula *n.* ventral exoskeletal plate protecting the pedicle in spiders.

planar cell polarity (PCP) structural and molecular polarity of cells in the plane of a tissue, such as an epithelium, giving a directionality to the cells such that, e.g., hairs produced by the cells all point in the same direction.

planarian *n.* any member of the order Tricladida, free-living flatworms (platyhelminths) living in streams, ponds, lakes and the sea. They have a broad, flattened body, well-developed sense organs at the anterior end, and an intestine with three main branches.

planation *n.* the flattening of branched structures, as has occurred e.g. in the evolution of fronds of ferns.

***Planctomyces* group** group of the Bacteria distinguished on DNA sequence grounds, containing aerobic organisms that reproduce by budding and lack peptidoglycan in their cell walls. They are primarily aquatic. Examples are *Planctomyces* and *Pirella*.

planetism *n.* the condition of having motile or swarm cell stages. *a.* **planetic**.

planidium *n.* active migratory larva of certain insects.

planiform *a.* with nearly flat surface, *appl.* certain articulating surfaces.

plankter *n.* an individual planktonic organism.

planktohyponeuston *n.* aquatic organisms which gather near the surface at night but spend their days in the main water mass.

plankton *n.* the small marine or freshwater photosynthetic organisms (phytoplankton) and animals (zooplankton) drifting with the surrounding water. *a.* **planktic.**

Planktosphaeroidea *n.* class of hemichordates known only from a free-swimming pelagic larval form that resembles that of pterobranchs.

planktotrophic *a.* feeding on plankton.

planoblast *n.* the free-swimming medusa form of a hydrozoan.

planoconidium *n.* zoospore of certain fungi.

planocyte *n.* (1) a wandering or migratory cell; (2) planospore *q.v.*

planogamete *n.* a motile, usually flagellate gamete, esp. of algae and fungi.

planont *n.* (1) any motile spore, gamete or zygote; (2) the initial amoebula stage in some sporozoans; (3) a swarm spore produced in thick-walled or resting sporangia of certain lower fungi.

planospore *n.* a motile spore.

planozygote *n.* a motile zygote.

planta *n.* (1) the sole of the foot; (2) the 1st tarsal joint of insects.

Plantae, plants *n.*, *n.plu.* the plant kingdom, which contains all multicellular eukaryotic photosynthetic organisms along with, in some modern classifications, the green algae (Chlorophyta *q.v.*), the red algae (Rhodophyta *q.v.*), and the Glaucophyta (*q.v.*), on molecular evidence that they all derive from one endosymbiotic event in which an engulfed cyanobacterium-type cell became the chloroplast. Other photosynthetic eukaryotes are placed in other groups (e.g. the brown algae are included in the Chromista/Stramenopila (*q.v.*)). The multicellular plants comprise the bryophytes (mosses and liverworts), seedless vascular plants (ferns, club mosses, horsetails) and the seed plants (gymnosperms and angiosperms). Some authorities restrict the kingdom Plantae to the multicellular land plants and the Chlorophyta. *see* Appendix 1.

plantar *a. pert.* sole of the foot.

plantaris *n.* a muscle of the lower leg.

plant defence response the response of a plant to infection or wounding, which aims to localize damage and prevent the spread of infection, and which may include the production of phytoalexins, tissue death, lignification and callus formation.

plant genetic resources the genetic diversity present within plants, usually *appl.* crop plants and their wild relatives, and which is of potential use in plant breeding.

plant growth regulator, plant hormone a signal molecule produced by a plant that regulates the plant's growth and development. Plant hormones include auxins, ethylene, gibberellins, abscisic acid.

plantigrade *a.* walking with the whole sole of the foot touching the ground.

Planctomycetes *n.* phylum of the Bacteria distinguished on DNA sequence criteria, containing aerobic organisms that reproduce by budding and lack peptidoglycan in their cell walls. They are primarily aquatic (e.g. *Planctomyces*) but also contain the obligately parasitic chlamydiae, which are human pathogens.

plant pathogen any agent (bacterium, fungus, virus or mycoplasma) causing disease in plants.

plant pathology the study of the diseases of plants.

plant reoviruses *see* Reoviridae.

plant rhabdoviruses *see* Rhabdoviridae.

plantula *n.* a pulvillus-like adhesive pad on the tarsal joints of some insects.

plant viruses viruses that infect plants.

planula *n.* the ovoid free-swimming ciliated larva of coelenterates.

planum *n.* a plane or area, *appl.* some cranial bone surfaces.

plaque *n.* a clear area in a continuous sheet of cultured cells or bacterial growth, which indicates destruction of infected cells by a virus or phage.

plaque assay an assay for the presence and concentration of infectious virus in a sample, by counting the number of clear plaques in a continuous sheet of cultured susceptible cells infected with the virus sample. Each plaque is due to a single virus and its progeny destroying a group of contiguous cells.

plaque-forming unit (PFU) quantitative measure of the number of infectious virus particles in a given sample, as each infectious virus can give rise to a single clear plaque on infection of a continuous 'lawn' of bacteria. *see* plaque assay.

plasma *n.* the liquid part of body fluids, such as blood, milk or lymph, as opposed to suspended material such as cells and fat globules.

plasmablast *n.* B cell in lymph node that shows some features of a plasma cell.

plasma cell antibody-secreting cell of the immune system. It differentiates from a B lymphocyte after that cell is stimulated by encounter with specific antigen.

plasmacyte plasma cell *q.v.*

plasmacytoid dendritic cell *see* dendritic cell.

plasmagel *n.* the more solid part of the cytoplasm, which usually may be reversibly converted to the more fluid plasmasol.

plasmagene *n.* any gene other than those carried in the nucleus of a eukaryotic cell, such as the genes of mitochondria and chloroplasts.

plasmalemma *n.* plasma membrane, esp. in plants.

plasmalemmasome *n.* membranous structure formed in plant cells, consisting of tubules, cisternae and vesicles, between the plasma membrane and the cell wall.

plasma lipoproteins lipid transport proteins in blood, carrying cholesterol, triacylglycerols and phospholipids complexed with a protein (apolipoprotein). There are three main types: high-density lipoproteins (HDL), carrying free cholesterol and phospholipids, low-density lipoproteins (LDL), carrying mainly cholesteryl esters, and very low-density lipoproteins (VLDL) rich in triacylglycerols.

plasmalogen *n.* any of a class of phospholipids found chiefly in animal heart and brain and in myelin, and including e.g. phosphatidalethanolamine, phosphatidalcholine. They differ from the corresponding phosphoglycerides in having one of the fatty acid chains replaced by a fatty alcohol attached through an ether linkage.

plasma membrane membrane bounding the surface of all living cells, formed of a fluid lipid bilayer in which proteins carrying out the functions of enzymes, ion pumps, transport proteins and receptors are embedded. It regulates the entry and exit of most solutes and ions, few substances being able to diffuse through the lipid bilayer unaided. *alt.* cell membrane, plasmalemma. *see* Figs 7 (p. 105) and 26 (p. 392).

plasmasol *n.* the more fluid part of the cytoplasm, which may be reversibly converted to the more solid plasmagel.

plasmaspore *n.* an adhesive spore.

plasma thromboplastin antecedent (PTA) factor XI *q.v.*

plasmatic *a. pert.* blood plasma.

plasmatoparous *a.* developing directly into a mycelium upon germination, instead of into zoospores, *appl.* spores of some fungi, as downy mildew of grapes (*Plasmopora viticola*).

plasmid *n.* small DNA replicating independently of the chromosome in bacteria and unicellular eukaryotes such as yeasts. Plasmids are maintained at a characteristic stable number from generation to generation. They typically carry genes for antibiotic resistance, colicin production or the breakdown of unusual compounds. They are widely used in genetic engineering as vectors into which foreign genes are inserted for cloning or expression in bacterial or yeast cells.

plasmid immunity *see* immunity.

plasmid incompatibility the inability of a plasmid to be maintained in a bacterial cell containing another plasmid of the same type or same compatibility group.

plasmin *n.* an enzyme in blood plasma that degrades fibrin. EC 3.4.21.7. *alt.* fibrinolysin.

plasminogen *n.* the inactive precursor of the proteolytic enzyme plasmin.

plasminogen activator *see* tissue plasminogen activator.

plasmodesmata *n.plu.* cytoplasmic threads running transversely through plant cell walls and connecting the cytoplasm of adjacent cells. *sing.* **plasmodesma.** *see* Fig. 7 (p. 105).

plasmodial *n.* growing as, or *pert.* a plasmodium.

plasmodial slime moulds widely distributed eukaryotic heterotrophic soil microorganisms, classified as members of the

Mycetozoa, in which unicellular amoebo-flagellate swarm cells aggregate to form a plasmodium – a naked, creeping, often reticulate, multinucleate mass of protoplasm that can cover up to several square metres. This develops multicellular sporangia. *alt.* myxogastrid slime moulds, Myxomycetes, true slime moulds.

plasmodiocarp *n.* fruiting body of some plasmodial slime moulds in which the whole plasmodium develops into a sporangium.

Plasmodiophoromycota, plasmodiophorans *n.*, *n.plu.* phylum of protists, mainly plant pathogens (e.g. the causal organism of clubroot, *Plasmodiophora brassicae*), which have a multinucleate plasmodial vegetative stage and uninucleate motile zoospores. Formerly classified in the fungi.

plasmodium *n.* (1) in the plasmodial slime moulds, a multinucleate, fan-shaped mass of reticulate streaming protoplasm without a cell wall which may cover several square metres, and which forms the non-reproductive stage of the organism; (2) a genus of parasitic protozoans including the causal agents of malaria, e.g. *Plasmodium falciparum*. *a.* **plasmodial**.

plasmogamy *n.* (1) in protozoans, fusion of several individuals into a multinucleate mass; (2) fusion of cytoplasm without nuclear fusion.

plasmolysis *n.* the withdrawal of water from a plant cell by osmosis if placed in a strong salt or sugar solution, resulting in contraction of cytoplasm away from the cell walls.

plasmoptysis *n.* (1) emission of cytoplasm from tips of hyphae in host cells, in certain endomycorrhizas; (2) localized extrusion of cell contents through cell wall of bacteria.

plasmotomy *n.* division of a plasmodium by cleavage into multinucleate parts.

plastic *a.* (1) formative; (2) capable of change, *appl.* e.g. connections in brain.

plasticity *n.* the capacity for change under the influence of stimuli.

plastidome *n.* the plastids of a cell collectively.

plastids *n.plu.* family of plant cell organelles with double membranes that develop differently in different plant cell types but which in any given species all contain the same small genome. They develop from proplastids. Examples are amyloplasts, chloroplasts and leucoplasts.

plastochron *n.* the time interval between the successive similar developmental events at the shoot apex, e.g. the initiation of successive leaf primordia.

plastocyanin *n.* blue copper-containing protein in chloroplasts, a component of the photosynthetic electron-transport chain.

plastodeme *n.* a deme that differs from others phenotypically but not genotypically.

plastogamy *n.* union of distinct unicellular individuals with fusion of cytoplasm but not of nuclei.

plastogenes *n.plu.* cytoplasmic factors, controlled by or interacting with the nucleus, which determine the differentiation of plastids, now known to be genes carried in plastids.

plastolysis *n.* dissolution of mitochondria.

plastome *n.* the genome of a plastid.

plastoquinone (Q) *n.* any of various quinones found in chloroplasts as components of the photosynthetic electron-transport chain.

plastorhexis *n.* the breaking up of mitochondria into granules.

plastron *n.* (1) bony plate on underside of body of turtles and tortoises; (2) thin film of air trapped on the surface of the bodies of some aquatic insects. *a.* **plastral**.

plate count a technique for counting the number of bacteria in a sample by spreading a diluted sample over the surface of a solid agar plate and then counting the number of colonies that grow.

platelet *n.* non-nucleated small disc-shaped cell fragments present in blood, produced by fragmentation of megakaryocytes. They are involved in blood clotting, gathering at sites of damage and releasing clotting factor X and other active products.

platelet-derived growth factor (PDGF) glycoprotein produced by platelets and other cells, which stimulates the proliferation of cells of mesenchymal origin and which has been implicated in repair of the vascular system.

plate meristem a ground meristem in plant parts with a flat form such as a leaf, consisting of parallel layers of cells dividing only anticlinally in relation to surface.

platy *a. appl.* soil crumbs in which the vertical axis is shorter than the horizontal.

platybasic *a. appl.* the primitive chondrocranium with wide hypophysial fenestra.

platydactyl *a.* with flattened out fingers and toes, as certain amphibians.

Platyhelminthes, platyhelminths *n., n.plu.* a phylum of multicellular, acoelomate animals, commonly called flatworms. They are flattened dorsoventrally and are bilaterally symmetrical, and have the epidermis (ectoderm) and gut separated by a solid mass of tissue. They include the free-living Turbellaria and the parasitic Monogea (skin and gill flukes), Trematoda (gut, liver and blood flukes), and Cestoda (tapeworms).

platyhieric *a.* having a sacral index above 100.

platymyarian *a.* having flat muscle cells, *appl.* some nematode worms.

platypus *n.* the duck-billed platypus of Australia. *see* Monotremata.

platyrrhines *n.plu.* the New World monkeys.

platysma *n.* broad sheet of muscle between superficial fascia of neck.

platyspermic *a.* having seed which is flattened in transverse section.

play *n.* behaviour exhibited esp. by young animals in which they explore the environment and learn by trial and error during the time when life is fairly easy for them.

β-pleated sheet regular periodic secondary structure common in proteins, in which fully extended polypeptide chains lying adjacent to each other are held together by hydrogen bonding to form a sheet structure. In a parallel β-sheet the adjacent chains all run in the same direction, in an antiparallel β-sheet adjacent chains run in opposite directions.

pleckstrin homology domain (PH domain) protein domain present in some intracellular signalling proteins which serves to recruit them to the cell membrane by specifically binding to inositol phospholipids.

Plecoptera *n.* order of insects commonly called stoneflies, similar in many respects to mayflies (Ephemoptera), but with two 'tails' and hindwings larger than forewings.

plectenchyma *n.* tissue of interwoven cell filaments or hyphae in algae or fungi. *a.* **plectenchymatous**.

plectoderm *n.* outer tissue of fungal fruit body when composed of densely interwoven branched hyphae.

Plectomycetes *n.* a group of ascomycete fungi, commonly called the blue, green and black moulds from the colour of their conidia, and generally bearing asci in closed ascocarps (cleistothecia). The asexual (imperfect) stages of many plectomycetes are similar to those of *Aspergillus* and *Penicillium*.

plectonemic *a. appl.* double spiral having the two strands interlocked at each twist, as in the structure of DNA.

plectonephridia, plectonephria *n.plu.* nephridia formed of networks of fine excretory tubules lying on body wall and septa of certain oligochaete worms.

plectostele *n.* modified type of actinostele found in some club mosses of the genus *Lycopodium*, being deeply fissured in cross-section.

plectron *n.* hammer-like form of certain bacilli during sporulation.

plectrum *n.* styloid process of temporal bone. *alt.* malleus.

pleioblastic *a.* having several buds germinating at several points, as spores of certain lichens.

pleiochasium *n.* axis of a cymose inflorescence bearing more than two lateral branches. *alt.* **pleiochasial cyme**.

pleiocotyl *a.* having more than two cotyledons. *n.* **pleiocotyledony**.

pleiocyclic *a.* living through more than one cycle of activity, as a perennial plant.

pleiomerous *a.* having more than the usual number of parts, as of sepals and petals in a whorl. *n.* **pleiomery**.

pleiomorphic pleomorphic *q.v.*

pleiopetalous *a.* (1) having more than the normal number of petals; (2) having double flowers.

pleiophyllous *a.* having more than the normal number of leaves or leaflets.

pleiosporous polysporous *q.v.*

pleiotaxy *n.* a multiplication in the number of whorls, as in double flowers.

pleiotropy, pleiotropism *n.* multiple effects of a single gene which affects more than one phenotypic character. *a.* **pleiotropic**, *appl.* genes.

pleioxenous *a*. living on more than one host during life cycle, *appl.* parasites.

Pleistocene *n*. glacial and postglacial epoch of the Quaternary sub-era, following the Pliocene and preceding the Holocene and lasting from *ca.* 2.5 million years ago to 10,000 years ago. Plants and animals included many modern genera and species, as well as extinct species such as mastodons and sabre-toothed cats. The earliest known fossils of anatomically modern humans, *Homo sapiens*, date from *ca.* 155,000 years ago and agriculture developed *ca.* 10,000–12,000 years ago.

pleochromatic *a*. exhibiting different colours under different environmental orphysiological conditions.

pleogamy *n*. maturation, and therefore pollination, at different times, as of flowers of one plant.

pleometrosis *n*. colony foundation by more than one female as in some social hymenopterans. *a*. **pleometrotic**.

pleomorphic *a*. (1) being able to change shape; (2) existing in different shapes at different stages of the life cycle. *see also* polymorphic. *n*. **pleomorphism**.

pleon *n*. the abdominal region in crustaceans.

pleopod swimmeret *q.v.*

pleotropy pleiotropy *q.v.*

plerocercoid *a*. a solid elongated metacestode of certain tapeworms, esp. when found in fish muscle. *alt.* plerocestoid.

plerocestoid (1) metacestoid *q.v.*; (2) plerocercoid *q.v.*

plerome *n*. the core or central part of an apical meristem.

plerotic *a*. (1) completely filling a space; (2) *appl.* oospore filling oogonium.

plesiobiosis *n*. the close proximity of two or more nests of social insects, accompanied by little or no direct communication between the colonies.

plesiobiotic *a*. (1) living in close proximity, *appl.* colonies of ants of different species; (2) building contiguous nests, *appl.* ants and termites.

plesiometacarpal *a. appl.* condition of retaining the proximal elements of metacarpals, as in many cervids.

plesiomorphic *a*. in cladistics, *appl.* the original pre-existing member of a pair of homologous characters.

plesiomorphous *a*. having a similar form.

Plesiosauria, plesiosaurs *n*., *n.plu.* extinct order of Mesozoic reptiles that were fully aquatic with a barrel-shaped body and paddle-shaped limbs.

pleura *n*. membrane lining thoracic cavity (chest) and extending to cover surface of lungs, delimiting the pleural cavity.

pleural *a*. (1) *pert.* a pleura or pleuron, as pleural ganglia; (2) *appl.* recesses: spaces within pleural sac not occupied by lung; (3) *appl.* costal plates of carapace of turtles and tortoises.

pleural cavity body cavity occupied by lung(s).

pleurapophysis *n*. a lateral vertebral projection or true rib.

pleurethmoid *n*. the compound ethmoid and prefrontal of some fishes.

pleurilignosa *n*. rain forest.

pleurite *n*. a sclerite of the lateral piece of a body segment of arthropods.

pleuroblastic *a*. producing, having, or *pert.* lateral buds or outgrowths.

pleurobranchiae *n.plu.* gills arising from lateral walls of thorax of some arthropods, esp. crustaceans.

pleurocarpic *a*. with lateral fructifications.

pleuroccipital exoccipital *q.v.*

pleurocentrum *n*. a lateral element of vertebral centrum in many fishes and fossil amphibians.

pleurocerebral *a. pert.* pleural and cerebral ganglia, in molluscs.

pleurocystidium *n*. sterile hair in hymenium, in agaric fungi.

pleurodont *a. appl.* teeth that are attached to the inside surface of jaw, as opposed to the outer edge (acrodont) or in sockets (thecodont).

pleurogenous *a*. originating or growing from the side or sides.

pleuron *n*. one of the external pieces on side of body segments of arthropods.

pleuropedal *a. pert.* pleural and pedal ganglia, in molluscs.

pleuroperitoneum *n*. pleura and peritoneum combined, a membrane lining the body cavity in vertebrates without a diaphragm.

pleuropneumonia-like organism (PPLO) *see* mycoplasma.

pleuropodium *n*. glandular process on abdomen of some insect embryos.

pleuropophysis *n.* a lateral vertebral projection or true rib.

pleurosphenoid sphenolateral *q.v.*

pleurospore *n.* spore formed on sides of basidium.

pleurosteon *n.* lateral projection from sternum of young birds, later becoming costal process.

pleurosternal *a.* connecting or *pert.* pleuron and sternum, *appl.* thoracic muscles in insects.

pleurotribe *a. appl.* flowers whose anthers and stigma are so placed as to rub the sides of insects entering them, a device for cross-pollination.

pleurovisceral *a. pert.* pleural and visceral ganglia, of molluscs.

pleurum pleuron *q.v.*

pleuston *n.* free-floating organisms, esp. those possessing a gas-filled bladder or float.

plexiform *a.* (1) entangled or complicated; (2) like a network, *appl.* two of the layers of the vertebrate retina, the outer and inner plexiform layers, containing synaptic connections between photoreceptors, horizontal and bipolar cells and between bipolar, amacrine and ganglion cells respectively. Also *appl.* outermost layer of grey matter of cerebral cortex.

plexus *n.* network of interlacing vessels, nerves or fibres.

plexus myentericus Auerbach's plexus *q.v.*

plica *n.* (1) fold of skin, membrane or lamella; (2) corrugation of brachiopod shell.

plicate *a.* (1) pleated, folded like a fan, *appl.* leaf; (2) folded or ridged.

pliciform *n.* (1) resembling a fold; (2) disposed in folds.

Pliocene *n.* the earliest geological epoch of the Neogene period, following the Miocene and preceding the Pleistocene epochs, lasting from *ca.* 5 million years ago to 2.5 million years ago. Ice accumulated at the poles as the climate cooled. Grazing mammals diversified and became dominant as grasslands and savannas spread. All existing families, orders, and some genera of mammals had evolved by the end of the epoch. The australopithecines and early species of *Homo* evolved in Africa.

ploidy *n.* number of complete sets of chromosomes in a cell or organelle, e.g. haploid cells have one set of chromosomes, diploid cells have two.

plotophyte *n.* plant adapted for floating.

PLP pyridoxal phosphate *q.v.*

pluma *n.* contour feather of birds.

plumate *a.* like a plume.

Plumbaginales *n.* order of dicot herbs or small shrubs and comprising the family Plumbaginaceae (leadwort).

plume *n.* (1) feather or feather-like structure; (2) feathery structure at 'head' end of vestimentiferan tubeworms through which carbon dioxide, oxygen and sulphide are absorbed from sea water.

plumicone *n.* a spicule with plume-like tufts.

plumicorn *n.* horn-like tuft of feathers on bird's head.

plumigerous *a.* feathered.

plumiped *n.* bird with feathered feet.

plumose *a.* (1) feathery; (2) having feathers; (3) feather-like; (4) *appl.* feathers without tiny hooks on barbules, i.e. down feathers.

plumula plumule *q.v.*

plumulaceous plumulate *q.v.*

plumular *a. pert.* a plumule.

plumulate *a.* with a covering of down.

plumule *n.* (1) (*bot.*) the developing shoot of a plant embryo, comprising the epicotyl and young leaves; (2) (*zool.*) down feather of adult birds.

plumule sheath coleoptile *q.v.*

pluriascal *a. pert.* or containing several asci.

pluriaxial *a.* having flowers developed on secondary shoots.

plurilocular *a.* having several compartments.

pluriparous *a.* giving birth to, or having given birth to, a number of offspring.

pluripartite *a.* with many lobes or partitions.

pluripolar *a.* with several poles.

pluripotent *a. appl.* cells capable of developing into many different cell types, e.g. embryonic stem cells or the cells of the inner cell mass of the mammalian blastocyst, which give rise to all the cells of the body and some extraembryonic membranes, but which have already lost the ability to contribute to the placenta. *cf.* multipotent, totipotent.

plurisegmental *a. pert.* or involving a number of segments, *appl.* nerve conduction, reflexes.

pluriseptate *a.* with multiple septa.

pluriserial *a.* arranged in two or more rows.

pluristratose *a.* (1) arranged in a number of layers; (2) much stratified.

plurivalent multivalent *q.v.*

plurivorous *a.* feeding on several substrates or hosts.

plus end of a microtubule or actin filament, the end at which monomers are being added most rapidly.

pluteus *n.* free-swimming larval stage of sea urchins and brittle stars, characterized by long processes stiffened by tiny spicules.

PMN polymorphonuclear leukocyte *q.v.*

PMS pregnant mare's serum.

POMC pro-opiomelanocortin *q.v.*

PNA peptide nucleic acid *q.v.*

pneumathode *n.* an aerial or respiratory root.

pneumatic *a. appl.* bones penetrated by air-filled cavities connecting with lungs in birds.

pneumaticity *n.* the condition of having air cavities, as bones of flying birds.

pneumatized *a.* having air cavities.

pneumatocyst *n.* (1) the air bladder or swim float of fishes; (2) air cavity used as a float; (3) air bladder of bladderwrack.

pneumatophore *n.* (*bot.*) (1) air bladder of marsh or shore plants; (2) aerating outgrowth in some ferns; (3) a root rising above level of water or soil and acting as a respiratory organ in some trees; (4) (*zool.*) the air sac or float of siphonophores.

pneumatopyle *n.* a pore of a pneumatophore, opening above to exterior in certain siphonophores.

pneumatotaxis pneumotaxis *q.v.*

pneumococcus *n.* colloquial name for the bacterium *Streptococcus pneumoniae*, forms of which cause pneumonia in humans and some mammals, and in which the phenomenon of bacterial transformation was first discovered. *plu.* **pneumococci**. *a.* **pneumococcal**.

pneumocyte *n.* epithelial cell lining the air spaces of lungs.

pneumogastric *a. appl.* 10th cranial or vagus nerve, supplying pharynx, larynx, heart, lungs and viscera.

pneumostome *n.* aperture through which air passes in and out of the mantle cavity in land snails.

pneumotaxis *n.* reaction to stimulus consisting of a gas, esp. carbon dioxide in solution.

P-nucleotides nucleotides not encoded in the genome which are added at the junctions between gene segments during the rearrangement of immunoglobulin and T-cell receptor genes. They contribute to antigen receptor diversity.

poad *n.* a meadow plant.

Poales *n.* order of herbaceous monocots, the grasses, comprising the family Poaceae (Graminae).

poculiform *a.* cup-shaped or goblet-shaped.

pod *n.* (1) legume *q.v.*; (2) a husk; (3) the cocoon in which eggs are laid in locusts; (4) a school of fish in which the bodies of the individuals actually touch; (5) a group of whales.

podal *a.* pertaining to feet. *alt.* pedal.

podeon *n.* the slender middle part of abdomen of hymenopterans (e.g. ants, bees and wasps), uniting propodeon and metapodeon. *alt.* petiole.

podetiiform *a.* resembling a podetium.

podetium *n.* a stalk-like elevation, an outgrowth of thallus in certain lichens that bears an apothecium.

podeum podeon *q.v.*

podia *plu.* of podium.

Podicipediformes *n.* order of small water birds without webbed feet, including the grebes.

podite *n.* a crustacean walking leg.

podium *n.* foot or foot-like structure, such as a tube foot of echinoderm. *plu.* **podia**. *a.* **podial**.

podobranchiae, podobranchs *n.plu.* gills arising from the basal segments of thoracic appendages of certain arthropods. *alt.* foot-gills.

podocephalous *a.* having a head of flowers on a long stalk.

podoconus *n.* conical mass of endoplasm connecting the central capsule with the disc of rhizopod protozoans.

podocyst *n.* sinus in the foot or a small sac at tail end of some gastropod molluscs.

podocyte *n.* epithelial cell of Bowman's capsule of kidney, having numerous processes which rest on the basement membrane.

pododerm *n.* dermal layer of hoof, within the horny layer.

podogynium *n.* stalk supporting gynoecium.

podomere *n.* limb segment in arthropods.

podophthalmite *n.* in crustaceans, eye-stalk segment farthest from the head.

podosoma *n.* the body region of mites and ticks, bearing the four pairs of walking legs.

Podostemales *n.* order of dicots living in fast-flowing water and on rocks in rivers, with often filamentous or ribbon-like leaves and stems. Comprises the family Podostemaceae. *alt.* **Podostemonales**.

podotheca *n.* (1) a foot-covering as of reptiles or birds; (2) leg sheath of a pupa.

podsol, podzol *n.* grey forest soil, the soil type of cold temperate regions, and formed on heathlands and under coniferous forest.

poecilogyny *n.* in some parasitic nematodes, a process where two different forms of females are produced.

pogonion *n.* most prominent point on chin as represented on mandible (lower jaw).

Pogonophora, pogonophorans *n.*, *n.plu.* phylum of marine sessile worm-like invertebrates, with similarities to hemichordates, which live in chitin tubes. *alt.* beard worms. *see also* Vestimentifera.

poikilochlorophyllous *a.* completely losing and regaining their chlorophyll in response to changes in environmental conditions, *appl.* some angiosperms.

poikilocyte *n.* a distorted form of erythrocyte present in certain pathological conditions.

poikilohydrous *a.* becoming dormant in the dry season after losing most of their water, *appl.* some angiosperms.

poikilosmotic *a.* having internal osmotic pressure varying with that of the surrounding medium, as with salinity.

poikilothermic *a.* *appl.* animals whose body temperature varies with that of the surrounding medium. Although most poikilotherms are also ectothermic, the two terms are not synonyms and describe different aspects of thermoregulation. *n.* **poikilothermy**. *cf.* homoiothermic.

point accepted mutation percent accepted mutation *q.v.*

point mutation a mutation involving a change at a single base-pair in DNA.

poiser haltere *q.v.*

Poisson distribution a probability distribution that characterizes the probability of a variable number (0, 1, 2, 3, etc.) of discrete events occurring (independently of each other) in a given time.

pokeweed mitogen (PWM) lectin isolated from *Phytolacca americana*.

pol I, II, III eukaryotic RNA polymerases, *see* RNA polymerase I, II, III.

polar *a.* (1) situated at one end of cell or structure; (2) *appl.* flagella, one or a small group situated at one end of the cell; (3) *appl.* molecule, chemical group, or covalent bond in which the electrons are unevenly distributed, thereby making one end of the bond, group, or molecule electrically negative and the other positive. *cf.* nonpolar.

polar body the smaller of the products of meiotic division in human oocytes, one each being produced at 1st and 2nd meiotic divisions. They lie just outside the membrane of the larger cell, which will become the egg.

polar capsules capsules of spores containing coiled extrusible filaments, in certain sporozoans.

polar cartilage posterior portion of cartilages surrounding hypophysis in embryo, or independent cartilage in that region.

polar fibres polar microtubules *q.v.*

polar granule (1) centromere *q.v.*; (2) ribonucleoprotein granule in polar plasm of insect egg.

polarilocular *a.* *appl.* a cask-shaped spore with two cells separated by a partition having a perforation, of certain lichens.

polarity *n.* (1) having one end morphologically and/or functionally distinct from the other, e.g. some cells, fertilized eggs, and DNA and protein chains, *see also* planar cell polarity; (2) tendency to develop from opposite poles (as of plants); (3) having one face of a different composition and functionally distinct from the other, *pert.* epithelial cells, membranes; (4) existence of opposite qualities; (5) differential distribution or gradation along an axis; (6) in some bacterial transcription units, the fact that a nonsense mutation in an early part of the unit can prevent the expression of subsequent genes in the unit.

polarization *n.* (1) the development of polarity in a cell, organ or organism; (2) the setting up of an electrical potential difference across a membrane.

polarizing regions small parts of developing embryo which direct the development of various structures, esp. limbs. *alt.* organizing regions.

polar lobe cytoplasmic protrusion that appears during the early cleavages of many gastropod embryos.

polar microtubules microtubules arising from poles of spindle and which are not attached to kinetochores of chromosomes. *alt.* polar fibres.

polar nuclei nuclei at each end of angiosperm embryo, which later form secondary nucleus.

polar organ the cluster of pole cells at tail of early embryo in insects.

polar plasm granular cytoplasm at posterior end of fertilized insect egg, from which the pole cells are formed.

polar plates (1) two narrow ciliated areas produced in transverse plane, part of equilibrium apparatus of certain coelenterates; (2) areas of cytoplasm without centrosomes beyond poles of spindle in dividing nucleus of certain protozoans.

polar rings two ring-shaped masses of cytoplasm formed near poles of ovum after fertilization.

polar translocation or transport movement of materials through tissues or cells in one direction only.

pole cell (1) large cell that buds off rows of smaller cells, as in annelid and mollusc embryos; (2) any of a group of large cells distinguishable at posterior end of embryos of some insects in the earliest stages, and from which the germ cells are produced.

Polemoniales *n.* order of dicot trees, shrubs, vines and herbs, including the families Boraginaceae (borage), Convolvulaceae (morning glory), Cuscutaceae (dodder) and others.

pole plasm cytoplasm at the posterior end of eggs of some insects which is involved in specifying the germ cells.

pole plates end plates or masses of cytoplasm at spindle poles in some protozoan mitosis.

Polian vesicles interradial vesicles opening into ring vessel of ambulacral system of most starfish and sea cucumbers.

poliovirus *n.* small icosahedral nonenveloped single-stranded RNA virus of the genus *Enterovirus* of the family Picornaviridae, which causes poliomyelitis.

pollakanthic *a.* having several flowering periods.

pollarding *n.* method of tree management in which trees are cut back at some distance above the ground, producing a crown of long straight shoots on a relatively short single trunk. *cf.* coppicing.

pollen *n.* fine powder produced by anthers and male cones of seed plants, composed of pollen grains which each enclose a developing male gamete.

pollen analysis quantitative and qualitative determination of pollen grains preserved in deposits such as peat, from which the former vegetation of the area can be reconstructed.

pollen basket the pollen-carrying hairs at back of tibia on worker bees.

pollen brush enlarged hairy tarsal joint of bee leg, which brushes pollen from anther.

pollen case *see* theca.

pollen chamber pit formed at apex of nucellus in ovule below micropyle, into which the germinating pollen tube grows.

pollen comb comb-like structure of bristles on leg of bee below pollen basket, serving to sweep pollen into basket.

pollen flower a flower without nectar, which attracts pollen-feeding insects.

pollen grain the two-celled haploid microspore of a seed plant, produced from anthers in angiosperms. It contains a tube cell, which forms the pollen tube on germination, and a generative cell that divides to form two sperm.

pollen profile vertical distribution of preserved pollen grains in a deposit such as peat, giving an indication of the former vegetation at different periods.

pollen sac the cavity in anther in which pollen is produced, being the microsporangium of seed plants.

pollen spectrum the relative numerical distribution or percentage of pollen grains of different species preserved in a deposit.

pollen tube tube that develops from a pollen grain after attachment to stigma. It grows down through the style towards an ovule, entering the ovule at the micropyle and delivering male gametes (sperm) to fuse with the female gamete.

pollen tube nucleus the vegetative nucleus in the pollen grain.

pollex *n.* the thumb, or corresponding innermost digit of the normal five of a forelimb.

pollinarium *n.* in orchids, the mass of pollen (pollinium) and its stalk and adhesive disc.

pollination *n.* transfer of pollen from anther (in angiosperms) or male cone (gymnosperms) to stigma or female cone respectively.

pollination drop mucilaginous drop exuded from micropyle and which detains pollen grains in gymnosperms.

pollinator *n.* any insect or other animal that carries pollen from one flower to another, or from anther to carpel in the same flower.

polliniferous, pollinigerous *a.* (1) pollen-bearing; (2) adapted for transferring pollen.

pollinium *n.* an agglutinated mass of pollen in orchids and some other flowers.

pollution *n.* any harmful or undesirable change in the physical, chemical or biological quality of air, water or soil as a result of the release of e.g. chemicals, radioactivity, heat, large amounts of organic matter (as in sewage). Usually *appl.* changes arising from human activity although natural pollutants, e.g. volcanic dust, sea salt, are known.

polospore *n.* fossil pollen.

polster *n.* a low compact perennial or cushion plant.

poly- prefix derived from Gk *polys*, many.

poly(A) polyribonucleotide composed exclusively of adenylate residues. Poly(dA) is the corresponding polydeoxyribonucleotide. *see also* poly(A) tail.

poly(A)⁺ denotes a messenger RNA carrying a poly(A) tail.

polyacrylamide gel electrophoresis (PAGE) widely used electrophoretic technique for separating proteins or nucleic acid fragments on a polyacrylamide gel, separation being primarily on the basis of size. *alt.* gel electrophoresis. *see also* SDS-PAGE.

polyadelphous *a.* having stamens united by filaments into more than two bundles.

polyadenylation *n.* the addition of a poly(A) tail to eukaryotic mRNA precursors in the nucleus.

polyamine *n.* any of a group of compounds composed of one or more basic units of two amino groups joined by a short hydrocarbon chain, formed from amino acids (in some cases during breakdown of protein by bacteria) and including putrescine, cadaverine and spermidine.

polyandrous *a.* (1) mating with more than one male at a time; (2) having 20 or more stamens. *n.* **polyandry**.

polyarch *a. appl.* roots and stems in which the primary xylem of the stele is made up of numerous alternating bundles of phloem and xylem.

polyarthric multiarticulate *q.v.*

poly(A) tail stretch of polyadenylate residues found at the 3′ ends of many eukaryotic messenger RNAs. It is added in the nucleus by the enzyme poly(A) polymerase after transcription and may be involved in stabilizing the mRNA.

polyaxon *n.* type of spicule laid down along numerous axes.

polyblastic *a.* having spores divided by a number of septa, *appl.* lichens.

poly(C) polyribonucleotide composed exclusively of cytidylate residues, poly(dC) being the corresponding polydeoxynucleotide.

polycarp *n.* gonad of some ascidians, on inner surface of mantle.

polycarpellary *a.* with a compound gynoecium consisting of many carpels.

polycarpic, polycarpous *a.* (1) with numerous, usually free carpels; (2) producing seed season after season.

polycentric *a.* (1) having several centromeres, *appl.* a chromosome; (2) with several centres of growth and reproduction, *appl.* fungal thallus.

polycercous, polycercoid *a. appl.* bladderworms developing several cysts, each with a head, as in an echinococcus.

Polychaeta, polychaetes *n., n.plu.* the bristle worms, a class of mainly marine annelid worms, e.g. ragworms and lugworms. They possess parapodia bearing numerous chaetae which are used for crawling, and have a pronounced head bearing tentacles, palps and often eyes.

polychasium *n.* cymose branch system when three or more branches arise about the same point.

polychlorinated biphenyls (PCBs) large group of toxic synthetic lipid-soluble chlorinated hydrocarbons, which are used in various industrial processes and which have become persistent and ubiquitous environmental contaminants which can be concentrated in food chains.

polychromatic *a.* with several colours, as pigmented areas.

polychromatophil *a.* (1) having a staining reaction characterized by various colours; (2) *appl.* erythrocytes with small haemoglobin content.

polycistronic *a. appl.* mRNA containing more than one polypeptide-coding sequence.

polyclad *n.* member of the order Polycladida, marine turbellarian flatworms with a very broad flattened leaf-shaped body, and large numbers of eyes at the anterior end.

polyclimax *n.* climax community consisting of several different climax associations, none of which shows a tendency to give way to any other.

polyclonal *a.* (1) *appl.* tissue or structure derived from a number of founder cells; (2) derived from several cell clones, *appl.* antigen-specific antibodies that have been obtained by immunization of an animal and therefore represent the products of different clones of antibody-producing cells specific for that antigen. *cf.* monoclonal antibody.

polyclonal mitogens substances that stimulate lymphocytes to proliferate without regard for their antigen specificity.

polyclone *n.* discrete area of tissue derived from several clones of cells.

polycotyledon *n.* plant with more than two cotyledons.

polycotyledonary *a.* having placenta in many divisions.

polycotyledony *n.* a great increase in the number of cotyledons.

polycrotism *n.* condition of having several secondary elevations in pulse curve.

polycyclic *a.* (1) having many whorls (of flowers); (2) with vascular system forming several concentric cylinders. *appl.* plant stems; (3) (*chem.*) having more than one aromatic ring in its chemical structure, *appl.* organic compounds;.

polycyclic aromatic hydrocarbon (PAH) any of several multi-ringed aromatic compounds that are found in e.g. soot, coal tar, cigarette smoke and barbecued meat. Examples include pyrene and benzo(a)pyrene. Some of these compounds are known carcinogens.

polycystid *a.* partitioned off.

polydactyly *n.* condition of having more than five fingers or toes.

polydelphic *a.* having more than one set of ovaries, oviducts and uteri, *appl.* nematodes.

polyderm *n.* protective tissue in plants made up of alternating layers of endoderm and parenchyma cells.

polydesmic *a. appl.* a type of fish scale growing by apposition at margin, and made up of several units of fused small bony teeth covered with dentine.

Polydnaviridae *n.* family of enveloped insect DNA viruses. Each particle contains many double-stranded DNAs of variable molecular weight.

polydomous *a. pert.* single colonies (of social insects) that occupy more than one nest.

polyembryony *n.* (1) development of several embryos within a single ovule; (2) the case where a zygote gives rise to more than one embryo, as in identical twins.

polyethism *n.* division of labour amongst members of an animal society, as in the social insects.

polyethylene glycol (PEG) chemical used to induce cell fusion in the formation of somatic cell hybrids.

poly(G) polyribonucleotide composed exclusively of guanidylate residues, poly(dG) being the corresponding polydeoxynucleotide.

polygalacturonan *n.* polysaccharide made of galacturonic acid residues linked together. It is a component of plant cell walls, found as part of the complex polysaccharide of the pectin fraction. It is hydrolysed to galacturonic acid by the enzyme polygalacturonase. *alt.* galacturonan.

polygalacturonase *n.* enzyme that degrades polygalacturonans to galacturonic acid in plant cell walls during fruit ripening.

Polygales *n.* order of dicot herbs, shrubs and small trees, and including the families Polygalaceae (milkwort) and others.

polygamy *n.* (1) having more than one mate at a time; (2) (*bot.*) having male, female or hermaphrodite flowers on the same plant. *a.* **polygamous**.

polygenes *n.plu.* genes which each have a small effect and which collectively produce a multifactorial or polygenic phenotypic trait.

polygenesis *n.* (1) derivation from more than one source; (2) origin of a new type at more than one place or time.

polygenic *a.* (1) *appl.* phenotypic characters (such as height in humans) that are determined by the collective effects of a number of different genes. *see also* quantitative variation; (2) *appl.* gene cluster containing several genes encoding related but differing proteins of similar function.

polygenic inheritance inheritance of phenotypic characters (such as height and eye colour in humans) that are determined by the collective effects of several or many different genes. *alt.* multifactorial inheritance, *see also* complex trait.

polyglucosan *n.* polymer of glucose units. *alt.* glucan.

Polygonales *n.* order of herbaceous dicots, rarely trees, and comprising the family Polygonaceae (buckwheat).

polygoneutic *a.* rearing more than one brood in a season.

polygynoecial *a.* having multiple fruits formed by united gynoecia.

polygynous *a.* (1) consorting with more than one female at a time; (2) *appl.* flowers with numerous styles. *n.* **polygyny**.

polyhaline *a. appl.* brackish water of salinity 18–30 parts per thousand, approaching that of sea water.

polyhaploid (1) *a. pert.* the gametic chromosome number of a polyploid organism; (2) *n.* a haploid organism derived from a normally polyploid species.

polyhybrid *n.* a hybrid heterozygous for many genes.

polyhydric *a. appl.* alcohol or acid with three, four or more hydroxyl groups.

poly-Ig receptor receptor on the basal membrane of epithelial cells which binds dimeric IgA (and to a lesser extent multimeric IgM) and carries it across the epithelium. Part of the receptor becomes secretory component.

polyisomeres *n.plu.* parts all homologous with each other, as leaves of plants of the same species.

polyisoprenoid *n.* any of a variety of compounds containing long-chain polymers of isoprene. *see also* isoprenoid.

polykaryocyte (1) megakaryocyte *q.v.*; (2) multinucleate cell, produced in some viral infections by virus-induced fusion of infected cells.

polykaryon *n.* nucleus with more than one centriole.

polykont multiflagellate *q.v.*

polylecithal *a. appl.* ova containing relatively large amounts of yolk.

polylepidous *a.* having many scales.

polymastia polymastism *q.v.*

polymastigote multiflagellate *q.v.*

polymastism *n.* occurrence of more than the normal number of mammae.

polymegaly *n.* occurrence of more than two sizes of sperm in one animal.

polymeniscous *a.* having many lenses, as compound eye.

polymer *n.* large organic molecule made up of repeating subunits covalently linked together.

polymerase *see* DNA polymerase, RNA polymerase.

polymerase chain reaction (PCR) technique for selectively and rapidly replicating a particular stretch of DNA *in vitro* to produce a large amount of a particular DNA sequence. Uncloned genomic DNA can be used as the starting material, thus obviating the need for a DNA cloning step in microorganisms.

polymeric *a.* (1) *appl.* system of independently segregating genes that are additive in their effects on the same phenotypic character; (2) composed of multiple subunits, *appl.* proteins.

polymerization *n.* formation of a polymer from smaller subunits.

polymerous *a.* consisting of many parts or members.

polymitosis *n.* excessive cell division.

polymorph polymorphonuclear leukocyte *q.v.*

polymorphic *a.* (1) existing in two or more different forms within a species or population, *appl.* genes, phenotypic traits, individuals of different morphological form; (2) showing a marked degree of variation in body form during the life cycle or within the species; (3) *pert.* or containing variously shaped units (i.e. cells, or individuals in a colony).

polymorphic locus genetic locus with two or more alleles, conventionally defined as a locus at which the most common homozygote has a frequency of less than 90% in a given population, or at which the variant allele has a frequency greater than 0.01.

polymorphism *n.* (1) the existence within a species or a population of different forms of individuals; (2) occurrence of different forms of, or different forms of organs in, the same individual at different periods of life. *see also* enzyme polymorphism, genetic polymorphism.

polymorphonuclear leukocyte type of phagocytic white blood cell characterized by irregularly shaped nucleus, principally refers to the neutrophil.

polymyarian *n.* having more than five longitudinal rows of muscle cells, *appl.* some nematodes.

polynemic *a.* consisting of several strands.

Polynesian region part of the Palaeotropical floristic kingdom, consisting of the Pacific islands east of Melanesia, except for the Fijian and Hawaiian islands.

Polynesian subkingdom a major division of the Palaeotropical floristic kingdom, consisting of the Hawaiian, New Caledonian, Melanesian and Micronesian, and Polynesian regions.

Polynesian subregion zoogeographical area, part of the Australian region, consisting of the islands of Polynesia and Melanesia.

polynuclear, polynucleate *a.* with several or many nuclei. *alt.* multinucleate.

polynucleotide *n.* unbranched chain of nucleotides linked through phosphodiester bonds between the sugar phosphates. *see* Fig. 31 (p. 501).

polynucleotide kinase enzyme that adds a phosphoryl group to the 5′-end of a DNA strand, used to label ^{32}P-DNA for sequencing by Maxam and Gilbert's method. EC 2.7.1.78, *r.n.* polynucleotide 5′-hydroxy-kinase.

polynucleotide phosphorylase enzyme catalysing the synthesis of polyribonucleotides from ribonucleoside diphosphates without the need for a DNA or RNA template. EC 2.7.7.8, *r.n.* polyribonucleotide nucleotidyltransferase.

polyoestrous *a.* having a succession of periods of oestrus in one sexual season. *cf.* monoestrous.

Polyomaviridae *n.* family of small, non-enveloped, double-stranded DNA viruses that includes polyoma and SV40 (simian virus 40). Polyoma and SV40 are oncogenic in the cells of certain species.

polyp *n.* (1) sedentary individual or zooid of a colonial animal; (2) in coelenterates, an individual having a tubular body, usually with a mouth and ring of tentacles on top, like a miniature sea anemone; (3) small stalked outgrowth of tissue from a mucous surface such as the intestine, usually benign but may become malignant.

polyparium, polypary *n.* the common base and connecting tissue of a colony of polyps.

polypeptide *n.* chain of amino acids linked together by peptide bonds. A polypeptide chain is the basic structural unit of a protein, some proteins consisting of one, some of several polypeptides. The polypeptide chains of proteins are synthesized on the ribosomes, using messenger RNA as a template. *see also* genetic code, protein.

polypeptide backbone the continuous chain of covalently linked repeating carbon and nitrogen atoms in a protein, from which the amino-acid side groups project.

polypetalous *a.* having separate, free or distinct petals.

polyphagous *a.* (1) eating various kinds of food; (2) of insects, using many different food plants.

polyphenism *n.* the occurrence in a population of several phenotypes which are not genetically controlled.

525

polyphenol oxidase any of several enzymes which catalyse the oxidation of tyrosine to dioxyphenylalanine using molecular oxygen. EC 1.14.18.1, *r.n.* monophenol monooxygenase.

polypheny pleiotropy *q.v.*

polyphyletic *a. appl.* a taxonomic group that includes several different lines of descent.

polyphyllous *a.* many-leaved.

polyphyodont *a.* having many successive sets of teeth.

Polyplacophora *n.* class of molluscs that includes the chitons *q.v.*

polyplanetic *a.* having several motile phases with intervening non-motile or resting stages.

polyplastic *a.* capable of assuming many forms.

polyploid (1) *a.* having more than two sets of chromosomes, e.g. triploid with 3, tetraploid with 4; (2) *n.* an organism with more than two chromosome sets per somatic cell.

polyploidy *n.* the polyploid condition, which may be the normal state of the somatic tissues of the whole organism, or a reduplication of chromosome number found in only some tissues or cells. Polyploidy can be induced artificially by chemicals such as colchicine or β-naphthol.

polypneustic *a. appl.* lateral lobes bearing multiple spiracle pores, in some insects.

polypoid *a.* resembling a polyp.

polypores *n.plu.* group of basidiomycete fungi including those commonly known as the bracket fungi, in which the fruit body is typically leathery, papery or woody with tubes or gills on the underside. Many bracket fungi grow on wood. This group also includes fungi with smooth, ridged, warty or spiny surfaces bearing the hymenium, and includes the coral fungi and cantharelles. *cf.* agarics.

polyprotein *n.* long polypeptide chain which is subsequently cleaved into separate functional proteins or peptides.

polyprotodont *a.* with four or five incisors on each side of upper jaw, and one or two fewer in lower.

polyrhizal multiradicate *q.v.*

polyribosome polysome *q.v.*

polysaccharide *n.* any of a diverse class of high-molecular-weight carbohydrates formed by the covalent linking together by condensation of monosaccharide, or monosaccharide derivative, units into linear or branched chains. They include homopolysaccharides (composed of one type of monosaccharide only) and heteropolysaccharides (composed of a mixture of different monosaccharides). Found as storage products (e.g. starch and glycogen) and structural components of cell walls (e.g. cellulose, xylans and arabinans), and as components of glycoconjugates. *see also* proteoglycan. *alt.* glycan.

polysaprobic *a.* (1) *appl.* category in the saprobic classification of river organisms comprising those that can live in water heavily polluted with organic pollutants, in which decomposition is mainly anaerobic, e.g. sewage fungus, bloodworms and the rat-tailed maggot (*Eristalis tenax*); (2) *appl.* aquatic habitats with heavy pollution by organic matter with little or no dissolved oxygen, the formation of sulphides, abundant bacteria, but few animals feeding on them or on the decaying matter. *cf.* α-mesosaprobic, β-mesosaprobic, oligosaprobic.

polysepalous *a.* having free or distinct sepals.

polysiphoneous, polysiphonic *a.* (1) consisting of several rows of cells, *appl.* thallus of red or brown algae; (2) consisting of several tubes.

polysome *n.* aggregate of ribosomes on messenger RNA during protein synthesis. *alt.* polyribosome.

polysomic *a. appl.* cells or organisms carrying more than the usual number of any particular chromosome.

polysomitic *a.* (1) having many body segments; (2) formed from fusion of primitive body segments.

polysomy *n.* condition in which more than two copies of any particular chromosome are present in normally diploid cells.

polyspecificity *n.* the ability to bind several different antigens, *appl.* antibodies.

polyspermous *a.* having many seeds.

polyspermy *n.* entry of several sperms into one ovum.

polyspondyly *n.* the condition in which the vertebrae have more than two centra, found in some fishes.

polysporocystid *a. appl.* oocyst of sporozoans when more than four sporocysts are present.

polysporous *a.* many-seeded or many-spored.

polystachyous *a.* with numerous spikes.

polystely *n.* arrangement of vascular tissue in several steles, each containing more than one vascular bundle. *a.* **polystelic**.

polystemonous *a.* having stamens more than double the number of petals or sepals.

polystichous *a.* arranged in numerous rows or series.

polystomatous *a.* having many pores, openings or mouths.

polystomium *n.* a suctorial mouth in certain jellyfish.

polystylar *a.* having many styles.

polysymmetrical *a.* divisible through several planes into bilaterally symmetrical portions.

poly(T) polynucleotide composed exclusively of thymidylate residues.

polytene chromosome type of giant chromosome formed in some tissues (e.g. salivary gland) of fly larvae by successive replications of paired homologous chromosomes without separation of the new DNA. It bears a characteristic pattern of dark and light bands visible in the light microscope.

polyteny *n.* the duplication of haploid chromosome content in a polytene chromosome, the degree of polyteny being the number of haploid chromosomes contained in each giant chromosome.

polythalamous *a.* (1) aggregate or collective, as *appl.* to fruits; (2) *appl.* shells made up of many chambers formed successively.

polythelia *n.* occurrence of supernumerary nipples.

polythermic *a.* tolerating relatively high temperatures.

polythetic *a. appl.* a classification based on many characteristics, not all of which are necessarily shown by every member of the group.

polytocous, polytokous *a.* (1) producing several young at a birth; (2) fruiting repeatedly.

polytomous *a.* (1) having more than two secondary branches; (2) with a number of branches originating in one place.

polytopic *a.* occurring or originating in several places.

polytrichous *a.* (1) having the body covered with an even coat of cilia, as some ciliate protozoans; (2) having many hair-like outgrowths.

polytrochal *a.* having several circlets of cilia between mouth and posterior end.

polytrophic *a.* (1) *appl.* ovariole in which nutritive cells are enclosed in oocyte follicles; (2) nourished by more than one organism or substance; (3) obtaining food from many sources; (4) eutrophic *q.v.*

polytropic pantropic *q.v.*

polytypic *a.* (1) having or *pert.* many types; (2) *appl.* species having geographical subspecies; (3) *appl.* genus having several species.

poly(U) polyribonucleotide composed exclusively of uridylate residues.

polyunsaturated *a. appl.* fatty acids with more than one C=C bond in their hydrocarbon chain.

polyvoltine *a.* producing several broods in one season.

polyxenous *a.* adapted to life in many different hosts, *appl.* parasites.

polyxylic *a.* having many strands of xylem and several concentric vascular rings.

Polyzoa, polyzoan *see* Bryozoa.

polyzoic *a. appl.* a colony of many zooids.

polyzooid *n.* an individual in a polyzoan colony.

pome *n.* fruit derived from a compound inferior ovary in which the fleshy portion is largely the enlarged base of the perianth or receptacle, e.g. apples, pears.

ponderal *a.* (1) *pert.* weight; (2) *appl.* growth by increase in mass.

poneroid complex one of two major taxonomic groups of ants, exemplified by the subfamily Ponerinae.

pongid *n.* any great ape other than the gibbons or siamang, i.e. chimpanzee, gorilla and orangutan.

pons *n.* (1) structure connecting two parts; (2) broad band of nerve fibres, pons Varolii, in the mammalian brain, connecting the two sides of the cerebellum and medulla oblongata.

pontic, pontile *a. pert.* a pons.

pontine *a. pert.* pons Varolii.

pooid *a. appl.* grasses of the genus *Poa.*

popliteal *a. pert.* region behind and above knee joint.

population *n.* a group of individuals of the same species living in a certain defined area.

population crash sharp reduction in the population of a species when its numbers exceed the carrying capacity of the habitat. *alt.* dieback.

population density number of individuals, usually with reference to a given species, living in a specified area.

population dispersion the distribution of the members of a population throughout its habitat.

population distribution the variation in population density over a given area.

population dynamics the changes in the structure of a population over time, i.e. the changes in the relative numbers of individuals of particular ages, different sexes, or different forms.

population ecology the study of factors influencing the numbers and structure of a given population.

population genetics the study of how genetic principles apply to groups of interbreeding individuals (a population) as a whole.

porcellanous *a.* resembling porcelain, white and opaque, *appl.* calcareous shells, as of foraminiferans and some molluscs.

pore *n.* a minute opening or passage.

pore canals minute spiral tubules passing through the cuticle, but not the epicuticle, of insects.

pore cell in sponges, cells of the outer layer that are perforated by a pore through which water enters.

pore chains in wood, extensive radial, tangential or oblique groups of xylem vessels, as seen in wood cross-sections.

pore complex complex structure present at pores in the nuclear membrane.

pore organ structure surrounding canal for excretion of mucilage through pores, in desmids.

pore space in soil, the spaces between particles of soil collectively.

poricidal *a.* dehiscing by valves or pores.

Porifera *n.* phylum of simple multicellular animals, commonly called sponges, with a simple body enclosing a single central cavity (in the simple sponges) or penetrated by numerous interconnected cavities. The body wall consists of an outer layer of epithelium separated from an inner layer of ciliated choanocytes (feeding cells) by a mesogloeal layer. There are no nerve or muscle cells. Water is drawn into the internal cavities through pores (ostia), food particles are taken up by the choanocytes, and the water flows out through a large pore (the osculum). There are three classes: the Calcarea, the calcareous sponges (e.g. *Leucosolenia*), which have spicules of calcium carbonate embedded in the mesogloea and projecting to the outside; the Hexactinellida, the glass sponges (e.g. *Euplectella*, Venus's flower basket), with silica spicules; and the Demospongia, which includes some species with silica spicules and some species without, and which often have the body wall strengthened by a tangled mass of fibres (e.g. the bath sponge *Spongia*).

poriferous *a.* furnished with numerous openings.

poriform *see* poroid.

porin *n.* protein which forms large pores in outer membrane of Gram-negative bacteria, and which enables small polar molecules to diffuse through the membrane.

porocyte *n.* phagocytic cell lining a pore in sponges.

porogamy *n.* entrance of a pollen tube into ovule by micropyle.

poroid *a.* (1) like a pore or pores; (2) having pore-like depressions, *alt.* poriform; (3) *n.* minute depression in theca of dinoflagellates and diatoms.

porophyllous *a.* having or *appl.* leaves with numerous transparent spots.

porose *a.* having or containing pores.

porphyrin ring *n.* structure formed of four 5-sided carbon and nitrogen rings (pyrrole rings) linked together into a square shape and which is the basis of porphyrin derivatives such as haem and chlorophyll.

porphyrophore *n.* a reddish-purple pigment-bearing cell.

porphyropsin *n.* light-sensitive visual pigment (rhodopsin) found in freshwater fish and amphibian tadpoles.

porpoise *n.* marine mammal, a member of the family Phocoenidae of the suborder

Odontoceti (toothed whales) of the Cetacea (*q.v.*). Porpoises are smaller and dumpier than dolphins and lack the typical dolphin 'beak'. They also cannot leap completely out of the water like a dolphin.

porrect *a.* extended outwards.

porta *n.* any gate-like structure, such as the transverse fissure of liver.

portal *a.* (1) *appl.* veins leading from alimentary canal, spleen and pancreas to liver; (2) also a vascular system to kidney in lower vertebrates.

ports, porters membrane transport proteins *q.v.*

positional cloning the isolation of a gene starting from a knowledge of its position on the chromosome.

positional information in embryonic development, the developmental signals a cell receives by virtue of its position in the embryo, and which direct its further development. Positional information may be given to the cell by means of gradients of a chemical signal emanating from a particular source, or by cell-surface molecules on neighbouring cells, or by the composition of the extracellular matrix.

position effect the influence that a gene's location on the chromosome has on its activity. Position effect can be demonstrated by the changes in gene expression that occur when genes are moved to different positions on the chromosome. The effect on gene expression is due to differences in the state of the chromatin at different positions on the chromosome.

position-effect variegation variegated appearance of the *Drosophila* eye and yeast colonies due to mutation that moves a pigment gene into heterochromatin during development and silences its expression in some cells.

position-specific score matrix (PSSM) a matrix that gives the probability of a given amino acid being found at a particular position in a set of aligned sequences. It is used to construct a profile of a protein family that can be used to search databases for other members of the family.

positive control type of control of gene expression in which the regulatory protein promotes the binding of RNA polymerase and transcription. *cf.* negative control.

positive cooperativity the situation when the binding of one molecule to a protein tends to promote the binding of further molecules at other sites on the protein. *cf.* negative cooperativity.

positive feedback type of control mechanism in which the end-product of a particular pathway or process activates the first step in the pathway or in other pathways that contribute to the overall process.

positive interference the effect that the occurrence of one chiasma decreases the likelihood of another forming nearby.

positive reinforcement a stimulus, or series of stimuli, which is pleasant to an animal and increases its response.

positive selection selection process undergone by developing T lymphocytes in thymus. Only cells whose T-cell receptors can recognize the MHC molecules of the body are selected for survival and further development.

positive-strand RNA viruses RNA viruses (e.g. poliovirus) in which the genome has the same base sequence as the messenger RNA. During virus replication a complementary copy of the genome RNA is made first, which then acts as a template for mRNA synthesis.

positive supercoiling twisting of a circular DNA molecule in the same direction as that of the right-handed double helix.

positive taxis, positive tropism tendency to move or grow towards the source of the stimulus.

positron emission tomography *see* PET scan.

possum *n.* any of around 60 species of small arboreal marsupial mammals, native to Australia and New Guinea. The brush-tailed possum (*Trichosurus vulpecula*) was introduced into New Zealand in the 19th century, where it has become a major pest and a threat to native wildlife and vegetation. The South American opossums (*q.v.*) are often called possums, but are only distant relatives of the Australasian possums.

post- prefix derived from L. *post*, after, signifying situated behind, the hindmost part of an organ or structure, or occurring after.

postabdomen *n*. (1) in scorpions, the metasoma or posterior narrower five segments of abdomen; (2) anal tubercle in spiders.

postalar *a*. situated behind the wings.

postanal *a*. behind the anus.

postantennal *a*. situated behind antennae.

postarticular *a*. posterior process of surangular, behind articulation with quadrate.

postaxial *a*. on posterior side of the axis, as e.g. on fibular side of leg.

postbacillary *a*. having nuclei behind sensory part of retinal cells.

postbranchial *a*. behind the gill clefts; *appl*. bodies: ultimobranchial bodies *q.v.*

postcapillary venule small blood vessel that joins the capillary bed to a system of increasingly larger venules and, finally, veins.

postcardinal *a*. behind the region of the heart.

postcava, postcaval vein the vein bringing blood to heart from posterior part of body. *alt*. posterior vena cava.

postcentral *a*. (1) behind central region; (2) *appl*. region of cerebral cortex immediately posterior to central sulcus, *appl*. gyrus; (3) *appl*. part of the intraparietal sulcus.

postcentrum *n*. the posterior part of centrum of vertebrae, in some vertebrates.

postcerebral *a*. posterior to the brain, *appl*. salivary glands in the head of hymenopterans.

postcingular *a*. posterior to cingulum.

postclavicle *n*. a membrane bone present in pectoral girdle of some teleost fishes.

postclimax *a*. *appl*. stable plant community whose composition reflects climatic conditions which are more favourable (e.g. moister, cooler) than usual for the region and which therefore differs from the usual climax vegetation for the region.

postclisere *n*. a series of vegetative formations that arise when the climate becomes wetter.

postclitellian *a*. situated behind clitellum.

postclival *a*. situated behind the clivus of cerebellum, *appl*. a fissure.

postclypeus *n*. the posterior part of clypeus of an insect.

postcolon *n*. part of gut between colon and rectum in certain mites.

postcornual *a*. *appl*. glands situated behind horns, as in chamois.

postcranial *a*. *appl*. skeleton of an animal other than the head.

postdicrotic *a*. *appl*. secondary wave of a pulse, or that succeeding the dicrotic.

postembryonic *a*. *pert*. the age or states succeeding the embryonic.

posterior *a*. (1) situated behind; (2) toward the tail end; (3) dorsal, in human anatomy; (4) behind the axis; (5) superior or next to the axis. *cf*. anterior.

posterior commissure tract of fibres connecting the two cerebral hemispheres at posterior end of corpus callosum.

posterior horn of spinal cord, that part of grey matter containing sensory cells. *cf*. anterior horn.

posterior lobe of pituitary gland the pars intermedia and the pars nervosa (neurohypophysis).

posterior marginal zone dense region of cells at the edge of the blastoderm of chick embryo, which gives rise to the primitive streak.

posterior vena cava postcava *q.v.*

posterolateral *a*. placed posteriorly and towards the side, *appl*. arteries.

posteromedial *a*. placed posteriorly and towards the middle, *appl*. arteries.

postesophageal postoesophageal *q.v.*

postestrum metoestrus *q.v.*

postflagellate *a*. *appl*. to forms of trypanosomes intermediate between flagellates and cyst.

postfrons *n*. portion of frons of insect head posterior to antennary base line.

postfrontal *a*. *appl*. a bone present behind the orbit of eye in some vertebrates.

postfurca *n*. forked process from sternum or an apodeme of metathorax in insects.

postganglionic *a*. *appl*. autonomic nerve fibres issuing from ganglia.

postgena *n*. the posterior part of gena (side) of insect head.

postgenital *a*. situated behind the genital segment, in arthropods and other segmented animals.

postglacial *appl*. Holocene *q.v.*

postglenoid *a*. situated behind the glenoid fossa, *appl*. a process or tubercle.

posthepatic *a*. *appl*. latter part of alimentary canal, that from liver to end.

posthypophysis postpituitary *q.v.*

postical, posticous *a.* (1) an outer or posterior surface; (2) *appl.* lower or under-surface of a thallus, leaf or stem, esp. in liverworts.

postischium *n.* lateral process on the hinder side of ischium in some reptiles.

postlabrum *n.* posterior portion of insect labrum, where differentiated.

postmeiotic segregation the segregation of the two strands of a heteroduplex DNA (formed during genetic recombination) in a subsequent round of DNA replication that succeeds meiosis.

postmentum *n.* the united sclerites constituting the base of labium of insects.

postminimus *n.* a rudimentary additional digit occasionally present in amphibians and reptiles.

postmitotic *a. appl.* a cell which, once it has been formed by mitosis, does not undergo another mitosis and cell division before its death.

postneural *a.* situated at the end of the back, *appl.* plates of shell of tortoises and turtles.

postnodular *a. appl.* a cerebellar fissure between nodule and uvula.

postnotum *n.* posterior portion of insect notum.

postocular *a.* behind the eye, *appl.* scales.

postoesophageal *a.* (1) *appl.* nerve fibres connecting ganglia serving antennae in crustaceans; (2) *appl.* nerve fibres connecting cerebral ganglia in various invertebrates.

postoestrus metoestrus *q.v.*

postoral *a.* behind the mouth.

postorbital *a.* (1) behind the eye socket, *appl.* bone forming posterior wall of eye socket; (2) *appl.* luminescent organ in certain fishes; (3) *n.* bone of vertebrate skull lying immediately behind orbit of eye.

postotic *a.* behind the ear.

postparietal *a. appl.* paired skull bones sometimes occurring between parietal and interparietal.

postpatagium *n.* in birds, small fold of skin extending between upper arm and trunk.

postpermanent *a. appl.* traces of a dentition succeeding the permanent.

postpetiole *n.* in ants, the 2nd segment of abdominal stalk.

postphragma *n.* a phragma developed in relation to a postnotum in insects.

postpituitary *a. pert.* or secreted by the posterior lobe of the pituitary gland (the pars intermedia and the neurohypophysis).

postpubic *a.* at posterior end of pubis, *appl.* processes of pubis parallel to ischium.

postpyramidal *a.* behind the pyramid of cerebellum, *appl.* a fissure.

postreduction *n.* halving the chromosome number in the 2nd meiotic division instead of the 1st.

postretinal *a.* (1) situated behind the retina; (2) *appl.* nerve fibres connecting peri-opticon and inner ends of ommatidia in a compound eye.

postscutellum *n.* (1) a projection under the scutellar lobe of mesothorax in insects; (2) a sclerite behind the scutellum.

postsegmental *a.* posterior to body segments or somites.

postsphenoid *n.* the posterior part of sphenoid in skull.

poststernellum *n.* most posterior portion of an insect sternite.

poststernite *n.* posterior sternal sclerite of insects.

postsynaptic *a. appl.* cell or part of cell (e.g. dendrite membrane or muscle cell membrane) on the receiving side of a synapse in nervous system or at a neuromuscular junction.

postsynaptic potential membrane potential generated in a receiving neuron by the action of a neurotransmitter at a synapse.

post-temporal *a.* behind the temporal bone, *appl.* bone and fossa.

post-transcriptional *a. appl.* processes occurring after transcription, such as capping and poly(A) addition to eukaryotic RNAs; *appl.* control: the regulation of gene expression that is exerted at any stage after transcription has begun.

post-translational *a. appl.* processes occurring after translation, e.g. modifications made to proteins such as glycosylation, phosphorylation, cleavage of preproteins, cleavage of signal sequences; *appl.* control: the regulation of protein function that is exerted at any stage after translation has begun.

post-trematic *a*. behind an opening such as a gill cleft, *appl*. nerves running in posterior wall of 1st gill cleft in pharynx.

postzygapophyses *n.plu*. the posterior zygapophyses of a vertebra, articulating with the prezygapophyses of the vertebra immediately succeeding it.

potamobenthos *n*. the bottom-living organisms in a river or other fresh water.

potamodromous *a*. migrating only in fresh water.

Potamogetonales Najadales *q.v*.

potamoplankton *n*. the plankton of rivers, streams and their backwaters.

potassium (K) *n*. metallic element, an essential nutrient for living organisms. As the ion K$^+$ it is involved in maintaining intracellular ion balance, in generating the membrane potential, and in producing electrical signals in neurons. One of the major elements required for plant growth.

potassium channel (K$^+$ channel) ion channel that is selective for potassium ions (K$^+$). Potassium channels are composed of four identical membrane-spanning protein subunits that enclose an aqueous pore through the lipid bilayer. Ion selectivity is conferred by a structural 'selectivity filter' towards the exit of the channel, through which a K$^+$ ion can pass, but not the similar Na$^+$ ion. Voltage-gated potassium channels are involved in generating action potentials in nerve and muscle cells.

potassium leak channel ungated ion channel through which potassium ions (K$^+$) continually leak out of the cell.

potato *n*. *Solanum tuberosum*, a crop plant originating in South America and now grown widely in the cooler, wetter parts of the world, whose tubers are eaten as food. *cf*. sweet potato.

potential *a*. latent or not expressed, as *appl*. characteristics. *see also* action potential, membrane potential.

Potyviridae *n*. family of single-stranded, positive-strand RNA plant viruses with flexuous rod-shaped particles, e.g. potato virus Y.

Poupart's ligament the ligament of the groin.

powder-down feathers those which do not develop beyond an early stage.

powdery mildews parasitic fungi of the order Erysiphales in the Pyrenomycetes, the powdery appearance being due to the large numbers of spores formed on the surface of the host tissue.

Poxviridae, poxviruses *n*., *n.plu*. family of large, double-stranded DNA viruses that includes vaccinia and smallpox, fowl pox, sheep pox and myxoma.

PP primary production *q.v*.

PPAR peroxisome proliferator-activated receptor *q.v*.

p.p.b., ppb parts per billion (10^9).

ppGpp guanosine tetraphosphate *q.v*.

PP$_i$ pyrophosphate *q.v*.

PPI peptidyl prolyl *cis-trans* isomerase *q.v*.

PPLO pleuropneumonia-like organisms *q.v*.

p.p.m., ppm parts per million (10^6).

PPP oxidative pentose phosphate pathway. *see* pentose phosphate pathway.

pppGpp guanosine pentaphosphate *q.v*.

praeabdomen *n*. the anterior, broader part of abdomen in scorpions.

praecoces *n.plu*. newly hatched birds able to take care of themselves. *a*. **praecocial**.

praeputium prepuce *q.v*.

prairie *n*. in North America, the natural grassland covering the middle of the continent in the mid-latitudes. It consists of tallgrass prairie in the cooler moister areas, most of which has now been converted into agricultural land, and short-grass prairie in the dryer areas.

pratal *a*. (1) *pert*. meadows; (2) *appl*. flora of rich humid grasslands.

Prausnitz–Kästner reaction specific allergy produced in non-allergic individual after injection of serum from allergic individual.

pre- prefix derived from L. *prae*, before, signifying situated before or occurring before.

preadaptation *n*. any previously existing anatomical structure, physiological process or behaviour pattern that makes new forms of evolutionary adaptation more likely.

preanal *a*. anterior to the anus.

preantenna *n*. one of the pair of feelers on the 1st segment in onychophorans.

pre-auricular *a*. *appl*. a groove at anterior part of auricular surface of hip-bone.

preaxial *a*. (1) in front of the axis; (2) on anterior border or surface.

prebacillary *a.* having nuclei distal to sensory portion of retinal cells.

pre-B cells an early stage in development of B lymphocytes, in which the immunoglobulin heavy-chain gene has been rearranged and the cells express heavy chains of isotype μ in the cytoplasm.

prebiotic *a.* before life appeared on earth.

prebranchial pretrematic *q.v.*

Precambrian *n.* geological time preceding the Cambrian period, reckoned generally as lasting from the earliest formation of Earth's crust until *ca.* 542 million years ago. It is divided into two eons, the Proterozoic and the earlier Archean. The Precambrian saw the origin of life, the evolution of living cells and the evolution of the eukaryotic cell. The first multicellular animals arose towards the end of the Precambrian.

precapillary *a. appl.* arterioles having an incomplete muscular layer. *n.* a small vessel conducting blood from arteriole to capillary.

precartilage *n.* type of cartilage preceding other kinds, or persisting as in fin-rays of certain fishes.

precava, precaval vein the vein bringing blood to the heart from the anterior part of the body. *alt.* anterior vena cava.

precentral *a.* (1) situated anterior to the centre; (2) *appl.* region of cerebral cortex immediately anterior to central sulcus, *appl.* gyrus; (3) *appl.* sulcus anterior to and parallel with the central sulcus.

precentrum *n.* the anterior part of centrum of vertebrae of certain vertebrates.

precheliceral *a.* anterior to chelicerae, *appl.* segment of mouth region in arachnids.

prechordal *a.* (1) anterior to notochord or spinal chord, *appl.* to part of base of skull; (2) *appl.* plate, a small thickened area of endoderm just anterior to the notochord in vertebrate embryos that eventually gives rise to part of the pharynx and contributes to the mesenchyme of the head.

precingular *a.* anterior to cingulum.

precipitin *n.* specific antibody that forms a precipitate with its corresponding antigen.

precipitin reaction assay for measuring the amount of antigen or antibody in a sample, based on the precipitation of antigen–antibody complexes from solution.

preclavia *n.* an element of the pectoral girdle.

preclimacteric *a.* a period before the climacteric or time of ripening.

preclimax *n.* the plant community immediately preceding the climax community.

preclival *a. appl.* fissure in front of clivus of cerebellum.

precocial *a. appl.* young that are able to move around and forage at a very early stage, esp. in birds. *cf.* altricial.

preconnubia *n.* gatherings of animals before the mating season.

precoracoid *a.* an anterior ventral bone of pectoral girdle in amphibians and reptiles.

precostal *a. appl.* short spurs on basal portion of hind wing of lepidopterans.

precoxa, precoxal subcoxa *q.v.*, subcoxal *q.v.*

precrural *a.* on anterior side of leg or thigh.

precuneus *n.* the medial surface of parietal lobe of cerebral hemisphere.

precursor *n.* (1) cell from which other cells will develop; (2) protein from which e.g. an enzyme or hormone will be produced by further modification; (3) any substance that precedes and is involved in the formation of a compound.

precystic *a. appl.* small forms appearing before the encystment stage in some protozoans.

predaceous *alt.* spelling of predacious.

predacious *a. appl.* fungi of the family Zoopagaceae, which trap and feed on protozoans and nematode worms.

predator *n.* any organism that catches and kills other organisms for food.

predator chain food chain that starts from plants and passes from herbivores to carnivores. *cf.* parasite chain, saprophyte chain.

predator–prey relationship interaction between two organisms of different species in which one (the predator) captures and feeds on the other (the prey).

predentine *n.* immature dentine which is not yet calcified but made mainly of fibrils.

prednisone *n.* anti-inflammatory steroid.

pre-embryo *n.* name sometimes given to the fertilized mammalian ovum and its cleavage stages up to blastocyst formation.

preen gland oil gland *q.v.*

pre-epistome *n.* plate covering base of epistome in arachnids.

pre-erythrocytic *a. appl.* phase of the malarial plasmodium life cycle in which it lives in the tissues of the human host, developing large schizonts that produce merozoites that infect erythrocytes.

prefemur *n.* second trochanter, as in walking legs of pycnogonids.

preferential species species that are present in several different communities, but are more common or thriving in one particular community.

preflagellate *a. appl.* forms of trypanosomes intermediate between cyst and elongated flagellates.

prefloration *n.* the form and arrangement of floral leaves in the flower bud.

prefoliation *n.* the form and arrangement of foliage leaves in the bud.

preformation theory theory current up to the 18th century, according to which it was supposed that a sperm contained a miniature adult (homunculus) from which the embryo developed.

prefrontal *a. appl.* a bone anterior to frontal bone in certain vertebrates.

preganglionic *a.* nerve fibres running from spinal cord and ending in synapses in sympathetic ganglia.

pregenital *a.* (1) situated anterior to genital opening; (2) *appl.* segment behind 4th pair of walking legs in arachnids.

pregnenolone *n.* cholesterol derivative, precursor of all steroid hormones. Its synthesis is stimulated by ACTH.

prehallux *n.* rudimentary additional digit on hindlimb.

prehaustorium *n.* rudimentary root-like sucker.

prehensile *a.* adapted for grasping and holding.

prehepatic *a. appl.* part of digestive tract anterior to liver.

prehyoid *a.* (1) *pert.* mandible and hyoid; (2) *appl.* cleft between mandible and ventral parts of hyoid arch.

prehypophysis prepituitary *q.v.*

preimaginal *a.* preceding the imaginal or adult stage.

preimaginal conditioning in insects a response learnt in a larva which is retained by the adult.

preimplantation *a. appl.* mammalian embryo at the blastocyst stage before it has become implanted in the ovarian wall.

preinterparietal *a. appl.* one of two small upper membranous centres of formation of supraoccipital bone.

prelacteal *a. pert.* a dentition that may occur previous to the milk teeth.

premandibular *a.* (1) anterior to mandible; (2) *appl.* somites of amphioxus; (3) *appl.* a bone of certain reptiles.

premaxilla *n.* a paired bone or cartilage anterior to the maxilla of jaw in most vertebrates. In terrestrial vertebrates it bears the front teeth. In most teleost fish it can be protruded independently of the maxilla. *plu.* **premaxillae.**

premaxillary *a.* (1) anterior to maxilla; (2) *pert.* premaxilla.

premedian *a.* (1) anterior to middle of body or part; (2) *appl.* vein in front of median vein in certain insect wings.

prementum *n.* the united stipites bearing ligula and labial palps of insects.

pre-messenger RNA pre-mRNA *q.v.*

premetaphase prometaphase *q.v.*

premolars *n.plu.* teeth located between canines and molars in mammalian dentition.

premorse *a.* with irregular and abrupt termination, as if end were bitten off.

premotor *a.* part of motor cortex lying anterior to the primary motor cortex.

pre-mRNA the transcript of a protein-coding gene before removal of introns and other processing.

premyoblast *n.* cell that gives rise to a myoblast.

prenasal *a. appl.* bone developed in septum in front of mesethmoid in skulls of certain vertebrates.

prenatal *a.* before birth, *appl.* tests for genetic defects performed on a foetus in the womb.

prenylation *n.* post-translational covalent attachment of an isoprenoid lipid group (either a farnesyl or a geranylgeranyl group) to a protein, which can then be linked to a membrane via the lipid group.

preoccipital *a. appl.* an indentation or notch in front of posterior end of cerebral hemispheres.

preocular *a.* in front of the eye, as antennae, scales, etc.

preopercular *a.* (1) anterior to gill cover; (2) *appl.* luminescent organ in certain fishes; (3) *appl.* bone: preoperculum *q.v.*

preoperculum *n.* anterior membrane bone of gill cover of fishes.

preoptic area area in the hypothalamus that contains a sexually dimorphic nucleus.

preoptic nerve cranial nerve associated with vomeronasal organ and ending in nasal mucosa.

preoral *a.* situated in front of the mouth.

preorbital *a.* (1) anterior to orbit of eye, *appl.* a bone in teleost fishes; (2) *appl.* glands in ruminants.

prepatagium *n.* (1) fold of skin extending between upper arm and forearm of birds; (2) part of the patagium between neck and fore-feet in flying lemurs.

prepatellar *a. appl.* a pouch between lower part of knee-cap (patella) and skin.

prepattern *n.* a system of spatially organized molecular developmental cues already laid down in the fertilized egg and which guide development.

prepenna *n.* down feather of nestling.

prepeptide *n.* signal peptide that is cleaved off many polypeptides after they enter the endoplasmic reiculum.

prepharynx *n.* narrow thin-walled structure connecting oral sucker and pharynx, in trematodes.

prephragma *n.* a phragma developed in relation to the notum in insects.

prepituitary *n.* anterior lobe of pituitary gland, the adenohypophysis.

preplacental *a.* occurring before placenta formation or development.

preplumula *n.* a nestling down feather which is succeeded by adult down feather.

prepollex *n.* rudimentary additional digit sometimes present before the thumb in some amphibians and reptiles.

prepotency *n.* (1) the fertilization of a flower by pollen from another flower in preference to pollen from its own stamens, when both are offered simultaneously; (2) capacity of one parent to transmit more characteristics to offspring than other parent.

prepotent *a.* (1) transmitting the majority of characteristics, *appl.* one of the parents in certain genetic crosses; (2) *appl.* a flower exhibiting a preference for cross-pollination; (3) having priority, as one reflex among other reflexes.

preprohormone *n.* precursor to a polypeptide prohormone, representing the original polypeptide chain synthesized from mRNA, and which is later cleaved enzymatically to produce first a prohormone and then an active hormone.

preprophase band band of microtubules and actin filaments that forms around the circumference of a plant cell, under the plasma membrane, before mitosis and cell division.

preproprotein *n.* the initial form of some secreted proteins from which the signal sequence (the prepeptide) and, subsequently, a propeptide are cleaved, leaving the functional protein.

preprotein *n.* transient protein precursor bearing terminal regions which are rapidly cleaved to produce a stable proprotein (such as proinsulin from preproinsulin) or a functional protein (as in many membrane proteins).

prepubertal *a. pert.* age or state before puberty or sexual maturity.

prepubic *a.* (1) *pert.* prepubis; (2) *appl.* projections from pelvic arch, in certain fishes; (3) on anterior part of pubis; (4) *appl.* elongated projections on pubis of certain vertebrates.

prepubis *n.* part of pelvic girdle of certain reptiles and fishes.

prepuce *n.* foreskin, part of the outer covering of the penis which leaves the surface at the neck of penis and is folded upon itself.

prepupa *n.* quiescent stage preceding the pupal in some insects.

preputial *a. pert.* the prepuce or foreskin.

prepygidial *a.* anterior to pygidium, *appl.* growth zone in polychaete worms.

prepyloric *a. appl.* ossicle hinged to pyloric ossicle in gastric mill of crustaceans.

prepyramidal *a.* (1) anterior to the pyramid of cerebellum, *appl.* a fissure; (2) *appl.* tract: the rubrospinal tract *q.v.*

prereduction *n.* halving the chromosome number in the 1st meiotic division.

prescutum *n.* anterior sclerite of insect notum.

presegmental *a.* anterior to body segments or somites.

presentation time minimum duration of continuous stimulus necessary for production of a response.

presequence *n.* sequence of amino acids at the N-terminal end of newly synthesized proteins destined to enter the mitochondria, and which is required for their transport across the mitochondrial membranes from their site of synthesis in the cytoplasm.

prespermatid *n.* secondary spermatocyte.

presphenoid *n.* (1) in many vertebrates, a bone of skull anterior to the basisphenoid; (2) the anterior part of the sphenoid bone.

prespiracular pretrematic *q.v.*

pressor *a.* causing a rise in arterial pressure.

pressure-flow hypothesis generally accepted theory that sugars and other solutes are transported via the phloem in plants along a gradient of hydrostatic pressure developed osmotically. Assimilates are actively transported into the phloem at their source, lowering the water potential in the phloem. Water enters by osmosis, and passively carries the assimilates to a sink, such as a storage root, where they are actively transported out of the phloem, causing an increase in water potential within the phloem, and causing water to leave the sieve tube by osmosis.

presternal *a.* (1) situated in front of sternum; (2) *pert.* anterior part of sternum; (3) *appl.* jugular notch, on superior border of sternum.

presternum *n.* (1) the manubrium or anterior part of sternum; (2) anterior sclerite of insect sternum.

prestomium *n.* in biting and sucking insects the aperture between tips of the mouthparts serving for the intake of food.

presumptive *a. appl.* embryonic cells or tissues that will, in the normal course of development, give rise to a particular tissue, e.g. presumptive endoderm, presumptive mesoderm. The designation does not imply that they are irreversibly committed to this course of development.

presynaptic *a.* (1) *appl.* neuron or part of neuron, e.g. membrane of axon terminal, on the transmitting side of a synapse; (2) *appl.* vesicles liberating neurotransmitter at axon terminals.

presynaptic facilitation enhanced transmission between one neuron and another at a synapse as the result of the action of a third neuron synapsing on the axon terminal of the transmitting neuron.

presynaptic inhibition action of an inhibitory neuron exerted on the axon terminal of another.

pretarsus *n.* terminal part or outgrowth on leg or claw of insects and spiders.

pretrematic *a.* anterior to an opening such as a gill cleft or spiracle.

prevalence *n.* the degree to which something (e.g. a disease) is present in a population. *a.* **prevalent** widespread.

prevenule *n.* small vessel conducting blood from capillaries to venule.

prevernal *a. pert.*, or appearing in, early spring.

prevertebral *a.* (1) *pert.* or situated in region in front of vertebral column; (2) *appl.* portion of base of skull; (3) *appl.* the ganglia of sympathetic nervous system.

prevomer *n.* (1) bone anterior to pterygoid in skull of some vertebrates; (2) the vomer of non-mammalian vertebrates; (3) in monotremes, a membrane bone in floor of nasal cavities.

prey *n.* organism that is actively captured and used as food by another organism (the predator).

prezygapophyses *n.plu.* the anterior zygapophyses of a vertebra, articulating with the postzygapophyses of the vertebra immediately preceding.

Priapulida, priapulids *n., n.plu.* phylum of burrowing and marine worm-like pseudocoelomate animals with a warty and superficially ringed body, with spines around the mouth.

Pribnow box short sequence of bases in the promoter regions of prokaryotic genes which appears to be the key recognition site for RNA polymerase. The consensus sequence is TATAATG, centred *ca.* 10 base pairs before the start of transcription.

prickle cells cells of the deeper layers of stratified squamous epithelium, connected by bundles of keratin fibres, which in isolated cells under the microscope project in tufts ('prickles') from the surface.

primaquine *n.* anti-malarial drug, which can cause haemolysis in people with glucose-6-phosphate dehydrogenase deficiency.

primary *a.* (1) first; (2) principal; (3) original; (4) *n.* one type of main flight feather in bird's wing, attached in the region of the posterior edge of the hand bones. *plu.* **primaries**.

primary cambium procambium *q.v.*

primary cell wall in plant cells, the first cell wall, laid down before and while the cell is growing and formed of cellulose microfibrils in a matrix of hemicelluloses, glycoprotein and pectic substances. It is relatively plastic and is the only cell wall present in most actively dividing cells or those engaged in photosynthesis or secretion. *cf.* secondary cell wall.

primary centre a part of central nervous system, especially of sensory cortex, which is the first to receive the input from a sensory organ.

primary consumer herbivore *q.v.*

primary culture culture of cells prepared directly from tissues.

primary ecological succession primary succession *q.v.*

primary endosperm nucleus triploid nucleus that derives from fusion of one generative nucleus with two polar nuclei in the embryo sac of flowering plants. The cell containing this nucleus divides by mitosis to produce the endosperm of the seed.

primary feathers primaries *q.v.*

primary focus collection of proliferating B and T cells in lymph node that is formed when B cells are stimulated by antigen in the presence of their corresponding T cells.

primary follicle *see* follicle (3).

primary germ layers *see* germ layers.

primary growth growth of roots and shoots from the time of initiation at the apical meristem in plant embryo to when their expansion and differentiation is completed. *cf.* secondary growth.

primary host host in which a parasite lives for much of its life cycle and in which it becomes sexually mature.

primary immune response the immune response made on contact with an antigen that has not been encountered before. It takes up to a week to develop and a week or two to reach maximum. *cf.* secondary immune response.

primary induction neural induction *q.v.*

primary lymphoid organs in mammals, the bone marrow and thymus, which produce B and T lymphocytes respectively, but which do not participate directly in immune reactions.

primary meristems the meristematic tissue in developing plant embryo. They consist of the outer protoderm (which gives rise to epidermis), surrounding the ground meristem (which gives rise to highly vacuolated loose ground tissue), which encloses the central procambium or procambial strand from which the vascular bundles arise.

primary metabolism the metabolic reactions essentials to keeping a cell or organism alive.

primary mycelium haploid mycelium originating from a basidiospore.

primary oocyte diploid female germ cell at the point of entering the first meiotic division of oogenesis. It is a large cell with a large nucleus (germinal vesicle).

primary organizer *see* organizer.

primary phloem collectively, the protophloem and metaphloem, the phloem derived from the primary cambium during primary growth.

primary pit fields areas of plant cell wall containing many plasmodesmata, and in which pits are clustered.

primary plant body the plant body formed from growth at the apical meristems.

primary producer autotroph *q.v.*

primary production (PP) the assimilation and fixation of inorganic carbon and other inorganic nutrients into organic matter by autotrophs, which are therefore called primary producers.

primary productivity the amount of organic matter fixed by the autotrophic organisms in an ecosystem per unit time.

primary root root that develops as a continuation of the radicle in plant seedling.

primary sere natural plant succession on an area previously without vegetation, starting from bare ground and ending in a climax community.

primary sexual characters differences between the sexes relating to the reproductive organs and gametes.

primary spermatocyte *see* spermatocyte.

primary structure in proteins, the amino acid sequence. In nucleic acids, the nucleotide sequence.

primary succession plant succession that begins on bare ground.

primary transcript in eukaryotic cells, an RNA as it is transcribed from a gene and which has not yet been modified by e.g. splicing, capping or polyadenylation.

primase *n.* RNA polymerase that synthesizes the RNA primer for DNA synthesis during DNA replication.

Primates *n.* order of mammals known from the Paleocene and including tree shrews, lemurs, monkeys, apes and humans. They are largely arboreal with limbs modified for climbing, leaping or brachiating (swinging), large brains in relation to body size, a shortening of the snout and elaboration of the visual apparatus, often with stereoscopic vision.

prime movers the ultimate factors that determine the direction of evolutionary change. They are of two kinds: basic genetic mechanisms, preadaptations and constraints imposed by an organism's existing developmental programme on the one hand, and the set of all environmental influences that constitute the agents of natural selection on the other.

primer *n.* (1) short RNA which must be synthesized on a DNA template before DNA polymerase can start elongation of a new DNA chain. It is subsequently removed and the gap filled in with DNA; (2) in the polymerase chain reaction, a pair of synthetic oligonucleotides complementary to flanking regions of the gene to be copied, which are bound to the DNA before the reaction commences to ensure that DNA replication is initiated at the required points.

priming *n.* (1) of T cells, their activation to effector status after encounter with antigen on a professional antigen-presenting cell; (2) in learning, the change that occurs in the processing of stimulus, e.g. a word or picture, as a result of previous exposure to the same stimulus or related stimuli.

primite *n.* the first of any pair of individuals in the chain-like colonies of gregarine protozoans, in which the front end of one (the satellite) becomes attached to the posterior end of another (the primite).

primitive *a.* (1) of earliest origin; (2) not differentiated or specialized; (3) *appl.* traits that appeared first in evolution and which give rise to other, more advanced, traits. They are often, but not always, less complex than the advanced ones.

primitive ectoderm that part of inner cell mass of mammalian blastocyst that gives rise to the embryo proper. *alt.* **epiblast**.

primitive endoderm that part of inner cell mass of mammalian blastocyst that gives rise to extra-embryonic membranes.

primitive node area of proliferating cells in which the primitive streak begins.

primitive pit enclosure at anterior end of the confluent folds of the primitive streak.

primitive streak in the early embryos of reptiles, birds and mammals, the two parallel longitudinal folds that develop on the epiblast and which represent the region at which cells are moving into the interior of the embryo to form the notochord. The anterior end of the primitive streak is an organizer region corresponding to the dorsal lip of the blastopore in amphibian gastrulas.

primordial *a.* (1) primitive; (2) original, first begun; (3) first formed; (4) *appl.* to embryonic cells, e.g. primordial germ cells, which will develop into particular cell types or tissues.

primordial cell initial *q.v.*

primordial follicle immature human ovarian follicle in the earliest stages of development, in which the oocyte is arrested in prophase of the first meiotic division and is surrounded by a single layer of follicle cells.

primordial germ cell in the early sexually undifferentiated embryo, cells whose descendants will eventually give rise to eggs or sperm.

primordium *n.* (1) original form; (2) a developing structure at the stage at which it starts to assume a form, *alt.* anlage; (3) (*bot.*) group of immature cells that will form a particular structure, e.g. leaf or flower primordium. *plu.* **primordia**.

primosome *n.* assembly of proteins concerned with the initiation of RNA primer formation in DNA replication.

Primulales *n.* order of dicot herbs, shrubs and trees comprising the woody tropical families Myrsinaceae and Theophrastaceae and the temperate family Primulaceae (primrose).

Principes Arecales *q.v.*, the palms.

principle of antithesis the principle, first formulated by Charles Darwin, that animals often convey signals with opposite meanings by expressions or postures that are opposites.

priodont *a.* (1) saw-toothed; (2) *appl.* stag beetles with smallest development of mandible projections.

prion *n.* proteinaceous infectious particle, which is the cause of the transmissible spongiform encephalopathies such as scrapie in sheep, bovine spongiform encephalitis (BSE) in cattle, and Creutzfeldt–Jakob disease (CJD) in humans. It contains no nucleic acid. The pathogenic prion protein is formed when a normal mammalian cell-surface protein, PrP, folds into a different three-dimensional conformation that forms aggregates that are toxic to the nerve cell and resistant to proteolytic digestion. The misfolded form can induce the normal PrP to adopt the pathogenic conformation (PrP*), thus accounting for the infectivity of prions and their ability to propagate themselves.

prisere primary sere *q.v.*

prismatic *a.* (1) like a prism, *appl.* cells, leaves; (2) consisting of prisms, as the prismatic layer of shells; (3) *appl.* soil crumbs in which the vertical axis is longer than the horizontal.

private *a. appl.* antigenic determinants unique to a particular haplotype in serological analysis of histocompatibility antigens. *cf.* public.

Pro proline *q.v.*

pro- prefix derived from Gk *pro*, before, denoting previous to, in front of, the precursor of, *or* from L. *pro*, forward, for.

pro-acrosome *n.* structure in spermatids which develops into the acrosome.

proamnion *n.* an area of blastoderm in front of head of early embryos of reptiles, birds and mammals.

proandry *n.* meroandry with retention of anterior pair of testes only.

proangiosperm *n.* a fossil type of angiosperm.

proatlas *n.* a median bone intercalated between atlas and skull in certain reptiles.

proband *n.* an individual affected by a genetic disease who is crucial in enabling a pedigree for the disease to be deduced. *alt.* propositus.

probasidium *n.* (1) thick-walled resting spore in Heterobasidiomycetes, which germinates to form a basidium (promycelium); (2) an immature basidium, before basidiospore formation.

pro-B cells developing B cells in bone marrow that display B-cell-specific surface proteins but which have not completed heavy-chain rearrangement.

probe *n.* (1) a defined, labelled fragment of DNA used to detect and identify corresponding sequences in nucleic acids by selectively hybridizing with them. *see also* DNA hybridization; (2) labelled antibody used to detect and identify proteins.

Proboscidea *n.* order of herbivorous placental mammals, known from the Eocene to the present, including the elephants and the extinct mammoths and mastodons. They are of great size, having a massive skeleton, stout legs, trunks and incisors modified as tusks.

proboscidiform *a.* like a proboscis.

proboscis *n.* (1) trunk-like projection of head, as in insects, annelid and nemertean worms, used for feeding; (2) in mammals, an elongated nose, such as the trunk of an elephant, or snout generally.

proboscis worms common name for the Nemertea *q.v.*

probud *n.* larval bud from the stolon of certain tunicates, which moves by pseudopodia to a specialized outgrowth and there divides to form definitive buds.

procambium *n.* meristematic tissue from which vascular bundles are developed. *alt.* procambial strand.

procarp *n.* female organ of red algae, consisting of carpogonium, trichogyne and auxiliary cells.

procartilage *n.* early stage of cartilage formation.

procary- *see* prokary-.

procaspase the inactive form of a caspase, which is activated by cleavage.

procedural memory non-declarative memory *q.v.*

Procellariiformes *n.* order of ocean birds with external tubular nostrils and hooked beaks, including the albatrosses, shearwaters and petrels. *alt.* tubenoses.

procercoid *n.* early larval form of certain cestodes in 1st intermediate host.

procerebrum *n.* the forebrain developed in the preantennary region of insects.

procerus *n.* pyramidal muscle of the nose.

process *n.* (1) a reaction or procedure; (2) an elongated portion of a cell, such as the axon and dendrites of nerve cells; (3) an elongated projection from any structure.

processed pseudogene pseudogene that seems to be derived from a reverse transcript of an mRNA that has become inserted into the genome.

processing *see* RNA processing.

processive enzyme enzyme that does not dissociate from its substrate between repeated catalytic events.

prochirality *n.* property of molecules lacking handedness in their chemical structure (i.e. their mirror images can be superimposed on each other) and which are optically inactive. *a.* **prochiral**. *cf.* chirality.

Prochlorophyta, prochlorophytes *n.*, *n.plu.* photosynthetic prokaryotes containing chlorophyll *a* and chlorophyll *b* but lacking phycobiliproteins, and therefore resembling plant chloroplasts rather than cyanobacteria. They include both ecto-symbiotic and free-living species.

prochorion *n.* an enveloping structure of embryo preceding formation of chorion.

proclimax *n.* stage in a sere appearing instead of usual climatic climax and not determined by climate. *alt.* subclimax.

procoelous *a.* with concave anterior face, *appl.* vertebral centra.

procollagen *n.* soluble precursor to the fibrous protein collagen. It forms tropocollagen after cleavage of terminal peptides.

proconvertin Factor VII *q.v.*

procoracoid *n.* an anteriorly directed projection from the glenoid fossa of urodele amphibians.

procruscula *n.plu.* a pair of blunt locomotory outgrowths on posterior half of a redia.

procrypsis *n.* (1) shape, pattern, colour or behaviour tending to make animals less conspicuous in their normal environment; (2) camouflage. *a.* **procryptic**.

proctal *a.* anal, *appl.* fish fins.

proctodaeum *n.* most posterior portion of the alimentary canal, lined by ectodermally derived epithelium. *alt.* proctodeum.

proctodone *n.* insect hormone thought to be secreted by anterior part of intestine and which ends diapause.

proctolin *n.* neuromodulatory neuropeptide hormone active in nervous system of certain crustaceans.

procumbent *a.* (1) trailing on the ground; (2) lying loosely along a surface; (3) *appl.* ray cells elongated in radial direction.

procuticle *n.* the colourless cuticle of insects, composed of protein and chitin, before differentiation into endocuticle and exocuticle.

procyclic *a. appl.* developmental stage in the life cycle of parasitic kinetoplastid flagellates in the invertebrate host.

prodeltidium *n.* a plate which develops into a pseudodeltidium.

prodentine *n.* a layer of uncalcified matrix capping tooth cusps before formation of dentine.

prodrome *n.* a preliminary process, indication or symptom. *a.* **prodromal**.

prodrug *n.* an inactive compound that is processed into the active drug in the body.

producer *n.* an autotrophic organism, e.g. photosynthetic green plant, which synthesizes organic matter from inorganic materials and is the first stage in a food chain. *alt.* primary producer.

production *n.* in ecology, the assimilation of nutrients into biomass. *see* net primary production, primary production.

productive infection of a virus, an infection in which new infectious viral particles are produced.

productivity *n.* the amount of organic matter fixed by an ecosystem per unit time. *see* primary production.

proecdysis *n.* in arthropods, the period of preparation for moulting. A new cuticle is laid down and the older one detaches from it.

proelastin *see* elastin.

proembryo *n.* (1) in plant embryos, stage before differentiation of embryo into embryo proper and the stalk-like suspensor; (2) an embryonic structure preceding true embryo.

proenzyme zymogen *q.v.*

proerythrocyte reticulocyte *q.v.*

proestrum pro-oestrus *q.v.*

professional antigen-presenting cells dendritic cells, macrophages and B cells, which are able to present antigen to and activate naive T cells. *see also* antigen-presenting cell.

profile-based threading method for the prediction of protein structure from sequence, in which the protein sequence is fitted to all known types of protein fold to find the nearest match.

profile transect a profile of vegetation along a predetermined line, drawn to scale and intended to show the heights of plant shoots. *alt.* stratum transect.

profilin *n.* a ubiquitous cytoplasmic protein in mammalian cells, to which unpolymerized actin subunits are bound.

proflavin *n.* a mutagenic acridine dye.

profunda (1) *a.* deep-seated, *appl.* a branch of brachial, femoral or costocervical artery, to the ranine artery, terminal part of lingual artery, and to a vein of femur; (2) *n.* a deep artery or vein.

profundal *a. appl.* or *pert.* the zone of deep water and bottom below compensation depth (point at which photosynthesis ceases) in lakes.

profundal zone the zone of a lake lying below the compensation point, comprising the deep water and the lake bottom.

progamete *n.* a structure giving rise to gametes by abstriction, in certain fungi.

progenesis *n.* the maturation of gametes before completion of body growth.

progenitor cells cells that give rise to other types of cell.

progeotropism *n.* positive geotropism.

progeric *a.* causing earlier onset of ageing and a reduced lifespan.

progestagens *n.plu.* a group of steroids that have effects like progesterone.

progestational *a.* (1) *appl.* phase of the oestrous cycle during luteal and endometrial activity; (2) *appl.* hormones controlling uterine cycle and preparing uterus for implantation of conceptus.

progesterone *n.* steroid hormone secreted by the female mammalian corpus luteum and placenta, which prepares the uterus to receive the fertilized egg, maintains the uterus during pregnancy, inhibits ovulation during pregnancy and is the precursor to several other hormones. It is a com-

ponent of some types of contraceptive pill. *alt.* progestin, progestone. Formerly also known as luteal hormone or luteosterone.

progestin, progestone forms of progesterone *q.v.*

progestogen *n.* any compound with progesterone-like effects in a female mammal.

proglottid, proglottis *n.* individual segment of an adult tapeworm, containing a set of reproductive organs. Eggs are formed in the posterior proglottids, are shed in the faeces and then enter the alternative host where they hatch. *plu.* proglottids, proglottides.

prognathous *a.* (1) having prominent or projecting jaws; (2) with mouthparts projecting downwards, *appl.* insects.

progonal *a. appl.* sterile anterior portion of genital ridge.

progoneate *a.* having the genital aperture anteriorly, as in some arthropods.

programmed cell death cell death that occurs as a normal part of the development or physiology of an organism. Cells dying in this way undergo apoptosis and not necrosis.

progressive provisioning the feeding of a larva in repeated meals. *cf.* mass provisioning.

progress zone in the developing vertebrate limb, a population of actively dividing mesodermal cells immediately underneath the apical ectodermal ridge of the limb bud. This zone provides the cells of the limb.

Progymnophyta, progymnophytes, progymnosperms *n., n.plu., n.plu.* division of extinct spore-bearing woody plants and trees, with secondary xylem similar to that of gymnosperms, and which are believed to be possible ancestors of the gymnosperms.

prohaptor *n.* anterior adhesive organ in trematodes.

prohormone *n.* an inactive precursor of a hormone, esp. of a polypeptide or peptide hormone, many of which are produced by enzymatic cleavage of an active portion from a longer polypeptide prohormone.

prohydrotropism *n.* positive hydrotropism.

proinflammatory *a.* promoting inflammation, *appl.* e.g. histamine, leukotrienes, prostaglandins.

proiospory *n.* the premature development of spores.

projectile *a.* (1) protrusible; (2) *appl.* structures that can be thrust forwards.

projection *n.* the perception of external sensation as external despite the fact that the stimuli are interpreted in the brain.

projection neurons neurons in the brain that have their cell bodies in one region and their axon terminals in another. *cf.* local-circuit neurons.

projicient *a. appl.* sense organs reacting to distant stimuli such as light or sound.

Prokarya, Prokaryota, prokaryotes *n., n., n.plu.* the Bacteria and the Archaea collectively, unicellular microorganisms whose small simple cells lack a membrane-bounded nucleus, mitochondria, chloroplasts and other membrane-bounded organelles typical of plant, animal, fungal, protozoan or algal cells. Their DNA is usually in the form of a single circular molecule and is not complexed with histones. They are divided into two domains or superkingdoms, the Bacteria and the Archaea, on the basis of biochemical differences and DNA sequence data. *a.* **prokaryotic.** *cf.* eukaryote. *see* Appendices 5 and 6.

prokaryon *n.* former term for the nucleoid of a prokaryotic cell.

prolabium *n.* middle part of upper lip.

prolactin *n.* glycoprotein hormone secreted by the anterior pituitary which stimulates the production of milk in mammals and of crop milk in some birds. It assists in maintaining the corpus luteum in mammals and has a range of effects in lower vertebrates.

prolactin-releasing hormones hypothalamic peptides that promote the release of prolactin from the anterior pituitary.

prolamellar body paracrystalline arrangement of membranes in etioplasts.

prolamines *n.plu.* simple proteins found in seeds of cereals, soluble in ethanol, and including gliadin from wheat, zein from maize, hordein from barley.

prolarva *n.* a newly hatched larva during the first few days when it feeds on its supply of embryonic yolk, as in some fish.

proleg *n.* an unjointed abdominal appendage of larvae of Lepidoptera and some other arthropods.

proleukocyte *n.* in insects, a small leukocyte with basophil cytoplasm and large nuclei, and developing into a macronucleocyte.

proliferation *n.* (1) increase in numbers by frequent and repeated reproduction; (2) *appl.* cell: increase in number by cell division.

proliferous *a.* (1) multiplying quickly, *appl.* bud-bearing leaves; (2) developing supernumerary parts abnormally.

prolification *n.* shoot development from a normally terminal structure.

proline (Pro, P) *n.* a cyclic amino acid (strictly an imino acid) with a non-polar hydrocarbon side chain, constituent of proteins. *see also* hydroxyproline.

proline-rich motif amino acid sequence motif present in many proteins in intracellular signalling pathways. It is recognized and bound by SH3 domains.

promeristem *n.* the part of the apical meristem consisting of the actively dividing cells (initials) and their most recent derivatives. *alt.* protomeristem.

prometaphase *n.* stage between prophase and metaphase in mitosis and meiosis. It is characterized by breakdown of the nuclear envelope (in most organisms) and attachment of chromosomes to the spindle.

promitochondria *n.plu.* abnormal mitochondria which are found in yeast grown anaerobically and which lack some cristae and cytochromes.

promonocyte *n.* a cell developing into a monocyte.

promonostelic *a. appl.* stem or root with a protostele or central cylinder of vascular tissue.

promontory *n.* prominence or projection, as of cochlea and sacrum.

promoter *n.* (1) DNA region involved in and necessary for initiation of transcription, and including the binding site for RNA polymerase, the startpoint of transcription and various other sites at which gene regulatory proteins can bind; (2) in carcinogenesis, any agent that hastens the process of carcinogenesis while not being a carcinogen on its own; (3) a protractor muscle.

promycelium *n.* short hypha that germinates from teleutospore in rust and smut fungi and on which basidiospores develop.

promyelocyte *n.* bone marrow cell that develops either into a granulocyte or a monocyte.

pronate *a.* prone or inclined.

pronation *n.* movement by which palm of hand is turned downwards by means of pronator muscles.

pronephros *n.* the kidney that develops first in embryonic or larval life, later replaced by mesonephros and metanephros. *a.* **pronephric.**

proneural clusters small clusters of cells within the fruit fly neurectoderm, in which one cell will give rise to a neuron.

pronograde *a.* walking with the long axis of the body parallel to the ground, as quadrupeds do.

pronotum *n.* the dorsal part of the prothorax (1st segment of thorax) of insects.

pronucleus *n.* haploid nucleus of egg or sperm after fertilization but before nuclear fusion and the first mitotic division.

pronymph *n.* in insect metamorphosis, the stage preceding the nymph.

pro-oestrus *n.* (1) the phase before oestrus or heat; (2) period of preparation for pregnancy.

proofreading *n.* (1) in DNA synthesis, the ability of DNA polymerase to recognize mismatched bases. *see also* editing; (2) mechanism for ensuring that the correct amino acid is inserted in a protein chain during translation.

pro-opiomelanocortin (POMC) *n.* polypeptide synthesized in the pituitary and which is differentially processed to produce β-lipotropin and ACTH in the anterior lobe, and α-MSH and β-endorphin in the intermediate lobe.

pro-otic (1) *n.* the anterior bone of otic capsule of vertebrates; (2) *a. pert.* a centre of ossification of petromastoid part of temporal bone.

propagate *v.* (1) to travel along a nerve fibre without losing strength, *pert.* electrical impulses; (2) to multiply, as of plants.

propagative *a.* reproductive.

propagule *n.* any spore, seed, fruit or other part of a plant or microorganism capable of producing a new plant and used as a means of dispersal. *alt.* diaspore.

propatagium *n.* part of bat wing membrane anterior to arm.

propeptide *n.* peptide that is cleaved off an inactive proprotein to give a functional protein.

properdin *n.* serum protein involved in the alternative pathway of complement activation. It stabilizes the C3/C5 convertase deposited on bacterial cell surfaces. *alt.* factor P.

properithecium *n.* young perithecium which contains a single zygote, giving rise ultimately to ascospores.

propes proleg *q.v.*

prophage *n.* bacteriophage DNA integrated into and replicating with the bacterial chromosome.

prophase *n.* the first stage in mitosis and meiosis. The replicated chromosomes condense and the two sister chromatids in each chromosome become visible. The spindle starts to assemble outside the nucleus.

prophialide sporocladium *q.v.*

prophloem protophloem *q.v.*

prophototropism *n.* positive phototropism.

prophyll(um) *n.* (1) a small bract; (2) first foliage leaf.

proplastid *n.* an immature plastid, as found in meristem cells.

propleuron *n.* lateral plate of exoskeleton of prothorax in insects.

propneustic *a.* with only prothoracic spiracles open for respiration.

propodeon, propodeum *n.* the first segment of abdomen fused to thorax in some hymenopterans.

propodite *n.* (1) foot segment of some crustacean limbs; (2) the tibia in arachnids.

propodium *n.* small anterior portion of a molluscan foot.

propodosoma *n.* region of body bearing 1st and 2nd legs in mites and ticks.

propolis *n.* resinous substance from buds of certain trees, used by worker bees to fasten comb portions and fill up crevices.

propons *n.* delicate bands of white matter crossing anterior end of pyramid of cerebellum below pons Varolii.

propositus proband *q.v.*

propranolol *n.* a drug that blocks β-adrenergic receptors.

proprioception *n.* reception of stimuli originating within the organism.

proprioceptor *n.* an internal sensory receptor sensitive to stimuli originating within the body, such as stretch receptors in muscle, and by which an animal receives information on its position and movements. *alt.* proprioreceptor.

propriogenic *a. appl.* effectors other than muscle, or organs which are both receptors and effectors.

proprioreceptor proprioceptor *q.v.*

propriospinal *a.* (1) *pert.* wholly to the spinal cord; (2) *appl.* nerve fibres confined to spinal cord.

prop roots adventitious aerial roots growing downwards from stem, as in mangrove and maize, and helping to support stem.

propterygium *n.* the foremost of three basal cartilages supporting pectoral fin of elasmobranch fishes.

propupa prepupa *q.v.*

propus propodite *q.v.*

propygidium *n.* dorsal plate anterior to pygidium in Coleoptera and some other insects.

proral *a.* from front backwards, *appl.* jaw movement in rodents.

prorennin *n.* precursor of rennin, secreted by the peptic cells and converted to active rennin by the action of hydrochloric acid.

proscapula clavicle *q.v.*

proscolex *n.* cysticercus, or the inverted scolex inside a cysticercus.

prosencephalization *n.* the progressive shifting of controlling centres towards the forebrain and the increasing complexity of the cerebral cortex in the course of vertebrate evolution.

prosencephalon forebrain *q.v.*

prosenchyma *n.* (1) elongated pointed cells, with thick or thin walls, as in mechanical and vascular tissues of plants; (2) fungal tissue formed of loosely woven hyphae. *a.* **prosenchymatous**.

prosequence *n.* part of a protein that assists in its folding but which is removed from the mature protein molecule.

prosethmoid *n.* an anterior bone of cranium of teleost fishes.

prosimian *n.* any primate, such as the lemurs and tarsiers, belonging to the primitive suborder Prosimii.

prosiphon endosiphuncle *q.v.*

prosocoel *n.* (1) a narrow cavity in epistome of molluscs, the 1st main part of coelom;

(2) median cavity between 3rd and lateral ventricles of brain.

prosoma *n.* the anterior part of the body in arachnids and some other invertebrates, corresponding to a cephalothorax.

prosome *n.* small ribonucleoprotein particle associated with mRNA, composed of >20 different proteins and one of several small prosomal RNAs.

prosopagnosia *n.* inability to recognize faces, including one's own.

prosoplectenchyma *n.* a false tissue of fungal hyphae where the cells are oriented parallel and the walls are indistinct, as in some lichens.

prosopyle *n.* the aperture of communication between adjacent incurrent and flagellate canals in some sponges.

prosorus *n.* thick-walled cell that develops into a sorus, in some lower fungi.

prospory *n.* seed production in plants which are not fully developed.

prostacyclins *n.plu.* compounds derived from arachidonic and other fatty acids which are produced e.g. in endothelial cells in response to damage and various other stimuli. They have vasodilator activity as a result of their relaxation of smooth muscle.

prostaglandins *n.plu.* any of a group of compounds formed from C_2 fatty acids and containing a five-carbon ring. There are four main classes, PGA, PGB, PGE and PGF. They are present in many mammalian tissues. They modify the effects of other hormones, stimulate contraction of smooth muscle of uterus, inducing labour or abortion, and are also involved in inflammatory reactions. The anti-inflammatory effect of aspirin is due to its inhibition of the enzyme prostaglandin synthetase, which catalyses the formation of prostaglandins from eicosatrienoate.

prostanoid *n.* compound derived from prostanoic acid, e.g. prostacyclins and prostaglandins.

prostate *n.* (1) the prostate gland in male mammals, a muscular and glandular organ around the beginning of the urethra in the pelvic cavity; (2) the spermiducal gland in annelids. *a.* **prostatic**.

prosternum *n.* (1) ventral part of prothorax of insects; (2) ventral part of cheliceral segment in some arachnids.

prostheca *n.* movable inner lobe of mandibles in certain insect larvae.

prosthetic group non-protein chemical group (e.g. haem, flavin or a metal atom) bound to a protein and essential for its biological activity.

prosthion *n.* the middle point of the upper alveolar arch.

prosthomere *n.* most anterior or preoral somite.

prostigmine neostigmine *q.v.*

prostomiate *a.* having a portion of head in front of the mouth.

prostomium *n.* in some annelids and molluscs, that part of the head anterior to the mouth. *a.* **prostomial**.

prostrate *a.* trailing on the ground or lying closely along a surface.

Protacanthopterygii *n.* group of generalized and fairly primitive teleost fishes existing from Cretaceous times to the present day, including the salmonids, pike and stickleback.

protamine *n.* any of a group of small highly basic arginine-rich proteins associated with DNA in fish sperm in a similar manner to histones in other eukaryotic DNAs.

protandry *n.* condition of hermaphrodite plants and animals where male gametes mature and are shed before female gametes mature. *a.* **protandrous**. *alt.* **protandria, protandrism** (in zoology).

pro-T cells early stage in T lymphocyte development in the thymus. They are immature thymocytes in which the T-cell receptor genes have not yet begun rearrangement.

Proteales *n.* order of xerophytic shrubs and trees with entire or much divided leaves with thick cuticle and hairs, and showy flowers, consisting of the family Proteaceae (protea).

protease *n.* enzyme that degrades proteins by hydrolysing internal peptide bonds to produce peptides. EC 3.4.21–24. *alt.* endopeptidase.

protease nexin a proteinase inhibitor of the serpin family.

proteasome *n.* large protein complex with proteolytic activity, present in cytosol of all cells, which degrades misfolded or unwanted proteins that are ubiquitinated or otherwise tagged for destruction, cleaving them into short peptides.

protective immunity immunity *q.v.*

protein *n.* one of the chief constituents of living matter, any one of a vast group of large polymeric organic molecules containing chiefly C, H, O, N, S. Proteins are essential in living organisms as enzymes, structural components of cells and tissues, and in the control of gene expression. An individual protein molecule consists of one or more unbranched polypeptide chains constructed from amino acids linked covalently together by peptide bonds. The chains are folded into three-dimensional structures which differ from one type of protein to another. The polypeptide chains are sometimes associated with non-protein compounds (e.g. haem, flavin) termed prosthetic groups. Each protein chain has a unique, genetically determined, amino acid sequence, which dictates its threedimensional structure and thus its function. Polypeptide chains are synthesized by translation of mRNA at the ribosomes. *see* Fig. 40 (p. 682).

proteinaceous *a. pert.* or composed of protein.

protein aggregate potentially damaging complex formed by misfolded proteins in cells. Protein aggregates are resistant to proteolytic destruction and often form fibrils. Cytotoxic aggregates of particular misfolded proteins are characteristic of some diseases, such as Alzheimer's disease.

protein array, protein chip a means of simultaneously detecting large numbers of different proteins, or molecules that bind to proteins, in a sample. Tens to thousands of 'spots' of different proteins or synthetic peptides are arranged in a predetermined pattern on a glass slide (microarray) or other suitable surface. When a sample is applied to the array, any proteins or other ligands (such as enzyme substrates) that bind to the array can then be detected by various means. Protein arrays composed of antibodies, e.g., are used as immunoassays that can simultaneously detect many different proteins in a sample.

proteinase *n.* any enzyme that degrades proteins by hydrolysing internal peptide bonds to produce peptides. EC 3.4.21–24. *alt.* endopeptidase, protease.

proteinase inhibitors small proteins, such as antitrypsin, which inhibit various proteinase enzymes.

protein bodies membrane-bounded granules composed of storage proteins, formed via the endoplasmic reticulum and Golgi apparatus in seeds and other plant tissues.

protein chip protein array *q.v.*

Protein Data Bank (PDB) the main publicly accessible database for experimentally determined three-dimensional protein structure.

protein degradation (1) the breakdown of proteins into their constituent amino acids by proteolysis; (2) *pert.* esp. to the cellular processes by which proteins are specifically targeted for destruction by the cell and to the degradative part of the normal turnover of cellular proteins.

protein disulphide isomerase (PDI) enzyme involved in the rearrangement of disulphide bonds during protein folding. EC 5.3.4.1.

protein domain *see* domain (2).

protein dynamics the study of the folding and internal movements of proteins.

protein engineering the alteration of the sequence, structure and function of a protein through the deliberate modification of the DNA that encodes it.

protein family group of proteins of related sequence and function, and which arise by the duplication and divergence of their genes from a single ancestral gene.

protein fingerprinting the identification of proteins from the pattern of peptide fragments obtained on two-dimensional gel electrophoresis of the protein digested with trypsin or other proteinases, which is characteristic for each protein. Peptide fragments can be eluted from the gel, or separated directly from the digest by liquid chromatography, and then further analysed by mass spectrometry.

protein folding the folding up of a newly synthesized polypeptide chain into a three-dimensional structure. The final conformation is determined by the amino acid sequence of the protein. Protein folding is a spontaneous process, although folding into the correct conformation may be aided by other proteins (molecular chaperones).

protein glycosylation *see* glycosylation.

protein histidine kinase enzyme that adds a phosphate group to specific histidine residues in a protein. This class of kinases is present in plants, fungi and bacteria, where they are part of two-component systems (*q.v.*), but are not found in animals.

protein intron intein *q.v.*

protein kinase enzyme that adds a phosphate group to specific amino acid residues in a protein. Different kinds of protein kinase phosphorylate serine and threonine, or tyrosine or histidine residues (the last in prokaryotes, plants and fungi only). They are an important component of intracellular pathways in all cells, regulating enzyme activity and relaying extracellular signals to the intracellular response machinery. *see* MAP kinase, protein kinase A, protein kinase B, protein kinase C, protein kinase G, protein phosphorylation, receptor tyrosine kinase, two-component systems.

protein kinase A (PKA) serine/threonine protein kinase activated by cyclic AMP in the intracellular signalling pathways of eukaryotic cells. *alt.* cyclic AMPdependent kinase.

protein kinase B (PKB) serine/threonine protein kinase involved in intracellular signalling pathways promoting cell survival. *alt.* Akt.

protein kinase C (PKC) ubiquitous serine/threonine protein kinase of mammalian and other eukaryotic cells which occurs in a variety of different types. It phosphorylates and alters the activity of a wide range of substrates *in vitro* and *in vivo* and is implicated in the modification of e.g. enzyme activity and ion channel conductivity in response to external signals.

protein kinase G (PKG) a serine/threonine protein kinase that is activated by the binding of cyclic GMP. It is part of signalling pathways leading from receptor guanylate cyclases.

protein microarray *see* protein array.

protein module any small protein domain that is found in different proteins, e.g. an SH2 domain.

protein motif short sequence, or pattern of particular amino acids, or pattern of secondary structure, that occurs in different proteins and is characteristic of a protein family and/or biochemical function.

proteinoid *n.* molecule like a protein, produced when trying to mimic primeval conditions by heating or other treatment of an amino acid mixture.

proteinoplast *n.* a storage plastid containing protein.

protein phosphatase enzyme that removes a phosphate group from a protein, often reversing the effect of a previous phosphorylation by a protein kinase. *alt.* phosphoprotein phosphatase.

protein phosphorylation *see* phosphorylation.

protein pump membrane protein or protein complex that pumps ions (e.g. Na^+, K^+, Cl^-) or simple organic compounds in or out of the cell against their concentration gradients in an energy-requiring reaction.

protein quality the nutritional value of a protein, which is determined both by its digestibility and by whether it contains adequate amounts of the essential amino acids that animals cannot synthesize for themselves.

protein sequence the order of amino acids in a protein chain.

protein serine/threonine kinase a protein kinase that adds phosphate groups to specific serine or threonine side chains on its target proteins. Many intracellular kinases are of this type.

protein serine/threonine phosphatase protein phosphatase that removes phosphate groups from phosphorylated serine and threonine residues in proteins. EC 3.1.3.16. *alt.* phosphoprotein phosphatase.

protein sorting the various mechanisms in eukaryotic cells that direct newly synthesized proteins from the cytosol to their final destinations.

protein structure *see* primary structure, quaternary structure, secondary structure, tertiary structure.

protein subunit a single polypeptide chain in a protein composed of several polypeptides.

protein superfamily group of proteins descended from a common ancestral protein but which have subsequently diverged considerably and acquired different functions.

protein synthesis the biosynthesis of a protein from amino acids. In a cell, protein synthesis occurs on the ribosomes using messenger RNA as a template to guide the order in which amino acids are added. *see also* genetic code, transfer RNA, translation.

protein targeting the process of transporting proteins synthesized in the cytoplasm to their correct destinations in other parts of a cell.

protein translocation the movement of certain proteins from the cytosol where they are synthesized into organelles or out of the cell, which involves their transport (translocation) across a membrane.

protein translocator a protein that mediates the movement of another protein across a membrane in a cell.

protein tyrosine kinases class of protein kinases that add phosphate groups to specific tyrosine residues in proteins. Some are dual-specificity kinases that can phosphorylate both tyrosine and serine/threonine residues. Protein tyrosine kinases are involved in the transmission of signals from cell-surface receptors, and some receptors have an intrinsic tyrosine kinase activity.

protein tyrosine phosphatase (PTP) a large family of the protein phosphatases. They hydrolyse the removal of phosphate from phosphotyrosine residues in proteins.

proteinuria *n.* the presence of proteins in the urine.

proteism *n.* the ability to change shape, as of amoeba and other cells. *a.* **protean**.

Proteobacteria one of the major phyla of the Bacteria distinguished on the evidence of DNA sequence data. The largest and most physiologically diverse of the main bacterial lineages, and containing most human pathogens. There are five major divisions. *see* alpha-proteobacteria, beta-proteobacteria, gamma-proteobacteria, delta-proteobacteria, epsilon-proteobacteria. *alt.* purple bacteria. *see* Appendix 5.

proteoclastic proteolytic *q.v.*

proteoglycan *n.* complex macromolecule consisting of polysaccharide (95%) and protein (5%) units. Proteoglycans form the ground substance of connective tissue.

proteolipid *n.* complex macromolecule composed of protein and lipid and which has the solubility properties of a lipid. *cf.* lipoprotein.

proteolysis *n.* breakdown of proteins and peptides into their constituent amino acids by enzymatic or chemical hydrolysis of peptide bonds. *a.* **proteolytic.**

proteolytic cascade a series of proteases that successively activate each other by cleavage of an inactive proenzyme. Such cascades are involve in blood clotting, where many of the clotting factors are proteases of this sort, and in the activation of the complement system.

proteome *n.* the proteins encoded by an organism's genome, considered collectively.

proteomics *n.* the study of the proteome. The large-scale identification and functional analysis of cellular proteins, usually by means of techniques that enable large numbers of different proteins to be analysed simultaneously.

proter *n.* the anterior individual produced when a protozoan divides transversely.

proterandry protandry *q.v.*

proteranthous *a.* flowering before foliage leaves appear.

proterogenesis *n.* the foreshadowing of adult or later forms by youthful or earlier forms.

proteroglyph *a.* with specialized canine teeth in upper jaw.

proterogyny protogyny *q.v.*

proterosoma *n.* (1) body region comprising mouth and the region bearing first and second legs in mites and ticks; (2) prosoma *q.v.*

Proterozoic *n.* the later division or eon of the Precambrian, lasting from *ca.* 2500 million years ago to 542 million years ago, immediately preceding the Cambrian, and whose rocks contain few fossils, mainly cyanobacteria and soft-bodied animals of problematical identity.

prothallial *a.* (1) *pert.* a prothallus; (2) *appl.* cell in microspore of gymnosperms and some pteridophytes, considered as a vestige of a prothallus.

prothalloid *a.* like a prothallus.

prothallus *n.* (1) the hyphae of lichens during the initial growth stages; (2) a small haploid gametophyte as in algae, ferns and some gymnosperms, bearing antheridia or archegonia or both, and developing from a spore.

protheca *n.* basal part of coral calicle, which is formed first.

prothecium *n.* a primary perithecium of many fungi.

protherians *n.plu.* egg-laying mammals, including the extinct triconodonts and multituberculates, and the extant monotremes (duck-billed platypus).

prothetely *n.* the development or manifestation of pupal or of imaginal characters in insect larvae.

prothoracic *a.* (1) *pert.* prothorax; (2) *appl.* glands secreting ecdysone; (2) *appl.* anterior lobe of pronotum.

prothorax *n.* the first segment of insect thorax.

prothrombin *n.* blood plasma protein, involved in blood clotting, formed in liver in the presence of vitamin K, and which is converted to the proteinase thrombin in the blood at site of trauma by a proteolytic enzyme, Factor Xa, in the presence of calcium ions. *alt.* thrombinogen.

proticity *n.* term coined to denote flow of protons (by analogy with electricity). *see* chemiosmotic hypothesis.

protist *n.* (1) any unicellular eukaryotic organism; (2) a member of the Protoctista *q.v.*

Protista Protoctista *q.v.*

proto- prefix derived from Gk *protos*, first.

proto-aecidium a cell mass surrounded by layers of hyphae, containing cells eventually producing aecidiospores and disjunctor cells.

Protoavis a possible fossil bird from the late Triassic, some 75 million years earlier than *Archaeopteryx*, but whose identification as a bird is still controversial.

protoaxis *n.* the primordial filament or axis in evolution of plant stem.

protobasidium *n.* a basidium of four cells, from each of which a basidiospore is developed by abstriction.

protoblast *n.* (1) a blastomere that develops into a definite organ or part; (2) an internal bud stage in life history of some sporozoans.

protobranch *a. appl.* gills of bivalve molluscs having flat non-reflected filaments.

protocell *n.* a simple self-replicating structure proposed as the origin of cellular life, composed of a self-replicating RNA genome surrounded by a simple membrane.

protocephalic *a. appl.* or *pert.* primary head region of insect embryo.

protocephalon *n.* (1) head part of cephalothorax in crabs, lobsters and similar crustaceans; (2) the first six body segments, which are fused to form the head of an insect.

protocercal *a.* with a caudal fin in which the vertebral column runs straight to tip, thereby dividing the fin symmetrically.

protocerebrum *n.* anterior part of 'brain' of insects and other arthropods.

protochlorophyll *n.* yellowish pigment in chloroplasts of plants grown in darkness, which is converted to chlorophyll when the plant is placed in the light.

protochordates *n.plu.* group of animals comprising the hemichordates, urochordates and cephalochordates, having gill slits, a dorsal hollow central nervous system, a persistent notochord and a postanal tail.

protocnemes *n.plu.* the six primary pairs of mesenteries of sea anemones and their relatives.

protocoel *n.* the front portion of the coelomic cavity.

protoconch *n.* the larval shell of molluscs, indicated by a scar on adult shell.

protocone *n.* inner cusp of upper molar tooth.

protoconid *n.* external cusp of lower molar tooth.

protoconidium *n.* a rounded or club-shaped cell at the tip of a filament, giving rise to conidia, as in dermatophytes.

protoconule *n.* anterior intermediate cusp of upper molar tooth.

protocorm *n.* (1) swelling of rhizophore, preceding root formation as in certain club mosses that have mycorrhizal fungi in the early stages; (2) in orchids, structure produced usually underground from germinating seedling and heavily infected with mycorrhizal fungus; (3) undifferentiated cell mass of archegonium in Ginkgoales; (4) (*zool.*) the posterior portion of germ band, which gives rise to trunk segments in insects.

protocranium *n.* posterior part of insect head.

Protoctista, protoctists *n., n.plu.* in some modern classifications a kingdom comprised of eukaryotic unicellular, colonial and simple multicellular organisms that do not fall easily into either the plant or animal kingdoms. The Protoctista are usually held to comprise the algae, including the multicellular seaweeds and other macroalgae, diatoms, protozoa, the water moulds and the cellular and acellular slime moulds. *alt.* protists. *see* Appendix 4.

protoderm *n.* (1) embryonic epidermis in plant embryos; (2) the outer layer of cells of apical meristem.

protoepiphyte *n.* a plant which is an epiphyte all its life and does not start life rooted to the ground or come to root in the ground later.

protofilament *n.* chain of protein monomers that associates laterally with other chains to form a protein filament, e.g. in cytoskeletal structures such as microtubules and intermediate filaments.

protogenic *a.* persistent from beginning of development.

protogyny *n.* the condition of hermaphrodite plants and animals in which female gametes mature and are shed before maturation of male gametes. *a.* **protogynous**.

protologue *n.* the printed matter accompanying the first description of a name.

protoloph *n.* anterior transverse crest of molar tooth.

protomala *n.* a mandible of millipedes.

protomer *n.* the asymmetric repeating unit (or units) that form the subunits of an oligomeric protein.

protomeristem promeristem *q.v.*

protomerite *n.* anterior part of medullary protoplasm of adult gregarines.

protomitosis *n.* a primitive form of mitosis as in some slime moulds.

protomorphic *a.* (1) first-formed; (2) primordial; (3) primitive.

proton *n.* subatomic particle with a mass of 1.0078 atomic units and an electric charge designated +1. Protons are part of the nuclei of all atoms. The alternative name for the hydrogen ion, H^+. *see also* proton gradient, protonmotive force, proton pump.

protonema *n.* (1) early filamentous stage in development of a fern prothallus; (2) filamentous stage in the development of some algae; (3) filamentous structure from which a moss plant buds. *plu.* **protonemata**.

protonematoid *a.* like a protonema.

protonephridial system excretory system in some simple invertebrates such as flatworms, nematodes, annelids and rotifers. It consists of a system of branching ducts (protonephridia) each closed at its internal end by a flame cell, and opening into a central duct or to the surface through pores. Flame cells bear a large bundle of flagella which project into the duct, and whose motion gives a flickering appearance under the light microscope similar to a flame.

protonephridium *n.* branching excretory duct of the protonephridial system. *plu.* **protonephridia**.

proton gradient electrochemical gradient of H⁺ set up across a membrane by active pumping of protons from one side to the other. This may occur by the agency of membrane H⁺-ATPases, as in plant and bacterial cell membranes, or as a result of movement of electrons along electron-transport chains in the membranes of mitochondria, bacteria and chloroplasts. The energy contained in the H⁺ gradient is used by cells to power many processes such as ATP synthesis, bacterial flagellar movement, and nutrient transport. *see also* protonmotive force.

protonmotive force the energy contained in an electrochemical gradient of protons formed e.g. across mitochondrial and chloroplast membranes during the passage of electrons down the electron-transport chains during respiration and photosynthesis, respectively. The movement of protons back across the membrane through the ATP synthases powers ATP synthesis. *see* chemiosmotic theory.

proton pump *n.* membrane protein which mediates the active transport of hydrogen ions (H⁺) across a membrane.

proto-oncogene the normal cellular gene from which an oncogene has been derived.

protopathic *a. appl.* stimuli and sensory systems involved in the perception of pain and of marked variations in temperature.

protopectin *n.* water-insoluble forms of pectin found in plant tissues that are hydrolyzed to pectin and pectic acid.

protopectinase *n.* enzyme that hydrolyzes protopectin.

protoperithecium *n.* primary haploid perithecium, as in some Pyrenomycetes.

protophloem *n.* first-formed phloem cells.

protoplasm *n.* living matter, the total substance of a living cell.

protoplasmic *a.* (1) *pert.* protoplasm; (2) *appl.* astrocytes: astrocytes found in grey matter and having thick branched processes similar to pseudopodia.

protoplasmodium *n.* microscopic plasmodium formed by some of the true slime moulds, which shows little streaming and usually gives rise to a single sporangium.

protoplast *n.* (1) plant cell with cell wall removed; (2) the living component of a cell, i.e. the protoplasm not including any cell wall.

protopod *a.* with feet or legs on anterior segments, not on abdomen, *appl.* insect larvae.

protopodite *n.* basal segment of arthropod limb.

protoporphyrin *n.* a precursor of porphyrin without the metal ion.

protoptile *n.* the first type of nestling down feather that develops.

protosoma, protosome prosoma *q.v.*

protospore *n.* fungal spore germinating to produce mycelium.

protosporophyte *n.* the first sporophyte stage in life cycle of red algae, the filament produced by the fertilized female cell.

protostele *n.* type of stele in which the vascular tissue forms a solid central strand, with the phloem either surrounding the xylem or interspersed within it. This arrangement is regarded as the most primitive type and is present in most roots and in stems of club mosses and some other groups.

protostelids slime moulds *n.plu.* a diverse group of largely microscopic slime moulds, which have a combination of characteristics of plasmodial and cellular slime moulds. *cf.* plasmodial, cellular.

protosterigma *n.* the basal portion of a sterigma.

protosternum *n.* sternite of cheliceral segment of prosoma in mites and ticks.

protostigmata *n.plu.* two primary gill slits of embryo. *sing.* **prostigma**.

protostoma blastopore *q.v.*

Protostomia, protostomes *n., n.plu.* collectively all animals with a true coelom, spiral cleavage of the egg, and in which the blastopore becomes the mouth (molluscs, annelids, arthropods, phoronids, bryozoans and brachiopods). *cf.* deuterostomes.

protostylic *a.* having lower jaw connected with cranium by original dorsal end of arch. *n.* **protostyly**.

protothallus *n.* the first-formed structure which gives rise to a thallus.

prototheca *n.* a skeletal cup-shaped plate at aboral end of coral embryo, the 1st skeletal formation.

Prototheria, prototherians *n., n.plu.* subclass of primitive egg-laying mammals which includes the orders Triconodonta, Multituberculata and Monotremata, of which only the monotremes (duck-billed platypus and spiny anteater) are extant.

prototroch *n.* preoral circlet of cilia of a trochosphere larva.

prototroph *n.* nutritionally independent, wild-type strain of bacterium or fungus that has no special nutritional requirements. *cf.* auxotroph.

prototrophic *a.* (1) nourished from one supply or in one manner only; (2) gaining nutrients and energy from inorganic matter.

prototype *n.* (1) an original type species or example; (2) an ancestral form.

protovertebrae *n.plu.* a series of primitive mesodermal segments in a vertebrate embryo.

protoxylem *n.* the first-formed primary xylem elements of a plant body or organ.

protozoa *n.* term generally applied to a large heterogeneous group of heterotrophic, non-photosynthetic, aquatic unicellular eukaryotes, lacking cell walls. Examples are the non-photosynthetic flagellates, the amoebas and foraminiferans, the ciliates, and parasitic protozoa such as *Eimeria*, which causes coccidiosis, and *Plasmodium*,

the malaria parasite. *a.* **protozoan**. *see* Actinopoda, Apicomplexa, Caryoblastea, Ciliophora, Cnidosporidia, Foraminifera, Rhizopoda, Zoomastigina.

protozoology *n.* that branch of biology dealing with protozoans.

protozoon *n.* individual protozoan cell.

protractor *n.* muscle that extends a part or draws it out from the body.

protriaene *n.* a trident-shaped sponge spicule with prongs of trident directed anteriorly.

Protura *n.* order of insects including the bark-lice, minute insects with 12 segments in the abdomen, no antenna or compound eyes and very small legs. Found under the bark of trees, in turf and in soil.

proventriculus *n.* (1) gizzard of insects and crustaceans; (2) in annelid worms, the portion of alimentary canal anterior to gizzard; (3) in birds, the glandular stomach anterior to the gizzard.

provinculum *n.* a primitive hinge on shells of young stages of certain lamellibranch molluscs.

provirus *n.* viral DNA that has become integrated into a host cell's chromosome and is carried from one cell generation to the next in the chromosome, not producing infective virus particles.

provisioning *n.* providing food for young, as in mass provisioning, nest provisioning, progressive provisioning, *all q.v.*

provitamin *n.* precursor to a vitamin.

proxi- prefix derived from L. *proximus*, next.

proximad *adv.* towards, or placed nearest the body or base of attachment.

proximal *a.* (1) nearest to the body, or centre or place of attachment; (2) *appl.* region of a gene close to the promoter. *cf.* distal.

proximate *a.* (1) nearest to, next to; (2) *appl.* cause: direct immediate cause.

proximate analysis of a food rough estimate of the nutritive value of a food made by first determining the total nitrogen and multiplying by 6.25 to get an approximate value for total protein, then determining the fat content by ether extraction, and finally determining the carbohydrate content by the difference between the above two values added together and the total dry weight of the sample.

proximoceptor *n.* a sensory receptor which reacts only to nearby stimuli, as a touch receptor.

proximo-distal axis (1) axis running from thumb to little finger on hand; (2) in limb or arthropod appendage, the axis running from point of attachment to body to tip.

prozonite *n.* the anterior part of a body segment consisting of two distinct parts.

prozymogen *n.* a precursor to a zymogen.

PRPP 5-phosphoribosyl-1-pyrophosphate, an activated donor of a ribose group.

PRR pattern recognition receptor *q.v.*

pruinose *a.* (1) covered with whitish particles or globules; (2) with a bloom on the surface.

Prymnesiophyta Haptophyta *q.v.*

PSI photosystem I *q.v.*

PSII photosystem II *q.v.*

PSSM position-specific score matrix *q.v.*

psalterium *n.* (1) omasum *q.v.*; (2) (*neurobiol.*) a thin triangular plate joining lateral portions of the fornix.

psammo- prefix derived from Gk *psammos*, sand.

psammon *n.* the organisms living between sand grains on freshwater and marine shores.

psammophilous arenicolous *q.v.*

psammophore *n.* one of rows of hairs under mandibles and sides of head in desert ants, used for removal of sand grains.

psammophyte *n.* plant growing in sandy or gravelly ground.

psammosere *n.* a plant succession originating in a sandy area, as on dunes.

pseud- prefix derived from Gk *pseudes*, false.

pseudambulacrum *n.* the lancet plate with adhering side plates and covering plates of certain echinoderms.

pseudannual *n.* plant that completes its growth in one year but provides a bulb or other means of surviving the winter.

pseudanthium *n.* flowerhead condensed to such an extent that it looks like a single flower.

pseudapogamy *n.* fusion of a pair of vegetative nuclei, as in certain fungi and fern prothalli.

pseudaposematic *a.* imitating warning coloration or other protective features of harmful or distasteful animals, i.e. showing Batesian mimicry.

pseudapospory *n.* spore formation without meiosis, resulting in a diploid spore that gives rise to the gametophyte.

pseudaxis *n.* an apparent main axis that really consists of a number of lateral branches running parallel.

pseudepipodite *n.* a flattened outer region of the 2nd maxilla and trunk appendages in certain crustaceans.

pseudepisematic *a.* having false coloration or markings, as in protective mimicry or for allurement or aggressive purposes.

pseudergate *n.* juvenile worker form in some termites, which can mature to a winged reproductive adult.

pseudholoptic *a. appl.* insect eyes intermediate between holoptic (where eyes meet in a coadapted line of union on top of the head) and dichoptic (where eyes are completely separate).

pseudimago *n.* a stage between pupa and imago in life history of some insects.

pseudoacrorhagus *n.* a tubercle near the margin of certain sea anemones containing ordinary epidermal nematocysts (stinging cells). *cf.* acrorhagus.

pseudoaethalium *n.* dense aggregation of distinct sporangia, as in some slime moulds. *cf.* aethalium.

pseudoalleles *n.plu.* subdivisions of a compound locus that can be distinguished by recombinational analysis.

pseudoallelic *a. appl.* two or more mutations that behave as alleles of the same locus in a complementation test but which can be separated by crossing-over, and which indicate the presence of a complex locus.

pseudoaposematic pseudaposematic *q.v.*

pseudoaquatic *a.* thriving in wet ground.

pseudoarticulation *n.* incomplete subdivision of a segment, or groove having the appearance of a joint, as in limbs of arthropods.

pseudoautosomal region small region in the mammalian X-chromosome which is homologous with a region at the tip of the short arm of the Y-chromosome, and which is involved in pairing of the sex chromosomes at meiosis.

pseudobrachium *n.* locomotory appendage formed from elongated ray of pectoral fin, used by anglerfish and batfish to 'walk' on the bottom.

pseudobulb *n.* a thickened internode of orchids and some other plants, for storage of water and food reserves.

pseudobulbil *n.* an outgrowth of some ferns, that substitutes for sporangia.

pseudobulbous *a.* adapted to hot dry conditions through development of pseudobulbs.

pseudocarp false fruit *q.v.*

pseudocartilage *n.* a cartilage-like substance serving as skeletal support in some invertebrates.

pseudocellus *n.* one of the scattered sense organs of unknown function in insects.

pseudocentrous *a. appl.* vertebra composed of two pairs of small cartilages which meet and form a suture later.

pseudocilia *n.plu.* threads of cytoplasm projecting from cell through surrounding sheath of mucilage in certain unicellular green algae.

pseudocoel *a.* fluid-filled cavity between epidermis and internal organs in rotifers, nematodes and some other invertebrates lacking a true coelom. It is derived from the embryonic blastocoel.

pseudocoelomate *a. appl.* animals whose body cavity is a pseudocoel and not a true coelom.

pseudoconch *n.* (1) a structure developed above and behind the true sphenoidal bone in crocodilians; (2) in some gastropod molluscs, a non-spiral shell.

pseudoconditioning *n.* situation where an unconditional response comes to be elicited by stimuli other than the unconditional stimulus even though there is no contingent relationship between them.

pseudocone *a. appl.* insect compound eye having ommatidia filled with transparent gelatinous material.

pseudoconidium *n.* one of the spores formed on lateral projections of pseudomycelium of certain yeasts.

pseudoconjugation *n.* conjugation in certain sporozoans, in which two individuals, temporarily and without fusion, join end to end, or side to side.

pseudocopulation *n.* in orchids, the case where the resemblance of the orchid flower to a female insect (as in bee orchids) leads

to an attempt by the male to copulate with it and so effect pollination.

pseudocortex *n.* a cortex composed of gelatinous hyphae, as in some lichens.

pseudocostate *a.* false-veined, having a marginal vein uniting all others, *appl.* leaves, insect wings.

pseudoculus *n.* an oval area on either side of head of certain millipedes, possibly a receptor for mechanical vibrations.

pseudocyphella *n.* a structure in lichens similar to a cyphella but smaller, also thought to be used in aeration of the thallus.

pseudocyst *n.* a residual mass of protoplasm which swells and ruptures, liberating spores of sporozoans.

pseudodeltidium *n.* a plate partly or entirely closing deltidial fissure in ventral valve of some brachiopods.

pseudoderm *n.* a skin-like covering of certain compact sponges.

pseudodominance *n.* expression of a recessive allele in the absence of the dominant allele.

pseudodont *n.* having horny pads or ridges instead of teeth, as monotremes.

pseudoelater *n.* one of the chains of cells in sporogonium of some liverworts, probably functioning as a true elater.

pseudofoliaceous *a.* with expansions resembling leaves.

pseudogamy *n.* (1) union of hyphae from different thalli; (2) activation of ovum by sperm which plays no part in further development; (3) pseudomixis *q.v.*

pseudogaster *n.* an apparent gastral cavity of certain sponges, opening to exterior by a pseudo-osculum and having no true oscula opening into itself.

pseudogastrula *n.* the stage in the development of certain sponges in which the archaeocytes become completely enclosed by flagellate cells.

pseudogene *n.* a stretch of DNA related in sequence to a functional gene but which is non-functional as a result of the mutations it has accumulated.

pseudohaptor *n.* a large discoidal organ in some trematodes, with a ventral armature of spines arranged in radial rows.

pseudoheart *n.* (1) the axial organ in echinoderms; (2) one of the contractile vessels pumping blood from dorsal to ventral vessel in annelids.

pseudoidium *n.* a separate hyphal cell which may germinate.

pseudoknot *n.* a common type of tertiary structure in RNA, in which the chain is folded back on itself to make a double fold with short stretches of base pairing.

pseudolamina *n.* expanded apical portion of a phyllode.

pseudometamerism *n.* (1) apparent serial segmentation; (2) an approximation to metamerism, as in certain cestodes.

pseudomitotic diaschistic *q.v.*

pseudomixis *n.* sexual reproduction by fusion of vegetative cells instead of gametes, leading to zygote formation.

pseudomonads *n.plu.* Gram-negative bacteria, family Pseudomonadaceae, widely distributed in soil and water. Some are plant pathogens and some are opportunistic pathogens of humans, esp. in hospital-acquired infections. They are typically aerobic or facultatively anaerobic heterotrophs with rod-shaped cells bearing polar flagella. Some species contain blue or green fluorescent pigments.

pseudomonocarpous *a.* with seeds retained in leaf bases until liberated, as in cycads.

pseudomonocotyledonous *a.* with two cotyledons coalescing to appear as one.

pseudomorph *n.* a structure having an indefinite form.

pseudomycelium *n.* an assemblage of chains of single cells, as in some yeasts.

pseudomycorrhiza *n.* mild pathological fungal infection of plant roots, superficially resembling mycorrhiza.

pseudonavicella *n.* a small boat-shaped spore containing sporozoites, in sporozoans.

pseudonotum postscutellum *q.v.*

pseudonychium *n.* a lobe or process between claws of insects.

pseudo-osculum *n.* the exterior opening of a pseudogaster in sponges.

pseudopallium *n.* in some gastropod molluscs parasitic on echinoderms, a ring-like fold of skin developing at the base of the proboscis and eventually extending like a sac over the whole parasite.

pseudoparaphysis basidiolum *q.v.*

pseudoparasitism *n.* accidental entry of a free-living organism into the body and its survival there.

pseudoparenchyma *n.* fungal or algal tissue formed of a tightly woven mass of hyphae, in which the hyphae have lost their individuality and which superficially resembles parenchyma tissue of plants.

pseudopenis *n.* (1) the protruded evaginated portion of male deferent duct, in some oligochaete worms; (2) copulatory structure in Orthoptera.

pseudoperculum *n.* a structure resembling an operculum or closing membrane.

pseudoperianth *n.* an envelope investing the archegonium in certain liverworts.

pseudoperidium *n.* the aecidiospore envelope in certain fungi.

pseudoplasmodium *n.* an aggregation of myxamoebae of a cellular slime mould, in which cells do not fuse.

pseudopod *n.* protrusion of cytoplasm put out by a cell, particularly by amoeboid protozoa and phagocytic cells of animals and plants where it serves for locomotion and feeding, *plu.* **pseudopodia**, *alt.* **pseudopodium**; (2) (*bot.*) the stalk supporting the sporangium in some mosses; (3) slender branch of the gametophyte that bears gemmae in some mosses.

pseudopregnancy *n.* condition of development of accessory reproductive organs simulating true pregnancy although fertilization has not taken place.

pseudopregnant *a. appl.* mice and other mammals which have been treated with hormones so that the uterus is receptive to *in vitro* fertilized ova and blastocysts which implant and develop normally.

pseudopupa coarctate pupa *q.v.*

pseudoramose *a.* having false branches.

pseudorhiza *n.* root-like structure connecting mycelium in the soil with the fruit body of a fungus.

pseudosacral *a. appl.* sacral vertebra attached to pelvis by transverse process and not by sacral rib.

pseudoscolex *n.* modified anterior proglottids of certain cestodes where the true scolex is absent.

Pseudoscorpiones, pseudoscorpions *n.*, *n.plu.* order of small arachnids, commonly called false scorpions, which resemble scorpions but whose opisthosoma is not divided into two regions.

pseudosematic pseudepisematic *q.v.*

pseudoseptate *a.* apparently but not morphologically septate, having a perforate or incomplete septum.

pseudoseptum *n.* (1) a septum with pores, as in certain fungi; (2) septum-like structure deposited at intervals in hyphae of some chytrids.

pseudosessile *a. appl.* abdomen of petiolate insects when petiole is so short that abdomen is close to thorax.

pseudosperm *n.* a small indehiscent fruit resembling a seed.

pseudospore *n.* an encysted resting myxamoeba.

pseudostele *n.* an apparently stelar structure, as midrib of leaf.

pseudostigma *n.* a cup-like pit in integument, as the socket of a sensory seta in ticks and mites.

pseudostiole, pseudo-ostiole *n.* a small opening formed by breakdown of cell walls or tissues, in certain fungi without perithecia.

pseudostipe *n.* a stem-like structure formed by presumptive spore-forming tissue, as in gasteromycete fungi.

pseudostipula, pseudostipule *n.* part of lamina at the base of a leaf stalk, which resembles a stipule.

pseudostoma *n.* a temporary mouth or mouth-like opening.

pseudostroma *n.* a mass of mixed fungal and host cells.

pseudosuckers *n.plu.* powerful organs of attachment, with gland cells, on either side of the oral sucker of trematodes.

pseudosymmetry *n.* approximate, but not exact, symmetry.

pseudothecium *n.* fruiting body resembling a perithecium.

pseudotrachea *n.* (1) a trachea-like structure; (2) one of the trachea-like food channels of labellum, as in dipterans.

pseudotroch *n.* inner ring of cilia around mouth of some rotifers.

pseudotrophi *a. appl.* mycorrhiza when the fungus is parasitic.

pseudouracil (ΨU) unusual pyrimidine base found in tRNA, formed by exchange of a carbon and nitrogen atom at positions 1 and 5 in the uracil ring. The nucleotide residue is called **pseudouridine**.

pseudovacuole gas vacuole *q.v.*

pseudovarium *n.* an ovary producing pseudova, ova that can develop without fertilization.

pseudovelum *n.* (1) in jellyfish, a velum without muscular or nerve cells; (2) in agaric fungi, a structure formed by outgrowths from cap and stalk, protecting immature hymenium. *alt.* pseudoveil.

Pseudoviridae *n.* family of DNA viruses with a reverse transcription stage in replication, including the Ty1 virus of *Saccharomyces cerevisiae* and the copia virus of *Drosophila*.

pseudovitellus *n.* a mass of fatty cells in abdomen of aphid.

pseudovum *n.* ovum that can develop parthenogenetically.

Psilophyta *n.* one of the four major divisions of extant seedless vascular plants, represented by only two living genera. They are tropical plants of simple structure, having a rootless sporophyte, dichotomously branching rhizomes, and aerial branches with small scale-like appendages (*Psilotum*) or larger bract-like outgrowths (*Tmesipteris*). *alt.* Psilopsida.

P-site site on the ribosome occupied by the last mRNA codon to have been read and by the peptidyl-tRNA.

psi torsion angle (ψ) the angle of rotation of the C–C$_\alpha$ bond to the adjacent peptide bond in the backbone of a polypeptide chain.

Psittaciformes *n.* an order of birds including the parrots.

psoas *n.* either of two muscles in the loin: psoas major and psoas minor.

Psocoptera, psocids *n.*, *n.plu.* order of small insects, commonly called book lice and bark lice, having incomplete metamorphosis, a globular abdomen and often no wings.

PSSM position-specific score matrix *q.v.*

psychoactive *a. appl.* any substance that influences the mind or behaviour.

psychogenetic *a.* (1) *pert.* mental development; (2) caused by the mind.

psychogenic *a.* of mental origin, *appl.* physiological and somatic changes.

psychophysiology *n.* physiology in relation to mental processes.

psychosomatic *a.* (1) *pert.* relationship between mind and body; (2) *pert.* or having bodily reactions to mental stimuli.

psychrophile *n.* organism that thrives at low temperatures. For bacteria, those growing best at temperatures below 15°C and with a maximum temperature for growth of below 20°C are considered to be psychrophiles. *a.* **psychrophilic.** *alt.* **psychrophil.**

psychrotolerant *a. appl.* organism that can grow at low temperatures but whose growth temperature optimum is above 20°C.

PT pertussis toxin *q.v.*

PTA (1) phosphotungstic acid, a staining agent for electron microscopy; (2) plasma thromboplastin antecedent *q.v.*

PtdInsP phosphatidylinositol phosphates *q.v.*

ptera-, ptero- prefixes derived from Gk *pteryx*, wing.

pteralia *n.plu.* axillary sclerites forming articulation of wing with the process of the mesonotum in insects.

pterate *a.* winged.

pterergate *n.* worker or soldier ant with vestigial wings.

pteridine *n.* organic compound composed of two fused six-membered rings of nitrogen and carbon with various substituents, which is a constituent of many natural compounds such as pterins (leucopterin, xanthopterin) and folic acid.

pteridology *n.* the branch of botany dealing with ferns.

pteridophytes *n.plu.* major group of spore-bearing vascular plants: the ferns, club mosses, horsetails and the Psilophyta, sometimes treated as a division, Pteridophyta. *see* Lycophyta, Psilophyta, Pterophyta, Sphenophyta.

Pteridospermophyta, pteridosperms *n., n.plu.* an extinct division of seed-bearing vascular plants, the seed ferns.

pterin *n.* any of a group of pigments, derivatives of pteridine, widespread in insects as eye pigments and in wings, and which are also found in vertebrates and plants.

pterion *n.* (1) the point of junction of parietal, frontal and great wing of sphenoid; (2) *appl.* ossicle, a sutural bone.

Pterobranchia, pterobranchs *n., n.plu.* class of small hemichordates with a vase-shaped body, U-shaped digestive tract, and a collar surrounding the mouth extended into pairs of hollow arms bearing ciliated tentacles. They are mostly colonial and some species are enclosed in a secreted tube.

pterocardiac *a. appl.* ossicles with curved ends in gastric mill of crustaceans.

pterocarpous *a.* with winged fruit.

pterodactyls *n.plu.* the common name for the pterosaurs *q.v.*

pterodium *n.* a winged fruit or samara.

pteroic acid *n.* a compound of a pterin and *p*-aminobenzoic acid, which combines with glutamine to form folic acid.

pteroid *a.* (1) resembling a wing; (2) like a fern.

pteropaedes *n.plu.* birds able to fly when newly hatched.

pteropegum *n.* an insect's wing socket.

Pterophyta *n.* one of the four major divisions of the spore-bearing vascular plants, commonly called the ferns. The sporophyte has roots, stems and large leaves (fronds) which are megaphylls and bear the sporangia.

pteropleura *n.* thoracic sclerite between wing insertion and mesopleura in dipterans.

pteropodium *n.* a winged foot as in certain bats.

pteropods *n.plu.* group of marine gastropod molluscs with wing-like extensions to the foot, commonly called sea butterflies.

Pteropsida *n.* plant classification which has been used in different ways, as an alternative to Filicophyta (ferns), or for a larger grouping containing the ferns and seed plants.

Pterosauria, pterosaurs *n., n.plu.* an order of Jurassic and Cretaceous archosaurs, flying reptiles commonly called pterodactyls, which have a membranous wing supported by a greatly elongated fourth finger.

pterospermous *a.* with winged seeds.

pterostigma *n.* an opaque cell on wing of insect.

pterote *a.* winged, having wing-like outgrowths.

pterotheca *n.* the wing-case of a pupa.

pterothorax *n.* fused mesothoracic and metathoracic segments, in dragonflies.

pterotic *n.* a cranial bone overlying horizontal semicircular canal of ear. *a. appl.* the bone between pro-otic and epiotic.

pterygia *plu.* of pterygium *q.v.*

pterygial *a.* (1) *pert.* a wing or fin; (2) *appl.* a bone or cartilage supporting a fin-ray.

pterygiobranchiate *a.* having feathery or spreading gills, as certain crustaceans.

pterygiophore *n.* (1) cartilaginous rod forming ray of fins of e.g. sharks and dogfishes; (2) similar bony ray in fins of bony fishes.

pterygium *n.* (1) prothoracic process in weevils; (2) small lobe at base of under-wings in Lepidoptera; (3) vertebrate limb. *plu.* **pterygia**.

pterygoda *n.plu.* the tegulae of an insect.

pterygoid *a.* (1) wing-like; (2) *appl.* a wing-like process of sphenoid bone; (3) *n.* a cranial bone, forming part of roof of mouth.

pterygoideus *n.* muscles causing protrusion and raising of mandible (lower jaw), pterygoideus externus and internus.

pterygomandibular *a.* (1) *pert.* pterygoid and mandible; (2) *appl.* a band of tendon or raphe of buccopharyngeal muscle.

pterygomaxillary *a. appl.* a fissure between maxilla and pterygoid process of sphenoid.

pterygopalatine *a.* (1) *pert.* a region of pterygoid and palatal cranial bones, *appl.* canal, fossa, groove; (2) *appl.* ganglion: sphenopalatine ganglion *q.v.*

pterygoquadrate *n.* a cartilage constituting dorsal half of mandibular arch of some fishes.

pterygospinous *a. appl.* a ligament between lateral pterygoid plate and spinous process of sphenoid.

pterygote, pterygotous *a. appl.* a large group of insects, the subclass Pterygota, which contains all the winged insects (and some wingless ones such as fleas and lice, considered to be descended from winged forms), and whose members undergo some form of metamorphosis and have no organs of locomotion on the abdominal segments. Applies to all insect orders except Thysanura, Diplura, Protura and Collembola.

pterylae *n.plu.* feather tracts.

pterylosis *n.* arrangement of feather tracts in birds.

PTH parathyroid hormone *q.v.*

ptilinum *n.* head vesicle or bladder-like expansion of head of a fly emerging from pupa.

ptilopaedic *a.* covered with down when hatched.

ptomaine *n.* any of a group of amino compounds, usually poisonous, produced during protein breakdown during putrefaction of dead animal matter and by some plants, and including choline, putrescine, cadaverine and muscarine.

ptosis *n.* drooping of the eyelids, congenital ptosis in humans usually being due to inheritance of a simple Mendelian dominant allele.

PTP protein tyrosine phosphatase.

ptyalin *n.* salivary amylase, found in man and some herbivores, which digests broken starch grains. EC 3.2.1.1, *r.n.* α-amylase.

ptyophagous *a. appl.* digestion, by host cells, of the cytoplasm emitted from tips of mycorrhizal hyphae invading the cells.

P-type ATPases type of ion-transport proteins with ATPase activity in which autophosphorylation is part of the transport cycle. They are single-chain proteins with 10 transmembrane helices. Examples are the Na^+-K^+ pump in the plasma membrane of animal cells and the Ca^{2+} pump in the sarcoplasmic reticulum of skeletal muscle cells.

ptyxis *n.* the form in which young leaves are folded or rolled on themselves in the bud.

puberty *n.* the beginning of sexual maturation.

puberulent *a.* covered with down or fine hair.

pubes *n.* the lower portion of hypogastric region of abdomen, the pubic region.

pubescent *a.* covered with soft hair or down.

pubic *a.* (1) in the region of pubes; (2) *pert.* pubis.

pubis *n.* in each half of pelvic girdle of vertebrates except fishes, bone forming anteroventral part of girdle, fused with ischium to form hip-bone.

public *a.* (1) *appl.* antigenic determinants shared by other haplotypes in serological analysis of histocompatibility antigens; (2) *appl.* any antigenic determinant shared by different antigens.

puboischium *n.* fused pubis and ischium, bearing acetabulum and ilium on each side.

pudendum *n.* vulva, or external female genitalia. *plu.* **pudenda.**

puffball *n.* common name for fungi of the order Lycoperdales in the Gasteromycetes *q.v.*

puffer fish *Takifugu (Fugu) rubripes,* a fish that has an unusually small genome for a vertebrate, due to very small introns and a lack of repetitive DNA.

puffs *n.plu.* temporarily swollen regions visible on polytene chromosomes at which chromosomal material is extruded and DNA is being actively transcribed.

pullulation *n.* reproduction by vegetative budding.

pulmo- prefix from L. *pulmo,* lung, usually denoting lung-like, to do with the lungs or breathing.

pulmobranch *n.* (1) a gill-like organ adapted to air-breathing conditions; (2) lung book, as of spiders. *alt.* pulmobranchia.

pulmogastric *a. pert.* lungs and stomach.

pulmonary *a. pert.* lungs.

pulmonary cavity, pulmonary sac the mantle cavity, modified as a lung, in pulmonate molluscs (slugs and snails).

pulmonates *n.plu.* molluscs of the subclass Pulmonata, the snails and slugs, characterized by lack of ctenidia and in which the mantle cavity is used as a lung.

pulmones *n.plu.* lungs. *sing.* **pulmo.**

pulp *n.* internal cavity of vertebrate tooth, containing connective tissue, nerves and blood vessels.

pulsating or pulsatile vacuole contractile vacuole *q.v.*

pulse *n.* the seed of a legume, e.g. peas, beans, lentils.

pulse-chase experiments experiments in which cells are very briefly labelled with a radioactive precursor of a particular molecule or pathway and the fate of the label is followed during subsequent incubation with non-labelled precursor.

pulsed-field gel electrophoresis electrophoretic technique for separating large pieces of DNA, e.g. chromosomes, in which an electric field is applied first in one direction and then in a direction at an angle to the first.

pulse wave a wave of increased pressure over the arterial system, started by ventricular systole.

pulverulent *a.* (1) powdery; (2) powdered.

pulvillar *a. pert.* or at a pulvillus.

pulvilliform *a.* like a small cushion.

pulvillus *n.* (1) pad, process or membrane on foot or between claws, sometimes serving as an adhesive organ, in insects; (2) lobe beneath each claw on feet of some mammals. *a.* **pulvillar.**

pulvinar *a.* (1) cushion-like; (2) *pert.* a pulvinus; (3) *n.* an angular prominence on thalamus in brain.

pulvinate *a.* (1) cushion-like, *appl.* a defensive or offensive gland in ants; (2) having a pulvinus.

pulvinoid *a.* resembling a pulvinus, *appl.* modified petiole.

pulvinus *n.* cellular swelling at junction of axis and leaf stalk, which plays a part in leaf or leaflet movement.

pulviplume powder-down feather *q.v.*

pump *n.* active transport mechanism for ions or small molecules in cell membrane.

punctae *n.plu.* small pores, holes or dots on a surface, esp. the markings on valves of diatoms.

punctate *a.* (1) dotted; (2) having surface covered with small holes, pores or dots; (3) having a dot-like appearance.

punctiform *a.* (1) having a dot-like appearance; (2) *appl.* distribution, as of cold, warm and pain spots on skin.

punctuated equilibrium the view that the course of evolution has been marked by long periods of little or no evolutionary change (stasis) punctuated by short periods of rapid evolution. The view is based on an interpretation of the fossil record which in some cases appears to show such a pattern. *cf.* phyletic gradualism.

punctulate *a.* covered with very small dots.

punctum *n.* a minute dot, point or orifice.

puncture *n.* (1) small round surface depression; (2) perforation.

pungent *a.* (1) producing a prickling sensation, *appl.* stimuli affecting chemoreceptors; (2) bearing a sharp point, *appl.* apex of leaf or leaflet.

punishment negative reinforcement *q.v.*

Punnett square a conventional representation used to calculate the proportions of different genotypes in progeny of a genetic cross, e.g. for parents *Aa* and *aa*:

	A	*a*
a	*Aa*	*aa*
a	*Aa*	*aa*

pupa *n.* in insects with complete metamorphosis, a resting stage in the life cycle where the larval insect is enclosed in a protective case, within which tissues are reorganized and metamorphosis into a new form, usually the adult, occurs.

puparium *n.* (1) the casing of a coarctate pupa, formed from the last larval skin, esp. in Diptera; (2) a larval instar.

pupate *v.* to pass into the pupal stage.

pupiform *a.* resembling a pupa in shape.

pupigerous *a.* containing a pupa.

pupil *n.* (1) central aperture of iris in vertebrate eye through which light enters, and whose size is varied by contraction of the ciliary body around the iris; (2) central spot of an eye-spot (ocellus).

pupillary *a.* (1) *pert.* pupil of an eye; (2) *appl.* reflex: variation in aperture of pupil due to change in illumination, closing in bright light, opening in dim light.

pupillate *a. appl.* eye-like markings with a differently coloured central spot.

pupiparous *a.* bringing forth young already developed to the pupa stage, as in certain parasitic insects.

pupoid pupiform *q.v.*

pure-breeding true-breeding *q.v.*

purifying selection type of selection that eliminates individuals carrying deleterious mutations that interfere with protein function and regulation.

purine *n.* a type of nitrogenous organic base, of which adenine and guanine are most common in cells.

purinergic *a.* secreting purines, *appl.* neurons that secrete purine neurotransmitters (e.g. adenosine and ATP).

purinoceptor *n.* cell-surface receptor for purines (e.g. adenosine and ATP), found on some neurons and other cells.

Purkinje cell, Purkinje neuron type of neuron found in cerebellum, whose cell bodies constitute a distinct layer in cerebellar cortex between the molecular and granular (nuclear) layers, and whose axons carry the output from cerebellum.

Purkinje fibres muscle fibres in the band of muscle and nerve fibres connecting auricles and ventricles of heart (His' bundle). They differ from typical cardiac fibres, esp. in a higher rate of conduction of contractile impulse.

pure line organisms originating from a single homozygous ancestor or identical homozygous ancestors, and which are therefore themselves homozygous for a given character or characters.

puromycin *n.* antibiotic which becomes incorporated into a polypeptide chain during protein synthesis, causing the release of the incomplete chain from the ribosome.

purple bacteria (1) Proteobacteria *q.v.*; (2) a group of photoautotrophic members of the proteobacteria. They contain bacteriochlorophylls *a* and *b* and purple (or brown or pink) carotenoid pigments, and carry out anoxygenic photosynthesis using one photosystem only.

purple membrane membrane in certain archaea, e.g. the halophile *Halobacterium*, which contains large amounts of the purple protein bacteriorhodopsin, a light-driven transmembrane proton pump.

purple sulphur bacteria group of mainly aquatic photoautotrophic bacteria, included in the gamma-proteobacteria, which oxidize sulphide to sulphur.

purposive behaviour goal-related behaviour.

pus *n.* creamy material formed by dead and dying cells, esp. neutrophils, in infected wounds.

pustule *n.* a fluid-filled blister-like prominence.

pusule *n.* (1) non-contractile vacuole containing watery fluid, filling or emptying by a duct, present in many dinoflagellates; (2) a contractile vacuole in some algae.

putamen *n.* lateral part of lentiform nucleus in cerebral hemispheres.

putrefaction *n.* decomposition of organic material, esp. the anaerobic breakdown of proteins by microorganisms, resulting in incompletely oxidized, ill-smelling compounds such as mercaptans, alkaloids and polyamines.

putrescine *n.* a foul-smelling polyamine, formed by decarboxylation of ornithine and often produced during breakdown of protein by bacteria, e.g. in putrefying meat, also found in small amounts in various mammalian tissues such as liver, pancreas, semen.

p-**value** a statistical value that gives the probability that a value equal to or more than a given value *x* could have occurred by chance.

PWM pokeweed mitogen *q.v.*

pycnidia *plu.* of pycnidium.

pycnidiophore *n.* a conidiophore producing pycnidia.

pycnidiospore *n.* a spore produced in a pycnidium.

pycnidium *n.* small hollow spherical or flask-shaped structure enclosing conidiophores in rust and smut fungi. *a.* **pycnidial**. *plu.* **pycnidia**.

pycnium (1) pycnidium *q.v.*; (2) spermogonium *q.v.*

Pycnogonida, pycnogonids *n., n.plu.* class of chelicerate marine arthropods commonly known as sea spiders. They have a long slender body consisting of an anterior cephalon, a trunk with four pairs of long walking legs, and a short segmented abdomen. Some species bear chelicerae and feelers, others have neither.

pycnosis *n.* cell degeneration including condensation of nuclear contents and formation of an intensely staining clump of chromosomes.

pycnospore pycnidiospore *q.v.*

pycnotic *a.* characterized by or *pert.* pycnosis, *appl.* small irregular nucleus of degenerate cells.

pycnoxylic *a.* having compact wood.

pygal *a.* (1) situated at or *pert.* posterior end of back; (2) *appl.* certain plates of shell of tortoises and turtles.

pygidium *n.* an exoskeletal shield covering tail region of some arthropods, and vari-ous structures in the same region in other insects. *a.* **pygidial**.

pygmy male a purely male form, usually small, found living close to the ordinary hermaphrodite form in certain animals, as in some polychaete worms and barnacles.

pygostyle *n.* a compressed upturned bone at end of vertebral column in birds, composed of fused vertebrae.

pykn- *see* pycn-.

pylangium *n.* proximal portion of amphibian or foetal heart, through which blood is driven to the ventricles.

pylome *n.* in some radiolarians, an aperture for emission of pseudopodia and intake of food.

pyloric *a.* (1) *pert.* or in the region of the pylorus; (2) *appl.* sphincter between mid-gut and hind-gut of insects; (3) *appl.* posterior region of gizzard in crustaceans.

pylorus *n.* the lower opening of stomach, into duodenum.

pyogenic *a.* pus-producing, *appl.* bacteria that cause the accumulation of pus at the site of infection.

pyramidal *a.* conical, like a pyramid, *appl.* leaves, a carpal bone, tract of nerve fibres in brain.

pyramidal cell a type of neuron found in the cerebral cortex, of characteristic shape.

pyramidal tract tract of motor nerve fibres running from cerebral motor cortex of brain to the anterior horn cells of spinal cord at all levels, concerned with voluntary motion. *alt.* corticospinal tract.

pyramid of biomass a representation of the total biomass at each level of a food chain, which forms a pyramid, the biomass at ·lower levels (e.g. primary producers) being greater than that at higher levels (e.g. carnivores).

pyramid of energy a representation of the energy available per unit time at each trophic level in an ecosystem, usually expressed in kilocalories per square metre per year.

pyramid of numbers a representation of the numbers of organisms at different levels of a food chain, which forms a pyramid, greater numbers of organisms being present at the lower levels (e.g. primary producers) than at higher levels (e.g. carnivores).

pyranose *n.* a monosaccharide in the form of a six-membered ring of five carbons and one oxygen. *cf.* furanose.

pyrene *n.* a seed surrounded by a hard body forming a fruit stone or kernel, often with several in one fruit.

pyrenocarp (1) perithecium *q.v.*; (2) drupe *q.v.*

pyrenoid *n.* in algae and certain liverworts, a protein body in the chloroplast which is the centre of starch formation.

pyrenolichen *n.* lichen in which the fungal component is a loculoascomycete.

Pyrenomycetes *n.* group of ascomycete fungi, commonly called the flask fungi, in which the generally club-shaped asci are usually borne in a hymenial layer in flask-shaped or spherical ascocarps (perithecia) which open in a terminal pore. They include the plant parasitic powder mildews, the saprophytic pink bread mould *Neurospora*, the agents of several plant cankers (including the coral-spot fungus), anthracnoses, and leaf-spot diseases.

pyretic *a.* (1) increasing heat production; (2) causing a rise in body temperature.

pyrexia *n.* an increase in body temperature above normal, often as a result of infection.

pyridine nucleotides nicotinamide adenine dinucleotide (*q.v.*), and nicotinamide adenine dinucleotide phosphate (*q.v.*).

pyridoxal *see* pyridoxal phosphate.

pyridoxal phosphate (PLP) a coenzyme derived from pyridoxine (vitamin B_6), the prosthetic group of many enzymes including all transaminases.

pyridoxamine *see* pyridoxine.

pyridoxine *n.* a form of vitamin B_6, a phenolic alcohol derived from pyridine. It can be converted to pyridoxal and pyridoxamine in the body and is a precursor to the coenzyme pyridoxal phosphate. *see also* vitamin B_6.

pyriform *a.* pear-shaped.

pyrimidine *n.* a type of nitrogenous organic base, of which cytosine, uracil and thymine are most common in living cells. They are present chiefly in nucleic acids and nucleotides.

pyrimidine dimer structure produced in DNA by ultraviolet light in which adjacent pyrimidines on the same strand become covalently linked, blocking DNA replication and transcription.

pyrogen *n.* a substance that can cause fever.

pyrophosphatase *n.* enzyme hydrolysing inorganic pyrophosphate to orthophosphate. EC 3.6.1.1. *r.n.* inorganic diphosphatase.

pyrophosphate (PPi) $HP_2O_7^{3-}$.

pyrophyte *n.* plant that likes to grow on burnt ground.

pyrosequencing *n.* a method of replication-based DNA sequencing in which the incorporation of nucleotides during the sequencing reactions generate flashes of light, which are digitally recorded and analysed to give the sequence of the DNA.

pyrotheres *n.* a group of extinct South American placental mammals of the Eocene to Oligocene with tusk-like teeth and tending to large size.

pyrrole ring five-membered ring containing four carbons and one nitrogen, which is a basic unit of the porphyrin ring (tetrapyrrole) of haem.

Pyrrophyta *n.* the dinoflagellates, a group of largely unicellular biflagellate eukaryotic microorganisms sometimes known as whirling whips. They include both photosynthetic and heterotrophic forms and are important members of both marine and freshwater plankton. A feature of many dinoflagellates is the plates of cellulose immediately under the plasma membrane, which form a sculptured wall (theca) around the cell.

pyruvate *see* pyruvic acid.

pyruvate carboxylase enzyme that catalyses the addition of a carboxyl group to pyruvate to produce oxaloacetate, using the cofactor biotin as a carrier of high-energy carboxyl groups. EC 6.4.1.1.

pyruvate decarboxylase key enzyme in alcoholic fermentation, catalysing the irreversible decarboxylation of pyruvate to yield acetaldehyde and CO_2. EC 4.1.1.1.

pyruvate dehydrogenase complex an assembly of three enzymes catalysing the oxidative decarboxylation of pyruvate to form acetyl-CoA for the tricarboxylic acid cycle. It is composed of pyruvate dehydrogenase (EC 1.2.4.1), dihydrolipoamide transacetylase (EC 2.3.1.12) and dihydrolipoamide dehydrogenase (EC 1.8.1.4).

pyruvate kinase enzyme catalysing the transfer of phosphate from phosphoenolpyruvate to ADP with formation of pyruvate and ATP (and vice versa) in glycolysis and elsewhere. EC 2.7.1.40.

pyruvic acid, pyruvate *n*. three-carbon organic acid ($CH_3COCOOH$) produced during glycolysis, which is an important intermediate in many metabolic pathways. It is converted to acetyl-CoA, in which form it is the starting-point of the tricarboxylic acid cycle of aerobic respiration.

pyxidiate *a*. (1) opening like a box by transverse dehiscence; (2) *pert*., or like a pyxidium.

pyxidium *n*. a capsular fruit which dehisces transversely, the top coming off as a lid.

pyxis pyxidium *q.v.*

Q

Q (1) glutamine *q.v.*; (2) ubiquinone *q.v.*

Q$_{CO_2}$ the volume of carbon dioxide (in microlitres of gas at normal temperature and pressure) given out per hour per milligram dry weight.

Q$_{O_2}$ oxygen quotient *q.v.*

Q$_n$ ubiquinone *q.v.*

Q$_{10}$ temperature coefficient *q.v.*

Q bands characteristic pattern of fluorescent bands on mitotic chromosomes stained with quinacrine, used in identification of individual chromosomes.

Q-cycle proposed recycling of ubiquinone through the cytochrome b-c_1 complex, leading to two H^+ being pumped for every electron transferred.

qPCR quantitative PCR *q.v.*

QTLs quantitative trait loci, *see* quantitative trait.

quadrangular *a.* (1) *appl.* four-cornered stems; (2) *appl.* lobes of cerebellar hemispheres, connected by the largest part of the superior vermis of cerebellum.

quadrant *n.* all the cells derived from one of the first four blastomeres.

quadrat *n.* a sample area enclosed within a frame, usually a square, within which a plant community, or sometimes an animal community, is analysed.

quadrate bone part of the hyomandibular in vertebrate skull, with which lower jaw articulates in birds, reptiles, amphibians and fishes.

quadratojugal *n.* membrane bone connecting quadrate and jugal bones. *alt.* **quadratomaxillary**.

quadratus *n.* the name of several rectangular muscles, such as quadratus femoris.

quadri- prefix derived from L. *quattuor*, four, signifying having four of, divided into four, in four parts.

quadricarpellary *a.* containing four carpels.

quadriceps *n.* muscle in front of thigh, extending lower leg and divided into four portions at the upper end.

quadrifarious *a.* in four rows, *appl.* leaves.

quadrifid *a.* cleft into four parts.

quadrifoliate *a.* (1) four-leaved; (2) *appl.* compound palmate leaf, with four leaflets arising at a common point.

quadrigeminal bodies corpora quadrigemina *q.v.*

quadrihybrid (1) *n.* a cross whose parents differ in four distinct characters; (2) *a.* heterozygous for four pairs of alleles.

quadrijugate *a. appl.* pinnate leaf having four pairs of leaflets.

quadrilateral *n.* the discal cell in wing of dragonflies, a large cell at base of wing completely enclosed by veins.

quadrilobate *a.* four-lobed.

quadrilocular *a.* having four loculi or chambers, as ovary, or anthers of certain flowers.

quadrimaculate *a.* having four spots.

quadripennate *a.* having four wings.

quadripinnate *a.* divided pinnately four times.

quadriradiate *a.* having four rays.

quadriseriate *a.* arranged in four rows or series. *alt.* tetrastichous.

quadritubercular *a. appl.* teeth with four cusps.

quadrivalent *n.* (1) association of four chromosomes that forms during meiosis in an individual carrying a heterozygous reciprocal translocation. This type of quadrivalent is formed of four complete chromosomes, the two chromosomes carrying the translocation and their untranslocated homologues, and has four centromeres; (2) the term is also sometimes used for the pair of duplicated homologous chromosomes, made up of four chromatids, that are linked by chiasmata during meiosis. This type of quadrivalent has only two centromeres.

quadrivoltine *a.* having four broods in a year.

quadrumanous *a.* having hind-feet as well as fore-feet constructed as hands, as in most primates, except humans.

quadrupedal *a.* walking on four legs.

quadruplex *a. appl.* polyploids having four dominant genes.

quaking bog type of bog covered with a floating mass of vegetation (often sphagnum moss) that moves when walked on.

qualitative *a.* concerned only with the nature of a property under investigation. *cf.* quantitative.

qualitative inheritance the inheritance of phenotypic characters that occur in two or more distinct states within a population and do not grade into each other. The states represent combinations of different alleles at a single locus. *alt.* simple Mendelian inheritance. *cf.* quantitative or polygenic inheritance.

quantitative *a.* concerned with the amounts, as well as the nature, of organisms or substances under investigation. *cf.* qualitative.

quantitative PCR (qPCR) type of polymerase chain reaction in which the starting DNA is quantified as well as amplified by measuring the amount after each round of amplification. *alt.* real-time PCR, **quantitative RT-PCR**, RTQ-PCR.

quantitative inheritance the inheritance of characters determined by many different genes acting independently, and which appear as continuously variable characters within a population. *cf.* qualitative inheritance.

quantitative trait in genetics, a phenotypic character determined by the effects of many genes (quantitative trait loci, QTLs), which shows a continuously graded spectrum of variation within a population that can only be measured quantitatively, such as height or weight. *see also* complex trait.

quantitative variation continuous variation *q.v.*

quantosomes *n.plu.* particles on the thylakoid membranes of chloroplasts, which contain the photosynthetic apparatus.

quantum *n.* (1) the smallest amount in which a neurotransmitter is normally secreted; (2) a unit of light. *plu.* **quanta**. *a.* **quantal**. *see* photon.

quantum dots semiconductor nanocrystals that can be made to fluoresce in many different colours, and which can be used to stain cells and tissues for imaging experiments.

quartet *n.* (1) a group of four nuclei or four cells resulting from meiosis; (2) four cells derived from a segmenting ovum during cleavage.

quasidiploid *a. appl.* cells with the diploid number of chromosomes but with an abnormal genetic makeup, as e.g. three copies of one chromosome and only one of another, as many cell lines.

quasisocial insects those social insects in which there is cooperative care of the brood but each female still lays eggs at some time.

quasi-species (1) member of a population of RNA or DNA molecules that differ from each other in nucleotide sequence and have been produced by mutation and selection *in vitro*; (2) variant of a highly mutable virus, such as the human immunodeficiency virus (HIV), that arises during the course of infection and differs in nucleotide sequence from the original infecting virus.

quaternary *a.* (1) fourth, or *pert.* to a fourth level of organization; (2) *appl.* flower symmetry when there are four parts in each whorl.

Quaternary *n.* geological period lasting from *ca.* 2 million years ago to present, comprising the Pleistocene and Holocene epochs.

quaternary structure the relationships of the various subunits to each other in a protein composed of several polypeptide chains (protein subunits).

quaternate *a.* (1) in sets of four; (2) *appl.* leaves growing in fours from one point.

queen *n.* a member of the reproductive caste in eusocial and semisocial insects, sometimes, but not always, morphologically different from the workers.

queen substance the set of pheromones by which a queen honeybee attracts workers and controls their reproductive activities. It generally denotes trans-9-keto-2-decenoic acid, the most powerful of the components.

queuosine (Q) *n.* unusual nucleoside found only in tRNA, a modified form of guanosine with an aromatic ring substituent at position 7 and which can pair with U as well as C.

quiescence *n.* temporary cessation of development or other cavity, owing to an unfavourable environment.

quiescent centre group of cells located in the centre of the root apical meristem that undergo mitosis very infrequently. They help keep those meristem cells in contact with them in a self-renewing state and prevent their premature differentiation.

quill *n.* (1) the central shaft of a feather; (2) a hollow horny spine, as of porcupine.

quill feathers feathers of wings (remiges) and tail (rectrices) of birds.

quill knobs tubercles or exostoses on ulna of birds, for attachment of fibrous ligaments connecting with follicle of feather.

quillworts *n.plu.* an order, the Isoetales, of vascular, non-seed-bearing plants of the division Lycophyta, having linear leaves and a 'corm' with complex secondary thickening.

quinary *a. appl.* flower symmetry when there are five parts in a whorl.

quinate *a. appl.* five leaflets growing from one point.

quincunx *n.* arrangement of five structures of which four are at corners of a square with the fifth in centre. *a.* **quincuncial**.

quinine *n.* alkaloid extracted from bark of the South American tree *Cinchona officinalis* and which has been used medicinally as an antimalarial drug and a febrifuge.

quinones *n.plu.* hydrophobic aromatic hydrocarbon molecules that act as mobile electron carriers in membrane-based electron-transport systems in living organisms. They act as hydrogen acceptors and electron donors and because of their lipid solubility can diffuse within the membrane.

quinque- prefix derived from L. *quinque*, five, signifying having five of, divided into five.

quinquecostate *a.* having five ribs on the leaf.

quinquefid *a.* cleft in five parts.

quinquefoliate *a.* with five leaves.

quisqualate *n.* structural analogue of glutamate, used to define a class of glutamate receptors.

quorum sensing the phenomenon whereby individual bacteria are able to sense the numbers within the population by sensing the concentration of signalling molecules secreted by the bacteria.

R

r (1) coefficient of relationship *q.v.*; (2) intrinsic rate of increase *q.v.*

r, R roentgen *q.v.*

R arginine *q.v.*

*R*₀ net reproductive rate *q.v.*

rabbit-fishes a small group of marine fishes, class Holocephalii, with long slender tails and large pectoral fins.

Rabl orientation the V-shape adopted by a chromosome during mitosis, in which the centromere leads the way, trailing the chromosome arms behind it.

Rab proteins a family of small GTPases that are involved in the specificity of vesicle transport between intracellular compartments.

Rac *see* Rho protein family.

race *n.* (1) group of individuals within a species which forms a permanent and genetically distinguishable variety; (2) rhizome, as of ginger.

racemase *n.* any of a group of enzymes catalysing the conversion of one optically active enantiomer into the other, or L-isomers to D-isomers, esp. of amino acids, classified in the isomerases in EC 5.1.

racemate *n.* a mixture of two optical isomers of a compound or molecule, dextrorotatory (+) and laevorotatory (−), whose steric formulae are mirror images of each other and not superimposable. *a.* **racemic**.

racemation *n.* a cluster, as of grapes.

raceme *n.* flowerhead having a common axis bearing stalked flowers arranged spirally around it, the bottom flowers opening first, as in hyacinth.

racemiferous *a.* bearing racemes.

racemiform *a.* in the form of a raceme.

racemose *a.* (1) *appl.* a flowerhead whose growing points continue to add to the head and in which there are no terminal flowers, individual flowers or side branches being arranged spirally or alternately along a single main axis, with the lowest flowers opening first, *alt.* indefinite; (2) *appl.* any structure resembling a raceme.

racemule *n.* a small raceme.

racemulose *n.* in small clusters.

rachial *a. pert.* a rachis.

rachides *plu.* of rachis.

rachidial rachial *q.v.*

rachidian *a.* (1) placed at or near a rachis; (2) *appl.* median tooth in row of teeth of radula.

rachiform *a.* in the form of a rachis.

rachiglossate *a.* having a radula with pointed teeth, as whelks.

rachilla *n.* (1) axis bearing the florets in a grass spikelet; (2) small or secondary rachis.

rachiodont *a. appl.* egg-eating snakes with well-developed hypophyses of anterior thoracic vertebrae, which function as teeth.

rachiostichous *a.* having a succession of somactids as axis of fin skeleton, as in lungfish.

rachis *n.* (1) the shaft of a feather; (2) a stalk or axis. *plu.* **rachides**.

rachitomous temnospondylous *q.v.*

racket mycelium raquet mycelium *q.v.*

rad unit formerly used to measure the amount of ionizing radiation absorbed by living tissue, 1 rad being equal to 100 erg per gram tissue. It has been replaced by the gray (Gy), with 1 rad = 10^{-2} Gy.

radial *a.* (1) *pert.* the radius; (2) growing out like rays from a centre; (3) *pert.* ray of an echinoderm; (4) *appl.* leaves or flowers growing out like rays from a centre; *n.* (5) cross-vein of an insect wing; (6) supporting skeleton of fin-ray.

radial apophysis a process on palp of male arachnids, inserted into groove of female epigynum during mating.

radial cleavage *appl.* type of cleavage in which the first divisions of the fertilized egg occur at right angles to each other, resulting in four upper blastomeres sitting directly on top of four lower blastomeres.

radiale *n.* a carpal bone in line with radius.

radial fibres fibrous tissue supporting the retina.

radial glial cells glial cells in the developing neural tube which span the wall from the lumen to the outer surface, and guide migrating neurons. They disappear from the brain and spinal cord towards the end of development, possibly differentiating into astrocytes.

radial notch lesser sigmoid cavity of coronoid process of ulna.

radial symmetry having a plane of symmetry about each diameter. Many flowers have this type of symmetry, as do animals such as sea anemones and starfish.

radiant *a.* (1) emitting rays; (2) radiating; (3) *pert.* radiants; (4) *pert.* radiation; (5) *n.* an organism or group of organisms dispersed from an original geographical location.

radiate *a.* (1) radially symmetrical; (2) radiating; (3) stellate; *v.* (4) to diverge or spread from a point; (5) to emit rays.

radiate-veined veined in a palmate manner.

radiatiform *a.* with radiating marginal florets.

radiation *n.* (1) emission of radiant energy in the form of waves or particles; (2) energy radiated in the form of waves or particles, e.g. electromagnetic radiation (radio waves, infrared, visible light, ultraviolet, X-rays and gamma rays) or emissions from radioactive sources (e.g. beta rays). Often refers esp. to that radiation potentially harmful to living organisms, e.g. ionizing radiation (e.g. X- and gamma rays and streams of α- and β-particles, emitted from radioactive elements); (3) (*evol.*) the relatively rapid increase in numbers of new species of a particular type of animal or plant and their diversification and spread into many new habitats, e.g. the mammalian radiation

that occurred after the end of the Cretaceous period to fill ecological niches left empty by the extinction of the dinosaurs.

radiation biology the study of the effects of potentially damaging radiation on living organisms.

radiation chimaera experimental animal, usually a mouse, whose immune system has been destroyed by irradiation and in which a new immune system has been reconsituted by bone marrow transplantation from a genetically different individual. *alt.* **radiation bone marrow chimaera**.

radiation dose equivalent dose equivalent *q.v.*

radiation ecology the study of radiation as affecting the relationship between living organisms and environment, and of the ecological effects and destination of radioactive elements.

radiation hybrid (RH) somatic cell hybrid made by fusion of a cell from one species with a cell from another species in which the chromosomes have been fragmented by radiation treatment. Used in radiation mapping.

radiation mapping genetic mapping technique involving irradiated somatic cell hybrids (radiation hybrids). It is based on the fact that the probability that two gene loci will be separated by a radiation-induced break and be carried on different chromosomal fragments, and thus be likely to be lost separately from the cell, should be proportional to their distance apart.

radical *a.* (1) arising from root close to ground, as basal leaves and flower stems; (2) a chemical entity possessing an unpaired electron, e.g. Cl•. *alt.* free radical.

radicant *a.* with roots arising from the stem.

radicate *a.* (1) rooted; (2) possessing root-like structures; (3) fixed to substrate as if rooted.

radication *n.* the rooting pattern of a plant.

radicel *n.* small root or rootlet.

radicicolous, radiciferous *a.* having roots.

radiciflorous *a.* with flowers arising at extreme base of stem, so apparently arising from root.

radiciform, radicine *a.* resembling a root.

radicivorous *a.* root-eating.

radicle *n.* embryonic plant root, developing at the lower end of the hypocotyl.

radicle sheath coleorhiza *q.v.*

radicolous *a.* living in or on roots.

radicose *a.* with a large root.

radicular *a. pert.* a radicle.

radicule *n.* rootlet.

radiculose *a.* having many rootlets or rhizoids.

radii *plu.* of radius.

radioactivity *n.* the disposition of some elements to undergo spontaneous disintegration of their nuclei associated with the emission of ionizing particles and electromagnetic radiation, as α-particles or β-particles and gamma radiation.

radioautography autoradiography *q.v.*

radiobiology *n.* study of the effects of radiation, esp. potentially harmful ionizing radiation such as X-rays, on living cells and organisms.

radiocarbon *n.* radioactive isotope of carbon, ^{14}C, occurring naturally in small amounts, used in biochemical and physiological research and as an indicator for dating in archaeology.

radiocarbon dating the use of the differential uptake of the rare radioactive isotope of carbon, ^{14}C and the much more abundant isotope ^{12}C, during carbon fixation by plants to date the remains of organic material in archaeology. The radiocarbon method can be used to date material between 3000 and 40,000 years old.

radiocarpal *a. pert.* radius and wrist.

radioecology radiation ecology *q.v.*

radioimmunoassay (RIA) *n.* very sensitive method for the detection and measurement of substances using radioactively labelled specific antibodies or antigens.

radioiodine *n.* radioactive isotope of iodine, ^{131}I, used for studying the thyroid and in treatment of thyroid cancers.

radioisotope *n.* radioactive isotope of an element, such as tritium (^{3}H), ^{32}P, radiocarbon (^{14}C) and radioiodine (^{131}I), widely used in experimental biology to label tracer compounds and biological molecules.

radiolabelling *n.* incorporation of a radioactive isotope into a compound, so that it can be traced in the body or the cell.

radiolarians *n.plu.* group of marine planktonic protists of the phylum Actinopoda (formerly classified as protozoans of the class Sarcodina), characterized by a symmetrical skeleton of silicaceous spicules.

radiole *n.* a spine of sea urchins.

radioligand *n.* radioactively labelled ligand (e.g. hormone, neurotransmitter) used for receptor-binding experiments.

radiomedial *n.* cross-vein between radius and medius of insect wing.

radiomimetic *a.* resembling the effects of radiation, *appl.* chemicals causing mutations.

radionuclide *n.* an unstable atomic nucleus, which undergoes spontaneous radioactive decay, emitting radiation and changing from one element into another.

radiophosphorus *n.* radioactive isotope of phosphorus, ^{32}P, used widely in biochemical and physiological research, and therapeutically.

radioreceptor *n.* a sensory receptor for receiving light or temperature stimuli.

radioresistant *a.* offering a relatively high resistance to the effects of radiation, esp. ionizing radiation such as X-rays.

radiosensitive *a.* sensitive to the effects of radiation, esp. ionizing radiation such as X-rays.

radiospermic *a.* (1) having seeds which are circular in transverse section; (2) *appl.* plants, esp. fossils, having such seeds.

radiosymmetrical *a.* having similar parts similarly arranged around a central axis.

radiotherapy *n.* the treatment of disease, such as cancer, by means of X-rays or radioactive substances.

radioulna *n.* radius and ulna combined as a single bone.

radius *n.* (1) bone of vertebrate forelimb between humerus and carpals; (2) main vein of insect wing. *plu.* **radii**.

radix *n.* (1) root; (2) point of origin of a structure.

radula *n.* short, broad organ with rows of chitinous teeth in mouth of most gastropod molluscs. Used for feeding.

radulate, raduliferous *a.* having a radula.

raduliform *a.* resembling a radula or a flexible file.

raffinose *n.* trisaccharide found in sugar beet, cereals and some fungi, giving glucose,

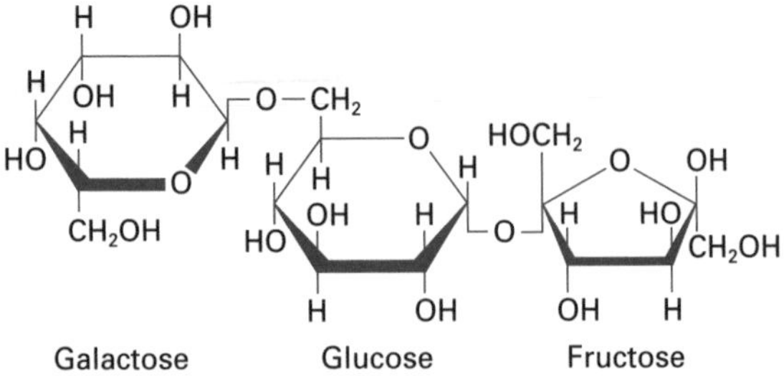

Galactose   Glucose   Fructose

Fig. 32 Raffinose.

fructose and galactose on hydrolysis. *see* Fig. 32.

Rafflesiales *n.* order of plant parasitic dicots in which the body is reduced to a simple thallus, and which comprises two families, Hydnoraceae and Rafflesiaceae.

Rainey's corpuscles spores of *Sarcocystis*, an elongated sporozoan parasite found in skeletal muscle.

Rainey's tubes elongated sacs found in voluntary muscle which are the adult stages of certain sporozoan parasites.

rain forest forest biomes that develop in areas with an annual rainfall of more than 254 cm. *see* monsoon rain forest, temperate rain forest, tropical rain forest.

raised bog, raised mire convex lens-shaped acid peatland developed in fen basins or river flood plains in wet climates.

Ramachandran plot a plot of the observed pairs of phi(ϕ)/psi(ψ) rotational angles in a protein structure, which shows that the range of possible angles is in fact fairly limited. *see* phi torsion angle, psi torsion angle.

ramal *a.* (1) belonging to branches; (2) originating on a branch.

ramapithecids, ramapithecines *n.plu.* a group of Miocene ape-like fossils from Asia and Africa, including the genus *Ramapithecus*, which show hominid-like features in the teeth.

ramate *a.* branched.

ramellose *a.* having small branches.

ramenta *n.plu.* (1) brown scales present on stems, leaves and petioles of ferns; (2) long epidermal hairs, of leaves. *sing.* **rament, ramentum.**

ramentaceous *a.* covered with ramenta.

rameous *a.* (1) branched; (2) *pert.* a branch.

ramet *n.* an individual member of a clone, e.g. an offshoot of a plant reproducing by stolons.

rami *plu.* of ramus.

rami communicantes nerve fibres connecting sympathetic ganglia and spinal nerves. *sing.* **ramus communicans.**

ramicorn *a.* having branched antennae, as some insects.

ramification *n.* (1) branching; (2) branch of e.g. tree, nerve or artery.

ramiflory *n.* the state of flowering from the branches. *a.* **ramiflorous.**

ramiform *a.* branch-like.

ramiparous *a.* producing branches.

ramose *a.* much branched.

ramule, ramulus *n.* a small branch, a twig.

ramuliferous *a.* bearing small branches.

ramulose *a.* with many small branches.

ramulus ramule *q.v.*

ramus *n.* (1) any branch-like structure; (2) barb of a feather. *plu.* **rami.**

ramus communicans *see* rami communicantes.

ramuscule ramule *q.v.*

Ran small GTPase involved in providing energy for nuclear import and export processes.

random genetic drift (1) random accumulation of changes in nucleotide sequence of a gene occurring over long periods of time and which do not appear to be due to any selective forces; (2) the random changes in gene frequency that can occur in a small population over time as a result of sampling of gametes in each generation. *alt.* genetic drift.

range *n.* (1) the area within which an animal or group of animals seeks food; (2) an area of unenclosed, unintensively managed grassland on which livestock are allowed to graze freely. *alt.* **rangeland**.

ranine *a. pert.* under surface of the tongue.

ranivorous *a.* feeding on frogs.

ranunculaceous *a. pert.* a member of the dicot flower family Ranunculaceae, the buttercups and their relatives.

Ranunculales *n.* order of herbaceous and woody plants, climbers or shrubs, and including the families Berberidaceae (barberry), Ranunculaceae (buttercup) and others.

Ranvier's nodes nodes of Ranvier *q.v.*

rapamycin *n.* an immunosuppressive drug.

RAPD randomly amplified polymorphic DNA. DNA representing sequence polymorphisms between individuals, which can be used for genome mapping.

raphe *n.* (1) seam-like suture, such as the junction of some fruits; (2) line of fusion of funicle and anatropous ovule; (3) slit-like line in diatom valves; (4) line, or ridge, of e.g. perineum, scrotum, hard palate, medulla oblongata.

raphe nucleus any of a cluster of nuclei in midbrain in which lie cell bodies of serotoninergic neurons.

raphide *n.* needle-like crystal of calcium oxalate, produced as a metabolic byproduct in many plant cells.

raphidiferous *a.* containing raphides.

rapid eye movement sleep REM sleep *q.v.*

raptatory *a.* preying.

raptorial *a.* adapted for snatching or robbing, *appl.* birds of prey.

raptors *n.* birds of prey, e.g. hawks, eagles, owls, etc.

raquet mycelium hyphae enlarged at one end of each segment, small and large ends alternating.

rare *a.* IUCN definition *appl.* species or larger taxa that have small populations, and, although not at present considered endangered or vulnerable, are at risk e.g. because of their highly restricted distribution within a habitat, or because they are thinly spread over a very large area, as some large carnivores. *see also* endangered, rarity, vulnerable.

rarity *n.* categories of rarity of plant and animal species have been defined by the International Union for Conservation of Nature and Natural Resources (IUCN). *see* endangered, rare, vulnerable.

Ras small GTP-binding protein which is part of many intracellular signalling pathways, where it activates the MAP kinase cascade. The *ras* gene is disrupted in many cancers.

rasiRNA repeat-associated small interfering RNA *q.v.*

rasorial *a.* adapted for scratching or scraping the ground, as fowls.

rastellus *n.* group of teeth on paturon of arachnid chelicera.

Rathke's pouch a diverticulum of ectoderm from the mouth cavity in vertebrate embryos that eventually forms the pituitary gland.

ratite *a.* having a sternum with no keel, *appl.* birds.

ratites *n.plu.* a group of flightless birds comprising the ostrich, emus, rheas, cassowaries and kiwis. They have rudimentary wings, a breast bone without a keel and fluffy feathers with no barbs.

rattle *n.* the series of horny joints at the end of a rattlesnake's tail, which produces the rattling sound.

Raunkiaer's life forms classification of plants by the type of perennating organs they possess and the position of these organs in relation to soil or water level. *see* chamaephyte, cryptophyte, geophyte, helophyte, hemicryptophyte, hydrophyte, phanerophyte, therophyte.

ray *n.* (1) (*bot.*) band of parenchyma tissue penetrating from cortex towards centre of stem into secondary xylem and phloem, or from pith outwards to edge of vascular tissue (medullary ray); (2) the stalk of a group of flowers in an umbel; (3) (*zool.*) one of the bony or cartilaginous spines supporting fins in fishes; (4) a division of a radially symmetrical animal, such as arm of starfish; (5) one of the straight urine-carrying tubules passing from medulla through cortex of kidney.

ray-finned fishes the common name for the Actinopterygii *q.v.*

ray florets large outer florets of some inflorescences, esp. of some Compositae, which are often sterile or carpellate. *cf.* disc florets.

ray initial in vascular cambium of plant stem, a cell that gives rise to medullary rays.

RCA family of lectins isolated from the castor oil bean, *Ricinis comunis*.

rcp reciprocal translocation *q.v.*

rDNA (1) DNA specifying ribosomal RNA *q.v.*; (2) recombinant DNA *q.v.*

re- prefix derived from L. *re*, again.

reaction centre the protein–chlorophyll complex within the photosystems in chloroplast and other photosynthetic membranes that receives electrons from the light-capturing systems and generates high-energy electrons for transfer to the photosynthetic electron-transfer chain. A 'special pair' of chlorophyll molecules near one side of the membrane acts as the electron donor and a quinone (in photosynthetic purple bacteria and photosystem II) or iron–sulphur centre (photosystem I and green sulphur bacteria) on the opposite side acts as acceptor. *alt.* photochemical reaction center.

reaction-diffusion mechanisms mechanisms proposed to account for some types of regular patterning seen in development. Such mechanisms depend on the generation of self-organizing spatial patterns of molecules.

reaction time (RT) the time interval between completion of presentation of a stimulus and the beginning of the response. *alt.* latent period.

reaction wood wood modified by bending of stem or branches, apparently trying to restore the original position, and including compression wood in conifers and tension wood in dicotyledons.

read abomasum *q.v.*

reading frame starting point on DNA or messenger RNA from which the base sequence is read off in triplet codons, the correct reading frame specifying the amino acid sequence of the corresponding protein.

readthrough the continuation of transcription or translation past a termination signal in DNA or mRNA respectively.

reafferent *a. appl.* stimulation that occurs as a result of an animal's bodily movements.

reaginic antibody immunoglobulin E (IgE) *q.v.*

realized niche the actual place and role in an ecosystem occupied by an organism or species, as opposed to its niche under ideal conditions.

real-time PCR quantitative PCR *q.v.*

reassociation *n.* of DNA, the pairing of complementary single strands to form a double helix. *alt.* **reannealing**.

recalcitrant *a.* non-biodegradable, *appl.* organic, usually man-made compounds in the soil.

recapitulation theory the theory, due largely to the 19th century biologist E. Haeckel, that ontogeny tends to repeat phylogeny, and the similar theory due to von Bauer that the individual's life history reproduces certain stages in the evolutionary history of the species. Now considered to be a misinterpretation of the similarities in early embryonic development of e.g. vertebrate embryos due to their common evolutionary ancestry.

Recent *n.* geological epoch following Pleistocene and lasting until the present day. *alt.* Holocene.

receptacle *n.* of a flower, the structure on which floral organs such as ovary, anthers and petals are borne. *alt.* floral axis. *see* Fig. 18 (p. 239).

receptacular *a.* (1) *pert.* a receptacle of any kind; (2) largely composed of the receptacle, as certain fruits.

receptaculum *n.* a receptacle of any kind.

receptaculum ovarum an internal sac in which ova are collected in oligochaetes such as earthworms.

receptive field of a neuron, the area on the corresponding sensory organ within which a stimulus will influence the firing of that neuron.

receptive hyphae female sex organs of rust fungi.

receptor *n.* (1) specialized tissue or cell sensitive to a specific stimulus; (2) sensory organ; (3) sensory nerve ending; (4) protein to which a signalling molecule such as a neurotransmitter, hormone, drug or metabolite binds specifically and stimulates a particular response by the cell; (5) protein to which an infectious agent such as a virus binds to gain entry into the cell.

α-receptor, β-receptor *see* adrenergic receptor.

αβ-receptor highly variable antigen receptor found on the majority of T lymphocytes. *see* T-cell receptor. *cf.* γδ-receptor.

γδ-receptor antigen receptor found on a subclass of T lymphocytes.

receptor-associated tyrosine kinases cytoplasmic protein tyrosine kinases that become associated with the cytoplasmic portions of cell-surface receptors when the receptors are stimulated.

receptor editing replacement of a light chain of self-reactive antigen receptors on immature B cells with a light chain that does not confer self-reactivity.

receptor guanylate cyclase a class of single-pass transmembrane proteins with an extracellular binding site for a ligand and an intracellular guanylate cyclase domain. Activation of the receptor results in the synthesis of the second messenger cyclic GMP. Natriuretic peptides act through such receptors.

receptor-like tyrosine phosphatase a class of transmembrane proteins with intracellular protein tyrosine phosphatase domains.

receptor-mediated endocytosis the internalization of substances by endocytosis after they have bound to cell-surface receptors.

receptor potential graded local depolarization of sensory nerve terminal membrane as a result of stimulation.

receptor serine/threonine kinases a class of transmembrane proteins with an external ligand-binding domain and an intracellular protein serine/threonine kinase domain. The transforming growth factor-β family of signal molecules act through such receptors. *see also* Smads.

receptor tyrosine kinases large family of transmembrane receptors with an extracellular ligand-binding portion and an intracellular portion with tyrosine kinase activity. Many growth factors, such as EGF, insulin and NGF, act through such receptors.

recess *n.* a fossa, sinus, cleft, or hollow space.

recessive allele allele that is not reflected in the phenotype when present as one member of a heterozygous pair, only determining the phenotype when present in the homozygous state.

recessive phenotype for a given gene, the phenotype expressed when the recessive allele is in the homozygous state. It is not expressed in the heterozygote.

recessivity *n.* the property displayed by a recessive allele.

reciprocal altruism social behaviour in which altruistic acts by one individual towards another are reciprocated. It is rare in most animals.

reciprocal cross two crosses between the same pair of genotypes or phenotypes in which the sources of the gametes are reversed in one cross.

reciprocal feeding trophallaxis *q.v.*

reciprocal hybrids two hybrids, one descended from a male of one species and a female of the other, the other from a female of the first species and a male of the second, such as the mule and the hinny.

reciprocal recombination genetic recombination in which exactly corresponding parts of the two DNAs undergoing recombination are exchanged. *cf.* integration, non-reciprocal recombination.

reciprocal translocation interchange of parts between non-homologous chromosomes.

reclinate *a.* curved downwards from apex to base.

reclining *a.* leaning over.

recognition sites in molecular genetics, conserved regions within promoters and other DNA sequences presumed to be recognized by proteins that bind to DNA such as RNA polymerase and regulatory proteins.

recombinant *a.* (1) *appl.* genotypes, phenotypes, gametes, cells or organisms produced as a result of natural genetic recombination; (2) *appl.* DNA. *see* recombinant DNA; (3) *appl.* proteins: proteins produced from cells containing recombinant DNA directing their synthesis; (4) *appl.* cell or organism into which recombinant DNA has been introduced; (5) *n.* any chromosome, cell or organism which is the result of recombination, either natural or artificial.

recombinant DNA (1) DNA produced by joining together *in vitro* genes from different sources, or DNA that has in some way been modified *in vitro* to introduce novel genetic information; (2) DNA produced as a result of natural genetic recombination.

recombinant DNA technology the techniques used to produce recombinant DNAs and artificially genetically modified organisms, cells and microorganisms, and their applications in biotechnology.

recombinant fraction, recombinant frequency recombination fraction *q.v.*

recombinant protein any protein produced from a recombinant DNA template.

recombinase *n.* enzyme activity that is responsible for catalyzing recombination reactions.

recombination *n.* (1) the process in sexually reproducing organisms by which DNA is exchanged between homologous chromosomes (reciprocal or balanced recombination) by chromosome pairing and crossing-over at meiosis during gamete formation. This produces gametes containing alleles from both parents on the same chromosome, *alt.* general or generalized recombination; (2) any exchange between, or integration of, one DNA molecule into another, which may be reciprocal or non-reciprocal. *see also* site-specific recombination.

recombination-activating genes *RAG-1* and *RAG-2*, which are required for gene rearrangement in developing lymphocytes. They encode the two subunits of a recombinase.

recombination fraction (RF) the proportion of the total number of offspring from a cross that do not have a parental combination of alleles for a given gene or genes. RF = (no. of recombinant offspring)/(no. of recombinant offspring + no. of non-recombinant offspring). *alt.* recombinant fraction, recombinant frequency.

recombination nodules spherical or cylindrical structures seen lying across the synaptonemal complex in fungi and insects and which may be sites of recombination.

recombination repair type of DNA repair in which gaps left in one DNA strand opposite damaged sites after replication are filled in by recombination with a normal DNA strand, the consequent gap in this strand being filled in by DNA synthesis.

recombination signal sequence (RSS) sequence that indicates a site at which recombination and joining of individual gene segments can take place during immunoglobulin and T-cell receptor gene rearrangement.

recombinogenic *a.* tending to cause recombination.

reconstitution *n.* reassembly of artificially separated differentiated cells to form a new tissue or even an individual, as can be done for sponges.

recrudescence *a.* state of breaking out into renewed growth.

recruitment *n.* (1) entry of new individuals into a population by reproduction or immigration; (2) (*neurol.*) activation of additional motor neurons, causing an increased reflex when stimulus of the same intensity is continued.

rectal *a.* (1) *pert.* rectum; (2) *appl.* gland: small vascular sac of unknown function near end of gut in some fishes.

rectal columns longitudinal folds of epithelium lining the rectum.

recti- prefix derived from L. *rectus*, straight.

rectigradation *n.* (1) adaptive evolutionary tendency; (2) structure showing an adaptive trend or sequence in evolution.

rectinerved *a.* with veins straight or parallel, *appl.* leaves.

rectirostral *a.* straight-beaked.

rectiserial *a.* arranged in straight or vertical rows.

rectivenous *a.* with straight veins, *appl.* leaves.

recto- prefix indicating *pert.*, or connecting with, the rectum.

rectogenital *a. pert.* rectum and genital organs.

rectovesical *a. pert.* rectum and bladder.

recto-uterine *a. appl.* posterior ligaments of uterus.

rectrices *n.plu.* stiff tail feathers of a bird, used in steering. *sing.* **rectrix**. *a.* **rectricial**.

rectum *n.* the posterior part of the alimentary canal, leading to the anus, in which water and inorganic ions are absorbed from the gut contents before they are passed out as faeces. *a.* **rectal**.

rectus *n.* a name for a rectilinear muscle, such as the rectus femoris.

recurrent *a.* (1) returning or reascending towards origin; (2) reappearing at intervals.

recurrent sensibility sensibility shown by motor roots of spinal cord due to sensory fibres of sensory roots.

recurved *a.* bent backwards.

recurvirostral *a.* with beak bent upwards.

recutite *a.* apparently devoid of epidermis.

red algae common name for the Rhodophyta *q.v.*

red blood cell, red blood corpuscle erythrocyte *q.v.*

red body, gland or spot rete mirabile *q.v.*

red-cell ghosts empty red-cell membranes obtained by haemolysis and used in investigations of cell membranes.

red drop the phenomenon that the quantum yield of photosynthesis falls sharply when the wavelength of light is greater than 680 nm. This is due to the fact that only photosystem I can be driven by light of longer wavelength.

red fibres *see* red muscle.

redia *n.* a larval stage of some endoparasitic flukes in the snail host. It is produced asexually from the sporocyst, has a mouth and gut and reproduces asexually to produce a further generation of rediae or cercariae.

redifferentiation *n.* in a cell or tissue, a reversal of the differentiated state and then differentiation into another type. *see also* transdifferentiation.

red light light of wavelength 620–680 nm.

red muscle type of muscle in fish, containing myoglobin, and with chiefly aerobic respiration, involved in slow swimming. *cf.* white muscle.

red muscle fibres skeletal muscle fibres rich in myoglobin, which give the slow-twitch response and are capable of sustained activity. *cf.* white muscle fibres.

red nucleus area in midbrain involved in relaying signals from brain to motor neurons in spinal cord.

redox *a. pert.* mutual oxidation and reduction.

redox potential (E_0') an electrochemical measure (in volts) of affinity for electrons relative to hydrogen in standard conditions (the redox potential of the $H^+:H_2$ couple is defined as 0 V). It is applicable to any substance that can exist in oxidized and reduced forms. A negative redox poten-

tial indicates a strong reducing agent, a positive redox potential a strong oxidizing agent. *alt.* oxidation–reduction potential.

redox reaction a chemical reaction in which one compound is reduced and another is oxidized.

red pulp erythroid tissue of spleen.

Red Queen hypothesis idea that each evolutionary advance by one species is detrimental to other species, so that all species must evolve as fast as possible simply to survive.

red tide a bloom of red dinoflagellates (e.g. *Gonyaulax polyedra*) which colours the sea red. Toxins contained in some of these microorganisms are concentrated in the shellfish that feed on them and can cause sometimes fatal poisoning in humans who eat the shellfish.

reduced *a.* in an anatomical context *appl.* structures that are smaller than in ancestral forms.

reducer organism decomposer *q.v.*

reducing power general term for the presence in cells of compounds such as NADH and NADPH, which are donors of hydrogen and electrons in metabolic reduction reactions.

reducing sugar sugar with a free aldehyde or ketone group that can act as a reducing agent in solution, as most monosaccharides and disaccharides.

reductase *n.* any enzyme that catalyses the reduction of a compound, esp. where hydrogen transfer from the donor is not readily demonstrable, classified among the oxidoreductases in EC class 1,

reduction *n.* (1) the halving of the number of chromosomes that occurs at meiosis; (2) structural and functional development in a species that is less complex than that of its ancestors. *cf.* amplification; (3) (*biochem.*) adding electrons to an atom, as occurs when hydrogen is added to a molecule, or removing oxygen from the molecule. *cf.* oxidation.

reduction division 1st meiotic division, sometimes used for meiosis as a whole.

reductionism *n.* the idea that breaking a system down into increasingly smaller parts will enable one to understand it completely. *adj.* reductionist.

reductive pentose phosphate pathway Calvin cycle *q.v.*

redundancy *n.* refers generally to the existence of back-up systems for many physiological and developmental processes. If one component is missing, another can substitute.

reduplicate *a. appl.* the arrangement of petals, etc. in flower bud, in which margins of bud sepals or petals turn outwards at points of contact.

reduviid *a. appl.* eggs of certain insects, protected by micropyle apparatus with porches.

reed abomasum *q.v.*

refection *n.* reingestion of incompletely digested food by some animals, such as eating faecal pellets, or in rumination.

referred *a. appl.* sensation in a part of the body remote from that to which the stimulus was applied.

reflected *a.* turned or folded back on itself.

reflector layer (1) layer of cells on inner surface of a light-emitting organ, as in fireflies; (2) silvery reflecting plates in skin of fishes above the argenteum.

reflex *a.* (1) involuntary, *appl.* reaction to stimulus; (2) turned or folded back on itself. *alt.* reflected; (3) *n.* involuntary movement (reflex action) or other response elicited by a stimulus at the periphery which is transmitted to the central nervous system and reflected back to a peripheral effector organ.

reflex arc unit of function in nervous system, consisting of a sensory receptor, an effector organ (muscle or gland) and a pathway of nerve cells conveying the sensory information to the central nervous system and carrying motor or other commands from the central nervous system to the effector organ.

reflexed *a.* turned or curved back on itself.

refracted *a.* bent backwards at an acute angle.

refractory *a.* (1) unresponsive; (2) *appl.* phase, the period after excitation of neuron during which repetition of stimulus fails to induce a response.

refuge, refugium *n.* an area that has remained unaffected by environmental changes to the surrounding area, such as a mountain area that was not covered with ice during the Pleistocene, and in which, therefore, the previous flora and fauna has survived.

regeneration *n.* (1) renewal of a portion of body which has been injured or lost; (2) reconstitution of a compound after dissociation, as e.g. of rhodopsin.

regma *n.* dry dehiscent fruit whose valves open by elastic movement, as in *Geranium* species.

regosols *n.plu.* soils which are developed on fairly deep unconsolidated parent material such as dune sands or volcanic ash.

regression *n.* (1) reversal in the apparent direction of evolutionary change, with simpler forms appearing; (2) replacement of a climax ecosystem with a previous stage in the succession, as e.g. the replacement of forest by grassland after felling.

regression coefficient the slope of a straight line that most closely relates two correlated variables.

regular *a. appl.* any structure, such as flower showing radial symmetry. *alt.* actinomorphic.

regular distribution a distribution in which the variance is less than the mean. *alt.* uniform distribution.

regulation *n.* in embryogenesis, the ability to compensate for the addition or removal of cells.

regulative *a. appl.* embryos which can compensate for the removal or addition of cells. In such embryos the fate of cells at the beginning of any developmental process is not yet irrevocably determined, and they can be influenced by developmental signals to follow another course if required.

regulator *n.* an animal that maintains its internal environment in a state that is largely independent of external conditions.

regulatory DNA any DNA that contains control sequences at which gene regulatory proteins bind to control gene expression.

regulatory genes genes that direct the production of proteins that regulate the activity of other genes, or which represent control sites in DNA at which gene expression is regulated.

regulatory protein *see* gene regulatory protein.

regulatory T cell any of several types of CD4 T cell that can inhibit T-cell responses.

regulome *n.* all the gene regulatory networks operating in a given cell, tissue or stage of development.

regulon *n.* all the genes whose expression is controlled by a given gene regulatory protein.

Reichert's membrane connective tissue membrane (basement membrane) covering mural trophectoderm in the mouse embryo.

reinforcement *n.* an event that alters an animal's response to a stimulus. Positive reinforcement rewards and increases the response, negative reinforcement is disagreeable or painful and suppresses the response.

reinforcing selection operation of selection pressures on two or more levels of organization, such as population, family and individual, in such a way that certain genes are favoured at all levels and their spread through the population is accelerated.

Reissner's membrane the vestibular membrane of cochlea, stretching from inner to outer wall of cochlea above basilar membrane and separating the scala vestibuli from the scala media.

rejungant *a.* coming together again, *appl.* related but hitherto separate taxa when in the course of time their ranges come to rejoin.

rejuvescence *n.* (1) renewal of youth; (2) regrowth from injured or old parts.

relationship, coefficient of coefficient of relationship *q.v.*

relative molecular mass (Mᵣ) the ratio of the mass of one molecule of a substance to one-twelfth the mass of an atom of ^{12}C. It is a ratio and therefore dimensionless. *alt.* molecular weight. *see also* molecular mass.

relaxation time the period during which excitation subsides after removal of a stimulus.

relaxed DNA a circular double-helical DNA molecule without any superhelical turns.

relaxed mutants mutant bacteria that do not show the stringent response in conditions of amino acid starvation.

relaxin *n.* hormone produced by the corpus luteum which produces relaxation of the pelvic ligaments during pregnancy.

relay cell interneuron *q.v.*

release channel calcium channel in sarcoplasmic reticulum which opens on muscle stimulation and releases calcium into the muscle cell.

release factor any of several proteins (e.g. RF-1, RF-2 in *E. coli*) which recognize termination codons in mRNA during translation and are involved in releasing the completed polypeptide chain from the ribosome. Proteins with the same function occur in eukaryotes and their names are prefixed by the letter e, e.g. eRF.

releaser *n.* a stimulus or group of stimuli that activates an inborn tendency or pattern of behaviour.

releasing hormone any of a number of small peptides produced by the hypothalamus and which act on the pituitary to induce the release of hormones such as growth hormone, thyroid-stimulating hormone, follicle-stimulating hormone and luteinizing hormone.

relict *a.* (1) not now functional, but originally adaptive, *appl.* structures; (2) surviving in an area isolated from the main area of distribution owing to intervention of environmental events such as glaciation, *appl.* species, populations. *alt.* **relic**.

rem roentgen equivalent man, the unit dose of ionizing radiation that gives the same biological effect as that due to 1 roentgen of X-rays. The rem has been replaced by the sievert, with 1 sievert = 100 rem.

remiges *n.plu.* primary wing feathers of bird. *sing.* **remex**.

remiped *a.* having feet adapted for rowing motion.

remotor *n.* a retractor muscle.

REM sleep rapid-eye-movement sleep, or paradoxical sleep, during which people are dreaming.

renal *a. pert.* kidneys.

renal columns cortical tissue between the pyramids of the medulla of kidneys.

renal corpuscle Malpighian corpuscle *q.v.*

renal portal *appl.* a circulation system in which some blood returning to heart passes through kidneys.

renal tubule fine tubule in kidney that carries waste material from glomerulus to collecting duct, and in which water resorption and ion resorption takes place.

renaturation *n.* the return of a denatured macromolecule such as a protein or nucleic acid to its original configuration.

rendzina *n.* any of a group of rich, dark, greyish-brown, limy soils of humid or subhumid grasslands, having a brown upper layer and yellowish-grey lower layers.

reniculus *n.* lobe of kidney comprising papillae, pyramid and surrounding part of cortex.

reniform *a.* kidney-shaped.

renin *n.* peptidase enzyme secreted by kidney and which converts angiotensinogen in the blood to angiotensin I. EC 3.4.23.15.

Renner complex a group of chromosomes that passes from generation to generation as a unit, as in the evening primrose *Oenothera*.

rennin chymosin *q.v.*

renopericardial *a. appl.* a ciliated canal connecting kidney and pericardium in higher molluscs.

Renshaw cell type of interneuron in spinal cord, involved in regulation of motor neurons supplying a single muscle.

Reoviridae, reoviruses *n.*, *n.plu.* family of icosahedral, non-enveloped, double-stranded RNA viruses of animals and plants, including the plant fijiviruses and the human rotaviruses that cause diarrhoea in children.

repand *a.* (1) with undulated margin, *appl.* leaf; (2) wrinkled, *appl.* bacterial colony.

repandodentate *a.* varying between undulate and toothed, *appl.* margin of leaf.

repandous *a.* curved convexly.

reparative *a.* restorative, *appl.* buds that develop after injury to a leaf.

repeat *n.* duplication, or additional serial repetition of a stretch of DNA as a result of unequal crossing-over or other molecular events. *see also* inverted repeats, tandem repeats.

repeat-associated small interfering RNA (rasiRNA) type of small non-coding RNA that targets retrotransposons and other repeated elements and acts through proteins of the Piwi/Argonaute family. Its best known function is the silencing of transposons in the *Drosophila* germline.

repent *a.* creeping along the ground.

repertoire *n.* in immunology, all the different antigen receptor specificities that can be produced by a single individual.

repetitive DNA repeated DNA sequences present in most eukaryotes, which may be from tens to thousands of bases long and be present in up to hundreds of thousands of copies. Most repetitive DNA seems to have no function and does not encode protein, although some is transcribed. *see also* satellite DNA, selfish DNA. *cf.* single-copy DNA.

replacement-level fertility birth rate that keeps a population constant, exactly replacing deaths by births.

replacement name scientific name adopted as substitute for one found invalid under the rule of the International Codes of Nomenclature.

replacement vector a cloning vector in which some of the vector DNA can be replaced by foreign DNA.

repletes *n.plu.* workers with distensible crops for storing and regurgitating honeydew and nectar, and constituting a physiological caste of honey ants.

replica *n.* in electron microscopy, a thin shell of electron-dense metal atoms that follows the surface contours of the specimen and which shows up details of the surface structure. It is produced by vaporizing metal atoms onto the surface.

replica plating production of an exact replica of a plate containing bacterial colonies by transfer of bacteria to a new plate by 'blotting' with a velvet pad or filter paper, which retains the exact positions of the colonies relative to each other. Destructive identification procedures can be carried out on the replica plate leaving the master plate as an untouched source of bacteria or DNA.

replicase *n.* general term for the enzyme activity involved in the replication of nucleic acids, as in DNA synthesis and the replication of RNA in some viruses.

replicate (1) *a.* doubled back over itself; *v.* (2) to duplicate; (3) to copy itself, like DNA.

replicatile *a. appl.* wings folded back on themselves when at rest.

replication *n.* (1) duplication, as of DNA, by making a new copy of an existing molecule; (2) duplication of organelles such as mitochondria, chloroplasts and nuclei, and of cells.

replication bubble, replication eye structure formed in replicating DNA when replication proceeds in both directions from an origin of replication.

replication-defective virus a virus which has lost the ability to multiply, usually because of loss or replacement of some of its genome.

replication fork site of simultaneous unwinding of double-stranded parental DNA and synthesis of new DNA. It is seen in electron micrographs of replicating DNA as a Y-shaped structure.

replication origin site on a bacterial, viral or eukaryotic chromosome at which DNA replication is initiation.

replicative form (RF) double-stranded DNA produced by certain single-stranded DNA viruses after infection, and which directs mRNA synthesis.

replicative transposition, replicative recombination movement of a DNA-only transposon to a new site on the chromosome by the synthesis of a copy of the transposon and integration of the copy into the new site, the old copy remaining at the original site.

replicon *n.* a unit length of DNA that replicates sequentially and which contains an origin of replication. Examples are bacterial plasmids and bacterial chromosomes, or regions of eukaryotic chromatin between two origins of replication.

replisome *n.* multiprotein assembly containing the various activities needed for DNA replication, formed in association with DNA.

replum *n.* a wall, not the carpellary wall, formed from ingrowths from the placenta and dividing a fruit into sections.

reporter gene a marker gene whose phenotypic expression can be monitored and thus can 'report' the activity of a control region attached to it.

repressed *a.* not being expressed as a result of interaction with a repressor or other regulatory mechanism, *appl.* genes.

repression *n.* (1) the specific shut-down of gene expression and consequent inhibition of protein synthesis that occurs when e.g. an enzyme substrate is not available, or a nutrient being synthesized by a microorganism becomes available in the medium;

(2) more generally, refers to the blocking of transcription of a gene by a repressor protein binding to a control region in DNA, or to the inhibition of translation of mRNA by a repressor protein binding to a specific site on the mRNA.

repressor *n.* (1) protein that binds to a site in DNA to shut down transcription; (2) protein which inhibits translation of mRNA by binding to mRNA.

reproduction *n.* formation of new individuals.

reproduction curve a plot giving the relationship between the number of individuals at a particular stage in one generation and the numbers at that stage in a previous generation.

reproductive hormones hormones such as oestrogen and progesterone that are involved in reproduction.

reproductive isolation the inability of two populations to interbreed because they are geographically isolated, or isolated from each other by differences of behaviour, mating time (or in plants maturation times of male and female sex organs), or genital morphology. This is a phase in the development of new species.

reproductive value the relative number of female offspring remaining to be born to each female of age x, symbolized by v_x.

reptant *a.* (1) creeping, *appl.* a polyzoan colony with zooecia lying on substrate; (2) *appl.* gastropod molluscs.

Reptilia, reptiles *n., n.plu.* class of amniote, air-breathing, poikilothermic ('cold-blooded') tetrapod vertebrates, mostly terrestrial, having dry horny skin with scales, plates or scutes, functional lungs throughout life, one occipital condyle and a four-chambered heart. Most reptiles lay eggs with a leathery shell but some are ovoviviparous. Reptiles include the tortoises and turtles, the tuatara, lizards and snakes, crocodiles, and many extinct forms, such as dinosaurs, pterosaurs. *see* Appendix 3.

reptiloid *a.* having the characteristics of a reptile.

repugnatorial *a.* defensive or offensive, *appl.* glands and other structures.

repulsion *n.* the case when the diploid genotype at two linked loci A and B is *Ab/aB*, where *A* is dominant over *a* and *B* is dominant over *b*.

RER rough endoplasmic reticulum. *see* endoplasmic reticulum.

RES reticuloendothelial system *q.v.*

reserve cellulose cellulose found in plant storage tissue and subsequently used for nutrition after germination.

reservoir *n.* host which carries a pathogen but is unharmed and acts as a source of infection to others.

residual bodies (1) anucleate portions of spermatids left after separation of spermatozoa; (2) secondary lysosomes in cytoplasm of eukaryotic cell which contain the indigestible remains of material taken in by phagocytosis.

residual volume volume of air remaining in lungs after strongest possible breathing out.

residue *n.* in chemistry and biochemistry, a compound such as a monosaccharide, nucleotide or amino acid when it is part of a larger molecule.

resilience *n.* ability of a living system to restore itself to its original condition after being disturbed.

resilifer, resiliophore *n.* the projection of shell carrying the flexible hinge in bivalve shells.

resilin *n.* protein present as cross-linked aggregates with rubber-like properties in some insect wing muscles and attachments of wing to thorax. It is involved in taking up the kinetic energy of the wingbeat at the end of each stroke.

resilium *n.* the flexible horny hinge of a bivalve shell.

resin *n.* any of various high molecular weight substances, including resin acids, esters and terpenes, which are found in mixtures in plants and often exuded from wounds. They harden to glassy amorphous solids and may protect against insect and fungal attack.

resin canals ducts in bark or wood, esp. in conifers, lined with glandular epithelium secreting essential oils, e.g. terpenes, which form resin oxidation products.

resiniferous *a.* producing resin.

resinous *a. appl.* a class of odours.

resistance factors R-factors *q.v.*

resistance transfer factor (RTF) the part of a transmissible drug-resistance plasmid that induces conjugation and plasmid transfer to another bacterium.

resolution *n.* (1) regeneration of normal duplex DNA molecules from structures formed during recombination; (2) regeneration of two replicons from a cointegrate by recombination between two transposons; (3) in microscopy, the size of object that can be viewed clearly. *alt.* resolving power.

resolvase *n.* enzyme activity involved in site-specific recombination between two transposons present in a cointegrate, resulting in regeneration of two replicons both containing copies of the transposon sequence.

resource *n.* anything provided by the environment to satisfy the requirements of a living organism, e.g. food, living space.

resource-holding potential the ability of an animal to gain and maintain possession of essential resources by fighting.

resource partitioning the division of scarce resources in an ecosystem so that species with similar requirements use the same resources at different times, in different ways, or in different places.

respiration *n.* any or all of the processes used by organisms to generate metabolically usable energy, chiefly in the form of ATP, from the oxidative breakdown of foodstuffs. May refer to processes ranging from the exchange of oxygen and carbon dioxide between organism and environment, to the biochemical processes generating ATP at the cellular level. *a.* **respiratory**, *pert.* or involved in respiration. *see also* aerobic respiration, anaerobic respiration, glycolysis, oxidative phosphorylation.

respiratory burst oxygen-requiring production of toxic oxygen metabolites triggered in macrophages and neutrophils by phagocytosis of bacteria and opsonized particles.

respiratory centres nuclei in medulla oblongata involved in the involuntary control of breathing.

respiratory chain series of electron carriers (e.g. quinones, flavoproteins, cytochromes) located in inner membrane of mitochondria. Electrons derived from respiratory metabolites are transferred along the chain in a series of redox reactions, which result eventually in the reduction of oxygen to water and the formation of ATP and NAD. *alt.* electron transport chain, electron transfer chain. *see also* chemiosmosis, oxidative phosphorylation.

respiratory control regulation of the rate of oxidative phosphorylation by the cellular ADP level. Low ADP inhibits, and high ADP stimulates oxygen consumption and ATP synthesis.

respiratory heart a name given to the auricle and ventricle of right side of heart in animals where there is no direct communication between right and left sides. *cf.* systemic heart.

respiratory index the amount of carbon dioxide produced per unit of dry weight per hour.

respiratory movements any movements connected with the supply of oxygen to respiratory surfaces and the removal of carbon dioxide, such as the movements of the thorax and diaphragm in mammals.

respiratory pigments (1) pigments such as haemoglobin and other haem proteins, which form an association with oxygen and carry it from the respiratory surfaces to tissue cells; (2) pigments concerned with cellular respiration, components of the respiratory chain such as cytochromes.

respiratory quotient (RQ) ratio of volume of carbon dioxide produced to volume of oxygen used in respiration.

respiratory sac a backward extension of the suprabranchial cavity, in certain air-breathing bony fishes.

respiratory substrate any substance that can be broken down by living organisms during respiration to yield energy.

respiratory surface surface at which gas exchange occurs between the environment and the body, e.g. gill lamellae or alveoli of lungs.

respiratory tract throat, bronchi and lungs.

respiratory trees in echinoderms, a respiratory system consisting of a series of tubules arising just within the anus into which water can be drawn and expelled.

respondent behaviour animal behaviour performed in response to an obvious stimulus.

response *n.* (1) the activity of a cell or organism in terms of e.g. movement, hormone or other secretion, enzyme pro-duction, changes in gene expression, as a result of a stimulus; (2) the behaviour of an organism as a result of fluctuations in the environment.

response latency the time interval between a stimulus and the response.

response regulator protein one of the members of a two-component system in bacteria, which is phosphorylated by the sensor protein.

restibrachium *n.* the restiform body or inferior peduncle of the cerebellum.

restiform *a.* having the appearance of a rope, *appl.* two bodies of nerve fibres on medulla oblongata, the inferior cerebellar peduncles.

resting cell or nucleus one that is not undergoing mitosis or meiosis.

resting metabolic rate (RMR) the metabolic rate, esp. of an animal, measured at rest. *cf.* basal metabolic rate.

resting potential the potential difference across a nerve or muscle cell membrane when it is not being stimulated, and which is usually about -70 millivolts (mV) in a nerve cell. *cf.* action potential.

Restionales *n.* order of xeromorphic tufted or climbing monocots including the family Restionaceae and others.

restitution *n.* (1) the reformation of a tissue or body by union of separated parts; (2) the reunion of breaks in chromosomes.

restriction *n.* (1) in bacteria, the breakdown of incoming DNA of another bacterial strain, phage or a foreign plasmid by particular types of endonucleases called restriction enzymes. *see also* modification, restriction endonuclease; (2) (*immunol.*) MHC restriction *q.v.*

restriction analysis the determination of the identity or internal structure of a gene or other piece of DNA by techniques that depend on the use of restriction enzymes to cut DNA into identifiable pieces.

restriction enzyme any of a large group of endonucleases produced by bacteria, which recognize short palindromic base sequences in DNA, cutting the double helix at a particular point within the sequence. Different restriction endonucleases recognize a different DNA sequence. Now used widely in genetic engineering, DNA cloning and gene mapping. *alt.*

restriction endonuclease, site-specific endonuclease. *see also* restriction–modification system.

restriction fragment DNA fragment generated by treatment of DNA with a restriction enzyme or combination of restriction enzymes.

restriction fragment length polymorphism (RFLP) the presence or absence of a cutting site for a particular restriction enzyme in DNA of different individuals, which can be detected by comparison of the DNA fragment lengths generated by that enzyme. RFLP can be used e.g. to detect the presence of a defective gene.

restriction mapping the characterization of a region of DNA by determining the number and relative positions of the sites at which particular restriction enzymes cut the DNA.

restriction–modification system in bacteria, a dual-function enzymatic system which both destroys incoming 'foreign' DNA and makes chemical modifications to the bacterium's own DNA which protect it against degradation. *see also* DNA methylation, modification, restriction endonuclease.

restriction point of eukaryotic cell cycle, the point of no return, occurring in the middle of G_1, and after which the cell is committed to completing its cycle.

restriction site a site of defined sequence in DNA at which a particular restriction enzyme cuts.

restriction-site polymorphism variation in the presence of a specific restriction site between individuals in a population. Such polymorphisms are used e.g. in gene mapping. *alt.* restriction fragment length polymorphism.

restrictive conditions conditions in which a conditional mutant organism shows the mutant phenotype.

resupinate *a.* twisted in such a way that parts are upside down.

resupination *n.* inversion.

resurgent *a. appl.* infectious disease thought to be under control but which starts to recur at high incidence.

rete *n.* (1) mesh of tissue; (2) network of interlaced vessels, nerves or fibres.

reteform, retiform *a.* in form of a network.

rete Malpighii Malpighian layer *q.v.*

rete mirabile small dense network of mainly arterial blood vessels in various organs of some vertebrates.

rete mucosum Malpighian layer *q.v.*

retention signal salvage sequences *q.v.*

retial *a. pert.* a rete.

retiary *a.* (1) making or having a net-like structure; (2) constructing a web.

reticular *a.* (1) like a meshwork; (2) netted; (3) *pert.* a reticulum.

reticular cells (1) fibroblasts that produce the reticulin fibres of connective tissues; (2) reticuloendothelial system *q.v.*

reticular formation area in brain stem involved in maintaining consciousness and level of arousal. Many sensory pathways from the body have branches that terminate in the reticular formation.

reticulate *a.* like a network; (1) *appl.* venation of leaf or insect wing; (2) *appl.* a pattern of lignin deposition in plant cell walls.

reticulocyte *n.* a precursor cell from which a mature red blood cell develops, having a reticular appearance when stained.

reticuloendothelial system (RES) system of phagocytic cells, including fibroblasts, macrophage precursors and specialized endothelial cells, that lines sinusoids of liver, spleen and bone marrow and is also present in lymph nodes. It is involved in the uptake and clearance of foreign particulate matter from the blood, in providing support for cells in lymphoid tissue and as a source of antigen-presenting macrophages.

reticulopodia *n.plu.* anastomosing threadlike pseudopodia, as of foraminiferans.

reticulose *a.* formed like a network.

reticulospinal *a. appl.* tracts of nerve fibres connecting reticular formation in brain with the spinal cord.

reticulum *n.* (1) network, esp. the framework of fibrous tissue in many organs; (2) in ruminants, the 2nd chamber of the stomach, in which water is stored; (3) (*bot.*) cross-fibres about base of petioles in palms.

retiform *a.* in the form of a network.

retina *n.* the inner light-sensitive layer of the eye. In vertebrates it comprises a double layer of epithelium lining the back of the eyeball. It is composed of a neural retina containing the photoreceptor cells (rods and cones) and neurons, and an inner pigmented epithelium containing melanin granules.

retinaculum *n.* (1) (*bot.*) a small glandular mass to which orchid pollinium adheres on dehiscence; (2) (*zool.*) a fibrous band that holds parts together; (3) a structure linking together fore- and hindwings of some insects. *plu.* **retinacula.**

retinaculum tendinum ligament around wrist or ankle.

retinal (1) *n.* aldehyde of retinol, part of the visual pigment rhodopsin, from which it is split off by the action of light; (2) *a. pert.* the retina.

retinal ganglion cells neurons in retina that transmit visual signals to brain via optic nerve.

retinella *n.* neurofibrillary network of phaeosome.

retinene retinal *q.v.*

retinerved *a.* having a network of veins, *appl.* leaf or insect wing.

retinitis pigmentosa genetically determined disorder characterized by night blindness, constriction of the visual fields, and changes in the fundus of the retina.

retinoblast *n.* retinal epithelial cells that give rise to neuroblasts and neuroglial precursor cells.

retinoblastoma *n.* rare tumour of retinal cells, usually occurring only in very young children. Inheritable familial retinoblastoma is due to a defective allele of the *Rb* gene, a tumour suppressor gene.

retinoic acid vitamin A derivative, used experimentally to induce differentiation in tumour cells. It is also a natural morphogen in vertebrate development.

retinoid *n.* analogue or metabolite of retinol (vitamin A). Retinoids have a role in embryonic development, as shown by the deleterious effects of vitamin A deficiency, the teratogenic effects of very large doses of vitamin A, and the influence of retinoid treatment on various developmental processes, e.g. pattern formation in the vertebrate limb.

retinoid receptor any of the receptors for retinoic acid and other retinoids, which belong to the nuclear receptor superfamily.

retinol vitamin A_1. *see* vitamin A.

retinotectal *a. pert.* retina and tectum, *appl.* projection: the point-to-point correspond-

ence of positions on the retina and on the optic tectum.

retinula *n.* group of elongated sensory photoreceptor cells, the innermost element of ommatidium of a compound eye.

retisolution *n.* dissolution of the Golgi apparatus in a cell.

retispersion *n.* peripheral distribution of Golgi bodies.

retractile *a. appl.* a part of an organ that may be drawn into the body.

retraction fibres fine cellular processes adhering to the substratum, which extend from the rear of a cell moving over a substrate as it moves forward.

retractor *n.* a muscle which on contraction withdraws that part attached to it, bringing it towards the body.

retrahens *n.* a muscle which draws a part backwards.

retraherence *n.* escape behaviour by which animals survive in conditions of extreme heat or cold, e.g. burrowing.

retral *a.* (1) backward; (2) posterior.

retro- prefix derived from L. *retro*, backwards.

retroarcuate *a.* curving backwards.

retrobulbar *a.* (1) posterior to eyeball; (2) on dorsal side of medulla oblongata.

retrocerebral *a.* situated behind the cerebral ganglion, in invertebrates.

retrocurved recurved *q.v.*

retroelement a general term for retrotransposons (*q.v.*) and retroviruses (*q.v.*).

retrofract *a.* bent backwards at an angle.

retrograde *a.* (1) *appl.* transport or movement of material in axons of neurons towards cell body; (2) *appl.* degeneration: destruction of neuronal cell body that occurs after its axon is damaged.

retrogressive *n. appl.* evolutionary trends towards more primitive rather than more complex forms.

retrolingual *a.* behind the tongue, *appl.* a gland.

retromandibular *a. appl.* posterior or temperomaxillary vein.

retroperitoneal *a.* (1) behind peritoneum; (2) *appl.* space between peritoneum and spinal column.

retropharyngeal *a.* behind the pharynx.

retroposon retrotransposon *q.v.*

retropubic *a. appl.* a pad or mass of fatty tissue behind pubic symphysis.

retrorse *a.* turned or directed backward.

retroserrulate *a.* with small backward-pointing teeth.

retrosiphonate *a.* with septal necks directed backwards.

retrotranslocation *n.* the movement of peptides and misfolded proteins from the endoplasmic reticulum back to the cytosol for degradation.

retrotransposition *n.* transposition of a transposable element by means of transcription of the element into RNA, reverse transcription of the RNA into DNA, and reinsertion of the DNA copy into the genome.

retrotransposon *n.* a transposon that moves by retrotransposition, i.e. via an RNA intermediate. Some are derived from retroviruses, e.g. *copia* of *Drosophila* and the THE-1 element in humans. Some, e.g. the L1 elements of mammalian genomes, appear to be derived from cellular RNAs. *alt.* retroposon.

retro-uterine *a.* behind the uterus.

retroverse retrorse *q.v.*

retroversion *n.* the state of being reversed or turned backwards.

retroviral *a.* (1) *pert.* or produced by retroviruses; (2) *appl.* vectors for genetic engineering derived from retroviruses.

retroviral-like retrotransposon transposable element that is derived from a retrovirus. Like a retrovirus it possesses flanking long terminal repeats (LTRs) and encodes a reverse transcriptase, but does not produce infectious viral particles. *see* retrotransposition.

Retroviridae, retroviruses *n., n.plu.* family of enveloped, single-stranded RNA viruses including the RNA tumour viruses and HIV (human immunodeficiency virus) as well as many apparently harmless, non-oncogenic viruses. They have a unique life history, copying their RNA into DNA by means of the viral enzyme reverse transcriptase. The DNA then enters a host cell chromosome, where it may continue to direct the production of virus particles, or may remain quiescent for many cell generations. The integrated DNA (the provirus) is passed on to all the cell's progeny. Vertebrates appear to carry a number of so-called endogenous proviruses permanently in their genomes without any ill effects.

retuse *a.* obtuse with a broad shallow notch in middle, *appl.* leaves, molluscan shells.

Retzius' striae growth markings in enamel of teeth, representing successive surfaces of enamel formation.

reuptake *n.* the uptake of neurotransmitter molecules by the presynaptic neuron that released them, which limits the strength and duration of neurotransmitter signalling at the synapse.

revehent *a.* in renal portal system, *appl.* vessels carrying blood back from excretory organs.

reversed *a.* inverted, *appl.* a spiral shell whose turns are left-handed.

reverse genetics the strategy that starts with a cloned DNA or a known protein sequence, and uses this to find the corresponding gene in the genome and introduce mutations into it to determine its function.

reverse mutation back mutation *q.v.*

reverse transcriptase (RT) DNA polymerase found in retroviruses (a class of RNA viruses including the RNA tumour viruses) which synthesizes DNA on a viral RNA template. *alt.* RNA-directed DNA polymerase.

reverse transcription (RT) the synthesis of DNA on an RNA template, catalysed by the enzyme reverse transcriptase.

reversion back mutation *q.v.*

revertant *n.* mutant organism or cell in which a back mutation to the original wild-type phenotype has occurred.

revolute *a.* rolled backwards from margin upon undersurface, as some leaves.

reward positive reinforcement *q.v.*

RF (1) general designation for protein synthesis release factors *q.v.*; (2) replicative form *q.v.*; (3) recombination frequency, recombination fraction *q.v.*

R-factor R plasmid *q.v.*

RFLP restriction fragment length polymorphism *q.v.*

R form the relaxed form of an allosteric protein, the form having a greater affinity for the substrate.

RH radiation hybrid *q.v.*

Rh rhesus factor *q.v.*

rhabd-, rhabdo- prefixes derived from Gk *rhabdos*, a rod.

rhabdacanth *n.* in certain roses, a compound thorn, consisting of small rod-like trabeculae wrapped around with lamellar tissue.

rhabdite *n.* short, rod-like body in epidermal cell of turbellarians, which is discharged if the worm is injured and swells up to form a gelatinous covering.

rhabditiform *a. appl.* larvae of nematodes with short straight oesophagus, with double bulb.

rhabditis *n.* larva of certain nematodes.

rhabdocoel *n.* member of the order Rhabdocoela, small turbellarian flatworms whose intestine is simple and sac-like.

rhabdoid *a.* rod-shaped.

rhabdolith *n.* calcareous rod found in some protozoans, strengthening the wall.

rhabdome *n.* pigmented rod formed by microvilli of retinula cells of ommatidium of a compound eye.

rhabdomere *n.* individual element of rhabdome of retinula in ommatidium of a compound eye.

rhabdomyosarcoma *n.* a tumour of muscle.

rhabdopod *n.* element of clasper in some male insects.

rhabdosphere *n.* aggregated rhabdoliths found in deep-sea calcareous oozes.

Rhabdoviridae, rhabdoviruses *n., n.plu.* family of bullet-shaped, enveloped, single-stranded, negative-strand RNA viruses including the vertebrate viruses rabies and vesicular stomatitis virus, insect viruses and plant viruses that are considered to be insect viruses becoming adapted to plants.

rhagon *n.* a bun-shaped type of sponge with an apical osculum and large gastral cavity.

Rhamnales *n.* order of trees, shrubs or woody climbers and including Rhamnaceae (buckthorn) and Vitaceae (grape).

rhamnose *n.* six-carbon (hexose) sugar found in the lipopolysaccharide outer membrane of some Gram-negative bacteria, and in plant cell wall polysaccharides.

rhamphoid *a.* beak-shaped.

rhamphotheca *n.* the horny sheath of a bird's beak.

Rheiformes *n.* an order of flightless birds, including the rheas.

rheobasis *n.* the minimal electrical stimulus that will produce a response.

rheophile, rheophilic *a.* preferring to live in running water. *n.* **rheophily**.

rheophyte *n.* plant that lives in running water.

rheoplankton *n.* the plankton of running waters.

rheoreceptors *n.plu.* cutaneous sense organs of fishes and some amphibians, receiving stimulus of water current, e.g. lateral line organs, ampullae of Lorenzini.

rheotaxis *n.* a taxis in response to the stimulus of a current, usually a water current.

rheotropism *n.* a growth curvature in response to a water or air current. *a.* **rheotropic**.

rhesus factor (Rh factor) blood group antigen found on red cells of rhesus monkey and a proportion of the human population, determined by a dominant gene. It is of medical importance because a Rh-negative mother carrying a Rh-positive foetus will produce antibodies against the foetal Rh antigen. In subsequent pregnancies these can lead to haemolytic disease of the newborn in any Rh-positive offspring (routine preventive treatment is now given to mothers at risk).

rheumatoid factor IgM antibodies found in the serum of individuals with rheumatoid arthritis. They react with IgG antibodies to form immune complexes.

rhexigenous *a.* resulting from rupture or tearing.

rhexilysis *n.* the separation of parts, or production of openings or cavities, by rupture of tissues.

rhexis *n.* fragmentation of chromosomes, caused by physical or chemical agents.

rhigosis *n.* sensation of cold.

rhin-, rhino- prefixes derived from Gk *rhis*, nose.

rhinarium *n.* (1) the muzzle or external nasal area of mammals; (2) nostril area; (3) part of nasus of some insects.

rhinencephalon *n.* the part of the forebrain forming most of the hemispheres in fishes, amphibians and reptiles, and comprising in humans the olfactory lobe, uncus, the supracallosal, subcallosal and dentate gyri, fornix and hippocampus.

rhinion *n.* the most prominent point at which the nasal bones touch.

rhinocaul *n.* narrowed portion of brain which bears the olfactory lobe.

rhinocoel *n.* cavity in olfactory lobe of brain.

rhinopharynx nasopharynx *q.v.*

rhinophore *n.* in some molluscs, an organ of sensory epithelium, sometimes borne in a pit, usually found on the tentacles and thought to have an olfactory function.

rhinotheca *n.* the sheath of upper jaw of a bird.

rhinoviruses *n. plu.* a numerous group of RNA viruses of the family Picornaviridae, the cause of the common cold and similar minor respiratory ailments in humans.

rhipidate *a.* fan-shaped.

rhipidistians *n.plu.* group of extinct crossopterygian fishes existing from Devonian to Permian times, and believed to include the ancestors of land vertebrates.

rhipidium *n.* (1) fan-shaped cymose inflorescence; (2) fan-shaped colony of zooids.

rhipidoglossate *a.* having a radula with numerous teeth in a fan-shaped arrangement, as in ear shells.

rhipidostichous *a. appl.* fan-shaped fins.

rhiptoglossate *a.* having a long prehensile tongue, as a chameleon.

rhiz-, rhizo- prefixes derived from Gk *rhiza*, a root.

rhizanthous *a.* having flowers arising so low down on a much reduced stem that they appear as if arising from root.

Rhizaria *n.* proposed major monophyletic taxonomic grouping, based on molecular phylogenetics, which includes amoeboid protozoa with fine thread-like pseudopodia such as the Foraminifera, Radiolaria and Cercozoa.

rhizine *n.* fine projection from lower fungal cortex of a lichen that attaches it to the substrate.

rhizobacteria *n.plu.* soil bacteria associated with root surfaces.

rhizobia *n.plu.* soil bacteria of the genus *Rhizobium* and related genera, Gram-negative rods that form nodules on the roots of leguminous plants, in which they carry out symbiotic nitrogen fixation.

rhizocarp *n.* (1) perennial herbaceous plant whose stems die down each winter, so that it persists by underground organs only; (2) plant producing underground flowers. *a.* **rhizocarpic, rhizocarpous.**

rhizocaul hydrorhiza *q.v.*

rhizocorm *n.* (1) underground stem like a single-jointed rhizome; (2) bulb or corm.

rhizodermis *n.* the outermost layer of tissue in roots, which may be the piliferous layer, or the exodermis in an older root where the piliferous layer has worn away.

rhizogenesis *n.* differentiation and development of roots.

rhizogenic *a.* (1) root-producing; (2) arising from endodermic cells, not developed from pericycle; (3) *pert.*, or stimulating, root formation.

rhizoid (1) *n.* filamentous outgrowth from prothallus that functions like a root; (2) *a.* root-like, *appl.* form of a bacterial colony.

rhizomatous *a.* resembling a rhizome, *appl.* fungal mycelium within a substratum or host.

rhizome *n.* thick horizontal plant stem, usually underground, bearing buds and scale leaves, sending out shoots above and roots below.

rhizomorph *n.* a root-like or bootlace-like structure formed from interwoven hyphae of some basidiomycete fungi.

rhizomorphic rhizomorphous *q.v.*

rhizomorphoid *a.* (1) resembling a rhizomorph; (2) branching like a root.

rhizomorphous *a.* in the form of a root.

rhizomycelium *n.* a many-branched system of hypha-like filaments, usually lacking nuclei, that anchor some chytrids to their substratum.

rhizophagous *a.* root-eating.

rhizophore *n.* a naked outgrowth of thallus, which grows down into soil and develops roots at the apex, as in club mosses.

rhizophorous *a.* root-bearing.

rhizoplane *n.* part of the rhizosphere immediately adjacent to the root surface, comprising a layer *ca.* 1 μm thick.

rhizoplast *n.* contractile structure in some green algae, connected to the basal bodies of the flagella.

Rhizopoda, rhizopods *n.*, *n.plu.* in protist classification, a phylum of mainly freeliving unicellular non-photosynthetic microorganisms, the amoebas, found in freshwater and marine habitats, characterized by the formation of pseudopodia and no flagella or cilia. In some groups, the body is surrounded by a casing or test, sometimes calcified.

Rhizopodea, rhizopods *n.*, *n.plu.* in older classifications, a class of protozoans consisting of the foraminiferans and the amoebas.

rhizosphere *n.* area of soil immediately surrounding and influenced by plant roots.

rhizotaxis *n.* root arrangement.

Rho *see* Rho protein family.

rhod-, rhodo- prefixes derived from the Gk *rhodon*, rose, signifying reddish.

rhodamine *n.* compound that fluoresces red, used to visualize cell structures by immunofluorescence techniques.

rhodogenesis *n.* formation or reconstitution after bleaching, of rhodopsin.

rhodophane *n.* a red chromophane in retinal cones of some fishes, reptiles and birds.

Rhodophyceae Rhodophyta *q.v.*

Rhodophyta *n.* the red algae, a group of largely multicellular, structurally complex photosynthetic organisms classified either in the plant kingdom or as a phylum of the Protocista. They are composed of close-packed filaments. The red colour is due to water-soluble phycobilin pigments. The storage carbohydrate is floridean starch, resembling amylopectin. The red algae are largely marine and there are many tropical species, and they usually grow attached to rocks or other substrates. Their cells have no flagella at any stage.

rhodoplast *n.* a reddish plastid in red algae.

rhodopsin *n.* rose-purple, light-sensitive pigment found in rod cells of the vertebrate retina (and in invertebrates). It is a conjugate of a protein (opsin), a phospholipid, and the vitamin A aldehyde retinal. It becomes colourless on exposure to light as a result of a reversible chemical change in retinal. *alt.* visual purple.

rhodosporous *a.* with pink spores.

rhodoxanthin *n.* a carotenoid pigment, found in aril of yew.

rho factor (ρ) protein factor regulating the termination of transcription of some bacterial genes.

rhombencephalon hindbrain *q.v.*

rhombic *a. appl.* lips and grooves of brain at rhomboid fossa.

rhombocoel *n.* dilation of the central canal of medulla spinalis near its terminal end, the terminal vesicle.

rhombogen *n.* stage in the life cycle of dicyemid parasites.

rhomboid *a.* having the shape of a rhombus, i.e. a diamond in a pack of playing cards.

rhomboideum *n.* the rhomboid or costo-clavicular ligament.

rhomboideus *n.* major and minor, parallel muscles connecting scapula with thoracic vertebrae.

rhomboid-ovate *a.* between rhomboid and oval in shape, *appl.* leaves.

rhombomeres *n.plu.* the seven distinct segments into which the vertebrate hindbrain is divided at an early stage in its development.

Rhombozoa *n.* class of the invertebrate phylum Mesozoa that includes the dicyemids and heterocyemids, parasites of the kidneys of marine invertebrates such as cephalopods, other molluscs, flatworms and annelids.

rhopalium *n.* a marginal sense organ in some jellyfish.

Rho protein family family of small GTPases that are involved in signalling pathways involving rearrangement of the actin cytoskeleton in response to external signals.

Rhynchocephalia, rhynchocephalians *n.*, *n.plu.* order of mainly extinct reptiles, with one living member, the tuatara (*Sphenodon punctatus*), a lizard-like animal confined to a few islands off New Zealand. *alt.* Sphenodonta.

rhynchocoel *n.* cylindrical cavity in body of nemertean worms that houses the proboscis when not extended.

rhynchodont *a.* with a toothed beak.

rhynchoporous *a.* beaked, *appl.* weevils.

rhynchostome *n.* anterior terminal pore through which proboscis is everted in nemertean worms.

α-rhythm spontaneous rhythmic fluctuations of electrical potential of cerebral cortex during mental inactivity.

β-rhythm spontaneous rhythmic fluctuations of electrical potential of cerebral cortex during mental activity.

rhytidome

rhytidome *n.* the outer bark consisting of the periderm and tissues isolated by it.

RIA radioimmunoassay *q.v.*

rib *n.* (1) in tetrapod vertebrates, curved thin bone of the thorax articulating with spine at one end and either free (floating ribs) or fixed to sternum (breast bone) at other; (2) central vein of leaf; (3) any elongated protrusion. *alt.* costa.

ribbon worms Nemertina *q.v.*

ribitol *n.* sugar alcohol derived from ribose, a constituent of the teichoic acids of bacterial cell walls.

rib meristem a meristem in which cells divide perpendicular to the longitudinal axis, producing a complex of parallel files or ribs of cells.

riboflavin *n.* vitamin B₂, consisting of ribose linked to the nitrogenous base dimethylisoalloxazine. It is synthesized by all green plants and most microorganisms, occurring free in milk and in some tissues of higher organisms and green plants, and in all living cells as a component of the coenzymes flavin adenine dinucleotide (FAD) and flavin mononucleotide (FMN). Liver, yeast and green vegetables are particularly rich sources. Deficiency causes skin cracking and lesions (ariboflavinosis).

riboflavin phosphate flavin mononucleotide *q.v.*

ribonuclease (RNase) *n.* any of various enzymes that cleave RNA into shorter oligonucleotides or degrade it completely into its constituent ribonucleotide subunits. *alt.* nuclease. *see also* RNase H, RNase P, RNase III.

ribonucleic acid *see* RNA.

ribonucleoprotein (RNP) *n.* any complex of RNA and protein.

ribonucleoside *see* nucleoside.

ribonucleotide *n.* a nucleotide containing the sugar ribose.

ribonucleotide reductase either of two enzymes that reduce ribonucleotides to deoxyribonucleotides, the enzyme from animals acting on ribonucleoside diphosphates (*r.n.* ribonucleoside-diphosphate reductase, EC 1.17.4.1), and that from some bacteria acting on ribonucleoside triphosphates (*r.n.* nucleoside-triphosphate reductase, EC 1.17.4.2).

Fig. 33 Ribose.

ribophorin *n.* protein on the cytoplasmic face of the endoplasmic reticulum, to which ribosomes are anchored.

riboprobe *n.* a short piece of RNA used as a probe for gene isolation and identification.

ribose *n.* a pentose sugar, the sugar in RNA and also an intermediate in the Calvin cycle of photosynthesis. *see* Fig. 33.

ribosomal DNA (rDNA) the DNA encoding ribosomal RNAs. In many eukaryotes it is present in many copies and clustered in chromosomal regions which form the nucleolar organizers. *alt.* **ribosomal genes**.

ribosomal protein *see* ribosome.

ribosomal RNA (rRNA) major component of ribosomes and the most abundant RNA species in cells. In eukaryotes it is synthesized in the nucleolus from rRNA genes tandemly repeated many times in the chromosomes. There are several different types, denoted by their sedimentation coefficients, e.g. 23S, 16S and 5S RNAs in eukaryotic ribosomes.

ribosome *n.* ribonucleoprotein particle found in large numbers in all cells, both free in cytoplasm and attached to the endoplasmic reticulum in eukaryotic cells. Ribosomes are the sites at which translation of messenger RNA and protein synthesis take place. Each consists of two subunits that differ in size (50S and 30S in *E. coli*, 60S and 40S in a typical eukaryotic cell). Each subunit contains a major RNA species complexed with proteins (52 in the *E. coli* ribosome, *ca.* 80 in eukaryotes), as well as minor rRNAs in the large subunit. The ribosome provides a framework for holding mRNA in position for translation and also contains the peptidyltransferase activity that catalyses formation of peptide bonds. *see also* microsomes, polysome, translation.

588

ribosome-binding site region in bacterial mRNA preceding the coding region, to which the ribosome binds to start translation. *alt.* Shine–Dalgarno sequence.

riboswitch *n.* a gene-regulatory RNA, or a sequence within an RNA that affects its transcription or translation.

ribothymidylate *n.* unusual nucleotide found in tRNA which is formed by modification of a uridylate residue after transcription and which is the ribonucleotide of thymine.

ribozyme *n.* an RNA molecule with enzymatic activity, such as the self-splicing introns of some RNAs, which can excise themselves from the molecule without the agency of protein enzymes.

ribulose (Ru) *n.* five-carbon ketose sugar which, as the phosphate and bisphosphate, is involved in carbon dioxide fixation in photosynthesis and in other metabolic pathways such as the pentose phosphate pathway.

ribulose 1,5-bisphosphate (RuBP) five-carbon sugar phosphate which is the primary carbon dioxide acceptor in photosynthesis. *alt.* (formerly) ribulose 1,5-diphosphate (RuDP).

ribulosebisphosphate carboxylase/oxygenase (RuBPc/o, Rubisco) enzyme found in chloroplasts of all green plants and in photosynthetic bacteria. It catalyses the fixation of carbon dioxide into carbohydrate via ribulose-1,5-bisphosphate as acceptor, and also has oxygenase activity, which is involved in photorespiration. EC 4.1.1.39, *alt.* ribulose-1,5-diphosphate carboxylase.

rickets *n.* inadequate calcification of bone in children, caused by deficiency of vitamin D.

rickettsiae, rickettsias *n.plu.* obligate intracellular prokaryotic parasites, members of the alpha group of the proteobacteria. They cause arthropod-transmitted disease in humans and other animals. Their simple cells lack cell walls and are obligate intracellular parasites of mammalian cells. Diseases caused by louse-borne rickettsias include typhus fever and trench fever, and tick-borne rickettsias cause Rocky Mountain spotted fever and other tick-borne typhuses. Scrub typhus is transmitted by mites. *sing.* **rickettsia**.

rickettsia-like organism (RLO) any of a group of organisms which resemble rickettsiae but are found in plants.

rictal *a. pert.* mouth and gape of birds.

rictus *n.* (1) the opening or throat of calyx; (2) the gape of a bird's beak.

rifampicin *n.* semisynthetic derivative of the antibiotic rifamycin.

rifamycin *n.* antibiotic from *Streptomyces* which specifically inhibits the initiation of RNA synthesis in bacterial cells.

riffle *n.* shallow broken water in a stream running over a stony bed.

RIG-I-like receptors (RLRs) family of intracellular proteins that recognize foreign nucleic acids and initiate antiviral defences such as the production of interferon.

rigor *n.* (1) the rigid state of plants when not sensitive to stimuli; (2) contraction and loss of irritability of muscle on heating or after death, or in some states such as shock or fever.

rigor mortis stiffening of body after death due to temporary rigidity of muscles.

rimate *a.* having fissures.

rimiform *n.* in the shape of a narrow fissure.

rimose *a.* having many clefts and fissures.

rimulose *a.* having many small clefts.

ring bark bark peeling off in rings, as in some cherries, etc. *cf.* scale bark.

ring canal (1) circular canal around margin of bell in medusae, connecting via radial canals to body cavity; (2) circular vessel around gullet in echinoderms.

ring cell thick-walled cell of annulus of sporangium in ferns.

ring centriole disc at end of body or middle portion of sperm, perforated for axial filament of flagellum.

ring chromosomes chromosomes formed by fusion of the ends of a chromosome fragment containing the centromere.

ringed worms common name for the annelids *q.v.*

ringent *a.* (1) having lips, as of corolla, or valves, separated by a distinct gap; (2) with upper lip arched; (3) gaping.

ring-form form assumed by immature trophozoite of the malaria parasite in red blood cells.

ring gland a glandular structure round aorta in insects, composed of various elements such as corpus allatum, corpus cardiacum, pericardial gland and hypocerebral ganglion.

ring-porous *a. appl.* wood in which the vessels formed early in the season are clearly larger than those formed later, producing a clear ring in cross-section.

ring species two species which overlap in range and behave as true species with no interbreeding, but are connected by a series (the ring) of interbreeding subspecies so that no true specific separation can be made.

ring vessel structure in head of cestodes which unites the four longitudinal excretory trunks.

ringworm *n.* fungal disease of skin, caused by species of *Microsporum* or *Trichophyton*.

ripa *n.* a line of ependymal fold over a plexus or tela.

riparian *a.* frequenting, growing on, or living on the banks of streams or rivers.

Ri plasmid a plasmid carried by *Agrobacterium rhizogenes*, the cause of hairy root in various dicotyledonous plants. The plasmid becomes stably integrated into the chromosomes of infected tissue.

risorius *n.* a cheek muscle stretching from over masseter muscle to corner of mouth.

riverine *a.* living in rivers.

rivinian *a.* (1) *appl.* sublingual glands and ducts; (2) *appl.* notch in ring of bone surrounding tympanic membrane.

rivose *a.* marked with irregularly winding furrows or channels.

rivulose *a.* marked with sinuous narrow lines or furrows.

RLO rickettsia-like organism *q.v.*

RLRs RIG-I-like receptors *q.v.*

R-loop a structure that can be formed when a piece of duplex DNA is being transcribed, composed of an RNA-DNA hybrid strand and the displaced DNA strand.

RMR resting metabolic rate *q.v.*

RNA *n.* ribonucleic acid, large linear molecule made up of a single chain of ribonucleotide subunits, containing the bases uracil, guanine, cytosine and adenine. Found in all cells as transfer RNA, ribosomal RNA and messenger RNA, all cellular RNAs being synthesized by transcription of chromosomal DNA acting as a template. It is the primary genetic material in some viruses and in some cases can be synthesized using viral RNA as a template. *see also* small cytoplasmic RNAs, small nuclear RNAs. *cf.* DNA.

5S RNA minor rRNA species in the large ribosomal subunit.

5.8S RNA minor rRNA species in the large ribosomal subunit of eukaryotic cells.

16S RNA rRNA component of the small subunit of bacterial ribosomes.

18S RNA rRNA component of the small subunit of a typical eukaryotic ribosome.

23S RNA major rRNA component of the large subunit of bacterial ribosomes.

28S RNA major rRNA component of the large subunit of a typical eukaryotic ribosome.

RNA blot Northern blot *q.v.*

RNA capping *see* cap.

RNA-dependent DNA polymerase reverse transcriptase *q.v.* *alt.* **RNA-directed DNA polymerase.**

RNA-dependent RNA polymerase RNA replicase *q.v.*

RNA editing the addition, deletion and conversion of nucleotides at specific sites in primary RNA transcripts after synthesis to form a functional mRNA. It occurs in some mitochondrial mRNAs of the protozoan *Trypanosoma* and other simple organisms, in some plant chloroplast and mitochondrial mRNAs and in one nuclear transcript in mammals. Separate 'guide RNAs' carrying the information for editing bind to the transcript and direct the changes to be made.

RNA helicase protein that unwinds double-stranded RNAs.

RNA interference (RNAi) the suppression of specific gene expression by an RNA complementary to part of the mRNA. This guides RNA-cleaving enzymes to the mRNA, leading to its destruction. *see also* small interfering RNA.

RNA ligase enzyme which catalyses the rejoining of exons in the splicing of certain mRNAs.

RNA maturase general name given to some enzymes that process RNA, especially ribosomal RNA, into its final functional form.

RNA plasmid linear or circular RNAs found in some plant mitochondria.

RNA polymerase general name for enzymes that catalyse the synthesis of RNA from a DNA template by the process of transcription (*q.v.*). *see also* RNA replicase.

RNA polymerase I eukaryotic RNA polymerase present in nucleolus, responsible for transcribing the rRNA genes.

RNA polymerase II eukaryotic RNA polymerase responsible for transcribing most protein-coding genes.

RNA polymerase III eukaryotic RNA polymerase responsible for transcribing tRNAs, 5S RNA, and some other small RNAs.

RNA primase *see* primase.

RNA primer *see* primer.

RNA processing any or all of the processes that result in the generation of a functional tRNA, rRNA or mRNA from a primary RNA transcript. It includes trimming the ends, removing introns (in eukaryotic RNAs), capping (in eukaryotic mRNAs), and cutting out individual rRNAs from their precursor transcripts. RNA processing in eukaryotes occurs in the nucleus.

RNA replicase RNA polymerase which synthesizes RNA using an RNA template, the means by which some RNA viruses replicate their genomes. *alt.* RNA synthetase, replicase.

RNase, RNAse ribonuclease *q.v.*

RNase H endonuclease with specificity for RNA in the form of an RNA–DNA hybrid.

RNase III endoribonuclease from *E. coli* involved in processing of rRNA precursors.

RNase P endoribonuclease from *E. coli* involved in processing of tRNA precursors and which has a protein and an RNA component, neither of which is active alone.

RNA splicing the process by which introns are removed from the primary RNA transcripts of eukaryotic genes. Introns are cut out at precisely defined splice points and the ends of the remaining RNA are rejoined to form a continuous mRNA, rRNA or tRNA. *alt.* splicing. *see also* spliceosome.

RNA synthetase RNA replicase *q.v.*

RNA transcript RNA produced by transcription of a DNA sequence.

RNA tumour viruses members of the retroviruses that can cause tumours in animals.

RNA viruses viruses that have RNA as their genetic material. They include the Reoviridae, Togaviridae, Coronaviridae, Picornaviridae, Caliciviridae, Rhabdoviridae, Paramyxoviridae, Orthomyxoviridae, Arenaviridae, Bunyaviridae and Retroviridae amongst vertebrate viruses, and most plant viruses. *cf.* DNA viruses. *see* Appendix 7.

RNA world a proposed stage in the evolution of life, pre-dating the appearance of protein synthesis and DNA. In the RNA world self-replicating RNA molecules were both the genetic material and 'enzymes' of very primitive cells. *see also* ribozyme.

RNP ribonucleoprotein *q.v.*

Robertsonian translocation type of chromosome abnormality in humans involving exchange between the long and short arms of two non-homologous acrocentric chromosomes. It produces one long chromosome and a tiny fragment, which is usually lost. *alt.* chromosomal fusion.

robust *a.* heavily built, *appl.* australopithecines: *Australopithecus robustus.*

robustness *n.* the property of self-regulation of biological systems that enables them to function reliably in the presence of variations in input, environmental coditions, random variation in the rate of biochemical reactions, and so on.

rock mosses granite mosses *q.v.*

Rocky mountain subregion zoogeographical area, part of the Nearctic region, consisting of the Rocky Mountains of North America.

rod *n.* (1) rod-shaped, light-sensitive sensory cell type in retina, containing the light-sensitive pigment rhodopsin, which is responsible for non-colour vision and vision in poor light, *cf.* cone; (2) common name for straight or slightly curved cylindrical bacterial cell.

Rodentia, rodents *n., n.plu.* the largest order of placental mammals, known from the Paleocene and including rats, mice, voles, hamsters, porcupines, beavers and squirrels. They are omnivorous and/or herbivorous, and have continuously growing chisel-like incisors adapted for gnawing, and no canines.

rod fibre nerve process that synapses with rod cell of retina.

roding *n.* patrolling flight of birds defending territory.

rod vision vision in dim light. Dark-adapted or 'night' vision.

roentgen (r, R) unit of ionizing radiation corresponding to an amount of ionizing radiation sufficient to produce two ionizations per cubic micrometre of water or living tissue.

rolling circle mode of replication of some circular DNAs, as in certain phages. Replication of one strand only is initiated at the origin, the newly synthesized strand displacing the other parental strand.

roof plate area of non-neuronal cells in the dorsal midline of neural tube, involved in patterning the neural tube.

root *n.* (1) descending portion of plant, fixing it in soil, and absorbing water and minerals, and having a characteristic arrangement of vascular tissues; (2) radix *q.v.*; (3) embedded portion of tooth, hair, nail or other structure; (4) pulmonary veins and artery joining lung to heart and trachea; (5) pedicle of vertebra; (6) efferent and afferent fibres of a spinal nerve, leaving or entering the spinal cord.

root apical meristem group of undifferentiated dividing cells at the tip of a root from which root growth is produced.

root cap protective cap of tissue at tip of root. *alt.* calyptra, pileorhiza.

root cell clear, colourless base of an alga, attaching thallus to substratum.

root climber plant which climbs by roots developed from the stem.

rooted *appl.* a phylogenetic tree that converges onto a common ancestor, or root.

root hairs unicellular outgrowths from epidermal cells of roots, concerned with uptake of water and solutes from soil.

rootlet *n.* an ultimate branch of a root.

root nodule structure formed on the roots of leguminous and some non-leguminous plants and which contains nitrogen-fixing bacteria.

root parasitism condition shown by semi-parasitic plants, the roots of which penetrate the roots of neighbouring plants and draw nutrients from them.

root pocket a sheath containing a root, especially of aquatic plants.

root pressure positive hydrostatic pressure in xylem which forces water and dissolved ions up the xylem. It is developed when transpiration is low, and is created by ion movements and osmosis in root cells.

root process branched structure attaching an algal thallus to the substratum.

root sheath *n.* (1) (*bot.*) a protective sheath surrounding the developing radicle of some flowering plants such as grasses, *alt.* coleorhiza; (2) velamen of orchid; (3) (*zool.*) that part of hair follicle continuous with epidermis.

root stalk (1) rootstock or rhizome; (2) root-like horizontal portion of certain hydrozoan colonies.

rootstock *n.* (1) more-or-less underground part of stem; (2) rhizome *q.v.*

root tubers swollen food-storing roots of certain plants.

roridous *a.* covered with droplets.

ROS rod outer segment, the photoreceptor portion of rod cells of vertebrate retina.

rosaceous *a.* (1) resembling a rose; (2) *pert.* the Rosaceae family of dicot flowers, which includes roses, cherries, mountain ash, hawthorn and many other trees and shrubs.

Rosales *n.* order of mainly woody dicots, including the families Chrysobalanaceae (coco plum), Neuradaceae and Rosaceae (rose, apple, etc.)

rosellate *a.* arranged in rosettes.

rosette *n.* (*bot.*) (1) a cluster of leaves arising in close circles from a central axis; (2) plant disease due to deficiency of boron or zinc; (3) cluster of crystals as in some plant cells; (*zool.*) (4) a swirl or vortex of hair in pelt of animal; (5) small cluster of blood cells; (6) group of spiracular channels in exocuticle of some aquatic insects.

rosette organ in some ascidians, a ventral complex stolon from which buds are constricted off.

rosin *n.* a resin obtained from pine.

rostellate *a.* furnished with a rostellum.

rostelliform *a.* shaped like a small beak.

rostellum *n.* a small beak or beak-like structure, e.g. the tubular mouthparts of some insects, the rounded hooked prominence on scolex of tapeworm. *a.* **rostellar**.

rostral *a.* (1) towards the anterior end of the body; (2) *pert.* a beak or rostrum.

rostrate *a.* beaked.

rostro-caudal axis antero-posterior axis, i.e. head-to-tail axis, of the animal body.

rostrum *n.* (1) beak of birds; (2) beak-like structures on heads of arthropods; (3) median ventral plate of shell of barnacles; (4) (*neurobiol.*) the prolongation of the anterior end of corpus callosum which curves under and backwards.

rotate *a.* shaped like a wheel.

rotation *n.* (1) turning as on a pivot, as limbs; (2) circulation, as of cell sap.

rotator *n.* a muscle that allows circular motion.

rotatores spinae paired muscles, one on each side of thoracic vertebra, each arising from transverse process and inserted into vertebra next above.

rotatorium *n.* a trochoid joint or pivot joint.

Rotifera, rotifers *n., n.plu.* phylum of microscopic, multicellular, pseudocoelomate animals, living mostly in fresh water. They are generally cone-shaped with a crown of cilia at the widest end surrounding the mouth. The epidermis is separated from the internal organs by a fluid-filled space, the pseudocoel. Formerly called wheel animals, as the crown of beating cilia looks as though it is rotating.

rotula *n.* (1) one of five radially parallel bars bounding circular aperture of oesophagus of sea urchin; (2) patella or knee cap. *a.* **rotular.**

rotuliform *n.* shaped like a small wheel.

rotundifolious *a.* with rounded leaves.

rough endoplasmic reticulum *see* endoplasmic reticulum.

rouleaux *n.* formations like piles of coins into which red blood cells tend to aggregate.

round dance type of repeated circular dance in bees that alerts other bees to the existence of a food source near the hive. *see also* waggle dance.

round window the lower of two membrane-covered openings in the bony wall between the tympanic cavity (middle ear) and vestibule of inner ear. It moves in response to movement of fluid within the cochlea. *alt.* fenestra rotunda.

roundworms common name for the Nematoda *q.v.*

royal cell (1) in honey bees the large, pitted waxen cell constructed by the workers to rear queen larvae; (2) in termites, the special cell in which the queen is housed.

royal jelly material supplied by workers to female larvae in royal cells, which is necessary for transformation of larvae into queens.

R plasmid plasmid carrying genes for resistance to various commonly used antibiotics, present in many of the enterobacteria. Some R plasmids are transmissible to other bacteria of the same and other related species. *alt.* drug-resistance plasmid, R-factor.

rpm, r.p.m. revolutions per minute, used for speed of a centrifuge.

R point restriction point *q.v.*

RPPP reductive pentose phosphate pathway. *see* Calvin cycle.

RQ respiratory quotient *q.v.*

rRNA ribosomal RNA *q.v.*

r-selected species species typical of variable or unpredictable environments, characterized by small body size and rapid rate of increase. *alt.* opportunist.

r selection selection favouring rapid rates of population increase, esp. prominent in species that colonize short-lived environments or undergo large fluctuations in population size. *cf. K* selection.

RSS recombination signal sequence *q.v.*

R-state in an allosteric protein, the relaxed state, in which the protein is more active, induced by the binding of an activating ligand.

RSV (1) Rous sarcoma virus, an avian RNA tumour virus; (2) respiratory syncytial virus.

RT (1) reaction time *q.v.*; (2) reverse transcriptase *q.v.*; (3) reverse transcription *q.v.*

RTF resistance transfer factor *q.v.*

RT-PCR reverse transcription and polymerase chain reaction. A combined technique in which cellular RNA is transcribed into cDNA and then selectively amplified using the polymerase chain reaction.

RTQ-PCR quantitative RT-PCR, in which the RT-PCR technique is modified to enable measurement of the amount of DNA produced at each round of amplification. *see* quantitative PCR.

Ru ribulose *q.v.*

rubber *n.* the coagulated latex of several trees, mainly *Hevea* species, being long-chain polymers of isoprene and hydrocarbons.

rubiaceous *a. pert.* a member of the dicot family Rubiaceae (the blackberries, raspberries, etc.).

rubiginose, rubiginous *a.* of a brownish-red tint, rust-coloured.

Rubisco ribulose bisphosphate carboxylase *q.v.*

RuBP ribulose 1,5-bisphosphate *q.v.*

rubrospinal *a. appl.* a descending tract of nerve fibres from red nucleus, in the ventrolateral column of the spinal cord. *alt.* prepyramidal tract.

ruderal *a.* (1) growing among rubbish or debris; (2) growing by the roadside or in disused fields.

rudiment *n.* an initial, or primordial group of cells which gives rise to a structure.

rudimentary *a.* (1) in an imperfectly developed condition; (2) at an early stage of development; (3) arrested at an early stage.

ruff *n.* fringe of fur or feathers around neck.

Ruffini endings pressure receptors in the dermis of vertebrates.

Ruffini's organs heat receptors in subcutaneous tissue of fingers.

ruffling *n.* the backward movement of unattached lamellipodia over the surface of a moving animal cell, which gives the leading edge of the cell a ruffled appearance.

rufine *n.* a reddish pigment in the mucous glands of slugs.

rufinism *n.* red pigmentation due to inhibition of formation of dark pigment. *alt.* rutilism.

ruga *n.* fold or wrinkle, as of skin or mucosal membranes in some animals.

rugate *a.* wrinkled, ridged.

Rugosa *n.* extinct group of colonial and solitary corals characterized by a horn shape and wrinkled surface, which existed from the middle Ordovician to the late Permian. *alt.* **rugose corals**, horn corals.

rugose *a.* with many wrinkles or ridges.

ruling reptiles archosaurs *q.v.*

rumen *n.* the first stomach in ruminants (cud-chewing mammals), in which food is digested by microorganisms and from which it can be regurgitated into the mouth for further chewing.

rumen flora microorganisms (bacteria and protozoa) present in the rumen and which help in the breakdown of plant material, esp. cellulose.

ruminants *n.plu.* herbivorous mammals such as cows, sheep, goats, deer, antelopes and giraffes, that chew the cud and have complex, usually four-chambered, stomachs containing microorganisms that break down the cellulose in plant material.

rumination *n.* the act of ruminant animals in returning partly digested food from the first stomach to mouth in small quantities for thorough mastication. *alt.* chewing the cud. **ruminate** *v.* to chew the cud.

runcinate *a. appl.* pinnatifid leaf when divisions point towards the base, as in dandelions.

runner *n.* a specialized stolon consisting of a prostrate stem rooting at the node and forming a new plant which eventually becomes detached from the parent, as strawberry.

runoff *n.* the drainage of water from waterlogged or impermeable soil.

rupestrine, rupicoline, rupicolous *a.* growing or living among rocks.

ruptile *a.* bursting in an irregular manner.

rush *see* Juncales.

rust fungi common name for a group of basidiomycete fungi, the Uredinales, many of which are serious and widespread plant pathogens of numerous important crops. They cause rust disease, which appears as small black, orange or brown pustules on the stem or leaf surface. They have complex life histories on two different hosts, e.g. the black stem rust of cereals (*Puccinia graminis*) uses the barberry (*Berberis vulgaris*) as an intermediate host.

rut *n.* period of sexual heat in male animals, when they often fight for females and defend territory before mating.

Rutales *n.* order of dicot trees and shrubs, rarely herbs, with leaves often dotted with glands, and including the families Anacardiaceae (cashew), Meliaceae (mahogany), Rutaceae (citrus, rue), Simaroubaceae, and others.

rutilant *a.* of a bright bronze-red colour.

rutilism rufinism *q.v.*

ryanodine receptor a calcium-activated Ca^{2+} release channel in sarcoplasmic reticulum, which is bound by the alkaloid ryanodine.

S

σ (1) sigma, symbol for 0.001 seconds; (2) symbol for standard deviation.

S (1) serine *q.v.*; (2) Svedberg unit *q.v.*

s (1) sedimentation coefficient *q.v.*; (2) coefficient of selection *q.v.*

S₁ 1st selfing generation, the offspring of a self-cross.

sabuline *a.* (1) sandy; (2) growing in sand, especially coarse sand.

sabulose, sabulous *a.* sandy.

saccade *n.* brief rapid movements of the eyes that occur when the direction of gaze is shifted. *a.* **saccadic.**

saccate *n.* pouched.

saccharide *see* monosaccharide, disaccharide, polysaccharide, oligosaccharide.

saccharobiose sucrose *q.v.*

Saccharomyces cerevisiae the bread and brewing yeast, a budding yeast that is widely used as a model organism in cell biology and genetics.

Saccharomycetes class of unicellular ascomycete fungi that includes the budding yeasts, e.g. *Saccharomyces cerevisiae. alt.* Hemiascomycetae.

saccharose sucrose *q.v.*

sacciferous *a.* furnished with a sac.

sacciform *a.* like a sac or pouch.

sacculate *a.* provided with sacculi.

sacculation *n.* (1) the formation of sacs or saccules; (2) series of sacs, as of haustra of colon.

saccule, sacculus *n.* (1) small sac or pouch; (2) part of the membranous labyrinth of inner ear that together with utricle forms vestibule of inner ear. *plu.* **sacculi.**

sacculus rotundus in rabbits and hares, a dilation of intestine between ileum and caecum.

sac fungi common name for the Ascomycota *q.v.*

sacral *a.* (1) *pert.* the sacrum; (2) *appl.* area of lower part of back and spinal cord.

sacral index one hundred times the breadth of sacrum at base, divided by anterior length.

sacralization *n.* fusion of sacral and lumbar vertebrae.

sacral ribs elements of sacrum joining true sacral vertebrae to pelvis.

sacrocaudal *a. pert.* sacrum and tail region.

sacrococcygeal *a. pert.* sacrum and coccyx.

sacro-iliac *a. pert.* sacrum and ilium (dorsal bone of pelvic girdle).

sacrolumbar *a. pert.* region of loins and termination of vertebral column (sacrum).

sacrospinal *a.* (1) *pert.* sacral region and spine; (2) *appl.* to ligament between sacrum and spine of ischium (ventral and posterior bone of each half of pelvic girdle).

sacrovertebral *a. pert.* sacrum and vertebrae.

sacrum *n.* (1) bone forming termination of vertebral column, usually composed of several fused vertebrae. *alt.* os sacrum; (2) vertebra or vertebrae to which the pelvic girdle is attached.

saddle clitellum *q.v.*

S-adenosylmethionine (SAM) *n.* high-energy compound of adenosine and methionine, a major donor of methyl groups in biosynthetic reactions.

Saefftigen's pouch internal sac near the posterior of male acanthocephalans that aids in copulation.

safe concentration the maximum concentration of a toxic substance that has no observable effect on a species after long-term exposure over one or more generations.

SAGE serial analysis of gene expression *q.v.*

sagitta *n.* (1) elongated otolith in sacculus of teleost fishes; (2) genus of arrow worms.

sagittae *n.plu.* the inner genital valves in hymenopterans.

sagittal *a.* (1) section or division in median longitudinal plane, bisecting a structure into right and left halves; (2) *appl.* sinus running between the two hemispheres of the brain; (3) *appl.* the suture between the parietal bones of the skull which forms a ridge or crest on top of skull in some primates.

sago palm *see* Cycadophyta.

salamanders *see* Urodela.

Salicales *n.* order of dicot trees and shrubs comprising the family Salicaceae (willow).

Salientia *n.* in some classifications the name given to the order of amphibians comprising the frogs and toads. *alt.* Anura.

salinization *n.* the deposition of excessive amounts of soluble mineral salts in the soil, making it unfit for cultivation. It is caused by high surface evaporation often exacerbated by artificial irrigation over long periods.

saliva *n.* secretion produced by salivary glands which open into or near the mouth in many vertebrates and invertebrates. In humans it contains mucoproteins and the starch-digesting enzyme α-amylase, in insects various digestive enzymes depending on diet, and in blood-sucking invertebrates various anticoagulants.

salivarium *n.* recess of preoral food cavity, with opening of the salivary duct, in insects.

salivary *a.* (1) *pert.* saliva; (2) *appl.* glands opening into or near the mouth which secrete saliva.

salmonellae *n.plu.* common name for bacteria of the genus *Salmonella*, which includes the causal agent of typhoid fever, *S. typhi*, and species causing food poisoning. *sing.* **salmonella**.

Salmonidae, salmonids *n., n.plu.* a family of the Salmoniformes which includes the genus *Salmo* (salmon, rainbow and brown trout).

Salmoniformes *n.* large order of fairly primitive marine and freshwater teleost fishes, including trout, salmon and pike.

salpingian *a. pert.* Eustachian or Fallopian tube.

salpingopalatine *a. pert.* Eustachian tubes and palate.

salpinx *n.* (1) Eustachian tube *q.v.*; (2) Fallopian tube *q.v.*; (3) any of various trumpet-shaped structures.

salps Thaliacea *q.v.*

salsuginous *a.* growing in soil impregnated with salts, as in a salt marsh.

saltation *n.* (1) jumping movement; (2) (*evol.*) the idea that major evolutionary changes can take place within a single generation through 'macromutations' – mutations of large effect. *cf.* phyletic gradualism.

saltatorial *a.* adapted for, or used in, leaping, *appl.* limbs of jumping insects.

saltatorians *n.plu.* crickets and grasshoppers, members of the insect order Orthoptera (called Saltatoria in some classifications).

saltatory (1) saltatorial *q.v.*; (2) *appl.* jerky movements of particles and organelles within cells.

saltatory conduction mode of impulse propagation in myelinated nerve fibres, where the impulse 'jumps' from node to node.

salt bridge hydrogen bond in which both donor and acceptor atoms are fully charged.

salt gland (1) organ near eye in marine reptiles and birds for excretion of excess sodium chloride; (2) similar structure in gills of fishes; (3) (*bot.*) an epidermal gland exuding salts in certain leaves.

saltigrade *a.* moving by leaps, as some insects and spiders.

salt linkage salt bridge *q.v.*

salt marsh the intertidal area on sandy mud in sheltered coastal areas and in estuaries, supporting characteristic plant and animal communities.

salvage pathway metabolic pathway that uses preformed compounds as precursors for new biosyntheses.

salvage sequences sequences present in proteins destined to remain in the endoplasmic reticulum.

SAM *S*-adenosylmethionine *q.v.*

samara *n.* winged fruit typical of elm and ash, a single winged achene.

SAN sinuatrial node *q.v.*

sand *n.* a soil having most particles between 2 mm and 0.02 mm in size, composed usually of silica, and being well drained and aerated.

sand dollars common name for sea urchins of the order Clypeasteroidea, having flattened tests.

sanguicolous *a.* living in blood.

sanguivorous *a.* feeding on blood.

Santales *n.* order of woody dicots, often parasitic on other angiosperms or, rarely, on gymnosperms, and including the families Santalaceae (sandalwood), Viscaceae (mistletoe) and others.

sap (1) *see* cell sap; (2) sugary fluid carried by phloem.

saphena *n.* a conspicuous vein of leg, extending from foot to femur.

saphenous *a. pert.* internal or external saphena, *appl.* a branch of the femoral nerve.

Sapindales *n.* order of dicot trees, shrubs, lianas, rarely herbs, and including the Aceraceae (maple), Hippocastanaceae (horse chestnut), Sapindaceae (soapberry) and others.

sapogenin *n.* the non-sugar part of a saponin, usually obtained from saponin by hydrolysis.

saponin *n.* any of various steroid glycosides present in many plants, such as soapwort and soapbark, and which produce a soapy solution in water.

saprobe *n.* a saprobic organism.

saprobic *a.* (1) living on decaying matter; (2) saprophytic or saprozoic.

saprobic classification, Saprobien system a biotic index for assessing the degree of organic pollution of a body of water. It is based on recognition of four stages in the oxidation of organic mattter, each characterized by the presence and relative abundance of certain groupings (saprobic groupings) of indicator species. *see* α-mesosaprobic, β-mesosaprobic, oligosaprobic, polysaprobic.

saprobiont *n.* saprobe *q.v.*

saprogenic *a.* (1) causing decay; (2) resulting from decay.

sapropelic *a.* living among debris of bottom ooze.

saprophage saprobe. *see* saprobic.

saprophyte *n.* plant, fungus or bacterium that gains its nourishment directly from dead or decaying organic matter. *a.* **saprophytic**. *see also* saprotroph.

saprophyte chain a food chain starting with dead organic matter and passing to saprophytic microorganisms.

saprotroph *n.* any organism that feeds on dead organic matter. *a.* **saprotrophic**. *alt.* saprophyte, esp. *appl.* fungi and bacteria.

saprozoic *a. appl.* animal that lives on dead or decaying organic matter. *n.* **saprozoite**.

sapwood *n.* the more superficial, younger, paler, softer wood of trees, which is water-conducting and contains living cells. *alt.* splintwood, alburnum. *cf.* heartwood.

SAR scaffold attachment region *q.v.*

sarciniform *a.* arranged in more-or-less cubical clumps, *appl.* cocci.

sarcocarp *n.* the fleshy or pulpy part of a fruit.

sarcocyte *n.* the middle layer of ectoplasm in some protozoans such as gregarines.

sarcoderm *n.* fleshy layer between a seed and its outer covering.

sarcodictyum *n.* the 2nd or network protoplasmic zone of radiolarians.

Sarcodina *n.* class or superclass of protozoa containing the Rhizopodea (amoebae and foraminiferans) and Actinopodea (heliozoans and radiolarians), which have pseudopodia or actinopodia, little internal differentiation of the cell, and no flagella at any stage.

sarcogenic *a.* flesh-producing.

sarcoid *a.* fleshy, as sponge tissue.

sarcolemma *n.* membranous sheath around an individual muscle fibre.

sarcoma *n.* a tumour of connective tissue, e.g. of fibrous tissue (fibrosarcoma) or bone (osteosarcoma). *cf.* carcinoma.

Sarcomastigophorea *n.* in some classifications, a phylum of protozoa comprising those that move by flagella, pseudopodia or both. It includes the classes (in some classifications considered as phyla) Zoomastigophorea (flagellates, including trypanosomes), Opalinatea, Lobosea (amoebae), Apicomplexa (sporozoans) and Ciliophora (ciliates).

sarcomere *n.* the portion of a striated muscle fibre between two Z-discs, comprising a complete contractile unit. *see* Fig. 28 (p. 425).

sarcophagous *a.* flesh-eating.

sarcoplasm *n.* the cytoplasm between fibrils of muscle cells.

sarcoplasmic reticulum membrane forming a network of fine channels around myofibrils in striated muscle cells, and which acts as an intracellular store of Ca^{2+}, which is needed for muscle contraction.

Sarcopterygii, sarcopterygians *n., n.plu.* a group of mostly extinct bony fishes having fleshy fins and nostrils opening into the mouth, comprising the lung-fishes (Dipnoi) and the crossopterygians (Crossopterygii).

sarcosoma *n.* the fleshy, as opposed to the skeletal, portion of an animal's body.

sarcotesta *n.* softer fleshy outer portion of testa.

sarcotheca *n.* the sheath of a hydrozoan polyp.

sarcotubular system the sarcoplasmic reticulum and T-tubules that form a series of membrane-bounded channels around the myofibrils of muscle fibres.

sarcous *a. pert.* flesh or muscle tissue.

sargasterol *n.* sterol present in some algae.

sarmentaceous, sarmentose, sarmentous *a.* having slender prostrate stems or runners.

sarmentum *n.* the slender stem of a climbing or creeping plant.

Sarraceniales *n.* order of carnivorous herbaceous dicots, typically with pitcher-like leaves for trapping insects. It comprises a single family, the Sarraceniaceae (pitcher plants).

SARS severe acute respiratory syndrome *q.v.*

sartorius *n.* strap-like muscle in thigh which helps to flex both hip and knee and enables legs to be moved inwards.

satellite *n.* (1) small piece of chromosomal material attached to the short arm of a chromosome by a slender thread; (2) the 2nd of any pair of pseudoconjugant individuals in colonies of gregarine protozoa.

satellite cells (1) cells in close physical association with another type of cell, such as neuroglial cells with neurons in the central nervous system or Schwann cells with peripheral neurons; (2) stem cells lying within the basal lamina of mature skeletal muscle fibres and from which new skeletal muscle cells are produced when required.

satellite chromosomes chromosomes that appear to be additional to the normal genome.

satellite DNA DNA composed of highly repeated short sequences, found in centromere regions of eukaryotic chromosomes.

satellite RNA encapsidated small plant pathogenic RNAs which require coinfection with a specific virus, a helper virus, for replication and encapsidation. *alt.* virusoid, **satellite virus**.

satiety *n.* feeling of fulfilment or satisfaction, as when hunger is satisfied.

saturated *a. appl.* hydrocarbons with a fully hydrogenated carbon backbone, *appl.* fatty acids.

saturation *n.* one of the three basic dimensions of perception of visible light by humans. Refers to the gradient of richness of colour from full colour to grey. *see also* brightness, hue.

saturnine *a.* forming, having, or *pert.* an equatorial ring.

saurian *a. appl.*, *pert.*, or resembling a lizard.

Saurischia *n.* large order of Mesozoic archosaurs, commonly called lizard-hipped dinosaurs. They included both bipedal carnivores (the theropods) and very large quadrupedal herbivores (e.g. the sauropods).

saurognathous *a.* with a lizard-like arrangement of jaw-bones.

sauropods *n.plu.* group of gigantic, herbivorous lizard-hipped dinosaurs which included *Diplodocus*, *Apatosaurus* (*Brontosaurus*) and *Brachiosaurus*.

Savageau's demand theory theory formulated by M.A. Savageau to explain why bacteria often exhibit positive gene regulation for resources they commonly encounter and negative gene regulation for resources they rarely encounter. It postulates that for resources in high demand, mutations causing loss of expression are more strongly selected against than those causing constitutive expression, and *vice versa* for resources in low demand.

savanna *n.* (1) subtropical or tropical dry grassland with drought-resistant vegetation and scattered trees; (2) transitional zone between dry grassland or semi-desert and tropical rain forest. *alt.* savannah.

sawflies *see* Symphyta.

saxatile, saxicoline, saxicolous *a.* living in, on, or among rocks.

Saxifragales *n.* order of dicot trees, shrubs, lianas and herbs, including the families Crassulaceae (orpine), Escalloniaceae (escallonia), Grossulariaceae (gooseberry), Hydrangeaceae (hydrangea), Saxifragaceae (saxifrage) and others.

saxitoxin (STX) *n.* poison produced by marine dinoflagellates and isolated from shellfish that feed on them. It selectively blocks the voltage-gated sodium channel in neurons, thereby blocking the generation of nerve impulses. It is used as a neurotoxin in experimental neurophysiology.

SBA soybean agglutinin, a lectin isolated from *Glycine max*.

SBMV Southern bean mosaic virus.

SC synaptonemal complex *q.v.*

scaberulous *a.* somewhat rough.

scabrate, scabrous *a.* rough, with a covering of stiff hairs, scales or points.

scaffold *n.* (1) of a chromosome, the proteinaceous structure in the shape of a sister chromatid pair which is generated when chromosomes are depleted of histones; (2) nuclear scaffold *q.v.*

scaffold attachment region (SAR) site in chromosomal DNA that attaches the chromosome to the proposed nuclear scaffold or matrix.

scaffold protein protein in an intracellular signalling pathway that binds groups of interacting signalling proteins.

scala *n.* any of the three fluid-filled canals separated by membranes and running the length of the cochlea of inner ear. The scala tympani lies below the basilar membrane, the scala media is delimited by the organ of Corti and Reissner's membrane, and the scala vestibuli lies on the other side of Reissner's membrane. The scala tympani and vestibuli contain perilymph, the scala media contains endolymph.

scalariform *a.* ladder-like, *appl.* structures having bars like the rungs of a ladder, such as the walls of some xylem vessels.

scale *n.* a flat, small, plate-like external structure. In plants it is formed from epidermis, in animals it may be made of chitin (in invertebrates), bone or horn (keratin) (in vertebrates).

scale bark bark flaking off in irregular sheets or patches. *cf.* ring bark.

scale insect member of the family Coccidae of the Hemiptera (bugs), feeding on plants, in which the wingless females remain fixed to the food plant and are covered with a waxy covering, the 'scale'. Some are serious plant parasites, others yield shellac and cochineal. *alt.* mealy bug.

scale leaf a small dry or hard leaf.

scalene *a.* (1) *pert.* the scalenus muscles; (2) *appl.* tubercle on 1st rib for attachment of scalenus anticus or anterior.

scalenus *n.* one of three neck muscles: scalenus posticus, scalenus medius, scalenus anticus.

scalids *n.plu.* spines arranged in a series of rings around the mouth in Kinorhyncha.

scaliform *a.* ladder-shaped. *see* scalariform.

scalpella *n.plu.* paired pointed processes, parts of maxillae of dipteran flies.

scalpriform *a.* chisel-shaped, *appl.* incisors of rodents.

scalprum *n.* the cutting edge of an incisor.

scandent *a.* (1) climbing by stem roots or tendrils; (2) trailing.

scanning electron microscope (SEM) electron microscope that produces a 'three-dimensional' image from electrons reflected from the surface of a specimen. *cf.* transmission electron microscope.

scanning probe microscope type of 'microscope' in which a molecular picture of the surface of an object is built up by a probe that travels over and interacts with the atoms in the surface.

scanning-tunnelling electron microscope (STEM) type of scanning probe electron microscope that can produce images down to almost atomic scale.

scansorial *a.* (1) adapted for climbing; (2) habitually climbing.

scape *n.* (1) flower stalk arising at or under ground level; (2) basal part of antenna in some flies.

scapha *n.* narrow curved groove between helix and anthelix of ear.

scaphocephalic *a.* with a narrow elongated skull.

scaphocerite *n.* a boat-shaped exopodite of 2nd antenna of decapod crustaceans.

scaphognathite *n.* process on 2nd maxilla of some decapod crustaceans that pumps water over the gills by a paddle-like action.

scaphoid *a.* shaped like a boat, *appl.* certain bones of wrists and ankles.

scapholunar *a. pert.* scaphoid and lunar carpal bones, or those bones fused. *n.* **scapholunatum.**

Scaphopoda *n.* class of marine molluscs, commonly called tusk shells or elephant-tooth shells, which have a tubular shell, a reduced foot and no ctenidia.

scapiform, scapigerous, scapoid, scapose *a.* resembling or consisting of a scape.

scapula *n.* in vertebrates, the shoulder blade, i.e. the dorsal part of pectoral girdle. *a.* **scapular.**

scapulars *n.plu.* feathers covering the ejunction of wing with body in bird.

scapulus *n.* modified submarginal region in certain sea anemones.

scarab(a)eiform *a. appl.* the C-shaped larval type of certain beetles.

scarious *a.* thin, dry, scurfy or scaly.

Scarpa's fascia deep layer of superficial abdominal fascia.

Scarpa's foramina two openings, for nasopalatine nerves, in middle line of palatine process of maxilla.

Scarpa's ganglion vestibular ganglion in internal ear.

Scarpa's triangle the femoral triangle formed by the adductor longus, sartorius and inguinal ligament.

Scatchard analysis, Scatchard plot graphical method of analysing the result of equilibrium binding experiments of receptors and ligands, which gives the association constant of binding and the number of binding sites per molecule.

scatophagous *a.* dung-eating.

scatter factor hepatocyte growth factor, a protein expressed in the liver of patients with fulminant hepatic failure, and which causes certain epithelial cells to scatter and to adopt fibroblast-like morphology.

scavenger *n.* an animal feeding on animals that have been killed by other predators, or have died naturally, or on organic refuse.

scavenger receptor receptor on liver cells that binds and removes damaged glycoproteins from blood.

SCE sister-chromatid exchange *q.v.*

scent scales androconia *q.v.*

Schild plot graphical method of analysing the results of experiments comparing the potency of competitive antagonists on the responses to an agonist drug.

schindylesis *n.* articulation in which a thin plate of bone fits into a cleft or fissure, as that between vomer and palatine bones.

Schisandrales Illicidales *q.v.*

schistocytes *n.* erythrocytes undergoing fragmentation and the resulting hollow fragments.

schistosome *n.* parasitic digenean blood fluke (Trematoda) infesting mammals, e.g. *Schistosoma mansoni*, which causes schistosomiasis (bilharzia) in humans in tropical regions. The larvae develop in certain freshwater snails.

schistosomule *n.* the migratory stage of a schistosome in the vertebrate host.

schizocarp *n.* fruit derived from a compound ovary but which splits into two or more one-seeded portions at maturity, e.g. the double 'keys' of sycamore and maple.

schizocarpic *a. appl.* dry fruits which split into two or more mericarps. Examples are carcerulus, cremocarp, lomentum, regma, compound samara.

schizocoel *n.* coelom formed by splitting of mesoderm into layers. *alt.* **schizocele.**

schizogamy *n.* fission into a sexual and non-sexual zooid in some polychaetes.

schizogenesis *n.* reproduction by fission.

schizogenetic *a.* (1) reproducing or formed by fission; (2) *appl.* intercellular spaces or glands in plants formed by separation of cell walls along middle lamella. *alt.* **schizogenous.** *n.* **schizogeny.**

schizogony *n.* reproduction by fission into many cells, in protozoans.

schizokinete *a.* a motile worm-like stage in the life history of some sporozoan blood parasites.

schizolysigenous *a.* formed schizogenously, by separation, and enlarged lysigenously, by breakdown, such as glands and cavities in pericarp, e.g. of citrus fruits.

schizolysis *n.* fragmentation.

schizont *n.* in some protozoans, esp. the sporozoan parasites, the stage in the life cycle following the trophozoite and reproducing by multiple fission.

schizontoblast agametoblast *q.v.*

schizontocytes *n.plu.* cells into which a schizont divides and which themselves divide into clusters of merozoites.

schizopelmous *a.* with two separate flexor tendons connected with toes, as some birds.

schizopod stage that stage in development of decapod crustacean larva when it resembles an adult *Mysis* in having an exopodite and endopodite to all thoracic limbs.

Schizosaccharomyces pombe species of fission yeast, i.e. yeasts whose cells divide equally at mitosis, which is used as a model organism in cell biology and genetics.

schizostele *n.* one of a number of strands formed by division of initial apical meristem of stem.

schizostely *n.* condition of stem in which apical meristem gives rise to a number of strands, each composed of one vascular bundle.

schizothecal *a.* having scale-like horny tarsal plates.

schizozoite *n.* (1) in sporozoan parasites, the stage in the life cycle produced by schizogony; (2) small cell produced by multiple fission of a schizont.

school *n.* a group of fish or marine mammals, such as porpoises, that swim together in an organized fashion.

Schwann cell glial cell which forms the fatty sheath around myelinated nerve fibres in the peripheral nervous system.

Schwann sheath neurilemma *q.v.*

sciatic *a. pert.* hip region, *appl.* e.g. artery, nerve, veins.

SCID severe combined immunodeficiency *q.v.*

scientific method the rational formulation of hypotheses, collection of data, and testing of hypotheses against observations or experimental results that is the basis of the scientific approach to explaining natural phenomena.

scion *n.* a branch or shoot which is to be grafted on to another plant.

scissile *a.* (1) cleavable; (2) splitting, as into layers.

Scitamineae Zingiberales *q.v.*

sclera *n.* the tough, opaque, fibrous coat of the eyeball, the white of the vertebrate eye.

scleractinians *n.plu.* order (the Scleractinia or Madreporina) of usually colonial Zoantharia. They are known as true corals, having a compact calcareous skeleton and polyps with no siphonoglyph.

scleratogenous layer strand of fused sclerotomes formed along the neural tube, later surrounding the notochord.

sclere *n.* (1) small skeletal structure; (2) sponge spicule.

sclereid *n.* type of sclerenchyma cell with a thick lignified wall, making up some seed coats, nutshells, the stone or endocarp of stone fruits, and which gives the flesh of pears its gritty texture.

sclerenchyma *n.* (1) plant tissue with thickened, usually lignified cell walls, which acts as a supporting tissue; (2) hard tissue of coral. *a.* **sclerenchymatous**.

sclerid sclereid *q.v.*

sclerification *n.* the process of becoming sclerenchyma.

sclerins scleroproteins *q.v.*

sclerite *n.* hard plate or spicule which may be of keratin, calcium carbonate or chitin, and is a skeletal or supporting element in invertebrates. *a.* **scleritic**.

sclerobasidium *n.* a thick-walled resting body or encysted probasidium of rust and smut fungi.

scleroblast *n.* (1) sponge cell from which a sclere or spicule develops; (2) immature sclereid.

scleroblastema *n.* embryonic tissue involved in development of skeleton.

scleroblastic *a. appl.* skeleton-forming tissues.

sclerocarp *n.* the hard seed coat or stone, usually the endocarp, in fruit such as plums and cherries.

sclerocauly *n.* condition of excessive skeletal structure in a stem.

sclerocorneal *a. pert.* sclera and cornea.

scleroderm *n.* (1) hard integument; (2) the hard skeletal part of corals.

sclerodermatous *a.* (1) with an external skeletal structure; (2) with horny, bony or calcareous plates in the skin.

sclerodermite *n.* the part of exoskeleton over one arthropod segment.

sclerogen *n.* wood-producing cells, i.e. the vascular cambium.

sclerogenic, sclerogenous *a.* producing lignin.

scleroid, sclerous *a.* (1) hard; (2) skeletal.

sclerophyll *n.* (1) plant with tough evergreen leaves; (2) one of the leaves of such a plant.

sclerophyllous *a.* hard-leaved, *appl.* leaves that are resistant to drought through having a thick cuticle, much sclerenchymatous tissue and reduced intercellular air spaces.

sclerophylly *n.* condition of excessive skeletal structure in leaves.

scleroprotein *n.* any of a group of proteins occurring in connective, skeletal and epidermal tissues, such as collagen, chondrin, elastin, keratin.

scleroseptum *n.* a radial vertical wall of calcium carbonate in scleractinian corals.

sclerosis *n.* hardening by an increase in the amount of connective tissue (in animals) or lignin (in plants).

sclerospermous *a.* having the seeds covered by a hard coat.

sclerotal sclerotic *q.v.*

sclerotesta *n.* the hard lignified inner layer of testa (seed coat).

sclerotic *a.* (1) hard; (2) containing lignin; (3) *pert.* sclerosis; (4) *pert.* sclera; having undergone sclerosis; (5) *n.* the sclera of eye.

sclerotin *n.* a highly resistant, stable, quinone-tanned protein occurring in insect cuticle and among structural proteins in many groups of invertebrates and vertebrates.

sclerotioid, sclerotiform *a.* like or *pert.* a sclerotium.

sclerotium *n.* resting or dormant stage of some fungi when they become a mass of hardened or mummified tissue. *plu.* **sclerotia**.

sclerotization *n.* the process of hardening and darkening the new exoskeleton which occurs in insects after moulting. *a.* **sclerotized**.

sclerotome *n.* part of somite of vertebrate embryo that develops into the cartilage of the developing vertebrae, and in most vertebrates later into bone.

sclerous *a.* (1) hard; (2) sclerotic *q.v.*

SCN suprachiasmatic nucleus *q.v.*

SCNT somatic cell nuclear transfer *q.v.*

scobina *n.* a spikelet of grasses.

scobinate *a.* having a rasp-like surface.

scobiscular, scobisculate, scobiform *a.* (1) granulated; (2) resembling sawdust.

scoleces *plu.* of scolex.

scolecid *a. pert.* to a scolex.

scoleciform *a.* like a scolex.

scolecite *n.* (1) worm-shaped body branching from mycelium of discomycete fungi; (2) Woronin hypha, a hypha inside coil of perithecial hyphae and giving rise to ascogonia.

scolecoid *a.* like a scolex.

scolespore *n.* a worm-like or thread-like spore.

scolex *n.* region at anterior end of tapeworm containing minute hooks and a sucker, by which it attaches itself to the gut wall. *plu.* **scoleces**.

scolite *n.* a fossil worm burrow.

scolpidium chordotonal sensilla *q.v.*

scolus *n.* a thorny process of some insect larvae.

scombrids *n.plu.* fish of the mackerel and tuna family (Scombridae).

scopa pollen brush *q.v.*

scopate, scopiform, scopiferous, scopulate *a.* like a brush.

scopula *n.* (1) small tuft of hairs; (2) in climbing spiders, an adhesive tuft of club-like hairs on each foot, replacing 3rd claw.

scopuliferous *a.* having a small brush-like structure.

scopuliform *a.* resembling a small brush.

scorpioid *a.* (1) resembling a scorpion; (2) (*bot.*) rolled up along the axis like a young fern frond, *appl.* inflorescence.

scorpioid cyme cymose inflorescence with one axis at each branching, and in which daughter axes are developed right and left alternately.

scorpion *see* Scorpiones.

Scorpiones, Scorpionoidea *n.* order of arachnids including the scorpions. They have a dorsal carapace on the prosoma, and an opisthosoma divided into two regions, with the posterior segments forming a flexible tail bearing a terminal poisonous sting that is used to paralyse prey. They are viviparous.

scorpion flies common name for the Mecoptera *q.v.*

scorteal *a. appl.* or *pert.* a tough cortex, as of certain fungi.

Scotobacteria *n.* in some classifications the large class of bacteria containing all heterotrophic Gram-negative bacteria and some other heterotrophic groups, the name signifying 'bacteria indifferent to light'.

scotoma *n.* blind spot, the point at which vision is absent within visual field.

scotopia *n.* adaptation of the eye to darkness.

scotopic vision vision at low intensities of light, in shades of grey, involving the rod cells of the vertebrate retina.

scotopsin opsin *q.v.*

scototaxis *n.* a positive taxis towards darkness, not a negative phototaxis.

SCP single-cell protein *q.v.*

scramblase *n.* protein that catalyses the general equilibration of membrane phospholipids to the two leaflets of the endoplasmic reticulum membrane after new phospholipids are added to the cytosolic layer.

scramble competition situation where a resource is shared equally between competitors.

scrapie *n.* a neurodegenerative disease of sheep, one of the transmissible spongiform encephalopathies, caused by a prion.

screening *n.* the analysis of large numbers of samples, cells or organisms, to find those with the desired property or mutation.

scRNA small cytoplasmic RNAs *q.v.*

scRNPs scyrps *q.v.*

scrobe *n.* a groove on either side of rostrum of beetles.

scrobicula, scrobicule *n.* the smooth area around the boss of an echinoderm test.

scrobiculate *a.* marked with little pits or depressions.

scrobiculus, scrobicule *n.* a small pit or depression.

Scrophulariales *n.* an order of dicot trees, shrubs, herbs and vines, including the families Acanthaceae (acanthus), Bignoniaceae (jacaranda), Buddleiaceae (buddleia), Solanaceae (nightshade, potato) and others.

scrotal *a. pert.* or in the region of the scrotum.

scrotum *n.* (1) external sac or sacs containing testicles in mammals; (2) covering of testis in insects.

scrounger *n. pert.* animal behaviour, an animal that waits for another animal of a different species to catch its prey, and then takes the prey from it. *see also* kleptoparasitism.

scrub *n.* a plant community dominated by shrubs.

scurf *n.* (1) scaly skin; (2) dried outer skin peeling off in scales; (3) scaly epidermal covering of some leaves.

scurvy *n.* deficiency disease caused by a lack of vitamin C (ascorbic acid), which among other symptoms prevents formation of effective collagen fibres, leading to skin lesions and blood vessel fragility.

scuta *plu.* of scutum.

scutate *a.* protected by large scales or horny plates.

scute *n.* (1) an external scale, as of reptile, fish or scaly insect; (2) plate of shell of turtles and tortoises. *see also* scutum.

scutella *plu.* of scutellum.

scutellar *a. pert.* a scutellum.

scutellate *a.* shaped like a small shield.

scutellation *n.* arrangement of scales, as on leg of bird.

scutelliform *a.* shaped like a small shield.

scutelligerous *a.* furnished with scutella or a scutellum.

scutelliplantar *a.* having tarsus covered with small plates or scutella.

scutellum *n.* (1) (*bot.*) development of part of cotyledon that separates embryo from endosperm in seed of grasses; (2) (*zool.*) any small shield-shaped structure; (3) posterior part of insect notum; (4) scale on tarsus of birds. *plu.* **scutella**.

scutiferous *a.* having scutella or a scutellum.

scutiform *a.* shaped like a shield.

scutigerous *a.* bearing a shield-like structure.

scutiped *a.* having foot, or part of it, covered by scutella.

scutum *n.* (1) (*bot.*) broad apex of style as in dicots of family Asclepiadaceae; (2) (*zool.*) a shield-like plate, horny, bony or chitinous, formed in the outer covering of an animal (as scales); (3) middle portion of insect notum; (4) dorsal shield of ticks; (5) one of the pair of anterior valves in goose barnacles. *alt.* **scute**, shield. *plu.* **scuta**.

scyllitol *n.* a sweet alcohol, $C_6H_6(OH)_6$, related to inositol.

scyphi *plu.* of scyphus.

scyphiferous *a.* bearing scyphi, as some lichens.

scyphiform, scyphoid, scyphose *a.* shaped like a cup.

scyphistoma *n.* inconspicuous asexual polyp stage of jellyfish (Scyphozoa).

Scyphozoa, scyphozoans *n.*, *n.plu.* the jellyfish, a class of marine coelenterates of the phylum Cnidaria, with a dominant medusa stage which is free-swimming or attached by an aboral stalk, and has no velum. *cf.* Hydrozoa.

scyphula scyphistoma *q.v.*

scyphulus *n.* small cup-shaped structure.

scyphus *n.* (1) funnel-shaped corolla of daffodil; (2) a cup-shaped outgrowth bearing apothecium in some lichens. *plu.* **scyphi**.

scyrps colloquial term for small cytoplasmic ribonucleoprotein particles, from the abbreviation scRNPs.

SDS sodium dodecyl sulphate *q.v.*

SDS-PAGE technique for separating e.g. membrane proteins, in which the membrane complex is first solubilized in the detergent sodium dodecyl sulphate and then subjected to electrophoresis in a polyacrylamide gel to separate the constituent proteins.

sea anemones common name for an order (Actiniaria) of coelenterates of the Zoantharia, which are generally solitary and have no skeleton. They have a hollow cylindrical body, often anchored to rocks, opening at one end in a small mouth surrounded by a ring of tentacles often numbering multiples of six.

sea butterflies pteropods *q.v.*

sea combs common name for the Ctenophora *q.v.*

sea cows the Sirenia, including the dugong and manatee, marine placental mammals highly specialized for an aquatic life with a naked body and front limbs modified as paddles.

sea cucumbers a common name for the Holothuroidea *q.v.*

sea fans gorgonians *q.v.*

sea gooseberries common name for the Ctenophora *q.v.*

sea lilies common name for the Crinoidea *q.v.*

sea mouse common name for the polychaete *Aphrodite*, which has a broad stout oval shape.

sea pens a group of corals of the subclass Alcyonaria, which form stalked colonies markedly resembling quill pens, composed of two different kinds of polyp.

search(ing) image a transitory filtering of external visual stimuli that enables an animal to focus its attention on finding e.g. a prey item of a particular colour or shape.

seashore *n.* the ground bordering the sea, between the highest high-water and lowest low-water marks, also including the splash zone above high-water mark. *see also* intertidal, littoral, lower shore, middle shore, splash zone, sublittoral, upper shore. *see* Fig. 34 for shore zonation.

sea slugs shell-less marine molluscs of the class Gastropoda, subclass Opisthobranchia.

sea spiders Pycnogonida *q.v.*

sea squirts common name for the Ascidiacea *q.v.*

sea stars a common name for the Asteroidea *q.v.*

sea urchins a common name for the Echinoidea *q.v.*

sea wasps box-jellies *q.v.*

seaweed *n.* marine multicellular algae belonging to various groups. *see* Chlorophyta, Phaeophyta, Rhodophyta.

sebaceous *a.* secreting or containing oils or fats, *appl.* glands of the skin secreting sebum.

sebiferous *a.* conveying fatty material.

sebific *a.* (1) sebaceous *q.v.*; (2) *appl.* gland: colleterium *q.v.*

sebum *n.* material rich in lipids (oils and fats) secreted by sebaceous glands of skin.

secodont *a.* furnished with teeth adapted for cutting.

secondaries *n.plu.* one type of main flight feather in bird's wing, attached in region of posterior edge of the ulna.

secondary *a.* (1) second in importance or in position; (2) arising not from a growing point but from other tissue.

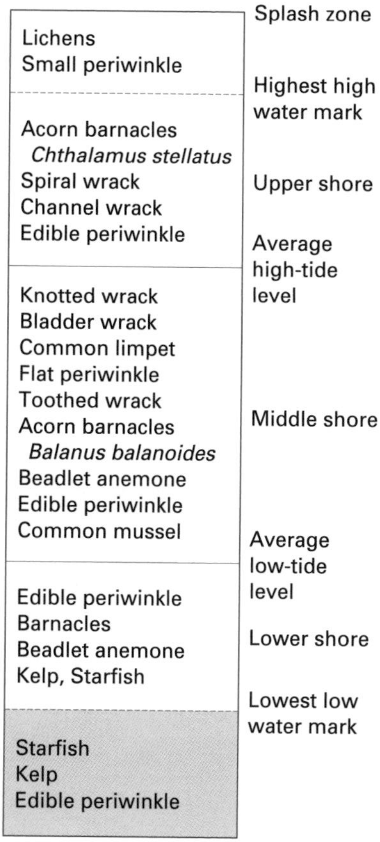

Fig. 34 Typical zonation for a sheltered Atlantic European shore.

secondary antibody heterologous anti-immunoglobulin *q.v.*

secondary bud an axillary bud, accessory to normal one.

secondary cell culture culture originating from cells taken from a primary cell culture.

secondary cell wall in many plant cells, material laid down on inner surface of primary wall, usually after the cell has stopped growing. It is rich in cellulose but lacks pectin or glycoproteins and is consequently more rigid than the primary

wall. Thick laminated secondary walls are found esp. in cells specialized for support and water conduction.

secondary constriction a non-staining constricted region of a chromosome, other than the centromere, which does not attach to spindle at metaphase.

secondary consumer carnivore that eats herbivores.

secondary cortex phelloderm *q.v.*

secondary ecological succession secondary succession *q.v.*

secondary follicle lymphoid follicle that arises from a primary follicle in secondary lymphoid tissue after B cells encounter antigen there and are activated. It forms a germinal centre of activated proliferating B cells.

secondary forest, secondary woodland forest or woodland that has developed as a result of secondary succession after complete clearance of pre-existing forest, or which has been planted.

secondary growth in plants, growth bringing about an increase in the thickness of stem and root, as opposed to extension of plant body at the apices of shoots and roots, and which is most marked in trees and shrubs. It is initiated at lateral meristems, which are the vascular cambium, a layer of tissue encircling root and stem between phloem and xylem, producing new xylem and phloem (secondary xylem and phloem), and the cork cambium, which contributes to the bark.

secondary host intermediate host *q.v.*

secondary immune response immune response made on a second or subsequent exposure to an antigen, which is faster in onset, stronger and more specific than the first, or primary, response. *cf.* primary immune response.

secondary lymphoid tissues in mammals, the lymph nodes and spleen, together with mucosa-associated lymphoid tissues such as tonsils, Peyer's patches, adenoids and appendix. They contain mature T and B lymphocytes, and are the sites at which lymphocytes encounter foreign antigen and initiate immune reactions.

secondary meristem cork cambium *q.v.*

secondary metabolites compounds produced by plants and microbes that are not essential to the growth of the organism, e.g. antibiotics, plant alkaloids, flower pigments.

secondary palate bony plate separating mouth cavity from nasal cavities in mammals and crocodiles.

secondary phloem phloem tissue formed from the vascular cambium during secondary growth, sometimes also called the inner bark.

secondary plant body the plant body formed from growth from lateral meristems, i.e. from the vascular and cork cambiums.

secondary production in an ecosystem, the yield due to primary consumers, i.e. herbivores.

secondary roots (1) branches of the primary root, arising within its tissue and in turn giving rise to tertiary roots; (2) roots arising at other than normal points of origin.

secondary sexual characteristics features characteristic of a particular sex other than the gonads and genitalia, usually developing under the influence of androgens and oestrogens, and including growth of a beard in men, antlers in stags, and enlarged breasts in women.

secondary structure distinct elements of substructure within proteins – α-helix, β-strand, β-sheet, β-turn – which are formed by folding of the protein chain by formation of regular patterns of hydrogen bonds between N–H and C=O groups of the chain backbone.

secondary succession a plant succession following the interruption of the normal or primary succession.

secondary thickening, secondary wood, secondary xylem *see* secondary growth.

second messengers small molecules such as cyclic AMP, diacylglycerol and inositol trisphosphate, which are formed intracellularly as a result of stimulation of cell-surface receptors and are involved in activating the cell's specific response to the stimulus.

second-order conditioning type of classical conditioning in which a second stimulus is associated with the conditional stimulus so that the conditioned animal comes to respond to the second conditional stimulus alone.

secrete *v.* to release material or fluid from a cell or tissue.

secretin *n.* polypeptide hormone produced by duodenum during digestion and which stimulates pancreas to produce pancreatic juice containing digestive enzymes.

secretion *n.* (1) material or fluid which is produced and released from a cell or gland; (2) the release of a substance from a cell. *a.* **secreted**.

secretion systems in bacteria, systems for secreting proteins across the cell envelope have been classed into five main types, I–V. Type I is based on ABC transporters, type II is a multiprotein complex through which folded proteins are secreted, type III is a syringe-like apparatus through which some bacteria inject their toxins into the cytoplasm of a host cell, type IV is related to the components of the bacterial conjugation machinery, type V involves secretion through a pore formed by a β-barrel protein in the outer membrane.

secretor *n.* person who secretes blood group antigens in saliva and other body fluids, a genetically determined trait.

secretory *a.* (1) *appl.* cells and tissues that secrete substances such as digestive enzymes, polypeptide hormones, neurotransmitters or complex material such as mucus and slime; (2) *appl.* proteins and other material that are secreted.

secretory component, secretory piece small protein component of the dimeric IgA found in mucous secretions but not present on monomeric serum IgA.

secretory granule small membrane-bounded vesicle in cytoplasm of eukaryotic cells, derived from the Golgi apparatus, and containing material to be secreted. *alt.* **secretory vesicle**.

sectile *a.* cut into small partitions or compartments.

section *n.* (1) thin slice of tissue prepared for microscopy; (2) a taxonomic group, often used as a subdivision of a genus, but used in different ways by different authors and never precisely defined.

sector *n.* area of colony of microorganisms, or of tissue of plant or animal, that is phenotypically different from the surrounding area.

secular *a.* long term, over a long period of time.

secundiflorous *a.* having flowers on one side of stem only.

secundine *n.* the internal integument of ovule.

sedentaria *n.plu.* sessile or sedentary organisms.

sedentary *a.* not free-living, *appl.* animals that live attached to some substratum.

sedimentation coefficient the rate of sedimentation of a particle in a centrifuge. It is abbreviated *s*, and has units of seconds. A Svedberg unit (S) is defined as a sedimentation coefficient of 1×10^{-13}.

sedoheptulose *n.* seven-carbon ketose sugar, which, as the phosphate and bisphosphate, is involved in carbon dioxide fixation in photosynthesis.

seed *n.* (1) reproductive unit formed from a fertilized ovule, and consisting of an embryo, food store and protective coat. Produced by gymnosperms and angiosperms; (2) semen *q.v.*; (3) *v.* to introduce microorganisms into a culture medium.

seed bank a conservation collection of seeds of wild plant species and cultivated varieties, usually of important crop plants, kept in long-term storage, usually freeze-dried under liquid nitrogen.

seed coat thin outer coat of mature seed, which may be dry and papery or hard and highly impermeable to water, which develops from the integuments of ovule. *alt.* testa.

seed ferns the Pteridospermophyta, a division of fossil seed-bearing vascular plants from the late Devonian and Carboniferous, which had fern-like leaves bearing seeds. They included both small shrubby and tall tree-like forms. The large pinnately compound leaves were borne at the top of the trunk.

seed leaf cotyledon *q.v.*

seed plants all seed-bearing plants, i.e. the gymnosperms and angiosperms, sometimes collectively termed Spermatophyta.

seed stalk *see* funicle.

seed storage proteins simple proteins produced in large quantities within seeds where they act as nitrogen storage compounds. They are broken down during germination and seedling growth and the amino acids utilized.

seed vessel fruit, esp. a dry fruit.

segment *n.* (1) part of an animal, esp. where the body has regular transverse divisions, as in annelid worms and arthropods, or of a jointed appendage; (2) division of a leaf cleft nearly to base. *a.* **segmental**.

segmental arteries diverticula from dorsal aorta arising in spaces between successive somites.

segmental duct an embryonic nephridial duct which gives rise to Wolffian or Müllerian duct.

segmental interchange (1) exchange of non-homologous segments, as between two chromosomes; (2) reciprocal translocation.

segmental organ an embryonic excretory organ. *alt.* nephridium.

segmental papillae conspicuous pigment spots by which true segments can be recognized in leeches.

segmental reflex a reflex involving a single segment of spinal cord, i.e. not involving additional input from brain or other parts of nervous system.

segmentation *n.* (1) division into segments or portions; (2) the repetition of a series of essentially similar segments along the length of the body of the animal, as seen esp. in arthropods and annelids. In annelids, each segment has a similar pattern of blood vessels, nerves, muscles, but in adult arthropods the internal structure is less obviously segmented; (3) the process in the embryonic development of insects and other segmented animals by which the segmented structure is established.

segmentation nucleus body formed by union of male and female pronuclei at fertilization of an ovum.

segmented *a. appl.* the genomes of double-stranded and some single-stranded RNA viruses, which consist of two or more separate RNA molecules each carrying a different gene or genes.

segment-polarity genes class of developmental genes in *Drosophila* that are involved in patterning the parasegments and segments.

segregation *n.* (1) separation of the two chromatids of a duplicated mitotic chromosome and the allocation of each to a different daughter nucleus; (2) separation of the paired homologous chromosomes at 1st division of meiosis, and the allocation of each to a different daughter nucleus. *see also* segregation of alleles.

segregation of alleles Mendel's first law, which describes the fact that the two alleles of any gene in a diploid cell each pass into a different gamete at the formation of the next generation. The outcome is that for any given gene, half the gametes will carry one allele and half the other.

seiroderm *n.* dense outer tissue, composed of parallel chains of hyphae, in certain fungi.

seirospore *n.* one of spores arranged like a chain.

seismathesia *n.* perception of mechanical vibration.

seismonasty *n.* plant movements in response to mechanical shock or vibration. *a.* **seismonastic**.

seismotaxis *n.* a taxis in response to mechanical vibrations.

sejugate, sejugous *a.* with six pairs of leaflets.

Selachii, selachians *n.*, *n.plu.* a class (or in some classifications an order) of cartilaginous fishes containing the sharks, dogfishes, skates and rays, having claspers and fins with a constricted base, existing from the Devonian to the present day.

selectable marker any characteristic by which a cell with a particular property can be specifically selected during an experiment. One common type of selectable marker is a gene, e.g. for antibiotic resistance, that is placed on a vector so that any cell receiving a recombinant DNA molecule can be selected by survival in medium containing the antibiotic.

selectins *n.plu.* family of structurally related cell-surface proteins: L-selectin on white blood cells, P-selectin on blood platelets and endothelial cells activated by inflammation, and E-selectin on activated endothelial cells. They have a lectin-like extracellular domain that recognizes an oligosaccharide ligand and mediate weak adhesion between leukocytes and endothelial cells during immune responses and inflammation.

selection *n.* (1) non-random differential reproduction of different genotypes; (2) natural selection *q.v.*, *see also* directional

selection, disruptive selection, sexual selection, stabilizing selection; (3) deliberate selection of mutants or recombinant microorganisms by culturing them in conditions in which only those with the desired genotype can grow.

selection, coefficient of coefficient of selection *q.v.*

selection differential for any given quantitative trait, the difference between the mean of a population and the mean of the individuals selected to be parents of the next generation.

selection pressure the effect of any feature of the environment that results in natural selection, e.g. food shortage, predator activity, competition from members of the same or other species.

selective advantage *pert.* any character that gives an organism a greater chance of surviving to reproductive age, breeding and rearing viable offspring.

selectivity filter part of an ion channel that determines which types of ion the channel can transport.

selector genes genes involved in selecting alternative states in development, as e.g. the homoeotic genes of *Drosophila*.

selenodont *a.* (1) *appl.* molars lengthened out anteroposteriorly and curved; (2) *appl.* or having molar teeth with crescent-shaped ridges on the grinding surface, as in artiodactyls.

selenoid *a.* shaped like a crescent.

selenocysteine *n.* form of cysteine in which the sulphur atom is replaced by a selenium atom. It is formed enzymatically from serine attached to a tRNA, which then inserts the selenocysteine at a UGA codon. It is present in a number of enzymes (selenoproteins) in which the selenium is necessary for catalytic activity.

selenophyte *n.* plant tolerating quite high levels of selenium in the soil.

selenotropism *n.* tendency to turn towards the moon's rays.

self-antigen any antigenic molecule produced by one's own body.

self-assembly *n.* the capacity of multi-subunit proteins and large macromolecular complexes such as a virus particle to assemble without the agency of other proteins. In many cases correct self-assembly may be facilitated *in vivo* by proteins known as molecular chaperones.

self-fertile *a.* capable of being fertilized by its own male gametes, *appl.* hermaphrodite animals and flowers. *alt.* **self-compatible.**

self-fertilization the fusion of male and female gametes from the same individual.

self-incompatible self-sterile *q.v.*

selfing *n.* self-pollination or self-fertilization.

selfish DNA DNA without any apparent function in its host cell. It maintains itself in the genome by virtue of its own intrinsic characteristics, including the ability to transpose and form multiple repeated copies.

selfish gene the idea that genes are primarily concerned with replicating themselves and passing on more copies of themselves to future generations, the organism in which they are carried being their means for doing this.

selfishness *n.* in sociobiology, behaviour that benefits the individual in terms of genetic fitness at the expense of the genetic fitness of other members of the same species.

self-MHC an individual's own MHC molecules.

self-pollination transfer of pollen grains from anthers to stigma of the same flower.

self-replicating *appl.* or *pert.* any object or structure that is able to make a copy of itself.

self-splicing introns introns in some RNAs that are able to excise themselves, leaving the exons correctly joined together, without the agency of protein enzymes.

self-sterile *a.* incapable of being fertilized by its own male gametes, *appl.* hermaphrodite animals and flowers.

self-tolerance the inability to mount an immune response against the body's own antigenic components.

sellaeform *a.* saddle-shaped. *a.* **selliform**.

sella turcica a deep depression in upper surface of sphenoidal bone, lodging the pituitary body.

selva *n.* the tropical rain forest.

SEM scanning electron microscope *q.v.*

semantic memory generalized memory, e.g. knowing the meaning of a word without remembering where or when it was learned.

semantide *n.* a molecule that carries information, such as DNA and RNA.

sematectonic *a. appl.* communication by means of objects.

sematic *a.* functioning as a danger signal, e.g. warning colours or odours, *appl.* warning and recognition markings. *see also* episematic, parasematic.

semelparity *n.* production of offspring by an organism all at one time. *a.* **semelparous.** *cf.* iteroparity.

semen *n.* fluid secreted from testes and accessory glands and containing sperm.

semi- prefix derived from L. *semi*, half, signifying half or partly.

semiamplexicaul *a.* partly surrounding the stem.

semianatropous *a.* with half-inverted ovule.

semiarid regions dry regions with sufficient rainfall (280–400 mm per annum) to support steppe or savanna grassland and some agriculture.

semiautonomous *a. appl.* organelles such as chloroplasts and mitochondria which contain DNA directing the synthesis of some of their own proteins and are self-replicating.

semicaudate *a.* with a rudimentary tail.

semicells *n.plu.* the two halves of a cell which are interconnected by an isthmus in some green algae.

semicircular canals the semicircular membranous tubes filled with endolymph in the inner ear, which are concerned with perception of body position and balance. In mammals there are three: one vertical and two more-or-less horizontal at right angles to each other, all interconnected at their bases in the sac-like utriculus. Each canal has a swollen base, the ampulla, containing sensory cells that detect the movement of endolymph with body movements. Other sensory cells detect the movement of small calcareous particles under gravity.

semiclasp *n.* one of two apophyses which may combine to form the clasper in certain male insects.

semicomplete *a.* incomplete, *appl.* metamorphosis.

semiconservative replication the way in which double-stranded DNA replicates itself, each strand of the double helix serving as a template for synthesis of a new strand.

semicylindrical *a.* round on one side, flat on the other, *appl.* leaves.

semidominant co-dominant *q.v.*

semifloret, semifloscule *n.* strap-shaped floret in flowers of Compositae.

semiherbaceous *a.* having lower part of stem woody and upper part herbaceous.

semilethal *a.* (1) not wholly lethal; (2) *appl.* genes causing mortality of more than 50% of the individuals that carry them, or permitting survival until reproduction has been effected.

semiligneous *a.* partially lignified, with stem woody only near the base.

semilocular *a. appl.* ovary with incomplete loculi.

semilunar *a.* (1) half-moon shaped, *appl.* branches of carotid artery, fibrocartilages of knee, ganglia, fascia, lobules of cerebellum, valves; (2) *appl.* notch: greater sigmoid cavity between olecranon and coronoid process of ulna; (3) *n.* a carpal bone: lunar bone *q.v.*

semimembranosus *n.* a thigh muscle with flat membrane-like tendon at upper extremity.

semimetamorphosis *n.* partial, or semicomplete metamorphosis.

seminal *a.* (*zool.*) *pert.* semen.

seminal fluid fluid component of semen, secreted by the seminal vesicles and prostate gland in mammals.

seminal funnel internal opening of vasa deferentia in oligochaete worms.

seminal receptacle spermatheca *q.v.*

seminal root the first formed root, developed from the radicle of the seed.

seminal vesicle (1) one of pair of glands of male mammals, which secrete the alkaline fluid component of semen; (2) organ for storing sperm in some invertebrates and lower vertebrates.

semination *n.* (1) dispersal of seeds; (2) discharge of sperm.

seminiferous *a.* (1) secreting or conducting semen; (2) bearing seed.

seminiferous tubules long tightly coiled tubules in the mammalian testis in which the sperm are produced.

seminude *a.* with ovules or seeds exposed.

seminymph *n.* stage in development of insects approaching complete metamorphosis.

semiorbicular *a.* half-rounded or hemispherical.

semiotics *n.* the study of communication.

semiovate *a.* half-oval or somewhat oval.

semioviparous *a.* between oviparous and viviparous, such as a marsupial whose young are imperfectly developed when born.

semiovoid *a.* somewhat ovoid in shape.

semipalmate *a.* having toes webbed halfway down.

semiparasite *n.* plant which only derives part of its nutrients from host and has some photosynthetic capacity.

semipenniform *a. appl.* certain muscles bearing some resemblance to the lateral half of a feather.

semipermeable *a.* (1) partially permeable, allowing some substances through but not others; (2) *appl.* membrane permeable to a solvent, esp. water, but not to solutes.

semiplacenta *n.* non-deciduate placenta.

semiplume *n.* feather with ordinary shaft but with a downy web.

semipupa coarctate pupa *q.v.*

semirecondite *a.* half-concealed, as insect head by thorax.

semisagittate *a.* shaped like a half arrowhead.

semisaprophyte *n.* a plant partially saprophytic.

semisocial insects those social insects in which there is cooperative care of the brood, a separate sterile worker caste but in which there is no overlap of generations caring for the brood. *cf.* eusocial insects.

semispecies *n.* taxonomic group intermediate between a species and subspecies, esp. if the result of geographical isolation.

semispinalis *n.* muscle of back and neck, on each side of spinal column.

semistreptostylic *a.* (1) between monimostylic and streptostylic; (2) with slightly movable quadrate.

semitendinosus *n.* dorsal muscle of the thigh stretching from the tuberosity of the ischium to the tibia.

semitendinous *a.* half-tendinous.

semituberous *a.* having somewhat tuberous roots.

Sendai virus an enveloped RNA virus of the family Paramyxoviridae, which enters cells by fusion of viral envelope with cell membrane and which also may cause cells to fuse.

senescence *n.* (1) advancing age; (2) the complex of ageing processes that eventually lead to death. *a.* **senescent**.

senile plaque area of abnormal deposition of the complex proteinaceous material amyloid, and destruction of neurons, in brains of patients with Alzheimer's disease.

senility *n.* degeneration due to old age.

sense organ an organ receptive to external stimuli, such as eye or ear. *alt.* receptor.

sense strand in double-stranded DNA, the DNA strand that carries the same base sequence as the RNA transcribed from the DNA. *alt.* coding strand.

sensile *a.* capable of affecting a sense.

sensilla, sensillum *n.* a small sense organ in invertebrates, usually in the form of a spine, hair or peg-like structure. *plu.* **sensillae, sensilla**.

sensitive *a.* (1) capable of receiving impressions from external objects; (2) reacting to a stimulus.

sensitive period period during development in which an organism can be permanently altered by a particular experience or treatment.

sensitization *n.* (1) condition in which an animal or human produces an enhanced immune response, or in some cases an allergic response, on a second encounter with an antigen. *see also* hypersensitivity; (2) condition in which the body shows an enhanced response to a drug after repeated doses; (3) form of non-associative learning in which an organism becomes more responsive to most stimuli after being exposed to unusually strong or painful stimuli.

sensor protein one of the members of two-component signalling systems in bacteria. It is a protein kinase that phosphorylates itself in response to an extracellular signal and then transfers the phosphoryl group to the response regulator protein.

sensory *a.* (1) involved in the reception, or processing, of stimuli from the external or internal environment; (2) of or concerned with sensation or the senses.

sensory cortex those parts of the cortex in the brain devoted to the reception and initial analysis of sensory signals, e.g. visual cortex, auditory cortex, olfactory cortex, somatosensory cortex.

sensory neuron nerve cell concerned with carrying impulses from a sense organ or sensory receptor to the central nervous system.

sensory receptor any cell or part of a cell specialized to respond to stimuli such as light, vibration, mechanical deformation or heat, and to convey signals to the central nervous system.

sensory transduction the transmission of a stimulus from a sensory receptor to the central nervous system.

sensu lato in a broad sense.

sensu stricto in a restricted sense.

sentient *a.* (1) capable of perceiving through the senses; (2) conscious.

sepal *n.* modified sterile leaf, often green and with others forming the calyx, or outer series of perianth segments, of a flower, sometimes the same colour as and resembling the petals. *see* Fig. 18 (p. 239).

sepaline, sepaloid *a.* like a sepal.

sepalled *a.* having sepals.

sepalody *n.* conversion of other parts of the flower into sepals.

sepalous *a.* having sepals.

separation layer abscission layer *q.v.*

sepia *n.* the brown 'ink' released by the cuttlefish *Sepia* to distract predators when threatened.

sepiapterin *n.* yellow pteridine pigment, a presumed intermediate in the formation of the eye pigment drosopterin in certain insects.

sepicolous *a.* living in hedges.

sepiment *n.* a partition.

sepion *n.* calcareous shell of cuttlefish.

sepsis *n.* infection of the bloodstream.

septa *plu.* of septum *q.v.*

septal *a.* (1) *pert.* a septum; (2) *pert.* hedgerows, *appl.* flora.

septal complex region of the brain that provides subcortical input to the hippocampal formation.

septal neck in the cephalopod *Nautilus*, a shelly tube continuous for some distance beyond each septum as support for the siphuncle.

septate *a.* divided by partitions (septa).

septate junction intercellular junction in insect cells, which has a ladder-like or honeycomb appearance in the electron microscope.

septempartite *a. appl.* leaf with seven divisions extending nearly to the base.

septenate *a.* with parts in sevens.

septibranch *a. appl.* gills of bivalve molluscs which are small and transformed into a transverse muscular pumping partition.

septic *a. pert.* sepsis.

septicaemia *n.* infection of the bloodstream.

septicidal *a.* (1) dividing through middle of ovary septa; (2) dehiscing at septa, *appl.* fruits.

septic shock systemic shock reaction produced by infection of the bloodstream, which triggers systemic activation of TNF-α.

septiferous *a.* having septa.

septifolious *a.* with seven leaves or leaflets.

septiform *a.* in the shape of a septum.

septifragal *a.* with slits as in septicidal dehiscence, but with septa broken and placentae and seeds left in middle.

septomaxillary *a.* (1) *pert.* maxilla and nasal septum; (2) *appl.* a small bone in many amphibians and reptiles and in certain birds.

septonasal *a. pert.* the internal partition between the nostrils.

septulum *n.* a small or secondary septum. *a.* **septulate.**

septum *n.* a partition separating two cavities or masses of tissue, as in fruits, chambered shells, fungal hyphae, corals, heart, nose, tongue. *plu.* **septa.**

septum lucidum thin inner walls of cerebral hemispheres, between corpus callosum and fornix. *alt.* **septum pallucidum.**

septum narium the partition between nostrils.

septum transversum (1) foetal diaphragm; (2) ridge within ampulla of semicircular canal.

sequenator *n.* machine for the automated sequencing of proteins.

sequence *see* DNA sequence, protein sequence.

sequence alignment *see* alignment.

sequence-tagged site (STS) short DNA sequence whose location in the genome is known, and which serves as a landmark on the developing map of a genome before it is fully sequenced.

Ser serine *q.v.*

SER smooth endoplasmic reticulum. *see* endoplasmic reticulum.

sera *plu.* of serum.

seral *a.* (1) *pert.* sere; (2) *appl.* a plant community before reaching equilibrium or climax; (3) *pert.* blood serum.

sere *n.* (1) successional series of plant communities; (2) stage in a succession.

serial analysis of gene expression (SAGE) an alternative method to DNA microarrays for determining the presence of multiple RNA transcripts.

serial endosymbiosis theory the idea, now generally accepted, that mitochondria and chloroplasts, and possibly some other organelles of eukaryotic cells, originated as endosymbiotic microorganisms. Mitochondria were acquired first, chloroplasts acquired later only in the line or lines that led to algae and plants.

seriate *a.* arranged in a row or series.

sericate, sericeous *a.* (1) covered with fine close-pressed hairs; (2) silky.

serine (Ser, S) *n.* β-hydroxyalanine, an amino acid with an uncharged, polar, aliphatic hydroxyl side chain, nonessential in human diet, a constituent of protein, and is also an important intermediate in phosphatide synthesis.

serine peptidase, serine protease, serine proteinase protease with a histidine and a serine residue involved in catalysis at the active site, e.g. chymotrypsin, elastase, subtilisin.

serine/threonine protein kinase protein serine/threonine kinase *q.v.*

serine/threonine protein phosphatase protein serine/threonine phosphatase *q.v.*

seroconversion *n.* the time during an infection when antibodies against the infecting agent first become detectable in the blood.

serodeme *n.* a deme, e.g. of parasites, differing from others on immunological criteria, i.e. possessing different surface antigens.

serological *n.* (1) *pert.* serology; (2) *appl.* use of specific antisera to characterize e.g. bacteria, viruses, histocompatibility antigens, as opposed to direct biochemical, genetic or molecular biological investigations.

serology *n.* the study of immune sera and the use of antisera to characterize pathogens, antigens or cells.

serosa *n.* (1) lining of peritoneal, pleural and pericardial cavities, also extending as a fine membrane over internal organs; (2) false amnion or outer layer of amniotic fold; (3) outer larval membrane of insects. *a.* **serosal**.

serosity *n.* (1) watery part of animal fluid; (2) condition of being serous.

serotinal, serotinous *a.* (1) appearing or blooming late in the season; (2) *pert.* late summer; (3) flying late in the evening, *appl.* bats.

serotonergic *a. appl.* neurons releasing the neurotransmitter serotonin (5-hydroxytryptamine).

serotonin 5-hydroxytryptamine *q.v.*

serotype *n.* a subdivision of a species, esp. of bacteria or viruses, characterized by its antigenic character.

serous *a.* (1) watery; (2) *pert.* serum or other watery fluid.

serous membrane a thin membrane of connective tissue, lining some closed cavity of the body, and reflected over viscera, as a mesentery.

Serpentes *n.* suborder of reptiles comprising the snakes.

serpin *n.* any of a structurally related family of proteins, including α_1-antitrypsin, which inhibit proteases.

serra *n.* any saw-like structure.

serrate *a.* saw-toothed.

serrate-ciliate *a.* with hairs fringing toothed edges, *appl.* leaves.

serrate-dentate *a.* with serrate edges themselves toothed, *appl.* leaves.

serratiform *a.* like a saw.

serration *n.* series of saw-like notches, e.g. on edge of leaf.

serratirostral *a.* with serrated beak, *appl.* birds.

serratodenticulate *a.* with many-toothed serrations.

serratulate serrulate *q.v.*

serrature *n.* (1) saw-like notch; (2) serration.

serratus anterior, serratus magnus a muscle stretching from upper ribs to scapula.

serratus posterior superior and inferior: two thin muscles of the chest aiding in respiration, spreading respectively backward to anterior ribs and forward to posterior ribs.

serriferous *a.* having a saw-like organ or part.

serriform *a.* like a saw.

serriped *a.* with notched feet.

serrula *n.* (1) comb-like ridge on chelicerae of some arachnids; (2) free-swimming larva of alcyonarians, preceding planula.

serrulate *a.* finely notched.

serrulation *n.* fine notches.

Sertoli cell large cell of the epithelium lining in testes, connected with group of developing spermatozoa, acts as a nurse cell.

serule *n.* (1) a minor sere; (2) a succession of minor organisms.

serum *n.* (1) fluid component of blood after removal of cells and fibrinogen; (2) antiserum *q.v. alt.* **sera**.

serum albumin small protein abundant in blood plasma where it is involved in osmotic regulation and transport of metabolic products.

serum-responsive element *see* SRE.

sesamoid *a. appl.* a bone developed within a tendon and near a joint, as patella, radial or ulnar sesamoid, fabella.

sesamoidal *a. pert.* a sesamoid bone.

sessile *a.* (1) sitting directly on base without a stalk or pedicel, *appl.* e.g. flowers, leaves; (2) attached or stationary, as opposed to free-living or motile, *appl.* e.g. certain protozoans.

seston *n.* (1) microplankton *q.v.*; (2) all bodies, living and non-living, floating or swimming in water.

seta *n.* (1) chitinous hair or bristle, arising from epidermis of many invertebrates, e.g. polychaete and oligochaete annelid worms and insects, *alt.* chaeta; (2) chitinous thread extruded by certain diatoms that holds individual diatoms together in chains; (3) the stalk bearing the capsule in mosses and liverworts; (4) bristle-like hair in or on fruiting bodies of some fungi. *plu.* **setae**.

setaceous *a.* (1) bristle-like; (2) set with bristles.

setiferous, setigerous *a.* bearing setae or bristles.

setiform *a.* bristle-shaped, *appl.* teeth when very fine and closely set.

setigerous sac a sac in which is lodged a bundle of chaeta, formed by invagination of epidermis in parapodium of Chaetopoda.

setiparous *a.* producing setae or bristles.

setirostral *a. appl.* birds with bristles on beak.

setobranchia *n.* a tuft of setae attached to gills of certain decapod crustaceans, being coxopodite setae.

set of chromosomes the basic haploid set of chromosomes of any organism.

setose *a.* bristly.

setula, setule *n.* a thread-like hair or bristle. *a.* **setuliform**.

setulose *a.* set with small bristles.

severe acute respiratory syndrome (SARS) potentially fatal respiratory disease caused by a coronavirus, which first appeared in humans in China in 2002 and was rapidly spread by air travel to other parts of the world.

severe combined immunodeficiency deficiency (SCID) a potentially fatal deficiency of the immune system, generally genetically based, in which both T-cell and B-cell immune responses are absent.

sewage fungus pale slimy growth that often occurs in water with heavy organic pollution, either as a slime or a fluffy fungoid growth with streamers, attached to the bed of the river or pond. It consists of a characteristic community of microorganisms dominated by filamentous and zoogloeal bacteria, but also including fungi and protozoa.

Sewall Wright effect genetic drift *q.v.*

sex *n.* the sum characteristics, structures, features and functions by which a plant or animal is classed as male or female. In some animals sex is entirely genetically determined, in others the sex may change in response to environmental conditions.

sex cell gamete *q.v.*

sex-chromatin body Barr body *q.v.*

sex chromosome chromosomes such as X and Y in humans, which form non-homologous pairs in one of the sexes. Their presence, absence or particular form may determine sex.

sex comb a row of bristles on tarsus of first leg of male in certain insects.

sex-conditioned trait sex-influenced trait *q.v.*

sex-controlled genes genes present and expressed in both sexes, but manifesting themselves differently in males and females.

sex cords proliferations from germinal epithelium of developing gonads which give rise either to seminiferous tubules or to medullary cords of ovary.

sex determination any of various ways in which the sex of an animal is determined. In many animals sex is determined genetically, i.e. by which sex chromosomes are carried, or, as in bees, by whether the organism is haploid or diploid. In other animals, however, such as some fishes, sex is environmentally determined by e.g. temperature.

sex differentiation (1) differentiation of gametes into male and female; (2) differentiation of organisms into kinds with different sexual organs.

sexdigitate *a.* having six fingers or toes.

sexduction *n.* in bacteria, the transfer of genes from one bacterium to another mediated by their attachment to a transferable piece of extrachromosomal DNA, the fertility factor (F).

sex factors a class of plasmids in bacteria which can induce conjugation and their own and chromosomal gene transfer, including the F-factor, some colicin plasmids and some R-factors.

sexfid *a.* cleft into six segments, as calyx.

sexfoil *n.* a group of six leaves or leaflets around one axis.

sex gland gonad *q.v.*

sex hormones chiefly the oestrogens, androgens and gonadotropins.

sex-influenced trait phenotypic character whose expression is influenced by the sex of the individual, such as the simple Mendelian trait of baldness in humans in which the allele is dominant in males but recessive in females. *alt.* sex-conditioned trait.

sex-limited trait phenotypic character expressed only in one sex.

sex linkage the case where a gene is carried on a sex chromosome.

sex-linked *a.* (1) *appl.* genes carried on one of the sex chromosomes (e.g. either X or Y), and which therefore show a different pattern of inheritance in crosses where the male carries the gene from those in which the female is the carrier; (2) *appl.* an inherited trait that is manifest in only one sex.

sex mosaic (1) an intersex *q.v.*; (2) gynandromorph *q.v.*

sex pilus F pilus *q.v.*

sex ratio the ratio of males to females in a population. May be given as proportion of male births, number of males per 100 females or per 100 births, or as percentage males in the population.

sex reversal, sex transformation a change of an individual from one sex to the other, natural, pathological or artificially induced.

sexual *a.* (1) *pert.* sex; (2) *pert.* reproduction: any kind of reproduction that involves the fusion of gametes to form a zygote.

sexual cell gamete (*q.v.*), or cells giving rise to gametes.

sexual coloration colours displayed during the breeding season but not at other times, often different in the two sexes.

sexual cycle (1) menstrual cycle *q.v.*; (2) oestrus cycle *q.v.*

sexual dimorphism marked differences, in shape, size, morphology, colour, etc., between male and female of a species.

sexual imprinting imprinting (*q.v.*) influencing mating preference as adult.

sexual reproduction reproduction involving the formation and fusion of two different kinds of gametes to form a zygote, usually resulting in progeny with a somewhat different genetic constitution from either parental type and from each other.

sexual selection the difference in the ability of individuals of different genetic types to acquire mates, and therefore the differential transmission of certain characteristics to the next generation. It is made up of the choices made between males and females on the basis of some outward characteristic such as bright plumage or length of tail in birds, and competition between members of the same sex.

SH2 domain Src homology region 2, a protein domain present in many intracellular signalling proteins. It binds to short amino-acid sequences containing phosphorylated tyrosine residues.

SH3 domain Src homology region 3, a protein domain present in many intracellular signalling proteins. It binds to short amino-acid sequences rich in proline residues.

shadowing *n.* technique for preparing specimens for electron microscopy in which electron-dense metal atoms are evaporated onto the specimen at an angle, throwing surface features into relief.

shaft *n.* (1) stalk of feather or hair; (2) straight cylindrical part of a long bone.

sheath *n.* (1) a protective covering; (2) theca *q.v.*; (3) lower part of leaf enveloping a stem or culm; (4) insect wing cover, esp. elytron.

β-sheet secondary structure element common in many proteins, composed of a folding of the polypeptide chain so that extended strands lie alongside each other, held together by regular hydrogen bonding between backbone atoms. *alt.* β-pleated sheet. *see also* antiparallel β-sheet; parallel β-sheet.

sheet erosion the removal of topsoil over a wide area after heavy rain.

shell *n.* (1) the hard outer covering of some eggs, animals or fruits; (2) calcareous, siliceous, bony, horny or chitinous covering.

shell gland, shell sac organ in which material for forming a shell is secreted.

shield *n.* (1) carapace *q.v.*; (2) clypeus *q.v.*; (3) scutellum *q.v.*; (4) scutum *q.v.*; (5) (*bot.*) disc-like ascocarp or apothecium borne on thallus of lichens.

shikimic acid, shikimate aromatic carboxylic acid, intermediate in synthesis of aromatic amino acids in microorganisms.

Shine–Dalgarno sequence a sequence present in bacterial mRNAs a few bases before the initiation codon and which forms the site at which the mRNA binds to a ribosome.

shipworm *n.* bivalve mollusc which is worm-like in form, with a much-reduced shell at the foot and two chalky plates closing the siphon, and which bores into wood.

shivering *n.* mechanism of heat production found in many animals, used by birds and mammals as an emergency protection to maintain body temperature in cold conditions.

shoot *n.* (1) the part of a vascular plant derived from the plumule, being the stem and usually the leaves; (2) a sprouted part, branch or offshoot of a plant.

shoot apical meristem group of undifferentiated dividing cells at the tip of a shoot from which shoot growth is produced.

shore *see* seashore.

short-day plants plants that will only flower if the daily period of light is shorter than some critical length: they usually flower in early spring or autumn. The critical factor is the period of continuous darkness they are exposed to. *cf.* long-day plants, day-neutral plants.

short-germ development type of development in some insects in which most of the segments are formed sequentially by growth, the blastoderm itself only giving rise to the anterior segments.

short-grass community type of grassland that develops on poorer soils in dry regions, e.g. short-grass prairie, short-grass steppe, and which consists of grasses no more than 60 cm tall and small herbaceous plants.

short-horned flies Brachycera *q.v.*

short-horned grasshoppers Acrididae *q.v.*

short interspersed element (SINE) short sequence repeated many times in the mammalian genome. An example is the Alu sequence. SINEs are considered to be retrotransposons.

short period interspersion type of sequence arrangement within eukaryotic genomes in which moderately repetitive DNA sequences of average length 300 bp alternate with non-repetitive sequences ranging from *ca.* 800 to 1500 bp.

short-term memory (STM) (1) a memory process underlying behaviour, available for seconds or a few minutes after exposure to information; (2) the actual behaviour of recalling such information. *cf.* long-term memory.

shotgun cloning the random fragmentation of an entire genome and cloning of the individual fragments.

shotgun sequencing strategy for DNA sequencing in which the DNA is fragmented at random, the fragments sequenced, and the sequence then assembled in the correct order.

shrub *n.* low-growing woody plant, usually less than 6 m high, that does not have a main trunk and which branches from the base.

shrub layer the horizontal ecological stratum of a plant community composed of shrubs, which is higher than the field, or herb, layer and lower than the tree layer. *alt.* bush layer.

shuttle vector cloning vector that can be propagated in two different organisms, e.g. *E. coli* and yeast.

sialic *a. pert.* saliva.

sialic acid, sialate any of a class of acidic sugars such as *N*-acetylneuraminic acid, found in many glycoproteins and in gangliosides.

sialidase *n.* enzyme that removes terminal sialic acid residues in polysaccharides, glycoproteins, glycolipids and other molecules, by hydrolysing a glycosidic bond. *see also* neuraminidase. EC 3.2.1.18.

sialoid *a.* like saliva.

Siberian subregion zoogeographical area, part of the Palaearctic region, consisting of Asia west to the Urals, south to the Himalayas and east to western China.

siblings *n.plu.* offspring of the same parents. *alt.* sibs.

sibling species true species which do not interbreed but are difficult to separate on morphological grounds alone.

sibmating *n.* mating between siblings.

sibs siblings *q.v.*

sibship *n.* collectively, the siblings of one family.

siccicolous *a.* drought-resistant.

siccous *a.* dry, with little or no juice.

sickle-cell anaemia disease caused when a person is homozygous for the sickle-cell mutation in the gene for β-globin. An abnormal haemoglobin is produced which causes deformation or 'sickling' of red blood cells, severe anaemia and other symptoms, leading to premature death if untreated.

sickle-cell haemoglobin (HbS) defective haemoglobin present in people with the genetically determined conditions of sickle-cell trait and sickle-cell anaemia. A glutamate has been replaced by a valine in the β chain of the molecule, a change that causes 'sickling' of red blood cells, reducing their capacity for oxygen transport.

sickle-cell trait phenotype produced by the allele responsible for sickle-cell anaemia when in the heterozygous state. Some sickle-cell haemoglobin is produced, there is a mild anaemia and the condition apparently confers resistance to falciparum malaria.

sickle dance a dance of bees where the bee performs a sickle-like semicircular movement and the axis of the semicircle indicates the direction of the food source.

side chain in an amino acid, that part of the molecule not involved in forming the peptide bond. The side chain gives each amino acid its characteristic chemical and physical properties.

siderocyte *n*. a red blood cell containing free iron not utilized in haemoglobin formation.

siderophilic *a*. (1) staining deeply with iron-containing stains; (2) tending to absorb iron; (3) thriving in the presence of iron. *alt*. **siderophil**. *n*. **siderophile**.

siderophore *n*. compound that chelates iron, found in many bacteria.

sierozem *n*. grey soil, containing little humus, of middle-latitude continental desert regions.

sieve area region in the cell wall of a sieve tube element, sieve cell or parenchyma cell, having pores (sieve pores) through which cytoplasmic connections pass to adjacent cells.

sieve cell conducting cell of phloem of ferns and gymnosperms. It is tapering and elongated and is distinguished from the sieve tube elements of angiosperms by having small pores only and no sieve plate.

sieve disc sieve plate. *see* sieve tube.

sieve elements the conducting parts of the phloem.

sieve field sieve area *q.v.*

sieve pit a primary pit giving rise to a sieve pore. *see* sieve area.

sieve plate (1) *see* sieve tube; (2) madreporite *q.v.*; ( ) area of pedipalp of spiders, with openings of salivary glands.

sieve pore *see* sieve area.

sievert (Sv) the SI unit of radiation dose equivalent, equal to 1 joule of energy per kilogram of absorbing tissue, and which replaces the rem, with 1 Sv = 100 rem.

sieve tube long tube in phloem of angiosperms, whose function is the translocation of nutrients, esp. sugars. Sieve tubes are made up of sieve tube elements, which are elongated cells with no nucleus at maturity, joined end to end and communicating through clusters of pores in the walls. The end walls have clusters of larger pores forming a sieve plate.

sieve tube element *see* sieve tube.

sigla *n*. name formed from letters or other characters taken from the words in a compound term.

sigma (σ) (1) subunit of bacterial RNA polymerase involved in recognizing and selecting promoters for initiating transcription and in opening the template DNA double helix; (2) symbol for 0.001 seconds; (3) symbol for standard deviation.

sigmoid *a*. (1) curved like the letter sigma, in two directions, *appl*. arteries, cavities, valves; (2) *appl*. curve, an S-shaped curve.

sigmoid flexure an S-shaped bend.

signalling *n*., *a*. general term referring to any aspect of the production and reception of chemical and physical signals by cells.

signalling pathways intracellular signalling pathways *q.v.*

signal molecule any molecule that triggers the response of a cell to external conditions or the behaviour of other cells.

signal patch a protein-sorting signal (*q.v.*) that consists of a specific structure on the surface of a protein, as opposed to a contiguous short sequence.

signal peptidase enzyme resident in the endoplasmic reticulum, which splits off the signal sequence from membrane and secretory proteins.

signal peptide *see* signal sequence.

signal recognition particle (SRP) ribonucleoprotein particle that binds to the signal sequence region of secretory and membrane proteins as they are being synthesized, and guides them to the endoplasmic reticulum.

signal sequence a hydrophobic sequence of around 15 amino acids at the N-terminal end of secretory proteins and certain membrane proteins, which is essential for their passage across or into the cell membrane (in bacteria) or the endoplasmic reticulum membrane (in eukaryotic cells). It is subsequently cleaved off. *alt.* signal peptide. *see also* KDEL, nuclear localization sequence.

signal sequence receptor protein in the membranes of the endoplasmic reticulum to which signal sequences bind and which is involved in the transport of newly synthesized proteins from the cytosol into the endoplasmic reticulum.

signal transduction the conversion of one type of signal into another. In cell biology it refers to the process of converting the signal produced by activation of a cell-surface receptor into intracellular biochemical signals and hence into an intracellular response.

sign stimulus an environmental stimulus which acts as a releaser of species-specific behaviour.

silencer *n.* a control site in DNA which is required for maintaining a neighbouring gene in an inactive state.

silencing gene silencing *q.v.*

silent mutation a mutation which does not affect function or production of gene product and therefore has no effect on phenotype. They are usually point mutations which change one codon into another specifying the same amino acid or a substitute amino acid which does not affect protein function (*alt.* neutral mutations), or mutations in a DNA region which has no genetic function.

silicified *a.* being impregnated with silica, as the walls of diatoms.

silicole *a.* plant thriving in markedly siliceous soil.

silicula, silicule *n.* broad flat capsule divided into two by a false septum and found in members of the Cruciferae, such as honesty. *cf.* siliqua.

siliqua *n.* (1) long thin capsule divided into two by a false septum, found in members of the Cruciferae such as wallflower. *cf.* silicula; (2) a pod-shaped group of fibres around the olive in mammalian brain.

silk *n.* very strong fine protein fibre produced by various insects and spiders. It is extruded as a fluid from specialized glands and hardens on contact with the air. It is composed of the fibrous protein fibroin and other proteins such as sericin, and is used to make e.g. spiders' webs and egg and pupal cocoons. Commercial silk is produced by the silkworm (the larva of the moth *Bombyx mori*).

silk moth the moth *Bombyx mori*, whose larva (the silkworm) spins a cocoon of silk around itself.

silkworm *n.* the larva of the silk moth *Bombyx mori*, which spins a cocoon of silk around itself when it pupates, from which silk fibre is gathered commercially.

silt *n.* a soil intermediate between sands and clays in size of the particles.

Silurian *a. pert.* or *appl.* geological period lasting from *ca.* 444 to 416 million years ago, and in which the first land plants arose.

silva *alt.* spelling of sylva *q.v.*

silverfish common name for many members of the Thysanura *q.v.*

silvicolous *a.* inhabiting or growing in woodlands.

silviculture *n.* the cultivation of trees and management of forests and woodland for timber.

simian immunodeficiency virus (SIV) retrovirus closely related to the human immunodeficiency virus (HIV), which infects monkeys. Numerous strains exist that do not usually cause disease in their natural hosts but can cause disease if they infect another species.

simians *n.plu.* general term for monkeys and apes, usually excluding humans. *a.* **simian** possessing characteristics of, or *pert.*, monkeys and apes, as opposed to humans.

simple eyes (1) single ocelli that occur either on their own or in addition to compound eyes in adults of many insects. They are usually the only eyes possessed by the larvae; (2) eyes with only one lens.

simple fruits fruits developing from a single carpel or several united carpels, e.g. berries, apples, plums.

simple Mendelian trait trait determined by a single gene.

simple tissues tissues composed of one type of cell.

Simpson's index of floristic resemblance an index of the similarity between two communities, which is calculated by dividing the number of species common to the two communities by the total number of species in the smaller of the two communities. *cf.* Gleason's index, Jaccard index, Kulezinski index, Morisita's similarity index, Sørensen similarity index.

Simpson diversity index (D) a measure of the species diversity in a community. It is based on the probability of randomly picking two individuals of different species from the community and is calculated from the formula $D = (N(N - 1))/\Sigma\, n(n - 1)$, where N = total number of individuals of all species counted, and n = number of individuals of individual species. The higher the value of D the more diverse the site.

Simpson dominance index (C) a measure of the dominance in a community based on the probability of randomly picking two individuals of the same species from a community.

simulation *n.* assumption of features or structures intended to deceive enemies, e.g. forms of leaf and stick insects, and all varieties of protective coloration.

Sinanthropus see Pekin man.

sinciput *n.* upper or fore-part of head. *a.* **sincipital**.

SINE short interspersed element *q.v.*

single-cell protein (SCP) protein derived from unicellular microorganisms (e.g. bacteria or fungi) grown on a feedstock of crude oil, carbohydrate raw materials, or wastes from food processing as a carbon source, and which may be used for human or animal consumption.

single-copy DNA that portion of the genomic DNA comprising gene sequences present in only one or a few copies per haploid genome, and which is taken to represent the protein-coding genes. *alt.* unique DNA. *cf.* repetitive DNA.

single-copy plasmid plasmid present in a bacterial cell as a single copy, replicating only once each cell cycle.

single-nucleotide polymorphism (SNP) any position in the genome at which there is a difference in a single nucleotide between two unrelated members of the same species. It has been estimated that there are about 10 million SNPs in the human population.

single-pass transmembrane protein membrane protein in which the protein chain passes only once through the membrane.

single-positive thymocyte mature T cell that carries either CD4 or CD8, not both, while it is still in the thymus.

single-strand DNA-binding protein (ssDNA-binding protein) protein that binds to single-stranded DNA that becomes exposed e.g. while DNA is being replicated or repaired, thus preventing reformation of the original helix. *alt.* helix-destabilizing protein.

singleton *n.* a single offspring.

sinigrin *n.* glycoside present in plants of the family Cruciferae and their relatives, whose breakdown products give the pungent taste to e.g. capers and mustard.

sinistral *a.* on or *pert.* the left.

sinistrorse *a.* growing in a spiral which twines from right to left, anticlockwise, as in most gastropod shells.

sink *n.* any cell, tissue or organism that is a net importer and end-user of a metabolite or other resource. A storage root of a plant is, for example, a sink for sugars synthesized in the leaves, converting them into storage polysaccharides.

Sino-Japanese region part of the Boreal floristic kingdom consisting of Japan, northern and eastern China, and the northern Himalayas.

sinuate, sinuous *a.* having a wavy indented margin, *appl.* leaves, *appl.* gills of an agaric.

sinuatrial node (SAN) group of specialized cardiac muscle cells in the wall of right atrium near opening of superior or anterior vena cava, in which the heartbeat is initiated. *alt.* pacemaker, sinuauricular node.

sinuatrial valve valve between sinus venosus and atrium of heart.

sinus *n.* (1) cavity, depression, recess or dilatation; (2) groove or indentation. *alt.* lacuna.

sinuses of Valsalva dilations of pulmonary artery and of aorta, opposite aortic semilunar and pulmonary valves of heart.

sinus glands endocrine glands in eye-stalks of decapod crustaceans.

sinusoid *n.* (1) a minute blood-filled space or channel in the tissues of an organ, as in liver; (2) blood-filled space of irregular shape connecting arterial and venous capillaries.

sinus venosus (1) posterior chamber of tubular heart of embryo; (2) in lower vertebrates, a corresponding structure receiving venous blood and opening into auricle; (3) cavity of auricle.

sinus venosus sclerae circular canal near sclerocorneal junction and joining with the anterior chamber of eye and anterior ciliary veins.

siphon *n.* of aquatic molluscs, funnel-shaped structure from mantle cavity to exterior, through which water is drawn in and out, and which in some molluscs can be used as a means of jet propulsion.

siphonaceous, siphoneous *a.* tubular, *appl.* algae consisting of a more-or-less elaborate multinucleate thallus, not divided into separate filaments.

Siphonaptera *n.* order of wingless insects with large hind legs, commonly called fleas, which as adults are ectoparasites of skin of birds and mammals. They drink blood through piercing mouthparts, have a worm-like larval stage and a pupa.

siphonate *a.* having a siphon.

siphonet *n.* the honey-dew tube of aphids.

siphonium *n.* membranous tube connecting air passages of quadrate bone with air space in mandible.

siphonocladial *a. appl.* filaments with tubular segments, as in green algae.

siphonogamy *n.* fertilization by means of pollen tube through which contents of pollen grain pass to the embryo sac.

siphonoglyph *n.* one of (usually) two longitudinal grooves in pharynx (gullet) of some anthozoans.

Siphonophora, siphonophores *n., n.plu.* a group of pelagic hydrozoans that form colonies consisting of both polyps and medusoid forms. Some individuals of the colony are modified as a float for swimming as in the Portuguese Man o' War. Siphonophores are sometimes mistakenly called jellyfish.

siphonoplax *n.* calcareous plate connected with siphon in certain molluscs.

Siphonopoda Cephalopoda *q.v.*

siphonostele *n.* type of tubular stele in which the vascular tissue surrounds a central core of parenchyma, the pith. The phloem may either be external to the xylem or both internal and external. Present in stems of many ferns and of some gymnosperms and angiosperms.

siphonostomatous *a.* (1) with a tubular mouth; (2) having front margin of shell notched for emission of the siphon.

siphonous *a. appl.* coenocytic green algae of the class Chlorophyceae, which form hollow tubular colonies.

siphonozoid, siphonozooid *n.* small modified polyp without tentacles, serving to draw water through canal system of some soft corals. It possesses reproductive organs and cannot feed.

siphuncle *n.* (1) thin strand of living tissue running in a calcareous tube through all the compartments of a nautiloid shell; (2) the honey dew tube of an aphid.

Siphunculata *n.* the Anoplura (*q.v.*), the sucking lice.

siphunculate *a.* (1) having a siphuncle; (2) having mouth parts modified for sucking, as certain lice.

Sipuncula, sipunculids *n., n.plu.* phylum of marine unsegmented coelomate worms, which have the anterior end of the body introverted and used as a proboscis and tentacles round the mouth.

Sirenia, sirenians *n., n.plu.* an order of placental mammals commonly called sea cows, including the dugong and manatee, which are highly modified for aquatic life, with a naked body and front limbs modified as paddles.

sirenin *n.* substance secreted by some water moulds during development of female gametes and facilitating fertilization.

sirens *n.plu.* family of eel-like amphibians that live in muddy pools in south-east USA and northern Mexico.

siRNA small interfering RNA *q.v.*

sister cell, sister nucleus one of a pair of cells or nuclei produced by division of an existing cell or nucleus.

sister-chromatid exchange (SCE) the exchange of segments of chromosome between sister chromatids during mitosis.

sister chromatids the two copies of a replicated chromosome held together at the centromere, seen in the prophase and metaphase of mitosis and meiosis.

sister species species that are each other's closest relatives.

site-directed mutagenesis techniques of mutation *in vitro* in which a mutation is made at a specific predetermined site in DNA.

site-specific endonuclease restriction endonuclease *q.v.*

site-specific mutation a general term covering various techniques by which specific changes can be introduced into isolated DNA *in vitro*.

site-specific recombination genetic recombination occurring between two specific but not necessarily homologous short DNA sequences, as in the integration or excision of phage from a bacterial chromosome. *alt.* conservative recombination. *see also* transposition.

sitosterol *n.* complex mixtures of sterols in fatty or oily tissue of higher plants, esp. in corn or wheat-germ oil and in certain algae.

situs inversus condition in which the left–right symmetry of the positioning of internal organs such as the heart, stomach and spleen in the mammalian body is reversed.

SIV simian immunodeficiency virus, a retrovirus closely related to HIV.

skeletal *a. pert.* the skeleton.

skeletal muscle striated muscle *q.v.*

skeletogenous *a.* (1) *appl.* embryonic structures or parts which later become parts of skeleton; (2) in sponges, *appl.* the mesogloeal layer of scattered cells and jelly in which the spicules arise and are embedded.

skeleton *n.* hard framework, internal or external, which supports and protects softer parts of plant, animal or unicellular organism, and to which muscles usually attach in animals. *see also* endoskeleton, exoskeleton.

Skene's glands racemose mucous glands of the female urethra. *alt.* para-urethral glands, Guerin's glands.

skin *n.* the outermost covering of an animal, plant, fruit or seed. *see also* epidermis. In mammals the skin is composed of two layers, an outer epidermis derived from ectoderm which becomes cornified at the surface and is continuously being renewed,

and an inner dermis, derived from mesoderm, composed of connective, vascular and muscular tissues. *alt.* cutis. *see* Fig. 35 (p. 622).

Skinner box apparatus developed by the behaviourist psychologist B.F. Skinner for use in operant conditioning experiments, in which a rat must press a lever to gain a reward or avoid punishment.

skull *n.* hard cartilaginous or bony part of head of vertebrate, containing the brain, and including the jaws.

slavery dulosis *q.v.*

S-layer paracrystalline outer wall layer composed of protein and glycoprotein, present in many prokaryotes.

SLE systemic lupus erythematosus *q.v.*

sleep movements change in position of leaves, petals, etc., at night, which may be brought about by external stimuli of light and temperature changes, or may be an endogenous circadian rhythm.

sliding filament mechanism of muscle contraction, the shortening of a muscle fibre brought about by the sliding of interdigitating actin filaments between each other. The sliding is brought about by filaments of myosin molecules whose heads form cross-bridges to two opposing actin filaments. ATPase hydrolysis powers the movement of the myosin head along the actin filament, resulting in sliding of actin filaments towards each other, and contraction of the muscle. *see* Fig. 28 (p. 425).

sliding growth of plant tissue, when new part of cell wall slides over walls of cells with which it comes in contact.

slime *n.* viscous substance secreted by plants, fungi and animals and rich in glycans (fungal and plant slimes) or glycoproteins (most animal slimes).

slime bacteria myxobacteria *q.v.*

slime layer a polysaccharide-rich sheath secreted by some bacteria.

slime moulds common name for members of the protist group Mycetozoa, comprising the plasmodial slime moulds, cellular slime moulds and the protostelid slime moulds. *all q.v.*

slime net Labyrinthulomycota *q.v.*

slime pits cavities in plant body of hornworts (Anthoceratales) filled with mucilage and sometimes colonized by algae.

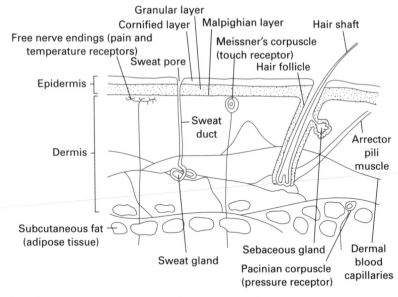

Fig. 35 Schematic cross-section of mammalian skin.

slime spore myxospore *q.v.*

slime tubes Cuvierian organs *q.v.*

slow muscle fibres type of striated muscle that contracts slowly but does not become readily fatigued. *alt.* slow-twitch fibres.

slow-twitch *appl.* muscle fibres incapable of very rapid contraction. *see* SO fibres.

slow viruses viruses that only show effects a long time after infection.

slug *n.* (1) shell-less terrestrial gastropod mollusc; (2) migrating pseudoplasmodium of a cellular slime mould.

Smads cytoplasmic proteins in animal cells that after phosphorylation by receptor serine/threonine kinases become active transcription factors and migrate to the nucleus to initiate transcription of target genes.

small cytoplasmic RNAs (scRNA) small RNA species ranging in size from 100 to 300 bases, found in ribonucleoprotein particles (scRNP) in the cytoplasm of eukaryotic cells.

small G proteins, small GTPases single-subunit guanine-nucleotide-binding proteins, which are involved in a variety of intracellular processes, including intracellular signalling and cytoskeletal rearrangement. Examples are Ras and Rac.

small interfering RNA (siRNA) double-stranded RNA of around 20–25 nucleotides which is complementary in sequence to part of an mRNA and which, when introduced into a cell, will form a complex with a ribonuclease and direct the enzyme specifically to the mRNA, resulting in cleavage of the mRNA and destruction of its function. siRNAs are used experimentally to suppress the expression of specific genes in mammalian cells. *see also* RNA interference.

small intestine collectively the duodenum, jejunum and ileum.

small molecule general name given to any natural or synthetic non-polymeric (i.e. non-protein, non-nucleic acid) organic molecule, e.g. metabolites, antibiotics and most therapeutic drugs.

small nuclear RNAs (snRNA) small RNA species ranging in length from 100 to 300 bases, found in ribonucleoprotein particles (snRNP) in the nuclei of eukaryotic cells, involved in RNA splicing.

small RNA (sRNA) general name for the many different types of small non-coding RNAs. *see* miRNAs, piRNAs, rasiRNA, siRNA.

smegma *n.* secretion of preputial glands or of clitoris glands.

smooth endoplasmic reticulum (SER) *see* endoplasmic reticulum.

smooth muscle non-striated muscle, e.g. of the walls of the alimentary canal, arteries and also found in many other organs. It is composed of spindle-shaped uninucleate cells and is not under voluntary control.

smut fungi common name for members of the Ustilaginales, a group of parasitic heterobasidiomycetes, some of which are serious pathogens of important crops. They form black, dusty masses of thick-walled teleutospores, resembling soot or smut.

snail *n.* generally refers to a member of the large group of terrestrial and freshwater gastropod molluscs of the subclass Pulmonata, which have a helically coiled shell, no ctenidia, and in which the mantle cavity is used as a lung.

snakes *see* Squamata.

SNAP receptors (SNAREs) membrane proteins of intracellular and plasma membranes involved in targeting the membrane of e.g. a transport vesicle to the membrane with which it is to fuse. The targeted fusion involves different SNAP receptors on vesicle and target membranes.

SNAPs soluble NSF attachment proteins. Proteins present in the membranes of transport vesicles and of intracellular compartments. They form a complex with the *N*-ethylmaleimide-sensitive fusion protein (NSF) and bind to intracellular membranes and plasma membrane by means of specific SNAP receptors. They are involved in mediating the specific targeting and fusion of transport vesicles to cellular compartments.

SNAREs SNAP receptors *q.v.*

SNP single-nucleotide polymorphism *q.v.*

snRNA small nuclear RNAs *q.v.*

snRNPs snurps *q.v.*

S1 nuclease deoxyribonuclease that degrades single-stranded DNA only.

snurps colloquial term for small nuclear ribonucleoproteins. *see* small nuclear RNAs.

social facilitation the effect of the actions of one animal in a group on the activity of others.

social insects the species of insect that live in organized groups and colonies, with division of labour between different morphological forms and which include many bees, ants and wasps (Hymenoptera) and the termites (Isoptera).

sociation *n.* a minor unit of vegetation, or microassociation.

socies *n.* an association of plants representing a stage in the process of succession.

society *n.* (1) a number of organisms forming a community; (2) in animals, a true society is characterized by cooperative behaviour between individuals, a division of labour and, in the social insects, morphological distinctions between members carrying out different functions; (3) a community of plants other than dominants, within an association or consociation.

sociobiology *n.* the study of the biological and genetic basis of social organization and social behaviour and their evolution in animals, a field of study which has caused controversy when applied to human social behaviour and organization.

sociohormone pheromone *q.v.*

SOD superoxide dismutase *q.v.*

sodium (Na) essential macronutrient, which as the ion, Na$^+$, is involved in generating the membrane potential in animal cells and is also required for the generation of nerve impulses.

sodium–calcium exchanger (Na^{2+}-Ca^{2+} exchanger) membrane transport protein that helps maintain a steep Ca^{2+} gradient across the plasma membrane of animal cells. It moves Ca^{2+} out of the cell powered by Na$^+$ moving into the cell down its electrochemical gradient.

sodium channel (Na$^+$) ion channel in cellular membranes that allows the passage of sodium ions.

sodium dodecyl sulphate (SDS) detergent widely used in experimental biology for solubilizing membrane protein assemblies for analysis as it disrupts most protein–protein and lipid–protein interactions.

sodium pump, sodium–potassium pump Na⁺-K⁺ ATPase *q.v.*

SO fibres slow oxidative muscle fibres in muscles of mammalian limbs, which contract slowly but do not readily become fatigued. They are muscle fibres adapted mainly for aerobic metabolism and are used when the animal is walking or running. *cf.* FG fibres.

soft corals Octocorallia *q.v.*

soft-rayed *appl.* fish having jointed fin-rays.

softwoods *n.plu.* conifers and their woods.

soil horizon a horizontal zone of soil of distinct composition and texture. A mature soil is made up of several distinct soil horizons. *see* soil profile.

soil profile a vertical section through soil showing the different layers from surface to underlying bedrock.

soil structure the texture of a soil, which is determined by the size and type of the soil particles and how they clump together.

soil water water that fills the spaces between soil particles and pores in rocks above the level of the water table. *cf.* groundwater.

sola *plu.* of solum *q.v.*

solaeus soleus *q.v.*

solarization *n.* retardation or inhibition of photosynthesis due to prolonged exposure to intense light.

solar plexus network of sympathetic nerves and ganglia situated behind stomach and supplying abdominal viscera.

soleaform *a.* slipper-shaped.

solenia *n.plu.* endoderm-lined canals, diverticula from coelenteron of a zooid colony.

solenocyte *n.* cell resembling flame cell of protonephridial system but with a single flagellum.

solenostele *n.* a stage after siphonostele in fern development, having phloem both internal and external to xylem.

soleus *n.* a flat muscle in calf of leg.

solfatara *n.* hot, sulphur-rich environment.

Solifuga, solifugids *n., n.plu.* order of arachnids commonly called false spiders or sun spiders, having very hairy bodies, large chelicerae and a segmented prosoma and opisthosoma. *alt.* Solpugida, solpugids.

soligenous *a. appl.* wet habitats that are supplied by groundwater rather than by precipitation, e.g. fens.

solitaria phase in locusts, the phase during which they live separately and do not aggregate into swarms.

solitary cells large pyramidal cells of brain, with axons terminating in superior colliculus or in midbrain.

solitary glands or follicles lymphoid nodules occurring singly on intestines and constituting Peyer's patches when aggregated.

solonchak *n.* light-coloured alkali soil with a high salt content found in poorly drained semidesert regions. It is infertile and low in organic matter.

solonetz *n.* dark-coloured alkali soil found in semidesert regions where the more soluble salts have been leached out. It supports some vegetation but is relatively infertile.

solpugids Solifuga *q.v.*

soluble NSF attachment proteins SNAPs *q.v.*

solum *n.* (1) floor, as of cavity; (2) soil between source material and topsoil. *plu.* **sola.**

solute *n.* any substance dissolved in a liquid (the solvent).

solvent *n.* any substance in which another substance is dissolved.

soma *n.* (1) the animal or plant body as a whole with the exception of the sex cells; (2) cell body of neuron. *plu.* **somata.**

somaclonal variation genetic variation arising from mutations in somatic plant cells undergoing regeneration in culture.

somactids *n.plu.* endoskeletal supports of dermal fin-rays.

somaesthesis *n.* sensation due to stimuli from skin, muscle and internal organs.

somata *plu.* of soma.

somatic *a.* (1) *pert.* purely bodily part of plant or animal, as opposed to germ cells; (2) *appl.* cell: body cell which will not produce gametes, as opposed to cells of the germ line; (3) *appl.* mutation: mutation occurring in a body cell; (4) *appl.* number: basic number of chromosomes in somatic cells.

somatic cell genetics the study of genetics using cultured somatic cells.

somatic cell hybrid a hybrid cell constructed from two somatic cells, often of two different species, by induced cell fusion.

somatic cell nuclear transfer (SCNT) a method of animal cloning in which the nucleus of a somatic cell is transferred into an enucleated unfertilized ovum *in vitro*. The resulting zygote is subsequently transferred into the uterus of a surrogate mother.

somatic gene therapy the therapeutic introduction of genes into somatic cells to correct a defect.

somatic hypermutation mutation that occurs in B cells after immunoglobulin gene rearrangement and after stimulation with antigen, and which introduces further changes into the variable regions of heavy- and light-chain genes.

somatic mutation mutation occurring in a somatic cell.

somatic nervous system the parts of the peripheral nervous system that carry sensory information from the periphery to the central nervous system and motor commands from the central nervous system to the skeletal muscles. *alt.* somatomotor system.

somatic polyploidy having polyploid somatic cells.

somatic recombination recombination that occurs in somatic cells, e.g. during gene rearrangement in developing lymphocytes.

somatic segregation unequal segregation of maternally and paternally derived organelle genes (as opposed to nuclear genes) in different tissues of the same organism.

somatic sensory fibres nerve fibres conveying impulses from sensory organs (other than those in head) involved in sensing external stimuli. They run from the periphery to the spinal cord and have cell bodies in dorsal root ganglia.

somatocoels *n.plu.* a pair of sacs which bud off from the primary coelomic sacs in the larval stage of echinoderms and later form the main coleom on the adult.

somatocyst *n.* cavity in pneumatophore of a siphonophore containing air or an oil droplet.

somatogamy *n.* sexual reproduction by fusion of two compatible somatic cells or hyphae, eventually giving rise to a zygote. *alt.* pseudomixis.

somatogenic *a.* (1) developing from somatic cells; (2) *appl.* variation and adaptation arising from external stimuli.

somatoliberin *n.* small polypeptide produced by hypothalamic neurons which regulates the synthesis of growth hormone in the pituitary.

somatome somite *q.v.*

somatomedin C insulin-like growth factor I *q.v.*

somatomotor system somatic nervous system *q.v.*

somatopleure *n.* dorsal sheet of mesoderm that forms the outer wall of the coelom.

somatosensory *a. appl.* sensation due to stimuli from skin, muscle or internal organs.

somatosensory cortex area of cerebral cortex in brain concerned with analysis of sensory information from body surface and internal organs.

somatostatin *n.* 14-amino acid peptide hormone, produced by hypothalamus, inhibiting the secretion of growth hormone, insulin and glucagon.

somatotrophic *a.* (1) stimulating nutrition and growth; (2) *appl.* hormone: somatotropin *q.v. alt.* **somatotropic.**

somatotropin *see* growth hormone.

somatotype *n.* body type or conformation.

somesthetic system the system of internal receptors that is responsible for bodily sensations in animals.

somite *n.* (1) one of a series of paired blocks of mesoderm on either side of the notochord in a vertebrate embryo. Each somite corresponds to one unit in the final sequence of articulated elements, e.g. vertebra and associated bones and muscles; (2) a segment of a metamerically segmented animal.

somitic *a. pert.* or giving rise to somites, *appl.* mesoderm.

songbirds Passeriformes *q.v.*

sonic *a. pert.* or produced by sound.

soral *a. pert.* a sorus.

soralium *n.* a well-defined group of soredia surrounded by a distinct margin on the thallus of a lichen.

sorbitol *n.* a faintly sweet alcohol isomeric with mannitol.

sorediate *a.* (1) bearing soredia; (2) with patches on the surface.

soredium *n*. a small round or scale-like body on the thallus of some lichens, which contains both algal or cyanobacterial cells and fungal hyphae, and by which the lichen is propagated. *plu.* **soredia**.

Sørensen similarity index a measure of the similarity of species composition between two plant communities. It is calculated by doubling the number of species common to the two communities and dividing by the sum of the total number of species in the two communities. *see also* Gleason's index, Jaccard index, Kulezinski index, Morisita's similarity index, Simpson's index of floristic resemblance.

sori *plu.* of sorus.

soriferous *a*. bearing sori.

sorocarp *n*. the fruiting body of the cellular slime moulds, composed of a stalk and spore head(s).

sorogen *n*. the cells or tissue that develops into a sorus.

sorophore *n*. thallus or stalk bearing a sorus or sorocarp.

sorosis *n*. composite fruit formed by fusion of fleshy axis and flowers, as pineapple.

sorption *n*. retention of material at surface by adsorption or absorption.

sorus *n*. cluster of sporangia in ferns, borne in large numbers on underside of frond, and similar masses of sporangia in other organisms. *plu.* **sori**.

SOS repair an error-prone DNA repair process in which replication is allowed to proceed through an area of chromosomal damage.

SOS response the response of *Escherichia coli* and other bacteria to treatments that cause extensive damage to DNA and inhibit replication. These include inhibition of cell division and stimulation of various DNA repair and recombination processes.

South African kingdom floristic kingdom consisting of the extreme southern tip of Africa.

South African region the part of the Palaeotropical floristic kingdom consisting of temperate and desert Southern Africa except for the extreme tip.

South African subregion zoogeographical area, part of the Ethiopian region, consisting of the temperate grasslands of Southern Africa.

South Brazilian region part of the Neotropical floristic kingdom consisting of Brazil south of the Amazon rain forest.

Southern blotting technique in which DNA fragments separated by gel electrophoresis are transferred by 'blotting' the gel onto a nitrocellulose membrane for subsequent hybridization with radioactively labelled nucleic acid probes for identification and isolation of sequences of interest.

South Temperate Ocean Islands region part of the Antarctic floristic kingdom consisting of the islands of the Southern Ocean south of latitude 50°, e.g. South Georgia and the South Sandwich Islands.

Southwest Australian region part of the Australian floristic kingdom, consisting of the extreme south-western corner of Australia.

SP suction pressure *q.v.*

sp. species *q.v.*

space-filling model a model of protein structure in which each atom is depicted as a solid sphere in proportion to its van der Waals radius.

spacer DNA the DNA found between genes.

spadiceous *a*. arranged like a spadix.

spadiciform, spadicose *a*. resembling a spadix.

spadix *n*. (1) (*bot*.) a branched inflorescence with an elongated axis, sessile flowers and an enveloping spathe, as of wild arum; (2) a spike, in a succulent; (3) (*zool*.) rudiment of developing manubrium in some coelenterates; (4) amalgamation of internal lobes of tentacles in the cephalopod, *Nautilus*.

spangle gall small reddish gall found on the underside of oak leaves and caused by gall wasp larvae.

spat *n*. the spawn or young of bivalve molluscs.

spathaceous, spathal *a*. resembling or bearing a spathe.

spathe *n*. large, enveloping leaf-like structure, green or coloured, protecting a spadix.

spathella *n*. a small spathe surrounding division of palm spadix.

Spathiflorae Arales *q.v.*

spathose *a*. with or like a spathe.

spatia zonularia a canal surrounding marginal circumference of lens of eye.

spatulate *a.* spoon-shaped.

spawn *n.* (1) collection of eggs deposited by e.g. bivalve molluscs, fishes, frogs; (2) mycelium of certain fungi; (3) *v.* to deposit eggs, e.g. by fishes.

Special Creation *see* Creationism.

specialist, specialist species species which can survive and thrive only within a narrow range of habitat and/or climatic conditions, or which can use only a very limited range of food, and is therefore usually less able to adapt to changing environmental conditions.

specialization *n.* adaptation to a particular mode of life or habitat in the course of evolution.

specialized transduction type of transduction (*q.v.*) in which only DNA from a particular region of the host chromosome is incorporated into the transducing phage particle. *cf.* generalized transduction.

speciation *n.* the evolution of new species.

species *n.* in sexually reproducing organisms, a group of interbreeding individuals not normally able to interbreed with other such groups. A species is given two names in binomial nomenclature (e.g. *Homo sapiens*), the generic name and specific epithet (italicized in the scientific literature), similar and related species being grouped into genera. Species can be subdivided into subspecies, geographic races, and varieties.

species-abundance curve graph illustrating the relationship between the number of species (plotted on the *y*-axis) and the number of individuals per species (on the *x*-axis).

species aggregate a group of very closely related species which have more in common with each other than with other species of the genus.

species-area curve graph illustrating the number of species (plotted on the *y*-axis) found in a given area (increase in area plotted on the *x*-axis). The shape of the curve provides information on the species diversity and species richness of an area and is helpful in determining the most efficient size of plot for sampling.

species-area effect the general rule that the number of species present in a given area

of a particular habitat is less than the total number of species that are supported by the habitat as a whole. Thus fragmentation of even a species-rich habitat will lead to loss of species.

species composition the different species in a given area or ecosystem.

species diversity the number and abundance of different species within a given area, which is one measure of biological diversity, a diverse environment having relatively small numbers of many different species.

species diversity index the extent to which different species are represented within a community.

species flock group of numerous species, endemic to a particular small area and ecologically diverse, such as the cichlid fish of some African lakes, and which are thought to have evolved from a single ancestor species.

species pair sibling species *q.v.*

species richness the number of different species within a given community or area.

species selection the view that selection can act at the level of the species as well as the individual, which is advanced as an explanation of some long-term evolutionary trends.

species-specific behaviour behaviour patterns that are inborn in a species and are performed by all members under the same conditions and which are not modified by learning.

specific *a.* (1) peculiar to a species; (2) *pert.* a species; (3) *appl.* characteristics distinguishing a species; (4) restricted to interaction with a particular substrate, *appl.* e.g. enzymes, antibodies.

specific activity (1) of an enzyme, the catalytic activity per milligram of protein; (2) of a radioisotope, the relative amount of radioactive atoms in a sample, expressed as becquerels per gram.

specification *n.* commitment of cells to follow their normal course of development if no external factors intervene. *a.* **specified.** *cf.* determination.

specific epithet the second name in a Latin binomial.

specific hunger the preference for certain foods containing an essential mineral or vitamin shown by animals on a diet deficient in the substance.

specificity *n*. (1) being limited to a species; (2) restriction of parasites, bacteria and viruses to particular hosts; (3) restriction of enzymes to certain substrates, and restriction of antibody to interactions with particular antigens.

speckles in cell biology, clusters of granules interspersed among chromatin in the nucleus.

spectrin *n*. a fibrous protein located on the cytoplasmic face of the red blood cell membrane and a constituent of the membrane cytoskeleton.

speculum *n*. (1) ocellus *q.v.*; (2) a wing bar with a metallic sheen, as in mallard drakes.

speleology *n*. the study of caves and cave life.

Spemann organizer the mesoderm of the dorsal lip of the blastopore in amphibian embryos, which provides signals for the further development of the body axis. When transplanted to another site on the embryo it is capable of generating a complete new body axis.

sperm *n*. (1) a spermatozoan; (2) male gamete (in plants and animals); (3) semen.

spermagglutination *n*. agglutination of spermatozoa, such as is brought about by some myxoviruses.

spermangium *n*. an organ producing male spore-like cells, in ascomycetes.

Spermaphyta Spermatophyta *q.v.*

spermaphytic *a*. seed-bearing.

spermary, spermarium *n*. any organ in which male gametes are produced.

spermatangium *n*. male sex organ of red algae, each developing into one non-motile gamete (spermatium).

spermatheca *n*. in female or hermaphrodite invertebrates, a sac for storing spermatozoa received on copulation. *plu.* **spermathecae.**

spermatia *plu.* of spermatium.

spermatic *a*. (1) *pert.* spermatozoa; (2) *pert.* testis.

spermatid *n*. haploid cell arising by division of secondary spermatocyte in testes and which becomes a sperm.

spermatiferous *a*. bearing spermatia.

spermatiform *a*. like a spermatium.

spermatiophore *n*. a hypha or hyphal outgrowth bearing spermatia.

spermatium *n*. (1) small non-motile cell that functions as a male sex cell in some ascomycete and basidiomycete fungi, and which fuses with a receptive female sex organ; (2) non-motile male gamete produced by red algae. *plu.* **spermatia.**

spermatoblast *n*. (1) spermatid *q.v.*; (2) Sertoli cell *q.v.*

spermatoblastic *a*. sperm-producing.

spermatocyst *n*. a seminal sac.

spermatocyte *n*. male cell in which meiosis occurs to form sperm. Primary spermatocytes are diploid cells developing from spermatogonia and which undergo a first meiotic division to form two secondary spermatocytes, each of which completes meiosis, producing four haploid spermatids from each original diploid spermatocyte.

spermatogenesis *n*. sperm formation, from spermatogonium through primary and secondary spermatocytes and spermatid to spermatozoon.

spermatogenic *a*. (1) *pert.* sperm formation; (2) sperm-producing.

spermatogenous *a*. giving rise to sperm.

spermatogonium *n*. primordial male germ cell, diploid stem cell producing the male germ cells, making up the inner layer of the lining of the seminiferous tubules, and which gives rise to spermatocytes. *see also* spermogonium. *plu.* **spermatogonia.**

spermatoid *a*. like a sperm.

spermatophore *n*. a number of sperms enclosed in a sheath of gelatinous material, the form in which sperm is released by many invertebrates.

Spermatophyta, spermatophytes *n*., *n.plu.* in some classifications a major division of plants containing all seed-bearing plants, i.e. gymnosperms (e.g. cycads, conifers, *Ginkgo*, *Ephedra* and quillworts) and angiosperms (flowering plants).

spermatoplasm *n*. cytoplasm of sperm.

spermatoplast *n*. a male gamete.

spermatostrate *a*. spread by means of seeds.

spermatozeugma *n*. union by conjugation of two or more spermatozoa, as in the vas deferens of some insects.

spermatozooid antherozooid *q.v.*

spermatozoon *n.* mature motile male gamete in animals, typically consisting of a head containing the nucleus, and a tail consisting of a single flagellum by which the cell moves. *alt.* sperm. *plu.* **spermatozoa.**

sperm competition type of sexual competition in animals in which there is competition not for access to females but for fertilization.

spermidine *n.* a widely distributed polyamine (C7H19N3), constituent of ribosomes, and acting generally as a membrane-stabilizing agent.

spermiducal *a.* (1) *appl.* glands into or near which sperm ducts open, in many vertebrates; (2) *appl.* glands associated with male ducts in oligochaete worms.

spermiduct *n.* duct for conveying sperm from testis to exterior.

spermine *n.* a widely distributed polyamine (C10H26N4), a constituent of ribosomes, and acting generally as a membrane-stabilizing agent.

spermiocyte primary spermatocyte. *see* spermatocyte.

spermiogenesis *n.* development of spermatozoon from spermatid.

sperm nucleus male pronucleus *q.v.*

spermo- *see also* spermato-.

spermodochium *n.* a group of spermatiophores derived from a single cell and lacking a capsule, in fungi.

spermogenesis spermatogenesis *q.v.*

spermogoniferous *a.* having spermogonia.

spermogonium *n.* the structure containing the sex organs of rust fungi and some lichen fungi, in which non-motile male gametes (spermatia) are formed. *plu.* **spermogonia.**

spermogonous *a.* like or *pert.* a spermogonium.

spermospore *n.* a male gamete produced in a spermangium.

spermotheca spermatheca *q.v.*

spermozeugma *n.* mass of regularly aggregated spermatozoa, for delivery into a spermatheca.

SPF specified pathogen-free.

sphacelate *a.* decayed and withered.

sphacelia *n.* the conidial stage of a fungus that produces a sclerotium or ergot.

sphaerenchyma *n.* a tissue composed of spherical cells.

sphaeridia *n.plu.* small rounded bodies, possibly balancing organs or other type of sense organ, on the surface of some echinoderms.

sphaerite *n.* rounded mass of calcium oxalate or starch forming crystals inside plant cells.

sphaerocyst *n.* large globose or oval cell in trama of *Russula* and *Lactarius* species.

sphaeroid *a.* (1) globular, ellipsoidal or cylindrical; (2) *appl.* an aggregate of individual protozoans; (3) *appl.* a dilated hyphal cell containing oil droplets, in lichens.

Sphaeropsida *n.* form class of deuteromycete fungi that reproduce by conidia borne in pycnidia, and which includes numerous plant pathogens (e.g. *Phyllosticta, Dendrophoma, Septoria*). *alt.* **Sphaeropsidales.**

sphagnicolous *a.* inhabiting sphagnum peat moss.

Sphagnidae sphagnum mosses *q.v.*

sphagniherbosa *n.* plant community on peat containing large amounts of *Sphagnum* moss.

sphagnous *a. pert.* peat moss.

sphagnum mosses mosses of the class Sphagnidae, also called peat mosses or bog mosses, having the gametophore with branches in whorls, many dead water-absorbing cells, and comprising one genus, *Sphagnum*. They are distinguished from other mosses by leaves lacking midribs and a plant body with no rhizoids when mature. The spore capsule lacks a peristome and the protonema germinating from the spore is plate-like rather than filamentous as in the true mosses.

S phase the phase of DNA replication in the eukaryotic cell cycle. *see* Fig. 8 (p. 106).

sphecology *n.* the study of wasps.

sphenethmoid *n.* single bone replacing the orbitosphenoids in anuran amphibians.

sphenic *a.* like a wedge.

Sphenisciformes *n.* an order of birds including the penguins.

spheno- prefix, *pert.* or involving the sphenoid, as sphenomandibular, *appl.* ligament connecting sphenoid and mandible.

Sphenodonta *see* Rhynchocephalia.

spheno-ethmoidal *a. pert.* or in the region of sphenoid and ethmoid, *appl.* a recess above superior nasal concha and a suture.

sphenofrontal *a. pert.* sphenoid and frontal bones, *appl.* a suture.

sphenoid *n.* a basal bone of skull in some vertebrates, composed of several fused bones. *alt.* butterfly bone.

sphenoidal *a.* (1) wedge-shaped; (2) *pert.* or in region of sphenoid.

sphenolateral *n.* one of dorsal pairs of cartilages parallel to the trabeculae embracing the hypophysis in embryo.

sphenomaxillary *a. pert.* sphenoid and maxilla, *appl.* fissure and (pterygopalatine) fossa.

sphenopalatine ganglion autonomic ganglion on the maxillary nerve in pterygopalatine fossa.

Sphenophyta *n.* one of the four main divisions of extant seedless vascular plants, commonly called horsetails and represented by a single living genus *Equisetum*. They have a sporophyte with roots, jointed stems and leaves in whorls, and have strobili of reflexed sporangia borne on sporangiophores.

sphenopterygoid *a.* (1) *pert.* sphenoid and pterygoid; (2) *appl.* mucous pharyngeal glands, near openings of Eustachian tubes, as in birds.

sphenosquamosal *a. appl.* cranial suture between spheroid and squamosal.

sphenotic *a.* postfrontal cranial bone in many fishes.

sphenozygomatic *a. appl.* cranial suture between sphenoid and zygomatic.

spher- *see also* sphaer-.

spheraster *n.* a many-rayed globular spicule.

spheroidal *a.* globular, but not perfectly spherical, *appl.* glandular epithelium (from shape of cells).

spheroidocyte *n.* type of haemocyte in insects.

spherome *n.* (1) cell inclusions producing oil or fat globules; (2) intracellular fatty globules as a whole.

spheromere *n.* segment of a radially symmetrical animal.

spheroplast *n.* bacterial or yeast cell from which the wall has been removed, leaving a protoplast.

spherosome *n.* organelle derived from endoplasmic reticulum and bounded by a single membrane leaflet, and which synthesizes lipids.

spherula *n.* a small sphere: a small spherical spicule.

spherulate *a.* covered with small spheres.

sphincter *n.* a muscle that closes or contracts an orifice, as of bladder or anus.

sphingolipid *n.* any of a group of complex lipids, glycolipids and phospholipids, which contain the amino alcohol sphingosine and not glycerol, and including sphingomyelin, cerebrosides, gangliosides and ceramide.

sphingomyelin *n.* any of a group of phospholipids, found esp. in the myelin sheath of nerve cells, and containing the amino alcohol sphingosine, fatty acids and phosphorylcholine.

sphingosine *n.* an amino alcohol, containing a long unsaturated hydrocarbon chain, and which is found in gangliosides, cerebrosides, sphingomyelin and ceramide.

sphragis *n.* a structure sealing the bursa copulatrix on female abdomen of certain lepidopterans after pairing, and consisting of hardened sphragidal fluid secreted by male.

sphygmic *a. pert.* the pulse, *appl.* 2nd phase of systole.

sphygmoid *a.* (1) pulsating; (2) like a pulse.

spica *n.* a spike.

spicate *a.* (1) having a flowerhead in the form of a spike; (2) bearing spikes.

spiciform *a.* spike-shaped.

spicula, spicule *n.* (1) minute needle-like body, siliceous or calcareous, found in invertebrates; (2) any minute pointed process.

spicular *a. pert.* or like a spicule.

spiculate *a.* (1) set with spicules; (2) divided into small spikes.

spiculiferous, spiculigerous, spiculose *a.* furnished with or protected by spicules.

spiculiform *a.* spicule-shaped.

spiculum *n.* the dart of a snail.

spider cell fibrous astrocyte. *see* astrocyte.

spiders order of arachnids, the Araneida, having spinning glands on the opisthosoma and poison glands on the chelicerae.

spike *n.* a flowerhead with stalkless flowers or secondary small spikes (spikelets) of flowers borne alternately along a single axis.

spikelet *n.* (1) one of the units of the flower-head of grasses, consisting of several florets along a thin stalk, at the base of which are two bracts (glumes) marking the end of the spikelet; (2) any small spike of flowers.

spinal *a. pert.* backbone, spinal cord, *appl.* foramen, ganglia and nerves.

spinal canal canal formed by vertebrae which encloses the spinal cord.

spinal column backbone, vertebral column *q.v.*

spinal cord column of nervous tissue running from brain along the back in vertebrates, enclosed in the spinal canal formed by vertebrae. Together with the brain it forms the central nervous system.

spinalis name given to muscles connecting vertebrae.

spinate *a.* (1) bearing spines; (2) spine-shaped.

spination *n.* the occurrence, development or arrangement of spines.

spindle *n.* (1) the structure formed by microtubules stretching between opposite poles of the cell during mitosis or meiosis and which guides the movement of the chromosomes. *see also* meiosis, mitosis; (2) muscle spindle *q.v.*; (3) tree of the genus *Euonymus*.

spindle cell (1) small spindle-shaped neuron; (2) a spindle-shaped coelomocyte.

spindle fibre a bundle of microtubules in the mitotic or meiotic spindle that forms a fibre thick enough to be visible in the light microscope.

spindle pole one end of a mitotic or meiotic spindle.

spindle pole body structure found at the poles of the mitotic spindles of many fungi, and which serves the role of a centrosome in organizing the spindle microtubules.

spine *n.* (1) sharp-pointed outgrowth as on leaves, stems, bones, echinoids; (2) sharp pointed modified hair as in hedgehogs, porcupines; (3) the backbone or vertebral column; (4) pointed process of vertebra; (5) scapular ridge; (6) fin-ray; (7) (*neurobiol.*) dendritic: small protuberances stiffened by actin filaments, at the ends of fine branches of dendrites of neurons, where many neurotransmitter receptors are concentrated.

spiniferous, spinigerous *a.* spine-bearing, *appl.* pads on ventral side of leg in the onychophoran *Peripatus*.

spiniform *a.* in the shape of a spine.

spinneret *n.* (1) organ perforated with tubes connecting to the silk glands in spiders, and from which the liquid silk is released to form webs; (2) similar organ from which cocoon is spun in some insects.

spinnerule *n.* a tube discharging silk secretion of spiders.

spinning glands glands that secrete silky material in arthropods, as for webs in spiders and cocoons in insect caterpillars.

spinocaudal *a. pert.* trunk of vertebrates.

spino-occipital *a. appl.* nerves arising in trunk somites which later form part of the skull.

spinose *a.* bearing many spines.

spinothalamic tracts tracts of nerve fibres connecting spinal cord and thalamus in brain, involved in relaying pain and temperature information.

spinous *a.* (1) spiny; (2) spine-like.

spinous process (1) median dorsal spine-like projection of vertebra; (2) a projection of sphenoid; (3) projection between articular surfaces of proximal end of tibia.

spinulate, spinulose, spinulous *a.* covered with small spines.

spinulation *n.* a defensive spiny covering.

spinule *a.* a small spine.

spinuliferous *a.* bearing small spines.

spiny-finned *a.* bearing fins with spiny rays for support.

spiny-headed worms Acanthocephala *q.v.*

spiny-rayed *a. appl.* fins supported by spiny rays.

spiracle, spiraculum *n.* (1) hole in the sides of thoracic and abdominal segments of insects, and of myriapods, through which the tracheal respiratory system connects with the exterior, and which can be opened and closed; (2) small round opening, a vestigial gill slit, immediately anterior to the hyomandibular cartilage in elasmobranch fishes; (3) various other exterior openings connected with breathing or respiration in other animals.

spiraculate, spiraculiferous *a.* having spiracles.

spiraculiform *a.* spiracle-shaped.

spiral *a.* winding like a screw, *appl.* leaves alternately placed up a stem, *appl.* flowers with spirally inserted parts, *appl.* pattern of lignified thickening in cell wall of xylem vessels and tracheids.

spiral cleavage mode of cleavage in which blastomeres divide obliquely so that at the eight-cell stage the four upper cells are not directly above the four lower cells, as in radial cleavage, but in the grooves between them. Further divisions are also oblique, alternating between right and left. Found in turbellarians, annelids, molluscs, nemertean worms and some other groups, and associated with a determinate type of development.

spiralia *n.plu.* coiled structures supported by crura, in some brachiopods.

spiral valve (1) in certain more primitive fishes, such as elasmobranchs, ganoids and dipnoans, a spiral infolding of the intestine wall; (2) of Heister, folds of mucous membrane in neck of gallbladder; (3) of heart, incomplete partition in conus arteriosus in dipnoans, preventing complete mixing of oxygen-rich and oxygen-poor blood.

spiral vessels first xylem elements of a stele, spiral fibres coiled up inside tubes and so adapted for rapid elongation.

spiranthy *n.* displacement of flower parts through twisting.

spire *n.* the whorled part of a spiral shell.

spiriferous *a.* having a spiral structure.

spirilla *n.plu.* group of β-Proteobacteria with helically curved or twisted thread-like cells. Some are flagellate. Some are free-living in freshwater or marine environments, others saprophytic or parasitic, including some human pathogens such as *Spirillum minor*, the cause of rat-bite fever. *sing.* **spirillum**.

spirivalve *n.* a gastropod with spiral shell.

Spirochaetes *n.* phylum of the Bacteria distinguished on DNA criteria, consisting of slender, helically coiled bacteria, commonly called spirochaetes, with flexible cells and no rigid cell wall. They include free-living, commensal and parasitic forms including some human pathogens such as *Treponema pallidum*, the causal organism of syphilis. *alt.* **spirochetes**.

spiroid *a.* spirally formed.

spiroplasms *n.plu.* very small motile helical prokaryotes lacking a cell wall, which cause plant disease. *alt.* **spiroplasma**.

Spirotrichia, spirotrichans *n.*, *n.plu.* group of ciliate protozoans having a well-defined gullet surrounded by a ring of composite cilia, the undulating membrane, e.g. *Stentor*.

spite *n.* in animal behaviour, the name given to a behaviour by which an animal reduces its own fitness in the process of harming another animal.

splanchnic *a.* (1) *pert.* internal organs; (2) *appl.* mesoderm, the mesoderm lying laterally to the central axis that gives rise to internal organs. *cf.* axial, paraxial.

splanchnocoel *n.* the cavity of lateral somites of embryo, persisting as the visceral cavity in adults.

splanchnopleure *n.* ventral sheet of mesoderm that forms the inner wall of the coelom.

splash zone zone of a seashore above the high-tide mark but which may be wetted by sea spray at high tide, and which supports some seaweeds (e.g. *Pelvetia caniculata*, channel wrack), small periwinkles and lichens. *see* Fig. 34 (p. 605).

spleen *n.* secondary lymphoid organ and vascular organ in vertebrates in which immune reactions are initiated and red blood cells destroyed.

splenetic splenic *q.v.*

splenial *a.* (1) *pert.* the splenium of corpus callosum; (2) *pert.* the splenial bone in vertebrate lower jaw; (3) *pert.* the splenius muscle of upper dorsal region and back of neck.

splenic *a. pert.* the spleen. *alt.* splenetic.

splenium *n.* posterior border of corpus callosum in brain.

splenius *n.* muscle of upper dorsal region and back of neck.

splenomegaly *n.* enlargement of the spleen, e.g. as a result of parasitic infection.

splenophrenic *a. pert.* spleen and diaphragm.

spliceosome *n.* multisubunit complex of small ribonucleoproteins (U1, U2, etc.) which assembles on RNA and carries out RNA splicing.

3′ splice site acceptor splice site *q.v.*

5′ splice site donor splice site *q.v.*

splicing *n.* (1) RNA splicing *q.v.*; (2) joining of two different pieces of DNA to form a recombinant DNA in genetic engineering.

splicing junctions the junctions between intron and exon in a primary transcript from a eukaryotic gene, which are the points at which introns are excised from the transcript and adjacent exons rejoined.

splint bone rudiments of metacarpals and some metatarsals in horses.

splintwood sapwood *q.v.*

spondyle, spondylus vertebra *q.v.*

spondylous vertebral *q.v.*

sponges *n.plu.* common name for the Porifera *q.v.*

spongicolous *a.* living in sponges.

spongin *n.* fibrous protein component of the horny sponges, such as the bath sponge.

spongioblast *n.* embryonic epithelial cell that gives rise to neuroglial cells and fibres radiating to periphery of spinal cord.

spongiocoel *n.* the cavity, or system of cavities in sponges.

spongiocyte *n.* a vacuolated cell of the zona fasciculata of the adrenal cortex.

spongiose *a.* of spongy texture.

spongophyll *n.* leaf having spongy parenchymatous tissue, without palisade tissue, between upper and lower epidermis, as in some aquatic plants.

spongy *a.* of open texture, containing air spaces.

spontaneous generation the idea that organisms could arise spontaneously from non-living material or from unlike living matter, widely held before the 19th century. It was shown to be untrue by Louis Pasteur, who demonstrated that if air was excluded from a sterilized flask of hay in water, no living organisms materialized.

spontaneous mutations mutations occurring as a result of normal cellular processes and random interaction with the environment. *cf.* induced mutations.

spools *n.plu.* minute tubes of spinnerets of spiders, from which the silk thread emerges.

spoon *n.* (1) small sclerite at base of balancers in Diptera; (2) pinion of tegula.

spoon worms Echiura *q.v.*

sporabola *n.* the trajectory of a spore discharged from a sterigma.

sporadic *a.* (1) *appl.* plants confined to limited localities; (2) *appl.* scattered individual cases of a disease; (3) *appl.* cases of spontaneously arising cancers, as opposed to familial cancers of the same type.

sporangia *plu.* of sporangium.

sporangial *a. pert.* a sporangium.

sporangiferous *a.* bearing sporangia.

sporangiform, sporangioid *a.* like a sporangium.

sporangiocarp *n.* (1) an enclosed collection of sporangia; (2) structure of asci and sterile hyphae surrounded by a peridium.

sporangiocyst *n.* (1) membrane enclosing a sporangium; (2) thick-walled resistant sporangium. *alt.* **sporangiole**.

sporangiolum *n.* small sporangium containing only one or a few spores. *plu.* **sporangiola**.

sporangiophore *n.* stalk-like structure bearing sporangia.

sporangiosorus *n.* compact group of sporangia.

sporangiospore *n.* spore, esp. if non-motile, formed in a sporangium.

sporangium *a.* cell or multicellular structure in which asexual non-motile spores are produced in fungi, algae, mosses, ferns. *plu.* **sporangia**.

spore *n.* (1) asexual spore: small, usually unicellular, reproductive body from which a new organism arises by cell division. Unlike a gamete, a spore can develop further without fusion with another cell; (2) sexual spore: haploid cell produced by meiosis in plants and fungi. *cf.* gamete.

spore case *see* theca.

spore coat envelope of a bacterial spore, external to cortex and surrounded by exosporium.

spore mother cell diploid cell which by meiosis gives rise to four haploid cells.

sporidium *n.* alternative name for basidiospore in smut and rust fungi. *plu.* **sporidia**.

sporiferous *a.* spore-bearing.

sporification *n.* formation of spores.

sporo-, -spore, -sporous word elements from Gk *sporos*, a seed, *pert.* spores and the structures that produce them.

sporoblast *n*. (1) the meristematic founder cell of a sporangium; (2) spore mother cell *q.v.*

sporocarp *n*. structure inside which spores are produced.

sporocladium *n*. a hyphal branch bearing sporangia or sporangiola, in some fungi.

sporocyst *n*. (1) one of the larval stages in the life cycle of endoparasitic flukes, which develops from the miracidium in the snail host. It has no mouth or gut and reproduces asexually to produce rediae or cercariae; (2) protective envelope of a spore in protozoans; (3) stage in spore formation preceding liberation of spores.

sporocystid *a. appl.* oocyst of sporozoans when the zygote forms sporocysts.

sporocyte *n*. spore mother cell, cell which gives rise to spores.

sporodochium *n*. mass of conidiophores and brightly coloured spores that erupts from the bark of trees and shrubs infected with coral spot fungi.

sporogenesis *n*. spore formation.

sporogenous *a*. spore-producing.

sporogonium *n*. the sporophyte generation in bryophytes. It consists of capsule and seta, develops from the fertilized ovum in the archegonium, and gives rise to spores. *a*. **sporogonial**.

sporogony *n*. (1) spore formation; (2) the formation of sporozoites or spores from a sporont in protozoans; (3) the formation of gametes from a sporont, their fusion and subsequent formation of spores and sporozoites from the zygote (sporont). *alt.* gamogony, in protozoans.

sporoid *a*. like a spore.

sporokinete *n*. a motile spore developing from the oocyst in certain protozoan blood parasites.

sporonine *n*. terpene-like substance found in the walls of spores and pollen grains.

sporont *n*. the individual or generation in sporozoan parasites which gives rise to a generation of sporozoites.

sporophore *n*. (1) a spore-bearing structure in fungi. It may be a simple sporangiophore or a complex structure such as the fruit-body of a mushroom or toadstool; (2) part of plasmodium producing spores on its free surface, in slime moulds; (3) an inflorescence.

sporophyll *n*. a leaf, or structure derived from a leaf, that bears a sporangium. It may be much modified, e.g. the stamens and carpels of a flower.

sporophyte *n*. the diploid or asexual phase in the alternation of generations in plants, in which meiosis occurs to produce a haploid gametophyte that gives rise to gametes. *cf.* gametophyte.

sporoplasm *n*. in some sporozoans, the cell released from cyst and forming an amoebula.

sporopollenin *n*. an alcohol found in the walls of spores and pollen grains, which is related to suberin and cutin but is much more durable, resulting in spores and pollen grains surviving for millions of years.

sporothallus *n*. thallus that produces spores.

sporotheca *n*. a membrane enclosing sporozoites.

Sporozoa, sporozoans *n., n.plu.* subphylum of parasitic protozoans containing many that cause disease in humans and domestic animals, and in other vertebrates and invertebrates. They include *Plasmodium*, the causal agent of human malaria, and *Eimeria*, the agent of coccidiosis in cattle, sheep and poultry. They usually have no feeding or locomotory organelles. Sporozoans have a complex life cycle, with asexual and sexual generations, sometimes in two hosts. The stage that infects new cells is a haploid sporozoite.

sporozoid *n*. a motile spore, a zoospore *q.v.*

sporozoite *n*. spore released from sporocyst of sporozoan protozoa. In malaria it is the stage in the salivary glands of the mosquito host and is transmitted to humans.

sport *n*. a somatic mutation in a plant, producing a plant or part of a plant with altered characteristics, and which can only be propagated vegetatively.

sporula, sporule *n*. small spore.

sporulation *n*. (1) the process of spore formation; (2) liberation of spores; (3) in bacteria, the segregation of the DNA to one part of the cell where it is surrounded by a spore coat, forming an endospore; (4) brood formation by multiple cell fission (in some protozoans).

spot desmosome *see* desmosome.

spp. *plu.* of species.

springtails the common name for the Collembola *q.v.*

spumaviruses *n.plu.* a subfamily of non-oncogenic retroviruses which produce a characteristic foamy appearance of the cytoplasm in the cells they infect.

spuriae *n.plu.* feathers of bastard wing.

spurious *a. appl.* a structure that appears to be something (e.g. teeth or fruit) but is not.

squalene *n.* a C_{30} hydrocarbon consisting of six isoprene units, an intermediate in cholesterol biosynthesis.

Squamata *n.* order of reptiles consisting of the snakes, lizards and amphisbaenians.

squamate *a.* scaly.

squamation *n.* the arrangement of scales on the surface of a lizard or snake.

squame *n.* (1) flattened cell of the outermost layers of the skin, consisting largely of keratin, and which eventually flakes off; (2) a scale; (3) a part arranged like a scale.

squamella *n.* a small scale or bract.

squamellate, squamelliferous *a.* having small scales or bracts.

squamelliform *n.* resembling a squamella.

squamid (1) *a.* scaly; (2) *n.* a member of the order Squamata: the lizards, amphisbaenians and snakes.

squamiferous, squamigerous *a.* bearing scales.

squamiform *a.* scale-like.

squamosal *n.* a membrane bone of vertebrate skull forming part of posterior side wall.

squamose *a.* covered in scales.

squamous *a.* (1) consisting of scales; (2) *appl.* simple epithelium of flat nucleated cells; (3) squamose *q.v.*

squamula, squamule *n.* a small scale.

squamulate *a.* having minute scales. *alt.* squamulose.

squamulose *a.* (1) squamulate *q.v.*; (2) *appl.* lichens having a foliose growth form with many loosely attached lobes to the thallus.

squarrose *a.* rough with projecting scales or rigid leaves.

squarrulose *a.* tending to become squarrose.

Src family family of tyrosine protein kinases that associate with the cytoplasmic portions of some receptor proteins and phosphorylate the stimulated receptor and other proteins. They help in transmitting the signal onwards from the receptor.

SRE serum-responsive element, a control site in various mammalian genes that is responsible for the induction of such genes in response to the stimulation of the cell by growth factors (or serum containing growth factors) and phorbol esters.

SRP signal recognition particle *q.v.*

ss-binding protein (SSB) *see* single-strand DNA-binding protein. *alt.* helix-destabilizing protein (HD protein).

ssDNA single-stranded DNA.

S-shaped curve a common type of dose–response curve, in which low concentrations of e.g. a drug have little effect until a threshold concentration is reached, after which increase in concentration produces a correspondingly increased response until another threshold concentration is reached, after which any further increase in dose produces little or no increase in response.

S-shaped growth curve type of growth curve (*q.v.*) in which the rate of population growth after an initial lag phase is exponential and then slows down as conditions for growth become progressively less favourable. Eventually an equilibrium is reached where birth rate equals death rate (stationary phase). In some circumstances this may be followed by a stage where death rate exceeds birth rate as conditions continue to deteriorate (decline phase). *alt.* sigmoid growth curve. *cf.* J-shaped growth curve.

ssp. subspecies *q.v.*

ssRNA single-stranded RNA.

stabilate *n.* population of microorganisms preserved in a stable and viable condition.

stability *n.* ability of a community or ecosystem to withstand or recover from changes or stresses imposed from outside. *see* constancy, inertia, resilience.

stabilizing selection selection that operates against the extremes of variation in a population and therefore tends to stabilize the population around the mean.

stachyose *n.* tetrasaccharide present in certain plant roots and other sources, made up of two galactose residues, one glucose and one fructose.

stachysporous *a.* bearing sporangia on the axis.

stade *n.* (1) stage in development or life history of a plant or animal; (2) interval between two successive moults. *alt.* **stadium**.

stag-horned *a.* having large branched mandibles, as the stag-beetles.

stagnicolous *a.* living or growing in stagnant water.

stalk cell (1) the barren cell of the two into which the antheridial cell of gymnosperms divides; (2) basal cell of crozier in discomycete fungi.

stamen *n.* male reproductive organ (microsporophyll) of a flower, consisting of a stalk or filament bearing an anther in which pollen is produced. *see* Fig. 18 (p. 239).

staminal *a. pert.* or derived from a stamen.

staminate *a. appl.* a flower containing stamens but not carpels.

staminiferous, staminigerous *a.* bearing stamens.

staminode *n.* (1) a leaf-like structure in some flowers, derived from a metamorphosed stamen; (2) a rudimentary, imperfect or sterile stamen.

staminody *n.* the conversion of any floral structures into stamens.

staminose *a. appl.* flowers having very obvious stamens.

stand *n.* aggregation of plants of uniform species and age, distinguishable from the adjacent vegetation.

standard *n.* (1) petal standing up at the back of a papilionaceous flower, such as pea, bean, vetch, which helps to make the flower conspicuous; (2) upstanding petal in flower of iris; (3) a unit of measurement or a material used in calibration.

standard deviation (σ, s.d., s) statistical measure of the variation around the mean $\bar{x}$ in a given set of n numbers. It is equal to the square root of the variance. In a normal distribution curve, a distance of 1 standard deviation on either side of the mean will include 68% of the cases.

standard free energy change (ΔG°) the gain or loss of free energy as one mole of reactant is converted to one mole of product under standard conditions of temperature and pressure.

standard metabolic rate the metabolic rate measured under a set of given conditions.

standard nutritional unit the unit expressing the energy available at a certain trophic level for the next level in the food chain, usually expressed as 10^6 kilocalories per hectare per year.

standing crop the biomass of a particular area or ecosystem at any specified time.

stapedius *n.* the muscle pulling the head of the stapes, one of the small bones of the middle ear.

stapes *n.* (1) stirrup-shaped innermost bone connecting incus and oval window of middle ear in mammals; (2) bone connecting eardrum and fenestra ovalis of middle ear in amphibians and reptiles.

staphylococci *n.plu.* common name for bacteria of the genus *Staphylococcus*, which are small spherical cells (cocci) often arranged in irregular clusters. Pathogenic species cause skin lesions, wound infections, food poisoning and occasionally more disseminated disease. *sing.* **staphylococcus**.

star cells Kupffer cells *q.v.*

starch *n.* polysaccharide made up of a long chain of glucose units joined by α-1,4 linkages, either unbranched (amylose) or branched (amylopectin) at a α-1,6 linkage. It is the storage carbohydrate in plants, occurring as starch granules in amyloplasts, and is hydrolysed during digestion in animals by amylases, maltase and dextrinases to glucose via dextrins and maltose. *see* Fig. 36.

starch gums dextrins *q.v.*

starch sheath an endodermis with starchy grains.

starch sugar glucose *q.v.*

starfish common name for the Asteroidea *q.v.*

star navigation learned method of navigation apparently used by many migrating songbirds, which have been shown to orient themselves according to the stars.

start codon the RNA triplet signalling the start of translation of a polypeptide chain, usually AUG.

startpoint of transcription the base pair in DNA corresponding to the first nucleotide incorporated into RNA.

star tree an initial tree made in some methods of phylogenetic tree building, in

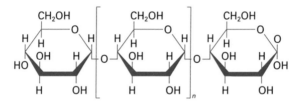

Fig. 36 Starch.

which all sequences arise from the same node, and which is subsequently refined.

stasimorphy *n.* deviation in form due to arrested growth.

stasis *n.* (1) in evolution, the apparent lack of major evolutionary change over long periods of time in any given lineage, as seen in the fossil record. *see* punctuated equilibrium, stabilizing selection; (2) stoppage or retardation in growth or development, or of movement of animal fluids.

static *a.* (1) *pert.* system at rest or in equilibrium, *appl.* postural reactions; (2) *appl.* receptors: proprioceptors, as otoliths and semicircular canals.

stationary phase the third stage in growth of a bacterial colony when multiplication slows down and virtually ceases, due to exhaustion of nutrients.

stato-acoustic *a. pert.* sense of balance and of hearing, *appl.* 8th cranial or cochlear nerves.

statoblast *n.* specialized asexual bud or 'winter egg' of some Bryozoa, enclosed in a chitinous shell. It remains dormant during winter and develops into a new colony in spring.

statocone *n.* a minute structure contained in a statocyst.

statocyst *n.* (1) vesicle lined with sensory cells and containing minute calcareous particles, either free in fluid or enclosed in cells, that move under gravity. Present in many invertebrates and concerned with perception of gravity; (2) statocyte *q.v.*

statocyte *n.* a cell containing statoliths and probably acting as a georeceptor, such as a root cap cell containing granules such as starch grains. *alt.* statocyst.

statokinetic *a. pert.* maintenance of equilibrium and associated movements.

statolith *n.* (1) particle of inorganic matter in fluid of statocysts (organs of balance and gravity detection) whose movement under gravity is detected by sensory cells. *alt.* otolith; (2) a cell inclusion, as oil droplet, starch grain, or crystal, which changes its intracellular position under the influence of gravity.

STATs signal transducers and activators of transcription. Cytoplasmic proteins that associate with the tails of some cytokine receptors and form transcription factors when the receptor is activated.

staurophyll *n.* a leaf having palisade or other compact tissue throughout.

steapsin *n.* a lipolytic enzyme of the digestive juice of many animals, including the pancreatic juice of vertebrates. EC 3.1.1.1, *r.n.* triacylglycerol lipase.

stearic acid, stearate widely distributed, saturated fatty acid, esp. in animals.

steatogenesis *n.* production of lipid material.

steganopodous *a.* having feet completely webbed.

stegocarpic, stegocarpous *a.* having a capsule with operculum and periostome, *appl.* mosses.

stelar *a. pert.* stele.

stelar parenchyma pith *q.v.*

stele *n.* column of primary vascular tissues – primary phloem and primary xylem – and pith if present, extending throughout the primary plant body.

stellar stellate *q.v.*

stellate *a.* star-shaped, radiating. *alt.* stellar, asteroid.

stellate cell (1) type of interneuron found in cerebellum with cell body in the molecular layer, acts on Purkinje cells; (2) Kupffer cell *q.v.*

Stelleroidea *n.* class of echinoderms containing the starfish (Asteroidea) and brittle stars (Ophiuroidea).

stelliform stellate *q.v.*

STEM scanning-tunnelling electron microscope *q.v.*

stem *n.* the main axis of a vascular plant, bearing buds and leaves or scale leaves, and reproductive structures (e.g. flowers), usually borne above ground (but *see* rhizome), and having a characteristic arrangement of vascular tissue.

stem cell undifferentiated cell in embryo or adult of an animal or plant which can both undergo unlimited division and can also give rise to one or several different cell types. In e.g. adult mammals, some tissues, e.g. blood and skin, are continually being renewed by the differentiation of stem cells. Stem cells are also present in skeletal muscle and in a few parts of the brain. *see also* embryonic stem cells, multipotent, totipotent.

stem-cell niche the cellular environment in e.g. skin or bone marrow, which keeps stem cells in the self-renewing state and guides their differentiation into specialized cells.

stemma *n.* (1) ocellus of arthropods; (2) a tubercle bearing an antenna. *plu.* **stemmata, stemmas**.

stenobaric *a. appl.* animals adaptable only to small differences in pressure or altitude.

stenobathic *a.* having a narrow vertical range of distribution.

stenobenthic *a. pert.*, or living within a narrow range of depth of the sea bottom. *cf.* eurybenthic.

stenochoric *a.* having a narrow range of distribution.

stenocyst *n.* one of the auxiliary cells in leaves of certain mosses.

stenoecious *a.* having a narrow range of habitat selection.

stenohaline *a. appl.* organisms adaptable to a narrow range of salinity only.

stenohygric *a. appl.* organisms adaptable to a narrow variation in atmospheric humidity.

Stenolaemata *n.* class of Bryozoa.

stenomorphic *a.* (1) dwarfed; (2) smaller than typical form, owing to cramped habitat.

stenonatal *a.* with a very small thorax, as worker insect.

stenopetalous *a.* with narrow petals.

stenophagous, stenophagic *a.* subsisting on a limited variety of food.

stenophyllous *a.* narrow-leaved.

stenopodium *n.* a crustacean limb in which the protopodite bears distally both exopodite and endopodite.

stenosepalous *a.* with narrow sepals.

stenosis *n.* narrowing or constriction of a tubular structure, as of a pore, duct or vessel.

stenostomatous *a.* narrow-mouthed.

stenothermic *a. appl.* organisms adaptable to only slight variations in temperature.

stenotopic *a.* having a restricted range of geographical distribution.

stenotropic *a.* having a very limited adaptation to varied conditions.

stephanion *n.* point on skull where superior temporal ridge is crossed by coronal suture.

steppe *n.* dry and treeless grassland covering extensive areas of Asia.

stercome *n.* excreted waste material of Sarcodina, in masses of brown granules.

stercoral *a. appl.* a dorsal pocket or sac of proctodaeum in spiders.

stereid *n.* (1) a lignified parenchyma cell with pit canals; (2) stone cell *q.v.*

stereid bundles bands or bundles of sclerenchymatous fibres.

stereocilium *n.* (1) projection on hair cell of cochlea, stiffened by a permanent cytoskeleton and which is involved in sensing vibration caused by sound waves; (2) similar non-motile projections on other cells. *plu.* **stereocilia**.

stereognosis *n.* the ability to recognize three-dimensional shapes by touch, the sense that appreciates size, weight or shape of an object. *a.* **stereognostic**.

stereoisomer *n.* any of two or more compounds with the same atomic composition but differing in their structural configuration.

stereokinesis *n.* movement, or inhibition of movement, in response to contact stimuli.

stereome *n.* (1) sclerenchymatous and collenchymatous masses along with hardened parts of vascular bundles forming supporting tissue in plants; (2) the thick-walled elongated cells of the central cylinder in mosses.

stereopsis *n.* the ability to perceive depth and see objects as three dimensional, using the differences in information received from the two eyes. *alt.* stereoscopic vision.

stereospecificity *n.* of enzymes, specificity for only one of several possible stereoisomers of a substrate.

stereospondylous *a.* having vertebrae each fused into one piece.

stereotaxis, stereotaxy *n.* the response of an organism to the stimulus of contact with a solid, such as the tendency of some organisms to attach themselves to solid objects or to live in crannies or tunnels.

stereotropism *n.* growth movement in plants associated with contact with a solid object, as the tendency for stems and tendrils of climbers to twine around a support.

stereotyped behaviour *see* fixation.

sterigma *n.* a short outgrowth arising from basidium in basidiomycete fungi and which develops the basidiospore at its tip. *plu.* **sterigmata**.

sterile *a.* (1) incapable of reproduction; (2) free of living organisms, esp. microorganisms, aseptic; (3) axenic *q.v.*

sterilize *v.* (1) to render incapable of reproduction, of animals; (2) to render incapable of conveying infection, of material containing microorganisms.

sternal *a. pert.* sternum, or sternite.

sternebrae *n.plu.* divisions of a segmented breast bone or sternum.

sternellum *n.* a sternal sclerite of insects, esp. sclerite behind antesternite.

sternite *n.* a ventral plate of an arthropod segment.

sternobranchial *a. appl.* vessel conveying blood to gills, in some crustaceans.

sternoclavicular *a. appl.* and *pert.* articulation between sternum and clavicle.

sternohyoid *a. appl.* a muscle between back of manubrium of sternum and hyoid.

sternokleidomastoid *a. appl.* an oblique neck muscle stretching from sternum to mastoid process.

sternopericardial *a. appl.* ligament connecting dorsal surface of sternum and fibrous pericardium.

sternopleurite *n.* a thoracic sternite formed by union of episternum and sternum in insects.

sternoscapular *a. appl.* a muscle connecting sternum and scapula.

sternothyroid *a. appl.* muscle connecting manubrium of sternum and thyroid cartilage.

sternotribe *a. appl.* flowers with fertilizing elements so placed as to be brushed by sternites of visiting insects.

sternoxiphoid *a. appl.* plane through junction of sternum and xiphoid cartilage.

sternum *n.* (1) breastbone of vertebrates; (2) the ventral plate of exoskeleton of typical arthropod segment; (3) the ventral plates of exoskeleton of a thoracic segment in insects.

steroid *n.* any of a large group of complex polycyclic lipids with a hydrocarbon nucleus and various substituents, synthesized from acetyl-CoA via isoprene, squalene and cholesterol, and which include bile acids, sterols, various hormones, cardiac glycosides and saponins.

steroid hormones oestrogens, testosterone and its derivatives, glucocorticoids, mineralocorticoids, and the insect hormone ecdysterone.

steroid receptor *see* nuclear receptor.

sterol *n.* any steroid alcohol, ubiquitous in plants, animals and fungi as components of the cell membranes and including ergosterol (a typical fungal sterol), cholesterol (in animal cells), and phytosterol (in plants).

STH somatotropin (growth hormone) *q.v.*

stichic *a.* in a row parallel to long axis.

stichidium *n.* a tetraspore receptacle in some red algae.

stick insects common name for many of the Phasmida *q.v.*

sticky ends cohesive ends *q.v.*

stigma *n.* (1) (*bot.*) the upper portion of the carpel which receives the pollen and which is usually connected to the ovary by an elongated structure, the style. *plu.* **stigmas**. *see* Fig. 18 (p. 239); (2) (*zool.*) various pigmented spots and markings, such as the coloured wing spot of some butterflies and other insects. *plu.* **stigmata**; (3) a pigmented spot near base of flagellum in photosynthetic euglenoids, involved in photoreception and phototaxis.

stigmasterol *n.* a plant sterol, also present in milk. A deficiency in the diet causes muscular atrophy and calcium phosphate deposits in muscles and joints.

stigmata *plu.* of stigma.

stigmatic *a. pert.* a stigma.

stilt roots buttress roots *q.v.*

stimulus *n.* an agent that causes a reaction or change in an organism or any of its parts.

stimulus–response theory theoretical explanation of classical conditioning which proposes that conditional responses occur because of reinforcement by being followed by a reward.

stimulus substitution theory theory proposed by Pavlov to explain classical conditioning, in which it is held that the animal comes to associate the unconditional and conditional stimuli and the conditional stimulus becomes a substitute for the unconditional stimulus. *cf.* stimulus–response theory.

stinging cell nematocyst *q.v.*

stinkhorns common name for fungi of the Phallales, an order of Gasteromycetes having a foetid odour to the gleba and a mature fruit body resembling a phallus.

stipe *n.* (1) stalk, esp. of mushrooms and other stalked fungi; (2) stalk of seaweeds; (3) stem of a fern frond.

stipella *n.* the stipule of a leaflet in a compound leaf.

stipes *n.* (1) the stalk of a stalked eye, as in crustaceans; (2) part of first segment of 1st maxilla of insects; (3) distal portion of embolus in spiders.

stipiform *a.* resembling a stalk or stem.

stipitate *a.* stalked.

stipites *n.plu.* small exoskeletal plates anterior to mentum forming part of maxilla in some arthropods.

stipitiform stipiform *q.v.*

stipular *a.* like, *pert.*, or growing in place of stipules.

stipulate *a.* having stipules.

stipule *n.* (1) one of two leaf-like or bract-like outgrowths at the base of a leaf-stalk, sometimes modified as spines or tendrils; (2) paraphyll *q.v.*; (3) pin feather *q.v.*

stipuliferous *a.* having stipules.

stipuliform *a.* in the form of a stipule.

STM short-term memory *q.v.*

stochastic *a. appl.* a process in which there is an element of chance or randomness.

stock *n.* (1) one or a group of individuals initiating a line of descent; (2) (*bot.*) stem of tree or bush receiving bud or scion in grafting; (3) the perennial part of a herbaceous plant; (4) (*zool.*) an asexual zooid which produces sexual zooids of one sex by gemmation, as in polychaetes; (5) livestock.

stocking rate the number of a particular kind of grazing animal feeding on a given area of grassland.

stoichiometry *n.* description of a metabolic reaction or cycle in terms of net proportions of molecules of each reactant consumed and produced. *a.* **stoichiometric**.

stolon *n.* (1) creeping plant stem or runner capable of developing rootlets and stem and ultimately forming a new individual; (2) hypha connecting two bunches of rhizoids in fungi; (3) similar structures in other organisms.

stolonate *a.* having, resembling, or developing from a stolon.

stoloniferous *a.* bearing a stolon or stolons.

stoma *n.* (1) *sing.* of stomata *q.v.*; (2) part of alimentary canal between mouth opening and oesophagus in nematodes.

stomach *n.* the large pouch of the intestine between oesophagus and intestines in vertebrates, and the corresponding part or entire digestive cavity in invertebrates.

stomal, stomatal, stomatic *a. pert.* or like a stoma.

stomata *n.plu.* (1) minute openings in epidermis of aerial parts of plants, esp. on undersides of leaves, through which air and water vapour enters the intercellular spaces, and through which carbon dioxide and water vapour from respiration is released. Stomata can be opened or closed by changes in turgor of the two guard cells that surround the central pore; (2) any small openings or pores in various structures. *a.* **stomatal**. *sing.* **stoma**.

stomatal index the ratio between number of stomata and number of epidermal cells per unit area.

stomate, stomatiferous *a.* possessing stomata.

stomatogastric *a.* (1) *pert.* mouth and stomach; (2) *appl.* visceral system of nerves supplying anterior part of alimentary canal; (3) *appl.* recurrent nerve from frontal to stomachic ganglion, in insects.

stomatogastric ganglion nerve centre in crustaceans situated on surface of stomach and controlling the movements of the teeth of the gastric mill.

stomatogenesis *n.* the formation of a mouth, as in ciliates.

stomions *n.plu.* pores or ostia in body wall of developing sponge.

stomium *n.* (1) group of thin-walled cells in fern sporangium where rupture of mature capsule takes place; (2) slit of dehiscing anther.

stomodaeal canal in Ctenophora, a canal given off by each perradial canal and situated parallel to the stomodaeum.

stomodaeum *n.* anterior portion of the alimentary canal, lined with ectodermally derived epithelium. *alt.* foregut.

stone canal cylindrical canal extending from madreporite to near mouth border in echinoderms.

stone cells short, isodiametric sclereids, as found in the flesh of a pear, giving it a gritty texture.

stoneflies *n.plu.* common name for the Plecoptera *q.v.*

stone fruit drupe *q.v.*

stoneworts *n.plu.* common name for the Charophyta *q.v.*

stony corals a group of colonial coelenterates, typified by the reef-building corals, in which individual polyps are embedded in a matrix of calcium carbonate and connected by living tissue.

stop codon termination codon *q.v.*

stop-transfer sequence amino-acid sequence present in membrane proteins that stops translocation of the polypeptide chain across the membrane of the endoplasmic reticulum.

storey *n.* layer of a given height in a plant community.

storied *a.* (1) arranged in horizontal rows on tangential surfaces, *appl.* axial cells and ray cells of wood cambium; (2) *appl.* cork in monocotyledons with suberized cells in radial rows.

stotting *n.* warning behaviour in some gazelles, which bound away with a stiff-legged gait and tails raised, displaying a white rump.

strain *n.* (1) pure-breeding variant line of a species of domesticated animal or cultivated plant; (2) subspecific group whose members are not sufficiently different genetically from the rest of the species to form a variety; (3) (*microbiol.*) in bacteria and other microorganisms, a population of genetically identical individuals with some characteristic differentiating them from other strains of the same species.

Stramenopila, stramenopiles *n.*, *n.plu.* proposed kingdom containing the phyla Hyphochytriomycota (hyphochytrids), Labyrinthulomycota (slime nets), Oomycota (oomycetes), Phaeophyta (brown algae), Chrysophyta (golden algae) and Bacillariophyta (diatoms) (and Haptophyta (Prymnesiophyta) and Xanthophyta in some classifications). A very similar grouping has also been named the Chromista (*q.v.*).

β-strand one of the polypeptide strands in a β-sheet.

strangulated *a.* (1) constricted in places; (2) contracted and expanded irregularly.

strata *plu.* of stratum.

stratification *n.* (1) arrangement in layers; (2) superimposition of layers of epithelial cells; (3) vertical grouping within a community or ecosystem; (4) differentiation of horizontal layers of soil.

stratified epithelium epithelium several cell layers thick.

stratiform *a.* layered.

stratum *n.* (1) a layer, as of cells, or of tissue; (2) a group of organisms inhabiting a vertical division of an area; (3) vegetation of similar height in a plant community, as trees, shrubs, herbs, mosses; (4) a layer of rock. *plu.* **strata**.

stratum corneum cornified layer *q.v.*

stratum germinativum basal layer of epidermis, with actively dividing cells.

stratum granulosum layer of small cells developing from the basal layer or stratum germinativum of mammalian epidermis.

stratum lucidum layer of cells becoming keratinized in epidermis of skin.

stratum Malpighii Malpighian layer *q.v.*

stratum spinosum layer of prickle cells in epidermis.

stratum transect a profile of vegetation, drawn to scale and intended to show the heights of plant shoots. *alt.* profile transect.

Strepsiptera *n.* an order of small insects with incomplete metamorphosis, commonly called stylopids, whose larvae and females are parasites of other insects and the males free-living. Forewings are halteres, hindwings are fan-shaped.

streptococci *n.plu.* any bacteria of the genus *Streptococcus*, which are small spherical cells (cocci) forming long chains. Many are harmless commensals living in the throat and gut, but some are human and animal pathogens, causing tonsilitis, scarlet fever and tissue destruction. *sing.* **streptococcus**.

streptolydigins *n.plu.* class of antibiotics, which inhibit bacterial transcription by interacting with the β subunit of RNA polymerase.

streptomycetes *n.plu.* the family Streptomycetaceae of the Gram-positive Bacteria, filamentous prokaryotic microorganisms widespread in soil and water, characterized by formation of a permanent mycelium and reproduction by means of conidia. Some species produce antibiotics, including streptomycin, chloramphenicol and the tetracyclines.

streptomycin *n.* trisaccharide antibiotic synthesized by the streptomycete *Streptomyces griseus*, which inhibits bacterial protein synthesis by interfering with the binding of formylmethionyl-tRNA to ribosomes, and also by causing misreading of mRNA. It is used to treat tuberculosis in humans, and downy mildew on hops.

Streptoneura prosobranch *q.v.*

streptoneurous *a.* having visceral cord twisted, forming a figure of eight, as certain gastropods.

streptonigrin *n.* antibiotic synthesized by the streptomycete *Streptomyces flocculus*, which causes chromosome breakage.

streptostylic *a.* having quadrate in movable articulation with squamosal.

stress-activated channel ion channel that is opened in response to mechanical force.

stress fibres bundles of actin filaments in cultured cells, lying parallel with the substrate surface.

stress responses the various responses made by organisms or cells to heat shock, cold shock and other stresses, such as drought.

stretch receptor sensory structure that monitors the degree of stretch of a muscle, e.g. muscle spindle in mammals, consisting of a specialized, modified muscle cell innervated by sensory neurons.

stretch reflex the contraction of a muscle in response to stretching of the muscle.

stria *n.* a narrow line, band, groove, streak or channel.

striae of Retzius Retzius' striae *q.v.*

striate, striated *a.* marked by narrow parallel lines or grooves.

striate cortex also known as visual area I or area 17, part of visual cortex concerned with the initial processing of visual information.

striated muscle muscle tissue composed of transversely striped (striated) fibres formed from the fusion of many individual muscle cells, and which makes up the muscles attached to the skeleton. It is under the control of the voluntary nervous system.

striate region the region of the brain that contains the basal ganglia, so-called because of its striped appearance.

striatum corpus striatum *q.v.*

stridulating organs special structures on various parts of the body of certain insects such as grasshoppers, crickets and cicadas, which produce the characteristic 'song' of these insects.

stridulation *n.* the characteristic sound made by grasshoppers, crickets and cicadas.

striga *n.* (1) band of stiff upright hairs or bristles; (2) bristle-like scale. *plu.* **strigae**.

strigate *a.* bearing strigae.

Strigiformes *n.* the owls, an order of mainly nocturnal, short-necked, large-headed birds of prey.

strigilis *n.* a structure for cleaning antennae, at junction of tibia and tarsus on 1st leg of certain insects.

strigillose *a.* minutely strigose.

strigose *a.* (1) covered with stiff hairs; (2) ridged; (3) marked by small furrows.

stringency *n.* in DNA hybridization reactions, refers to the degree to which DNAs of differing sequence will form duplexes, conditions of low stringency (e.g. low temperature) allowing duplex formation between non-identical related DNAs, conditions of high stringency (e.g. high temperature) allowing duplex formation only between identical DNAs.

stringent factor enzyme in bacteria which catalyses the formation of ppGpp and pppGpp in conditions of amino acid starvation and is involved in the stringent response.

stringent response phenomenon seen in bacterial cells under conditions of amino acid starvation, when many cellular functions such as general RNA synthesis are shut down.

striola *n.* a fine narrow line or streak.

striolate *a.* finely striped.

striped muscle striated muscle *q.v.*

strobila chain of proglottids of tapeworms.

strobilaceous *a. pert.* or having strobili.

strobilation, strobilization *n.* reproduction by separating off successive segments of the body, as in some jellyfish and in tapeworms.

strobile strobilus *q.v.*

strobili *plu.* of strobilus.

strobiliferous *a.* producing strobili.

strobiliform, strobiloid *a.* resembling or shaped like a strobilus or cone.

strobilus *n.* (1) (*bot.*) cone-shaped assemblage of sporophylls in horse-tails, club mosses and gymnosperms. *alt.* cone; (2) in flowering plants, a spike formed by persistent membranous bracts, each having a pistillate flower; (3) (*zool.*) stage in development of some jellyfish, a sessile polyp-like form that separates off a succession of disc-like embryos by segmentation. *plu.* **strobili**.

stroma *n.* (1) in ovary, soft vascular framework in which ovarian follicles are embedded; (2) (*bot.*) in chloroplasts, the colourless material enclosed by the inner membrane and in which the grana are embedded, and in which carbon dioxide fixation takes place during photosynthesis; (3) the non-pigmented part of other plastids; (4) (*mycol.*) tissue of hyphae, or of fungal cells and host tissue, in or upon

which a spore-bearing structure may be produced *plu.* **stromata**.

stromate *a.* having, or being within or upon, a stroma, *appl.* fruit bodies of fungi.

stromatic *a. pert.*, like, in form or nature of, a stroma.

stromatolite *n.* layered structure, sometimes of considerable size, formed in certain warm shallow waters by mats of cyanobacteria mixed with other microorganisms and deposits of calcium carbonate. Fossils of similar structure have been found in Archaean rocks, indicating the presence of life at that time.

strombuliferous *a.* having spirally coiled organs or structures.

strombuliform *a.* spirally coiled.

strongyle *n.* a type of nematode larva.

strophanthidin *n.* a digitalis glycoside whose effect on the heart is mediated by inhibition of Na^+-K^+ ATPase.

strophanthin *n.* a glycoside with effects on the nervous system, obtained from various plants of the family Apocynaceae and used as a tropical arrow poison.

strophiolate *a.* having excrescences around hilum, *appl.* seeds.

strophiole *n.* one of the small excrescences arising from various parts of seed testor, arising after fertilization.

strophotaxis *n.* twisting movement or tendency, in response to an external stimulus.

structural colours colours of fish skin, insect wings, etc., that are not due to pigment but to surface structure, e.g. reflecting layers, plates of guanine crystals.

structural gene a gene that codes for an enzyme or other protein required for a cell's structure or metabolism, or for tRNA or rRNA. *cf.* regulatory genes.

θ-structure intermediate structure formed in the replication of circular DNA molecules.

struma *n.* a swelling on a plant organ.

strumiform *a.* cushion-like.

strumose, strumulose *a.* having small cushion-like swellings.

Struthioniformes *n.* an order of flightless birds including the ostriches.

strut roots buttress roots *q.v.*

strychnine *n.* an alkaloid produced from seeds of *Strychnos* species and some other plants, a mammalian poison because of its effects on the nervous system.

Stuart factor Factor X *q.v.*

stupeous, stupose *a.* (1) like tow; (2) having a tuft of matted filaments.

STX saxitoxin *q.v.*

stylar *a. pert.* style.

stylate *a.* having a style.

style *n.* (1) (*bot.*) portion of carpel connecting stigma and ovary, often slender and elongated. *see* Fig. 18 (p. 239); (2) (*zool.*) translucent revolving rod of protein and carbohydrate in stomach of bivalve molluscs which contains digestive enzymes; (3) bristle-like process on clasper of male insect; (4) arista *q.v.*; (5) embolus of spider.

style sac a tubular gland in some molluscs which secretes the crystalline style.

stylet *n.* (1) slender, hollow mouthpart, present in two pairs in aphids, through which they suck sap; (2) any of various small, sharp appendages, used for stinging or piercing prey.

stylifer *n.* portion of clasper that carries the style.

styliferous *a.* (1) bearing a style; (2) having bristly appendages.

styliform *a.* prickle- or bristle-shaped.

styloconic *a.* having terminal peg on conical base, *appl.* type of olfactory sensilla in insects.

styloglossus *n.* a muscle connecting styloid process and side of tongue.

stylohyal *n.* (1) distal part of styloid process of temporal bone; (2) a small interhyal between hyal and hyomandibular.

stylohyoid *a.* (1) *appl.* a ligament attached to styloid process and lesser cornu of hyoid; (2) *appl.* a muscle; (3) *appl.* a branch of facial nerve.

styloid (1) *a.* pillar-like, *appl.* processes of temporal bone, fibula, radius, ulna; (2) *n.* a columnar crystal.

stylomandibular *a. appl.* ligamentous band extending from styloid process of temporal bone to angle of lower jaw.

stylomastoid *a. appl.* foramen between styloid and mastoid processes, also an artery entering that foramen.

stylopharyngeus *n.* a muscle extending from the base of styloid process downwards along side of pharynx to thyroid cartilage.

stylopids common name for the Strepsiptera *q.v.*

stylosome *n.* tube-like structure used for feeding by the parasitic larvae of trombiculid mites.

stylospore *n.* a spore borne on a stalk.

stylostegium *n.* the inner corolla of milkweed plants.

stylus *n.* (1) style *q.v.*; (2) stylet *q.v.*; (3) simple pointed spicule; (4) molar cusp; (5) pointed process.

sub- prefix from L. *sub*, under, signifying under, below (as in anatomical terms, e.g. subauricular), less than (e.g. subthreshold), not quite, nearly, or somewhat (esp. in descriptions of plant and animal parts, e.g. subdentate, slightly toothed (of leaves), subcarinate, somewhat keel-shaped). In classification it indicates a group just below the status of the taxon following it, as in subclass. Terms whose meanings can be unambiguously obtained by looking up the main part of the word are not given below; e.g. for subreniform (somewhat kidney shaped) look up reniform (kidney-shaped).

subabdominal *a.* nearly in the abdominal region.

subaerial *a.* growing just above the surface of the ground.

subalpine *a. appl.* ecological zone just below the timber line in high mountains, and to the plants and animals that live there.

subanconeous *n.* small muscle extending from triceps to elbow.

subapical *a.* nearly at the apex.

subarachnoid space cavity filled with cerebrospinal fluid between arachnoid membrane and pia mater surrounding brain and spinal cord.

subarborescent *a.* somewhat like a tree.

subarcuate *a. appl.* a blind fossa which extends backwards under superior semicircular canal in infant skull.

subatrial *a.* below the atrium, *appl.* longitudinal ridges on inner side of metapleural folds, uniting to form ventral part of atrium, in development of amphioxus.

subauricular *a.* below the ear.

subaxillary *a. appl.* outgrowths just beneath the axil.

sub-basal *a.* situated near the base.

sub-branchial *a.* under the gills.

sub-bronchial *a.* below the bronchial tubes.

subcalcarine *a.* under the calcarine fissure, *appl.* lingual gyrus of brain.

subcallosal *a. appl.* a gyrus below corpus callosum.

subcapsular *a.* under a capsule.

subcardinal *a. appl.* a pair of veins between mesonephroi.

subcaudal *a.* beneath or on the ventral side of the tail.

subcaudate *a.* having a small tail-like process.

subcaulescent *a.* borne on a very short stem.

subcellular *a. appl.* functional units or organelles within a cell.

subchela *n.* in some arthropods, a prehensile claw of which the last joint folds back on the preceding one.

subchelate *a.* (1) having subchelae; (2) having imperfect chelae.

subchordal *a.* situated under the notochord.

subcingulum *n.* the lower lip part of cingulum or girdle of rotifers.

subclass *n.* taxonomic grouping between class and order, e.g. Theria (mammals that have live-born young) in the class Mammalia.

subclavian *a.* below the clavicle, *appl.* artery, vein, nerve.

subclavius *n.* a small muscle connecting 1st rib to clavicle.

subclimax *n.* stage in plant succession preceding the climax, which persists because of some arresting factor such as fire or human activity.

subclone *n.* part of a cloned DNA recloned into another vector.

subcoracoid *a.* below the coracoid.

subcorneous *a.* (1) under a horny layer; (2) slightly horny.

subcortical *a.* beneath the cortex or cortical layer.

subcosta *n.* an auxiliary vein joining costa of insect wing.

subcostal *a.* beneath ribs.

subcoxa *n.* basal ring of arthropod segment, which articulates with coxa of leg.

subcrureus *n.* muscle extending from lower femur to knee.

subcubical *a. appl.* cells not quite so long as broad.

subcutaneous *a.* (1) under the skin; (2) *appl.* parasites living just under skin; (3) *appl.* fat under the skin.

subcuticula *n.* epidermis beneath cuticle, as in nematodes.

subcuticular *a.* under the cuticle, epidermis or outer skin.

subcutis *n.* (1) a loose layer of connective tissue between corium and deeper tissues of dermis; (2) (*mycol.*) inner layer of cutis of agaric fungi, under the epicutis.

subdermal *a.* (1) beneath the skin; (2) beneath the dermis.

subdominant *n.* species that may seem more abundant than the true dominant species in a climax plant community at particular times of the year, or which is more abundant than the dominant species but occurs at a lower frequency.

subdorsal *a.* situated almost on the dorsal surface.

subdural *a. appl.* the space separating the spinal dura mater from arachnoid.

subepicardial *a. appl.* areolar tissue attaching visceral layer of pericardium to muscular wall of heart.

subepithelial *a.* (1) below epithelium, *appl.* a plexus of cornea; (2) *appl.* endothelium: Débove's membrane *q.v.*

suber *n.* cork tissue.

subereous *a.* of corky texture.

suberic *a. pert.* or derived from cork.

suberiferous *a.* cork-producing.

suberification *n.* conversion into cork tissue.

suberin *n.* waxy substance developed in thickened plant cell wall, characteristic of cork tissues.

suberization *n.* modification of plant cell walls due to suberin deposition.

suberose *a.* (1) with corky waterproof texture; (2) somewhat gnawed.

subfornical organ one of the circumventricular organs, lying under the fornix, that mediate communication between cerebrospinal fluid and the brain.

subgalea *n.* part of maxilla at base of stipes, of insects.

subgenital *a.* below the reproductive organs.

subgenomic RNA viral RNA produced by transcription of part of a genomic RNA, and which serves as an mRNA.

subgerminal *a.* below the germinal disc, *appl.* cavity.

subglenoid *a.* below the glenoid cavity.

subglossal *a.* beneath the tongue.

subhyaloid *a.* beneath hyaloid membrane or fossa of eye.

subhymenium *n.* layer of small cells between trama and hymenium in gill of agarics.

subhyoid *a.* below hyoid bone(s) at base of tongue in mammals.

subiculum *n.* brain region underlying the hippocampus, and through which come the main inputs to the hippocampus from the entorhinal cortex. It is considered as part of the hippocampal formation. *alt.* hippocampal gyrus.

subimago *n.* winged stage between pupa and full adult (imago) in some insects, e.g. mayflies and other Ephemoptera, which undergoes a further moult.

subinguinal *a.* situated below a horizontal line at level of great saphenous vein termination, *appl.* lymph glands.

subjugal *a.* below jugal or cheek bone.

subjugular *a. appl.* a ventral fish fin nearly far enough forward to be jugular.

sublethal *a.* not causing death directly, but having cumulative deleterious effects.

subliminal *a. appl.* stimuli that are not strong enough to evoke a sensation.

sublingua *n.* a double projection or fold beneath tongue in some mammals.

sublingual *a.* beneath the tongue.

sublittoral *a.* (1) below littoral, *appl.* the shallow water zone of the sea from the extreme low-tide level to a depth of around 200 m. *see* Fig. 34 (p. 605); (2) zone of a lake too deep for rooted plants to grow.

sublobular *a. appl.* veins at base of lobules in liver.

sublocus *n.* a part of a complex genetic locus that acts as an individual locus in some genetic tests.

submandibular submaxillary *q.v.*

submarginal *a.* placed nearly at margin.

submarginate *a. appl.* a bordering structure near a margin.

submaxilla mandible *q.v.*

submaxillary *a.* beneath lower jaw, *appl.* duct, ganglion, gland, triangle.

submedian *a. appl.* tooth or vein next to median.

submental *a.* (1) beneath the chin, *appl.* artery, glands, triangle, vibrissae; (2) *pert.* submentum.

submentum *n.* part of labium of insects.

submersed *a. appl.* plants growing entirely under water.

submetacentric (1) *a. appl.* chromosomes with the centromere nearer one end than the other giving arms of unequal length; (2) *n.* a submetacentric chromosome.

submission *n.* the behaviour of a losing animal in a conflict where it takes up a submissive posture to prevent further attack.

submitochondrial particles small particles produced by sonication of mitochondria, having the inner membrane on the outside, used to study respiratory chain function and arrangement in the membrane.

submucosa *n.* a layer of gut wall between the mucosa and external muscular coat, composed of connective tissue and accommodating blood vessels, nerves, Meissner's plexus and some glands.

subneural *a.* (1) *appl.* blood vessel in annelid worms; (2) *appl.* gland and ganglion of nervous system in tunicates.

subnotochord *n.* a rod ventral to true notochord.

suboccipital *a. appl.* muscles, nerve, triangle, under occipitals of skull.

suboesophageal *a.* below the gullet, *appl.* anterior ganglion of ventral nerve cord in invertebrates.

subopercular *a.* under operculum of fishes, or shell of molluscs.

suboperculum *n.* a membrane bone of operculum in fishes.

suboptic *a.* below the eye.

suboral *a.* below the mouth.

suborbital *a.* below the orbit of the eye.

subparietal *a.* beneath parietals, *appl.* sulcus which is lower boundary of parietal lobe of brain.

subpedunculate *a.* resting on a very short stalk.

subperitoneal *a. appl.* connective tissue under the peritoneum.

subpessular *a.* below the pessulus of syrinx, *appl.* air sac.

subpetiolar, subpetiolate *a.* (1) within petiole, *appl.* bud so concealed; (2) almost sessile.

subpharyngeal *a.* (1) below the throat; (2) *appl.* gland or endostyle beneath pharynx, with cells containing iodine in ammocoetes.

subphrenic *a.* below the diaphragm.

subphylum *n.* taxonomic grouping between phylum and class, e.g. Vertebrata (in the phylum Chordata).

subpial *a.* under the pia mater.

subpleural *a.* beneath inner lining of thoracic wall.

subpubic *a.* below the pubic region, *appl.* arcuate ligament.

subpulmonary *a.* below the lungs.

subradicate *a.* to have a slight downward extension at base, as of stipe.

subradius *n.* in radially symmetrical animals, a radius of the 4th order, that between adradius and perradius or between adradius and interradius.

subradular *a.* *appl.* organ containing nerve endings situated at anterior end of odontophore.

subretinal *a.* beneath the retina.

subrostral *a.* below the beak or rostrum, *appl.* a cerebral fissure.

subsartorial *a.* *appl.* plexus under sartorius muscle of thigh.

subscapular *a.* beneath the scapula, *appl.* artery, muscles, nerves.

subsclerotic *a.* (1) beneath sclera; (2) between sclerotic and choroid layers of eye.

subscutal *a.* under a scutum.

subsere *n.* plant succession on a denuded area, secondary succession.

subsong *n.* the first attempts at song by a young bird, which resembles the adult song but is imprecise and lacks some elements, with ill-defined phrasing and lack of tonal purity.

subspecies (ssp.) *n.* a taxonomic term usually meaning a group consisting of individuals within a species having certain distinguishing characteristics separating them from other members and forming a breeding group, but which can still interbreed with other members of the species. *alt.* variety.

subspinous *a.* tending to become spiny.

substance P peptide found in gut tissue and in central nervous system, thought to act as a neurotransmitter and be involved in pain pathways.

substantia adamantina enamel of teeth.

substantia alba white matter of brain and spinal cord.

substantia eburnea dentine *q.v.*

substantia gelatinosa the gelatinous neuroglia in spinal cord.

substantia grisea grey matter of spinal cord.

substantia nigra semicircular layer of grey matter in brainstem, contains cell bodies of dopaminergic neurons.

substantia ossa cement of teeth.

substantia reticularis reticular formation *q.v.*

substantia spongiosa cancellous tissue of bone.

substantive variation change in actual constitution or substance of parts. *cf.* meristic variation.

substernal *a.* below the sternum or breast bone.

substitution *n.* a mutation that results in the replacement of one nucleotide for another. *see also* non-synonymous, synonymous.

substitution matrix a probability matrix derived from the analysis of large numbers of related protein sequences that gives the probability of one type of amino acid being replaced by another during evolution. Such matrixes are used to assign quantitative scores to sequence alignments.

substrate *n.* (1) the substance on which an enzyme acts in a biochemical reaction; (2) respiratory substrate: substance undergoing oxidation during respiration; (3) any material used by microorganisms as a source of food; (4) inert substance containing or receiving a nutrient solution on which microorganisms grow; (5) the base to which a sedentary animal or plant is fixed. *alt.* substratum.

substrate cycle pair of irreversible metabolic reactions catalysed by two different enzymes in which one compound is converted into another and back again. It may amplify metabolic signals and possibly generate heat, as in the flight muscle of bumble bees. *alt.* futile cycle.

substrate-level phosphorylation formation of energy-rich phosphate compounds such as ATP by transfer of phosphate from a metabolic substrate to ADP directly with no respiratory chain involvement, as occurs in glycolysis. *cf.* oxidative phosphorylation, photophosphorylation.

substratose *a.* slightly or indistinctly stratified.

substratum *n.* the base to which a sedentary animal or plant is fixed. *see also* substrate.

subtalar *a. appl.* joint: the articulation between talus and calcaneus.

subtectal *a.* lying under a roof, esp. of skull.

subtegminal *a.* under the tegmen or inner coat of a seed.

subtentacular canals two prolongations of echinoderm coelom.

subterminal *a.* situated near the end.

subthalamus *n.* part of hypothalamus excluding optic chiasma and region of the mamillary bodies.

subthoracic *a.* not quite so far forward as to be called thoracic, *appl.* certain fish fins.

subtilisin *n.* proteolytic enzyme produced by *Bacillus subtilis.* Included in EC 3.4.21.14.

subtruncate *a.* terminating abruptly.

subtypical *a.* deviating slightly from type.

subulate *a.* awl-shaped, i.e. narrow and tapering from base to a fine point, *appl.* leaves, as of onion.

subumbellate *a.* tending to an umbellate arrangement with peduncles arising from a common centre.

subumbonal *a.* beneath or anterior to the umbo of a bivalve shell.

subumbonate *a.* (1) slightly convex; (2) having a low rounded protuberance.

subumbrella *n.* the concave inner surface of the bell of a medusa.

subungual *a.* under a nail, claw or hoof.

subunguis *n.* the ventral scale of a claw or nail.

subunit *see* protein subunit.

subunit vaccine vaccine made from purified protein components of viruses, bacteria and other parasites, rather than the complete organism.

subvaginal *a.* within or under a sheath.

subventricular zone region around ventricle in brain that continues to produce glial cells after birth.

subvertebral *a.* under the vertebral column.

subzonal *a. appl.* layer of cells internal to zona radiata.

subzygomatic *a.* below the cheek bone.

succate *a.* containing juice, juicy.

succession *n.* (1) a geological, ecological or seasonal sequence of species; (2) the sequence of different communities developing over time in the same area, leading to a dynamic steady state or climax community (used esp. of plant or microbial communities); (3) the occurrence of different species over time in a given area.

successive percentage mortality apparent mortality *q.v.*

succiferous *a.* conveying sap.

succinate dehydrogenase flavoprotein enzyme catalysing the oxidation of succinate to fumarate in the tricarboxylic acid cycle. EC 1.3.99.1.

succinic acid, succinate *n.* four-carbon dicarboxylic acid of the tricarboxylic acid cycle, converted to fumarate by succinate dehydrogenase.

succinyl-CoA 'energy-rich' compound formed from succinate and CoA during the tricarboxylic acid cycle and other metabolic cycles, and which also provides carbon skeleton for porphyrin synthesis.

succise *a.* (1) abrupt; (2) appearing as if part were cut off.

succubous *a.* with each leaf covering part of that under it.

succulent *a.* (1) full of juice or sap; (2) *appl.* fruit having a fleshy pericarp, as berries; (3) *appl.* plants adapted to dry and desert conditions, with swollen water-storing stems and leaves.

succus entericus the secretions of the epithelium lining the small intestine, containing peptidases, maltase, sucrase, lactase, lipase, nucleases, nucleotidases.

sucker *n.* (1) (*bot.*) a branch of stem, at first running underground and then emerging, and which may eventually form an independent plant; (2) haustorium *q.v.*; (3) (*zool.*) an organ adapted to attach to a surface by creating a vacuum, in some animals for the purpose of feeding, in others to assist locomotion or attachment.

sucking disc a disc assisting in attachment, as at end of echinoderm tube-foot.

sucking lice common name for the Anoplura *q.v.*

sucrase *n.* digestive enzyme hydrolysing the disaccharides sucrose and maltose to their component monosaccharides (glucose and fructose, and glucose respectively) by an α-D-glucosidase action, found in intestinal mucosa. EC 3.2.1.48, *r.n.* sucrose-α-D-glucohydrolase.

sucrose *n.* a non-reducing disaccharide present in many green plants and hydrolysed by the enzymes invertase or sucrase or by dilute acids to glucose and fructose. *alt.* cane sugar, invert sugar, beet sugar, saccharose, saccharobiose. *see* Fig. 37.

suction pressure (SP) the capacity of a plant cell to take up water by osmosis, being the difference between the osmotic pressure of the cell sap causing water to enter and the back pressure exerted by the cell wall (turgor pressure). When a cell is turgid it has no suction pressure.

Suctoria *n.* group of predatory ciliate protozoans which usually lose their cilia as adults and possess one or more suctorial tentacles.

suctorial *a.* adapted for sucking.

Sudanese Park Steppe region phytogeographical area, part of the Palaeotropical kingdom, consisting of a semi-arid belt south of the Sahara from the west coast to the Rift valley.

sudation *n.* (1) discharge of water and other substances in solution, as through pores; (2) sweating.

sudden correction model proposed mechanism for maintenance of fidelity of multiple repeated DNA sequences (such as the rRNA genes) in which the entire gene cluster is replaced from time to time with a set of copies derived from one or a few original sets of genes.

sudoriferous *a.* conveying, producing or secreting sweat, *appl.* glands and their ducts.

sudorific *a.* causing or *pert.* secretion of sweat.

sudoriparous sudoriferous *q.v.*

suffrescent *a.* slightly shrubby, *appl.* plants that are woody at the base but herbaceous above and that do not die back to ground level in winter.

suffrutex *n.* an under-shrub. *plu.* **suffrutices**.

suffruticose *a.* somewhat shrubby.

sugar *n.* (1) the general name for any mono-, di- or trisaccharide; (2) sucrose *q.v.*

sugar nucleotide a nucleotide covalently linked to a sugar, e.g. UDP-glucose, which is the activated form of sugars for glycan synthesis.

sugent suctorial *q.v.*

suids *n.plu.* members of the mammalian family Suidae: the pigs.

suines *n.plu.* members of the mammalian suborder Suina, which includes the non-ruminant artiodactyls: the hippopotamuses, pigs, peccaries, and a number of extinct groups.

sulcate *a.* grooved or furrowed.

sulcation *n.* (1) fluting; (2) formation of ridges and furrows.

sulcus *n.* (1) a groove; (2) a groove between two convolutions on the surface of the brain. *plu.* **sulci**.

sulf- *see* sulph-.

sulphanilamide *n.* an antibacterial compound, a sulphur-containing aromatic amide and its derivatives which prevent bacterial growth by inhibiting purine synthesis. *alt.* **sulfanilamide**.

sulphatase *n.* enzyme catalysing the hydrolysis of sulphuric esters, e.g. arylsulphatase (EC 3.1.6.1), which hydrolyses a phenol sulphate to phenol plus sulphate ion.

sulphatide *n.* any of several sulphur-containing glycolipids, derivatives of ceramide, found in animal brain and other tissues.

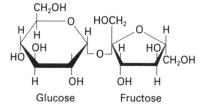

Glucose   Fructose

Fig. 37 Sucrose.

sulpholipid *n.* sulphur-containing glycolipid, occurring in chloroplasts in green plants and chromatophores of photosynthetic bacteria.

sulphur bacteria a group of unrelated bacteria which can variously utilize sulphur or sulphide as a respiratory substrate or electron acceptor in photosynthesis or reduce sulphate to sulphide. They comprise the photosynthetic green sulphur bacteria and purple suphur bacteria, which oxidize sulphide to sulphur (and which can also fix nitrogen), the colourless non-photosynthetic sulphur bacteria (e.g. *Thiobacillus*) which oxidize sulphide to sulphur and sulphate, and the heterotrophic sulphate-reducing bacteria which reduce sulphate to sulphide.

sulphur cycle a cycle of biological processes by which sulphur circulates within the biosphere. It includes assimilation of sulphur by plants from soil sulphate, its incorporation into plant and animal protein, and putrefaction of dead organic matter by bacteria which releases sulphide. This can be converted to elemental sulphur, to sulphate, and back to sulphide by a heterogeneous group of sulphur bacteria.

sulphydryl group –SH group. *alt.* thiol.

summation *n.* combined action of either simultaneous or successive nerve impulses, subliminal stimuli or subthreshold potentials which produces an excitatory or inhibitory response.

summer wood late wood *q.v.*

sumoylation *n.* the post-translational modification of a protein by the covalent attachment of a small protein called SUMO.

sun-basking *n.* behaviour shown by many poikilothermic animals to control body temperature.

sun spiders common name for the Solifuga *q.v.*

super- prefix derived from L. *super*, over. In classification, a group just above the status of the taxon following it, as in superclass.

superantigen *n.* antigen which provokes a massive T-cell response by nonspecifically activating all T cells with receptors containing a particular type of V segment.

supercarpal (1) *a.* above the carpus; (2) *n.* an upper carpal.

superciliary *a.* (1) *pert.* eyebrows; (2) above orbit of eye.

superciliary arches two arched elevations on forehead below frontal eminences.

superclass *n.* taxonomic grouping between subphylum and class.

supercoiled DNA DNA in which the double helix is further twisted on itself. *alt.* **superhelical DNA, supertwisted DNA**.

superdominance overdominance *q.v.*

superfamily (1) *see* gene superfamily; (2) protein superfamily: a group of proteins with different functions but all having related sequences and presumed to be encoded by genes derived from a common ancestral gene.

superfemale metafemale *q.v.*

superfetation *alt.* spelling of superfoetation *q.v.*

superficial *a.* (1) on or near the surface; (2) *appl.* placentation in ovary of flower in which ovules are scattered over inner surface of ovary wall.

superfluent *n.* an animal species of the same importance in an ecosystem as a subdominant plant species in a succession.

superfoetation *n.* (1) fertilization of ovules of an ovary with more than one kind of pollen; (2) successive fertilization of two or more ova of different oestrous periods, in the same uterus.

supergene *n.* region of chromosome which contains a number of genes but in which crossing-over does not occur, so that the genes are transmitted together from generation to generation.

superhelical DNA supercoiled DNA *q.v.*

superior *a.* (1) upper; (2) higher; (3) anterior; (4) growing or arising above another organ; (5) *appl.* ovary having perianth inserted around the base; (6) *appl.* sepals, petals or stamens attached to the receptacle of a flower below the ovary; (7) *appl.* vena cava: precava *q.v.*

superior colliculi pair of small prominences on dorsal surface of midbrain that receive visual information. *cf.* inferior colliculi.

superior olive structure in brainstem that receives auditory information from right and left ears. *alt.* **superior olivary nucleus**.

superinfection *n.* phage infection of a cell that harbours that phage as a prophage.

superlinguae *n.plu.* paired lobes of hypopharynx in certain insects.

supermale metamale *q.v.*

supernumerary chromosomes extra heterochromatic chromosomes present in some plants above the normal number for the species, such as B chromosomes.

superordinate circuit neural circuit that is hierarchically superior to other circuits.

superorganism *n.* any society, such as a colony of a eusocial insect species, possessing features of organization analogous to the properties of a single organism.

superovulation *n.* the production of an unusually large number of eggs at any one time.

superoxide anion O_2^-, a highly reactive oxygen derivative, toxic to cells.

superoxide dismutase (SOD) widely distributed enzyme which destroys superoxide anions (O_2^-) with the formation of hydrogen peroxide and molecular oxygen. *alt.* erythrocuprein, haemocuprein, cytocuprein.

superparasite hyperparasite *q.v.*

super-regeneration *n.* the development of additional or superfluous parts in the process of regeneration.

super-repressed *a. appl.* mutant genes that cannot be derepressed. *alt.* uninducible.

supersacral *a.* above the sacrum.

supersecondary structure level of protein structure, such as a β-barrel, in which secondary structure elements are combined into a more complex discrete structural element.

supersonic ultrasonic *q.v.*

superspecies *n.* (*bot.*) complex of morphologically very similar semispecies that replace each other geographically. *alt.* coenospecies.

supersphenoidal *a.* above sphenoid bone.

supertwisted DNA supercoiled DNA *q.v.*

supervolute *a.* having a plaited and rolled arrangement in the bud.

supinate *a.* inclining or leaning backwards.

supination *n.* movement of arm by which palm of hand is turned upwards.

supinator brevis, supinator longus two forearm muscles used in turning the palm upwards.

supplemental air volume of air that can be expelled from the lungs after normal breathing out.

supplementary motor cortex region of non-primary motor cortex that receives input from basal ganglia and modulates activity of primary motor cortex.

supporting tissue in plants, tissue made of cells with thickened walls such as collenchyma and sclerenchyma, adding strength to plant body; in animals, skeletal tissue forming endo- or exoskeleton.

suppression *n.* (1) non-development of an organ or part; (2) the cancelling out of the effects of one mutation by another, usually in another gene (if in the same gene it is termed intragenic suppression), the second mutation being called a suppressor.

suppressor mutation mutation that cancels out the effects of another mutation elsewhere in the genome. *see* suppression.

suppressor T lymphocyte, suppressor T cell regulatory T cell *q.v.*

suppressor tRNA mutant tRNA with altered codon specificity, which can correct a nonsense mutation elsewhere in the genome by inserting the correct amino acid at the mutant codon.

supra- prefix derived from L. *supra*, above, signifying situated above.

supra-acromial *a.* above the acromion of shoulder blade.

supra-anal *a.* above anus or anal region.

supra-angular surangular *q.v.*

supra-auricular *a.* above the auricle of the ear, *appl.* feathers.

suprabranchial *a.* above the gills.

suprabuccal *a.* above cheek and mouth.

suprabulbar *a. appl.* region between hair bulb and fibrillar region of hair.

supracallosal *a. appl.* a gyrus on the upper surface of corpus callosum of brain.

supracaudal *a.* above the tail or caudal region.

supracellular *a.* (1) *appl.* structures originating from many cells; (2) *appl.* level of organization above cellular level, as of tissues, organs.

supracerebral *a. appl.* lateral pharyngeal glands, as in hymenopterans.

suprachiasmatic nucleus (SCN) one of the nuclei in the hypothalamus, in humans it is involved in regulating diurnal rhythms such as the sleep-waking cycle, body temperature and hormone release.

suprachoroid *a.* (1) over the choroid; (2) between choroid and sclera of eye.

supraclavicle *n.* a bone of pectoral girdle of fishes. *alt.* supracleithrum.

supraclavicular *a.* above or over the clavicle, *appl.* nerves.

supracondylar *a.* above a condyle, *appl.* ridge and process.

supracoracoideus *n.* a flight muscle in birds, running indirectly from breast bone to humerus and responsible for raising the wing.

supracostal *a.* over or external to the ribs.

supracranial *a.* over or above the skull.

supradorsal *a.* (1) on or over the dorsal surface; (2) *appl.* small cartilaginous elements in connection with primitive vertebral column.

supra-episternum *n.* upper sclerite of episternum in some insects.

supraethmoid dermethmoid *q.v.*

supraesophageal supraoesophageal *q.v.*

supraglenoid *a.* (1) above the glenoid cavity; (2) *appl.* tuberosity at apex of glenoid cavity.

suprahyoid *a.* over or above the hyoid bone(s) lying at base of tongue in mammals.

supralabial *a.* on the lip, *appl.* scutes or scales.

supraliminal *a.* above the threshold of sensation, *appl.* stimuli.

supralittoral *a. pert.* seashore above high-water mark, or spray zone.

supraloral *a.* above the loral region, as in birds, snakes.

supramastoid crest ridge at upper boundary of mastoid region of temporal bone.

supramaxillary *a. pert.* upper jaw.

suprameatal *a. appl.* triangle and spine over external auditory meatus.

supranasal *a.* over nasal bone or nose.

supraoccipital *n.* a large bone in the middle of the back of the skull.

supraocular *a.* over or above the eye.

supraoesophageal *a.* above or over the gullet.

supraorbital *a.* above eye orbits.

suprapatellar *a. appl.* pouch between upper part of patella and femur.

suprapharyngeal *a.* above or over the pharynx.

suprapopulation *n.* population that includes all the developmental phases of a species at a given time.

suprapubic *a.* above the pubic bone.

suprapygal *a.* above the pygal bone.

suprarenal *a.* (1) situated above the kidneys; (2) *appl.* bodies, glands: adrenals *q.v.*

suprarostral *a. appl.* a cartilaginous plate anterior to trabeculae of skull in amphibians.

suprascapula *n.* (1) cartilage of dorsal part of pectoral girdle in certain cartilaginous fishes; (2) incompletely ossified extension of scapula of amphibians and some reptiles.

suprascapular *a.* above the shoulder blade, *appl.* ligament, nerve.

supraseptal *a. appl.* two plates diverging from interorbital septum of skull.

suprasphenoid *n.* membrane bone dorsal to sphenoid cartilage.

suprasphenoidal *a.* above sphenoid bone of skull.

supraspinal *a.* (1) above and over the spinal column; (2) above ventral nerve cord in insects.

supraspinatous *a. appl.* scapular fossa and fascia for origin of supraspinatus.

supraspinatus *n.* shoulder muscle inserted into proximal part of greater tubercle of humerus.

suprastapedial *n.* the part of columella of ear above stapes, homologous with mammalian incus.

suprasternal *a.* over or above breast bone, *appl.* a slit-like space in cervical muscle, *appl.* supernumerary sternal elements in some mammals, *appl.* body plane.

suprastigmal *a.* above a stigma or breathing pore of insects.

supratemporal *a.* (1) *pert.* upper temporal region of skull, *appl.* bone, arch, fossa; (2) pterotic of teleosts.

suprathoracic *a.* above thoracic region.

supratidal *a.* above high-water mark, *appl.* the spray zone and organisms living there.

supratonsillar *a. appl.* a small depression in lymphoid mass of palatine tonsil.

supratrochlear *a.* over trochlear surface, *appl.* nerve, foramen, lymph glands.

supratympanic *a.* above the ear drum.

sural *a. pert.* calf of leg.

surangular *n.* a bone of lower jaw of some fishes, reptiles and birds.

surculose *a.* of plants, bearing suckers.

surcurrent *a.* proceeding or prolonged up a stem.

surface exclusion the inability of a plasmid to enter a bacterial cell already carrying a plasmid of the same type, and which is mediated at the surface of the bacterium.

surface immunoglobulin membrane-bound immunoglobulin on B cells which acts as the antigen receptor.

surface water water from rain or other precipitation that does not sink into the ground or evaporate, and which runs off the land surface, flowing eventually into streams and rivers.

surrogate genetics the study of gene expression by making defined alterations in an isolated DNA template and looking at the effects of these changes on its transcription in *in vitro* or other transcription systems, i.e. bypassing the need to induce or search for suitable mutations in the living organism.

survivorship schedule demographic data giving the number of individuals surviving to each particular age in a population.

suspensor *n.* (1) chain of cells developing from the zygote in angiosperms, attaching embryo to embryo sac; (2) similar structure in other plants; (3) modified portion of a hypha from which a gametangium or a zygospore is suspended.

suspensorium *n.* skeletal element forming the side wall of mouth cavity in bony fishes and other vertebrates.

suspensory *a.* (1) *pert.* a suspensorium; (2) serving for suspension, *appl.* various ligaments.

sustainable yield highest rate at which a renewable resource can be used without reducing its supply.

sustentacular *a.* (1) supporting, *appl.* cells; (2) *appl.* connective tissue acting as a supporting framework for an organ.

sutural *a.* (1) *pert.* a suture; (2) *appl.* dehiscence taking place along a suture.

sutural bones irregular isolated bones occurring along sutures.

suture *n.* (1) line of junction of two parts immovably connected, as between bones of skull, sclerites of exoskeleton covering an arthropod segment; (2) line of seed capsule along which dehiscence occurs.

Sv sievert *q.v.*

Svedberg unit (S) unit in which the rate of sedimentation of a particle in the ultracentrifuge is expressed ($1 S = 10^{-13} s$ under standard conditions) and which is an indirect measurement of size and molecular weight.

swamp *n.* wet ground, saturated or periodically flooded, dominated by woody plants and with no surface accumulation of peat.

swarm cells motile zoospores or amoebae in fungi, algae and slime moulds.

sweat glands specialized glands in the skin of some mammals, through which water and salts are exuded to aid evaporative cooling of the body. They are of two types: apocrine sweat glands and eccrine sweat glands.

sweating *n.* exudation of water from the body surface through sweat glands, which is used as a cooling mechanism by some mammals, evaporation of the water from the surface having a cooling effect.

sweet potato *Ipomoea batatus*, a crop plant originating in Central America, now grown as a staple food worldwide, a member of the family Convolvulaceae (order Passiflores), whose tubers are eaten.

swimbladder *n.* a gas-filled sac in body cavity of most teleost fishes, developed as an outgrowth of the alimentary canal, and which is an aid to buoyancy.

swimmeret *n.* small paired appendage of crustaceans, present on up to five abdominal segments, possibly involved partly in swimming.

swimming bell nectocalyx *q.v.*

swimming funnel the siphon in some cephalopods through which water is expelled from mantle cavity, producing a means of jet propulsion.

swimming ovaries groups of ripe ova in some acanthocephalans, detached from ovary and floating in body cavity.

swimming plates *see* Ctenophora.

switch gene a gene such as the homeotic genes, which when mutated cause development to switch from one pathway to another, and therefore are believed normally to act as genetic master switches, selecting a particular developmental pathway.

switch plant a xerophyte which produces normal leaves when young, then sheds them, and photosynthesis is taken over by another structure such as a cladode or phyllode.

switch region in immunoglobulin loci, a site found to the 5′ side of each C gene at which recombination occurs during immunoglobulin class switching, resulting in production of antibodies with the same antigen specificity but different class during an immune response.

syconium, syconus *n.* a composite fruit consisting mainly of an enlarged succulent receptacle, as a fig.

sylva *n.* (1) forest of a region; (2) forest trees collectively.

Sylvian fissure deep fissure that demarcates the temporal lobe of brain.

sym-, syn- prefixes from the Gk *syn*, with.

symbiont *a.* one of the partners in a symbiosis.

symbiosis *n.* (1) close and usually obligatory association of two organisms of different species living together, not necessarily to their mutual benefit; (2) often used exclusively for an association in which both partners benefit, which is more properly called mutualism. *plu.* **symbioses**. *a.* **symbiotic**.

symmetrical *a. pert.* symmetry.

symmetrodonts *n.plu.* an order of Mesozoic trituberculate mammals having molar teeth with three or more cusps in a triangle.

symmetry *n.* regularity of form. *see* bilateral symmetry, radial symmetry.

symparasitism *n.* the development of several competing species of parasite within or on one host.

sympathetic *a.* (1) *appl.* components of sympathetic nervous system; (2) *appl.* segmental nerves supplying spiracles in insects; (3) *appl.* coloration in imitation of surroundings.

sympathetic chain chain of sympathetic ganglia that runs along each side of the spinal column. *see* sympathetic nervous system.

sympathetic nervous system part of the autonomic nervous system comprising nerve fibres that leave the spinal cord in the thoracic and lumbar regions and supply viscera and blood vessels by way of a chain of sympathetic ganglia running on each side of the spinal column. These communicate with the central nervous system via a branch to a corresponding spinal nerve. The sympathetic nervous system controls movements and secretions from viscera and monitors their physiological state. Stimulation of the sympathetic system induces e.g. the contraction of gut sphincters, heart muscle and the muscle of artery walls, and the relaxation of gut smooth muscle and the circular muscles of the iris. The chief neurotransmitter in the sympathetic system is adrenaline, which is liberated in heart, visceral muscle, glands and internal vessels. It acts as a neurotransmitter at ganglionic synapses and at sympathetic terminals in skin and smooth muscle of blood vessels. The actions of the sympathetic system tend to be antagonistic to those of the parasympathetic system.

sympathomimetic *a. appl.* substances that produce effects similar to those produced by stimulation of the sympathetic nervous system.

sympatric *a. appl.* species inhabiting the same or overlapping geographic areas. *cf.* allopatric.

sympatric speciation speciation that occurs due to intrinsic factors within the population, in the absence of geographical isolation.

sympetalous *a.* having petals joined into a tube, at least at base.

symphily *n.* the situation where one species of insect lives as a guest (symphile) in the nest of a social insect which feeds and protects it in return for its secretions which are used as food. Examples are certain beetles in the nests of ants and termites.

symphoresis *n.* movement or conveyance collectively, as movement of spermatid group to a Sertoli cell.

symphygenesis *n.* the development of an organ from the union of two others.

symphyllodium *n.* (1) structure formed by coalescence of external coats of two or more ovules; (2) a compound ovuliferous scale.

symphysis *n.* (1) the line of junction at which two bones, e.g. left and right halves of jaw, are fused; (2) slightly movable articulation of two bones connected by fibrocartilage; (3) the growing together of parts which are separate in early development. *a.* **symphyseal, symphysial.**

Symphyla *n.* class of arthropods allied to the myriapods. They have 14 body segments and 6 pairs of walking legs.

Symphyta *n.* the sawflies, a suborder of hymenopteran insects having no well-defined 'waist', and considered more primitive than members of the suborder Apocrita (bees, wasps, ants, ichneumon flies). The ovipositor is serrated like a saw.

symplast *n.* the interconnected protoplasm of plant tissue, the protoplasm of individual cells connected by plasmodesmata through the cell walls. *a.* **symplastic.**

symplastic transport in plant tissue, the movement of ions and other material from cell to cell via plasmodesmata (cytoplasmic bridges in the cell wall).

symplesiomorphy *n.* in the cladistic method of classification, the case where a homologous character state shared between two or more taxa is believed to have originated as a novelty in a common ancestor earlier than the most recent common ancestor. *a.* **symplesiomorphic, symplesiomorphous.** *cf.* homoplasy, synapomorphy.

sympodial *a. pert.* or resembling a sympodium in mode of branching.

sympodite protopodite *q.v.*

sympodium *n.* (1) a plant or part of a plant whose main axis arises not from growth of an apical bud but from that of a lateral branch which also stops growing after a while, growth continuing from a lateral bud near the apex, and then another, e.g. many orchids; (2) a stem vascular bundle and its associated leaf traces. *plu.* **sympodia.**

symport *n.* membrane protein that transports a solute across the membrane in one direction, transport depending on the sequential or simultaneous transport of another solute in the same direction.

synacme, synacmy *n.* conditions when pistils and stamens mature simultaneously.

synaesthesia *n.* the accompaniment of a sensation due to stimulation of the appropriate receptor, such as sound, by a sensation characteristic of another sense, such as colour.

synandrium *n.* the cohesion of anthers in male flowers of some aroids (e.g. arum lilies).

synandry *n.* condition where stamens normally separated are united.

synangium *n.* (1) compound sporangium in which sporangia are coherent, as in some ferns; (2) most anterior portion of the foetal or amphibian heart, through which blood is driven from the ventricle; (3) any arterial trunk from which arteries arise.

Synanthae Cyclanthales *q.v.*

synantherous *a.* having anthers united to form a tube.

synanthous *a.* (1) having flowers and leaves appearing simultaneously; (2) having flowers united together as in Compositae.

synanthropic *a.* associated with humans or their dwellings.

synanthy *n.* adhesion of flowers usually separate.

synapomorphy *n.* in cladistic phylogenetics denotes a homologous character common to two or more taxa and thought to have originated in their most recent common ancestor. *a.* **synapomorphous.** *cf.* apomorphous.

synaporium *n.* an animal association owing to unfavourable environmental conditions or disease.

synaposematic *a.* having warning colours in common, *appl.* mimcry of a more powerful or dangerous species as a means of defence.

synapse (1) *n.* the point of communication between one nerve cell and another or between nerve cell and a target cell such as muscle. The two main types of synapses are chemical synapses and electrical synapses. Synapses are commonly made between the axon terminals of the transmitting cell and the dendrites or cell body of another, but may also be made by an axon terminal on the axon of another neuron. At chemical synapses the signal is transmitted across a narrow gap between the neuron and the target cell by neurotransmitter molecules released from the pre-synaptic cell. These stimulate (or inhibit) the postsynaptic cell. Electrical synapses are formed by gap junctions between two neurons, through which electrical current can flow directly; (2) *v.* come into contact with each other, *appl.* nerve endings, homologous chromosomes during meiosis.

synapsid *a.* having skull with one ventral temporal fenestra on each side.

Synapsida, synapsids *n.*, *n.plu.* subclass of reptiles living from the Carboniferous to Triassic, the mammal-like reptiles, with synapsid skulls, some forms of which gave rise to the mammals, and including the pelycosaurs and therapsids.

synapsis *n.* association of two homologous chromosomes at the start of meiosis to form a bivalent.

synaptic *a. pert.* or occurring at a synapse.

synaptic bouton bouton *q.v.*

synaptic cleft narrow fluid-filled gap between two opposing cell membranes at a chemical synapse in nervous tissue or at a neuromuscular junction.

synaptic potential electrical potential difference produced across the postsynaptic membrane by the action of neurotransmitter at a single synapse. Individual synaptic potentials are summed by the receiving neuron to produce a grand postsynaptic potential, which may be inhibitory or excitatory.

synaptic vesicles vesicles containing neurotransmitter which aggregate in the cytoplasm at the tip of axon terminals and which liberate their contents into the synaptic cleft. *alt.* presynaptic vesicles.

synaptogenesis *n.* formation of a synapse as axon terminal and dendrite grow towards each other.

synaptonemal complex proteinaceous ladder-like structure linking two paired homologous chromosomes in meiosis, seen in the electron microscope.

synaptosomes *n.plu.* membrane-bounded structures, representing pinched off nerve terminals and postsynaptic membrane, formed when brain tissue is homogenized.

synaptospermous *a.* having seeds germinating close to the parent plant.

synaptospore *n.* aggregate spore.

synarthrosis *n.* a joint in which bone surfaces are in almost direct contact, fastened together by connective tissue or hyaline cartilage, with no appreciable capability for movement.

syncarp *n.* an aggregate fruit with united carpels.

syncarpy *n.* condition of having carpels united to form a compound gynaecium. *a.* **syncarpous.**

syncerebrum *n.* a secondary brain formed by union with brain of one or more ventral cord ganglia, in some arthropods.

synchondrosis *n.* a synarthrosis where the connecting medium is cartilage.

synchorology *n.* (1) study of the distribution of plant and animal associations; (2) geographical distribution of communities.

synchronic *a.* (1) contemporary; (2) existing at the same time, *appl.* e.g. species. *cf.* allochronic.

synchronizer *n.* some environmental factor, such as light or temperature, that interacts with an endogenous circadian rhythm, causing it to be precisely synchronized (entrained) to a 24-hour cycle rather than free-running. *alt.* zeitgeber.

syncladous *a.* with offshoots or branchlets in tufts, *appl.* some mosses.

synconium, synconus syconium *q.v.*

syncraniate *a.* with vertebral elements fused with skull.

syncranterian *a.* with teeth in a continuous row.

syncryptic *a. appl.* animals which appear alike, although unrelated, due to common protective resemblance to the same surroundings.

syncytial blastoderm multinucleate blastoderm of insect embryo before formation of individual cells.

syncytium *n.* a multinucleate mass of protoplasm which is not divided into separate cells. *a.* **syncytial.** *cf.* coenocyte.

syndactylism, syndactyly *n.* whole or part fusion of two or more digits. *a.* **syndactyl.**

syndesmology *n.* that branch of anatomy dealing with ligaments and articulations.

syndesmosis *n.* a slightly movable joint, with bone surfaces connected by a ligament.

syndetocheilic *a. appl.* type of stomata found in gymnosperms in which the subsidiary cells are derived from the same cell as the guard mother cell.

syndrome *n.* a group of concomitant symptoms, characteristic of a particular condition.

synecology *n.* the ecology of plant or animal communities. *cf.* autecology.

synema synnema *q.v.*

synencephalon *n.* the part of embryonic brain between diencephalon and mesencephalon.

synenchyma *n.* fungal tissue composed of laterally joined hyphae.

syneresis *n.* (1) contraction of a gel with expression of fluid; (2) contraction of clotting blood and expression of serum.

synergic, synergistic *a.* acting together, often to produce an effect greater than the sum of the two agents acting separately. *n.* **synergism**.

synergid *n.* each of two cells without cell walls lying beside ovum at micropylar end of ovule.

synesthesis synaesthesia *q.v.*

syngameon *n.* a group of species or semispecies linked by frequent or occasional hybridization in nature.

syngamy *n.* the complete fusion (cytoplasmic and nuclear) of two gametes to form a zygote. *alt.* fertilization.

syngeneic *a.* genetically identical.

syngenetic *n.* descended from the same ancestors.

syngraft homograft *q.v.*

syngynous epigynous *q.v.*

synkaryon *n.* a zygote nucleus resulting from fusion of gamete nuclei.

synkaryotic *a.* diploid, *appl.* nucleus.

synnema *n.* (1) in some fungi, a group of conidiophores cemented together by their stalks; (2) in some flowers, a bundle of stamens united by their filaments. *plu.* **synnemata**.

synochreate, synocreate *a.* with stipules united and enclosing stem in a sheath.

synoecious *a.* having antheridia and archegonia on same receptacle, or stamens and pistils on same flower, or male and female flowers on same capitulum.

synonym *n.* (1) in molecular biology, any of two or more codons specifying the same amino acid; (2) in classification, an alternative Latin name.

synonymous *a. appl.* codons that encode the same amino acid; *appl.* substitutions: the mutation of a codon into another codon that encodes the same amino acid.

synosteosis, synostosis *n.* ossification from two or more centres in the same bone, as from diaphysis and epiphyses in long bones.

synotic tectum in higher vertebrates, a cartilaginous arch between otic capsules representing cartilaginous roof or tegmentum of cranium in lower vertebrates.

synovia *n.* mucopolysaccharide and mucoprotein secretion from cells lining joint cavities, serving to lubricate the joint.

synovial cell epithelial cell lining joint cavities, secreting hyaluronic acid.

synovial fluid fluid contained in joint cavity, secreted by synovial cells, serving to lubricate the joint.

synovial membrane the inner layer of capsule around a movable joint, made of connective tissue and secreting material to lubricate the joint.

synoviparous *a.* secreting synovia.

synpelmous *a.* having two tendons united before they go to separate digits.

synpolydesmic *a. appl.* scales growing by apposition at margin, and which are made up of fused bony teeth covered with a continuous layer of dentine.

synsacrum *n.* mass of fused vertebra supporting the pelvic girdle of birds.

synsepalous *n.* having a calyx composed of fused sepals. *n.* **synsepaly**.

synspermous *a.* with seeds united.

synsporous *a.* propagating by cell conjugation, as in algae.

syntagma *see* tagma.

syntelome *n.* a compound telome.

syntenic *a.* (1) *appl.* genetic loci that lie on the same chromosome; (2) *appl.* to blocks of genetic loci that are in the same order on a chromosome in different species. *n.* **synteny**.

syntenosis *n.* articulation of bones by means of tendons.

synthetase *see* ligase.

synthetic *n. appl.* molecules produced by chemical synthesis, not in a living organism.

synthetic biology the re-engineering of biological components to provide new, designed functions. One aim of synthetic biology is to build a working cell.

synthetic lethal a combination of two mutations that renders cells non-viable, although cells with either of the mutations singly are viable.

synthetic theory of evolution *see* neo-Darwinism.

syntopic *a.* sharing the same habitat within the same geographical range, *appl.* different species or to phenotypic variants within a species.

syntrophy *n.* nutritional interdependence, the phenomenon in which a bacterial species can only grow in association with another species because the action of the second is required to couple an energetically unfavourable reaction (e.g. ethanol fermentation) in the first bacterium with an energetically favourable reaction (e.g. methanogenesis) in the second to provide a positive overall energy yield. *a.* **syntrophic.**

syntropic *a.* turning or arranged in the same direction.

syntype *n.* any one specimen of a series used to designate a species when holotype and paratypes have not been selected. *alt.* cotype.

synusia *n.* a plant community of relatively uniform composition, living in a particular environment and forming part of the larger community of that environment.

syringeal *a. pert.* the syrinx.

syringes *plu.* of syrinx.

syringium *n.* a syringe-like organ through which some insects eject a disagreeable fluid.

syringograde *a.* jet-propelled, moving by alternate suction and ejection of water through siphons, as squids and salps.

syringyl sinapyl alcohol. *see* lignin.

syrinx *n.* sound-producing organ in birds, situated at junction of windpipe with bronchi. It consists of two patches of thin membrane (tympaniform membrane) in the wall of each bronchus. Sound is produced when the tympanic membranes are pushed inwards partially blocking the bronchi. Air pushed out of the lungs makes the membranes vibrate and produces a sound. *plu.* **syringes, syrinxes.**

systaltic *a.* (1) contractile; (2) alternately contracting and dilating.

systematics *n.* the study of the identification, taxonomy and nomenclature of organisms, including the classification of living things with regard to their natural relationships and the study of variation and the evolution of taxa.

systemic *a.* throughout the body, involving the whole body.

systemic anaphylaxis potentially fatal type of immediate hypersensitivity reaction. Antigen in the bloodstream triggers the activation of mast cells throughout the body, causing inflammatory reactions that lead to vasodilation and tissue swelling.

systemic arteries arteries leading from the heart to the aorta in reptiles, carrying blood to the body (as opposed to the lungs).

systemic autoimmunity an autoimmune reaction that involves common cellular constituents and is not restricted to a particular organ or location.

systemic circulation in vertebrates, the course of blood from ventricle through body to auricle, as opposed to the pulmonary circulation.

systemic heart the heart of invertebrates, and the left auricle and ventricle in higher vertebrates.

systemic lupus erythematosus (SLE) autoimmune disease involving antibodies directed against nucleic acids (amongst other cellular constituents).

systems biology the study of biological processes in e.g. a cell as an integrated system of dynamic networks of interacting elements, defined by one of its leading practitioners as 'the search for the syntax of biological information'.

systole *n.* (1) contraction of heart causing circulation of blood; (2) contraction of any contractile cavity. *cf.* diastole.

systrophe *n.* aggregation of starch grains in chloroplasts, induced by illumination.

systylous *a.* (1) with coherent styles; (2) with fixed columella lid, as in mosses.

syzygium *n.* a group of associated gregarines.

syzygy *n.* (1) a close suture of two adjacent arms, found in crinoids; (2) a number of individuals, from two to five, adhering in strings in association of gregarines; (3) reunion of chromosome fragments at meiosis.

T

θ- *for all entries with prefix* θ-, *refer to theta- or to the headword itself.*

T (1) threonine *q.v.*; (2) thymine *q.v.*

*T*ₘ melting temperature, the temperature at which double-stranded DNA is 50% separated into single strands. It varies for different DNAs depending on their base composition.

tables *n.plu.* outer and inner layers of flat compact bones, esp. of skull.

tabula *n.* a horizontal layer.

tabular *a.* (1) arranged in a flat surface; (2) flattened, as certain cells.

tabulare *n.* skull bone posterior to parietals in some vertebrates.

Tabulata *n.* extinct group of colonial stony corals organized as horizontal layers. *alt.* **tabulate corals**.

tabulate (1) *a.* composed of horizontal layers; (2) *v.* to list systematically or to arrange data in the form of a table.

tachyblastic *a.* with cleavage immediately following oviposition, *appl.* quickly hatching eggs.

tachycardia *n.* excessive heartbeat rate of >100 per minute.

tachygenesis *n.* development with omission of some embryonic stages as in certain crustaceans, or of nymphal stages in some insects, or of tadpole stages in some amphibians.

tachykinins *n.plu.* small group of neuroactive peptides including substance P.

tachytelic *a.* evolving at a rate faster than the standard rate.

tachyzoite *n.* fast-developing merozoite stage in the acute phase of a *Toxoplasma* infection.

tacrolimus *n.* an immunosuppressive drug. *alt.* FK506.

tactic *a. pert.* a taxis.

tactile *a.* serving the sense of touch, *appl.* sensory hairs, cells, cones.

tactor (1) touch receptor *q.v.*; (2) any sensory receptor receptive to touch.

tactual *a. pert.* sense of touch.

taenia *n.* a band, as of nerve or muscle.

taeniate *a.* ribbon-like.

taenioid *a.* ribbon-shaped, like a tapeworm.

TAF tumour angiogenesis factors *q.v.*

tagma *n.* a region of the body of a metamerically segmented animal formed by fusion of metameres (segments), e.g. head of insects. *plu.* **tagmata**.

tagmatism, tagmosis *n.* the specialization of groups of body segments to form different functional regions of the body, as in arthropods.

taiga *n.* northern coniferous forest zone, esp. in Siberia, adjacent to tundra.

tail fan in decapod crustaceans, the uropods and telson.

tali *plu.* of talus.

talin *n.* protein associated with the cytoskeleton in many animal cells, providing a connection between cell-surface molecules that bind to the extracellular matrix and the cell's internal cytoskeleton.

tall-grass community type of grassland with grasses up to 2 m or more high that develops in parts of the tropics (tall-grass savanna), and in temperate regions (tall-grass prairie, tall-grass steppe).

talocalcaneal *a. pert.* talus and calcaneus, *appl.* articulation, ligaments.

talocrural *a.* (1) *pert.* ankle and shank bones; (2) *appl.* articulation: the ankle joint.

talon *n.* posterior heel of upper molar teeth.

talonid *a. appl.* hollow (talonid basin) and cusps, at posterior heel of lower molar teeth of therian mammals.

taloscaphoid *a. pert.* talus and scaphoid bone.

talus *n.* (1) the second largest tarsal bone in humans, the astragalus; (2) a tarsal bone in vertebrates. *plu.* **tali.**

Tamaricales *n.* order of dicot trees, shrubs or rarely herbs, with small scale-like or heather-like leaves, including the families Fouquieriaceae (ocotilla), Frankeniaceae (sea heath) and Tamaricaceae (tamarisk).

tandem mass spectrometry (MS/MS) *see* mass spectrometry.

tandem repeat(s) two or more adjacent copies of a gene or other nucleotide sequence, all arranged in the same orientation. *cf.* inverted repeats.

tangential *a.* at right angles to a radius of a cylindrical structure, *appl.* section of e.g. a stem.

tannase *n.* enzyme which hydrolyses the tannin digallate to gallate, but will also hydrolyse certain ester linkages in other tannins. EC 3.1.1.20.

tannins *n.plu.* complex aromatic compounds some of which are glucosides, occurring in the bark of various trees. They possibly give protection against insect or fungal attack or are concerned with pigment formation. They are used in tanning and dyeing.

tanyblastic *a.* with a long germ band.

TAP transporter associated with antigen processing. A membrane transport protein complex (TAP-1 and TAP-2) that transports peptides from the cytoplasm into the endoplasmic reticulum.

tapetum *n.* (1) (*zool.*) tapetum lucidum *q.v.*; (2) main body of fibres of corpus callosum in brain; (3) (*bot.*) special nutritive layer investing sporogenous tissue of a sporangium; (4) layer of cells lining the cavity of anther and absorbed as pollen grains mature.

tapeworms common name for the Cestoda *q.v.*

taphonomy *n.* the study of the conditions of the burial of fossils.

taphrophyte *n.* a ditch-dwelling plant.

taproot *n.* long straight main root formed from radicle of embryo in gymnosperms and dicotyledons. In plants such as carrot and turnip it forms a swollen food-storage organ.

taraxanthin *n.* a xanthophyll carotenoid pigment found in red algae.

Tardigrada, tardigrades *n., n.plu.* (1) phylum of small animals, commonly called water bears, that have some features similiar to arthropods and are sometimes included in the Arachnida; (2) an infraorder of Edentata, including the tree sloths.

target organ the end-organ upon which a hormone or nerve acts.

tarsal (1) *a. pert.* tarsus, of foot and eyelid; (2) *n.* ankle bone. *plu.* **tarsalia.**

tarsal glands modified sebaceous glands of the eyelids, the ducts opening on the free margins. *alt.* Meibomian glands.

tarsi (1) *plu.* of tarsus; (2) two elongated plates of dense connective tissue helping to support the eyelid.

tarsomeres *n.plu.* the two parts of the dactylopodite in spiders, basitarsus and telotarsus.

tarsometatarsal *a. pert.* an articulation of the tarsus with the metatarsus.

tarsometatarsus *n.* a short straight bone of bird's leg formed by fusion of distal row of tarsals with 2nd to 5th metatarsals.

tarsophalangeal *a. pert.* tarsus and phalanges.

tarsus *n.* (1) ankle bones; (2) segment of insect leg beyond the tibia, which bears the terminal claw and pulvillus; (3) fibrous connective tissue plate of eyelid. *plu.* **tarsi.**

tartareous *a.* having a rough and crumbling surface.

tartaric acid organic acid that gives acid taste to grapes.

tassel *n.* male inflorescence of maize plant.

tassel-finned fishes a common name for the crossopterygians *q.v.*

taste bud sense organ consisting of a small flask-shaped group of cells, numbers of which are found chiefly on the upper surface of the tongue, and by which the sensations of sweetness, sourness, bitterness and saltiness are perceived.

taste pore orifice in epithelium leading to terminal hairs of sensory cells in a taste bud.

TATA box conserved sequence preceding the startpoint of transcription in many eukaryotic genes, involved in the binding of the complex of transcription factor proteins and RNA polymerase required for initiation of transcription. *alt.* Hogness box.

TATA box binding factor/protein (TBF, TBP) protein subunit of the transcription factor TFIID that binds to the TATA box in the promoters of many eukaryotic genes, and on which the transcription initiation complex is formed.

tau *n.* a microtubule-associated protein present in normal cells, and also in the abnormal paired helical filaments in neurons in brains of patients with Alzheimer's disease.

taurine *n.* amino acid containing sulphonic acid which is a possible neurotransmitter in some invertebrate nervous systems.

taurocholic acid bile acid hydrolysed to taurine and cholic acid.

tautomerism *n.* type of isomerism in which the two isomers (**tautomers**) are readily interconvertible and exist in equilibrium with each other, e.g. H–O–C(CH₃)= CH–CO₂Et (enol) $\leftrightarrow$ (CH₃)C(=O)–CH₂–CO₂Et (keto). *a.* **tautomeric.**

tautonym *n.* the same name given to a genus and one of its species or subspecies.

taxa *plu.* of taxon.

Taxales *n.* order of gymnosperms, being evergreen shrubs or small trees with small linear leaves, and ovules solitary and surrounded by an aril, such as the yews (Taxaceae).

taxes *plu.* of taxis.

taxis *n.* (1) movement of a freely motile cell, such as a bacterium or protozoan, or of part of an organism, towards (positive taxis) or away from (negative taxis) a source of stimulation, such as light (phototaxis) or chemicals (chemotaxis); (2) orientation behaviour related to a directional stimulus. *a.* **tactic.** *plu.* **taxes.**

taxol *n.* anticancer drug obtained from the Pacific yew tree, which inhibits cell division by binding to microtubules and preventing their normal growth and disassociation.

taxon *n.* the members of any particular taxonomic group e.g. a particular species, genus, family. *plu.* **taxa.**

taxonometrics numerical taxonomy *q.v.*

taxonomic category a category used in the classification of living organisms, e.g. phylum, class, order, family, genus or species.

taxonomy *n.* the analysis of an organism's characteristics for the purpose of classification. *see also* systematics.

taxospecies, taxonomic species species defined by similarities in morphological characters only, and which do not necessarily correspond to biological species.

taxy taxis *q.v.*

Tay–Sachs disease an inherited condition characterized by blindness, seizures, degeneration of mental and motor function and early death in childhood in homozygotes. It is caused by a recessive defect in the gene for hexosaminidase A, which leads to an abnormal accumulation of ganglioside GM₂ in the central nervous system in homozygotes.

TBF, TBP TATA box binding factor/protein *q.v.*

TBSV tomato bushy stunt virus.

TC₅₀ a measure of infectivity for viruses, the dose at which 50% of tissue cultures become infected and show degeneration.

TCA cycle tricarboxylic acid cycle *q.v.*

T cell (1) (*immunol.*) T lymphocyte *q.v. see also* CD4 T cells, CD8 T cells, effector T cell, memory cells, regulatory T cell, Th1 cell, Th2 cell, Th17 cell; (2) (*zool.*) sensory cell responding to touch in leech segmental ganglion.

T-cell co-receptor *see* CD4, CD8.

T-cell help stimulation of antibody production in B cells by activated helper T cells.

T-cell receptor (TCR) the antigen receptor of T lymphocytes (*q.v.*), a highly variable antigen-specific heterodimeric glycoprotein present on the surface of T lymphocytes, which recognizes and binds antigen in the form of a peptide bound to an MHC molecule on the surface of another cell. Like antibodies, the receptor consists of an antigen-specific variable portion and a constant region. In conjunction with other signals, activation of the T-cell receptor on a naive T lymphocyte by binding antigen leads to activation of the T cell and its proliferation and differentiation into effector T cells. Effector T cells in turn recognize antigen on their target cells via their T-cell receptors, which stimulates the effector T cells to carry out their various functions. Two types of T-cell receptor have been identified in mammals: αβ, made up of an α chain and a β chain, and γδ, made up of a γ chain and a δ chain. Like antibody genes, T-cell receptor genes undergo re-arrangement before they can be expressed. *see also* B lymphocyte, CD4 T cells, CD8 T cells, cytotoxic T lymphocyte, effector T cell, helper T cells, immune response, memory cell, regulatory T cell, Th1 cell, Th2 cell, Th17 cell.

TDN total digestible nutrients *q.v.*

T-DNA part of the Ti plasmid of *Agrobacterium* that is transferred into the host plant cell genome.

TDP (1) thiamine diphosphate *q.v.*

TDP, dTDP thymidine diphosphate *q.v.*

TdT terminal deoxynucleotidyltransferase *q.v.*

TEA tetraethylammonium *q.v.*

tear pit dacryocyst *q.v.*

tectal *a.* of or *pert.* tectum.

tectorial membrane a shelf of tissue running the length of the cochlea above the hair cells, which have their stereocilia (hairs) embedded in it. Movement of the tectorial membrane in response to sound-induced vibrations in the fluid of the inner ear and movements of the cochlear partitions stimulates the hair cells, which transmit a signal to the auditory nerve.

tectorium *n.* (1) tectorial membrane *q.v.*; (2) coverts of birds.

tectospondylic *a.* having vertebrae with several concentric rings of calcification, as in some elasmobranchs.

tectostracum *n.* thin waxy outer covering of exoskeleton in ticks and mites.

tectrices *n.plu.* small feathers covering bases of remiges. *alt.* wing coverts.

tectum *n.* (1) structure in brain of amphibians and birds to which the visual nerve projects; (2) a roof-like structure, such as the corpora quadrigemina in brain which forms the roof of the mesencephalon (midbrain).

teeth *n.plu.* (1) hard, bony, outgrowths from jaws of mammals, each tooth composed of a core of soft tissue (pulp) supplied with nerves and blood vessels, surrounded by a layer of dentine and covered with an outer layer of enamel. The adult set generally comprises molars or grinding teeth, premolars, canines and incisors or biting teeth; (2) similar structures in jaws or throat of reptiles and fish; (3) any similarly shaped structures in invertebrates used

for rasping, seizing or grinding food, and which are generally composed of chitin or keratin; (4) (*bot.*) the pointed outgrowths on edges of leaves, petals or calyx.

tegmen *n.* (1) (*bot.*) the integument or inner seed coat; (2) (*zool.*) thin, hardened forewing of grasshoppers, stick insects and cockroaches. *plu.* **tegmina**.

tegmentum *n.* (1) tegmen *q.v.*; (2) (*bot.*) a protective bud scale; (3) (*neurobiol.*) dorsal part of cerebral peduncles. *plu.* **tegmenta**.

tegula *n.* a tile-shaped structure.

tegument *n.* (1) integument *q.v.*; (2) type of epidermis present in flukes and tapeworms.

teichoic acid polymer of glycerol or ribitol, often also containing *N*-acetylglucosamine, *N*-acetylgalactosamine and alanine, forming the cell wall of Gram-positive bacteria.

tela *n.* (1) a web-like tissue; (2) folds of the pia mater forming roof of 3rd and 4th ventricles in brain; (3) interlacing fibrillar or hyphal tissue in fungi.

telarian *a.* web-spinning.

teleceptor distance receptor *q.v.*

telegamic *a.* attracting females from a distance, *appl.* scent apparatus of lepidopterans.

telencephalon *n.* the anterior part of forebrain, including the cerebral hemispheres, lateral ventricles, optic part of hypothalamus and anterior portion of 3rd ventricle. *a.* **telencephalic**.

teleodont *a. appl.* forms of stag beetle with the largest mandible development.

teleology *n.* the invalid view that evolutionary developments are due to the purpose or design that is served by them. *a.* **teleological**.

teleomorph *n.* the sexual reproductive stage of a fungus. *cf.* anamorph, holomorph.

teleonomy *n.* the idea that if a structure or process exists in an organism it must have conferred an evolutionary advantage. *a.* **teleonomic**.

teleoptile *n.* a feather of definitive plumage of adult bird.

Teleostei, teleosts *n., n.plu.* group of fish including all modern bony fishes except lungfishes, holosteans and crossopterygians. They have thin bony scales covered by an epidermis, a homocercal tail, a hydrostatic air bladder (swimbladder), no spiracle and no spiral valve in the gut.

telereceptor distance receptor *q.v.*

telescopiform *a.* having joints that telescope into each other.

teleutosorus *n.* in rust fungi, a group of developing teleutospores forming a pustule on the host. *alt.* **telium, teliosorus**.

teleutospore *n.* thick-walled spore in rust and smut fungi, known as smut spore in the latter.

teleutostage *n.* the stage in life cycle of a rust fungus when teleutospores are produced. *alt.* **telial stage, teliostage**.

telia *plu.* of telium *q.v.* **telial** *a. pert.* or having telia.

Teliomycetes *n.* group of heterobasidiomycete fungi containing the rusts (Uredinales) and smuts (Ustilaginales).

telium *n.* group of cells that produce teleutospores in rust fungi. *plu.* **telia**.

telmophage *n.* a blood-sucking insect which feeds from a blood pool produced by laceration of blood vessels.

teloblasts *n.plu.* in leeches and other annelid embryos, cells that give rise to the structures of the segments.

telocentric (1) *a.* with a terminal centromere; (2) *n.* a telocentric chromosome.

telocoel *n.* 1st or 2nd ventricle in brain.

telogen *n.* resting stage in the hair growth cycle.

teloglia *n.plu.* glial cells around endings of axon at a neuromuscular junction.

telokinesis *n.* (1) telophase *q.v.*; (2) changes in cell after telophase, where this is considered only as a cytoplasmic division, resulting in the reconstitution of daughter nuclei.

telolecithal *a.* having yolk accumulated in one hemisphere, as in eggs of amphibians.

telolemma *n.* capsule containing a nerve fibre termination, in neuromuscular spindles. *alt.* **end-sheath**.

telome *n.* (1) morphological unit in a vascular plant which is either a terminal branch bearing a sporangium and a vascular supply, or the simplest part of the plant body, whether terminal or not; (2) the sterile and fertile axes of certain fossil leafless primitive vascular plants.

telomerase *n.* RNA-containing enzyme responsible for replicating telomeres.

telomere *n.* region of highly repetitive DNA at the extreme ends of a eukaryotic chromosome, which 'seals' the end of the chromosome and prevents it becoming joined to another chromosome. Telomeres comprise inverted repeated DNA sequences to which various proteins are bound. They become shortened at each round of DNA replication and in some cells are restored by the enzyme telomerase.

telophase *n.* the stage of mitosis and meiosis in which sister chromatids or homologous chromosomes (telophase I in 1st meiotic division) reach the poles of the spindle, cytoplasmic division takes place, and nuclei are reformed.

teloplasm *n.* cytoplasm in eggs of annelids which is involved in the specification of the teloblasts.

telopod *n.* male copulatory appendage in some millipedes.

telotaxis *n.* movement along line between animal and source of stimulus.

telson *n.* (1) the unpaired terminal abdominal segment of crustaceans and the king crab *Limulus*; (2) the 12th abdominal segment in some insect larvae, and in Protura.

telum *n.* last abdominal segment of insects.

TEM transmission electron microscope *q.v.*

Temnospondyli, temnospondyls *n., n.plu.* extinct order of labyrinthodont amphibians, found in the Carboniferous.

temnospondylous *a.* with individual vertebrae not fused but in articulated pieces. *n.* **temnospondyly**.

temperate *a.* (1) *appl.* bacteriophage capable of integrating its DNA into the bacterial chromosome where it lies dormant for many generations, instead of causing a lytic infection; (2) *appl.* climate, moderate, having long warm summers and short cold winters.

temperate rain forest type of forest that develops in some temperate coastal areas with high rainfall or continual moisture from ocean fogs, as the coastal forest of western North America.

temperature coefficient (Q_{10}) quotient of two reaction rates at temperatures differing by 10°C.

temperature-sensitive mutations mutations that only produce an effect when the cell or organism carrying them is kept at a certain temperature. They are generally mutations that cause a small change in the amino acid sequence of the protein product that does not affect the protein's function at one temperature, but is sufficient to disrupt its normal structure and function at another (usually higher) temperature.

template *n.* (1) blueprint or pattern from which a copy can be made, *appl.* role of DNA and RNA in transcription and translation, respectively; (2) in animal behavioural development, a hypothetical internal representation of the final version of a particular behaviour against which the animal measures its current performance.

template strand the strand in a DNA double helix that is acting as the template for DNA synthesis or for transcription.

temporal *a.* (1) *pert.* time; (2) in the region of the temples; (3) *appl.* compound bone on side of mammalian skull whose formation includes fusion of petrosal and squamosal.

temporalis *n.* broad radiating muscle arising from whole temporal fossa and extending to coronoid process of mandible.

temporal isolation prevention of interbreeding between different species as a result of time differences in events such as shedding of pollen or mating.

temporal lobe of brain, lateral lobe of cerebral hemisphere behind the main sulcus.

temporomalar *a. appl.* branch of maxillary nerve supplying the cheek.

temporomandibular *a.* (1) *appl.* articulation: the hinge of the jaws; (2) *appl.* external lateral ligament between zygomatic process of temporal bone and neck of mandible.

tenacle, tenaculum *n.* (1) holdfast of algae; (2) ectodermal area modified for adhesion of sand grains, in certain sea anemones; (3) in Collembola, paired appendages of 3rd abdominal segment modified to retain furcula; (4) in teleost fishes, a fibrous band stretching from the eyeball to skull.

tendines *plu.* of tendo. *see* tendon.

tendinous *a.* (1) of the nature of a tendon; (2) having tendons.

tendon *n.* a band or cord of white fibrous connective tissue connecting a muscle with a movable structure such as a bone. *alt.* **tendo**.

tendon cells fibroblasts in white connective tissue, with wing-like processes extending between bundles of collagen fibres.

tendon organ type of sensory receptor within muscle, found near tendon–muscle junctions, sensitive to contraction of nearby muscle fibres. *alt.* Golgi tendon organ.

tendon reflex contraction of muscles in a state of slight tension by a tap on their tendons.

tendril *n.* specialized twining stem, leaf, petiole or inflorescence by which climbing plants support themselves.

tendril fibres cerebellar nerve fibres with branches communicating with dendrites of Purkinje neurons.

tendrillar *a.* (1) acting as a tendril; (2) twining.

teneral *a.* (1) immature; (2) *appl.* stage of some insects on emergence from the nymphal covering.

Tenericutes *n.* one of the main divisions of the prokaryotic kingdom in some classifications, which includes the various obligate intracellular parasitic, often wall-less, prokaryotes such as rickettsias, mycoplasmas and chlamydias.

tenia *alt.* spelling of taenia *q.v.*

Tenon's capsule the fibroelastic membrane surrounding the eyeball from optic nerve to ciliary region.

tenoreceptor *n.* a proprioceptor in tendon, reacting to contraction.

ten per cent law the generalization that 90% of energy at one trophic level in a food chain is lost as respiration when being transformed into the energy of the next trophic level.

tension wood reaction wood of dicotyledons, having little lignification and many gelatinous fibres, and produced on the upper side of bent branches.

tensor *a. appl.* muscles that stretch parts of the body.

tentacle *n.* (1) slender flexible organs on head of many animals, used for feeling, exploration, grasping or attachment; (2) adhesive structure of insectivorous plants, e.g. sundew. *alt.* tentacula.

tentacular *a.* (1) *pert.* tentacles; (2) *appl.* canal branching from perradial canal to tentacle base in ctenophores.

tentaculiform *a.* like a tentacle in shape or structure.

tentaculocyst *n.* in some coelenterate medusae, a sense organ consisting of a modified club-shaped tentacle on the margin of the bell, containing one or more lithites.

tentaculozooids *n.plu.* long, slender tentacular individuals at outskirts of hydrozoan colony.

tentaculum tentacle *q.v. plu.* **tentacula.**

tentilla, tentillum *n.* a branch of a tentacle.

tentorium *n.* (1) chitinous framework supporting brain in insects; (2) transverse fold of dura mater, ossified in some mammals, that separates cerebellum and occipital lobes of brain.

tenuissimus *a. appl.* a slender muscle beneath biceps femoris in some mammals.

Tenuivirus *n.* genus of negative-strand RNA plant viruses with filamentous particles, occasionally branched.

teosinte *n.* any of several species of large grasses resembling maize that occur in Central America.

tepal *n.* a perianth segment which is not differentiated into petal or sepal.

teratocarcinoma *n.* a cancer originating from a teratoma.

teratogen *n.* any agent that can cause malformations during embryonic development, e.g. the drug thalidomide. *a.* **teratogenic.**

teratogenesis *n.* abnormal development, resulting in a foetus with congenital deformities.

teratology *n.* the study of abnormal embryonic development and congenital malformations.

teratoma *n.* abnormal growth of an oocyte or testis germ cell *in situ*, resulting in a disorganized mass of cells of many different types and undifferentiated stem cells that continue to divide.

terebra *n.* ovipositor modified for boring, sawing or stinging as in certain bees and wasps.

terebrate *a.* adapted for boring.

teres *n.* (1) the round ligament of liver; (2) two muscles, teres major and minor, extending from scapula to humerus.

terete, teretial *a.* nearly cylindrical in section, as stems.

tergite *n.* dorsal plate of exoskeleton of a typical arthropod skeleton.

tergum *n.* (1) dorsal part of typical arthropod segment; (2) the notum (*q.v.*) in insects; (3) dorsal plate of barnacles. *plu.* **terga.** *a.* **tergal.**

terminal *a.* (1) *pert.* or situated at the end, as terminal bud at end of twig; (2) final, *appl.* the last stage of cell differentiation, after which a cell cannot differentiate further.

terminal complement components complement proteins C5, C6, C7, C8, C9, which form a pore in cell membrane, leading to lysis of the cell.

terminal deoxynucleotidyl transferase (TdT) enzyme that adds nucleotides to the 3′ ends of DNA. In T cells and B cells it adds non-templated nucleotides at the junctions between V, J and D gene segments during gene rearrangement. *alt.* terminal transferase.

terminalia *n.plu.* external genitalia in dipterans.

terminalization *n.* movement of chiasmata towards ends of chromosomes during diplotene and telekinesis stages of meiosis.

terminal oxidase an oxidase which reacts with oxygen to form water at the end of an electron-transport chain, e.g. cytochrome *c* oxidase in the respiratory chain.

termination codon any of three codons that can signal the end of a protein-coding sequence in DNA or RNA, and at which polypeptide synthesis stops, and which are (in mRNA) UAA, UAG or UGA. *alt.* nonsense codon, stop codon.

terminator *n.* region of DNA involved in and necessary for correct termination of transcription and release of the RNA transcript.

termitarium *n.* elaborately constructed nest of a termite colony.

termites common name for the Isoptera *q.v.*

termiticole *n.* organism that lives in a termite nest, e.g. some fungi and insects.

ternary, ternate *a.* (1) arranged in threes; (2) having three leaflets to a leaf; (3) trilateral, *appl.* symmetry; (4) *appl.* crosses involving three genera.

ternatopinnate *a.* having three pinnate leaflets to each compound leaf.

terpenes, terpenoids *n.plu.* large class of natural plant products based on the isoprenoid unit. They include volatile aromatic compounds such as menthol, pigments such as β-carotene, and gibberellins.

terrestrial *a.* (1) *appl., pert.* or living on Earth, as opposed to any other planet or in space, *cf.* extraterrestrial; (2) living or found on land, as opposed to in rivers, lakes or oceans or in the atmosphere.

terricolous *a.* inhabiting the soil.

terriherbosa *n.* terrestrial herbaceous vegetation.

territoriality *n.* a social system in which an animal establishes a territory which it defends against other members of the same species.

territory *n.* (1) area defended by an animal or group of animals, mainly against other members of the same species; (2) an area sufficient for food requirements of an animal or aggregation of animals; (3) foraging area.

tertiary *a.* third, or *pert.* a third level of organization.

Tertiary *n.* geological sub-era (sometimes considered as a period) lasting from *ca.* 65 million years ago to 2.5 million years ago. It followed a mass extinction event at the end of the preceding Cretaceous period, which saw the extinction of the dinosaurs. During the Tertiary, mammals diversified and the ancestors of humans evolved.

tertiary consumer a carnivore that eats other carnivores.

tertiary feathers third row of flight feathers in bird's wing, attached to humerus. *alt.* scapulars, **tertials**.

tertiary parasite an organism parasitic on a hyperparasite.

tertiary roots roots produced by secondary roots.

tertiary structure the overall three-dimensional conformation of the folded polypeptide chain in a protein.

tessellate *a.* (1) chequered, *appl.* markings or colours arranged in squares; (2) *appl.* epithelium formed of cuboid cells.

test, testa *n.* (1) shell or hard outer covering; (2) (*bot.*) seed coat.

testaceous *a.* (1) protected by a shell-like outer covering; (2) made of shell or shell-like material.

test cross the mating of an organism to a double recessive in order to determine whether it is homozygous or heterozygous for a character under consideration.

tester *n.* individual homozygous for one or more recessive alleles, used in a test cross.

testes *plu.* of testis.

testicle testis *q.v.*

testicular *a.* (1) *pert.* testis; (2) testicle-shaped; (3) (*bot.*) having two oblong tubercles, as in some orchids.

testicular-feminization syndrome inherited X-linked condition in humans (and other mammals) in which genetic (XY) males show an immature female human phenotype as a result of the unresponsiveness of their tissues to the male sex hormone testosterone.

testis *n.* male reproductive gland producing spermatozoa. *alt.* testicle. *plu.* **testes**.

testis-determining region small region of the mammalian Y chromosome that is required for development as a male.

testosterone *n.* steroid hormone produced chiefly by the testis in vertebrates. It is the main male sex hormone in mammals. It is secreted from an early stage in mammalian development by the gonads of male embryos and stimulates the differentiation of male sexual organs.

testudinate *a.* having a hard protective shell, as of tortoise.

tetanic *a. pert.* tetanus.

tetaniform, tetanoid *a.* like tetanus.

tetanus *n.* (1) state of a muscle undergoing a continuous series of contractions due to electrical stimulation; (2) disease characterized by progressive paralysis, caused by infection with the bacterium *Clostridium tetani*, which produces a powerful protein neurotoxin, tetanus toxin, that blocks the release of neurotransmitter at nerve endings.

Tethys *n.* the sea between Laurasia and Gondwana. *alt.* **Tethyian sea**.

tetra- prefix derived from Gk *tetras*, four, signifying having four of, arranged in fours, divided into four parts.

tetra-allelic *a. appl.* a polyploid with four different alleles at a locus.

tetrabranchiate *a.* having four gills.

tetracarpellary *a.* having four carpels.

tetracerous *a.* four-horned.

tetrachaenium *n.* fruit composed of four adherent achenes, as in members of the mint family (Labiatae).

tetrachotomous *a.* divided up into fours.

tetracoccus *n.* any microorganism found in groups of four.

tetracotyledonous *a.* having four cotyledons.

tetractinal *a.* with four rays.

tetracyclic *a.* with four whorls, *appl.* flowers.

tetracycline *n.* antibiotic produced by *Streptomyces*, which inhibits the binding of tRNA to bacterial ribosomes, and which can be used against both Gram-negative and Gram-positive bacterial infections.

tetracytic *a. appl.* plant stomata accompanied by four subsidiary cells.

tetrad *n.* (1) a group of four, esp. the four spores formed by 1st and 2nd meiotic divisions of spore mother cell, esp. in fungi; (2) group of four chromatids at meiosis; (3) pair of homologous chromosomes separating at mitosis, forming a quadrangular shape.

tetradactyl *a.* having four digits.

tetrad analysis the use of tetrads (1) to study the sequence of events in meiosis.

tetraethylammonium (TEA) quaternary ammonium compound, selectively blocks potassium conductance channels in neurons, thereby blocking the generation of nerve impulses.

tetragenic *a.* controlled by four genes, *appl.* characters.

tetragonal *a.* having four angles.

tetrahedral *a.* (1) having four triangular sides; (2) *appl.* apical cell in plants having a unicellular growing point.

tetrahydrofolate *n.* tetrahydropteroyl-glutamate, which is a donor of one-carbon units in many biosyntheses, e.g. in the synthesis of dTMP and other nucleotides, and an acceptor of one-carbon units in degradative reactions. It is required in the diet of mammals, e.g. as the vitamin folic acid.

tetra-iodothyronine thyroxine *q.v.*

tetralophodont *a. appl.* molar teeth with four ridges.

tetramer *n.* (1) molecule formed of four subunits; (2) a sequence of four nucleotides in a nucleic acid. *a.* **tetrameric.**

tetramorphic *a.* having four forms.

tetrandrous *a.* having four stamens.

tetraparental *a.* having four parents, *appl.* mice produced from fusion of blastomeres from two different embryos.

tetrapetalous *a.* having four petals.

tetraphyllous quadrifoliate *q.v.*

tetraploid (1) *a.* having four sets of chromosomes per somatic cell; (2) *n.* a tetraploid organism or cell.

tetrapod *n.* a four-footed animal.

tetrapterous *a.* having four wings or wing-like processes.

tetraquetrous *a.* having four angles, as some stems.

tetraradiate *a. appl.* pelvic girdle consisting of pubis, prepubis, ilium and ischium.

tetrarch *a. appl.* roots and stems in which the primary xylem of the stele forms a four-lobed cylinder (in cross-section somewhat like a Maltese cross), with four alternating bundles of phloem.

tetrasaccharides *n.plu.* carbohydrates made up of four monosaccharide units, e.g. stachyose.

tetraselenodont *a.* having four crescentic ridges on molar teeth.

tetrasepalous *a.* having four sepals.

tetraseriate *a.* arranged in four rows.

tetrasomic (1) *a. pert.* or having four homologous chromosomes; (2) *n.* an organism with four chromosomes of one type.

tetraspan *a. appl.* a superfamily of membrane proteins with four membrane-spanning segments.

tetraspermous *a.* having four seeds.

tetrasporangium *n.* sporangium produced on tetrasporophyte phase of red algae, each producing four haploid tetraspores from which the male and female gametophytes develop.

tetraspore *n.* (1) one of a group of four non-motile haploid spores produced by tetrasporangium of red algae; (2) one of four basidial spores, in certain fungi.

tetrasporic, tetrasporous *a.* four-spored.

tetrasporocystid *a. appl.* oocyst of sporozoans when four sporocysts are present.

tetrasporophyte *n.* diploid vegetative phase in life cycle of red algae which bears tetrasporangia in which meiosis occurs.

tetraster *n.* a mitotic spindle having four poles rather than two, found in zygotes produced by polyspermy.

tetrastichous *a.* arranged in four rows.

tetrathecal *a.* having four compartments or chambers.

tetratype *n.* a tetrad type containing four different genotypes, two parental and two recombinant.

Tetraviridae *n.* family of insect RNA viruses.

tetrazoic *a.* having four sporozoites.

tetrodotoxin (TTX) *n.* poison obtained from puffer fish which selectively blocks the voltage-gated sodium channel in neurons and muscle fibres, thereby blocking the generation of nerve impulses and muscle contraction.

tetrose *n.* any monosaccharide having the formula $(CH_2O)_4$, such as erythrose.

TF abbreviation for transcription factor, esp. the general transcription factors of the RNA polymerase II transcription initiation complex, e.g. TFIIA, TFIIB, TFIID, TFIIG, TFIIH.

T form the 'tense' form of an allosteric protein, the form having a lower affinity for the substrate.

TGF-β transforming growth factor-β *q.v.*

TGMV tomato golden mosaic virus.

TGN *trans* Golgi network *q.v.*

Th1 cell type of effector CD4 T cell that can activate macrophages to kill pathogenic bacteria that have taken up residence inside the macrophage. Th1 cells can also interact with activated B cells specific for the same antigen recognized by the T cell, 'helping' the B cells to differentiate into antibody-producing plasma cells. *cf.* Th2 cell, Th17 cell.

Th2 cell type of effector CD4 T cell that interacts with activated B cells specific for the same antigen as the T cell, 'helping' the B cells to differentiate into antibody-producing plasma cells. *alt.* helper T cell. *cf.* Th1 cell, Th17 cell.

Th17 cell type of effector CD4 T cell that acts early in an adaptive immune response to help recruit neutrophils to a site of infection. It has also been implicated in some types of autoimmune disease.

thalamencephalon *n.* (1) the part of the forebrain comprising thalamus, corpora geniculata and pineal body; (2) diencephalon *q.v.*

thalamomamillary *a. appl.* fasciculus: Vicq-d'Azyr, bundles of *q.v.*

thalamus *n.* (1) large paired egg-shaped masses of grey matter in the diencephalon of vertebrate forebrain, below lateral ventricles and surrounding 3rd ventricle. Many sensory pathways converge on the thalamus before being passed on to the cerebral cortex; (2) (*bot.*) the receptacle of a flower.

thalassaemia *n.* any of a group of hereditary anaemias in which various defects in haemoglobin production are present as a result of defective globin genes. α-Thalassaemia is characterized by non-production of α-globin, β-thalassaemia by the non-production of functional β-globin. Thalassaemia minor is the mild or asymptomatic condition in heterozygotes for a defective gene, thalassaemia major the severe and potentially fatal anaemia that develops in some homozygotes. *alt.* **thalassemia**.

thalassoid *a. pert.* freshwater organisms resembling, or originally, marine forms.

thalassoplankton *n.* marine plankton.

Thaliacea *n.* class of tunicates (phylum Hemicordata), which are small, filter-feeding, free-swimming, barrel-shaped marine organisms. Some are colonial.

thalliform thalloid *q.v.*

thalline *a.* consisting of a thallus.

Thallobacteria *n.* in some classifications, a class of prokaryotes containing the actinomycetes and related forms.

thalloid, thallose *n.* growing as a thallus or resembling a thallus.

thallome *n.* a thallus or thallus-like structure.

thallophyte *n.* member of the plant kingdom in which the plant body is not divided into root, stem and leaves, e.g. an alga.

thallus *n.* (1) simple plant body not differentiated into leaf and stem, as of lichens, multicellular algae and some liverworts; (2) the entire body of a fungus.

thamnium *n.* branched and shrub-like thallus of certain lichens.

thanatoid *a.* (1) deadly poisonous; (2) resembling death.

thanatosis *n.* (1) habit or act of feigning death; (2) death of a part.

thaumatin *n.* protein with an intensely sweet taste obtained from the tropical plant *Thaumatococcus*.

Theales *n.* order of dicot trees, shrubs and woody climbers, rarely herbaceous, with evergreen leaves. Includes many families e.g. Dipterocarpaceae, Hypericaceae (St John's Wort) and Theaceae (tea).

thebesian valve *n.* valve in the coronary sinus in right atrium.

theca *n.* (1) protective covering encapsulating various structures, organs or organisms; (2) theca interna *q.v.*

thecacyst *n.* sperm envelope or spermatophore formed by spermatheca.

theca interna internal layer of covering of ovarian follicle which contains oestrogen-secreting cells.

thecal *a.* (1) surrounded by a protective membrane or tissue; (2) *pert.* a theca; (3) *pert.* an ascus.

thecate *a.* covered or protected by a theca.

thecial *a.* within or *pert.* a thecium.

thecium *n.* that part of fungus or lichen bearing spores.

thecodont *a.* having teeth in sockets.

thecodonts *n.plu.* group of primitive reptiles of Permian to Triassic age, including bipedal or crocodile-like forms, and thought to be ancestral to several groups.

thelygenic *a.* producing offspring preponderantly or entirely female.

thelyotoky *n.* parthenogenesis where females only are produced.

thenal *a. pert.* the palm of the hand.

thenar *n.* the muscular mass forming ball of thumb.

theobromine *n.* 3,7-dimethylxanthine, a purine similar to caffeine, found in coffee, tea and chocolate, which is a stimulant of the central nervous sytem and a diuretic.

theophylline *n.* plant alkaloid closely related to theobromine and obtained originally from tea leaves. It inhibits the hydrolysis of cyclic AMP by phosphodiesterase, and is a diuretic and heart stimulant.

theory *n.* in scientific usage, a plausible or scientifically acceptable general principle or body of principles offered to explain phenomena, such as Darwin's theory of evolution. *cf.* hypothesis.

Therapsida, therapsids *n., n.plu.* order of extinct mammal-like reptiles, living from the Permian to the Triassic, believed to be ancestral to mammals.

Theria, therians *n.plu.* a subclass of mammals, including all living mammals except monotremes, with molar teeth bearing a triangle of cusps (tribosphenoid), and ribs fused to vertebrae, and a spiral cochlea.

thermaesthesia *n.* sensitivity to temperature stimuli.

thermal panting method of body cooling employed by some birds and mammals, and a few reptiles, in which heat is lost from the body by panting, which increases the evaporation of water from the moist surface tissues of the respiratory tract.

thermal pollution discharge of heat into the environment from industrial processes (e.g. the release of cooling water from a power station into a river), raising the temperature of the environment. The rise in temperature may affect living organisms directly, and the warmer water also holds less dissolved oxygen.

thermium *n.* plant community in warm or hot springs.

thermoacidophilic bacteria group of Archaea (e.g. *Sulfolobus*, *Thermococcus*, *Thermoproteus*) adapted for life in acidic hot springs, thriving at temperatures of 70–75°C and pH range 1–3. *alt.* thermoacidophiles.

thermobiology *n.* study of the effects of thermal energy on all types of living organisms and biological molecules.

thermocleistogamy *n.* self-pollination of flowers when unopened owing to unfavourable temperature.

thermocline *n.* (1) the upper layer of water of rapidly changing temperature in lakes and seas in summer; (2) zone between warm surface water and colder deep water, in which the temperature decreases sharply, in lake, reservoir or ocean.

Thermodesulfobacteria *n.* phylum of the Bacteria distinguished on DNA sequence criteria, containing some thermophilic bacteria (e.g. *Thermodesulfobacterium*).

thermoduric *a.* resistant to relatively high temperatures (70–80°C), *appl.* microorganisms.

thermogenesis *n.* the generation of heat by metabolism, carried out by some animals (e.g. mammals) and some plants (e.g. the *Arum* spadix).

thermolysis *n.* chemical dissociation as a result of heating.

Thermomicrobia *n.* phylum of the Bacteria distinguished on DNA sequence criteria, currently containing a single genus of thermophilic bacteria, *Thermomicrobium*.

thermonasty *n.* nastic movement in plants in response to variations of temperature. *a.* **thermonastic**.

thermoneutral zone for an endothermic animal, the temperature range it can tolerate without change in its metabolic rate.

thermoperiodicity, thermoperiodism *n.* the response of living organisms to regular changes of temperature, either with day or night or season to season.

thermophase *n.* first developmental stage in some annual and perennial plants, and which can be partly or entirely completed during seed ripening if temperature and humidity are favourable. *alt.* vernalization phase.

thermophile *n.* organism with an optimum growth temperature between 45 and 80°C. *a.* **thermophilic.** *cf.* hyperthermophile.

thermophilous *a.* heat-loving, *appl.* plants.

thermophobic *a.* able to live or thrive only at relatively low temperatures.

thermophylactic *a.* (1) heat-resistant; (2) tolerating heat, as certain bacteria.

thermophyte *n.* a heat-tolerant plant.

thermoreceptor *n.* a sense organ that responds to temperature stimuli.

thermoregulation *n.* control of body temperature, either by metabolic or behavioural means, so that a more-or-less constant temperature is maintained.

thermoscopic *a.* adapted for recognizing changes of temperature, e.g. sense organs of certain cephalopods.

thermotaxis *n.* movement in response to temperature stimulus. *a.* **thermotactic.**

Thermotogae *n.* phylum of the Bacteria distinguished on DNA sequence criteria, containing the hyperthermophilic chemoorganotroph *Thermotoga*.

thermotropism *n.* curvature in plants in response to temperature stimulus.

theromorphs *n.plu.* an extinct order of primitive mammal-like reptiles of the Carboniferous to Permian, having a primitive sprawling gait and including the sail-back lizards.

therophyllous *a.* (1) having leaves in summer; (2) with deciduous leaves.

therophyte *n.* a plant which completes its life cycle within a single season, being dormant as seed during unfavourable period, i.e. an annual.

theropod *a. appl.* a group of small carnivorous bipedal dinosaurs, a suborder of the Saurischia, the lizard-hipped dinosaurs.

thiamine *n.* a water-soluble vitamin, a member of the vitamin B complex, found especially in seed embryos and yeast, its absence causing beriberi in humans, or polyneuritis, and which is a precursor of the coenzyme thiamine diphosphate (thiamine pyrophosphate) required for carbohydrate metabolism. *alt.* thiamine, vitamin B_1.

thiamine diphosphate, thiamine pyrophosphate (TDP, TPP) cofactor for key enzymes in the pentose phosphate pathway, Calvin cycle of photosynthesis and some decarboxylation reactions.

thick filament filament made of the protein myosin, which is one of the two types of interdigitating protein filaments in muscle cell myofibrils. *cf.* thin filament. *see* Fig. 28 (p. 425).

thigmaesthesia *n.* the sense of touch.

thigmokinesis stereokinesis *q.v.*

thigmotaxis stereotaxis *q.v.*

thigmotropism *n.* (1) growth curvature in reponse to a contact stimulus found in clinging plant organs such as stems, tendrils; (2) response of sessile organisms to stimulus of contact. *a.* **thigmotropic.**

thin filament one of two types of interdigitating protein filaments in muscle cell myofibrils, containing actin, tropomyosin and troponin. *cf.* thick filament. *see* Fig. 28 (p. 425).

thinophyte *n.* plant of sand dunes.

thiobacilli *n.plu.* non-filamentous, chemolithotrophic Bacteria of the genus *Thiobacillus*, characterized by their use of elemental sulphur or other inorganic sulphur compounds.

thioester *n.* high-energy −S–O− bond formed by a condensation reaction between an acyl group and a sulphydryl (−SH) group. Present in acetyl-CoA and in many enzyme–substrate complexes.

thiogenic *a.* sulphur-producing, *appl.* bacteria utilizing sulphur compounds.

thiol *n.* organic compound that contains an −SH group.

thiolytic *a. appl.* reactions or enzymes that break S–C bonds.

thiophilic *a.* an organism thriving in the presence of sulphur compounds, as certain bacteria.

Thiopneutes *n.* in some classifications the name for the sulphate-reducing bacteria (e.g. *Desulfovibrio, Desulfuromonas*).

thioredoxin *n.* protein cofactor in some enzyme reactions, undergoes redox reactions via the –SH group of active-site cysteines and is a carrier of reducing power.

thoracic *a. pert.* or in region of thorax.

thoracic duct large lymphatic vessel conveying lymph from most of the body to left subclavian vein.

thoracolumbar *a.* (1) *pert.* thoracic and lumbar region of spine; (2) *appl.* nerves: the sympathetic system.

thoracopod *n.* thoracic leg of crabs and lobsters.

thorax *n.* (1) in higher vertebrates, that part of trunk between neck and abdomen, the chest in humans, containing heart and lungs; (2) the region behind head in other animals; (3) in insects, first three segments behind head, bearing legs and wings.

thorny-headed worms Acanthocephala *q.v.*

Thr threonine *q.v.*

thread cells (1) stinging cells of coelenterates; (2) in the skin of hagfishes, cells whose long threads form a network in which mucous secretions of ordinary gland cells is entangled.

threading *see* profile-based threading.

threadworms (1) Nematomorpha *q.v.*; (2) small nematode worms, commonly also called pin worms, common inhabitants of human bowel.

threatened *a. appl.* wild species that is still abundant in its natural range but is likely to become endangered because of declining numbers. *see also* endangered, rare, vulnerable.

threonine (Thr, T) *n.* an amino acid, aminohydroxybutyric acid, with an uncharged polar side chain. It is a constituent of protein and essential in human diet.

threshold *n.* (1) value of membrane potential that is just sufficient to trigger an action potential in a neuron; (2) minimal stimulus required for producing sensation; (3) level or value which must be reached before an event occurs.

threshold effect the harmful effect of a small change in environment which exceeds the limit of tolerance of an organism or population, and which becomes evident e.g. as a sudden and dramatic decrease in population size.

threshold trait in genetics, a phenotypic character that cannot be measured quantitatively, such as a predisposition to certain illnesses with a genetic component.

thrips *n.plu.* common name for the insect order Thysanoptera *q.v.*

thrombin *n.* serine protease produced from prothrombin at a wound, which converts the soluble protein fibrinogen to insoluble fibrin in blood clot formation. EC 3.4.21.5.

thrombinogen prothrombin *q.v.*

thrombocyte *n.* (1) platelet *q.v.*, in mammals; (2) in non-mammalian vertebrates, spindle-shaped nucleated cells involved in blood clotting.

thrombocytopenia *n.* lack of platelets in the blood.

thrombokinase Factor X *q.v.*

thrombokinesis *n.* the process of blood clotting.

thrombomodulin *n.* protein produced by endothelial cells which is involved in limiting the extent of blood clotting. It forms a complex with thrombin, which interacts with one component of the coagulation pathway – protein C – to downregulate clotting.

thromboplastid *n.* former name for a blood platelet.

thromboplastin Factor X *q.v.*

thrombosis *n.* blood clot formation, esp. when it causes blockage of a blood vessel.

thromboxanes *n.plu.* compounds derived from arachidonic acid, and which are involved in inflammation, acting to constrict blood vessels and aggregate platelets.

thrum-eyed short-styled, with long stamens extending to mouth of tubular corolla. *cf.* pin-eyed.

thylacogens *n.plu.* substances produced by parasites and causing reactive hypertrophy of the host's tissue at the site of infection.

thylakoids *n.plu.* (1) membrane-bounded structures in the stroma of chloroplast, in which the light reactions of photosynthesis are carried out. The thylakoid membrane, across which the protonmotive force is developed, contains chlorophyll, photosystems I and II, the photosynthetic electron-transport chain and the ATP-synthesizing complex. The membrane encloses a space, the thylakoid space, functionally analogous to the space bounded by the inner membrane of mitochondria;

(2) invagination of the cell membrane in cyanobacteria forming lamellae on which the photosynthetic pigments are borne.

thymectomy *n.* removal of the thymus.

Thymelaeales *n.* order of dicot shrubs comprising the family Thymelaeaceae (mezereon).

thymic *a. pert.* thymus.

thymidine *n.* nucleoside made up of the pyrimidine base thymine linked to deoxyribose. *alt.* deoxythymidine.

thymidine diphosphate (TDP, dTDP) thymidine nucleotide containing a diphosphate group, forms part of some activated sugars.

thymidine kinase (TK) enzyme catalysing the conversion of thymidine to thymidine-5′-monophosphate in the minor pathway of DNA biosynthesis. EC 2.7.1.21.

thymidine monophosphate (TMP, dTMP) nucleotide composed of thymine, deoxyribose and a phosphate group, a product of partial hydrolysis of DNA, synthesized *in vivo* by reduction of UMP to dUMP and methylation of the uracil ring. *alt.* deoxythymidine monophosphate, deoxythymidylate, thymidylate, thymidylic acid, thymidine 5′-phosphate. *cf.* ribothymidylate.

thymidine triphosphate (TTP, dTTP) thymidine nucleotide containing a triphosphate group, one of the four deoxyribonucleotides needed for DNA synthesis, also acts in some metabolic reactions in a manner analogous to ATP.

thymidylate thymidine monophosphate *q.v.*

thymidylyltransferase *see* nucleotidyltransferase.

thymine (T) *n.* a pyrimidine base, one of the four nitrogenous bases in DNA, in which it pairs with thymine (A). It is the base in the nucleoside thymidine. *see* Fig. 5 (p. 69).

thymine dimer *see* pyrimidine dimer.

thymocyte *n.* a developing T cell while resident in the thymus.

thymovidin *n.* a hormone produced by thymus in birds, which influences egg albumen and shell formation.

thymus *n.* a primary lymphoid organ, located in humans in upper chest, in which T lymphocytes develop. In the thymus, self-reactive T cells are removed from the population, and T cells are positively selected for their ability to recognize antigen in complex with the MHC molecules of the body. In humans the thymus begins to atrophy after puberty.

thymus-dependent antigen (TD antigen) antigen that elicits an antibody response only in the presence of T cells. A B-cell response to such an antigen requires T-cell help. Almost all protein antigens are thymus-dependent antigens.

thymus-independent antigen (TI antigen) antigen that can induce an antibody response in the absence of T cells.

thyridium *n.* hairless whitish area on certain insect wings.

thyro-arytaenoid *n.* a muscle of larynx.

thyrocalcitonin calcitonin *q.v.*

thyroepiglottic *a. appl.* ligament connecting epiglottis stem and angle of thyroid cartilage.

thyroglobulin *n.* a glycoprotein containing iodine, the iodine of the thyroid gland being stored chiefly as this protein.

thyroglossal *a.* (1) *pert.* thyroid and tongue; (2) *appl.* an embryonic duct from which the thyroid gland develops.

thyrohyals *n.plu.* greater cornua of hyoid bone.

thyrohyoid *a. appl.* muscle extending from thyroid cartilage to cornu of hyoid bone.

thyroid *a.* (1) shield-shaped, peltate; (2) *pert.* or produced by the thyroid gland.

thyroid gland, thyroid shield-shaped gland in neck of vertebrates which secretes the iodine-containing hormones thyroxine and tri-iodothyronine. These regulate oxidative metabolism in the body, and initiate metamorphosis in tadpoles. In mammals the thyroid also secretes the hormone calcitonin. Its function is essential for proper growth.

thyroid hormone receptor intracellular protein of the steroid receptor superfamily that is the receptor protein for thyroid hormones (e.g. thyroxine).

thyroid hormones the hormones secreted by the thyroid gland, chiefly thyroxine and tri-iodothyronine.

thyroid-stimulating hormone (TSH) glycoprotein hormone secreted by the anterior pituitary which regulates the growth of the thyroid gland and the synthesis and secretion of its hormones. *alt.* **thyrotropin.**

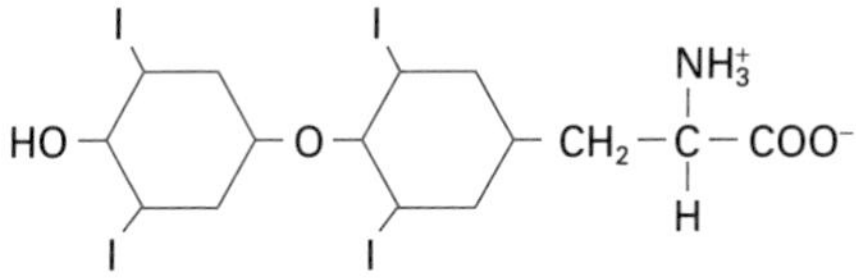

Fig. 38 Thyroxine (tetraiodothyronine).

thyroid-stimulating hormone-releasing hormone (TRH) tripeptide secreted by hypothalamus which causes the secretion and release of thyroid-stimulating hormone from the anterior pituitary.

thyrotropic, thyrotrophic *a.* influencing the activity of the thyroid gland, *appl.* hormone: thyroid-stimulating hormone.

thyrotropin thyroid-stimulating hormone *q.v.*

thyroxine *n.* tetraiodothyronine, iodine-containing hormone produced by the thyroid, derived from the amino acid tyrosine. It initiates metamorphosis in tadpoles, and is essential for normal growth and development in mammals, where it is required throughout life. *see* Fig. 38.

thyrsoid *a.* resembling a thyrsus in shape.

thyrsus *n.* a mixed inflorescence with main axis racemose, later axes cymose, with cluster shaped almost like a double cone. *alt.* **thyrse**.

Thysanoptera *n.* order of small slender sap-sucking insects, commonly called thrips, with piercing mouthparts and no or very narrow wings fringed with long setae, which can become pests. They have an incomplete metamorphosis.

Thysanura, thysanurans *n.*, *n.plu.* order of insects containing the silverfish and the firebrat, small primitive wingless insects with small leg-like appendages on the abdomen and three long processes at the posterior end.

Ti the Ti plasmid (*q.v.*) of *Agrobacterium tumefaciens*.

tibia *n.* (1) shin bone, inner and larger of leg bones between knee and ankle; (2) 4th segment oflegs of insects, spiders and some myriapods.

tibial *a. pert.* or in region of tibia.

tibiale *n.* a sesamoid bone in tendon of posterior tibial muscle.

tibialis *n.* anterior and posterior, tibial muscles acting on ankles and intertarsal joints.

tibiofibula *n.* bone formed of fused tibia and fibula. *a.* **tibiofibular**.

tibiotarsus *n.* tibial bone to which proximal tarsals (ankle-bones) are fused, as in birds. *a.* **tibiotarsal**.

ticks common name for many of the Acari *q.v.*

tidal *a.* (1) *pert.* tides, ebbing and flowing; (2) *appl.* air: volume of air normally inhaled and exhaled at each breath; (3) *appl.* wave: main flow of blood during systole.

tidal rhythms endogenous physiological or behavioural rhythms governed by the bimonthly tidal cycle. They are seen in many marine and shore animals, which synchronize their daily behavioural rhythms with the changes in high- and low-tide levels.

tigellum *n.* the central embryonic axis of a plant, consisting of radicle and plumule.

tight junction area of closely apposed plasma membrane of two adjacent animal cells. Tight junctions prevent lateral diffusion of proteins within the lipid bilayer and are found especially in epithelial cells where they 'seal' the sheet of epithelium, preventing the movement of large and small molecules between the cells.

tiller *n.* flowering stem of a grass.

tiling microarray type of DNA microarray in which the probes are composed of overlapping sequences, usually of a complete genome.

TIM triose phosphate isomerase *q.v.*

timbal *n.* sound-producing organ in cicadas.

timberline *n.* the altitude (in mountains) and latitude above which trees are unable to grow.

time–energy budget in behavioural ecology, the quantitative evaluation of the amount of time and energy an animal expends on various tasks.

Timofeev's corpuscles specialized sensory nerve endings in submucosa of urethra and in prostatic capsule.

Tinamiformes *n.* order of birds including the tinamous.

tinctorial *a.* producing dyestuff.

tinsel *a. appl.* flagella which are branched and feathery. *cf.* whiplash.

tip cell the uninucleate ultimate cell of a hyphal crozier, distal to the dome cell, and directed towards the basal cell.

Ti plasmid a plasmid carried by the crown-gall bacterium *Agrobacterium tumefaciens*, part of which (T-DNA) becomes integrated into the chromosomes of infected tissue. Normally encodes unusual amino acids that are used by the bacterium as a food source. Foreign genes artificially spliced into T-DNA are also stably integrated into the plant cell chromosomes and *Agrobacterium* and its T-DNA is one of the most important vectors in present use for introducing foreign DNA into plant cells to produce transgenic plants. Modified T-DNA that retains its capacity for integration but does not cause tumours is used.

tipophyte *n.* a pond plant.

tissue *n.* an organized aggregate of cells of a particular type or types, e.g. nervous tissue, connective tissue.

tissue culture the culture *in vitro* of cells taken from an animal or plant. *see also* cell culture.

tissue factor non-enzyme lipoprotein released from damaged blood vessels, which together with Factor VIIa initiates blood clotting.

tissue fluid interstitial fluid *q.v.*

tissue graft tissue such as skin transplanted from one individual to another, or from one site on the body to another.

tissue plasminogen activator (t-PA, TPA) a serine proteinase that converts plasminogen to plasmin by limited proteolytic cleavage.

tissue-specific *a. appl.* genes expressed only in a particular cell type or tissue.

tissue type the set of MHC molecules displayed by an individual on the surface of their body cells.

titer *alt.* spelling of titre.

titin *n.* giant protein in a sarcomere of a muscle fibre that extends from the Z disc to the M line along with a thick filament, and acts as a spring to keep the thick filament positioned in the middle of the sarcomere.

titre *n.* the concentration of specific antibodies, antigens or virus particles in a sample. It is often presented in terms of the maximum dilution at which the sample still gives a reaction in an immunological test with a standard preparation of an appropriate antigen or antibody or other assay.

TK thymidine kinase *q.v.*

T lymphocyte, T cell type of lymphocyte that goes through a developmental phase in the thymus. It is responsible for cell-mediated immunity and aids the production of antibodies by B cells. Mature T cells are found in large numbers in the blood and lymphoid tissues. T lymphocytes bear antigen-specific T-cell receptors that interact with a complex of peptide antigen with MHC molecules on surfaces of other cells. After an initial activation by antigen, T cells differentiate into several subtypes of effector T lymphocytes with different functions in the immune response. *see also* CD4 T cells, CD8 T cells, cytotoxic T lymphocyte, memory cell, regulatory T cell, Th1 cell, Th2 cell, Th17 cell. *cf.* B lymphocyte.

TLRs Toll-like receptors *q.v.*

TMP, dTMP thymidine monophosphate *q.v.*

TMV tobacco mosaic virus, a rod-shaped RNA virus causing mottling (mosaic) disease of tobacco and other plants.

TnC troponin C, the calcium-binding subunit of the troponin complex of striated muscle.

TNF tumour necrosis factor *q.v.*

TnI troponin I, the ATPase-inhibiting subunit of the troponin complex of striated muscle.

TnT troponin T, the tropomyosin-binding subunit of the troponin complex of striated muscle.

toad *n.* common name for those members of the amphibian order Anura that are stout-bodied, with a warty skin, and live in damp places, returning only to the water to spawn. There is no biological or evolutionary distinction between frogs and toads and many anuran families contain species of both types. The term frog is generally used to cover all anurans.

toadstool *n.* common name for agaric fungi other than edible mushrooms of the genus *Agaricus*, but which has no biological significance.

Tobamovirus *n.* genus of positive-strand RNA plant viruses with rigid rod-shaped particles, e.g. tobacco mosaic virus.

Tobravirus *n.* genus of rigid rod-shaped single-stranded RNA viruses, e.g. tobacco rattle virus. They are multicomponent viruses in which two genomic RNAs are encapsidated in two different virus particles.

tocopherol vitamin E *q.v.*

Togaviridae *n.* family of small, enveloped, single-stranded RNA viruses that includes Sindbis virus and rubella (German measles). Many are arthropod-borne viruses (arboviruses) such as Semliki Forest virus, Eastern encephalomyelitis virus and Ross River virus.

token stimulus a stimulus which operates indirectly by having become linked through experience with an action or object. For example, colour is the token stimulus attracting bees to some flowers, although they actually want to eat the nectar.

tokology *n.* the study of reproductive biology.

tolerance *n.* (1) the ability to survive and grow in the presence of a normally toxic substance, e.g. heavy metals; (2) (*immunol.*) situation in which an animal does not mount an immune response against an antigen, as e.g. against the antigens carried on its own cells; (3) situation in which an individual becomes less responsive to a drug after repeated doses. *a.* **tolerant**.

Toll-like receptors (TLRs) family of receptor proteins in animals (named after the first-discovered family member, Toll, in the fruit fly) that recognize various general characteristic features of bacteria, viruses, and other microbial pathogens, alerting the innate immune system to the presence of infection, and, in vertebrates, helping to activate the adaptive immune system. Toll itself also has a role in fruit-fly embryonic development.

Tombusviridae *n.* family of positive-strand RNA plant viruses with isometric particles, e.g. tomato bushy stunt virus.

tomentose *a.* covered closely with matted hairs or fibres.

tomentum *n.* (1) the closely matted hair on leaves or stems, or filaments on cap and stem in fungi; (2) other matted structures.

tomite *n.* free-swimming, non-feeding stage following protomite stage in life cycle of certain ciliates.

tomium *n.* the sharp edge of a bird's beak.

tomography *see* computerized axial tomography, PET scan.

tomont *n.* stage in life cycle of holotrichan ciliates when body divides, usually inside a cyst.

tone *n.* (1) sound, especially with reference to pitch, quality and strength; (2) tonus *q.v.*; (3) the condition of tension found in living animal tissue, especially muscle. *alt.* **tonicity**.

tongue *n.* (1) usually movable and protrusible organ on floor of mouth; (2) any tongue-like structure, e.g. radula, ligula; (3) hypopharynx, in some insects.

tongue worms Pentastomida *q.v.*

tonic *a.* producing tension. *see also* tonus.

tonicity (1) osmolarity; (2) *see* tone, tonus.

tonofilament *n.* keratin filament found in and connecting epithelial cells through spot desmosomes.

tonoplast *n.* the membrane delimiting the central vacuole of plant cells.

tonotaxis *n.* a taxis in response to a change in density.

tonotopic *a. appl.* spatially organized mapping of input from auditory nerves onto the auditory cortex, so that each point on the cochlea has a corresponding point on the cortex.

tonsilla *n.* (1) tonsil *q.v.*; (2) posterior lobule at side of cerebellar hemisphere.

tonsillar ring ring of lymphoid tissue formed by the palatine, pharyngeal and lingual tonsils. *alt.* Waldeyer's tonsillar ring.

tonsils *n.plu.* paired lymphoid tissues in pharynx or near base of tongue.

tonus *n.* (1) condition of persistent partial excitation, which in muscles results in a state of partial contraction; (2) in certain nerve centres, the state in which motor impulses are given out continuously without any sensory impulses from the receptors. *alt.* tonicity, tone. *a.* **tonic**.

tool use the manipulation of an object by an animal to achieve some end which it could not achieve without it.

tooth *see* teeth.

toothed whales Odontoceti *q.v.*

toothplate *n.* a flexible plate of cartilage carrying horny teeth in hagfishes, used to tear pieces of prey off and pull them into the mouth.

tooth shells Scaphopoda *q.v.*

topaestheia, topaesthesis *n.* appreciation of the location of a tactile sensation.

topobiology *n.* the concept that topologically positioned molecular activities and cell behaviour can generate morphological form.

topochemical *a. appl.* sense: the perception of odours in relation to track or place, as in ants.

topocline *n.* a geographical variation, not always related to an ecological gradient, but to other factors such as topography and climate.

topodeme *n.* a deme occupying a particular geographical area.

topogamodeme *n.* individuals occupying a precise locality which form a reproductive or breeding unit.

topogenous mire a mire that develops in places where there is a permanently high water table.

topographical maps in cerebral cortex, the point-to-point mapping of points in the sensory field to points in the cortex, resulting in an internal map. The relationship between neighbouring points in the sensory field is maintained in the cortex, albeit often in a distorted, fragmented and duplicated form.

topoinhibition *n.* inhibition of a cellular process as a result of the proximity of other cells.

topoisomerases *see* DNA topoisomerases.

toponym *n.* (1) the name of a place or region; (2) a name designating the place of origin of a plant or animal.

topotaxis *n.* movement induced by spatial differences in stimulation intensity, and orientation in relation to sources of stimuli.

topotropism *n.* orientation towards the source of a stimulus. *a.* **topotropic.**

topotype *n.* a specimen from locality of original type.

toral *a. pert.* a torus.

torcular *n.* occipital junction of venous sinuses with dura mater.

tori *plu.* of torus.

tornaria *n.* the free larval stage in development of some Enteropneusta *q.v.*

tornate *a.* (1) with blunt extremities, as a spicule; (2) rounded off.

torose *a.* having fleshy swellings.

Toroviridae *n.* family of enveloped single-stranded RNA viruses that cause enteric infections in humans and other mammals. The elongated virus particles sometimes bend into the shape of a torus.

torpor *n.* state of complete inactivity accompanied by decreased body temperature and greatly reduced metabolic rate shown by some animals. It may occur on a daily basis, or seasonally, when it is more usually known as hibernation.

torques *n.* a necklace-like arrangement of fur or feathers.

torsion *n.* (1) spiral bending; (2) the twisting around of a gastropod body as it develops. *a.* **torsive.**

torsion angle the angle of rotation of a group at either end of a peptide bond in a protein chain. *see* phi torsion angle, psi torsion angle.

torticone *n.* a turreted, spirally twisted shell.

tortoises Chelonia *q.v.*

torula *n.* (1) a small torus; (2) a small round protuberance; (3) a yeast cell.

torulose *a.* (1) with small swellings; (2) beaded.

torulus antennifer *q.v.*

torus *n.* (1) receptacle of a flower; (2) firm prominence, or marginal fold or ridge; (3) any of the pads on feet of various animals, as of cat. *plu.* **tori.**

total digestible nutrients (TDN) a measure of the nutrient value of a food which combines chemical analysis with estimates of digestibility obtained from analysis of the composition of the faeces.

total range the entire area covered by an individual animal in its lifetime.

totipotent *a. appl.* cells capable of forming any cell type. In mammals, the only genuinely totipotent cells are those of the embryonic morula stage, which give rise to some placental tissue, the extraembryonic membranes and the embryo itself. *n.* **totipotency.** *cf.* multipotent, pluripotent.

touch receptors swellings of the terminals of sensory cutaneous nerves in epidermis of skin which transmit sensation of touch.

toxa toxon *q.v.*

toxic *a.* (1) *pert.*, caused by, or of the nature of a poison; (2) poisonous.

toxicity *n.* the harmful properties of a poison.

toxicogenic toxigenic *q.v.*

toxicogenomics *n.* the study of toxic compounds in terms of the genome-wide gene-expression responses they elicit.

toxicology *n.* the study of poisons and their effects on living organisms.

toxicophorous, toxiferous *a.* holding or carrying poison.

toxic shock syndrome systemic shock induced by toxins secreted by certain bacteria infecting the bloodstream, esp. some strains of *Staphylococcus aureus.* The toxins act as superantigens, provoking a massive and non-specific T-cell response, with the release of cytokines that cause the shock syndrome.

toxigenic *a.* producing a poison.

toxiglossate *a.* having hollow radula teeth conveying poisonous secretion of salivary glands, as certain carnivorous marine gastropods.

toxin *n.* any poison (esp. a toxic protein) derived from a plant, animal or microorganism. *see also* phytotoxin, zootoxin, mycotoxin.

toxognaths *n.plu.* first pair of limbs, with opening of poison duct, in centipedes.

toxoid *n.* a toxin which has been inactivated but which can still stimulate the production of antibodies effective at neutralizing the original toxin.

toxon *n.* a bow-shaped spicule. *alt.* **toxa.**

toxophore *n.* the part of a molecule responsible for its toxic properties.

TP turgor pressure *q.v.*

tPA tissue plasminogen activator *q.v.*

TPA (1) tissue plasminogen activator *q.v.*; (2) tetraphorbol acetate.

TPP thiamine pyrophosphate *q.v.*

trabecula *n.* (1) columnar structure bridging a space, such as row of cells spanning a cavity, small fibrous band forming part of imperfect septa or framework of organs, muscular column projecting from inner surface of ventricles of heart; (2) (*bot.*) rod-like

part of cell wall extending across lumen of cell, plate of sterile cells extending across sporangium of pteridophytes, primordial gill of agarics. *plu.* **trabeculae.**

trabecular *a.* (1) having a cross-barred framework; (2) *pert.* or formed of trabeculae. *alt.* **trabeculate.**

trace elements those chemical elements occurring in minute quantities in living tissue and required in only small amounts as nutrients by living organisms. They are required principally as prosthetic groups for certain proteins. They are chromium, cobalt, copper, manganese, molybdenum, nickel, selenium, tungsten, vanadium, zinc and iron. Not all these elements are required by all living organisms. *alt.* micronutrients. *cf.* macronutrient.

tracer *n.* molecule or atom that has been labelled chemically or radioactively and can thus be followed in a biochemical reaction, during metabolism, or during passage through the body.

trachea *n.* (1) (*zool.*) the windpipe in vertebrates; (2) in insects and other arthropods one of the air-filled tubules of the respiratory system, which opens to the exterior through openings (spiracles) in the sides of thorax and abdomen; (3) (*bot.*) element of xylem tissue of plants with spiral or annular thickenings of the wall. *plu.* **tracheae.** *a.* **tracheal, tracheary.**

tracheal gills small wing-like respiratory outgrowths from the abdomen of some aquatic insect larvae.

tracheary elements the water-conducting cells of the xylem.

tracheate *a.* having tracheae.

tracheid *n.* type of water-conducting xylem cell present in all vascular plants, with lignified secondary cell wall usually containing spiral thickening or bordered pits, but which does not contain perforations through cell wall. *cf.* vessel element.

tracheidal cells cells of the pericycle resembling tracheids.

trachelate *a.* narrowed as in neck formation.

trachelomastoid *a.* (1) *pert.* neck region and mastoid process; (2) *appl.* muscle, longissimus capitis.

trachenchyma *n.* tracheal vascular tissue.

tracheobronchial *a.* (1) *pert.* trachea and bronchi, *appl.* lymph glands; (2) *appl.* a syrinx formed of lower end of trachea and upper bronchi.

tracheole *n.* fine branch of the tracheal or respiratory system in insects and other arthropods, directly supplying the tissues.

Tracheophyta, tracheophytes vascular plants *q.v.*

trachyglossate *a.* with rasping or toothed tongue.

tract *n.* (1) region or area or system considered as a whole, as alimentary tract; (2) distinguishable band or bundle of nerve fibres within the central nervous system.

tractellum *n.* flagellum at anterior end of Mastigophora, or of zoospores, which causes rotatory motion.

tragus *n.* small pointed eminence in front of concha of ear, well-developed in many bats.

trait *n.* a distinct phenotypic character, which may be either heritable or environmentally determined or both. Examples of heritable traits include white and other flower colours, white or red eyes in *Drosophila*, common baldness in humans, and sickle-cell anaemia and other genetic diseases.

trama *n.* the inner core of interwoven hyphae of the gill of agaric fungi.

trans (1) beyond, or on the other side; (2) in genetics, two different mutations at the same locus on a homologous pair of chromosomes are said to be in *trans* if one is on one chromosome and one on the other. *cf. cis*; (3) *appl.* molecular configuration, one of two configurations of a molecule caused by the limitation of rotation around a double bond, the alternative configuration being the *cis*-configuration.

transacetylase acetyltransferase *q.v.*

trans-acting, trans-regulatory *a. appl.* genes or their protein products produced by one chromosome and cooperating with or acting on genes elsewhere.

transacylase acyltransferase *q.v.*

transad *n.* (1) closely related species that have become separated by a geographical or environmental barrier; (2) *adv. appl.* organisms of the same or closely related species which have become separated by an environmental barrier, as European and American reindeer.

transaldolase *n.* widely distributed enzyme catalysing the transfer of a three-carbon (aldehyde) unit from a ketose to an aldose, e.g. the conversion of sedoheptulose 7-phosphate + glyceraldehyde 3-phosphate into erythrose 4-phosphate + fructose 6-phosphate in the pentose phosphate pathway. EC 2.2.1.2.

transalpine *a.* situated to the north of the Alps.

transaminase aminotransferase *q.v.*

transamination *n.* transfer of amino (NH_2) groups from one molecule to another.

transcript *n.* the RNA that is synthesized by RNA polymerase on a DNA or RNA template. *see also* RNA editing, RNA processing, RNA replicase, RNA splicing, transcription.

transcriptase *n.* (1) any enzyme catalysing transcription. *alt.* RNA polymerase, DNA-dependent RNA polymerase.

transcription *n.* the copying of a DNA strand nucleotide by nucleotide, following the base-pairing rules, by an RNA polymerase to produce a complementary RNA copy. In eukaryotes it occurs in the nucleus. *see* Fig. 39.

transcriptional control control of gene expression exerted at the level of initiation of transcription.

transcriptional regulators regulatory proteins that bind to DNA at specific control sites to initiate or prevent transcription and gene expression. Eukaryotic genes typically have complex control sites to which a combination of regulatory proteins must bind to attain a maximum level of transcription.

transcriptional terminator short sequence in a gene that signals the end of transcription.

transcription complex a complex of transcription factors and RNA polymerase that is formed on the promoter of a gene and initiates transcription.

transcription factor general term for any protein that is directly involved in regulating the initiation of transcription. *alt.* gene regulatory protein. *see also* general transcription factor.

transcription unit stretch of chromosome (or DNA *in vitro*) transcribed into a continuous length of RNA. It may comprise one or more genes.

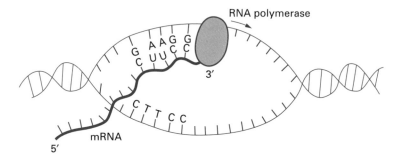

RNA polymerase

3′

mRNA

5′

Fig. 39 Transcription of RNA from DNA.

transcriptome *n.* all the genes that are actually expressed in a particular cell or organism under a particular condition or stage in development. It sometimes refers to all the genes that can be expressed from a given genome.

transcytosis *n.* the passage of materials or solutes across an epithelium by endocytosis at one face of a cell, transport in vesicles and release at the other face.

transdetermination *n.* the switch of a cell from one state of determination to another as seen in some cultured *Drosophila* imaginal discs.

transdifferentiation *n.* the change of a cell from one fully differentiated state to another. *alt.* metaplasia.

transduce *see* transduction.

transducer *n.* any structure that converts a signal from one form into another, e.g. a sensory receptor that converts a touch into an electrical impulse.

transducin (G$_t$) *n.* heterotrimeric G protein in the rod and cone cells of the retina that is involved in the cells' conversion of signals from light-activated rhodopsin into electrical signals for transmission to retinal neurons.

transducing phage a bacteriophage that carries genes picked up from one bacterium and which can transfer them to a new host bacterium.

transduction *n.* (1) the process of conveying or carrying over of information. *see* sensory transduction, signal transduction; (2) (*genet.*) the transfer of genes from

one bacterium to another by carriage as part of bacteriophage DNA. *v.* **transduce**.

transect *n.* a line (line transect *q.v.*), strip (belt transect *q.v.*), or profile (profile transect *q.v.*), chosen to sample the organisms present within a particular habitat and to gain an idea of their distribution. *see also* quadrat, transect sampling.

transection *n.* section across a longitudinal axis. *alt.* cross-section, transverse section.

transect sampling sampling technique in which samples are taken at regular intervals along a line (line transect) or within a strip between two parallel transect lines (belt transect).

transeptate *a.* having transverse septa.

transfection *n.* the genetic modification of cells by DNA added to the culture medium. The DNA enters the cells, which have been made transiently permeable to DNA, and in some cases is stably incorporated into the genome. *v.* **transfect**.

transfer cell type of parenchyma cell in many plant tissues, with ingrowths of cell wall greatly increasing the surface area of plasma membrane. It is probably involved in transport of solutes over short distances.

transfer RNA (tRNA) type of small RNA found in all cells, which carries amino acids to the ribosomes for protein synthesis. Each tRNA is specific for one kind of amino acid only (e.g. tRNA^{Cys}, tRNA^{Lys}). tRNAs enable the addition of the correct amino acid to a growing polypeptide chain by specific base pairing of a three-base anticodon on each tRNA with a complementary codon in messenger RNA.

transferase *n.* any enzyme that catalyses the transfer of a group of atoms from one molecule to another. EC 2.

transferrin *n.* iron-transporting protein found in blood plasma.

transformation *n.* (1) genetic modification of a bacterium by DNA added to the culture and which enters the bacterial cell. Transformation can occur naturally, during mixed infections, and is also used in experimental bacterial genetics and to produce bacteria containing recombinant DNA; (2) changes that occur in cultured cells after infection with tumour viruses, introduction of an oncogene, or treatment with carcinogens. The cells become able to divide indefinitely and undergo various other changes characteristic of a potential cancer cell. In some cases a transformed cell has become fully malignant and can produce a tumour if transplanted into an animal. *alt.* neoplastic transformation.

transformation series *see* evolutionary transformation series.

transformed cell *see* transformation.

transforming *a. appl.* viruses, proteins or genes that can induce neoplastic transformation if introduced into susceptible cultured cells.

transforming growth factor-β (TGF-β) cytokine produced by some CD4 T cells, esp. regulatory T cells, and which tends to inhibit T-cell responses and macrophage activation. Other members of the TGF-β family of proteins (e.g. bone morphogenetic proteins, Nodal and Nodal-related proteins) have multiple key roles as intercellular signals in vertebrate embryonic development.

transforming principle name originally given to the substance responsible for the transformation of avirulent (R) pneumococci to the virulent (S) form, and which was later identified as DNA.

transfusion tissue tissue of gymnosperm leaves consisting of parenchymatous and tracheidal cells.

transgene *n.* any gene transferred into an animal or plant artificially by the techniques of genetic engineering, such organisms being known as transgenic animals

or plants. The transferred gene is usually from another species, but may also be a mutated version of a gene in the organism's own genome.

transgenesis *n.* the deliberate introduction of a gene from one species into another species by genetic engineering techniques.

transgenic *a. appl.* animals and plants into which genes from another species, or mutated versions of their own genes, have been deliberately introduced by genetic engineering.

trans **Golgi network (TGN)** system of interconnected tubules and cisternae at the *trans* face (the face nearest the cell membrane) of the Golgi aparatus. *cf. cis* Golgi network.

transgressive *a. appl.* a species that overlaps two adjacent communities.

transhydrogenase *n.* formerly, one of a group of enzymes which catalyse the transfer of hydrogen from one molecule to another, now placed in EC group 1.

transient *a.* (1) passing; (2) of short duration.

transient diploid the stage of the life cycle of predominantly haploid fungi and algae during which meiosis occurs.

transient polymorphism the existence of two or more distinct types of individuals in the same breeding population for only a short while, one type eventually replacing the other.

transilient *a. appl.* nerve fibres connecting non-adjacent parts of the cerebral cortex.

transit-amplifying cell type of cell found in the basal layer of the epidermis and gut epithelia which is produced by a stem cell and divides and differentiates into differentiated cells.

transit sequence a sequence of amino acids, generally at the N-terminal end, present in newly synthesized proteins destined for chloroplasts. It is required for their transport across the chloroplast membranes from their site of synthesis in the cytoplasm.

transition *n.* mutation in DNA in which a pyrimidine or purine is substituted for another pyrimidine or purine, respectively.

transitional *a.* (1) between one state and another; (2) *appl.* epithelium occurring in ureters and urinary bladder renewing itself by mitotic division of 3rd and inner layer; (2) *appl.* inflorescence intermediate between racemose and cymose; (3) *appl.* endoplasmic reticulum, areas that are partly smooth and partly rough.

transition state the transient chemical species of highest free energy in a chemical reaction, in which chemical bonds are in the process of being made or broken. It occurs at the top of the activation-energy barrier. Enzymes have a much higher affinity for the transition state than for the substrate. *see also* activation energy.

transition state analogue compound similar in structure to the transition state of a chemical reaction, but which inhibits an enzyme-catalysed reaction as it binds to the enzyme but does not react further.

transketolase *n.* widely distributed enzyme that catalyses the transfer of a two-carbon unit from a ketose to an aldose, e.g. the conversion of ribose 5-phosphate + xylulose 5-phosphate into sedoheptulose 7-phosphate + glyceraldehyde 3-phosphate in photosynthesis and in the pentose phosphate pathway. EC 2.2.1.1.

translation *n.* the process by which the genetic information encoded in messenger RNA directs the synthesis of proteins. Messenger RNA provides a template for the step-by-step synthesis of a polypeptide chain. Amino acids are incorporated in the correct order by the matching of codons in messenger RNA and complementary anticodons on transfer RNAs carrying the amino acids. Translation takes place on the ribosomes in the cytoplasm. *see* Fig. 40 (p. 682). *See also* genetic code.

translational control control of gene expression at the level of translation, when mRNAs are produced but not translated, sometimes being stored for future use, as in eggs and early embryos.

translational frameshifting in translation, the bypassing of a normal stop signal and the continuation of translation in a different reading frame. It occurs e.g. in retroviruses to produce the Gag–Pol fusion protein.

translational recoding the insertion of selenocysteine into a protein chain during translation by the pairing of a specialized selenocysteine-carrying tRNA that pairs with the UGA codon that normally signifies a stop signal.

translational research research intended to convert experimental laboratory findings into a medically useful therapy or procedure.

translocase *n.* (1) any enzyme mediating the facilitated diffusion of a substance across a membrane permeability barrier; (2) the elongation factor (*q.v.*) EF-G.

translocation *n.* (1) movement or removal to a different place or habitat; (2) movement of material in solution within an organism, esp. in phloem of plant; (3) chromosomal rearrangement in which part of a chromosome breaks off and is rejoined to a non-homologous chromosome. *see also* reciprocal translocation; (4) *appl.* protein, movement of a protein across a membrane.

translocation factor elongation factor EF-G.

translocation quotient ratio of content of a particular substance in shoot to that in root, a measure of a substance's mobility or relative translocation.

translocator *n.* membrane protein mediating the transfer of a substance across a membrane.

transmedian *a. pert.* or crossing the median line.

transmembrane *a.* across a membrane, as in transmembrane potential, transmembrane protein.

transmembrane protein a protein that completely spans a biological membrane. The orientation of transmembrane proteins with one transmembrane domain has been classified as follows. Type I have a signal sequence which is cleaved, the N-terminus is extracellular, the C-terminus is cytoplasmic. Type II have an N-terminal signal sequence which is not cleaved and acts as the membrane-spanning domain (signal anchor), the N-terminus is cytoplasmic, the C-terminus is extracellular. Type III have no signal sequence, extracellular N-terminus and cytoplasmic C-terminus. Type IV have a signal anchor in the C-terminal region, cytoplasmic N-terminus, extracellular C-terminus. *see* Fig. 26 (p. 392). *see also* membrane protein, membrane transport, membrane transport protein.

transmembrane potential membrane potential *q.v.*

transmethylase

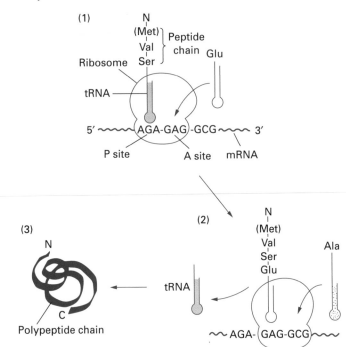

Fig. 40 Stages in translation of messenger RNA to produce a protein. (1) A transfer RNA bearing the part of the polypeptide chain that has already been synthesized occupies the P site on a ribosome. The appropriate aminoacyl-tRNA lines up beside it at the A site, matching its anticodon to the codon in the mRNA. (2) The peptide chain is transferred to the amino acid carried by the incoming tRNA by the formation of a peptide bond. In this way a new amino acid is added to the growing polypeptide. At the same time, the ribosome moves one place along the mRNA, leaving the A site free for the next aminoacyl-tRNA to bind. Having transferred its peptide chain, a tRNA exits from the ribosome. (3) The outcome of translation is a complete protein chain. After synthesis it folds up into a three-dimensional conformation characteristic for each protein.

transmethylase *alt.* methyltransferase.

transmissible *a. appl.* diseases that can be transmitted from one individual to another by contact, or by air, water, food or insect vectors, i.e. diseases caused by bacteria, viruses, fungi and other parasites. *cf.* nontransmissible.

transmissible spongiform encephalopathies group of neurodegenerative diseases of humans and animals in which

the brain has a characteristic spongelike appearance after death. They include scrapie in sheep, bovine spongiform encephalopathy (BSE) in cattle, and Creutzfeld–Jakob disease (CJD) in humans. The transmissible agent, called a prion, and the mechanism of the transmission, are not yet completely understood.

transmission electron microscope (TEM) electron microscope that produces an image from the diffraction of electrons after passage through the specimen. *cf.* scanning electron microscope.

transmission genetics genetics studied from the point of view of the mechanisms involved in the passage of a gene from one generation to the next.

transmitter neurotransmitter *q.v.*

transpalatine *n.* a cranial bone of crocodiles, connecting pterygoid with jugal and maxilla.

transpeptidation *n.* formation of peptide bonds between the short peptides in peptidoglycans of bacterial cell wall.

transpinalis *n.* a muscle connecting transverse processes of vertebrae.

transpiration *n.* the evaporation of water through stomata of plant leaves and stem.

transpirational pull the continuous loss of water through leaves by transpiration, causing the flow of water through xylem from roots to leaves.

transpiration stream the movement of water and inorganic solutes upwards from roots to leaves through the xylem.

transplant (1) *v.* to transfer tissue from one part to another part of the body of the same individual, or to transfer tissue from one individual to another; (2) *n.* tissue transferred in this way. *alt.* graft.

transplantation antigen histocompatibility antigen *q.v.*

transport *see* active transport, facilitated diffusion, passive transport, translocation.

transport protein (1) any protein that binds and transports a particular substance within the body or within a cell; (2) protein that transports a substance or ion across a biological membrane. *alt.* transporters. *see also* membrane transport protein.

transport vesicle small membrane-bounded vesicle which carries material from one intracellular compartment to another within eukaryotic cells, budding off one compartment and fusing with the membrane of another.

transporter *see* transport protein.

transposable genetic elements DNA sequences that possess the property of inserting themselves elsewhere on the chromosomes by a process called transposition. They are present in both eukaryotic and prokaryotic genomes. The

simplest types of transposable element are bacterial insertion sequences, which carry only the genes necessary for their own transposition. Larger bacterial transposons also carry genes for other functions, such as antibiotic resistance. *alt.* mobile DNA elements, transposons. *see also* controlling element, drug-resistance plasmids, insertion sequence, retrotransposon, transposition.

transposase *n.* protein(s) specified by a transposable genetic element and responsible for its transposition.

transposition *n.* (1) the movement of a DNA sequence to another position on the same or a different DNA molecule. This may occur by excision and re-insertion elsewhere, or by replication of the sequence and insertion of a copy elsewhere. DNA sequences that can do this are known as transposable genetic elements or transposons, and most encode all the functions necessary for their transposition. Transposition may result in gene inactivation or changes in gene expression at the site of insertion. *alt.* **transpositional site-specific recombination**. *see also* retrotransposon; (2) translocation (definition) *q.v.*

transposon *n.* transposable genetic element *q.v.*

transposon mutagenesis the use of transposon insertion into a gene to produce a mutation whose location on the chromosome can be determined by tracking the transposon.

transpyloric plane upper of imaginary horizontal planes dividing abdomen into artificial regions.

***trans*-splicing** the generation of a mature RNA by the joining of two separate transcripts, as opposed to the more usual process of RNA splicing *q.v.*

transtubercular *a. appl.* plane of body through tubercles of iliac crests.

transudate *n.* any substance which has oozed out through a membrane or pores.

transvection *n.* the case where a chromosomal rearrangement that prevents synapsis between homologous regions changes the phenotype although the overall genotype remains unchanged. This is due to the ability of a mutation at or near one allele to influence the other allele only when they are synapsed.

transversal *a.* lying across or between, as a transversal wall.

transverse *a.* lying across or between, *appl.* section, a cross-section.

transverse process projection from each side of the neural arch of vertebra, with which a rib articulates.

transverse tubules ingrowths of plasma membrane which form interconnected tubules surrounding each myofibril in a skeletal muscle fibre.

transversion *n.* in nucleic acids, the substitution of a purine for a pyrimidine base, or vice versa.

transversum *n.* in many reptiles, a cranial bone extending from pterygoid to maxilla.

transversus *n.* a transverse muscle, as of abdomen, thorax, tongue, foot.

trapeziform *a.* trapezium-shaped.

trapezium *n.* (1) the 1st carpal bone at base of 1st metacarpal; (2) a portion of the pons Varolii in brain.

trapezius *n.* a broad, flat, triangular muscle of neck and shoulders.

trapezoid *a.* shaped like a trapezium, *appl.* various muscles, bones.

traumatic *a. pert.* or caused by, a wound or other injury.

traumatonasty *n.* a movement in response to wounding.

traumatotropism *n.* a growth curvature in plants in response to wounds.

treadmilling *n.* the process by which subunits are added to one end of an actin filament and removed from the other end at the same rate, resulting in a polymer of constant length.

tree *n.* (1) a woody perennial plant which has a single main trunk at least 7.5 cm in diameter at 1.3 m height, a definitely formed crown of foliage, and a height of at least 4 m; (2) phylogenetic tree *q.v.*

tree ferns a group of tropical and subtropical ferns (*Cyathea*) with stems several metres high.

tree layer the highest horizontal layer in a plant community, comprising the tree canopy.

tree line line marking the northern, southern or altitude limit beyond which trees do not grow.

tree-ring dating dendrochronology *q.v.*

tree rings growth rings *q.v.*

trefoil *n.* a flower or leaf with three lobes.

trehalose *n.* a disaccharide composed of two glucose units, esp. abundant in some lichens and in insects, where it is one of the principal fuels for the flight muscles.

Trematoda, trematodes *n., n.plu.* class of parasitic flatworms, including the digenean gut, liver and blood flukes, such as *Fasciola*, the liver fluke of sheep and cattle, and *Schistosoma* spp. which cause the serious disease schistosomiasis in humans, with tissue damage and inflammation.

tremelloid, tremellose *a.* gelatinous in substance or appearance.

TRH thyroid-stimulating hormone releasing hormone *q.v.*

triactinal *a.* three-rayed.

triacylglycerol *n.* uncharged fatty acid ester of the three-carbon sugar alcohol glycerol, containing three fatty acids per molecule. Triacylglycerols are the storage form of fatty acids and the main components of plant and animal fats and oils. *alt.* neutral fat, triglyceride.

triadelphous *a.* with stamens united by their filaments into three bundles.

triaene *n.* a somewhat trident-shaped spicule.

trial-and-error conditioning or learning a kind of learning in which a random and spontaneous response becomes associated with a particular stimulus, because that response has always produced a reward whereas other responses have not done so. *alt.* instrumental conditioning, operant conditioning.

triallelic *a. appl.* polyploid with three different alleles at a locus.

triandrous *a.* having three stamens.

triangularis *n.* (1) muscle from mandible to lower lip, which pulls down corner of mouth; (2) muscle and tendinous fibres between dorsal surface of sternum and costal cartilages, transversus thoracis, which assists breathing out.

trianthous *a.* having three flowers.

triarch *n.* refers to roots, stems in which the primary xylem of the stele forms a three-lobed cylinder, with three alternating bundles of phloem.

triarticulate *a.* three-jointed.

Triassic *n.* geological period lasting from *ca.* 251 million years ago to 200 million years ago, the first period in the Mesozoic era and followed the mass extinction at the Permian–Triassic boundary, when *ca.* 90–95% of marine species are estimated to have gone extinct. Animal groups first appearing during the Triassic include modern corals, marine reptiles (ichthyosaurs), dinosaurs and mammals (late Triassic).

triaxon *n.* a sponge spicule with three axes.

tribe *n.* subdivision of a family in taxonomy. Tribal names generally have the ending -eae in plant taxonomy and -ini in animal taxonomy.

triboloid *a.* (1) like a burr; (2) prickly.

triboluminescence *n.* luminescence produced by friction.

tribosphenic *a.* trituberculate, *appl.* teeth of therian mammals.

tribracteate *a.* with three bracts.

tricarboxylic acid organic acid bearing three COOH groups, e.g. citric acid and aconitic acid.

tricarboxylic acid cycle (TCA cycle) a key series of metabolic reactions in aerobic cellular respiration. In animals and plants it occurs in the matrix of the mitochondria. Acetyl-CoA formed from pyruvate produced by glycolysis is completely oxidized to CO_2 via interconversions of various carboxylic acids (oxaloacetate, citrate, ketoglutarate, succinate, fumarate, malate). It results in the reduction of NAD to NADH and FAD to $FADH_2$. Their reducing power is then used indirectly in the synthesis of ATP by oxidative phosphorylation. The TCA cycle also produces substrates for many other metabolic pathways. *alt.* citric acid cycle, Krebs cycle.

tricarpellary *a.* having three carpels.

tricentric *a.* having three centromeres.

triceps *n. appl.* a muscle with three heads or insertions.

trichilium *n.* a pad of matted hairs at base of certain leaf petioles.

trichoblast *n.* plant epidermal cell that develops into a root hair.

trichocarpous *a.* with hairy fruits.

trichocyst *n.* oval or spindle-shaped protrusible body in ectoplasm of ciliates and dinoflagellates.

trichoderm *n.* filamentous outer layer of cap and stipe in agamic fungi.

trichogen *n.* hair-producing cell.

trichogyne *n.* elongated hair-like receptive cell at end of female sex organ in some fungi, lichens, red or green algae, which may receive the male gamete.

trichoid *a.* hair-like.

trichome *n.* (1) any of various outgrowths of the epidermis in plants, including branched and unbranched hairs, vesicles, hooks, spines and stinging hairs; (2) a hair tuft; (3) (*bact.*) strand of vegetative cells ensheathed in a hollow tubular structure, forming the filament of filamentous cyanobacteria and some filamentous sulphur bacteria.

Trichomycetes *n.* class of endo- and exoparasitic fungi living in or on arthropods, having a simple or branched mycelium, and reproducing by non-motile spores.

trichophore *n.* (1) group of cells bearing a trichogyne; (2) chaetigerous sac of annelid worms.

trichopore *n.* opening for emerging hair or bristle, as in spiders.

Trichoptera *n.* order of insects known commonly as caddis flies, whose aquatic larvae (caddis worms) often build elaborate protective cases incorporating pieces of sand and other debris, or nets of silk in which food is trapped. They have a complete metamorphosis. The adults have two pairs of long slender wings and much reduced mouthparts and rarely feed.

trichosclereid *n.* a sclereid with thin hairlike branches extending into intercellular spaces.

trichosiderin *n.* iron-containing red pigment isolated from human hair.

trichosis *n.* (1) distribution of hair; (2) abnormal hair growth.

trichothallic *a.* (1) having a filamentous thallus; (2) *appl.* growth of a filamentous thallus in certain algae by division of intercalary meristematic cells at the base of a terminal hair.

trichotomous *a.* divided into three branches. *n.* **trichotomy**, a three-way branch.

trichroic *a.* showing three different colours.

trichromatic *a. pert.*, or able to perceive, the three primary colours.

trichromatic theory theory developed in 19th century that three different pigments must be present in retinal cones to account for colour vision, since proved to be true. *alt.* Young–Helmholtz trichromatic theory.

tricipital *a.* (1) having three heads or insertions, as triceps muscle; (2) *pert.* triceps.

triclads *n.plu.* order of elongated turbellarians, the Tricladida, commonly called planarians, having an intestine with three main branches and well-developed sense organs.

Tricoccae Euphorbiales *q.v.*

tricoccus *a. appl.* a fruit consisting of three carpels.

triconodont *a. appl.* tooth with three crown prominences in a line parallel to jaw axis.

Triconodonta, triconodonts *n., n.plu.* order containing the earliest known mammals, living from the late Triassic to the Jurassic, somewhat resembling shrews, and with molar teeth with three cusps in a straight line.

tricostate *a.* with three ribs.

tricotyledonous *a.* with three cotyledons.

tricrotic *a.* having a triple beat in the arterial pulse.

tricrural *a.* with three branches.

tricuspid *a.* with three cusps, *appl.* triangular valve of heart.

tricuspidate *a.* having three points, *appl.* leaf.

tridactyl *a.* having three digits.

tridentate *a.* having three tooth-like divisions.

trifacial *a. appl.* 5th cranial nerve, the trigeminal.

trifarious *a.* (1) in groups of three; (2) of three kinds; (3) in three rows; (4) having three surfaces.

trifid *a.* cleft into three lobes.

triflagellate *a.* having three flagella.

trifoliate *a.* having three leaves growing from the same point.

trifoliolate *a.* with three leaflets growing from the same point.

trifurcate *a.* with three forks or branches.

trigamous *a. appl.* flowerhead with staminate, pistillate and hermaphrodite flowers.

trigeminal *a.* (1) consisting of, or *pert.*, three structures; (2) *appl.* nerve: 5th cranial nerve whose root is divided into three branches conveying sensation from eye orbit, forehead and front of scalp (ophthalmic), upper jaw, teeth and overlying

skin (maxillary), and lower jaw, teeth and overlying skin and sensation other than taste from tongue and mouth (mandibular). *alt.* facial nerve.

trigeneric *a. pert.* or derived from three genera.

trigenetic *a.* requiring three different hosts in the course of a life cycle.

trigenic *a. pert.* or controlled by three genes.

triglyceride triacylglycerol *q.v.*

trigon *n.* the triangle of cusps on molar teeth of upper jaw.

trigonal *a.* (1) ternary or triangular when *appl.* symmetry with three parts to a floral whorl; (2) triangular in cross-section, *appl.* stems.

trigone *n.* a small triangular space.

trigoneutic *a.* producing three broods in each breeding season.

trigonid *n.* triangle of cusps of lower molar teeth.

trigonum *n.* posterior process of talus forming a separate ossicle.

trigynous *a.* (1) having three pistils or styles; (2) having three carpels to the gynoecium.

triheterozygote *n.* an organism heterozygous for three genes.

trihybrid (1) *n.* a cross whose parents differ in three distinct characters; (2) *a.* heterozygous for three pairs of alleles.

tri-iodothyronine *n.* iodine-containing hormone derived from tyrosine produced by the thyroid gland.

trijugate *a.* having three pairs of leaflets.

trilabiate *a.* having three lips, with lip of corolla divided into three.

trilobate *a.* having three lobes.

Trilobita, Trilobitomorpha, trilobites *n., n., n.plu.* group of fossil arthropods (*q.v.*) found throughout the Paleozoic era. They had a body divided into a head and a posterior region of numerous segments, a single pair of antennae, and numerous similar biramous appendages.

trilocular *a.* having three compartments or loculi.

trilophodont *a.* having teeth with three crests.

trimer *n.* (1) protein composed of three subunits; (2) a sequence of three nucleotides in a nucleic acid. *a.* **trimeric**.

trimeric GTP-binding proteins *see* heterotrimeric G proteins.

Trimerophyta *n.* group of primitive vascular plants, known from the mid-Devonian and now extinct, thought to represent the ancestors of the ferns and progymnosperms. The main axis was branched, with some branches bearing sporangia, and they lacked leaves.

trimerous *a.* composed of three or multiples of three, as parts of flower.

trimonoecious *a.* with male, female and hermaphrodite flowers on the same plant.

trimorphic *a.* (1) having three different forms, *appl.* species; (2) having three different types of individual; (3) with stamens and pistils of three different length.

trimorphism *n.* occurrence of three distinct forms or forms of organs in one life cycle or one species.

trinervate *a.* having three veins or ribs running from base to margin, as leaf.

trinomial *a.* consisting of three Latin names, such as names of subspecies.

trinucleotide repeats consecutive repeats of a three-nucleotide sequence, which are found in some genes. In Huntington's disease and some other inherited diseases, affected individuals have an increased number of trinucleotide repeats in the disease gene.

trioecious *a.* producing male, female and hermaphrodite flowers on different plants.

trionym trinomial name.

triose *n.* any three-carbon monosaccharide $(CH_2O)_3$, such as glyceraldehyde, dihydroxyacetone phosphate.

triose-phosphate isomerase enzyme that catalyses the isomerization of dihydroxyacetone phosphate to glyceraldehyde3-phosphate for use in glycolysis. EC 5.3.1.1.

triosseum *a. appl.* foramen, the opening between coracoid, clavicle and scapula.

triovulate *a.* having three ovules.

tripartite *a.* divided into three parts, as leaf.

tripetalous *a.* having three petals.

triphyllous *a.* having three leaves.

tripinnate *a.* divided pinnately three times.

tripinnatifid *a.* divided three times in a pinnatifid manner.

tripinnatisect *a.* three times lobed, with divisions nearly to midrib, *appl.* leaves.

triplet *n.* three consecutive bases in DNA or RNA, encoding an amino acid.

triplet code genetic code *q.v.*

triplet expansion trinucleotide repeats *q.v.*

triplicostate *a.* having three ribs.

triploblastic *a. appl.* embryos with three primary germ layers.

triplocaulescent *a.* having axes of the third order, i.e. having a main stem with branches which are branched.

triploid (1) *a.* having three sets of chromosomes in the somatic cells; (2) *n.* a triploid organism or cell.

triplostichous *a.* (1) arranged in three rows, as of cortical cells on small branches of green algae of the genus *Chara*; (2) *appl.* eyes with preretinal, retinal and postretinal layers, as of larval scorpion.

tripolite diatomaceous earth *q.v.*

tripton *n.* non-living material difting in plankton.

triquetral, triquetrous *a.* (1) *appl.* stem with three angles and three concave faces; (2) *appl.* three-cornered or wedge-shaped bone: one of the carpal bones.

triquetrum *n.* the cuneiform carpal bone.

triquinate *a.* divided into three, with each lobe again divided into five.

triradial *a.* having three branches arising from one centre.

triradiate *a. appl.* pelvic girdle consisting of pubis, ilium and ischium.

triramous, triramose *a.* divided into three branches.

trisaccharide *n.* a carbohydrate made up of three monosaccharide units, e.g. raffinose.

trisepalous *a.* having three sepals.

triseptate *a.* having three septa.

triserial *a.* (1) arranged in three rows; (2) having three whorls.

triskelion *n.* three-legged protein assembly.

trisomic *a. pert.* or having three homologous chromosomes or genetic loci. *n.* **trisomy.**

trispermous *a.* having three seeds.

trisporic, trisporous *a.* having three spores.

tristachyous *a.* with three spikes.

tristichous *a.* arranged in three vertical rows.

tristyly *n.* the condition of having short, medium-length and long styles. *a.* **tristylic.**

triternate *n.* thrice ternately divided.

tritibial *n.* compound ankle bone formed when centrale unites with talus.

tritium *n.* radioactive isotope of hydrogen, ^{3}H, widely used in experimental biology to label tracer compounds and biological molecules.

tritocone *n.* a cusp of premolar tooth.

trituberculate *a.* having three cusps to the molar teeth.

trituberculates *n.plu.* group of Mesozoic therian mammals known mainly from remains of jaws and teeth, the molar teeth having the characteristic triangle of cusps, and which are forerunners of the living therians. *alt.* tribosphaenids.

Triuridales *n.* order of saprophytic monocots with scale leaves. They consist of the family Triuridaceae.

trivalent (1) *n.* association of three chromosomes held together by chiasmata between diplotene and metaphase of 1st division in meiosis; (2) *a. appl.* antigen that can bind three antibody molecules.

trivoltine *a.* having three generations of broods in a year.

trixenous *a.* of a parasite, having three hosts.

trizoic *a. appl.* protozoan spore containing three sporozoites.

tRNA transfer RNA *q.v.*

tRNAfMet initiator tRNA in bacteria, which carries formylmethionine, the initial amino acid in most newly synthesized bacterial polypeptide chains.

tRNAiMet initiator tRNA in eukaryotes, which carries methionine and is used solely for initiation of translation.

troch *n.* a circlet or segmental band of cilia of a trochophore.

trochal *a.* wheel-shaped.

trochantellus *n.* segment of leg between trochanter and femur, in some insects.

trochanter *n.* (1) prominences at upper end of thigh bone; (2) small 2nd segment of leg between coxa and femur in insects and spiders.

trochanteric fossa deep depression on medial surface of neck of femur.

trochate *a.* (1) having a wheel-like structure; (2) wheel-shaped.

trochlea *n.* a pulley-shaped structure, esp. one through which a tendon passes, as of femur, humerus.

trochlear *a.* (1) shaped like a pulley; (2) *pert.* trochlea; (3) *appl.* nerve: 4th cranial nerve to superior oblique muscle of eye.

Trochodendrales *n.* order of dicot shrubs and trees including the families Tetracen-

traceae (e.g. the rare Chinese endemic *Tetracentron sinensis*) and Trochodendraceae. (e.g. *Trochodendron aralioides*, the wheel tree).

trochoid *a.* (1) wheel-shaped; (2) capable of rotating motion, as a pivot joint.

trochophore *n.* free-swimming top-shaped pelagic larval stage of annelids, bryozoans, and some molluscs, forming part of the zooplankton. It has a ring of cilia around the rim and a terminal ring or tuft of cilia in front of the mouth. *alt.* trochosphere.

Trochozoa *n.* animal taxonomic group comprising the Nemertina, Mollusca, Sipuncula, Echiura, Pogonophora and Annelida (*all q.v.*).

troglobiont *n.* any organism living only in caves.

Trogoniformes *n.* an order of very softplumaged birds including the trogons.

tropeic *a.* keel-shaped.

trophallaxis *n.* in social insects, the exchange of alimentary liquid among colony members and guest organisms, either mutually or unilaterally.

trophamnion *n.* sheath around developing egg of some insects, and passing nourishment to the embryo.

trophectoderm *n.* outer layer of mammalian blastocyst, which will form chorion. *alt.* **trophoblast.**

trophi (1) *n.plu.* hard jaw-like structures in the pharynx of rotifers, used to grind food; (2) the mouthparts of arthropods, esp. insects.

trophic *a.* (1) *pert.* or connected with nutrition and feeding; (2) *appl.* hormones influencing the activity of endocrine glands and growth, such as those secreted by the anterior lobe of the pituitary.

trophic factor protein that is required for the survival and growth, and sometimes repair, of a particular tissue or cell type.

trophic level (1) a level in a food chain defined by the method of obtaining food, and in which all organisms are the same number of energy transfers away from the original source of the energy (e.g. photosynthesis) entering the ecosystem. *see* autotroph, herbivore, secondary consumer, tertiary consumer; (2) the nutrient status of a body of water. *see* eutrophic, oligotrophic.

trophidium *n.* 1st larval stage of certain ants.

trophifer, trophiger *n.* posterolateral region of insect head with which mouthparts articulate.

trophobiont *n.* organism that lives in a symbiosis where each partner feeds the other, as ants and aphids. *see* trophobiosis.

trophobiosis *n.* the life of ants in relation to their nutritive organisms, as to fungi and insects.

trophoblast *n.* the outer layer of cells of epiblast or of morula. *alt.* trophectoderm.

trophocytes *n.plu.* cells providing nourishment for other cells, as nurse cells to oocyte.

trophoderm *n.* (1) outer layer of chorion; (2) trophectoderm with a mesodermal cell layer.

trophogenic *a.* due to food or feeding, *appl.* features of social hymenopterans.

trophogone *n.* a nutritive organ in ascomycetes.

trophonemata *n.plu.* uterine villi or hair-like projections which transfer nourishment to embryo.

trophont *n.* stage of vegetative growth in holotrichan ciliates.

trophophase *n.* growth phase in secondary metabolism. *cf.* idiophase.

trophophore *n.* in sponges, an internal bud or group of cells destined to become a gemmule.

trophophyll *n.* a sterile or foliage leaf of certain pteridophytes.

trophosome *n.* (1) polyp involved in feeding in a hydroid colony; (2) vascularized organ in vestimentiferan tubeworms, which contains symbiotic sulphur-oxidizing CO_2-fixing chemoautotrophic bacteria which supply fixed carbon to their host.

trophospongia *n.* spongy vascular layer of mucous membrane between uterine wall and trophoblast.

trophotaxis *n.* response to stimulation by an agent which may serve as food.

trophotropism *n.* tendency of a plant organ to turn towards food, or of an organism to turn towards food supply.

trophozoite *n.* the adult stage of a sporozoan.

trophozooid *n.* in some tunicates, lateral buds that collect food for the colony.

tropic *a.* (1) *pert.* tropism, *appl.* movement or curvature in response to a directional or unilateral stimulus; (2) *appl.* pituitary hormones that act on endocrine glands; (3) tropical, *appl.* regions.

tropical *a. appl.* climate characterized by high temperature, humidity and rainfall, found in a belt between latitudes 23°27′N and 23°27′S, and to the flora and fauna of these regions.

tropical rain forest evergreen broadleaf forest that develops in areas near the Equator with a climate of high temperature, humidity and rainfall and no marked seasons, and which is characterized by a high biological diversity and productivity. Tropical rain forest is found in the Amazon basin, parts of Central America, central West Africa, parts of the southeastern African coast and Madagascar, South-east Asia and Indonesia, New Guinea, and the northern tip of Australia. *see also* rain forest.

tropism *n.* a plant or sessile animal growth movement, usually curvature towards (positive) or away from (negative) the source of stimulus.

tropocollagen *n.* basic structural unit of the fibrous protein collagen, consisting of three long polypeptide chains wound round each other in a helical conformation.

tropoelastin *n.* precursor to the protein elastin.

tropomyosin *n.* protein in the thin filaments of striated muscle, which in the absence of Ca^{2+} prevents the interaction of actin and myosin, and thus prevents contraction.

troponin *n.* protein complex in thin filaments of striated muscle, consisting of three components, one of which (TnC) binds Ca^{2+}. In this state it interacts with tropomyosin, allowing interaction of actin and myosin and muscle contraction.

tropophil(ous) *a.* (1) tolerating alternating periods of cold and warmth, or of moisture or dryness; (2) adapted to seasonal changes.

tropophyte *n.* (1) a plant which adapts to the changing seasons, being more or less hygrophilous in summer and xerophilous in winter; (2) a plant growing in the tropics.

tropotaxis *n.* a taxis in which an animal orients itself in relation to source of stimulus by simultaneously comparing the amount of stimulus on either side of it by symmetrically placed sense organs.

Trp tryptophan *q.v.*

true-breeding *a. appl.* organisms that are homozygous for any given genotype and therefore pass it on to all their progeny in a cross with a similar homozygote. *alt.* pure-breeding.

true corals common name for the Scleractinia, an order of mainly colonial hydrozoans, having a compact calcareous skeleton and no siphonoglyph.

true flies Diptera *q.v.*

true mosses the Bryophyta *q.v.*

true ribs ribs connected directly with the sternum (breast bone).

trumpet hyphae large trumpet-shaped sieve cells in some brown algae.

truncate *a.* terminating abruptly, as if tapering end were cut off. *alt.* abrupt.

truncus arteriosus the most anterior region of foetal, or of amphibian, heart, through which blood is driven to ventricles.

trunk *n.* (1) the main stem of tree; (2) body, exclusive of head and limbs; (3) main stem of a vessel or nerve; (4) elongated proboscis, as of elephant.

tryma *n.* a type of drupe with a separable rind and two-halved endocarp with spurious dissepiment, such as the fruit of the walnut tree. *plu.* **trymata.**

trypanomonad *a. appl.* phase of development of trypanosome while in its invertebrate host.

trypanosome *n.* member of a genus of parasitic flagellate protozoans, *Trypanosoma*, which includes the organisms causing trypanosomiasis, or sleeping sickness, in Africa (e.g. *T. brucei*) and Chagas' disease in South America (*T. cruzi*). *T. brucei* is transmitted by the bite of infected tsetse flies and *T. cruzi* via the faeces of blood-sucking bugs.

trypsin *n.* proteolytic digestive enzyme found in pancreatic juice of mammals (and similar enzymes found in other animals and in plants). It is a serine peptidase, and is formed from an inactive precursor, trypsinogen, by enzymatic cleavage by enteropeptidase in the small intestine. EC 3.4.21.4.

trypsin inhibitors various proteins that inhibit the enzyme trypsin by binding

tightly to the active site. Some belong to the serpin family (e.g. α_1-antitrypsin, a deficiency of which is responsible for lung damage due to unrestrained protease activity, resulting in the clinical condition emphysema) and some to another family of proteinase inhibitors (e.g. pancreatic trypsin inhibitor).

trypsinogen *see* trypsin.

tryptic *a.* produced by or *pert.* trypsin.

tryptophan (Trp, W) *n.* β-indolealanine, a non-polar amino acid with an aromatic side chain, constituent of protein, essential in human diet and a precursor of the auxin indoleacetic acid in plants.

tryptophanase *n.* enzyme catalysing the breakdown of tryptophan to ammonia, pyruvic acid and indole, present esp. in some colon bacteria. EC 4.1.99.1.

tryptophan synthase enzyme that catalyses the synthesis of tryptophan from indoleglycerol phosphate and serine. EC 4.2.1.20.

tsetse fly *Glossina* spp., the vector of African trypanosomiasis.

TSH thyroid-stimulating hormone *q.v.*

T-state in an allosteric protein, the tense state, in which the protein is less active.

TTP, dTTP thymidine triphosphate *q.v.*

T tubules transverse tubules *q.v.*

TTX tetrodotoxin *q.v.*

tuatara *see* Rhyncocephalia.

tuba *n.* (1) a salpinx or tube, as tuba acustica or auditiva, the Eustachian tube; (2) tuba uterina: Fallopian tube.

tubal *a. pert.* the Eustachian tube or Fallopian tubes.

tubar *a.* consisting of an arrangement of tubes, or forming a tube, as *appl.* system and skeleton of sponges.

tubate *a.* tube-shaped, tubular.

tube feet projections from the body wall in echinoderms which are connected to the water vascular system and generally used for locomotion. They may also be modified to serve respiratory, food-catching and sensory functions.

tubenoses Procellariformes *q.v.*

tuber *n.* thickened, fleshy, food-storing underground root, or similar underground stem with surface buds.

tubercle *n.* (1) small rounded protuberance; (2) root swelling or nodule; (3) dorsal articular knob on a rib; (4) cusp of a tooth.

tubercle bacillus *Mycobacterium tuberculosis*, the cause of tuberculosis.

tubercular, tuberculate *a. pert.*, resembling, or having tubercles.

tuberculin test immunological test for previous infection with *Mycobacterium tuberculosis*, which depends on a reaction to tuberculin, an extract of the bacterium.

tuberculose *a.* having many tubercles.

tuberculosis *n.* disease caused by the bacterium *Mycobacterium tuberculosis*, most commonly affecting the lungs.

tuberiferous *a.* bearing or producing tubers.

tuberoid *a.* shaped like a tuber.

tuberose, tuberous *a.* covered with or having many tubers.

tuberosity *n.* a rounded eminence on a bone, usually for muscle attachment.

tuber vermis part of superior vermis of cerebellum, continuous laterally with inferior semilunar lobules.

tubeworms *see* Annelida, Echiura, Phoronida, Pogonophora, Pterobranchia, Vestimentifera.

tubicolous *a.* inhabiting a tube.

tubicorn *a.* with hollow horns.

tubifacient *a.* tube-making, as some polychaete worms.

tubificid worms freshwater worms of the genus *Tubifex*, which are tolerant of heavy organic pollution.

tubiflorous tubuliflorous *q.v.*

tubiform *a.* having the form of a tube or tubule.

tubilingual *a.* having a hollow tongue, adapted for sucking.

tubiparous *a.* secreting tube-forming material, *appl.* glands.

tubo-ovarian *a. pert.* oviduct and ovary.

tubotympanic *a. appl.* recess between 1st and 3rd visceral arches, from which are derived the tympanic cavity and the Eustachian tubes.

tubular, tubulate *a.* in the form of a tube, having tubes, consisting of tubes.

tubule *n.* a small tube.

Tubulidentata *n.* order of placental mammals known from the Miocene, or possibly the Eocene, whose only living member is the African ant-eater or aardvark (*Orycteropus*), which possesses unique peg-like teeth with tubular canals in the dentine, is ant-eating and has powerful digging forelimbs.

tubuliferous *a.* having tubules.

tubuliflorous *a.* having florets with a tubular corolla.

tubuliform *a.* (1) tube-shaped; (2) *appl.* type of spinning glands in spiders.

tubulin *n.* protein subunit of microtubules. There are two types: α-tubulin and β-tubulin.

tubulose *a.* (1) having, or composed of, tubular structures; (2) hollow and cylindrical.

tubulus *n.* tubule.

tumescence *n.* swelling.

tumid *a.* swollen or turgid.

tumor *alt.* spelling of tumour *q.v.*

tumorigenesis *n.* the development of a tumour.

tumorigenic *a.* (1) *appl.* any agent that can cause a tumour, such as a tumour virus or certain chemicals; (2) *appl.* a cell that can give rise to a tumour.

tumour *n.* a growth resulting from the abnormal proliferation of cells. It may be self-limiting or non-invasive, when it is called a benign tumour, or continue proliferating indefinitely and invade underlying tissues and metastasize, when it is called a malignant tumour or cancer.

tumour angiogenesis factors (TAF) cytokines released from malignant tumours which induce the formation of a network of blood capillaries invading the tumour. *alt.* angiogenin.

tumour antigen novel cell-surface protein that appears on tumour cells, and which is not present on the normal cell.

tumour necrosis factor-α (TNF-α) cytokine produced by activated T cells, NK cells and macrophages during an immune response. It induces inflammation and, if activated systemically, is responsible for inducing systemic shock. It also has anti-tumour activity and disturbs lipid metabolism, producing the cachexia (wasting) seen in patients with some types of cancer. Sometimes now called simply tumour necrosis factor (TNF). *alt.* cachectin.

tumour necrosis factor-β (TNF-β) lymphotoxin *q.v.*

tumour necrosis factor family (TNF family) family of cytokines and membrane-associated proteins important in immune responses. As well as TNF-α it also includes CD40 ligand and Fas ligand.

tumour promoters compounds that are not carcinogenic in themselves, but which hasten the effects of a carcinogen.

tumour suppressor genes genes involved in various aspects of regulation of cell proliferation. Individuals inheriting one defective copy of such a gene show a much greater predisposition than normal to develop various types of cancer. The development of the cancer is due to subsequent loss or inactivation of the other, 'good', copy of the gene. An example is the retinoblastoma gene (Rb).

tumour viruses *see* DNA tumour viruses, RNA tumour viruses.

tundra *n*. treeless Polar region with permanently frozen subsoil, bare of vegetation or may support mosses, lichens, herbaceous plants and dwarf shrubs.

TUNEL assay assay used to identify apoptotic cells *in situ*.

tunic *n*. (1) membrane or tissue enclosing or surrounding a structure; (2) the body wall or test of a tunicate.

tunica *n*. (1) body wall or outer covering; (2) investing membrane or tissue or outer wall of organ.

tunica albuginea perididymis *q.v.*

tunica–corpus type of cellular organization, e.g. of plant apical meristem, in which the region is differentiated into two parts distinguished by their plane of cell division. The outer or tunica layer has mainly anticlinal divisions and the inner corpus has divisions in various planes.

tunica intima innermost layer of a blood vessel wall.

tunica muscularis layer of smooth muscle that forms the outer layer of the mucosa of the digestive, respiratory, urinary and genital tracts.

tunicamycin *n*. antibiotic that inhibits the glycosylation of glycoproteins in eukaryotic cells.

Tunicata, tunicates *n*., *n.plu*. subphylum of chordates containing the classes Ascidiacea (the sessile sea squirts), the free-swimming tadpole-like Larvacea, and the Thaliacea (free-swimming salps). Chordate features (i.e. notochord and nerve cord) are found only in the larva

and are generally lost in the adult. The adult secretes a tough cellulose sac (tunic) in which the animal is embedded. *alt.* Urochordata, urochordates.

tunicate *a*. provided with a tunic or test.

tunicin *n*. a polysaccharide related to cellulose, found in the tunic of ascidians.

tunicle *n*. a natural covering or integument.

tunnelling electron microscope type of electron microscope that gives atomic-level resolution of certain types of structure or surfaces, including biological macromolecules.

Turbellaria, turbellarians *n*., *n.plu*. class of free-living flatworms with a leaf-like shape and a ciliated epithelium.

turbid plaque type of plaque produced by the growth of certain bacteriophages on a lawn of bacteria, which has a cloudy appearance due to the survival and growth of some bacteria within the area.

turbinal (1) *a*. spirally rolled or coiled, as bone or cartilage; (2) *n*. one of the nasal bones in vertebrates, supporting the olfactory tissues. *alt.* turbinate bone.

turbinate *a*. top-shaped, spirally rolled or coiled, *appl.* shells.

turbinate bones fragile scrolled bones forming the side walls of nasal cavities.

turbinulate *a*. shaped like a small top, *appl.* certain apothecia.

turgescence, turgidity, turgor *n*. (1) the swelling of a plant cell due to the internal pressure of vacuolar contents; (2) distention of any living tissue due to internal pressures. *a.* **turgid, turgescent**.

turgor pressure (TP) the pressure set up inside a plant cell due to the hydrostatic pressure of the vacuole contents pressing on the rigid cell wall. It provides mechanical support to non-woody plant stems, and changes in turgor pressure due to osmosis are responsible for the opening and closing of stomata and some seismonastic movements.

turio *n*. (1) young scaly shoot budded off from underground stem; (2) detachable winter bud used for perennation in many water plants. *alt.* **turion**.

Turkish saddle sella turcica *q.v.*

β-turn structure common in proteins whereby polypeptide chains make a sharp bend by hydrogen bonding between the CO group of one amino acid residue and the NH group of the next third residue. *alt.* β-bend.

turnover *n.* (1) in an ecosystem, the ratio of productive energy flow to the biomass; (2) the fraction of a population which is exchanged per unit time through loss by death or emigration and replacement by reproduction and immigration; (3) in cells, the continual process of the degradation and replacement of proteins.

turnover number in enzyme reactions, the number of substrate molecules converted into product per second when the enzyme is fully saturated with substrate.

turnover time (1) time taken to complete a biological cycle; (2) time from birth to death of an organism.

turtles *see* Chelonia.

tusk shells common name for the Scaphopoda *q.v.*

twins *see* dizygotic, monozygotic.

two-component system a signal transduction system found in bacteria, plants and fungi, but not known in animals so far, consisting of a sensor protein (a receptor histidine kinase), which senses an external signal, is activated, and acts on an intracellular response regulator protein (e.g. a transcription factor), which transmits the signal to other components of the cell.

two-dimensional polyacrylamide gel electrophoresis (2D-PAGE) technique for the separation of large numbers of proteins, in which the mixture is first separated by isoelectric focusing or electrophoresis in one direction, and then subjected to electrophoresis in a direction at right angles to the first.

two-hybrid system *see* yeast two-hybrid system.

tycholimnetic *a.* temporarily attached to bed of lake and at other times floating.

tychoplankton *n.* (1) drifting or floating organisms which have been detached from their previous habitat, as in plankton of the Sargasso Sea; (2) inshore plankton.

tychopotamic *a.* thriving only in backwaters.

tylopods *n.plu.* the mammalian suborder Tylopoda, of the Artiodactyla, which includes the camel family and a number of extinct groups.

tylosis *n.* (1) development of irregular cells in a cavity; (2) intrusion of parenchyma cells into a xylem vessel, esp. of secondary xylem, through pits; (3) a callosity; (4) callus formation. *a.* **tylotic.**

tylosoid *a.* resembling a tylosis, e.g. a resin duct filled with parenchymatous cells.

tymbal *alt.* spelling of timbal *q.v.*

Tymovirus *n.* genus of positive-strand RNA plant viruses with isometric particles, e.g. turnip yellow mosaic virus.

tympanic *a. pert.* tympanum.

tympanic canal one of the three main canals running the length of the cochlea.

tympanic membrane eardrum, thin membrane at the internal end of the external channel of the ear via which sound vibrations are transmitted to the ossicles of the middle ear.

tympaniform membrane *see* syrinx.

tympanohyal (1) *a. pert.* tympanum and hyoid bone; (2) *n.* part of hyoid arch embedded in petromastoid.

tympanoid *a.* shaped like a flat drum, *appl.* certain diatoms.

tympanum *n.* (1) the drum-like cavity constituting the middle ear; (2) the eardrum or tympanic membrane; (3) membrane of auditory organ borne on tibia, metathorax or abdomen of insect; (4) an inflatable air sac on the neck of birds of the grouse family; (5) (*bot.*) the membrane closing capsule in some mosses. *a.* **tympanic.**

TYMV turnip yellow mosaic virus.

type *n.* (1) sum of characteristics common to a large number of individuals, e.g. of a species, and serving as the basis for classification; (2) a primary model, the actual specimen described as the original of a new genus or species. *alt.* holotype.

type 1 diabetes, type I diabetes type of diabetes mellitus due to autoimmune reactions that destroy insulin-producing pancreatic β-cells, leading to non-production of insulin. It usually appears in childhood or young adulthood and can only be controlled by life-long insulin injections. *alt.* insulin-dependent diabetes mellitus, juvenile-onset diabetes.

type 2 diabetes, type II diabetes type of diabetes mellitus that usually appears later in life than type 1 diabetes and is caused by deficient production of insulin and/or resistance of cells to the effects of insulin. It can be treated by careful management of the diet, drugs that reduce blood sugar levels, and, sometimes, insulin injections. *alt.* adult-onset diabetes.

type I hypersensitivity reactions allergic reactions involving IgE.

type II hypersensitivity reactions immunological hypersensitivity reactions involving IgG antibodies produced against cell-surface or extracellular matrix antigens.

type III hypersensitivity reactions immunological hypersensitivity reactions involving formation of immune complexes and their deposition in tissues.

type IV hypersensitivity reactions immunological hypersensitivity reactions involving T cells.

type I, type II, type III, type IV membrane proteins *see* transmembrane protein.

type I, type II, type III, type IV secretion systems *see* secretion systems.

type locality locality in which the holotype or other type used for designation of a species was found.

type number the most frequently occurring chromosome number in a taxonomic group.

type specimen the single specimen chosen for the designation and description of a new species.

Typhales *n.* order of marsh or aquatic monocots with rhizomes and linear leaves, comprising the families Sparaginiaceae (bur reed) and Typhaceae (cat-tail).

typhlosole *n.* median longitudinal fold of the intestine projecting into lumen of gut in some vertebrates and in cyclostomes.

typical *a.* (1) *appl.* specimen conforming to type or primary example; (2) exhibiting in marked degree the characteristics of species or genus.

typogenesis *n.* (1) phase of rapid type formation in phylogenesis; (2) quantitative or explosive evolution.

typonym *n.* a name designating or based on type specimen or type species.

Tyr tyrosine *q.v.*

tyramine *n.* phenolic amine formed by decarboxylation of tyrosine, produced in small amounts in animal liver and which causes a rise in arterial blood pressure. It is also produced by bacterial action on tyrosine-rich substrates, secreted by cephalopods, and found in various plants such as mistletoe and ergot of rye.

tyrocidin *n.* cyclic decapeptide antibiotic produced by *Bacillus brevis*.

tyrosinase *n.* general name for a group of copper-containing enzymes catalysing the oxidation of tyrosine with the formation of dopa (dihydroxyphenylalanine). *r.n.* monophenol monooxygenase, EC 1.14.18.1; catechol oxidase, EC 1.10.3.1.

tyrosine (Tyr, Y) *n.* amino acid with polar, uncharged, aromatic side chain, constituent of protein, also important as a precursor of adrenaline and noradrenaline, and of thyroxine and melanin.

tyrosine hydroxylase tyrosinase *q.v.*

tyrosine kinase *see* protein tyrosine kinases.

tyrosine phosphatase *see* protein tyrosine phosphatase.

tyvelose *n.* a 3,6-dideoxyhexose sugar found in the lipopolysaccharide outer membrane of some enteric bacteria.

U

U uracil *q.v.*

U1, U2, U4, U6 and others small nuclear ribonucleoproteins involved in RNA splicing. They form the spliceosome.

ψU pseudouridine *q.v.*

UAS upstream activating sequence *q.v.*

ubiquinone *n.* quinone derivative with a tail of isoprenoid units (the number varying with species), a mobile electron carrier between flavoproteins and cytochromes of the respiratory electron-transport chain.

ubiquitin *n.* small acidic protein which is involved in targeting proteins for degradation in proteasomes. The enzyme ubiquitin ligase attaches it to lysine residues in the protein to be degraded, and the ubiquitin is later cleaved off and recycled during proteolysis.

ubiquitination *n.* addition of ubiquitin to a protein.

UCR unconditional response or reflex *q.v.*

UCS unconditional stimulus *q.v.*

UDP uridine diphosphate *q.v.*

UDPG UDP-glucose *q.v.*

UDP-(sugar) activated form of monosaccharide attached to the nucleotide UDP, formed from UTP and the sugar. UDP-sugars are donors of monosaccharide residues in polysaccharide synthesis, interconversion of sugars, etc., and include UDP-glucose, UDP-galactose.

ula *n.* the gums (of teeth).

uletic *a. pert.* the gums.

uliginose, uliginous *a.* (1) swampy; (2) growing in mud or swampy soil.

ulna *n.* one of the long bones of vertebrate forearm, parallel with radius and in some vertebrates combined with it to form a single bone. *a.* **ulnar**.

ulnar nervure radiating or cross-vein in insect wing.

ulnare *n.* one of the wrist bones, lying at the far end of the ulna.

ulnocarpal *a. pert.* ulna and carpus.

ulnoradial *a. pert.* ulna and radius.

uloid *a.* resembling a scar.

ulotrichous *a.* having woolly or curly hair.

ultimate *a. appl.* factor thought to be the fundamental cause of some biological phenomenon. *cf.* proximate.

ultimobranchial bodies pair of gland rudiments derived from the endoderm of the 5th pharyngeal pouches, which secrete calcitonin and later degenerate and disappear.

ultra-abyssal hadral *q.v.*

ultracentrifuge *n.* instrument in which extracts of broken cells can be separated into their different components by spinning at various speeds (up to 150,000*g*), the different organelles sedimenting at different speeds, and which is also used to separate large molecules of different molecular weights.

ultradian rhythm biological rhythm with a periodicity greater than 24 hours.

ultrametric tree in phylogenetic analysis, a rooted phylogenetic tree in which one axis of the tree is directly proportional to time, as the same and constant rate of mutation is assumed along all branches.

ultramicroscopic *a. appl.* structures or organisms too small to be visible under the light microscope but which can be seen in the electron microscope.

ultramicrotome *n.* machine with fine glass or diamond knife-blade for slicing ultrathin tissue sections for electron microscopy.

ultrastructure *n.* the fine structure of cells as seen in the electron microscope.

ultraviolet light (UV) electromagnetic radiation of wavelengths between those of the violet end of the visible light spectrum and X-rays. Although invisible to the human eye, it can be captured on photographic film. The ultraviolet (UV) spectrum is subdivided by wavelength into A (400–320 nm), B (320–280 nm) and C (280–10 nm) bands, of which UV-B is most harmful to living organisms. Much of the solar UV radiation, esp. the shorter wavelengths, is absorbed by the stratospheric ozone layer before reaching Earth's surface.

umbel *n.* (1) flowerhead in which each flower or cluster of flowers arises from a common centre, forming a flat-topped or rounded cluster; (2) various structures in other organisms resembling an umbel.

Umbellales Cornales *q.v.*

umbella (1) umbel *q.v.*; (2) umbrella of jellyfish.

umbellate *a.* arranged in umbels.

umbellifer *n.* member of the large dicot family Umbelliferae, e.g. cow-parsley, carrot, having small flowers borne in umbels and much-divided leaves. They include many plants grown for food, but also highly poisonous species (e.g. *Conium maculatum*, called hemlock in the UK).

umbelliferous *a.* having flowerheads in umbels, as the umbellifers.

umbelliform *a.* shaped like an umbel.

umbellula *n.* large cluster of polyps at tip of elongated stalk. *alt.* **umbellule**.

umbellule *n.* (1) small secondary umbel; (2) umbellula *q.v.*

umber opal *q.v.*

umbilical *a. pert.* navel or umbilical cord.

umbilical cord *n.* cord of tissue connecting embryo with placenta.

umbilicate *a.* (1) having a central depression; (2) like a navel.

umbilicus *n.* (1) the navel, place of attachment of the umbilical cord; (2) hilum *q.v.*; (3) basal depression of certain spiral shells; (4) an opening near base of feather.

umbo *n.* (1) protuberance like a boss on a shield; (2) (*bot.*) swollen part of cone scale; (3) (*zool.*) beak or older part of shells of bivalve molluscs. *plu.* **umbones**. *a.* **umbonal**.

umbonate *a.* having, *pert.*, or resembling an umbo.

umbraticolous *a.* growing in a shaded habitat.

umbrella *n.* the upper surface of the bell of a jellyfish medusa.

UMP uridine monophosphate *q.v.*

Umwelt *n.* the total sensory input of an animal, distinctive for each species.

unbalanced *a.* in genetics, *appl.* chromosomal rearrangements that result in duplications and deficiencies and the production of aneuploid cells and gametes.

uncate *a.* hooked.

unciferous *a.* bearing hooks.

unciform (1) *a.* shaped like a hook or barb; (2) *n.* unciform bone of wrist.

uncinate *a.* (1) hook-like; (2) *appl.* apex, as of a leaf.

uncinate fasciculus band of fibres connecting temporal and frontal lobes of brain.

uncinus *n.* (1) small hooked or hook-like structure; (2) crotchet *q.v.*; (3) marginal tooth of radula in gastropods.

unconditional reflex (UCR) an inborn reflex, produced involuntarily in response to a stimulus. *alt.* **unconditioned reflex**.

unconditional response (UCR) *see* classical conditioning. *alt.* **unconditioned response**.

unconditional stimulus (UCS) a stimulus that produces a simple reflex response. *alt.* **unconditioned stimulus**.

uncoupler, uncoupling agent any molecule, such as 2,4-dinitrophenol, that blocks ATP generation without blocking the preceding steps of respiration and electron flow down the electron-transport chain. Uncoupling agents carry protons across membranes, thus dissipating the proton-motive force.

uncus *n.* hook-shaped extremity of hippocampus in brain.

undate *a.* wavy, undulating.

undecaprenyl phosphate, undecaprenyl pyrophosphate phosphate or pyrophosphate derivatives of the very long-chain isoprenyl lipid undecaprenol, carrier of activated oligosaccharides in the biosynthesis of glycans in eubacteria.

underdominance *n.* condition in which the phenotypic expression of the heterozygote is less than that of either homozygote.

undershoot afterpotentials *q.v.*

understorey *n.* vegetation layer between tree canopy and the ground cover in a forest or wood. It is composed of shrubs and small trees.

under-wing one of the posterior wings of any insect.

undifferentiated *a.* (1) not differentiated, in immature state, *appl.* embryonic cells that have not yet acquired a specialized structure and function; (2) *appl.* meristematic and stem cells.

undose *a.* having undulating and nearly parallel depressions which run into each other and resemble ripple marks.

undulate *a.* having wave-like undulations, *appl.* leaves.

undulating membrane (1) a membrane in some flagellates which attaches part of the flagellum along the length of the cell; (2) similar structure in tail of spermatozoon.

undulipodium *n.* flagellum or cilium of a eukaryotic cell. *plu.* **undulipodia.**

unequal crossing-over type of recombination that sometimes occurs between chromosomes carrying stretches of identical repeats or clusters of similar genes. Misalignment between identical sequences, followed by recombination, produces recombinants of unequal length, one containing fewer and the other more copies of the sequence than normal. *alt.* nonreciprocal recombination.

unguiculate *a.* (1) clawed; (2) (*bot.*) *appl.* petals with a claw-like stalk.

unguiculus *a.* a small nail or claw.

unguis *n.* (1) nail or claw; (2) the fang of an arachnid chelicera, through which the poison gland opens; (3) (*bot.*) narrow stalk of some petals. *a.* **ungual.**

ungula *n.* hoof.

ungulate *a.* hoofed or hoof-like.

ungulates *n.plu.* hoofed mammals. *see* Artiodactyla, Perissodactyla.

unguliform *a.* hoof-shaped.

unguligrade *a.* walking upon hoofs, which are formed from the tips of the digits.

uni- prefix from L. *unus*, one, generally meaning having one of.

uniascal *a.* containing a single ascus, *appl.* locules.

uniaxial *a.* (1) with one axis; (2) *appl.* movement only in one plane, as a hinge joint.

unibranchiate *a.* having one gill.

unicamerate *a.* one-chambered.

unicell *n.* single-celled organism.

unicellular *a.* having only one cell, or consisting of one cell.

unicolour *a.* having only one colour, of the same colour throughout.

unicorn *a.* having a single horn.

unicostate *a.* having a single prominent midrib, as certain leaves.

unicuspid *a.* having one tapering point, as tooth.

unidactyl *a.* having one digit.

unidirectional replication DNA replication that proceeds in one direction only from the origin.

unifacial *a.* (1) having one face or chief surface; (2) having similar structure on both sides.

unifactorial *a. pert.* or controlled by a single gene.

uniflagellate *a.* having only one flagellum.

uniflorous *a.* having one flower.

unifoliate *a.* (1) with one leaf; (2) with a single layer of zooecia, *appl.* polyzoan colony.

uniforate *a.* having only one opening.

uniformity *n.* in ecology, the tendency of the component species of an association to be uniformly distributed within it.

unijugate *a. appl.* a pinnate leaf having one pair of leaflets.

unilabiate *a.* with one lip or labium.

unilacunar *a.* (1) with one lacuna; (2) having one leaf gap, *appl.* nodes.

unilaminate *a.* having one layer only.

unilateral *a.* arranged on one side only.

unilocular *a.* (1) having a single compartment or cell; (2) containing a single oil droplet, as cells in white fat.

unimodal *a.* having only one mode, *appl.* frequency distribution with a single maximum.

unimucronate *a.* having a single sharp point at tip.

uninducible *a. appl.* mutant, normally inducible genes in a permanent state of repression and which cannot be induced. *alt.* super-repressed.

uninemal *a.* single-stranded.

uninuclear, uninucleate *a.* having one nucleus.

uniovular *a.* (1) *pert.* a single ovum; (2) *appl.* twinning, twins produced from a single egg.

uniparental *a.* arising from a single parent.

uniparental disomy condition in which an embryo inherits two copies of a chromosome or part of a chromosome from one parent.

uniparental inheritance transmission of a phenotype from one parental type to all the progeny, usually due to organelle genes.

uniparous *a.* (1) producing one offspring at a birth; (2) (*bot.*) having a cymose inflorescence with one axis of branching.

unipennate *a. appl.* muscle having its tendon of insertion extending along one side only.

unipetalous *a.* having one petal.

unipolar *a.* having one pole only, *appl.* nerve cells with a single process.

uniport, uniporter *n.* membrane protein that transports a solute across a membrane in one direction only.

unipotent(ial) *a. appl.* cells that can develop into only one type of cell. *cf.* totipotent.

unique sequences single-copy DNA *q.v.*

uniradiate *a.* one-rayed.

Uniramia *n.* in some classifications a group of arthropods containing the insects and myriapods. *alt.* Atelocerata.

uniramous *a.* (1) having one branch; (2) *appl.* crustacean appendage lacking an exopodite; (3) *appl.* antennule.

unisepalous *a.* having a single sepal.

uniseptate *a.* having one septum or dividing partition.

uniserial *a.* (1) arranged in one row or series; (2) *appl.* certain ascospores; (3) *appl.* fins with radials on one side of basilia; (4) *appl.* medullary rays.

uniseriate *a.* occurring in a single row, or layer.

uniserrate *a.* with one row of serrations along edge.

unisetose *a.* having one bristle.

unisexual *a.* (1) of one or other sex, distinctly male or female; (2) (*zool.*) sometimes *appl.* animal producing both sperm and eggs; (3) (*bot.*) *appl.* plants and flowers having stamens and carpels in separate flowers.

unistrate, unistratose *a.* having one layer.

unit membrane any cell membrane with a lipid bilayer structure.

unitunicate *a. appl.* asci in which both inner and outer wall are rigid and inelastic.

univalent *a. appl.* a single unpaired chromosome at meiosis.

univalve *n.* a shell in one piece, as of a gastropod mollusc.

universal donor *see* blood groups.

universal genetic code *see* genetic code.

universal recipient *see* blood groups.

universal veil tissue that completely encloses the developing toadstool in some fungi such as *Amanita*, and which tears as the toadstool grows to leave a cup-shaped remnant (volva) around base of stalk.

univoltine *a.* producing only one brood in a season.

unken reflex defensive posture adopted by newts in which they hold themselves rigidly immobile, tail upheld, showing brightly coloured underparts.

unmyelinated *a. appl.* axons that lack a myelin sheath.

unpaired *a.* (1) situated in median line of body, consequently single; (2) *appl.* nucleotide in DNA lacking a complementary nucleotide on opposite strand.

unrooted *appl.* a phylogenetic tree which shows the relative degrees of difference (or similarity) between a set of present-day DNA or amino-acid sequences, but in such a way that the direction of time cannot be determined for any branch, and thus the tree does not converge onto a common ancestor, or root. *cf.* rooted.

unsaturated *a. appl.* fatty acids and other long-chain hydrocarbons with one or more double bonds, C=C, in their carbon chain.

unscheduled DNA synthesis DNA synthesis occurring outside of the S phase during the cell cycle.

unstable mutation mutation that has a high frequency of reversion.

unstriped muscle smooth muscle *q.v.*

3′ untranslated region, 5′ untranslated region *see* 3′ UTR, 5′ UTR.

unweighted pair group method using arithmetic mean UPGMA *q.v.*

unwindase, unwinding enzyme helicase enzyme that unwinds the DNA double helix in front of the DNA polymerase during DNA replication.

u-PA urokinase *q.v.*

UPGMA unweighted pair group method using arithmetic mean. A method of building phylogenetic trees from evolutionary distance data by first subclustering the most similar sequences, assuming a constant molecular clock. It produces an ultrametric rooted tree.

up mutation mutation in which transcription of a particular gene(s) is much enhanced, usually due to a mutation in the promoter controlling that gene(s).

upper shore zone of seashore between the average high-tide level and the highest high-water mark, which supports only a few species.

upregulation *n.* increase in number, as of receptors on a cell surface, or in rate of production, as of mRNA.

upstream *a.* (1) *appl.* sequences to the 5′ side of any given point on the DNA coding strand or in the RNA corresponding to it; (2) *appl.* events preceding a given stage in a metabolic or signalling pathway. *cf.* downstream.

upstream activating sequence (UAS) general name for positive regulatory sequences located on the 5′ side of the startpoint of transcription.

urachus *n.* (1) the median umbilical ligament; (2) fibrous cord extending from apex of bladder to umbilicus.

uracil (U) *n.* a pyrimidine base, one of the four bases in RNA, in which it pairs with thymine (A). It is the base in the nucleoside uridine. *see* Fig. 41.

uracil-DNA glycosylase an enzyme that removes uracil from uracil-containing nucleotides in DNA, which are then excised and replaced by other enzymes. It thus prevents mutations caused by the spontaneous deamination of cytosine to uracil and its subsequent mispairing with adenine. *alt.* DNA-uracil glycosidase.

urate oxidase enzyme found in liver and kidney and in some fungi, responsible for

the oxidation of urate to allantoin which occurs in animals other than primates. EC 1.7.3.3. *alt.* uricase.

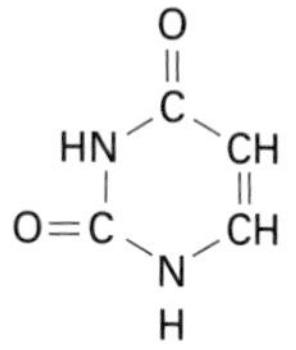

Urbilateria *n.* name given to the hypothetical last common ancestor of all bilaterally symmetrical animals (the Bilateria).

urceolate *a.* urn- or pitcher-shaped.

urceolus *n.* any pitcher-shaped structure.

urea *n.* carbamide, NH_2CONH_2, soluble waste product of the breakdown of proteins and amino acids in mammals and some other animals. It is the chief nitrogenous constituent of the urine, and is also found in some fungi and higher plants.

urea cycle metabolic cycle principally involving arginine, citrulline and ornithine, and found in all terrestrial vertebrates except reptiles and birds, in which ammonium ion formed during aminoacid breakdown is converted to urea for excretion. *alt.* arginine–urea cycle, Krebs–Henseleit cycle, ornithine cycle.

urease *n.* enzyme which catalyses hydrolysis of urea into ammonia and carbon dioxide. EC 3.5.1.5.

Uredinales *n.* order of fungi including the rust fungi, plant pathogenic fungi that produce characteristic masses of brown, black or yellow spores. They have a complex life cycle that usually involves two different host plants.

uredinium, uredium *n.* in macrocyclic rust fungi, a structure resembling an acervulus, in which binucleate uredospores, or summer spores, are produced. *plu.* **uredinia, uredia.**

uredo *n.* summer stage of rust fungi.

uredospore *n.* in macrocyclic rust fungi, the summer spore, the main propagative phase of the rust life cycle, a binucleate spore which germinates to form a mycelium from which new uredospores are produced, and so on.

ureide *n.* nitrogen-containing compound formed during amino acid metabolism (e.g. citrulline) or by the oxidation of purines (e.g. allantoin and allantoic acid). It is used as a nitrogen transport compound in some plants, and as an excretion product in some animals.

ureotelic *a.* excreting nitrogen as urea, as adult amphibia, elasmobranch fishes, mammals. *cf.* ammonotelic, uricotelic.

Fig. 41 Uracil.

ureter *n.* duct conveying urine from kidney to bladder or cloaca.

ureteric bud embryonic diverticulum of metanephros that gives rise to ureters.

urethra *n.* duct conveying urine from the bladder to the outside.

uric acid, urate 2,6,8-trioxypurine, an almost insoluble end-product of purine metabolism in certain mammals. It is the product of the breakdown of nucleic acids and proteins, is excreted in urine in primates, is the main nitrogenous excretion product in birds, reptiles and some invertebrates esp. insects.

uricase urate oxidase *q.v.*

uricotelic *a.* excreting nitrogen as uric acid, as insects, birds, reptiles. *cf.* ureotelic.

uridine *n.* nucleoside made of the pyrimidine base uracil linked to ribose.

uridine diphosphate (UDP) uridine nucleotide containing a diphosphate group, forms part of activated sugars in many metabolic reactions.

uridine monophosphate (UMP) nucleotide consisting of uracil, ribose and a phosphate group, product of the partial hydrolysis of RNA, synthesized *in vivo* from orotidylate by decarboxylation. *alt.* uridylate, uridylic acid, uridine 5′-phosphate.

uridine triphosphate (UTP) uridine nucleotide containing a triphosphate group, one of the four ribonucleotides needed for synthesis of RNA, takes part in many metabolic reactions in a manner analogous to ATP, esp. forming activated UDP-sugars.

uridylic acid, uridylate uridine monophosphate *q.v.*

uridylyltransferase *see* nucleotidyltransferase.

urinary *a. pert.* urine, *appl.* organs including kidneys, ureters, bladder and urethra.

urine *n.* (1) fluid excretion from kidney in mammals, containing waste nitrogen chiefly as urea; (2) solid or semisolid excretion in birds and reptiles, containing waste nitrogen chiefly as uric acid.

uriniferous *a.* (1) urine-producing; (2) *appl.* tubules of nephron leading from Bowman's capsules to collecting ducts.

urinogenital *a. pert.* urinary and genital systems.

urinogenital ridge a paired ridge in mammalian embryo from which urinary and genital systems are developed.

urinogenital sinus bladder or pouch in connection with the urinary and genital systems in many animals.

urn *n.* (1) theca or capsule of mosses; (2) urn-shaped structure; (3) one of the ciliated bodies floating in coelomic fluid of annulates.

urobilin *n.* brown pigment in urine.

urobilinogen *n.* colourless compound derived from bilirubin, oxidized to urobilin and excreted in urine.

urocardiac ossicle stout short bar forming part of gastric mill in certain crustaceans.

urochord *n.* the notochord when confined to tail region, as in tunicates.

Urochordata, urochordates Tunicata *q.v.*

urochrome *n.* yellowish pigment which gives urine its normal colour.

urocoel *n.* an excretory organ in molluscs.

urocyst *n.* the urinary bladder.

urodaeum *alt.* spelling of urodeum.

Urodela, urodeles *n., n.plu.* one of the three orders of extant amphibians, containing the newts and salamanders, amphibians with well-developed tails and two pairs of more-or-less equal legs. Called the Caudata in some classifications. *cf.* Anura.

urodelous *a.* with a persistent tail.

urodeum *n.* the part or chamber of the cloaca into which ureters and genital ducts open. *alt.* **urodaeum.**

urogastric *a. pert.* posterior part of gastric region in some crustaceans.

urogastrone epidermal growth factor *q.v.*

urogenital urinogenital *q.v.*

urohyal *n.* median bony element in the hyoid arch.

uroid *n.* in some amoebae, a region of posterior gel-like cytoplasm.

urokinase (u-PA) *n.* proteolytic enzyme present in mammalian urine, closely related to tissue plasminogen activator, and which can convert plasminogen to plasmin. *alt.* urinary plasminogen activator.

uromere *n.* abdominal segment in arthropods.

uromorphic *a.* like a tail.

uroneme *n.* tail-like structure of some ciliate protozoans.

uronic acid any of a group of acids formed as oxidation products of sugars, and found in urine and as subunits of some polysaccharides.

uropatagium *n.* (1) membrane stretching from one femur to the other in bats; (2) part of patagium extending between hind-feet and tail in flying lemurs.

urophan *n.* any ingested substance found chemically unchanged in urine. *a.* **urophanic**.

urophysis *n.* in fishes, a concentration of neurosecretory nerve endings at the end of the spinal cord, resembling the neurohypophysis of mammals.

uropod *n.* fan-shaped paired appendage on the penultimate segment of crustaceans, used for swimming.

uropore *n.* opening of excretory duct in certain invertebrates.

uroporphyrin *n.* a brownish-red product of haem metabolism, lacking iron, a pigment of urine.

uropygium *n.* the hump at the end of a bird's body, containing vertebrae and supporting tail feathers. *a.* **uropygial**. *alt.* **uropyge**.

uropyloric *a. pert.* posterior region of crustacean stomach.

urorectal *a. appl.* embryonic septum which ultimately divides intestine into anal and urinogenital parts.

urorubin *n.* red pigment of urine.

urosacral *a. pert.* caudal and sacral regions of the vertebral column.

urosome *n.* (1) the tail region of fish; (2) abdomen of some arthropods.

urostege, urostegite *n.* ventral tail plate of snakes.

urosternite *n.* ventral plate of arthropod abdominal segment.

urosthenic *a.* having the tail strongly developed for propulsion.

urostyle *n.* an unsegmented bone forming the posterior part of vertebral column of frogs and toads.

Urticales *n.* an order of dicots including herbs, vines, shrubs and trees, and including the families Cannabaceae (hemp), Moraceae (mulberry), Ulmaceae (elm) and Urticaceae (nettle).

urticant *a.* stinging, irritating.

urticaria *n.* local inflammatory swellings on skin as a result of allergic reactions. *alt.* hives.

urticarial *a.* producing a rash as of stinging nettle, *appl.* hairs of some caterpillars.

Ustilaginales *n.* order of fungi that includes the smut fungi, plant pathogenic fungi which produce masses of black resting spores that look like soot, giving the group its name.

uterine *a. pert.* uterus.

uterine cervix the neck of the uterus where it adjoins the vagina.

uterine crypts depressions in uterine mucosa, for accommodation of chorionic villi.

uteroabdominal *a. pert.* uterus and abdominal region.

uterosacral *a. appl.* two ligaments of sacrogenital folds attached to sacrum.

uterovaginal *a. pert.* uterus and vagina.

uterovesical *a. pert.* uterus and urinary bladder.

uterus *n.* (1) in female mammals, the organ in which the embryo develops and is nourished before birth; (2) in other animals, an enlarged portion of oviduct modified to serve as a place for development of young or of eggs.

UTP uridine triphosphate *q.v.*

3′ UTR 3′ untranslated region, region in mRNA (and DNA) following the coding region, which is transcribed but is not translated.

5′ UTR 5′ untranslated region, *see* leader sequence.

utricle *n.* small fluid-filled sac in the inner ear, which responds to movement of the position of the head. It is part of the membranous labyrinth, and together with the sacculus forms vestibule of inner ear. *alt.* utriculus. *q.v.*

utricular, utriculate *a.* (1) *pert.* utricle; (2) shaped like a utricle; (3) containing vessels like small bags.

utriculiform *a.* shaped like a utricle or small bladder.

utriculus utricle *q.v.*

utriform *a.* bladder-shaped, with a shallow constriction.

UV ultraviolet light *q.v.*

uva *n.* berry formed from a superior ovary and with a central placenta, such as a grape.

uvea *n.* pigmented epithelium covering posterior surface of iris.

uvula *n.* conical flap of soft tissue hanging from soft palate.

V

V (1) valine *q.v.*; (2) symbol for the chemical element vanadium *q.v.*; (3) SI symbol for volt *q.v.*

V$_H$ variable region of antibody heavy chain. *see* V region.

V$_L$ variable region of antibody light chain. *see* V region.

V$_{max}$ *see* Michaelis–Menten kinetics *q.v.*

Val valine *q.v.*

v–a mycorrhiza vesicular–arbuscular mycorrhiza *q.v.*

vaccination *n.* administration of a preparation of an avirulent or killed pathogen that results in an immune response against the pathogen without accompanying illness, and which results in protective immunity against the pathogen. *alt.* immunization, inoculation. *see also* vaccine. *v.* **vaccinate.**

vaccine *n.* a preparation of microorganisms or their antigenic components which can induce protective immunity against the appropriate pathogenic bacterium or virus but which does not itself cause disease. Vaccines may be composed of killed pathogenic microorganisms (killed vaccines), or live non-pathogenic strains of virus or bacterium (live vaccines), or isolated protein antigens (subunit vaccines).

vacuolar *a. pert.* or resembling a vacuole.

vacuolated *a.* containing vacuoles.

vacuolation *n.* (1) formation of vacuoles; (2) appearance or formation of drops of clear fluid in growing or ageing cells. *alt.* **vacuolization.**

vacuole *n.* any large fluid-filled membrane-bounded cavity in the cytoplasm of eukaryotic cells. A vacuole is a prominent feature of plant and algal cells, where it is known as the central vacuole and takes up a large part of the cellular volume.

see also contractile vacuole. *see* Fig. 7 (p. 105).

vacuum activity a fixed-action pattern of activity carried out by an animal in the absence of any external stimulus.

vagal *pert.* the vagus.

vagiform *a.* (1) having an indeterminate form; (2) amorphous.

vagile *a.* (1) freely motile; (2) able to migrate. *n.* **vagility.**

vagina *n.* (1) canal leading from uterus to external opening of genital canal; (2) sheath or sheath-like tube.

vaginal *a. pert.* or supplying the vagina, e.g. arteries, nerves.

vaginal process (1) projecting plate on inferior surface of petrous portion of temporal bone; (2) a plate on sphenoid.

vaginate *a.* sheathed.

vaginervose *a.* with irregularly arranged veins.

vaginicolous *a.* building and inhabiting a sheath or case.

vaginipennate *a.* having wings protected by a sheath.

vagus *n.* (1) 10th cranial nerve, supplies many internal organs including stomach and duodenum, liver, spleen and kidneys, forms cardiac and pulmonary plexuses, supplies muscles of oesophagus and bronchi and their glands, and pharynx and larynx; (2) visceral accessory nervous system in insects.

valence *n.* (1) the combining power of an atom, which is given a number equal to the number of hydrogen atoms (valence 1) the atom could combine with. A carbon atom has a valence of 4, and an oxygen atom a valence of 2; (2) (*immunol.*) of an antibody, the number of antigen-binding sites per molecule; (3) (*immunol.*) of an antigen, the number of epitopes that antibodies can bind. *alt.* valency.

valine (Val, V) *n.* amino acid with a non-polar hydrocarbon side chain, constituent of protein, essential in the diet of humans and some animals.

valinomycin *n.* antibiotic which increases the permeability of cell membranes to potassium ions.

vallate *a.* with a rim surrounding a depression, *appl.* papillae with taste buds on back of tongue.

vallecula *n.* a depression or groove.

valleculate *a.* grooved.

Valsalva's sinuses sinuses of Valsalva *q.v.*

valvate *a.* (1) hinged at margin only; (2) meeting at edges; (3) opening or furnished with valves.

valve *n.* (1) structure in heart and blood vessels that permits blood flow in one direction only, preventing backward flow; (2) each half of the hinged two-part shells of e.g. clams and cockles; (3) each part of the silica case of diatoms; (*bot.*) (4) any of the sections formed by a seed capsule on dehiscence; (5) lid-like structure of certain anthers.

valvula *n.* a small valve in the coronary sinus in right atrium.

valvulae conniventes circular, spiral or bifurcated folds of mucous membrane found in alimentary canal from duodenum to ileum, affording an increased area for secretion and absorption.

valvular *a. pert.* or like a valve, *appl.* dehiscence of certain capsules and anthers which split into several sections.

valvule *n.* (1) upper palea of grasses; (2) valvula *q.v.*

VAM fungi members of the zygomycete order *Glomales*, which form vesicular–arbuscular mycorrhizae.

vanadium (V) *n.* metallic element which is a prosthetic group of some bacterial nitrogenases.

vanadocyte *n.* blood cell containing a vanadium compound, in certain ascidians.

vancomycin *n.* antibiotic which blocks cell wall formation in bacteria.

van der Waals forces weak non-covalent interatomic attractive forces, of importance in forming and maintaining the three-dimensional structure of proteins and in interactions between proteins.

vane *n.* the web of a feather, consisting of the thread-like barbs.

vannus *n.* fan-like posterior lobe of hindwing of some insects. *a.* **vannal**.

variability plot Wu and Kabat plot *q.v.*

variable expressivity of a genetic trait, expression to a different degree in different individuals.

variable lymphocyte receptor (VLR) type of variable antigen receptor expressed by lymphocyte-like cells in the lamprey. VLRs are not immunoglobulins, but are generated by a process of somatic gene rearrangement.

variable number tandem repeats (VNTR) minisatellite DNA *q.v.*

variable region (V region) a region in the light and heavy chains of an immunoglobulin molecule and in the two chains of a T-cell receptor that varies considerably in amino acid sequence between molecules of different antigen specificities. It forms a distinct domain at the amino-terminal end of each chain. The variable regions are the parts of the molecule that contribute to the antigen-binding site. *see also* constant region, V gene segment.

variance *n.* (1) the condition of being varied; (2) the mean of the squares of individual deviations from the mean.

variant *n.* an individual or species deviating in some character or characters from type.

variate *n.* the numerical value of a variable.

variation *n.* (1) divergence from type in certain characteristics; (2) the phenotypic differences that exist between individuals of the same species (other than those due to age), and which reflect both genetic differences and the influence of the environment. *see also* continuous variation, discontinuous variation, genetic variation; (3) (*stat.*) deviation from the mean.

varicellate *a. appl.* shells with small and indistinct ridges.

varicella-zoster virus the virus that causes chickenpox and shingles, a member of the Herpesviridae.

varices *n.plu.* prominent ridges across whorls of univalve shells, showing previous position of the outer lip. *sing.* **varix**.

varicose *a.* unusually swollen and dilated.

variegation *n.* (1) patchiness of pigmentation of leaves or other plant organs. It may be caused by viral interference, non-development of chloroplasts in some cell lineages, or somatic mutations leading to a change of phenotype during development; (2) any discontinuous phenotype caused by a change in genotype in somatic cells during development.

variety *n.* a taxonomic group below the species level.

variola major smallpox.

variola minor cowpox.

variole *n.* small pit-like marking on various parts, in insects.

varix *sing.* of varices.

vas *n.* a small vessel, duct, canal or blind tube. *plu.* **vasa**.

vasa afferentia *plu.* of vas afferens.

vasa deferentia *plu.* of vas deferens.

vasa efferentia *see* ductulus efferens.

vas afferens lymphatic vessel entering lymph node. *plu.* **vasa afferentia**.

vasal *a. pert.* or connected with a vessel.

vasa vasorum blood vessels supplying the larger arteries and veins and found in their coats.

vascular *a. pert.*, consisting of, or containing vessels adapted for the carriage or circulation of fluid. In animals it refers to the blood and lymphatic systems, in plants to the xylem and phloem.

vascular addressins cell adhesion molecules on vascular endothelial cells. They bind to cell adhesion molecules on leukocytes and are involved in the selective migration of leukocytes to different sites in the body. Vascular addressins belong to the immunoglobulin superfamily.

vascular bundle discrete strand of xylem and phloem cells, sometimes separated by a strip of cambium, in stems of some plants.

vascular cambium *see* cambium.

vascular cylinder the xylem and phloem of stem and root.

vascular endothelial growth factor (VEGF) a protein growth factor secreted by tissues where oxygen supply is low, and which stimulates the growth of blood vessels.

vascularized *a.* infiltrated with blood capillaries.

vascular plants common name for all plants containing xylem and phloem and which include the club mosses, ferns, cycads, gymnosperms and angiosperms.

vascular strand fine strand of primary xylem and primary phloem in stems of some plants, leaves and flowers.

vascular system (1) (*bot.*) the plant tissues that carry water and solutes around the plant, i.e. the xylem and phloem; (2) (*zool.*) the blood system.

vascular tissue in plants, xylem and phloem.

vasculogenesis *n.* the formation of blood vessels.

vasculum *n.* (1) a pitcher-shaped leaf; (2) a small blood vessel.

vas deferens duct leading from testes to penis, for carriage of sperm. *plu.* **vasa deferentia**.

vasoactive *a.* affecting blood vessels, causing them either to dilate or contract.

vasoactive intestinal peptide (VIP) a biologically active peptide occurring in both gut and nervous tissue.

vasoconstriction *n.* constriction of blood vessel with reduction in the size of the lumen, as a result of contraction of smooth muscle in vessel wall. It causes an increase in blood pressure.

vasoconstrictor *a.* causing constriction of blood vessels.

vasodentine *n.* a variety of dentine permeated with blood vessels.

vasodilation *n.* enlargement of the lumen of blood vessel, by relaxation of smooth muscle in the vessel wall. It causes a reduction in blood pressure.

vasodilator *a.* causing enlargement of lumen of blood vessels.

vasoganglion *n.* dense network of blood vessels such as a rete mirabile.

vasoinhibitory vasodilator *q.v.*

vasomotor *a. appl.* nerves supplying muscles of walls of blood vessels and regulating the size of the lumen of the vessel by vasodilation and vasoconstriction.

vasopressin *n.* peptide hormone produced by the neurohypophysis (the posterior lobe of the pituitary). It stimulates smooth muscle contraction, causes constriction of small blood vessels and raises blood pressure, and has an antidiuretic effect by causing water resorption in kidney tubules. *alt.* antidiuretic hormone.

vasotocin *n.* peptide hormone with similar properties to oxytocin and vasopressin. It is secreted by the posterior lobe of the pituitary gland in lower vertebrates, and by foetus and pineal in mammals.

vastus *n.* a division of the quadriceps muscle of thigh.

vCJD variant Creutzfeldt–Jakob disease, *see* Creutzfeldt–Jakob disease.

V(D)J recombinase enzyme composed of RAG-1 and RAG-2 subunits that carries out the recombination reaction in the somatic recombination of immunoglobulin and T-cell receptor genes. *see* V(D)J recombination.

V(D)J recombination the DNA rearrangement that occurs at immunoglobulin and T-cell receptor loci in B cells and T cells, respectively, to produce the DNA sequence encoding the variable region of these molecules. At immunoglobulin light-chain loci and T-cell receptor α-loci, a V gene segment and a J gene segment are joined together, whereas at immunoglobulin heavy-chain loci and T-cell receptor β-loci, V, D and J gene segments are joined.

vector *n.* (1) any agent (living or inanimate) that acts as an intermediate carrier or alternative host for a pathogenic organism and transmits it to a susceptible host; (2) phage, plasmid or virus DNA into which another DNA is inserted for introduction into bacterial or other cells for amplification (DNA cloning) or studies of gene expression.

vegetable starch starch, as compared with glycogen.

vegetalized *a.* in experimental embryology, *appl.* embryo that develops an increased amount of endoderm at the expense of ectoderm, as a result of some treatment. *alt. n.* **vegetalization**.

vegetal pole, vegetal region (1) that part of a yolky egg that cleaves more slowly than the rest, due to the presence of the yolk. *cf.* animal pole; (2) the end of a blastula at which the larger cleavage products (megameres) collect.

vegetation *n.* the plant cover of an area, considered generally, and not taxonomically.

vegetative *a.* (1) *appl.* stage of growth in plants when reproduction does not occur;

(2) *appl.* foliage shoots on which flowers are not formed; (3) *appl.* the assimilative phase in fungi, when mycelium is being produced; (4) *appl.* reproduction by bud formation or other asexual method in plants and animals.

vegetative apomixis asexual reproduction in plants by e.g. rhizomes, stolons and bulbils.

vegetative nucleus (1) macronucleus *q.v.*; (2) pollen tube nucleus.

VEGF vascular endothelial growth factor *q.v.*

vehicle *n.* non-living environmental source of pathogens, e.g. food or water.

veil *n.* in fungi, a sheet of fine tissue attached to the stalk and stretching over the cap in some fruit bodies. It is ruptured as the fruit body develops, remaining as a ring on the stalk and sometimes as patches on the cap. *alt.* velum.

vein *n.* (1) branched vessel that conveys blood to heart; (2) in insect wing, fine extension of the tracheal system; (3) in leaf, branching strand of vascular tissue.

vela *plu.* of velum.

velamen *n.* the multiple-layered epidermis of aerial orchid roots, providing mechanical protection, reducing water loss, and possibly specialized for water absorption.

velaminous *a.* having a velamen.

velar *a. pert.* or situated near a velum.

velate *a.* (1) veiled; (2) covered by a velum.

veld *n.* the open temperate grasslands of southern Africa.

veliger *n.* second larval stage in some molluscs, developing from the trochophore.

velum *n.* (1) membrane or structure similar to a veil; (2) in medusa of Hydrozoa, a ring of tissue projecting inwards from margin of bell; (3) in lampreys and some other vertebrates, a flap of muscular tissue in the buccal cavity; (4) (*mycol.*) veil, a sheet of tissue stretching from stipe over top of cap in some basidiomycete fungi. *plu.* **vela**.

velvet *n.* soft vascular skin which covers the antlers of deer during their growth and is rubbed off as the antlers mature.

velvet worms Onychophora *q.v.*

vena *n.* a vein, esp. large blood vessel carrying blood to heart.

vena cava one of the main veins that carries blood to the right auricle of heart.

vena comitantes vein accompanying or alongside an artery or nerve.

venation *n.* the arrangement of veins of leaf or insect wing.

Vendian *n.* period in the late Precambrian lasting from around 650 million years ago to 543 million years ago and characterized by fossils of small and large soft-bodied organisms. *see also* Ediacaran fauna.

Venezuelan region part of the Neotropical floristic kingdom consisting of the area north of the Amazon river to central Venezuela.

venin *n.* a toxic substance of snake venom.

venomous *a.* having poison glands and able to inflict a poisonous wound.

venose *a.* having many and prominent veins.

venous *a.* (1) *pert.* veins; (2) *appl.* blood returning to heart after circulation in body.

vent *n.* (1) the anus; (2) cloacal or anal aperture in lower vertebrates.

vent community *see* hydrothermal vent community.

venter *n.* (1) (*zool.*) abdomen; (2) lower abdominal surface; (3) (*bot.*) the swollen lower portion of archegonium of bryophytes, containing a single egg.

ventrad *adv.* towards lower or abdominal surface.

ventral *a.* (1) *pert.* or nearer the belly or underside of an animal; (2) *pert.* the undersurface of leaf or wing. *cf.* dorsal.

ventral aorta large artery in fish and in amniote embryos running forward from ventricle of heart.

ventralized *n.* in experimental embryology, *appl.* embryo that develops an increased ventral region at the expense of the dorsal region, as a result of some treatment. *n.* **ventralization**.

ventral root (1) of cranial nerve, a nerve root with some sensory fibres; (2) of spinal nerve, a nerve root with some motor fibres.

ventricle *n.* (1) cavity or chamber; (2) one of several fluid-filled cavities in centre of brain; (3) one of pair of lower chambers in heart.

ventricose *a.* swelling out towards the middle, or unequally.

ventricular *a. pert.* a ventricle, *appl.* ligaments and folds of larynx, *appl.* septum and valves in heart.

ventricular membrane basement membrane underlying epithelial tissues of retina.

ventricular proliferative zone layer of dividing cells lining the lumen of the embryonic neural tube, from which neurons and glial cells are formed.

ventrobronchus *n.* one of a number of tubes in lungs of birds which branch off the bronchi and are connected with the anterior air sacs.

ventrodorsal *a.* extending from ventral to dorsal.

ventrolateral *a.* (1) at the side of the ventral region; (2) central and lateral.

ventromedial hypothalamus region of hypothalamus involved in inhibition of eating, amongst other functions.

venule *n.* (1) small vessel conducting venous blood from capillaries to a vein; (2) small vein of leaf or insect wing.

veratridine *n.* alkaloid poison acting on nervous system.

vermicular *a.* resembling a worm in appearance or movement.

vermiculate *a.* marked with numerous sinuate fine lines or bands of colour or by irregular depressed lines.

vermiform *a.* shaped like a worm, *appl.* embryo: stage in life cycle of dicyemid parasite that attaches to host kidney.

vermiform appendix a remnant of the caecum present in some mammals, in humans being a worm-like blind tube extending from the gut.

vermis *n.* the median portion of the cerebellum, distinguished from the cerebellar cortex.

vernal *a. pert.* or appearing in mid or late spring.

vernalization *n.* the exposure of certain plants or their seeds to a period of cold which is necessary either to cause them to flower at all or to make them flower earlier than usual, and is used esp. on cereals such as winter varieties of wheat, oats and rye.

vernation *n.* the arrangement of leaves within a bud.

vernicose *a.* (1) having a varnished appearance; (2) glossy.

verruca *n.* a wart-like projection.

verruciform *a.* wart-shaped.

Verrucomicrobia *n.* phylum of the Bacteria distinguished on DNA sequence criteria.

verrucose *a.* covered with wart-like projections.

versatile *a.* (1) swinging freely. *appl.* anthers; (2) capable of turning backwards and forwards, *appl.* bird's toe.

versical *a. pert.* or in relation to bladder.

versicoloured *a.* (1) variegated in colour; (2) capable of changing colour.

versiform *a.* (1) changing shape; (2) having different forms.

vertebra *n.* any of the bony or cartilaginous segments that make up a backbone, having a central hole for the passage of the spinal cord. *plu.* **vertebrae.**

vertebral *a.* (1) *pert.* backbone; (2) *appl.* various structures situated near or connected with backbone, or with any structure like a backbone; (3) *appl.* arteries that ascend alongside the spinal cord, enter the skull and join together to form the basilar artery.

vertebral column the series of vertebrae running from head to tail along the back of vertebrates, and which encloses the spinal cord. *alt.* backbone, spinal column.

vertebrarterial canal canal formed by foramina in transverse process of cervical vertebrae or between cervical rib and vertebra.

Vertebrata, vertebrates *n.*, *n.plu.* subphylum of the Chordata, animals characterized by the possession of a brain enclosed in a skull, ears, kidneys and other organs, and a well-formed bony or cartilaginous vertebral column or backbone enclosing the spinal cord. The Vertebrata includes the classes Agnatha (lampreys and hagfish), Holocephalii (rabbit fish), Selachii (sharks, dogfishes and rays), Osteichthyes (bony fish), Amphibia, Reptilia, Aves (birds) and Mammalia. *see* Appendix 3.

vertebration *n.* division into segments or parts resembling vertebra.

vertex *n.* (1) top of the head; (2) region between compound eyes of insect.

vertical *a.* (1) standing upright; (2) lengthwise; (3) in direction of axis; (4) *pert.* vertex.

verticil whorl *q.v.*

verticillaster *n.* a much condensed cyme with appearance of whorl but really arising in axils of opposite leaves.

Verticillatae Casuarinales *q.v.*

verticillate *a.* arranged in whorls or verticils.

very low density lipoprotein (VLDL) a group of plasma lipoproteins, synthesized by the liver, which transport triacylglycerols from liver to adipose tissue.

vesica *n.* bladder, esp. urinary bladder.

vesical *a. pert.* or in relation to bladder, *appl.* arteries.

vesicle *n.* (1) general term for small membrane-bounded sacs in eukaryotic cells, which are derived mainly from plasma membrane, endoplasmic reticulum and Golgi apparatus and carry materials from one cellular compartment to another or for secretion. *see* secretory granule, synaptic vesicles, transport vesicle; (2) small spherical air space in tissues; (3) small cavity or sac usually containing fluid; (4) one of three primary cavities in the human brain; (5) (*mycol.*) hyphal swelling as in mycorrhiza; (6) (*zool.*) hollow prominence on shell or coral; (7) base of postanal segment in scorpions. *alt.* **vesicula.**

vesicle-mediated transport the intracellular transport of material enclosed in a membrane vesicle from one cellular compartment to another. *see also* transport vesicle.

vesicula *n.* (1) small bladder-like cyst or sac; (2) vesicle *q.v.*

vesicular *a.* (1) composed of or marked by the presence of vesicle-like cavities; (2) bladder-like.

vesicular–arbuscular mycorrhiza (v–a mycorrhiza) common type of endomycorrhiza characterized by the occurrence of vesicles (swellings on invading hyphae) and arbuscules (discrete masses of branched hyphae) in infected tissues, the lack of a fungal sheath around the roots, and hyphae ramifying within and between the cells of the root cortex.

vesicular gland gland in tissue underlying epidermis in plants and containing essential oils.

vesicula seminalis seminal vesicle *q.v.*

vespertine *a.* blossoming or active in the evening.

vespoid *a.* wasp-like.

vessel *n.* any tube with properly defined walls in which fluids such as blood, lymph and sap move. *see also* xylem.

vessel element one type of water-conducting cell in xylem of angiosperms, with heavily lignified secondary cell walls and large perforations through the cell wall, especially the end walls. Joined end to end with similar cells to form a long hollow tube or xylem vessel.

vestibular *a. pert.* a vestibule.

vestibular apparatus sensory apparatus in inner ear that responds to forces such as gravity and acceleration and is concerned with maintaining balance. It consists of the semicircular canals, utriculus and sacculus. *alt.* vestibular organs, vestibular system.

vestibular canal one of three main canals running the length of the cochlea.

vestibular nerve branch of auditory nerve.

vestibular nuclei nuclei in brainstem that receive input from the vestibular apparatus through the 8th cranial nerve.

vestibular organs, vestibular system vestibular apparatus *q.v.*

vestibulate *a.* (1) in the form of a passage between two channels; (2) resembling, or having, a vestibule.

vestibule, vestibulum *n.* (1) cavity leading into another cavity or passage; (2) portion of ventricle directly below opening of aortic arch in heart; (3) cavity leading to larynx; (4) nasal cavity; (5) posterior chamber of bird's cloaca; (6) (*bot.*) pit leading to stoma of leaf.

vestibulocochlear *a. appl.* nerve: the auditory nerve which innervates the inner ear.

vestibulo-ocular reflex (VOR) the reflex eye movements that ensure that the eyes remain stably pointing in one direction when the head turns, so that the image of the visual field on the retina does not become blurred.

vestige *n.* a small degenerate or imperfectly developed organ or part which may have been complete and functional in some ancestor.

vestigial *a.* (1) of smaller and simpler structure than corresponding part in an ancestral species; (2) small and imperfectly developed.

Vestimentifera, vestimentiferans *n.*, *n.plu.* phylum proposed to include certain genera of the Pogonophora (e.g. *Riftia*, *Lamellibrachia*) which are sessile deep-sea worms that produce fixed chitin tubes

in bottom sediments or on decaying wood on the sea floor. They carry symbiotic sulphide-oxidizing chemoautotrophic bacteria contained within a structure called a trophosome. Their haemoglobin can carry hydrogen sulphide as well as oxygen.

vestiture *n.* a body covering, as of scales, hair or feathers.

vexilla *plu.* of vexillum.

vexillar(y) *a.* (1) *pert.* a vexillum; (2) *appl.* type of imbricate aestivation in which upper petal is folded over others.

vexillate *a.* bearing a vexillum.

vexillum *n.* (1) the upper petal standing at the back of a papilionaceous flower such as pea and bean, which helps to make the flower conspicuous. *alt.* banner, standard; (2) the vane of a feather. *alt.* web. *plu.* **vexilla**.

V gene segment DNA sequence coding for most of the variable region of the polypeptide chains of an immunoglobulin or a T-cell receptor. Each locus contains many different variants arranged in series. To produce the variable region of an immunoglobulin gene, one V gene segment is joined at random to a J segment and, in heavy-chain genes, to a D segment also. *alt.* **V segment.** *see also* D gene segment, gene rearrangement, J gene segment, variable region.

via *n.* a way or passage.

viable *a.* (1) capable of living; (2) capable of developing and surviving parturition, *appl.* foetus. *n.* **viability.**

viable count measurement of concentration of live cells in a microbial population.

viatical *a. appl.* plants growing by the roadside.

vibraculum *n.* whip-like cell modified for defence in Ectoprocta.

vibratile *a.* (1) oscillating; (2) *appl.* antennae of insects.

vibratile corpuscles cells closely resembling sperms present in coelomic fluid of starfish.

vibratile membrane structure formed by fused cilia for wafting food to mouth in ciliate protozoans.

vibrio *n.* any of a group of bacteria with short curved cells, appearing comma-shaped under the microscope, esp. the cholera bacillus, *Vibrio cholerae* and related organisms.

vibrissa *n.* (1) stiff hair growing on nostril or face of animal, as whiskers of cat or mouse, often acting as a tactile organ; (2) feather at base of bill or around eye; (3) one of paired bristles near upper angles of mouth cavity in Diptera; (4) one of the sensitive hairs of an insectivorous plant, as of *Dionaea*, Venus' fly trap. *plu.* **vibrissae.**

vicariance, vicariation *n.* the separate occurrence of corresponding species, e.g. reindeer and caribou, in corresponding but separate environments, divided by a natural barrier.

vicariant speciation the formation of new species resulting from the geographical isolation and separation of an ancestral population into two or more isolated populations. *alt.* geographical isolation.

vicarious *a. appl.* species belonging to a closely related group that are equivalent in ecological terms but which live in separate regions divided by environmental barriers.

vicilin *n.* a seed storage protein of legumes.

vicinism *n.* tendency to variation due to proximity of related forms.

Vicq-d'Azyr, bundle of a bundle of nerve fibres running from corpora mamillaria to the thalamus. *alt.* mamillothalamic tract, thalamomamillary fasciculus.

villi *plu.* of villus.

villiform *a.* having form and appearance of velvet, *appl.* dentition.

villin *n.* protein component of the cytoskeleton of intestinal microvilli.

villose, villous *a.* (1) shaggy; (2) having villi or covered with villi.

villus *n.* (1) one of the small vascularized projections on the lining of the small intestine, which increases the area for nutrient absorption; (2) one of the processes on chorion through which nourishment passes to the embryo; (3) invagination of a synovial membrane into joint cavity; (4) a fine straight process on epidermis of plants. *plu.* **villi.**

vimen *n.* long slender shoot or branch. *plu.* **vimina.**

vimentin *n.* protein component of intermediate filaments in many cell types.

vinblastine, vincristine plant alkaloids derived from *Vinca* species and used as anti-cancer drugs. They inhibit microtubule formation and kill rapidly dividing cells by disrupting the mitotic spindle.

vinculin *n.* a protein associated with the cytoskeleton in some mammalian cells.

vinculum *n.* (1) slender band of tendon; (2) band uniting two main tendons in bird's foot; (3) part of sternum bearing claspers in male insects. *plu.* **vincula.**

Violales *n.* order of dicot shrubs or small trees, less often herbs, and including the families Cistaceae (rock rose) and Violaceae (violet) and others.

violaxanthin *see* xanthophyll.

VIP vasoactive intestinal peptide *q.v.*

viral *a. pert.*, belonging to, consisting of, or due to, a virus.

viral interference the case where the multiplication of one virus in a cell is inhibited when the cell is also infected by another type of virus.

virescence *n.* production of green colouring in petals instead of usual colour.

virescent *a.* turning greenish or green.

virgalium *n.* a series of rod-like elements forming petaloid rays of an ambulacral plate, as in some Asteroidea.

virgate *a.* (1) rod-shaped; (2) striped.

virgin lymphocytes immature lymphocytes that have acquired antigen specificity but have not yet encountered antigen. *alt.* naive lymphocytes.

virgula *n.* (1) small rod; (2) paired or bilobed structure or organ at oral sucker in some trematodes.

virgulate *a.* (1) with or like a small rod or twig; (2) having minute strips.

viridant *a.* becoming or being green.

virilization *n.* (1) masculinization of genetic females caused by disturbances of sex hormone metabolism, due to various causes; (2) precocious sexual development in genetic males.

virion *n.* mature virus particle consisting of nucleic acid core and protein coat, and in some types an outer lipid envelope.

viroids *n.plu.* small circles of RNA which cause various diseases in plants, replicated entirely by host cell enzymes and not coding for any proteins. Transmitted from plant to plant by insect vectors.

virology *n.* the study of viruses.

virose, virous *a.* containing a virus.

virulence *n.* the ability to cause disease. *a.* **virulent**, *appl.* bacteria and viruses.

virulence factor any aspect of a pathogenic microorganism that enables it to give rise to disease.

virulence gene gene in a pathogenic microorganism which is responsible for its ability to cause disease.

virulence plasmid bacterial plasmid that carries genes, e.g. toxin genes, that render its bacterial host pathogenic.

virulent bacteriophage a bacteriophage which can only enter the lytic cycle of infection, causing lysis of the host.

virus *n.* minute, intracellular obligate parasite, visible only under the electron microscope. A virus particle consists of a core of nucleic acid, which may be DNA or RNA, surrounded by a protein coat, and in some viruses a further lipid/glycoprotein envelope. It is unable to multiply or express its genes outside a host cell as it requires host cell enzymes to aid DNA replication, transcription and translation. Viruses cause many diseases of man, animals and plants. Viruses infecting bacteria are called bacteriophages. *see also* DNA viruses, RNA viruses. *see* Appendix 7.

virusoid satellite RNA *q.v.*

virus receptor protein on surface of host cell to which virus particles bind and which aids their entry into the cell.

viscera *n.plu.* the internal organs collectively. *a.* **visceral**.

visceral afferent fibres nerve fibres conveying impulses from sensory receptors in internal organs to the spinal cord. Their cell bodies are in dorsal root ganglia.

visceral arches branchial arches *q.v.*

visceral clefts gill slits *q.v.*

visceral efferent fibres nerve fibres of the sympathetic and parasympathetic systems of the autonomic (involuntary) nervous system, conveying impulses from brain and spinal cord to smooth muscle and glandular tissue of internal organs.

visceral endoderm endoderm derived from the primitive endoderm that develops on the egg cylinder of mammalian blastocyst.

visceral hump, visceral mass in molluscs, a central concentration of viscera covered by a soft skin, the mantle.

viscerocranium *n.* jaws and visceral arches.

visceromotor *a. appl.* nerves carrying motor impulses to internal organs.

viscerotropic *a.* having a predilection to colonize sites in visceral tissues, *appl.* white blood cells, parasites.

viscid *a.* sticky.

viscidium *n.* in orchid flower, a sticky disc at end of the stalk of pollinium, by which it is attached to an insect's head.

viscid silk a highly extensible and sticky type of silk produced by spiders, which forms the spiral of a typical orb web. *see also* frame silk.

viscin *n.* sticky substance obtained from various plants, esp. from berries of mistletoe.

visible light electromagnetic radiation of wavelength 380–780 nm, which is perceptible by the human eye.

visual acuity sharpness of vision.

visual agnosia inability to identify familiar objects by sight.

visual axis the straight line between the point to which the focused eye is directed and the fovea.

visual cortex that part of cerebral cortex in brain concerned with visual perception.

visual field everything that can be seen without moving the eyes or head.

visual pigments (1) pigments used for light detection or vision in animals; (2) rhodopsin and the cone pigments, which are the photoreceptor pigments of the vertebrate eye.

visual transduction the conversion of the light signal received by the photoreceptor pigments in the eye into a nerve impulse.

vital capacity of lungs, the sum of complemental, tidal and supplemental air.

vitalism *n.* a belief that phenomena exhibited in living organisms are due to a special force distinct from physical and chemical forces.

vital staining the staining of living cells or tissues with non-toxic dyes.

vitamin *n.* any of various organic compounds needed in minute amounts for various metabolic processes and synthesized by plants and some lower animals, but which must be supplied in the diet of higher animals. The lack of the appropriate vitamin causes a deficiency disease. *see individual entries.*

vitamin A fat-soluble vitamin derived from carotenes, found in liver oils of certain fish and in milk and eggs. It is a precursor of retinal, the light-sensitive pigment of the rods and cones of the eye. Its deficiency retards growth, causes night blindness and keratinization of epithelia.

vitamin B complex group of water-soluble vitamins obtained from yeast, wheat-germ and liver, and given separate B numbers, now usually replaced by specific names. *see below.*

vitamin B$_c$ folic acid *q.v.*

vitamin B$_1$ thiamine *q.v.*

vitamin B$_2$ riboflavin *q.v.*

vitamin B$_3$ pantothenic acid *q.v.*

vitamin B$_4$ biotin *q.v.*

vitamin B$_5$ pantothenic acid *q.v.*

vitamin B$_6$ any or all of three interconvertible compounds, pyridoxine, pyridoxal and pyridoxamine, found in eggs, milk, meat, whole grains, fresh vegetables and yeast.

vitamin B$_7$ (1) niacin *q.v.*; (2) nicotinamide or nicotinic acid *q.v.*

vitamin B$_{12}$ *see* cobalamin.

vitamin C ascorbic acid, a water-soluble vitamin found in fresh fruit and vegetables, esp. citrus fruit and blackcurrants. It is required in the diet of primates and some other animals, its deficiency causing scurvy. Its precise role in metabolism is not yet known but it is believed to act as a cofactor in oxidation–reduction reactions, and is involved in the synthesis of bone, cartilage and dentine.

vitamin D any or all of several fat-soluble sterols, found esp. in fish liver oils, egg yolk and milk, or formed from precursors in skin exposed to ultraviolet light (sunlight). A deficiency in children causes rickets due to deficient and abnormal bone growth. D vitamins are necessary for normal bone and tooth structure, increasing calcium and phosphate absorption from the gut. Vitamin D$_2$ is calciferol, Vitamin D$_3$ is cholecalciferol and vitamin D$_4$ is dihydrotachysterol.

vitamin E any of several fat-soluble vitamins, the most common being α-tocopherol (vitamin E or E$_1$), β-tocopherol (vitamin E$_2$), and γ-tocopherol (vitamin E$_3$). They occur in leaves of various plants and in oils of some seed germs. Their absence leads to sterility in some animals, and they are possibly necessary for reproduction in all mammals. They have strong antioxidant properties and may be necessary for stabilizing membranes and preventing oxidation in cells.

vitamin G riboflavin *q.v.*

vitamin H biotin *q.v.*

vitamin K a group of vitamins necessary for blood clotting as they are concerned in the production of prothrombin and other coagulation factors in the liver. They are obtained from green leaves, putrefying fish, or are synthesized by bacteria in the gut. Vitamin K$_1$ is α-phylloquinone or phytonadione, vitamin K$_2$ is β-phylloquinone (farnoquinone), vitamin K$_3$ is menadione.

vitamin P citrin (*q.v.*) with its active constituent hesperidin, which affects the permeability and fragility of blood capillaries.

vitellarium yolk gland *q.v.*

vitellin *n.* (1) abundant phosphoprotein in egg yolk; (2) similar or related substance in seeds.

vitelline *a.* (1) *pert.* yolk or yolk-producing organ; (2) *appl.* membrane: zona pellucida *q.v.*

vitelline duct duct conveying yolk from yolk gland to oviduct.

vitelline layer, vitelline membrane thick transparent extracellular layer surrounding plasma membrane of ovum of vertebrates and invertebrates.

vitellogenesis *n.* yolk formation.

vitellogenic hormone juvenile hormone *q.v.*

vitellogenin *n.* protein produced in the liver of certain female amphibians, and which is converted into yolk protein.

vitellus *n.* yolk of ovum or egg.

vitrella *n.* cell of an ommatidium which secretes the crystalline cone.

vitreodentine *n.* a very hard variety of dentine.

vitreous *a.* hyaline or transparent.

vitreous body, vitreous humour clear jelly-like substance in inner chamber of eye.

vitreous membrane (1) innermost layer of dermal coat of hair follicle; (2) innermost layer of cornea.

vitreum *n.* the vitreous humor of eye.

vitronectin *n.* a glycoprotein of the extracellular matrix.

vitta *n.* (1) one of the resinous canals in pericarp of dicots of the Umbelliferae and some other families; (2) longitudinal ridge on diatoms; (3) a band of colour. *plu.* **vittae.**

vittate *a.* having lengthwise ridges, bands or stripes.

viverrids *n.plu.* members of the Viverridae, a family of carnivores including the genet.

viviparous *a.* (1) producing young alive rather than laying eggs, *appl.* all mammals except monotremes, and some animals in other groups; (2) *appl.* plants, having seeds that germinate while still attached to the parent plant, e.g. mangrove. *n.* **viviparity.**

VLDL very low density lipoprotein *q.v.*

VLR variable lymphocyte receptor *q.v.*

VNTR variable number tandem repeats. *see* minisatellite DNA.

vocal chords folds of mucous membrane that project into larynx and whose vibration produces sound.

vocal sac extension of mouth cavity in some amphibians that is involved in sound production.

voice box larynx *q.v.*

volant *a.* adapted for flying or gliding.

volar *a. pert.* palm of hand or sole of foot.

Volkmann's canals *see* Haversian canals.

volt (V) the derived SI unit of electric potential, defined as the potential difference between two points on a conducting wire carrying a current of 1 ampere, when the power dissipated between the points is 1 watt.

voltage clamp apparatus for studying membrane conductance changes in response to changes in membrane potential. It prevents the conductance changes from influencing the membrane potential by 'clamping' membrane potential at a level determined by the experimenter.

voltage-gated channel ion channel in cell membrane whose opening (or closing) depends on a certain threshold membrane potential bei g reached. *alt.* **voltage-sensitive.**

voltine *a. pert.* number of broods in a year.

voltinism *n.* a polymorphism in some insect species, where some individuals enter diapause and some do not.

voluble *a.* twining spirally.

voluntary *a.* subject to or regulated by the will.

voluntary muscle striated muscle *q.v.*

volute *a.* (1) rolled up; (2) spirally twisted.

volutin granules polyphosphate granules found in some microorganisms.

volva *n.* cup-shaped remnant of the universal veil that remains around base of stalk in some fungi such as *Amanita. plu.* **volvae.**

volvate *a.* having a volva.

vomer *n.* paired bone forming floor of nasal cavity. *alt.* ploughshare bone.

vomerine *a.* (1) *pert.* vomer; (2) *appl.* teeth borne on vomers.

vomeronasal organ Jacobson's organ *q.v.*

vomeropalatine *n.* fused vomer and palatine in some fishes and amphibians.

von Baer's law *see* recapitulation theory.

von Ebner's gland gland in tongue that secretes a watery fluid.

von Willebrand factor glycoprotein involved in blood clotting which is required for the adhesion of platelets to damaged regions of blood vessels, and stabilization of Factor VIII. A significant deficiency (von Willebrand disease) causes a bleeding disorder.

VOR vestibulo-ocular reflex *q.v.*

VPL ventroposterolateral nucleus of thalamus.

V region variable region *q.v.*

VSV vesicular stomatitis virus, a rhabdovirus, a mild pathogen of cattle.

V-type ATPase type of proton-transporting ATPase, found e.g. in the membranes of plant vacuoles and lysosomes.

vulnerable *a.* IUCN definition *appl.* species or larger taxa (1) thought likely to move into the endangered category in the near future if circumstances do not change, as most or all of their populations are decreasing through e.g. loss of habitat or over-exploitation, (2) whose populations are seriously depleted and whose security is not assured, and (3) which are at present abundant but are threatened by major adverse factors throughout their range. *see also* endangered, rare, rarity.

vulva *n.* (1) external female genitalia in mammals; (2) opening of ovary to exterior in nematodes.

W

W tryptophan *q.v.*

waggle dance the sequence of movements by which honeybees communicate the location and distance of food sources and new nest sites. It comprises a repeated figure-of-eight movement, made up of a straight run, a loop back to right (or left), then another straight run, then a loop back in the opposite direction and so on, the straight run containing information on the direction and distance away of the target.

Wallace's line an imaginary line separating the Australian and Oriental zoogeographical regions, between Bali and Lombok, between Celebes and Borneo, and then eastwards of the Philippines.

Wallerian degeneration degeneration of a nerve fibre distally to the site of injury.

wall pressure (WP) turgor pressure *q.v.*

Warburg–Dickens pathway pentose phosphate pathway *q.v.*

Warburg manometer apparatus for measuring changes in the amount of a gaseous substrate or product of a reaction by the resulting pressure changes in a fixed volume.

Warburg's factor former name for cytochrome oxidase *q.v.*

warning coloration bright and distinctive coloration, such as the yellow and black stripes of a wasp, that warns a potential predator that it is unpleasant-tasting or dangerous. *alt.* aposematic coloration.

wart *n.* (1) small dry benign growth on skin, *alt.* papilloma; (2) glandular protuberance.

wasp *n.* insect of the superfamily Vespoidea (true wasps) of the order Hymenoptera, generally with a smooth shiny body and a well-defined waist between thorax and abdomen. Some species are social and some solitary. All feed their young on small insects, larvae, etc. Social wasps live in colonies with a queen, males and workers, in a cellular nest made from paper produced from chewed wood pulp. *see also* digger wasps.

water balance the balance between the water intake of an organism directly, in food, and as metabolic water, and the water lost by excretion and evaporation.

water bears the common name for the tardigrades *q.v.*

water channel a water-transporting pore in the plasma membrane, formed by proteins of the aquaporin family.

water cycle the processes by which the Earth's supply of water is converted from one physical state to another by evaporation and condensation, and is moved between the land, oceans and atmosphere, and is passed through the bodies of plants and animals.

water fleas *see* Branchiopoda.

water gland a structure in leaf mesophyll that regulates exudation of water through pores (hydathodes) in the leaf.

water meadow grassland bordering a river which regularly floods in winter.

water moulds common name for the Oomycota *q.v.* and the Hyphochytridiomycota *q.v.*

water pore (1) in various invertebrates, a pore by which water tubes connect to exterior, esp. a pore in echinoderms from which the madreporite is derived; (2) hydathode *q.v.*

water potential (WP) suction pressure *q.v.*

water slater small freshwater isopod crustacean, resembling a terrestrial woodlouse, with no carapace, and carrying young in a brood pouch under the hind part of the body.

water-splitting enzyme the enzymatic activity in photosystem II of green plants that withdraws electrons from water to replace the electrons lost from the reaction centre by the action of light, splitting water to form oxygen and hydrogen in the process.

water table upper surface of the zone of saturation, below which all available pores in soil and rock are filled with water.

water vascular system system of vessels through which water circulates, characteristic of echinoderms, extending into the arms and opening to the exterior through the madreporite.

Watson–Crick base pairing the normal A to T (or U) and C to G pairing that occurs in double-helical DNA or RNA.

Watson–Crick helix *see* DNA.

wattle *n.* fleshy excrescence under throat of cock or turkey or of some reptiles.

wax cells modified leukocytes charged with wax, as in certain insects.

waxes *n.plu.* esters of fatty acids with long chain monohydric alcohols, insoluble in water and difficult to hydrolyse. They are found as protective waterproof coatings on leaves, stems, fruits, animal fur and integument of insects, and include beeswax and lanolin.

wax hair a filament of wax extruded through pore of a wax gland, as in certain scale insects.

wax pocket one of the paired wax-secreting glands on abdomen of honeybee.

W chromosome the X chromosome in animals in which the female is the heterogametic sex.

weathering *n.* the action of external factors such as rain, frost, snow, sun or wind on rocks, altering their texture and composition and converting them to soil.

web *n.* (1) membrane stretching from toe to toe as in frog and swimming birds; (2) network of threads spun by spiders. *see* orb web.

Weberian ossicles chain of four small bones that connect the swimbladder to the ear in teleost fish of the superorder Ostariophysi (cyprinid fish and their relatives). *alt.* **Weberian apparatus**.

Weber's line imaginary line separating islands in the Indonesian archipelago with a preponderant Indo-Malayan fauna from those with a preponderant Papuan fauna.

wedge bones small infravertebral ossifications at junction of two vertebrae, often present in lizards.

weed *n.* a plant growing where it is not wanted.

weevils *n.plu.* family of beetles, the Curculionidae, which have the head prolonged into a beak.

Weil's disease leptospirosis *q.v.*

Weismannism *n.* the concepts of evolution and heredity put forward by the German biologist A. Weismann in the late 19th century, and which deal chiefly with the continuity of germ plasm and the nontransmissibility of acquired characteristics.

Welwitschiales *n.* an order of Gnetophyta including the single living genus *Welwitschia*, having a mainly subterranean stem and two thick long leaves surviving throughout the plant's life.

Wernicke's aphasia language impairment resulting in fluent but meaningless speech, due to injury to Wernicke's area.

Wernicke's area region in temporal lobe of left cerebral cortex involved in language comprehension. *alt.* area 22.

West African subregion zoogeographical area, part of the Ethiopian region, consisting of the equatorial rain forest and the area immediately surrounding it.

West African Rainforest region part of the Palaeotropical floristic kingdom consisting of the equatorial rain forest of West Africa.

Western and Central Asiatic region part of the Palaeotropical floristic kingdom consisting of Asia between *ca.* 30°N and 50°N, from the Caspian Sea in the west to Manchuria in the east.

Western blot the transfer of proteins from a gel on which they have been separated to a surface on which they can be identified by the addition of antibodies specific for the proteins of interest.

wetland *n.* area habitually saturated with water. It may be partly or wholly covered permanently, occasionally or periodically, by fresh or salt water up to a depth of 6 metres. Wetlands include bogs, fens, flood meadows, marshland and salt marshes, shallow ponds, river estuaries and intertidal mud flats, but exclude rivers, streams, lakes and oceans.

WGA wheat-germ agglutinin *q.v.*

whales *see* cetaceans.

Wharton's duct the duct of the submaxillary gland.

Wharton's jelly the gelatinous core of the umbilical cord.

wheal-and-flare reaction positive reaction in a skin test for an allergy to a particular substance.

wheat-germ agglutinin (WGA) protein from wheat-germ, a lectin which binds specifically to terminal *N*-acetylgalactosamine residues in carbohydrate chains.

wheel animals *n.plu.* common name for the Rotifera *q.v. alt.* **wheel animalcules**.

whiplash *a. appl.* flagella that are unbranched and not feathery.

whip scorpions common name for members of the Uropygi, an order of arachnids having the last segment bearing a long jointed flagellum.

whisk ferns common name for the Psilophyta *q.v.*

white adipose tissue *see* white fat.

white blood cell leukocyte *q.v.*

white body so-called optic gland of molluscs, a large soft body of unknown function.

white commissure a transverse band of white fibres forming floor of median ventral tissue of spinal cord.

white fat type of adipose tissue containing cells with large lipid globules of triacylglycerols, almost filling the cytoplasm. *alt.* white adipose tissue.

white fibres (1) myelinated nerve fibres; (2) white muscle fibres *q.v.*; (3) unbranched inelastic fibres of connective tissue, made of collagen and occurring in wavy bundles.

white matter regions of brain and spinal cord that appear white, consisting chiefly of myelinated axons (nerve fibres).

white muscle type of muscle in fish, with chiefly anaerobic respiration, involved in bursts of fast swimming. *cf.* red muscle.

white muscle fibres skeletal muscle fibres which give the fast-twitch response and are involved in rapid intermittent movement. *cf.* red muscle fibres.

white pulp the lymphoid tissue in spleen.

whorl *n.* (1) (*bot.*) circle of flowers, parts of a flower, or leaves arising from one point;

(2) (*zool.*) the spiral turn of a univalve shell; (3) the concentric arrangement of ridges of skin on fingers.

wilderness *n.* land that has never been permanently occupied by humans, or exploited by mining, agriculture, logging in any way.

wildlife corridor in urban and suburban areas, narrow continuous areas of favourable habitat which connect built-up areas with the country and allow the movement of animals, birds and plants along them. *alt.* green corridor.

wild type *n.* the organism carrying the 'normal' form of a given gene or genes. *cf.* mutant. *a.* **wild-type**, may refer to genotype or phenotype.

wilting *n.* loss of turgidity in plant cells, due to inadequate water absorption.

wilting coefficient percentage of moisture in soil when wilting takes place.

wing *n.* (1) vertebrate forelimb modified for flying, in birds, bats (and the extinct pterodactyls); (2) epidermal structure modified for flying, in insects; (3) large lateral process on sphenoid bone; (4) (*bot.*) one of two lateral petals in a papilionaceous flower; (5) lateral expansion on many fruits and seeds; (6) any broad membranous expansion.

wing cells distally rounded polyhedral cells in epithelium of cornea, proximally with extensions between heads of basal cells.

wing coverts tectrices *q.v.*

wing disc small undifferentiated sac of epithelium in dipteran larva that develops into wing at metamorphosis. *alt.* **wing imaginal disc**.

winged insects common name for the Pterygota *q.v.*

winged stem stem having expansions of photosynthetic tissue, as some vetches.

wingless insects common name for the Apterygota *q.v.*

Wingless signalling pathway *see* Wnt signalling pathway.

wing pad undeveloped wing of insect pupa.

wing quill remex *q.v.*

wing sheath elytron *q.v.*

Winslow's foramen opening of bursa omentalis and large sac of peritoneum.

winter bud dormant bud, protected by hard scales during winter.

winter egg egg of many freshwater invertebrates, provided with a thick shell which preserves it as it lies quiescent over the winter, and which hatches in spring.

wisdom teeth four back molar teeth which complete the permanent set in humans, erupting late.

witches' broom twiggy growth occurring on some trees, caused by infection with fungi or mites.

Wnt signalling pathways important signalling pathway in embryonic development and in adult multicellular animals in which secreted Wnt proteins (e.g. Wingless in *Drosophila*, Wnts in mammals) stimulate cell-surface receptors to activate an intracellular signalling system. In the classical Wnt pathway, this results in the protein β-catenin acting as a co-transcription factor to activate target genes, resulting in localized cell growth and cell proliferation, among other responses. Defects in this pathway, due to defects in one of the components, the APC protein, are very common in cases of colorectal cancer.

wobble the recognition by some anticodons of a wider range of codons than is strictly specified by the genetic code. The base-pairing rules between the 3rd base of the codon and 1st base of the anticodon are relaxed.

Wolffian ducts a pair of ducts developing in the early mammalian embryo which represent a primitive kidney and which will give rise to the internal sex organs in males.

Wolffian ridges ridges which appear on either side of the middle line of early embryo, and upon which limb buds are formed.

wood *n.* (1) secondary xylem *q.v.*; (2) the hard, generally non-living part of a tree.

woodlice *see* Isopoda.

wood vessel xylem vessel *q.v.*

woody plant any perennial plant, e.g. tree or shrub, having secondary lignified xylem in stem.

worker *n.* a member of the labouring, non-reproductive caste of semisocial and eusocial insect species.

worm *n.* (1) general name for any member of the phyla Platyhelminthes (flatworms), Nemertina (ribbon worms), Acanthocephala (thorny-headed worms), Nematoda (roundworms), Nematomorpha (horse-hair worms), Phoronida (tubeworms), Annelida (ringed worms) and some other phyla; (2) in developmental biology, genetics and molecular biology, refers usually to the nematode model organism *Caenorhabditis elegans*.

Wormian bones sutural bones *q.v.*

Woronin hypha a hypha inside coil of perithecial hyphae and giving rise to ascogonia.

wound cambium cambium forming protective tissue at site of injury in plants.

wound hormones substances produced by wounded plant cells, which stimulate renewed growth of tissue near wound.

wound response *see* plant defence response.

WP (1) wall pressure. *see* turgor pressure; (2) water potential. *see* suction pressure.

Wu and Kabat plot a plot of the sequence differences, at each amino-acid position, in different variants of a protein, which provides a measure of the protein's variability.

wyosine (Y) *n.* unusual nucleoside found only in tRNA, an extensively modified form of guanosine which can pair with U as well as C.

writhe (W) number of turns of superhelix in a supercoiled DNA molecule. *alt.* writhing number.

X

xanth- prefix from the Gk *xanthos*, yellow, indicating yellow colouring.

xanthein *n.* water-soluble yellow pigment in cell sap.

xanthin *n.* yellow carotenoid pigment in flowers.

xanthine *n.* a purine, 2,6-dioxypurine, found especially in animal tissues such as muscle, liver, pancreas, spleen and in urine, and also in certain plants, and which is a breakdown product of AMP and guanine, and is oxidized to urate (uric acid).

xanthine monophosphate (xanthylate, xanthylic acid) (XMP) ribonucleotide containing the purine base xanthine, a biosynthetic precursor of guanosine monophosphate (GMP).

xanthine oxidase enzyme which oxidizes hypoxanthine to xanthine and then to urate, using molecular oxygen. EC 1.2.3.2.

xanthism *n.* colour variation in which the normal colour is replaced almost entirely by yellow.

xanthocarpous *a.* having yellow fruits.

xanthochroic *a.* having a yellow or yellowish skin.

xanthodermic *a.* having a yellowish skin.

xanthommatin *n.* brown ommatochrome pigment of eyes, ocelli, larval Malpighian tubules in certain insects.

xanthophore *n.* yellow chromatophore.

xanthophyll *n.* any of a group of widely distributed yellow or brown carotenoid pigments, oxygenated derivatives of carotenes, and including lutein, violaxanthin, neoxanthin, cryptoxanthin. *alt.* phylloxanthin.

Xanthophyta *n.* phylum of yellow-green photosynthetic protists, formerly considered as algae and now included in the kingdom Stramenopila or Chromista, which have two unequal flagella, the longer hairy and the shorter smooth. They contain the chlorophylls *a*, *c*, c_2 and *e* and have xanthin pigments. Storage products are oils. Common in fresh water, many xanthophytes have multicellular and syncytial forms. *alt.* yellow-green algae.

xanthoplast *n.* yellow plastid or chromatophore.

xanthopous *a.* having a yellow stem.

xanthopterin *n.* yellow pigment, a pterin, found esp. in wings of yellow butterflies and the yellow bands of wasps and other insects, and also in mammalian urine, and which can be oxidized to leucopterin and converted to folic acid by microorganisms.

xanthospermous *a.* having yellow seeds.

X cell type of retinal ganglion cell that responds to stationary spots and lines of light, and, in animals with good colour vision, to particular wavelengths of light.

X chromosome (1) the female sex chromosome in mammals, two copies of which are present in each somatic cell of females with one copy being permanently inactivated, and one (active) copy being present in males; (2) in general, the sex chromosome present in two copies in the homogametic sex, and in one copy in the heterogametic sex.

X-chromosome inactivation the inactivation of all but one copy of the X chromosome in the somatic cells of female mammals. *alt.* **X inactivation**. *see also* dosage compensation.

xenarthral *a.* having additional articular facets on dorsolumbar vertebrae.

xenic *n. pert.* a culture containing one or more unidentified microorganisms.

xeno- prefix derived from Gk *xenos*, strange.

xenoantibody heteroantibody *q.v.*

xenoantigen heteroantigen *q.v.*

xenobiosis *n.* the condition where colonies of one species of social insect live in the nests of another species and move freely among them, obtaining food from them by various means but keeping their broods separate.

xenobiotic *a.* (1) foreign to a living organism; (2) *appl.* foreign substances such as drugs; (3) *appl.* a completely synthetic chemical, not occurring naturally on Earth.

xenodeme *n.* a deme of parasites differing from others in host specificity.

xenoecic *a.* living in the empty shell of another organism.

xenogamy cross-fertilization *q.v.*

xenogeneic *a.* (1) *pert.* or *appl.* different species; (2) *appl.* immunization of one animal with antibodies of another of a different species, *alt.* heterologous; (3) *appl.* grafting of tissue between animals of different species.

xenogenous *a.* (1) originating outside the organism; (2) caused by external stimuli.

xenograft *n.* a graft of tissue from one species to another.

xenomorphosis heteromorphosis *q.v.*

xenoplastic *a. appl.* graft established in a different host.

Xenopus laevis a species of frog, commonly known as the African clawed toad, widely used in research in developmental biology because of the size and robustness of its eggs and their amenability to surgical manipulation. Its oocytes are also used to study the expression and function of isolated foreign genes and RNAs injected into them.

xenotransplantation *n.* the transplantation of tissue or organs from one species into another, most commonly used to refer to transplantation of organs from an animal species into humans.

xerad xerophyte *q.v.*

xerantic *a.* (1) drying up; (2) withering, parched.

xerarch *a. appl.* seres progressing from xeric towards mesic conditions.

xeric *a.* (1) dry; (2) arid; (3) tolerating, or adapted to, arid conditions.

xero- prefix derived from the Gk *xeros*, dry.

xerochasy *n.* dehiscence of fruits when induced by drying. *a.* **xerochastic**.

xeroderma pigmentosum rare heritable skin disease in humans, in which skin is abnormally sensitive to ultraviolet or sunlight. It is characterized by parchment skin, ulceration of the cornea and a predisposition to skin cancer. It is caused by a defect in a DNA repair enzyme responsible for excising pyrimidine dimers formed in DNA on exposure to sunlight.

xeromorphic *a.* structurally modified so as to withstand drought, *appl.* desert plants such as cacti. *n.* **xeromorphy**.

xerophile *n.* plant or fungus adapted to a limited water supply and able to withstand drought. *a.* **xerophilous**.

xerophyte *n.* a plant adapted to arid conditions, either having xeromorphic characteristics, or being a mesophyte growing only in a wet period.

xerosere *n.* stages in a plant succession that begins on a dry site and develops towards moister conditions.

xerothermic *a. appl.* organisms thriving in hot dry conditions.

X-gal synthetic substrate for the enzyme β-galactosidase, which gives a blue product when hydrolysed.

X inactivation X-chromosome inactivation *q.v.*

X-inactivation centre the site in the middle of the X chromosome at which X-chromosome inactivation begins. Heterochromatin formation proceeds in both directions from the centre. It also encodes a short RNA, XIST RNA, which eventually coats the entire inactive X chromosome.

xiphihumeralis *n.* a muscle extending from xiphoid cartilage to humerus.

xiphisternum *n.* the posterior portion of the sternum, usually cartilaginous.

xiphoid *a.* sword-shaped.

xiphoid process (1) xiphisternum *q.v.*; (2) tail or telson of the king crab, *Limulus*.

Xiphosura *n.* order of aquatic arthropods, commonly called king or horseshoe crabs, in the class Merostomata, having a heavily chitinized body with the prosoma covered with a horseshoe-shaped carapace.

XIST RNA *see* X-inactivation centre.

X-linked agammaglobulinaemia genetically determined immunodeficiency in which antibody synthesis is deficient as a result of arrest of B-cell development.

X-linked gene any gene carried on the X chromosome. *alt.* sex-linked gene.

X-linked inheritance *see* sex-linked (1).

X-linked spinal and bulbar muscular atrophy Kennedy's disease *q.v.*

XMP xanthine monophosphate *q.v.*

X-organ small compact sac-like neurosecretory organ in eye-stalk of some crustaceans.

XP xeroderma pigmentosum *q.v.*

X-ray crystallography technique for determining the three-dimensional structures of molecules such as proteins from the diffraction patterns of X-rays passed through a crystal of the molecule.

xylan *n.* any of a group of polysaccharides composed of a central chain of linked xylose residues, with other monosaccharides attached as single units or side chains, found in the cell walls of many angiosperms.

xylary *a.* (1) *pert.* xylem, *appl.* fibres; (2) *appl.* procambium which gives rise to xylem.

xylem *n.* the main water-conducting tissue in vascular plants which extends throughout the body of the plant and is also involved in transport of minerals, food storage and support. Primary xylem is derived from the procambium, secondary xylem (e.g. the wood of trees and shrubs) from the vascular cambium. Xylem is composed of tracheary elements: tracheids and (in angiosperms) vessel elements. Both are elongated hollow cells, with thickened, usually heavily lignified walls, and lacking protoplasts when mature. They are joined end to end to form a continuous conducting tube. *see also* vascular bundle, phloem.

xylem canal narrow tubular space replacing central xylem in demersed stem of some aquatic plants.

xylem parenchyma short lignified cells surrounding conducting elements or produced with other xylem cells towards the end of the growing season.

xylem ray (1) ray or plate of xylem between two medullary rays; (2) part of a ray of parenchyma found in secondary xylem.

xylocarp *n.* hard woody fruit.

xylochrome *n.* a pigment of tannin, produced before death of xylem cells and giving colour to heartwood.

xylogen *n.* the forming xylem in a vascular bundle.

xylogenesis *n.* the formation of xylem.

xyloglucan *n.* a polysaccharide found in the cell walls of most dicotyledonous plants, composed of a central chain of linked glucose residues with xylose units attached as single units or as part of a complex side chain.

xyloic xylary *q.v.*

xyloid ligneous *q.v.*

xyloma *n.* (1) hardened mass of mycelium which gives rise to spore-bearing structures in certain fungi; (2) tumour of woody plants.

xylophagous *a.* wood-eating, as certain termites and beetles.

xylophilous *a.* (1) preferring wood; (2) growing on wood.

xylophyte *n.* a woody plant.

xylose *n.* five-carbon aldose sugar, a constituent of polysaccharides, esp. in the cell walls of some plants.

xylostroma *n.* the felt-like mycelium of some wood-destroying fungi.

xylotomous *a.* able to bore or cut wood.

xylulose *n.* five-carbon ketose sugar. Xylulose 5-phosphate is involved in carbon dioxide fixation in photosynthesis.

Y

Y (1) tyrosine *q.v.*; (2) wyosine *q.v.*; (3) male chromosome in mammals.

Y2H yeast two-hybrid system *q.v.*

YAC yeast artificial chromosome *q.v.*

Yamanaka cells *see* induced pluripotent stem cells.

yarovization vernalization *q.v.*

Y-cartilage cartilage joining ilium, ischium and pubis in the acetabulum.

Y cell type of retinal ganglion cell that responds to changes in illumination or moving stimuli.

Y chromosome (1) the male sex chromosome in mammals, smaller than and nonhomologous with the X chromosome, being absent from cells of females and present in one copy in the somatic cells of males; (2) in general, the sex chromosome that pairs with the X chromosome in the heterogametic sex.

yeast artificial chromosome (YAC) an artificial chromosome containing telomeres and centromeres from yeast chromosomes. The rest of the DNA can be from any source. It can replicate in yeast cells and is used as a recombinant DNA vector for large fragments of DNA.

yeast mating-type locus *see* mating-type locus.

yeasts general name for a diverse group of unicellular ascomycete fungi that includes the brewer's or baker's yeast *Saccharomyces cerevisiae* and the fission yeast *Schizosaccharomyces pombe*, both of which are used as model organisms in cell biology and genetics. Yeasts are ubiquitous inhabitants of the soil and plant surfaces esp. on sugary substrates, which they ferment when growing anaerobically, producing alcohols and carbon dioxide. This is the basis of their use in wine making, brewing and baking. Some yeasts are human pathogens, such as *Candida*.

yeast two-hybrid assay (Y2H) technique for determining which proteins a given protein can interact with. It involves transfecting the genes for both proteins into yeast in specialized vectors that are designed so that if the proteins expressed from these vectors physically interact, the protein complex turns on a reporter gene in the yeast cell. *alt.* **yeast two-hybrid system**.

yellow body corpus luteum *q.v.*

yellow cartilage a cartilage with matrix pervaded by yellow elastic connective tissue fibres.

yellow cells chloragogen cells surrounding gut of annelid worms.

yellow fibres elastic connective tissue fibres composed largely of elastin.

yellow-green algae *see* Chrysophyta.

yellow spot macula lutea *q.v.*

Y-ligament iliofemoral ligament *q.v.*

Y linkage the inheritance pattern of genes found only on the Y chromosome.

Y-linked gene any gene carried on the Y chromosome.

yolk nutrient material rich in protein and fats, forming a large part of the ova of many egg-laying animals (e.g. amphibians, reptiles and birds) and which nourishes the developing embryo.

yolk duct vitelline duct *q.v.*

yolk gland gland associated with the reproductive system in most animals and which produces yolk cells filled with nutrients, which in most animals become incorporated into the egg as the yolk.

yolk plug mass of yolky cells filling up the blastopore in some gastrulating amphibian eggs.

yolk sac membranous sac rich in blood vessels which develops around the yolk in the eggs of vertebrates and which is attached to the embryo and through which nutrients pass from the yolk. In some viviparous lower vertebrates the yolk sac forms a placenta with the uterine wall. In mammals the yolk sac is empty of yolk and in the marsupials forms the main, and short-lived, placenta.

yolk sac placenta placenta formed by embryonic yolk sac and chorion, the only type of placenta in most marsupials. Viviparous selachians, lizards and snakes also develop a yolk sac placenta.

Y-organs in crabs and lobsters, a pair of glands in the antennary or maxillary segment, resembling the prothoracic glands of insects, and which secrete ecdysone.

ypsiloid *a.* U-shaped, *appl.* cartilage anterior to pubis in salamanders for attachment of muscles used in breathing.

Z

Z either glutamine or glutamic acid in the single-letter code for amino acids.

zalambdodont *a. appl.* insectivores with very narrow molar teeth with V-shaped transverse ridges.

Z chromosome the Y chromosome when female is the heterogametic sex.

Z-disc, Z-line a dark line seen separating sarcomeres of muscle myofibrils under the microscope and which represents the membrane to which the actin filaments of each sarcomere are anchored. *see* Fig. 28 (p. 425).

Z-DNA *see* deoxyribonucleic acid.

zeatin *n.* a natural cytokinin isolated from maize (*Zea mays*), a derivative of adenine.

zeaxanthin *n.* yellow carotenoid pigment found in many plants, including maize, in some classes of algae and in egg yolk and which is an isomer of lutein.

zebrafish *Danio rerio*, a small subtropical freshwater fish widely used for studies in developmental biology.

zein *n.* a simple protein in seeds of maize, lacks tryptophan and lysine.

zeitgeber *n.* a synchronizing agent, e.g. environmental cues responsible for keeping circadian rhythms of plants in tune with the daily 24-hour light–dark cycle.

zeugopodium *n.* (1) forearm; (2) shank.

zidovudine AZT *q.v.*

zinc (Zn) metallic element, an essential micronutrient for plants.

zinc finger a structural feature shared by various proteins that bind to DNA and act as transcriptional regulators, and which is involved in DNA binding. Each zinc finger is a hairpin fold of amino-acid chain held together by a Zn atom.

Zingiberales *n.* order of tropical monocot herbs with rhizomes, and including Cannaceae (canna), Musaceae (banana), Strelitziaceae (bird of paradise flower), Zingiberaceae (ginger) and others.

Zinjanthropus *see* australopithecines.

Zoantharia *n.* subclass of Anthozoa, including the stony corals and sea anemones, which are solitary or colonial with paired mesenteries usually in multiples of six, and having the skeleton, if present, external and not made of spicules.

zoarium *n.* all the individuals of a polyzoan colony.

zoëa *n.* early larval form of certain decapod crustaceans.

zoecial, zoecium zooecial, zooecium *q.v.*

zoic *a.* (1) containing remains of organisms and their products; (2) *pert.* animals or animal life.

zoid zoospore *q.v.*

zona *n.* a zone, band or area.

zona fasciculata radially arranged columnar cells in adrenal cortex below zona glomerulosa, secreting glucocorticoids.

zona glomerulosa groups of cells forming external layer of adrenal cortex beneath capsule, secreting glucocorticoids and mineralocorticoids.

zona granulosa granular zone in a Graafian follicle, the mass of cells in which the ovum is embedded.

zonal *a.* of or *pert.* a zone.

zona orbicularis circular fibres of capsule of hip joint, around the neck of femur.

zona pellucida thick transparent layer surrounding plasma membrane of mammalian ovum.

zona reticularis or reticulata inner layer of adrenal cortex.

zonary *a. appl.* placenta with villi arranged in a band or girdle.

zona striata zona pellucida *q.v.*

zonate *a.* (1) zoned or marked with rings; (2) arranged in a single row, as some tetraspores.

zonation *n.* arrangement or distribution in zones.

zone *n.* (1) an area characterized by similar fauna or flora; (2) a belt or area to which certain species are limited; (3) stratum or set of beds characterized by typical fossil or set of fossils; (4) an area or region of the body. *a.* **zonal.**

zone fossil index fossil *q.v.*

zone of polarizing activity (ZPA) small area of tissue in vertebrate limb bud which directs development along the anterior–posterior axis (thumb–little finger axis).

zonite *n.* a body segment of Diplopoda.

zonoid *a.* like a zone.

zonolimnetic *a.* of or *pert.* a certain zone in depth.

zonula zonule *q.v.*

zonula adh(a)erens belt desmosome, intermediate junction. *see* desmosome.

zonula ciliaris the hyaloid membrane forming suspensory ligament of lens of eye.

zonula occludens tight junction *q.v.*

zonule *n.* a small zone, belt or girdle.

zoo-, -zooid, -zoite word elements derived from Gk *zoos*, animal.

zooanthellae *n.plu.* cryptomonads symbiotic with certain marine protozoans.

zoobenthos *n.* the fauna of the sea bottom, or of the bottom of inland waters.

zoobiotic *a.* parasitic on, or living on an animal.

zoo blot comparison of DNAs from different animals, by digestion of DNA with the same restriction enzymes and electrophoresis of the resulting DNA fragments.

zoochlorellae *n.plu.* symbiotic green algae living in the cells of various animals.

zoochoric *a. appl.* plants dispersed by animals. *n.* **zoochory.**

zoocoenocyte *n.* a coenocyte bearing cilia, in certain algae.

zoocoenosis *n.* an animal community.

zoocyst sporocyst *q.v.*

zooecial *a. pert.* or resembling a zooecium.

zooecium *n.* chamber or sac enclosing zooid in bryozoan colony.

zooerythrin *n.* red pigment found in plumage of various birds.

zoofulvin *n.* yellow pigment found in plumage of various birds.

zoogamete *n.* a motile gamete.

zoogamous *a.* having motile gametes.

zoogamy *n.* sexual reproduction in animals.

zoogenesis *n.* the origin of animals, *appl.* usually to phylogenetic origin.

zoogenetics *n.* animal genetics.

zoogenic *a.* arising from the activity of animals.

zoogenous *a.* produced or caused by animals.

zoogeographical regions large areas of the world with distinct natural faunas. In the classification proposed by Alfred Russel Wallace, the terrestrial world is divided into six large regions: Australian, Ethiopian, Nearctic, Neotropical, Oriental, Palaearctic.

zoogeography *n.* the geographical distribution of animal species.

zoogloea *n.* a mass of bacteria embedded in a mucilaginous matrix and frequently forming an iridescent film on surface of water.

zooid *n.* an individual in a colonial animal, such as an individual polyp in a coral, or an individual still linked to others in asexual reproduction in some annelid worms.

zoology *n.* the science dealing with the structure, function, behaviour, history, classification and distribution of animals.

Zoomastigina *n.* heterogeneous phylum of non-photosynthetic, mainly unicellular, protists bearing from one to many thousands of flagella, commonly known as the animal flagellates, zooflagellates or zoomastigotes. Some (e.g. *Naegleria*) can change from a flagellated to an amoeboid form. They include intestinal parasites of fish and amphibians (e.g. the opalinids), the kinetoplastids, typified by the genus *Trypanosoma*, which causes sleeping sickness in humans, wood-digesting protozoa that live in the gut of termites (e.g. *Staurojenia*), the choanoflagellates, which resemble the body cells of sponges, and several other groups. They correspond to the Zoomastigophora in protozoan classifications.

zoomorphic *a.* having the form of an animal.

zoonomy *n.* physiology, esp. animal physiology.

zoonosis *n.* a disease of animals that can be transmitted to humans. *plu.* **zoonoses**.

Zoopagales *n.* an order (in some classifications considered as a class) of zygomycete fungi parasitic on small soil animals. They are known as the predaceous fungi, or animal traps, and capture and parasitize soil amoebae, rhizopods and nematodes by attaching to them and growing within or on them.

zooparasite *n.* any parasitic animal.

zoophilous *a. appl.* plants adapted for pollination by animals other than insects.

zoophobic *a. appl.* plants shunned by animals because they are protected by spines or hairs.

zoophyte *n.* an animal resembling a plant in appearance, as some colonial hydrozoans.

zooplankton *n.* animal plankton.

zoosis *n.* any disease caused by animals.

zoosphere *n.* biciliate zoospore of algae.

zoosporangiophore *n.* structure bearing zoosporangia.

zoosporangium *n.* a sporangium in which zoospores are produced. *plu.* **zoosporangia**.

zoospore *n.* motile, flagellated asexual reproductive cell in protozoans, algae and fungi.

zoothecium *n.* in certain ciliates, the common gelatinous and often branched matrix.

zootic climax any stable climax community dependent for its maintenance on animal activity such as grazing.

zootoxin *n.* any toxin produced by animals.

zootrophic heterotrophic *q.v.*

zootype *n.* (1) representative type of animal; (2) pattern of expression of Hox genes and some other genes in embryo which is considered to be characteristic of all metozoan embryos.

zooxanthellae *n.plu.* parasitic or symbiotic yellow or brown algae living in various marine invertebrates.

zooxanthin *n.* yellow pigment found in the plumage of certain birds.

Zuckerkandl's bodies aortic bodies *q.v.*

zwitterion *n.* an ion with both positive and negative charges, such as all amino acids.

zygantrum *n.* fossa on posterior surfaces of neural arch of vertebrae of snakes and certain lizards.

zygapophysis *n.* one of the processes of a vertebra by which it articulates with adjacent vertebrae.

zygobranchiate *a.* having gills symmetrically placed and renal organs paired, *appl.* certain gastropods.

zygocyst oocyst *q.v.*

zygodactyl(ous) *a.* having two toes pointing forward, two backwards, as in parrots.

zygodont *a.* having molar teeth in which the four cusps are united in pairs.

zygogenetic, zygogenic *a.* produced by fertilization.

zygoid *a.* diploid, *appl.* parthenogenesis.

zygoma *n.* the bony arch of the cheek.

zygomatic *a. pert.* or in the area of the cheekbone.

zygomatic arch the bony arch of the cheek, formed by the jugal and squamosal bones, that runs along the lower edge of the temporal fenestra from the end of the jaw to the orbit of the eye.

zygomatic gland the infraorbital salivary gland.

zygomaticofacial *a.* (1) *appl.* foramen on malar surface of zygomatic bone for passage of nerve and vessels; (2) *appl.* branch of zygomatic or temporomalar nerve.

zygomaticotemporal *a. appl.* e.g. suture, foramen or nerve at temporal surface of zygomatic bone.

zygomaticus *n.* muscle from zygomatic bone to angle of mouth.

zygomelous *a.* having paired appendages, *appl.* fins.

zygomorphic *a. appl.* e.g. flowers that are bilaterally, rather than radially, symmetrical.

Zygomycota, Zygomycotina, zygomycetes *n., n., n.plu.* diverse group of terrestrial fungi including the bread moulds, fly fungi and animal traps (predacious fungi). They are characterized by sexual reproduction by fusion of gametangia, the production of a resting sexual spore (zygospore) and asexual reproduction by non-motile spores.

zygoneury *n.* in certain gastropods, the condition of having a connective band of nerve fibres between the pleural ganglion and a ganglion on visceral branch of opposite side.

zygophore *n.* hyphal branch bearing zygospores.

zygophyte *n.* plant with two similar reproductive cells which unite in fertilization.

zygopleural *a.* bilaterally symmetrical.

zygopodium *n.* forearm or shank.

zygopterans *n.plu.* damselflies, members of the suborder Zygoptera, of the order Odonata.

zygosis *n.* (1) conjugation; (2) union of gametes.

zygosphene *n.* an articular process on anterior surface of neural arch of vertebrae of snakes and certain lizards, which fits into zygantrum.

zygosphere *n.* gamete which unites with a similar one to form a zygospore.

zygospore *n.* thick-walled sexual spore resulting from fusion of gametangia in zygomycete fungi.

zygotaxis *n.* mutual attraction between male and female gametes.

zygote *n.* diploid cell formed from the union of two gametes, and from which a new individual develops.

zygotene *n.* stage of prophase of meiosis at which homologous chromosomes pair.

zygotic *a. pert.* a zygote.

zygotic gene a gene from the zygote genome that is expressed in the zygote or early embryo. *cf.* maternal-effect gene.

zygotic meiosis meiosis that occurs in a zygote, giving rise to a haploid phase of the life cycle. *cf.* gametic meiosis.

zygotic mutation a mutation occurring in the zygote, immediately after fertilization.

zygotic number the diploid number of chromosomes.

zymodeme *n.* strain of organisms distinguished from another strain on the basis of the pattern of their proteins on gel electrophoresis.

zymogen *n.* functionally inactive precursor of certain enzymes, the active form being produced by specific cleavage of the polypeptide chain. *alt.* proenzyme.

zymogen granules in pancreatic cells secreting digestive enzymes, small dense vesicles containing enzyme precursors (zymogens) for secretion from the cell.

zymogenic *a. pert.* or causing fermentation.

zymogenous *a. appl.* microflora in soil normally present in the resting state and only becoming active when a fresh supply of organic material is added.

zymosan *n.* a material derived from yeast cell walls.

Appendix 1
AN OUTLINE OF THE PLANT KINGDOM
(DOMAIN EUKARYA)

The following outline is intended as a brief summary of the different types of plants that exist and does not represent a rigorous taxonomic classification. The names of divisions, classes and orders, and the rank assigned to the various groups, differ considerably from authority to authority, and no attempt has been made to give all alternatives. The nomenclature of the main groups largely follows that used in P.H. Raven, R.F. Evert and H. Curtis, *Biology of Plants*, 5th edn (Worth, New York, 1992) and L. Margulis and K.V. Schwartz, *Five Kingdoms*, 2nd edn (Freeman, 1988). The nomenclature and arrangement of the flowering plants follows that proposed by Takhatajan and Cronquist rather than the older systems of Engler and Bentham and Hooker (see S. Holmes, *An Outline of Plant Classification*, Longman, Harlow, 1983). The divisions of algae are now considered as members of either the Protoctista or the Stramenopila (see Appendix 4), but those divisions that were traditionally included in the plant kingdom are listed here. For fungi and lichens, see Appendix 2.

All divisions and classes named here, and some orders and additional common names, have an entry in the body of the dictionary.

Algae

DIVISION CHRYSOPHYTA (diatoms and golden algae)
DIVISION PYRROPHYTA (dinoflagellates)
DIVISION EUGLENOPHYTA (euglenoids)
DIVISION RHODOPHYTA (red algae)
DIVISION PHAEOPHYTA (brown algae)
DIVISION CHLOROPHYTA (green algae)

Bryophytes

DIVISION HEPATOPHYTA (liverworts)
DIVISION ANTHOCEROPHYTA (hornworts)
DIVISION BRYOPHYTA
    class Sphagnidae (sphagnum or peat mosses)
    class Andreaeidae (granite or rock mosses)
    class Bryidae (true mosses)

Early vascular plants, now extinct

DIVISION RHYNIOPHYTA
DIVISION ZOSTEROPHYLLOPHYTA
DIVISION TRIMEROPHYTA (possibly the progenitor of the ferns, horsetails and progymnophytes)

Pteridophytes: seedless vascular plants

DIVISION PSILOPHYTA (PSILOTOPHYTA) (whisk ferns, only two living genera, *Psilotum* and *Tmesipteris*)

Appendix 1: The Plant Kingdom

DIVISION LYCOPHYTA (lycophytes)
      order Lycopodiales (lycopods or club mosses)
            Lepidodendrales (tree lycophytes, extinct)
            Selaginellales (one living genus, *Selaginella*)
            Isoetales (quillworts)
            and other orders (extinct)
DIVISION SPHENOPHYTA (horsetails)
      order Equisetales (one living genus, *Equisetum*)
            and other orders (extinct)
DIVISION PTEROPHYTA (FILICOPHYTA) (ferns)
      order Marattiales (giant ferns)
            Ophioglossales (e.g. *Ophioglossum*, adder's tongue)
            Osmundales (e.g. *Osmunda*)
            Filicales (most living ferns, e.g. maidenhair fern, filmy fern,
               hart's tongue, bracken)
            Marsileales (water ferns, e.g. *Pilularia*, pillwort)
            Salviniales (water ferns, e.g. *Azolla*)
DIVISION PROGYMNOPHYTA (progymnosperms, extinct, e.g. *Archaepteris*)

Spermatophytes: seed plants

Gymnosperms

DIVISION PTERIDOSPERMOPHYTA (seed ferns, extinct)
DIVISION CYCADEOIDOPHYTA (BENNETTITALES) (cycadeoids, extinct)
DIVISION CYCADOPHYTA (cycads)
DIVISION GINKGOPHYTA (one species, *Ginkgo*)
DIVISION CONIFEROPHYTA (conifers and their allies)
      order Coniferales (most living conifers, e.g. firs, monkey puzzle,
               cedars, cypresses, junipers, pines, redwoods)
            Cordaitales (extinct)
            Voltziales (extinct)
            Taxales (yews, *Torreya*)
DIVISION GNETOPHYTA
      order Welwitschiales (one species, *Welwitschia*)
            Ephedrales (one genus, *Ephedra*)
            Gnetales (one genus, *Gnetum*)

Angiosperms

DIVISION ANTHOPHYTA (MAGNOLIOPHYTA) (the common names in
    brackets after orders refer to families)
      class Dicotyledones (Magnoliopsida)
        subclass Magnoliidae
          order Magnoliales (e.g. magnolia, nutmeg)
               Laurales (e.g. laurel, calycanthus)
               Piperales (e.g. pepper (the spice))
               Aristolochiales (birthwort)
               Rafflesiales (rafflesia)
               Nymphaeales (e.g. water lily)

subclass Ranunculidae
   order Illiciales (e.g. star anise)
      Nelumbonales (Indian lotus)
      Ranunculales (e.g. buttercup, barberry)
      Papaverales (poppy, fumitory)
      Sarraceniales (pitcher plant (fam. Sarraceniaceae))
subclass Hamamelididae
   order Trochodendrales
      Hamamelidales (e.g. witch hazel, plane)
      Eucommiales
      Urticales (e.g. elm, nettle, mulberry)
      Casuarinales (she oak)
      Fagales (beech (incl. oaks), birch, hazel)
      Myricales (sweet gale)
      Leitneriales
      Juglandales (e.g. walnut)
subclass Caryophyllidae
   order Caryophyllales (e.g. pink, amaranth, goosefoot)
      Cactaceae (cacti)
      Polygonales (polygonum)
      and other orders
subclass Dilleniidae
   order Paeoniales (peonies)
      Theales (e.g. tea, St John's Wort)
      Violales (e.g. violet, rock rose)
      Passiflores (passion flower, pawpaw)
      Cucurbitales (cucurbits: e.g. marrow, squash, gourd)
      Datiscales (e.g. begonia)
      Capparales (e.g. crucifers (brassicas), caper, mignonette)
      Tamaricales (e.g. tamarisk)
      Salicales (willow)
      Ericales (e.g. heather, wintergreen)
      Primulales (e.g. primrose)
      Malvales (e.g. mallow, cocoa, lime-tree)
      Euphorbiales (e.g. spurge, box-tree, jojoba)
      Thymelaeales (daphne (mezereon))
subclass Rosidae
   order Saxifragales (e.g. gooseberry, saxifrage, hydrangea)
      Rosales (e.g. rose (rose, apple, hawthorn, etc.), coco plum)
      Fabales (Leguminosae) (e.g. peas, beans, etc., mimosa)
      Nepenthales (sundews, pitcher plant (fam. Nepenthaceae))
      Myrtiflorae (e.g. myrtle, mangrove, pomegranate, evening
        primrose, loosestrife)
      Hippuridales (mare's tail, gunnera)
      Rutales (e.g. rue, citrus fruits, mahogany)
      Sapindales (Acerales) (e.g. maple, horse chestnut)
      Geraniales (e.g. balsam, geranium, nasturtium)
      Cornales (e.g. dogwood, umbellifers, *Davidia*, ginseng)

           Celastrales (e.g. holly)
           Rhamnales (e.g. grape, buckthorn)
           Oleales (Ligustrales) (olive, privet)
           Elaeagnales (oleaster)
           and other orders
     subclass Asteridae
       order Dipsacales (e.g. honeysuckles, valerian)
           Gentianales (e.g. gentian, bog bean)
           Polemoniales (e.g. borage, convolvulus, dodder, phlox)
           Scrophulariales (personatae) (e.g. acanthus, buddleia, African
              violet, nightshade)
           Lamiales (e.g. verbena, mint)
           Campanulales (e.g. bellflower, lobelia)
           Asterales (composites (e.g. daisy, cornflower, thistle, dandelion))
  class Monocotyledones (Liliopsida, Liliatae)
    subclass Alismidae
      order Alismales (e.g. flowering rush)
           Potamogetonales (e.g. pondweed, eel-grass)
    subclass Liliidae
      order Liliales (e.g. agave, lily, daffodil, onion, yam)
           Iridales (e.g. iris)
           Zingiberales (e.g. canna, banana, ginger)
           Orchidales (orchid)
    subclass Commelinidae
      order Juncales (e.g. rush)
           Cyperales (sedge)
           Bromeliales (pineapple)
           Commelinales (e.g. tradescantia, xyris)
           Eriocaulales (pipe-worts)
           Restionales
           Poales (grasses)
    subclass Arecidae
      order Arecales (palms)
           Arales (e.g. arum)
           Pandanales (screw pine)
           Typhales (bur reed, cat-tail)
           and other orders

Appendix 2

AN OUTLINE OF THE KINGDOM FUNGI
(DOMAIN EUKARYA)

The following outline is intended as a brief summary of the different types of fungi that exist and does not represent a rigorous taxonomic classification. The outline follows that given in C.J. Alexopoulos, C.W. Mims and M. Blackwell, *Introductory Mycology*, 4th edn (Wiley, New York, 1996). The main groups distinguished within each phylum are given. The Oomycota, Hyphochytriomycota and the Labyrinthulomycota are now generally considered taxonomically as members of the kingdom Stramenopila (see Appendix 4). Other groups often considered along with the fungi (e.g. acellular and cellular slime moulds) are now considered as phylogenetically quite separate (see Appendix 4). The lichens are a symbiotic association, in which one partner is a member of the Fungi and the other is an alga or cyanobacterium.

See entries in the body of the dictionary for more information.

Kingdom Fungi

PHYLUM CHYTRIDIOMYCOTA

Chytridiomycetes (chytrids or water moulds, e.g. *Synchytrium endobioticum*, black wart disease of potato)

PHYLUM ZYGOMYCOTA

Zygomycetes (moulds, e.g. *Mucor, Rhizopus, Pilobolus*, and the predaceous fungi or animal traps)

Trichomycetes (fly fungi: fungi associated with living arthropods, e.g. Entomophthorales)

PHYLUM ASCOMYCOTA

Archiascomycetes (yeasts and yeast-like fungi: *Schizosaccharomyces, Taphrina, Protomyces, Saitoella, Pneumocystis*)

Saccharomycetales (yeasts, e.g. *Saccharomyces cerevisiae, Candida albicans*)

Filamentous ascomycetes

    Plectomycetes (*Eurotiales*; the black moulds and blue moulds, e.g. *Penicillium, Aspergillus*; and other families)

    Pyrenomycetes (flask fungi, e.g. *Neurospora, Nectria* (coral spot))

    Discomycetes (cup fungi and others, e.g. *Botrytis* (grey mould), *Monilinia* (brown rot of peach), *Rhytisma acerinum* (tar spot of maples), morels, truffles, and many lichen fungi)

    Loculoascomycetes (e.g. *Alternaria, Venturia*, and many lichen fungi)

    Other groups allied to discomycetes and loculoascomycetes (e.g. powdery mildews, *Laboulbeniales*)

PHYLUM BASIDIOMYCOTA

Hymenomycetes

    Aphyllophorales (polypores, chanterelles, tooth fungi, coral fungi and corticioids)

    Agaricales (all gilled fungi and the boletes)

    Gasteromycetes (puffballs, earthstars, stinkhorns and bird's nest fungi)

Ustilaginales (smut fungi, e.g. *Ustilago*)

Uredinales (rust fungi, e.g. *Puccinia*, *Uromyces*)

DEUTEROMYCOTA (FUNGI IMPERFECTI) (an artificial grouping of fungi with no known sexual stage. The following common names are still in use.)

Coelomycetes (conidia produced in pycnidia or acervuli)

Hyphomycetes (conidia not produced within specialized structures)

Mycelia Sterilia (no conidia produced)

Lichens

Ascolichens (where the fungal partner is an ascomycete)

Discolichens (where the fungal partner is a discomycete)

Pyrenolichens (where the fungal partner is a loculoascomycete)

Basidiolichens (where the fungal partner is a basidiomycete)

Deuterolichens (where the fungal partner is a deuteromycete)

Appendix 3

AN OUTLINE OF THE ANIMAL KINGDOM
(DOMAIN EUKARYA)

This outline is intended simply to give an overall view of the different types of animals that exist, and is not a rigorous or comprehensive taxonomic classification. Only extant phyla are included, but some extinct groups within phyla are indicated. The names of phyla, classes and orders and the rank given to the various groups differ from authority to authority, and no attempt has been made to give all the alternatives. The Metazoa, the multicellular animals, are considered to include all vertebrates and invertebrates with the exception of the phyla Mesozoa and Porifera. The arrangement and nomenclature of phyla used here generally follows L. Margulis and K.V. Schwartz, *Five Kingdoms*, 2nd edn (Freeman, 1988). A traditional zoological classification of protozoa is included here for historical reference; a more modern treatment of the protozoa, in which the group is divided into numerous separate phyla, is included in Appendix 4.

See entries in the body of the dictionary for more information.

PHYLUM PROTOZOA (unicellular nonphotosynthetic organisms)
  subphylum Sarcomastigophora
    superclass Mastigophora
      class Phytomastigophora (photosynthetic flagellates e.g. *Chlamydomonas, Euglena, Gymnopdinium*)
      class Zoomastigophora (non-photosynthetic flagellates, e.g. *Trypanosoma* (parasitic), choanoflagellates)
    superclass Opalinata (multiflagellate protozoans inhabiting the gut of some amphibians)
    superclass Sarcodina
      class Rhizopodea (amoebas (including the parasitic amoebas) and foraminiferans)
      class Actinopodea (radiolarians and heliozoans)
  subphylum Ciliophora (the ciliates)
    class Ciliatea
      subclass Holotrichia (e.g. *Paramecium, Tetrahyena*)
          Peritrichia (e.g. *Vorticella*)
          Suctoria (stalked sessile cells, lacking cilia when mature, e.g. *Podophyra*)
      subclass Spirotrichia (e.g. *Stentor*)
  subphylum Sporozoa (exclusively parasitic protozoa)
    class Telosporea
      subclass Gregarinia (gregarines: endoparasites of invertebrates)
          Coccidia (e.g. *Eimeria*, the cause of coccidioidosis, and *Plasmodium*, the agent of malaria)
    class Piroplasmea (e.g. *Babesia*)

subphylum Cnidospora (exclusively parasitic protozoa)
    class Myxosporidea (e.g. *Myxobolus*) (now considered to have metazoan
        affinities and to be related to the cnidarians)
    Microsporidea

In modern classifications protozoa are included in the separate kingdom Protoctista along with other groups of eukaryotic microorganisms (see Appendix 4).

Invertebrates

PHYLUM MYXOZOA (unicellular parasitic microorganisms with multinucleate
    spores, e.g. *Myxobolus*)
PHYLUM PLACOZOA one species known, *Trichoplax adhaerens*
PHYLUM PORIFERA (sponges)
    class Calcarea (calcareous sponges)
            Hexactinellidea (glass sponges, e.g. *Euplectella*, Venus's flower
                basket)
            Demospongia (e.g. *Spongia*, the bath sponge)
PHYLUM CNIDARIA
    class Anthozoa (corals and sea anemones)
        subclass Octocorallia (Alcyonaria) (soft corals, sea fans, sea pens)
                Zoantharia (sea anemones, stony corals)
    class Hydrozoa (milleporine corals, solitary hydroids, e.g. *Hydra*, and
            colonial hydroids, e.g. *Bouganvillea*, and the siphonophores, e.g.
            Physalia, the Portuguese Man o' War)
    class Scyphozoa (true jellyfishes)
            Cubozoa (box-jellies)
PHYLUM CTENOPHORA (sea gooseberries or comb jellies)
    class Tentacula
            Nuda
(The Cnidaria and Ctenophora collectively are the coelenterates.)
PHYLUM MESOZOA
    class Rhombozoa (dicyemids, heterocyemids)
    class Orthonectidea (orthonectids)
PHYLUM PLATYHELMINTHES (flatworms)
    class Turbellaria (free-living flatworms)
        order Acoela
                Rhabdocoela
                Tricladida
                Polycladida
                and other orders
    class Monogenea (flukes)
        order Monopisthocotylea (skin flukes)
                Polyopisthocotylea (gill flukes)
    class Trematoda (flukes)
        order Aspidobothrea
                Digenea (gut, liver and blood flukes)
    class Cestoda (tapeworms)

PHYLUM NEMERTINA (NEMERTEA, RHYNCHOCOELA) (nemertine worms, ribbon worms)
    class Anopla
        Enopla
PHYLUM GNATHOSTOMULIDA
PHYLUM GASTROTRICHA
PHYLUM ROTIFERA (rotifers)
    class Seisonacea
        Bdelloidea
        Monogononta
PHYLUM KINORHYNCHA
PHYLUM LORICIFERA (only 10 species known)
PHYLUM ACANTHOCEPHALA (thorny-headed worms)
PHYLUM ENTOPROCTA (moss animals)
PHYLUM NEMATODA (nematodes or roundworms)
    class Aphasmidia (Adenophorea)
        Phasmidia (Secernentea)
PHYLUM NEMATOMORPHA (gordian worms, horse-hair worms)
PHYLUM ECTOPROCTA (moss animals)
(The Entoprocta and Ectoprocta were formerly grouped together as the Bryozoa.)
PHYLUM PHORONIDA (tubeworms, *ca.* 15 species known)
PHYLUM BRACHIOPODA (lamp shells)
    class Inarticulata
        Articulata
PHYLUM MOLLUSCA (molluscs)
    class Monoplacophora (*Vema*, *Neopilina*)
        Aplacophora (solenogasters)
        Caudofoveata
        Polyplacophora (chitons)
    class Pelecypoda (Bivalvia) (clams, etc.)
      subclass Protobranchia (*Nucula*)
          Lamellibranchia (most other genera)
    class Gastropoda
      subclass Prosobranchia (winkles, etc.)
Opisthobranchia (sea slugs, sea hares)
Pulmonata (whelks, snails and land slugs)
    class Scaphopoda (tusk shells)
    class Cephalopoda
      subclass Nautiloidea (pearly nautilus)
          Ammonoidea (ammonites, extinct)
          Coleoidea (octopus, squid, cuttlefish)
PHYLUM PRIAPULIDA
PHYLUM SIPUNCULA (peanut worms)
PHYLUM ECHIURA (spoon worms)
PHYLUM ANNELIDA (ringed worms)
    class Polychaeta (e.g. ragworms, lugworms, myzostomarians)
        Oligochaeta (earthworms, etc.)
        Hirudinea (leeches)

Appendix 3: The Animal Kingdom

PHYLUM TARDIGRADA (water bears)
PHYLUM PENTASTOMIDA (PENTASTOMA) (tongue worms)
PHYLUM ONYCHOPHORA (velvet worms)
PHYLUM ARTHROPODA (arthropods)
  subphylum Trilobitomorpha (extinct)
      class Trilobita
  subphylum Chelicerata
      class Merostomata (horseshoe crabs)
          Arachnida
        order Scorpiones (scorpions)
              Pseudoscorpiones (false scorpions)
              Araneae (Araneida) (spiders)
              Palpigrada (palpigrades)
              Solifuga (Solpugida) (solifugids)
              Opiliones (harvestmen)
              Acari (Acarina) (ticks and mites)
      class Pycnogonida (sea spiders)
  subphylum Crustacea
      class Cephalocarida
          Branchiopoda (water fleas, etc.)
          Ostracoda (ostracods)
          Copepoda (copepods)
          Mystacocarida
          Branchiura (fish lice)
          Cirripedia (barnacles)
          Malacostraca (crabs, lobsters, shrimps, woodlice)
  subphylum Atelocerata
      class Diplopoda (millipedes)
          Chilopoda (centipedes)
          Pauropoda
          Symphyla
          Insecta
        order Diplura
              Thysanura (silverfish)
              Collembola (springtails)
              Protura (bark lice)
              Odonata (dragonflies)
              Ephemeroptera (mayflies)
              Plecoptera (stoneflies)
              Dictyoptera (cockroaches)
              Dermaptera (earwigs)
              Embioptera (web spinners)
              Isoptera (termites)
              Psocoptera (book-lice and their allies)
              Anoplura (biting and sucking lice)
              Orthoptera (locusts, grasshoppers and crickets)
              Thysanoptera (thrips)
              Hemiptera (bugs)

          Neuroptera (lace-wings, ant-lions)
          Mecoptera (scorpion flies)
          Trichoptera (caddis-flies)
          Lepidoptera (moths and butterflies)
          Coleoptera (beetles)
          Diptera (house flies, mosquito, tsetse fly, crane-flies)
          Siphonaptera (fleas)
          Strepsiptera (stylopids)
          Hymenoptera (ants, wasps, bees, saw-flies, ichneumon flies, gall-wasps)
PHYLUM POGONOPHORA (beard worms, tubeworms)
PHYLUM VESTIMENTIFERA (beard worms, tubeworms)
PHYLUM ECHINODERMATA (echinoderms)
  subphylum Pelmatozoa
     class Crinoidea (sea lilies and feather stars)
  subphylum Eleutherozoa
     class Stelleroidea (starfish, brittle stars)
          Echinoidea (sea urchins, sand dollars)
          Holothuroidea (sea cucumbers)
     and many extinct classes
PHYLUM CHAETOGNATHA (arrow worms)
PHYLUM HEMICHORDATA (hemichordates)
     class Enteropneusta (acorn worms, tongue worms)
          Pterobranchia (pterobranchs)
          Planktosphaeroidea (known from a larval form only)
          Graptolita (graptolites, extinct)
PHYLUM CHORDATA (the chordates, including tunicates (urochordates), cephalochordates and vertebrates)
  subphylum Tunicata (Urochordata)
     class Ascidiacea (sea squirts)
          Larvacea
          Thaliacea (salps)
  subphylum Cephalocordata (lancelets)
     class Leptocardii

Vertebrates

  subphylum Agnatha (vertebrates without jaws)
     class Cyclostomata (lampreys, hagfishes, slime eels)
     class Ostracodermi (extinct) (ostracoderms: cephalaspids, anaspids, pteraspids and thelodonts)
  subphylum Gnathostomata (jawed vertebrates)
  superclass Pisces
     class Chondrichthyes (cartilaginous fishes)
       subclass Elasmobranchii
         order Heterodontiformes (the Port Jackson shark)
            Hexanchiformes
            Lamniformes (most dogfishes and sharks)
            Raiiformes (rays and skates)

               Torpediniformes (electric rays)

               and extinct orders

class Osteichthyes (bony fishes)

  subclass Actinopterygii

    infraclass Palaeoniscoidei (extinct except *Polypterus* and *Erpetoichthys*)

    infraclass Chondrostei (sturgeons)

          Holostei (extinct except *Amia* and *Lepisosteus*)

          Teleostei (most living bony fishes)

    superorder Elopomorpha (eels, tarpons, etc.)

    superorder Clupeomorpha (herrings, etc.)

    superorder Osteoglossomorpha (mooneyes, knife-fish and bony tongues)

    superorder Protacanthopterygii

      order Salmoniformes (salmon)

      order Gonorhynchiformes (millifishes and deep-sea lantern fishes) and extinct orders

    superorder Ostariophysi

      order Cypriniformes (characins, American knife-fishes, carps and minnows)

         Siluriformes (catfishes)

    superorder Paracanthopterygii

      order Gadiformes (cod, etc.)

      and other orders

    superorder Acanthopterygii

      order Atheriniformes (tooth-carps, etc.)

         Perciformes (perches, mackerel, tuna, etc.)

         Pleuronectiformes (flatfishes)

      and other orders

 class Sarcopterygii

  subclass Crossopterygii

    order Rhipidistia (extinct)

        Actinistia (extinct except for the coelacanth *Latimeria*)

  subclass Dipnoi (lungfishes)

      and extinct classes acanthodians and placoderms

superclass Tetrapoda

 class Amphibia

  subclass Labyrinthodontia (extinct)

  subclass Lepospondyli (extinct)

  subclass Lissamphibia (includes all living amphibians)

    order Anura (Salientia) (frogs and toads)

       Urodela (newts and salamanders)

       Apoda (caecilians)

 class Reptilia

  subclass Anapsida

    order Cotylosauria (Captorhinida) (extinct)

       Mesosauria (extinct)

subclass Testudinata
  order Chelonia (tortoises and turtles)
subclass Lepidosauria
  order Sphenodonta (extinct except the tuatara, *Sphenodon*)
      Eosuchia (extinct)
      Squamata
  suborder Lacertilia (lizards)
        Amphisbaenia (amphisbaenids)
        Serpentes (snakes)
subclass Archosauria
  order Thecodontia (extinct)
      Saurischia (lizard-hipped dinosaurs, extinct)
      Ornithischia (bird-hipped dinosaurs, extinct)
      Pterosauria (pterosaurs, extinct)
      Crocodylia (crocodiles and alligators)
      and other extinct orders
subclass uncertain
  order Nothosauria (extinct)
      Placodontia (extinct)
      Plesiosauria (extinct)
      Ichthyosauria (extinct)
      and other extinct orders
subclass Synapsida (mammal-like reptiles, extinct)
  order Pelycosauria
      Therapsida
class Aves (birds)
subclass Archaeornithes (*Archaeopteryx* only, extinct)
subclass Odontornithes (extinct)
  order Hesperornithiformes
      Ichthyorniformes
subclass Neornithes (all other birds)
  order Tinamiformes (tinamous)
      Rheiformes (rheas)
      Struthiorniformes (ostriches)
      Casuariiformes (cassowaries and emus)
      Aepyornithiformes (e.g. *Aepyornis*, extinct)
      Dinornithiformes (moas and kiwis)
      Podicipediformes (Colymbiformes) (grebes)
      Procellariformes (albatrosses, shearwaters and petrels)
      Sphenisciformes (penguins)
      Pelecaniformes (pelicans and allies)
      Anseriformes (waterfowl, ducks, geese, swans)
      Phoenicopteriformes (flamingoes and allies)
      Ciconiiformes (herons and allies)
      Falconiformes (falcons, hawks, eagles, buzzards and other
        birds of prey)
      Galliformes (grouse, pheasants, partridges, etc.)

Gruiformes (hemipodes, cranes, bustards, rails, coots, etc.)
Charadriiformes (shorebirds, waders, gulls, auks)
Gaviiformes (divers)
Columbiformes (doves, pigeons, sandgrouse)
Psittaciformes (parrots, etc.)
Cuculiformes (cuckoos and others)
Strigiformes (owls)
Caprimulgiformes (nightjars)
Apodiformes (swifts)
Coliiformes (colies)
Trogoniformes (trogons)
Coraciiformes (kingfishers, bee-eaters, hoopoes)
Piciformes (woodpeckers)
Passeriformes (perching birds and songbirds, a very large
    order, including finches, crows, warblers, sparrows and
    weavers, etc.)
class Mammalia
  subclass Prototheria
    order Monotremata (duck-billed platypus)
    and several extinct orders
  subclass Theria
  infraclass Metatheria (marsupials)
    order Didelphiformes (opossums)
    Peramelina (bandicoots)
    Diprotodonta (kangaroos, etc.)
  infraclass Eutheria (placental mammals)
    order Insectivora (shrews, moles, etc.)
    Chiroptera (bats)
    Dermoptera (flying lemurs)
    Fissipedia (dogs, cats, bears, mustelids and other specialized
    carnivores)
    Pinnipedia (seals)
    Cetacea (whales and dolphins)
    suborder Odontoceti (toothed whales and dolphins)
    Mysticeti (baleen whales)
    order Rodentia (rodents)
    Lagomorpha (rabbits, hares, etc.)
    Artiodactyla (even-toed ungulates)
    suborder Suina (pigs, hippopotamus, etc.)
    Tylopoda (camels, etc.)
    Ruminantia (deer, antelope, etc.)
    order Perissodactyla (odd-toed ungulates)
    suborder Hippomorpha (horses)
    Ceratomorpha (rhinos and tapirs)
    order Proboscidea (elephants)
    Hyracoidea (hyraxes)
    Tubulidentata (anteaters)

Sirenia (dugongs, manatees)
Pholidota (pangolins)
Primates
suborder Prosimii (tree shrews, lemurs, etc.)
Anthropoidea (monkeys, apes and humans)
and many extinct groups

Appendix 4

AN OUTLINE OF THE PROTISTS
(DOMAIN EUKARYA)

The multitude of unicellular and simple multicellular eukaryotic organisms do not sit happily in the animal, plant or fungal kingdoms, and until recently were put in a Kingdom of their own as the Protoctista (Protista) (L. Margulis and K.V. Schwartz, *Five Kingdoms*, 2nd edn, Freeman, 1988). However, the phylogeny and taxonomy of the protists and their relation to other eukaryotes are currently undergoing major revision as a result of the increasing amounts of genomic and other molecular data available. One current view (Adl, S.M. *et al. J. Eukaryot. Microbiol.* 2005, **52**:399–451) places all eukaryotic organisms into six monophyletic groups (i.e. all organisms within a group are thought to be descended from a single common ancestor). These groups are given first here, followed by the older names for large groups of protists that may be encountered in the scientific literature. The names given here are not comprehensive.

See entries in the body of the dictionary for more information.

The six 'Kingdoms' of eukaryotes

OPISTHOKONTA (animals, fungi, the choanoflagellates, myxozoa, microsporidia, and some other non-photosynthetic protozoa)

AMOEBOZOA (the classical amoebae, cellular slime moulds and plasmodial slime moulds)

PLANTAE (land plants, green algae, red algae and glaucophytes)

CHROMALVEOLATA (the Chromista/Stramenopila (see below) and the Alveolata (ciliates, apicomplexans and dinoflagellates))

RHIZARIA (amoeboid protozoa with mineral shells through which protrude fine thread-like pseudopodia: they include the foraminifera, radiolarians and cercozoans)

EXCAVATA (heterotrophic, mostly flagellate, unicellular eukaryotes such as diplomonads, trypanosomes and the sometimes photosynthetic euglenids, and also some amoeba-like protozoa and the acrasid slime moulds)

Phyla in the Kingdom Protoctista

Commonly known as algae

DINOFLAGELLATA (DINOMASTIGOTA or PYRROPHYTA) (dinoflagellates, e.g. *Gymnodinium*, *Gonyaulax*)

EUGLENOPHYTA (euglenoids, e.g. *Euglena*)

CRYPTOPHYTA (cryptomonads)

EUSTIGMATOPHYTA (e.g. *Pleurochloris*)

RHODOPHYTA (red algae, e.g. *Porphyra*, *Corallina*, *Chondrus*)

GAMOPHYTA (conjugating green algae, e.g. *Mougeotia*, *Spirogyra*)

CHLOROPHYTA (green algae, e.g. *Chlamydomonas*, *Chlorococcus*, *Chlorella*, *Volvox*, *Ulva*, *Oedogonium*, *Stigeoclonium*, *Chaetomorpha*, *Acetabularia*, *Nitella*, *Platymonas*)

Commonly known as protozoa

CARYOBLASTEA (one species, *Pelomyxa palustris*)

PARABASALA (lacking mitochondria and containing parabasal bodies, e.g. *Trichomonas*)

CERCOZOA (includes the foraminifers (e.g. *Globigerina*) and other similar amoeboid protozoa with fine pseudopodia extending through a shell)

RADIOLARIA (amoeboid protozoa with ornate silica shells or strontium sulphate shells though which fine stiff pseudopodia extend like the spokes of a wheel)

AMEOBOZOA (classical amoebas with finger-like pseudopodia, e.g. *Acanthamoeba, Amoeba, Difflugia*)

ALVEOLATA (ciliates (e.g. *Didinium*), apicomplexans (e.g. *Plasmodium*), and dinoflagellates (e.g. *Gonyaulax*)

DIPLOMONADIDA (flagellates lacking mitochondria, and with two equal-sized nuclei (e.g. *Giardia*))

EUGLENOZOA (euglenids, which are flagellate and sometimes photosynthetic (e.g. *Euglena*), and the kinetoplastids, characterized by a single large mitochondrion called the kinetoplast (e.g. *Trypanosoma* and *Leishmania*))

Formerly classified in the Fungi

ACRASIOMYCOTA (acrasids)

MYCETOZOA (acellular plasmodial slime moulds (myxogastrids, e.g. *Physarum*), the cellular slime moulds (dictyostelids, e.g. *Dictyostelium*), and the protostelid slime moulds (e.g. *Planoprotostelium aurantium*))

PLASMODIOPHOROMYCOTA (plasmodiophorans, e.g. *Plasmodiophora brassicae* (club-root))

Kingdom Chromista (or Stramenopila)

OOMYCOTA (oomycetes – water moulds, white rusts and downy mildews, e.g. *Phytophthora infestans* (potato blight), *Peronospora* (downy mildew), parasitic water moulds)

HYPHOCHYTRIOMYCOTA (only 23 species known) (hyphochytrids or water moulds, e.g. *Rhizidiomyces*)

LABYRINTHULOMYCOTA (slime nets, e.g. *Labyrinthula*)

CHRYSOPHYTA (golden algae, e.g. *Ochromonas*)

BACILLARIOPHYTA (diatoms, e.g. *Asterionella, Navicula, Thalassiosira*)

HAPTOPHYTA (PRYMNESIOPHYTA) (haptomonads, coccolithophorids, e.g. *Prymnesium*)

PHAEOPHYTA (brown algae, e.g. *Fucus, Laminaria*)

XANTHOPHYTA (yellow-green algae, e.g. *Vaucheria*)

Appendix 5

AN OUTLINE OF THE DOMAIN BACTERIA

Prokaryotes are divided on the basis of biochemical and DNA sequence data into two distinct Domains – the Archaea (formerly known as the Archaebacteria, see Appendix 6) and the Bacteria. The Bacteria are distinguished from the Archaea by having peptidoglycan cell walls, ester-linked membrane lipids only, and by their ribosome structure and rRNA sequences. The classification of the prokaryotes is being revised to agree with relationships discerned on the basis of 16S rRNA sequences.

The outline below lists the recognized bacterial phyla according to Garrity, G.M., Bell, J.A., & Lilburn, T.G. Taxonomic Outline of the Prokaryotes. *Bergey's Manual of Systematic Bacteriology*, Second Edition, Release 5.0, Springer-Verlag, New York, 2004. DOI: 10.1007/bergeysoutline200405

Bacterial phyla

Aquificae (e.g. *Hydrogenobacter, Aquifex*)
Thermotogae (e.g. *Fervidobacterium, Thermotoga*)
Thermodesulfobacteria (e.g. *Thermodesulfobacterium*)
Deinococcus-Thermus (e.g. *Thermus, Deinococcus*)
Chrysiogenetes (*Chrysiogenes*)
Chloroflexi (e.g. *Chloroflexus, Herpetosiphon*)
Thermomicrobia (*Thermomicrobium*)
Nitrospira (e.g. *Leptospirillum, Thermodesulfovibrio*)
Deferribacteres (e.g. *Deferribacter*)
Cyanobacteria oxygenic photoautotrophic bacteria; some also have nitrogen-fixing capacity (e.g. *Synechococcus, Microcystis, Prochloron, Anabaena*)
Chlorobi (green sulphur bacteria, e.g. *Chlorobium*)
Proteobacteria (purple bacteria) large phylum of Gram-negative bacteria with a wide range of morphologies and physiologies, and including many pathogens of humans. There are five Classes:
  alpha-proteobacteria (e.g. *Rhodospirillum, Acetobacter, Rickettsia, Anaplasma, Rhodobacter, Sphingomonas, Caulobacter, Rhizobium, Bartonella, Brucella, Phyllobacterium, Methylocystis, Beijerinckia, Hyphomicrobium, Methylobacterium*)
  beta-proteobacteria (e.g. *Burkholderia, Alcaligenes, Bordetella, Hydrogenophilus, Neisseria, Nitromonas, Spirillum, Thiobacillus*)
  gamma-proteobacteria (e.g. *Chromatium, Thiospirillum, Halothiobacillus, Xanthomonas, Beggiatoa, Leucothrix, Pseudomonas, Francisella, Legionella, Coxiella, Methylococcus, Oceanospirillum, Azotobacter, Moraxella, Vibrio, Aeromonas, Escherichia, Shigella, Salmonella, Klebsiella, Yersinia*)
  delta-proteobacteria *Desulfurella, Desulfovibrio, Desulfuromonas, Bdellovibrio, Myxococcus*
  epsilon-proteobacteria (e.g. *Campylobacter, Helicobacter, Thiovulvum, Wolinella*)

Firmicutes large phylum of Gram-positive bacteria, of very diverse physiology and morphology, and including many pathogens of humans. The main Classes distinguished are:

    Clostridia (e.g. *Clostridium, Sarcina, Butyrivibrio, Peptostreptococcus, Heliobacterium*)

    Mollicutes (obligate intracellular parasites lacking cell walls, e.g. *Mycoplasma, Spiroplasma, Acholeplasma*)

    Bacilli (e.g. *Bacillus, Listeria, Planococcus, Staphylococcus, Thermoactinomyces, Lactobacillus, Enterococcus, Leuconostoc, Streptococcus, Lactococcus*)

Actinobacteria (e.g. *Actinomyces, Micrococcus, Myobacterium, Corynebacterium*)

Planctomycetes (e.g. *Planctomyces, Chlamydia*)

Spirochaetes (e.g. *Treponema, Borrelia*)

Fibrobacteres (*Fibrobacter*)

Acidobacteria (e.g. *Acidobacterium, Geothrix*)

Bacteroidetes (e.g. *Bacteroides, Flavobacterium, Sphingobacterium*)

Fusobacteria (e.g. *Fusobacterium*)

Verrucomicrobia (e.g. *Verrucomicrobium*)

Dictyoglomi (*Dictyoglomus*)

Gemmatimonadetes (*Gemmatimonas*)

Appendix 6

AN OUTLINE OF THE DOMAIN ARCHAEA

Prokaryotes are divided on the basis of DNA sequence and biochemical data into two distinct Domains or Superkingdoms – the Archaea (formerly known as the Archaebacteria) and the Bacteria (see Appendix 5). The Archaea are prokaryotes whose cell walls lack peptidoglycan, and which also have ether-linked membrane lipids and distinctive ribosomes and rRNA sequences. As with the Bacteria, the phylogenetic classification of the Archaea is still far from resolved. The following taxa are generally distinguished on grounds of morphology, biochemistry and DNA sequence data (outline based on Garrity, G.M., Bell, J.A., & Lilburn, T.G. Taxonomic Outline of the Prokaryotes. *Bergey's Manual of Systematic Bacteriology*, Second Edition, Release 5.0, Springer-Verlag, New York, 2004. DOI: 10.1007/bergeysoutline200405).

Crenarchaeota

Extremely thermophilic sulphur metabolizers

Class Thermoprotei
  Orders Thermoproteales (e.g. *Thermoproteus, Desulfurococcus*)
        Caldisphaerales (*Caldisphaera lagunensis*)
        Desulfurococcales (e.g. *Aeropyrum*)
        Sulfolobales (*Sulfolobus, Acidianus*)

Euryarchaeota

Methanogens
Class Methanobacteria
  Order Methanobacteriales (e.g. *Methanobacterium, Methanothermus*)
Class Methanococci
  Order Methanococcales (*Methanococcus*)
Class Methanomicrobia
  Orders Methanomicrobiales (*Methanomicrobium, Methanospirillum*)
        Methanosarcinales (e.g. *Methanosarcina*)
Class Methanopyri
  Order Methanopyrales (e.g. *Methanopyrus*)

Extreme thermophiles and sulphate reducers
Class Thermococci
  Order Thermococcales (e.g. *Thermococcus, Pyrococcus*)
Class Archaeoglobi
  Order Archaeoglobales (e.g. *Archaeoglobus, Geoglobus*)

Extreme halophiles
Class Halobacteria
  Order Halobacteriales (e.g. *Halobacterium*)

Archaea lacking cell walls
Class Thermoplasmata
  Order Thermoplasmatales (e.g. *Thermoplasma*)

Appendix 7

VIRUS FAMILIES

Listed below are those virus families recognized by the International Committee on Taxonomy of Viruses. This list is not comprehensive, as many well-known plant viruses, such as tobacco mosaic virus, have not yet been assigned a family and are classified only to 'genus' level. Some of these genera are included in the body of the dictionary, e.g. Capillovirus, Carlavirus, Hordeivirus, Marafivirus, Tenuivirus, Tobravirus, Tobamovirus.

a. DNA virus families
Double-stranded DNA

Bacteriophages	Corticoviridae (e.g. PM2)
	Fuselloviridae (e.g. SSV1)
	Lipothrixviridae (e.g. TTV1)
	Myoviridae (T4 and the T-even phages)
	Plasmaviridae (e.g. MVL2)
	Podoviridae (T7 and the T-odd phages)
	Rudiviridae (e.g. SIRV1)
	Siphoviridae (e.g. lambda and P22)
	Tectiviridae (e.g. PRD1)
Animal viruses	Adenoviridae (e.g. adenovirus)
	Ascoviridae
	Asfarviridae (e.g. African swine fever virus)
	Baculoviridae (e.g. insect baculovirus)
	Herpesviridae (e.g. herpesviruses, chickenpox)
	Iridoviridae (e.g. insect iridescent viruses)
	Papillomaviridae (e.g. papillomaviruses)
	Polydnaviridae
	Polyomaviridae (e.g. SV40)
	Poxviridae (e.g. vaccinia, smallpox)
Algal viruses	Phycodnaviridae

Single-stranded DNA

Bacteriophages	Inoviridae (fd)
	Microviridae (e.g. φX174, G4)
Plant viruses	Geminiviridae (e.g. maize streak virus)
	Circoviridae (e.g. chicken anaemia virus)
Animal viruses	Parvoviridae (e.g. canine distemper virus)

b. Reverse-transcribing DNA and RNA viruses

Plant viruses	Caulimoviridae (e.g. cauliflower mosaic virus)
Animal viruses	Hepadnaviridae (e.g. hepatitis B)
	Retroviridae (e.g. HIV, animal tumour viruses)
Others	Metaviridae (yeast, plants and invertebrates)
	Pseudoviridae (yeast, plants and invertebrates)

Appendix 7: Virus Families

c. RNA virus families
Double-stranded RNA

Bacteriophages	Cystoviridae (f6)
Plant viruses	Partitiviridae
	Reoviridae (e.g. fijivirus)
Animal viruses	Birnaviridae (e.g. infectious panreatic necrosis virus of fish)
	Reoviridae (e.g. human rotaviruses)
Other	Chrysoviridae (fungal viruses)
	Hypoviridae (fungal viruses)
	Totiviridae (viruses of fungi and protozoa)

Single-stranded RNA (negative strand)

Plant viruses	Bunyaviridae (e.g. tomato spotted wilt virus)
	Neoviridae (e.g. subterranean clover stunt virus)
	Rhabdoviridae (e.g. potato yellow dwarf virus)
Animal viruses	Arenaviridae (e.g. lymphocytic choriomeningitis virus)
	Bornaviridae
	Bunyaviridae (e.g. Rift Valley fever)
	Filoviridae (e.g. Ebola virus)
	Orthomyxoviridae (e.g. influenza)
	Paramyxoviridae (e.g. measles, mumps)
	Rhabdoviridae (e.g. rabies)

Single-stranded RNA (positive strand)

Bacteriophages	Leviviridae (MS2)
Plant viruses	Barnaviridae (e.g. maize rayado fino virus)
	Bromoviridae (e.g. alfalfa mosaic virus)
	Closteroviridae (e.g. beet yellow virus)
	Comoviridae (e.g. cowpea mosaic virus)
	Flexiviridae (e.g. potato virus X)
	Luteoviridae (e.g. barley yellow dwarf virus)
	Potyviridae (e.g. potato virus Y)
	Sequiviridae (e.g. parsnip yellow fleck virus)
	Tetraviridae (e.g. Southern bean mosaic virus)
	Tombusviridae (e.g. tomato bushy stunt virus)
Animal viruses	Arteriviridae (e.g. equine arteritis virus)
	Astroviridae (e.g. human astrovirus)
	Caliciviridae (e.g. swine vesicular exanthema virus)
	Coronaviridae (e.g. human coronaviruses)
	Dicistroviridae (insect viruses)
	Flaviviridae (e.g. yellow fever virus)
	Nodaviridae (e.g. Nodamura virus)
	Picornaviridae (e.g. poliovirus)
	Togaviridae (e.g. Sindbis virus, rubella)
Others	Narnaviridae (yeast viruses)
	Marnaviridae (algal viruses)

Appendix 8

GEOLOGICAL TIMESCALE (NOT DRAWN TO SCALE)
IN MILLIONS OF YEARS BEFORE PRESENT

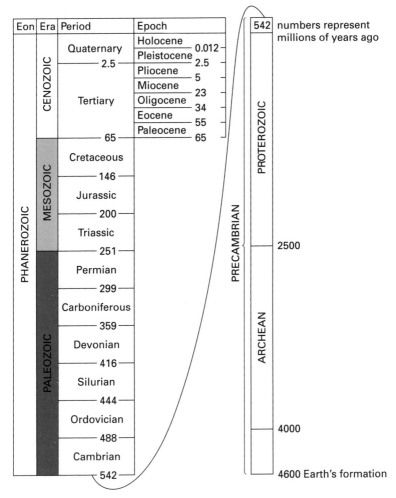

Source: From Lawrence, E., Jackson, R. W. and Jackson, J. M. (1998) *Longman Dictionary of Environmental Science*, Pearson Education, page 463. Reprinted with permission. This chart has been updated on the basis of the 2009 International Stratigraphic Chart produced by the International Commission on Stratigraphy.

Appendix 9

ETYMOLOGICAL ORIGINS OF SOME COMMON WORD ELEMENTS IN BIOLOGY

The word element as it appears in English is in **bold** type. The Greek (Gk) or Latin (L.) word from which it is derived is shown in italics and is followed by its original meaning.

a- *a* (Gk), not.
ab- *ab* (L.), from.
absci- *abscidere* (L.), to cut off.
abyss- *abyssos* (Gk), unfathomed.
acanac- *akanos* (Gk), thistle.
acanth- *akantha* (Gk), thorn.
acer- *acer* (L.), sharp.
acid- *acidus* (L.), sour.
acra- *akros* (Gk), tip.
actin- *aktis* (Gk), ray.
ad- *ad* (L.), to, towards.
adeno- *aden* (Gk), gland.
adipo- *adeps* (L.), fat.
aeolian *Aeolus* (Gk), god of the winds.
aer- *aer* (L.), *aēr* (Gk), air.
-aesthesia, -aesthetic *aisthēsis* (Gk), sensation.
agrost- *agrōstis* (Gk), grass.
albo-, albu-, albino *albus* (L.), white.
alga-, algo- *alga* (Gk), seaweed.
allele, allelo- *allēlōn* (Gk), one another.
allo- *allos* (Gk), other.
ambi- *ambo* (L.), both.
amoeba *amoibē* (Gk), change.
amphi- *amphi* (Gk), both.
amylo- *amylum* (Gk), starch.
an- *an* (Gk), not.
ana- *ana* (Gk), up, again.
andr-, andro- *anēr* (Gk), male, *andrikos* (Gk), masculine.
anemo- *anemos* (Gk), wind.
angio-, -angium *anggeion* (Gk), vessel.
aniso- *anisos* (Gk), unequal.
ankylo- *agkylos* (Gk), crooked.
anlage *Anlage* (Ger.) predisposition.
annelid, annulate *annulus* (L.), a ring.
ano- *anus* (L.), anus.
anomalo- *anomalos* (Gk), uneven.
anomo- *anomos* (Gk), lawless, irregular.
ante- *ante* (L.), before.
antha-, antho-, -anthous, -anthy *anthos* (Gk), flower.
anthero- *anthēros* (Gk), flowering.

anthropo- *anthrōpos* (Gk), man.
anti- *anti* (Gk), against, opposite.
-apl'oid *aploos* (Gk), onefold, and *eidos* (Gk), form.
apo- *apo* (Gk), from.
arachni-, arachno- *arachnē* (Gk), spider, cobweb.
arbor- *arbor* (L.), tree.
archaeo- *archaios* (Gk), primitive, ancient.
arche- *archē* (Gk), beginning.
archi- *archi* (Gk), first.
archo- *archon* (Gk), ruler.
arci- *arcus* (L.), bow.
argent- *argentum* (L.), silver.
argyro- *argyros* (Gk), silver.
arthro- *arthron* (Gk), a joint.
artio- *artios* (Gk), even (numbered).
-asci, -ascus, asco- *askos, askidion* (Gk), bag, little bag.
astra-, astro-, -aster *astra* (Gk), star.
-atomy, -otomy *tomē* (Gk), cutting.
auri-, auricul- *auris, auricula* (L.), ear, small ear.
auto- *autos* (Gk), self.
auxi-, auxo- *auxein* (Gk), to increase.
avi- *avis* (L.), bird.
axill- *axilla* (L.), armpit.
axis, axial *axis* (L.), axle.
axo- *axon* (Gk), axis.
barb-, barba *barba* (L.), beard.
baro- *baros* (Gk), pressure, weight.
bathy- *bathys* (Gk), deep.
batrach- *batrachos* (Gk), frog.
benthos, benthic *benthos* (Gk), depths of sea.
bi- *bis* (L.), twice.
bio-, -biotic *bios* (Gk), life, *biosis*, living, *biōtikos*, pert. life.
blast- *blastos* (Gk), bud.
botany *botanē* (Gk), pasture.
bothr- *bothros* (Gk), pit.
botry- *botrys* (Gk), bunch of grapes.

brachia- *brachium* (L.), arm.
brachy- *brachys* (Gk), short.
brady- *bradys* (Gk), slow.
branchi- *branchiae* (L.), gills, or
brangchia (Gk), gills.
brevi- *brevis* (L.), short.
bryo- *bryon* (Gk), moss.
bucco- *bucca* (L.), cheek.
caeno- *kainos* (Gk), recent.
calci- *calx* (Gk), lime.
calyptr- *kalyptra* (Gk), covering.
cambium, cambio- *cambium* (L.),
change.
capit- *caput* (L.), head, *capitellum*, small
head.
capsid, capso-, capsul- *capsa* (L.), box,
capsula, little box.
carbo- *carbo* (L.), coal.
carcino- *karkinos* (Gk), a crab.
cardia-, cardio- *kardia* (Gk), heart,
stomach.
-carp, -carpous *karpos* (Gk), fruit.
carpa- *carpal* (L.), wrist.
cata- *katalysis* (Gk), dissolving.
cata- *kata* (Gk), down.
cauda- *cauda* (L.), tail.
caul-, cauli- *caulis* (L.), stalk or *kaulos*
(Gk), stalk.
cell, cellular *cellula* (L.), small room.
centro-, -centric *kentron* (Gk), centre.
cephal- *kephalē* (Gk), head.
-ceptor, -ceptive *capere* (L.), to take.
cerca-, cerco- *kerkos* (Gk), tail.
cerebr- *cerebrum* (L.), brain.
-cerous *keras* (Gk), horn.
cervic- *cervix* (L.), neck.
ceta-, ceto- *cetus* (L.), whale.
-chaene-, -chene *chainein* (Gk),
to gape.
chaet- *chaitē* (Gk), hair.
chela- *chēlē* (Gk), claw.
chemi-, chemistry *chēmeia* (Gk),
transmutation.
chiasm- *chiasma* (Gk), cross.
chitin *chiton* (Gk), tunic.
chlamy- *chlamys* (Gk), cloak, mantle.
chloro- *chlōros* (Gk), yellow, green, pale.
chondro- *chondros* (Gk), cartilage.
-chord, chorda-, chordo- *chorde* (Gk),
string.
-chore *chōros* (Gk), place.
chroma-, chromo-, -chrome *chrōma*
(Gk), colour.

chrono- *chronos* (Gk), time.
-cidal *caedere* (L.), to kill.
clade, -cladous *klados* (Gk), branch.
clav- *clava* (L.), a club.
cleisto- *kleistos* (Gk), closed.
clino-, -cline, -clinous *klinē* (Gk), bed.
-clinous *klinein* (Gk), to bend.
clype- *clypeus* (L.), shield.
cocc-, cocco- *koccos* (Gk), berry.
cochli- *kochlias* (Gk), a snail.
coel- *koilos* (Gk), hollow.
coen-, -coenosis *koinos* (Gk), shared in
common.
coleo- *koleos* (Gk), sheath.
conch- *concha* (L.), shell, *kongchē* (Gk),
shell.
cono-, -cone *kōnos* (Gk), a cone.
copro- *kopros* (Gk), dung.
-corn, corne-, corni- *cornus* (L.), horn.
corona- *corona* (L.), crown.
corp-, corpor- *corpus* (L.), body.
cortex, cortic- *cortex* (L.), bark.
costa- *costa* (L.), rib.
cotyl- *kotylē* (Gk), cup.
coxa-, coxo- *coxa* (L.), hip.
crani-, crania- *kranion* (Gk), skull.
-crine *krinein* (Gk), to separate.
cruci- *crux* (L.), cross.
cryo- *kryos* (Gk), frost.
crypto- *kryptos* (Gk), hidden.
cyano- *kyanos* (Gk), dark blue.
cyath- *kyathus* (Gk), a cup.
cyclo-, -cyclic *kyklos* (Gk), circle.
cyst- *kystis* (Gk), bladder.
-cyte, cyto- *kytos* (Gk), hollow.
-dactyl *daktylos* (Gk), finger.
de- *de* (L.), away.
deme, demo- *dēmos* (Gk), people.
demi- *dimidius* (L.), half.
dendr- *dendron* (Gk), tree.
dent- *dens* (L.), tooth.
derma-, dermo-, -derm *derma* (Gk),
skin.
desm- *desmos* (Gk), bond.
di- *dis* (Gk), twice.
dia- *dia* (Gk), asunder.
dicho- *dicha* (Gk), in two.
dictyo- *dictyon* (Gk), net.
digito- *digitus* (L.), finger.
dino- *dinos* (Gk), rotation.
dino- *deinos* (Gk), terrible.
diplo- *diploos* (Gk), double.
dors- *dorsum* (L.), back.

Appendix 9: Some Etymological Origins

-drome, -dromic, -dromous *dramein* (Gk), to run, *drōmos* (Gk), running.
-duct *ducere* (L.), to lead.
duplico- *duplex* (L.), double.
dynamo- *dynamis* (Gk), power.
dys- *dys* (Gk), mis-.
e- *ex* (L.), out of.
ec- *ek* (Gk), out of.
echino- *echinos* (Gk), spine.
eco- *oikos* (Gk), house, household.
ect- *ektos* (Gk), without, outside.
-ectomy *ektomē* (Gk), a cutting out.
elaeo-, elaio- *elaion* (Gk), oil.
electro- *elektron* (Gk), amber (a fossil resin which produces static electricity when rubbed).
embryo- *embryon* (Gk), embryo.
endo- *endon* (Gk), within, inside.
-ennial *annus* (L.), year.
entero- *enteron* (Gk), gut.
entomo- *entomon* (Gk), insect.
epi- *epi* (Gk), upon.
equi- *aequus* (L.), equal.
erg-, -ergic, -ergy *ergon* (Gk), activity, work.
erythro- *erythros* (Gk), red.
eu- *eu* (Gk), well.
eury- *eurys* (Gk), wide.
exo- *exō* (Gk), outside.
extra- *extra* (L.), beyond.
-farious *fariam* (L.), in rows.
fauna- *faunus* (L.), god of woods.
ferre-, ferri-, ferro- *ferrum* (L.), iron.
fibrino- *fibra* (L.), a band.
-fid *findere* (L.), to split.
fili- *filum* (L.), a thread.
flavo- *flavus* (L.), yellow.
flor- *flos* (L.), flower.
folia- *folium* (L.), leaf.
fronto- *frons* (L.), forehead.
fuco- *fucus* (L.), seaweed.
-fugal, -fuge *fugere* (L.), to flee.
galacto- *gala* (Gk), milk.
gamete, gameto- *gametes* (Gk), spouse.
gamo-, -gamy, -gamous *gamos* (Gk), marriage.
ganglio- *ganglion* (Gk), swelling.
gastro- *gaster* (Gk), stomach.
-geminal *geminus* (L.), double.
-gen, -genous *genos* (Gk), descent.
gene-, -genetic *genesis* (Gk), birth, descent, origin.

-genic, -genous *gennaein* (Gk), to produce.
genito- *gignere* (L.), to beget.
geno- *genos* (Gk), race.
geo- *gē*, or *gaia* (Gk), earth.
germ-, germin- *germen* (L.), a bud.
geronto- *gerōn* (Gk), old man.
glia-, -gloea *gloia* (Gk), glue.
-globin, -globulin *globus* (L.), a sphere.
glosso- *glossa* (L.), tongue.
gluco-, glyco- *glykys* (Gk), sweet.
gnatho-, -gnath, -gnathous *gnathos* (Gk), jaw.
gonad-, -gone, -gonic *gonē* (Gk), seed.
-gone, goni- *gonos* (Gk), offspring.
-grade *gradus* (L.), step.
-gram, -graphy *graphein* (Gk), to write.
gyn-, -gynous *gynē* (Gk), female.
haem-, haema- *haima* (Gk), blood.
halo- *hals* (Gk), sea, salt.
haplo- *haploos* (Gk), simple.
hapto- *haptos* (Gk), touch.
helio- *helios* (Gk), sun.
hemi- *hēmi* (Gk), half.
hepa-, hepatico- *hepar* (Gk), liver.
hepta- *hepta* (Gk), seven.
hetero- *heteros* (Gk), other.
hex- *hex* (Gk), six.
histio- *histion* (Gk), tissue.
histo- *histos* (Gk), tissue.
holo- *holos* (Gk), whole.
homeo- *homoios* (Gk), alike.
homo- *homos* (Gk), the same.
hormone *hormaein* (Gk), to excite.
hyalo- *hyalos* (Gk), glass.
hydr-, hydro- *hydor* (Gk), water.
hygro- *hygros* (Gk), wet.
hyper- *hyper* (Gk), above.
hypo- *hypo* (Gk), under.
ichthy- *ichthys* (Gk), fish.
-icole, -icolous *colere* (L.), to dwell.
-iferous *ferre* (L.), to carry.
-ific, -ification *facere* (L.), to make.
-igen *generare* (L.), to beget.
-igerous *gerere* (L.), to bear.
im-, in- *in* (L.), not.
immuno- *immunis* (Gk), free.
in- *in* (L.), into.
infero- *inferus* (L.), beneath.
infra- *infra* (L.), below.
inter- *inter* (L.), between.
intra- *intra* (L.), within.
iso- *isos* (Gk), equal.

-jugate *jugare* (L.), to join.
jugo- *jugum* (L.), yoke.
juxta- *juxta* (L.), close to.
kary- *karyon* (Gk), nucleus, nut.
kera- *keras* (Gk), horn.
-kinesis, -kinetic *kinesis* (Gk), movement, *kinein* (Gk), to move.
labia-, labio- *labium* (L.), lip.
lacto- *lac* (L.), milk.
lati-, latero- *latus* (L.), wide.
lepido- *lepidotos* (Gk), scaly.
lepto- *leptos* (Gk), slender.
leuco-, leuko- *leukos* (Gk), white.
limn- *limne* (Gk), marsh.
lipo- *lipos* (Gk), fat.
litho-, -lith *lithos* (Gk), stone.
lopho- *lophos* (Gk), crest.
luci- *lux* (L.), light.
luteo- *luteus* (L.), orange-yellow.
-lysin, -lysis, lyso-, -lytic *lysis* (Gk), loosening, *lyein* (Gk), to dissolve.
macro- *makros* (Gk), large.
masto- *mastos* (Gk), breast.
matro- *mater* (L.), mother.
medi- *medius* (L.), middle.
mega- *megas* (Gk), large.
megalo- *megalon* (Gk), great.
meio- *meion* (Gk), smaller.
meiosis- *meiosis* (Gk), diminution.
-mere, mero- *meros* (Gk), a part.
meso- *mesos* (Gk), middle.
meta- *meta* (Gk), after.
metabolism, metabolic *metabole* (Gk), change.
-metric, -metry *metron* (Gk), measure, *metreo* (Gk), to count.
micro- *mikros* (Gk), small.
mito-, mitosis *miton* (Gk), a thread.
mono- *monos* (Gk), alone, single.
morpho-, -morph, -morphism, -morph *morphe* (Gk), shape, form, *morphosis* (Gk), form.
multi- *multus* (L.), many.
mutate, muta- *mutare* (L.), to change.
myco-, -mycin *mykes* (Gk), fungus.
myelo- *myelos* (Gk), marrow.
myo- *mys* (Gk), muscle.
myrme- *myrmēx* (Gk), ant.
myxo- *myxa* (Gk), slime.
nano- *nanos* (Gk), dwarf.
necro- *nekros* (Gk), dead.
nema-, nemato-, -neme *nēma* (Gk), a thread.

neo- *neos* (Gk), new.
nephr- *nephros* (Gk), kidney.
neuro- *neuron* (Gk), nerve.
nexus, -nexed *nectare* (L.), to bind.
nigro- *niger* (L.), black.
nitro- *nitron* (Gk), soda.
noci- *nocere* (L.), to hurt.
-nomics, -nomy *nomos* (Gk), law.
nomin-, -nomial *nomen* (L.), name.
noto- *noton* (Gk), back.
nucleus *nucleus* (Gk), kernel.
ob- *ob* (L.), against, reversely.
occipi- *occiput* (L.), back of head.
octa-, octo- *okta* (Gk), *octo* (L.), eight.
-odont, odonto- *odous* (Gk), tooth.
-oecious, -oecium *oikos* (Gk), house.
oestro-, oestrus *oistros* (Gk), gadfly.
-ogen *genos* (Gk), birth.
-ogony *gonos* (Gk), generation.
oi-, -oo- *ōon* (Gk), egg.
-oid *eidos* (Gk), form.
olei-, oleo- *oleum* (L.), oil.
oligo- *oligos* (Gk), few.
-ology *logos* (Gk), discourse.
onco- *onkos* (Gk), bulk, mass.
onto- *on* (Gk), being.
-oo- *ōon* (Gk), egg.
ophthal- *ophthalmos* (Gk), eye.
opsi-, -opsin, -opsy, opto- *opsis* (Gk), eye.
ora-, oro- *os, oris* (L.), mouth.
organ *organon* (Gk), instrument.
orni- *ornis* (Gk), bird.
ortho- *orthos* (Gk), straight.
osmo- *ōsmos* (Gk), impulse.
ost-, osteo- *osteon* (Gk), bone.
ostraco- *ostrakon* (Gk), shell.
oto- *ous* (Gk), ear.
-otomy *temnein* (Gk), to cut.
ova-, ovi-, ovo- *ovum* (L.), egg.
oxy- *oxys* (Gk), sharp.
pachy- *pachys* (Gk), thick.
palaeo- *palaios* (Gk), ancient.
palpi- *palpare* (L.), to stroke.
pan-, panto- *pan* (Gk), all.
para- *para* (Gk), beside.
-parous *parere* (L.), to produce.
patho-, -pathy *pathos* (Gk), suffering.
patri- *pater* (L.), father.
-patric *patria* (L.), native land.
ped-, -pedal *pes* (L.), foot.
-pelagic *pelagos* (Gk), sea.
penta-, pento- *pente* (Gk), five.

Appendix 9: Some Etymological Origins

per- *per* (L.), through.
peri- *peri* (Gk), around.
perisso- *perissos* (Gk), odd (numbered).
petalo- *petalon* (Gk), leaf.
petro- *petros* (Gk), stone.
-phage, phago-, -phagous *phagein* (Gk), to eat.
-phase *phasis* (Gk), aspect, appearance.
-phene, pheno- *phainein* (Gk), to appear.
-phil, -phile, -phily *philein* (Gk), to love.
philo- *philos* (Gk), loving.
-phobe, -phobic *phobos* (Gk), fear.
phono- *phōnē* (Gk), sound.
-phore *phorein* (Gk), to carry.
phospho- *phosphoros* (Gk), bringing light.
photo- *phos* (Gk), light.
phragmo- *phragmos* (Gk), fence.
-phylactic *phylaktikos* (Gk), fit for preserving.
-phyll, -phyllous *phyllon* (Gk), leaf.
physi-, -physis *physis* (Gk), growth.
physics *physis* (Gk), nature.
phyt-, -phyte *phyton* (Gk), plant.
pinna-, penna- *penna* (L.), feather.
pisci- *piscis* (L.), fish.
placo- *plax* (Gk), plate.
plana- *planatus* (L.), flattened.
-planetic, planeto- *planetes* (Gk), wanderer.
plano- *planos* (Gk), wandering.
-plasia *plasis* (Gk), a moulding, *plassein* (Gk), to form.
plasma-, -plasm *plasma* (Gk), form.
-plast, -plastic, plastid *plastos* (Gk), formed.
plasti-, plasto- *plastos* (Gk), formed.
platy- *platys* (Gk), flat.
pleio- *pleion* (Gk), more.
plero- *plērēs* (Gk), full.
plesio- *plēsios* (Gk), near.
pleura-, -pleurite *pleuros* (Gk), side.
-plicate *plicare* (L.), to fold.
pluri- *plus* (L.), more.
pneu- *pnein* (Gk), to breathe.
-pod, -podite *pous* (Gk), foot.
-poiesis, -poietic *poiesis* (Gk), making.
poly- *polys* (Gk), many.
poro-, -pore *poros* (Gk), channel.
porphyr- *porphyra* (Gk), purple.
post- *post* (L.), after.
pre- *prae* (L.), before.
primo- *primus* (L.), first.

pro- *pro* (Gk), before.
proto- *prōtos* (Gk), first.
pseudo- *pseudes* (Gk), false.
psycho- *psychē* (Gk), mind.
-pter- *pteron* (Gk), wing.
pterido- *pteris* (Gk), fern.
ptero- *pteron* (Gk), wing.
pteryg- *pterygion* (Gk), little wing.
pteryg- *pterygion* (Gk), fin.
-ptile *ptilon* (Gk), feather.
pulmo- *pulmo* (L.), lung.
pycno- *pyknos* (Gk), dense.
pygo- *pygē* (Gk), rump.
-pyle *pyle* (Gk), gate.
pyreno- *pyrēn* (Gk), fruit stone.
pyri- *pyrum* (L.), pear.
pyrro- *pyrrhos* (Gk), tawny-red.
quadrato- *quadratus* (L.), squared.
quadri-, quadru- *quattuor* (L.), four.
quin-, quinque- *quinque* (L.), five.
racem- *racemus* (L.), bunch.
rachi- *rachis* (Gk), spine.
radic- *radix* (L.), a root.
radio- *radius* (L.), a ray.
-ramous *ramus* (L.), branch.
rani- *rana* (L.), frog.
re- *re* (L.), back.
rena-, reni- *renes* (Gk), kidney.
reti- *rete* (L.), net.
reticulo- *reticulum* (L.), small net.
retro- *retro* (L.), backwards.
rhabdo- *rhabdos* (Gk), rod.
rheo- *rheein* (Gk), to flow.
rhin- *rhis* (Gk), nose.
rhiza-, rhizo- *rhiza* (Gk), root.
rhodo- *rhodon* (Gk), rose.
rhyncho- *rhyngchos* (Gk), snout.
ribo- *ribes* (L.), currant.
rostra- *rostrum* (L.), beak.
rubi- *ruber* (L.), red.
rubro- *ruber* (L.), red.
sacchar- *sakchar* (Gk), sugar.
salpingo- *salpingx* (Gk), trumpet.
sangui- *sanguis* (L.), blood.
sapo- *sapo* (L.), soap.
sapro- *sapros* (Gk), decayed.
-sarc, sarco- *sarx, sarkōdēs* (Gk), flesh, fleshy.
-saur *sauros* (Gk), lizard.
schisto-, -schist *schistos* (Gk), split.
schizo- *schizein* (Gk), to cleave.
-scopic *skopein* (Gk), to view.
scoto- *skotos* (Gk), dark.

scute, scutum *scutum* (L.), shield.
seismo- *seismos* (Gk), a shaking.
seleno- *selēnē* (Gk), moon.
-sematic *sema* (Gk), signal.
semi- *semi* (L.), half.
septa-, septi-, septo- *septum* (L.),
   partition.
septi- *septum* (L.), seven.
-sere *serere* (L.), to put in a row.
serum *serum* (L.), whey.
seta-, seti-, seto- *seta* (L.), bristle.
siali-, sialo- *sialon* (Gk), saliva.
sidero- *sidēros* (Gk), iron.
soma-, somato-, -some *sōma* (Gk),
   body.
sora-, sori-, soro- *sōros* (Gk), heap.
speleo- *spelaion* (Gk), cave.
sperma-, -sperm *sperma* (Gk), seed.
sphaero- *sphaira* (Gk), globe.
spheno- *sphen* (Gk), a wedge.
sphero- *sphaira* (Gk), globe.
spondyl- *sphondylos* (Gk), vertebra.
spor-, -spore *sporos* (Gk), seed.
squame, squama- *squama* (L.), scale.
stachy- *stachys* (Gk), ear of corn.
stamin-, -stemonous *stemon* (Gk), spun
   thread.
-stat, -static *stare* (L.), to stand.
stato- *statos* (Gk), stationary, standing.
stega-, stegi- *stega* (Gk), roof.
-stelic, -stely *stele* (Gk), pillar.
stereo-, -steric *stereos* (Gk), solid.
stern- *sternum* (L.), breast.
steroid, -sterone *stear* (Gk), suet.
stoma-, stomato-, -stome *stoma* (Gk),
   mouth.
-strate *stratum* (L.), layer.
strepto- *streptos* (Gk), twisted, pliant.
strobilus *strobilos* (Gk), fir-cone.
strom-, stroma *stroma* (Gk), bedding.
-stylic *stylos* (Gk), pillar.
sub- *sub* (L.), under.
super- *super* (L.), over.
supra- *supra* (L.), above.
sylv- *sylva* (L.), forest.
sym- *syn* (Gk), with.
syn- *syn* (Gk), with.
synaps-, synapto- *synapsis, synaptos*
   (Gk), union, joined.
synthesis *synthesis* (Gk), composition.
tachy- *tachys* (Gk), quick.
talo- *talus* (L.), ankle.
tarso- *tarsos* (Gk), sole of foot.

tauto- *tautos* (Gk), the same.
-taxis, -taxy *taxis* (Gk), arrangement.
taxo- *taxis* (Gk), arrangement.
tect- *tectum* (L.), roof.
tele- *tēle* (Gk), far.
teleo- *teleos* (Gk), complete.
telo-, telio- *telos* (Gk), end.
tempor- *tempora* (L.), temples.
-tene *tainia* (Gk), a band.
tensin- *tonos* (Gk), tension.
terga- *tergum* (L.), back.
ternato- *terni* (L.), three.
terr- *terra* (L.), earth.
tetra- *tetras* (Gk), four.
thallo- *thallos* (Gk), branch.
-theca, -thecium *theke* (Gk), box.
-theria *therion* (Gk), small animal.
thermo- *thermē* (Gk), heat.
thero- *theros* (Gk), summer.
thigmo- *thigēma* (Gk), touch.
thio- *theion* (Gk), sulphur.
thrombo- *thrombos* (Gk), clot.
thylako- *thylakos* (Gk), pouch.
thyro- *thyra* (Gk), door.
-tocin *tokos* (Gk), birth.
-tope, -topic, topo- *topos* (Gk), place.
toti- *totus* (L.), all.
toxico-, -toxin *toxikon* (Gk), poison.
tracheo- *trachia* (L.), windpipe.
trachy- *trachys* (Gk), rough.
trans- *trans* (L.), across.
trauma- *trauma* (Gk), wound.
tri- *tria* (Gk), three, *tres* (L.), three.
trich- *thrix* (Gk), hair.
trocho- *trochos* (Gk), hoop.
-troph, -trophic, tropho-, -trophy
   trophē (Gk), maintenance, nourishment.
-tropic, -tropism *tropē* (Gk), turn.
tubi-, tubo- *tubus* (L.), pipe.
tympano- *tympanon* (Gk), drum.
-type *typos* (Gk), pattern.
ulna-, ulno- *ulna* (L.), elbow.
ultra- *ultra* (L.), beyond.
umbell- *umbella* (L.), a sunshade.
umbona- *umbo* (L.), shield boss.
unci- *uncus* (L.), hook.
uni- *unus* (L.), one.
uredo- *urēdo* (L.), blight.
uro- *ouron* (Gk), urine or *oura* (Gk), tail.
vacuol- *vacuus* (L.), empty.
vagini- *vagina* (L.), sheath.
valv- *valvae* (L.), folding doors.
vasa-, vaso- *vas* (L.), vessel.

Appendix 9: Some Etymological Origins

ventr- *venter* (L.), belly.
vermi- *vermis* (L.), worm.
versi- *versare* (L.), to turn.
vesicul- *vesicula* (L.), small bladder.
viro- *virus* (L.), poison.
vitello- *vitellus* (L.), yolk.
vitreo- *vitreus* (L.), glassy.
vivi- *vivus* (L.), living.
-vorous *vorare* (L.), to devour.

xantho- *xanthos* (Gk), yellow.
xeno-, -xenous *xenos* (Gk), host or strange.
xero- *xeros* (Gk), dry.
zo-, zoo- *zōon* (Gk), animal.
-zoic *zoikos* (Gk), pertaining to life.
zygo-, zygote *zygon* (Gk), yoke, *zygotos* (Gk), yoked.
zymo-, -zyme *zymē* (Gk), leaven.